MILLER'S

ANATOMY OF THE DOG

Second Edition

HOWARD E. EVANS, Ph.D.
Professor and Chairman,
Department of Anatomy,
New York State College of Veterinary Medicine,
Cornell University, Ithaca, New York

GEORGE C. CHRISTENSEN, D.V.M., M.S., Ph.D.
Vice President for Academic Affairs
and Professor of Veterinary Anatomy,
Iowa State University,
Ames, Iowa

W. B. SAUNDERS COMPANY
Philadelphia London Toronto Mexico City Rio de Janeiro Sydney Tokyo

W. B. SAUNDERS COMPANY
Harcourt Brace Jovanovich, Inc.

The Curtis Center
Independence Square West
Philadelphia, PA 19106

Library of Congress Cataloging in Publication Data

Miller, Malcolm Eugene.

Miller's anatomy of the dog.

Includes index.

1. Dogs—Anatomy.   I. Evans, Howard E., 1922–
   II. Christensen, George C.   III. Title.   IV. Title:
   Anatomy of the dog.

SF767.D6M54 1979   636.7'08'91   78–20728

ISBN 0–7216–3438–9

Listed here is the latest translated edition of this book together with
the language of the translation and the publisher.

Japanese (*1st Edition*)—Gakutosha, Ltd., Tokyo, Japan

Miller's Anatomy of the Dog   ISBN   0-7216-3438-9

The majority of the illustrations used in this volume were prepared at the Department of Anatomy
at the New York State College of Veterinary Medicine at Cornell University and are used by permission of and remain the property of Cornell University. © Assigned to Cornell University 1964.
Designated executor, H. E. Evans.

printing, last digit: 10

MALCOLM E. MILLER
B.S., D.V.M., M.S., PH.D.

1909–1960

DR. MALCOLM E. MILLER was born on a farm in Durrell, Pennsylvania, studied for two years at Pennsylvania State University, and then earned his B.S. and D.V.M. (1934), M.S. (1936), and Ph.D. (1940) degrees from Cornell University. He was appointed Instructor in 1935, and at the time of his death was Professor and Head of the Department of Anatomy and Secretary of the New York State Veterinary College at Cornell University. His zest for life, devotion to his family, and enjoyment of teaching and research sustained his spirit through several operations which provided only temporary relief.

This volume was envisioned by Dr. Miller in 1944 as a comprehensive treatise documenting the morphology of the dog. His efforts were aided considerably by the encouragement of Dean W. A. Hagan, whose initiative resulted in the appointment of a Medical Illustrator in 1946. Preliminary work resulted in the preparation in 1947 of a *Guide to the Dissection of the Dog*, which now appears as *Miller's Guide to the Dissection of the Dog*, 2nd Edition, by Evans and deLahunta, published by the W. B. Saunders Company.

FOREWORD

This text is a monument to the memory of the author, Malcolm E. Miller, who labored faithfully and long to produce it. During many of the years while it was being written the author worked under the handicap of ill health. The struggle for health was a losing one that culminated in his death in 1960 while he was in his fiftieth year.

From the beginning it was the author's intent to produce a wholly original work. He was not content to accept and use descriptions of others. His original goal, to which I think he adhered to the end, was to base each of his anatomical descriptions and illustrations on not less than five original dissections. This work was meticulously done. It was slow, and it was interrupted by several hospitalizations and other periods when he was incapable of working. Altogether more than 15 years elapsed between the time the work was begun and when the task had to be relinquished. By the time some of the later sections had been finished it was necessary to revise parts that had been finished years earlier.

Unfortunately time ran out on him before he was able to complete the manuscript, and the work had to be finished by two of his friends, former students and colleagues, who have done it as a labor of love.

Since I am not an anatomist, I cannot adequately judge of the excellence of the work. Since I saw the manuscript in the making, know of the devotion of the author to his specialty, and know of the large amount of conscientious labor that went into it, I am led to believe that the volume will be a fitting memory to "Mac" Miller, an excellent and well-loved teacher, a devoted veterinary anatomist, and a long-time colleague and friend.

The generous spirit of George C. Christensen and Howard E. Evans, the two friends who assumed the task of completing the manuscript and preparing it for publication, deserves mention here. Without their efforts the volume could not have been published.

WILLIAM A. HAGAN

Late Professor Emeritus and former Dean,
New York State Veterinary College,
Cornell University, Ithaca, New York

Late Director, National Animal Disease Laboratory,
U.S. Department of Agriculture, Ames, Iowa

PREFACE

It has been fifteen years since this book first appeared, and during this time a veterinary anatomical nomenclature (Nomina Anatomica Veterinaria), patterned after Nomina Anatomica for humans, was developed and adopted by the World Association of Veterinary Anatomists. The major effort in this present revision was to update the anatomical nomenclature in accordance with Nomina Anatomica Veterinaria (N.A.V.), incorporate recent literature, and make corrections. All of the illustrations have been relabeled where required to eliminate obsolete terms and enter substitutions. On several plates, structural changes have been made or entire figures replaced. Added in this revision are new chapters or sections on the classification of dogs, reproduction and development, the endocrine system, tongue, spinal cord, and eye accompanied by several new illustrations and radiographs.

Anatomy of the Dog attempts to meet the varied needs of the veterinary student, the clinician, the experimentalist, and the anatomist. The correspondence received over the intervening years is evidence that the book has served well as a reference text. The illustrations have been favorably received, and requests for reuse stand as a tribute to the dissectors and illustrators whose combined efforts produced them. In the present revision a basic anatomical understanding on the part of the reader is assumed, and an attempt is made to describe the specific morphology of the dog with minimal reference to other species.

The terms used for structures of the body are numerous, and therefore it is necessary that they be clear and precise. With this in mind, anatomical nomenclature committees have eliminated synonyms and eponyms to promote understanding and facilitate learning.

In 1950 an International Committee on Veterinary Anatomical Nomenclature was established by the World Association of Veterinary Anatomists. In 1968 they issued Nomina Anatomica Veterinaria, which has since been revised in a second edition (1973) and updated by emendation in 1975 and 1978.

The terminology in this text follows the list of approved terms in N.A.V.

The following principles served as guidelines in the work of the Committee:

1. Each anatomical concept should be designated by a single term.
2. Each term should be in Latin.
3. Each term should be as short and simple as possible.
4. The terms should be easy to remember and should have instructive and descriptive value.
5. Structures that are closely related topographically should have similar names, as *Arteria femoralis, Vena femoralis, Nervus femoralis.*
6. Differentiating adjectives should generally be opposites, as *major* and *minor.*
7. Terms derived from proper names (eponyms) should not be used.

The directional terms cranial and caudal apply to the neck, trunk, and tail as well as to the limbs as far distally as the end of the antebrachium and crus. The terms dorsal and palmar are used on the manus; dorsal and plantar on the pes. On the head the terms rostral, caudal, dorsal, and ventral are preferred. Only in a few locations such as the jaws, eye, and inner ear are such terms as anterior, posterior, superior, or inferior used. Medialis and lateralis apply to the whole body except on the digits, where axialis and abaxialis refer to the sides of the digit toward the axis of the limb or away from the axis of the limb, respectively. The axis of the limb passes through the third digit. Structures are generally designated by the anglicized forms in common use. Each term, when introduced for the first time in the text, is followed by its Latin equivalent.

ACKNOWLEDGMENTS

The completion of this revision at the present time would not have been possible without the cooperation of the following contributing authors:

F. KAREEM AL-BAGDADI, B.V.M.S., M.S., Ph.D., Assistant Professor, Department of Veterinary Anatomy and Fine Structure, Louisiana State University, Baton Rouge, Louisiana: "Integument."

GREGORY A. CHIBUZO, D.V.M., M.S., Associate Professor, Department of Anatomy, School of Veterinary Medicine, Tuskegee Institute, and Visiting Lecturer, Department of Anatomy, College of Veterinary Medicine, Cornell University, Ithaca, New York: "The Tongue."

THOMAS F. FLETCHER, D.V.M., Ph.D., Professor, Department of Veterinary Biology, College of Veterinary Medicine, University of Minnesota, St. Paul, Minnesota: "The Spinal Cord and Meninges."

ROBERT GETTY, D.V.M., M.S., Ph.D., Late Professor and Head, Department of Veterinary Anatomy, Iowa State University, Ames, Iowa: "The Ear."

RONALD L. HULLINGER, D.V.M., Ph.D., Associate Professor, Department of Anatomy, School of Veterinary Medicine, Purdue University, West Lafayette, Indiana: "The Endocrines."

JAMES LOVELL, D.V.M., M.S., Ph.D., Retired Professor and Head, Department of Veterinary Anatomy, College of Veterinary Medicine, University of Illinois, Urbana, Illinois: "Integument."

ROBERT C. MCCLURE, D.V.M., Ph.D., Professor, Department of Veterinary Anatomy-Physiology, College of Veterinary Medicine, University of Missouri, Columbia, Missouri: "The Cranial Nerves."

HERMANN MEYER, Dr. Med. Vet., Ph.D., Professor, Department of Veterinary Anatomy-Physiology, College of Veterinary Medicine, University of Missouri, Columbia, Missouri: "The Brain."

Roy V. H. Pollock, D.V.M., Lecturer, Department of Anatomy, College of Veterinary Medicine, Cornell University, Ithaca, New York: "The Eye."

Victor Rendano, V.M.D., Assistant Professor of Radiology, Department of Clinical Sciences, College of Veterinary Medicine, Cornell University, Ithaca, New York: Radiographs for various chapters.

Melvin W. Stromberg, D.V.M., Ph.D., Professor and Head, Department of Anatomy, School of Veterinary Medicine, Purdue University, West Lafayette, Indiana: "Autonomic Nervous System."

Most of the illustrations have been prepared by Marion Newson, Medical Illustrator in the Department of Anatomy from 1951 to 1972. Her skill as an illustrator, her knowledge of anatomy, and her concern for accuracy proved invaluable. Several drawings were made by other illustrators in the Department of Anatomy, including Pat Barrow (1947–50), Lewis Sadler (1973–76), and William Hamilton IV, the present Medical Illustrator. There are also illustrations by Robert R. Billar, Dan Hillman, Santiago B. Plurad, Algernon Allen, and Dave Williams. Karen Allaben Confer and Barbara Pollock aided in lettering the plates. Permission to use illustrations which appeared in *Das Lymphgefasssystem des Hundes* by Baum (1918) was kindly granted by Springer Verlag. Paragraphs in the text on "Muscles" have been freely translated and adapted from Baum-Zietzschmann's *Anatomie des Hundes* (1936) by permission of Paul Parey Verlag.

Many students and colleagues have made suggestions for corrections and additions which are appreciated, particularly Jon McCurdy, Robert Worthman, George G. Stott, and Michael Shively. Special thanks are due Alexander deLahunta and John Cummings, whose advice I frequently sought. The faculty and the staff of the New York State College of Veterinary Medicine have cooperated to provide specimens and literature when required. Edward C. Melby, Jr., Dean of the College, has supported my efforts by permitting time and facilities to be devoted to the research and writing required. I am pleased to have the moral support of Mrs. Mary Miller Ewing.

My coauthor, Dr. George C. Christensen, with whom I shared the work on the first edition, has encouraged my efforts in this revision, although his duties as Vice President for Academic Affairs at Iowa State University have not allowed time for active participation in anatomical matters. Dr. Christensen, a former graduate student of Dr. Miller, was Associate Professor of Veterinary Anatomy at Iowa State University, Head of Veterinary Anatomy at Purdue University, and then Dean of the College of Veterinary Medicine at Iowa State University before assuming his present position.

I have enjoyed the constant encouragement and assistance of my wife, Erica, whose patient support and help with translations have expedited completion.

My relations with the publishers have been most cordial, and I wish to acknowledge their prompt advice and cooperation, particularly in coordinating this revision with the revision of *Miller's Guide to the Dissection of the*

Dog which shares some of the illustrations. Mr. Carroll Cann, of the W. B. Saunders Company, graciously pursued me during periods of inertia and prepared the necessary groundwork for production schedules with Laura Tarves and Ray Kersey. I have also had the assistance of Robert Reinhardt and copy editor Bonnie Breme, for which I am grateful.

HOWARD E. EVANS

CONTENTS

Chapter 1

CLASSIFICATION AND NATURAL HISTORY OF THE DOG

By HOWARD E. EVANS

This text is based upon anatomical studies of many dogs of diverse and usually unknown ancestry commonly referred to as mongrels. The degree of anatomical similarity between domestic and wild canids, and particularly among mongrels is considerable, although their external morphology differs.

Taxonomically the dog is in the

Order Carnivora
  Family Canidae
    Genus Canis
      Species familiaris

THE ORDER CARNIVORA

The order Carnivora, whose natural history and behavior is discussed so well by Ruth Ewer in *The Carnivores* (1973, Cornell University Press), includes an assemblage of intelligent, flesh-eating mammals with prominent canine teeth, molars adapted for crushing and cutting, and a relatively short alimentary canal. Members of the order have toes provided with claws and behavioral characteristics which identify them as predators with strong family ties, devoted to the care of their young. Many of the species adapt readily to domestication.

The Carnivora are worldwide in distribution and presumably evolved from stock represented by the late Cretaeceous genus *Cimolestes* 70 million years ago. Later relatives, the Creodonts, appeared in the Eocene 40 million years ago, and by Oligocene time 20 million years ago the first canids evolved from Miacoid ancestors like *Cynodictis* (Colbert 1953, McKenna 1969).

The seal, sea lion, and walrus (Pinnipedia) are no longer considered a formal taxon. It is believed (Mitchell and Tedford 1973) that the Pinnipedia are polyphyletically derived from ursids and mustelids and that the aquatic families should be arranged with their terrestrial relatives within the Canoidea.

The living species of Carnivores can be arranged in two superfamilies: Canoidea — the dog group (also called Arctoidea) — and Feloidea — the cat group (also called Aeluroidea) (Simpson 1945) (Fig. 1–1).

1

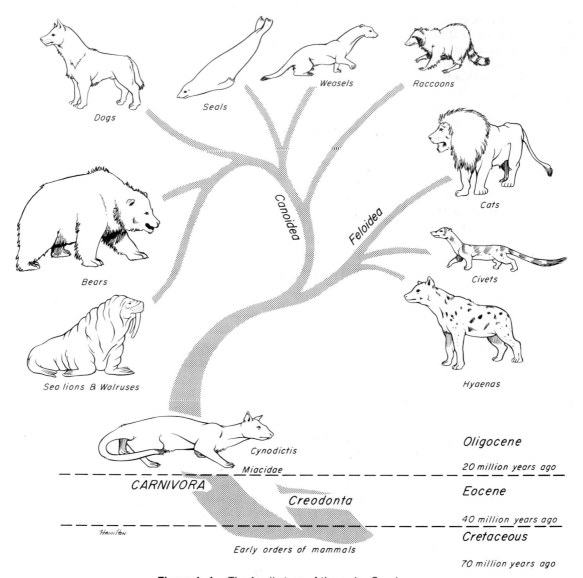

Figure 1–1. The family tree of the order Carnivora.

Canoidea

Canidae	Dogs
Ursidae	Bears
Otariidae	Sea lion and walrus
Mustelidae	Weasels
Phocidae	Seals
Procyonidae	Raccoons

Feloidea

Felidae	Cats
Viverridae	Civets
Hyaenidae	Hyaenas

THE FAMILY CANIDAE

The family Canidae is a distinct group of dog- and foxlike animals distributed throughout the world. (Mivart 1890, Bueler 1974, Fox 1975, Fiennes 1976). Although there is general agreement about which genera are included in the family, there has been disagreement regarding the generic status of several species and their grouping into subfamilies. Clutton-Brock, Corbet, and Hills (1976) reviewed the family Canidae using numerical methods and suggested a classification at the generic level based upon 90 characters of the skeleton, pelage, internal anatomy, and behavior. Their results indicated that the three largest genera, *Canis* — dog, wolf, coyote, jackal; *Vulpes* — foxes; and *Dusicyon* — South American foxes and foxlike animals, are all closely related but merit the separate designations they now have. Confirmed by this study were the isolated taxonomic positions of the monospecific genera of the African hunting dog *(Lycaon)*, the South American bush dog *(Speothos)*, and the Asiatic red dog or Dhole *(Cuon)*. They concluded that the monospecific genus of the Arctic fox *(Alopex)*, although still deserving of separate status, is clearly related to other foxes of the genus *Vulpes*. They suggested that the gray fox *(Urocyon)* and fennec fox *(Fennecus)* should be included in *Vulpes,* whereas the common zorro *(Cerdocyon)* and small-eared zorro *(Atelocynus)* should be included in the South American fox genus *Dusicyon*. The classification of the family Canidae by Clutton-Brock et al (1976) included 37 species in 10 genera. The dental formula for almost all species of the family Canidae is $I\frac{3}{3} C\frac{1}{1} P\frac{4}{4} M\frac{2}{3}$. Exceptions are the Bat-eared fox *(Otocyon)* $\frac{3}{3}\frac{1}{1}\frac{4}{4}\frac{3}{4}$, the South American bush dog *(Speothos)* $\frac{3}{3}\frac{1}{1}\frac{4}{4}\frac{1}{2}$, and the Asiatic dhole *(Cuon)* $\frac{3}{3}\frac{1}{1}\frac{4}{4}\frac{2}{2}$.

A subsequent study of the classification of the family Canidae by Van Gelder (1978) suggests a compromise between the classifications of Langguth (1975) and Clutton-Brock et al. (1976). Van Gelder's arrangement recognizes six monotypic genera and one polytypic genus, *Canis*. The six genera with only one species are:

Chrysocyon	Maned wolf (South America)
Speothos	Bush dog (South America)
Nyctereutes	Raccoon dog (Asia)
Cuon	Dhole (Asia)
Lycaon	Hunting dog (Africa)
Otocyon	Bat-eared fox (Africa)

The genus *Canis* includes eight subgenera and all of the remaining species of the family Canidae. The subgenera recognized are:

Canis (Canis) spp.	Dog, wolf, coyote, jackal
Canis (Dusicyon) sp.	Falkland Island wolf
Canis (Pseudalopex) spp.	Foxes
Canis (Lycalopex) sp.	Hoary fox
Canis (Cerdocyon) sp.	Common zorro
Canis (Atelocynus) sp.	Small-eared zorro
Canis (Vulpes) spp.	Foxes
Canis (Alopex) sp.	Arctic fox

The domestic dog *Canis familiaris* is taxonomically conspecific with the wolf *Canis lupus* and is not considered a separate species by some workers (Groves 1971, Clutton-Brock et al 1976). *Canis familiaris* is the type species of the genus designated by Linnaeus in 1758, although it is generally recognized that *Canis lupus,* the wolf, is indistinguishable anatomically and as the wild stock would more properly serve as the type for the genus. *Canis dingo,* a native of Australia, is generally regarded as a distinct feral domestic dog. Hybrids occur in the wild between several species (wolf and dog, dog and coyote) and between several genera or subgenera such as *Vulpes* and *Alopex, Vulpes* and *Canis, Pseudalopex* and *Canis* (Van Gelder 1977).

The subgenus *Canis (Canis)* includes the wolf, coyote, jackal, and domestic dog. All interbreed and may raise hybrid pups which can backcross with either of the parent stocks. Van Gelder (1978) recognizes eight species in the subgenus Canis:

Canis (Canis) lupus	Wolf
Canis (Canis) laterans	Coyote
Canis (Canis) rufus	Red wolf
Canis (Canis) aureus	Golden jackal
Canis (Canis) adustus	Side-striped jackal
Canis (Canis) mesomelas	Black-backed jackal
Canis (Canis) simensis	Ethiopian jackal

The skull in all of the canids is strongly built, often with a temporal crest, especially in the male, and bears robust canines, medium-sized incisors, and large upper fourth premolar teeth. Their limbs are adapted for endurance running, and their behavior is that of an alert, active predator with strong family or pack ties, marked sociality, and domesticability.

Domestic dogs are present in all parts of the world, including oceanic islands, and have been continuously associated with human culture. The dog of the American Indians was introduced from Asia either across the Bering Strait or via rafts with ancient man. It was formerly thought that the domestic dog of America arrived about 1000 B.C. (Haag 1948), but recent finds in Illinois (6000 B.C.) and Idaho (8400 B.C.) indicate that it was much earlier. Epstein (1971), after studying the circumglobal as well as the African origins of the dog, concluded that domestication of the dog took place in southwest Asia in early Neolithic times, about 10,000 B.C.

Several domestic dog breeds of today are very similar to those of 3500 B.C. as we know them from paintings, sculptures, and ancient writings. Human interest in dogs stemmed from a variety of circumstances which made the association of humans and dogs beneficial to one or the other or both: hunting needs, companionship propensities, and guarding against predators were common to both; when dogs were eaten, sacrificed, or used as rat killers, it was a one-sided relationship. It was to the dog's advantage when a warm place to sleep and excess food were available in exchange for sociality.

The relative geographical isolation of early human populations led to the development of many local breeds utilized for particular purposes. When improved transportation allowed more population mobility and voyages of discovery, dogs were exchanged more frequently and although many were bred for their characteristic features, it is likely that the mongrel population increased tremendously, especially in large cities, allowing the gene pool of every geographical area to incorporate the features of many domestic breeds. When Darwin wrote about animals under domestication in 1868, he recognized about 100 breeds of dogs and wondered whether the domesticated varieties were descended from a single species or several. His remarks on various native dogs encountered on the voyage of the *Beagle* are still pertinent today.

Borgannkar et al (1968) found no significant differences in the number of chromosomes, 78, or their morphology among the European beagle, Shetland sheep dog, African basenji, and Malayan telomian. Epstein (1971) concluded that the pariah dogs of Africa trace their genealogy to several types of dogs introduced into Africa in prehistoric times, in addition to the indigenous Greyhound. For a history of the domestic dog see Ash (1927), Zeuner (1963), and Epstein (1971).

Since all domestic dogs are interfertile, many intermediate physiognomies which are unassignable to any particular breed are seen in mongrels. Even first-generation hybrids of purebred parents may exhibit bizarre combinations of body characteristics. Charles Stockard in 1937 undertook an investigation of the genetic and endocrinic basis for differences in form and behavior in dogs. He made crosses between many breeds and between their F_1 and F_2 generations. The monograph documenting this study (Stockard 1941) contains photographs of the physiognomy of many hybrid litters and numerical comparisons of their skeletal and body proportions.

An extensive collection of reference materials on dogs (over 3000 titles) is housed in the Library of the College of William and Mary in Williamsburg, Virginia, as the Peter Chapin Collection of Books on Dogs (Chapin 1938).

THE BREEDS OF DOGS

There are about 300 breeds of dogs in the world, and each country has its own distinctive dogs, as well as introduced forms. The Chinese were probably the earliest breeders of purebred dogs, several of which are still popular. Often there are several names for the same breed, since the name may change when the dog is introduced into another country; thus the Borzoi is also known as the Russian Wolfhound, the German Shepherd as an Alsatian, the Vizsla as the Hungarian Pointer, the Scottish Deerhound as the Irish Wolfhound, the Great Dane as the German Mastiff, the Chow as the Canton dog, and so on.

Several breeds are well known in one country and almost unheard of in another, such as the Chinese Crested dog, the Finnish Hound, the Danish Pointer, the Chinese Shar-Pei, and the Spanish Greyhound.

A breed is any group of animals derived
Text continued on page 11.

GROUP I.   SPORTING DOGS

Retriever

Setter

Pointer

Spaniel

Figure 1-2.   There are 24 breeds of sporting dogs.

Retrievers: Labrador, Golden, Flat-coated, Curly-coated, Chesapeake Bay
Setters: Irish, Gordon, English
Pointers: English, German wirehaired, German shorthaired
Spaniels: Brittany, Clumber, Cocker, English Cocker, English Springer, Field, Sussex, Welsh Springer, American Water, Irish Water
Vizsla
Weimaraner
Wirehaired Pointing Griffon

GROUP II.   HOUNDS

Figure 1–3.   There are 19 breeds of hounds.

Afghan	Greyhound
Basset	Harrier
Beagle	Irish Wolfhound
Basenji	Norwegian Elkhound
Black and Tan Coonhound	Otterhound
Bloodhound	Rhodesian Ridgeback
Borzoi	Saluki
Dachshund	Scottish Deerhound
American Foxhound	Whippet
English Foxhound	

GROUP III.   WORKING DOGS

Collie

Malamute

St. Bernard

German Shepherd

Corgi

Doberman

English Sheepdog

Hamilton

Figure 1–4. There are 30 breeds of working dogs.

Akita	Collie	Old English Sheepdog
Alaskan Malamute	Doberman Pinscher	Puli
Belgian Malinois	German Shepherd	Rottweiler
Belgian Sheepdog	Giant Schnauzer	St. Bernard
Belgian Tervuren	Great Dane	Samoyed
Bernese Mountain dog	Great Pyrenees	Shetland Sheepdog
Bouvier des Flandres	Komondor	Siberian Husky
Boxer	Kuvasy	Standard Schnauzer
Briard	Mastiff	Cardigan Welsh Corgi
Bull mastiff	Newfoundland	Pembroke Welsh Corgi

GROUP IV.   TERRIERS

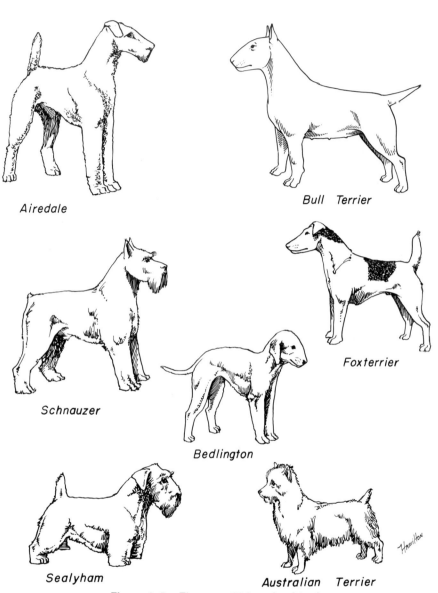

Figure 1–5.   There are 22 breeds of terriers.

Airedale	Lakeland
American Staffordshire	Manchester
Australian	Miniature Schnauzer
Bedlington	Norwich
Border	Scottish
Bull	Sealyham
Cairn	Skye
Dandie Dinmont	Soft-coated Wheaten
Fox	Staffordshire Bull
Irish	Welsh
Kerry Blue	West Highland White

GROUP V. TOY DOGS

Pug

Poodle

Pekingese

Chihuahua

Maltese

Figure 1-6. There are 16 toy breeds.

Affenpinscher	Maltese	Pug
Brussels Griffon	Miniature Pinscher	Shih Tzu
Chihuahua	Papillon	Silky Terrier
English Toy Spaniel	Pekingese	Yorkshire Terrier
Italian Greyhound	Pomeranian	
Japanese Spaniel	Poodle (miniature)	

GROUP VI.   NON-SPORTING DOGS

Chow Chow

Dalmation

English Bulldog

Schipperke

Figure 1–7.   There are 11 breeds of non-sporting dogs.

Bichon Frise	Keeshond
Boston Terrier	Lhasa Apso
Bulldog	Standard Poodle
Chow Chow	Schipperke
Dalmation	Tibetan Terrier
English Bulldog	

from a common stock and bred for their distinctive features, which are codified as the standard for the breed by those willing to recognize the breed.

The English Kennel Club recognizes only 84 breeds in two groups: sporting dogs (52) and non-sporting dogs (32). In the United States more than half of the breeds are of English ancestry, and the American Kennel Club recognizes six groups of dogs plus a miscellaneous class for a total of 130 breeds.

The six categories of dog breeds recognized by the American Kennel Club (A.K.C.) are grouped for show purposes and do not necessarily represent genetic closeness or ancestry. A few of the breeds have been moved from one group to another at the request of breed clubs. Each club originates its own standards and revisions and submits them to the A.K.C. for approval. The standard for each breed is therefore a composite of the desired features for each breed. Rarely are all the desired features found in one dog. There may be several varieties within a breed, and often features acceptable in one breed are unacceptable in another.

For the many breeds not officially recognized by the American Kennel Club the first step for registry is acceptance into the "Miscellaneous Class." This requires proof that a substantial nationwide interest in the breed exists, as shown by an active club, breed registry, and expanding breeding activity. When the American Kennel Club is satisfied that a breed is continuing to grow in number and sponsorship, it may admit the breed to registration in the Stud Book and grant the opportunity to compete in regular classes. The following eight breeds are presently recognized in the miscellaneous class:

Australian Cattle Dog
Australian Kelpie
Bearded Collie
Border Collie
Cavalier King Charles Spaniel
Ibizan Hound
Miniature Bull Terrier
Spinone Italiano

The relative popularity of the breed may vary from time to time and from country to country. As a result, the mongrel population will also change owing to the prevalence of a particular breed's acting as a sire or dam, although the change is gradual. Since there are specific characteristics associated with each breed, one can expect to see subtle changes in the anatomical features of the mongrel population. As an example one may cite the tendency toward reduction of the stylohyoideus muscle in the beagle (Evans 1959) and the loss of teeth in brachycephalic breeds. There are several specific pathologies more prevalent in some breeds than in others. For reference to these conditions see Earl (1975), de Lahunta (1977), and Kalter (1968).

The six groups of dogs recognized by the American Kennel Club and the standards for these dogs appear in *The Complete Dog Book,* 15th edition (1975). Examples of each group were chosen randomly for illustration.

BIBLIOGRAPHY

American Kennel Club. 1975. The Complete Dog Book: The Photograph, History and Official Standard of Every Breed Admitted to AKC Registration, and the Selection, Training, Breeding, Care and Feeding of Pure-bred Dogs. 15th ed. 672 pp.

Ash, E. C. 1927. Dogs: Their History and Development. 2 vols. London. (Reprinted 1972, New York, Benjamin Blom, Inc.)

Bueler, L. E. 1973. Wild Dogs of the World. New York, Stein & Day.

Chapin, H. M. 1938. The Peter Chapin Collection of Books on Dogs. Bull. Coll. William and Mary 32(7).

Clutton-Brock, J., G. B. Corbet, and M. Hills. 1976. A review of the family Canidae with a classification by numerical methods. Bull. Brit. Museum Zool. 29(3): 1–99.

Colbert, E. H. 1953. The Origin of the Dog. New York, American Museum of Natural History, Man and Nature Publications. 14 pp.

Darwin, C. 1868. The Variation of Animals and Plants Under Domestication, Vol. 1. 2nd Ed., 1900. New York, Appleton Co.

deLahunta, A. 1977. Veterinary Neuro-anatomy and Clinical Neurology. Philadelphia, W. B. Saunders Co. 439 pp.

Epstein, H. 1971. The Origin of the Domestic Animals of Africa, Vol. 1. New York, Africana Pub. Corp. 719 pp.

Evans, H. E. 1959. Hyoid muscle anomalies in the dog. *(Canis familiaris).* Anat. Rec. *133*:145–162.

Ewer, R. 1973. The Carnivores. Ithaca, N.Y., Cornell Univ. Press.

Fiennes, R., and A. Fiennes. 1968. The Natural History of the Dog. London, Weidenfeld and Nicolson.

Fiennes, R. 1976. The Order of Wolves. Indianapolis, Bobbs-Merill. 206 pp.

Fox, M. W. (ed.) 1975. The Wild Canids, Their Systematics, Behavioral Ecology and Evolution. New York, Van Nostrand Reinhold Co.

Groves, C. P. 1071. Request for a declaration modifying article so as to exclude names proposed for domestic animals from zoological nomenclature. Bull. Zool. Nom. 27:269–272.

Haag, W. G. 1948. An osteometric analysis of some aboriginal dogs. Univ. Ky. Rept. Anthropol. 7(3): 107–264.

Kalter, H. 1968. Teratology of the Central Nervous System. Chicago, Univ. Chicago Press. 483 pp.

Kleiman, D. G. 1967. Some aspects of social behavior in the Canidae. Amer. Zool. 7(2):365–372.

Langguth, A. 1975. Ecology and evolution in the South American canids. In Fox, M. W. (ed.) The Wild Canids, Their Systematics, Behavioral Ecology and Evolution. New York, Van Nostrand Reinhold Co. pp. 192–206.

Lawrence, B., and W. H. Bossert. 1976. Multiple character analysis of *Canis lupus, laterans,* and *familiaris,* with a discussion of the relationship of *Canis niger.* Amer. Zool. 7(2): 223–232.

McKenna, M. C. 1969. The origin and early differentiation of Therian mammals. Ann. N.Y. Acad Sci. *167*:217–240.

Mitchell, E., and R. M. Tedford. 1973. The enaliarctinae, a new group of extinct aquatic carnivora and a consideration of the otoriidae. Bull. Amer. Mus. Nat. Hist. *151*(3):203–284.

Mivart, St. G. 1890. A Monograph of the Canidae. London.

Simpson, G. G. 1945. The principles of classification and a classification of mammals. Bull. Amer. Mus. Nat. Hist. 85:1–350.

Titcomb, M. 1969. Dog and Man in the Ancient Pacific. Bernice P. Bishop Museum Special Publ. (Honolulu) 59:1–91.

Van Gelder, R. G. 1977. Mammalian hybrids and generic limits. Amer. Mus. Novitates *2635*:1–25.

Van Gelder, R. G. 1978. A review of canid classification. Amer. Mus. Novitates *2646*:1–10.

Walker, E. P. 1975. Mammals of the World, Vols. 1 and 2. 3rd Ed. Baltimore, Johns Hopkins Press.

Zeuner, F. E. 1963. A History of Domesticated Animals. New York, Harper & Row.

Chapter 2

REPRODUCTION AND PRENATAL DEVELOPMENT

By HOWARD E. EVANS

ESTROUS CYCLE

The normal dog has two estrous or "heat" periods a year, which may occur in any month and may vary in time of occurrence from year to year.

Sokolowski et al. (1977) found that estrous activity occurred in all seasons for seven breeds: Toy Poodle, Cocker spaniel, Basset hound, Boston terrier, German shepherd, Pekingese, and Beagle. The Basset hounds and Cocker spaniels had mean interestrous intervals of about five months, whereas the German shepherd had the shortest interestrous interval, 149 ± 28.5 days. The mean occurrence of estrus for the German shepherd was 2.4 times per year, and for the other breeds it was 1.5 times per year. Neither natural nor artificial light had any effect on estrus. These periods of sexual activity were early characterized by behavioral patterns, later by cytological changes in the vaginal epithelium, and most recently by hormonal levels in the circulating blood correlated with vaginal and ovarian cytology.

The changes through which the nonpregnant uterus passes were designated by Heape (1900) as proestrum, estrum, metestrum and anestrum. These terms were used by Evans and Cole (1931), Griffiths and Amoroso (1939), and other early investigators of the dog's estrous cycle. Later workers suggested that the phase of estrus during which the reproductive organs are mainly under the influence of progesterone should be designated as diestrus, leaving the term metestrus for the short transition stage between estrus and diestrus when the corpora lutea of the ovary are becoming functional.

Investigations by Holst and Phemister (1974), Concannon, Hansel and Visek (1975), and Concannon, Hansel and McEntee (1977) have facilitated a redefinition of the estrous cycle.

Proestrus — one to two weeks with a mean of about nine days. Characterized by swelling of the vulva, sanguineous discharge, and heightened estrogen secretion. This period ends at the time of first acceptance (coitus) of the male when luteinizing hormone (LH) is at its peak. A vaginal smear would show a predominance of red blood cells (Table 2–1).

Estrus — one to two weeks with a mean of about nine days. Characterized by declining plasma estrogens, increased plasma progesterones, and acceptance of the male. Ovulation occurs during this period, usually on the second or third day of estrus (40 to 44

13

TABLE 2–1. CELLS SEEN ON A VAGINAL SMEAR DURING ACTIVE PERIODS OF THE ESTROUS CYCLE

	Red Blood Cells	Leucocytes	Cornified Cells	Nucleated Epithelial Cells
Proestrus	++++	−	+	+++
Estrus	±	−	++++	+
Diestrus	−	++++	−	++

+ = present; − = absent.

hours after the LH peak, according to Concannon et al. 1977); or six days before the onset of diestrus, as shown by Phemister et al. (1973). For practical purposes dogs can be bred with success by one mating 24 hours after first acceptance or at any time during the first week of estrus. The causes of reproductive failure in the bitch have been reviewed by Phemister (1974).

Metestrus — a three- to five-day period during which the corpora lutea of the ovary are becoming functional. Progesterone concentration continues to increase, and the bitch still accepts the male.

Diestrus — a period of about three days characterized by a marked shift in the epithelial cell type seen in the vaginal smear. On the first day of diestrus there is a decrease by at least 20 per cent in the number of superficial (cornified) cell types and an increase of more than 10 per cent (often more than 50 per cent) in intermediate and parabasal (non-cornified) cells from the deeper layers of the vaginal epithelium (Holst and Phemister 1974). The corpora lutea are fully functional, and progesterone is dominant. The bitch may still accept the male. The onset of this period can be used as a fixed point to backdate early stages of embryonic development (Holst and Phemister 1971). Diestrus minus one day (D–1) is the stage of a two-cell cleavage; D minus three = fertilization; D minus four = formation of secondary oocyte; D minus five = beginning of meiosis; D minus six = ovulation.

Subsequent studies of the diestrus period by Holst and Phemister (1974) have shown that it is closely correlated with conception rate, litter size, and the length of gestation. They found that if the day of fertilization (diestrus minus three days) is counted as day 1, then the Beagle has a gestation period

of 60 days, and whelping takes place on day 57 of diestrus. Concannon et al. (1977) found that plasma progesterone levels in pregnant bitches decline rapidly 36 to 48 hours prior to parturition and that no pups are born until levels are below 2 ng/ml. They suggest that a prepartum decline in plasma progesterone plays an important role in parturition.

Anestrus — a period of quiescence of the reproductive organs, with a fall in circulating hormone levels and an absence of sexual behavior.

EARLY DEVELOPMENT

Ovulation in the dog occurs spontaneously early in estrus, as was shown by Bischoff in 1844. The germ cell of the dog at this time (unlike most other mammals) is a primary oocyte, since the first and second polar bodies are not formed until after fertilization. A recently ovulated oocyte with its covering of corona radiata cells can be seen with the unaided eye and is about 230 μm. in diameter. Photographs of preimplantation ova have been published by Holst and Phemister (1971).

At the time of ovulation the fimbria of the infundibulum are swollen and aid to engulf ovulated oocytes. They effectively block the slitlike opening of the ovarian bursa and thus prevent transperitoneal migration of oocytes or their loss into the peritoneal cavity.

Passage of spermatozoa through the uterus and tubouterine junction and up the uterine tube can be rapid in the dog. Bischoff (1844) noted that it takes 6, 18, or 20 hours for sperm to enter the uterine tube. Doak et al. (1967) found motile spermatozoa in high concentration in all parts of the dog's uterus for 6 days after copulation, and some were present for as long as 11 days. Holst and Phemister (1974) confirmed at least a seven-day life span of dog spermatozoa. It is thought that the fertile life of a sperm is about half of its motile life, and this long fertile period may be responsible for the high conception rate in dogs.

The nuclear chromosomes of the sperm become the male pronucleus which fuses with the female pronucleus of the ovum to determine the genetic constitution of the zygote. The dog has a great number of small chromosomes, which makes it difficult to demonstrate all of them in a single preparation even after cell culture. Moore and Lambert (1963) illustrated and described the karyotype of the Beagle from an analysis of several metaphase plates taken from cultured kidney cells. They confirmed the findings of Minouchi (1928) and Ahmed (1941) that the diploid number in somatic cells of the dog was 78 in both sexes. The chromosomes appear to be aligned into 38 homologous pairs of autosomes and two sex chromosomes. All of the autosomes have acrocentric or terminal centromeres. The X chromosome is one of the largest, while the Y chromosome is equal to the smallest. The two X chromosomes in the female have metacentric centromeres. For a discussion of genetics in the dog, see Asdell (1966).

PRENATAL PERIODS

The successive stages of cleavage, gastrulation, implantation, and early somite formation are completed before day 20 of gestation. The length of time required for the entire development of a fertilized oocyte into a newborn puppy is approximately 60 days, and it is no wonder that marked changes in development can be seen at daily intervals. Phemister (1974) divided prenatal development in the dog into three periods: (1) *the period of the ovum* following fertilization is characterized by a zygote which lies free in the uterine tube and migrates to the uterus (day 2 to 17); (2) *the period of the embryo* begins with implantation of the blastocyst and ends with the completion of major organogenesis (day 19 to 35); (3) *the period of the fetus* is the time during which the characteristic features of the dog appear and most of the growth occurs (day 35 to birth).

THE OVUM

Fertilization takes place in the upper part of the uterine tube (oviduct), and the fertilized oocyte (zygote) begins to divide within a few hours. Tubal transport of ova takes longer in the dog (7 to 10 days) than in

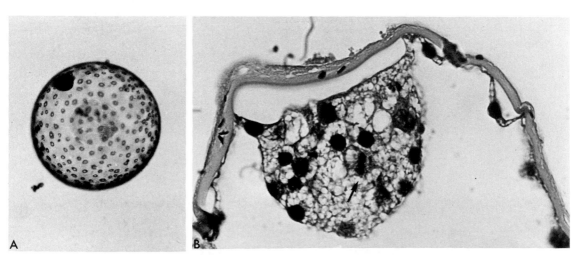

Figure 2–1. *A.* Dog blastocyst about 600 μm. in diameter, 11 days after breeding. The inner cell mass is dense and globular in form. Note the trophoblast cells. Unstained, ×90. (From Holst and Phemister 1971.)

B. Section of a 500 μm. dog blastocyst showing the inner cell mass. Note mitotic figure (at the *arrow*) and several spermatozoan heads embedded in the zona pellucida. Eleven days after breeding. Stained with hematoxylin and eosin, ×460. (From Holst and Phemister 1971.)

most other mammals (3 to 4 days), although some embryos reach the region of the tubouterine junction in about 5 days. Holst and Phemister (1971) found that developing embryos may enter the uterus as early as the 16-cell stage but more commonly as morulae or even early blastocysts. This agrees with the observations of Anderson (1927) and Van der Stricht (1923). Estrogen inhibits the transport of developing eggs into the uterus, whereas progesterone enhances it.

Within the uterus the morula develops into the rather spherical blastocyst with an inner cell mass and a thin surface trophoblast surrounded by a zona pellucida (Fig. 2–1). Unattached blastocysts range in size from 250 μm. to 3 mm. as they pass along the uterine lumen prior to implantation on about day 17 after breeding. The free-floating blastocyst stage lasts about seven days (Gier 1950; Tietz and Selinger, 1967; Holst and Phemister, 1971). There is usually an even spacing of blastocysts in each uterine horn, but there is often a marked difference in the number of oocytes ovulated from each ovary, as ascertained by the number of corpora lutea seen. Thus oocytes ovulated by one ovary and fertilized in the upper end of the uterine tube on that side may migrate through the body of the uterus and ascend the opposite uterine horn to implant.

The path taken by ovulated oocytes from one ovary to the opposite uterine horn in the mammal depends upon the morphology of the genital tract. External, or transperitoneal, migration is possible in those species which lack an ovarian bursa and have ovaries in close proximity to one another. The process involves ovulation into the peritoneal cavity in close proximity to the ostia of both uterine tubes, thus allowing either to engulf and transport the oocyte. The distance traveled within the peritoneal cavity is short and aided by ciliary action in the region of the infundibulum. Internal or transuterine migration is the more common phenomenon in mammals. It frequently occurs even in species which are anatomically capable of transperitoneal migration and always occurs in those species which have an ovarian bursa surrounding the ovary. In animals with duplex uteri, such as the rabbit, where each uterine horn enters the vagina separately, transuterine migration is not possible.

The early studies of Bischoff (1844) on the dog called attention to the phenomenon of transuterine migration, as did later studies in the cat by Hill and Tribe (1924) and other carnivores by Boyd, Hamilton, and Hammond (1944). Evans (1956) examined the ovaries and uterine horns of 13 bred Beagles for evidence of transuterine migration

TABLE 2–2. TRANSUTERINE MIGRATION OF BLASTOCYSTS IN 13 BEAGLES

Number of Oocytes Ovulated from Each Ovary		Number of Implantations in Each Uterine Horn		Probable Number That Migrated	Number of Blastocysts or Early Embryos Lost
Left	Right	Left	Right		
7	1	4	3	2 L to R	1
1	6	4	3	3 R to L	0
7	3	4	3	None	3
3	6	4	3	1 R to L	2
4	2	2	4	2 L to R	0
2	4	3	3	1 R to L	0
3	4	2	4	None	1
4	6	6	4	2 R to L	0
5	3	3	5	2 L to R	0
3	6	3	6	None	0
0	7	3	4	3 R to L	0
2	3	2	1	None	2
3	4	3	4	None	0
44	55	43	47		9
99		90		6 L to R 10 R to L	

and found that in 8 dogs a total of 16 morulae or blastocysts had crossed from one uterine horn to the other. There was no way of telling how many others had changed sides without affecting the final balance, since no marker was used to distinguish right from left ovulations (Table 2–2).

The spacing of blastocysts along the uterine horn is the result of ciliary action, muscular contraction, and unoccupied sites of uterine receptivity. Irregularities in spacing caused by the early death of a blastocyst or embryo are adjusted by differential hypertrophy of the uterus and resorption of the dead conceptus.

Summary of Morphological Events During the First Trimester of Development

Day 1–7 Formation of the zygote and cleavage in the uterine tube.
8 8- to 16-cell morula.
9 16- to 32-cell morula.
11 32- to 64-cell morula.
12–16 Blastocyst enters the uterus and increases in size while migrating to a receptive site. Inner cell mass distinct.
17–18 Implantation by the trophoblast, zona pellucida shed, formation of the neural plate and primitive streak.
19–20 Formation of the first somites.

THE EMBRYO

The implanting blastocyst on day 18 of gestation lies in a pear-shaped cavity of the uterus that is hardly discernible upon external examination (Fig. 2–2). The lumen between successive implantations is very narrow, and, therefore, implantation sites can be readily identified upon dissection. Fluid which collects within the blastocyst causes it to expand and fill the small uterine cavity. The attachment process, referred to as central implantation, is a superficial apposition of trophoblast to the antimesometrial surface of the uterine endometrium. The trophoblast cells are responsible for initiating formation of the placenta, and in the dog, they elaborate the enzymes which erode the maternal tissues to form the definitive endotheliochorial deciduate zonary placenta.

Barrau et al. (1975) studied early implantation in the dog and found the invasive

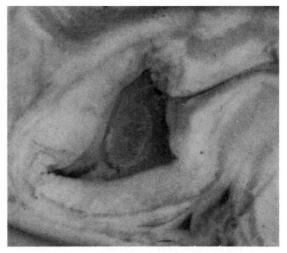

Figure 2–2. Blastocyst within the implantation chamber on the eighteenth day in the Beagle.

form of trophoblast to be syncytial, as did Schoenfeld (1903), rather than cellular, as reported by Amoroso (1952). Three types of trophoblast are ultimately recognizable:

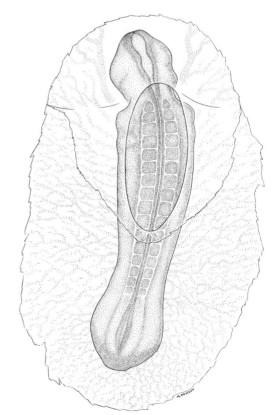

Figure 2–3. Dorsal view of a 16-somite dog embryo being enclosed by the amniotic fold.

(1) syncytial trophoblast around maternal blood vessels; (2) cytotrophoblast in the necrotic zone; and (3) phagocytic cytotrophoblast around the hematomata. In addition to the variety of absorptive cells present in the dog's placenta, there are occasional decidual cells associated with the trophoblast of the labyrinth (Anderson 1969). In some species the trophoblast is primarily nutritive and hormone-producing, while in others it is very invasive. There is evidence of tissue-specific antigens in the trophoblast (see Billington 1971).

The embryo develops in a cephalocaudal sequence, beginning with head folds, then neural tube closure, followed by somite formation, appearance of branchial arches, lens placode, otic placode, cardiac bulge, and the growth of limb buds.

Several days after implantation, the developing embryo is enveloped by amniotic folds which grow across its dorsal surface (Fig. 2–3). At the 21-day stage (16 somites, 4.5 mm.), the embryo is crescent-shaped and the yolk sac, which is continuous with the midgut, rapidly fills the chorionic vesicle (Fig. 2–4). Where the trophoblastic ectoderm and yolk sac endoderm are in apposition on the mesometrial wall of the uterus, it is called the bilaminar omphalopleure. Closer to the embryo, where the extraembryonic mesoderm and its vitelline vessels extend between the trophoblast and the yolk sac endoderm, a trilaminar omphalopleure or choriovitelline membrane is formed which serves as a transitory placenta.

The embryonic membranes of the dog include a chorion, yolk sac, amnion, and allantois which grow and change their relationships to each other during the early embryonic period.

The amnion of the dog develops by outfolding after the embryo is well organized. This is unlike the human, in whom the amnion develops by cavitation of the inner cell

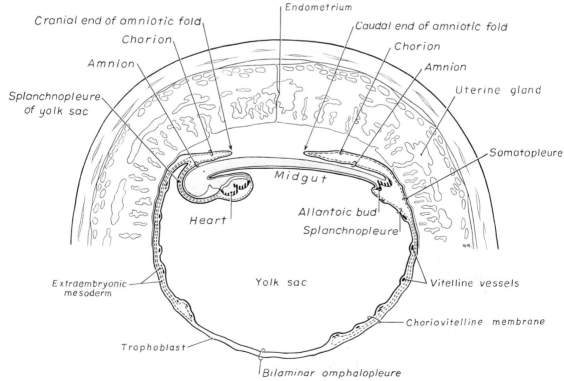

Figure 2–4. Early somite dog embryo, approximately 16 somites, 4.5 mm., 21 days. A diagrammatic transverse section of a portion of the uterus and placenta which passes through the longitudinal axis of the embryo. The large yolk sac fills the uterine cavity, and extensions of extraembryonic mesoderm are seen between the yolk sac and the chorion. The amnion forms by the folding of somatopleure over the dorsum of the embryo, and the allantoic bud in the hindgut region makes its appearance.

mass in the presomite embryo prior to the establishment of germ layers. By day 21 in the dog, the amniotic folds are well developed over the dorsum of the embryo, and the vitelline vessels are extensive (Fig. 2–5). The amnion fills with fluid as the embryo grows, and in later fetal stages becomes a sizable sac. By day 23, when the

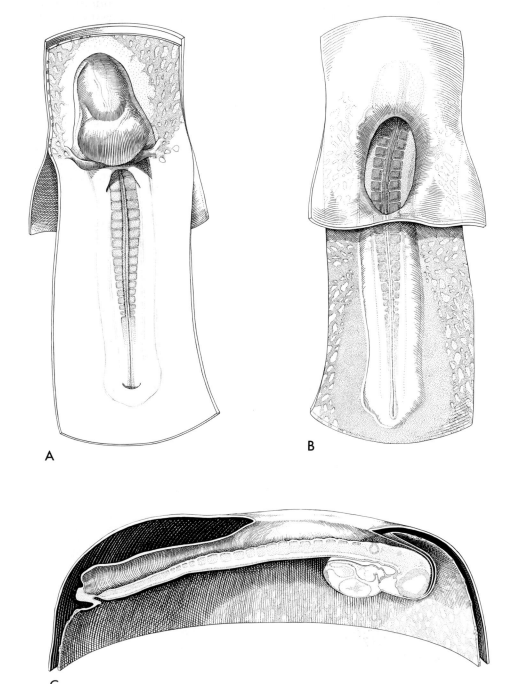

A

B

C

Figure 2–5. A 19-somite dog embryo showing closure of the amniotic fold and development of the vitelline circulation.

    A. Ventral view.
    B. Dorsal view.
    C. Sagittal section.

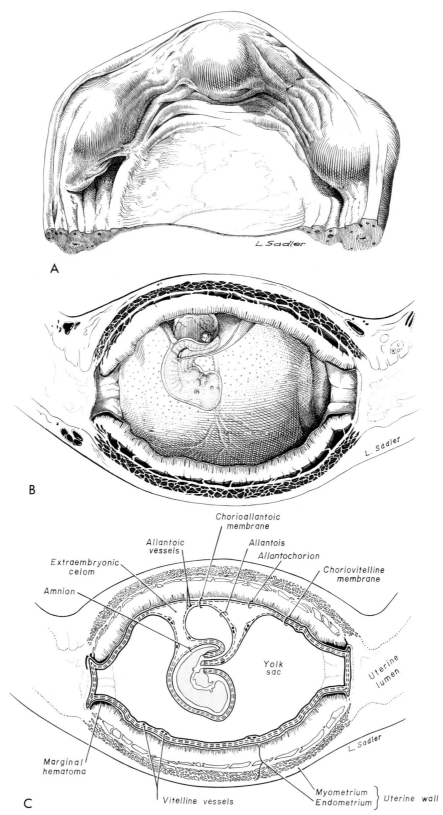

A

B

C

Figure 2–6. Uterus and early limb-bud embryo at 23 days of gestation.
   A. External view.
   B. Longitudinal section with a 32-somite embryo in situ.
   C. Schematic section of B showing fetal membranes and placenta.

20

uterine swellings are almost spherical (Fig. 2–6A), the embryo has about 32 somites and is about 5 mm. long. The yolk sac is very extensive, closely applied to the chorion, and well vascularized (Fig. 2–6B). The vas-cularity, as well as the size of the yolk sac, increases throughout gestation, although it is superseded functionally by the allantois on about day 25 of gestation.

The allantois, a diverticulum of the hind-

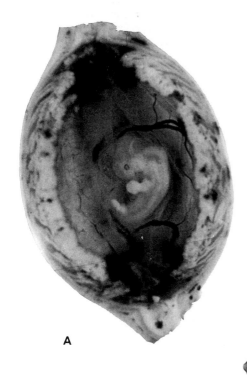

A

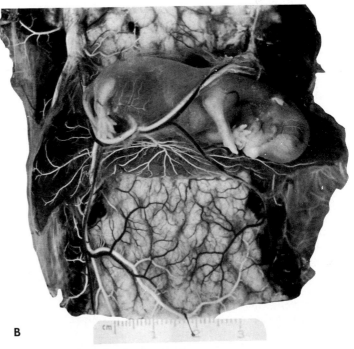

B

Figure 2–7. Placental vessels and marginal hematomata.
A. Longitudinal section of a uterine swelling on the twenty-fifth day of gestation.
B. A mongrel fetus and placenta of about 35 days of gestation.

gut, is first visible as a bud on day 21 and by day 23 it is a spherical sac located beneath the caudal end of the embryo and projecting into the exocelom (Fig. 2–6C). As it progressively expands, the allantois contacts the chorion (somatopleure) over most of the exocelomic surface, and allantoic blood vessels can be seen on its surface. By day 25 the allantois has intercalated itself between the chorion and the body of the yolk sac, thus completing the formation of the definitive chorioallantoic membrane. The enveloped yolk sac is separated from the chorion except at the poles, where it remains attached.

The choriovitelline placenta that exists from day 21 to 23 has simple villi which project into the openings of the endometrial glands. The villi are avascular and thus likely serve a histiotrophic function, as suggested by Wimsatt (1974), who studied the 23-somite black bear.

As the placenta grows, the uterine swelling changes from pear-shaped (20 days) (Fig. 2–2) to spherical (25 days) to ovoid (30 days) (Fig. 2–7A). During this time the thickening of the uterine glands and the rapid expansion of the allantoic sac result in a change from a primary choriovitelline (yolk sac) placentation to the definitive chorioallantoic placentation. Embryos appear most vulnerable to physiologic stress during this period of rapid growth and placental change. Pressures within the uterus are greatest when the allantoic sac has expanded and the uterine swellings are spherical or beginning to elongate on day 25 of gestation. Death of the embryo prior to day 25 usually results in complete resorption with little or no visual evidence at term of the placental site. Death of early embryos is manifest by a reduced litter size but can be confirmed only by examination of the interior of the uterus.

Most carnivores have marginal or central blood sinuses or hematomata associated with the placenta. The dog has a marginal "green band" which can first be recognized in early limb-bud stages as a slightly thickened region between the choriovitelline membrane and the endometrium at the constricted ends of each uterine loculus (Fig. 2–6C). This incipient marginal hematoma thus appears simultaneously with the expansion of the allantoic sac but prior to allantoic contact in the marginal region. The green band at each end of the zonular placenta increases in size throughout gestation (Fig. 2–7A and B). Occasionally, there are isolated hematomatous patches in the normal dog placenta. These structures have been called hemophagous organs by Creed and Biggers (1963). Hematomas form by necrosis of both the maternal and fetal tissues. Maternal vessels hemorrhage into the necrotic area and form pools of blood with a greenish color. These hematomata are surrounded by trophoblast rather than being between cytotrophoblast and maternal endometrium (Barrau et al. 1975).

Summary of Morphological Events During the Second Trimester of Development

17–18 days
Bitch: Uterus same size as in a pseudopregnant dog.
Embryo: Blastocysts evenly spaced from one another; trophoblastic attachment; zona pellucida shed; primitive streak visible on neural plate.

20 days
Bitch: Implantation chamber of the uterus is a pear-shaped cavity.
Embryo: 8 somites; 4 mm. long; neural tube closing.

21 days
Bitch: Uterus slightly enlarged at placental sites.
Embryo: 16 somites, 5 mm. long and crescent-shaped; longitudinal axis of embryo transverse to uterine horn; head flexed invaginating yolk sac; cardiac bulge prominent; allantoic bud developing; amniotic folds closing; yolk sac fills uterine cavity and is widely confluent with the midgut; branchial arches I and II present.

23 days
Bitch: Uterine swellings distinct; vascular and glandular layers of the uterus hypertrophied (Fig. 2–6A, B, and C).
Embryo: 32 somites; 10 mm. long; twisted with cranial end invaginated into yolk sac; forelimb bud promi-

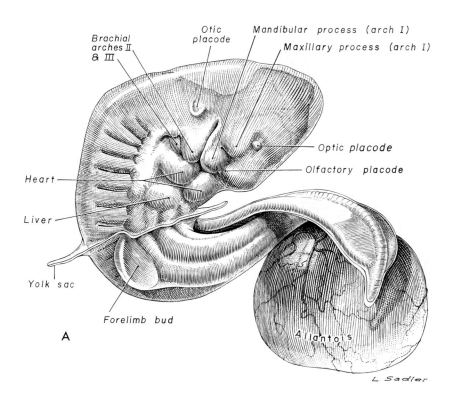

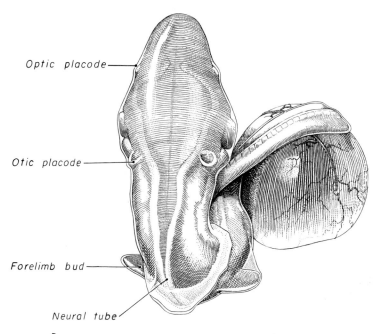

Figure 2–8.   Early limb-bud embryo at 23 days of gestation.
A. Lateral view.
B. Dorsal view.

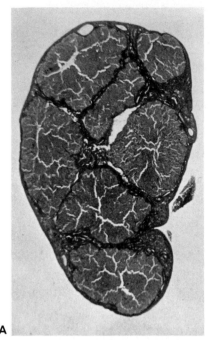

Figure 2–9. A. Ovary with seven corpora lutea. B. Beagle embryo at 25 days of gestation.

nent; otic placode and lens placode present; mandibular and maxillary processes distinct; branchial arches I, II and III present; liver bulge marked; allantoic sac spherical and beneath tail (Fig. 2–8).

25 Days
Bitch: Corpora lutea completely fill the ovary (Fig. 2–9A); uterine swellings almost spherical, 30 × 35 mm.; width of placenta about 29 mm.

Embryo: 14 mm.; cephalic flexure prominent; otic pit and eye well formed; mammary ridge present (Fig. 2–9B); limb buds at plate stage; vertebral elements condrify; dental lamina forms.

28 Days
Embryo: 17 mm.; first ossification seen in mandible, maxilla, frontal, and clavicle.

30 Days
Bitch: Uterine swellings 33 × 50 mm.; width of placenta is greater than length of embryo (Fig. 2–10A).
Embryo: 19 mm.; eyelids and external ear forming; sensory hairs on snout, chin, and eyebrow; intestine

herniated into umbilical stalk; five pairs of nipples; digits on forelimbs distinct; genital tubercle prominent (Fig. 2–10B).

33 Days
Embryo: 27 mm.; ossification of nasal, incisive, palatine, zygomatic, and parietal bones; midshaft of ribs 4 through 10; midshaft of humerus, radius, and ulna; femur, tibia, and fibula; canine teeth in early cap stage; palatal shelves fuse; digits on hindpaws distinct (Fig. 2–10C).

AGE DETERMINATION

Various external and internal features have been used as criteria for determining chronological age. External features include somite count; branchial arch development; eye, ear, and nose formation; limb-bud development; and linear measurements. Internal features include maturation of various organs, stages of tooth development, and, most commonly, the appearance of ossification centers.

Streeter (1951) divided the human embryonic period into 23 "horizons" based on both external and internal features. Following the twenty-third horizon was the fetal period, characterized in the human by the

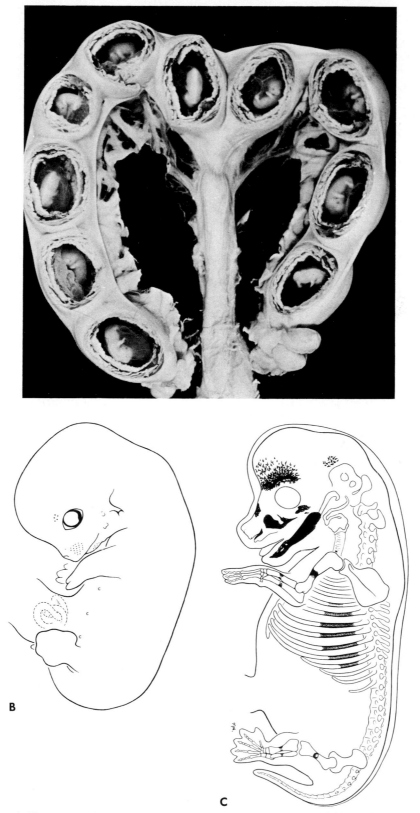

Figure 2–10. A. The uterus of a nine-month-old Beagle with 10 embryos on the thirtieth day of gestation.
B. External features of a 30-day Beagle embryo.
C. Ossifications in a 33-day, 27-mm. embryo.

TABLE 2–3. CROWN-RUMP LENGTH ESTIMATES FOR EACH DAY OF GESTATION IN THE BEAGLE

Day	Total Length	Day	C–R Length	Day	C–R Length
18	1–2 mm.	32	25 mm.	47	95 mm.
19	2–3 mm.	33	30 mm.	48	97 mm.
20	4 mm.	34	32 mm.	49	100 mm.
21	4.5 mm.	35	35 mm.	50	107 mm.
22	4.5–5 mm.	36	41 mm.	51	120 mm.
23	5–6 mm.	37	47 mm.	52	130 mm.
24	10 mm.	38	53 mm.	53	138 mm.
25	14 mm.	39	59 mm.	54	142 mm.
26	15 mm.	40	65 mm.	55	144 mm.
27	16 mm.	41	67 mm.	56	145 mm.
28	17 mm.	42	70 mm.	57	150 mm.
29	18 mm.	43	75 mm.	58	155 mm.
30	19 mm.	44	78 mm.	59	157 mm.
31	20 mm.	45	86 mm.	60	158 mm.
		46	90 mm.	63	165 mm.

appearance of bone marrow in the humerus. This occurred in the 28- to 30-mm. embryo at approximately seven weeks of age. During the fetal period, linear dimensions are used for the assessment of age, the commonest being crown-rump, or sitting height, and crown-heel, or total length, for man.

In domestic animal embryos and fetuses the most widely used measurement is the straight line crown-rump (C–R) length, which allows for continuous measurement during most of the prenatal period (Table 2–3). Evans and Sack (1973) plotted prenatal growth in several domestic animals, including the dog, and considered the fetal stage of the dog to begin on day 35, when the digits on both limbs were fully formed and the external features were clearly those of a dog.

The weights of individual embryos or fetuses within a litter often vary considerably, and there may be a great disparity between the smallest and the largest individual within an age class or within a litter (Fig. 2–11). When the litter is small, each individual tends to be heavier in the later stages of gestation, and, conversely, in large litters individual weights are often below average for their age class.

Measurements and Growth Plots

For recording standard linear measurements of fetal growth it is necessary to take straight line as well as curvilinear measurements, as indicated on Figure 2–12. For a discussion of the coefficients of variation for various fetal measurements, see Cloete (1939), who worked with sheep.

C–R The crown-rump length is a straight line measurement from the most rostral point of the crown to the base of the tail. In early somite embryos total length is measured.

HL The head length is a straight line measurement from the tip of the snout to a point on the midline which represents the extension of a line through the medial margin of the base of the ear. The width of the head is measured by calipers at the widest point between the zygomatic arches.

LV The vertebral column length is taken by a string, along the body contour from the midline point where head length (HL) intersects the occiput to the tip of the tail.

G The chest girth is a string measurement of the greatest circumference behind the thoracic limb.

Successive removals of embryos and fetuses during gestation allows comparisons to be made between litters and provides data for calculating incremental growth (Evans 1974). The average increase in length between days 35 and 40 is approximately 30.5 mm., which indicates growth of 6.1 mm. per day for the Beagle fetus (Fig. 2–13).

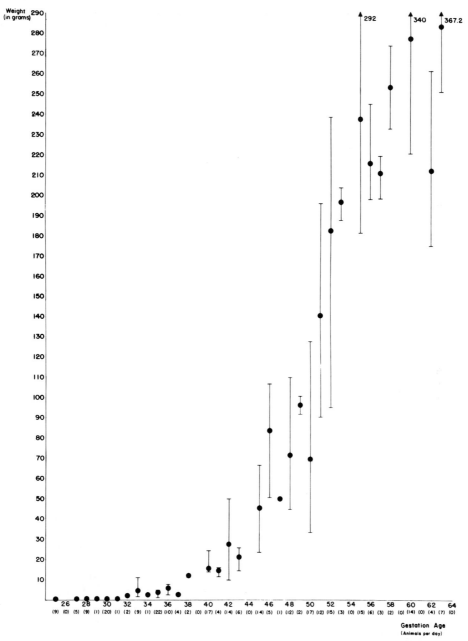

Figure 2–11. Weight in grams versus gestation age in days. The vertical bar indicates the minimum and maximum weights of individuals within the age class (Evans 1974).

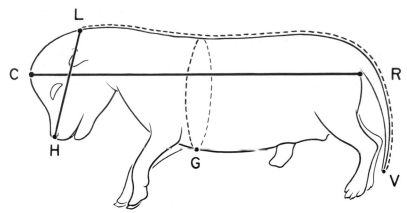

Figure 2–12. Standard measurements taken for plotting fetal growth.

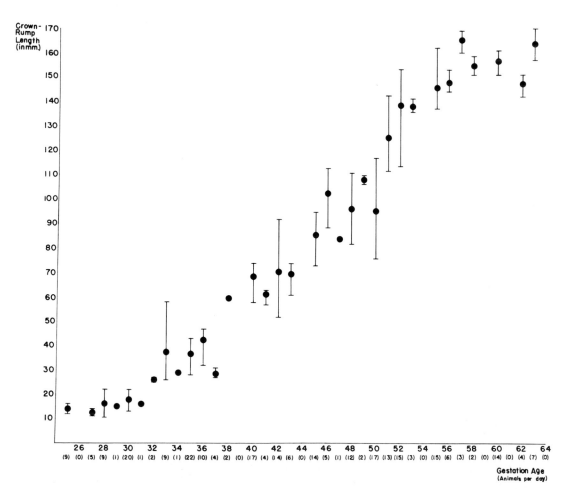

Figure 2–13. Crown-rump length in millimeters versus gestation age in days for the Beagle. The smallest and largest fetus for each age is indicated by the vertical bar, while the spot indicates the mean crown-rump length. (Evans 1974).

Size Index

The size index is a combined figure which is an average of the sum of the five most reliable measurements: crown-rump length, head length, head width, vertebral column length, and chest girth (Fig. 2–14).

THE FETUS

By the time the dog embryo is a little more than halfway through gestation (35 days of the 60-day mean), it has completed the embryonic period of major organ formation. The external features enable one to recognize it as a dog, and from this point on it can be referred to as a fetus. A Beagle fetus of 35 days postcoitus is about 35 mm. in crown-rump length. Larger dogs would have somewhat larger fetuses, but in either case the length of the fetus is equal to the width of the placental band.

Successive removals of fetuses at intervals from the same dog midway through gestation show the gradual increase in the size of the uterus and the marked increase in the size of the embryo and fetus. Figures 2–15A, B, C, and D show changes at five-day intervals.

External features characteristic of fetal stages are development of pigmentation, growth of hair and claws, closure and fusion of the eyelids, growth of the external ear, elongation of the trunk, and sexual differentiation. As the genital area develops in the male, the mammary primordia in the vicinity involute and regress as the prepuce forms around the growing phallus. In the female, the vulva grows caudally as it envelops the genital tubercle, which becomes the clitoris (Figs. 2–16 to 2–22).

Text continued on page 36.

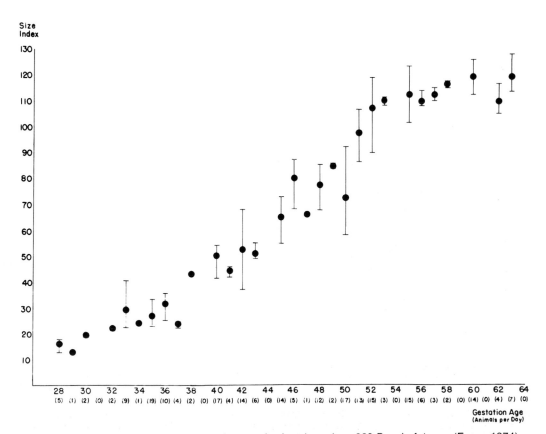

Figure 2–14. Size index versus gestation age in days based on 229 Beagle fetuses (Evans 1974).

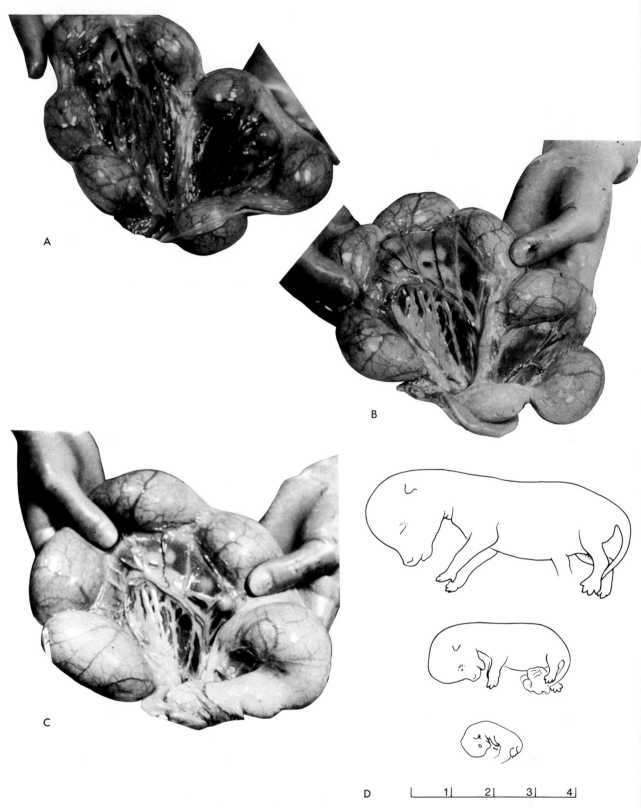

Figure 2–15. A. Uterus at 28 days.
B. Uterus at 33 days.
C. Uterus at 38 days.
D. 28-, 33-, and 38-day embryos removed from A, B, and C.

30

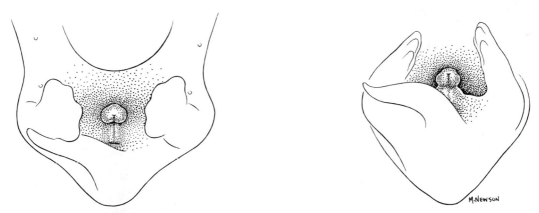

Figure 2–16. Undifferentiated phallic tubercle of a 30-day Beagle embryo.

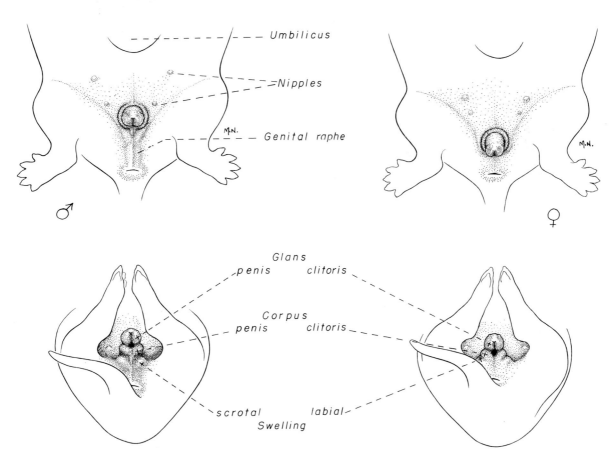

Figure 2–17. Differentiated male and female external genitalia at 35 days of gestation in the Beagle.

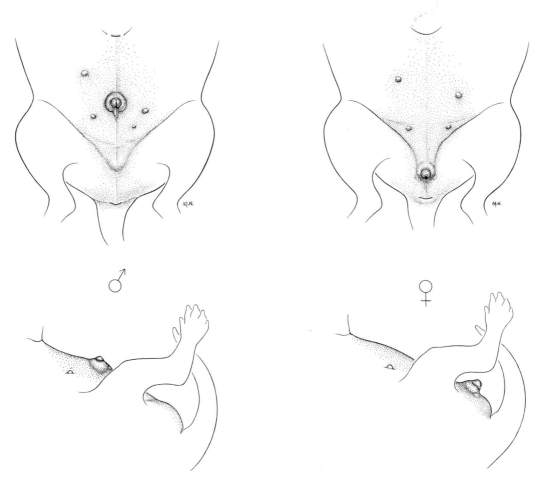

Figure 2–18. External genitalia of 40-day fetal Beagles.

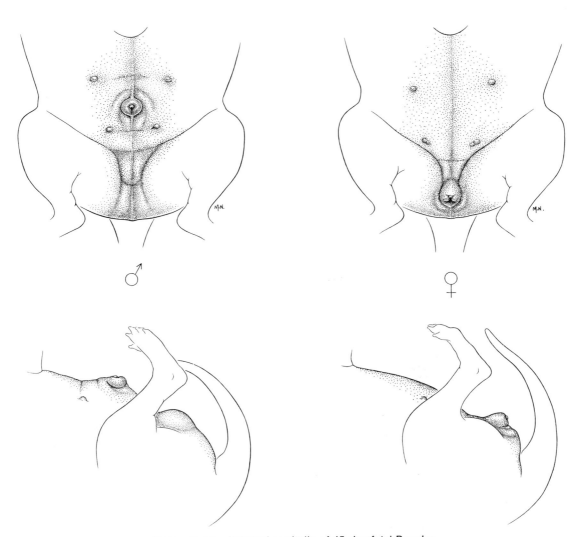

Figure 2–19. External genitalia of 45-day fetal Beagles.

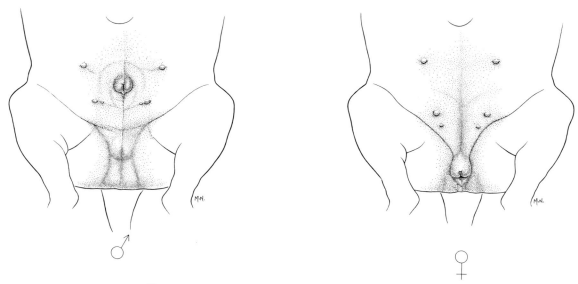

Figure 2–20. External genitalia of 50-day fetal Beagles.

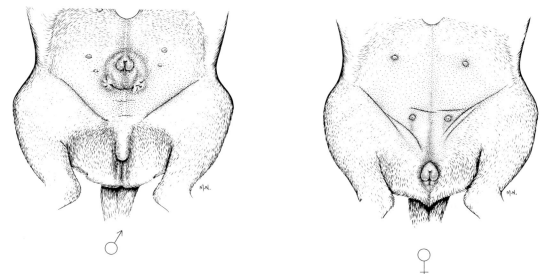

Figure 2–21. External genitalia of 55-day Beagles.

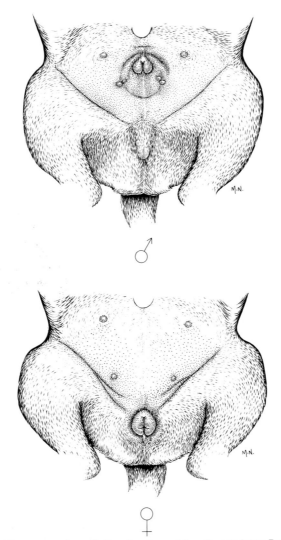

Figure 2–22. The external genitalia of male and female newborn Beagles (60 days).

Skeletal Age Criteria

Although the weight or size of embryos and fetuses may vary within a litter, the sequence of ossification is rather uniform within the litter.

To see the cartilaginous structures and incipient ossifications in three dimensions, it is possible to stain the cartilage and bone preferentially, clear the entire embryo in hydroxide and glycerine, and examine it under a dissecting microscope. For staining embryonic cartilage, in toto, the alcian blue technique is ideal and can be applied to fixed or fresh material (Simons and Van Horn 1971, Watson 1977). When calcified cartilage or bone is present, it can be stained with alizarin red S. Both techniques may be combined (Dingerkus and Uhler 1977), resulting in a differentially stained specimen which leaves no doubt as to the morphological stage of the cartilage or bone in question.

Ossification centers can be seen at an earlier stage using alizarin-stained clearings than by using radiographic techniques. When staining and clearing is preceded by the injection of contrast media into the vessels or organs (Evans 1948), the resulting specimen is ideal for studying developing vessels and organs in relation to the skeleton (Fig. 2–23).

Skull

The earliest indication of the skull is seen as parachordal, trabecular, and branchial cartilages which are probably of neural crest origin. A well-illustrated study of the development of the vertebrate skull was made by de Beer (1937). As the parachordal and trabecular cartilages enlarge and fuse with one another beneath the brain, they incorporate the sense capsules (olfactory, optic, and otic) and form a troughlike cartilaginous skull called the **chondrocranium** (Fig. 2–24). This cartilaginous forerunner of the definitive skull, which has been studied in the dog by Olmstead (1911) and Schliemann (1966), is destined to ossify endochondrally beneath the brain (Fig. 2–25) and combine with the membrane-formed bone of the **desmocranium** above the brain (Fig. 2–26). Drews (1933), using mongrel dog fetuses, and Schliemann (1966), studying Whippet embryos and fetuses, have described several stages of skull ossification. The cartilages of the jaws, hyoid, and middle ear, constituting the **viscerocranium** (Fig. 2–27), form bone in both membrane and cartilage which is incorporated or attached to the adult skull, or **cranium.**

Regardless of whether the bones are formed in cartilage or membrane, there may be one or more centers of ossification, and

Text continued on page 43.

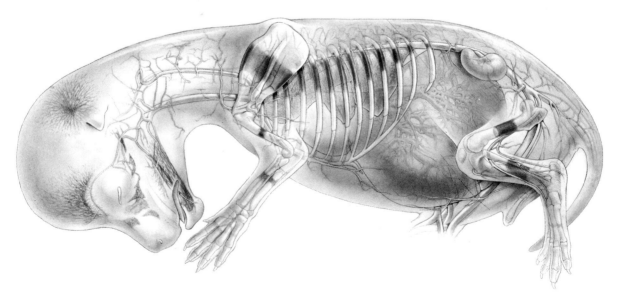

Figure 2–23.   Developing vessels and organs in relation to the skeleton of a 40-day Beagle fetus. (Drawn by Lewis Sadler.)

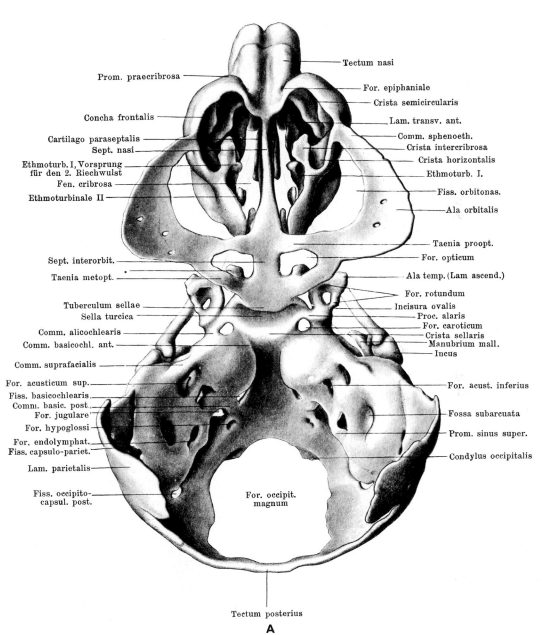

Prom. praecribrosa

Concha frontalis

Cartilago paraseptalis
Sept. nasi

Ethmoturb. I, Vorsprung
für den 2. Riechwulst
Fen. cribrosa
Ethmoturbinale II

Sept. interorbit.
Taenia metopt.

Tuberculum sellae
Sella turcica

Comm. alicochlearis
Comm. basicochl. ant.

Comm. suprafacialis

For. acusticum sup.
Fiss. basicochlearis
Comm. basic. post.
For. jugulare
For. hypoglossi

For. endolymphat.
Fiss. capsulo-pariet.

Lam. parietalis

Fiss. occipito-
capsul. post.

Tectum nasi

For. epiphaniale
Crista semicircularis
Lam. transv. ant.
Comm. sphenoeth.
Crista intercribrosa
Crista horizontalis
Ethmoturb. I.
Fiss. orbitonas.
Ala orbitalis

Taenia proopt.
For. opticum
Ala temp. (Lam ascend.)

For. rotundum
Incisura ovalis
Proc. alaris
For. caroticum
Crista sellaris
Manubrium mall.
Incus

For. acust. inferius

Fossa subarcuata

Prom. sinus super.

Condylus occipitalis

For. occipit.
magnum

Tectum posterius

A

Figure 2–24.   The chondrocranium of the dog. (From Olmstead 1911.)
A. Dorsal view.
B. Ventral view.
C. Lateral view.

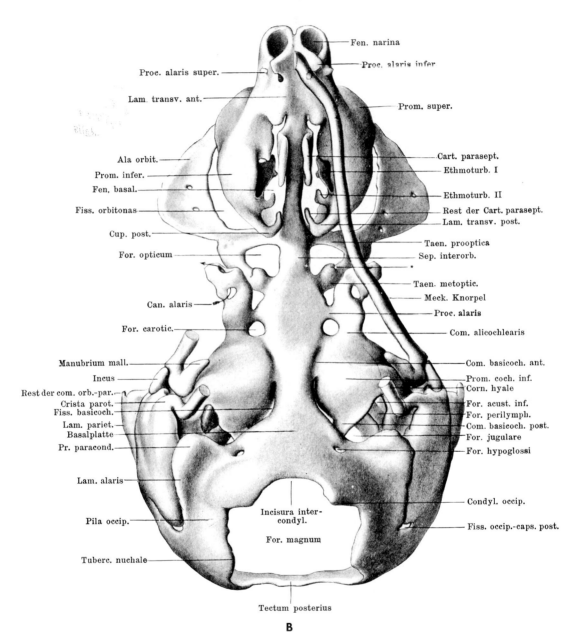

B

Figure 2–24. *Continued*

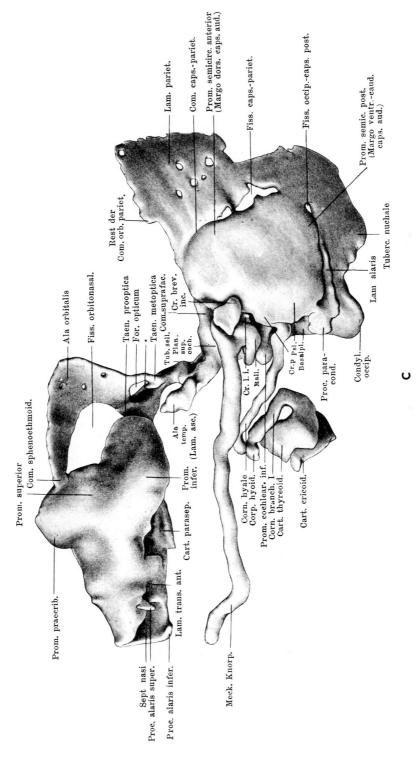

Figure 2-24. *Continued.*

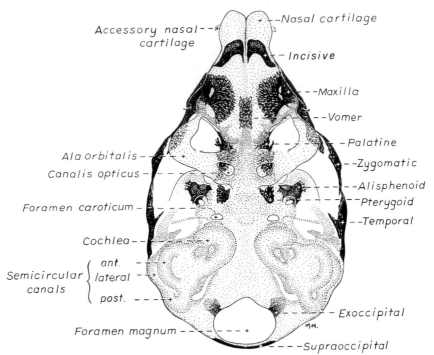

Figure 2–25. Dorsal view of the skull of a 40-day Beagle fetus with the roof removed to show the cartilaginous chondrocranium. Several membrane bones of the jaws, palate, and face are ossified, as are the membrane bones of the roof. (See Fig. 2–26.)

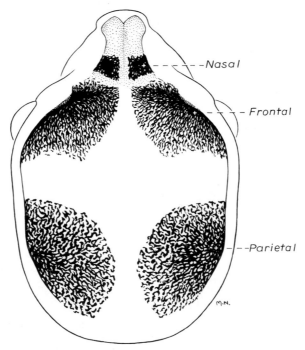

Figure 2–26. Ossification of the desmocranium in a 40-day Beagle fetus.

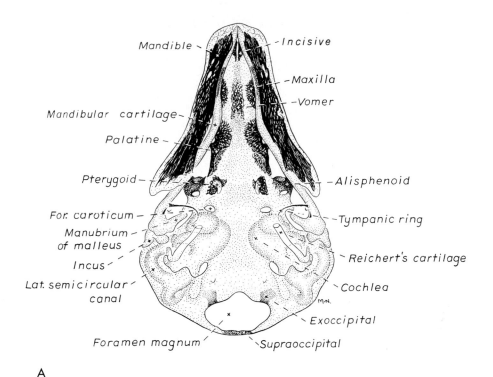

A

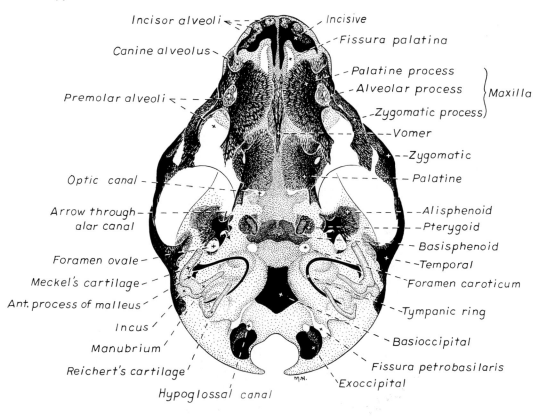

B

Figure 2–27. A. The skull of a 40-day Beagle fetus in ventral view to show parts of the viscerocranium: mandibular cartilages, middle ear cartilages, and hyoid apparatus (Reichert's cartilage).

B. Ventral view of the skull of a 45-day Beagle fetus with the lower jaw removed to show the extent of the palate and alveolar surface.

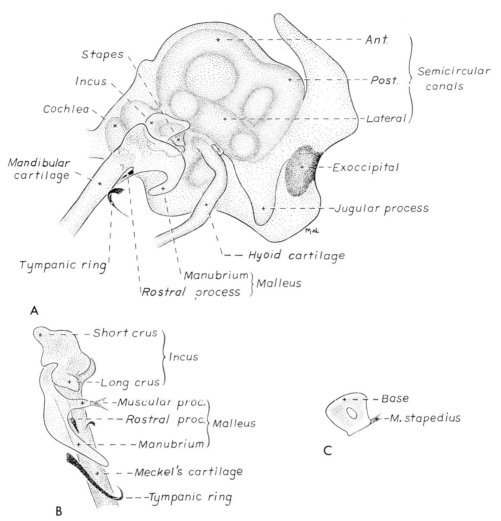

Figure 2–28.   The otic region.
A. The labyrinth of the inner ear, the developing middle ear ossicles, and the attachment of the hyoid apparatus in lateral view.
B. Ventral view of the middle ear cartilages.
C. The stapes prior to ossification.

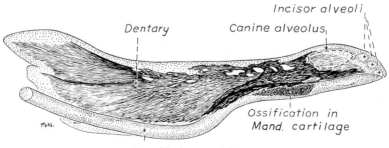

Figure 2–29.   Medial view of the lower jaw of a 40-day Beagle fetus. Note the ossification within the mandibular cartilage in the region of the canine alveolus and the lack of any articular condyle at this stage.

the structure of the definitive bone is the same. Bones developed in membrane ossify earlier than those developed in cartilage. Thus the sheathing bone of the mandibular cartilage (dentary) and the sheathing bone of the palatoquadrate cartilage (maxilla) begin to ossify on day 28 in the 19-mm. embryo, at the same time as the frontal bone of the skull roof and the membranous portion of the clavicle. Other membrane-formed bones of the skull, which develop early, are the nasal, incisive, palatine, zygomatic, and parietal, which form by day 32 in the 27-mm. embryo.

The branchial arches give rise to the cartilages of the jaws, the auditory ossicles, the hyoid apparatus, and the larynx. The lower jaw, which is part of the first branchial arch, is present as a rodlike mandibular cartilage (Meckel's cartilage) on each side by day 25 of gestation, when the embryo is 13 to 16 mm. in length. The mandibular cartilages join rostrally, whereas caudally each lies within a middle ear cavity, where its hook-shaped articular cartilage is destined to become the malleus of the middle ear (Fig. 2–28). By day 28 of gestation (19 mm.) membrane bone begins to form around the mandibular cartilage as the dentary. The cartilage itself at a later stage will undergo endochondral ossification (Fig. 2–29) in the region of the canine tooth, and thus the lower jaw is formed by both perichondral membrane bone and endochondral cartilage bone. On the dorsocaudal border of the dentary bone a condyle develops as secondary cartilage, which then ossifies and fuses with the body of the mandible to become part of the temporomandibular joint.

Later develpmental stages of the skull show that the sequence of primary ossification centers shifts from facial and calvarial centers to basicranial and otic centers, followed by hyoid centers (Fig. 2–30). In some instances bone is being formed in both membrane and cartilage simultaneously, as in the supraoccipital, mandible, pterygoid, and temporal bones.

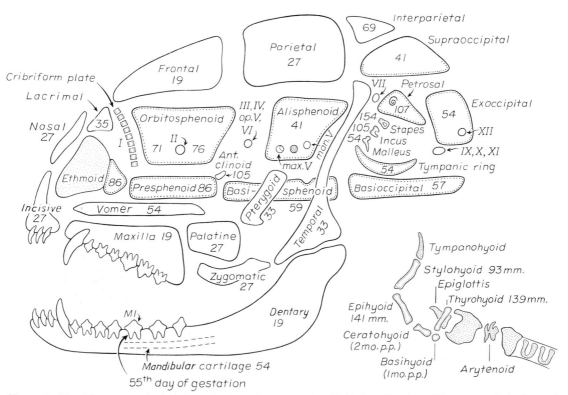

Figure 2–30. Diagrammatic dog skull showing the sequence of initial ossifications. The numerals indicate the fetal size in millimeters at which time each of the bones begins to ossify. Bones with a dotted outline are formed in cartilage (Evans 1974).

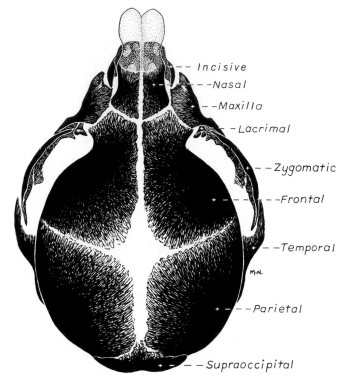

Figure 2–31.   Dorsal view of the skull in a 45-day Beagle fetus. The fontanel between the frontal and parietal bones will close before birth.

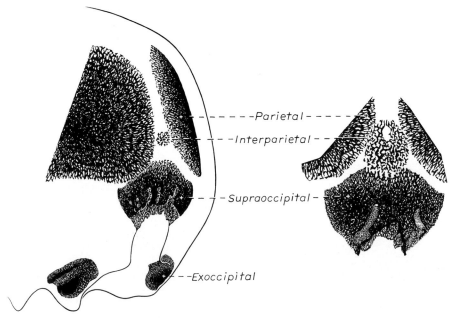

Figure 2–32.   The interparietal bone makes a transitory appearance on about the forty-fifth day of gestation. It soon fuses with the supraoccipital bone and becomes indistinguishable from it. On rare occasions it remains as a separate bone in the adult. Note that its position is more superficial than that of the supraoccipital and parietal.

CALVARIAL CENTERS

The skull roof, or calvaria (Fig. 2–31), is composed of bones which develop in membrane as paired frontal and parietal centers. Each bone shows a central trabecular network (Fig. 2–23) which spreads to cover the brain. In addition to these primary roofing bones, there is an unpaired interparietal bone which fuses with the supraoccipital bone on day 45 of gestation and only rarely maintains its separate identity (Fig. 2–32).

BASICRANIAL CENTERS

The chondrocranium ossifies to form the bones of the floor, walls, and sense capsules of the braincase. The process is gradual, as can be seen from Figures 2–33 to 2–40, and involves endochondral loci which spread to predetermined borders of homogeneous cartilaginous anlagen. Thus the basicranial axis is formed by a central ossification for the basioccipital bone, a similar one for the basisphenoid bone, and paired ossifications for the presphenoid and ethmoid bones. These median elements fuse with their lateral component wings to form a sphenoidal complex. The presphenoid fuses with the orbitosphenoids on each side to form a presphenoid with orbital wings, whereas the basisphenoid fuses with the alisphenoids on each side to form a basisphenoid with temporal wings. Both of these sphenoids fuse with each other (never completely in the dog) to form the sphenoidal complex. On the rostral end of this complex the lateral ethmoids fuse to form the sphenethmoids below the cribriform plate, and the mesethmoid forms a median perpendicular plate.

The joint between the basisphenoid and the basioccipital is known as the spheno-occipital synchondrosis, and that between the basisphenoid and presphenoid is the intersphenoidal synchondrosis. Premature fusion of either of these joints, as is seen in brachycephalic animals, limits growth and shortens the basicranial axis.

OTIC CENTERS

The otic region consists of a cartilaginous **otic capsule** containing the **membranous labyrinth** (Fig. 2–28A). Numerous ossification centers develop around the inner ear

Text continued on page 52.

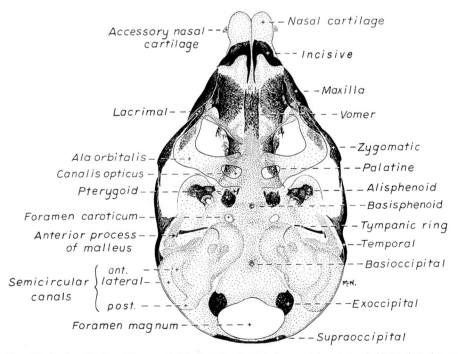

Figure 2–33. Skull of a 40-day, 59-mm. fetal Beagle, dorsal view, roof removed. Note that the earliest ossifications of the basicranial axis are in the basisphenoid and basioccipital bones.

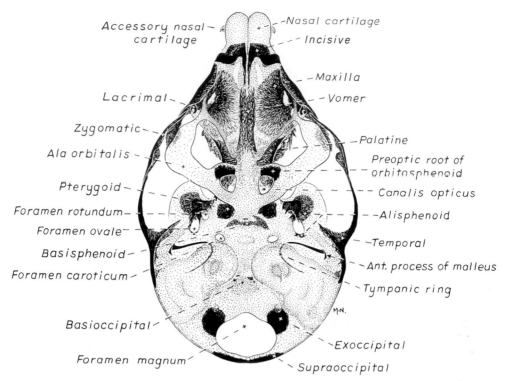

Figure 2–34. Skull of a 40-day, 71-mm. fetal Beagle, dorsal view, roof removed. Note that the preoptic root of the orbitosphenoid wing has ossified.

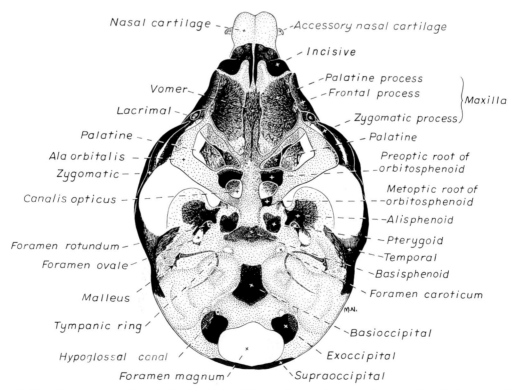

Figure 2–35. Dorsal view of the skull of a 45-day, 73 mm. Beagle fetus, roof removed. Note the metoptic root of the orbitosphenoid on one side.

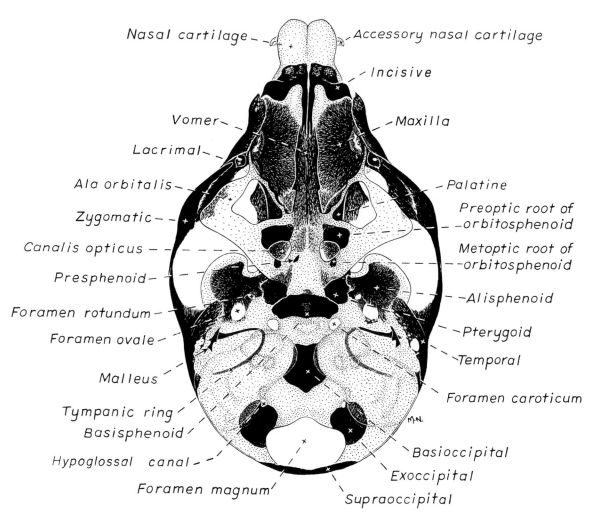

Figure 2–36. Dorsal view of the skull of a 42-day, 88-mm. Beagle fetus, roof removed. Note the presphenoid on the left side.

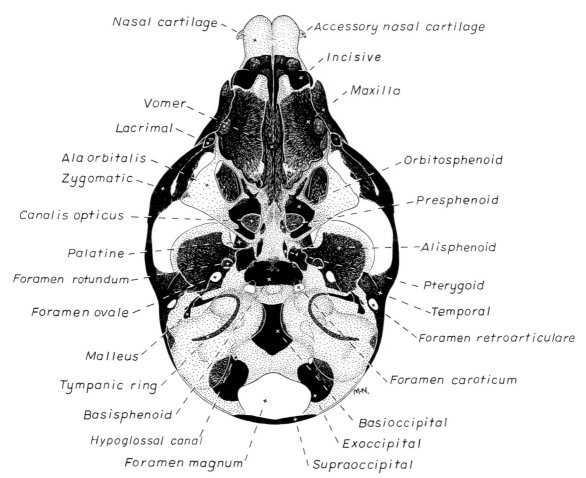

Figure 2–37. Dorsal view of the skull of a 45-day, 92-mm. Beagle fetus, roof removed. Note that the preoptic and metoptic roots have fused, that both presphenoids are present, and that each pterygoid has two ossification centers.

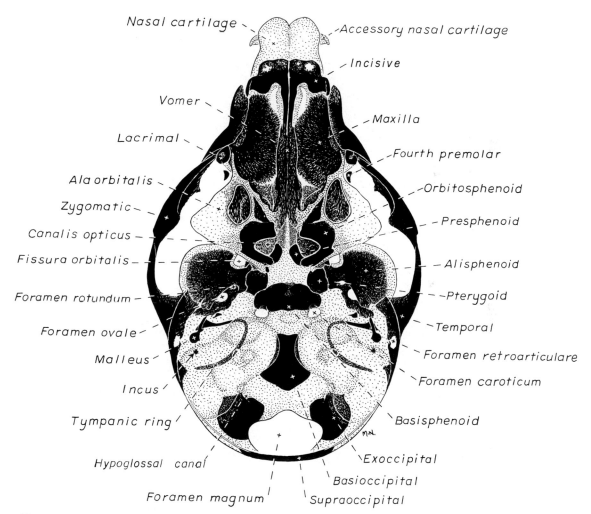

Figure 2–38. Dorsal view of the skull of a 50-day, 105-mm. Beagle fetus, roof removed. Note that the incus and fourth upper premolar are ossified.

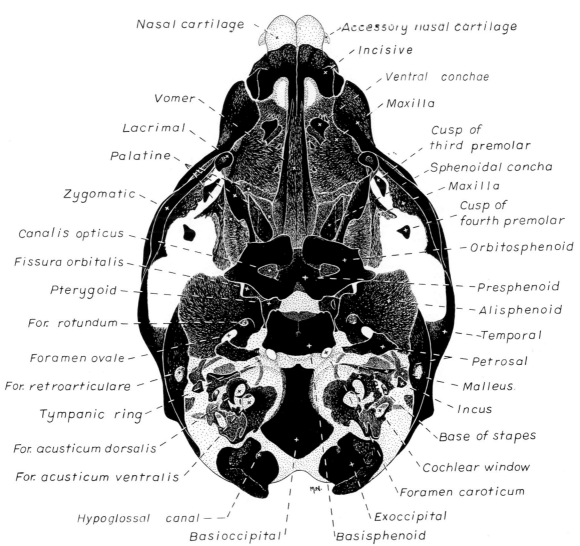

Figure 2–39. Dorsal view of the skull of a 55-day Beagle fetus, roof removed. Note the ossifications of the inner ear region. The ventral nasal concha is beginning to ossify.

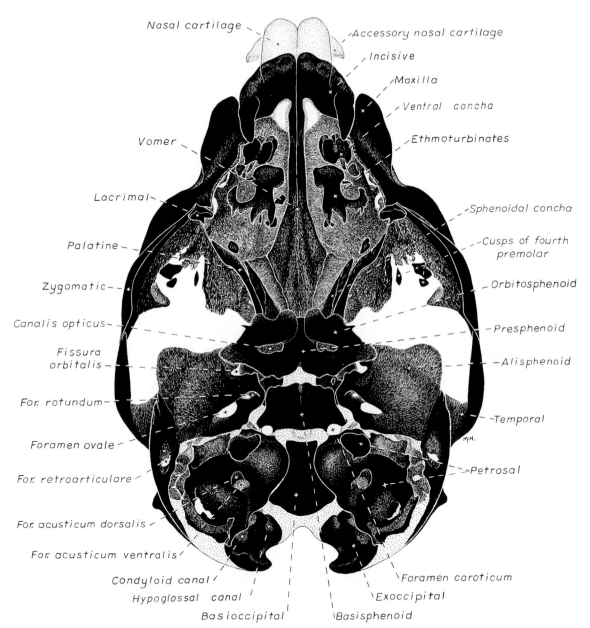

Figure 2–40. Dorsal view of the skull of a newborn Beagle (60 days), roof removed. The intersphenoidal synchondrosis and the spheno-occipital synchondrosis are very prominent.

structures to form the petrosal bone. The latter fuses with the squamous temporal bone and the tympanic ring to form the temporal complex. As the tympanic cavity develops, the **middle ear ossicles** (of branchial arch origin) are enclosed in the tympanic bulla of tympanic ring origin. The sequence of ossification for the ossicles is malleus (38 days), incus (50 days), and stapes (55 days).

HYOID APPARATUS

The **stylohyoid** cartilage (Fig. 2–41) ossifies by 47 days (93 mm.) followed by the **thyrohyoid** at 53 days (139 mm.) and the **epihyoid** on day 54 (141 mm.). The **basihyoid** ossifies one month after birth and the **ceratohyoid** two months after birth. The **tympanohyoid** does not ossify.

TEETH

The development of the dentition has been studied by Williams (1961), who found that calcification of all deciduous teeth was initiated by day 55 of gestation and completed by day 20 postpartum for the crowns or day 45 postpartum for the roots. The only tooth of the permanent dentition to calcify prenatally is the lower first molar, which appears on day 55 of gestation.

Williams and Evans (1978) examined sectioned dog embryos and fetuses as well as wholemounts to determine the time sequence of standard morphological stages of dental development. The **dental lamina** first

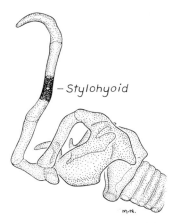

Figure 2–41. The larynx and hyoid apparatus of a 47-day Beagle fetus.

appeared at 25 days of gestation (14 mm.), and by 30 days the right and left laminae joined across the midline, forming a continuous dental arch. The **vestibular lamina,** which separates the gums from the lips and cheeks, first arises as a distinct invagination lateral to the dental lamina. It is recognizable in the incisor region of the 30-day embryo, best developed on the lower jaw.

Differentiation of the deciduous enamel organs in the dog begins on about day 30 of gestation and exhibits the following typical sequence: bud, early cap, cap, advanced cap, early bell, and advanced bell stages. The relationship of the enamel organ to the dentary bone and mandibular cartilage can be seen in Figure 2–42.

The sequence of calcification of the teeth in the Beagle is:

Third lower premolars Fourth lower premolars	42 days (70 mm.)
Upper and lower canines Third upper premolars	45 days (86 mm.)
Upper incisors	45 to 48 days (86–97 mm.)
Lower incisors Fourth upper premolars Second upper and lower premolars	49 to 51 days (100–120 mm.)

By day 55 (141 mm.) all deciduous teeth show calcification, as does the lower first molar of the permanent dentition.

Williams and Evans (1978) were unable to determine for certain whether the first premolar tooth is a member of the deciduous or permanent dentition. Tooth buds for the first premolars did not appear until 47 days (95 mm.), which is long after all of the other deciduous tooth primordia have made their appearance.

Vertebral Column

The notochord is the forerunner of the vertebral column, and in the early embryo it is present as a solid rod of cells surrounded by a sheath of paraxial mesoderm. Condensations within the somites form the sclerotomes which surround the notochord and partially enclose the neural tube. When resegmentation and chondrification take place

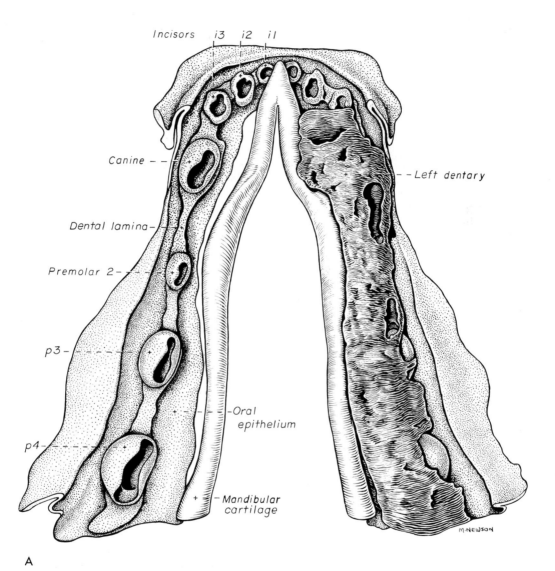

A

Figure 2–42. The relationship of the enamel organs to the dentary bone and mandibular cartilage. Based on a reconstruction of a 71-mm. Beagle fetus.
     A. Ventral view—dentary bone removed on the right side.
     B. Lateral view.
     C. Transection.
(From Williams and Evans 1978. Zbl. Vet. Med. C).

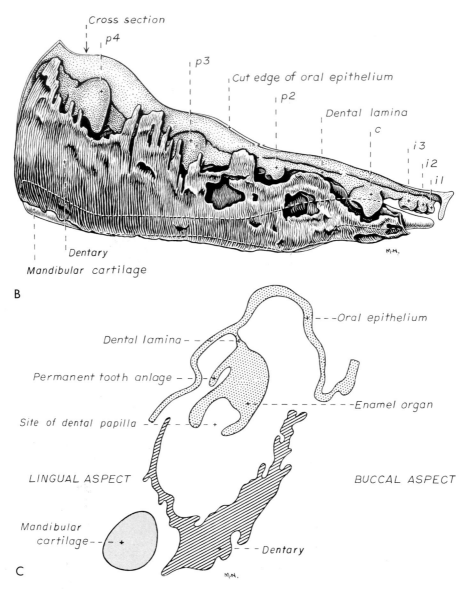

Cross section
p4
p3
Cut edge of oral epithelium
p2
Dental lamina
c
i3
i2
i1
Dentary
Mandibular cartilage

B

Dental lamina
Oral epithelium
Permanent tooth anlage
Enamel organ
Site of dental papilla
LINGUAL ASPECT
BUCCAL ASPECT
Mandibular cartilage
Dentary

C

Figure 2–42. *Continued.*

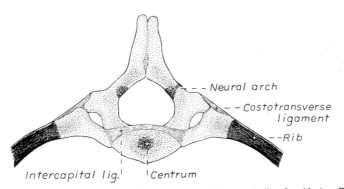

Neural arch
Costotransverse ligament
Rib
Intercapital lig.
Centrum

Figure 2–43. Cranial view of the eighth thoracic vertebra and rib of a 40-day Beagle fetus.

in the sclerotome, the notochord is almost completely obliterated within the centrum. The notochord persists as a soft central core, the nucleus pulposus, within the intervertebral disc. The intervertebral disc is a remnant of the intercentral portion of the early chondrification.

The lateral portions of the vertebral condensation chondrify and grow dorsally to form the neural arches and ventrolaterally to form the transverse and costal processes and part of the body. The spinous process of the neural arch develops after the arches meet and fuse. Failure of the neural arches to close dorsally results in rachischisis or spina bifida.

By day 25 of gestation the vertebral column consists of individual chondrified elements resembling definitive vertebrae. Each vertebra (except the atlas and axis) is ossified from three primary centers, one for the centrum and one for each neural arch. The centra first appear as endochondral nodular condensations, whereas the neural arches form as perichondral collars around the base of the cartilaginous arch (Fig. 2–43). Although the manner of formation of centra and neural arches differs in the early stages, later growth and ossification involves simultaneous perichondral as well as endochondral ossification.

CENTRA

The first endochondral ossifications of the Beagle vertebral column appear at 38 days (54 mm.) in thoracic and lumbar regions. These are closely followed by ossifications of C2, C6, and T4 to T6. Intervening centra are ossified rapidly in both directions, resulting in a continuous series from C2 through L6 by 55 mm. (Fig. 2–44).

The body of the atlas (intercentrum I) first appears as an ossification in the 42-day (73 mm.) fetus and is always present after 46 days (92 mm.). The earliest appearance of an ossified dens is on day 42 (71 mm.), although it may still be lacking in some individuals as old as 46 days (89 mm.). Between the dens and the centrum of the axis after birth there is an ossified intercentrum II. All elements of the axis fuse with one another by the fourth month post partum.

Lumbar centra and the first sacral centrum are all present by day 40 (59 mm.).

Sacral vertebrae are slow to ossify, but by 43 days (73 mm.) all three sacral centra are present. Additional ossifications in the sacral region are located lateral to S1 and S2 and represent sacral "ribs" of ancestral forms. Subsequent growth and fusion result in a combined sacrum dominated by S1, with its large auricular surface for articulation with the ilium.

Caudal centra are the last to appear. They exhibit typical endochondral ossifications from Cd1 to Cd4 but have perichondral plus endochondral ossifications from Cd5 to Cd20. The number of caudal vertebrae was a constant 20 in the Beagles studied.

NEURAL ARCHES

Paired perichondral neural arch ossifications first appear in the cervical region and increase in a craniocaudal sequence. Their earliest appearance at 38 days (54 mm.) shows them to be present from C1 through T7. Rarely do the right or left ossification centers differ in their time of formation. By day 42 (72 mm.) the sequence of neural arch ossifications becomes discontinuous because caudal neural arches 5 through 8 show premature ossification before sacrals 1 and 2 or caudals 1 to 4 ossify (Fig. 2–44). The precocious development of these caudal vertebrae in fetuses between 42 and 45 days may be a response to the developing rectococcygeus muscle, which attaches to the fifth and sixth caudal vertebrae.

The extent of vertebral ossification at 40 days and at birth is shown for each of the regions in Figures 2–45 to 2–50.

Ribs

All ribs are cartilaginous until day 31, when ribs 3 through 9 begin to ossify at midshaft as perichondral collars. By 40 days the shafts of all ribs are ossified dorsally (See Fig. 2–63B).

Sternum

The sternum consists of eight sternebrae with intervening cartilages to which the ribs attach. In the early embryo (15 mm.) a longitudinal mesodermal bar develops on each

Text continued on page 60.

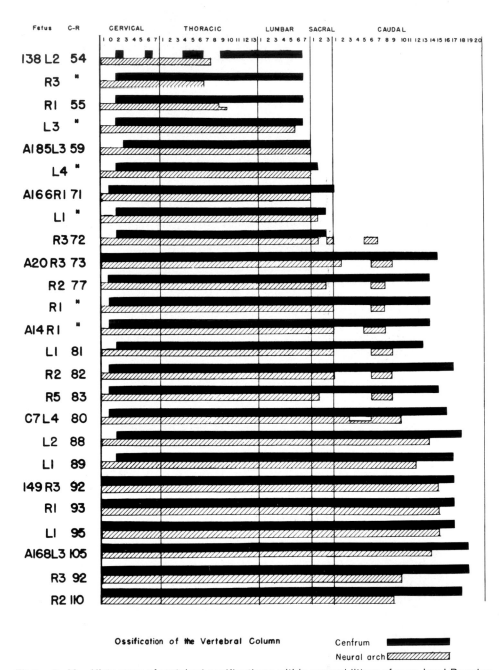

Figure 2-44.   Histogram of vertebral ossifications within several litters of pure-bred Beagles.

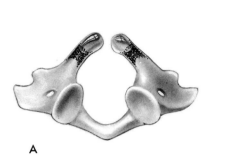
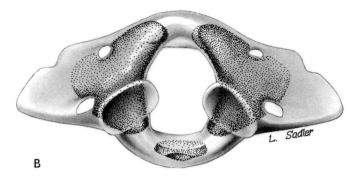

A

B

Figure 2–45. Atlas, caudal aspect.
A. At 40 days of gestation.
B. At birth.

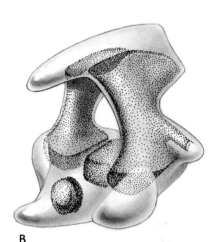

A

B

Figure 2–46. Axis, craniolateral aspect.
A. At 40 days of gestation.
B. At birth.

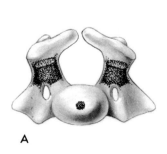
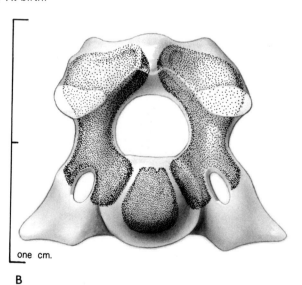

A

one cm.

B

Figure 2–47. Third cervical vertebra, cranial aspect.
A. At 40 days of gestation.
B. At birth.

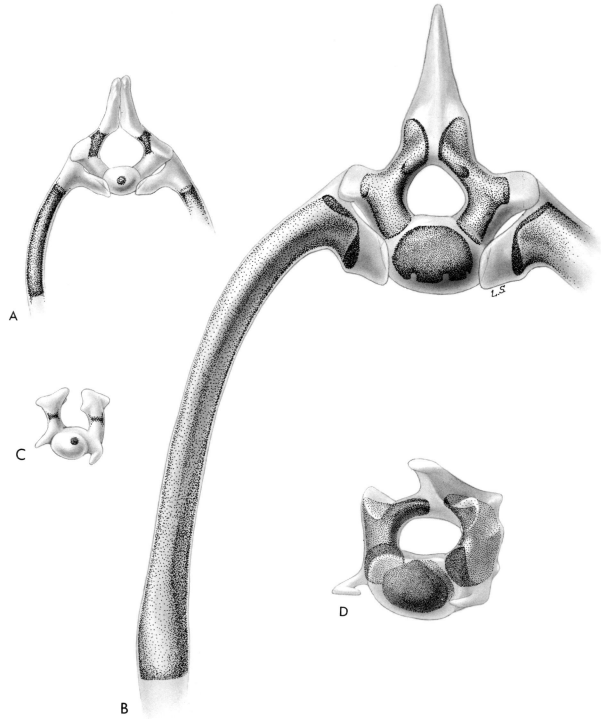

Figure 2–48. Cranial aspect of fourth thoracic vertebra.
A. At 40 days of gestation.
B. At birth.

Craniolateral aspect of fourth lumbar vertebra.
C. At 40 days.
D. At birth.

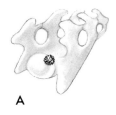

A

B

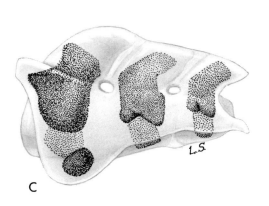

C

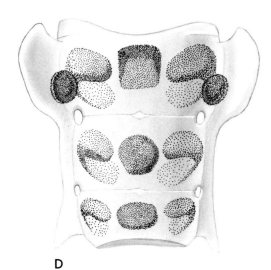

D

Figure 2–49. The sacrum at 40 days.
A. Craniolateral aspect.
B. Left lateral aspect.

The sacrum at birth.
C. Left lateral aspect.
D. Ventral aspect.

A

B

C

D

E

Figure 2–50.   Caudal vertebrae.
A. First caudal vertebra, 40 days, lateral aspect.
B. First caudal vertebra, 40 days, dorsal aspect.
C. First caudal vertebra at birth, craniolateral aspect.
D. Fifth caudal vertebra, 40 days, craniolateral aspect.
E. Tenth caudal vertebra, at birth, craniolateral aspect.

side in the lateral body wall independently of the ribs and clavicle. As the body wall grows and encloses the pericardial sac, the sternal bars migrate toward the ventral midline, followed closely by the ventral ends of the ribs. By 18 mm. (25 days) the sternal bars almost meet at the manubrium but are widely separated caudally (Fig. 2–51). Sub-sequent growth and fusion of the sternal bars is followed by hypertrophy of the chondroblasts and initiation of ossification in sternal regions between the attachments of the costal cartilages.

The ventral ends of ribs 2 through 7 fuse homogeneously with the sternal bars, while the first, eighth, and ninth ribs remain inde-

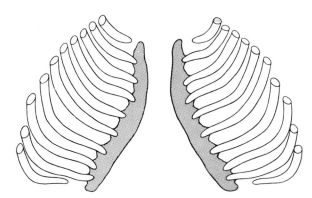

Figure 2–51.   Sternal bars and ribs of a 25-day Beagle fetus. Dorsal view.

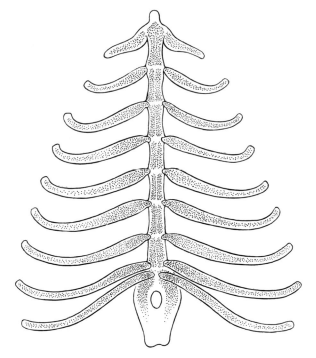

Figure 2–52. Evidence of the bilateral origin of the sternum can still be seen after fusion of right and left sternal bars.

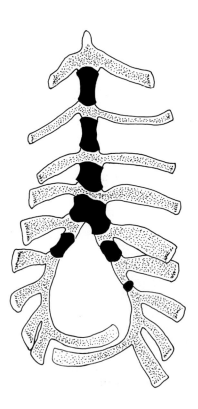

Figure 2–53. Widely separated ends of the nonunited sternal bars can result in an anomalous infracostal arch.

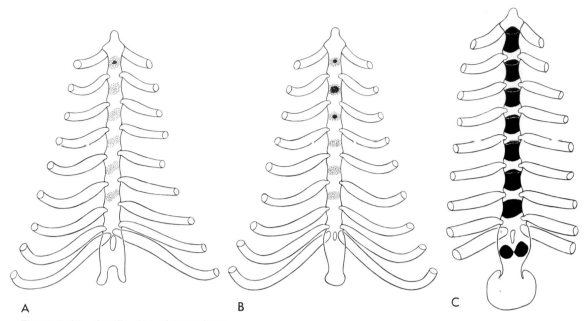

Figure 2–54. Ossification of sternebrae.

A and B. Littermates showing early endochondral ossifications of sternebrae.

C. A more advanced stage of perichondral ossifications. Note differences in the terminal segment or xiphisternum, a frequent occurrence within and between litters. In A the sternebral condensations are oblique owing to uneven placement of ribs.

pendent of the sternum for a short period. Growth of the body wall allows the sternal bars to unite progressively caudalward with each other on the midline (Fig. 2–52). All evidence of the bilateral nature of the sternal bars is usually lost after fusion of the cartilages occurs and prior to ossification. However, non-union of the sternal bars in the xiphoid region can result in the presence of a xiphisternal foramen (Figs. 2–52 and 2–54) or widely separated terminal ends (Fig. 2–53).

The earliest ossifications of the sternum are seen in 40-day (73 mm.) fetuses as endochondral centers. Later ossifications may be either endochondral or perichondral (Fig. 2–54). There is considerable interlitter and intralitter variation in both the manner of ossification and the number of sternebrae ossified at any particular time, even in closely related dogs. Occasionally, the ossification center for the first sternebra is lacking, although the second through sixth are present. The seventh and eighth sternebrae often exhibit eccentric ossification centers in early development. The xiphisternum is the most variable element and

may have a foramen, a fissure, or a cleft, depending upon the manner of final fusion of the sternal bars.

Chen (1953) demonstrated experimentally that hypertrophy of the chondroblasts and subsequent ossification are inhibited where sternal ribs are attached. Evidence of this phenomenon can be seen in Figure 2–55, which represents atypical development of

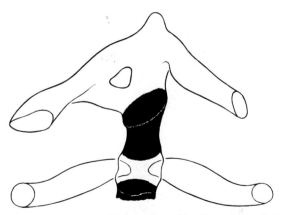

Figure 2–55. Inhibition of ossification of the first sternebra due to an anomalous first rib.

the first sternebra due to the inhibitory effect of an atypical sternal rib. All ossification centers of the sternebrae bear a constant relationship to the attachment of ribs, although a very wide sternum may minimize the inhibitory effect and result in variable fusions of adjacent sternebrae to form a sternal bar or irregular plates.

Limbs and Girdles

The thoracic limb-bud is the first to develop, on about day 23 of gestation, when the embryo is about 5 mm. long. It appears as an oval paddle closely attached at midbody (Fig. 2–6). The pelvic limb bud develops about a day later.

On day 25 (14 mm.) the thoracic limb shows incipient digit formation in the form of a crenulated margin of the apical ridge (Fig. 2–9B). On day 28 the clavicle ossifies as one of the first four bones to appear in the embryo. (The others are the mandible, maxilla, and frontal.)

By day 30 the thoracic limb has lengthened and pronated, and its digits are distinct. The pelvic limb at 30 days resembles the thoracic limb of five days earlier (Fig. 2–10B). A further increase in the size and length of the limbs by day 35 (35 mm.) delimits the joints and produces well-formed digits with developing claws. At this time perichondral ossification of the long bones begins at midshaft in both limbs and results in primary bone collars without endochondral involvement (Fig. 2–56 A to H).

The **scapula** at 35 days has three perichondral ossification centers: (1) a triangular area on the cranial margin of the supraspinous fossa, (2) a short bar at the midpoint of the scapula spine, and (3) a plaque in the central area of the infraspinous fossa (Fig. 2–56A). All three ossification centers join by day 40 and form a continuous perichondral collar around the scapula, although there is a distinct triangular region on the leading edge of the supraspinous fossa that persists until birth.

The **clavicle** originates as a comma-shaped "membrane" bone in the tendinous intersection of the brachiocephalicus muscle on day 28 and increases in size by the addition of bone formed in secondary cartilage. It continues to grow in size after birth.

The humerus, radius, and ulna do not form epiphyses prior to birth. Although all of the metacarpals and phalanges are ossified by the end of gestation, none of the carpals ossify prior to birth (Fig. 2–57).

The pelvic girdle is completely cartilaginous until day 40, when a perichondral bone collar develops around the ilium (Fig. 2–58). Several days later (day 45) the ischium ossifies (Fig. 2–59,), and shortly before or at birth (day 55 to 60) pubic centers appear (Fig. 2–60). The acetabular bone does not appear until several weeks after birth.

The femur, tibia, and fibula ossify perichondrally at first, as do the metacarpals and phalanges. A cartilaginous patella is present in the tendon of the quadriceps muscle throughout the second half of gestation. Only the talus and calcaneus ossify before birth in the tarsus.

The forepaw and hindpaw show intra- and interlitter variations of the metapodials as to their presence, duplication, and time of ossification. Metapodials 3 and 4 are followed closely by 2, 5, and 1. All of the phalanges ossify in the typical mammalian sequence of distal, then proximal, then middle phalanx (Figs. 2–59 and 2–61). Digits 2 and 3 are the first to ossify.

Summary of Morphological Events During the Third Trimester of Development

35 Days (Fig. 2–62)

Bitch: Uterine swellings 43 × 74 mm., width of placenta equal to length of fetus.

Fetus: 35 mm.; eyelids developing so that eye is almost closed; pinna covers ear opening; sex determination possible externally; sternal bars united on midline; ossification of temporal, pterygoid, and lacrimal bones; scapula and ribs 2 through 13 are ossified at midshaft.

37 Days

Fetus: 47 mm.; ossification of supraoccipital and temporal wing of basisphenoid; first rib neural arches C1 through C4; metacarpals 2, 3, and 4; ilium at midshaft.

Text continued on page 70.

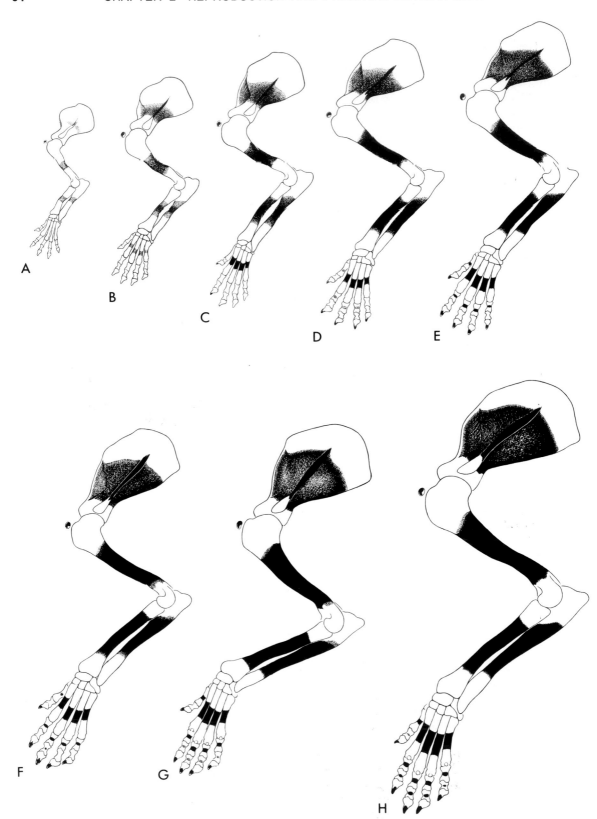

Figure 2–56.   A to H. Progressive ossification of the thoracic limb and girdle.

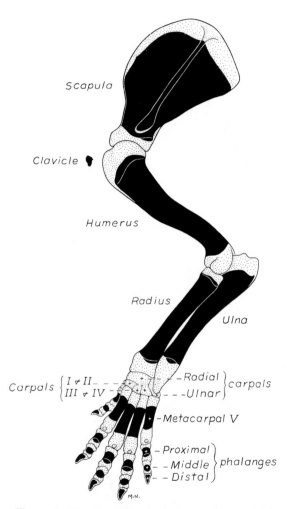

Figure 2–57. The thoracic limb and girdle at birth (60 days).

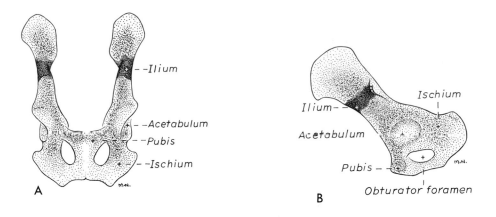

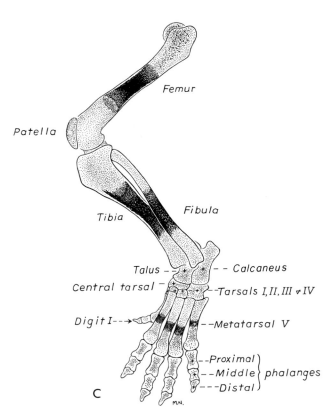

Figure 2–58. Ossification of the pelvic limb and girdle on the fortieth day of gestation in the Beagle.
A. Ventral view of pelvis.
B. Lateral view of pelvis.
C. Laterodorsal view of pelvic limb.

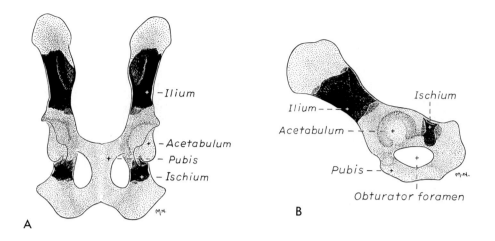

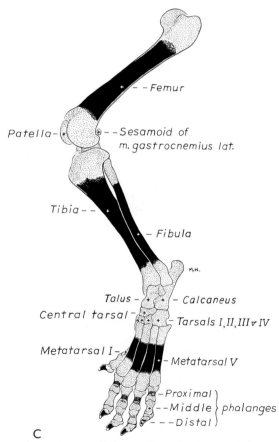

Figure 2–59. Ossification of the pelvic limb and girdle on the forty-fifth day of gestation in the Beagle.
A. Ventral view of pelvis.
B. Lateral view of pelvis.
C. Laterodorsal view of pelvic limb.

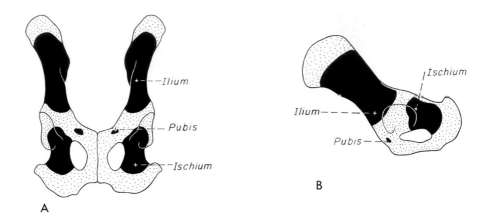

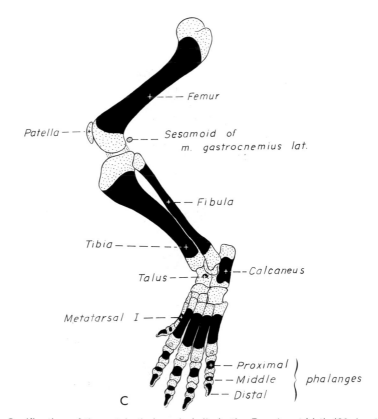

Figure 2–60. Ossification of the pelvic limb and girdle in the Beagle, at birth (60 days).
A. Ventral view of the pelvis.
B. Lateral view of the pelvis.
C. Laterodorsal view of the pelvic limb.

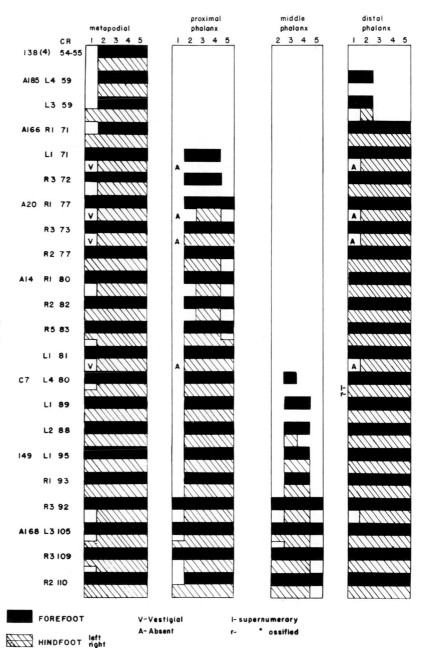

Figure 2–61. Histogram of the sequence of ossification in the metapodials and phalanges.

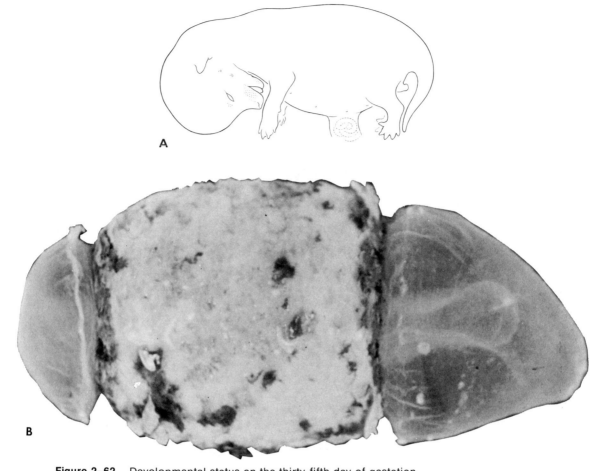

Figure 2–62.   Developmental status on the thirty-fifth day of gestation.
A. External features.
B. Fetus within its membranes and placenta. Note the faint outline of the yolk sac.

38 Days
Fetus: 53 mm.; ossification of exoccipital, vomer, tympanic ring, malleus, and midportion of mandibular cartilage; vertebral centra C2 through L6; vertebral arches C1 through T8; metacarpals 1 through 5 and metatarsals 2 through 5.

39 Days
Fetus: 60 mm.; ossification of orbital wing of presphenoid; basisphenoid; and basioccipital; centra C2 through S1, arches C1 through S1; distal phalanges of digits 1 and 2 of forepaw and digit 2 of hindpaw.

40 Days (Fig. 2–63)
Bitch: Uterine swellings 54 × 81 mm.; width of placenta about equal to the length of fetus; a firm cervical plug is formed.
Fetus: 65 mm.; eyes closed and lids fused; umbilical hernia eliminated; claws formed on all digits.

42 Days
Fetus: 70 mm.; ossification of dens of axis; caudal vertebral centra 6 through 11; sternebrae 1 through 5; distal phalanges 1 through 5 in both paws; proximal phalanges 3 and 4 in both paws.

43 Days
Fetus: 76 mm.; ossification of body of atlas (intercentrum); caudal vertebral centra 1 through 14, arches 6 through 8; sternebrae 1 through 7;

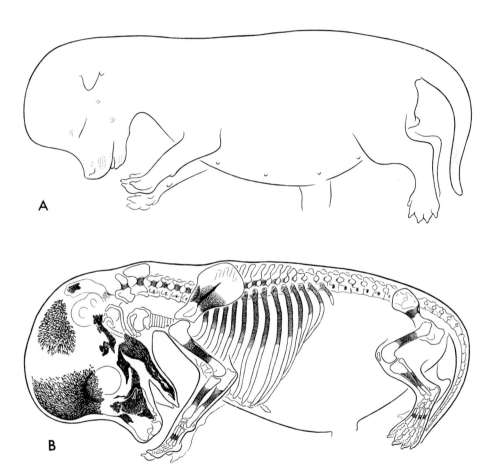

Figure 2–63. Developmental status on the fortieth day of gestation.
A. External features.
B. The skeleton.

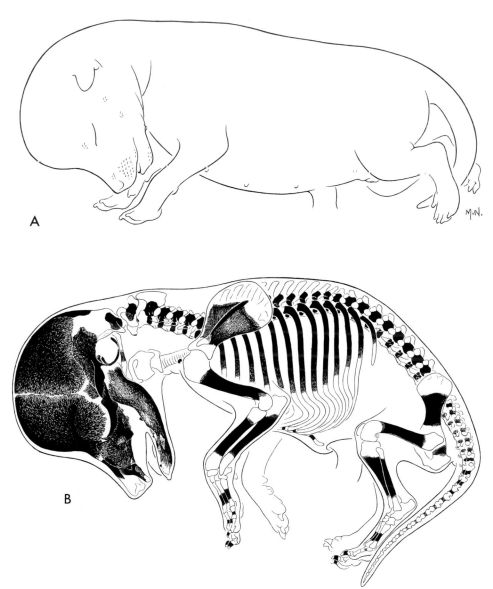

Figure 2–64. Developmental status on the forty-fifth day of gestation.
A. External features.
B. The skeleton.

proximal phalanges 1 through 5 in forepaw, 2 through 5 in hindpaw.

45 Days (Fig. 2–64)
Bitch: Uterine swellings less distinct from one another; uterus bent upon itself to conform with available space in the abdomen; width of placenta less than the length of fetus.
Fetus: 86 mm.; color markings appear, and body hair begins to grow; scrotal swellings are large, and labia are prominent; calcification of lower premolars; interparietal bone ossifies independently and then fuses with the supraoccipital; ossification of presphenoid; vertebral centra C1 through Cd17, arches C1 through S3, Cd 2 through Cd4, Cd9 through Cd13; middle phalanges 3 and 4 of forepaw; phalanx 3 of hindpaw; ischium.

Figure 2–65. Developmental status on the fiftieth day of gestation.
A. Exterior of the uterus.
B. Placental bands, myometrium removed.
C. External features of the fetus.

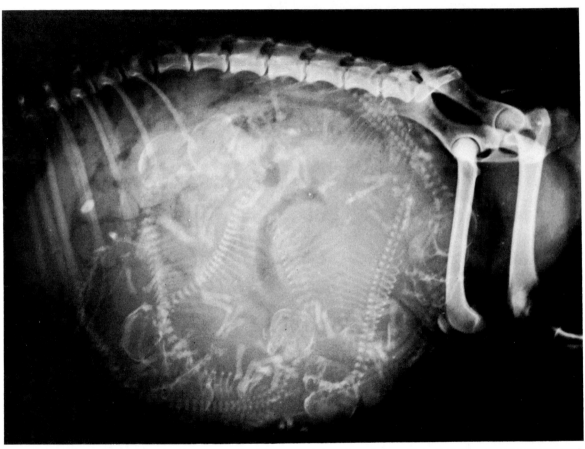

Figure 2–66. Radiograph of a Beagle on the fifty-second day of gestation. Note that the pups are distributed throughout the abdominal cavity, indicating the folded nature of the enlarged uterine horns (Evans 1956).

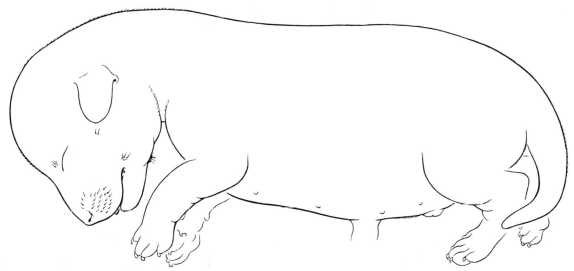

Figure 2–67.   External features on the fifty-fifth day.

50 Days (Fig. 2–65)

Bitch: The uterus has enlarged to such a degree that individual swellings are no longer apparent; adjacent fetuses are in contact.

Fetus: 107 mm.; body well proportioned; caudal nipples of male involute as penile structures grow cranially; ossification of lateral ethmoid, petrosal, incus, stylohyoid; vertebral centra and arches C1 through Cd18; sternebrae 1 through 8; all metacarpals and phalanges of forepaw.

55 Day (Fig. 2–67)

Bitch: The uterus is very large, and the fetuses can move freely within the placental band.

Fetus: 144 mm.; all deciduous teeth show calcification, and lower first molar calcifies (a permanent tooth); ossification of thyrohyoid and epihyoid; sacral "ribs" 1 and 2; all vertebral arches and centra; all metapodials and phalanges; pubis; calcaneus.

57 Days (Fig. 2–68)

Fetus: 150 mm.; ossification of basihyoid; sacral wing of S1; talus.

60–63 Days — Whelping

Bitch: Restless, prepares bed, cervix dilates, fetal movement apparent.

Pup: 158 to 175 mm. long; well haired; eyelids closed; carpals not ossified, tarsals not ossified except for calcaneus and talus.

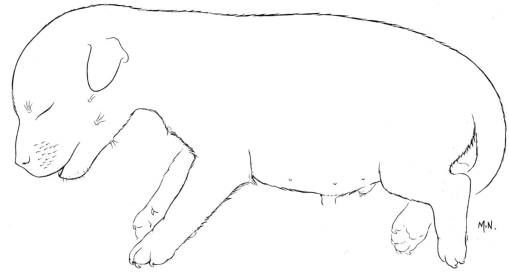

Figure 2–68.   External features on the sixtieth day, at the time of birth.

BIBLIOGRAPHY

Ahmed, I. A. 1941. Cytological analysis of chromosome behavior in three breeds of dogs. Proc. Roy. Soc. Edinburgh 61:107–118.

Amoroso, E. C. 1952. Placentation, In Marshall's Physiology of Reproduction, edited by A. S. Parkes. Boston, Little, Brown & Co., Vol. 2, pp. 127–311.

Anderson, D. 1927. The rate of passage of the mammalian ovum through various portions of the fallopian tube. Am. J. Physiol. 82:557–569.

Anderson, J. W. 1969. Ultrastructure of the placenta and fetal membranes of the dog. 1. The placental labyrinth. Anat. Rec. 165:15–36.

Asdell, S. A. 1966. Dog Breeding. Reproduction and Genetics. Boston, Little, Brown & Co.

Barrau, M. D., J. H. Abel, C. A. Torbit, and W. J. Tietz. 1975. Development of the implantation chamber in the pregnant bitch. Am. J. Anat. 143: 115–130.

Billington, W. D. 1971. Biology of the trophoblast. In Advances in Reproductive Physiology, Vol. V, edited by M. Bishop. New York, Academic Press.

Bischoff, Th. L. W. 1844. Beweis der von der Begattung unabhängigen periodischen Reifung und Loslösung der Eier der Säugetiere und des Menschen als die erste Bedingung ihrer Fortpflanzung. Giessen.

Boyd, J. D., W. J. Hamilton, and J. Hammond, Jr. 1944. Transuterine (internal) migration of the ovum in sheep and other mammals. J. Anat. 78:5–14.

Chen, J. M. 1953. Studies on the morphogenesis of the mouse sternum. III Experiments on the closure and segmentation of the sternal bars. J. Anat. 87:130–149.

Cloete, J. H. L. 1939. Prenatal growth in the Merino sheep. Onder. J. Vet. Sci. Anim. Indust. 13:417–564.

Concannon, P., W. Hansel, and K. McEntee. 1977. Changes in LH, progesterone and sexual behavior associated with preovulatory luteinization in the bitch. Biol. Reprod. 17:604–613.

Concannon, P. W., W. Hansel, and W. J. Visek. 1975. The ovarian cycle of the bitch: plasma estrogen, LH and progesterone. Biol. Reprod. 13:112–121.

Concannon, P. W., M. E. Powers, W. Holder, and W. Hansel. 1977. Pregnancy and parturition in the bitch. Biol. Reprod. 16:517–526.

Creed, R. F. S. 1963. Haemophagocytic structures in the placenta of some Carnivora. J. Physiol. 170:44–45.

Creed, R. F. S., and J. D. Biggers. 1963. Development of the raccoon placenta. Am. J. Anat., 113:417–445.

de Beer, G. R. 1937. The Development of the Vertebrate Skull. Oxford, Oxford Univ. Press.

Dingerkus, G., and L. D. Uhler. 1977. Enzyme clearing of alcian blue stained whole small vertebrates for demonstration of cartilage. Stain Technol. 52:229–232.

Doak, R. L., A. Hall, and H. E. Dale. 1967. Longevity of spermatozoa in the reproductive tract of the bitch. J. Reprod. Fertil. 13:51–58.

Drews, M. 1933. Uber Ossifikationsvorgange am Katzen und Hundeschadel. Gegenbaurs Morphol. Jahrb., 73:185–237.

Evans, H. E. 1948. Clearing and staining small vertebrates, in toto, for demonstrating ossification. Turtox News 26:42–47.

– – – – – . 1956. A dog comes into being. Gaines Dog Research Progress, Fall 1956. Gaines Dog Research Center.

– – – – –. 1974. Prenatal development of the dog. Gaines Vet. Symp. at Cornell Univ., 1974, pp. 18–28.

Evans, H. E., and W. O. Sack. 1973. Prenatal development of domestic and laboratory mammals. Zbl. Vet. Med. C. Anat. Histol. Embryol. 2:11–45.

Evans, H. M., and H. H. Cole. 1931. An introduction to the study of the oestrus cycle of the dog. Mem. Univ. Calif. 9:65–103.

Gier, H. T. 1950. Early embryology of the dog. Anat. Rec. 108:561–562.

Griffiths, W. F. B., and E. C. Amoroso. 1939. Prooestrus, oestrus, ovulation and mating in the Greyhound bitch. Vet. Rec. 51:1279–1284.

Heape, W. 1900. The sexual season of mammals. Q. J. Microscop. Sci. 44:1–70.

Hill, J. P., and M. Tribe. 1924. The early development of the cat. (Felis domestica). Q. J. Microscop. Sci. 68:513–602.

Holst, P. A., and R. D. Phemister. 1971. The prenatal development of the dog: preimplantation events. Biol. Reprod. 5:194–206.

– – – – – . 1974. Onset of diestrus in the Beagle bitch: definition and significance. Am. J. Vet. Res. 35:401–406.

Markee, J. E., and J. C. Hinsey. 1933. Internal migration of ova in the cat. Proc. Soc. Exp. Biol. Med. 31:267–270.

Minouchi, O. 1928. The spermatogenesis of the dog, with special reference to meiosis. Jap. J. Zool. 1:255–268.

Moore, W., Jr., and P. D. Lambert. 1963. The chromosomes of the Beagle dog. J. Hered. 54:273–276.

Olmstead, M. P. 1911. Das Primordialcranium eines Hundembryo. Anat. Hefte 130:339–375.

Phemister, R. D. 1974. Nonneurogenic reproductive failure in the bitch. Vet. Clin. North Am. 4:573–586.

Phemister, R. D., P. A. Holst, J. S. Spano, and M. L. Hopwood. 1973. Time of ovulation in the Beagle bitch. Biol. Reprod. 8:74–82.

Schliemann, H. 1966. Zur Morphologie und Entwicklung des Craniums von Canis lupus familiaris. Gegenbaurs Morphol. Jahrb. 109:501–603.

Schoenfeld, H. 1903. Contributions à l'étude de la fixation de l'oeuf des mammifères dans la cavité utérine des premiers stades de la placentation. Arch. Biol. 19:701–830.

Simons, E. V., and J. R. Van Horn. 1971. A new procedure for whole-mount alcian blue staining of the cartilaginous skeleton of chicken embryos, adapted to the clearing procedure in potassium hydroxide. Acta Morphol. Neerl. Scand. 8:281–292.

Sokolowski, J. H., D. G. Stover, and F. Van Ravenswaay. 1977. Seasonal incidence of estrus and interestrous interval for bitches of seven breeds. J. Am. Vet. Med. Assoc. 171:271–273.

Streeter, G. L. 1951. Developmental horizons in human embryos. Carnegie Institute Wash. Publ. 592. Contrib. Embryol. 34:165–196.

Tietz, W. J., and W. G. Selinger. 1967. Temporal relationship in early canine embryogenesis. Anat. Rec. 157:333–334.

Van der Stricht, O. 1923. The blastocyst of the dog. J. Anat. 58:52–53.

Watson, A. G. 1977. In toto alcian blue staining of the cartilaginous skeleton in mammalian embryos. Anat. Rec. *187*:743.

Williams, R. C. 1961. Observations on the Chronology of Deciduous Dental Development in the Dog. Thesis, Cornell Univ., Ithaca, N. Y.

Williams, R. C., and H. E. Evans. 1978. Prenatal dental development in the dog, *Canis familiaris*: chronolo-gy of tooth germ formation and calcification of decid-uous teeth. Zb1. Vet. Med. C. Anat. Histol. Embryol. 7:152–163.

Wimsatt, W. A. 1974. Morphogenesis of the fetal mem-branes and placenta of the black bear, *Ursus ameri-canus* (Pallas). Am. J. Anat. *140*:471–496.

Zietzschmann, O., and O. Krolling. 1955. Lehrbuch der Entwicklungsgeschichte der Haustiere. 2nd Ed. Ber-lin, Paul Parey.

Chapter 3

THE INTEGUMENT

By FAKHRI AL-BAGDADI AND JAMES LOVELL

The **common integument** *(integumentum commune)* includes the skin, hair, claws, pads, and associated glands, including the mammary glands.

The **skin** *(cutis)* performs many important functions. In addition to preventing desiccation and hydration, the integument informs the central nervous system of its contacts (as a sensory organ, the skin is the receptor for the perception of touch, pressure, vibration, heat, cold, and pain), prevents trauma, protects the body from the invasion of microorganisms and noxious chemicals, and regulates temperature change. In regard to heat regulation, however, the skin of the dog serves only a limited role via sweat glands and superficial capillary beds because of the alternate route of thermal panting. The skin also acts as the site of vitamin D synthesis, and the subcutaneous tissue serve as a reservoir for fat. Finally, secretions of the glands associated with the skin not only waterproof and lubricate it, but also function as pheromones for recognition and, in the case of the mammary glands, as nourishment for the young.

The skin is often subject to allergic reactions, dermatitis, and parasitic invasion. By being icteric, cyanotic, dry, edematous, and so on, it may reflect the general health of the animal.

In terms of structure, the skin consists of a connective tissue bed containing blood vessels, lymphatics, muscles, and nerve endings covered by a stratified squamous epithelium. It is continuous at the natural body openings with the mucous membrane of the digestive, respiratory, and urogenital tracts as well with the conjunctivae of the eyelids, the lacrimal duct, and the tympanic membrane.

The greater surface of the body is covered with hair. The **hair coat** *(pili)*, consisting of **cover hairs** *(capilli)* and **wool hairs** *(pili lanei)*, is most dense on the dorsal and lateral portions of the body, while the abdomen, the inside of the flanks, the inside of the ears, and the underside of the tail are sparsely haired. Some areas, such as the nasal surface and carpal, metacarpal, tarsal, metatarsal, and digital pads, are hairless. The **claws** *(unguicula)* are horny coverings of the third phalanges of the digits.

There are large tactile hairs *(pili tactiles)* on the muzzle *(pili labiales maxillares)*, mandible *(pili labiales mandibulares)*, and above the eyes *(pili supraorbitales)*. There are usually two genal tubercles on each side of the face from which long hairs grow. A tuft of inter-ramal hairs stands out from surrounding hair between the rami of the mandible. The specialized hairs of the eyelids *(pili palpebrae)* or eyelashes are stiff and larger than other hairs. The ventral body surface is characterized by the **median raphe of the linea alba** *(linea mediana ven-*

tralis), the hairless **umbilicus** *(regio umbilicalis)* and **nipples** *(papilla mammae),* and the sparsely haired **mammary glands** *(glandula mammaria),* which are arranged in two rows with four to six in each row. The hairy skin is thickest over the neck and back, where it is loosely attached.

All skin areas are composed of epidermis and dermis (corium). When the skin is removed during dissection, subcutaneous tissue and some thin, small cutaneous muscles often remain with it, although they are not components of the skin. The skin, hair, and subcutis of the newborn puppy represents approximately 24 per cent of the total body weight. Owing to differential growth of various body parts, this percentage is reduced to 12 per cent by six months of age. Additional information concerning canine skin may be found in: Blackburn (1965), Calhoun and Stinson (1976), Epling (1962), Lovell and Getty (1957), Muller and Kirk (1969), Warner and McFarland (1970), and Webb and Calhoun (1954).

EPIDERMIS

The **epidermis** overlying hairy skin varies from 25 to 40 micrometers (microns) in thickness. The thicker portions are found near the hair follicle orifices and the hairy margins of the mucocutaneous junctions. The epidermis of non-hairy skin varies in thickness owing to the system of ridges that occur at the dermal-epidermal junction, which is found in the non-hairy margin of the lip, eyelid, prepuce, vulva, and anus. The thickest epidermis occurs on the nasal skin and digital pads. The epidermis of the planum nasale is 200 micrometers at the time of birth and 600 micrometers at six months of age. The epidermis of the foot pads is 200 micrometers at birth and increases to 1800 micrometers at six months of age.

DERMIS

The thickness of the **dermis** or corium varies in different body areas and at different ages. The dermis of the planum nasale and foot pads is 300 micrometers at birth and increases to 800 micrometers at six months of age. The differences in thickness of the hairy skin that can be observed by comparing the dorsal area to the abdominal area is due to the difference in thickness of the dermis. The dermis of the skin of the dorsal region is 700 micrometers at birth and increases to 1500 micrometers at six months of age. The dermis in the abdominal region is 300 micrometers at birth and increases to 800 micrometers by six months of age.

Structure of the Dermis and Changes with Age

The dermis is composed of fibroblasts, fibers, and various structures such as blood vessels, nerves, cells of blood or tissue origin, and tissue fluid. Skin glands and hair follicles are embedded in the dermal tissue. The fibers of smaller diameter are found in the superficial dermis just under the epidermis. The larger collagenous fibers are in the deeper layer of the dermis. The size of the fibers, the population density of fibroblast nuclei, and the plasma cell content of the dermis undergo a developmental change with the increase in thickness that occurs between birth and six months of age.

At the time of birth, there are many reticular fibers throughout the dermis. By three weeks most of them have been replaced by collagen fibers, and only a few reticular fibers remain just under the epidermis, around hair follicles, and surrounding sebaceous and sweat glands. Collagen fiber bundles increase in size and number as the dermis thickens with age. At birth, the collagen fiber bundles measure 3 or 4 micrometers in diameter, and at six months they measure 19 to 20 micrometers. Concurrent with this gradual increase there is a corresponding decrease in the size of the spaces between the fiber bundles. There is also an increase in the size and number of elastic fibers. At birth, the elastic fibers are small, branching, filamentous fibers, less than ½ micrometer in diameter. At six months they are thicker undulating fibers of 1½ to 2 micrometers in diameter. Irwin (1966) illustrated the tension lines in dog skin that are determined by orientation of connective tissue, gravity, and physical forces.

Fibroblast nuclei in the dermis are more

densely distributed at birth than at six months of age. Per unit area, there are twice as many nuclei at birth as at six months. Mast cells are present in greatest numbers in the reticular layer of the dermis, with fewer in the papillary layer and subcutis. They are usually adjacent to small vessels and frequently surround sebaceous glands and apocrine or eccrine glands. They are most numerous in the skin of the ears and appear in decreasing numbers in the skin of the vulva, prepuce, medial thigh, foot pad, and external nares (Emerson and Crass 1965). This may be related to histamine levels, as studied by Copeman and Gaafar (1967). Sebaceous glands double or triple in diameter during the first six months, and hair follicles double in length during this time.

PIGMENTATION

Conroy and Beamer (1970) studied the development of melanoblasts and melanocytes in the skin of Labrador retriever fetuses. The earliest melanoblasts were demonstrable in the primordial dermis of the head, thorax, and abdomen of 29-day fetuses. The melanoblasts were most numerous in the lower two-thirds of the primordial dermis in contact with or near blood vessels. The frequent contact or close association of melanoblasts with blood vessels suggests that the cells migrate along blood vessels in their journey from the neural crest to their destination in the epidermis. Dendritic dermal melanocytes first appeared in the 37-day fetus, and scattered epidermal melanocytes appeared in 29-day fetuses. Their numerical distribution in various regions of the body conformed to a dorsoventral gradient. Melanocytes were present in primordial and differentiating hair follicles, eccrine sweat glands, and sebaceous glands.

Skin pigment was not visible in fetuses of less than 33 days. Retinal pigment was first detected in a 25-day embryo, while skin pigment was first readily observed in a 37-day embryo. In 40-day fetuses, cutaneous pigmentation was prominent on the muzzle, eyelids, and ears. A fetus of 46 days was heavily pigmented except for parts of the digital pads, the central part of the planum nasale, the metapodial pads, the median

upper and lower lip, the chin, the hard palate, and the tongue. Skin pigmentation was most prominent on the dorsal and lateral aspects of the head and body. A nearly full-term fetus of 55 days had abundant hair and was completely pigmented externally, including digital pads and claws, although the claws were somewhat less pigmented than the integument. The oral cavity was not pigmented except for the lips.

NASAL SKIN

The nasal skin is usually heavily pigmented, tough, and moist. On close examination of the surface of the *planum nasale*, polygonal, plaquelike areas are observed which give the nasal skin an irregular appearance (Fig. 3–1B).

On histological examination, no glands can be demonstrated in the epidermis or dermis of nasal skin (Figure 3–1C). The moisture that appears on the nasal surface is derived primarily from serous gland secretions of the lateral nasal gland and other glands that drain into the vestibule (Blatt, Taylor, and Habal 1972).

The **dermis** of nasal skin is composed of reticular, collagenous, and elastic fibers, together with fibroblasts, blood vessels, and nerves. The blood vessels and nerves are larger in the deeper layers of the dermis than in the more superficial layers. Directly under the epidermis, the dermal papillae interdigitate with epidermal projections to form an irregular line of attachment between the dermis and epidermis (Calhoun and Stinson 1976). The epidermal surface is marked by deep grooves which divide it into polygonal areas.

The **epidermis** of the nasal skin, which averages 630 micrometers in thickness in adult dogs, is composed of three layers: stratum basale, stratum spinosum, and stratum corneum. The *stratum basale* of the epidermis rests on the condensed, thickened superficial portion of the dermis and consists of one layer of cylindrical cells. The *stratum spinosum* is made up of 10 to 20 layers of diamond-shaped, dome-shaped, or flattened polygonal cells which have a lighter-straining cytoplasm than the cylindrical cells. In heavily pigmented nasal skin there are many pigment granules in the cyto-

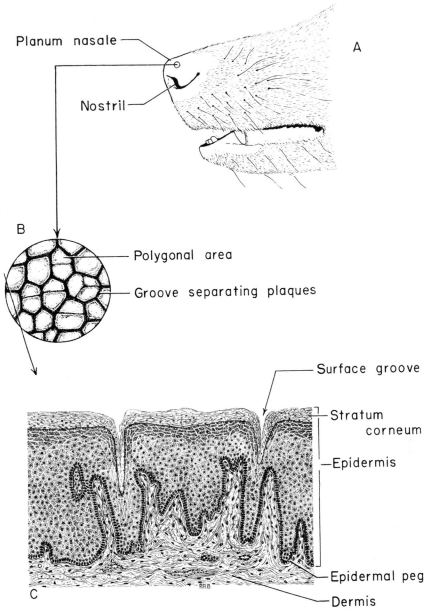

Figure 3-1. The planum nasale. (After Lovell and Getty 1957.)
A. Area from which enlargement was made.
B. Surface plaques.
C. Histological section.

plasm. There is no stratum granulosum or stratum lucidum in the epidermis of the nasal skin. The more peripheral spinosum cells apparently do not undergo keratinization, as they do in other regions of epidermis. Their cytoplasm becomes weakly acidophilic, and the nuclei become pyknotic, the cells flattening out into a squamous type. As they approach the surface, they remain as a thin, atypical nucleated *stratum corneum,* four to eight cell layers thick. The stratum corneum of the nasal skin is surprisingly thin for such an exposed and tough-appearing area of the body (Fig. 3–1C).

DIGITAL PADS

The skin of the digital pads is usually heavily pigmented and is the toughest region of canine skin. The surface of the pads is rough, owing to the presence of numerous conical papillae which are heavily keratinized and are readily seen with the naked eye (Fig. 3–2B). When dogs are kept on concrete or rough surfaces, the papillae sometimes become worn smooth so that they are rounded instead of conical in shape.

The **digital cushion** or base of the foot pad is made up of subcutaneous adipose tissue which is partitioned by reticular, collagenous, and elastic fibers (Fig. 3–2C). Many elastic fibers are present in the deeper layers. Eccrine sweat glands and lamellar corpuscles are embedded in the adipose tissue. The excretory ducts of the eccrine sweat glands are found in the dermis, through which they carry secretion to the surface of the epidermis. Directly under the epidermis, the connective tissue is dense

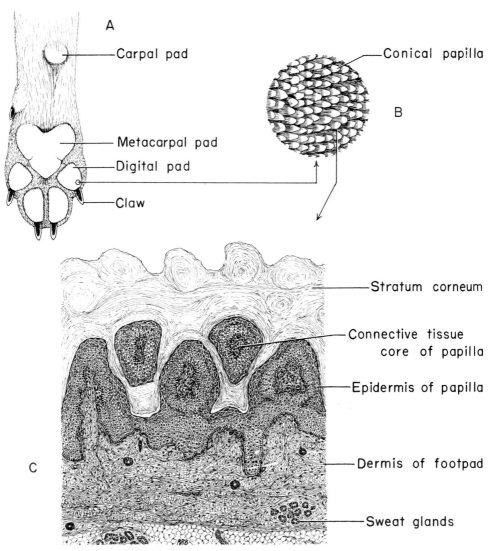

Figure 3–2. Surface contour and histology of foot pads. (After Lovell and Getty 1957.)
A. Gross appearance of pads.
B. Conical papillae arranged on surface of digital pad.
C. Histological section of foot pad.

and papillate, forming conical dermal cores for the epidermal papillae. There are also secondary dermal papillae within the conical structure.

The epidermis of the digital pad, which averages 1800 micrometers in thickness in the adult dog, is composed of five layers: stratum basale, stratum spinosum, stratum granulosum, stratum lucidum, and stratum corneum. The *stratum basale* is made up of a single layer of basal cells resting on the connective tissue of the dermis. The *stratum spinosum* is composed of 10 to 15 layers of diamond- or dome-shaped cells. In both the digital pads and the planum nasale, cell outlines and intercellular bridges (demosomal attachments) may be observed on the spinous cells. The *stratum granulosum* is made up of four or five layers of flattened cells which contain basophilic keratohyalin granules in their cytoplasm. The *stratum lucidum* appears as a shiny acidophilic layer of homogeneous substance with refractile droplets called eleidin (Bloom and Fawcett 1975). The *stratum corneum* of the digital pads consists of a thick layer of keratinized, non-nucleated material thicker than all the cellular layers combined (Fig. 3–2C). The excretory ducts of the eccrine sweat glands of the digital pad become continuous with the epidermis at the depths of the epidermal pegs, where their epithelium joins with the stratum basale of the epidermis. The lumen of the excretory duct then follows a tortuous path through the epidermal cells to the surface, where the glandular secretion is expelled.

HAIRY SKIN

The basic unit of hair production is the individual **hair follicle** *(folliculus pili)*. The follicle wall, which is continuous with the surface epithelium, is divided into two layers, the outer and inner root sheaths. The follicle attains its greatest diameter at the base, where it is dilated to form a bulb in which the hair-producing matrix is contained. Invaginating the bulb is the dermal papilla, which supplies the germinative epithelium by diffusion as long as the hair is growing. The hair shaft consists of a central medulla; a thick cortex, which forms the bulk of the hair; and a single-layered cuticle

on the outside. The keratin shaft of the hair is formed by the germinative epithelium, which is active only during the time of hair growth. There are periods during which the growth of the hair is arrested. At this time there is a regression of the hair root, and the dead hair is held in the follicle completely disconnected from the inactive germinal matrix. After a variable period of time, the dormant germinal cells become active and enter a period of organogenesis in which a new hair root is regenerated and production of hair is resumed. At this time the old dead hair will be shed and replaced by the new hair. Growing hair follicles are said to be in **anagen** and quiescent ones in **telogen**; the period of transition between the two is called **catagen**.

Growth Rate of the Hair Shaft

Differences may be observed in hair growth rates in various breeds and during certain seasons of the year. Al-Bagdadi (1975) found the average rate of daily hair growth in male beagle dogs was 0.40 mm. per day in the winter and 0.34 mm. per day in summer. Comben (1951) reported determinations from male Greyhounds to be 0.04 mm. per day in summer and 0.18 mm. per day in the fall. Although the two observers found widely different values, which may reflect breed difference, they agreed that the daily growth rate of the hair shaft was greater during the colder season than it was in the warmer time of the year.

Embryology of Hair Follicles

The terms pregerm, hair germ, hair peg, and bulbous peg were used by Pinkus (1958) to designate developmental stages of the canine hair follicle. In a study of the development of cutaneous pigment, Conroy and Beamer (1970) described the embryological development of the canine hair follicle.

The first hairs to appear in the 29-day fetal dog are in the region of the eyebrows, upper lip, and chin (see Chapter 2). These develop into the large hairs of the face, which will become specialized sinus hairs. The follicles of the general hairy skin appear in the

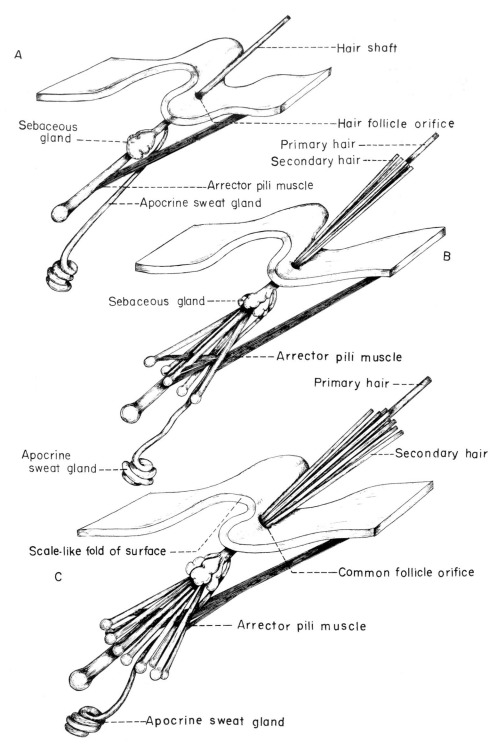

Figure 3–3.   Development of the postnatal hair follicle of the dog, schematic.
A. Simple hair follicle during the first postnatal week.
B. Compound hair follicle during the twelfth postnatal week.
C. More elaborate compound hair follicle during the twenty-eighth postnatal week.

pregerm stage on the head and neck as early as 30 days. They reach the hair-germ stage at 32 days and the hair-peg stage at 37 days of gestation. In the general development of the pelage, the hairs are farthest advanced cephalad, and the development spreads caudad and ventral. The primary hair germs form more or less simultaneously at fairly even distances. As the skin grows, increasing the surface area, new primary germs develop among the earlier ones. This results in two, three, or four groups of follicles being clustered together. The triad arrangement is most frequent, with a large central primary follicle bounded on either side by smaller lateral primaries. Later, the secondary germs develop close to the primary ones and form the compound follicle arrangement. This process starts before birth and is completed after full term (Fig. 3–3).

The first evidence of the follicle in the embryo is seen as a thickening of the epidermis (pregerm stage) at regular intervals. The pregerm stage passes rapidly into the hair-germ stage as the basal cells become higher and the entire structure protrudes downward into the dermis. From its point of origin, the hair germ grows obliquely downward into the mesenchyme in the form of a solid column. This is called the hair-peg stage. The advancing border enlarges, becomes bulbous, and envelops part of the mesenchymal material which was pushed down ahead of the invading epidermal cells, thus becoming the bulbous-peg stage. Later, the hair bulb and the dermal papilla become differentiated into the productive hair follicle complete with glandular and muscular accessories. As the dermis increases in thickness between birth and six months of age, the length of the hair follicles increases. In 40-day fetuses, all stages of follicle development are present. In 55-day fetuses, most primary hair follicles have developed to the bulbous-peg stage. The secondary hair follicles begin development before birth but usually have no external hair shaft until after full term.

Development of Compound Follicle

Observations of the development of the hair follicle in the dog have been reported by Lovell and Getty (1957), Baker (1966),

and Al-Bagdadi et al. (1977). It can be observed by examining the hair coat of a puppy during the first few days after birth that there is usually only a single hair emerging from each external follicle orifice of the skin. On microscopic examination it can be observed that secondary follicles are forming as strands of intensely basophilic cells running deeply into the dermis from their point of origin just below the sebaceous gland of the primary follicle (Figure 3–4B). These satellite or accessory hairs appear externally at three to four weeks of age, when each primary follicle can be seen to be giving rise to two or three secondary hairs. At 8 to 10 weeks of age the secondary follicles can be seen arranged in a crescent about the central and lateral primary hairs. They are on the same side as the apocrine gland. Subsequently, secondary follicle formation continues until puberty, when from 6 to 10 or more hairs may emerge from a single follicle orifice. The larger primary hairs have a better developed nerve and blood supply than do the satellite hair follicles. As a general rule, the coarser guard-hairs appear earlier than the satellite hairs. In the development of the dog, the nature of the puppy hair changes, and in the young adult, the fine, fluffy hair of the pup is replaced by a coarser hair. In a young adult dog, the hair growth is profuse and abundant. In old dogs the hair covering is thinner, the hairs are not as long, and frequently the coloring changes to gray. Also, the hairs are more brittle and are accompanied by loss of flexibility of the skin and subcutaneous tissue. Baker (1967) illustrated the atrophy of the epidermis that takes place in aging dogs.

Compound Hair Follicle

The hairy skin of an adult dog contains bundles of hairs which share common openings in the surface. These bundles are usually arranged in groups of three oriented into irregular rows. The typical bundle consists of a group of under-hairs and a single longer and stiffer cover-hair (Fig. 3–4A and B). The cover-hair of a three-bundle group is coarser than those in the lateral bundles.

The hair shafts that share a common open-

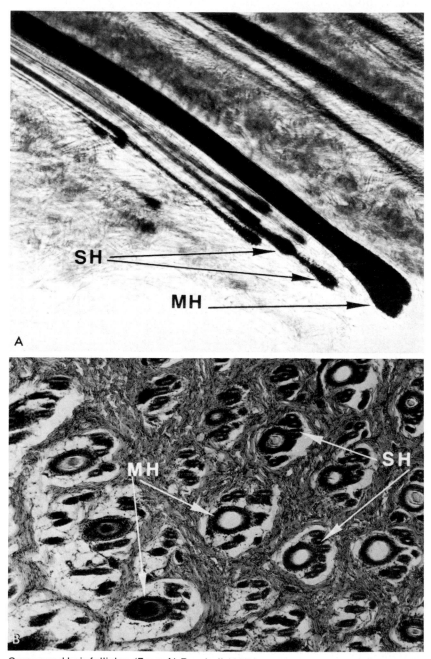

Figure 3—4.   Compound hair follicles. (From Al-Bagdadi 1975.)

A. Longitudinal section from a two-week-old Beagle. The bulbs of the hairs making up the complex extend to different levels of the dermis. The main hair (MH) extends into the adipose tissue, which is in the subcutis (175 ×).

B. Transverse section. The compound hair follicles are in groups of one main (MH) and six to eight secondary (SH) follicles (130 ×).

ing in the skin are enclosed in a common follicle down to the level of the sebaceous glands. Below this point the hair shafts branch away from one another into their own individual hair follicle and bulb (Figure 3–3). In this way, as many as 15 hairs may share a single external follicle orifice. The individual follicle and hair bulb of the cover-hair or guard-hair is larger and penetrates more deeply into the subcutaneous tissue than those of the subsidiary hairs (Fig. 3–4A). There are breed variations in the number of follicle bundle complexes per square centimeter and also in the number of hairs in each bundle. Brunsch (1956) found that Smooth-haired Dachshunds, Smooth-haired Terriers, and Toy Poodles have 400 to 600 bundles per square centimeter. German shepherds, Airedales, and Rottweilers had only 100 to 300 bundles per square centimenter. The other breeds were somewhere in between these figures. The number of hairs per bundle varied from 9 to 15 in the Rottweiler to two to five in the dachshund. In general, the hair of those breeds which have many bundles is finer than in those breeds which have a smaller number of hair bundles. Compared to the larger animals of the same breed, midget animals have a greater number of bundles with fewer and finer hairs.

The canine hair follicle could be defined as a pilosebaceous arrector muscle complex. The sebaceous glands of the individual hair follicles bunch together in clusters and sometimes fuse. The arrector pili muscles originate from the outer root sheath of each hair follicle and then join together into a common muscle bundle which is inserted in the dense dermal tissue beneath the epidermis. When the arrector pili muscle contracts, the entire complex of follicles is elevated and the sebaceous gland material empties into the upper common follicle structure, which is shared by all the hairs. A single apocrine gland is associated with each follicle complex. The coiled secretory tubule extends deep into the subcutaneous tissue. A direct extension of this tubule becomes the excretory duct for the apocrine secretion, which extends up along the follicle complex and empties into the common part of the follicle above the opening of the sebaceous glands. The apocrine glands are sweat glands but do not play a major role in the heat-regulating mechanism of the dog. They are comparable to the apocrine sweat glands associated with hair follicles of the axillary and pubic regions of man. The oily secretion from the glands associated with the hair follicles tends to keep the skin soft and pliable and spreads out over the hair shafts. This gives the coat a glossy sheen. During periods of sickness, malnutrition, or parasitism, the hair coat frequently becomes dull and dry as a result of inadequate functioning of the skin glands.

Odorous glandular secretions produced by one individual which evoke a response in another via olfactory receptors have been called "pheromones." Donovan (1969) suggests that canine anal glands produce several pheromones, including a sexual attractant or deterrent or both.

Hair Types in the Dog

There is a great deal of variability in hair length, color, diameter, and transverse contour among the various breeds of dogs and between individuals of the same breed (Fig. 3–5). Brunsch (1956) has classified canine hairs into six types:

1. Straight Hair. This is a bristly, firm hair, often deeply pigmented. It is sometimes called a protective hair, primary hair, or cover-hair. This is the strongest hair and is the chief hair in the follicle bundle. It is also usually the longest hair, and the shaft is either straight or bowed. It has a thick medulla and a thin cortex.

2. Bristle Hair. This is a bristle with a spinelike tip, but weaker and softer near the base. The distal third is similar to type 1, but the proximal two-thirds may be slightly wavy. In the hair coat it is difficult to distinguish from type 1. The medulla is slightly smaller than that of type 1. The bristle hair is shorter than the straight hair, but is regarded as an over-hair or protective hair. This type may be the chief hair in a bundle, but is usually a subsidiary hair to type 1.

3. Wavy Bristle Hair. This type is finer and shorter than type 2. It is wavy with a well-developed bristle. These are the larger subsidiary hairs, but are usually included with the cover-hairs or protective hairs. The medulla and cortex are smaller than in type 2, but the cortex is relatively heavier.

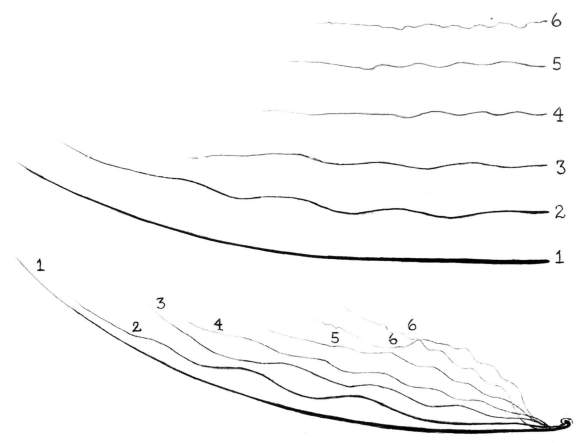

Figure 3–5. Hair types in the dog (see text for description).

4. Bristled Wavy Hair. This is a long, soft hair which is shorter and finer than type 3, with a poorly developed bristle and a smaller medulla. It is wavy in the lower two-thirds of the shaft. This type represents the largest hairs of the undercoat.

5. Large Wavy Hair. This type is shorter and finer than type 4, and the shaft is very wavy with a small bristle on the tip. The medulla is very small and may be discontinuous. The cortex is relatively thick. This type gives a furlike or wool-like feel to the undercoat.

6. Fine Wavy Hair. This type is shorter and finer than type 5 and is sometimes described as vellus hair, fuzz, down, or lanugo hair. The medulla is discontinuous or absent. This type represents the finest and smallest hairs of the undercoat and is usually wavy with a small and poorly developed bristle on the tip.

Variability in Hair Coat of the Dog

The formation of bristles at the tips of the hair shafts of greater diameter than the remaining parts of the hairs suggests that the early part of the hair growth cycle is the most productive. Bristle formation is greater in hair types 1, 2, and 3 than in types 4, 5, and 6. Brunsch (1956) explains the difference in growth intensity of different follicles on a physiological basis. Those follicles with a rich blood supply and source of metabolites will synthesize more hair shaft. Thus, smaller follicles with less blood supply produce smaller hairs.

Gair (1928) classified the coat of the dog into three groups on the basis of hair length. The normal coat, which resembles the hair covering of wild canidae (Wolf, Jackal), is typified by the German shepherd. The short-hair type is represented by the Boxer.

The long-hair type may be illustrated by the Chow. There are many variations among the long-haired types, such as wire hair, tight, curly hair, and flat, long hair.

The various coats observed in domestic breeds of dogs are made up of the six types of hair described by Brunsch (1956), with some exceptions. The wire-haired breeds, such as the Schnauzer, have a preponderance of bristle-type hairs, with a seventh type not found in other breeds. The Cocker spaniel and the Setter have fine, long silky hairs with less obvious bristle development. The Poodle has extremely long hair which resembles the wool hair type. The medullary canal of a Poodle hair is greatly reduced or absent. Bristle formation in this breed is characterized by a rhythmical pattern of differences in the thickness of the hair, thus suggesting continous growth with variations in growth intensity.

Coat Color in the Dog

The color of the hair shaft is produced by pigment cells in the bulb of the hair follicle. From these cells granules of pigment enter cortical and medullary cells during development. The granules may remain between the cells, as is the case in the medulla, but most of them are engulfed by the cells. The amount of pigment and variations in location produce different optical effects. The pigmentation may be uniform through the entire length of the hair, or it may vary. In the agouti type of hair, which is found in wolves and in some breeds of dog (German shepherd and Norwegian elkhound), the tip of the hair is white and the thick part of the bristle is heavily pigmented (black or dark brown), the proximal two-thirds of the hair having lighter pigmentation (yellow or red).

Despite the wide range of colors that are possible in the coat, microscopic examination has revealed only black, brown, and yellow pigment granules. The black-brown pigment is designated as "tyrosine-melanin," since it is formed by enzyme oxidation of tyrosine to melanin. The yellow-red pigment is designated as "pheomelanin." Its origin is unknown. Da Fonsica and Cabral (1945) have classified the dog's coat according to color and pattern into three

types: simple, compound, and mixed. The studies of inheritance and genetic control of color and coat patterns have been summarized by Burns (1966) and Little (1957).

Hair Length

The length of the hair is controlled to a large extent by the genetic make-up of the individual. A short coat is dominant to long; straight or wavy types are recessive or partially recessive to wire coat types. Temperature and climate also stimulate seasonal variation in hair length in most breeds of dogs.

There has been a general tendency in domestic dogs toward the reduction of undercoat, a tendency paralleling that in domestic sheep. The short-haired breeds show a definite reduction in the undercoat. This process has gone furthest in the Poodle, in which the outer coat has also been reduced, thus increasing the proportion of undercoat. In such a manner, selective breeding by man has succeeded in altering the characteristics of the coat of dogs from that found in foxes, wolves, and dingos.

Implantation of Hairs in the Skin of the Dog

Some of the differences that occur in the appearance of the coat of various types of dogs are due to the variation in the implantation angle of the hair follicles. The Chow, Airedale, and Scottish terrier have an implantation angle of 45 degrees. Other breeds, such as the Long-haired Dachshund, Cocker spaniel, and Irish setter, have an implantation angle of less than 30 degrees. The majority of all breeds examined by Brunsch (1956) had an angle between 30 and 40 degrees. There is a tendency for long-haired dogs to have a higher implantation angle. Generally, the hairs slant in a caudal direction from the nose toward the tip of the tail.

Niedoba (1917) has described the streams and convergent and divergent whirls and the lines where streams of different directions join. The patterns are subject to great variation. Some of the more obvious features that can be easily observed on short-haired dogs are the center of nasal divergence, cheek whirls, ear center, ventral cervical stream, neck diverging line, diverging breast whirls, ventral center line (division of hair cover on both sides of the body), thoracic whirls from the ventral cervical stream, whirl in the region of the elbow, and rump whirls.

The Hair Follicle Cycle and Seasonal Shedding

The process of shedding is gradual, and the coat of one season merges into that of the next, so that the dog is normally never without a protective covering. Shedding is genetically controlled to some extent, but

environment is certainly a factor in expression of genetic potential. It has been observed by dog owners that short-haired breeds of house dogs may shed a little all year round and that long-haired outdoor dogs may be seasonal shedders twice a year. Blackburn (1965) found in confined dogs with hair of normal length that there is shedding in spring and autumn. In the spring the shedding of the hair in a dog that is groomed daily lasts about five weeks. During the first 10 to 14 days, the majority of the hairs shed are bristle hairs and bristle-lanugo hairs, and after this it is mainly the lead hairs and lanugo hairs that are shed.

Al-Bagdadi et al. (1977) related the stages of the hair follicle cycle as observed from microscopic examination of monthly skin

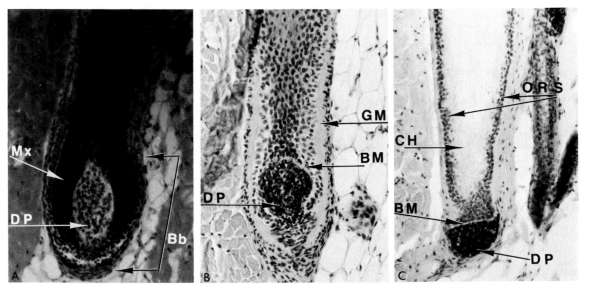

Figure 3–6. Stages of the hair follicle cycle. (From Al-Bagdadi 1975.)
A. Anagen. Longitudinal section of a main hair follicle from the saddle region of a six-month-old Beagle dog. This is an example of the anagen stage, and it illustrates a well-developed dermal papilla (DP), which is completely bordered by the matrix cells (Mx). The bulb of the hair follicle is labeled (Bb). Magnification 240 ×. Stained with hematoxylin and eosin.
B. Catagen. Longitudinal section of a main hair follicle from the saddle region of a two-week-old Beagle dog. This is an example of a catagen hair follicle. It has a rounded dermal papilla (DP). The glassy membrane (GM) is thick and somewhat irregular above the bulb region. The basement membrane (BM) can be observed. Magnification 265 ×. Stained with hematoxylin and eosin.
C. Telogen. Longitudinal section of a main hair follicle from a nine-month-old Beagle dog. This is an example of a telogen hair follicle. The dermal papilla (DP) is outside the bulb, separated from the matrix cells by a basement membrane (BM). The outer root sheath (ORS) borders the club hair (CH) directly owing to lack of inner root sheath. Magnification 190 ×. Stained with hematoxylin and eosin.

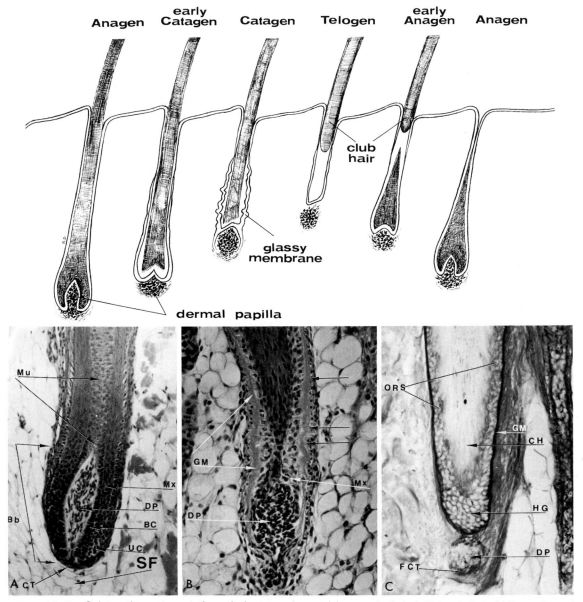

Figure 3-7. Schematic representation of changes observed in a main hair follicle of the saddle region of Beagle dogs during the hair cycle. (From Al-Bagdadi 1975).

    A. Anagen hair follicle. Longitudinal section of a secondary hair follicle in the anagen stage from the saddle region of a 28-month-old Beagle. The bulb (Bb) extends into the subcutaneous fat (SF). The spindle-shaped dermal papilla (DP) extends toward the medulla of the hair (Mu), and the base of the dermal papilla is continuous with the connective tissue (CT) of the hair follicle. The dermal papilla is surrounded by the matrix cells (Mx) of the bulb (Bb). The basal cells or the matrix are columnar (BC). The lower part of the bulb contains undifferentiated matrix cells (UC). Stained with hematoxylin and eosin. Magnification 350 ×.

    B. Catagen. Longitudinal section of a hair follicle in the catagen stage from the saddle region of a two-week-old Beagle. The dermal papilla (DP) is oval in shape. The nuclei are crowded closely together, and the matrix cells (Mx) that border the dermal papilla have lost their orientation. The glassy membrane (GM) is thick and straight at the upper part of the hair follicle (single black unlabeled arrow in the upper part of the picture), while above the bulb the glassy membrane is undulating (two black unlabeled arrows in the lower part of the picture). Stained with hematoxylin and eosin. Magnification 395 ×.

    C. Telogen. Longitudinal section of a main hair follicle in the telogen stage from the saddle region of a three-month-old Beagle. The dermal papilla (DP) is separated from the matrix cells of the hair follicle. It is surrounded by fibrous connective tissue (FCT) and appears to contact the base of the follicle at one point. The hair germ cells (HG) are located at the base of the club hair (CH). The cells of the outer root sheath (ORS) lack glycogen granules. The glassy membrane (GM) is thick and PAS–positive. The hair follicle at this stage is surrounded by connective tissue which separates the follicle from the adipose tissue. Prepared with Periodic Acid Schiff reaction of McManus (1968) without diastase treatment. Magnification 400 ×.

biopsies to the mean weight of monthly hair samples collected by combing male Beagle dogs each week. A definite correlation was found between the highest percentage of telogen hair follicles and the greatest amount of shedding which occurred in the spring and autumn. The following is a description of the stages of the hair follicle cycle as observed in the Beagle by Al-Bagdadi, Titkemeyer, and Lovell (1977).

The **anagen** stage is characterized by a well-developed single-shaped dermal papilla which is completely capped by the hair matrix, forming the bulb of the hair follicle (Figs. 3–6A and 3–7A). Ultrastructural studies of the basement membrane between the matrix cells and dermal papilla cells suggests that granular ground substance diffuses from the dermal papilla cells to the matrix cells to furnish metabolites and materials that are needed for the rapid synthesis of keratin (Al-Bagdadi et al. 1977–B). Hair follicles of the anagen stage are the longest, with the bulb extending deep into the dermis or even into the hypodermis, where they are surrounded by adipose tissue. In the larger main hair follicles, blood vessels have been demonstrated entering the dermal papilla of the hair bulb, (Al-Bagdadi et al. 1978). The smaller hair follicles of lanugo hair seem to have dermal papillae that are devoid of blood vessels.

The **catagen** stage is identified by the presence of a thick, glassy membrane on the outside of the follicle (Figs. 3–6B and 3–7B). This glassy membrane is irregular and has an undulated appearance in the lower third of the hair follicle. Thickening of the glassy membrane is accompanied by a thickening of the basement membrane between the dermal papilla and the matrix. The follicle bulb becomes smaller and the dermal papilla more rounded. The entire follicle becomes shorter, and the rounded bulb is not as deep in the dermis as the spindle-shaped bulb of the anagen hair follicle.

The hair follicles during the **telogen stage** have a smaller dermal papilla, which is separated from the bulb and is no longer capped by matrix cells, which have decreased in number (Figs. 3–6C and 3–7C). The hair follicle of the telogen stage is very short; it contains a club hair, and the inner root sheath disappears.

Hairs mature and are shed according to the metabolic state of the skin. The rate of growth varies in different follicles and in different regions of the body. A hair which has been shed naturally is differentiated from one which has been broken or shorn by the slightly bulbous proximal end, which is frayed out into fibrillae. When club hairs are plucked during the resting phase (telogen), new hairs begin to grow at once, whereas new hair growth occurs much later if the resting hair is allowed to shed naturally. This may influence the development of coats of Wire-haired Fox terriers, which are customarily plucked when being groomed for show purposes. When a growing hair is plucked during anagen, nearly all of the lower half of the follicle is pulled out with it.

Dogs shed more cover-hairs in the spring than in the summer, and the number of hairs in each bundle increases in the winter. There is a great deal of variation in the manner in which dogs shed their hair, even among individuals of the same breed kept under similar environmental condtions and fed a like diet.

Surface Contour of Hairy Skin and Histology of Epidermis

The surface of the hairy skin is irregular because of scalelike folds which form depressions into which the hair follicles invaginate. The pattern of skin folds is occasionally interrupted by the presence of knoblike enlargements, 0.33 to 0.35 mm. in diameter, the **tactile elevations** (Fig. 3–8A and B). Various terms have been used to describe these structures, such as epidermal papillae (Lovell and Getty 1957), integumentary papillae (Strickland and Calhoun 1963), Haarscheiben (Mann 1965, Smith 1967, and Straile 1961), toruli tactiles or touch spots (Iggo 1962). Iggo and Muir (1963) described distinct dome-shaped swellings that can be seen in the hairy skin of cats. These domes are often associated with guard-hairs. The guard-hair that is associated with the Haarscheiben, or tactile elevations, is referred to as a **tylotrich hair** by many authors: Mann and Straile (1965), Straile (1960 and 1961), Mann (1965), and Smith (1967). The Haarscheibe, which is more pedunculated in the dog and cat than

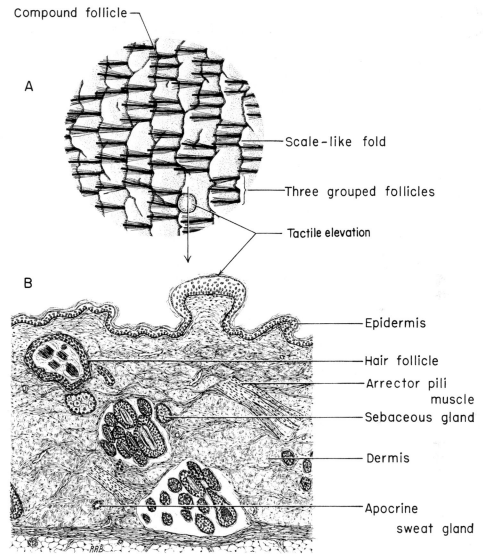

Figure 3–8. Surface contour, hair arrangement, and histology of the hair skin. (After Lovell and Getty 1957.)
A. View of scalelike folds and arrangement of hair follicles.
B. Histological section of hairy skin.

in other species, may lie medial, lateral, cephalad, or caudal to the tylotrich. This positional relationship is of importance in the sensory function. Rarely, no tylotrich is seen adjacent to a Haarscheibe in cats. Montagna (1967) states that all animals have Haarscheiben. Adams, Calhoun, Smith, and Stinson (1970) illustrate the epidermal pad

of the tylotrich follicle in the dog. It has been observed by Lovell and Getty (1957) that the epidermal papilla (tactile elevation) is a dome-shaped enlargement of the epidermis at birth and becomes more pedunculated by six months of age.

Histological study shows that the epidermis of the hairy skin, varying in thickness from 25 to 40 micrometers, usually consists of three layers: *stratum basale, stratum spinosum,* and *stratum corneum.* In a few areas the *stratum granulosum* and *stratum lucidum* are evident, but these are infrequent and are in areas where keratinization is retarded (i.e., around hair follicle orifices). The number of layers of cells varies between three and six. In regions where the stratum granulosum and stratum lucidum are evident, there are as many as eight layers of cells. Cells of the stratum basale usually have their nuclei oriented with the long axis perpendicular to the surface of the skin. The tactile elevations are covered by a thickened epidermis which is usually 6 to 12 cell layers thick, about twice as thick as the surrounding epidermis (Figures 3–8B). The dermis under the tactile elevations is composed of very fine, closely packed connective tissue fibers which extend under the thickened epidermis to form the elevation.

MUSCLES OF THE SKIN

The *arrector pili* muscles are best developed on the dorsal line of the neck, back, and tail. They are very small or absent in the ventral surface of the body. During the first eight weeks of life in pups, the *arrector pili* muscles of the interscapular area were found to range from 10 to 40 micrometers in diameter. At the ages of four to six months of age they ranged from 30 to 40 micrometers in the same region (Lovell 1955).

Cutaneous muscle fibers occur in the superficial fascia beneath the skin, closely associated with the dermis. In the cranial region they consist of the *sphincter colli superficialis, platysma,* and *sphincter colli profundus.* The fibers of these muscles are very irregular and tend to run transversely. Around the muzzle region some fibers are associated with the sinus hair follicles. A large cutaneous muscle called the *cutaneus trunci* occurs over a great portion of the thorax and abdomen. It extends from the gluteal region craniad and ventrad to the pectoral region. Some fibers from the cutaneus trunci make up the *preputialis* muscle in the male and the *supramammarius* in the female (St. Clair 1975). The cutaneous muscles are attached to the dermis of the skin

and are anchored to the subcutaneous fascia rather than to bone. Contraction of cutaneous muscles tends to cause movement of the skin or skin structures like the sinus hair follicles, prepuce, or mammary gland.

GLANDS OF THE CANINE SKIN

Eccrine sweat glands are found only in the foot pads (Nielsen 1953). They are placed deeply in the fat and fibrous tissue under the thick foot pads. They are small (25 to 35 micrometers in diameter), tightly coiled, tubular glands, with minute lumina which are lined with cuboidal cells. They contain coarse granules scattered in the clear cytoplasm. Myoepithelial cells may be demonstrated peripheral to the secretory tubules. The excretory ducts follow a tortuous path through the dermis and epidermis and empty in the crevices between the conical papillae of the foot pads. The eccrine secretion is watery.

Apocrine sweat glands are found mainly in connection with hair follicles (Speed 1941). The secretory parts of the glandular tubules are situated in the dermal and subcutaneous layers of the skin. The excretory duct passes up through the dermis and empties into the hair follicles above the ducts of the sebaceous glands. The tubules and individual cells attain various sizes, 30 to 90 micrometers, depending upon the secretory phase. In some sections, huge, dilated, cyst-like tubules, 90 micrometers in diameter, lined with flattened, elongated cells, are found. In others the tubules are small, with high, cylindrical epithelium 30 to 45 micrometers in diameter. The highly developed mammary glands have a similar structure.

Sebaceous glands have the holocrine type of secretion. Distributed over the integument in connection with the hair follicles, they are largest along the dorsal part of the neck, back, and tail, particularly in the specialized tail gland area. The meibomian glands, or **tarsal glands of the eyelids,** are also specialized sebaceous glands (Fig. 20–21). The size of the sebaceous glands in the skin of the back at birth is 30 to 50 micrometers in diameter. There was a gradual increase to 80 to 250 micrometers at six months of age. The largest sebaceous glands are present at the mucocutaneous junctions of the lips, vulva, and eyelids.

The **glands of the ear canal** (*gll. ceruminosae*) are apocrine and sebaceous. Cerumen is a product of both glandular types and appears as a fairly dry, dark brownish substance. Fernando (1966) reported that long-haired breeds have more sebaceous and apocrine glandular tissue in the external auditory canal than short-haired breeds.

The **circumanal glands** (*gll. circumanales*) are most numerous in the vicinity of the anal orifice (Parks 1950). They consist of upper sebaceous portions with open ducts to hair follicles and of deeper, nonsebaceous portions. The nonsebaceous lobules are solid masses of large polygonal cells. It is believed that the ducts to this portion of the glands are solid and have no secretory activity.

The **anal sac** (*sinus paranalis*) of the dog is spherical and averages about a centimeter in diameter. They are paired and lie on each side of the anal canal between internal and external anal sphincter muscles. Each sac opens onto the lateral margin of the anus by a single duct. The sacs form pockets which function as a reservoir into which apocrine and sebaceous glands open (Montagna and Parks 1948). They are lined by a thin, stratified squamous epithelium. Under the connective tissue supporting the epithelium there are many sebaceous and apocrine glands (*gll. sinus paranalis*). The sebaceous glands tend to line the neck of the sac, while the apocrine glands are concentrated in the fundus. The combined secretions of the tubules of the anal sac and the sebaceous glands associated with its excretory duct form a viscous, putrescent liquid or paste. Gerisch and Neurand (1973) found only tubular apocrine glands in the sac. Therapeutic administration of female steroids has been reported to affect anal sac secretions (Donovan 1969).

Tail Gland Area

An oval to diamond-shaped area, 2.5 to 5 cm. long, is located on the dorsal aspect of the tail, at the level of the seventh and ninth caudal vertebrae (Fig. 3–9A). The hair shafts are larger in diameter and differ in appearance from the surrounding hair. The hairs of this area emerge from the hair follicles singly (Fig. 3–9B), whereas surrounding hair is of the complex follicle type, supporting 6 to 11 hairs. The single hairs of the specialized area are very stiff and coarse, and the surface of the skin has a yellow, waxy appearance probably due to an abundance of sebaceous secretion. The sebaceous glands and apocrine glands of the area are large, extending deep into the dermis and subcutaneous tissue (Lovell and Getty 1957). Hildebrand (1952) suggests that secretions of the tail gland in wild canids function in species recognition and identification. Meyer (1971) studied the dorsal tail gland organ of 135 pedigreed dogs and found that it was similar to that described in wild canidae. Meyer and Wilkens (1971) observed a definite seasonal dimorphism of this organ in both sexes and concluded that it is significant as an organ associated with estrus.

BLOOD SUPPLY TO THE SKIN

The arteries to the skin include simple cutaneous arteries, which reach the skin by running between muscles while supplying small branches to the muscles, and mixed cutaneous arteries, which run through muscles and supply large muscular branches before terminating in the skin. The arteries are roughly arranged in a segmental distribution which is not as regular as that of the nervous system. Hughes and Dransfield (1959) have listed 23 mixed cutaneous arteries and 16 simple cutaneous arteries. For additional information on regional blood supply to the skin refer to Chapters 11 and 12. The vessels anastomose extensively with one another. The plexuses which are formed in the subcutaneous fascia are especially evident over bony prominences and large muscle masses.

Microscopic examination has revealed that the arterial supply to the skin of the dog is divided into three distinct plexuses, all lying parallel to the surface. These are the deep or subcutaneous plexus, the middle or cutaneous plexus, and the superficial or subpapillary plexus (Fig. 3–10).

The **subcutaneous plexus** is made up of the terminal branches of the cutaneous arteries. Branches from this plexus form the

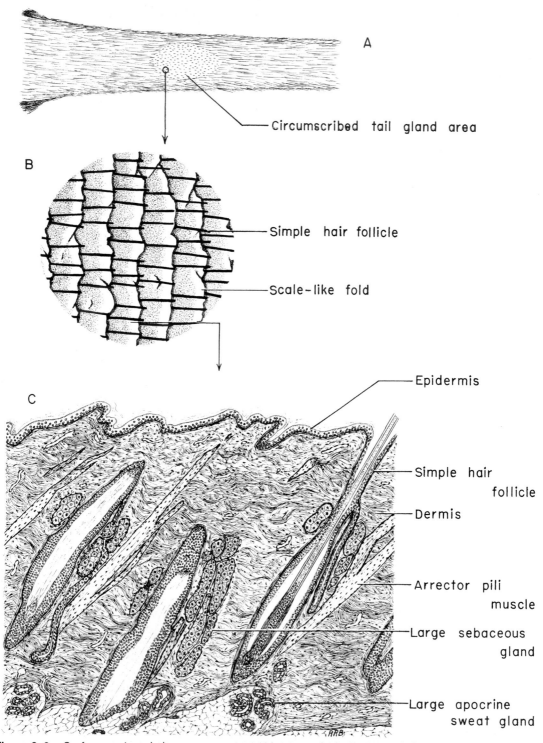

Figure 3–9. Surface contour, hair arrangement, and histology of tail gland area. (After Lovell and Getty 1957.)

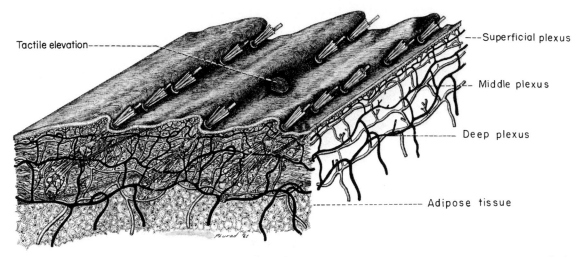

Figure 3–10. Schematic section of the skin of the dog showing tactile elevation and blood vessels (veins in black).

cutaneous plexus, which is associated with the hair follicles and glands.

The **subpapillary plexus** is formed by the union of small vessels arising from the middle plexus. The papillary body contains numerous capillary loops which come from the superficial plexus. In general the veins and arteries parallel one another. Arteriovenous anastomoses have been observed in the deeper layers. Variations in the circulatory pattern have been noted in the various modified skin areas.

The lymphatics arise from capillary nets which lie in the superficial part of the dermis or surround the hair follicles and glands. The vessels arising from these nets drain into a subcutaneous lymphatic plexus. Baum (1917) has described the lymph vessels and nodes associated with the skin of the dog.

NERVE SUPPLY TO THE SKIN

Small nerve branches are generally distributed in a segmental arrangement to the hypodermis in all areas of the body. These originate from cutaneous components of the trigeminal and facial nerves in the cranial area, the dorsal and ventral branches of the cervical, thoracic, lumbar, sacral, and caudal spinal nerves. The segmental order is altered somewhat in the region of the limbs, where the cutaneous nerves arise from the axillary, radial, ulnar, and median nerves in the brachial plexus; and from the gluteal, sciatic, tibial, and femoral nerves in the lumbosacral plexus. For more detailed information on nerve distribution to the skin, refer to Chapters 15 to 18.

Microscopic examination reveals that large nerve trunks enter the dermis from the hypodermis, where they branch and give rise to nerves which ramify alongside the blood vessels, forming a branching plexus which supplies the blood vessels, hair follicles, skin glands, and epidermis (Fig. 3–11). Nerve fibers have not been demonstrated in the apocrine sweat glands of the general body surface of the dog. Iwabuchi (1967) presented evidence that catecholamines of the adrenal medulla provoke general sweating on the hairy skin of dogs and suggests that these sweat glands receive adrenergic innervation from the sympathetic nerve. The eccrine sweat glands of the foot pad have been found to be innervated. Nerve fibers have been found around and apparently ending on the cells of the sebaceous glands of both primary and secondary hairs. Cutaneous arteries are supplied by a network of nerve fibers. The veins have a

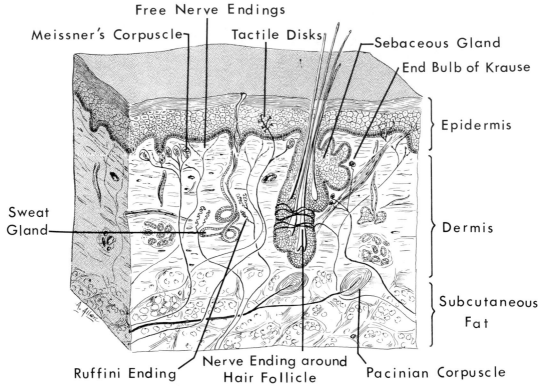

Figure 3–11. Schematic representation of the nerve supply to the skin, illustrating receptor morphology. (From Kitchell 1964. Modified after Woolard et al. 1940 and Gardner 1963.)

less complex innervation and are supplied by single nerve fibers. In the dog there are connections between the nerves associated with the hair follicle and those in the arrector pili muscle. Iggo (1977) has classified the afferent nerve units in the hairy skin that innervate receptors in the hair follicles and the epidermis.

The largest axons (type 7), 9 to 16 micrometers in diameter, supply large guard-hairs that project well above the general hairy coat. Each axon runs to 3 to 15 large guard-hairs that emerge from Strailer's tylotrich follicle. Sinus hairs of the muzzle of the dog are large examples of this type.

A group of thinner axons (type G) supplies as many as 50 guard-hairs each. These hairs are usually shorter than those supplied by type T units. These would be the larger three types of hair that make up the cover-hairs.

Type D units have fine myelinated afferent fibers which innervate both the numerous down or lanugo hair follicles as well as the guard-hair follicles.

A palisade of nerve fibers both circular and longitudinal to the hair follicle has been observed around both primary and secondary hairs. The nerve fibers supplying these palisade endings encircle the hair cluster and each of the three groups within it, sometimes giving the impression of a palisade ending around the cluster. Some of the circular and longitudinal fibers within the palisade ending reacted for specific cholinesterase and others for monamine oxidase, according to Jenkinson, McEwan, and Blackburn (1968). The palisade endings around the secondary hairs were seemingly interconnected by fine nerve fibers, and a nerve fiber from the ending around a primary hair was sometimes seen disappearing amid a complex nerve palisade of a neighboring secondary hair. By various means, the hair follicles are innervated in a characteristic manner so that they are excited by hair movement.

Another type of mechanoreceptor that is separate from hair follicle innervation is present in the hairy skin of animals. It has been well illustrated by Iggo (1977). This is a conspicuous elevation of the epidermis which appears among the hairs of the general body surface. These structures have been referred to as cutaneous touch spots by Iggo and Muir (1963). They are formed by dome-shaped enlargements (tactile elevations) of the epidermis, within the basal layer of which the myelinated afferent nerve fiber ends in a number of expanded discs (Merkel's discs). Movement of the adjacent hairs does not excite these receptors unless the hair presses down on top of the receptor. These structures have been studied by Smith (1967) in the cat and were described as "Haarscheiben" which were located caudally, rostrally, medially, or laterally from a tylotrich (large guard-hair), so that a slowly adapting response could be evoked by various movements of the hairs. Each hair has a specific movement to which the tactile elevation responded (that motion which compressed the dome). Movements of the hair in other directions elicited little or no response from its tactile elevation.

Another type of nerve ending is the pacinian corpuscle in the dermis. These occur in the foot pad of the dog and also have been described in the hairy skin of monkeys and cats (Iggo 1977). These corpuscles are excited by mechanical stimuli, such as stretching of the skin.

Jenkinson, McEwan, and Blackburn (1968) observed that the epidermis of the hairy skin of the dog was supplied by numerous single nerve fibers, reactive for specific cholinesterase, which end freely in it (epidermis). In two dogs, a Labrador and a Collie, encapsulated nerve endings were observed in the dermis of the hairy skin.

SKIN GRAFTING

Autogenous skin grafting (Jensen 1959), homographs (Puza and Gombos 1958), and allografts (Rehfeld et al. 1970) have been performed on dogs.

Histopathological studies of transplants indicate that degenerative changes involve the epidermis and the upper layers of the dermis during the first 8 to 10 postoperative days, at which time regenerative processes equalize the degenerative changes. The blood supply to the transplant is adequate by the twelfth day and completely normal by the twenty-fourth day.

THE CLAW

The superficial layers of the epidermis are modified to form the **horny claw** (*unguicula*). Gross examination shows that the claw consists of a sole, two walls, and a central dorsal ridge. The claw is frequently strongly pigmented and is curved and compressed laterally. The dorsal ridge is made up of thicker horny material than the walls and sole, which maintain the pointed appearance of the claw. The coronary border (*vallum*) of the claw fits into the space under the ungual crest of the third phalanx. This relationship is hidden by the skin of the claw fold. Dorsally, this fold is a modification of the hairy skin which is free from hair on one side and fused to the horn of the claw. As the horny material is produced and grows out, it is covered by a thin *stratum tectorium* which adheres to the proximal part of the claw. A furrow along the palmar or plantar surface of the claw separates it from the foot pad in a similar manner.

The periosteum of the third phalanx and the dermis of the claw are continuous and fill the space between the bony and epidermal structures. The vascularity of this tissue is well demonstrated by the hemorrhage which follows trimming the canine toenail into the connective tissue. On microscopic examination, the dermis of the coronary and dorsal ridge areas has been described as having a papillate structure (Trautman and Fiebiger 1957).

The *stratum basale,* which is the epidermal layer supported by the dermis, is most active in the coronary and dorsal ridge areas, where most of the horny claw is formed. The inner surface of the claw wall bears small epidermal lamellae. The epidermis of the claw is largely composed of the horny *stratum corneum,* which consists of flat, cornified epidermal cells. The epidermis of the sole has a well-developed *stratum granulosum* and *stratum lucidum.*

The claw grows at a rapid rate and, if not worn off or trimmed, may continue to grow in a circular fashion until the point of the claw approaches or invades the region of the volar furrow between the base of the claw and the foot pad.

BIBLIOGRAPHY

Adams, W. S., M. L. Calhoun, E. M. Smith, and A. W. Stinson. 1970. Microscopic Anatomy of the Dog. Integumentary System. Springfield, Ill., Charles C Thomas, Chapter 1, pp. 3–29.

Al-Bagdadi, F. A., 1975. The Hair Cycle in Male Beagle Dogs. Ph.D. Thesis, Univ. Ill., Champaign, Ill.

Al-Bagdadi, F. A., C. W. Titkemeyer, and J. E. Lovell. 1977–A. Hair follicle cycle and shedding in male Beagle dogs. Am. J. Vet. Res. 38:611–616.

———. 1977–B. Ultrastructural morphology of the anagen stage hair follicle of male Beagle dogs. Thirty-fifth Ann. Proc. Electron Microscopy Soc. Amer. 35:652–653. Boston, Massachusetts.

———. 1978. Alkaline phosphatase reaction in hair follicles of male Beagle dogs during hair cycle stages. Zbl. Vet. Med. C. Anat. Hist. Embryol. 7:245–252.

Baker, K. P. 1966. Postnatal development of the dog's skin. Br. Vet. J. 122:344–347.

———. 1967. Senile changes of dog skin. J. Small Anim. Pract. 8:49–54.

Baum, H. 1917. Die Lymphgefässe der Haut des Hundes. Anat. Anz. 50:1–15.

Blackburn, P. S. 1965. The hair of cattle, horse, dog and cat. In Comparative Physiology and Pathology of Skin, edited by A. J. Rook and G. Ş. Walton. Philadelphia, F. A. Davis Co., pp. 201–210.

Blatt, C. M., C. R. Taylor, and M. B. Habal. 1972. Thermal panting in dogs: the lateral nasal gland, a source of water for evaporative cooling. Science 177:804–805.

Bloom, W., and D. W. Fawcett. 1975. 10th Ed. Textbook of Histology. Philadelphia, W. B. Saunders Co., Chap. 23, pp. 563–597.

Brunsch, A. 1956. Vergleichende Untersuchungen am Haarkleid von Wildcaniden und Haushunden. Z. Tierzuchtung und Zuchtungs. Biologie 67:205–240.

Burns, M. 1966. Genetics of the Dog — Inheritance of Color and Hair Type. Philadelphia, J. B. Lippincott Co., Chap. 4, pp. 38–63.

Calhoun, M. L., and A. W. Stinson. 1976. Integument. In Textbook of Veterinary Histology, edited by H.-D. Dellmann and E. M. Brown. Philadelphia, Lea & Febiger, Chap. 18, pp. 459–493.

Comben, N. 1951. Observations on the mode of growth of the hair of the dog. Br. Vet. J. 107:231–235.

Conroy, J. D., and P. D. Beamer. 1970. The development of cutaneous and oral pigmentation in Labrador Retriever fetuses (Canis familiaris). J. Invest. Dermatol. 54:304–315.

Copeman, D. B., and S. M. Gaafar. 1967. Skin histamine in dogs. Am. J. Vet. Res. 28:309–312.

Da Fonsica, P., and A. Cabral. 1945. Pelagnes dos caes [classification of coat types in dogs]. Rev. Med. Vet. 40:187–191.

Donovan, C. A. 1969. Canine anal glands and chemical signals (pheromones). J. Am. Vet. Med. Assoc. 155:1995–1996.

Emerson, J. L., and R. F. Crass. 1965. The distribution of mast cells in normal canine skin. Am. J. Vet. Res. 26:1379–1382.

Epling, G. P. 1962. The anatomy of the skin. In Canine Medicine. 2nd Ed. Evanston, Ill., American Veterinary Publications, Inc., Chapter 10, pp. 405–410.

Fernando, S. D. A. 1966. A histological and histochemical study of the glands of the external auditory canal of the dog. Res. Vet. Sci. 7:116–119.

Gair, R. 1928. Die Wuchsformen des Haarkleides bei Haustieren nach Untersuchungen beim Hunde. Z. Tiersucht u. Zuchtungsbiol 11 (1):57–58.

Gerish, D., and K. Neurand. 1973. Topographie und Histologie der Drusen der Regio analis des Hundes. Anat. Hist. Embryol. 2:280–294.

Ham, A. W. 1969. Histology. 6th Ed. Philadelphia, J. B. Lippincott Co.

Hildebrand, M. 1952. The integument in canidae. J. Mammal. 33:419–428.

Hughes, H. V., and J. W. Dransfield. 1959. The blood supply to the skin of the dog. Br. Vet. J. 115:1–12.

Humphries, A. L., W. S. Harms, and W. H. Moretz. 1961. Skin homographs in dogs deficient in pyridoxine. J.A.M.A. 178:490.

Iggo, A. 1962. New specific sensory structures in hairy skin. Acta Neuroregatativa 24:175–180.

———. 1977. Somesthetic sensory mechanisms. In Dukes' Physiology of Domestic Animals, 9th Ed., edited by M. J. Swenson. Ithaca, N.Y., Cornell Univ. Press, Chap. 43, pp. 589–591.

Iggo, A., and A. R. Muir. 1963. A cutaneous sense organ in the hairy skin of cats. J. Anat. 97:151.

Irwin. 1966. Tension lines in the skin of the dog. J. Small Anim. Pract. 7:593–598.

Iwabuchi, T. 1967. General sweating on the hairy skin of the dog and its mechanisms. J. Invest. Dermatol. 49:61–70.

Jenkinson, D. McEwan, and P. S. Blackburn. 1968. The distribution of nerves, monoamine oxidase and cholinesterase in the skin of the cat and dog. Res. Vet. Sci. 9:521–528.

Jensen, E. C. 1959. Canine autogenous skin grafting. Am. J. Vet. Res. 20:898–908.

Little, C. C. 1957. The Inheritance of Coat Color in Dogs. Ithaca, N.Y., Comstock Publishing Associates.

Lovell, J. E. 1955. Histological and Histochemical Studies of Canine Skin. Masters Thesis, Iowa State Univ., Ames, Iowa.

Lovell, J. E., and R. Getty. 1957. The hair follicle, epidermis, dermis, and skin glands of the dog. Am. J. Vet. Res. 18:873–885.

Machida, H., L. Giacometti, and E. Perkins. 1966. Histochemical and pharmacological properties of the sweat glands of the dog. Am. J. Vet. Res. 27:566–573.

Mann, S. J. 1965. Haarscheiben in the skin of sheep. Nature 205:1228–1229.

Mann, S. J., and W. E. Straile. 1965. Tylotrich (hair) follicle. Association with a slowly adapting tactile receptor in the cat. Science 147:1043–1045.

Meyer, P. 1971. Das dorsale Schwanzorgan des Hundes (Canis familiaris). Zbl. Vet. Med. 18:541–557.

Meyer, P., and H. Wilkens. 1971. Die "Viole" des Rot Fuchses (Vulpes vulpes L.) Zbl. Vet. Med. 18:353–364.

Montagna, W. 1967. Comparative anatomy and physiology of the skin. Arch. Dermatol. 96:357–363.

Montagna, W., and H. F. Parks. 1948. A histochemical study of the glands of the anal sac of the dog. Anat. Rec. 100:297–318.

Muller, G. H., and R. W. Kirk. 1969. Small Animal Dermatology. Anatomy of the skin, pp. 3–36. Philadelphia, W. B. Saunders Co.

Mykytowycz, R. 1968. Territorial marking by rabbits. Sci. Am. *218*:116–126.

Niedoba, T. 1917. Untersuchungen über die Haarrichtung der Haussäugetiere. Anat. Anz. *50*:178–192, 204–216.

Nielsen, S. W. 1953. Glands of canine skin; morphology and distribution. Am. J. Vet. Res. *14*:448–454.

Parks, A. S. and H. M. Bruce. 1961. Olfactory stimuli in mammalian reproduction. Science *134*:1049–1054.

Parks, H. 1950. Morphological and cytochemical observations on the circumanal glands of the dog. Ph.D. thesis, Cornell Univ., Ithaca, N.Y.

Pinkus, H. 1958. Embryology of hair, *in* The Biology of Hair Growth, edited by W. Montagna and R. A. Ellis. New York, Academic Press, pp. 1–32.

Puza, A., and A. Gombos. 1958. Immunological tolerance to skin homographs in dogs. Transplantation Bull. *5*:66.

Rehfeld, C. E., G. J. Dammin, and W. J. Hester. 1970. Skin graft survival in partially inbred Beagles. Am. J. Vet. Res. *31*:733–745.

Smith, K. R. 1967. The structure and function of the Haarscheibe. J. Comp. Neurol. *131*:459–474.

Speed, J. G. 1941. Sweat glands of the dog. Vet. J. *97*:252–256.

St. Clair, L. E. 1975. Carnivore myology, *in* Sisson and Grossman's Anatomy of the Domestic Animal, edited by R. Getty. 5th Ed. Philadelphia, W. B. Saunders Co., Vol. 2, Chap. 50, pp. 1507.

Straile, W. E. 1960. Sensory hair follicles in mammalian skin: the tylotrich follicle. Am. J. Anat. *106*:133–147.

———. 1961. Morphology of tylotrich follicles in the skin of the rabbit. Am. J. Anat. *109*:1–13.

Strickland, J. H., and M. L. Calhoun. 1963. The integumentary system of the cat. Am. J. Vet. Res. *24*:1018–1029.

Trautmann, A., and J. Fiebiger, 1957. Fundamentals of the Histology of Domestic Animals. Translated and revised by R. E. Habel and E. L. Biberstein. Ithaca, N.Y., Comstock Publishing Association, Chap. 17, pp. 334–369.

Warner, R. L., and L. Z. McFarland. 1970. Integument. *In* The Beagle as an Experimental Dog, edited by A. Anderson and L. S. Good. Ames, Iowa, Iowa State Univ. Press, pp. 126–148.

Webb, A. J., and M. L. Calhoun. 1954. The microscopic anatomy of the skin of mongrel dogs. Am. J. Vet. Res. *15*:274–280.

THE MAMMAE

By GEORGE C. CHRISTENSEN

Each *mamma* consists of a glandular complex and its associated papilla. The mammary gland is an accessory gland of the skin and resembles a sweat gland in its mode of development. Mammae are characteristic of mammals and provide nourishment to the newborn. In the male they remain rudimentary throughout life, but in the female they are subject to conspicuous changes during pregnancy and during and after lactation.

The mammary glands are typically arranged in two bilaterally symmetrical rows extending from the ventral thoracic to the inguinal region (Fig. 3–12). The teats (*papillae mammae*) indicate the position of the glands in the male or in the non-lactating female. The number of glands varies from 8 to 12, with 4 to 6 gland complexes on each side of the midline. Most commonly, there are a total of 10 glands, and 8 are more frequently seen than 12. In a few cases, there are more glands on one side than on the other. Four pairs of glands are more commonly found in the smaller breeds (Kitt 1882). When 10 glands are present, the relatively small cranial 4 are the thoracic mammae, the following 4 are the abdominal mammae (median in size), and the relatively large caudal 2 are the inguinal, or pubic, mammae. In some instances, especially during involution of the glands following lactation, the relative sizes of the glands may digress from the typical pattern. Under these circumstances, the caudal abdominal mammae may rarely be slightly larger than the inguinal glands. Turner and Gomez (1934) examined 20 dogs and found 10 glands in each of 16 dogs, 9 in 3, and 8 in 1. Supernumerary glands are found in both thoracic and abdominal regions. Wakuri (1966) found the number of mammary gland teats in mongrel dogs (29 fetuses, 9 newborn, and 40 adults) to vary between 6 and 10, with a mean of 9.4 in fetuses, 8.8 in newborns, and 8.1 in adults. There appears to be a loss of mammae as the dog matures, particularly in the cranial pectoral region. Study of the mammary glands of the dog however, has been comparatively neglected.

Structure

The **mammary gland** (*glandula mammaria*) consists of epithelial glandular tissue (*lobuli glandulae mammariae*), connective tissue, and the covering skin.

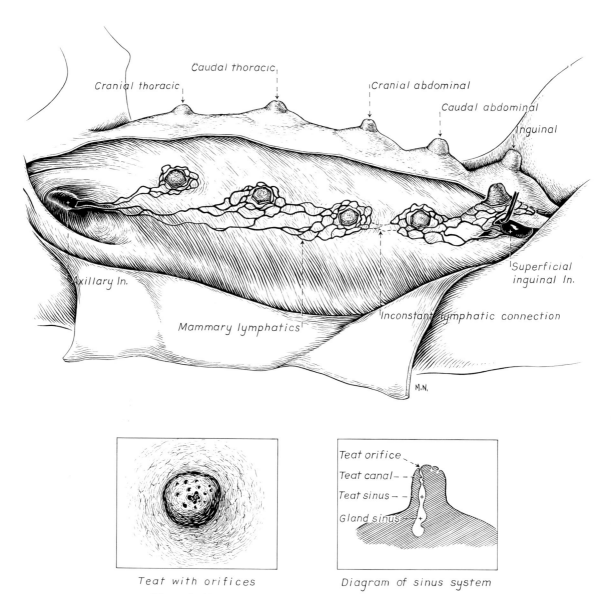

Teat with orifices   Diagram of sinus system

Figure 3–12. Mammary glands, topography and structure.

The secretory tissue is present to a significant degree only during pregnancy, pseudopregnancy, the period of lactation when pups are nursing, and for 40 to 50 days following weaning. After this postpartum period, the alveoli and lobules are reduced to a shrunken system of ducts with relatively few remnants of lobules. For a general discussion of lactation structures, see Larson (1974), Falconer (1971), and Cowie and Tindal (1971).

The number of ducts opening on a teat varies. In a study of the external teat openings of nine mature dogs of different breeds, as few as 7 and as many as 16 ducts were observed on a teat. The duct openings are located on the blunt end of the teat in an irregular, sievelike pattern. The peripheral ducts tend to form a circle, whereas the centrally placed ones form an irregular design. Turner (1939) described 8 to 14 ducts in one case and 12 to 22 in another. Martin (1910) found from 8 to 12 ducts.

Each **papillary duct** or **teat canal** (*ductus papillares*) occupies approximately one-third of the length of the teat. It is lined by stratified squamous epithelium. The epithelium usually lies in folds near the margin of the teat sinus.

The **teat sinus** or **teat cistern** (*sinus lactiferi*) extends from the teat canal into the parenchyma of the gland. In large dogs, the sinus system may be seen upon gross examination of the sectioned gland. The epithelium of the teat canal changes gradually, in the teat sinus, from stratified squamous to columnar.

In the middle portion of the teat there is an intermingling of diversely running smooth muscle fibers and connective tissue elements. The circular musculature radiates from the axis of the central zone of the teat into the area among the teat canals. The musculature encircles the canals and joins or condenses into the **sphincters of the canals** (*m. sphincter papillae*). Elastic fibers radiate among the teat canals, forming an extensive network. Essentially, the tunica propria of the teat is composed of bands of loose connective tissue, blood vessels, and smooth muscle and elastic fibers.

According to Turner (1939), the teat and gland sinuses unfold when filled with milk and show no constrictions or circular folds between the two divisions of the sinuses. Each **gland sinus** is separated from surrounding sinuses by connective tissue septa and has a distinct glandular area composed largely of minute alveoli (Käppeli 1918).

The skin is very thin over the distal tip of the teat. It increases in thickness near the base. The corium contains elastic elements, smooth muscle fibers, and blood vessels. The epidermis may be pigmented, in which case the pigment is present in the germinal layer. Although the distal blunt end of the teat is bare, the rest of it is covered by very fine hairs which are accompanied by sebaceous glands. Turner (1939) found sweat glands at the base of the teat.

Vessels and Nerves

The mammary glands are highly vascular. Veins are more extensive than arteries. The thoracic mammae receive their arterial blood supply from the perforating sternal branches of the internal thoracic arteries. These penetrate the thoracic wall through the intercostal spaces. Intercostal and lateral thoracic arteries may also contribute blood to the thoracic glands. Abdominal and inguinal mammae are supplied by mammary branches of the epigastric arteries. The cranial superficial epigastric artery arises from the cranial epigastric artery, a branch of the internal thoracic. It penetrates the abdominal wall approximately 2 to 4 cm. from the midventral line, mesial to the costal arch. It sends mammary branches to the cranial abdominal gland and anastomoses with the caudal superficial epigastric artery. The latter artery, a branch of the external pudendal, runs cranially, deep to the inguinal mamma, which it supplies. The artery continues forward to supply the abdominal mammae and terminates in numerous superficial branches which anastomose with the end branches of the cranial superficial epigastric artery.

The veins of the mammae in the dog parallel the course of the arteries to a large degree. The cranial and caudal superficial epigastric veins are the major veins of the glands. The abdominal and inguinal glands drain into the caudal superficial epigastric veins. The thoracic mammae drain into the cranial superficial epigastric veins, as well as directly into the internal thoracic veins as far cranially as the fifth intercostal space.

Lymphatic drainage of the mammary glands is also bilaterally symmetrical. By studies involving injection of solutions of Berlin blue and India ink into the glandular tissue of each mamma in live, lactating bitches and observation of the flow of dye in the lymphatic vessels, the findings of Stalker and Schlotthauer (1936) and of Baum (1918) were substantiated. Each gland has its own plexus of lymphatic channels which anastomose and surround the area of the teat (Fig. 3–12). Lymphatic networks in the teat anastomose to form an irregular channel encircling the base of the teat. Lymphatic plexuses are found in the parenchyma, subcutis, and teat. Usually, one to three main channels leave each glandular plexus and pass laterally and superficially to the nearest superficial lymph node. The cranial and caudal thoracic mammae drain directly, by separate lymphatics, to the axillary node. Typically, the caudal abdominal gland drains into the lymphatic meshwork of the inguinal mammary gland and also directly to the superficial inguinal lymph node. The inguinal mammary gland has an extensive interlocking lymphatic plexus which drains into the adjoining superficial inguinal lymph node. The drainage of the cranial abdominal mamma is inconsistent. Although draining toward the axillary lymph node in most instances, its lymphatics may join those of the caudal abdominal gland and drain toward the superficial inguinal lymph node. Schlotthauer (1952) postulates that, in addition, there may be direct connections between the lymphatics of the mammae and the vascular system, accounting for the direct internal metastasis of tumors. Turner (1939) was able to trace lymphatic ducts from the cranial abdominal gland into the thoracic cavity and directly into the sternal lymph nodes. This has not been verified by other investigators.

The axillary lymph nodes are drained by the sternal nodes within the thoracic cavity. The superficial inguinal lymph nodes drain into the iliac nodes through the lymphatics of the femoral canal.

The cranial thoracic mammary gland is innervated by branches of the fourth, fifth, and sixth ventral cutaneous nerves. The caudal thoracic gland receives its nerve supply from the sixth and seventh ventral cutaneous nerves. The abdominal and inguinal mammae are innervated by the inguinal nerve and the ventral superficial branches of the first three lumbar nerves: cranial iliohypogastric, caudal iliohypogastric, and ilioinguinal. Sympathetic fibers accompany the blood vessels to the mammae. Nerves are distributed to the parenchyma of the gland, to the blood vessels, to the smooth muscle of the teat, and to the skin. In addition to being subject to nervous control, secretion of the mammary glands is influenced by hormones from the hypophysis and other organs, brought to them by the blood.

Development

Although prenatal mammary development has been studied in carnivora as well as in many other species of mammals, the developing mammae of the fetal dog have not been extensively investigated. The mammary ridge is present at 25 days of gestation (Fig. 2–9B) and by the thirtieth day has differentiated into five pairs of nipples (Fig. 2–10). Involution of the mammae in the male fetus is illustrated in Chapter 2 (Figs. 2–17 to 2–22). Turner and Gomez (1934) have studied the development of the gland during the estrus cycle, pregnancy, and pseudopregnancy. A few workers have studied mammary growth in dogs as influenced by estrogen and progesterone (Kunde et al. 1930, Trentin et al. 1952).

Male mammae, and female mammae from birth until the approach of the first estrus, consist of small primary ducts extending a short distance below the base of the teat. During estrus, the duct system of the gland grows rapidly and the alveolar system develops. Marshall and Halnan (1917) reported that within a week after estrus slow growth of the tissues of the gland (a few ducts surrounding the teat) changes to a period of rapid development in the pregnant animal. The growth phase appears to be completed between day 30 and 40 after initiation of estrus. There is then a gradual increase in the size of the gland, owing to secretory activity of the alveolar epithelial cells.

Turner and Gomez (1934) made a detailed study of the gross and microscopic glandular changes during pregnancy. Ten days after conception the growth of the gland is grossly perceptible. At 20 days, the peripheral borders of adjoining glands in each row begin to unite and to extend toward the midventral line. The glandular systems always remain intrinsically separate, however. On microscopic examination, the connective tissue stroma is reduced, adipose cells are present, and the growth of the duct system is very marked. At 30 days, a typical duct and lobule system is present, as well as anlagen of alveoli. Individual alveoli with lumina are seen at 40 days, and for the next 20 days there is a gradual enlargement of the gland owing to initiation of secretion by the alveolar epithelial cells. Within one day after parturition, the alveoli become greatly reduced, compared with the total amount of parenchyma. Changes in the mammary gland during pseudopregnancy are essentially identical to those of pregnancy, except that secretory activity at 60 days is less well developed.

Approximately 10 days after parturition, the size of the mammae is greatly reduced, the lobule-alveolar structures being affected sooner than the duct system. By 40 days the lobule-alveolar system is largely degenerated, and the ducts are shrunken. After cessation of lactation, the mammary gland in the dog regresses to a simple duct system.

Many afflictions may involve the mammary glands of the dog, such as inflammation (mastitis), aplasia, hypoplasia, and tumors. Mastitis occurs relatively infrequently, compared with its incidence in the cow, but neoplasms of the mammary glands of the bitch are not uncommon, especially in aged animals. Hormone imbalance in some dogs of advanced age is the factor responsible for the development and growth of mammary tumors. Malignant mammary tumors may metastasize either by the blood stream or through the lymphatic system, the latter being the most common route. Consequently, the success of surgical removal of malignant mammary gland neoplasms is dependent, among other things, on the pattern of lymphatic drainage. The established pattern of lymphatic drainage indicates that a malignant tumor of the cranial thoracic gland may be corrected by removal of that one gland. If the caudal thoracic mamma contains a malignant neoplasm, both thoracic mammary glands on that side should be extirpated. When a tumor is present in the cranial abdominal gland, this gland should be removed, along with the thoracic glands. The relatively rare case of lymphatic anastomosis between the cranial and caudal abdominal mammae indicates the occasional advisability of removal of all of the glands on the affected side. If either the caudal abdominal or the inguinal mamma contains tumors, both should be removed. The superficial inguinal lymph node is then also removed. Schlotthauer (1952) ascertained the lack of necessity for removal of the axillary lymph node when the thoracic and cranial abdominal mammae contain tumors. Metastasis of mammary neoplasms to the axillary or accessory axillary lymph nodes is infrequent in the bitch. Spayed bitches rarely have mammary tumors.

BIBLIOGRAPHY

Baum, H. 1918. Das Lymphgefässsystem des Hundes. Berlin, Hirschwald.

Cowie, A. T., and J. S. Tindal. 1971. The Physiology of Lactation. London, Arnold Pub., 392 pp.

Falconer, I. R. 1971. Lactation. London, Butterworths, 476 pp.

Käppeli, F. 1918. Über Zitzen- und Zisternenverhältnisse der Haussäugetiere. Inaugural Dissertation, Universität Zürich.

Kitt, T. 1882. Zur Kenntniss der Melchdrüsenpapillen unserer Haustiere. Dtsch. Z. Thiermed. u. vergl. Pathol. 8:245–269.

Kunde, M. M., F. E. D'Amour, A. J. Carlson, and R. G. Gustavson. 1930. Studies on metabolism; the effect of estrin injections on the metabolism, uterine endometrium, lactation, mating, and maternal instincts in the adult dog. Am. J. Physiol. 95:630–640.

Larson, B. L. 1974. Lactation: A Comprehensive Treatise. 3 vols. Vol. I. The Mammary Gland: Development and Maintenance. New York, Academic Press.

Marshall, F. H. A., and E. T. Halnan. 1917. On the post-oestrous changes occurring in the generative organs and mammary glands of the non-pregnant dog. Proc. Roy. Soc. B 89:546–559.

Martin, P. 1910. Die Milchdrüse, in Handbuch der vergleichenden mikroskopischen Anatomie, edited by W. Ellenberger. Berlin, Springer-Verlag.

Schlotthauer, C. F. 1952. The mammary glands. Pp. 504–521 *in* Canine Surgery, edited by J. V. Lacroix and H. P. Hoskins. 3rd Ed. Evanston, Ill., American Veterinary Publications, Inc.

Stalker, L. K., and C. F. Schlotthauer. 1936. Neoplasms of the mammary gland in the dog. North Am. Vet. *17*:33–43.

Trentin, J. J., J. DeVita, and W. U. Gardner. 1952. Effect of moderate doses of estrogen and progesterone on mammary growth and hair growth in dogs. Anat. Rec. *113*:163–177.

Turner, C. W. 1939. The Comparative Anatomy of the Mammary Glands. Columbia, Mo., University Coop. Store.

—————. 1952. The Mammary Gland. I. The Anatomy of the Udder of Cattle and Domestic Animals. Columbia, Mo., Lucas Bros.

Turner, C. W., and E. T. Gomez. 1934. The Normal and Experimental Development of the Mammary Gland. II. The Male and Female Dog. Mo. Agr. Exp. Sta. Res. Bul. 207.

Wakuri, H. 1966. Embryological and anatomical studies on the mammary formula of mongrel dogs. On the newborn and adult dog. J. Mamm. Soc. Jap. 3:19–23.

Chapter 4

THE SKELETON

GENERAL

The skeleton serves for support and protection while providing levers for muscular action. It functions as a storehouse for minerals, and as a site for fat storage and blood cell formation. In the living body the skeleton is composed of a changing, actively metabolizing tissue which may be altered in shape, size, and position by mechanical or biochemical demands. The process of bone repair and the incorporation of heavy metals and rare earths (including radioisotopes) in the adult skeleton attest to its dynamic nature. Bone responds in a variety of ways to vitamin, mineral, and hormone deficiency or excess. Inherent in these responses are changes in the physiognomy, construction, and mechanical function of the body.

For a review of the history of the vertebrate skeleton and the bones which comprise it, reference may be made to comparative anatomy texts such as those by Romer and Parsons (1978) and Bolk et al. (1931–39) or to older works by Flower (1870) on mammals or Owen (1868) on all vertebrates.

Specific information on the skeleton of the dog is included in the veterinary anatomical texts of *Sisson and Grossman's Anatomy of the Domestic Animals* by Getty (1975); *Lehrbuch der Anatomie der Haustiere* by Nickel, Schummer, and Seiferle (Vol. I, 1977); and in the most detailed *Handbuch der Vergleichenden Anatomie*

der Haustiere by Ellenberger and Baum (1943).

For a discussion of the structure and function of bone in health and disease, reference may be made to *The Biochemistry and Physiology of Bone* by Bourne (Vols. I, II, III, 1972; IV, 1976); *The Biology of Bone* by Hancox (1972); *Biological Mineralization* by Zipkin (1973); *The Physiological and Cellular Basis of Metabolic Bone Disease* by Rasmussen and Bordier (1974); and *Bone: Fundamentals of the Physiology of Skeletal Tissue* by McLean and Urist (1968). Various aspects of skeletal morphology in the dog have been considered by Lumer (1940) — evolutionary allometry; Stockard (1941) — genetic and endocrine effects; Haag (1948) — osteometric analysis of aboriginal dogs; Hildebrand (1954) and Clutton-Brock et al. (1976) — comparative skeletal morphology in canids.

Classification of Skeletal Elements

Bones may be grouped according to shape, structure, function, origin, or position. The total average number of bones in each division of the skeletal system, as found in an adult dog (Fig. 4–1), is given in Table 4–1. In this enumeration, the bones of the dewclaw (the first digit of the hindpaw) are not included, because this digit is absent in many breeds of dogs, and in other breeds a single or double first

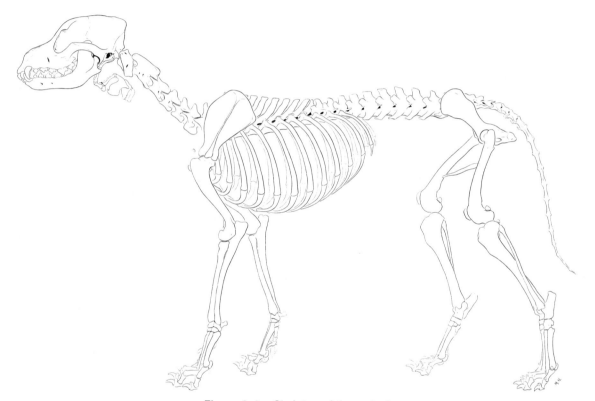

Figure 4–1.   Skeleton of the male dog.

TABLE 4–1. BONES OF SKELETAL SYSTEM

Division	Total Average Number
Axial Skeleton	
Vertebral column	50
Skull and hyoid	50
Ribs and sternum	34
Appendicular Skeleton	
Pectoral limbs	90
Pelvic limbs	96
Heterotopic Skeleton	
Os penis	1
Total	321

digit is required for showing purposes (American Kennel Club 1975). Sesamoid bones associated with the limbs are included.

Classification of Bones According to Shape

Bones may be classified in various ways. Anatomists have long grouped bones according to shape although borderline forms exist. For descriptive purposes five general divisions on this basis are recognized: long bones, short bones, sesamoid bones, flat bones, and irregular bones. Long, short, and sesamoid bones are found in the limbs, whereas the flat and irregular bones are characteristic of the skull and vertebral column. The terms are readily understandable, except possibly **sesamoid,** which is derived from the Greek word for a seed that is small, flat, and obovate. Sesamoid bones vary from tiny spheres to the slightly bent, ovoid patella (kneecap), which is 2 or more centimeters long in a large dog. Some sesamoid elements never ossify but remain as cartilages throughout life, such as those of the distal interphalangeal joints.

Long bones *(ossa longa)* are characteristic of the limbs. The bones of the thigh and arm, that is, the femur and humerus, are good examples. Typically a long bone, during its growth, possesses a shaft or **diaphysis,** and two ends, the **epiphyses.** During development each end is separated from the shaft by a plate of growing cartilage, the **epiphyseal cartilage** *(cartilago*

epiphysialis), or **plate** (referred to as the "physis" in some recent publications). At maturity the epiphyseal cartilage ceases to grow, and the epiphysis fuses with the shaft as both share in the bony replacement of the epiphyseal cartilage. Fractures sometimes occur at the epiphyseal plate. Usually after maturity no distinguishable division exists between epiphysis and diaphysis. The ends of most long bones enter into the formation of freely movable joints. Long bones form levers and possess great tensile strength. They are capable of resisting many times the stress to which they are normally subjected. The stress on long bones is both through their long axes, as in standing, and at angles to these axes, as exemplified by the pull of muscles which attach to them. Although bones appear to be rigid and not easily influenced by the soft tissues which surround them, soft tissues actually do contour the bones. Indentations in the form of grooves are produced by blood vessels, nerves, tendons, and ligaments that lie adjacent to them, whereas roughened elevations or depressions are produced by the attachments of tendons and ligaments. The ends of all long bones are enlarged and smooth. In life, these smooth surfaces are covered by a layer of hyaline cartilage, as they enter into the formation of joints. The enlargement of each extremity of a long bone serves a dual purpose. It diminishes the risk of dislocation and provides a large bearing surface for the articulation. The distal end of the terminal phalanx of each digit is an exception to the stated rule. Since it is covered by horn and is not articular, it is neither enlarged nor smooth.

Short bones *(ossa brevis)* are confined to the carpal (wrist) and tarsal (ankle) regions, which contain seven bones each. They vary in shape from the typical cuboidal shape with six surfaces to irregularly compressed rods with only one flat, articular surface. In those bones having many surfaces, at least one surface is nonarticular. This surface provides an area where ligaments may attach and blood vessels may enter and leave the bone.

Sesamoid bones *(ossa sesamoidea)* are present near freely moving joints. They are usually formed in tendons, but they may be developed in the ligamentous tissue over which tendons pass. They usually possess only one articular surface, which glides on a flat or convex surface of one or more of the long bones of the extremities. Their chief function is to protect tendons at the places where greatest friction is developed.

Flat bones (*ossa plana*) are found in the limb girdles, where they serve for muscle attachment, and in the head, where they surround and protect the sense organs and brain as well as serve for muscle attachment. The bones of the face are flat, providing maximum shielding without undue weight, and streamlining the head. Furthermore, the heads of all quadrupeds overhang their centers of gravity; a heavy head would be a handicap in locomotion. The flat bones of the cranium consist of outer and inner tables of compact bone and an intermediate uniting spongy bone, called *diploë*. In certain bones of the head the diploë is progressively invaded, during growth, by extensions from the nasal cavity which displace the diploë and cause a greater separation of the tables than would otherwise occur. The intraosseous air spaces of the skull formed in this way are known as the paranasal sinuses. Bones which contain air cavities are called **pneumatic bones** (*ossa pneumatica*).

Irregular bones (*ossa irregulata*) are those of the vertebral column, but the term also includes all bones of the skull not of the flat type, and the three parts of the hip bone (*os coxae*). Jutting processes are the characteristic features of irregular bones. Most of these processes are for muscular and ligamentous attachments; some are for articulation. The vertebrae of quadrupeds protect the spinal cord and furnish a relatively incompressible bony column through which the propelling force generated by the pelvic limbs is transmitted to the trunk. The vertebrae also partly support and protect the abdominal and thoracic viscera, and give rigidity and shape to the body in general. The amount of movement between any two vertebrae is small, but the combined movement permitted in all the intervertebral articulations is sufficient to allow considerable mobility of the whole body in any direction.

Development of Bone

Bone is about one-third organic material, which is both intracellular and extracellular in location. Within or around the bone cells, known as **osteoblasts,** the bone matrix is laid down. The osteoblasts later become the **osteocytes** of mature bone. The cells which direct the deposition of cartilage and bone are derived from mesenchyme, which forms the greater part of the middle germ layer, or mesoderm, of the embryo. Bone consists of cells in an intercellular matrix of mucopolysaccharide with collagen and reticular fibers, within which a mineral complex of calcium, phosphate, carbonate, and citrate is deposited. Each bone cell, or osteocyte, rests in a lacuna and has long, branching processes which extend through canaliculi in the mineralized matrix to the lacunae of neighboring cells.

The fetal skeleton (see Chapter 2) is characterized by bones formed in membrane (intramembranous) which precede or accompany bones formed in cartilage (endochondral). Both intramembranous bone and endochondral bone are remodeled during development and form lamellar bone with haversian systems indistinguishable from each other. The terms "membrane bone" and "cartilage bone" refer to the primary tissue being mineralized. Almost all so-called "cartilage bones" begin their ossification in a perichondral membrane followed by vascular invasion and endochondral ossification. Several "membrane bones" develop secondary cartilage after membranous ossification has begun. This secondary cartilage ossifies to form compact bone indistinguishable from the remainder of the structure.

For further information, see *Calcified Tissue Research,* an international journal founded in 1967 and devoted to the structure and function of bone and other mineralized systems, and *Developmental and Cellular Skeletal Biology* by Hall (1978).

The bones of the face and dorsum of the cranium develop in sheets of connective tissue, not in cartilage. This type of bone formation is known as **intramembranous ossification.** Osteoblasts and osteoclasts continue to be the laborers in this activity.

The compact bone formed by the periosteum is identical with membrane bone in its elaboration. Bony tissue of either type is capable of growing in any direction. The jaws and hyoid arches are preceded by cartilages, which are probably derived from the neural crest.

Structure of Bone

The gross structure of a dried, macerated bone is best revealed if the bone is sectioned in various planes. Two types of bone structure will be seen. One is compact, or dense, which forms the outer shell of all skeletal parts. The other is spongy, or cancellous, which occupies the interior of the extremities of all long bones and the entire interior of most other bones, except certain of the skull bones and the bones of the pectoral and pelvic girdles. Spongy bone is not found in the girdles, where the two compact plates are fused.

Compact bone (*substantia compacta* and *substantia corticalis*) is developed in direct ratio to the stress to which the bone is subjected. It is thicker in the shafts of long bones than in their extremities. It attains its greatest uniform thickness where the circumference of the bone is least. The maximum thickness of the compact bone found in the femur and humerus of an adult Great Dane was 3 mm. Local areas of increased thickness are present at places where there is increased tension from muscles or ligaments.

Spongy bone (*substantia spongiosa*) is elaborated in the extremities of long bones, forms the internal substance of short and irregular bones, and is interposed between the two compact layers of most flat bones. Spongy bone consists of a complicated maze of crossing and connecting osseous leaves and spicules which vary in shape and direction. The spongy bone of the skull is known as **diploë.**

The shafts of long bones in the adult are largely filled with **yellow bone marrow** (*medulla ossium flava*). This substance is chiefly fat. In the fetus and the newborn, **red bone marrow** (*medulla ossium rubra*) occupies this cavity and functions in forming red blood cells. No spongy bone is present in the middle of the shaft of a long bone, and the marrow-filled space thus formed is known as a **medullary cavity** (*cavum medullare*).

Spongy bone is developed where greatest stress occurs. The leaves or lamellae and bars are arranged in planes where pressure and tension are greatest, this structural development for functional purposes being best seen in the proximal end of the femur. The interstices between the leaves and bars of spongy bone are occupied by red marrow. The spongy bone of ribs and vertebrae and of many other short and flat bones is filled with red marrow throughout life. In the emaciated or the extremely aged, red marrow gives way to fatty infiltration.

The **periosteum** is an investing layer of connective tissue which covers the nonarticular surfaces of all bones in the fresh state. The connective tissue covering of cartilage, known as **perichondrium,** does not differ histologically from periosteum. Perichondrium covers only the articular margins of articular cartilages, but invests cartilages in all other locations. Periosteum blends imperceptibly with tendons and ligaments at their attachments. Muscles do not actually have the fleshy attachment to bone which they are said to have, since a certain amount of connective tissue, periosteum, intervenes between the two. At places where there are no tendinous or ligamentous attachments it is not difficult, when bone is in the fresh state, to scrape away the periosteum.

The **endosteum** is similar in structure to periosteum, but is thinner. It lines the large medullary cavities, being the condensed peripheral layer of the bone marrow. Both periosteum and endosteum, under emergency conditions, such as occur in fracture of bone, provide cells (osteoblasts) which aid in repair of the injury. Sometimes the broken part is overrepaired with bone of poor quality. Such osseous bulges at the site of injury are known as **exostoses.**

Mucoperiosteum is the name given to the covering of bones which participate in forming boundaries of the respiratory or digestive system. It lines all of the paranasal sinuses and contains mucous cells.

Physical Properties of Bone

Bone is about one-third organic and two-thirds inorganic material. The inorganic matrix of bone has a microcrystalline structure composed principally of calcium phosphate. The exact constitution of the crystal lattice is still under study, but it is generally agreed that bone mineral is largely a hydroxyapatite $3\ Ca_3(PO_4)_2 \cdot Ca(OH)_2$ with adsorbed carbonate. Some consider that it may exist as tricalcium phosphate hydrate $3\ Ca_3(PO_4)_2 \cdot 2\ H_2O$ with adsorbed calcium carbonate (Dixon and Perkins 1956). The organic framework of bone can be preserved while the inorganic part is dissolved. A 20 per cent aqueous solution of hydrochloric acid will decalcify any of the long bones of a dog in about one day. Such bones retain their shape but are pliable. A slender bone, such as the fibula, can be tied into a knot after decalcification. The organic material is essentially connective tissue, which upon boiling yields gelatin.

Surface Contour of Bone

Much can be learned about the role in life of a specific bone by studying its eminences and depressions. There is a functional, embryological, or pathological reason for the existence of almost every irregularity.

Most eminences serve for muscular and ligamentous attachments. Grooves and fossae in some instances serve a similar function. Facets are small articular surfaces which may be flat, concave, or convex. Trochleas and condyles are usually large articular features of bone. The roughened, enlarged parts which lie proximal to the condyles on the humerus and femur are known as epicondyles.

Vessels and Nerves of Bone

Bone, unlike cartilage, has both a nerve and a blood supply. Long bones and many flat and irregular bones have a conspicuous **nutrient** (medullary) **artery** and **vein** passing through the compact substance to serve the marrow within. Such arteries pass through a **nutrient foramen** (*foramen nutricium*) and **canal** (*canalis nutricius*) of a bone and, upon reaching the marrow cavity, divide into proximal and distal branches which repeatedly subdivide and supply the bone marrow and the adjacent cortical bone. In the long and short bones, terminal branches reach the epiphyseal plate of cartilage, where, in young animals, they end in capillaries. In adults it is likely that many twigs nearest the epiphyses anastomose with twigs arising from vessels in the periosteum. Nutrient veins pursue the reverse course. Not all of the blood supplied by the nutrient artery is returned by the nutrient vein or veins; much of it, after traversing the capillary bed, returns through veins which perforate the compact bone adjacent to the articular surfaces at the extremities of these bones. The **periosteal arteries** and **veins** are numerous but small; these arteries supply the extremities of long bones and much of the compact bone also. They enter minute canals which lead in from the surface, and ramify proximally and distally in the microscopic tubes which tunnel the compact and spongy bone. The arterioles of the nutrient artery anastomose with those of the periosteal arteries deep within the compact bone. It is chiefly through enlargement of the periosteal arteries and veins that an increased blood supply and increased drainage are obtained at the site of a fracture. Veins within bone are devoid of valves, the capillaries are large, and the endothelium from the arterial to the venous side is continuous. **Lymph vessels** are present in the periosteum as perivascular sheaths and probably also as unaccompanied vessels within the bone marrow. The **nerves** of bone are principally sensory. They serve as an inner defense against injury. The sensory nerves of the skin form the outer defense. Both carry impulses which result in pain. Kuntz and Richins (1945) state that both the afferent and sympathetic fibers probably play a role in reflex vasomotor responses in the bone marrow.

Function of Bone

The skeleton of the vertebrate body serves four functions.

1. Bone forms the supporting and in

many instances the protecting framework of the body.

2. Many bones serve as first-,second-, or third-class levers, owing to the action of different muscles at different times and to changes in the positions of force and fulcrum. Nearly all muscles act at a mechanical disadvantage. The speed at which the weight travels is in direct proportion to the shortness of the force arm, and this is determined by the distance of the insertion of the muscle from the joint, or fulcrum.

3. Bone serves as a storehouse for calcium and phosphorus and for many other elements in small amounts. The greatest drain occurs during pregnancy; conversely, the greatest deposition takes place during growth. In the large breeds, such as the Great Dane and St. Bernard, the skeleton is the system most likely to show the effects of a nutritional deficiency. Undermineralization of the skeleton is a common manifestation of underfeeding, improper feeding, or inability of the individual to assimilate food adequately. Overnutrition can result in a variety of skeletal diseases (Hedhammer et al. 1974). Excessive calcium deposition may lead to the destruction of vessels and nerves passing through bones, with a net result of hypertrophic osteodystrophy, osteochondrosis dissecans, "wobblers," enostosis, and so on.

4. Bone serves as a factory for red blood cells and for several kinds of white blood cells. In the normal adult it also stores fat.

AXIAL SKELETON

THE SKULL

The skull (cranium) is the most complex and specialized part of the skeleton. It lodges the brain and houses the sense organs for hearing, equilibrium, sight, smell, and taste while providing attachment for the teeth, tongue, larynx, and a host of muscles. It is basically divided into a facial plus palatal region, and a neural, or braincase, portion (Fig. 4–2).

The facial and palatal region, consisting of 36 bones, is specialized to provide a large surface area subserving respiratory

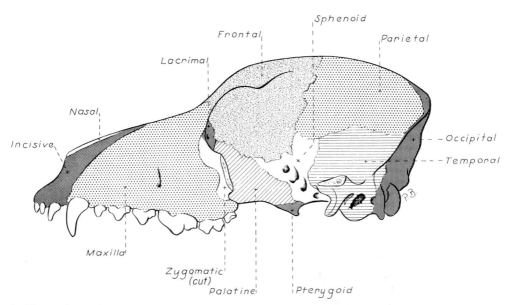

Figure 4–2. Bones of the skull, lateral aspect. (Zygomatic arch and mandible removed.)

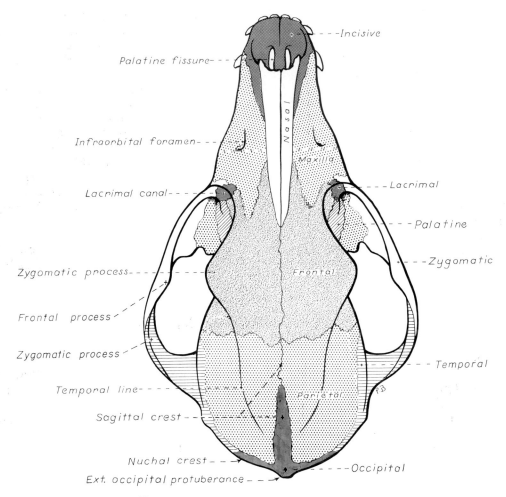

Figure 4–3.   Bones of the skull, dorsal aspect.

and olfactory functions and a long surface for the implantation of the teeth. This elongation results in a pointed rostral end, or apex, and a wide, deep base which imperceptibly blends with the braincase.

The braincase (Fig. 4–3) or **cranial cavity** *(cavum cranii)* is separated from the **cavity of the nose** *(cavum nasi)* by a perforated plate of bone, the cribriform plate. Caudally the large opening through the occipital region, the *foramen magnum,* allows for passage of the spinal cord and its associated vessels.

The ventral part of the cranium has a number of foramina and canals for the passage of nerves and blood vessels. At the junction of the facial and cranial parts, on each side, are the orbital cavities, in which

are located the globes of the eyes and accessory structures.

The bones of the ventral part (Fig. 4–4) of the cranium, or basicranial axis, are preformed in cartilage, whereas those of the dorsum, or calvaria, are formed in membrane. A classical treatment of the development of the vertebrate skull by de Beer (1937) considers the homologies of skull components, compares chondrocrania, and discusses modes of ossification.

Skulls differ more in size and shape among domestic dogs than in any other mammalian species. For this reason, craniometry in dogs takes on added significance when characterizing specific breeds and crosses. Certain points and landmarks on the skull are recognized in making lin-

Text continued on page 118.

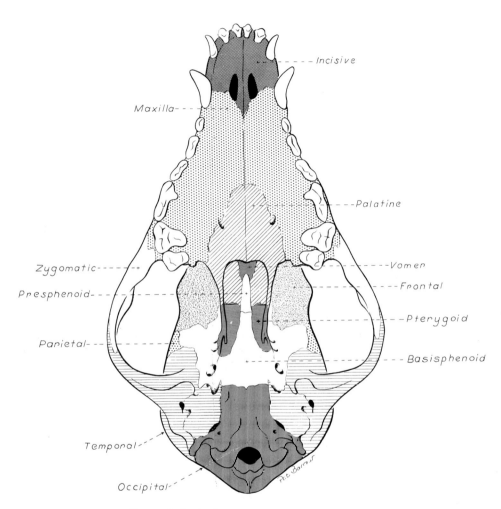

Figure 4-4. Bones of the skull, ventral aspect.

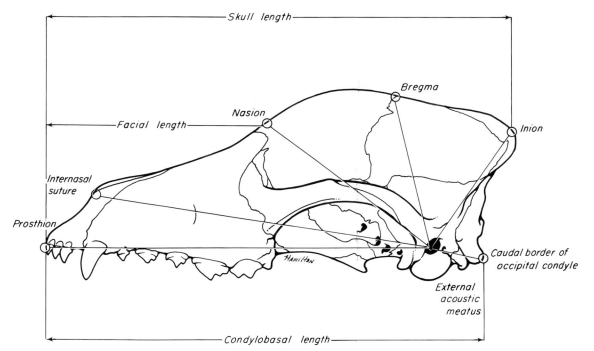

Figure 4–5.   Skull, lateral view showing craniometric points.

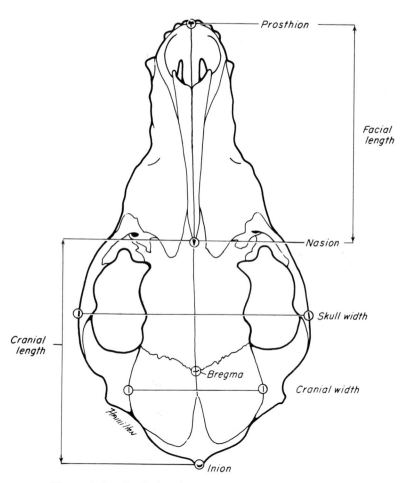

Figure 4–6. Skull, dorsal view showing craniometric points.

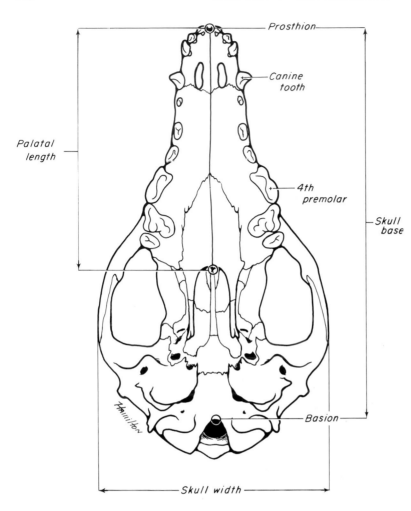

Figure 4–7. Skull, ventral view showing craniometric points.

ear measurements and have been used by Stockard (1941) and others. The more important of these are (Figs. 4–5, 4–6, 4–7):

Inion: Central surface point on the external occipital protuberance.

Bregma: Junction on the median plane of the right and left frontoparietal sutures, or the point of crossing of the coronal and sagittal sutures.

Nasion: Junction on the median plane of the right and left nasofrontal sutures.

Prosthion: Rostral end of the interincisive suture, located between the roots of the upper central incisor teeth.

Pogonion: Most rostral part of the mandible, at the symphysis, located between the roots of the lower central incisor teeth.

Basion: Middle of the ventral margin of the foramen magnum.

The center of the external acoustic meatus: Although unnamed, this spot also serves as a reference point.

Three terms are frequently used to designate head shapes (see Fig. 4–42):

Dolichocephalic, meaning long, narrow-headed. Breed examples: Collie, Russian wolfhound.

Mesaticephalic, meaning a head of medium proportions. Breed examples: German shepherd, Beagle, Setter.

Brachycephalic, meaning short, wide-headed. Breed examples: Boston terrier, Pekingese.

The face of the dog varies more in shape and size than does any other part of the skeleton. In brachycephalic breeds the facial skeleton is shortened and broadened. In some brachycephalic breeds, the English bulldog, for example, the lower jaw protrudes rostral to the upper jaw, pro-

TABLE 4–2. AVERAGE MEASUREMENTS OF THREE SKULL TYPES

	Measurement	Brachy-cephalic	Mesati-cephalic	Dolicho-cephalic
Facial length	Nasion to prosthion	48 mm.	89 mm.	114 mm.
Facial width	Widest interzygomatic distance	103 mm.	99 mm.	92 mm.
Cranial length	Inion to nasion	99 mm.	100 mm.	124 mm.
Cranial width	Widest interparietal distance	56 mm.	56 mm.	59 mm.
Cranial height	Middle of external acoustic meatus to bregma	54 mm.	60 mm.	61 mm.
Mandibular length	Caudal border of condyle to pogonion	85 mm.	134 mm.	163 mm.
Skull length	Inion to prosthion	127 mm.	189 mm.	238 mm.
Skull width	Widest interzygomatic distance	103 mm.	99 mm.	92 mm.
Skull base length	Basion to prosthion	107 mm.	170 mm.	216 mm.
Indices $\left(\dfrac{\text{width} \times 100}{\text{length}}\right)$				
Skull index		81	52	39
Cranial index		57	56	48
Facial index		215	111	81

ducing the undershot condition known as **prognathism** of the mandible. Most other breed types have **brachygnathic** mandibles, that is, receding lower jaws. Although brachygnathism of the mandibles is relative, both the Collie and the Dachshund frequently exemplify this condition to a marked extent. Stockard (1941) demonstrated that discrepancies in the pattern between the upper and lower jaws in the dog are inherited and developed as separate and independent characters. This can lead to marked disharmonies in facial features and dental occlusion, as was shown by the many crosses he made between purebred dogs. In the cross between the Basset hound and Saluki, two dogs with different skull proportions but without abnormally dissimilar jaws, some of the F_2 hybrids showed the independent inheritance of upper and lower jaw features. When one pup can inherit the muzzle and upper jaw of one parent and the lower jaw from the other, it can have serious effects on dental occlusion and thus mastication, tooth loss, prehension, and so on. Occasionally, breed-specific features are accentuated in the crossbred dog so that minor aberrations become major features. Photographs of a variety of crosses of purebred dogs can be found in Stockard's memoir (1941). Included are such crosses as Basset hound-Shepherd, Basset hound-Saluki, Basset hound-English bulldog, Dachshund-Boston terrier, Dachshund-French bulldog, Dachshund-Brussels griffon, Pekingese-Saluki, Dachshund-Basset hound, and Dachshund-Pekingese.

Table 4–2 shows average measurements in millimeters taken from randomly selected adult skulls of the three basic types. From these data it can be seen that the greatest variation in skull shape occurs in the facial part. In making comparisons of skull measurements it is essential that the overall size of the individuals measured is taken into consideration. As a rule the dolichocephalic breeds are larger than the brachycephalic, whereas the working breeds fall in the mesaticephalic group, and these as a division have the greatest body size. The only measurement in which the brachycephalic type exceeds the others, in the small sampling shown, is facial width. To obviate the size factor among the breed types, indices are computed (Table 4–3). These indicate relative size and are expressed by a single term representing a two-dimensional relationship. The cranial index is computed by multiplying the cranial width by 100 and dividing the product by the cranial length. Skull and facial indices are computed in the same manner. (Stockard (1941) found rather consistent differences between the sexes in most breeds, suggesting an endocrine influence for the differential structural expression.

Differences among the breeds in facial skeletal development are the most salient features revealed by craniometry. The face is not only short in the brachycephalic

TABLE 4–3. A COMPARISON OF INDICES*

	German shepherd	Saluki	English bulldog	Pekingese	Brussels griffon
Cranial index	51	64	69	84	84
Skull index	56	56	107	107	103
Palatal index	60	57	122	122	125
Snout index	60	53	171	179	183

Groups of breeds having similar types of skulls:

I. German shepherd	II. St. Bernard	III. English Bulldog
Foxhound	Great Dane	French Bulldog
Saluki	Dachshund	Boston Terrier

*From Stockard 1941.

breeds but it is also actually wider than it is in the heavier, longer-headed breeds. These data do not show that appreciable asymmetry exists, especially in the round-headed types. Even though the neurocranium varies least in size, it frequently develops asymmetrically. The caudal part of the skull is particularly prone to show uneven development. The further a breed digresses from the ancestral wolf type, the more likely are distortions to be found. This is particularly true of the round-headed breeds. The appearance of the English bulldog is produced by the prognathic condition of the lower jaw, as well as the brachygnathic condition of the upper jaw. This structural disharmony results in poor occlusion of the teeth. Stockard (1941) found that the formation of the bulldog type of skull results from a defective growth reaction of the basicranial epiphyseal cartilages. This defective growth is foreshadowed by a deficiency in the cartilaginous matrix which is the precursor of the basioccipital and basisphenoid bones themselves. An early ankylosis of these growth cartilages (chondrodystrophy) causes the shortening of the basicranial axis. On sagittal section (Fig. 4–8) the limits of the cranial and facial portions of the skull are clearly demarked by the cribriform plate of the ethmoid.

Cranial capacity may vary between breeds and can be measured by filling the

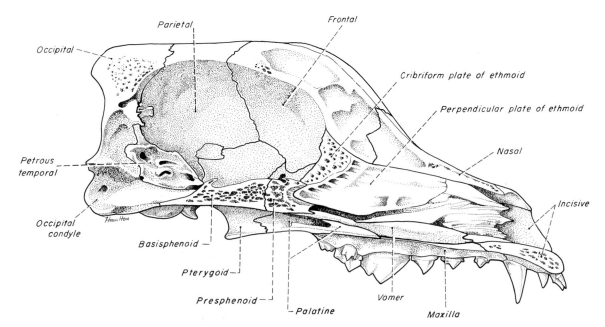

Figure 4–8. Bones of the skull, medial aspect of sagittal section.

TABLE 4-4.   INDIVIDUAL BONES OF THE SKULL

Bones of the braincase:

Paired:	1. Exoccipital	3. Frontal
	2. Parietal	4. Temporal

Unpaired:	1. Supraoccipital	4. Basisphenoid
	2. Interparietal	5. Presphenoid
	3. Basioccipital	6. Ethmoid

Bones of the face and palate:

Paired:	1. Premaxilla	6. Zygomatic
	2. Nasal	7. Palatine
	3. Maxilla	8. Lacrimal
	4. Dorsal concha	9. Pterygoid
	5. Ventral concha	10. Mandible

Unpaired:	1. Vomer

Bones of the hyoid apparatus and middle ear:

Paired:	1. Stylohyoid	5. Malleus
	2. Epihyoid	6. Incus
	3. Ceratohyoid	7. Stapes
	4. Thyrohyoid	

Unpaired:	1. Basihyoid

crania with mustard seed after the foramina had been closed with modeling clay, and then determining the volume of seed used. Average Boston terrier skulls held 82 cc. A sampling of skulls of medium size and medium length showed an average capacity of 92 cc.; the average skull capacity of the crania of the Russian wolfhound and of the Collie was 104 cc.

The names of the individual bones making up the 50 which compose the skull are listed in Table 4-4.

Bones of the Braincase

OCCIPITAL BONES

The **occipital bones** (Figs. 4-9 and 4-10) form a ring, the foramen magnum, around the entrance of the spinal cord. The ring develops from four centers: a squamous part dorsally, two lateral condylar parts, and a basilar part ventrally. A keyhole-shaped notch may be present dorsally (Fig. 4-11), resulting in what has been referred to as occipital dysplasia (Parker and Park 1974).

The **squamous part** (*pars squamosa*), or **supraoccipital bone,** is the largest division. Dorsorostrally it is wedged between the parietal bones to form the **interparietal process** (*processus interparietalis*). This process represents the the unpaired interpairetal bone which fuses prenatally with the supraoccipital. Occasionally an unfused interparietal bone is found in an adult dog. It may be more apparent inside the cranium than externally. From the interparietal process arises the mid-dorsal **external sagittal crest** (*crista sagittalis externa*), which, in some specimens, is confined to this bone. The rostral end of the interparietal process is narrower and thinner than the caudal part, which turns ventrally to form a part of the caudal surface of the skull. The **nuchal crest** (*crista nuchae*) marks the division between the dorsal and caudal surfaces of the skull. It is an unpaired sharp-edged crest of bone which reaches its most dorsal point at the external occipital protuberance. On each

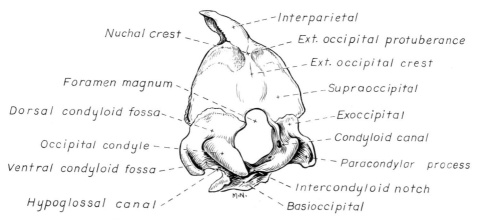

Figure 4-9.   Occipital bone, caudolateral aspect.

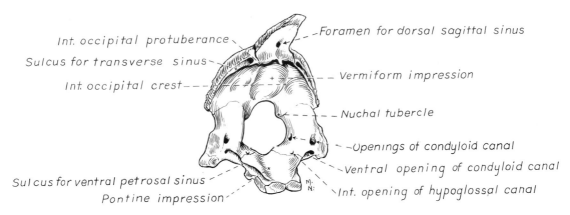

Figure 4–10.   Occipital bone, rostrolateral aspect.

side it arches ventrally before ending on a small eminence located dorsocaudal to the external acoustic meatus. The **external occipital protuberance** (*protuberantia occipitalis externa*) is the median, triangular projection forming the most dorsocaudal portion of the skull. The **external occipital crest** (*crista occipitalis externa*) is a smooth median ridge extending from the external occipital protuberance to the foramen magnum. It is poorly developed in some specimens.

Within the dorsal part of the occipital bone and opening bilaterally on the cerebral surface is the **transverse canal** (*canalis transversus*), which, in life, contains the venous transverse sinus. The transverse canal is continued laterally, on each side, by the **sulcus for the transverse sinus** (*sulcus sinus transversi*). Mid-dorsally or to one side, the dorsal sagittal sinus enters the transverse sinus via the *foramen impar*. Between the laterally located sulci the skull protrudes rostroventrally to form the **internal occipital protuberance** (*protuberantia occipitalis internus*). Extending rostrally from the internal occipital protuberance is the variably developed, usually paramedian, and always small **internal sagittal crest** (*crista sagittalis interna*). The **vermiform impression** (*impressio vermialis*), forming the thinnest part of the caudal wall of the skull, is an irregular excavation of the median portion on the cerebellar surface of the squamous part of the occipital bone which houses a part of the vermis of the cerebellum. The vermiform impression is bounded laterally by the paired **internal occipital crest** (*crista occipitalis interna*),

which is usually asymmetrical and convex laterally. Lateral to the internal occipital crest, as well as on the ventral surface of the interparietal process, there are elevations, *juga cerebralia et cerebellaria*, and depressions, *impressiones digitatae*. Ventrally the squamous part is either curved or notched to form the dorsal part of the foramen magnum. On either side the supraoccipital is fused with the exoccipital. This union represents the former articulation (*synchondrosis intraoccipitalis squamolateralis*) which extended from the foramen magnum to the temporal bone.

The **lateral parts** (*partes laterales*), or **exoccipital bones,** bear the **occipital condyles** (*condyli occipitales*), which are convex and, with the atlas, form the atlanto-occipital joints. The **jugular process** (*processus jugularis*) is located, one on either side, lateral to the condyle, and ends in a rounded knob ventrally, usually on a level with the bottom of the rostrally located tympanic bulla. Between the jugular process and the occipital condyle is the **ventral condyloid fossa** (*fossa condylaris ventralis*). On a ridge of bone rostral to this fossa is the **hypoglossal foramen** (*foramen hypoglossi*), which is the external opening of the **hypoglossal canal** (*canalis hypoglossi*), a direct passage through the ventral part of the occipital bone. The **dorsal condyloid fossa** (*fossa condylaris dorsalis*) is located dorsal to the occipital condyle. The rather large **condyloid canal** (*canalis condylaris*) runs through the medial part of the lateral occipital bone. There is an intraosseous passage between the condyloid canal and the hypoglossal canal. Usually there is also a small passage between the condyloid canal and the petrobasilar fissure.

The **basilar part** (*pars basilaris*), or **basioccipital bone,** is unpaired, and forms the posterior third of the cranial base. It is roughly rectangular, although caudally it tapers to a narrow, concave end which forms the central portion of the **intercondyloid notch** (*incisura intercondyloidea*). The central area of the basioccipital bears the **pontine impression** (*impressio pontina*). The adjacent occipital condyles on each side deepen the incisure as they contribute to its formation. The incisure bounds the ventral part of the foramen magnum. The *foramen magnum* is a large, transversely oval opening in the posteroventral portion of the skull, through which pass the spinal cord and its associated structures, the meninges, vertebral venous sinuses, the spinal portion of the accessory nerve, and the various arteries associated with the spinal cord. In brachycephalic breeds it is more circular than oval, and it is frequently asymmetrical or notched. The dorsal boundary of the foramen magnum is featured by the caudally flared ventral part of the supraoccipital bone. The caudal extension is increased by the paired **nuchal tubercles** (*tubercula nuchalia*). These projections are sufficiently prominent at times to make spinal punctures at this site difficult. The dorsal surface of the basioccipital bone is concave to form the *sulcus medullae oblongatae*. The lateral surfaces of the caudal half of the basioccipital bone fuse with the exoccipital bones along the former **ventral intraoccipital synchondrosis** (*synchondrosis intraoccipitalis basilateralis*). The ventral surface of the basioccipital bone adjacent to the petrotympanic synchondrosis possesses muscular tubercles (*tubercula muscularia*). These are rough, sagitally elongated areas, located medial to the smooth, rounded tympanic bullae. The **pharyngeal tubercle** (*tuberculum pharyngeum*) is a single triangular rough area rostral to the intercondyloid incisure. Laterally the basioccipital bone is grooved to form the ventral petrosal sulcus, which concurs with the pyramid of the temporal bone to form the **petro-occipital canal** (*canalis petro-occipitalis*).

Ventrally the rostral end of the basioccipital bone articulates with the body of the basisphenoid bone at the cartilaginous **spheno-occipital joint** (*synchondrosis spheno-occipitalis*). Ventrolaterally the occipital bone articulates with the tympanic part of the temporal bone to form the cartilaginous **occipitotympanic joint** (*sutura occipitotympanica*). Deep to this joint is the important **petro-occipital suture** (*sutura petro-occipitalis*), in which the foramen lacerum caudalis, or jugular foramen, opens. The joint between the petrosal and the occipital bones which forms the petro-occipital suture is the *synchondrosis petro-occipitalis*. Laterally, and proceeding dorsally, the occipital bone first articulates with the squamous temporal bone superficially, the **occipitosquamous suture** (*sutura occipitosquamosa*), and with the mastoid part of the petrous temporal bone deeply, the **occipitomastoid suture** (*sutura occipitomastoidea*); further dorsally it articulates with the parietal bone, the **lambdoid suture** (*sutura lambdoidea*). Where the squamous and lateral parts of the occipital bone articulate with each other and with the mastoid part of the temporal bone, the **mastoid foramen** (*foramen mastoideum*) is formed.

Variations in the occipital bone are nu-

Figure 4–11. Occipital region of mongrel mesocephalic dogs showing a "keyhole" notch compared with a rather circular foramen magnum.

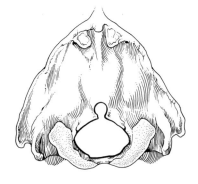
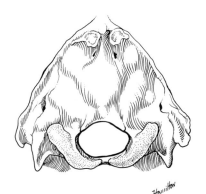

merous. The foramen magnum varies in shape and is not always bilaterally symmetrical (Fig. 4–11). The condyloid canal may be absent on one or both sides. Even when both canals are present, connections between the hypoglossal and condyloid canals may fail to develop. The jugular processes may extend several millimeters ventral to the tympanic bullae so that they will support a skull without the mandibles when it is placed on a horizontal surface; conversely, they may be short, retaining the embryonic condition. The vermiform impression may be deep, causing a caudomedian rounded, thin protuberance on the caudal surface of the skull. The foramen impar may be double. It is rarely median in position. A sutural bone may be present at the rostral end of the interparietal process.

PARIETAL BONE

The **parietal bone** (*os parietale*) (Fig. 4–12) is paired and forms most of the dorsolateral part of the cranial wall. It articulates dorsally with its fellow and with the interparietal process of the occipital bone. Each parietal bone lies directly rostral to the squamous occipital and dorsal to the squamous temporal. In the newborn no elevation is present at the sagittal interparietal suture or on the interparietal process, but soon thereafter in the heavily muscled breeds, particularly in the male, the mid-dorsal external sagittal crest is developed. This crest, which increases in size with age, forms the medial boundary of the **temporal fossa** (*fossa temporalis*), a large area on the **external surface** (*facies externa*) of the cranium from which the temporal muscle originates. In dolichocephalic breeds with heavy temporal muscles, the external sagittal crest may reach a height of more than 1 cm. and extend from the external occipital protuberance to the parietofrontal suture. Rostrally, it continues as the diverging frontal crests. In most brachycephalic skulls the external sagittal crest is confined to the interparietal part of the occipital bone and is continued rostrally as the diverging **temporal lines** (*lineae temporales*). The temporal lines at first are convex laterally, then become concave as they cross the parietofrontal, or coronal, suture and are continued as the external frontal crests to the zygomatic processes. The temporal lines replace the external sagittal crest in forming the medial boundaries of the temporal fossae in most brachycephalic skulls.

The **internal surface** (*facies interna*) of the parietal bone presents *digital impressions* and *intermediate ridges* corresponding, respectively, with the cerebral gyri and sulci. A well-defined vascular groove, the **sulcus for the middle meningeal artery** (*sulcus arteriae meningeae mediae*), starts at the ventrocaudal angle of the bone and arborizes over its internal surface. The groove runs toward the opposite angle of the bone, giving off smaller branched grooves along its course. A leaf of bone projects rostromedially from the dorsal part of the caudal border. This leaf concurs with its fellow and with the internal occipital protuberance to form the curved *tentorium ossium*. On the internal surface of the parietal bone near its caudal border is a portion of the **transverse sulcus**,

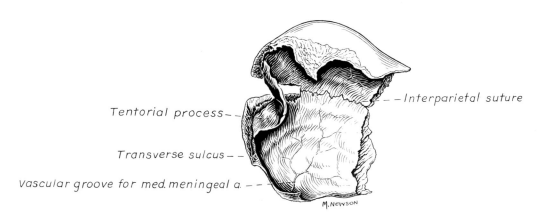

Tentorial process-

Transverse sulcus – –

Vascular groove for med. meningeal a. – –

– –Interparietal suture

M. NEWSON

Figure 4–12. Parietal bones, ventral lateral aspect.

which leads dorsally into the transverse canal of the occipital bone and ventrally into the temporal meatus.

The borders of the parietal bone are rostral, dorsal, and ventral in position, since the bone is essentially a curved, square plate. The rostral or **frontal border** (*margo frontalis*) overlaps the frontal bone, forming the frontoparietal or **coronal suture** (*sutura coronalis*). The caudal or **occipital border** (*margo occipitalis*) meets the occipital bone to form the **occipitoparietal suture** (*sutura occipitoparietalis*). The rostral half of the dorsal or **sagittal border** (*margo sagittalis*) articulates with its fellow on the midline to form the **sagittal suture** (*sutura sagittalis*). The caudal half of the dorsal border articulates with the interparietal process of the occipital bone to form the **parietointerparietal suture** (*sutura parietointerparietalis*). The ventral or **squamous border** (*margo squamosus*) is overlaid by the squamous temporal bone in forming the **squamous suture** (*sutura squamosa*). A small area of the squamous border at its rostral end articulates with the temporal wing of the sphenoid bone to form the **parietosphenoidal suture** (*sutura parietosphenoidalis*). Overlapping of the bones at the squamous and coronal sutures allows for cranial compression of the fetal skull during its passage through the pelvic canal.

FRONTAL BONE

The **frontal bone** (*os frontale*) (Figs. 4–13, 4–14) is irregular in shape, being broad caudally and somewhat narrower rostrally. Laterally, the rostral part is concave and forms the medial wall of the orbit. Caudal to this concavity, it flares laterally to form part of the temporal fossa. The **frontal sinus** (*sinus frontalis*) is an air cavity located between the inner and outer tables of the rostral end of the frontal bone and is divided into two or three compartments. It is discussed in greater detail under the heading Paranasal Sinuses.

For descriptive purposes the frontal bone is divided into orbital, temporal, frontal, and nasal parts.

The **orbital part** (*pars orbitalis*) is a segment of a cone with the apex located at the optic canal and the base forming the medial border of the **infraorbital margin** (*margo infraorbitalis*). Lateral to the most dorsal part of the **frontomaxillary suture** (*sutura frontomaxillaris*) the orbital margin is slightly flattened for the passage of the vena angularis oculi. Ventrally, a long, distinct, dorsally arched muscular line marks the approximate ventral boundary of the bone. The **ethmoidal foramina** (*foramina ethmoidalia*) are two small openings about 1 cm. rostral to the optic canal. The smaller opening is in the frontosphenoidal suture; the larger foramen, located dorsocaudal to the smaller, passes obliquely through the orbital part of the frontal bone. Sometimes the two ethmoidal foramina are confluent. At the orbital margin, the frontal and orbital surfaces meet, forming an acute angle. The **supraorbital** or **zygomatic process** (*processus zygomaticus*) is formed where the orbital margin meets the **external frontal crest** (*crista frontalis externa*), which curves rostrolaterally from the temporal line or saggital crest. On the orbital surface of the zygomatic process is a small foramen which is only large enough to admit a horse hair. Ventrorostral to this foramen in some adult skulls

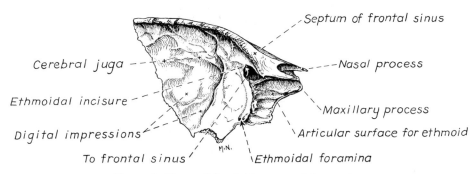

Figure 4–13. Left frontal bone, medial aspect.

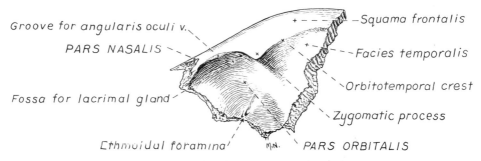

Groove for angularis oculi v.
PARS NASALIS
Fossa for lacrimal gland
Ethmoidal foramina
Squama frontalis
Facies temporalis
Orbitotemporal crest
Zygomatic process
PARS ORBITALIS

Figure 4–14. Left frontal bone, lateral aspect.

the **fossa for the small lacrimal gland** (*fossa glandulae lacrimalis*) can be seen.

The **temporal surface** (*facies temporalis*) forms that part of the frontal bone caudal to the orbital part. Dorsally the two tables of the frontal bone are separated to form the frontal sinus, whereas ventrally and caudally the two tables are fused or united by a small amount of diploë to form the braincase.

The **frontal squama** (*squama frontalis*), is roughly triangular, with its base facing medially, and articulating with that of the opposite bone. It is gently rounded externally and is largely subcutaneous in life. Its caudal boundary is the external frontal crest and the lateral part of its rostral boundary is the orbital margin.

The **nasal part** (*pars nasalis*) is the rostral extension of the frontal bone. Its sharp, pointed **nasal process** (*processus nasalis*) lies partly under and partly between the caudal parts of the nasal and maxillary bones. The **septum of the frontal sinus** (*septum sinuum frontalium*) is a vertical median partition which closely articulates with its fellow in separating right and left frontal sinuses. It is widest near its middle, which is opposite the cribriform plate. Rostrally it is continuous with the septal process of the nasal bone. The ventral part of the septum of the frontal sinus is the **internal frontal crest** (*crista frontalis interna*). The conjoined right and left crests articulate with the perpendicular plate of the ethmoid bone ventrally and with the conjoined right and left septal processes of the nasal bones rostrally. The **ethmoid incisure** (*incisura ethmoidalis*), which lies dorsal and lateral to the cribriform plate of the ethmoid bone, is formed by the smooth concave edge of the internal table of the nasal part of the frontal bone.

The **internal surface** (*facies interna*) of the frontal bone forms a part of the braincase caudally and a small portion of the nasal cavity rostrally. The salient ethmoidal notch separates the two parts. The caudal part is deeply concave and divided into many fossae by the **digital impressions** and the **cerebral juga.** Fine, dorsocaudally running vascular grooves indicate the position occupied in life by the rostral meningeal vessels. The large **aperture to the frontal sinus** is located dorsal to the ethmoidal incisure. The nasal part of the internal surface of the frontal bone is marked by many longitudinal lines of attachment for the ethmoturbinates.

The mid-dorsal articulation of the frontal bones forms the **frontal suture** (*sutura interfrontalis*). This suture is a forward continuation of the sagittal suture between the parietal bones. Caudally the frontal bone is overlapped by the parietal bone, forming the **frontoparietal suture** (*sutura frontoparietalis*). Ventrally the rather firm **sphenofrontal suture** (*sutura sphenofrontalis*) is formed. Rostrally the frontal bone articulates with the nasal, maxillary, and lacrimal bones to form the **frontonasal suture** (*sutura frontonasalis*), the **frontomaxillary suture** (*sutura frontomaxillaris*), and the **frontolacrimal suture** (*sutura frontolacrimalis*). Deep in the orbit, the frontal bone articulates with the palatine bone to form the **frontopalatine suture** (*sutura frontopalatina*). Medially, hidden from external view, the frontal bone articulates with the ethmoid bone in forming the **frontoethmoidal suture** (*sutura frontoethmoidalis*).

SPHENOID BONES

The **sphenoid bones** (*ossa sphenoidales*) (Figs. 4–15, 4–16, 4–17) form the rostral two-thirds of the base of the neurocranium, be-

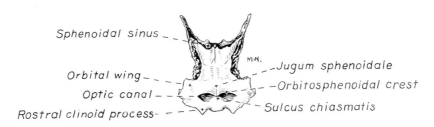

Sphenoidal sinus

Orbital wing

Optic canal

Rostral clinoid process

M.N.

Jugum sphenoidale

Orbitosphenoidal crest

Sulcus chiasmatis

Figure 4–15. Presphenoid, dorsal aspect.

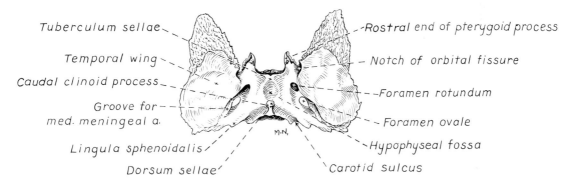

Tuberculum sellae

Temporal wing

Caudal clinoid process

Groove for
med. meningeal a.

Lingula sphenoidalis

Dorsum sellae

M.N.

Rostral end of pterygoid process

Notch of orbital fissure

Foramen rotundum

Foramen ovale

Hypophyseal fossa

Carotid sulcus

Figure 4–16. Basisphenoid, dorsal aspect.

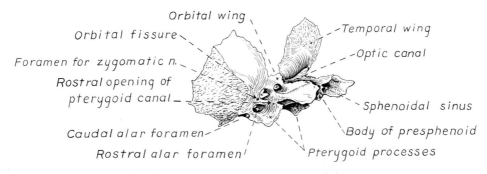

Orbital wing

Orbital fissure

Foramen for zygomatic n.

Rostral opening of
pterygoid canal

Caudal alar foramen

Rostral alar foramen

Temporal wing

Optic canal

Sphenoidal sinus

Body of presphenoid

Pterygoid processes

Figure 4–17. Presphenoid and basisphenoid, rostrolateral aspect.

tween the basioccipital caudally and the ethmoid rostrally. They each consist of a pair of wings and a median body. The more rostral bone with orbital wings is the **presphenoid** (*os presphenoidale*); the caudal bone with the larger wings is the **basisphenoid** (*os basisphenoidale*).

Presphenoid. The dorsal part of the body of the presphenoid is roofed over by the fusion of right and left **orbital wings** (*alae orbitales*) to form the **yoke** (*jugum sphenoidale*). The yoke forms the base of the rostral cranial fossa. A small median tubercle, the **rostrum** (*rostrum sphenoidale*), divided in the newborn, projects from the rostral border of the yoke. Caudally, the yoke forms a shelf, the **orbitosphenoidal crest** (*crista orbitosphenoidalis*), under which lie the diverging **optic canals** (*canales optici*). On each side of the caudal end of the presphenoid is a **rostral clinoid process** (*processus clinoideus rostralis*) which projects caudally. On the dorsum of the body, caudal to the optic canals, is the unpaired *sulcus chiasmatis*, in which lies the optic chiasma.

Basisphenoid. The body of the basisphenoid (Fig. 4–16) forms the base of the middle cranial fossa. The middle of its dorsal surface is slightly dished to form the oval **hypophyseal fossa** (*fossa hypophysialis*). The fossa is limited rostrally by the *tuberculum sellae*, an upward sloping ridge of bone formed at the junction of the presphenoid and basisphenoid. The hypophyseal fossa is limited caudally by a bony process, the *dorsum sellae*, which, in adult skulls, is flattened and expanded at its free end. Projecting rostrally on each side of the dorsum sellae is a **caudal clinoid process** (*processus clinoideus*). This complex of bony structures, consisting of the tuberculum sellae, the hypophyseal fossa, and the dorsum sellae with its two caudal clinoid processes, is called the *sella turcica*, or Turkish saddle. In life it contains the hypophysis. Occasionally the small **craniopharyngeal canal** (*canalis craniopharyngeus*) persists in the adult. This canal is a remnant of the pharyngeal diverticulum to the hypophyseal fossa from which the pars glandularis of the hypophysis develops.

The **orbital wings** (*alae orbitales*), also known as lesser wings or orbitosphenoids, leave each side of the presphenoid and roof across its body. Rostrally, at the junction of the wings and the body, the presphenoid is hollow and divided by a longitudinal septum to form the **sphenoidal fossae** (*fossae sphenoidales*), into which extend the ventrocaudal parts of the ethmoturbinates. The orbital wings articulate ventrally with the palatine and dorsally with the frontal bones. In the frontosphenoidal suture is located the **ethmoidal foramen** (*foramen ethmoidale*); a second, larger ethmoidal foramen is usually present in the frontal bone dorsocaudal to the one in the suture. These foramina may be confluent. The caudal parts of the orbital wings slope upward and outward and are thicker, but smaller, than the rostral parts. Their bases are perforated by the optic canals. Medially, in young specimens, the two elliptical optic canals are confluent across the midline. The **orbital fissures** (*fissurae orbitales*) are located lateral to the body of the sphenoid in the sutures between the orbital wings and the temporal wings. These large openings are at a lower level and are located slightly caudolateral to the optic canals.

The **temporal wings** (*alae temporales*), also known as great wings (Fig. 4–16) or alisphenoids, are larger than the orbital wing and curve outward and upward. Rostrally they extend to the lateral margin of each frontal bone to form the sphenofrontal suture. The caudal two-thirds of the temporal wings are covered laterally by the squamous temporal bone in forming the sphenotemporal suture. At the base of each wing, near its junction with the body, are a series of foramina. The **oval foramen** (*foramen ovale*) is a large opening which leads directly through the cranial wall. It is located about 0.5 cm. medial to the temporomandibular joint. A small notch or even a foramen, *foramen spinosum*, may be present in its caudolateral border for the transmission of the middle meningeal artery. The **alar canal** (*canalis alaris*) runs through the rostral part of the base of the temporal wing. Its smaller opening is the **caudal alar foramen** (*foramen alare caudalis*), and its larger one is the **rostral alar foramen** (*foramen alare rostralis*). Entering the canal from the cranium is the **round foramen** (*foramen rotundum*). It can be seen by viewing the medial wall of the alar canal through the rostral alar foramen. Dorsorostral to the alar canal is the **orbital fissure**. A small *foramen alare parvum* may be present as the dorsal opening of a small canal which leaves the alar canal. It is located on the ridge of bone separating the orbital fissure from the rostral alar foramen. When present it conducts the zygomatic nerve from the maxillary trunk. Two pairs of grooves are present on the basisphenoid bone. The extremely small **pterygoid groove** (*sulcus nervi pterygoidei*) leads into the minute **pterygoid canal** (*canalis pterygoideus*). It begins rostral to the small, pointed, muscular process of the temporal bone, where it is located in the suture between the pterygoid and basisphenoid bones. It ends in the caudal part of the pterygopalatine fossa. Probing with a horse hair will reveal that it runs medial to the pterygoid process of the sphenoid in the suture between this process and the pterygoid bone. The second groove of the basisphenoid is the **sulcus for the middle meningeal artery** (*sulcus arteriae meningeae mediae*). This groove runs obliquely dorsolaterally from the oval foramen on the cerebral surface of the temporal wing and continues mainly on the temporal and parietal bones. Two notches indent the caudal border of the temporal wing. The **medial notch** (*incisura carotica*) concurs with the temporal bone to form the foramen lacerum. The lateral notch, with its counterpart on the temporal bone, forms the short **musculotubal canal** (*canalis musculotubarius*), which transmits the tendon of the *m. tensor veli palatini* and the *tuba auditiva*. A low ridge of bone, the **lingula** (*lingula sphenoidalis*), ending in a process, separates the two openings.

The **pterygoid processes** (*processi pterygoidei*) are the only ventral projections of the basisphenoid. They are thin, sagittal plates about 1 cm. wide, 1 cm. long, and a little over 1 cm. apart. Attached to their medial surfaces are the caudally hooked, approximately square pterygoid bones. The processes and pterygoid bones separate the caudal parts of the pterygopalatine fossae from the nasal pharynx.

The body of the basisphenoid articulates caudally with the basioccipital, forming the **spheno-occipital synchondrosis** (*synchondrosis spheno-occipitalis*); and rostrally with the presphenoid, forming the **intersphenoidal synchondrosis** (*synchondrosis intersphenoidalis*). Rostrally, the presphenoid contacts the vomer, forming the **vomerosphenoidal suture** (*sutura vomerosphenoida-*

lis). The ethmoid also contacts the body of the presphenoid, forming the **sphenoethmoidal suture** (*sutura sphenoethmoidalis*). As the orbital wing of the presphenoid bone extends dorsorostrally, the **sphenopalatine suture** (*sutura sphenopalatina*) is formed ventrally, and the **sphenofrontal suture** (*sutura sphenofrontalis*), dorsally. Caudodorsally, the temporal wing is overlapped by the squamous temporal bone, forming the **sphenosquamous suture** (*sutura sphenosquamosa*). The dorsal end of the temporal wing overlaps the parietal bone, forming the **sphenoparietal suture** (*sutura sphenoparietalis*). The medial surface of the pterygoid process, with the pterygoid bone, forms the **pterygosphenoid suture** (*sutura pterygosphenoidalis*).

TEMPORAL BONE

The **temporal bone** (*os temporale*) (Figs. 4–18 to 4–21) forms a large part of the ventrolateral wall of the cranium. Its structure is intricate, owing to the presence of the cochlea and the semicircular canals, and an extension of the nasal pharynx into the middle ear. In a young skull the temporal bone can be separated into petrosal, tympanic, and squamous parts. The petrosal part has a mastoid process caudally, with an external surface. The pyramid houses the cochlea and the semicircular canals, and is the last to fuse with the other parts in development. It is located completely within the skull. The tympanic part, or bulla tympanica, is a sac-shaped protuberance, roughly as large as the end of one's finger, which lies ventral to the mastoid process. The squamous part consists of two basic divisions, an expanded plate which lies above the bulla, and the rostrally projecting zygomatic process which forms the caudal half of the zygomatic arch.

The **petrosal part** (*pars petrosa*), **pyramid**, or **petrosum** (Fig. 1–14), is fused around its periphery laterally to the medial surfaces of the tympanic and squamous parts; caudally, it may on rare occasions be united with the mastoid process. It is roughly pyramidal in shape, and is called the pyramid for this reason. The part immediately surrounding the membranous labyrinth ossifies first and is composed of very dense bone. The cartilage which surrounds the inner ear, known as the **otic capsule**, is a conspicuous feature

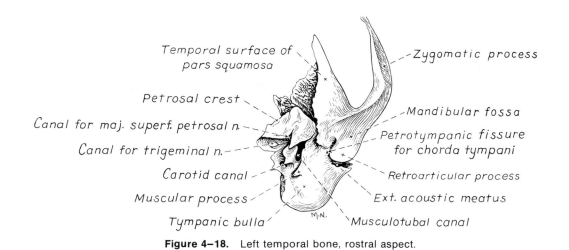

Temporal surface of
pars squamosa

Zygomatic process

Petrosal crest

Canal for maj. superf. petrosal n.

Canal for trigeminal n.

Mandibular fossa

Petrotympanic fissure
for chorda tympani

Carotid canal

Muscular process

Tympanic bulla

Retroarticular process

Ext. acoustic meatus

Musculotubal canal

Figure 4–18.   Left temporal bone, rostral aspect.

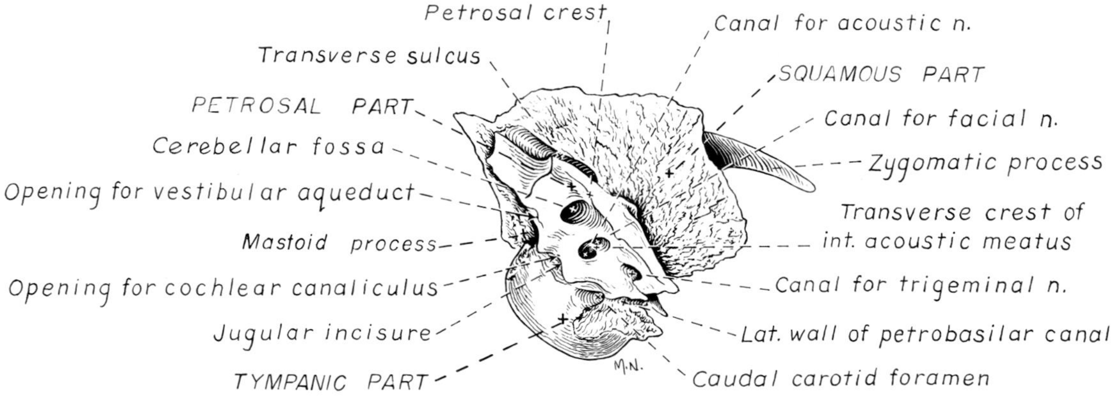

Petrosal crest

Transverse sulcus

Canal for acoustic n.

SQUAMOUS PART

PETROSAL   PART

Cerebellar fossa

Opening for vestibular aqueduct

Mastoid   process

Opening for cochlear canaliculus

Jugular incisure

TYMPANIC PART

Canal for facial n.

Zygomatic process

Transverse crest of
int. acoustic meatus

Canal for trigeminal n.

Lat. wall of petrobasilar canal

Caudal carotid foramen

Figure 4–19.   Left temporal bone, medial aspect.

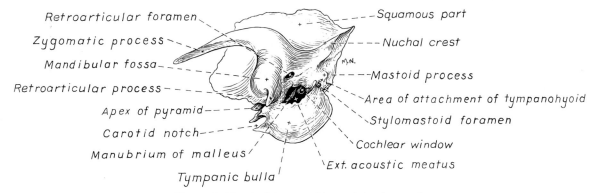

Retroarticular foramen
Zygomatic process
Mandibular fossa
Retroarticular process
Apex of pyramid
Carotid notch
Manubrium of malleus
Tympanic bulla

Squamous part
Nuchal crest
Mastoid process
Area of attachment of tympanohyoid
Stylomastoid foramen
Cochlear window
Ext. acoustic meatus

Figure 4–20. Left temporal bone, lateral aspect.

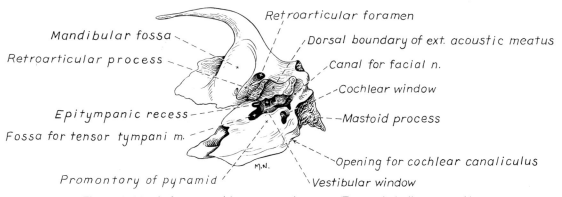

Mandibular fossa
Retroarticular process
Epitympanic recess
Fossa for tensor tympani m.
Promontory of pyramid

Retroarticular foramen
Dorsal boundary of ext. acoustic meatus
Canal for facial n.
Cochlear window
Mastoid process
Opening for cochlear canaliculus
Vestibular window

Figure 4–21. Left temporal bone, ventral aspect. (Tympanic bulla removed.)

of early embryos. Its sharp **petrosal crest** (*crista petrosa*) extends downward and forward; its axis forms an angle of about 45 degrees caudally with a longitudinal axis through the skull. It nearly meets the tentorium ossium dorsally to form a partial partition between the cerebral and cerebellar parts of the brain; rostrally it ends in a sharp point, the *apex pyramidalis*. Its **cerebral** and **cerebellar surface** (*facies encephalica*) is divided by the petrosal crest into rostrodorsal and caudomedial parts. The third surface, facing ventrolaterally, is the **tympanic surface** (*facies tympanica*).

The caudomedial surface presents several features. The most dorsal of these is the **cerebellar fossa** (*fossa cerebellaris*), which attains its greatest relative size in puppies and houses the paraflocculus of the cerebellum. Ventral to the cerebellar fossa is a very important recess, the **internal acoustic meatus** (*meatus acusticus internus*). The opening into this recess is the *porus acusticus internus*. The meatus is an irregularly ellipitcal depression which is divided deeply by the **transverse crest** (*crista transversa*). Dorsal to the crest is the opening of the facial canal, which contains the facial nerve as well as the cribriform **dorsal vestibular area** (*area vestibularis utriculo-ampullaris*) for the passage of nerve bundles from the membranous labyrinth. Ventral to the crest is the **ventral vestibular area** (*area vestibularis saccularis*), through which pass additional vestibular nerve bundles that come from a deep, minute depression, the *foramen singulare*. The **spiral cribriform tract** (*tractus spiralis foraminosus*) is formed by the wall of the hollow modiolus of the cochlea. The perforations are formed by the fascicles of the cochlear nerve which arise from the spiral ganglion on the outside of the modiolus. The cochlea and semicircular canals can be seen by removing a portion of the pyramid. Ventrorostral to the internal acoustic meatus is the short **canal** through the pyramid for the passage of the **trigeminal nerve** (*canalis trigemini*). The caudoventral part of the pyramid articulates with the occipital bone. On the cerebral surface, or on the border between the cerebral surface and the suture for the occipital bone, is the **external opening of the cochlear canaliculus** (*apertura externa canaliculi cochleae*). This opening is in the rostral edge of the jugular foramen and is large enough to be probed with a horse hair. The **jugular foramen** (*foramen jugulare*) is located between the pyramid and the occipital bone. A smaller opening for the vestibular aqueduct, the *apertura externa aqueductus vestibuli*, is located caudodorsal to the opening of the cochlear canaliculus in a small but deep cleft in the bone.

The rostrodorsal part of the cerebral surface of the pyramid is gently undulating, its only features being the digital impressions and elevations corresponding to the gyri and sulci of the cerebrum. Its lateral border is usually grooved by the small middle meningeal artery.

The **tympanic** or **ventral surface** of the pyramid forms much of the dorsal wall of the **tympanic cavity** (*cavum tympani*). At its periphery it articulates with the squamosum dorsally and the tympanicum ventrally. It can be seen from the outside through the external acoustic meatus. An eminence, two openings (windows), and three fossae are the prominent features of this surface. The barrel-shaped eminence, or **promontory** (*promontorium*), has at its larger caudolateral end the **cochlear (round) window** (*fenestra cochlea*), which, in life, is closed by the secondary tympanic membrane. Just rostral and slightly dorsolateral to the cochlear window is the **vestibular (oval) window** (*fenestra vestibuli*), which is occluded by the foot plate of the stapes. The fossae lie at the angles of a triangle located rostrolateral to the windows. The smallest fossa is a curved groove with its concavity facing the vestibular window; it is the open part of the canal for the facial nerve peripheral to the genu and rostral to the stylomastoid foramen. The largest is the **fossa for the tensor tympani muscle** (*fossa m. tensoris tympani*); it is a spherical depression which lies rostral to the vestibular window. A thin scale of bone with a point extending caudally forms part of its ventral wall. The **epitympanic recess** (*recessus epitympanicus*), the third fossa, lies caudolateral to the fossa m. tensoris tympani and at a higher level. The incus and the head of the malleus lie in this recess.

The petrosum contains the **osseous labyrinth**, which is divided into three parts: the cochlea, semicircular canals, and vestibule. The basal turn of the cochlea is located lateral to the ventral part of the internal acoustic meatus, its initial turn producing the bulk of the promontory. The **semicircular canals** (*canales semicirculares ossei*) are three in number, each located in a different plane caudal to the cochlea. The **bony vestibule** (*vestibulum*) is the osseous common chamber where the three semicircular canals and the cochlea join. The vestibular and cochlear windows of the vestibule communicate with the tympanic cavity in well-cleaned skulls. For details of the labyrinth, see Chapter 19, The Ear.

The **facial canal** (*canalis facialis*) carries the seventh cranial, or facial nerve. It enters the petrosum in the dorsal part of the internal acoustic meatus and, after pursuing a sigmoid course throught the temporal bone, emerges at the stylomastoid foramen. The initial 3 mm. of the canal, starting at the internal acoustic meatus, is straight. The canal makes its first turn on arriving at the thin medial wall of the fossa m. tensoris tympani. At this turn, or **genu of the facial canal** (*geniculum canalis facialis*), there is an indistinct enlargement for the sensory geniculate ganglion of the facial nerve. In the concavity of this bend is the rostral half of the vestibule. As the canal straightens after the first turn, and before the second turn begins, it opens into the cavity of the middle ear lateral to the vestibular window. The direction of the second bend of the canal is the reverse of that of the first, so that the whole passage is S-shaped, but does not lie in one plane. The **fossa for the stapedius muscle** (*fossa m. stapedius*) is located on the dorsal wall of the facial canal just before the canal opens into the middle ear cavity from the cranium. After completing its second arch the facial canal opens to the outside by the deeply placed **stylomastoid foramen** (*foramen stylomastoideum*). The small **petrosal canal** (*canalis petrosus*) leaves the facial canal at the genu by extending rostrally, dorsal to the fossa m. tensoris tympani. It runs rostroventrally just within the wall of the fossa to a small opening near the distal end of the petrosquamous suture and lateral to the canal for the trigeminal nerve. If a dark bristle is inserted in the canal, its path can be seen through the wall of the fossa. The greater superficial petrosal nerve passes through the petrosal canal. The *canaliculus chordae tympani* carries the chorda tympani nerve from the facial canal to the cavity of the

middle ear. It arises from the peripheral turn of the facial canal. After the nerve has crossed the medial surface of the handle of the malleus, it passes under a fine bridge of bone of the tympanic ring to continue in the direction of the auditory tube. The chorda tympani usually passes through a small canal in the rostrodorsal wall of the bulla tympanica and emerges through the **petrotympanic fissure** (*fissura petrotympanica*) by a small opening medial to the retroarticular process. When the canal fails to develop, the opening is through the rostrolateral wall of the tympanic bulla. Both the petrosal canal and the canaliculus chordae tympani leave the facial canal at an acute angle and course toward the brain.

Two minute canals run from the labyrinth to the cerebral surface of the pyramid. The **cochlear aqueduct** (*aqueductus cochleae*) runs ventrad from a point on the ventral wall of the scala tympani near the cochlear window to the border of the jugular foramen. It carries the perilymphatic duct to the subarachnoid space. The **vestibular aqueduct** (*aqueductus vestibuli*) from the vestibule passes posteroventrally to the posterior part of the cerebral surface of the pyramid about 3 mm. dorsorostral to the cochlear opening. This duct is too small to be probed easily. It carries the endolymphatic duct to the epidural space.

The **mastoid process** (*processus mastoideus*) of the petrosum is the only part to have an external surface. This surface lies between the **mastoid foramen** (*foramen mastoideum*) dorsally and the **stylomastoid foramen** (*foramen stylomastoideum*) ventrally, both of which it helps to form. It articulates with the exoccipital medially and the squamosum laterally. The ventral part is slightly enlarged, and serves for the attachment of the tympanohyoid cartilage. The facial canal, as it leaves the stylomastoid foramen, grooves the ventral surface of the pars mastoidea. The stylomastoid foramen is dorsal to the caudal part of the tympanic bulla.

The **tympanic part** (*pars tympanica*) of the temporal bone, or **tympanicum**, is the ventral portion and is easily identified by its largest component, the smooth bulbous enlargement, or *bulla tympanica*, which lies between the retroarticular and jugular processes. In puppies its walls are not thicker than the shell of a hen's egg. The cavity it encloses is the fundic part of the tympanic cavity, which is delimited from the dorsal part of the tympanic cavity proper by a thin edge of bone. In old animals this bony ledge has fine, knobbed spicules protruding from its free border. The osseous **external acoustic meatus** (*meatus acusticus externus*) is the canal from the external ear to the tympanic membrane. Its length increases with age but rarely exceeds 1 cm. even in old, large skulls. It is piriform, with its greatest dimension dorsoventrally and its smallest dimension transversely. In carefully prepared skulls the malleus can be seen through the meatus, somewhat displaced but articulated with the incus. All but the dorsal part of the external acoustic meatus is formed by the tympanicum. The **tympanic membrane** (*membrana tympani*), or ear drum, is a membranous diaphragm attached to the **tympanic ring** (*anulus tympanicus*). If planes are drawn through the ear drums, they meet at the rostral end of the braincase. At the rostral margin of the bulla, lateral to the occipitosphenoidal suture, there are paired notches and two large openings. The more medial of the two openings is the *foramen lacerum*, formerly called the external carotid foramen. It is flanked on the medial and lateral sides by sharp, pointed processes of bone from the bulla wall. The medial process meets the lingula of the sphenoid bone in separating the foramen from the lateral opening, the **musculotubal canal** (*canalis musculotubarius*). The musculotubal canal is continuous with the auditory tube. By means of the auditory (eustachian) tube, the tympanic cavity communicates with the nasal pharynx. The **carotid canal** (*canalis caroticus*) runs longitudinally through the medial wall of the osseous bulla where it articulates with the basioccipital bone. It begins at the **caudal carotid foramen** (*foramen caroticum caudalis*), which is hidden in the depths of the petrobasilar fissure. It runs rostrally, makes a ventral turn at a little more than a right angle, and opens to the outside at the foramen lacerum. At its sharp turn ventrad it concurs with the caudal part of the sphenoid bone which here forms not only the rostral boundary of the vertical parts of the carotid canal, but also the rostral boundary of an opening in the braincase, the **internal carotid foramen** (*foramen caroticum internum*). The carotid canal transmits the internal carotid artery. The lateral boundary of the **petrobasilar canal** (*canalis petrobasilaris*) is formed by the tympanic bulla and petrosum. Medially the basioccipital bone bounds it. The petrobasilar canal contains the ventral petrosal sinus which parallels the horizontal part of the carotid canal and lies medial to it.

The **tympanic cavity** (*cavum tympani*) is the cavity of the middle ear. It can be divided into three parts: the largest, most ventral part is located entirely within the tympanic bulla and is the fundic part. The smaller, middle compartment, which is located opposite the tympanic membrane, is the tympanic cavity proper, and its most dorsal extension, for the incus, part of the stapes, and head of the malleus, is the **epitympanic recess** (*recessus epitympanicus*).

The **squamous part** (*pars squamosa*) of the temporal bone possesses a long, curved, **zygomatic process** (*processus zygomaticus*) (Figs. 4–20, 4–21), which extends rostrolaterally and overlies the caudal half of the zygomatic bone in forming the **zygomatic arch** (*arcus zygomaticus*). The ventral part of the base of the zygomatic process expands to form a transversely elongated, smooth area, the **mandibular fossa** (*fossa mandibularis*), which receives the condyle of the mandible to form the **temporomandibular joint** (*articulatio temporomandibularis*). The **retroarticular process** (*processus retroarticulare*) is a ventral extension of the squamous temporal bone. Its rostral surface forms part of the mandibular fossa, and its caudal surface is grooved by an extension of the **retroarticular foramen** (*foramen retroarticulare*). The dorsal part of the squamosum is a laterally arched, convex plate of bone which articulates with the parietal bone dorsally, the temporal wing of the basisphenoid bone rostrally, the tympanicum ventrally, and the pars mastoidea and the supraoccipital caudally. Near the caudolateral border of the bone is the ventral part of the **nuchal crest.** This crest is continued rostrally dorsal to the external acoustic meatus as the **temporal crest** (*crista temporalis*) of the zygomatic process. The smooth, rounded outer surface above the root of the zygomatic process is the *facies temporalis*. The **temporal canal**, seen on the inner surface, between the squamous and petrous parts, forms a passage for the temporal sinus, which exits by means of the retroarticular foramen as the retroarticular vein.

The squamous part of the temporal bone overlaps the parietal bone, forming a **squamous suture** (*sutura squamosa*). It also extends over the caudal margin of the temporal wing of the sphenoid bone, forming the **sphenosquamosal suture** (*sutura sphenosquamosa*). Rostrally, the zygomatic process of the squamosum meets the zygomatic bone at the **temporozygomatic suture** (*sutura temporozygomatica*). Ventrally, the tympanic part of the temporal bone meets the basioccipital to form the rostral part of the **tympano-occipital fissure** (*fissura tympano-occipitalis*). Caudally, the tympanicum articulates with the jugular process of the exoccipital to form the caudal part of this joint. The **petro-occipital fissure** (*fissura petro-occipitalis*) is formed between these articulations. At the depth of this fissure the petrous temporal bone articulates with the occipital bone in forming the **petro-occipital synchondrosis** (*synchondrosis petro-occipitalis*).

ETHMOID BONE

The **ethmoid bone** (*os ethmoidale*) (Figs. 4–22 to 4–26) is located between the cranial and facial parts of the skull, both of which it helps to form. It is completely hidden from view in the intact skull. Its complicated structure is best studied from sections and disarticulated specimens. Although unpaired, it develops from paired anlagen. It is situated between the walls of the orbits, and is bounded dorsally by the frontal, laterally by the maxillary, and ventrally by the vomer and palatine bones. It consists of four parts: a median perpendicular plate, or lamina; two lateral masses covered by the external lami-

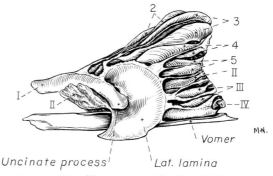

Figure 4–22. Vomer and left ethmoid, lateral aspect.
Roman numerals indicate endoturbinates.
Arabic numerals indicate ectoturbinates.

na; and a cribriform plate, to which the ethmoturbinates of the lateral masses attach.

The **perpendicular plate** (*lamina perpendicularis*), or **mesethmoid**, is a median vertical sheet of bone which, by articulating with the vomer below and the septal processes of the frontal and nasal bones above, forms the **osseous nasal septum** (*septum nasi osseum*). This bony septum is prolonged rostrally by the cartilaginous nasal septum. Caudally, it fuses with the cribriform plate, but usually does not extend through it to form a crista galli. It forms only the ventral half of the nasal septum as the septal plates of the frontal and nasal bones extend down halfway and fuse with it. The perpendicular plate is roughly rectangular in outline, with a rounded rostral border and an inclined caudal one, so that it is longer ventrally than it is dorsally. The turbinates of the lateral masses fill the nasal cavities so completely that an inappreciable **common nasal meatus** (*meatus nasi communis*) remains between each lateral mass and the lateral surface of the septum. The dorsal border does not fol-

Figure 4–23. Vomer and medial aspect of left ethmoid. (Perpendicular plate removed.)
Roman numerals indicate endoturbinates.
Arabic numerals indicate ectoturbinates.

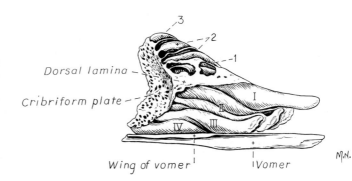

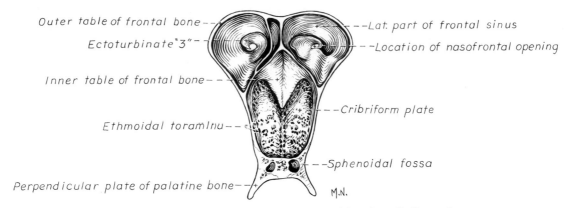

Outer table of frontal bone
Ectoturbinate"3"
Inner table of frontal bone
Ethmoidal toramlnu
Perpendicular plate of palatine bone

Lat. part of frontal sinus
Location of nasofrontal opening
Cribriform plate
Sphenoidal fossa

M.N.

Figure 4–24.   Cross section of the skull caudal to the cribriform plate.

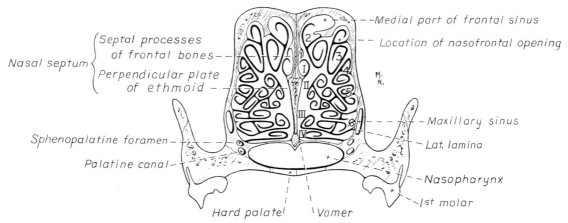

Nasal septum
{
Septal processes
  of frontal bones
Perpendicular plate
  of ethmoid
}

Sphenopalatine foramen
Palatine canal

Medial part of frontal sinus
Location of nasofrontal opening

M.N.

Maxillary sinus
Lat. lamina
Nasopharynx
1st molar

Hard palate   Vomer

Figure 4–25.   Scheme of the ethmoturbinates in cross section immediately rostral to cribriform plate.
Roman numerals indicate endoturbinates.
Arabic numerals indicate ectoturbinates.

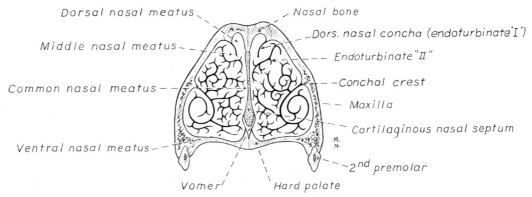

Dorsal nasal meatus
Middle nasal meatus
Common nasal meatus
Ventral nasal meatus

Nasal bone
Dors. nasal concha (endoturbinate'I')
Endoturbinate "II"
Conchal crest
Maxilla
Cartilaginous nasal septum
2nd premolar

M.N.

Vomer   Hard palate

Figure 4–26.   Scheme of the ventral concha in cross section.

low the contour of the face, but parallels the hard palate. The roof of the thin external lamina arises from the dorsal part of the perpendicular plate.

The **external lamina** (*lamina externa*), or **papyraceous lamina**, is developmentally the osseous lining of the nasal fundus. It is extremely thin, and in places it is deficient, as it coats the inner surfaces of the heavier bones that form this part of the face. This lamina is divided into dorsal, lateral, and ventral parts, commonly called the roof (*lamina tectoria*), side (*lamina orbitalis*), and floor (*lamina basalis*) plates, respectively, of the lateral masses. From its origin on the perpendicular plate the external lamina runs dorsally in contact with frontal and nasal parts of the septum, swings laterally over the top of the ethmoidal labyrinth, forming the **roof plates** (lamina tectoria), and down each lateral side (lamina orbitalis). It partly covers the side of the ethmoturbinates. This portion of the lamina is exceedingly thin, incomplete in places, and porous throughout. Its rostrodorsolateral part is channeled to form the **uncinate process** (*processus uncinatus*), which is a part of the first endoturbinate as well as of the orbital (lateral) lamina. The **uncinate notch** (*incisura uncinata*), in the meatus between the first two endoturbinates, is located dorsocaudal to the uncinate process. The depressed area of the orbital lamina, the **maxillary fossa** (*fossa maxillaris*), forms the medial wall of the **maxillary sinus** (*sinus maxillaris*). The external lamina is deficient caudally, occurring only as paper-thin, irregular plaques which remain attached to the basal laminae of the scrolls of bone. The individual turbinates arise from the roof and lateral portions of this delicate covering. The **ventral** or **transverse lamina** (lamina basalis), which forms the floor plate, can be isolated as a thin, smooth leaf fused to the medial surfaces of the maxillae. It continues from the ventral part of the orbital plate medially to the vomer in a transverse, dorsally convex arch. It is closely applied to the horizontal part of the vomer, the two conjoined sheets in this manner forming a partition which separates the ethmoturbinates in the nasal fundus from the nasopharyngeal duct.

The **cribriform plate** (*lamina cribrosa*) (Figs. 4–23, 4–24) is a deeply concave partition, protruding rostrally, which articulates with the ethmoidal notches of the frontal bones dorsally and with the presphenoid ventrally and laterally. It is the sievelike partition between the nasal and cranial cavities. Approximately 300 foramina, some as large as 1.5 mm. in diameter, perforate the plate and serve for the transmission of olfactory nerve bundles. These **cribriform foramina** (*foramina laminae cribrosae*) are grouped into tracts which surround the attachments of the turbinates, the larger foramina being adjacent to these attachments as well as around the periphery of the bone. Extending rostromedially from the middle of the lateral border is a slightly raised, foramen-free ridge of bone which is surrounded by large foramina. Caudal to this low ridge the ethmoid concurs with the presphenoid to form one of the double **ethmoidal foramina** (*foramina ethmoidalia*) on each side. These two foramina carry the ethmoidal vessel and nerve. A *crista galli*, dividing the cranial surface of the cribriform plate into right and left fossae for the olfactory bulbs of the brain, is present only in old specimens. The most rostral limit of the single **ethmoidal fossa** (*fossa ethmoidalis*) touches a transverse plane passing through the middle of the orbital openings. The cribriform plate is not transverse in position, the right and left halves lying in nearly sagittal planes, and meeting in front at an angle of about 45 degrees. Its cerebral surface forms the inside of a laterally compressed cone which is curved in all directions.

The **ethmoidal labyrinth** (*labyrinthus ethmoidalis*) forms the bulk of the lateral mass. It is largely composed of delicate bony scrolls, or **ethmoturbinates** (*ethmoturbinalia*), which attach to the external lamina by basal laminae and attach caudally to the cribriform plate. Since the cribriform plate does not extend to the body of the presphenoid, but only to its inner table, a caudally extending space on each side of the body is formed. Likewise, the cribriform plate attaches dorsally to the inner table of the frontal bone, which in old, long skulls is separated from the outer table by over 2 cm. Into these spaces extend the ethmoturbinates. So completely is the cavity of the presphenoid filled by the ethmoturbinates that the dog is usually regarded as not possessing a **sphenoidal sinus** (*sinus sphenoidalia*), although in every other respect a sinus does exist. Dorsally, the uppermost turbinates grow upward and backward from the cribriform plate into the cavity of the **frontal sinus**. Usually all compartments of the medial part of the frontal sinus have secondary linings formed by ethmoturbinates. The rostral end of the large lateral compartment contains the end of an ethmoturbinal scroll that is always open, allowing free interchange of air between the nasal fossa and the sinus. The ethmoturbinates are surprisingly alike in different specimens. They may be divided into four long, deeply lying **endoturbinates** (*endoturbinalia I to IV*) and six smaller, more superficially lying **ectoturbinates** (*ectoturbinalia 1 to 6*). The difference between these two groups of turbinates is in their location and not in their form. Each ethmoidal element (turbinate) possesses a basal leaf which attaches to the external lamina. Most of these scrolls come from the lateral part of this lamina, but some arise from the roof plate proper and others from the septal part. Most turbinates also attach to the cribriform plate caudally. Each ethmoturbinate is rolled into one or more delicate scrolls of 1½ to 2½ turns. Those turbinates with a single scroll turn ventrally, with the exception of the first endoturbinate, which turns dorsally. The elements with two scrolls usually turn toward each other, and thus toward their attachments. Variations are common, as the illustrations show. The endoturbinates nearly reach the nasal septum medially. The **first endoturbinate** is the longest and arises from the dorsal part of the cribriform plate caudally as well as from the medial part of the roof plate. In the region dorsal to the infraorbital foramen it passes from the roof plate to the medial surface to the maxilla. Further forward it attaches to the medial wall of the nasal bone as the **dorsal nasal concha** (formerly nasal turbinate). The uncinate process is formed at the attachment of the basal lamina to the nasal bone. This process is coextensive with the lateral lamina and extends caudoventrally into the maxillary sinus. The caudal part of the first endoturbinate is represented by a dorsomedially rolled plate. The small, ventrally infolded first endoturbinate is located dorsally. The **second endoturbinate** arises from its basal lamina near the middle of the lateral lamina. It divides into two or more scrolls, which become widened and flattened in a sagittal plane rostrally and rest against the

caudodorsal part of the **ventral nasal concha**, formerly maxilloturbinate. Viewed from the medial side the **third** and **fourth endoturbinates** have the same general form as the second. They are progressively shorter than the second, so that the wide rostral free end of the second overlaps the third as do shingles on a roof. The fourth element is the smallest, and lies dorsal to the wing of the vomer. Caudally it invades the sphenoidal fossa.

The **ectoturbinates** are squeezed in between the basal laminae of the endoturbinates and do not approach the nasal septum as closely as do the endoturbinates. The first two protrude through the floor of the frontal sinus. According to Maier (1928), the second ectoturbinate pushes up into the medial compartment of the frontal sinus, whereas the third ectoturbinate pushes up into the lateral compartment. Since the form of any one turbinate changes so drastically from level to level, these delicate bones can best be studied from sagittally sectioned heads which have been decalcified.

The **cribriform plate** of the ethmoid articulates ventrally with the presphenoid to form the **sphenoethmoid suture** (*sutura sphenoethmoidalis*) and with the vomer to form the **vomeroethmoid suture** (*sutura vomeroethmoidalis*). Laterally and dorsally the **frontoethmoidal suture** (*sutura frontoethmoidalis*) is formed by the union of the cribriform plate with the medial surface of the frontal bone. The transverse lamina and the rostral part of the lateral lamina attach to the maxilla, forming the **ethmoidomaxillary suture** (*sutura ethmoidomaxillaris*). The caudal part of the lateral lamina as it meets the transverse lamina attaches to the palatine bone, forming the **palatoethmoid suture** (*sutura palatoethmoidalis*). The lateral lamina attaches to the small lacrimal bone to form the **lacrimoethmoidal suture** (*sutura lacrimoethmoidalis*). Dorsally the dorsal lamina of the ethmoid articulates with the nasal bones to form the **nasoethmoidal suture** (*sutura nasoethmoidalis*). The external lamina intimately fuses with the bones against which it lies so that in a young, disarticulated skull lines and crests are present on the inner surfaces of the bones against which the lateral mass articulates. The more salient lines are for the attachment of the endoturbinates, and the smaller ones for the attach-

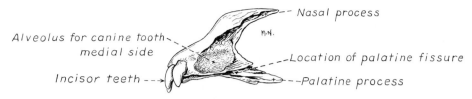

Figure 4–27. Left incisive bone (premaxilla), ventral lateral aspect.

ment of the exoturbinates, since the external lamina has largely fused to the bones against which it lies.

Bones of the Face and Palate

INCISIVE BONE

Each **incisive bone** (*os incisivum*), or **premaxilla** (Fig. 4–27), carries three upper incisor teeth. These teeth are anchored in the **body** (*corpus ossis incisivi*) by deep, conical **sockets** (*alveoli dentales*), which increase in size from the medial to the lateral position. The bony partitions between the alveoli are the **interalveolar septa** (*septa interalveolaria*). A laterally facing concavity on the caudal alveolar surface forms the rostromedial wall of the alveolus for the canine tooth. The dorsocaudal part of the incisive bone is the curved, tapering **nasal process** (*processus nasalis*), the free rostral border of which bounds the piriform aperture. A minute groove on the medial surface of each incisive bone concurs with its fellow to form the **incisive canal** (*canalis incisivus*) in the **interincisive suture** (*sutura interincisiva*). This canal varies in size and position and occasionally is absent. Extending caudally from the body is the laterally compressed, pointed **palatine process** (*processus palatinus*). This process, with that of the opposite bone, forms a **dorsal sulcus** (*sulcus septi nasi*), in which the rostral part of the cartilaginous nasal septum fits. The oval space me-

dial to the palatine process is the **palatine fissure** (*fissura palatina*), which is the only large opening in each half of the bony palate. The incisive bone articulates caudally with the maxilla to form the **incisivomaxillary suture** (*sutura incisivomaxillaris*). The caudodorsal parts of the right and left palatine processes form the **vomeroincisive suture** (*sutura vomeroincisiva*) as they articulate with the vomer. The medial surface of each nasal process articulates with the nasal bone to form the **nasoincisive suture** (*sutura nasoincisiva*).

NASAL BONE

The **nasal bone** (*os nasale*) (Fig. 4–28) is long, slender, and narrow caudally, but in large dogs is almost 1 cm. wide rostrally. The dorsal or **external surface** (*facies externa*) of the nasal bone varies in size and shape, depending on the breed. In brachycephalic types the nasal bone is very short, whereas in dolichocephalic breeds of the same weight its length may exceed its width by 15 times. The external surface usually presents a small foramen at its midlength for the transmission of a vein.

The ventral or **internal surface** (*facies interna*) in life, is covered by mucous membrane. It is deeply channeled throughout its rostral half, where it forms the **dorsal nasal meatus** (*meatus nasi dorsalis*) and bears the nasal turbinate. The caudal half of the nasal surface is widened to form the shallow **ethmoidal fossa**, which bounds the dorsal part of

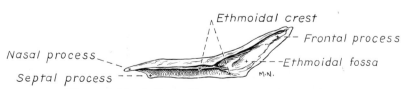

Figure 4–28. Left nasal bone, ventral lateral aspect.

the lateral mass of the ethmoid. The **nasoturbinate crest** (*crista nasoturbinalis*) is a thin shelf of bone which gives rise to the nasal turbinate throughout its rostral half and to the first endoturbinate in its caudal half. The division between the two parts of this turbinate is arbitrary. The nasal bone ends rostrally in a concave border which, with that of its fellow, forms the dorsal boundary of the piriform aperture. The lateral pointed part is more prominent than the medial and is called the **nasal process** (*processus nasalis*). The caudal extremity of the bone is usually pointed near the midsagittal plane and is known as the **frontal process** (*processus frontalis*). The nasal bone articulates extensively with its fellow on the median plane, forming the **internasal suture** (*sutura internasalis*) externally and the **nasoethmoidal suture** (*sutura nasoethmoidalis*) internally. Caudally it articulates with the frontal bone, forming the **frontonasal suture** (*suture frontonasalis*). Laterally the nasal bone articulates with the maxilla and the incisive bone, forming the **nasomaxillary suture** (*sutura nasomaxillaris*) and the **nasoincisive suture** (*sutura nasoincisiva*), respectively.

MAXILLA

The **maxilla** (Fig. 4–29) and the incisive bone of each side form the upper jaw. The maxilla is divided grossly into a body and four processes. It is the largest bone of the face, and bears all of the upper cheek teeth. It is roughly pyramidal in form, with its apex rostrally and its wide base caudally. Like the other facial bones, it shows great variation in size and form, depending on the skull type.

The smooth **external surface** (*facies facialis*) of the maxilla has as its most prominent feature an elliptical **infraorbital foramen** (*foramen infraorbitale*), for the passage of the infraorbital nerve and artery. The ventrolateral surface of the bone which bears the teeth is the **alveolar process** (*processus alveolaris*). The partitions between adjacent teeth are the **interalveolar septa** (*septa interalveolaria*), and the septa between the roots of an individual tooth are the **interradicular septa** (*septa interradicularia*). The smooth elevations on the lower facial surface of the maxilla caused by the roots of the teeth are the **alveolar juga** (*juga alveolaria*), the juga for the canine and the lateral roots of the shearing tooth (fourth upper premolar) being the most prominent. The alveolar process contains fifteen **sockets** (*alveoli dentales*) for the roots of the seven teeth that it contains. Where the teeth are far apart the spaces between them are known as **interdental spaces**, and the margin of the jaw at such places is called the **interalveolar margin**. Interdental spaces are found between each of the four premolar teeth and caudal to the canine tooth. The **lateral border** of the alveolar process (*margo alveolaris*) is scalloped as a result of the presence of the tooth sockets, with their interalveolar and interradicular septa. There are three alveoli for each of the last three cheek teeth, two each for the next two rostrally and one for the first cheek tooth. In addition to these alveoli the large caudally

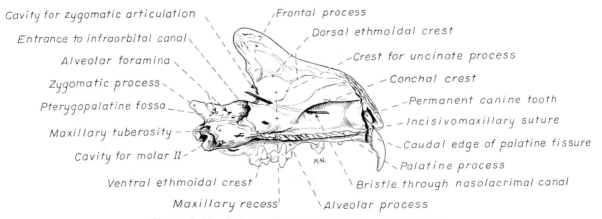

Cavity for zygomatic articulation
Entrance to infraorbital canal
Alveolar foramina
Zygomatic process
Pterygopalatine fossa
Maxillary tuberosity
Cavity for molar II
Ventral ethmoidal crest
Maxillary recess

Frontal process
Dorsal ethmoidal crest
Crest for uncinate process
Conchal crest
Permanent canine tooth
Incisivomaxillary suture
Caudal edge of palatine fissure
Palatine process
Bristle through nasolacrimal canal
Alveolar process

M.N.

Figure 4–29.   Left maxilla of a young dog, medial aspect.

curved alveolus for the canine tooth lies dorsal to those for the first two cheek teeth, or premolars I and II. Lying dorsal to the three alveoli for the shearing tooth is the short **infraorbital canal** (*canalis infraorbitalis*). This canal begins at the **maxillary foramen** (*foramen maxillare*). Leading from the infraorbital canal to the individual roots of the premolar teeth (first four cheek teeth) are the **alveolar canals** (*canales alveolares*), which open by numerous **alveolar foramina** (*foramina alveolaria*) at the apex of each alveolus. The special **incisivomaxillary canal** (*canalis maxilloincisivus*) carries the nerves and blood vessels to the first three premolar and the canine and incisor teeth. It leaves the medial wall of the infraorbital canal within the infraorbital foramen, passes dorsal to the apex of the canine alveolus with which it communicates, and enters the incisive bone. It continues rostrally and medially in the incisive bone, giving off branches to the incisor alveoli.

The **frontal process** (*processus frontalis*) arches dorsally between the nasal bone and orbit to overlap the frontal bone in a squamous suture. The **zygomatic process** (*processus zygomaticus*) is largely hidden, in an articulated skull, by the laterally lying zygomatic bone, which is mitered into the maxilla both above and below the bulk of the process. This type of articulation prevents dislocation at a place where injury frequently occurs. The **palatine process** (*processus palatinus*) is a transverse shelf of bone which, with its fellow, forms most of the **hard palate** (*palatum osseum*) and separates the respiratory from the digestive passageway. The dorsal surface of the palatine process forms part of the floor of the ventral nasal meatus. Its **ventral surface** (*facies palatina*) is grooved on each side by the **palatine sulcus** (*sulcus palatinus*) and forms part of the roof of the oral cavity. Each sulcus extends rostrally from the **major palatine foramen** (*foramen palatinum majus*), which is an oval, oblique opening in the suture between the palatine process of the maxilla and the palatine bone. In some specimens the palatine sulcus may reach the **palatine fissure** (*fissura palatina*), which is a large, sagittally directed oval opening formed caudally by the rostral border of the palatine process of the maxilla. The most caudal process of the maxilla is a small pointed spur, the **pterygoid process of**

the maxilla (*processus pterygoideus*), located caudomedial to the alveolus for the last cheek tooth. This process and the palatine bone form a notch, rarely a foramen, through which the minor palatine vessels pass.

The **nasal surface** (*facies nasalis*) of the maxilla is its medial surface, and bears several crests. The **crest of the ventral concha** (*crista conchalis*) begins at or near the incisivomaxillary suture, runs caudally, inclines ventrally, and terminates rostral to the opening of the maxillary sinus. The small **ventral ethmoidal crest** (*crista ethmoidalis ventralis*) is a sagittofrontal crest for the attachment of the floor plate or transverse lamina of the ethmoid bone. The **dorsal ethmoidal crest** (*crista ethmoidalis dorsalis*) limits the maxillary sinus dorsally and marks the line of attachment of the lateral lamina of the ethmoid to the maxilla. An oblique line passes from the nasoturbinate crest caudoventrally and laterally to the mouth of the maxillary sinus to which the uncinate process of the ethmoid is attached. The **lacrimal canal** (*canalis lacrimalis*) continues from the lacrimal bone into the maxilla, where it opens ventral to the nasoturbinate crest. The medial wall of the canal is thin and may be incomplete. The **maxillary sinus** or **recess** (*sinus maxillaris*) lies medial to the infraorbital and lacrimal canals, both of which protrude slightly into it. The lateral wall of the maxillary sinus is formed largely by the maxilla with the addition of the palatine bone caudally. The floor of the **pterygopalatine fossa** (*fossa pterygopalatina*) lies caudal to the maxillary foramen. The shelf of bone which forms it is thicker rostrally and contains many foramina which lead to the alveoli for the last two cheek teeth. The thin caudal part, barely thick enough to cover the roots of the last molar tooth, is the **maxillary tuberosity** (*tuber maxillae*).

The maxilla articulates with the incisive bone rostrally, forming the **incisivomaxillary suture** (*sutura incisivomaxillaris*). Dorsomedially, the nasal bone meets the maxilla at the **nasomaxillary suture** (*sutura nasomaxillaris*). Dorsocaudally, the maxilla articulates with the frontal bone, forming the **frontomaxillary suture** (*sutura frontomaxillaris*) at its dorsocaudal angle. Ventral to this suture the lacrimal bone and maxilla form the short **lacrimomaxillary suture** (*sutura lacrimomaxillaris*). The ventrolateral part of the maxilla forms the unusually stable **zygomaticomaxillary suture** (*sutura zygomaticomaxillaris*), as it articulates with the zygomatic bone. Ventrocaudally, the maxilla forms the extensive **palatomaxillary suture** (*sutura palatomaxillaris*) with the palatine bone. The **median palatine suture** (*sutura palatina mediana*) is formed by the two palatine processes. The transitory joint between the ethmoid and maxilla is the **ethmoideomaxillary suture** (*sutura ethmoideomaxillaris*). The **vomeromaxillary suture** (*sutura vomeromaxillaris*) is formed in the median plane, within the nasal cavity.

DORSAL NASAL CONCHA

The **dorsal nasal concha** (*concha nasalis dorsalis*) was formerly called the nasal turbinate. It is the continuation of endoturbinate I of the ethmoid, which attaches by means of an ethmoidal crest (Fig. 4–28) to the nasal bone. Baum and Zietzschmann (1936) regarded the first endoturbinate and the nasal turbinate as one structure. The uncinate process and the caudally extending scroll constitue endoturbinate I of the ethmoid. The dorsal nasal concha, unlike the ethmoturbinates and maxilloturbinate, is a simple, curved shelf of bone which is separated from the ventrally lying maxilloturbinate by a small cleft, the **middle nasal meatus** (*meatus nasi medius*). In life, the scroll is continued rostral to the ethmoidal crest by a plica of mucosa which diminishes and disappears in the vestibule of the nose.

VENTRAL NASAL CONCHA

The **ventral nasal concha** (*os conchae nasalis ventralis*) (see Fig. 4–26) was formerly called the maxilloturbinate. It is attached to the medial wall of the maxilla by a single basal lamina, the conchal crest. The **common nasal meatus** (*meatus nasi communis*) is a small sagittal space between the conchae and the nasal septum. The space dorsal to the conchae is the **middle nasal meatus** (*meatus nasi medius*), and the space ventral to it is the **ventral nasal meatus** (*meatus nasi ventralis*). The osseous plates are continued as soft tissue folds which converge anteriorly to form a single medially protruding ridge that ends in a clublike eminence in the vestibule. The direction of the bony scrolls is caudoventral. Usually five primary scrolls can be identified, and they are numbered, dorsoventrally, from 1 to 5. The first primary unit leaves the dorsal surface of the basal lamina and runs toward the dorsal concha. It is displaced laterally in its caudal part by endoturbinates I and II. The second primary unit arises several millimeters peripheral to the first, and some of its subsequent leaves reach nearly to the nasal septum. The third unit and its secondary and tertiary scrolls largely fill the space formed by the union of the nasal septum with the hard palate. The fourth unit at first runs ventrally nearly to the palate and then inclines medially, ventral to the third unit. The fifth, or terminal, unit curves dorsally as a simple caudally closed scroll which runs under the conchal crest. It has fewer secondary scrolls than do the others. The secondary scrolls divide further, so that the whole nasal fossa is nearly filled with a labyrinthine mass of delicately porous, bony plates. The larger the nasal cavity, the more numerous the bony scrolls.

ZYGOMATIC BONE

The **zygomatic bone** (*os zygomaticum*) (Fig. 4–30), sometimes called jugal or malar bone, forms the rostral half of the **zygomatic arch** (*arcus zygomaticus*). It is divided into two surfaces, four borders, and two processes. The **lateral surface** (*facies lateralis*) is convex longitudinally and transversely, although it is slightly

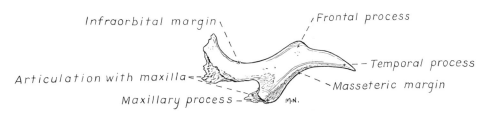

Figure 4–30. Left zygomatic bone, lateral aspect.

dished ventral to the orbit. Usually a nutrient foramen is present near its middle. The medial or **orbital surface** (*facies orbitalis*) is concave in all directions. The **maxillary border** articulates broadly with the maxilla and is recessed to form an unusually stable foliate type of sutural joint. At the middle of this articular border the zygomatic bone receives the zygomatic process of the maxilla, which it partly overlays. The **temporal border** forms a long harmonial suture with the zygomatic process of the temporal bone. This suture is one of the last to close. The **infraorbital border** (*margo infraorbitalis*) forms the ventral margin of the eye socket. It is thick and beveled medially. The **masseteric border** (*margo massetericus*) is also thick, but is beveled laterally. Both the thickness of the border and the degree to which it is beveled decrease caudally. This border provides the orgin for the strong masseter muscle. The caudoventral margin of the zygomatic bone is turned down and pointed; it is the **temporal process** (*processus temporalis*). The **frontal process** (*processus frontalis*), smaller than the others, is located between the orbital and temporal borders. It is joined to the zygomatic process of the frontal bone by the **orbital ligament**.

The zygomatic bone articulates with the maxilla in forming the mitered **zygomaticomaxillary suture** (*sutura zygomaticomaxillaris*). At the rostral edge of the orbit the **lacrimozygomatic suture** (*sutura lacrimozygomatica*) is formed by the zygomatic joining the lacrimal bone. The **temporozygomatic suture** (*sutura temporozygomatica*) is an oblique, late-closing suture between the zygomatic process of the temporal bone and the temporal process of the zygomatic bone.

PALATINE BONE

The **palatine bone** (*os palatinum*) (Fig. 4–31) is located caudomedial to the maxilla, where it forms the caudal part of the hard palate, the rostromedial wall of the pterygopalatine fossa, and the lateral wall of the nasopharyngeal duct. It is divided into horizontal and perpendicular laminae. The **horizontal lamina** (*lamina horizontalis*) forms, with its fellow, the posterior third of the **hard palate** (*palatum osseum*). Each horizontal lamina has a **palatine surface** (*facies palatina*), a **nasal surface** (*facies nasalis*), and a free concave caudal border. The nasal surface of the bone adjacent to the median palatine suture is raised to form the **nasal crest** (*crista nasalis*). The rostral part of this crest articulates with the vomer. The nasal crest ends caudally in the unpaired, but occasionally bifid, **nasal spine** (*spina nasalis*). Sometimes a notch in the lateral, sutural margin of the horizontal part concurs with a similar, but always deeper, notch in the maxilla to form the **major palatine foramen** (*foramen palatinum majus*), which opens on the hard palate. Caudal to this foramen there is usually one or, occasionally, two or more **minor palatine foramina** (*foramina palatina minora*). All of these openings lead into the **palatine canal** (*canalis palatinus*), which runs through the palatine bone from the pterygopalatine fossa. This canal transmits the major palatine artery, vein, and nerve.

The **perpendicular lamina** (*lamina perpendicularis*) of the palatine bone leaves the caudolateral border of the horizontal lamina at nearly a right angle. Medially it forms the lateral wall (*facies nasalis*) of the nasopharyngeal meatus, and laterally it forms the medial wall (*facies maxillaris*) of

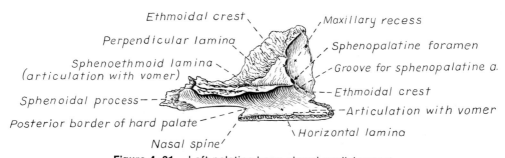

Ethmoidal crest

Perpendicular lamina

Sphenoethmoid lamina
(articulation with vomer)

Sphenoidal process

Posterior border of hard palate

Nasal spine

Maxillary recess

Sphenopalatine foramen

Groove for sphenopalatine a.

Ethmoidal crest

Articulation with vomer

Horizontal lamina

Figure 4–31.   Left palatine bone, dorsal medial aspect.

the pterygopalatine fossa. The nasal surface is partly divided by a frontally protruding shelf, the *lamina sphenoethmoidalis*. This shelf parallels the horizontal part of the bone as it lies dorsal and extends about half its length caudal to it. Dorsal to the rostral end of the sphenoethmoid lamina is the **sphenopalatine foramen** *(foramen spheno-palatinum)*, which lies dorsal to the caudal palatine foramen and extends from the pterygopalatine fossa to the nasal cavity. The nasal vessels and nerve of the spheno-palatine foramen groove the rostral end of the lamina sphenoethmoidalis. The area dorsal to this lamina is articular for the orbital wing of the sphenoid and the lateral lamina of the ethmoid bones. The rostral part of the nasal surface is excavated to form the caudolateral part of the *maxillary fossa*, which bounds part of the maxillary sinus laterally. The small ventral ethmoidal crest, located at the caudoventral margin of the maxillary fossa, marks the line along which the lateral lamina of the ethmoid articulates with the palatine bone to form the medial wall of the maxillary sinus. The part of the palatine bone ventral to the sphenoethmoid crest is smooth, slightly concave, and faces medially to form the rostral part of the lateral wall of the nasopharyngeal meatus. The caudal border of the hard palate provides attachment for the soft palate. The perpendicular lamina of the palatine bone has three processes. The caudal part between the pterygoid bone medially and the sphenoid bone laterally is the **sphenoidal process** *(processus sphenoidalis)*.

The **maxillary process** *(processus maxillaris)* of the palatine bone articulates with the maxilla at the rostroventrolateral extremity of the perpendicular part. The thin, irregularly convex border of the most dorsal part of the palatine bone is the **ethmoidal crest** *(crista ethmoidalis)*, which conceals the medially lying ethmoidal bone. The medial wall of the **pterygopalatine fossa** is formed by the palatine bone. The round dorsal opening is the **sphenopalatine foramen,** and the oblong ventral one, about 1 mm. distant, is the **caudal palatine foramen** *(foramen palatinum caudalis)*. The palatine bone articulates caudally with the sphenoid and pterygoid bones, rostrally with the maxilla and ethmoid, dorsolaterally with the lacrimal and frontal bones, dorsomedially with the vomer, and ventromedially with its fellow at the **median palatine suture** *(sutura palatina mediana)*. Rostrally the palatine bones articulate with the maxillae by a suture which crosses the midline, the **transverse palatine suture** *(sutura palatina transversa)*. Dorsally, at the rostal end of the median palatine suture, the vomer articulates with the palatine bones, forming the **vomeropalatine suture** *(sutura vomeropalatina)*. In the medial part of the pterygopalatine fossa the palatine bone articulates with the maxilla, forming the **palatomaxillary suture** *(sutura palatomaxillaris)*, which is a continuation of the transverse palatine suture. Where the palatine bone articulates with the pterygoid process of the sphenoid bone, as well as with its orbital wing, the **sphenopalatine suture** *(sutura sphenopalatina)* is formed. The **pterygopalatine suture** *(sutura pterygopalatina)* is formed by the small pterygoid bone articulating with the medial surfaces of the posterior part of the palatine bone as it unites with the pterygoid process of the sphenoid. On the medial side of the orbit the **frontopalatine suture** *(sutura frontopalatina)* runs dorsorostrally. The deep surface of the palatine bone joins rostrally with the ethmoid bone to form the **palatoethmoidal suture** *(sutura palatoethmoidalis)*.

LACRIMAL BONE

The **lacrimal bone** *(os lacrimale)* (Fig. 4-32), located in the rostral margin of the orbit, is roughly triangular in outline and pyramidal in shape. Its **orbital face** *(facies orbitalis)* is concave and free. Located in its center is the **fossa for the lacrimal sac** *(fossa sacci lacrimalis)*, which is about 6 mm. in diameter. (The two lacrimal ducts, one from each eyelid, unite in a slight dilatation to form the lacrimal sac. From the lacrimal sac the soft nasolacrimal duct courses to the vestibule of the nose.) The osseous **lacrimal canal** *(canalis lacrimalis)*, containing the nasolacrimal duct, begins in the lacrimal bone at the fossa for the lacrimal sac, runs ventrorostrally through the lacrimal bone, and leaves at the apex of the bone. It continues in a dorsally concave groove in the maxilla and opens ventral to the caudal end of the conchal crest. The **orbital crest** *(crista orbitalis)* is that part of the lacrimal bone

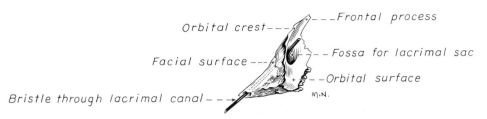

Figure 4–32. Left lacrimal bone, lateral aspect.

which forms the margin of the orbit. The **frontal process** *(processus frontalis)* is a narrow strip of the orbital margin which projects dorsally. The crest is formed by the **facial surface** *(facies facialis)* meeting the orbital surface at an acute angle. Only a small part of the facial surface is free; most of it is covered by the maxilla and zygomatic bones. In some specimens a free facial surface is lacking. The **nasal surface** *(facies nasalis)* forms a small portion of the nasal cavity.

The lacrimal bone articulates dorsocaudally with the frontal bone, forming the **frontolacrimal suture** *(sutura frontolacrimalis);* rostrally with the maxilla, forming the **lacrimomaxillary suture** *(sutura lacrimomaxillaris);* and rostroventrally with the zygomatic bone, forming the **zygomaticolacrimal suture** *(sutura zygomaticolacrimalis).* Caudoventrally the **palatolacrimal suture** *(sutura palatolacrimalis)* is formed by the articulation between the palatine and lacrimal bones. Medially, the ethmoid bone articulates with the lacrimal, forming the **lacrimoethmoidal suture** *(sutura lacrimoethmoidalis).*

PTERYGOID BONE

The **pterygoid bone** *(os pterygoideum)* (Fig. 4–33) is a small, thin, slightly curved, nearly four-sided plate of bone which articulates with the bodies of both the presphenoid and the basisphenoid bone, but particu-

larly with the medial surface of the pterygoid process of the basisphenoid. It extends ventrally beyond this process, to form the caudal part of the osseous lateral wall of the nasal pharynx, or the basipharyngeal canal. The caudoventral angle, *hamulus pterygoideus,* is in the form of a caudally protruding hook around which, in life, the tendon of the m. tensor veli palatini plays. The smooth concave medial surface is the *facies nasopharyngea.* Running in the suture between the pterygoid bone and the pterygoid process of the sphenoid is the minute **pterygoid canal** *(canalis pterygoideus),* which carries the autonomic nerve of the pterygoid canal. The pterygoid bone forms an extensive squamous suture with the pterygoid process of the sphenoid bone caudally, the **pterygosphenoid suture** *(sutura pterygosphenoidalis),* and with the palatine bone rostrally, the **pterygopalatine suture** *(sutura pterygopalatina).*

VOMER

The **vomer** (Fig. 4–34) is an unpaired bone which forms the caudoventral part of the nasal septum. It contributes to the roof of the choana. Since this bone runs obliquely from the base of the neurocranium to the nasal surface of the hard palate, the choanae are located in oblique planes in such a way that the ventral parts of the choanae are rostral to a transverse plane through the caudal border of the hard pal-

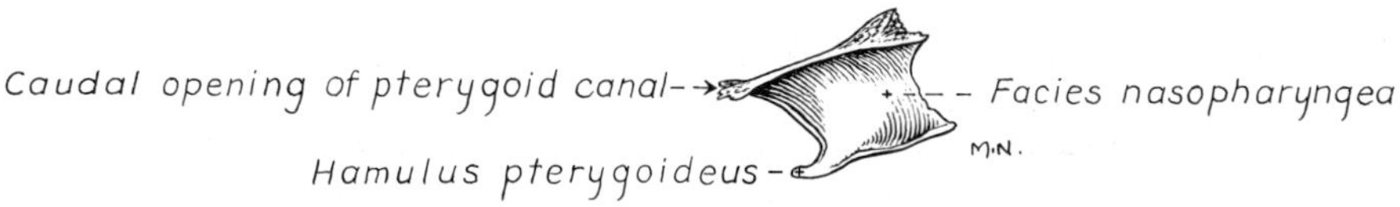

Figure 4–33. Left pterygoid bone, medial aspect.

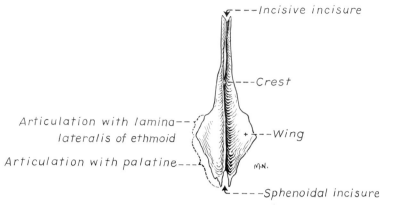

Figure 4–34. Vomer, ventral aspect.

ate. The choanae are the openings whereby the right and left nasopharyngeal meatuses are continued as the single basipharyngeal canal, the skeletal base of the nasal pharynx. The vomer has sagittal and horizontal parts.

The **sagittal part** is formed of two thin, bony leaves, *laminae laterales,* which unite ventrally to form a sulcus, *sulcus septi nasi,* which in turn receives the cartilaginous nasal septum rostrally and the bony nasal septum, or perpendicular plate of the ethmoid caudally. It articulates ventrally with the palatine processes of the maxillae, with the caudal parts of the palatine processes of the incisive bones, and with the rostral parts of the horizontal portions of the palatine bones. This caudal articulation is at the palatine suture and the ventrorostral half of the **base of the vomer.** The sagittal part of the vomer is sharply forked at each end. The rostral notch is the **incisive incisure** *(incisura incisiva);* the caudal is the **sphenoidal incisure** *(incisura sphenoidalis).*

The **horizontal part** of the vomer, constituting the **wings** *(alae vomeris),* is at right angles to the sagittal part, flaring laterally and articulating with the sphenoid, ethmoid, and palatine bones. The wings, with the transverse lamina of the ethmoid, form a thin septum that separates the dorsally lying nasal fundus, in which lie the ethmoturbinates, from the ventrally lying nasopharyngeal meatuses and the nasal pharynx.

The vomer articulates dorsally with the sphenoid bone, forming the **vomerosphenoid suture** *(sutura vomerosphenoidalis).* Rostral to this suture and hidden from external view is the **vomeroethmoid suture** *(su-*

tura vomerothemoidalis), for articulation with the ethmoid bone. Laterally the wings of the vomer articulate with the palatine bones, forming the **dorsal vomeropalatine suture** *(sutura vomeropalatina dorsalis).* The vomer articulates with the conjoined palatine crests to form the **ventral vomeropalatine suture** *(sutura vomeropalatina ventralis).* Rostral to this suture the vomer articulates with the palatine processes of the maxillae and incisive bones to form the **vomeromaxillary suture** *(sutura vomeromaxillaris)* and the **vomeroincisive suture** *(sutura vomeroincisiva),* respectively.

MANDIBLE

The **mandible** *(mandibula)* (Fig. 4–35) of the dog consists of right and left halves firmly united in life at the **mandibular symphysis** *(symphysis mandibulae),* which is a strong, rough-surfaced, fibrous joint. Each half is divided into a horizontal part, or body, and a vertical part, or ramus.

The **body of the mandible** *(corpus mandibulae)* can be further divided into the part that bears the incisor teeth *(pars incisiva)* and the part that contains the molar teeth *(pars molaris).* The sockets, or alveoli *(alveoli dentales),* which are conical cavities for the roots of the teeth, indent the **alveolar border** *(arcus alveolaris)* of the body of the mandible. There are single alveoli for the roots of the three incisor teeth, the canine, and the first and last cheek teeth. The five middle cheek teeth have two alveoli each, with those for the first molar, or fifth cheek tooth, being the largest, as this is the shearing tooth of the mandible. The free dorsal

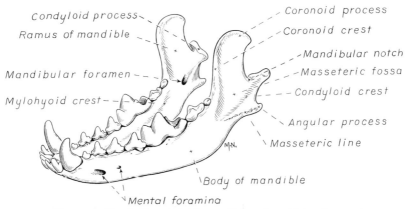

Condyloid process
Ramus of mandible
Mandibular foramen
Mylohyoid crest
Coronoid process
Coronoid crest
Mandibular notch
Masseteric fossa
Condyloid crest
Angular process
Masseteric line
Body of mandible
Mental foramina

Figure 4–35. Left and right mandibles, dorsal lateral aspect.

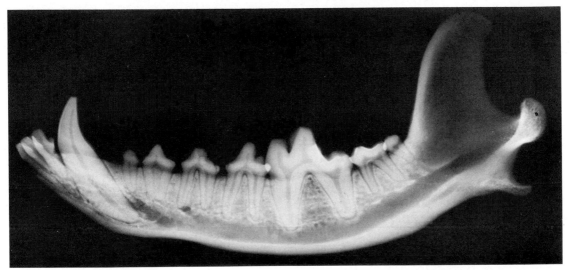

Figure 4–35. Radiograph. Lateromedial radiograph, left half of the mandible.

border of the mandible between the canine and first cheek tooth (first premolar) is larger than the others and is known as the **interalveolar margin** (*margo interalveolaris*). Similar but smaller spaces are usually present between adjacent premolar teeth, where the **interalveolar septa** (*septa interalveolaria*) end in narrow borders. From the symphysis, the bodies of each half of the mandible diverge from each other, forming the **mandibular space** (*spatium mandibulae*), in which lies the tongue. On each side the body of the mandible presents a **lateral surface,** which faces to the cheek (*facies buccalis*) and lips (*facies labialis*), and a **lingual surface** (*facies lingualis*), which faces the tongue. The lingual surface may present a wide, smooth, longitudinal ridge, the **mylo-**

hyoid line (*linea mylohyoidea*), for the attachment of the mylohyoid muscle. Rostrally, the lingual surface gives way to the **symphyseal surface,** which articulates extensively with its fellow. The lateral surface is long, smooth, and of a uniform width caudal to the symphysis. It ends in the thick, convex **ventral border,** with which the lateral and lingual surfaces are confluent. Rostrally it turns medially and presents a **mental foramen** (*foramen mentale*) near the symphysis, ventral to the alveolus of the central incisor tooth. The largest of the mental foramina, the **middle mental foramen,** is located ventral to the septum between the first two cheek teeth. A small mental foramen or several foramina are present caudal to the middle opening.

The **ramus of the mandible** *(ramus mandibulae)* is the caudal non–tooth-bearing, vertical part of the bone. It contains three salient processes. The **coronoid process** *(processus coronoideus)*, which forms the most dorsal part of the mandible, extends upward and outward. It is a large, thin plate of bone with a thickened rostral border. The **condyloid** or **articular process** *(processus condylaris)* is a transversely elongated, sagittally convex articular process which forms the temporomandibular joint by articulating with the temporal bone, The **mandibular notch** *(incisura mandibulae)* is located between the condyloid and coronoid processes. The **angle of the mandible** *(angulus mandibulae)* is the caudoventral part of the bone. It contains a salient hooked process in the dog, the **angular process** *(processus angularis)*, which serves for the attachment of the pterygoids medially and the masseter laterally. The lateral surface of the ramus contains a prominent, three-sided depression, the **masseteric fossa** *(fossa masseterica)*, for the insertion of the masseter muscle. This muscle attachment is limited by the **coronoid crest** *(crista coronoidea)* rostrally and by the **condyloid crest** *(crista condyloidea)* caudally. The medial surface of the ramus is slightly dished for the insertion of the temporal muscle. Directly ventral to this insertion is the **mandibular foramen** *(foramen mandibulae)*. It is the caudal opening of the **mandibular canal** *(canalis mandibulae)*, which opens rostrally by means of the mental foramina. The mandibular canal contains the mandibular artery and vein, and the mandibular alveolar nerve, which supply the lower teeth and jaw. The mandible articulates with the temporal bones.

Bones of the Hyoid Apparatus

The **hyoid apparatus** *(apparatus hyoideus)* (Figs. 4–36, 4–37) acts as a suspensory mechanism for the tongue and larynx. It attaches to the skull dorsally, and to the larynx and base of the tongue ventrally, suspending these structures in the caudal part of the mandibular space. The component parts, united by synchondroses, consist of the single basihyoid and the paired thyrohyoid, ceratohyoid, epihyoid, and stylohyoid bones, and the tympanohyoid cartilages.

BASIHYOID

The **basihyoid body** *(basihyoideum)* is a transverse unpaired element lying in the musculature of the base of the tongue as a ventrally bowed, dorsoventrally compressed rod. Its extremities articulate with both the thyrohyoid and the ceratohyoid bones.

THYROHYOID

The **thyrohyoid** *(thyrohyoideum)*, or cornu majus of man, is a laterally bowed, sagittally compressed, slender element which extends dorsocaudally from the basihyoid to articulate with the cranial cornu of the thyroid cartilage of the larynx.

CERATOHYOID

The **ceratohyoid** *(ceratohyoideum)* is a small, short, tapered rod having a distal extremity which is about twice as large as its proximal extremity. It articulates with the basihyoid and the thyrohyoid. The proximal extremity, which points nearly cranially in life, articulates with the epihyoid at a right angle.

EPIHYOID

The **epihyoid** *(epihyoideum)* is approximately parallel to the thyrohyoid bone. It articulates with the ceratohyoid at nearly a right angle distally and with the stylohyoid proximally.

STYLOHYOID

The **stylohyoid** *(stylohyoideum)* is slightly longer than the epihyoid, with which it articulates. It is flattened slightly craniocaudally and is distinctly bowed toward the median plane. It gradually increases in size from its proximal to its distal end. Both ends are slightly enlarged.

TYMPANOHYOID CARTILAGE

The **tympanohyoid cartilage** *(cartilago tympanohyoideum)* is a small cartilaginous

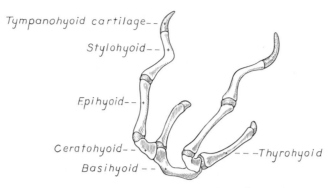

Figure 4-36. Hyoid bones, rostrolateral aspect.

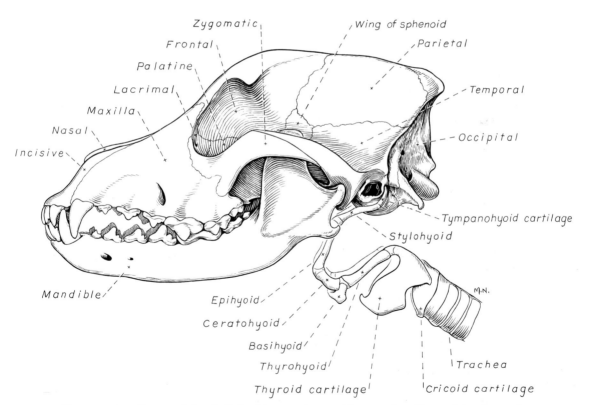

Figure 4-37. Bones of the skull, hyoid apparatus, and laryngeal cartilages, lateral aspect.

bar which continues the proximal end of the stylohyoid to the inconspicuous mastoid process of the skull.

The Skull as a Whole

DORSAL SURFACE OF SKULL (FIGS. 4–3, 4–6)

Cranial Part. The dorsal surface of the cranial or neural-part of the skull (neurocranium) is nearly hemispherical in the newborn and is devoid of prominent markings. On the other hand, a skull from a heavily muscled adult possesses a prominent **external sagittal crest,** a median longitudinal projection which is the most prominent feature of the dorsal surface of the skull. Caudally, the dorsal surface is limited by the **nuchal crest,** a transverse, variably developed crest which marks the transition between the dorsal and the caudal surface of the skull. The right and left **temporal lines** diverge from the sagittal crest and continue rostrally to the zygomatic processes of the frontal bones. The convex surface on each side of the dorsum of the skull is the **temporal fossa,** from which the temporal muscle arises. Each is bounded by the temporal line and the external sagittal crest medially in brachycephalic breeds, and by the dorsal nuchal line caudally in all breeds. This surface of the skull is the **parietal plane** (*planum parietale*).

Facial Part. The dorsal surface of the facial part of the skull is extremely variable, depending on the breed, and is greatly foreshortened in brachycephalic dogs. It is formed by the dorsal surfaces of the nasal, incisive, and maxillary bones, and the nasal processes of the frontal bones. Its most prominent feature is the unpaired external **nasal opening** or **piriform aperture** (*apertura nasi ossea*). In brachycephalic skulls this opening is not piriform, since its transverse dimension is greater than its dorsoventral one.

The **stop,** or **glabella,** prominent only in brachycephalic skulls, is a wide, smooth, transverse ridge which lies directly dorsal to the dish of the face or in a transverse plane through the caudodorsal parts of the frontomaxillary sutures. An unpaired midsagittal fossa, the frontal fossa, extends forward on the nasal bones from the frontal bones.

LATERAL SURFACE OF THE SKULL (FIGS. 4–37, 4–38)

Cranial Part. The salient features of the lateral surface of the cranial part of the skull are the prominent zygomatic arch and the orbit. The **zygomatic arch** is a heavy, laterodorsally convex bridge of bone located between the facial and neural parts of the skull; it is laterally compressed rostrally and laterally, and dorsoventrally compressed caudally. It is composed of the zygomatic bone and the zygomatic processes of the temporal and maxillary bones. It serves three important functions: to protect the eye, to give origin to the masseter and a part of the temporal muscle, and to provide an articulation for the mandible. The **external acoustic meatus** is the opening to which the external ear is attached. Ventral and medial to the external acoustic meatus is the bulla tympanica, which can be seen best from the ventral aspect. The **jugular process** is a sturdy ventral projection caudal to the bulla tympanica, and lateral to the occipital condyle.

The **orbital region** is formed by the **orbit** and the ventrally lying **pterygopalatine fossa.** The orbital opening faces rostrolaterally, and is nearly circular in the brachycephalic breeds and irregularly oval in the dolichocephalic. Approximately the caudal fourth or the orbital margin is formed by the **orbital ligament.** A line from the center of the optic canal to the center of the orbital opening is the axis of the orbit. The eyeball and its associated muscles, nerves, vessels, glands, and fascia are the structures of the orbit. Only the medial wall of the orbit is entirely osseous. Its caudal part is marked by three large openings which are named, from rostrodorsal to caudoventral, the **optic canal, orbital fissure,** and **rostral alar foramen.** In addition to these there are usually two **ethmoidal foramina,** which are located rostrodorsal to the optic canal. Within the rostral orbital margin is the **fossa for the lacrimal sac.** The **lacrimal canal** leaves the fossa and extends rostroventrally. Ventral to the medial surface of the orbit, and separated from it by the dorsally arched **ventral orbital crest** (*crista orbitalis ventralis*), is the **pterygopalatine fossa.** The rostral end of this fossa funnels down to the maxillary foramen which is located dorsal to the caudal end of the shearing tooth. A small part of the me-

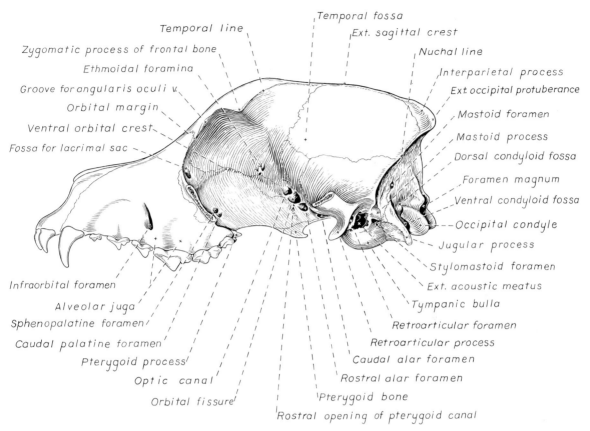

Figure 4–38. Skull, lateral aspect. (Zygomatic arch removed.)

dial wall of the fossa just caudal to the maxillary foramen frequently presents a defect. Still farther caudally are the more ventrally located **sphenopalatine foramen** and the **caudal palatine foramen.** The more dorsally located sphenopalatine foramen is separated from the caudal palatine foramen by a narrow septum of bone. The ventral orbital crest marks the dorsal boundary of the origin of the medial pterygoid muscle. The crest ends caudally in the septum between the orbital fissure and the rostral alar foramen. The caudal border of the pterygoid bone also forms the caudal border of the pterygopalatine fossa.

Facial Part. The lateral surface of the facial part of the skull is formed primarily by the maxilla. It is gently convex dorsoventrally, and has as its most prominent feature the vertically oval **infraorbital foramen,** which lies dorsal to the septum between the third and fourth cheek teeth. The **alveolar juga** of the shearing and canine teeth are features of this surface.

VENTRAL SURFACE OF THE SKULL (FIGS. 4–4, 4–7, 4–39)

Cranial Part. The ventral surface of the cranial part of the skull extends from the foramen magnum to the hard palate. Caudally, it presents the rounded **occipital condyles** with the intercondyloid notch and the median basioccipital which extends forward between the hemispherical **tympanic bullae.** The **muscular tubercles** are low, rough, sagittally elongated ridges of the basioccipital bone which articulate with the medial surfaces of the bullae. Between the bullae and the occipital condyle is the **ventral condyloid fossa,** in which opens the small circular **hypoglossal foramen.** Between this small, round opening and the tympanic bulla (in the petro-occipital suture) is the

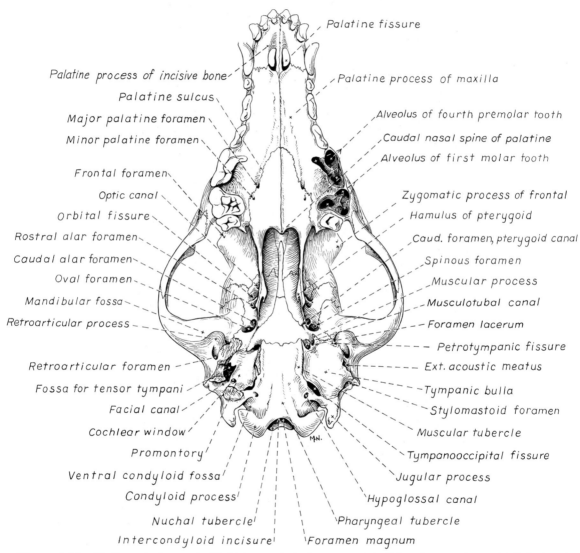

Palatine fissure

Palatine process of incisive bone

Palatine sulcus

Major palatine foramen

Minor palatine foramen

Frontal foramen

Optic canal

Orbital fissure

Rostral alar foramen

Caudal alar foramen

Oval foramen

Mandibular fossa

Retroarticular process

Retroarticular foramen

Fossa for tensor tympani

Facial canal

Cochlear window

Promontory

Ventral condyloid fossa

Condyloid process

Nuchal tubercle

Intercondyloid incisure

Palatine process of maxilla

Alveolus of fourth premolar tooth

Caudal nasal spine of palatine

Alveolus of first molar tooth

Zygomatic process of frontal

Hamulus of pterygoid

Caud. foramen, pterygoid canal

Spinous foramen

Muscular process

Musculotubal canal

Foramen lacerum

Petrotympanic fissure

Ext. acoustic meatus

Tympanic bulla

Stylomastoid foramen

Muscular tubercle

Tympanooccipital fissure

Jugular process

Hypoglossal canal

Pharyngeal tubercle

Foramen magnum

Figure 4–39. Skull, ventral aspect. (Right tympanic bulla removed. Left fourth premolar and left first molar removed.)

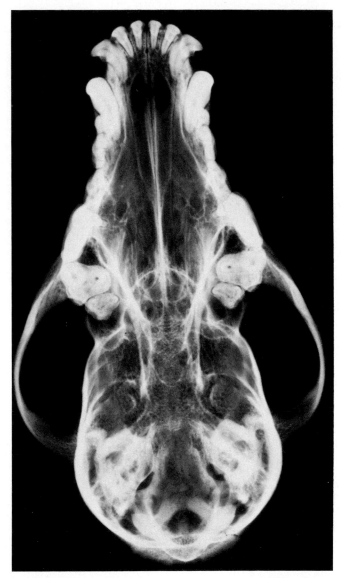

Figure 4–39. Radiograph. Dorsoventral radiograph, skull.

obliquely placed, oblong **petro-occipital fissure,** into which open the jugular and carotid foramina. Fused to the caudal surface of the bulla is the **jugular process.** Immediately rostral to the bulla and guarded ventrally by the sharp-pointed muscular process of the temporal bone is the **musculotubal canal of the auditory tube.** The *foramen lacerum* lies medial to the osseous auditory tube and lateral to the rostral part of the basioccipital, where it is flanked by small bony processes from the tympanic bulla. The largest foramen of this region is the **oval foramen,** which lies medial to the **mandibular fossa.**

The mandibular fossa is the smooth articular area on the transverse caudal part of the zygomatic arch. The minute opening medial to the retroarticular process is the **petrotympanic fissure,** through which passes the chorda tympani nerve. Caudal dislocation of the mandible, which articulates in the mandibular fossa, is prevented by the curved, spadelike **retroarticular process.** The caudal surface of this process contains a groove which helps to form the **retroarticular foramen.**

The **basipharyngeal canal** of Jayne (1898), or the **choanal region,** is the osseous part of the nasal pharynx which extends from the choanae to the caudal borders of the pterygoid bones. It is twice as long as it is wide, and its width approximates its depth. The palatine and pterygoid bones form its lateral walls and part of the roof. The median portion of its roof is formed by the vomer, presphenoid, and basisphenoid. In young skulls a small space exists between the presphenoid and vomer, which is later closed by a caudal growth of the vomer. In the living animal the soft palate completes the basipharyngeal canal and forms a tube, the nasal pharynx, which starts rostrally at the choanae and ends caudally at the pharyngeal isthmus.

At the junction of the temporal wing with the body of the basisphenoid is the short **alar canal.** Running in the suture between the pterygoid process of the sphenoid bone and the pterygoid bone is the **pterygoid canal.** The minute **pterygoid groove** leading to the caudal opening of the canal will be seen in large skulls lying dorsal to and in the same direction as the muscular process of the temporal bone. The rostral opening of the canal is in the caudal part of the pterygopalatine fossa in the vicinity of the septum between the orbital fissure and the optic canal. It conducts the nerve of the pterygoid canal.

Facial Part. The ventral surface of the facial part of the skull is largely formed by the horizontal parts of the palatine, maxillary, and incisive bones, which form the **hard palate.** Lateral to the hard palate on each side lie the teeth in their alveoli. There are three alveoli for each of the last three cheek teeth, two for each of the next two, anteriorly, and one for the first cheek tooth. The largest alveolus is at the rostral end of the maxilla, for the canine tooth. At the rostral end of the hard palate, in the incisive bones, are the six incisor teeth in individual alveoli. In the puppy skull only nine alveoli are present in each maxilla. There is one for the canine tooth, two for the first cheek tooth, and three for each of the last two deciduous molar teeth. The first permanent premolar has no deciduous predecessor. The medial alveoli for the last three cheek teeth diverge from the lateral ones, and the lateral alveoli of the shearing tooth diverge from each other.

The features of the hard palate vary with age. The palatine sulcus extends to the **palatine fissure** only in adult skulls. In old skulls, transverse ridges and depressions may be present on the hard palate. The **major palatine foramina** medial to the carnassial teeth lie rostral to the **minor palatine foramina.** The minor palatine foramina are usually two in number, located close together ventral to the palatine canal. The major palatine vessels and a nerve leave the palatine foramina, run forward in the palatine sulcus, and supply the hard palate and adjacent soft structures. The caudal border of the hard palate forms the **choanal border,** and the median eminence, or **caudal nasal spine,** may be inconspicuous. The lateral posterior part of the hard palate presents a distinct notch, which follows the palatomaxillary suture and is located between the palatine bone and the pterygoid process of the maxilla. The minor palatine vessels pass through it. The sagittal parts of the palatine bones and the pterygoid bones project ventrally to a frontal plane through the hard palate. The oval **palatine fissures** between the canine teeth are separated by the palatine processes of the premaxillary bones. Through them the palatine vessels anastomose with the infraorbital and nasal vessels. On the midline the two halves of the hard palate join to form the **palatine suture.** On the premaxillary part of this suture is located the small ventral opening of the **incisive canal.**

CAUDAL SURFACE OF THE SKULL

The caudal surface of the skull (*planum nuchale*) is three-sided and irregular. It is formed laterally by the **exoccipitals,** with their condyles and jugular processes, dorsally by the **supraoccipital,** and midventrally by the **basioccipital,** with its intracondyloid incisure. The lateral sides of the caudal surface are separated from the temporal fossae by the **nuchal crest.** The **external occipital protuberance** is the mid-dorsal caudal end of the external sagittal crest. Lateral to the external occipital protuberance is a rough area for the attachment of the m. semispinalis capitis. Between the external occipital protuberance and the foramen magnum is the **external occipital crest,** which is frequently bulged in its middle by the **vermiform fossa.** The **foramen magnum** is the large, frequently asymmetrical, ventral, median opening for the spinal cord and associated structures. Lateral to the foramen magnum are the smooth, convex occipital condyles. Each is separated from the jugular process by the ventral condyloid fossa, in the rostral part of which is the hypoglossal foramen. In the young skull the occipitomastoid suture is present lateral to the jugular process. This suture fails to close dorsally, forming the **mastoid foramen.** The mastoid process is that part of the temporal bone dorsal to the **stylomastoid foramen.**

APEX OF THE SKULL

The apex of the skull is formed by the rostral ends of the upper and lower jaw, each of which bears six incisor teeth. Its most prominent feature is the nearly circular **nasal** or **osseus aperture.**

Cavities of the Skull

CRANIAL CAVITY

The **cranial cavity** (*cavum cranii*) (Figs. 4–8, 4–40, 4–41) contains the brain, with its coverings and vessels. Its capacity varies more with body size than with head shape. The smallest crania have capacities of about 40 cc., and are known as microcephalic; the largest have capacities of approximately 140 cc., and are known as megacephalic. The boundaries of the cranial cavity may be considered as the roof, base, caudal wall, rostral wall, and the side walls. The roof of the skull (cranial vault or skull cap) is the **calvaria.** It is formed by the parietal and frontal bones, although caudally the interparietal process of the occipital bone contributes to its formation. The rostral two-thirds of the base of the cranium is formed by the sphenoid bones and the caudal third by the basioccipital. The caudal wall is formed by the occipitals, and the rostral wall by the cribriform plate of the ethmoid. The lateral wall on each side is formed by the temporal, parietal, and frontal bones, although ventrally the sphenoid and caudally the occipital bones contribute to its formation. The base of the cranial cavity is divided into rostral, middle, and caudal cranial fossae. The interior of the cranial cavity contains

smooth digital impressions bounded by irregular elevations, the cerebral and cerebellar juga. These markings are formed by the gyri and sulci, respectively, of the brain.

The **rostral cranial fossa** (*fossa cranii rostralis*) supports the olfactory bulbs and tracts and the remaining parts of the frontal lobes of the brain. It lies at a higher level and is much narrower than the part of the cranial floor caudal to it. It is continued rostrally by the deep **ethmoidal fossa.** Only in old dogs is a *crista galli* present, and this vertical median crest is confined usually to the ventral half of the ethmoidal fossa. In most specimens a line indicates the caudal edge of the perpendicular plate of the ethmoid, which takes the place of the crest. The **cribriform plate** is so deeply indented that its lateral walls are located more nearly in sagittal planes than in a transverse plane. It is perforated by the numerous **cribriform foramina.** At the junction of the ethmoid with the frontal and sphenoid bones are located the double ethmoidal foramina. The transversely concave **sphenoidal yoke** of the presphenoid bone forms most of the floor of this fossa. The right and left optic canals diverge as they run rostrally through the cranial floor. The *sulcus chiasmatis* lies caudal to the optic canals, and in young specimens its middle part forms a transverse groove connecting the internal portions of the two canals. The shelf of bone located above the rostral part of the sulcus chiasmatis is the **orbitosphenoidal crest.**

The **middle cranial fossa** (*fossa cranii media*) is situated at a lower level than the rostral fossa. The body of the basisphenoid forms its floor. Caudally, it is limited by the rostrodorsal surfaces of the pyramids, which end medially in the sharp petrosal crests. The **orbital fissures** are large, diverging openings on the lateral sides of the tuberculum sellae. Caudal and slightly lateral to the orbital fissures are the **round foramina,** which open into the alar canals. Caudolateral to the round foramina are the larger **oval foramina.** The complex of structures which surround the hypophysis is called the *sella turcica.* It consists of a rostral pommel, or *tuberculum sellae,* and a caudal elevation, or *dorsum sellae.* The **caudal clinoid processes,** which are irregular in outline, form the sides of the flat but irregular top of the dorsum sellae. The **hypophyseal fossa,**

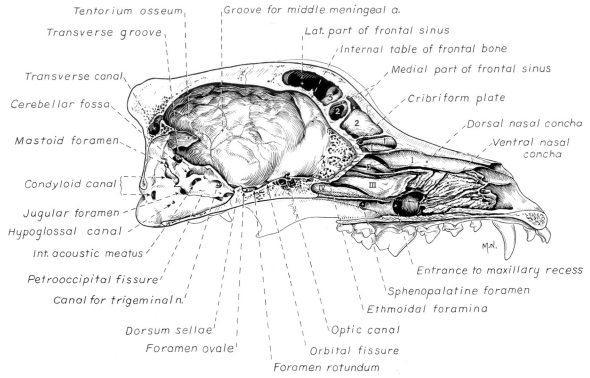

Tentorium osseum,
Transverse groove,
Transverse canal
Cerebellar fossa
Mastoid foramen
Condyloid canal
Jugular foramen
Hypoglossal canal
Int. acoustic meatus
Petrooccipital fissure
Canal for trigeminal n.
Dorsum sellae
Foramen ovale

Groove for middle meningeal a.
Lat. part of frontal sinus
Internal table of frontal bone
Medial part of frontal sinus
Cribriform plate
Dorsal nasal concha
Ventral nasal concha
Entrance to maxillary recess
Sphenopalatine foramen
Ethmoidal foramina
Optic canal
Orbital fissure
Foramen rotundum

M.N.

Figure 4–40. Sagittal section of skull. The position of the vomer is indicated by a dotted line.
Roman numerals indicate endoturbinates.
Arabic numerals indicate ectoturbinates.

in which the pituitary body lies, is a shallow oval excavation of the basisphenoid bone, located between the tuberculum sellae and the dorsum sellae. The temporal lobes of the brain largely fill the lateral parts of the middle cranial fossa.

The **caudal cranial fossa** *(fossa cranii caudalis)* is formed by the dorsal surface of the basioccipital bone and is located caudal to the middle cranial fossa. It is bounded in front by the dorsum sellae; caudally, it ends at the *foramen magnum.* Its dorsal surface is concave where the pons, medulla oblongata, and vessels rest upon it. Laterally, a considerable cleft exists between the apical part of the petrous temporal and the basioccipital bones. At the caudomedial part of this cleft is located the **petro-occipital foramen** *(foramen petro-occipitalis),* by means of which the petrobasilar canal opens rostrally. This foramen is continued toward the dorsum sellae by a groove. The foramen and canal conduct the ventral petrosal venous sinus. The **internal carotid foramen** is located rostrolateral to the petrobasilar foramen, which it resembles in size and shape.

It lies directly under the apex of the petrous temporal bone, where it is located ventral to the canal for the trigeminal nerve and dorsal to the external carotid foramen. It is the internal rostral opening of the carotid canal, which conducts the internal carotid artery and a vein. The **canal for the trigeminal nerve** runs almost directly rostrally and is nearly horizontal in direction as it perforates the petrous temporal bone.

Caudolateral to the canal for the trigeminal nerve is located the **internal acoustic pore,** which leads into the short **internal acoustic meatus.** Dorsolateral to the pore is the variably developed **cerebellar fossa.** In the petro-occipital suture, at the caudoventral angle of the pyramid, is the **jugular foramen.** Caudomedial to this opening is the small internal opening of the **hypoglossal canal.** Located within the medial part of the lateral occipital bone is the large **condyloid canal,** for the transmission of the condyloid vein. The rostral part of the canal frequently bends dorsally, so that its opening faces rostrodorsally.

The cranial fossae form the floor of the

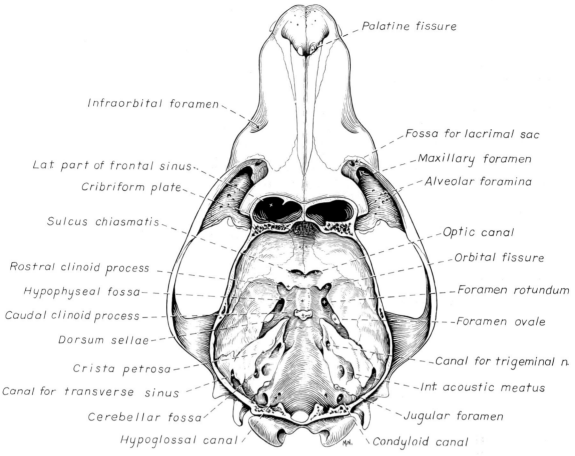

Figure 4–41. Skull with calvaria removed, dorsal aspect.

cranial cavity. The remaining portion of the cranium is marked internally by smooth depressions and elevations which are formed by the gyri and sulci of the brain. The impressions are called **digital impressions.** The elevations are formed by the sulci of the cerebellum as well as those of the cerebrum, but unless specifically indicated by the name cerebellar juga, they are all known by the more comprehensive term, cerebral juga. The **vascular groove** *(sulcus vasculosus arteriae meningeae mediae),* for the middle meningeal artery and vein, begins at the oval foramen and ramifies dorsally. Its branches vary greatly in their course and tortuosity, and in old specimens parts of the groove may be bridged by bone.

The edges of the **petrosal crests** and the *tentorium ossium* serve for the attachment of the *tentorium cerebelli,* which separates the cerebrum from the cerebellum. Extending from the tentorium ossium to the suture, between the petrosum and squamosum, is the groove of the transverse sinus. This **canal for the transverse sinus** *(canalis sinus transversi)* connects the **transverse canal** in the internal occipital protuberance with the **temporal canal,** which leads to the outside by the retroglenoid foramen. The *foramen impar* is usually a single opening, not necessarily median in position, which is located on the rostral surface of the internal occipital protuberance dorsal to the tentorium ossium. The small **internal sagittal crest** is a median, low, smooth ridge which runs a short distance forward from the internal occipital protuberance and provides attachment for the falx cerebri. No constant sulcus for the dorsal sagittal sinus exists. Ventral to the internal occipital protuberance is the vermiform impression for the vermis of the cerebellum. The divided **internal occipital crest** flanks it.

NASAL CAVITY

The **nasal cavity** (*cavum nasi*) is the facial part of the respiratory tract. It is composed of two symmetrical halves, the **nasal fossae** (*fossae nasales*), which are separated from each other by the **nasal septum** (*septum nasi*). This median partition is formed rostrally by the septal cartilage, and caudally by the septal processes of the frontal and nasal bones, the perpendicular plate of the ethmoid, and the sagittal portion of the vomer. The osseous nasal opening (*apertura nasi ossea*) was formerly known as the piriform aperture. Each nasal fossa is largely filled by the ventral nasal conchae rostrally and the ethmoturbinates caudally.

The **dorsal nasal concha** (*concha nasalis dorsalis*), formerly called the nasoturbinate (see Fig. 4–26), is a curved shelf of bone which protrudes medially from the ethmoidal crest into the dorsal part of the nasal fossa. It separates the relatively large unobstructed **dorsal nasal meatus** from the **middle nasal meatus,** which is located between the dorsal and ventral nasal conchae.

The **ventral nasal concha** (*concha nasalis ventralis*), formerly called the maxilloturbinate (see Fig. 4–26), protrude into the nasal fossa from a single leaf of attachment, the **conchal crest** (*crista conchalis*). The basal lamina of the maxilloturbinates curves inward and downward from this crest. From the convex surface of the lamina arise five or six accessory leaves which divide several times, forming a complicated but relatively constant pattern of delicate bony scrolls. The greatest number of subdivisions leave the first accessory leaf. Subsequent accessory leaves have fewer subdivisions. The free ends of the bony plates are flattened near the floor of the nasal septum and dorsal nasal concha.

In each nasal fossa the conchae divide the nasal cavity into four primary passages, known as meatuses.

The **dorsal nasal meatus** (*meatus nasi dorsalis*) is located between the dorsal turbinate and the nasal bone. The **middle nasal meatus** (*meatus nasi medius*) is located between the dorsal concha and the ventral concha. The **ventral nasal meatus** (*meatus nasi ventralis*) is located between the ventral concha and the dorsum of the hard palate. The **common nasal meatus** (*meatus nasi communis*) is the median longitudinal space located between the conchae and the nasal septum.

The **nasopharyngeal meatus** (*meatus nasopharyngeus*) is the air passage extending from the caudal ends of the middle, ventral, and common nasal meatuses to the choana. In the fresh state, it is continued by the nasal pharynx or by the basipharyngeal canal in the skull. It is bounded by the sagittal part of the vomer medially and by the maxillary and palatine bones laterally and ventrally. The dorsal part is bounded by the floor plate of the ethmoid bone. The entire mass of bony scrolls of the ventral conchae are so formed that numerous ventrocaudally directed air passages exist. The caudal portion of the ventral conchae is overlapped medially by endoturbinates II and III. Incoming air is directed by the conchae scrolls toward the maxillary sinus and the nasopharyngeal meatus.

The **ethmoidal labyrinth** (*labyrinthus ethmoidalis*) (see Figs. 4–22, 4–23) forms the scrolls which lie largely in the nasal fundus. Each ethmoidal labyrinth is composed of four medially lying **endoturbinates** and six smaller, laterally lying **ectoturbinates.** The ectoturbinates are interdigitated between the basal laminae of the endoturbinates. The endoturbinates attach caudally to the cribriform plate. By means of basal laminae both the endoturbinates and ectoturbinates attach to the **external lamina** of the ethmoid bone. The external lamina is a thin, imperfect, papyraceous osseous coating of the ethmoidal labyrinth. It is fused largely to adjacent bones around its periphery. The most ventrocaudal extension of the ethmoturbinates is endoturbinate IV, which fills the **sphenoidal fossa** so that what would otherwise be a **sphenoidal sinus** is largely obliterated. The most dorsocaudal extensions of the ethmoturbinates are the first two ectoturbinates, which invade the **frontal sinus,** completely lining the medial part and also, to some extent, the rostral portion of the lateral part. A caudoventrally running canal exists between the ventral conchae and ethmoturbinates. This canal lies against the maxilla and directs incoming air past the opening of the maxillary sinus into the nasopharyngeal meatus. The area occupied by the ethmoturbinates is the **fundus of the nasal fossa** *(fundus nasi)*. It is separated from the nasopharyngeal meatus by the floor plate of the ethmoid bone and the wings of the vomer. Rostrally, the floor of each nasal fossa contains the oblong palatine fissure. The nasolacrimal canal arises from the rostral part of the orbit and courses to the concavity of the conchal crest, where it opens. Its medial wall may be deficient in part. The **sphenopalatine foramen** is an opening into the nasopharyngeal meatus from the rostral part of the pterygopalatine fossa.

PARANASAL SINUSES

The **maxillary sinus** or recess *(sinus maxillaris)* (Fig. 4–42) is a large, lateral diverticulum of the nasal cavity. The opening into the sinus, or *aditus nasomaxillaris*, usually lies in a transverse plane through the rostral roots of the upper shearing tooth; the sinus runs caudally to a similar plane through the last cheek tooth. The caudal part of the sinus forms a rounded fundus by a convergence of its walls. The medial wall of the maxillary sinus is formed by the lateral lamina of the ethmoid bone, and the lateral wall is formed by the maxillary, palatine, and lacrimal bones. The medial and lateral walls meet dorsally and ventrally at acute angles.

Although this diverticulum of the nasal fossa may appear as a recess in the prepared dry skull, it is present as a sinus with a restricted opening in the fresh state. The lateral nasal gland lies against the lateral wall of the maxilla within the maxillary sinus.

The **frontal sinus** (Fig. 4–42) is located chiefly between the outer and inner tables of the frontal bone. It varies more in size than any other cavity of the skull. It is divided into lateral, medial, and rostral parts. The **lateral part** occupies the whole truncated enlargement of the frontal bone which forms the supraorbital process. It may be partly divided by osseous septa which extend into the cavity from its periphery. Rostrally an uneven transverse partition unites the two tables of the frontal bone. This par-

tition is deficient medially, resulting in formation of the **nasofrontal opening** *(apertura sinus frontalis)* into the nasal fossa. Through the opening extends the delicate scroll of ectoturbinate 3, the caudal extremity of which flares peripherally and ends as a delicate free end closely applied to the heavier frontal bone. Not only is the ectoturbinate covered by mucosa, but the whole sinus is also lined with mucosa, because it is an open cavity in free communication with the nasal fossa in and around ectoturbinate 3. The **medial part** of the frontal sinus is more irregular and subject to greater variations in size than is the lateral. The inner table of the frontal bone here is largely deficient, so that the ethmoturbinates completely invade this compartment. Ectoturbinates 1 and 2 are the scrolls which are located in this compartment. They are usually separated by a lateral shelf of bone, to which ectoturbinate 2 is attached in such a way that ectoturbinate 1 lies rostral to 2, although many variations occur. The rostral part of the frontal sinus is small. The size and form of the frontal sinus are dependent on skull form and age. In heavily muscled, dolichocephalic breeds, the lateral compartment is particularly large. In brachycephalic breeds, the medial compartment is much reduced in size or absent, and the lateral part is small. All paranasal sinuses enlarge with age, and only the largest definitive diverticula are present at birth.

The **sphenoid sinus** *(sinus sphenoidalis)* lies within the presphenoid bone and is largely occupied by the endoturbinate IV (Fig. 4–40).

THE VERTEBRAL COLUMN

The **vertebral column** *(columna vertebralis)*, or spine (see Fig. 4–1), consists of approximately 50 irregular bones, the **vertebrae**. (The three separate hemal arches to be described later are not included in this number.) The vertebrae are arranged in five groups: **cervical, thoracic, lumbar, sacral,** and **caudal** (or coccygeal). The first letter (or abbreviation) of the word designating each group, followed by a digit designating the number of vertebrae in the specific group, constitutes the vertebral formula. That of the dog is $C_7 T_{13} L_7 S_3 Cd_{20}$. The number 20

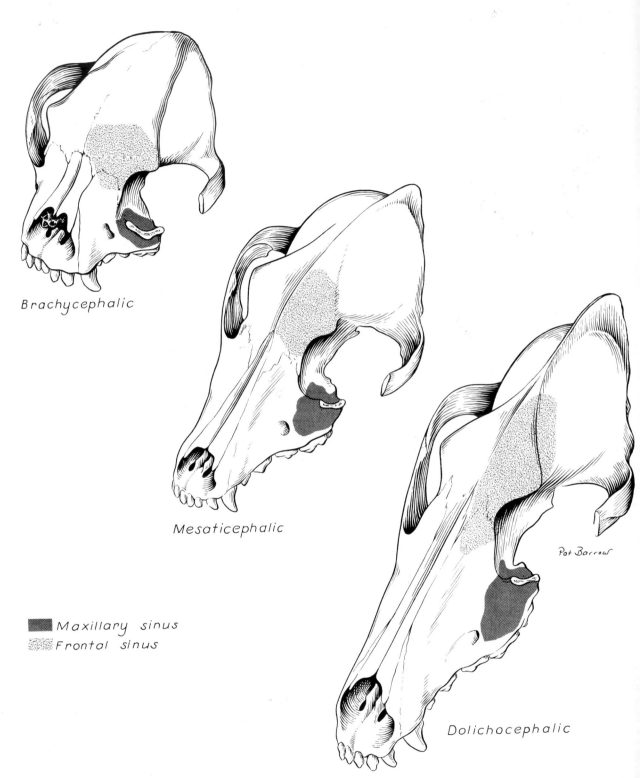

Brachycephalic

Mesaticephalic

Dolichocephalic

Pat Barrow

■ Maxillary sinus
▒ Frontal sinus

Figure 4–42. Paranasal sinuses in three types of skull.

for the caudal vertebrae may be rather constant for the beagle, but many dogs have fewer, and a few have more. All vertebrae except the sacral vertebrae remain separate and articulate with contiguous vertebrae in forming movable joints. The three sacral vertebrae are fused to form a single bone, the sacrum *(os sacrum)*. The vertebrae protect the spinal cord and roots of the spinal nerves, aid in the support of the head, and furnish attachment for the muscles governing body movements. Although the amount of movement between any two vertebrae is limited, the vertebral column as a whole possesses considerable flexibility (Slijper 1946, Badoux 1969 and 1975).

A typical vertebra consists of a **body** *(corpus vertebrae)*; a **vertebral arch** *(arcus vertebrae)*, consisting of right and left **pedicles** and **laminae;** and various processes for muscular or articular connections, which may include **transverse, spinous, articular, accessory,** and **mamillary processes.**

The **body** *(corpus vertebrae)* of a typical vertebra is constricted centrally. It has a slightly convex cranial articular surface and a centrally depressed caudal articular surface. In life, the **intervertebral fibrocartilage** or **disc** *(discus intervertebralis)* is located between adjacent vertebrae. Its center is composed of a **pulpy nucleus** *(nucleus pulposus)*, which bulges freely when the confining pressure of the outer portion, or **fibrous ring** *(annulus fibrosus)*, is released. The tough outer portion of the disc attaches firmly to adjacent vertebrae, forming a formidable retaining wall for the amorphous, gelatinous center (Hansen 1952).

The **vertebral arch** *(arcus vertebralis)*, or neural arch, consists of two **pedicles** *(pediculi arcus vertebrae)* and two **laminae** *(laminae arcus vertebrae)*. Together with the body, the arch forms a short tube, the **vertebral foramen** *(foramen vertebrale)*. All the **vertebral foramina** concur to form the **vertebral canal** *(canalis vertebralis)*. On each side the root of the vertebra extends dorsally from the dorsolateral surface of the body, presenting smooth-surfaced notches. The **cranial vertebral notch** *(incisura vertebralis cranialis)* is shallow; the **caudal vertebral notch** *(incisura vertebralis caudalis)* is deep. When the vertebral column is articulated in the natural state, the notches on either side of adjacent vertebrae, with the

intervening fibrocartilage, form the right and left **intervertebral foramina** *(foramina intervertebralia)*. Through these pass the spinal nerves, arteries, and veins. The dorsal part of the vertebral arch is composed of right and left laminae which unite at the mid-dorsal line to form a single **spine,** or **spinous process** *(processus spinosus)*, without leaving any trace of its paired origin. Most processes arise from the vertebral arch. Each typical vertebra has, in addition to the single, unpaired, dorsally located spinous process, on either side an irregularly shaped **transverse process** *(processus transversus)* which projects laterally from the region where the pedicle joins the vertebral body. At the root of each transverse process, in the cervical region, is the **transverse foramen** *(foramen transversarium)*, which divides the process into dorsal and ventral parts. The dorsal part is an intrinsic part of the transverse process. It is comparable to the whole transverse process found in a thoracic vertebra. The part ventral to the transverse foramen is serially homologous with a rib, a costal element which has become incorporated into the transverse process. It is not unusual in the dog for this costal element to be free from the seventh cervical vertebra on one or both sides. In such instances the separate bone is known as a cervical rib.

Paired **articular processes** are present at both the cranial and the caudal surface of a vertebra, at the junction of the root and lamina. The **cranial process** *(processus articularis cranialis)*, or prezygapophysis, faces craniodorsally or medially, whereas the **caudal process** *(processus articularis caudalis)*, or postzygapophysis, faces caudoventrally or laterally.

CERVICAL VERTEBRAE

The **cervical vertebrae** *(vertebrae cervicales)* (Figs. 4–43 to 4–47) are seven in number in most mammals. The first two, differing greatly from each other and also from all the other vertebrae, can be readily recognized. The third, fourth, and fifth differ only slightly, and are difficult to differentiate. The sixth and seventh cervical vertebrae present differences distinct enough to make their identification possible.

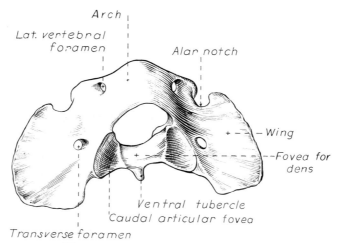

Figure 4–43. Atlas, caudodorsal aspect.

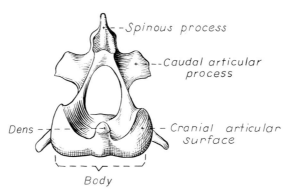

Figure 4–44. Axis, cranial aspect.

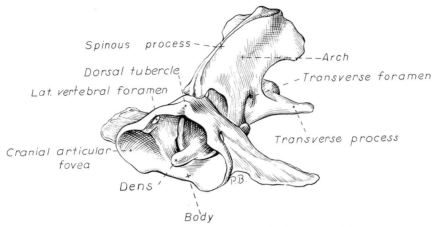

Spinous process

Dorsal tubercle

Lat. vertebral foramen

Arch

Transverse foramen

Transverse process

Cranial articular fovea

Dens

Body

P.B.

Figure 4–45. Atlas and axis articulated, cranial lateral aspect.

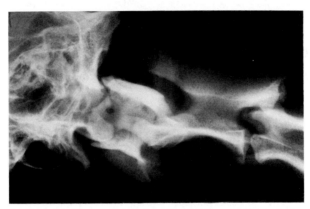

Figure 4–45. Radiograph. Occipitoatlantoaxial region. Occipital condyles and atlas have been rotated to show the dens.

The **atlas** (Fig. 4–43), or first cervical vertebra, is atypical in both structure and function. It articulates with the skull cranially, and with the axis caudally. Its chief peculiarities are the modified articular processes which "cup" the exoccipital condyles, the winglike lateral expansions, the lack of a spinous process, and reduction of its body. The thick lateral portions of the atlas are known as *massae laterales*. They unite the vertebral arch with the so-called "ventral arch," which is the body of the atlas formed by intercentrum I. The elliptical space between the dorsal arch of the atlas and the occipital bone is the *spatium interarcuale atlanto-occipitale*. The shelflike transverse processes, or **wings** *(alae atlantis)*, project from the lateral masses. Other eminences of the atlas are the **dorsal** and the **ventral tu-**

bercle *(tuberculum dorsale et ventrale)*, located on the vertebral arch and body, respectively. Frequently the dorsal tubercle is bifid, and the ventral tubercle may take the form of a conical process. The **cranial articular surface** *(fovea articularis cranialis)* consists of two cotyloid cavities which sometimes meet ventrally. They articulate with the occipital condyles of the skull, forming a joint of which the main movements are flexion and extension. Since the atlanto-occipital joint allows rather free up-and-down movement of the head, it may be remembered as the "yes joint." The **caudal articular surface** *(fovea articularis caudalis)* consists of two shallow glenoid cavities which form a freely movable articulation with the second cervical vertebra. This is sometimes spoken of as the "no joint," since rotary movement of the head occurs at this articulation. The dorsal surface of the body of the atlas contains the **fovea of the dens** (*fovea dentis*) (Fig. 4–43), which is concave from side to side and articulates with the dens of the second cervical vertebra. This articular area blends with the articular areas on the caudal surface of the lateral masses. Besides the large **vertebral foramen**, through which the spinal cord passes, there are two pairs of foramina in the atlas (Fig. 4–43). The **transverse foramen** is a short canal passing obliquely through the transverse process, or wing, of the atlas, whereas the **lateral vertebral foramen** (formerly intervertebral) perforates the craniodorsal part of the vertebral arch. Cranial and caudal notches are located at the origin of the transverse processes. The **atlantal fossae** (*fossae atlantis*) are depressions ventral to the wings. In some specimens there is an intraosseous canal running from the atlantal fossa into the lateral mass. The vertebral vein and artery traverse the atlantal fossa. The vein extends through the transverse foramen caudally and anastomoses with the internal jugular vein in the ventral condyloid fossa rostrally. A venous branch runs dorsally through the cranial notch in the wing and aids in forming the external vertebral venous plexus. The vertebral artery enters the vertebral canal through the lateral vertebral foramen, after first having run through the transverse foramen of the atlas.

The **axis** (Figs. 4–44, 4–45), or second cervical vertebra, presents an elongated, dorsal **spinous process,** which is bladelike cranially and expanded caudally. The spinous process overhangs the cranial and caudal articular surfaces of the vertebral body. The axis is further characterized by a cranioventral peglike eminence, the *dens,* or odontoid process. This process is morphologically the caudal part of the centrum of the atlas. The dens lies within the vertebral foramen of the atlas, held down by the transverse ligament. The cranial articular surfaces of the axis are located laterally on the expanded cranial end of the vertebral body. The caudal articular processes are ventrolateral extensions of the vertebral arch which face ventrally. Through the pedicles of the vertebra extend the short transverse canals. Two deep fossae, separated by a median crest,

mark the ventral surface of the body. The cranial vertebral notches concur on either side with those of the atlas to form the large intervertebral foramina for the transmission of the second pair of cervical nerves and the intervertebral vessels. The caudal notches concur with those of the third cervical vertebra to form the third pair of intervertebral foramina, through which pass the third pair of cervical nerves and the intervertebral vessels.

A review of the history of tetrapod vertebrae by Williams (1959) summarizes the developmental theories of the most influential workers in the field. He concludes that the resegmentation of the vertebral column (the union of the caudal half of one sclerotome with the cranial half of the next) is basic for all tetrapods. It appears that three elements — a neurapophysis, a pleurocentrum, and a hypocentrum (with its associated rib and ventral arch) — can be traced with clarity from the Paleozoic amphibia through reptiles to mammals. The pleurocentrum has become the centrum of mammals, and the hypocentrum has been reduced to a remnant intercentrum.

Watson and Evans (1976) studied the development of these elements in the dog and confirmed that the adult atlas develops from three bony elements: a pair of neural arches which become the vertebral arch, and a body which develops from intercentrum 1 (Fig. 4–46). The axis develops from seven bony elements: a pair of neural arches; centrum 2 and a caudal epiphysis; intercentrum 2; centrum 1 (from the atlas), which forms the dens; and an apical element on the dens which represents the centrum of the proatlas. The latter is distinct for only a short time in the newborn pup.

The **third, fourth, and fifth cervical vertebrae** (Fig. 4–47) differ slightly from each other. The spinous processes increase in length from the third to the fifth vertebra. The laminae are particularly strong on the third cervical vertebra, but gradually become shorter and narrower on the remaining vertebrae of the series. Tubercles are present on the caudal articular processes, decreasing in prominence from the third to seventh cervical segment. The transverse processes are two-pronged and slightly twisted in such a manner that the caudal

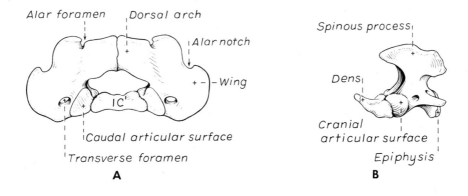

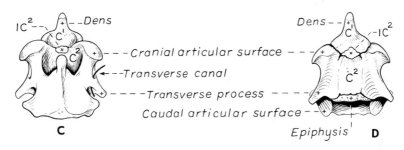

Figure 4–46.   Atlas and axis of a Beagle puppy (104 days old).
A, Atlas, dorsal view.
B, Axis, left lateral view.
C, Axis, dorsal view.
D, Axis, ventral view.

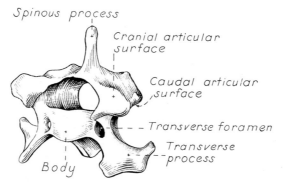

Figure 4–47.   Fifth cervical vertebra, cranial lateral aspect.

prong lies at a more dorsal level than the cranial. The transverse processes of the fifth cervical vertebra are the shortest. On each vertebra there is a pair of **transverse foramina,** which extend through the pedicles of the vertebral arches.

The **sixth cervical vertebra** possesses a higher spine than the third, fourth, or fifth, but its main peculiarity is the expanded platelike transverse processes. These plates, which extend downward and outward, represent only the caudal portion of the transverse processes. The remaining cranial portion is in the form of a conical projection ventrolateral to the transverse foramen.

In contrast to all other vertebrae, the first six cervical vertebrae are characterized by transverse foramina.

The **seventh, or last, cervical vertebra** (Fig. 4–48) lacks transverse foramina. Cervical ribs, when these are present, articulate with the ends of the single-pronged transverse processes of this vertebra. The spine of this vertebra is the highest of all those on the cervical vertebrae. Sometimes rib foveae appear caudoventral to the caudal vertebral notches. In these instances the heads of the first pair of true ribs articulate here.

The transverse processes of cervical vertebrae represent, in part, fused ribs and are sometimes referred to as pleurapophyses. Cave (1975) has reviewed the terms applied by various authors to the parts of a cervical transverse process. He redefines the mammalian cervical pleurapophysis to include: (1) the greater portion of the *caput costae*

(ventral bar and tubercle), (2) the greater part of the *tuberculum costae* (dorsal bar and tubercle), and (3) the intervening *collum costae* (intertubercular lamella). Hence the cervical transverse process, ventral, lateral, and dorsal to the outer half of the vertebroarterial foramen, is pleurapophyseal. The dorsomedian portion of the "fused rib" represents diapophysis or true processus transversus, whereas the ventromedial portion is the parapophysis.

THORACIC VERTEBRAE

There are 13 **thoracic vertebrae** (*vertebrae thoracicae*) (Figs. 4–49 to 4–51). The first nine are similar; the last four present minor differences from each other and from the preceding nine. The bodies of the thoracic vertebrae are shorter than those of the cervical or lumbar region. Although there are about twice as many thoracic as there are lumbar vertebrae, the thoracic region is slightly less than a third longer than the lumbar region. The body of each thoracic vertebra possesses a cranial and a caudal **costal fovea** or **demifacet** (*fovea costalis cranialis et caudalis*) on each side as far caudally as the eleventh. The body of the eleventh frequently lacks the caudal demifacets, and the twelfth and thirteenth thoracic vertebrae always have one complete fovea on each side. The foveae on the bodies of the thoracic vertebrae are for articulation with the heads of the ribs. The bodies of most of the thoracic vertebrae have a pair of nutrient foramina entering the middle of the ventral surface. All show paired vascular foramina on the flattened dorsal surface of the body. The **pedicles** of the vertebral arches are short. The caudal vertebral notches are deep, but the cranial notches are frequently absent. The **laminae** give rise to a **spinous process,** which is the most conspicuous feature of the first nine thoracic vertebrae. The spine of the first thoracic vertebra is more massive than the others, but is of about the same length. The massiveness gradually decreases with successive vertebrae, but there is little change in the length and direction of the spines until the seventh or eighth thoracic is reached. The spines then become progressively shorter and are inclined increasingly caudad through the ninth and tenth seg-

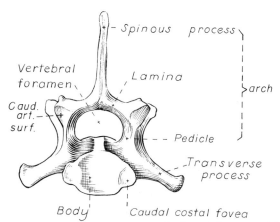

Figure 4–48. Seventh cervical vertebra, caudal aspect.

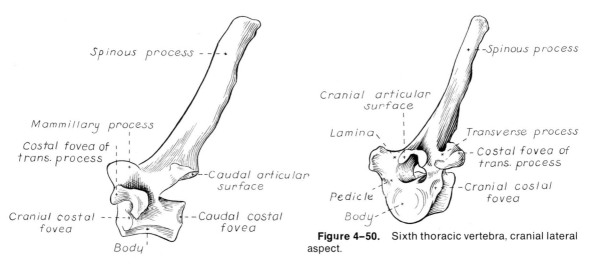

Figure 4–49. First thoracic vertebra, left lateral aspect.

Figure 4–50. Sixth thoracic vertebra, cranial lateral aspect.

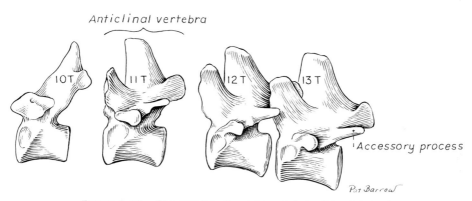

Figure 4–51. The last four thoracic vertebrae, lateral aspect.

ments. The spine of the eleventh thoracic vertebra is nearly perpendicular to the long axis of that bone. This vertebra, the **anticlinal vertebra** (*vertebra anticlinalis*), is the transitional segment of the thoracolumbar region. All spines caudal to those of the twelfth and thirteenth thoracic vertebrae are directed cranially, whereas the spines of all vertebrae cranial to the eleventh thoracic are directed caudally. In an articulated vertebral column the palpable tips of the spines of the sixth and seventh thoracic vertebrae lie dorsal to the cranial parts of the bodies of the eighth and ninth; the tips of the spines of the eighth to tenth thoracic vertebrae lie dorsal to the bodies of the vertebrae behind them.

The heads of the first pair of ribs articulate with the first thoracic and sometimes with the last cervical vertebra. The first ribs therefore articulate usually with the cranial part of the body of the first thoracic vertebra and with the fibrocartilage which forms the joint between the last cervical and the first thoracic segment. The tubercles of the ribs articulate with the transverse processes of the thoracic vertebrae of the same number in all instances. The last three thoracic vertebrae usually possess only one pair of costal fovea on their bodies, owing to a gradual caudal shifting of the heads of each successive pair of ribs.

The **transverse processes** are short, blunt, and irregular. All contain **foveae** (*foveae costales transversales*) for articulation with the tubercles of the ribs. These foveae decrease in size and convexity from the first to the last thoracic vertebra.

The **mamillary processes** (*processus mamillares*), or metapophyses, start at the second or third thoracic vertebra and continue as paired projections through the remaining part of the thoracic and through the lumbar, sacral, and coccygeal regions. They are small knoblike eminences which project dorsally from the transverse processes. At the eleventh thoracic vertebra they become associated with the cranial articular processes and continue as laterally compressed tubercles throughout the remaining vertebrae of the thoracic and those of the lumber region.

The **accessory processes** (*processus accessorii*), or anapophyses, appear first in the midthoracic region and are located on succeeding segments as far caudally as the fifth or sixth lumbar vertebra. They leave the caudal borders of the pedicles and, when well developed, form a notch lateral to the caudal articular process which receives the cranial articular process of the vertebra behind.

The **articular processes** are located at the junctions of the pedicles and the laminae. The cranial pairs of facets are widely separated on the first and second thoracic vertebrae, and nearly confluent at the median plane on thoracic vertebrae 3 to 10. On thoracic vertebrae 11, 12, and 13, the right and left facets face each other across the median plane and are located at the base of the mammillary processes. The cranial articular facets, with the exception of those on the last three thoracic vertebrae, face forward and upward. The caudal facets articulate with the cranial ones of the vertebra behind, are similar in shape, and face downward and backward on thoracic vertebrae 1 to 9. The joints between thoracic vertebrae 10 to 13 are conspicuously modified, since the caudal articular facets are located on the lateral surfaces of dorsocaudally projecting processes. This type of interlocking articulation allows flexion and extension of the caudal thoracic and the lumbar region, while limiting sagittal movement. Foveae on the transverse processes and demifacets on the centra for articulation with the ribs characterize the thoracic vertebrae.

LUMBAR VERTEBRAE

The **lumbar vertebrae** (*vertebrae lumbares*) (Figs. 4–52 to 4–54), seven in number, have longer bodies than do the thoracic vertebrae. They gradually increase in width throughout the series, and in length through the first five or six segments. The body of the seventh lumbar vertebra is approximately the same length as the first. The ventral foramina of each body are not always paired or present. The dorsal foramina are paired and resemble those of the thoracic vertebrae. Although longer and more massive, the pedicles and laminae of the lumbar vertebrae resemble those of typical vertebrae of the other regions.

The **spinous processes** are highest and most massive in the midlumbar region. The spines are about half as long, and the dorsal borders are approximately twice as wide as those of the vertebrae at the cranial end of the thoracic region.

The **transverse processes** are directed cranially and slightly ventrally. They are longest in the midlumbar region. In emaciated animals the broad extremities of the transverse processes can be palpated.

The **accessory processes** are well developed on the first three or four lumbar vertebrae, and absent on the fifth or sixth. They overlie the caudal vertebral notches and extend caudad lateral to the articular processes of the succeeding vertebrae.

The **articular processes** lie mainly in sagittal planes. The caudal processes lie between the cranial processes of succeeding vertebrae and restrict lateral flexion. All cranial articular processes bear **mamillary processes.**

There are 20 vertebrae in the thoracolumbar region. This number is quite constant. Iwanoff (1935) found only one specimen out of 300 with 21 thoracolumbar vertebrae; all of the remaining had 20. Among the specimens he studied the last lumbar segment was sacralized (fused to the sacrum) in three, and the first sacral vertebra was free in two.

SACRAL VERTEBRAE

The bodies and processes of the three **sacral vertebrae** (*vertebrae sacrales*) fuse in the adult to form the **sacrum** (*os sacrum*) (Figs. 4–55 to 4–58). The bulk of this four-sided, wedge-shaped complex lies between the ilia and articulates with them. The body of the first segment is larger than the bodies of the other two segments combined. The

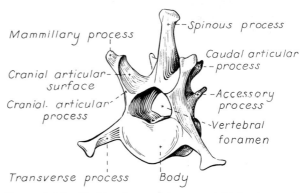

Mammillary process

Spinous process

Cranial articular surface

Caudal articular process

Cranial articular process

Accessory process

Vertebral foramen

Transverse process

Body

Figure 4–52. First lumbar vertebra, cranial lateral aspect.

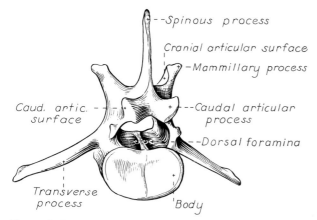

Spinous process

Cranial articular surface

Mammillary process

Caud. artic. surface

Caudal articular process

Dorsal foramina

Transverse process

Body

Figure 4–53. Fifth lumbar vertebra, caudal lateral aspect.

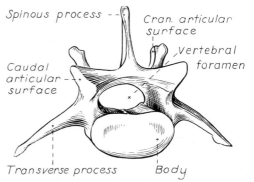

Spinous process

Cran. articular surface

Vertebral foramen

Caudal articular surface

Transverse process

Body

Figure 4–54. Seventh lumbar vertebra, caudal aspect.

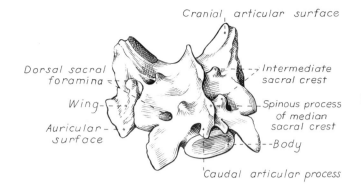

Cranial articular surface

Dorsal sacral foramina

Wing

Auricular surface

Intermediate sacral crest

Spinous process of median sacral crest

Body

Caudal articular process

Figure 4–55. Sacrum, caudal lateral aspect.

Figure 4–56. Sacrum, dorsal aspect.

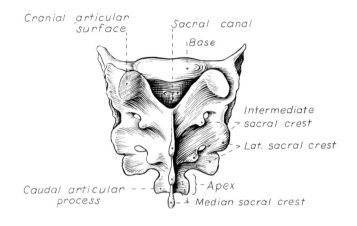

Cranial articular surface

Sacral canal

Base

Intermediate sacral crest

Lat. sacral crest

Caudal articular process

Apex

Median sacral crest

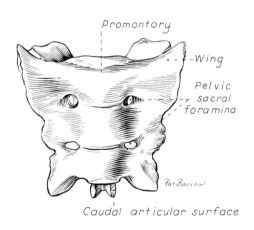

Promontory

Wing

Pelvic sacral foramina

Pat Barrow

Caudal articular surface

Figure 4–57. Sacrum, ventral aspect.

Figure 4–58. Sacrum and first caudal vertebra, lateral aspect.

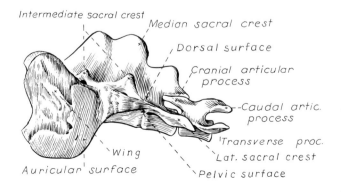

Intermediate sacral crest

Median sacral crest

Dorsal surface

Cranial articular process

Caudal artic. process

Transverse proc.

Lat. sacral crest

Pelvic surface

Wing

Auricular surface

three are united to form an arched, bony mass with a concave ventral, or pelvic surface, a feature of obstetrical importance.

The **dorsal surface** *(facies dorsalis)* (Fig. 4–56) presents the **median sacral crest** *(crista sacralis mediana)*, which represents the fusion of the three spinous processes. Two indentations on the crest indicate the areas of fusion. The dorsal surface also bears two pairs of **dorsal sacral foramina** *(foramina sacralia dorsalia)*, which transmit the dorsal divisions of the sacral nerves and vessels. Medial to these processes are low projections representing the fused mammilloarticular processes of adjacent segments. In some specimens the three mammilloarticular processes on each side are united by intervening ridges. The aggregate of the processes and the connecting ridges then forms the **intermediate sacral crest** *(crista sacralis intermedia)*. The **caudal articular processes** are small and articulate with the first caudal vertebra. The **cranial articular processes** are large, face dorsomedially, and form joints with the seventh lumbar vertebra.

The **pelvic surface** *(facies pelvina)* (Fig. 4–57) of the sacrum is variable in its degree of concavity. During the first six postnatal months two intervertebral fibrocartilages mark the separation of the vertebral bodies. These persist in the adult as two **transverse lines** *(lineae transversae)*. Two pairs of **pelvic sacral foramina** *(foramina sacralia pelvina)*, situated just lateral to the fused sacral bodies, are larger than the corresponding dorsal foramina. In addition to blood vessels, they transmit the ventral branches of the first two sacral nerves. Lateral to the pelvic sacral foramina are the fused transverse processes. Those of the first and part of the second segment are greatly enlarged and modified for articulation with the ilium. The transverse processes of the third segment and part of the second form the narrow, thin **lateral sacral crest** *(crista sacralis lateralis)*, which terminates caudally in a flattened, pointed process, the caudolateral angle. This angle frequently articulates with the adjacent transverse process of the first caudal vertebra.

The so-called **wing of the sacrum** *(ala ossis sacri)* is the enlarged **lateral part** *(pars lateralis)*, which has a large, rough semilunar facet, the **auricular surface** *(facies auricularis)*, which articulates with the ilium.

The **base of the sacrum** *(basis ossis sacri)* faces cranially. Above its slightly convex articular surface is the beginning of the wide **sacral canal** *(canalis sacralis)*, which traverses the bone and is formed by the coalescence of the three vertebral foramina. The dorsal and ventral parts of the base are clinically important. The cranioventral part of the base has a transverse ridge, the **promontory** *(promontorium)*. This slight ventral projection, along with the ilia, forms the dorsal boundary of the smallest part of the bony ring, or **pelvic inlet** *(inlet pelvina)*, through which the fetuses pass during birth. The angle formed at the sacrolumbar junction is known as the **sacrovertebral angle** *(angulus sacrovertebralis)*. The laminae above the entrance to the sacral canal do not extend to the median plane, but leave a concave caudal recession in the osseous dorsal wall of the sacral canal, which is covered only by soft tissue. Lumbar punctures are made through this osseous opening. The caudal extremity of the sacrum, although broad transversely, is known as the **apex** *(apex ossis sacri)* and articulates with the first caudal vertebra. Its base, in a similar manner, articulates with the last lumbar vertebra. Occasionally the first caudal vertebra is fused to the sacrum.

CAUDAL VERTEBRAE

The average number of **caudal vertebrae** *(vertebrae caudales)* (Figs. 4–59 to 4–64) is usually 20, although the number may vary from 6 to 23. The caudal vertebrae, also referred to as coccygeal vertebrae, are subject to greater variation than are the vertebrae of any other region, although they may be constant within a breed. The cranial members

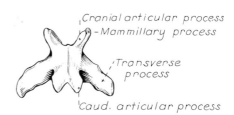

Figure 4–59. First caudal vertebra, dorsal aspect.

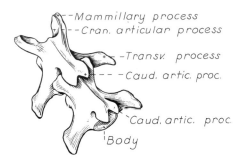

Figure 4–60. Second and third caudal vertebrae, dorsal lateral aspect.

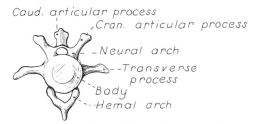

Figure 4–61. Fourth caudal vertebra, cranial aspect.

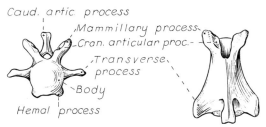

Figure 4–62. Fifth caudal vertebra, cranial and dorsal aspects.

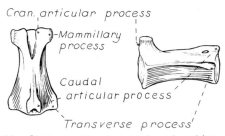

Figure 4–63. Sixth caudal vertebra, dorsal and lateral aspects.

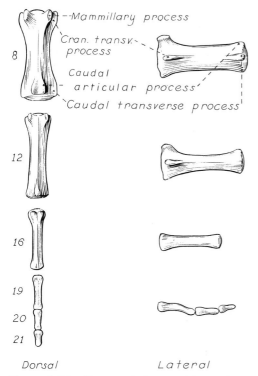

8
Mammillary process
Cran. transv. process
Caudal articular process'
Caudal transverse process'

12

16

19

20

21

Dorsal   Lateral

Figure 4–64. Representative caudal vertebrae.

of the series conform most typically to the representative type, whereas the caudal segments are gradually reduced to simple rods.

The **body** of the first caudal vertebra is as wide as it is long. Succeeding segments gradually lengthen, as far as the middle of the series, after which they become progressively shorter. The segments decrease in width from the sacrum caudally. The last segment is minute, and ends as a tapering process.

The **vertebral arch** is best developed in the first caudal segment. The lumen, which the consecutive arches enclose, becomes progressively smaller until in the sixth or seventh caudal vertebra only a groove remains to continue the vertebral canal. The caudal part of the vertebral canal contains the coccygeal nerves which supply the structures of the tail (the spinal cord usually ends at the articulation between the last two lumbar vertebrae). The **cranial articular processes** exist, although they have lost their articular function. Each vertebra bears a **mammillary process,** which persists cau-

dally in the series until all trace of the articular process has vanished. The **caudal articular processes** project from the caudal border of the arch and are frequently asymmetrical. They gradually disappear in a craniocaudal sequence, so that at the twelfth caudal vertebra both caudal and cranial articular processes are no longer present. The spinous processes are small and disappear early in the series—at about the seventh caudal vertebra. The first four or five pairs of transverse processes are well developed and typical. Caudal to the fifth caudal vertebra they are reduced in size, and they disappear at about the fifteenth segment.

Hemal arches *(arcus hemales)* (Fig. 4–61) are present as separate bones which articulate with the ventral surfaces of the caudal ends of the bodies of the fourth, fifth, and sixth caudal vertebrae. They slope caudally and are shaped like a **V** or **Y.** In life, they protect the median coccygeal artery, which passes through them. Caudal to the hemal arches, and in corresponding positions on succeeding vertebrae, are the paired **hemal processes** *(processus hemales).* Hemal processes are the last processes to disappear, and remnants of them can still be identified as far caudally as the seventeenth or eighteenth caudal vertebra.

The Vertebral Column as a Whole

The vertebral column protects, supports, and acts as a flexible, slightly compressible rod through which the propelling force generated by the pelvic limbs is transmitted to the rest of the body. It is also utilized by the axial and abdominal muscles in locomotion. The basic movements of the vertebral column are: flexion or dorsal arching of the spine; extension, straightening or ventral arching of the spine; lateral flexion; and rotation.

In the support of the viscera of the trunk, Slijper (1946) compares the vertebral column to a bow, and the abdominal muscles and linea alba to a string. As the string, the abdominal muscles, particularly the recti, do not attach to the ends of the bow, but at some distance from them. Cranially, the attachment is to the rib cage; caudally it is to the ventral cranial edge of the pelvis. This variance does not alter the aptness of the comparison, since the abdominal muscles and the vertebral column form a functional unit which is transported by the four legs. The intrinsic architecture of the vertebral column would not support the abdominal viscera without the powerful abdominal muscles, which act for this purpose as a complete elastic apron. Badoux (1975) explains the support of the body axis as a compromise between the requirements of meeting the forces of gravity and the requirements of propulsion for locomotion. The modern view is represented by a modified "bow and string" concept by which the vertebral column forms a bow with a variable curvature stabilized by its ligaments and muscles. Changes in the curvature are effected by the action of three muscular "strings" with adjustable tension: (1) a dorsal string of epaxial muscles, (2) an interrupted ventral string of hypaxial and psoas muscles, and (3) an uninterrupted abdominal muscle group.

In the fetus the vertebral column is uniformly dorsally arched from the head to the tip of the tail. In the adult standing position the head is elevated, resulting in a secondary cervical curvature, which extends the joints between the caudal cervical vertebrae. It is interesting to note that the greatest movement of the vertebral column takes place near one or both ends of the several regions into which it is divided: at both ends of the cervical region, near the caudal end of the thoracic region, at the lumbosacral junction, and in the cranial part of the caudal region.

The total length of a freshly isolated vertebral column of a shepherd-type, medium-proportioned mongrel dog weighing 45 pounds was found to be 109 cm. The lengths of the various regions as measured along the ventral surface of the articulated vertebral column are shown in Table 4–5.

The size of the vertebral canal reflects quite accurately the size and shape of the contained spinal cord, since there is only a small amount of epidural fat in the dog. The spinal cord is largest in the atlas, where its diameter is about 1 cm. It tapers to about half this size in the caudal end of the axis. The canal in the first three cervical vertebrae is nearly circular. In the fourth cervical vertebra the canal enlarges and becomes

TABLE 4–5. LENGTH OF VARIOUS REGIONS OF A FRESHLY ISOLATED VERTEBRAL COLUMN

Region	With Intervertebral Fibrocartilages	Without Intervertebral Fibrocartilages
Cervical	19 cm.	16.5 cm.
Thoracic	25.5 cm.	23.0 cm.
Lumbar	20.0 cm.	17.5 cm.
Sacrum	4.5 cm.	4.0 cm.
Caudal	40.0 cm.	36.0 cm.

slightly oval transversely. This shape and enlargement continues through the second thoracic vertebra. The increased size of the spinal cord in this region is caused by the issuance of the brachial plexus of nerves and accounts for the larger size and oval shape of the vertebral canal. From the second thoracic to the anticlinal segment, or eleventh thoracic vertebra, the vertebral canal is nearly circular in cross section and is of a uniform diameter. From the eleventh thoracic vertebra through the lumbar region the height of the canal remains about the same but the width increases, so that the canal becomes transversely oval. The shape of the canal does not grossly change in the last two lumbar vertebrae, where it is larger than in any other vertebra caudal to the first thoracic. The lumbar enlargement of the vertebral canal accommodates the lumbosacral enlargement of the spinal cord. Possibly the small lumbar subarachnoid cistern and epidural fat contribute to this enlargement, as the spinal cord usually ends opposite the fibrocartilage between the last two lumbar vertebrae.

THE RIBS

The **ribs** (*costae*) (Figs. 4–65, 4–66) form the thoracic skeleton, except for narrow mid-dorsal and mid-ventral strips formed by the vertebral column and the sternum, respectively. There are usually 13 pairs of ribs in the dog. Each rib is divided into a laterally and caudally convex dorsal bony part, the *os costale*, and a ventral cartilaginous part, the **costal cartilage** (*cartilago costalis*). The first nine ribs articulate with the sternum and are called the **sternal ribs** (*costae sternales*); the last four are called the **asternal**

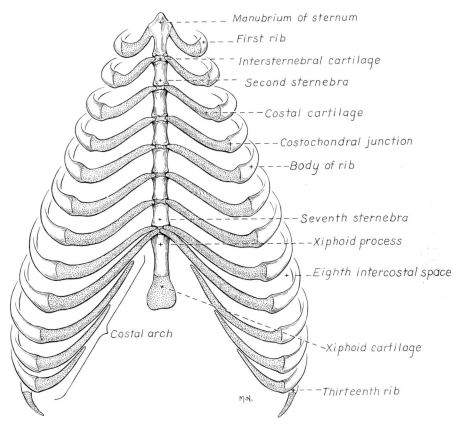

Manubrium of sternum

First rib

Intersternebral cartilage

Second sternebra

Costal cartilage

Costochondral junction

Body of rib

Seventh sternebra

Xiphoid process

Eighth intercostal space

Costal arch

Xiphoid cartilage

Thirteenth rib

M.N.

Figure 4–65. Ribs and sternum, ventral aspect.

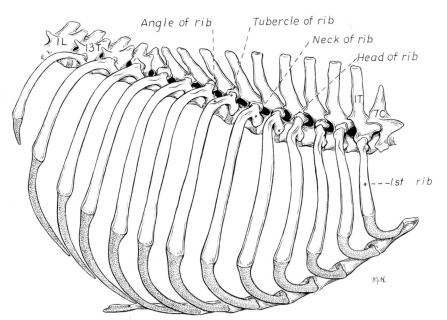

Angle of rib

Tubercle of rib

Neck of rib

Head of rib

1L

13T

1T

7C

1st rib

M.N.

Figure 4–66. Ribs and sternum, right lateral aspect.

ribs *(costae asternales)*. The costal cartilages of the tenth, eleventh, and twelfth ribs unite with the cartilage of the rib above to form the **costal arch** *(arcus costalis)* on each side. Since the cartilages of the last (thirteenth) pair of ribs end freely in the musculature, these ribs are sometimes called **floating ribs.** The ninth ribs are the longest, with the longest costal cartilages. Passing both caudally and cranially from the ninth rib, both the bony and the cartilaginous parts of the other ribs become progressively shorter. The costochondral junctions of the third through eighth ribs lie nearly in the same horizontal plane. Since the sternum and thoracic spine diverge from the thoracic inlet and the successive ribs become progressively more laterally arched, the caudal part of the rib cage is much more capacious than the cranial part. The space between adjacent ribs is known as the **intercostal space** *(spatium intercostale)*. These spaces are two or three times as wide as the adjacent ribs.

A typical **rib** (os costale) as exemplified by the seventh, presents a vertebral extremity, a sternal extremity, and an intermediate shaft, or body. The vertebral extremity consists of a **head** *(caput costae)*, a **neck** *(collum costae)*, and a **tubercle** *(tuberculum costae)*. The head of the rib has a wedge-shaped **articular surface** which articulates with adjacent costal foveae of contiguous vertebrae and the intervening fibrocartilage. The articular areas, corresponding to those of the vertebrae with which they articulate, are of about equal size and convex, and face cranially *(facies articularis capitis costae cranialis* and caudally *(facies articularis capitis costae caudalis)*. In the thoracic region T_1 to T_{10} the head of each rib (caput costae) articulates over the intervertebral disc in demifacets formed by parapophyses of adjacent vertebrae. At the eleventh or twelfth thoracic vertebra the caudal pair of demifacets or articular areas on the head of the rib disappear, as the last two or three ribs articulate only with their corresponding vertebrae. The heads of these ribs are modified accordingly, and each lacks the **crest** *(crista capitis costae)*, or transverse ridge, which separates the two articular areas when they are present. The tubercle of the rib bears an **articular facet** *(facies articularis tuberculi costae)* for articulation with the transverse process of the vertebra of the same number. The space between the neck and tubercle of the rib and the body of the vertebra is known as the **costotransverse foramen** *(foramen costotransversarium)*, which is homologous to the transverse foramen of a cervical vertebra. In the last two or three ribs the articular areas of the head and that of the tubercle become confluent, but the tubercle remains for muscular attachment.

The **body** of the rib *(corpus costae)*, in general, is cylindrical and slightly enlarged at the costochondral junction. The third, fourth, and fifth ribs show some lateral compression of the distal halves of the bony parts. In the large breeds the ribs are flatter than they are in the small breeds. In all breeds the vertebral portions of the ribs are slightly thicker from side to side than they are from front to back. The **angle** *(angulus costae)* is an indistinct lateral eminence about 2 cm. distal to the tubercle. The **costal groove** *(sulcus costae)* on the inner surface, for the intercostal vessels and nerve, is not distinct on any of the ribs.

The **costal cartilage** is the cartilaginous cylindrical distad continuation of the bony rib. It is smaller in diameter than the bony rib and, in mature dogs, may be calcified. Near the costochondral junctions the cartilages incline cranially. This is most marked in the first and twelfth ribs. The first rib articulates with the first sternebra, or manubrium sterni. Succeeding true rib cartilages articulate with successive intersternebral cartilages. However, the eighth and ninth costal cartilages articulate with the cartilage between the seventh sternebra and the last sternebra, or xiphoid process. The costal cartilages of the tenth, eleventh, and twelfth ribs are long, slender rods each joined to the one above by connective tissue to form the costal arch. The costal cartilage of the thirteenth rib, shorter and more rudimentary than those of the adjacent ribs, enters the musculature of the flank, in which it terminates.

THE STERNUM

The **sternum** (see Figs. 4–65, 4–66) is an unpaired segmental series of eight bones (sternebrae) which form the floor of the thorax. It is slightly turned up in front and turned down behind. The consecutive sternebrae are joined by short blocks of cartilage, the **intersternebral cartilages** (cartilago intersternebralis). The sternal ends of the ribs articulate with the intersternebral cartilages, with the exception of the first pair, which articulate with the first sternebra. The sternum of the dog is laterally compressed, so that its width is in a vertical plane and its thickness is in a horizontal one. The first and last sternebrae are specialized. The cranial half of the first sternebra is expanded and bears lateral projections for the attachment of the first costal cartilages. The first sternebra is longer than the others and is known as the **manubrium** (manubrium sterni).

The last sternebra, called the **xiphoid process** (processus xiphoideus), is wide horizontally and thin vertically. Its length is about three times its width. It is roughly rectangular, and may have an elliptical foramen in its caudal half. A thin cartilaginous plate (cartilago xiphoidea) prolongs the xiphoid process caudally. In rare instances the xiphoid cartilage may appear as a "fork" or a perforated plate because of a failure of the sternal bars to unite completely in the embryo.

The firm cartilaginous joints between the sternebrae (synchondroses sternales) may ossify in old individuals.

APPENDICULAR SKELETON

The development of the limbs and epiphyseal fusion in the dog has been studied by Schaeffer (1934), Pomriaskinsky-Kobozieff and Kobozieff (1954), Bressou et al. (1957), Hare (1961), Smith (1960), and Smith and Allcock (1960).

BONES OF THE THORACIC LIMB

Each pectoral or thoracic limb (membrum thoracicum) consists of its half of the pectoral girdle (cingulum membri thoracici), composed of the **clavicle** and **scapula;** the arm or brachium, represented by the **humerus;** the forearm or antebrachium, consisting of the **radius** and **ulna;** and the forepaw, or manus. The manus includes the wrist or **carpus,** with its digits, consisting of **metacarpals, phalanges,** and dorsal as well as palmar **sesamoid** bones.

CLAVICLE

The **clavicle** (clavicula) (Fig. 4–67) is not articulated with the skeleton. It is located at the tendinous intersection of the brachiocephalicus muscle, and its medial end is attached to the sternal fascia by a distinct ligamentous band. A large clavicle may be over 1 cm. long, and a third as wide. It is thin and slightly concave both longitudinally and transversely. Its medial half may be twice as wide as its lateral half. The clavicle is more closely united to the clavicular tendon between the m. cleidomastoideus and the m. cleidobrachialis than to the underlying axillary fascia to which it is related. The clavicle of the dog does not usually appear on radiographs.

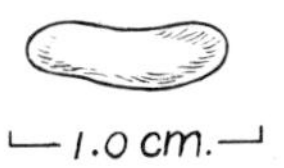

SCAPULA

The scapula (Figs. 4–68 to 4–70) is the large, flat bone of the shoulder. Its highest part lies just below the level of the free end of the spine of the first or second thoracic vertebra. Longitudinally, it extends from a transverse plane cranial to the manubrium sterni to one through the body of the fourth or fifth thoracic vertebra. Since the pectoral limb has no articulation with the axial skeleton and supports the trunk by muscles only, the normal position of the scapula may vary by the length of one vertebra. In outline it forms an imperfect triangle having two surfaces, three borders, and three angles.

The **lateral surface** (facies lateralis) Fig.

└─ 1.0 cm. ─┘

Figure 4–67. Left clavicle, cranial aspect.

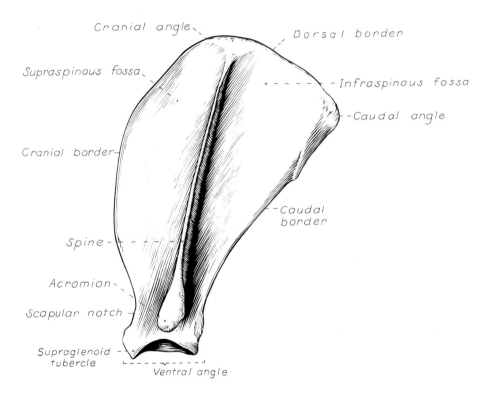

Figure 4–68.   Left scapula, lateral aspect.

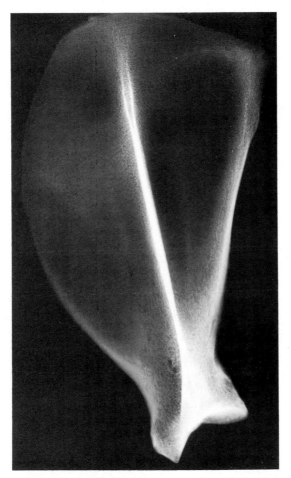

Figure 4–68. Radiograph. Lateromedial radiograph, left scapula.

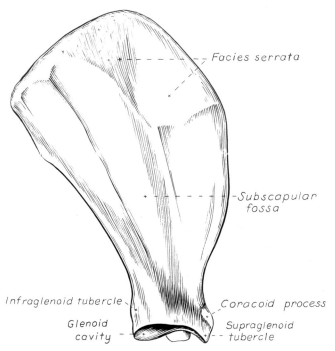

Figure 4-69. Left scapula, medial aspect.

4–68) is divided into two nearly equal fossae by a shelf of bone, the **spine of the scapula** *(spina scapulae)*. The spine is the most prominent feature of the lateral surface of the bone. It begins at the junction of the cranial and middle thirds of the vertebral border as a thick, low ridge, which gradually becomes wider but thinner as it is traced distally, so that it presents definite cranial and caudal surfaces throughout most of its length, and near its distal end there is a definite caudal protrusion. The free border or crest of the spine is slightly thickened and rolled caudally in heavily muscled specimens. The widened truncate distal end of the spine of the scapula is called the **acromion.** Its broadened superficial portion

is subcutaneous and easily palpated in the living animal. A nutrient foramen is frequently present at the junction of the ventral border of the spine and the scapula proper. The acromial part of the m. deltoideus arises from the acromion and extends distally. The m. omotransversarius arises from the distal end of the spine adjacent to the acromion and extends cranially. The m. trapezius inserts on, and the spinous part of the m. deltoideus arises from, the whole crest of the spine dorsal to the origin of the m. omotransversarius. (Fig. 6–45).

The **supraspinous fossa** *(fossa supraspinata)* is bounded by the cranial surface of the scapular spine and the adjacent lateral surface of the scapula. It is widest in the mid-

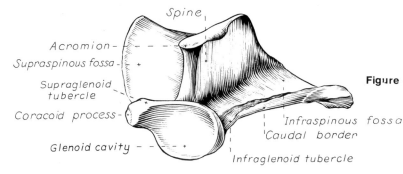

Figure 4-70. Left scapula, ventral aspect.

dle because the cranial border of the scapula extends in an arc from the scapular notch to the cranial angle. The whole thin plate of bone which forms the supraspinous fossa is sinuous, possessing at its greatest undulation a lateral projection involving the middle of the fossa. The m. supraspinatus arises from all but the distal part of the supraspinous fossa.

The **infraspinous fossa** *(fossa infraspinata)* is in general triangular. Since the caudal and vertebral borders are thick and the spine leaves the lateral surface at nearly a right angle, this fossa is well defined. The m. infraspinatus arises from the infraspinous fossa.

The medial or **costal surface** *(facies costalis)* (Fig. 4–69) of the scapula lies opposite the first five ribs and the adjacent four or five thoracic vertebrae. Two areas are recognized: a small dorsocranial rectangular area, *facies serrata*, from which arises the powerful m. serratus ventralis, and the large remaining part of the costal surface, or the **subscapular fossa** *(fossa subscapularis)*. It is nearly flat and usually presents three relatively straight muscular lines which converge toward the distal end of the bone. Between the lines the bone is smooth, and in some places it is concave. The largest concavity lies opposite the spine. From the whole subscapular fossa, and particularly from the muscular lines, arises the m. subscapularis.

The **cranial border** *(margo cranialis)* is thin except at its extremities. Distally, it forms a concavity, the **scapular notch** *(incisura scapulae)* which marks the position of the constricted part of the bone. The border undulates as it reflects the warped nature of the supraspinous fossa. In the working breeds, the cranial border forms an arc, whereas in dogs with slender extremities, the border is nearly straight. Distally, the border becomes smoother and thicker; proximally, it becomes rougher and thicker, as it runs into the vertebral border at the cranial angle.

The **dorsal border** *(margo dorsalis)*, sometimes called the base, extends between the cranial and caudal angles. In life it is capped by a narrow band of **scapular cartilage** *(cartilago scapulae)*, which represents the unossified part of the bone. The m.

rhomboideus attaches to the vertebral border of the scapula.

The **caudal border** *(margo caudalis)* (Fig. 4–70) is the thickest of the three borders and bears, just dorsal to the ventral angle, the **infraglenoid tubercle** *(tuberculum infraglenoidale)*. This tuberosity is much thicker than the border, and is located largely on the costal surface of the bone. Parts of the mm. triceps — caput longum and teres minor — arise from the infraglenoid tuberosity. The ventral third of the thick caudal border contains two muscular lines which diverge distally; the more cranially located line extends nearly to the lip of the glenoid cavity, and the more caudal one ends in the infraglenoid tubercle. The more cranial line and adjacent caudal border of the scapula give origin to the m. teres minor; the more caudal line and adjacent caudal border give origin to the m. triceps–caput longum. The middle third of the caudal border of the scapula is broad and smooth; the m. subscapularis curves laterally from the medial side and arises from it here. Approximately the dorsal fourth of the caudal border is surrounded by a lip in the heavily muscled breeds. From this part arises the m. teres major.

The **caudal angle** *(angulus caudalis)* is obtuse as it unites the adjacent thick caudal border with the thinner, rougher, gently convex vertebral border. The m. teres major arises from the caudal angle and the adjacent caudal border of the scapula.

The **cranial angle** *(angulus cranialis)* imperceptibly unites the thin, convex cranial border to the rough, convex, thick vertebral border. No muscles attach directly to the cranial angle.

The **ventral angle** *(angulus ventralis)* — formerly called the articular, glenoid, or lateral angle — forms the expanded distal end of the scapula. Clinically, the ventral angle is the most important part of the bone, since it contains the **glenoid cavity** *(cavitas glenoidalis)*, which receives the head of the humerus in forming the shoulder joint. The glenoid cavity is very shallow; its lateral border is flattened, and cranially it extends out on the tuber scapulae. The medial border forms a larger arc than does the caudal border.

The **supraglenoid tuberosity** *(tuberculum*

supraglenoidale) is the largest tuberosity of the scapula. For the most part it projects cranially, with a medial inclination. From it arises the single tendon of the m. biceps brachii. The small beaklike process which leaves the medial side of the scapular tuberosity is the **coracoid process** *(processus coracoideus),* from which the m. coracobrachialis arises. The coracoid process is a remnant of the coracoid bone, which is still distinct in monotremes. Although in dogs this osseous element is no longer a separate bone, it still retains its own center of ossification.

HUMERUS

The **humerus** (Figs. 4–71 to 4–73) is the bone of the arm, or brachium. Proximally it articulates with the scapula in forming the shoulder joint; distally it articulates with

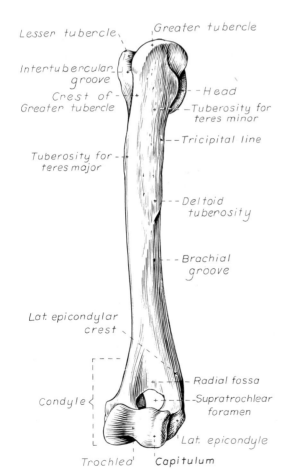

Figure 4–71. Left humerus, cranial lateral aspect.

the radius and ulna in forming the elbow joint. Developmentally it is divided into a shaft and two extremities; definitively it is divided into a head, neck, body, and condyle.

The **head** *(caput humeri)* is oval, being elongated in a sagittal plane. The articular area it presents is about twice the size of that of the glenoid cavity of the scapula with which it articulates. Although it is rounded in all planes, it does not form a perfect arc in any plane, as the cranial part is much flatter than the caudal part. The articular area of the head is continued distally by the **intertubercular groove** *(sulcus intertubercularis),* which ridges the craniomedial part of the proximal extremity of the bone. The extension of the shoulder joint capsule into the groove lubricates the bicipital tendon which lies in it (Fig. 5–9).

The **greater tubercle** *(tuberculum majus),* or large craniolateral projection of the proximal extremity of the humerus, has a smooth, convex summit which in most breeds extends higher than the head. It serves for the total insertion of the m. supraspinatus and the partial insertion of the m. pectoralis profundus. Between the head and greater tubercle are several small foramina for the transmission of veins. The relatively smooth facet distal to the summit of the greater tubercle serves for the insertion of the m. infraspinatus. The **lesser tubercle** *(tuberculum minus)* is a medially flattened enlargement of the proximal medial part of the humerus, the convex border of which does not extend as high as the head. To this convex border attaches the m. subscapularis. Planes through the lesser and the greater tubercle meet at about a right angle cranially. The two tubercles are separated craniomedially by the intertuberal groove, and caudolaterally by the head of the humerus.

The **neck** of the humerus *(collum humeri)* is distinct only caudally and laterally. It indicates the line along which the head and parts of the tubercles have fused with the shaft.

The **body** of the humerus *(corpus humeri),* or shaft, is the long, slightly sigmoid-shaped part of the humerus which unites the two extremities. It varies greatly in shape and size, depending on the breed. Usually it is laterally compressed, possessing medial and lateral surfaces, and cranial

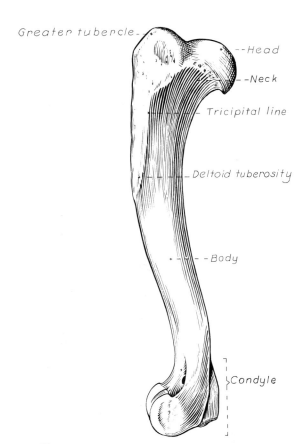

Greater tubercle

Head

Neck

Tricipital line

Deltoid tuberosity

Body

Condyle

Figure 4–72. Left humerus, lateral aspect.

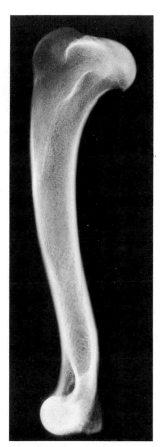

Figure 4–72. Radiograph. Lateromedial radiograph, left humerus.

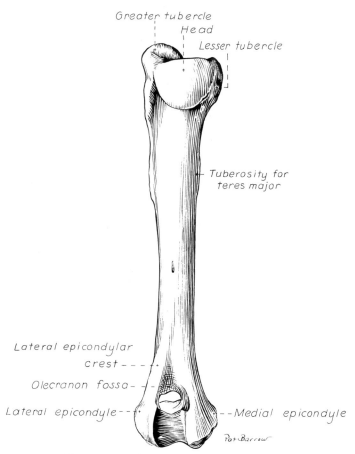

Greater tubercle

Head

Lesser tubercle

Tuberosity for teres major

Lateral epicondylar crest

Olecranon fossa

Lateral epicondyle

Medial epicondyle

Pat Barrow

Figure 4–73. Left humerus, caudal aspect.

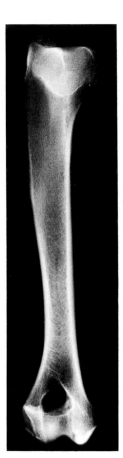

Figure 4–73. Radiograph. Caudocranial radiograph, left humerus.

and caudal borders which are not distinct throughout their length.

The **lateral surface** *(facies lateralis)* (Fig. 4–72) is marked proximally by the **tricipital line** *(linea m. tricipitis)*, formerly called the anconeal line, and a tuberosity which divides it into a narrow, slightly convex area, which faces craniolaterally, and a wider, smoother surface, which is slightly concave and faces caudolaterally. The tricipital line begins at the head of the humerus caudal to the greater tubercle, and in an uneven, cranially protruding arc extends distally to the elongated **deltoid tuberosity** *(tuberositas deltoides)*. On the crest, below the head, is a small enlargement for the insertion of the m. teres minor. The remaining distal part of the crest serves for the origin of the m. triceps–caput laterale. The deltoid tuberosity is the most prominent feature of the lateral surface of the humerus and serves for the insertion of the m. deltoideus. The **brachial groove** *(sulcus m. brachialis)*, or **musculospiral groove** (Hughes and Dransfield 1953), forms the smooth, flat to convex, lateral surface of most of the humerus. It begins at the neck caudally and extends laterally and finally cranially as it twists to the distal extremity of the bone. Although the m. brachialis lies in the whole groove, it arises from the proximal part only. Both the proximal and the distal part of the lateral surface incline cranially. The proximal part lies between the crest of the major tubercle medially and the tricipital line laterally.

The **medial surface** *(facies medialis)* is rounded transversely, except for a nearly flat triangular area in its proximal fourth. Caudally, this area is bounded by the **crest of the lesser tubercle** *(crista tuberculi minoris)*, which ends distally in an inconspicuous eminence, and the **tuberosity for the teres major,** which lies in the same transverse plane as the laterally located deltoid tuberosity. The m. coracobrachialis inserts on the crest of the lesser tubercle adjacent to the teres tuberosity. Cranial to this insertion the m. triceps–caput mediale arises from the crest by a small aponeurosis. The mm. teres major and the latissimus dorsi insert on the teres tuberosity. The medial surface of the humerus is loosely covered by the m. biceps brachii.

The **cranial surface** *(facies cranialis)* of the humerus begins proximally at the crest of the greater tubercle. This crest swings laterally just medial to the deltoid tuberosity, where it reaches the cranial edge of the spiral groove. The entire m. pectoralis superficialis attaches to the crest of the greater tubercle, and a portion of the m. pectoralis profundus attaches to its proximal part (Figs. 6–49 and 6–50).

The **caudal surface** *(facies caudalis)* (Fig. 4–73) begins at the neck of the humerus where the m. triceps–caput accessorium arises. As a transversely rounded margin, it extends to the distal fourth of the bone, where it is continued by the lateral epicondyloid crest. The caudal border is perforated below its middle by the distally directed **nutrient foramen.**

The sagittally rounded distal end of the humerus may be divided into a small, lateral articular area, the **capitulum humeri,** for articulation with the head of the radius, and the **trochlea humeri,** a much larger, medially located pulley-shaped part which extends proximally into the adjacent fossae (Fig. 4–71). The trochlea articulates extensively with the trochlear notch of the ulna in forming one of the most stable hinge joints in the body. The distal end of the humerus, including its articular areas and the adjacent fossae, may be regarded as the **humeral condyle** *(condylus humeri)* (Fig. 4–72).

The **lateral epicondyle** *(epicondylus lateralis)* (Figs. 4–71, 4–73) is the enlarged distolateral end of the humerus. It lies caudoproximal to the lateral articular margin of the capitulum. It gives origin to the mm. extensor digitorum communis, extensor digitorum lateralis, and the ulnaris lateralis. Functionally, it is known as the extensor epicondyle of the humerus. The proximal end of the lateral ligament of the elbow joint attaches to the articular margin and adjacent surface of the lateral epicondyle. The lateral epicondylar crest *(crista epicondyli lateralis)* extends proximally from the lateral epicondyle. It is a thick, rounded crest which ends by blending with the caudal border at the beginning of the distal fourth of the humerus. The m. brachioradialis arises from the proximal part of the crest, and the m. extensor carpi radialis arises from the remaining part.

The **medial epicondyle** *(epicondylus medialis)* (Fig. 4–73) is also known as the flexor epicondyle. Larger than the lateral epicondyle, it gives origin to the m. flexor digitorum superficialis and the humeral heads of the mm. flexor digitorum profundus, flexor carpi ulnaris, and flexor carpi radialis. The proximal end of the medial ligament of the elbow joint attaches to the articular margin and adjacent surface of the medial epicondyle.

The **olecranon fossa** *(fossa olecrani)* is a deep excavation of the caudal part of the distal extremity of the humerus. It receives the anconeal process *(processus anconeus)* of the ulna when the elbow joint is extended. The olecranon fossa, in life, is covered by the m. anconeus, which arises from its margin. Opposite the olecranon fossa is the **radial fossa** *(fossa radialis)*, which is also called the *coronoid fossa (fossa coronoidea)* by Getty 1975, Baum and Zietzschmann 1936, and Hughes and Dransfield 1953). The dog has no coronoid fossa, since only the head of the radius enters this depression when the elbow joint is flexed, and not the coronoid process of the ulna. The radial and olecranon fossae communicate with each other by means of the **supratrochlear foramen** *(foramen supratrochleare)*. The foramen may be absent when the humerus is small.

RADIUS

The **radius** (Figs. 4–74, 4–75, 4–76) is the main weight-supporting bone of the forearm; it is shorter than the ulna, which parallels it and serves primarily for muscle attachment. The radius articulates with the humerus proximally in forming the elbow joint and with the carpal bones distally in forming the main joint of the carpus. It also articulates with the ulna proximally by its caudal surface and distally by its lateral border. The radius, like the humerus, is divided into proximal and distal extremities, with an intervening shaft or body.

The **head** *(caput radii)* is irregularly oval in outline as it extends transversely across the proximal end of the bone. Its concave **articular fovea** *(fovea capitis radii)* articulates with the capitulum of the humerus, and bears practically all the weight transmitted from the arm to the forearm. The **articular circumference** *(circumferentia articularis s. articularis ulnaris radii proximalis)* is a caudal, smooth, osseous band on the head for articulation with the radial notch of the ulna. The articular circumference is longer than the corresponding notch in the ulna, so that a limited amount of rotation of the forearm is possible. The bulbous eminence on the lateral surface of the head does not serve for muscular attachment; the m. supinator plays over it in its course to a more distal attachment. A sesamoid bone is frequently present in this region.

The **neck** *(collum radii)* is the constricted segment of the radius which joins the head to the body. The constriction is more distinct laterally and cranially than it is elsewhere.

The **body** *(corpus radii)*, or shaft, is compressed so that it presents two surfaces and two borders. Its width is two or three times its thickness. The **radial tuberosity** *(tuberositas radii)* is a small projection which lies distal to the neck on the medial border and adjacent caudal surface of the bone. It is particularly variable in development, depending on the breed. This tuberosity serves for the lesser insertions of the mm. biceps brachii. A large eminence lies proximal to the radial tuberosity on the lateral border of the radius just distal to the neck and serves for the distal attachment of the cranial crus of the lateral ligament of the elbow joint (Fig. 5–13).

The **cranial surface** *(facies cranialis)* is convex both transversely and vertically. At the junction of the proximal and middle thirds, on the medial border, there frequently is an obliquely placed rough line or ridge to which the m. pronator teres attaches. The m. supinator attaches to most of the cranial surface of the radius proximal to the insertion of the m. pronator teres. In large specimens, starting at the middle of the lateral border, and continuing distally from this border, there are alternating smooth ridges and grooves which run across the cranial surface of the radius; these markings converge toward a short, but distinct, oblique groove on the medial part of the distal extremity of the bone. The m. abductor pollicis longus, which arises on the ulna, as it courses distally, crosses the cranial surface of the radius obliquely and accounts for these markings.

The **caudal surface** *(facies caudalis)* is divided into two flat to concave areas by the vertical **interosseous border** *(margo interos-*

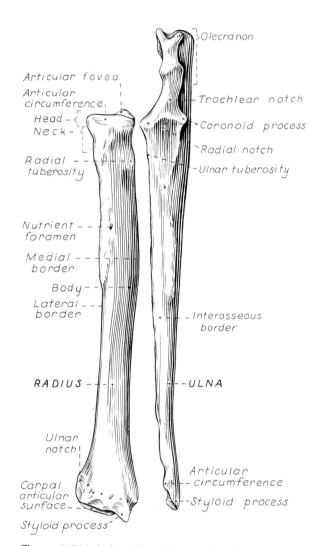

Olecranon

Articular fovea

Articular circumference

Head

Neck

Radial tuberosity

Trochlear notch

Coronoid process

Radial notch

Ulnar tuberosity

Nutrient foramen

Medial border

Body

Lateral border

Interosseous border

RADIUS

ULNA

Ulnar notch

Carpal articular surface

Styloid process

Articular circumference

Styloid process

Figure 4–74. Left radius, ulnar surface. Left ulna, radial surface.

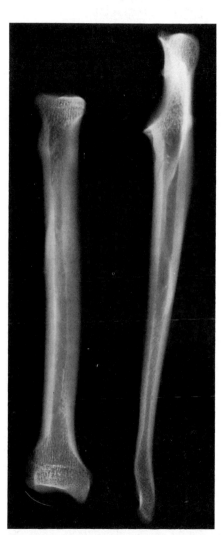

Figure 4–74. Radiograph. Craniocaudal radiograph, left radius and ulna disarticulated.

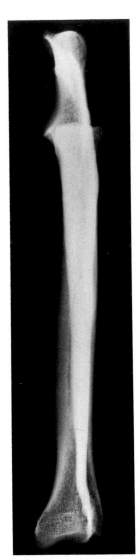

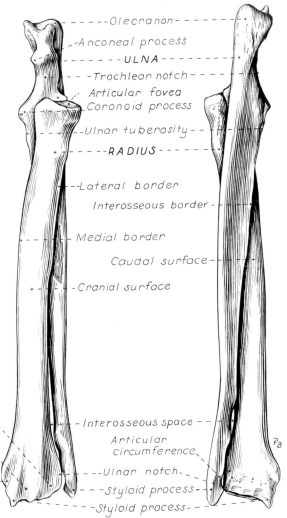

Olecranon

Anconeal process

ULNA

Trochlear notch

Articular fovea
Coronoid process

Ulnar tuberosity

RADIUS

Lateral border

Interosseous border

Medial border

Caudal surface

Cranial surface

Groove for
Ext. digitalis communis

Groove for
Ext. carpi radialis

Groove for
Abd. pollicis longus

Interosseous space

Articular
circumference

Ulnar notch

Styloid process

Styloid process

Figure 4–75. Radiograph. Craniocaudal radiograph, left radius and ulna articulated.

Figure 4–75. Left radius and ulna articulated, cranial aspect.

Figure 4–76. Left radius and ulna articulated, caudal aspect.

sea). The crest, which does not extend to either extremity of the radius, divides the surface into a medial two-thirds and a lateral one-third. The interosseous membrane attaches to it. The larger, flat, rough area medial to the border gives attachment to the m. pronator quadratus. A prominent rough area extends from the proximal part of the interosseous border distally to the lateral border. The heavy, short, interosseous ligament which unites the radius and ulna attaches to this raised, roughened area. Slightly above the middle of the caudal surface of the radius is the proximally directed nutrient foramen. Distally, the caudal surface becomes smoother, wider, and more convex, as it blends with the caudal surface of the distal extremity of the bone.

The **medial** and **lateral borders** of the radius present no special features. They are smooth and acutely rounded as they form the margins of the two surfaces of the bone. In large specimens, the rough area just above the middle of the bone on the caudal surface, for the attachment of the interosseous ligament, encroaches on the lateral border.

The **distal extremity** of the radius is the most massive part of the bone. Its distal surface *(facies articularis carpea)* articulates primarily with the radial carpal bone and to a lesser extent with the ulnar carpal. This surface is concave, both transversely and longitudinally, except for a caudomedial projection which lies in the groove of the radial carpal bone. The lateral surface of the distal extremity is slightly concave and lipped, forming the **ulnar notch** *(incisura ulnaris),* which articulates with a facet near the distal end of the ulna. Medially, the **styloid process** *(processus styloideus)* extends distal to the main carpal articular surface in the form of a sharp, wedge-shaped projection. The lateral surface of the styloid process enters into the formation of the carpal articular surface; the medial portion is somewhat flattened for the proximal attachment of the medial ligament of the carpal joint. The dorsal surface of the distal extremity of the radius presents three distinct grooves. The most medial groove, which is short, distinct, and obliquely placed, lodges the tendon of the m. abductor pollicis longus. The middle groove, which is the largest, contains the tendon of the m. exten-

sor carpi radialis. The most lateral groove, which is wider but occasionally less distinct than the others, contains the tendon of the m. extensor digitorum communis. The extensor retinaculum blends with the periosteum on the lip of the carpal articular surface. The caudal surface of the distal extremity is rough-ended and tuberculate. It contains many foramina for the passage of veins from the bone. The flexor retinaculum blends with the periosteum on this surface.

ULNA

The **ulna** (see Figs. 4–74, 4–75, 4–76), for descriptive purposes, is divided into a body, or shaft, and two extremities. Located largely in the postaxial part of the forearm, it exceeds the radius in length, and is, in fact, the longest bone in the body. Proximally it articulates with the humerus by the **trochlear notch** *(incisura* trochlearis) and with the articular circumference of the radius by the **radial notch** *(incisura radialis)* of the ulna. Distally it articulates with the ulnar notch of the radius and with the ulnar carpal and accessory carpal bones by means of two confluent facets on the knoblike distal end.

The **proximal extremity** of the ulna is the **olecranon,** which serves as a lever arm, or tension process, for the powerful extensor muscles of the elbow joint. It is four-sided, laterally compressed, and medially inclined; its proximal end is grooved cranially and enlarged and rounded caudally. The mm. triceps brachii, anconeus, and tensor fasciae antebrachii attach to the caudal part of the olecranon; the mm. flexor carpi ulnaris–caput ulnare and flexor digitorum profundus–caput ulnare arise from the medial surface of the olecranon (Figs. 6–55 and 6–56).

The trochlear notch (incisura trochlearis) is known, in some texts, as the semilunar notch. It is a smooth, vertical, half-moon–shaped concavity which faces cranially. A transverse plane through the middle of the trochlear notch separates the proximal extremity from the shaft of the ulna. The semilunar outline of this salient notch is formed by a sagittally placed ridge which divides its articular area into two nearly equal parts. The whole trochlear notch articulates with the trochlea of the humerus so that the sharp-edged, slightly hooked **anconeal process** *(processus anconeus),* at its proximal end, fits in the olecranon fossa of the humerus when the elbow is extended. The **coronoid process** *(processus coronoideus),* distal to the trochlear notch, is divided into a prominent, medial projection and a less prominent, lateral one. Both of these eminences are articular, facing cranially and proximally, where they articulate with the radius and humerus, respectively. They increase the surface area of the elbow joint without contributing materially to its weight-bearing function.

The **body** *(corpus ulnae),* or shaft, in the larger working breeds is typically compressed laterally in its proximal third, three-sided throughout its middle third, and cylindrical in its distal third. Great variation exists, however, and in long-limbed breeds the body is somewhat flattened throughout its length. The **cranial surface** *(facies cranialis)* is rough and convex, both longitudinally and transversely. Its most prominent feature is a slightly raised, oval, rough area on the middle third of the bone. It serves for the ulnar attachment of the short, but strong, interosseous ligament that attaches to the radius. The **interosseous border** extends proximally from the notch which separates the distal extremity from the body of the ulna. The interosseous membrane attaches to the interosseous border. Medial to the crest a faint vascular groove indicates the position, in life, of the palmar interosseous artery. The largest nutrient foramen is directed proximally and is usually located proximal to the rough area for the attachment of the interosseous ligament, near the interosseous crest. Other smaller nutrient foramina are located along the course of the vascular groove in the middle third of the body. The m. pronator quadratus attaches to the cranial surface of the ulna medial and adjacent to the interosseous crest. The mm. abductor pollicis longus and extensor pollicis longus et indicis proprius arise in that order from the cranial surface of the body of the ulna, progressing from the interosseous crest to the lateral border. The **caudal border** *(margo caudalis)* of the ulna, unlike the cranial surface, is smooth and concave throughout. It gradually tapers toward the distal end. The m. flexor digitorum profun-

dus–caput ulnare arises largely from this surface lateral to the radius. The mm. biceps brachii and brachialis insert mainly on the roughened **ulnar tuberosity** *(tuberositas ulnae),* which is located near the proximal end of the caudal surface just distal to the trochlear notch. **The medial border** *(margo medialis)* is sharper and straighter than the lateral one. The **lateral border** *(margo lateralis)* continues the wide, rounded, caudal border of the olecranon distally and laterally to the distal extremity of the bone. The foregoing description of the body of the ulna does not apply to some specimens, in which the middle third is more prismatic than flat. When the middle third is definitely three-sided, this feature continues distally, transforming the usually rodlike distal third to one which is three-sided.

The **distal extremity** of the ulna is separated from the body of the bone by a notch in its cranial border. An oval, slightly raised facet *(circumferentia articularis s. facies articularis radialis ulnae distalis)* is located in the distal part of the notch for articulation with the ulnar notch of the radius. The ulna of the dog possesses no head. The pointed, enlarged distal extremity of the ulna is the **styloid process** *(processus styloideus);* on its distomedial part there are two confluent facets. The one which faces cranially is concave and articulates with the ulnar carpal bone; the smaller, convex, medial facet articulates with the accessory carpal bone. The styloid process of the ulna projects slightly farther distally than the styloid process of the radius.

The Forepaw

The skeleton of the **forepaw** *(manus)* includes the carpus, metacarpus, phalanges, and certain sesamoid bones associated with them. The **carpus,** or wrist, is composed of seven bones arranged in two transverse rows, plus a small medial sesamoid bone. Articulating with the distal row of carpal bones are the five metacarpal bones which lie alongside one another and are enclosed in a common integument. Each of the lateral four metacarpal bones bears three phalanges which, with their associated sesamoid bones, form the skeleton of the four main digits. The small, medially located, first metacarpal bone bears only two, which form the skeleton of the rudimentary first digit. The bones of a typical tetrapod manus are serially homologous with those of the pes. In the lower vertebrate forms three groupings of the carpal and tarsal bones are made. The proximal grouping includes the radial, intermediate, and ulnar carpal bones for the manus, and the tibial, intermediate, and fibular tarsal bones for the pes. The middle grouping includes the central elements, of which there are three or four in each extremity. The distal grouping comprises a row of five small bones which articulate distally with the five metacarpal or metatarsal bones. There has been considerable modification of this primitive arrangement in mammals, with the fusion or loss of various elements.

CARPUS

The **carpus** (Figs. 4–77 to 4–79, 4–82, and 4–83), or wrist, includes the **carpal bones** (*ossa carpi*) and the associated sesamoid bones. The term carpus also designates the compound joint formed by these bones, as well as the region between the forearm and metacarpus. The carpal bones of the dog are arranged in a proximal and a distal row so that they form a transversely convex cranial outline and a concave caudal one. The bones of the proximal row are the radial,

ulnar, and accessory carpal bones. Those of the distal row are the first, second, third, and fourth carpal bones.

The **radial carpal bone** (*os carpi radiale*) or scaphoid, located on the medial side of the proximal row, is the largest of the carpal elements. It represents a fusion of the primitive radial carpal bone with the central and intermediate carpal bones. The proximal surface of the bone is largely articular for the distal end of the radius. The distal surface of the radial carpal bone articulates with all four distal carpal bones. Laterally it articulates extensively with the ulnar carpal. Its transverse dimension is about twice its width.

The **ulnar carpal bone** (*os carpi ulnare*), or triquetrum, is the lateral bone of the proximal row. It is shaped somewhat like the radial carpal, but is smaller. It articulates proximally with the ulna and radius, distally with the fourth carpal and the fifth metacarpal, medially with the radial carpal, and on the palmar side with the accessory carpal. It possesses a small lateral process and a larger palmar one for articulation with the accessory carpal and metacarpal V. This latter process is separated from the main part of the bone on the lateral side by a concave articular area for articulation with the styloid process of the ulna.

The **accessory carpal bone** (*os carpi accessorium*), or pisiforme, is a truncated rod

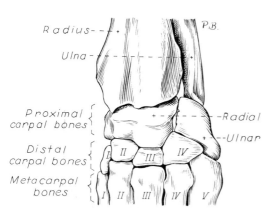

Figure 4–77. Left carpus, articulated, dorsal aspect.

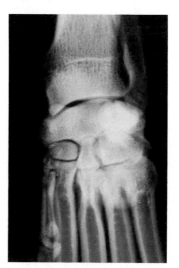

Figure 4–77. Radiograph. Dorsopalmar radiograph of left carpus, articulated.

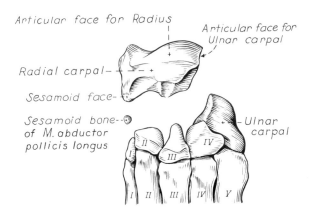

Figure 4–78. Left carpus, dorsal aspect. Radial carpal disarticulated.

of bone located on the caudal or palmar side of the ulnar carpal. Both ends of this bone are enlarged. The basal enlargement bears a slightly saddle-shaped articular surface for the ulnar carpal which is separated by an acute angle from a smaller, transversely concave, proximally directed articular area for the styloid process of the ulna. The free end is thickened and overhangs slightly. The accessory carpal bone is not a true carpal bone phylogenetically, but is rather a relatively new acquisition found in reptiles and mammals (Romer and Parsons 1978). The m. flexor carpi ulnaris inserts on it.

The **first carpal bone** *(os carpale primum)*, or trapezium, is the smallest carpal bone. It is somewhat flattened as it articulates with the palmaromedial surfaces of the second carpal and the base of metacarpal II. It artic-

ulates proximally with the radial carpal and distally with metacarpal I.

The **second carpal bone** *(os carpale secundum)*, or trapezoideum, is a small, wedge-shaped, proximodistally compressed bone which articulates proximally with the radial carpal, distally with metacarpal II, laterally with the third carpal, and medially with the first carpal.

The **third carpal bone** *(os carpale tertium)*, or capitatum, is larger than the second carpal. It has a large palmar projection, which articulates with the three middle metacarpal bones. It articulates medially with the second carpal, laterally with the fourth carpal, proximally with the radial carpal, and distally with metacarpal III.

The **fourth carpal bone** *(os carpale quartum)*, or hamatum, is the largest bone of the distal row. It presents a caudal enlargement and is wedge-shaped in both cranial and proximal views. It articulates distally with metacarpals IV and V, medially with the third carpal, and proximomedially with the radial carpal.

Each carpal element chondrifies independently before losing its identity. The intermediate carpal element fuses with the radial carpal, and then the two in turn fuse with the central carpal. The accessory carpal bone has an epiphyseal center of ossification which elaborates the cap of the enlarged caudal end of the bone.

The smallest bone of the carpus is a spherical sesamoid bone, about the size of a radish seed, which is located in the tendon of insertion of the m. abductor pollicis longus on the medial side of the proximal end of the first metacarpal. According to Baum and Zietzschmann (1936), on the palmar side of the carpus between the two rows of bones there are two flat bones similar in size to this small sesamoid bone.

METACARPUS

The term metacarpus refers to the region of the manus, or forepaw, located between the carpus and the digits. The **metacarpal bones** *(ossa metacarpalia I-V)* (Figs. 4–80, 4–84) are typically five in number in primitive mammals, although supernumerary metacarpal bones and digits may appear. In many mammals some of the metacarpal bones and their accompanying digits

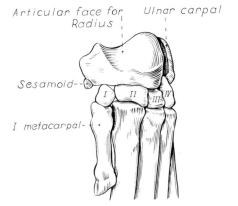

Figure 4–79. Left carpus, articulated, medial aspect.

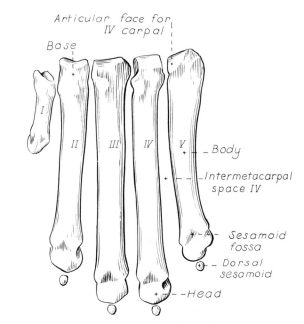

Figure 4–80. Left metacarpal and sesamoid bones, disarticulated, dorsal aspect.

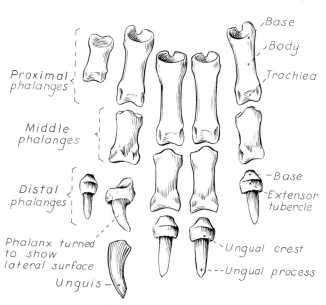

Figure 4–81. Phalanges, disarticulated, dorsal aspect.

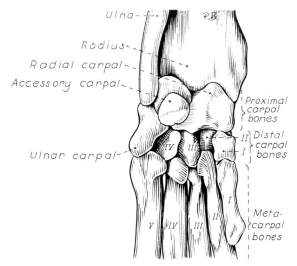

Figure 4–82.   Left carpus, articulated, palmar aspect.

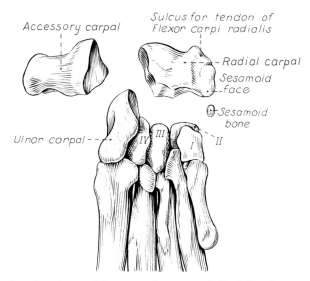

Figure 4–83.   Left carpal and metacarpal bones, palmar aspect. Radial and accessory carpals disarticulated.

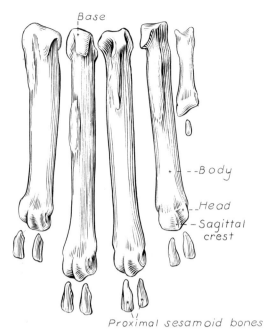
Base

_Body
_Head
_Sagittal crest

Proximal sesamoid bones

Figure 4-84. Left metacarpal and sesamoid bones, disarticulated, palmar aspect.

have been lost. Like the distal row of carpal bones, the metacarpal bones are numbered from the medial to the lateral side. The five metacarpal bones are each cylindrically shaped and enlarged at each end, proximally to form the **base,** and distally to form the **head.** The middle portion, or shaft, of each metacarpal bone is known as the **body.** Unlike the first metatarsal bone of the hind-paw, the **first metacarpal bone** of the fore-paw is usually present, although it is by far the shortest and most slender of the meta-carpal bones. It bears the first digit, which does not quite reach the level of the second metacarpophalangeal joint. Metacarpal I ar-ticulates proximally with the first carpal, and laterally with the second metacarpal. Distally, its laterally enlarged trochlea artic-ulates with the proximal phalanx of the first digit and a single palmar sesamoid bone.

Metacarpal bones II to **V** are the main metacarpal bones. They are irregular rods with a uniform diameter. Metacarpals II and V are shorter than III and IV and are four-sided, particularly at their proximal ends, whereas metacarpals III and IV are more triangular proximally. Distally, the bones diverge, forming the intermetacarpal spaces. The trochlea of the main metacarpal bones possess roller-like cranial parts which are undivided and are separated from the bodies dorsally by **sesamoid fossae** *(fossae sesamoidales).* Between the trochlea and bodies of the metacarpal bones on the palmer side are the **sesamoid impressions** *(impressiones sesamoidales).* The caudal parts of the heads possess prominent, sharp-edged **sagittal crests** *(cristae sagittales),* which effectively prevent lateral luxation of the two crescent-shaped sesamoid bones which articulate with these trochleae. The base of metacarpal II extends farther prox-imally than do the other metacarpal bones. It articulates with the first, second, and third carpals, as well as with metacarpals I and III. Besides articulating with adjacent meta-carpals, the base of metacarpal III articu-lates with the third and fourth carpals; the base of metacarpal IV articulates with the fourth carpal; the base of metacarpal V artic-ulates with the fourth carpal and the disto-caudal extension of the ulnar carpal. The interosseous muscles arise from the palmar surfaces of the bases of all of the main meta-carpal bones. The proximal palmar surfaces of the bodies of metacarpals II and III pro-vide insertion for the m. flexor carpi radialis, and the dorsal surfaces of the bases provide insertion for the m. extensor carpi radialis. The small m. adductor digiti quinti inserts on the medial surfaces of the distal parts of metacarpals IV and V, and on the lateral surface of metacarpal V near the base of the bone. The accessory carpal bone provides insertion for the m. extensor carpi ulnaris, the m. abductor pollicis longus inserts on the proximal medial part of metacarpal I, and the m. extensor pollicis longus inserts on the distal medial part of metacarpal I.

The middle parts of the bodies of the met-acarpal bones have particularly dense walls. These walls become thinner toward the ex-tremities, so that the articular cartilages lie on thin cortical bases. During development, the main metacarpal bones have only distal epiphyses. According to Schaeffer (1934), metacarpal I has only a proximal epiphysis. On the proximal third of the caudal surface of each of the four main metacarpal bones there is a nutrient foramen.

PHALANGES

The **digital skeleton** *(ossa digitorum manus)* (Figs 4–81, 4–85) of the forepaw consists of five units, of which four are fully developed and one is rudimentary. Each main digit consists of a proximal phalanx, middle phalanx, and distal phalanx, and two large palmar sesamoid bones at the metacarpophalangeal joint. A small osseous nodule is also located in the dorsal part of the joint capsule of each of the four main metacarpophalangeal joints, and a small cartilaginous nodule is located in a like place on each of the distal interphalangeal joints.

The **proximal** or **first phalanx** *(phalanx proximalis)* of each of the main digits, II to V, is a medium-length rod with enlarged extremities. Proximally, at its base, it bears a transversely concave articular surface with a sharp cranial border and a bituberculate palmar border. The palmar tubercles are separated by a deep groove which receives the sagittal crest of the head of a metacarpal bone when the joint is flexed. The palmar tubercles articulate with the distal end of the palmar sesamoid bones. The joint surface of the distal trochlea is saddle-shaped, sagittally convex, and transversely concave. It extends more proximally on the palmar surface than on the dorsal one. As if to prevent undue spreading of the main abaxial digits, the m. adductor digiti quinti inserts on the medial surface of the proximal phalanx of digit V, and the m. adductor digiti

secundi inserts on the lateral surface of the proximal phalanx of digit II. The proximal phalanx of digit I receives the insertions of mm. abductor pollicis brevis et opponens pollicis and adductor pollicis.

The **middle** or **second phalanx** *(phalanx media)* is present only in each of the main digits, there being none in digit I. Each middle phalanx is a rod about one-third shorter than the corresponding proximal phalanx with which it articulates. A palmar angle of about 135 degrees is formed by the proximal interphalangeal joint, whereas distally an obtuse palmar angle is formed as the distal phalanx butts against the middle phalanx, forming nearly a right angle dorsally. Each middle phalanx, like the proximal ones, is divided into a proximal **base**, a middle **body**, and a distal **head**. The base of each middle phalanx possesses an intermediate sagittal ridge, with palmar tubercles which are smaller and a palmar groove between these tubercles which is shallower than are those of the proximal phalanges. The m. flexor digitorum superficialis attaches to the palmar, proximal surfaces of the four middle phalanges by means of its four tendons of insertion.

The **distal** or **third phalanx** *(phalanx distalis)* is approximately the same size in all four main digits. The distal phalanx of digit I is similar to the others in form, but is smaller. The proximal part of the distal phalanx is enlarged. It has a shallow, sagitally concave articular area for contact with the

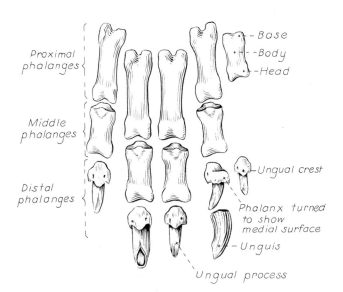

Figure 4–85. Phalanges, disarticulated, palmar aspect.

middle phalanx (proximal phalanx of digit I), to form the distal interphalangeal joint. A rounded, broad, low tubercle on the palmar side serves for the insertion of one of the five parts into which the tendon of the m. flexor digitorum profundus divides. Each side of this tubercle is perforated by a formen, the opening of a vascular canal which transversely perforates the bone. The dorsal part of the bone is also perforated by a vascular canal. The dorsal parts of the four main distal phalanges serve for the insertions of the four branches into which the tendon of the m. extensor digitorum communis divides. Joining the branches of the tendon of the m. extensor digitorum communis over the proximal phalanges is the tendon of the mm. extensor pollicis longus et indicis proprius and the tendons of the m. extensor digitorum lateralis to digits III, IV, and V. The distal part of the distal phalanx is a laterally compressed cone which is shielded by the horny claw. It is porous and has ridges on its proximal dorsal part which fade distally. The wall of the claw attaches to this surface. The sole of the claw attaches to the flattened palmar surface. The lateral and dorsal parts of the base of the cone are overhung by a crescent-shaped shelf of bone, the **ungual crest** (*crista unguicularis*), under which the root of the claw is located (Fig. 4–85).

SESAMOID BONES

On the palmar surface of each metacarpophalangeal joint of the main digits are two elongated, slightly curved **sesamoid bones** (*ossa sesamoidea*) (Fig. 4–84), which are located in the tendons of insertion of the interosseus muscles. They articulate primarily with the head of each metacarpal bone and secondarily with the palmar tubercles of each proximal phalanx. Their truncated distal ends articulate by small facets with the palmar tubercles of the corresponding proximal phalanges. Only a single osseous bead is located on the palmar side of the metacarpophalangeal joint of digit I.

Small bony nodules are located in the dorsal parts of the tendons of the four main digits at the metacarpophalangeal joints, (Fig. 4–80), whereas cartilaginous nodules are found at both the dorsal and palmar sides of the distal interphalangeal joints.

BONES OF THE PELVIC LIMB

Each pelvic limb (*membrum pelvinum*) consists of its half of the pelvic girdle (*cingulum membri pelvini*), composed of the **ilium, ischium, pubis,** and **acetabular bone** fused as the hip bone (*os coxae*); the thigh, represented by the **femur;** the stifle, or knee joint, with its menisci and sesamoids; the crus, or leg, consisting of the **tibia** and **fibula;** and the hindpaw, or pes. The pes includes the ankle, or **tarsus,** with its digits, consisting of **metatarsals, phalanges,** and the **sesamoid bones** associated with the phalanges.

The bony **pelvis** (Fig. 4–86) is formed by the ossa coxarum, the sacrum, and the first caudal vertebra.

OS COXAE

The *os coxae*, or hip bone (Figs. 4–87 to 4–89), is composed of four distinct bones developmentally. These are the ilium, ischium, pubis, and acetabular bone. They fuse during the twelfth postnatal week, forming the socket which receives the head of the femur in creation of the hip joint. This socket is a deep, cotyloid cavity, called the **acetabulum.** The acetabulum in a medium-sized dog is 1 cm. deep and 2 cm. in diameter. The **lunate surface** (*facies lunata*) is the smooth articular circumference which is deficient over the medial portion of the acetabulum. The cranial part of the lunate surface is widest as it extends from the acetabular margin three-fourths of the distance to the depth of the acetabulum. The lunate surface is narrowest midlaterally, being about one-half its maximum width. The cranial portion ends medially in a rounded border. Medially the acetabulum is indented by a notch, the *incisura acetabuli.* The caudal part of the acetabular margin or lip which forms the caudal boundary of the notch is indented by a fissure 2 to 4 mm. deep. The quadrangular, non-articular, thin, depressed area which extends laterally from the acetabular notch is the **acetabular fossa** (*fossa acetabuli*). During the seventh post-

Text continued on page 203.

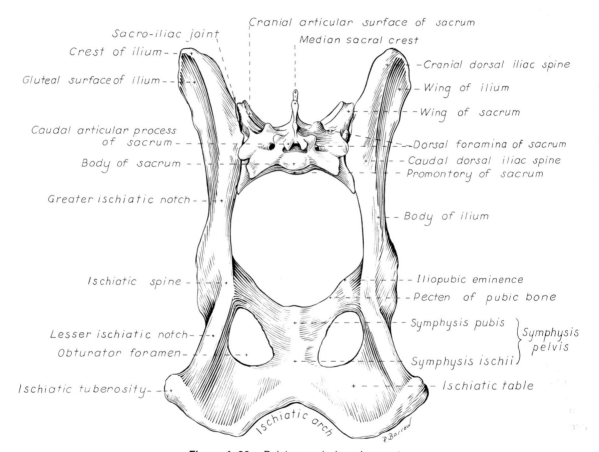

Figure 4–86. Pelvis, caudodorsal aspect.

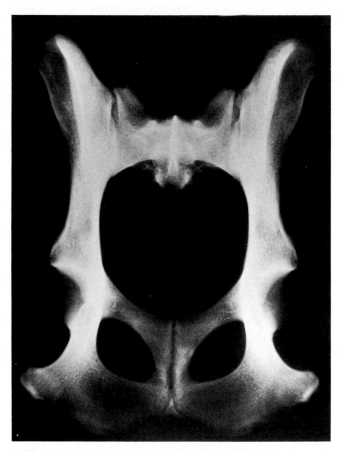

Figure 4–86. Radiograph. Ventrodorsal radiograph, ossa coxae and sacrum.

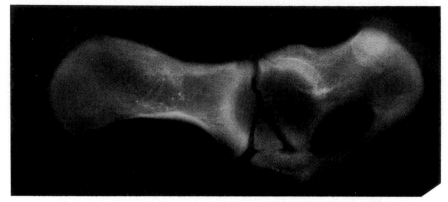

Figure 4–87. Radiograph. Lateromedial radiograph, left os coxa of young dog.

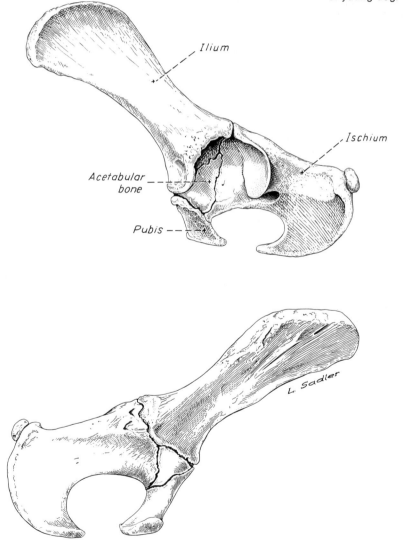

Ilium

Ischium

Acetabular bone

Pubis

L. Sadler

Figure 4–87. Left os coxa of a 15-week-old Beagle.

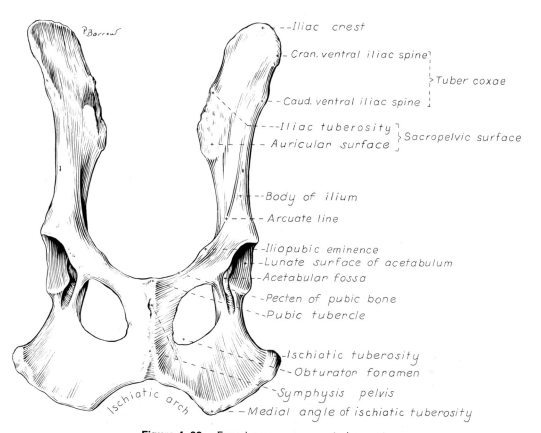

Figure 4–88. Fused ossa coxae, ventral aspect.

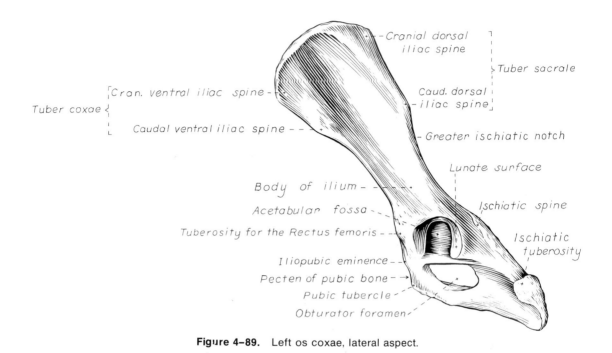

Cranial dorsal iliac spine

Tuber sacrale

Caud. dorsal iliac spine

Cran. ventral iliac spine

Tuber coxae

Caudal ventral iliac spine

Greater ischiatic notch

Lunate surface

Body of ilium

Acetabular fossa

Ischiatic spine

Tuberosity for the Rectus femoris

Ischiatic tuberosity

Iliopubic eminence

Pecten of pubic bone

Pubic tubercle

Obturator foramen

Figure 4–89. Left os coxae, lateral aspect.

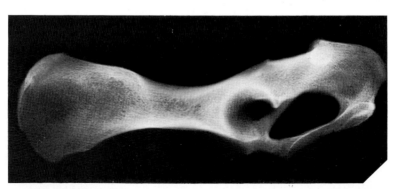

Figure 4–89. Radiograph. Lateromedial radiograph, left os coxa.

natal week, a small osseous element, the **acetabular bone** *(os acetabuli)* (Fig. 4–86), located in the floor of the acetabulum between the ilium and ischium, becomes incorporated with these larger bones. The **pelvic cavity** is of considerable obstetrical importance, since, for survival of the species in nature, it must be large enough to allow for the passage of the young during parturition. The **cranial pelvic aperture** *(apertura pelvis cranialis),* or **pelvic inlet,** is formed by the **promontory** of the sacrum dorsally, the **cranial border of the pubis** or **pecten** ventrally, and the **arcuate line** bilaterally. The arcuate line passes along the ventromedial edge of the sacropelvic surface of the ilium. It extends from the auricular surface to the **iliopubic eminence** *(eminentia iliopubica).*

The following conventional measurements of the pelvis are useful in obstetrics: the **transverse diameter** *(diameter transversa)* is the greatest transverse measurement of the bony pelvic cavity. According to Roberts (1956), only in the achondroplastic types of dogs like the Sealyham and Pekingese are the transverse diameters greater than the conjugate or sacropubic diameters. The **conjugate** *(conjugata)* measurement is the distance from the sacrovertebral angle or the sacral promontory to the cranial border of the symphysis pubis. The **oblique diameter** *(diameter obliqua)* is measured from the sacroiliac articulation of one side to the iliopubic eminence of the other. The **pelvic axis** *(axis pelvis)* is an imaginary, slightly curved line drawn through the middle of the pelvic cavity from the pelvic inlet to the pelvic outlet. The **caudal pelvic aperture** *(apertura pelvis caudalis),* or **pelvic outlet,** is bounded dorsally by the first caudal vertebra, bilaterally by the sacrotuberous ligament, and ventrally by the caudolateral border of the tuber ischiadicum on each side and the ischiatric arch located between them. The sacral part of the roof of the pelvic canal is about as long as its floor, but is offset to the extent that a transverse plane touching the caudal part of the sacrum also touches the cranial border of the pubis. The lateral osseous wall of the pelvic canal is formed largely by the body of the ilium and caudally, to a small extent, by the bodies of the ischium and pubis as these fuse to form the acetabulum. The floor of the

bony pelvis is formed by the sacropelvic surfaces of the rami of the pubes and ischii. Between these rami and the body of the ischium is the large, oval to triangular **obturator foramen** *(foramen obturatum).* The **symphysis pelvis** is the median synostosis formed by the right and left pubic and ischial bones. It is, therefore, composed of the **symphysis pubis** cranially and the **symphysis ischii** caudally. Occasionally, in young specimens, there is in the caudal part of the symphysis a separate triangular bone which is widest and thickest caudally.

The **ilium** *(os ilium)* is the largest and most cranial of the bones which compose the os coxae. It is basically divided into a cranial, nearly sagittal, laterally concave part, the **wing** *(ala ossis ilii)* and a narrow, more irregular caudal part, the **body** *(corpus ossis ilii).* The body, at its expanded caudal end, forms the cranial two-fifths of the acetabulum. In this cavity it fuses with the ischium caudally and the pubis medially. The **cranial border** is more commonly known as the **iliac crest** *(crista iliaca).* It forms a cranially protruding arc which is thin in its ventral half; the dorsal half gradually increases in thickness until it reaches a width of nearly 1 cm. dorsally in the large working breeds. The iliac crest, in heavily muscled breeds, presents a slight lateral eversion. The **dorsal border** is thicker in its cranial half than in its caudal half. The caudal half of the dorsal border is gently concave, forming the **greater ischiatic notch** *(incisura ischiadica major).* The dorsal border of the ilium is continuous with the dorsal border of the ischium as a slight convexity dorsal to the acetabulum. This is the **ischiatic spine.** The eminence located dorsal to the iliosacral joint between the thick parts of this border is the **caudal dorsal iliac spine** *(spina iliaca dorsalis caudalis).* The obtuse angle located between the cranial and dorsal borders is the **cranial dorsal iliac spine** *(spina iliaca dorsalis cranialis).* These two spines and the intermediate border constitute what is known as the *tuber sacrale* in the dog and in the large herbivores, in which it is more salient than it is in the dog. The ventral margin begins at the **cranial ventral iliac spine** *(spina iliaca ventralis cranialis).* About 1 cm. caudal to this spine is a small eminence on the thin ventral border which is known as the **caudal ventral iliac spine** *(spina iliaca ventralis caudalis).* These two spines and the connecting border constitute what is called the *tuber coxae.* Grooving the ventral border just caudal to the tuber coxae and extending on the lateral surface of the ilium in old specimens is the vascular groove for the iliolumbar artery and vein. After running caudal parallel to a plane through the lateral surface of the ilium, the prominent caudal half of the ventral border ends in the low, but prominent, iliopubic eminence (eminentia iliopubica). Slightly craniodorsal to the iliopubic eminence is the **tuberosity for the origin of the m. rectus femoris** *(area lateralis m. recti femoris).*

The **gluteal surface** *(facies glutea)* of the ilium faces laterally and slightly upward. It embodies the whole external surface of the bone. An intermediate fossa which parallels the axis of the bone divides the surface into a strong ridge dorsally and a triangular, moderately rough area ventrally. The medial or **sacropelvic surface** *(facies sacropelvina)* articulates with the wing of the sacrum by a synchondrosis which forms the **auricular surface** *(facies auricularis).* The **iliac tuberosity** *(tuberositas iliaca)* is the rough, slightly protruding eminence of the sacropelvic surface *(facies sacropelvina)* located dorsal to the auricular surface. The **iliac surface** *(facies iliaca)* is a nearly square, flat area cranial to the auricular surface. Between the ventral and dorsal margins of the body of the ilium lies the lateral part of the **arcuate line,** which extends from the auricular surface to the iliopubic eminence. It divides the sacropelvic surface of the body of the ilium into a medial two-thirds and a ventromedial one-third. The caudally directed nutrient foramen is located near the middle of this surface adjacent to the ventral border. The mm. sartorius and tensor fasciae latae arise from the tuber coxae. The iliac portion of the m. iliopsoas, the m. sacrospinalis, and portions of the mm. coccygeus and levator ani attach to the sacropelvic surface (Figs. 6–42 and 6–73). The mm. gluteus medius, gluteus profundus, and capsularis coxae arise from the gluteal surface of the ilium. The mm. psoas minor, rectus abdominis, rectus femoris, and pectineus attach to the eminence and the line adjacent to the eminence.

The **ischium** *(os ischii)* consists of a body, ramus, and tuberosity. It forms the caudal third of the os coxae and enters into the formation of the acetabulum, obturator foramen, and symphysis pelvis. The **ischiatic tuberosity** *(tuber ischiadicum)* is the caudolateral part of the bone. It gradually thickens, from the medial to the lateral side, where it ends in a pronounced rough hemispherical eminence. The caudal end of the sacrotuberous ligament attaches to the dorsal surface of this eminence.

The **body of the ischium** *(corpus ossis ischii)* is that part of the bone which lies lateral to the obturator foramen and at its cranial end forms about two-fifths of the acetabulum. Its thick dorsal border continues with the dorsal border of the ilium in a slight convexity, forming the **ischiatic spine** *(spina ischiadica).* Caudal to the spine the dorsal border is flattened and creased by about five shallow groves, in which lie the multiple tendons of the m. obturatorius internus. In life the **lesser ischiatic notch** *(incisura ischiadica minor)* is converted into a large opening, the lesser ischiatic foramen, by the sacrotuberous ligament. The **ramus of the ischium** *(ramus ossis ischii)* joins the body at a right angle. The ramus and the body are twisted at their junction in such a way that the ramus faces dorsally, forming the **ischiatic table** *(tabula ossis ischii),* and the body faces medially, forming the caudal part of the lateral boundary of the pelvic cavity. The m. obturatorius internus arises from the shallow fossa of the ischiatic table which lies cranial to the ischiatic tuberosity, as well as from the medial and cranial edges of the obturator foramen and the adjacent pelvic surface of the os coxae. The cranial border of the ramus forms the caudal boundary of the obturator foramen, and the caudal border meets its fellow in forming the deep **ischial arch** *(arcus ischiadicus),* which is formed by the joining of the caudomedial parts of the adjacent bones. The ventral surface of the ischiatic tuberosity gives rise to the most powerful muscles of the thigh, the hamstring muscles: mm. biceps femoris, semitendinosus, and semimembranosus. The adjacent ventral surface of the ramus gives rise to the m. quadratus femoris, and a zone next to the caudal and medial borders of the obturator foramen gives rise to the m. obturatorius externus. The m. adductor arises from the symphysis and the ventral surface of the ischium adjacent to it. The mm. gemelli arise from the lateral surface of the ischium ventral to the lesser ischiatic notch. The root of the penis, with its m. ischiocavernosus, attaches to the medial angle of the ischiatic tuberosity.

The **pubis** (*os pubis*) is a dorsoventrally compressed, curved bar of bone which extends from the ilium and ischium laterally to the symphysis pubis medially. Its caudal border bounds the cranial part of the obturator foramen, which is particularly smooth and partly grooved by the n. obturatorius and the ascending obturator vessels. It is divided into a body and two rami. The **body** (*corpus ossis pubis*) is the central part of the bone, forming the cranial border of the obturator foramen and fusing with the opposite side to form the symphysis pubis. The cranial ramus enters into the formation of the acetabulum. The iliopubic eminence (*eminetia iliopubica*) is located at the cranial end of the ramus as it joins the ilium. The caudal ramus extends to the ischium, where is fuses without demarcation. The **iliopectineal** eminence (*eminentia iliopectinea*), its largest process, is located on the cranial border of the bone as it joins the ilium. At its narrowest part the body gives way to the **ramus** (*ramus ossis pubis*), the flattened, triangular part of the pubis. Medially it fuses with its fellow, forming the pubic part. The ventral surface of the pubis gives origin to the mm. gracilis, adductor, and obturatorius externus. The dorsal or pelvic surface gives rise to the m. levator ani and a part of the m. obturatorius internus. The **pubic tubercle** (*tuberculum pubicum*) is located on the ventral surface of the pubic symphysis. The cranial border of the pubis, stretching from the iliopubic eminence to the symphysis pubis, is also called the *pecten ossis pubis*, or the medial part of the arcuate line. The pubic tubercle and pecten serve for the attachment of the prepubic tendon, whereby all of the abdominal muscles, except for the m. transversus abdominis, attach wholly or in part. The m. pectineus also arises here.

FEMUR

The **femur** (*os femoris*) (Figs. 4–90 to 4–92) is the heaviest bone in the skeleton. In well-proportioned breeds it is slightly shorter than the tibia and ulna, but is about one-fifth longer than the humerus. It articu-

lates with the os coxae proximally, forming a flexor angle of 110 degrees cranially. Distally, it articulates with the tibia, forming a flexor angle of 110 degrees caudally. Right and left femurs lie in parallel sagittal planes when the animal is standing. In fact, all the main bones of the pelvic limb are in about the same sagittal plane as those of the ipsilateral pectoral limb, but the flexor angles of the first two joints of each limb face in opposite directions.

The **proximal end** of the femur presents a smooth, nearly hemispherical **head** (*caput femoris*), supported by a neck on its proximolateral side and three processes, or trochanters. The head caps the dorsocaudal and medial parts of the neck. The *fovea capitis femoris* is a small rather indistinct, circular pit on the medial part of the head. Occasionally a depressed, moderately rough, nonarticular strip extends from the fovea to the nearest caudoventral nonarticular margin. The fovea serves for the attachment of the **ligament of the head of the femur** (*ligamentum capitis ossis femoris*), formerly called the round ligament. The **neck** (*collum femoris*) unites the head with the rest of the proximal extremity. It is about as long as the diameter of the head, slightly compressed craniocaudally, and it is reinforced by a ridge of bone which extends from the head to the large, laterally located greater trochanter.

The **greater trochanter** (*trochanter major*), the largest tuber of the proximal extremity of the bone, is located directly lateral to the head and neck. Its free, pyramid-shaped apex usually extends nearly to a frontal plane lying on the head. Between the femoral neck and the greater trochanter, caudal to the ridge of bone connecting the two, is the deep **trochanteric fossa** (*fossa trochanterica*). The mm. gluteus medius, gluteus profundus, and piriformis insert on the greater trochanter. The mm. gemelli, obturatorius internus, and obturatorius externus insert in the trochanteric fossa. The **lesser trochanter** (*trochanter minor*) is a distinct, pyramid-shaped eminence which projects from the caudomedial surface of the proximal extremity near its junction with the shaft. It is connected with the greater trochanter by a low but wide arciform crest, the **intertrochanteric crest** (*crista intertrochanterica*). The m. quadratus femoris in-

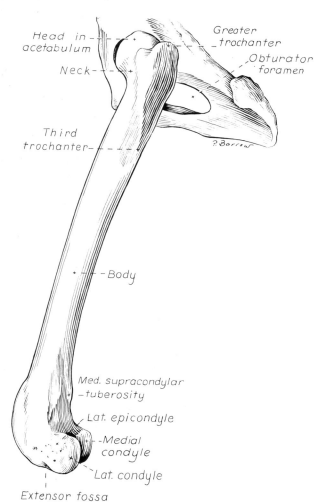

Head in acetabulum

Neck

Third trochanter

Greater trochanter

Obturator foramen

P. Barrow

Body

Med. supracondylar tuberosity

Lat. epicondyle

Medial condyle

Lat. condyle

Extensor fossa

Figure 4–90. Left femur and os coxae articulated, lateral aspect.

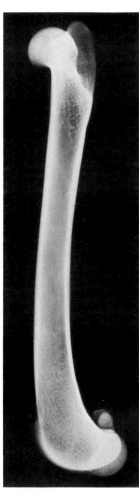

Figure 4–90. Radiograph. Lateromedial radiograph, left femur and fabellae.

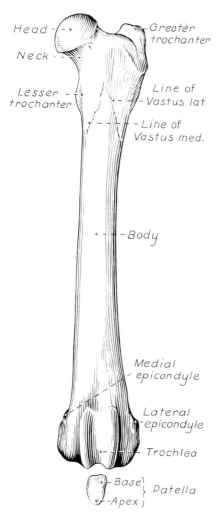

Figure 4–91. Left femur with patella, cranial aspect.

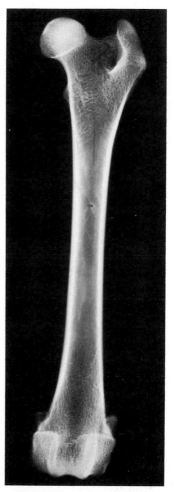

Figure 4–91. Radiograph. Craniocaudal radiograph, left femur with fabellae.

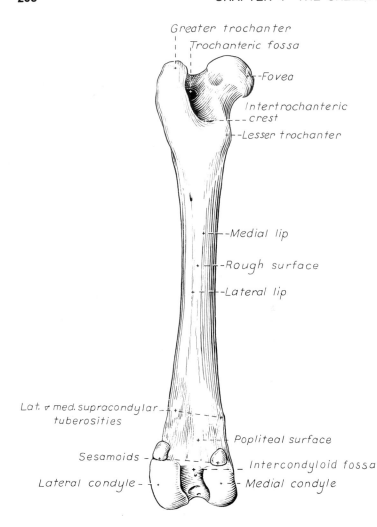

Greater trochanter
Trochanteric fossa
Fovea
Intertrochanteric crest
Lesser trochanter
Medial lip
Rough surface
Lateral lip
Lat. & med. supracondylar tuberosities
Popliteal surface
Sesamoids
Intercondyloid fossa
Lateral condyle
Medial condyle

Figure 4–92. Left femur with fabellae, caudal aspect.

serts distal to the intertrochanteric crest adjacent to the lesser trochanter. The most craniolateral eminence of the greater trochanter is called the cervical tubercle. On the line which arches distocaudally from this tubercle is the **third trochanter** *(trochanter tertius)*, which is about as large as the cervical tubercle. The m. gluteus superficialis inserts on the third trochanter. This lateral eminence is about 2 cm. distal to the apex of the greater trochanter. The **transverse line** *(linea transversa)* is dorsally arched and runs from the femoral head across the cranial surface of the intertrochanteric crest to the greater trochanter.

The shaft, or **body** *(corpus femoris)*, is nearly cylindrical, and is straight proximally and cranially arched distally. Its cranial, lateral, and medial surfaces are not demarcated from each other, but the caudal surface is flatter than the others. A small proximal nutrient foramen pierces the cranial surface of the cortex in a distal direction. Covering all but the caudal surface of the femur is the large m. quadriceps femoris. All except the rectus femoris division of this muscle arise from the proximal part of the body of the femur, where occasionally indistinct lines indicate the most proximal attachments for the m. vastus lateralis and the m. vastus medialis. The caudal surface is marked by a finely roughened surface, the *facies aspera,* which is narrow in the middle and wider at both ends. This slightly roughened face is bounded by the **medial and lateral lips** *(labium mediale et laterale)*, which diverge proximally, running into the lesser and greater trochanters, and, distally, becoming obscured in the medial and lateral epicondyles, respectively. The sagittally concave, transversely flat area enclosed distally by these lips is the **popliteal surface** *(facies poplitea)*. The relatively flat surface proximally, which is flanked by the diverging femoral lips, is called the **trochanteric surface** by Nickel et al. (1977). The largest nutrient foramen to enter the femur is found on the caudal surface at approximately the junction of the proximal and middle thirds of the bone. The m. adductor longus inserts on the lateral lip distal to the third trochanter, whereas the m. adductor magnus et brevis inserts on the whole lateral lip from the third trochanter to the popliteal surface. The m. pectineus inserts on the popliteal surface, and the m. semimembranosus inserts on the adjacent distal end of the medial lip.

The **distal end** of the femur is quadrangular and protrudes caudally. It contains three main articular areas. Two of these are on the medial and lateral condyles, and the third is an articular groove on the cranial surface. The **lateral condyle** *(condylus lateralis)* is convex in both the sagittal and the transverse plane. The **medial condyle** *(condylus medialis)* is smaller and less convex in both the transverse and the sagittal plane. Each condyle articulates directly with the tibia, but most extensively with the menisci of the tibia. They are separated by the **intercondylar fossa** *(fossa intercondylaris)*, which is slightly oblique in direction as the caudal part of the intercondyloid fossa is located farther laterally than is the cranial part. The articular surfaces of the condyles are continuous caudally with small facets on the adjacent epicondyles. The facet on the lateral epicondyle is larger than that on the medial one. These articulate with sesamoid bones in the tendons of origin of the m. gastrocnemius. The **femoral trochlea or patellar surface** *(trochlea femoris s. facies patellaris)* is the smooth, wide articular groove on the cranial surface of the distal extremity which is continuous with the articular surfaces of the condyles. Proximally, the limiting ridges diverge slightly. The medial ridge is somewhat thicker than the lateral one. The patella, or knee cap, articulates with the patellar surface of the femur.

Proximal and cranial to the medial and lateral condyles are the **medial and lateral epicondyles** *(epicondylus medialis et lateralis)*. These serve for the proximal attachments of the medial and lateral collateral ligaments of the stifle joint. The **extensor fossa** *(fossa extensoria)* is a small pit located at the junction of the lateral ridge of the patellar surface and the lateral epicondyle. From it arises the m. extensor digitorum longus. The m. popliteus arises under the lateral collateral ligament from the lateral condyle of the femur. On the caudal proximal surfaces of the medial and lateral condyles are facets for the articulation of the medial and lateral fabellae, the sesamoid bones located in the tendons of origin of the heads of the m. gastrocnemius. Proximal to these facets, at the proximal edge of the popliteal surface, are located tubercles which are known as the **medial and lateral supracondylar tuberosities** *(tuberositas supracondylaris medialis et lateralis)*. The m. gastrocnemius arises from both tuberosities. The m. flexor digitorum superficialis also arises from the lateral supracondylar tuberosity.

SESAMOID BONES OF THE STIFLE JOINT

The **patella** (Fig. 4–91), or knee cap, is the largest sesamoid bone in the body. It is ovate in shape and curved so as to articulate with the patellar surface of the femur. The **base** *(basis patellae)* is blunt and faces proximally. It may extend beyond the adjacent articular surface. The distally located **apex** *(apex patellae)* is slightly more pointed than the base and does not extend beyond the articular surface. The **articular surface** *(facies articularis)* is smooth, convex in all directions, and in some specimens shows longitudinal striations. Several nutrient foramina enter the bone from the medial side. The patella is an ossification in the tendon of insertion of the great extensor of the stifle, the m. quadriceps femoris. That part of the tendon between its insertion on the tibial tuberosity and the patella is also known as the patellar ligament. The patella alters the direction of pull of the tendon of the quadriceps; it protects the tendon, and it provides a greater bearing surface for the tendon to play on the trochlea of the femur than would be possible without it.

The cranial articular area of the stifle is greatly increased by the presence of two or three **parapatellar fibrocartilages.** These are grooved cartilages, one on each side of the patella, which articulate with the ridges of the patellar surface of the femur. Proximally, the two cartilages may extend far enough above the patella to curve toward each other and meet, or a third cartilage may be located at this site. For a more complete description of these cartilages refer to Chapter 5, Joints and Ligaments.

There are three sesamoid bones, or **fabellae,** in the stifle region. Two of these are located in the heads of the m. gastrocnemius caudal to the stifle joint on the medial and lateral condyles (fig. 4–92), and the third, intercalated in the tendon of the m. popliteus (Fig. 4–95). The sesamoid located in the lateral head of origin of the m. gastrocnemius is the largest. It is globular in shape, except for a truncated end, which faces distally and has a nearly flat articular surface for the facet on the caudal part of the lateral femoral condyle. The sesamoid in the medial head of origin of the m. gastrocnemius is smaller than the lateral one, and is angular in form. It may not leave a distinct facet on the medial condyle. The smallest sesamoid bone of the stifle region is the sesamoid located in the tendon of origin of the m. popliteus, adjacent to its muscle fibers. It articulates with the lateral condyle of the tibia.

TIBIA

The **tibia** (Figs. 4–93 to 4–95) is a long, strong bone which lies in the medial part of the crus, or true leg. It articulates proximally with the femur, distally with the tarsus, and on its lateral side both proximally and distally with the companion bone of the crus, the fibula. Its proximal half is triangular in cross section and more massive than its distal half, which is nearly cylindrical.

The **proximal end** of the tibia is relatively flat and triangular, with its apex cranial. Extending from the margin of the base on each side of a central elevation are articular areas which form the **proximal articular surface** (*facies articularis proximalis*). The divided proximal articular surface lies on the **lateral and medial condyles** (*condylus lateralis et medialis*). The articular areas of the condyles are separated by a sagittal, non-articular strip, and two eminences. Al-

though the surface area of the two is approximately the same, the medial condyle is oval and the lateral condyle is nearly circular. Both are convex in the sagittal plane, and concave transversely. In the fresh state they are covered by articular cartilage and have only a small area of contact with the articular cartilage of the femoral condyles. Functionally the medial and lateral tibial condyles are separated from the medial and lateral femoral condyles by the **medial and lateral menisci** (*meniscus medialis et lateralis*). These fibrocartilages are biconcave, incomplete discs, which are open toward the axis of the bone. The central edges of these C-shaped cartilages are thin and concave, and their peripheral margins are thick and convex. The **intercondylar eminence** (*eminentia intercondylaris*) is a low but stout divided eminence between the medial and lateral tibial condyles. The two spurs which are articular on their abaxial sides are known as the **medial and lateral intercondylar tubercles** (*tuberculum intercondylare mediale et laterale*). The oval, depressed area cranial to the intercondyloid eminence is the *area intercondylaris cranialis;* the smaller, depressed area caudal to it is the *area intercondylaris caudalis.* The meniscial ligaments attach to these areas. The condyles are more expansive than the articular areas located on their proximal surfaces. Between the condyles caudally is the large **popliteal notch** (*incisura poplitea*). The **extensor groove** of the tibia (*sulcus extensorius*) is a smaller notch which cuts into the lateral condyle as far as the articular area. On the caudolateral surface of the lateral condyle is the *facies articularis fibularis,* an obliquely placed facet for articulation with the head of the fibula. The m. semimembranosus inserts on the caudal part of the medial condyle, and the proximal part of the origin of the m. tibialis cranialis arises from the lateral condyle. The **tibial tuberosity** (*tuberositas tibiae*) is the large, quadrangular, proximocranial process which provides insertion for the powerful m. quadriceps femoris and parts of the mm. biceps femoris and sartorius. Extending distally from the tibial tuberosity is the **cranial border** of the tibia (*margo cranialis*) formerly called the tibial crest. To it insert the mm. gracilis, semitendinosus, and parts of the mm. sartorius and biceps femoris.

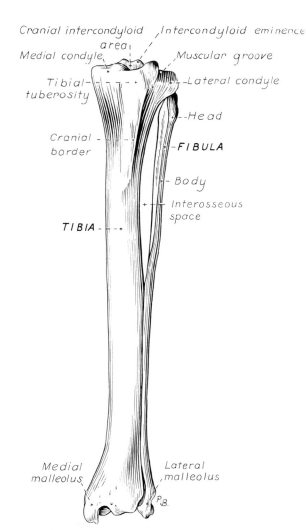

Cranial intercondyloid area

Intercondyloid eminence

Medial condyle

Muscular groove

Tibial tuberosity

Lateral condyle

Head

Cranial border

FIBULA

Body

Interosseous space

TIBIA

Medial malleolus

Lateral malleolus

P.B.

Figure 4–93. Left tibia and fibula articulated, cranial aspect.

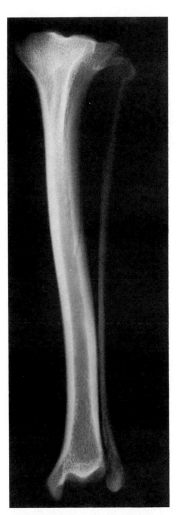

Figure 4–93. Radiograph. Craniocaudal radiograph of left tibia and fibula, articulated.

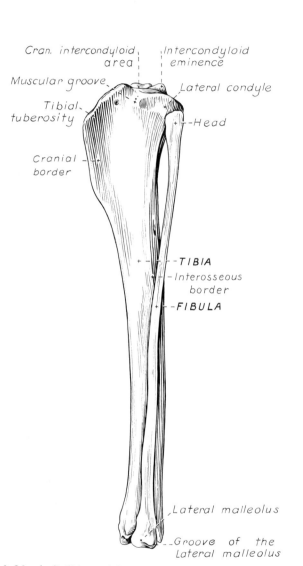

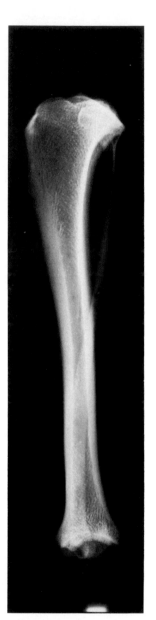

Figure 4–94. Left tibia and fibula articulated, lateral aspect.

Figure 4–94. Radiograph. Lateromedial radiograph of left tibia and fibula, articulated.

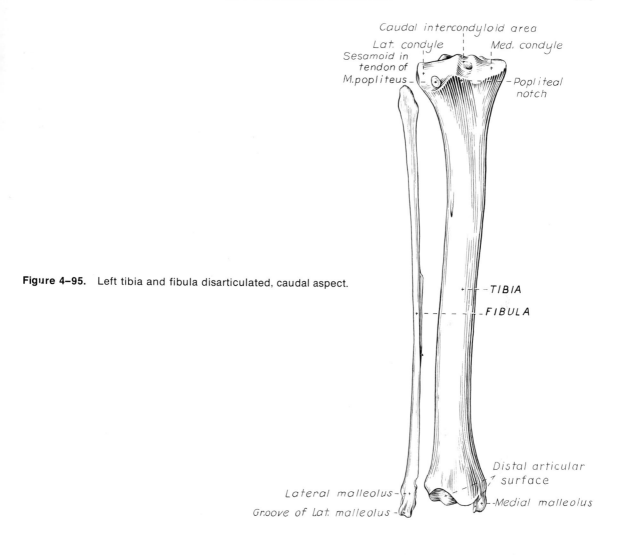

Caudal intercondyloid area
Lat. condyle   Med. condyle
Sesamoid in
tendon of
M.popliteus   Popliteal
                                       notch

-TIBIA

-FIBULA

Distal articular
surface

Lateral malleolus-   -Medial malleolus
Groove of Lat. malleolus-

Figure 4-95. Left tibia and fibula disarticulated, caudal aspect.

The **body** *(corpus tibiae)* is three-sided throughout its proximal half, whereas the distal half is essentially quadrilateral or cylindrical. Three surfaces and three borders are recognized in the proximal half of the tibia. These are the caudal, medial, and lateral surfaces, and the medial, interosseous, and cranial borders. The **interosseous border** *(margo interosseus)* is replaced in the distal half of the tibia by a narrow, flat surface apposed to the adjacent, closely lying fibula.

The **caudal surface** *(facies caudalis)* presents an oblique line which courses from the proximal part of the interosseous border to the middle of the medial border. At the junction of the proximal and middle thirds of the interosseous border is the distally

directed nutrient foramen of the bone. The m. popliteus inserts on the proximal medial part of the caudal surface, the proximal part of the medial border, and the adjacent medial surface of the tibia proximal to the oblique line. The mm. flexor hallucis longus, tibialis posterior, and flexor digitorum longus arise from the proximal half of the caudal surface in lateral to medial sequence. Running obliquely distolaterally across the lower part of the caudal surface may be a vascular groove which extends to the distal end of the bone adjacent to the lateral malleolus.

The **medial surface** *(facies medialis)* of the tibia is wide and nearly flat proximally, as it is partly formed by the tibial crest. Near the tibial crest in large specimens is a low, but wide, muscular line for the insertions of the mm. semitendinosus, gracilis, and sartorius. The medial surface of the tibia is relatively smooth throughout, as it is largely subcutaneous in life.

The **lateral surface** *(facies lateralis)* of the tibia is smooth, wide, and concave proximally, flat in the middle, and narrow and convex distally. Part of the m. biceps femoris inserts on the medial surface of the tibial crest, and just caudal to this attachment the m. tibialis cranialis arises. This muscle intimately covers the lateral surface of the tibia. The m. flexor hallucis longus arises from the proximal three-fourths of the lateral border of the tibia. The m. fibularis brevis arises from the lateral surfaces of the distal two-thirds of the fibula and tibia.

The **distal end** of the tibia is quadrilateral and slightly more massive than the adjacent part of the body. The **distal articular surface** is in the form of two nearly sagittal, arciform grooves, the **cochlea tibiae,** which receive the ridges of the proximal trochlea of the tibial tarsal bone. The grooves are separated by an intermediate ridge. A transversely located synovial fossa extends from one groove to the other across the intermediate ridge. The whole medial part of the distal extremity of the tibia is the **medial malleolus** *(malleolus medialis).* Its cranial part is formed by a stout pyramid-shaped process. Caudal to this is a semilunar notch. The small, but distinct, sulcus for the tendon of the m. flexor digitorum longus grooves the lip of the medial malleolus at the center of the semilunar notch. On the caudal side of the distal extremity is a much wider sulcus for the tendon of the m. flexor hallucis longus. The lateral surface of the distal extremity of the tibia is in an oblique plane as it slopes caudolaterally. It is slightly flattened by the fibula. At the distal end of the fibular surface is a small facet, the *facies articularis malleoli,* for articulation with the distal end of the fibula. No muscles attach to the distal half of the tibia.

FIBULA

The fibula (see Figs. 4–93 to 4–95) is a long, thin, laterally compressed bone located in the lateral part of the crus. It articulates with the caudolateral part of the lateral condyle of the tibia proximally and with the

tibia and tibial tarsal bone distally. It serves mainly for muscle attachment, as it supports little weight. It is divided into a head, or proximal extremity; body, or shaft; and lateral malleolus, or distal extremity.

The **head** *(caput fibulae)* is flattened transversely, being expanded beyond the planes through the borders of the body both cranially and caudally. A small tubercle, which is articular, projects from its medial surface, facing proximomedially. This small facet, the *facies articularis capitis fibulae,* articulates with a similar one on the caudolateral part of the lateral condyle of the tibia.

The **body of the fibula** *(corpus fibulae)* is slender and irregular. Its distal half is flattened transversely; its proximal half is also thin transversely, but is slightly concave facing medially. Near its middle it is roughly triangular in cross section. The proximal half of the body of the fibula is separated from the tibia by a considerable interosseous space. The cranial margin of the fibula is the **interosseous border** *(margo interosseus).* It runs straight distally and disappears at about the middle of the fibula, where the bone widens as it contacts the tibia. The interosseous membrane, which in life stretches across the interosseous space, attaches to this border or to the rounded ridge of bone which lies adjacent to it, facing the tibia. The proximal half of the fibula may be twisted; the distal half is wider, thinner, and more regular than the proximal half. The **medial surface** *(facies medialis)* is rough, as it lies closely applied to the tibia. A fine, proximally directed nutrient foramen pierces the middle of its medial surface. The **lateral surface** *(facies lateralis)* is smooth, as it lies embedded in the muscles of the crus. The distal end of the fibula is known as the **lateral malleolus.** Medially, it contains the articular surface, *facies articularis malleoli,* which slides on the lateral surface of the trochlea of the tibial tarsal bone. The distal border of the lateral malleolus is thin and flat. Its caudal angle contains a distinct groove, the *sulcus malleolaris lateralis,* through which run the tendons of the mm. extensor digitorum lateralis and fibularis brevis. The muscles which attach to various parts of the fibula include: head of fibula—m. flexor digitorum longus; the head and adjacent shaft—mm. extensor digitorum lateralis and fibularis longus; the medial part of the proximal end — m. tibialis, caudalis; caudal surface of proximal three-fifths — m. flexor hallucis longus; cranial border between proximal and middle thirds — m. extensor hallucis longus; distal two-thirds — m. fibularis brevis.

The Hindpaw

The skeleton of the hindpaw (*pes*) (Figs. 4–96 to 4–103) is composed of the tarsus, metatarsus, phalanges, and the sesamoid bones associated with the phalanges. The tarsus is composed of bones basically arranged in two transverse rows. Articulating with the distal surfaces of the most distally located tarsal bones are the four (sometimes five) metatarsal bones. Each of the four main metatarsal bones bears three phalanges which, with their associated sesamoid bones, form the skeleton of each of the four digits. The first digit, or *hallux*, is usually absent in the dog. When it is fully developed, as it is in some breeds, it contains only two phalanges. The first digit of the hindpaw is known as the **dewclaw**, regard-

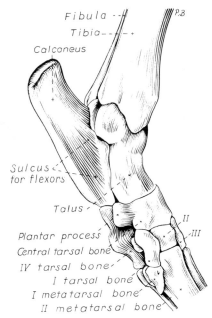

Figure 4–96. Left tarsus, articulated, medial aspect.

less of its degree of development. Except for the first digit, the skeleton of the hindpaw distal to the tarsus closely resembles the comparable part of the forepaw.

Text continued on page 220

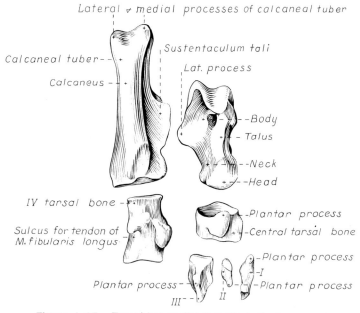

Figure 4–97. Tarsal bones disarticulated, plantar aspect.

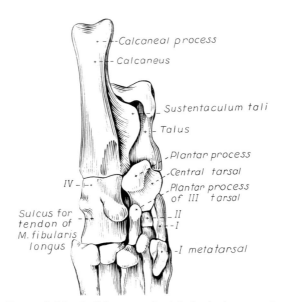

Figure 4-98. Left tarsus, articulated, plantar aspect.

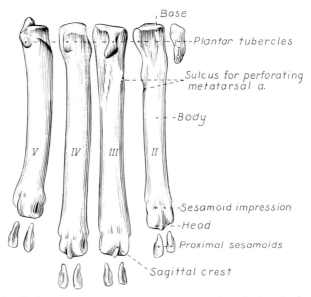

Figure 4-99. Left metatarsal and sesamoid bones, disarticulated, plantar aspect.

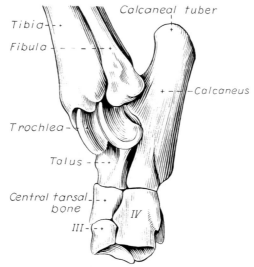

Figure 4–100. Left tarsus, articulated, lateral aspect.

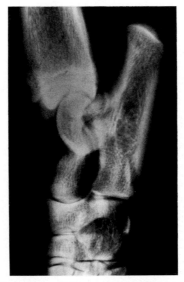

Figure 4–100. Radiograph.   Oblique radiograph of left tarsus, articulated.

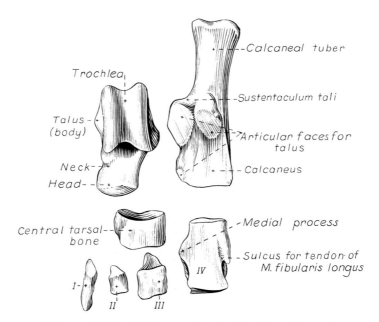

Figure 4–101.   Left tarsus, disarticulated, dorsal aspect.

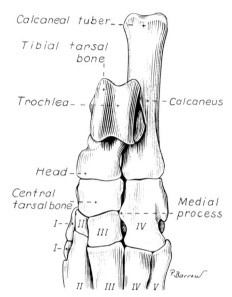

Calcaneal tuber

Tibial tarsal bone

Trochlea

Calcaneus

Head

Central tarsal bone

Medial process

I — II

III

IV

I

II III IV V

P. Barrow

Figure 4–102. Left tarsus, articulated, dorsal aspect.

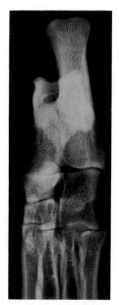

Figure 4–102. Radiograph. Dorsoplantar radiograph of left tarsus, articulated.

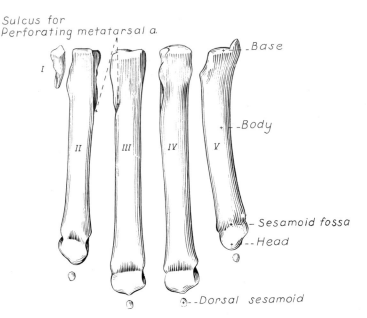

Sulcus for
Perforating metatarsal a.

I

II III IV V

Base

Body

Sesamoid fossa

Head

Dorsal sesamoid

Figure 4–103. Left metatarsal and sesamoid bones, disarticulated, dorsal aspect.

TARSUS

The tarsus, or hock, consists of seven **tarsal bones** *(ossa tarsi)*. The term also applies collectively to the several joints between the tarsal bones, as well as the region between the crus and the metatarsus. The tarsal bones of the dog are arranged in such a way that the tibia and fibula articulate with only the tibial tarsal bone. The tarsus is more than three times as long as the carpus, and the distance between its most proximal and its most distal articulation may be 9 cm. The long, laterally located fibular tarsal bone and the shorter, medially located tibial tarsal bone make up the proximal row. The distal row consists of four bones. Three small bones, the first, second, and third tarsal bones, are located side by side and are separated from the proximal row by the central tarsal bone. The large fourth tarsal bone, which completes the distal row laterally, is as long as the combined lengths of the third and central tarsal bones against which it lies.

The **talus** or tibial tarsal bone is the second largest of the tarsal bones. It articulates proximally with the tibia and fibula, distally with the central tarsal, and on the plantar side with the fibular tarsal. The tibial tarsal may be divided for descriptive purposes into a head, neck, and body. The **body** *(corpus tali)* forms the proximal half of the bone. The most prominent feature of the body is the **proximal trochlea** *(trochlea tali proximalis),* the surface which articulates with the sagittal grooves and the intermediate ridge of the distal articular surface of the tibia. The sides of the trochlea articulate with the medial and lateral malleoli and are known as the *facies malleolaris medialis* and *facies malleolaris lateralis,* respectively. The tibial tarsal articulates with the fibular tarsal by three distinct and separate facets. The large, concave, **proximal articular surface** of the tibial tarsal bone *(facies articularis calcanea proximalis)* is plantarolaterally located. The lateral part of this facet is located on a large right-angled process which is articular on three sides. It is the **lateral process of the talus** *(processus lateralis tali).* The oval **middle articular surface** *(facies articularis calcanea media)* is separated from the distal part of the dorsal articular surface by the deep but narrow *sulcus tali.* The smallest articular surface for the fibular tarsal bone is located on the extreme distolateral part of the tibial tarsal. It is the **distal articular surface** *(facies articularis calcanea distalis).* The **head** *(caput tali)* of the tibial tarsal bone is the transversely elongated distal extremity. The distal surface is rounded, and irregularly oval transversely, and contacts only the central tarsal to form the **articular surface for the** central tarsal (*facies articularis centralis*). The **neck** (*collum tali*) unites the large, proximally located body with the head. It is smooth and convex medially, and lies directly under the skin.

The **calcaneus,** also called the os calcis or fibular tarsal bone, is the largest and longest bone of the tarsus. The distal half of the bone is wide transversely and possesses three facets and two processes whereby it is mortised with the tibial tarsal bone to form a very stable joint. The *tuber calcanei,* or proximal half of the bone, is a sturdy traction process which serves for the insertion of the calcanean tendon. Its slightly bulbous free end contains the **medial and lateral processes,** which are separated by a wide groove. A jutting shelf, the *sustentaculum tali,* leaves the medial side of the bone. On the plantar side of this process is a wide, shallow groove over which the tendon of the m. flexor hallucis longus glides. On the dorsomedial side is a concave, oval facet, the *facies articularis talaris media,* for articulation with the middle articular surface of the tibial tarsal. The dorsal articular surface, *facies articularis talaris dorsalis,* is convex as it articulates with the comparable surface of the tibial tarsal. The most distal and the smallest articular surface on the dorsal part of the bone is the *facies articularis talaris distalis.* This surface is confluent with a small articular facet for the central tarsal on the distal surface. Between the middle and distal articular surfaces is the **calcanean sulcus** (*sulcus calcanei*). This sulcus concurs with a similar one of the tibial tarsal to form the **tarsal sinus** (*sinus tarsi*). On the distal end of the fibular tarsal is a large flat *facies articularis cuboidea,* for articulation mainly with the central tarsal and by a small facet with the tibial tarsal.

The **central tarsal bone** (*os tarsi centrale* or *os naviculare*) lies in the medial part of the tarsus between the proximal and distal rows. It articulates with all of the other tarsal bones. Proximally it articulates with the tibial tarsal by a large, concave, roughly oval area. On the proximal surface of the plantar process of the bone, *tuberositas plantaris,* is a small facet for articulation with the fibular tarsal. The central tarsal articulates distally with the first, second, and third tarsals, and laterally with the proximal half of the fourth tarsal.

The **first tarsal bone** (*os tarsale I,* or *os cuneiforme mediale*) varies greatly in development. When it does not exist as a separate bone, it is fused with the distally lying first metatarsal bone. It is always compressed transversely. When it is fused with the first metatarsal, it forms a rough, bent plate. The first tarsal bone normally articulates with the central tarsal, the second tarsal, and the first metatarsal. Occasionally the first tarsal bone articulates with the second metatarsal. Other possible variations are described in the discussion of the first digit of the hindpaw, under Phalanges.

The **second tarsal bone** (*os tarsale II,* or *os cuneiforme intermedium*) is the smallest of the tarsal bones. It is a wedge of bone which extends toward the plantar side only a short distance. It articulates with the central tarsal proximally, the third tarsal laterally, the first tarsal medially, and the second metatarsal distally. The joint with the second metatarsal is at a higher level than the similar joints lateral to it.

The **third tarsal bone** (*os tarsale III,* or *os cuneiforme laterale*) is nearly three times larger and two times longer than the second tarsal bone. It articulates proximally with the central tarsal, laterally with the fourth tarsal, distally with the third metatarsal, and medially with the second tarsal and metatarsal. On the plantar side it ends in a rounded plantar tuberosity which is embedded in the joint capsule.

The **fourth tarsal bone** (*os tarsale IV* or *os cuboideum*) is as long as the combined dimensions of the central and third tarsals, with which it articulates medially. The joint between the fourth and central tarsals slopes upward and outward, whereas the joint with the third tarsal slopes downward and inward. Proximally, the fourth tarsal articulates mainly with the fibular tarsal and slightly with the tibial tarsal on its dorsomedial edge. Medially, the fourth tarsal articulates with the central and third tarsals and distally with metatarsals IV and V. The distal half of the lateral surface is widely grooved for the tendon of the m. fibularis longus, forming the *sulcus tendinis m. fibularis longi.* Proximal to the sulcus is the salient **tuberosity of the fourth tarsal bone** (*tuberositas ossis tarsalis quarti*). Distally there are two indistinct rectangular areas, sometimes partly separated by a synovial fossa, for articulation with metatarsals IV and V. All tarsal bones of the distal row possess prominent plantar processes for the attachment of the heavy plantar portion of the joint capusle.

METATARSUS

The term metatarsus refers to the region of the pes, or hindpaw, located between the tarsus and the phalanges. The **metatarsal bones** *(ossa metatarsalia I-V)* resemble the corresponding metacarpal bones in general form. They are, however, longer. The shortest main metatarsal bone, metatarsal II, is about as long as the longest metacarpal bone. The metatarsus is compressed transversely, so that the dimensions of the bases of the individual bones are considerably greater sagitally than they are transversely. Furthermore, as a result of this lateral crowding the areas of contact between adjacent bones is greater and the intermetatarsal spaces are smaller. The whole skeleton of the hindpaw is longer and narrower than that of the forepaw.

The **first metatarsal bone** *(os metatarsale I)* is usually atypical and will be described with the phalanges of the first digit.

Metatarsal bones II, III, IV, and V *(ossa metatarsalia II-V)* are similar. A typical metatarsal bone consists of a proximal **base** *(basis)*, which is transversely compressed and irregular and a **shaft** or **body** *(corpus)*, which in general is triangular proximally, quadrangular at midshaft, and oval distally. Each body possesses one large and several small nutrient foramina which enter the proximal halves of the bones from either the contact or the plantar surface. Oblique grooves on the opposed surfaces of the proximal fourths of metatarsals II and III form a space through which passes the perforating metatarsal artery from the dorsal to the plantar side of the paw. The distal end of each main metatarsal bone, like each corresponding metacarpal bone, has a ball-shaped **head** *(caput)*, which is separated from the body dorsally by a deep transverse **sesamoid fossa** *(fossa sesamoidalis)*. On the plantar part of the head the articular surface is divided in such a way that the area nearer the axis of the paw is slightly narrower and less oblique transversely than the one on the abaxial side. The four mm. interossei arise

from the plantar side of the bases of the main metatarsal bones and intimately cover most of their plantar surfaces. The m. tibialis cranialis inserts on the medial side of the base of metatarsal II and the m. fibularis brevis inserts on the lateral side of the base of metatarsal V.

PHALANGES

The phalanges and sesamoid bones of the hindpaw are so similar to those of the forepaw that no separate description is necessary, except for the bones of digit 1.

The term "**dewclaw**" is applied to the variably developed first digit of the hindpaw of the dog. It should not be applied to the first digit of the forepaw because that appendage, although rudimentary, is always present. Some breeds are recognized by the American Kennel Club (1975) as normally possessing fully developed first digits on their hindpaws. Many individuals of the larger breeds of dogs possess "dewclaws" in various degrees of development (Kadletz 1932). In the most rudimentary condition an osseous element bearing a claw is attached only by skin to the medial surface of the tarsus. The proximal phalanx may be absent, and metatarsal I, much reduced in size, may or may not be fused with the first tarsal. Occasionally, two claws of equal size are present on the medial side of the hindpaw. These supernumerary digits probably have no phylogenetic significance. Complete duplication of the phalanges and metatarsal I is sometimes encountered. The first metatarsal may also be divided into a proximal and a distal portion. The distal metatarsal element is never fused to the proximal phalanx. It may be united to its proximal part by fibrous tissue, or a true joint may exist. Although the dewclaw may be lacking, a rudiment of the first metatarsal is occasionally seen as a small, flattened osseous plate which lies in the fibrous tissue on the medial side of the tarsus.

OS PENIS

The *os penis* (Fig. 4–104), or *baculum*, is always present in the male dog. In large dogs it is about 10 cm. long, 1.3 cm wide, and 1 cm. thick. The bone forms a rigid axis

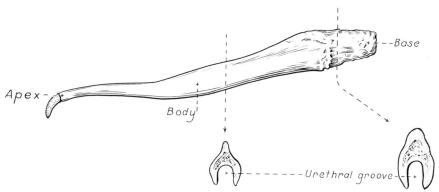

Figure 4–104. Os penis, lateral aspect with two transections.

of the glans penis, passing through the bulbus glandis. The caudal part, or **base** of the os penis, is truncate and attached to the tunic of each cavernous body, whereas the cranial part, or **apex** tapers gradually and ends in a cartilaginous tip, which is attached by a fibrous strand to the corona of the glands. The **body** is long and has as its most distinctive feature a **urethral groove** *(sulcus urethrae)*, which runs ventrally along the base and body of the bone. The urethral groove is of clinical importance because of its obstruction of calculi passing through the urethra. It begins caudally, where its width equals its depth. From the middle of the bone, where the depth of the groove is about 7 mm., it gradually becomes shallower, until it becomes the flattened cranial ventral surface of the bone. The lips of the groove are about 4 m.. apart over the base; then they converge, leaving a space about 2 mm. wide between them in the middle of the bone. They gradually diverge as the groove becomes shallower on the ventrally flattened cranial portion. The urethral groove contains the urethra and the corpus spongiosum which surrounds it. Throughout the length of the bone there are many foramina pitting its surface. These foramina are particularly large and numerous in the caudal half. The external surface of the os penis is rough and tuberculate on the base but smooth over the body. Some specimens show, on their lateral surfaces, dorsally arched vascular grooves which are formed by the dorsal veins of the penis.

Many female carnivores have a homologous bone, known as the **os clitoridis.** The dog, however, usually lacks this element. When present, it is often associated with an endocrine imbalance.

The morphology of the canid baculum has been described by Pohl (1911), Chaine (1926), and Hildebrand (1954). Its structure is extremely variable, being more massive in the large breeds.

BIBLIOGRAPHY

American Kennel Club. 1975. The Complete Dog Book: The Photograph, History and Official Standard of Every Breed Admitted to AKC registration, and the selection, training, breeding, care and feeding of pure-bred dogs. 15th ed. 672 pp.

Badoux, D. M. 1969. Biostatics of the cervical vertebrae in domesticated dogs. Proc. Kon. Nederl. Akad. Wet. Biol Med. 72:478–490.

– – – – –. 1975. General biostatics and biomechanics. *In* Sisson and Grossman's Anatomy of the Domestic Animals. 5th Ed., by R. Getty. Philadelphia, W. B. Saunders Co.

Barone, R. 1976. Anatomie Comparée des Mammiferes Domestiques. I, Osteologie. Paris, Vigot Freès, 580 pp.

Baum, H., and O. Zietzschmann. 1936. Handbuch der Anatomie des Hundes. Band I: Skelette- und Muskel-system. Berlin. Paul Parey.

de Beer, G. R. 1937. The Development of the Vertebrate Skull, London, Oxford University Press.

Bolk, L., et al. 1931–1939. Handbuch der vergleichenden Anatomie der Wirbeltiere. 6 vols. Berlin and Vienna, Urban and Schwartzenberg.

Bourdelle, E., and C. Bressou. 1953. Anatomie régionale des Animaux Domestiques. IV. Carnivores. Chien et Chat. Paris, J.-B. Baillière.

Bourne, G. H. 1972. The Biochemistry and Physiology of Bone. 2nd Ed. Vols. I, II, III. 1976, Vol IV, Calcification and Physiology. New York, Academic Press.

Bressou, C., N. Pomriaskinsky-Kobozieff, and N. Kobozieff. 1957. Étude radiologique de l'ossification du squelette du pied du chien aux divers stade de son évolution, de la naissance à l'âge adulte. Rec. Méd. Vet. Alfort *133*:449–464.

Cave, A. J. E. 1975. The morphology of the mammalian cervical pleurapophysis. J. Zool. London *177*:377–393.

Chaine, J. 1926. L'os pénien; étude descriptive et comparative. Actes Soc. Linné. Bordeaux *78*:1–195.

Chauveau, A. 1891. The Comparative Anatomy of the Domesticated Animals. Rev. by S. Arloing. 2nd English Ed. translated and edited by G. Fleming. New York, Appleton Co.

Clutton-Brock, J., G. B. Corbet, and M. Hills, 1976. A review of the family Canidae with a classification by numerical methods. Bull. Brit. Museum Zool. *29*(3): 1–99.

Dixon, T. F., and H. R. Perkins, 1956. The Chemistry of Calcification. Chap. X, pp. 287–317, *in* The Biochemistry and Physiology of Bone, edited by G. H. Bourne. New York, Academic Press.

Ellenberger, W., and H. Baum. 1943. Handbuch der vergleichenden Anatomie der Haustiere. 18th Ed. Berlin, Springer.

Flower, W. H. 1870. An Introduction to the Osteology of Mammalia, London, Globe.

Getty, R. 1975. Sisson and Grossman's Anatomy of the Domestic Animals. 5th Ed. 2 vols. Philadelphia, W. B. Saunders Co.

Haag, W. G. 1948. An osteometric analysis of some aboriginal dogs. Univ. of Kentucky Reports in Anthropology VII(3):107–264.

Haines, R. W. 1942. The evolution of epiphyses and of endochondral bone. Biol Rev. *17*:267–292.

Hall, B. K. 1978. Developmental and Cellular Skeletal Biology. New York, Academic Press.

Hancox, N. M. 1972. Biology of Bone, Cambridge, Cambridge Univ. Press.

Hansen, H. 1952. A pathologic-anatomical study on disc degeneration in the dog. Acta Orth. Scandinav. Suppl. *11*:1–117.

Hare, W. C. D. 1961. Radiographic anatomy of the cervical region of the canine vertebral column. J.A.V.M.A. *139*:209–220.

— — — — —. 1961. The ages at which the centers of ossification appear roentgenographically in the limb bones of the dog. Am. J. Vet. Res. *22*:825–835.

Hedhammer, A., et al. 1974. Overnutrition and skeletal disease. An experimental study in growing Great Dane dogs. Cor. Vet. *64*(2) (Suppl. 5):1–160.

Hildebrand, M. 1954. Comparative morphology of the body skeleton in recent canidae. Univ. of Calif. Pub. Zool. *52*:399–470.

Hughes, H. V., and J. W. Dransfield. 1953. McFadyean's Osteology and Arthrology of the Domesticated Animals. London, Baillière, Tindall & Cox.

Iwanoff, S. 1935. Variations in the ribs and vertebrae of the dog. Jb. Vet. Med. Fat. Sofia (Bulg.) *10*:461–497.

Jayne, H. 1898, Mammalian Anatomy; Part I. The Skeleton of the Cat. Philadelphia, J. B. Lippincott Co.

Kadletz, M. 1932. Anatomischer Atlas der Extremitätengelenke von Pferd und Hund. Berlin, Wien, Urban & Schwarzenberg.

Kuntz, A., and C. A. Richins, 1945. Innervation of the bone marrow. J. Comp. Neurol. *83*:213–222.

Lumer, H. 1940. Evolutionary allometry in the skeleton of the domesticated dog. Am. Nat. *74*:439–467.

Maier, V. 1928. Untersuchungen über die Pneumatizität des Hunde-schädels mit Berücksichtigung der Rassenunterschiede. Ztschr. Anat. u. Entw. *85*:251–286.

McLean, F. C., and M. R. Urist. 1968. Bone Fundamentals of the Physiology of Skeletal Tissue. 3rd Ed. Chicago, Univ. Chicago Press. 314 pp.

Murray, P. D. F. 1936. Bones; A Study of the Development and Structure of the Vertebrate Skeleton. Cambridge, Univ. Press.

Nickel, R., A. Schummer, E. Seiferle, J. Frewin, and K. H. Wille. 1977. Lehrbuch der Anatomie der Haustiere. Band 1: Bewegungsapparat. Berlin, Paul Parey.

Owen, R. 1866. On the Anatomy of Vertebrates. 3 vols. London, Longmans and Green.

Parker, A. J., and R. D. Park. 1974. Occipital dysplasia in the dog. J. Amer. Anim. Hosp. Assoc. *10*:520–525.

Pohl, L. 1911. Das Os penis der Carnivoren einschliesslich der Pinnipedia. Jen. Ztschr. Naturw. *47*:115–160.

Pomriaskinsky-Kobozieff, N., and N. Kobozieff. 1954. Étude radiologique de l'aspect du squelette normal de la main du chien aux divers stades de son évolution, de la naissance à l'âge adulte. Rec. Méd. Vet. Alfort *130*:617–646.

Rasmussen, H., and P. Bordier. 1974. The physiological and cellular basis of metabolic bone disease. Baltimore, The Williams & Wilkins Co.,

Reynolds, S. H. 1913. The Vertebrate Skeleton. 2nd Ed. Cambridge, Cambridge Univ. Press.

Roberts, S. 1956. Veterinary Obstetrics and Genital Diseases. Ithaca, N.Y., Pub. by Author.

Romer, A. S., and T. Parsons. 1978. The Vertebrate Body. 5th Ed. Philadelphia, W. B. Saunders Co.

Schaeffer, H. 1934. Die Ossifikationsvorgänge im Gliedmassenskelett des Hundes. Morph. Jahrb. *74*:472–512.

Schebitz, H., and H. Wilkins. 1977. Atlas of Radiographic Anatomy of the Dog and Cat. Berlin, Paul Parey.

Slijper, E. J. 1946. Comparative biologic-anatomical investigations on the vertebral column and spinal musculature of mammals. Kon. Ned. Akad. Wet., Verh. (Tweede Sectie) *42*(5):1–128.

Smith, R. N. 1960. Radiological observations of the limbs of young greyhounds. J. Small Anim. Pract. *1*(2):84–90.

Smith, R. N., and J. Allcock. 1960. Epiphysial fusion in the greyhound. Vet. Rec. *72*(5):75–79.

Stockard, C. R. 1941. The Genetic and Endocrinic Basis for Differences in Form and Behavior. Amer. Anat. Memoir 19. Philadelphia, Wistar Institute of Anatomy and biology.

Watson, A. G., and H. E. Evans. 1976. The development of the atlas-axis complex in the dog. Anat. Rec. *184*:558 (w/illustration).

Weinmann, J. P., and H. Sicher, 1947. Bone and Bones; Fundamentals of Bone Biology. St. Louis, The Mosby Co.

Williams, E. E. 1959. Gadows arcualia and the development of tetrapod vetebrae. Q. Rev. Biol. *34*:1–32.

Zipkin, I. 1973. Biological Mineralization. New York, John Wiley & Sons.

Chapter 5

JOINTS AND LIGAMENTS

GENERAL

Articulations or **joints** *(articulationes [juncturae ossium])* are formed when two or more bones are united by fibrous, elastic, or cartilaginous tissue, or by a combination of these tissues. Three main groups are recognized and named according to their most characteristic structural features. Where little movement is required, the union is short, direct, and often transitory. A **fibrous joint** *(junctura fibrosa),* formerly known as a synarthrosis, is one of this nature; such joints include syndesmoses, sutures, and gomphoses. A **cartilaginous joint** *(junctura cartilaginea),* formerly known as an amphiarthrosis, permits only limited movement, such as compression or stretching. A **synovial joint** *(junctura synovialis),* or true joint, formerly known as a diarthrosis, facilitates mobility. The studies of Kadletz (1932) provide detailed information on the arthrology of the dog, and the well-documented work of Barnett, Davies, and MacConaill (1961) discusses the structure and mechanics of synovial joints in considerable detail.

The term "Syndesmologia" was used in the Basel Nomina Anatomica (B.N.A.) of 1895 for the joints and ligaments of the body. This was changed to "Arthrology" in the Birmingham Revision (B.R.) of 1933 and back to the original in Paris (N.A.P.) in 1955. The latest revision of Nomina Anatomica in Tokyo in 1955 adopted "Arthrologia" as the most appropriate heading and "Articulatio" in place of "Junctura" (4th Ed., N.A., 1977). It should be noted that the similar-sounding term "Syndesmosis" is used to denote one type of fibrous joint.

Fibrous Joints

A **syndesmosis** is a fibrous joint with a considerable amount of intervening connective tissue. The attachment of the hyoid apparatus to the petrous temporal bone is an example of a syndesmosis.

A **suture** *(sutura)* is a fibrous joint of the type which is confined largely to the flat bones of the skull. Depending on the shape of the apposed edges, sutures are further divided into: (1) **serrate suture** *(sutura serrata),* one which articulates by means of reciprocally alternating processes and depressions; (2) **squamous suture** *(sutura squamosa),* one which articulates by overlapping of reciprocally beveled edges; (3) **plane suture** *(sutura plana),* one in which the bones meet at an essentially right-angled edge or surface; and (4) **foliate suture** *(sutura foliata),* one in which the edge of one bone fits into a fissure or recess of an adjacent bone. Serrate sutures are found where stable noncompressible joints are needed, such as the parieto-occipital and the interparietal union. Where a slight degree of compressibility is advantageous, such as is required in the fetal cranium at birth, squamous sutures are found. Similarly, the frontonasal and frontomaxillary squamous sutures allow enough movement to absorb the shock of a blow which might otherwise fracture the bones of the face. Examples of plane sutures are those of the ethmoid and those between most of the bones of the face. Where extreme stability is desirable, foliate sutures are formed. The best example of this type is the zygomaticomaxillary suture. The various fibrous sutures of the skull also permit growth to take place at the periphery of the bones.

The implantation of a tooth in its socket is by means of a fibrous union known as a **gomphosis.** This specialized type of fibrous joint is formed by the periodontal ligament, which attaches the cementum of the tooth to the alveolar bone of the socket and permits slight movement while at the same time it provides firm attachment.

Cartilaginous Joints

Many bones are united by cartilaginous joints, which are sometimes referred to as synchondroses. Unions of this type may be formed by hyaline cartilage, by fibrocartilage, or by a combination of the two, and they are subject to change with increasing age.

Hyaline cartilage joints, or primary joints, are usually temporary and represent persistent parts of the fetal skeleton or secondary cartilage of growing bones. The epiphysis of an immature long bone is united with the diaphysis by a cartilaginous epiphyseal plate. When adult stature is reached, osseous fusion occurs and a joint no longer exists, although a slight epiphyseal line may mark the union. This osseous union, in some anatomical works, is called a synostosis. Similar transitory hyaline cartilage joints are typical of the union of the shafts with the femoral trochanters or the humeral tubercles, and of the spheno-occipital synchondrosis. Some hyaline cartilage joints, such as the costochondral junctions, remain throughout life.

Fibrocartilaginous joints, or secondary joints, are sometimes referred to as amphiarthroses. The best examples of such joints are those of the pelvic symphysis, mandibular symphysis, sternebrae, and vertebral bodies. The fibrocartilage uniting these bones may have an intervening plate of hyaline cartilage at each end. Occasionally these joints may ossify, as do hyaline cartilage joints.

Synovial Joints

The true joints of the extremities permit the greatest degree of movement and are most commonly involved in dislocations. All **synovial joints** *(juncturae synoviales)* are characterized by a **joint cavity** *(cavum articulare),* a **joint capsule** *(capsula articularis)* including an outer fibrous and an inner synovial membrane, **synovial fluid** *(synovia),* and **articular cartilage** *(cartilago articularis).* A few of the synovial joints have modifications peculiar to the functions they perform and may possess intra-articular ligaments, menisci, fat pads, or synovial projections in the form of plicae or villi.

The blood supply of synovial joints is provided by an arterial and venous network from parent trunks in the vicinity of the joint. The vessels supply the capsule and also the epiphyses bordering the joint. Around the articular margins the blood vessels of the synovial membrane form anastomosing loops, referred to collectively as the *circulus articuli vasculosus.*

Lymphatic vessels are also present in synovial membrane and account for the rapid removal of some substances from the joint cavity (Bauer, Short, and Bennett 1933.)

The nerve supply of synovial joints is derived from peripheral or muscular branches in the vicinity of the joint. Included in these articular nerves are proprioceptive fibers, pain receptor fibers, and sympathetic fibers related to vasomotor or vasosensory functions. Some areas of the joint capsule are more richly innervated than others. Gardner (1950) reviewed the morphology and physiology of joints in the human, including their innervation, and cites over 500 references. Ansulayotin (1960) studied the nerves which supply the appendicular joints in the dog.

STRUCTURE OF SYNOVIAL JOINTS

The **joint capsule** is composed of an inner synovial membrane and an outer fibrous membrane. The **synovial membrane** *(membrana synovialis)* is a vascular connective tissue which lines the inner surface of the capsule and is responsible for the production of synovial fluid. The synovial membrane does not cover the articular cartilage but blends with the periosteum as it reflects onto the bone. Joint capsules may arise postnatally if the need exists, and thus false joints often form following unreduced fractures. Synovial membrane covers all structures within a synovial joint except the articular cartilage and the contact surfaces of fibrocartilaginous plates. Synovial membrane also forms sleeves around intra-articular ligaments and covers muscles, tendons, nerves, and vessels if these cross the joint closely. Adipose tissue often fills the irregularities between articulating bones, and in some instances it is aspirated into or squeezed out of the joint as the surfaces of the articulating bones part or come together during movement. Fat in such locations is covered by synovial membrane. A **synovial fold** *(plica synovialis)* is an extension of the synovial membrane; such folds usually contain fat. Around the periphery of some synovial joints the synovial membrane is in the form of numerous processes, or **synovial villi** *(villi synoviales)*. These are soft and velvety. The synovial membrane may extend beyond the fibrous layer and act as a bursa under a tendon or ligament, or even be in communication with a synovial sheath.

The **fibrous membrane** *(membrana fibrosa)* of a joint capsule is composed mainly of white fibrous tissue containing yellow elastic fibers. It is also known as the capsular ligament. In some synovial joints the true ligaments are quite separate from the fibrous capsular ligament, such as the patellar ligaments of the stifle joint, but in most joints the ligaments are thickenings of the fibrous portion of the joint capsule. In those joints where great movement occurs in a single plane the fibrous membrane is usually thin and loose on the flexor and extensor surfaces, and thick on the sides of the bone which move the least. Such thickenings of the fibrous layer are known as **collateral ligaments** *(ligg. collateralia)* and are present to a greater or lesser degree in all hinge joints. The fi-

brous membrane attaches at the margin of the articular cartilage, or at most 3 cm. from it, where it blends with the periosteum.

The **synovial fluid** *(synovia)* serves chiefly to lubricate the contact surfaces of synovial joints. In all cases these surfaces are hyaline cartilage or fibrocartilage. Fibrocartilage contains few blood vessels and nerves, and hyaline cartilage has neither. Therefore, the synovial fluid serves the additional function of transporting nutrient material to the hyaline cartilage and removing the waste metabolites from it. Synovia also enables the wandering leucocytes to circulate in the joint cavity and phagocytize the products of the wear and tear of the articular cartilage. In many joints of man there is little, if any, free synovia. The average volume in the stifle joint of adult dogs of various sizes varies from 0.2 ml. to 2 ml. The general health and condition of the dog has a marked influence on the amount of synovia present in the joints. Synovia is thought to be a dialysate, although mucin is probably produced by the fibroblasts of the synovial membrane (Davies 1944). The chemical composition of synovia closely resembles that of tissue fluid. In addition to mucin, it contains salts, albumin, fat droplets, and cellular debris. The quantitative composition of synovia is largely dependent on the type of tissue underlying the surface fibroblasts and the degree of vascularity of this tissue (Coggeshall et al. 1940).

The **articular cartilage** *(cartilago articularis)* is usually hyaline cartilage. It covers the articular surfaces of bones where its deepest part may be calcified. It contains no nerves or blood vessels, although it is capable of some regeneration after injury or partial removal (Bennett, Bauer, and Maddock 1932). It receives its nutrition from the synovia. Because of its mucin content, the synovia forms a viscous capillary film on the articular cartilage. The articular cartilage varies in thickness in different joints and in different parts of the same joint. It is thickest in young, healthy joints and in joints which bear considerable weight. Its thickness in any particular joint is in direct proportion to the weight borne by the joint, and it may atrophy from disuse. Healthy articular cartilage is translucent, with a bluish sheen. Elasticity and compressibility are necessary physical properties which it possesses. This resiliency guards against fracture of bone by absorbing shock.

A **meniscus** *(meniscus articularis),* or **disc** *(discus articularis),* is a complete or partial fibrocartilaginous plate which divides a joint cavity into two parts. The temporomandibular joint contains a thin, but complete, disc, and, because the capsular ligament attaches to the entire periphery of the disc, the joint cavity is completely divided into two parts. Two menisci are found in the stifle joint, and neither is complete, thus allowing all parts of the joint cavity to intercommunicate. Menisci have a small blood and nerve supply, and are capable of regeneration (Dieterich 1931). Their principal function, according to MacConaill (1932), is ". . . to bring about the formation of wedge-shaped films of synovia in relation to the weight-transmitting parts of joints in movement." An obvious function is the prevention of concussion. The stifle and temporomandibular joints are the only synovial joints in the dog which possess menisci, or discs.

A **ligament** *(ligamentum)* is a band or a cord of nearly pure collagenous tissue which unites two or more bones. The term has also been used to designate remnants of fetal structures and relatively avascular narrow serous membrane connections. Ligaments, as used in this chapter, unite bone with bone. Tendons unite muscle with bone. Ligaments may be intracapsular (stifle and hip joints) or extracapsular where they are developed within or in relation to the capsular ligament. They are heaviest on the side of joints where the margins of the bones do not separate but glide on each other. Hinge joints with the greatest radii of movement have the longest ligaments. The ligaments often widen at their attached ends, where they blend with the periosteum. Histologically, ligaments are largely composed of long parallel or spiral collagenous fibers, but all possess some yellow elastic fibers also. The integrity of most joints is ensured by the ligaments, but in some (shoulder and hip) the heavy muscles which traverse the joints play a more important part than do the ligaments. Such muscles and their tendons are sometimes spoken of as active ligaments. In hinge joints ligaments limit lateral mobility, and some (cruciate ligaments of the stifle joint) limit folding and opening of the joint as well. In certain ball-and-socket synovial joints the sockets are deepened by ridges of dense fibrocartilage, known as **glenoid lips** *(labia glenoidalia).*

PATHOLOGY

Articular separations are spoken of as **luxations** or dislocations. Although most luxations are due to injury or degenerative changes, there are also predisposing genetic factors (often breed-specific) which play an important role.

CLASSIFICATION OF SYNOVIAL JOINTS

Syovial joints may be classified according to (1) the number of articulating surfaces involved, (2) the shape or form of the articular surfaces, or (3) the function of the joint (Barnett, Davies, and MacConaill 1961).

According to the number of articulating surfaces a joint is either **simple** *(articulatio simplex)* or **compound** *(articulatio composita).* A simple joint is formed by two articular surfaces within an articular capsule. When more than two articular surfaces are enclosed within the same capsule, the joint is compound.

The classification of synovial joints (Nomina Anatomica Veterinaria 1973) is based on the shape or form of the articular surfaces. There are seven basic types:

A **plane joint** *(articulatio plana)* is one in which the articular surfaces are essentially flat. It permits a slight gliding movement. An example would be the costotransverse joint.

A **ball-and-socket joint** *(articulatio spheroidea)* is formed by a convex hemispherical head which fits into a shallow glenoid cavity (shoulder joint) or into a deep cotyloid cavity (hip joint).

An **ellipsoidal joint** *(articulatio ellipsoidea)* is similar to a spheroidal joint. It is characterized by an elongation of one surface at a right angle to the other, forming an ellipse. The reciprocal convex (male) and concave (female) elongated surfaces of the radiocarpal articulation form an ellipsoidal joint.

A **hinge joint** (*ginglymus*) permits flexion and extension with a limited degree of rotation. The most movable surface of a hinge joint is usually concave. An example would be the elbow joint.

A **condylar joint** (*articulatio condylaris*) resembles a hinge joint in its movement but differs in structure. The surfaces of such a joint include rounded prominences or condyles which fit into reciprocal depressions or condyles on the adjacent bone, resulting in two articular surfaces usually included in one articular capsule. Examples of condylar joints include the temporomandibular joint and the knee joint. The knee, or stifle, joint is best classified as a complex condylar joint, since it possesses an intra-articular fibrocartilage which partially subdivides the intra-articular cavity.

A **trochoid** (*articulatio trochoidea*), or **pivot joint,** is one in which the chief movement is around a longitudinal axis through the bones forming the joint. The median atlantoaxial joint and the proximal radioulnar joint are examples of trochoid joints.

A **saddle joint** (*articulatio sellaris*) is characterized by opposed surfaces each of which is convex in one direction and concave in the other, usually at right angles. When opposing joint surfaces are concavo-convex, the main movements are also in planes which meet at right angles. The interphalangeal joints are examples of this type of articulation.

MOVEMENTS OF SYNOVIAL JOINTS

Joint movements that are brought about by the contraction of muscles which cross the joints are known as active movements; those joint movements caused by gravity or secondarily by the movement of some other joint or by an external force are known as passive movements. Synovial joints are capable of diverse movements. **Flexion,** or folding, denotes moving two or more bones so that the angle between them becomes less than 180 degrees. **Extension,** or straightening, denotes movement by which the angle is increased to

180 degrees. It is readily seen that some joints such as the metacarpophalangeal and metatarsophalangeal joints are in a state of overextension. This is also called dorsal flexion. When an animal "humps up," it flexes its vertebral column. Some parts of the vertebral column are normally in a state of flexion (the joints between the first few caudal vertebrae), whereas others are in a state of overextension (the joints between the last few cervical vertebrae). Flexion and extension occur in the sagittal plane unless the movement is specifically stated to be otherwise (right or left lateral flexion of the vertebral column). **Adduction** is the term applied to moving an extremity toward the median plane or a digit toward the axis of the limb. **Abduction,** or taking away, is the opposite movement. **Circumduction** occurs when an extremity follows in the curved plane of the surface of a cone. **Rotation** is the movement of a part around its long axis.

LIGAMENTS AND JOINTS OF THE SKULL

TEMPOROMANDIBULAR JOINT

The **temporomandibular joint** (*articulatio temporomandibularis*) (Fig. 5–1) is a condylar joint which allows considerable sliding movement. The transversely elongated condyle of the mandible does not correspond entirely to the articular surface of the mandibular fossa of the temporal bone. A thin **articular disc** (*discus articularis*) lies between the cartilage-covered articular surface of the condyloid process of the mandible and the similarly covered mandibular fossa of the temporal bone. The loose **joint capsule** extends from the articular cartilage of one bone to that of the other. On the temporal bone the capsular ligament also attaches to the retroarticular process. It attaches to the entire edge of the disc as it passes between the two bones. The joint cavity is thus completely divided into a dorsal *compartment*, between the disc and temporal bone, and a ventral *compartment*, between the disc and mandible. Laterally the fibrous part of the joint capsule is strengthened by fibrous stands to form the **lateral ligament** (*lig. laterale*).

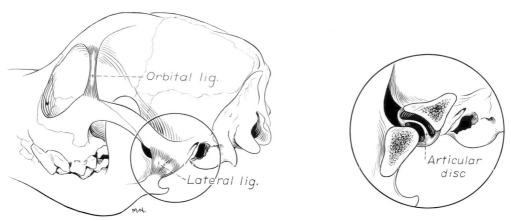

Figure 5–1. Temporomandibular joint, lateral aspect and sagittal section.

Figure 5–1. Radiograph. Lateral radiograph, temporomandibular joint.

The **symphysis of the mandible** *(articulatio intermandibularis)* is the median synchondrosis uniting right and left mandibular bodies. The opposed articular surfaces are interdigitated, and the fibrocartilage of the symphysis persists throughout life.

JOINTS OF AUDITORY OSSICLES

The **joints of the auditory ossicles** *(articulationes ossiculorum auditus)* allow for movement of the malleus, incus, and stapes. The head of the malleus articulates with the body of the incus via a synovial **incudomallear joint** *(articulatio incudo-* *mallearis)*. The lenticular process of the long crus of the incus, with the head of the stapes, likewise forms a synovial joint, which is called the **incudostapedial joint** *(articulatio incudostapedia)*. The footplate, or base, of the stapes articulates with the vestibular window *(fenestra vestibuli)* by means of a fibrous union *(syndesmosis tympanostapedia)*.

The **ligaments of the auditory ossicles** *(ligg. ossiculorum auditus)* function to hold the ossicles in place and to limit their movement. Associated with the malleus is a short **lateral ligament** between the lateral process of the malleus and the tympanic notch, a **dorsal ligament** joining the head of the malleus to the roof of the epitympanic recess, and a short **rostral ligament** connecting the rostral process of the malleus to the osseous tympanic ring. The body of the incus is attached to the roof of the epitympanic recess by a **dorsal ligament,** and the short crus of the incus is attached to the fossa incudis by a **caudal ligament.** The base of the stapes is attached to the margin of the vestibular window by an **annular ligament.**

JOINTS OF HYOID APPARATUS

The tympanohyoid cartilage articulates with the mastoid part of the petrous temporal bone, forming the *articulatio temporohyoidea.* This articulation is adjacent to the stylomastoid foramen. Except for the temporohyoid joint there are tightly fitting synovial cavities between all parts of the hyoid complex, as well as a small synovial cavity between the thyrohyoid bone and the cranial cornu of the thyroid cartilage.

INTERMANDIBULAR JOINT

The *articulatio intermandibularis,* or **mandibular symphysis,** is formed partly by cartilage but mostly by fibrous connective tissue.

SYNCHONDROSES OF THE SKULL

The **synchondroses of the skull** *(synchondroses cranii)* include the following:

Synchondrosis intraoccipitalis squamolateralis
Synchondrosis intraoccipitalis basilateralis
Synchondrosis spheno-occipitalis
Synchondrosis petro-occipitalis
Synchondrosis intersphenoidalis
Synchondrosis sphenopetrosa
Synchondrosis intermandibularis
Synchondrosis temporohyoidea

SUTURES OF THE SKULL

The **sutures of the skull** *(suturae capitis)* are described in the discussion of the individual bones of the skull in Chapter 4. The name of each bone in the following list is followed by the names of the sutures in which it participates.

Occipital Bone
Sutura occipitosquamosa
Sutura occipitomastoidea
Sutura occipitoparietalis

Parietal Bone
Sutura parietointerparietalis
Sutura occipitoparietalis
Sutura coronalis
Sutura squamosa
Sutura sagittalis
Sutura sphenoparietalis

Frontal Bone
Sutura interfrontalis
Sutura coronalis
Sutura sphenofrontalis
Sutura frontonasalis
Sutura frontomaxillaris
Sutura frontolacrimalis
Sutura frontopalatina
Sutura frontoethmoidalis

Sphenoid Bone
Sutura vomerosphenoidalis
Sutura sphenoethmoidalis
Sutura sphenopalatina
Sutura sphenofrontalis
Sutura sphenosquamosa
Sutura sphenoparietalis
Sutura pterygosphenoidalis

Temporal Bone
Sutura squamosa
Sutura sphenosquamosa
Sutura temporozygomatica

Ethmoid Bone
Sutura sphenoethmoidalis
Sutura vomeroethmoidalis
Sutura frontoethmoidalis
Sutura ethmoideomaxillaris
Sutura ethmolacrimalis
Sutura nasoethmoidalis
Sutura palatoethmoidalis

Incisive Bone
Sutura incisivomaxillaris
Sutura vomeroincisiva
Sutura nasoincisiva
Sutura interincisiva

Nasal Bone
Sutura internasalis
Sutura frontonasalis
Sutura nasomaxillaris
Sutura nasoincisiva
Sutura nasoethmoidalis

Maxilla
Sutura incisivomaxillaris
Sutura nasomaxillaris
Sutura frontomaxillaris
Sutura lacrimomaxillaris
Sutura zygomaticomaxillaris
Sutura palatomaxillaris
Sutura palatina mediana (s. intermaxillaris)
Sutura ethmoideomaxillaris
Sutura vomeromaxillaris

Zygomatic Bone
Sutura zygomaticomaxillaris
Sutura zygomaticolacrimalis
Sutura temporozygomatica

Palatine Bone
Sutura palatina mediana
Sutura palatina transversa
Sutura vomeropalatina dorsalis
Sutura vomeropalatina ventralis
Sutura palatomaxillaris
Sutura sphenopalatina
Sutura pterygopalatina
Sutura frontopalatina
Sutura palatoethmoidalis
Sutura palatolacrimalis

Lacrimal Bone
Sutura frontolacrimalis
Sutura lacrimomaxillaris
Sutura zygomaticolacrimalis
Sutura palatolacrimalis
Sutura ethmolacrimalis

Pterygoid Bone
Sutura pterygosphenoidalis
Sutura pterygopalatina

Vomer
Sutura vomerosphenoidalis
Sutura vomeroethmoidalis
Sutura vomeropalatina dorsalis
Sutura vomeropalatina ventralis
Sutura vomeromaxillaris
Sutura vomeroincisiva

LIGAMENTS AND JOINTS OF THE VERTEBRAL COLUMN

ATLANTO-OCCIPITAL ARTICULATION

The **atlanto-occipital joint** *(articulatio atlanto-occipitalis)* is formed by the dorsolaterally extending occipital condyles and the corresponding concavities of the atlas. The spacious **joint capsule** *(capsula articularis)* on each side attaches to the margins of the opposed articular surfaces. Ventromedially the two sides are joined so that an undivided U-shaped joint cavity is formed. The atlantooccipital joint cavity communicates with the atlantoaxial joint cavity. The dorsal and ventral atlanto-occipital membranes reinforce the joint capsule at their respective locations.

The **dorsal atlanto-occipital membrane** *(membrana atlanto-occipitalis dorsalis)* extends between the dorsal edge of the foramen magnum and the cranial border of the dorsal arch of the atlas. Two oblique straplike thickenings, about 8 mm. wide, arise on each side of the notch of the supraoccipital bone, diverge as they run caudally, and attach to the dorsolateral parts of the atlas. In the triangular space formed by these bands, punctures are made for the removal of cerebrospinal fluid from the cisterna magna.

The **ventral atlanto-occipital membrane** *(membrana atlanto-occipitalis ventralis)* and its synovial layer form the uniformly thin joint capsule located between the ventral edge of the foramen magnum and the ventral arch of the atlas.

The **lateral atlanto-occipital ligament** *(lig. atlanto-occipitalis lateralis)* (Fig. 5–2) runs from the lateral part of the dorsal arch of the atlas to the jugular process of the occipital bone. Its course is cranioventrolateral, and its caudal attachment is narrower than its cranial one. Another small ligament runs from each side of the inner surface of the lateral part of the ventral arch of the atlas to the lateral part of the foramen magnum. Ventral and medial to these ligaments the unpaired joint cavities between the skull and the atlas and between the atlas and the axis freely communicate.

ATLANTOAXIAL ARTICULATION

The **atlantoaxial joint** *(articulatio atlantoaxialis)* (Figs. 5–2, 5–3) is a pivot joint which permits the head and atlas to rotate around a longitudinal axis. The **joint capsule** is loose and uniformly thin as it extends from the dorsal part of the cranial articular surface of one side of the axis to a like place on the opposite side. Cranially it attaches to the caudal margins of the sides and ventral arch of the atlas. The fibrous layer of the joint capsule extends from right to left dorsally

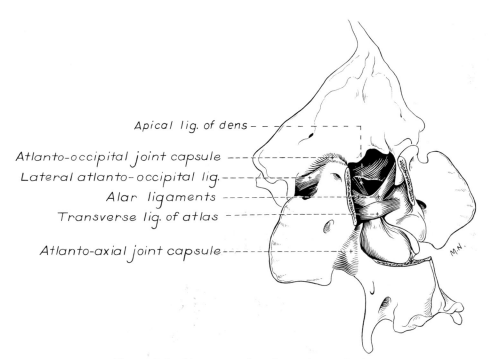

Apical lig. of dens

Atlanto-occipital joint capsule

Lateral atlanto-occipital lig.

Alar ligaments

Transverse lig. of atlas

Atlanto-axial joint capsule

Figure 5–2. Ligaments of occiput, atlas, and axis.

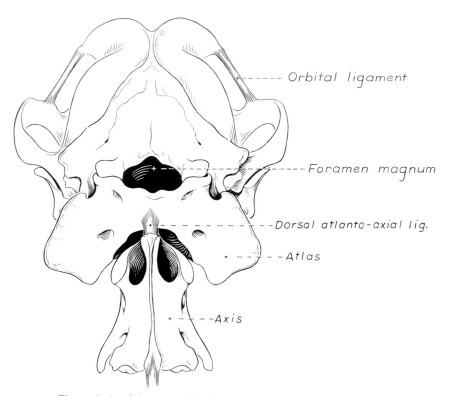

------ Orbital ligament

------ Foramen magnum

------ Dorsal atlanto-axial lig.

---Atlas

---Axis

Figure 5–3.   Atlanto-occipital space, head flexed, caudal aspect.

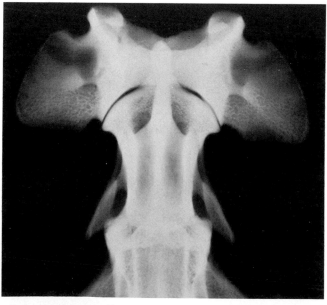

Figure 5–3.   Radiograph. Dorsoventral radiograph, atlanto-axial articulation.

between the dorsal arch of the atlas and the neural arch of the axis. This is the **dorsal atlantoaxial membrane**, or *membrana tectoria.*

The **apical ligament of the dens** *(lig. apicis dentis)* leaves the apex of the dens by three pillars. The middle one goes straight forward to the ventral part of the foramen magnum. According to Nomina Anatomica Veterinaria (p. 28, item 101), the apical ligament represents a remnant of the notochord. The lateral pillars *(ligg. alaria)* are wider and heavier than the middle one; they diverge from each other and attach to the occipital bone medial to the caudal parts of the occipital condyles. The **transverse atlantal ligament** *(lig. transversum atlantis)* is a strong ligament which connects one side of the ventral arch of the atlas to the other. It crosses dorsal to the dens and functions to hold this process against the ventral arch of the atlas. A spacious bursa exists between the ventral surface of the ligament and the dens.

Other Synovial Joints of the Vertebral Column

The synovial joints of the vertebral column caudal to the axis are those which appear in pairs between the articular processes of contiguous vertebrae *(juncturae zygapophyseales)* and the joints between the ribs and the vertebrae *(articulationes costovertebrales).* The articular capsules are most voluminous in the cervical region and at the base of the tail, where the greatest degrees of movement occur. In the lumbar region there is essentially a sagittal interlocking of the cranial and caudal articular processes. At the tenth thoracic vertebra the direction of the articular processes changes. The caudal articular processes of this segment face laterally, and the cranial articular processes face dorsally. The articular processes of all vertebrae cranial to the tenth thoracic are in nearly a frontal plane so that the cranial articular processes face dorsally and the caudal articular processes face ventrally.

Long Ligaments of the Vertebral Column

The **nuchal ligament** *(lig. nuchae)* (Fig. 5–4) is composed of longitudinal yellow elastic fibers which attach cranially to the caudal part of the heavy spinous process of the axis. It extends caudally to the tip of the spinous process of the first thoracic vertebra. It is a laterally compressed, paired band which lies between the medial surfaces of the mm. semispinales capiti. The yellow nature of the nuchal ligament can be traced in the supraspinous ligament to the tenth thoracic spinous process (Baum and Zietzschmann 1936).

The **supraspinous ligament** *(lig. supraspinale)* (see Fig. 5–7) extends from the spinous process of the first thoracic vertebra to the third coccygeal vertebra. It is a heavy band especially in the thoracic region, where it attaches to the apices of the spines as it passes from one to another. Bilaterally the dense collagenous lumbodorsal fascia imperceptibly blends with it throughout the thoracic and lumbar regions. The feeble interspinous ligaments send some strands to its ventral surface, but the supraspinous ligament more than the interspinous ligaments prevents abnormal separation of the spines during flexion of the vertebral column.

The **ventral longitudinal ligament** *(lig. longitudinale ventrale)* (Fig. 5–5) lies on the ventral surfaces of the bodies of the vertebrae. It can be traced from the axis to the sacrum, but it is best developed caudal to the middle of the thorax. The **dorsal longitudinal ligament** *(lig. longitudinale dorsale* (Fig. 5–6) lies on the dorsal surfaces of the bodies of the vertebrae. It therefore forms a part of the floor of the vertebral canal. It is narrowest at the middle of the vertebral bodies and widest over the intervertebral fibrocartilages. The dorsal longitudinal ligament attaches to the rough ridges on the dorsum of the vertebral bodies and to the intervertebral fibrocartilages. It extends from the dens of the axis to the end of the vertebral canal in the coccygeal region. The dorsal longitudinal ligament is heavier than the ventral longitudinal ligament.

Short Ligaments of the Vertebral Column

The **intervertebral discs** *(disci intervertebrales)* are interposed in every intervertebral space (except between C_1 and C_2), uniting the bodies of the adjacent vertebrae. In the sacrum of young specimens, transverse

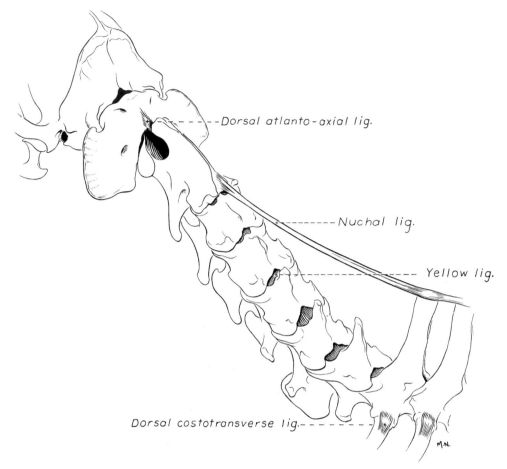

Figure 5-4. Nuchal ligament.

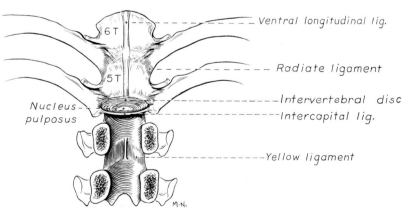

Figure 5-5. Ligaments of vertebral column and ribs, ventral aspect.

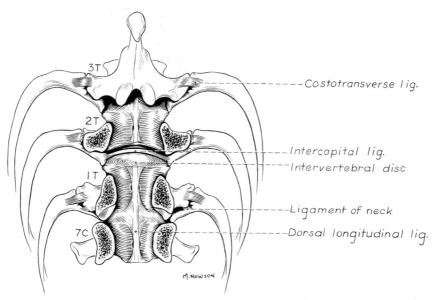

Figure 5–6. Ligaments of vertebral column and ribs, dorsal aspect.

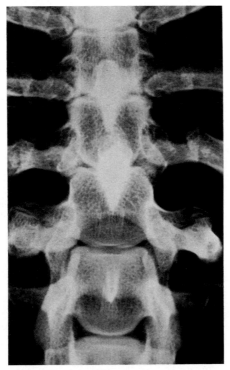

Figure 5–6. Radiograph. Dorsoventral radiograph, vertebral column and ribs; C6 through T4 are shown.

lines indicate the planes of fusion of the discs with the adjoining vertebral bodies. The thickness of the discs is greatest in the cervical and lumbar regions, the thickest ones being between the last few cervical vertebrae. The thinnest discs are in the caudal region, those between the last few segments being smaller in every way than any of the others. Each intervertebral disc consists of an outer laminated fibrous ring and a central, amorphous, gelatinous center, the pulpy nucleus. (Present N.A.V. terminology considers the disc to be an *anulus fibrosus disci intervertebralis* with a central nucleus pulposus.)

The **fibrous ring** (*anulus fibrosus*) consists of bands of parallel fibers which run oblique-

ly from one vertebral body to the next. They provide a means for the transmission of stresses and strains which are required by all lateral and upward movements. These bands of fibers cross each other in a lattice-like pattern and are over eight layers thick ventrally. Near the nucleus pulposus the anulus fibrosus loses its distinctive structure and form and becomes more cartilaginous and less fibrous.

The **pulpy nucleus** (*nucleus pulposus*) is a gelatinous remnant of the notochord. Its position and shape are indicated on each end of the vertebral body as a depressed area surrounded by a line. Since its consistency is semifluid, it bulges when the retaining fibrous ring ruptures or degenerates. Resul-

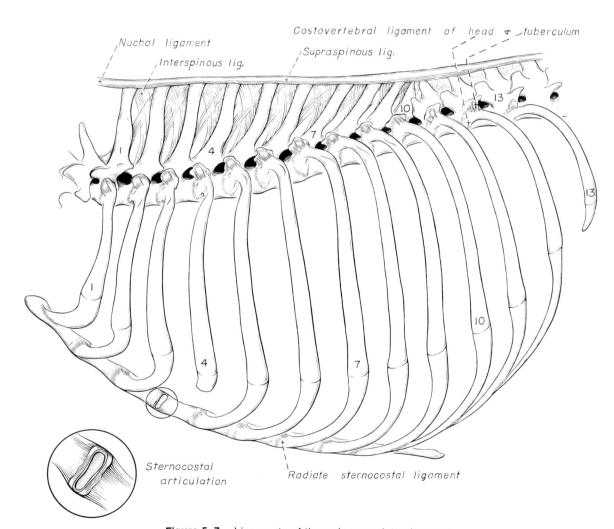

Figure 5–7. Ligaments of thoracic cage, lateral aspect.

tant pressure upon the spinal cord may cause pain or paralysis.

The **interspinous ligaments** *(ligg. interspinalia)* (Fig. 5–7) connect adjacent vertebral spines. They consist of laterally compressed bands of tissue interspersed with muscle bundles of the mm. interspinalis. The bands run from the bases and borders of adjacent spines and decussate as they insert on the opposed caudal and cranial borders of adjacent spines near their dorsal ends. The stronger fibers of the interspinous ligaments lie almost vertically. Some of their fibers blend dorsally with the supraspinous ligament. Great variation exists, and there seems to be no correlation with body type.

The **intertransverse ligaments** *(ligg. intertransversaria)* consist of bundles of fibers which unite the craniolaterally directed transverse processes of the lumbar vertebrae. They are not distinct in any of the other regions of the spine.

The **yellow ligaments** *(ligg. flava)* (Fig. 5–4), or interarcuate ligaments, are loose, thin elastic sheets between the arches of adjacent vertebrae. Laterally they blend with the articular capsules surrounding the articular processes. Ventral to this ligament is the epidural space, which separates the ligaments and the arches of the vertebrae from the dura covering the spinal cord.

LIGAMENTS AND JOINTS OF THE RIBS

Each typical rib articulates with the vertebral column by two synovial joints and with the sternum by one. There is usually a slightly enlarged synchondrosis between the rib and its costal cartilage.

The **costovertebral joints** *(articulationes costovertebrales)* are formed by the articulation of the capitulum of each rib *(articulatio capitis costae)* with the costal facets of the appropriate vertebrae, and the articulation of each tuberculum *(articulatio costotransversaria)* with the transverse process of the corresponding vertebra. The **articular capsules** of these joints are thin-walled synovial sacs which completely surround each joint and are associated with the four ligaments of the costovertebral articulation. These are the ligament of the head and the intercapital

ligament, both of the capitular joint, and the ligaments of the tubercle and neck of the tubercular joint.

The **ligament of the head** *(lig. capituli costae)* (see Fig. 5–5) is a small ligamentous band which passes from the head of the rib to the lateral part of the disc. The last three or four ribs are displaced caudally at their vertebral articulations; the ligament also shifts caudally and attaches to the body of the vertebra adjacent to the disc.

The **intercapital ligament** *(lig. intercapitale)* (see Figs. 5–5, 5–6) is probably homologous with the intra-articular ligaments of man. Because of its unique position and function it merits a separate description. It runs from the head of one rib over the dorsal part of the disc, but under the dorsal longitudinal ligament, to the head of the opposite rib. It grooves the dorsal part of the disc. A synovial membrane between the ligament and the disc joins the joint capsules of the opposite rib heads. The intercapital ligament is attached both cranially and caudally to the disc by a delicate membrane, and is attached dorsally to the dorsal longitudinal ligament and dura by areolar tissue. The ligament functions to hold the heads of opposite ribs tightly against their articular sockets and to prevent excessive cranial and caudal movements of the ribs. There is no intercapital ligament uniting the first pair or the last two pairs of ribs, and that connecting the heads of the eleventh pair of ribs is smaller than the others.

The dorsal costotransverse ligament (see Fig. 5–6), or the **ligament of the tubercle** *(lig. tuberculi costae)*, is the strongest single ligament uniting the rib to the vertebra. It attaches just distal to the articular capsule of the tubercle, crosses the capsule, and blends with the periosteum of the transverse process of the vertebra corresponding to the rib. The ligaments of the tubercles of the first five ribs lie cranial to the joints and run obliquely craniomedially from the tubercles to the transverse processes, the ligaments of the next three run almost directly medially to the transverse processes from the dorsal surfaces of the tubercles, and those of the last four incline increasingly caudally as they run from the rib tubercles to the transverse processes of the vertebrae. Great variation in size and position of these ligaments exists in different dogs. The ligaments of the tubercles are usually strongest on the last four ribs.

The **ligament of the neck** *(lig. colli costae)* (see Figs. 5–5, 5–6) consists of collagenous bundles which extend from the neck of the rib to the ventral surface of the transverse process and the adjacent lateral surface of the body of the vertebra. Running from the caudal surface of the neck of the rib to the body of the vertebra of the same number is a ligament; passing caudally, these ligaments progressively decrease in size. A ligament also runs from the cranial surface of the neck of each rib to the transverse process of the vertebra of the same number.

The **sternocostal joints** *(articulationes sternocostales)* (Figs. 5–7, 5–8) are synovial joints formed by the first eight costal cartilages articulating with the sternum. The second to seventh pairs of joints are typical, but the first and last pairs present special features. The first sternebra is widened cranially by the formation of lateral shelves of bone which articulate with the transversely compressed costal cartilages of the first ribs. These costal cartilages approach their sternal articulations at a more acute angle than do any of the other costal cartilages. The last sternocostal joints, typically, are formed by the ninth pair of cartilages joining each other and together articulating with the ventral surface of the fibrocartilage between the last two sternebrae, or with the sternebra cranial to the xiphoid process. The ends of the right and left ninth costal cartilages are united by an indistinct collagenous ligament. No synovial joint is found here, as the ninth costal cartilages lie closely applied to the eighth costal cartilages. A typical sternocostal joint is vertically elongated so that its length is double its width. Being in a vertical plane, it allows only forward and backward movements. The joint capsule is usually thin, except dorsally and ventrally, where the heavy perichondrium leaving the costal cartilages thickens and spreads out as it goes to the intersternebral fibrocartilages. These are the **dorsal and ventral sternocostal radiate ligaments** *(ligg. sternocostalia radiata dorsalia et ventralia).* The dorsal and ventral surfaces of the sternum are covered by white membranous sheets and bands of thickened periosteum, the **sternal membrane** (membrana sterni). The dorsal part is divided into two or more strands, whereas the ventral part consists of a single median band. The **costoxiphoid ligaments** *(ligg. costoxiphoidea)* are two flat cords which originate on the eighth costal cartilages. They cross ventral to the ninth costal cartilages, and converge and blend as they join the periosteum on the ventral surface of the caudal half of the xiphoid process.

The **costochondral joints** *(articulationes costochondrales)* are the joints between the ribs and the costal cartilages. Apparently, no synovial cavities ever develop here. In puppies these joints are slightly enlarged and appear as a longitudinal line of beads on the ventrolateral surface of the thorax.

LIGAMENTS AND JOINTS OF THE THORACIC LIMB

Shoulder Joint

The **shoulder joint** *(articulatio humeri)* (Figs. 5–9, 5–10) is the ball-and-socket joint between the glenoid cavity of the scapula and the head of the humerus. It is capable of

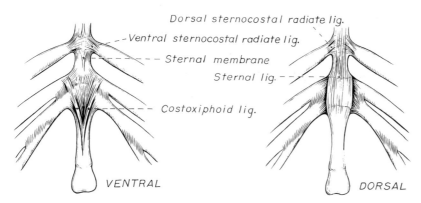

Dorsal sternocostal radiate lig.

--Ventral sternocostal radiate lig.

-- Sternal membrane

Sternal lig. --

-- Costoxiphoid lig.

VENTRAL

DORSAL

Figure 5–8. Ligaments of xiphoid region.

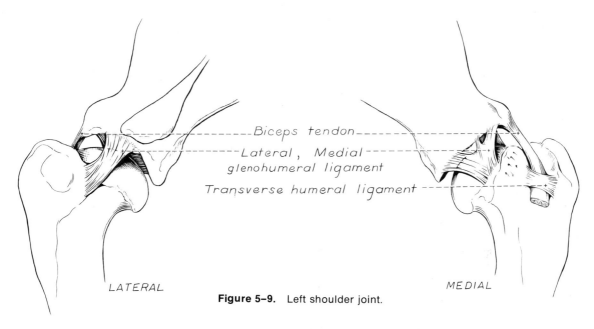

Biceps tendon

Lateral, Medial glenohumeral ligament

Transverse humeral ligament

LATERAL   MEDIAL

Figure 5–9.   Left shoulder joint.

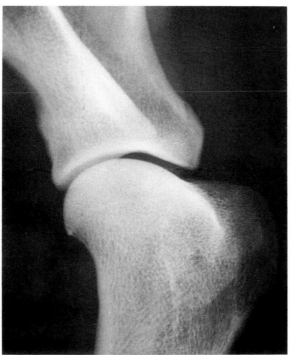

Figure 5–9.   Radiograph. Lateromedial radiograph, left shoulder joint.

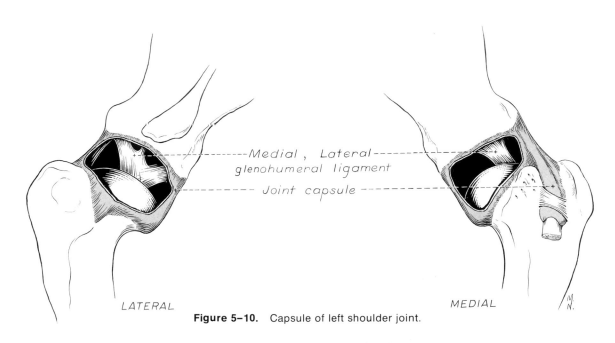

Figure 5-10. Capsule of left shoulder joint.

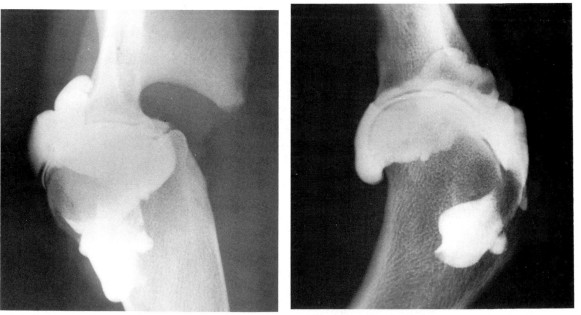

Figure 5-10. Radiograph. Left, Craniocaudal radiograph, left shoulder joint; positive contrast arthrogram performed to define the synovial space. Right, Lateromedial radiograph, left shoulder joint; positive contrast arthrogram performed to define the synovial space.

movement in any direction, but its chief movements are flexion and extension. The shallow, small glenoid cavity of the scapula is increased in size and deepened by the **glenoid lip** *(labrum glenoidale),* which extends 1 or 2 mm. beyond the edge of the cavity caudolaterally. The **articular capsule** *(capsula articularis)* forms a loose sleeve which attaches just peripheral to the glenoid lip proximally. In places the capsule attaches several millimeters distal to the articular part of the humeral head, where is blends with the periosteum on the neck of the humerus. A part of the joint capsules surrounds the tendon of origin of the m. biceps brachii and extends distally about 2 cm. in the intertubercular groove. The tendon with its synovial sheath is held in the groove by the **transverse humeral ligament** *(lig. transversum humerale).* The capsule blends with this ligament craniomedially and with the tendon of the m. subscapularis medially. Laterally the joint capsule blends with the tendons of the mm. supraspinatus and infraspinatus. Elsewhere, especially caudally, the articular capsule is thin and possesses a number of irregular pouches when it is distended. Medially and laterally the fibrosa of the capsule is irregularly thickened internally to form the **medial and lateral glenohumeral ligaments** *(ligg. glenohumeralia medialis et lateralis).* These reinforcing bands protrude appreciably into the joint cavity. The heavy tendons which cross the joint may be called active ligaments. They ensure its integrity, and do this so well that shoulder dislocation is practically unknown in domestic animals.

Suter and Carb (1969), using silastic injections and subsequent maceration, have demonstrated the extent of the shoulder joint capsule. In cranial view a lateral extension of the joint capsule is seen beneath the tendon of the supraspinatus, a large medial synovial sheath descends to surround the biceps tendon, and a large medial pouch can be found underlying the broad insertion of the subscapularis.

Elbow Joint

The **elbow joint** *(articulatio cubiti)* (Figs. 5–11 to 5–14) is a composite joint formed by the humeral condyle with the head of the radius, the **humeroradial joint** *(articulatio humeroradialis),* and with the semilunar notch of the ulna, the **humeroulnar joint** *(articulatio humeroulnaris). The* **proximal radioulnar joint** *(articulatio radioulnaris proximalis)* freely communicates with the main part of the elbow joint, and is regarded as a part of it. The humeroradial part of the elbow joint transmits most of the weight supported by the limb. The humeroulnar part stabilizes and restricts the movement of the joint to a sagittal plane, and the proximal radioulnar joint allows rotation of the antebrachium. Lateral movements of the elbow joint are minimal because of the strong collateral ligaments and the forward protrusion of the anconeal process of the ulna into the deep olecranon fossa of the humerus. Enough rotational movement occurs at the radioulnar and carpal joints so that the forepaws can be supinated about 90 degrees.

The **joint capsule** is common to all three articular parts. It is taut on the sides but expansive in front and behind. On the cranial or flexor surface, it attaches proximal to the supratrochlear foramen and encompasses most of the radial fossa. Caudally, or on the extensor surface, the joint capsule forms a loose, fat-covered synovial pouch which attaches distal to the supratrochlear foramen, so that there is no intercommunication between the extensor and flexor pouches through the supratrochlear foramen. The joint capsule dips down between the radial notch of the ulna and the articular circumference of the radius. Everywhere but cranially the synovial membrane attaches closely to the articular cartilage. Medially it sends a distal pouch under the m. biceps brachii, and similar extensions occur laterally under the mm. extensor carpi radialis and extensor digitorum communis. On the caudomedial side, extensions of the capsule occur under the mm. flexor carpi radialis and flexor digitorum profundus, caput humerale.

The **lateral (ulnar) collateral ligament** *(lig. collaterale laterale)* attaches proximally to the lateral epicondyle of the humerus. Distally it divides into two crura. The slightly larger cranial crus attaches to a small lateral eminence distal to the neck of the radius. The flatter caudal crus passes to the ulna. At the level of the articular circumference the

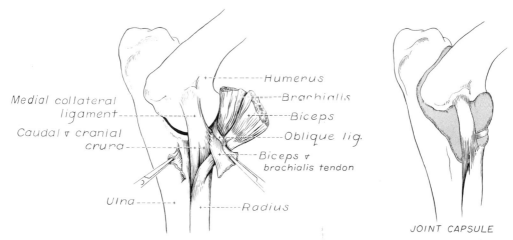

Figure 5–11.   Left elbow joint, medial aspect.

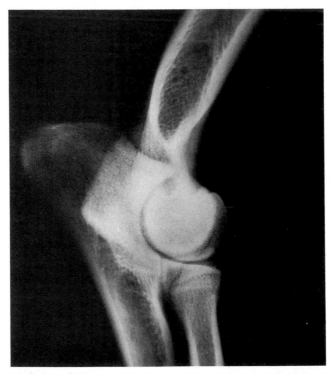

Figure 5–11.   Radiograph. Lateromedial radiograph, left elbow joint.

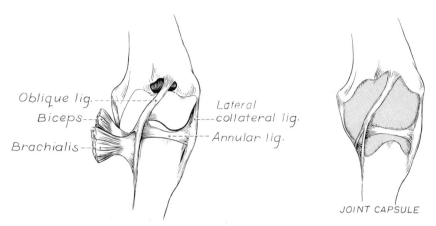

Oblique lig.
Biceps
Brachialis
Lateral collateral lig.
Annular lig.

JOINT CAPSULE

Figure 5–12.   Left elbow joint, cranial aspect.

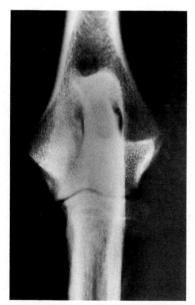

Figure 5–12. Radiograph.   Craniocaudal radiograph, left elbow.

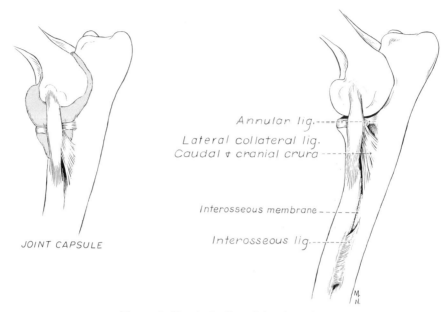

JOINT CAPSULE

Annular lig.
Lateral collateral lig.
Caudal & cranial crura
Interosseous membrane
Interosseous lig.

Figure 5–13. Left elbow joint, lateral aspect.

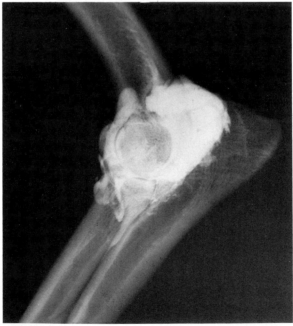

Figure 5–13. Radiograph. Lateromedial radiograph, left elbow joint; positive contrast arthrogram performed to define the synovial space.

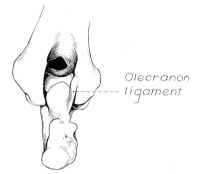
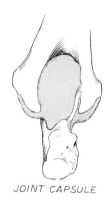

Olecranon
ligament

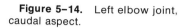

Figure 5–14. Left elbow joint, caudal aspect.

JOINT CAPSULE

ligament blends with the annular ligament and, according to Baum and Zietzschmann (1936), often contains a sesamoid bone.

The **medial (radial) collateral ligament** *(lig. collaterale mediale)* is weaker than the lateral collateral ligament, which it resembles. It attaches proximally to the medial epicondyle of the humerus, crosses the annular ligament distally, and divides into two crura. The weaker cranial crus attaches proximal to the radial tuberosity. The stronger caudal crus passes deeply into the interosseous space, where it attaches mainly on the ulna but also partly on the radius.

The **annular ligament of the radius** *(lig. anulare radii)* is a thin band which runs transversely around the radius. It attaches to the lateral and medial extremities of the radial incisure of the ulna. It lies under the collateral ligaments and is slightly blended with the ulnar collateral ligament. In conjunction with the ulna, it forms a ring in which the articular circumference of the radius turns when the forearm is rotated.

A small but distinct band of fibers in the joint capsule, sometimes called the oblique ligament, arises on the dorsal edge of the supratrochlear foramen and crosses the flexor surface of the elbow joint distomedial to the tendons of the mm. biceps brachii and brachialis. At the level of these tendons, directly peripheral to the annular ligament, it divides into two parts. The shorter part blends with the cranial crus of the radial collateral ligament. The longer branch ends on the medial border of the radius after looping around the tendons of the mm. biceps brachii and brachialis.

Radioulnar Joints

The radius and ulna are united by the proximal and distal radioulnar synovial joints and by the surprisingly heavy interosseous ligament and the narrow weak interosseous membrane, which extends both proximally and distally from the interosseous ligament.

The **proximal radioulnar joint** *(articulatio radioulnaris proximalis),* already mentioned as a part of the main elbow joint, extends distally between the articular circumference of the radius and the radial notch of the ulna to a depth of about 5 mm. The joint allows rotation of the radius in the radial notch of the ulna.

The **interosseous ligament of the antebrachium** *(lig. interossei antebrachii)* (Fig. 5–13) is a heavy but short collagenous ligament which extends across the interosseous space from the apposed rough areas on the radius and ulna. It is about 2 cm. long, 0.5 cm. wide, and 0.2 cm. thick. It extends distally slightly beyond the middle of the ulna but not quite to the middle of the radius, since this bone does not extend as far proximally as does the ulna. The long axis of the ligament is slightly oblique so that the distal part is more lateral than the proximal. It is wider distally and is separated from the interosseous membrane by a small fossa, which extends under it for about half its length. In the fornix of the fossa the interosseous membrane and ligament fuse. There appears to be no precedent for the name interosseous ligament of the antebrachium. Yet it is so much heavier than the interosseous membrane located both proximal and distal to it that it warrants separate treatment. The **interosseous membrane of the antebrachium** *(membrana interossea antebrachii)* (Fig. 5–13) is a narrow, thin septum which connects the radius and ulna both above and below the interosseous ligament. It attaches to the apposed interosseous crests of the radius and ulna. The membrane extends from the proximal to the distal radioulnar synovial joints but is perforated proximally for the passage of the common interosseous artery and vein and the interosseous nerve. Distally a smaller perforation in the membrane allows for the passage of the distal dorsal interosseous artery and vein from the palmar to the dorsal side. There are also, throughout the length of the interosseous membrane, small openings for the anastomotic vessels which course between the palmar interosseous and the dorsal interosseous vessels.

The **distal radioulnar joint** *(articulatio radioulnaris distalis)*, which extends between the radius and ulna distally, is a proximal extension of the antebrachiocarpal joint capsule. The distal end of the ulna bears a slight articular convexity, and the adjacent surface of the radius bears a shallow articular cavity. The fibrosa of the joint capsule is essentially a part of the interosseous membrane and is short and tight *(lig. radioulnare)*. It is the distal pivotal joint for the small amount of rotational movement permitted between the bones of the forearm.

Carpal, Metacarpal, and Phalangeal Joints (Articulationes Manus)

CARPAL JOINTS

The **carpal joints** *(articulationes carpi)* are composite articulations which include proximal, middle, distal, and intercarpal joint surfaces. The **antebrachiocarpal joint** *(articulatio antebrachiocarpea)* is located between the distal part of the radius and ulna and the proximal row of carpal bones. The **middle carpal joint** *(articulatio mediocarpea)* is located between the two rows of carpal bones. The **carpometacarpal joints** *(articulationes carpometacarpeae)* are located between the carpus and metacarpus. Joints between the individual carpal bones of each row constitute the **intercarpal joints**

(articulationes intercarpeae s. intraordinarii carpi). The carpal joint as a whole acts as a ginglymus, permitting flexion and extension with some lateral movement. Greatest movement occurs in the antebrachiocarpal and middle carpal joints. Considerably less movement takes place in the intercarpal and carpometacarpal joints.

There are no continuous collateral ligaments for the three main joints of the carpus. The dorsal and palmar parts of the joint capsule are much thicker than is usually the case on the extensor and flexor surfaces of hinge joints. Long collateral ligaments are lacking. Two superimposed sleeves of collagenous tissue, with tendons located between them, ensure the integrity of the carpus. The superficial sleeve is a modification of the deep carpal fascia, and the deep sleeve is the fibrous layer of the joint capsule. Laterally and medially, the two sleeves fuse and become specialized in part to form the short collateral ligaments.

The **flexor retinaculum** *(retinaculum flexorum)*, formerly called the transverse palmar carpal ligament (Fig. 5–15), is well developed in the dog; it is a modification of the caudal part of the carpal fascia. It attaches laterally to the medial part of the enlarged base of the accessory carpal bone, and widens as it passes medially to attach to the styloid process of the radius and on the palmar projections of the radial and first carpals. The transverse palmar carpal ligament is divided into two parts. One lies superficial and the other lies between the tendons of the superficial and deep digital flexors. The **carpal canal** *(canalis carpi)* on the palmar side of the carpus is formed superficially by the superficial part of the flexor retinaculum and deeply by the palmar part of the joint capsule. It is bounded laterally by the accessory carpal bone. It contains the tendons and synovial sheaths of the mm. flexor digitorum superficialis and flexor digitorum profundus, as well as the radial, ulnar, and palmar interosseous arteries and veins and the ulnar and median nerves.

The **palmar carpal fibrocartilage** *(fibrocartilago carpometacarpeum palmare)* (Fig. 5–15) is quite thick and sharply defined proximally. As it crosses the palmar surfaces of the carpal bones, it attaches to all except the accessory carpal bone. The thickest attachment on the accessory carpal bone is to its

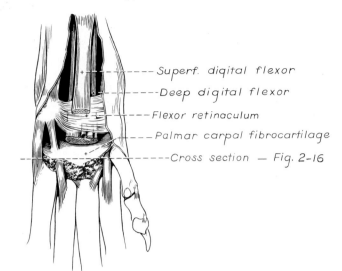

Figure 5–15. Superficial ligaments of left carpus, palmar aspect.

- - - - Superf. digital flexor
- - - Deep digital flexor
- - - Flexor retinaculum
- - - Palmar carpal fibrocartilage
- - - - Cross section — Fig. 2-16

dorsomedial border, just caudal to the articulation of the radial carpal with the ulnar carpal. The palmar carpal fibrocartilage is very heavy distally. This attaches to the palmar surfaces of the distal row of carpal bones and the adjacent surfaces of the proximal parts of metacarpals III, IV, and V. The palmar carpal fibrocartilage serves as the origin for most of the special muscles of digits 2 and 5, as well as furnishing part of the origin for the interosseous muscles. It flattens the palmar irregularities at the carpometacarpal joints and furnishes a smooth, deep surface for the carpal canal.

The **special ligaments of the carpus** will be treated briefly. Some of the smaller ones will not be described, but all are illustrated in Figures 5–17 to 5–21.

The **short radial collateral ligament** (*lig. collaterale radiale breve*) consists of a straight and an oblique part. The straight part runs from a tubercle above the styloid process to the most medial part of the radial carpal. The oblique part, after leaving the styloid process, runs obliquely to the palmaromedial surface of the radial carpal. The tendon of the m. abductor pollicis longus lies between the two parts as it crosses the medial surface of the carpus.

The **short ulnar collateral ligament** (*lig. collaterale ulnare breve*) extends from the styloid process of the ulna to the ulnar carpal.

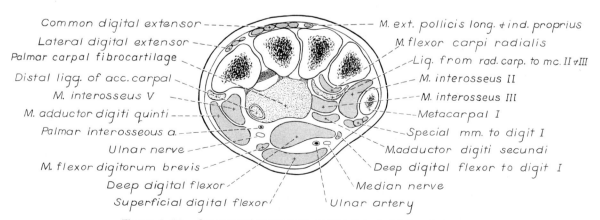

Common digital extensor - - - - -
Lateral digital extensor - - -
Palmar carpal fibrocartilage - - -
Distal ligg. of acc. carpal - -
M. interosseus V - - - -
M. adductor digiti quinti - - - -
Palmar interosseous a. - - -
Ulnar nerve - - -
M. flexor digitorum brevis - -
Deep digital flexor - -
Superficial digital flexor -

- - M. ext. pollicis long. + ind. proprius
M. flexor carpi radialis
- Lig. from rad. carp. to mc. II + III
- M. interosseus II
- - M. interosseus III
- - - Metacarpal I
- - - Special mm. to digit I
M. adductor digiti secundi
Deep digital flexor to digit I
Median nerve
Ulnar artery

Figure 5–16. Cross section through proximal end of left metacarpus.

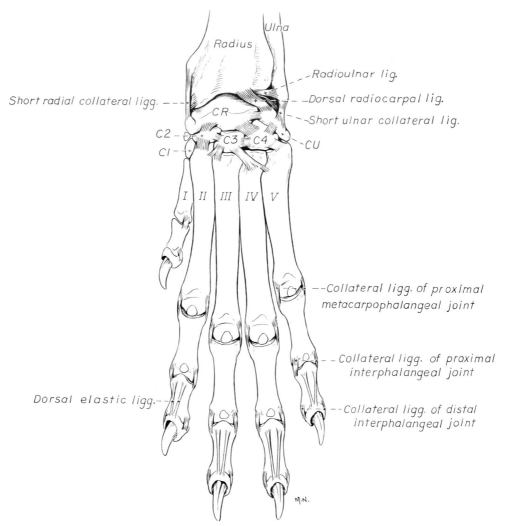

Figure 5–17.   Ligaments of left forepaw, dorsal aspect.
CR = radial carpal. CU = ulnar carpal.
C1 to C4 = first, second, third, fourth carpals.
I to V = metacarpals.

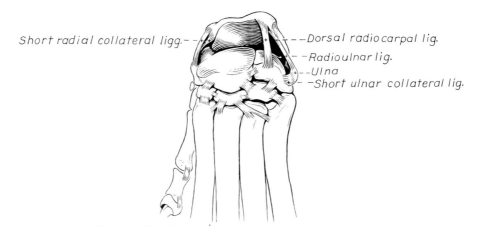

Figure 5–18.   Ligaments of flexed carpus, dorsal aspect.

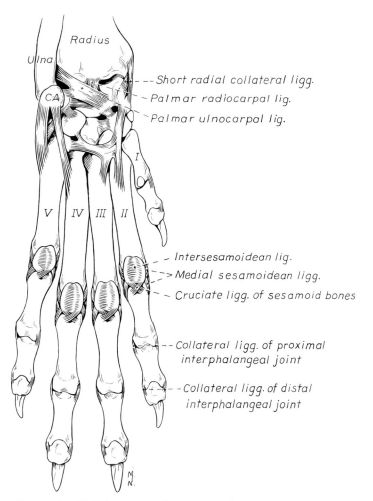

Figure 5–19. Deep ligaments of left forepaw, palmar aspect. CA = accessory carpal. I to V = metacarpals.

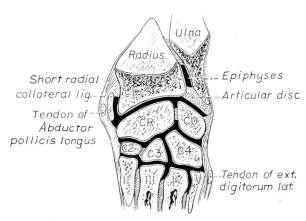

Figure 5–20. Schematic section of left carpus, showing articular cavities.
CR = radial carpal. CU = ulnar carpal.
C2, C3, C4 = second, third, fourth carpals. II to V = metacarpals.

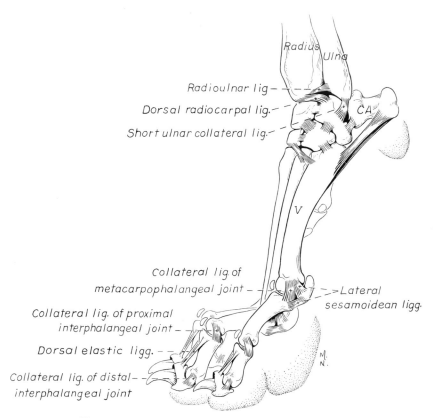

Figure 5–21. Ligaments of forepaw, lateral aspect.
CA = accessory carpal. V = metacarpal V.

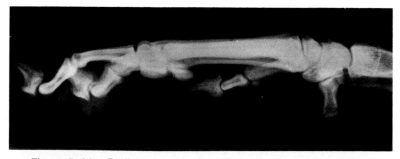

Figure 5–21. Radiograph. Lateromedial radiograph, left carpus.

In addition to the short collateral ligaments the cranial distal lip of the radius is attached to the cranial surface of the ulnar carpal by a strong ligament. These ligaments diverge as they run distally, thus allowing a free opening on the cranial surface of the antebrachiocarpal joint during flexion. The ulna is securely anchored to the palmar side of the radial carpal by an obliquely running ligament located just proximal to the accessory carpal bone. From the palmar surface of the radius, near its distal articular cartilage, a ligament runs to the palmar surface of the radial carpal. A short leaf of this ligament runs from the midpalmar surface of the radius to the radial carpal. A flat band nearly 1 cm. wide runs from the palmarolateral surface of the radius from within the distal part of the interosseous space to the lateral surface of the radial carpal adjacent to the ulnar carpal. The accessory carpal bone is secured distally by two ligaments which originate near its enlarged, rounded, free end. Distally one attaches to metacarpal V and the other to metacarpal IV. Many short ligaments unite the carpal bones transversely, holding them as units in the two rows.

METACARPAL JOINTS

The **intermetacarpal joints** *(articulationes intermetacarpeae)* are close-fitting joints between the proximal ends of adjacent metacarpal bones. The synovial membrane from the adjacent carpometacarpal joint extends a few millimeters between the metacarpal bones. Distal to the synovial part, the bones are united for variable distances by fibrous tissue, the **interosseous metacarpal ligaments** *(ligg. metacarpea interossea)*. Distal to these ligaments are the **interosseous spaces** of the metacarpus *(spatia interossea metacarpi)*.

The **metacarpophalangeal joints** *(articulationes metacarpophalangeae)* are the five joints formed by the distal ends of the metacarpal bones and the proximal ends of the proximal phalanges. To these are added in each of the four main joints the two palmar sesamoid bones. Each joint has a **joint capsule** that runs between the four bones which form the joint and the two **collateral ligaments** which unite the osseous parts. Each pair of palmar sesamoid bones of the four main joints are joined together by the **inter-**

sesamoidean ligament *(lig. intersesamoideum)*. This short, cartilaginous ligament consists of transverse fibers which unite the paired sesamoid bones and cover their palmar surfaces. The **lateral and medial sesamoidean ligaments** *(ligg. sesamoideum laterale et mediale)* are short, flat bands on each side of the metacarpophalangeal joint. The first part attaches the corresponding lateral and medial surfaces of the sesamoid bones to the distal surfaces of the metacarpal bone caudal to the proximal attachments of the collateral ligaments. The second part goes to the medial and lateral tubercles of the proximal phalanx. From the distal ends of each pair of sesamoid bones there is a thin, flat band which attaches to the palmar side of the proximal phalanx. It is called the **distal sesamoidean ligament** *(lig. sesamoideum distale)*. The **cruciate ligaments of the sesamoid bones** *(ligg. sesamoidea cruciata)* extend from the bases of the sesamoid bones to the diagonally opposite tubercles on the proximal end of the proximal phalanges. In the first digit there is usually only one sesamoid bone, and therefore only one ligament.

The dorsal sesamoid bones of the metacarpophalangeal joints are secured by delicate fibers from the tendons of the m. extensor digitorum communis and the mm. interossei proximally, and by a ligament to the dorsal surface of the middle phalanx distally.

PHALANGEAL JOINTS

The **proximal interphalangeal joints** *(articulationes interphalangeae proximales)* are formed by the heads of the proximal phalanges articulating with the fossae of the middle phalanges in each of the main digits, II to V. These are saddle-type joints. The **joint capsules** have dorsal walls which are thickened by a bead of cartilage. Here the capsules are intimately united with the extensor tendons so that the sesamoid cartilages appear to be intercalated in them. On the palmar side the joint capsules are intimately fused with the flexor tendons. The **collateral ligaments** are stout collagenous bands which do not parallel the axis through the digit but extend in vertical planes as the dog stands. They attach proximally to the fossae on the sides of the distal ends of the first phalanges and distally to the collateral tubercles on the proximal ends of the middle phalanges. In the first digit, which has only two phalanges, the collateral ligaments attach distally to the proximal end of the distal phalanx.

The **distal interphalangeal joints** *(articulationes interphalangeae distales)*, in the second to fifth digits, are formed by the heads of the middle phalanges articulating with the saddle-shaped fossae on the proximal ends of the distal phalanges. A single, small, spheroidal, sesamoid cartilage is located on the palmar side of the joint capsule. The **joint capsule** is thickened to form the **collateral ligaments,** which attach proximally to the shallow fossae on each side of the head on the middle phalanx and extend obliquely caudodistally to attach to the sides of the ungual crest of the third, or distal, phalanx. The **dorsal ligaments** are two elastic cords which extend across the dorsal part of the distal interphalangeal joint some distance from its surface. They attach proximally to the dorsal surface of the proximal part of the middle phalanx, where they are about 2 mm. apart. Distally they attach close together on the dorsal part of the ungual crest. They passively keep the claws retracted, so that the claws do not touch the substratum except when their tension is overcome by the m. flexor digitorum profundus.

INTERDIGITAL LIGAMENTS

The **interdigital ligaments** *(ligg. interdigitalia)* form a continuous superficial, V-shaped ligamentous structure which not only holds the digits together but also acts as a fastening mechanism for the large heart-shaped metacarpal pad. They originate bilaterally as small fibrous strands from the abaxial borders of the second and fifth tendons of the m. flexor digitorum superficialis. From their origin proximal to the metacarpophalangeal joints they extend distally to the proximal digital annular ligaments which cross the flexor tendons at these joints of digits II and V. The interdigital ligaments attach to the proximal digital annular ligaments of the second and fifth digits and, augmented in size, run distoaxially to the proximal digital annular ligaments of the third and fourth digits. They attach to these annular ligaments and again increase in size, reaching a maximum width of 4 mm. in large dogs. Continuing distally they unite in a single broad band located dorsal to the metacarpal pad. The conjoined interdigital ligaments continue to the integument of the pad and cover the flexor tendons opposite the proximal interphalangeal joints. This is the main supportive structure of the pad, but there are in addition several fibroelastic strands which pass radially into the substance of the pad from the interdigital ligaments as they cross and are fused to the annular ligaments. Proximal to the interdigital ligaments is a feeble collagenous strand which runs from the palmar surface of metacarpal II to a like place on metacarpal V. It is not present in the hindpaw, according to Baum and Zietzschmann (1936).

LIGAMENTS AND JOINTS OF THE PELVIC LIMB

Joints of Pelvic Girdle (Juncturae Cinguli Membri Pelvinae)

The right and the left os coxae, in young dogs, are united midventrally by cartilage, to form the **pelvic symphysis** *(symphysis pelvis).* The cranial half is formed by the **pubic symphysis** *(symphysis pubica)* and the caudal half by the **ischial symphysis** *(symphysis ischiadica).* In the adult, the joint ossifies, although the pubic symphysis is subject to periodic resorption as a result of advanced pregnancy.

SACROILIAC JOINT

The **sacroiliac joint** *(articulatio sacroiliaca)* is a combined synovial and cartilaginous joint. The apposed auricular surfaces on the wings of the sacrum and ilium are covered by cartilage, and their margins are united by a thin **joint capsule.** The fibrosa of the caudoventral part is so thin that the capsular wall is translucent. Dorsal to the crescent-shaped auricular surfaces, the wing of the sacrum and the wing of the ilium are rough and possess irregular projections and depressions which tend to interlock. In life this space is occupied by a plate of fibrocartilage (homologous to the ligamenta sacroiliaca interossea of man), which unites the two wings. When this joint is disarticulated by injury, or by force as in an autopsy procedure, the fibrocartilage usually remains at-

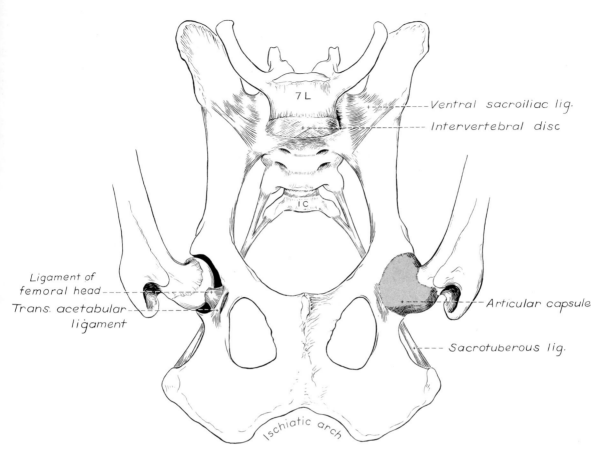

Figure 5–22.   Ligaments of pelvis, ventral aspect.

tached to the sacrum. Through the medium of this fibrocartilage, the ilium and sacrum are firmly united, to form the **sacroiliac synchondrosis** *(synchondrosis sacroiliaca).* The sacroiliac synchondrosis is located craniodorsal to the synovial portion of the joint.

The **ventral sacroiliac ligament** *(lig. sacroiliacum ventrale)* (Fig. 5–22) consists of many short, fibrous fascicles, which are arranged in two groups. Those of the cranial group run inward and backward; those of the shorter caudal group run inward and forward. The thin joint capsule appears between them.

The **dorsal sacroiliac ligaments** *(ligg. sacroiliacum dorsale breve et longum)* (Fig. 5–23) are more formidable than the ventral ones. They can be divided into a short and a long part. The short part consists of collagenous bands which extend obliquely caudomedially from the caudal dorsal iliac

spine to the cranial two-thirds of the lateral border of the sacrum. The long part is dorsocaudal to the short part, and is fused to it cranially. It is questionable whether or not a long dorsal sacroiliac ligament should be recognized, since it represents largely the attachment of the fasciae of the rump and tail. The long part of the ligament extends further caudally on the sacrum and may even reach the transverse process of the first caudal vertebra.

The **sacrotuberous ligament** *(lig. sacrotuberale)* (Fig. 5–23) is a fibrous cord which is flattened at both ends. It extends from the caudolateral part of the apex of the sacrum and the transverse process of the first vertebra to the lateral angle of the ischiatic tuberosity. In large dogs the middle part of the ligament may be 3 mm. thick and its flattened ends may be 1 cm. wide. The sacrotuberous ligament lies hidden mainly by

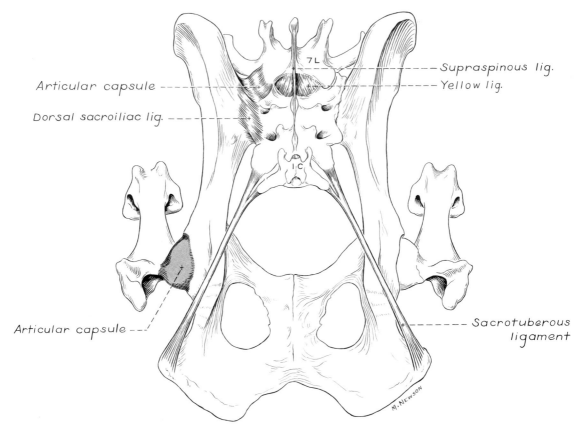

Figure 5–23. Ligaments of pelvis, dorsal aspect.

the m. gluteus superficialis. It forms the caudodorsal boundary of the lesser ischiadic foramen (foramen ischiadicum minus). The following muscles arise wholly or in part from it: mm. biceps femoris, gluteus superficialis, piriformis, and abductor cruris caudalis.

Hip Joint

The **hip joint** *(articulatio coxae)* (see Figs. 5–22, 5–23) is formed by the head of the femur articulating with the acetabulum, the cotyloid cavity of the os coxae. Axes through the femur and os coxae meet at the hip joint in a cranially open angle of about 95 degrees. Although flexion and extension are the chief movements of the joint, its ball-and-socket construction allows a great range of movement. The deep acetabulum is further deepened in life by a band of fibrocartilage, the **acetabular lip** *(labrum acetabu-*

lare), which is applied to the rim of the acetabulum. It extends across the acetabular notch as a free ligament, the **transverse acetabular ligament** *(lig. transversum acetabuli).* The **joint capsule** is very capacious. It attaches, medially, a few millimeters from the edge of the acetabular lip, and, laterally, on the neck of the femur, 1 or 2 cm. from the cartilage-covered head. The fibrous coat has various thickenings, but no definite ligaments. The most distinct thickening is in the dorsal part of the fibrosa. This causes a nearly horizontal bulging of the synovial membrane, known as the **orbicular zone** *(zona orbicularis).* As it arches from the cranial to the caudal border across the dorsal surface of the neck, it parallels both the dorsal part of the acetabular rim and the dorsal part of the head-neck junction. It presents no definite fiber pattern and appears as a white thickening in the joint capsule, measuring less than 1 mm. thick by 2 or 3 mm. wide.

The **ligament of the head of the femur** (*lig. capitis femoris*), formerly called the round ligament, is a rather heavy flattened cord which extends from the fovea in the head of the femur to the acetabular fossa. Since it is largely intracapsular and not weight-bearing it is covered by synovial membrane. In large dogs it is about 1.5 cm. long, and 5 mm. wide at its femoral attachment. Its acetabular attachment is wide as it blends with the periosteum of the acetabular fossa, and the transverse acetabular ligament. The pelvic attachment of the ligament of the femoral head is over 1 cm. wide in large dogs. In and peripheral to the rectangular acetabular fossa there is usually a small quantity of fat.

Hip dysplasia in the dog has a high incidence in some breeds. It is a progressive disparity between muscle mass and bone growth which results in degenerative joint disease. The evidence, according to Riser (1975), indicates that the soft tissues do not have sufficient strength to maintain the femoral head in the acetabulum.

Stifle Joint

The **stifle joint,** or **knee** (*articulatio genus* (Figs. 5–24 to 5–30), is a complex condylar synovial joint. The main spheroidal part is formed by the thick roller-like condyles of the femur articulating with the flattened condyles of the tibia to form the femorotibial or condyloid part of the joint (*articulatio femorotibialis*). Freely connected with this is the **femoropatellar joint** (*articulatio femoropatellaris*), located between the patella and the trochlea of the femur. The two joints are interdependent in that the patella is held to the tibia firmly by ligamentous tissue so that any movement between the femur and tibia also occurs between the patella and the femur. The incongruence which exists between the tibia and femur is occupied, in life, by two fibrocartilages, or **menisci,** one located between the adjacent medial condyles (*meniscus medialis*) and the other (*meniscus lateralis*) between the adjacent lateral condyles of the femur and tibia. The proximal tibiofibular joint is also a component of the stifle joint.

The **joint capsule** of the stifle joint is the largest in the body. It forms three sacs, all of which freely intercommunicate. Two of these are between the femoral and tibial condyles (*saccus medialis et lateralis*), and the third is beneath the patella. The patellar part of the joint capsule is very capacious. It attaches to the edges of the parapatellar fibrocartilages and extends beyond these in all directions. Proximally a sac protrudes 1.5 cm. under the tendon of the m. quadriceps femoris. Laterally and medially the patellar part of the joint capsule extends about 2 cm. from the crests of the trochlear ridges toward the femoral epicondyles. Distally, the patellar and femorotibial parts join without sharp demarcations. Distal to the patella the synovial and fibrous layers of the joint capsule are separated by a quantity of fat, the **infrapatellar fat body** (*corpus adiposum infrapatellare*), which increases in thickness distally. The femorotibial sacs are less roomy than the femoropatellar. Both femorotibial sacs are partly divided by the menisci into femoromeniscal and tibiomeniscal parts. The fibrosa of the capsule is firmly attached to the discs so that the two parts communicate only around their concave, sharp-edged axial borders, where the tibial and femoral condyles contact each other. A free transverse communication also exists between the lateral and medial condyloid parts of the joint. Both the lateral and medial condyloid parts extend between the caudal, proximal parts of the femoral condyles and the fabellae that articulate with them. The lateral femorotibial joint capsule has three other subpouches, in addition to the extension between the lateral fabella and femur. One of these extends laterally between the head of the fibula and the lateral condyle of the tibia to form the proximal tibiofibular joint capsule. This subpouch is tight-walled, since little movement is necessary here. Cranial to this articulation in the sulcus muscularis of the tibia another pouch extends distally about 2 cm. The tendon of the m. extensor digitorum longus at its origin from the extensor fossa of the femur (in the proximal part of the sulcus muscularis) is ensheathed by synovial membrane. The tendon of origin of the m. popliteus on the lateral epicondyle of the femur is never completely surrounded by synovial membrane, but it does possess, on its deep surface, a well-defined synovial pouch which acts as a bursa.

Text continued on page 263.

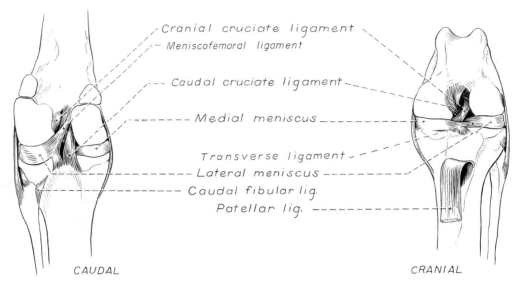

Cranial cruciate ligament
Meniscofemoral ligament
Caudal cruciate ligament
Medial meniscus
Transverse ligament
Lateral meniscus
Caudal fibular lig.
Patellar lig.

CAUDAL   CRANIAL

Figure 5–24. Ligaments of left stifle joint.

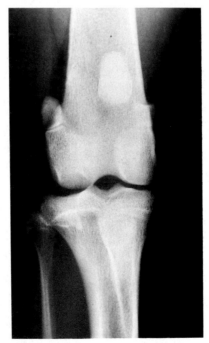

Figure 5–24. Radiograph. Craniocaudal radiograph, left stifle joint.

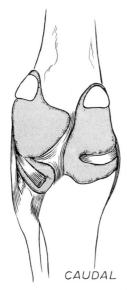 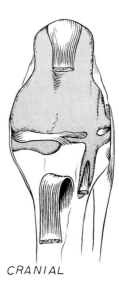

CAUDAL   CRANIAL

Figure 5–25.   Capsule of left stifle joint.

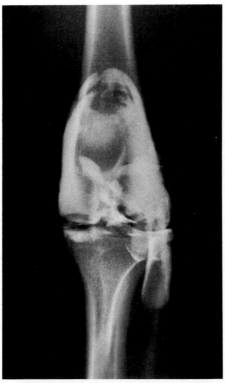

Figure 5–25.   Radiograph. Craniocaudal radiograph, left stifle joint; positive contrast arthrogram performed to define the synovial space.

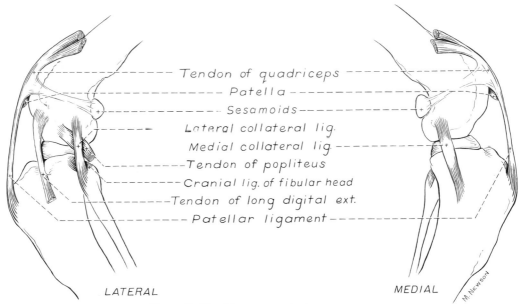

Tendon of quadriceps
Patella
Sesamoids
Lateral collateral lig.
Medial collateral lig.
Tendon of popliteus
Cranial lig. of fibular head
Tendon of long digital ext.
Patellar ligament

LATERAL   MEDIAL

Figure 5–26.   Ligaments of left stifle joint.

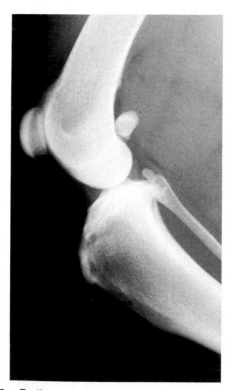

Figure 5–26.   Radiograph. Lateromedial radiograph, left stifle joint.

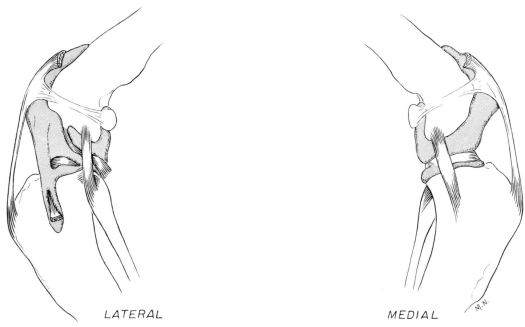

LATERAL   MEDIAL

Figure 5-27. Capsule of left stifle joint.

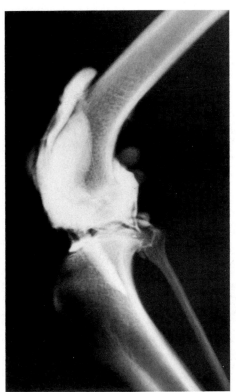

Figure 5-27. Radiograph. Lateromedial radiograph, left stifle joint; positive contrast arthrogram performed to define the synovial space.

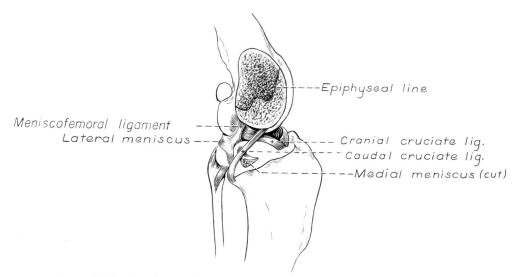

Figure 5–28. Cruciate and meniscal ligaments of left stifle joint, medial aspect.

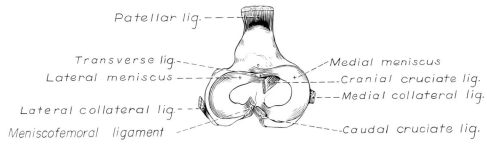

Figure 5–29. Menisci and ligaments of left stifle joint, dorsal aspect.

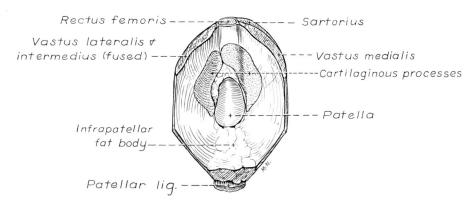

Figure 5–30. Patella, caudal aspect.

The **lateral and medial menisci** *(meniscus lateralis et medialis)* are semilunar fibrocartilaginous discs with sharp, deeply concave axial, and thick convex abaxial borders. The lateral meniscus is slightly thicker and forms a slightly greater arc than the medial one. In large dogs the peripheral border of the lateral meniscus measures about 8 mm. The lateral meniscus does not reach the border of the tibia caudolaterally, but allows the tendon of origin of the m. popliteus to pass over the tibial condyle.

LIGAMENTS OF STIFLE JOINT

The **meniscal ligaments** attach the menisci to the tibia and femur. Four of these, two from each meniscus, go to the tibia. Nomina Anatomica Veterinaria (1973) recognizes only two meniscal ligaments: one from the lateral meniscus to the femur and one transverse ligament between menisci.

The **cranial tibial ligament of the medial meniscus** goes from the cranial, axial angle of the medial meniscus to the cranial intercondyloid area of the tibia. This attachment is immediately cranial to the intermeniscal ligament, the cranial tibial attachment of the lateral meniscus, and the tibial attachment of the cranial cruciate ligament.

The **caudal tibial ligament of the medial meniscus** goes from the caudal axial angle of the medial meniscus to the caudal intercondyloid area of the tibia. This attachment is just cranial to the tibial attachment of the caudal cruciate ligament.

The **cranial tibial ligament of the lateral meniscus** goes to the cranial intercondyloid area of the tibia, where it attaches caudal to the cranial tibial attachment of the medial meniscus.

The **caudal tibial ligament of the lateral meniscus** goes from the caudal axial angle of the lateral meniscus to the popliteal notch of the tibia just caudal to the caudal intercondyloid area of the tibia.

The **femoral ligament of the lateral meniscus** *(lig. meniscofemorale)* is the only femoral attachment of the menisci. It passes from the caudal axial angle of the lateral meniscus dorsally to that part of the medial femoral condyle which faces the intercondyloid fossa.

The **transverse or intermeniscal ligament** *(lig. transversum genus)* is a small transverse fibrous band which leaves the caudal side of the cranial tibial ligament of the medial meniscus and goes to the cranial part of the cranial tibial ligament of the lateral meniscus.

The **femorotibial ligaments** are the collateral and the cruciate ligaments. The **cruciate ligaments of the stifle** *(ligg. cruciata genus)* are located within the joint cavity.

The **medial (tibial) collateral ligament** *(lig. collaterale mediale)* is a strong ligament which extends between the medial epicondyle of the femur and the medial border of the tibia about 2 cm. distal to the medial tibial condyle. As it passes over this condyle a bursa is interposed between the ligament and the bone. The total length of the ligament is over 4 cm. in medium-sized dogs. It fuses with the joint capsule and the medial meniscus.

The **lateral (fibular) collateral ligament** *(lig. collaterale laterale)* is similar to its fellow in size and length. As it crosses the joint cavity it passes over the tendon of origin of the m. popliteus. It ends distally on the head of the fibula, with a few fibers going to the adjacent lateral condyle of the tibia.

The **cranial (lateral) cruciate ligament** *(lig. cruciatum craniale)* runs from the caudomedial part of the lateral condyle of the femur somewhat diagonally across the intercondyloid fossa to the cranial intercondyloid area of the tibia.

The **caudal (medial) cruciate ligament** *(lig. cruciatum caudale)* runs from the lateral surface of the medial femoral condyle caudodistally to the lateral edge of the popliteal notch of the tibia. The caudal cruciate ligament is slightly heavier and definitely longer than the cranial one. As their name implies, the cruciate ligaments decussate, or cross each other. The caudal cruciate ligament lies medial to the cranial one. Being intra-articular, they are covered by synovial membrane, which, in fact, forms an imperfect sagittal septum in the joint. However, this is incomplete, allowing right and left parts to communicate.

Although the patella is a large sesamoid bone intercalated in the tendon of insertion of the m. quadriceps femoris, it is acceptable to regard the tendon from the patella to the tibial tuberosity as the **patellar ligament** *(lig. patellae).* The patellar ligament is separated from the joint capsule by a large quantity of fat, which is particularly thick distally. Between the distal part of the patellar ligament and the tibial tuberosity, just proximal to its attachment, there is frequently located a small synovial bursa. The patella is held in the trochlea of the femur mainly by the heavy lateral femoral fascia, or fascia lata, and the lighter medial femoral fascia. Aiding in this function are the delicate **medial and lateral femoropatellar ligaments** *(ligg. femoropatellare mediale et laterale).* They are narrow bands of loose fibers which partially blend with the overlying femoral fasciae. The lateral band can usually be traced from the lateral side of the patella to the fabella in the lateral head of the m. gastrocnemius. The medial ligament, weaker than the lateral, usually blends with the periosteum of the medial epicondyle of the femur. The sides of the patella are continued into the femoral fascia by the **medial and lateral parapatellar fibrocartilages** *(cartilago parapatellaris medialis et lateralis).* These usually meet dorsally. Baum and Zietzschmann (1936) mention a suprapatellar fibrocartilage being present in older dogs in the tendon of the m. rectus femoris. The lateral and medial cartilages ride on the crests of the femoral trochlea and tend to prevent dislocation of the patella.

Paatsama (1952) has shown that a ruptured cruciate ligament may be reconstituted by fascia lata, which will provide a new functional ligament. Free forward movement of the tibia, with the joint in extension, is positive evidence of a rupture of the cranial cruciate ligament (Schroeder and Schnelle 1941). The cranial cruciate ligament is the one most often torn or severed, as a result, usually, of trauma. Hyperflexion is more likely to cause it than other movements. In extreme instances the tibial collateral ligaments also may be ruptured. Schreiber (1947) has written an anatomical treatise on the canine stifle joint.

It has long been known that the removal of all or part of a meniscus results in a regenesis of the excised part. According to Smillie (1943) and Nilsson (1949), the regenerated portion consists entirely of connective tissue. By close attention to the radii of patellar movement the femoral trochlea, the femoropatellar ligaments can be successful-

ly reinforced to prevent chronic luxation of the patella.

Tibiofibular Joints

The fibula articulates with the tibia at each end by small synovial cavities and, in addition, possesses an extensive tibiofibular syndesmosis. Barnett and Napier (1953) studied the rotatory mobility of the fibula in eutherian mammals and concluded that in the dog no rotation could be demonstrated on passive movements of the foot.

The **proximal tibiofibular joint** *(articulatio tibiofibularis proximalis)* is small and tightly fitting; its synovial lining is a distal extension of the lining for the lateral femorotibial part of the stifle joint capsule. The fibrous layer is not well developed, although a recognizable spray of short fibers goes from the head of the fibula proximocaudally under the fibular collateral ligament to the adjacent lateral condyle of the tibia as the **ligament of the fibular head** *(lig. capitis fibulare).* **The interosseous membrane of the crus** *(membrana interossea cruris)* extends from the proximal to the distal tibiofibular articular capsule. The fibula has many muscles attaching to it, and many of these extend beyond their fibular attachment to the interosseous membrane, which, more than anything else, fastens the fibula to the tibia. The fibers which compose this fibrous sheet decussate, forming a lattice-like flat ligament. Proximally an opening exists in the ligament for the passage of the large cranial tibial artery and its small satellite vein.

The **distal tibiofibular joint** *(articulatio tibiofibularis distalis)* receives an extension of the synovial membrane from the lateral side of the talocrural joint. Like the proximal tibiofibular joint, the distal joint is hardly more than a synovial pocket between the lateral malleolus and the distal lateral surface of the tibia. Besides the strong connection to the calcaneus, fourth tarsal, and metatarsal V, by means of the lateral collateral ligament, the lateral malleolus of the fibula has a **cranial tibiofibular ligament** *(lig. tibiofibulare craniale)* which runs a short distance transversely from the cranial edge of the lateral malleolus to the adjacent lateral surface of the tibia. The lateral collateral ligament has short (deep) and long (superficial) parts (Fig 5–33).

Tarsal, Metatarsal, and Phalangeal Joints (Articulationes Pedis)

TARSAL JOINTS

The **tarsal joints** *(articulationes tarsi* (Figs. 5–31 to 5–33), like the carpal joints, are composite articulations. The **talocrural joint** *(articulatio talocruralis),* or ankle joint, permits the greatest degree of movement. The trochlea of the tibial tarsal, formed largely of two articular ridges, fits into reciprocal grooves which form the cochlea of the tibia. The grooves and ridges are not quite in sagittal planes but deviate laterally about 25 degrees so that the open angle faces cranially. This allows the hind-paws to be thrust past the forepaws on the outside when the dog gallops. Owing to the presence of the central tarsal bone in the medial half of the tarsus, the **intertarsal joint** *(articulatio intertarsea)* is divided into the **proximal intertarsal joint** *(articulatio intertarsea proximalis)* and the **distal intertarsal joint** *(articulatio intertarsea distalis).* The middle tarsal joint of the lateral side and the proximal middle tarsal joint of the medial side are coextensive; they are formed be-

tween the talus (tibial tarsal) and the calcaneus (fibular tarsa) proximally, and between the central and the fourth tarsal distally. Some side movement as well as flexion and extension are possible here as the slightly convex distal ends of the talus and calcaneus fit into glenoid cavities of the central and fourth tarsals. The four distal tarsal bones, the first to fourth, articulate with metatarsals I to V, forming the **tarsometatarsal joints** *(articulationes tarsometatarseae).* The vertical joints between the individual bones of the tarsus are the **intratarsal joints** *(articulationes intratarsea),* all of which are exceedingly rigid.

The fibrous part of the **tarsal joint capsule** extends from the periosteum proximal to the distal articular cartilage of the tibia and fibula to the proximal ends of the metatarsal bones. As the fibrous layer covers the individual tarsal bones and their ligaments, it fuses to the free surfaces of the bones and ligaments. Like the comparable carpal joint capsule, thickenings are found on both the dorsal and plantar surfaces. On the plantar surface the fibrous part is deeply concave transversely and thickened distally to form the deep wall of the **tarsal canal** *(canalis tarsi).* The tarsal canal contains the tendon and sheath of the m. flexor hallucis longus, the plantar branches of the saphenous artery

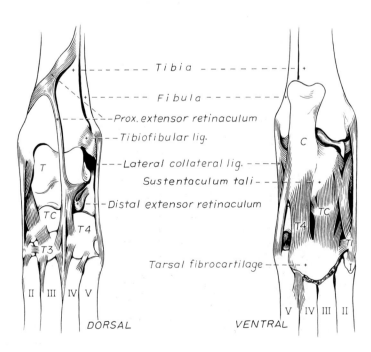

Figure 5–31. Ligaments of left tarsus. C = calcaneus. T = talus. T1, T3, T4 = first, third, fourth tarsals. TC = central tarsal. I to V = metatarsals.

Tibia

Fibula

Prox. extensor retinaculum

Tibiofibular lig.

Lateral collateral lig.

Sustentaculum tali

Distal extensor retinaculum

Tarsal fibrocartilage

DORSAL

VENTRAL

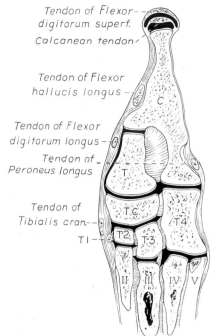

Figure 5–32. Schematic section of left tarsus showing articular cavities, dorsal aspect.

   C = calcaneus.

   T = talus.

   T2, T3, T4 = second, third, fourth tarsals.

   TC = central tarsal.

   II to V = metatarsals.

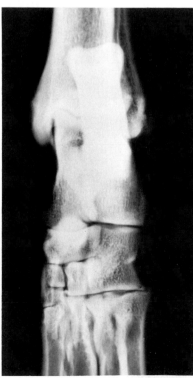

Figure 5–32. Radiograph. Dorsoplantar radiograph, left tarsus.

and vein, and the medial and lateral plantar nerves. The synovial layer of the joint capsule extends to the edges of the articular cartilages. There are three lateral and four medial joint sacs. Proximally the largest sac lines the most freely movable joint of the tarsus, the talocrural joint. Distal to this, in the medial part of the tarsus, are the proximal and distal middle tarsal sacs. Laterally, only a single middle tarsal sac exists between the calcaneus and the fourth tarsal. Between the tarsus and metatarsus is the tarsometatarsal sac enclosing the joint which extends between the distal row of tarsal bones, the first to fourth, and the bases of metatarsals I to V. According to Baum and Zietzschmann (1936), the talocrural and the proximal middle intertarsal sacs communicate with each other and these communicate with the synovial sheath surrounding the tendon of the m. flexor hallucis longus. These authors further state that the distal middle tarsal sac communicates with the tarsometatarsal sac, but that the two intercommunicating proximal sacs do not communicate with the two intercommunicating distal sacs.

The **medial collateral ligament** *(lig. collaterale mediale)* (Fig. 5–33) is divided into a long and a short part. The long, more superficial part is a rather strong band which runs from the medial (tibial) malleolus to attach firmly to the first tarsal and feebly to metatarsals I and II. As it crosses the tarsus it attaches weakly to the free surface of the talus and strongly to the free surface of the central. The short part, attaching craniodistal to the long part, divides as it passes under it. One part extends caudally to attach on the talus. The other part is longer and parallels the long part of the medial collateral ligament, caudal to which it lies, after passing under it. A few fibers may end distally on the sustentaculum tali of the calcaneus by a fascial connection. However, it attaches primarily to the first tarsal and metatarsal bones.

The **lateral collateral ligament** *(lig. collaterale laterale),* like the medial collateral ligament, is divided into a long and a short part. The long part passes from the lateral (fibular) malleolus to the base of metatarsal V, attaching along its course to the calcaneus and fourth tarsal. The short part lies under the long part proximally. From the lateral malleolus one band extends to the

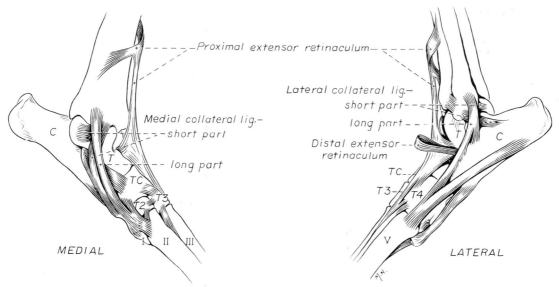

Figure 5–33.   Ligaments of left tarsus.
C = calcaneus.
T = talus. T2, T3, T4 = second, third, fourth tarsals.
TC = central tarsal. I to V = metatarsals.

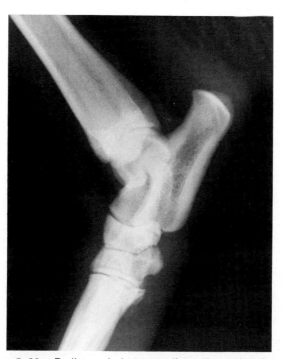

Figure 5–33.   Radiograph. Lateromedial radiograph, left tarsus.

tuber calcanei of the calcaneus; a second band goes to the more dorsally located talus. Both bands run at nearly right angles to the long part of the lateral collateral ligament.

On the dorsal surface of the tarsus there are various short dorsal ligaments. One prominent ligament unites the talus with the third and fourth tarsals. It blends proximally with the **proximal transverse ligament of the tarsus** (*lig. tarsi transversi proximalis*), or the **proximal extensor retinaculum,** which holds the tendons of the mm. extensor digitorum longus, extensor hallucis longus, and tibialis cranialis to the tibia. A small band connects the second and third tarsals. Oblique bands exist between the central and second tarsals, as well as between the central and third tarsals. The distal row of tarsal bones is joined to the proximal ends of the metatarsal bones by small vertical ligaments on the dorsal surface. A ligamentous loop which attaches to the calcaneus surrounds the tendon of the m. extensor digitorum longus. This is the so-called **distal transverse ligament of the tarsus** (*lig. tarsi transversi distalis*), or the **distal extensor retinaculum.** On the plantar surface of the tarsus the special plantar ligaments are heavier than those on the dorsal side. Most of these fuse distally with the thickened part of the joint capsule at the tarsometatarsal junction. Several of these ligaments are distinct. A ligament from the body of the calcaneus passes over and is attached to the fourth tarsal on its way to the bases of metatarsals IV and V. Another ligament leaves the plantar surface of the sustentaculum tali, and attaches to and passes over the central tarsal to end in the thickened tarsometatarsal joint capsule. Laterally, a conspicuous band leaves the caudolateral surface of the calcaneus and blends with the long fibular collateral ligament attached to the base of metatarsal V.

METATARSAL AND PHALANGEAL JOINTS

The joints and ligaments of the metatarsus and digits are similar to the comparable joints and ligaments of the forepaw.

BIBLIOGRAPHY

Ansulayotin, C. 1960. Nerve Supply to the Shoulder, Elbow, Carpal, Hip, Stifle and Tarsal Joints of the Dog as Determined by Gross Dissection. Thesis, Cornell University, Ithaca, New York.

Barnett, C. H., D. V. Davies, and M. A. MacConaill, 1961. Synovial Joints; Their Structure and Mechanics. Springfield, Ill., Charles C Thomas.

Barnett, C. H., and J. R. Napier. 1953. The rotatory mobility of the fibula in eutherian mammals. J. Anat. 87:11–21.

Bauer, W., C. L. Short, and G. A. Bennett. 1933. The manner of removal of proteins from normal joints. J. Exp. Med. 57:419–433.

Baum, H., and O. Zietzschmann. 1936. Handbuch der Anatomie des Hundes. 2nd Ed. Berlin, Paul Parey.

Bennett, G. A., W. Bauer, and S. J. Maddock. 1932. A study of the repair of articular cartilage and the reaction of normal joints of the adult dogs to surgically created defects of articular cartilage, "joint mice" and patellar displacements. Am. J. Path. 8:499–523.

Coggeshall, H. C., C. F. Warren, and W. Bauer. 1940. The cytology of normal human synovial fluid. Anat. Rec. 77:129–144.

Davies, D. V. 1944. Observations on the volume, viscosity, and nitrogen content of synovial fluid, with a note on the histological appearance of the synovial membrane. J. Anat. 78:68–78.

Dieterich, H. 1931. Die Regeneration des Meniscus. Dtsch. Ztschr. Chir. 230:251–260.

Gardner, E. 1950. Physiology of movable joints. Physiol. Rev. 30:127–176.

Kadletz, M. 1932. Anatomischer Atlas der Extremitätengelenke von Pferd und Hund. Berlin, Wien, Urban & Schwarzenberg.

MacConaill, M. A. 1932. The function of intra-articular fibrocartilages, with special reference to the knee and inferior radio-ulnar joints. J. Anat. 66:210–227.

Nilsson, F. 1949. Meniscal injuries in dogs. North Am. Vet. 30:509–516.

Nomina Anatomica. 1977. International Anatomical Nomenclature Committee. 4th Ed. Amsterdam, Excerpta Medica Foundation.

Nomina Anatomica Veterinaria. 1973. International Committee on Veterinary Anatomical Nomenclature 2nd Ed. Vienna, World Assoc. Vet. Anat.

Riser, W. H. 1975. The dog as a model for the study of hip dysplasia. Vet. Pathol. 12:229–334.

Schreiber, J. 1947. Beiträge zur vergleichenden Anatomie und zur Mechanik des Kniegelenkes. Wiener Tierärztl. Monatsschr. 34:725–744.

Schroeder, E. F., and G. B. Schnelle. 1941. Veterinary radiography; the stifle joint. North Am. Vet. 22:353–360.

Smillie, I. S. 1943. Observations on the regeneration of the semilunar cartilages in man. Brit. J. Surg. 31:398–401.

Suter, P. F., and A. V. Carb. 1969. Shoulder arthrography in dogs — radiographic anatomy and clinical application. J. Small Anim. Pract. 10:407–413.

Chapter 6

MUSCLES

The muscular system is composed of contractile units of varied morphology, energized by voluntary or involuntary nerve impulses or humoral substances. Muscles provide the necessary motive power for locomotion, respiration, circulation, alimentation, and a host of other functions, including the indication of emotional states via barking, facial expression, raising the hair, or wagging the tail. The functional cell unit is known as a **muscle fiber,** and it is customary to classify muscle fibers as smooth, cardiac, or skeletal.

Smooth muscle fibers are spindle-shaped in form, with a central nucleus. Like other muscle cells they possess myofibrils, but they are homogeneous and not striated. They are found in the walls of hollow organs, and in blood vessels, as well as in association with glands, and with the spleen, the eyeball, and hair follicles of the skin. Smooth muscle is innervated by the autonomic nervous system, and in many cases is also under humoral control. Other names that have been used for smooth muscle are: unstriated, plain, involuntary, white, or visceral muscle.

Cardiac muscle fibers form the bulk of the heart. The fibers are arranged in a network of individual cellular units with intercalated discs between the cell extremities. They exhibit cross striations, as do skeletal muscle fibers, and have centrally placed nuclei like smooth muscle fibers. Cardiac muscle is capable of rhythmic contractions and is under autonomic control. Specialized cardiac muscle fibers (Purkinje fibers) serve as a conducting system for impulses within the heart.

Skeletal muscle fibers are long, cylindrical, multinucleated cells organized into distinct bundles with connective tissue envelopes. Other names applied to skeletal muscle include striated, voluntary, or somatic muscle. The cells appear striated because the light and dark bands of adjacent myofibrils are in register with each other. Each **muscle fiber** is composed of several hundred or several thousand parallel **myofibrils,** which also exhibit cross striations. The myofibril is in turn composed of several hundred thick and thin **myofilaments,** which consist of the proteins myosin (thick) and actin (thin). These myofilaments alternate and interdigitate along the length of the myofibril and thus produce the characteristic alternation of the isotropic (I), or light, bands and the anisotropic (A), or dark, bands. The control of skeletal muscles is largely voluntary, although they are capable of involuntary contraction and reflex movements. There are some muscles, such as the retractor penis, which have both smooth and skeletal muscle fibers.

Another way to classify muscles is based on their developmental origin and innervation. Thus one can speak of somatic muscles with striated fibers and somatic motor innervation versus visceral muscles with smooth, striated, or cardiac muscle fibers and visceral motor innervation.

For a consideration of structural detail, the reader is referred to any of the standard histology texts; for muscles as functional units in regard to mechanics see Basmajian (1974); and for an overview of structure, biochemistry, physiology, pharmacology, and disease of muscle refer to Bourne (1972–73). The phylogenetic history of muscles as seen in lower vertebrates offers many insights for explaining observed anomalies, deficiencies, or excesses in mammals. For a general review of comparative aspects of the muscular system, see Romer and Parsons (1977).

SKELETAL MUSCLES

The present chapter is concerned primarily with the axial and appendicular muscles of the body. In mammalian species the skeletal muscles comprise approximately one-third to one-half of the total body weight. They range in size from the minute stapedius muscle of the middle ear to the large gluteus medius muscle of the rump.

Each muscle fiber is surrounded by a thin **sarcolemma** and a delicate connective tissue sheath known as the **endomysium.** When several fibers are grouped into a fasciculus they are enclosed by **perimysium.** The definitive muscle is composed of several fasciculi wrapped by an **epimysium,** which delimits one muscle from another or occasionally fuses with the intervening fascia. The size of an individual muscle fiber depends on the species as well as on the physical condition of the animal, since individual muscle fibers are capable of hypertrophy as well as atrophy.

Lockhart and Brandt (1938) found muscle fibers running the entire length of the sartorius muscle (5 cm.) in a human fetus, and were able to isolate fibers 34 cm. long in a 52-cm. sartorius muscle of an adult. Huber (1916) and Van Harreveld (1947), working with rabbit thigh muscles, found that many fibers do not extend from end to end. They concluded that, although the longer fasciculi have longer fibers, many fibers end intrafascicularly.

Muscles take diverse shapes and are usually named according to some structural or functional feature, although other criteria have also been used. The variations encountered in the muscular system within a species are numerous and may constitute a breed-specific feature. Huntington (1903) considered problems of gross myological re-

search and the significance and classification of muscular variations. The most complete account of the muscles in the dog is by Baum and Zietzschmann (1936).

Origin and Insertion

Most skeletal muscles are attached by connective tissue to a bone or cartilage. Some are attached to an organ (eye, tongue), to another muscle, or to the skin; others lie free beneath the skin and act as sphincters of orifices. The connective tissue attachment may be in the form of a cordlike **tendon** or a flat, sheetlike **aponeurosis.** Some muscles have no demonstrable tendons or aponeuroses but attach directly to the periosteum of bones. Such origins or insertions are spoken of as fleshy attachments. The more fixed point of muscle attachment is spoken of as the **origin;** the more movable point of attachment is called the **insertion** or **termination.** In the limb the insertion of a muscle is always considered to be distal to its origin, although functionally it may be the most fixed point at some phase of the stride. Certain muscles have equally fixed or mobile attachments, and the naming of an origin and an insertion is rather arbitrary.

The expanded fleshy portion of a muscle is its **belly,** the origin is a **head,** and the termination is a **tail.** Minor divisions of origin or termination are called **slips.** A muscle may have more than one belly (digastric) or more than one head (triceps) and several slips.

Function

Muscles which attach to long bones (the levers) and span one or more joints usually work at a mechanical disadvantage. When a muscle fiber contracts it does so at its maximum power, and it is capable of contracting to about half of its stretched length. The contraction is initiated by a nerve impulse traveling over a motor nerve fiber (axon) to the muscle fibers, or cells. Each axon supplies several muscle fibers. These neuromuscular units are known as **motor units,** and the number of motor units functioning at any one time determines the activity of the muscle. If a muscle has many motor units, each of which includes only a few muscle fibers, then the precision of movement is great (as in the extrinsic muscles of the eyeball).

It is important to study and experiment with muscles in the living body to appreciate the full significance of precise muscular movement and the value of such movement in a neurological examination for the determination of intact or defective nerve supply. Electromyography is an excellent technique for studying living muscles.

Of two muscles of equal size and shape, the muscle which contains the greater number of fibers is the stronger. Straplike and sheetlike muscles contract to a greater degree than do those of the extremities. Muscles possessing tendons throughout all of their length are known as pennate muscles. A muscle with a tendon running along one side is called **unipennate**; if there is a tendon on each side of the muscle, it is **bipennate**; when a muscle is invaded by tendons in several places, it is **multipennate**. Pennate muscles are stronger (exert more force) than straplike or sheetlike types because they are composed of many short oblique fibers which have an additive effect upon the insertion. Because of their elasticity, tendons protect muscles from sudden strains. Fleshy, rectangular muscles usually attach farther from the fulcrum than do pennate types; they also move their insertions farther, but with less speed.

Muscles which straighten bone alignment, or open a joint, are called **extensors;** those which angulate the bones, or bend the joint, are known as **flexors.** Flexion and extension are the primary movements necessary for locomotion. Accompanying movements include **adduction,** or the movement of an extremity toward the median plane; **abduction,** or movement away from the median plane (in the case of the digits the reference point is the axis of the limb; **circumduction,** or moving an extremity in a plane describing the surface of a cone; and **rotation,** or moving a part around its long axis. The pattern of movement resulting from muscle contractions, even for apparently simple movements, is brought about by the complex interactions of many muscles. The characteristic movement of a joint is produced by a muscle or muscles, called **prime movers,** or **agonists.** The muscles responsible for the opposite action are known as **antagonists,** although they actually aid the prime mover by relaxing in a controlled manner so that the movement will be smooth and precise. For the elbow joint, the prime mover in flexion is the biceps brachii; the antagonist is the triceps. Conversely, in bringing about extension, the prime mover is the triceps. **Fixation and articular muscles** are those which stabilize joints while the prime movers are acting. **Synergists** are fixation muscles which stabilize intermediate or proximal joints and enable the force of the prime mover to be exerted on a more distal joint.

Accessory Structures

Associated with muscles are accessory structures of great physiological and clinical importance, such as sesamoid bones, bursae, synovial tendon sheaths, and fascia.

Sesamoid bones are located in certain tendons or joint capsules as small, rounded nodules. Occasionally they develop in response to friction, but usually they form prenatally. The patella, or knee cap, is an example of a large sesamoid bone in the tendon of insertion of the quadriceps femoris muscle. Sesamoid bones serve three important functions: (1) they protect tendons which pass over bony prominences; (2) they increase the surface area for attachment of tendons over certain joints; and (3) they serve to redirect the pull of tendons so that greater effective force can be applied to the part being moved.

Bursae are simple connective tissue sacs containing a viscous fluid and serving to reduce friction. They are usually located between a tendon, ligament, or muscle, and a bony prominence. Occasionally they are located between tendons or between a bony prominence and the skin. Inconstant bursae may develop at various sites in response to undue friction, and, conversely, cellular proliferation due to infection or trauma may eliminate them.

Synovial tendon sheaths are double-layered, elongated sacs containing synovia which wrap tendons as they pass through osseous or fibrous grooves. The inner layer of the sheath, which is fused to the tendon, attaches to the wall of the passageway and is known as the **mesotendon.** Blood vessels and nerves enter the tendon via the mesotendon. The tendon sheath with its contained synovia serves for reducing friction during movement. The sleeve formed by the superficial digital flexor tendon around the deep digital flexor tendon at each metacarpo- and metatarsophalangeal joint is known as a *manica flexoria*.

Fascia is connective tissue which remains after the recognizable mesodermal structures have been differentiated in the fetus. It serves many important functions, and has considerable clinical significance. For descriptive purposes it is convenient to distinguish many fascial entities which envelop, separate, or connect muscles, vessels, and nerves. Fascial sheets provide routes for the passage of blood vessels, lymphatics, and nerves, as well as serving for the storage of fat. The superficial fascia beneath the skin is closely associated with the dermis, and often includes cutaneous muscle fibers. The deep fascia which covers and passes between the muscles is particularly tough and distinct in the limbs. It functions as a sleeve within which the muscles can operate, and often serves as an aponeurosis of origin or insertion. In certain locations fascia blends with the periosteum of bone, forming interosseous membranes or annular bands which confine tendons or redirect their force. Most commonly, distinct fascial septa separate groups of muscles from one another and result in fascial planes along which infection may spread or fluids drain.

Connective Tissue

The amount of connective tissue present is much greater in some muscles than in others. When the connective tissue content is high, the muscle has a high tensile strength and tends to be capable of more finely graded movements. Connective tissue elements include collagen fibers, elastic fibers, reticular fibers, fibroblasts, and histiocytes.

Blood and Nerve Supply

Muscles have a high metabolic rate and are well supplied with blood by twigs from neighboring blood vessels. The arteries supplying a muscle enter at rather definite places and often anastomose within the muscle. There is much constancy in arterial supply, although variations do occur. Lymphatics accompany the arteries and, like them, form capillary plexuses around the muscle fibers. Veins also accompany the arteries, and during muscular contraction blood is forced into the larger veins, which, as a rule, are more superficial than the arteries.

Nerves accompany the blood vessels and ramify within the muscle. About half of the fibers are motor and the other half sensory. Efferent nerve fibers form motor end-plates which are neuromuscular junctions on muscle fibers. Sensory receptors of a muscle include spindles, Golgi tendon endings, free nerve endings, and paciniform corpuscles, which discharge proprioceptive impulses in response to relaxation or contraction of the muscle, and modify the activities of motor neurons. Matthews (1972) has summarized findings related to primary and secondary endings, static and dynamic fibers, feedback loops, and possible mechanisms of muscle spindle operation.

Regeneration

Mammalian skeletal muscle fibers are capable of regeneration, although the success of the reparative process is variable. Surgical implants of minced muscle (Carlson 1972) regenerate to about 25 per cent of their former bulk, whereas transplanted whole muscle regains about 80 per cent of its volume and function. The inward progression of regeneration of an implant is correlated with its revascularization. The ability of minced muscle to survive vascular deprivation is one of the striking features of muscle regeneration. The regenerative process can be aborted by conditions which stimulate connective tissue formation, such as circulatory insufficiency, widening of the gap, infection, or the presence of foreign bodies. The ability of cardiac muscle to regenerate is denied by some authors and supported by others. Field (1960), after reviewing the literature, concluded that, although cardiac muscle has less regenerative capacity than skeletal muscle even under optimal conditions, it does at times exhibit appreciable regeneration.

MUSCLES OF THE HEAD

The muscles of the head are composed of six groups on the basis of their embryonic origin and their innervation: (1) the facial musculature, innervated by branches of the facial nerve; (2) the masticatory musculature, innervated by the mandibular branch of the trigeminal nerve; (3) the tongue musculature, supplied by the hypoglossal nerve; (4) the pharyngeal musculature, under the control of the glossopharyngeal and vagus nerves; (5) the laryngeal musculature, supplied for the most part by the vagus nerve; and (6) the eye musculature, innervated by the oculomotor, trochlear, and abducent nerves. The cranial muscles of many vertebrates have been described by Edgeworth (1935). The facial musculature of the dog has been described and illustrated by Huber (1922, 1923).

Superficial Muscles of the Face

MUSCLES OF THE CHEEK AND LIPS

The superficial muscles of the face are derived from three primary layers of the primitive sphincter colli. They include the m. sphincter colli superficialis, platysma, and m. sphincter colli profundus.

The *m. sphincter colli superficialis* (Fig. 6–1) is best developed in the laryngeal region beneath the skin. Its delicate transverse fibers span the ventral borders of the platysma muscles at the junction of the head and neck. Caudally the fibers of the loose sphincter colli superficialis blend with isolated bundles of fibers from the sphincter colli primitivus. Occasionally fibers of the sphincter colli superficialis reach the thorax, radiate over the shoulder joint, or blend with the cervical part of the platysma.

The *platysma* is a well-developed muscle sheet which takes origin from the mid-dorsal tendinous raphe of the neck and the skin. The two separate layers of origin fuse near the midline. In its longitudinal course it extends over the parotid and masseter re-gions to the cheek and commissure of the lips, where it radiates into the m. orbicularis oris. In the lips the platysma is designated as the m. cutaneus labiorum. At the ventral midline these bilateral cutaneous muscles, when they are well developed, approach each other and meet behind a transverse plane through the commissures of the lips. The platysma covers large portions of the m. sphincter colli profundus; its dorsal border, extending from the neck to the commissural portion of the upper lip, is united by many fiber bundles with the underlying sphincter colli profundus. The ventral border has a distinct boundary. Only rarely does the platysma have defects.

Action: To draw the commissure of the lips caudally.

Innervation: Rami buccalis dorsalis et ventralis, and ramus auricularis caudalis, n. facialis.

The *m. sphincter colli profundus* includes muscles of the cheek, lip, and external ear. It is divided into a pars oralis, pars palpebralis, pars intermedia, and pars auricularis.

Pars oralis. The pars oralis of the sphincter colli profundus includes the mm. orbicularis oris, incisivus superioris et inferioris, maxillonasolabialis, buccinator, and mentalis.

The *m. orbicularis oris* (Figs. 6–1, 6–2), the principal component of the lips, extends from the commissural region into the lips near their free borders. In the median segment of both lips, the muscle is interrupted. It lies between the skin and mucosa. The other muscles of the lips and the muscles of the cheeks (platysma, and mm. buccinator, zygomaticus, and levator nasolabialis) enter the m. orbicularis oris so that here these muscles blend with each other; the m. incisivus is also attached to it. The portion of the orbicularis oris lying in the upper lip is the stronger; separate fibers extend from it to the external naris.

Action: The muscle closes the mouth opening and is a pressor of the labial glands. Of the bundles extending to the lateral nasal cartilage on either side, the medial ones act to pull the entire nose downward (in sniffing), and the lateral bundles act to increase the diameter of the external nares. In strong contrac-

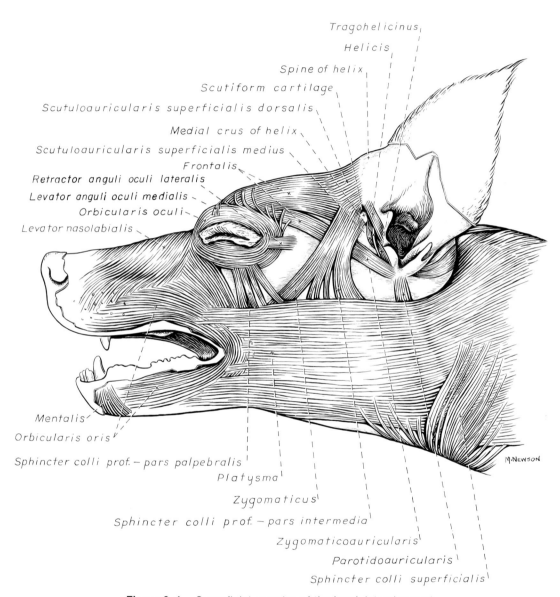

Tragohelicinus

Helicis

Spine of helix

Scutiform cartilage

Scutuloauricularis superficialis dorsalis

Medial crus of helix

Scutuloauricularis superficialis medius

Frontalis

Retractor anguli oculi lateralis

Levator anguli oculi medialis

Orbicularis oculi

Levator nasolabialis

Mentalis

Orbicularis oris

Sphincter colli prof. — pars palpebralis

Platysma

Zygomaticus

Sphincter colli prof. — pars intermedia

Zygomaticoauricularis

Parotidoauricularis

Sphincter colli superficialis

M·NEWSON

Figure 6–1. Superficial muscles of the head, lateral aspect.

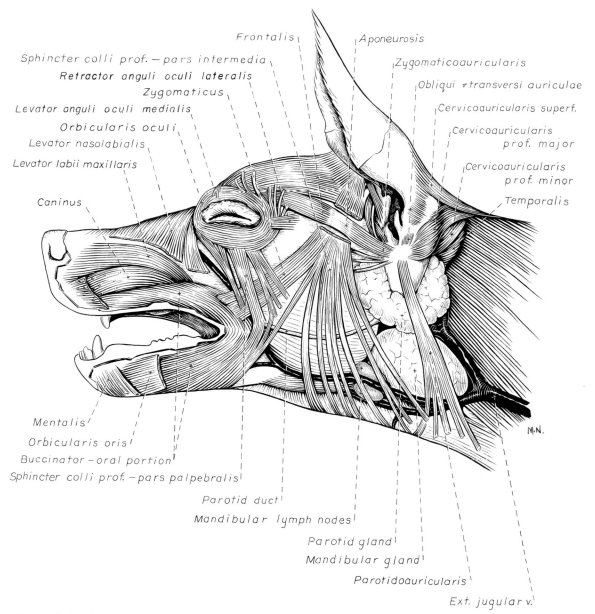

Frontalis
Aponeurosis
Sphincter colli prof. — pars intermedia
Zygomaticoauricularis
Retractor anguli oculi lateralis
Obliqui & transversi auriculae
Zygomaticus
Cervicoauricularis superf.
Levator anguli oculi medialis
Orbicularis oculi
Cervicoauricularis prof. major
Levator nasolabialis
Cervicoauricularis prof. minor
Levator labii maxillaris
Temporalis
Caninus

Mentalis
Orbicularis oris
Buccinator — oral portion
Sphincter colli prof. — pars palpebralis
Parotid duct
Mandibular lymph nodes
Parotid gland
Mandibular gland
Parotidoauricularis
Ext. jugular v.

M·N.

Figure 6–2. Superficial muscles of the head, lateral aspect. (Platysma and sphincter colli superficialis removed.)

tions both the medial and lateral fiber bundles function to dilate the external nares.

Innervation: Rami buccalis dorsalis et ventralis, n. facialis.

The *mm. incisivus superioris et inferioris* lie deep to the orbicularis oris. These are two thin muscles not clearly defined from the orbicularis and buccinator; they arise on the alveolar borders of the incisive bone and mandible as far as the corner incisor teeth, and are situated immediately beneath the mucosa of the lips. They extend to the orbicularis oris. According to Huber (1922), the lower one cannot be isolated as a separate muscle.

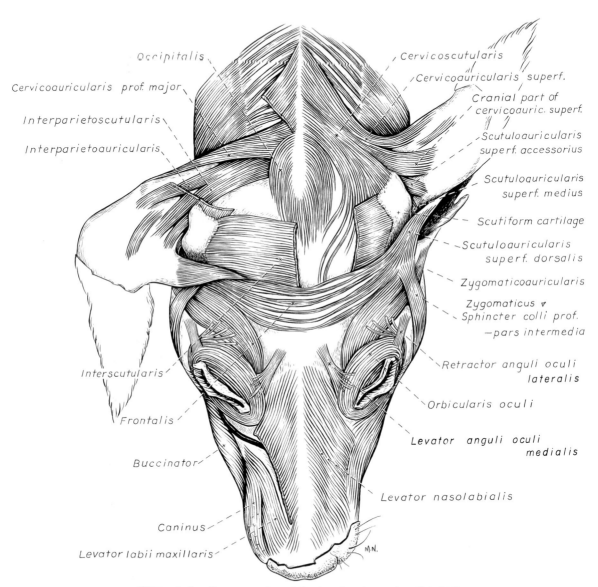

Figure 6–3. Deep muscles of the head and ear, dorsal aspect.

Action: The m. incisivus superioris raises the upper lip. The m. incisivus inferioris pulls the lower lip downward.

Innervation: Rami buccalis dorsalis et ventralis, n. facialis.

A **maxillonasolabial muscle** has developed from the upper lip portion of the orbicularis oris. It corresponds to the m. levator labii superioris alaeque nasi of the primates. In the dog it is distinctly separated into nasal and labial portions, which are both covered by the levator nasolabialis.

The nasal part, *m. levator labii superioris* (Figs. 6–2, 6–4), is a flat muscle which lies beneath the apical end of the levator nasolabialis on the maxilla and incisive bone. The nasal part arises caudoventral to the infraorbital foramen deep to the palpebral part of the sphincter colli profundus. The fibers of insertion spread out as they enter the nasal ala and the upper lip.

The labial part, *m. caninus* (Figs. 6–2, 6–4), is immediately ventral to the nasal portion and separates from it, in that it extends over the labial end of the levator nasolabialis into the upper lip.

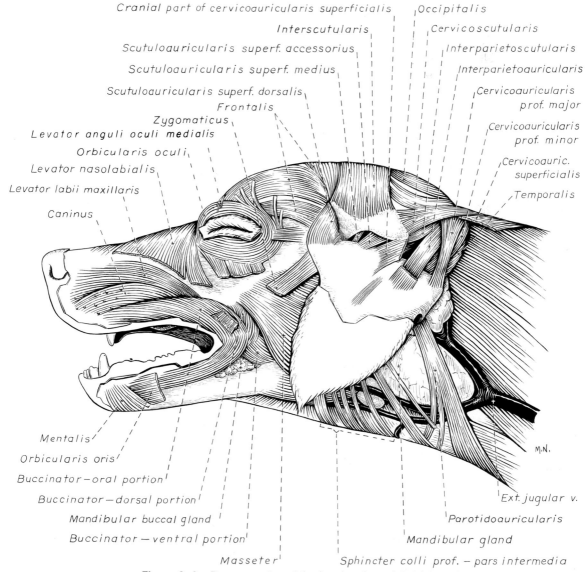

Figure 6–4. Deep muscles of the head and ear, lateral aspect.

Action: To increase the diameter of the external naris, and lift the apical portion of the upper lip.

Innervation: Ramus buccalis dorsalis, n. facialis.

The *m. buccinator* (Figs. 6–2, 6–3) has developed from the deep part of the orbicularis oris. It is a strong, flat, wide muscle which forms the foundation of the true cheek. It is composed of two portions, which extend caudally from the labial commissure. Only a superficial portion proceeds in an arch from one lip into the other, like the orbicularis. The upper portion has been called the m. buccalis, the lower the m. molaris.

The *pars buccalis* (formerly called pars dorsalis) is the somewhat stronger portion. It arises from the upper lip and secondarily from the region caudal to the infraorbital foramen on the alveolar border of the maxilla. It is superficial as far as the raphe. It proceeds caudoventrally and, except for a narrow posterior portion, crosses through the ventral portion. After running deeply, it ends in three or four distinct portions in such a way that the oral portion remains superficial and proceeds sphincter-like into the lower lip. This portion runs beneath and parallel to the orbicularis oris, and can be isolated in carefully prepared specimens. Between the two, the rostral end of the m. zygomaticus enters. The other portions of the pars dorsalis of the buccinator pierce the ventral part and cross beneath the latter in a caudoventral direction to end more or less independently on the mandible. The caudal branch runs almost horizontally and passes beneath the masseter. Furthermore, as already mentioned, a little muscle bundle separates from the uncrossed part of the caudal border; this arches caudally and in some animals extends superficially over the rostral border of the masseter.

The *pars molaris* (formerly called pars ventralis) is the weaker portion of the buccinator. It arises from the lower lip and, secondarily, in the region of the three lower molars, from the alveolar border of the mandible. It proceeds beyond the caudal edge of the dorsal portion. The rostral portions end on the maxilla, although most of the fibers pass beneath the masseter.

The fibers of the dorsal and ventral portions of the buccinator, disappearing beneath the edge of the masseter, fuse into a loose, delicate muscle lying directly upon the mucous membrane of the cheek. Closely attached to the buccal mucosa, they also attach on the maxilla and mandible peripheral to the edge of the mucosal covering. Functionally, this is the origin of the muscle. This attachment may be circumscribed by the following lines: the alveolar border of the mandible in the region of the last two cheek teeth, the rostral edge of the coronoid process to half its length, and the alveolar border of the maxilla in the region of the last two cheek teeth. The "free" portions of the buccinator lying rostral to the masseter also surround the ventral buccal glands with fibrous masses medially. At the level of the third maxillary cheek tooth, the parotid duct perforates the buccinator near its dorsal border.

Action: To return food from the vestibule to the masticatory surface of the teeth.

Innervation: Rami buccalis dorsalis et ventralis, n. facialis.

The *m. mentalis* is an incomplete division of the ventral portion of the m. buccinator. It arises from the alveolar border and body of the mandible near the corner teeth. The fibers unite with those of the opposite side and radiate into the lower lip, forming a strong, fat-infiltrated muscle.

Action: To stiffen the lower lip in the apical region.

Innervation: Ramus buccalis ventralis, n. facialis.

Pars palpebralis. The pars palpebralis of the sphincter colli profundus is a thin muscle consisting of separated, delicate muscular strands extending in front of the masseter (beneath the platysma) from the ventral midline to the lower lid. At the level of the dorsal border of the platysma, it is divided into two portions by the pressure of the platysma: (1) a ventral portion, covered by the platysma, and (2) the free dorsal segment. The muscle is found between the pars oralis and pars intermedia of the sphincter colli profundus, and covers the rostral end of the m. zygomaticus and the m. buccinator. In the lower eyelid it extends upon the orbicularis oculi.

Action: To depress the lower lid.

Innervation: Ramus buccalis ventralis, n. facialis.

Pars intermedia. The pars intermedia of

the sphincter colli profundus (Fig. 6–2) is the continuation of the palpebral portion and consists of loose muscular strands which arise in the intermandibular region behind those of the palpebral portion. There the two portions extend over to the other side and cross each other. Toward the rostroventral border of the scutulum on which they end, the strands are closer together. The superficial strands occasionally arise from the superficial surface of the platysma at its dorsal border. From the rostral margin of the pars intermedia the long, broad muscular band which proceeds toward the angle of the mouth is the m. zygomaticus (auriculolabialis).

The *m. zygomaticus* (Figs. 6–1 to 6–4), a derivative of the pars intermedia of the sphincter colli profundus, is usually intimately fused with the intermediate part at the scutulum; only exceptionally can both be completely isolated at this point. The straplike, long muscle extends from the rostral angle of the scutulum to the edge of the mouth, where it sinks into the orbicularis oris after crossing beneath the palpebral portion of the sphincter colli profundus. Its apical portion is deep and bears no relationship to the platysma. Proximally it is distinctly separated from the m. frontalis, which arises from a fascia at the deep surface of the scutulum; this portion of the m. zygomaticus is covered by the skin.

Action: To fix the angle of the mouth and draw it back, or to fix and draw the scutulum forward.

Innervation: Ramus auriculopalpebralis, n. facialis.

MUSCLES OF THE FOREHEAD AND THE DORSUM OF THE NOSE

Dorsal to the pars intermedia of the sphincter colli profundus is a large homogeneous muscle mass, the orbitofrontoauricular muscle leaf. This extends from the concha over the scutulum to the forehead and the eyelids; it sends further derivatives between the orbits to the nose and upper lip.

The rostroauricular muscles, the m. frontalis, the muscles of the lids, and the m. levator nasolabialis belong to the orbitofrontoauricular muscle complex (Figs. 6–1 to 6–4).

The *m. frontalis* is a thin muscle which lies on the temporalis. It arises in front of the rostral border of the scutulum, beneath the pars intermedia of the sphincter colli profundus, by means of a fascial leaf, and extends to the forehead and toward the upper eyelid. The frontal portion, unseparated caudally from the m. interscutularis, unites with that of the opposite side and rostrally joins the nasofrontal fascia by which it attaches to the zygomatic process. The palpebral portion passes beneath the orbicularis oculi and attaches to the orbital ligament. From the concha a considerable number of muscle strands of the m. scutuloauricularis superficialis dorsalis extend over the scutulum into the frontalis. It is seen from this relationship that these two muscle groups belong together. The m. frontalis has also been called the m. frontoscutularis. In the dog it has become attached to the frontal bone and the orbital ligament. It has been completely separated from the mm. retractor anguli oculi lateralis, levator anguli oculi medialis, and levator nasolabialis.

Action: To fix and pull the scutulum forward.

Innervation: Rami temporalis and auriculopalpebralis, n. facialis.

The *m. orbicularis oculi* surrounds the palpebral fissure. Portions of the muscle adjacent to the borders of the lids extend from the medial palpebral ligament over the upper lid, around the temporal angle of the lids, and along the lower lid back to the ligament. Thus in the dog, this muscle, which originally was divided into dorsal and ventral portions, has become one. Huber (1922) states that the ventral portion comes from the m. zygomaticus, and the dorsal portion comes from the m. frontalis.

Action: To close the palpebral fissure.

Innervation: Ramus auriculopalpebralis, n. facialis.

The *m. retractor anguli oculi lateralis*, as a division of the m. frontalis, arises beside the latter from the temporal fascia. It extends horizontally to the lateral palpebral angle, and, in so doing, it crosses the orbicularis oculi before it sinks into the fibers of the latter.

Action: To draw the lateral palpebral angle caudally.

Innervation: Ramus auriculopalpebralis, n. facialis.

The *m. levator anguli oculi medialis* is a small, strong muscle strand which arises from the median line on the frontal bone from the nasofrontal fascia. It extends over the orbicularis oculi into the medial half of the upper lid. It passes through portions of the upper lid which bear the hairs (pili supraorbitales) designated as the eyebrow.

Action: To lift the upper lid, especially its nasal portion, and erect the hairs of the eyebrow.

Innervation: Ramus auriculopalpebralis, n. facialis.

The *m. levator nasolabialis* is the most rostral extension of the original muscle sheet (pars intermedia of the sphincter colli profundus) that passes from the ear past the eye to the muzzle. It is a very flat, thin, and broad muscle (even in large dogs), lying immediately beneath the skin on the lateral surface of the nasal and maxillary bones. It arises in the frontal region between the orbits from the nasofrontal fascia, the nasal palpebral ligament, and the maxillary bone; occasionally a few additional fibers come from the lacrimal bone. Spreading out, it proceeds to the nose and upper lip to push beneath the orbicularis oris. The caudal portion inserts on the buccinator; the apical, larger portion passes beneath the orbicularis partly between both portions of the m. maxillonasolabialis, to end near the edge of the lip. The most dorsal and rostral fibers, however, force their way between the fibers of the nasal portion of the maxillonasolabialis and so break it up into separate leaflike layers in gaining attachment to the external naris. These fibers cannot be considered a special m. nasalis (Boas and Paulli 1908). The specific mm. nasales of other mammals (ungulates, proboscidae, primates) are divisions of the pars oralis of the sphincter colli profundus. Further divisions of the m. levator nasolabialis, which would serve for special nasal movements, do not exist in the dog. At most, one finds a flat, thin dilator of the naris which arises on the rostrodorsal portions of the nasal cartilage and extends into the lateral nasal ala. It is noticeable only in large dogs.

Action: To increase the diameter of the naris, and lift the apical portion of the upper lip.

Innervation: Ramus auriculopalpebralis, n. facialis.

Muscles of the External Ear

The muscles of the ear, from comparative and ontogenetic viewpoints, fall into three groups. The rostroauricular group is a derivative of the pars intermedia of the sphincter colli profundus; the ventroauricular group is represented by the pars auricularis of the sphincter colli profundus; and the caudoauricular group is a derivative of the deep portion of the platysma of the neck.

ROSTROAURICULAR MUSCLES

Because of the position of the scutiform cartilage or scutulum, the m. scutuloauricularis superficialis dorsalis, the zygomaticoauricularis, and the intrinsic muscles—the mm. tragohelicinus, tragotubohelicinus, and conchohelicinus—have all become completely separated from the orbitofrontoauricular muscle complex. The m. interscutularis and the m. scutuloauricularis profundus major are also distinct. All of these rostroauricular muscles are innervated by the ramus temporalis, n. facialis.

The *m. scutuloauricularis superficialis dorsalis* (m. auricularis anterior superior of Huber) (Figs. 6–1, 6–3, 6–4) is the dorsal inward rotator of the ear. It separates dorsocaudally from the m. frontalis, with which it is always partly united. It is a broad, strong muscle lying free over the rostral portion of the scutulum and coursing in a fold of skin to the medial border of the concha, where it attaches in the region of the spine of the helix (crus helicis distale).

Action: To turn the conchal fissure forward.

The *m. zygomaticoauricularis* (Figs. 6–1, 6–2, 6–3) is the external inward rotator that arises as a rather broad muscle from the tendinous leaf lying in front of the scutulum. It is continuous rostrally with the lower division of the frontalis. Caudally, it extends ventrally to the basal portion of the tragus.

Action: To turn the conchal fissure forward.

The *mm. tragohelicinus, tragotubohelicinus,* and *conchohelicinus* lie together in

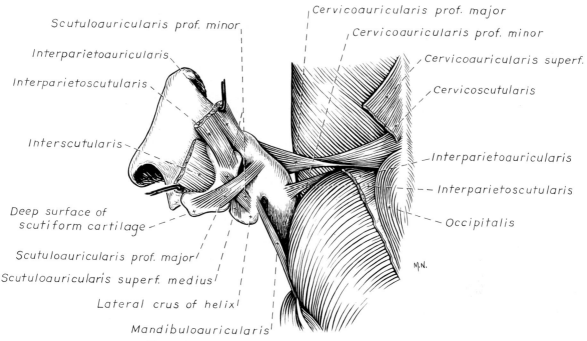

Scutuloauricularis prof. minor

Interparietoauricularis

Interparietoscutularis

Interscutularis

Deep surface of
scutiform cartilage

Scutuloauricularis prof. major

Scutuloauricularis superf. medius

Lateral crus of helix

Mandibuloauricularis

Cervicoauricularis prof. major

Cervicoauricularis prof. minor

Cervicoauricularis superf.

Cervicoscutularis

Interparietoauricularis

Interparietoscutularis

Occipitalis

M.N.

Figure 6–5. Muscles of the external ear, dorsal aspect.

one muscle complex, which bridges the space between the superimposed conchal edges of the conchal canal. This muscle aggregate passes from the deep surface of the lateral crus of the helix to the tragus. In certain cases all of these muscles are independent. See Leahy (1949) for further pictorial and descriptive treatment. The m. tragohelicinus arises from the external surface of the tragus, the m. tragotubohelicinus from the tragus and the conchal canal, and the m. conchohelicinus from the external surface of the concha.

Action: To narrow the entrance to the conchal canal and thus make the concha rigid.

The *m. interscutularis* (Fig. 6–3) is a thin muscle extending from one scutulum to the other, without attaching to the cranial bones. It has arisen from the fusion of bilateral portions. The origin is from the entire dorsomedial border of the scutulum. The caudal portion of the muscle has a distinct border and covers the m. occipitalis and the m. cervicoscutularis, both of which blend with the interscutularis. Rostrally it has no distinct border and encroaches upon the frontal portion of the m. frontalis. The inter-

scutularis is sometimes considered to belong to the dorsal auricular muscle group.

Action: Fixation of the scutulum.

The *m. scutuloauricularis profundus major*, or large rotator of the concha (Fig. 6–5), is completely separated from the m. frontalis to which it belongs. As a strong, well-defined muscle, it lies beneath the scutulum and arises on its deep surface to extend to the concha over the m. temporalis. The muscle has an almost sagittal course as it crosses the scutuloauricularis profundus minor on its deep surface.

Action: To turn the conchal fissure backward.

VENTROAURICULAR MUSCLE

The *m. parotideoauricularis* (Fig. 6–2) has also been called the depressor auriculae. It arises caudal to the laryngeal region, on or near the midline, where it blends with the cervical fascia. As a strong, well-defined band, it runs obliquely toward the concha, crossing the mandibular and parotid gland in its course. The muscle is almost completely covered by the platysma and inserts on the antitragus.

Action: To depress the ear.

Innervation: Ramus colli, n. facialis.

CAUDOAURICULAR MUSCLES

The caudoauricular muscles are derivatives of the platysma and are innervated by rami caudoauriculares of the facial nerve. The muscles forming this complex consist of the cervicoauricular musculature, the caudal conchal muscles, and the m. mandibuloauricularis.

The cervicoauricular musculature is a continuation of the deep layer of the platysma in the region of the neck. It is divided into three layers which, in the dog, are not completely separated.

The superficial layer, m. cervicoauriculo-occipitalis, forms a completely homogeneous muscle plate which comes from the dorsum of the neck and the occipital bone. It inserts on the caudal border of the scutulum as well as on the m. interscutularis and nasofrontal fascia. In this muscle complex are contained the m. cervicoauricularis superficialis, the m. occipitalis, and, as a connecting link between the two, the m. cervicoscutularis.

The *m. cervicoauricularis superficialis,* or long levator (Fig. 6–4), arises from the cervical midline and the external occipital protuberance. As a broad muscle mass it passes to the concha and ends by two branches on the dorsum of the ear. The rostral branch is made wider by fibers from the lateral border of the scutulum; these correspond to the m. scutuloauricularis superficialis accessorius, or short levator, of other animals. The caudal branch covers the auricular end of the interparietoauricularis.

Action: To raise the concha.

The *m. cervicoscutularis* (cervicointerscutularis of Huber) (Figs. 6–3, 6–4) is a narrow, intermediate portion of the muscle complex which is not clearly defined; it goes to the caudal border and the caudomedial angle of the scutulum, and is united with the deep surface of the interscutularis by means of a few fibers.

Action: To draw the scutulum downward, or fix it when the scutulum is drawn forward at the same time.

The *m. occipitalis* (Figs. 6–3, 6–4) is the third portion of the cervicoauriculo-occipital complex. From the external sagit-

tal crest, its fibers turn rostrally in bilaterally symmetrical arches in such a way that they form an unpaired, oval, thin membranous muscle which can be followed a short distance forward beneath the caudal portion of the m. interscutularis; there, on the frontal bone, they spread out into the nasofrontal fascia.

Action: To tense the nasofrontal fascia.

The middle and deep layers of the cervicoauricular musculature in the dog are divided into a number of individual muscles. Of these the m. cervicoauricularis profundus major, the m. interparietoscutularis, and the three parts of the m. scutuloauricularis belong to the middle layer, while the m. interparietoauricularis and m. cervicoauricularis profundus minor belong to the deep layer.

The *m. cervicoauricularis profundus major,* or long outward rotator (cervicoauricularis medius of Huber) (Fig. 6–4, 6–5), is a strong, relatively wide muscle which, covered partly by the cervicoauricularis superficialis, arises on the external sagittal crest, the external occipital protuberance and the neighboring attachment of the nuchal ligament. It extends to the base of the concha and finally ends on the root of the lateral conchal border (antitragus), where it lies next to the insertion of the depressor auriculae. This muscle covers a portion of the origin of the interparietoauricularis, the greater part of the cervicoauricularis profundus minor, and the m. temporalis of that region. Primitively, the auricular end of this muscle is simple. There are cases, however, in which this end is double.

Action: To turn the conchal fissure outward and backward.

The *m. interparietoscutularis* (Figs. 6–3, 6–4) is only exceptionally an independent muscle. It is described by Huber (1923) as the m. cervicoscutularis medius belonging to the middle layer. It arises from the interparietal portion of the external sagittal crest and inserts on the caudal border of the scutulum, which is completely covered by the superficial layer of this muscle complex. Ordinarily this muscle is united with the m. interparietoauricularis almost as far as the scutulum.

Action: With other scutular muscles, it aids in fixation of the scutulum.

The *m. scutuloauricularis profundus*

minor, or short rotator (Fig. 6–5), is considered by Huber (1922) to be a special branch of the scutuloauricularis medius. Owing to the presence of the scutulum, the short rotator is interrupted as a continuation of the m. interparietoscutularis. From the deep surface of the scutulum near the lateral angle it descends in a somewhat vertical direction to the concha. At the same time, this small, short rotator crosses the large rotator which belongs to the rostroauricular muscle group. It inserts on the external surface of the lateral crus of the helix opposite the attachment of the m. mandibuloauricularis.

Action: To turn the conchal fissure laterally.

The *m. scutuloauricularis superficialis accessorius,* or short levator (Figs. 6–3, 6–4), is the other part of Huber's m. scutuloauricularis medius, as the second indirect continuation of the m. interparietoscutularis. It arises from the caudal portion of the lateral scutular border with the rostral segment of the m. cervicoauricularis superficialis. It inserts basal to the long levator or cervicoauricularis superficialis.

Action: With the other levators, erection of the concha.

The *m. scutuloauricularis superficialis medius,* or middle inward rotator (Figs. 6–3, 6–4, 6–5), is, according to Huber (1922), a branch of the m. scutuloauricular medius and comes from the platysma. It extends from the deep surface of the caudal half of the lateral edge of the scutulum to the lateral crus of the helix. Rostral to the short levator, it lies deeply between the scutulum and the base of the concha.

Action: To turn the conchal fissure forward.

The *m. interparietoauricularis,* or middle levator (cervicoauricularis profundus anterior of Huber) (Fig. 6–4), is only seldom completely isolated. Indeed, it belongs to the deep layer, but usually fuses with the m. interparietoscutularis, which becomes separate only near the scutulum. In its entire course it is covered by the superficial layer of the caudoauricular musculature. It arises from the interparietal segment of the external sagittal crest and goes directly over to the dorsum of the concha, where it attaches under the caudal terminal branch of the m. cervicoauricularis superficialis basal to its insertion.

Action: To raise the concha.

The *m. cervicoauricularis profundus minor,* or short outward rotator (Figs. 6–3, 6–5), is a division of the deep layer of the caudoauricular musculature. At its origin, it is rather variable in that it can be divided into two to five clearly defined muscle bundles. Of these, the caudal one usually comes from the external occipital protuberance, whereas the other portions are more or less shortened and arise aponeurotically on the m. temporalis. Covered by the long outward rotator, the muscle runs toward the concha and beneath it to the extended lateral concha border; in extreme cases it can descend to the conchal canal itself.

Action: To turn the conchal fissure outward and backward.

The *mm. obliqui et transversi auriculae* (Fig. 6–2) are unevenly distributed muscle strands on the convex surface of the concha which, for the most part, proceed in a longitudinal direction. This layer probably belongs to the deep portion of the cervicoauricular musculature. A few of its divisions appear to be distinct: thus there are short longitudinal fibers in the transverse groove between the conchal fossa and the dorsum of the remaining distal part of the pinna. There is a pars sulci transversi, of which a particularly long segment extends over the end of the long levator. Directly lateral to this is the pars marginalis, located distal to the end of the long outward rotator in the region of the antitragus. It extends distally. There is also the m. helicis, or m. helicis retroauricularis of Huber, which proceeds quite independently from the deep surface of the lateral crus of the helix. It runs distally between the medial and lateral crura to the insertion of the m. scutuloauricularis superficialis dorsalis.

Action: Erection of the free portion of the pinna.

Innervation: The innervation by a branch of the ramus auricularis caudalis, n. facialis, which can be traced from behind the concha to the medial edge thereof, indicates that the m. helicis belongs to the caudoauricular muscular group. The same branch supplies the mm. obliqui et transversi, the short levator, and the m. mandibuloauricularis.

The *m. mandibuloauricularis* (Figs. 6–5, 6–13) is a muscle of the auditory canal. It is a long, narrow muscle, which also bears the name m. tragicus lateralis in descriptive nomenclature. It arises tendinously in the niche between the angular and condyloid processes of the mandible and extends dorsally to the lateral crus of the helix, covered by the parotid salivary gland. In its course it passes over the root of the zygomatic process of the temporal bone, extends along the rostral side of the cartilaginous auditory canal, and ends opposite the short rotator. This muscle may undergo great reduction, and in extreme cases may be represented only by tendinous remains. Often it is connected directly with the m. helicis, whose innervation it shares.

Deep Muscles of the Face

The deep facial muscles are completely separated from the superficial facial musculature and are innervated by deep branches of the n. facialis (Huber, 1923).

The *m. stapedius* (see Fig. 19–5) was originally associated with the hyomandibular bone of the primitive mandibular joint. During evolution the hyomandibular became the stapes and with its associated muscle was incorporated into the middle ear. The muscle is described along with the ear.

The *m. digastricus* (biventer mandibulae) (Fig. 6–6) runs from the processus jugularis of the occiput to the ventral border of the mandible. Althouth it appears as a single-bellied muscle in the dog, a tendinous intersection and an innervation by both the n. trigeminus and the n. facialis are evidence of its dual nature. Much has been written concerning this muscle in mammals (Rouviere 1906, Bijvoet 1908, and Chaine 1914). Only the caudal belly of the digastricus belongs to the deep facial muscle group.

The digastricus lies medial to the parotid and mandibular glands. After crossing the ventrocaudal edge of the insertion of the masseter, it has a fleshy ending on the ventromedial border of the mandible over a distance of about 2.5 cm., to the level of the canine tooth. Small muscle bundles extend far forward toward the chin.

Action: To open the jaws.

Innervation: N. facialis to caudal belly and n. trigeminus to rostral belly.

The *m. stylohyoideus* (Figs. 6–6, 6–13) is a narrow muscle bundle which proceeds from the tympanohyoid and proximal end of the stylohyoid obliquely across the lateral surface of the m. digastricus to the lateral end of the basihyoid. It inserts by means of a small terminal tendon or aponeurosis which is intimately related to the hyglossus or mylohyoideus, or both, cranially and the sternohyoideus caudally. Over much of its course the muscle is hidden from view by the mandibular gland. Occasionally the muscle divides into two bellies as it crosses the tendinous intersection of the digastricus. Huber (1923) noted the tendency toward reduction of the stylohyoideus and the possible elimination of this muscle in the dog. Evans (1959) described loss and anomaly of the hyoid muscles in mongrels and beagles which reflected the phylogenetic history of the muscles. The stylohyoideus was frequently absent bilaterally or unilaterally, particularly in beagles, and also exhibited secondary slips or fusions with the

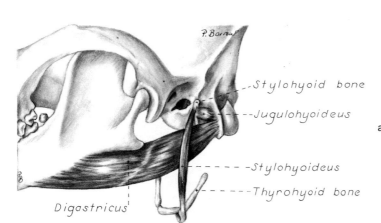

Figure 6–6. Superficial hyoid muscles and the digastricus, lateral aspect.

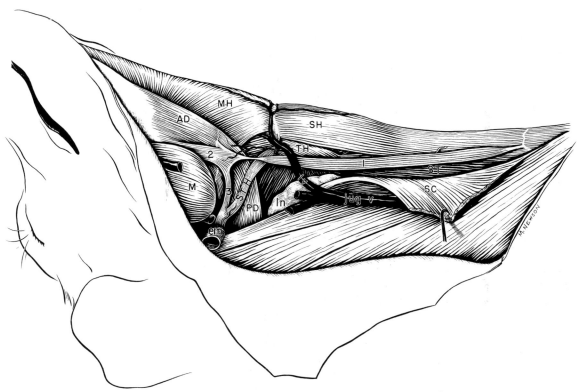

Figure 6–7.   Lateral view of anomalous slips of the sternohyoideus (1), rostral digastricus (2), and stylohyoid-eus (3). (From Evans, Anat. Rec. 1959.)

AD, rostral digastricus	lv, lingual vein	SH, sternohyoideus
em, external acoustic meatus	M, masseter	ST, sternothyroideus
jug v, external jugular vein	MH, mylohyoideus	STH, stylohyoideus
In, retropharyngeal lymph node	PD, caudal digastricus	TH, thyrohyoideus
	SC, sternocephalicus	

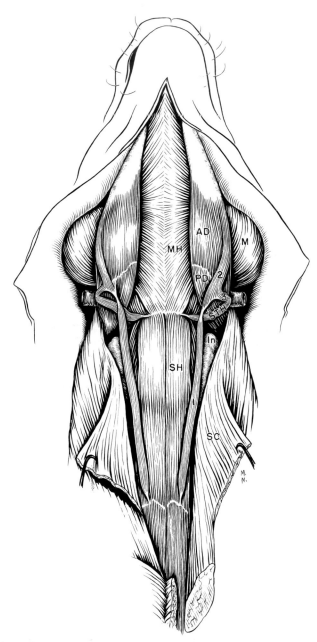

Figure 6–8.   Ventral view of anomalous slips of hyoid muscles: sternohyoideus (1), rostral digastricus (2), and stylohyoideus (3). (From Evans, Anat. Rec. 1959.)

AD, rostral digastricus
In, retropharyngeal lymph node
M, masseter
MH, mylohyoideus

PD, caudal digastricus
SC, sternocephalicus
SH, sternohyoideus
STH, stylohyoideus

digastricus, mylohyoideus, sternohyoideus, or subhyoidean septum (Figs. 6–7, 6–8).

Action: To raise the basihyoid bone.

Innervation: N. facialis.

The *m. jugulohyoideus* (jugulostyloideus of some authors) (Fig. 6–6) is a small rectangular muscle which extends from the jugular process to the occiput to the cartilaginous tympanohyoid and proximal end of the stylohyoid. The muscle is partly covered by the end of the m. sternomastoideus. The m. jugulohyoideus arises on the laterally projecting caudal border, whereas the m. digastricus attaches to the knobby end of the jugular process.

Action: To move the stylohyoid bone caudally.

Innervation: N. facialis.

Extrinsic Muscles of the Eyeball

There are seven extrinsic muscles of the eyeball: two oblique muscles, four recti muscles, and the retractor bulbi. Closely associated with these, but inserting in the upper eyelid, is the m. levator palpebrae. All of the extrinsic ocular muscles insert in the fibrous coat of the eyeball near its equator. The level of insertion of the recti muscles is nearer the corneoscleral junction than is that of the four parts of the retractor. In general, the oblique muscles insert in an intermediate zone between the insertions of the recti and retractor groups. All arise from the margin of the optic canal and orbital fissure, except the ventral oblique, which comes from the rostral part of the pterygopalatine fossa. Gilbert (1947) has investigated the origin and development of the extrinsic ocular muscles in the domestic cat. (See Chapter 20 for a more complete treatment of the eyeball.)

The *m. obliquus ventralis* (Fig. 6–9) arises from the rostrolateral margin of a variably sized opening in the palatine bone adjacent to the suture between the palatine, lacrimal, and maxillary bones. Frequently a groove harbors the origin of the muscle and extends caudally. As the ventral oblique muscle passes beneath the eyeball, it gradually widens, and crosses below the tendon of insertion of the ventral rectus. The ventral oblique divides as it reaches the ventral border of the lateral rectus. Part of its tendon crosses that of the lateral rectus superficially; the deep part goes underneath the lateral rectus, and ends in the sclera. The superficial part ends in the sclera lateral to the insertion of the dorsal rectus.

Action: To rotate the eyeball around an axis through its poles so that the ventral part is moved medially and dorsally.

Innervation: N. oculomotorius.

The *m. obliquus dorsalis*, or trochlearis (Figs. 6–9, 6–10), arises from the medial border of the optic canal. It ascends on the dorsomedial face of the periorbita to a cartilaginous pulley located on the medial wall of the orbit near the medial canthus of the eye. The pulley, or *trochlea* (Figs. 6–9A, and 6–10), is a disc of hyaline cartilage located dorsocaudal to the medial canthus of the lids on the medial wall of the orbit, less than 1 cm. from the orbital margin. It is spherical to oval in outline, with its long axis parallel to that of the head. It is about 1 cm. long by 1.5 mm. thick. It is closely related to the periorbita, in which its ligaments run. Three of these may be recognized. A long, distinct ligament runs from the anterior end of the trochlea, where the trochlear tendon bends around the cartilage, to the periosteum at the medial canthus. A short but wide thickening of the periorbita anchors the trochlea to the dorsal wall near its margin. The third ligament is a thickening in the periorbita which runs from the posterior pole of the trochlea to the periosteum on the ventral surface of the supraorbital process.

The slender tendon of the dorsal oblique muscle passes through a groove on the medial surface of the ventrorostral end of the trochlear cartilage, where it is held in place by a collagenous ligament. As it passes through this pulley, or trochlea, it bends at an angle of about 45 degrees. It passes dorsolaterally and under the tendon of the dorsal rectus, at the lateral edge of which it inserts in the sclera. It is the longest and slenderest muscle of the eyeball.

Action: To rotate the eyeball around an axis through its poles so that the dorsal part is pulled medially and ventrally.

Innervation: N. trochlearis.

The *mm. recti*, or straight muscles of the eyeball, include the mm. rectus lateralis, rectus medialis, rectus dorsalis, and rectus ventralis (Figs. 6–9, 6–10). They all arise from a poorly defined fibrous ring which is

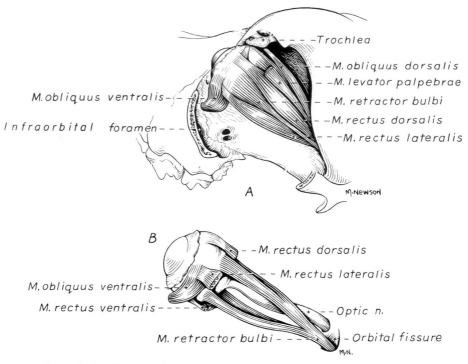

Figure 6–9.   Muscles of the eyeball.
A. Caudolateral aspect. (The eye is displaced slightly laterally
B. The m. retractor bulbi, lateral aspect.

attached around the optic canal and is continuous with the dural sheath of the optic nerve. The dorsal and medial recti arise farther peripherally from the optic canal than do the others. As the four muscles course rostrally from this small area of origin, they diverge and insert laterally, medially, dorsally, and ventrally on an imaginary line circling the eyeball, about 5 mm. from the margin of the cornea. The muscles are fusiform, with widened peripheral ends which give rise to delicate aponeuroses.

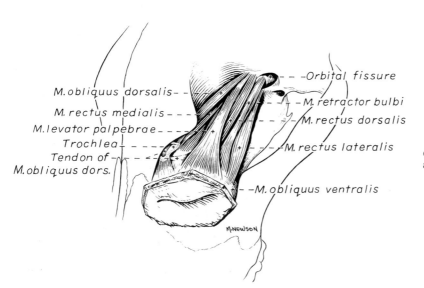

Figure 6–10.   Extrinsic muscles of the eyeball, dorsolateral aspect.

The muscles diverge from each other so that wedgelike spaces are formed between them. In the depths of these spaces the four segments of the m. retractor bulbi lie under the fascia and fat. The recti are longer and larger than the parts of the retractor with which they alternate. They therefore insert a greater distance from the caudal pole than do the parts of the retractor. The medial rectus is slightly larger than the others.

Action: The medial and lateral recti rotate the eyeball about a vertical axis through the equator; the dorsal and ventral recti rotate the eyeball about a horizontal axis through the equator.

Innervation: N. oculomotorius to the ventral, medial, and dorsal recti; n. abducens to the lateral rectus.

The *m. retractor bulbi* (Figs. 6–9, 6–10) arises deep to the mm. recti at the apex of the orbit, where they attach to the ventral end of the pterygoid crest and the adjacent orbital fissure. This places the initial part of the muscle lateral to the optic nerve. The four fasciculi of the m. retractor bulbi diverge as they run to the equator of the eye. The muscle fasciculi can be divided into dorsal and ventral pairs. The optic nerve, as it emerges from the optic canal, passes between the dorsal and ventral portions. The insertion of the several parts of the retractor on the globe of the eye is about 5 mm. caudal and deep to the recti. This muscle is sometimes spoken of as the choanid.

Action: To retract the eyeball. In addition, because of its essentially alternate attachments with the recti, it aids in bringing about oblique eye movements.

Innervation: N. abducens.

Muscles of the Eyelids

The mm. orbicularis oculi, retractor anguli oculi lateralis, and levator anguli oculi medialis have been described with the superficial muscles of the face.

The *m. levator palpebrae superioris* (Figs. 6–9, 6–10) is the main retractor of the upper eyelid. It arises dorsal to the optic canal between the dorsal rectus and the dorsal oblique muscles. It courses deep to the periorbita and superficial to the ocular muscles in reaching the upper eyelid. The leva-

tor inserts in the upper eyelid by means of a wide, flat tendon which passes between the fascicles of the m. orbicularis oculi.

Action: To lift the upper eyelid.

Innervation: N. oculomotorius.

There are also *smooth muscles* associated with the eyeball, orbit, and lids. Several of these, including the ventral and dorsal palpebral muscles, have been referred to in the past as muscles of Müller. Acheson (1938) described and illustrated the inferior and medial smooth muscles of the kitten's eye and showed their relationship to the eyelids and nictitating membrane. In the dog a delicate fan of muscle fibers arises from the trochlear cartilage and inserts in the upper lid. These fibers are nearly continuous at their insertion with the edge of the m. levator palpebrae superioris. Walls (1942) illustrates a muscle of Müller in man as an unstriped slip of the levator which inserts in the upper lid.

Muscles of the Hyoid Apparatus

The *m. sternohyoideus* (Figs. 6–11, 6–48) is a straplike muscle that arises from the deep surface of the manubrium sterni and the cranial edge of the first costal cartilage. It lies in contact with its fellow, and together they extend up the neck, covering the ventral surface of the trachea, to be inserted on the basihyoid bone. At its origin, and throughout its caudal third, the deep surface of the m. sternohyoideus is fused to the m. sternothyroideus. The caudal third of the m. sternohyoideus is covered by the m. sternocephalicus, and, in specimens in which there is a decussation of fibers between the sternocephalic muscles, the caudal two-thirds of the muscle will be covered by these crossed fasciculi. The cranial portion of the muscle which is not covered by the m. sternocephalicus is the most ventral muscle of that portion of the neck, except for the platysma. A transverse fibrous intersection separates the caudal third from the cranial two-thirds of the muscle.

Variations: Occasionally, muscle slips will arise from the transverse fibrous intersection of the m. sternohyoideus and pass up the neck to be inserted half in the stylohyoideus muscle, just lateral to the basihyoid bone, and half in the digastricus mus-

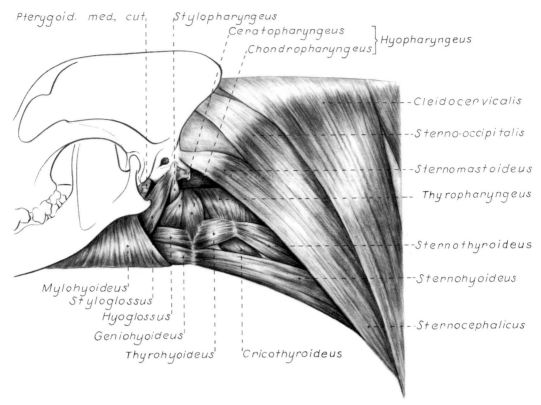

Figure 6–11. The hyoid muscles and muscles of the neck, lateral aspect. (Stylohyoideus and digastricus removed.)

cle at the angle of the jaw. The m. digastricus may be considerably smaller than normal and separated by a short intermediate tendon, as is the homologous muscle of man and horse. Leahy (1949) and Evans (1959) have described unilateral and bilateral anomalous slips in the dog.

Action: To pull the basihyoid bone and tongue caudally.

Innervation: Nn. cervicales et n. accessorius.

The **m. thyrohyoideus** (Figs. 6–11, 6–17) originates on the lamina of the thyroid cartilage. At the thyroid attachment it is bordered dorsally by the insertion of the m. thyropharyngeus and caudally by the mm. cricothyroideus and sternothyroideus. Its fibers pass obliquely forward and downward, over the surface of the thyroid lamina, to be inserted along most of the caudal border of the thyrohyoid bone.

Action: To draw the hyoid apparatus caudally and dorsally.

Innervation: Nn. cervicales et n. accessorius.

The **m. mylohyoideus** (Figs. 6–11, 6–12) lies most ventrally in the intermandibular space. Together with the muscle of the opposite side, it forms a sling for the tongue. It has a long origin from the medial side of the mandible. In most specimens the most cranial fibers are opposite the first lower premolar tooth and the most caudal fibers are slightly caudal to the last lower molar tooth. From its origin the muscle fibers extend medially, forming a thin plate which is largely inserted on a median fibrous raphe with its fellow of the opposite side. The most cranial fibers curve and insert on the midline farther forward than their point of origin. A few of the most caudal fibers curve and pass caudally, to be inserted on the basihyoid bone. The deep surface of the muscle is related to the m. geniohyoideus, the tongue, and the oral mucosa.

Action: To raise the floor of the mouth and draw the hyoid apparatus cranially.

Innervation: Ramus mandibularis, n. trigeminus.

The **m. ceratohyoideus** (see Figs. 6–18,

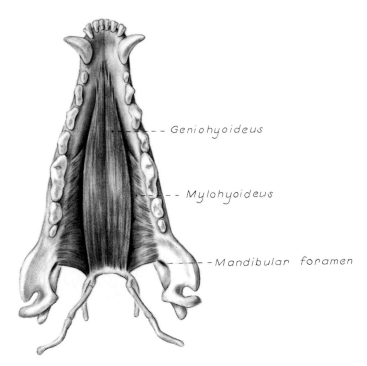

- - Geniohyoideus

- - Mylohyoideus

- -Mandibular foramen

Figure 6–12. Muscles of mandible and basihyoid, dorsal aspect.

6–22) is a small triangular plate of muscle, one side of which attaches to the anterior border of the thyrohyoid bone. The fibers run cranioventrally from the thyrohyoid bone to the ceratohyoid bone, to be attached along the dorsal border of the bone. In some specimens a few fibers attach to the ventral end of the epihyoid bone. The deep surface of the muscle is related to the root of the tongue and the oral mucosa, and the outer surface is related to the m. ceratopharyngeus.

Action: To decrease the angle formed by the thyrohyoid and ceratohyoid bones.

Innervation: N. glossopharyngeus.

The **m. geniohyoideus** (Fig. 6–12) is a fusiform muscle which extends from the chin, parallel to the midventral line, to the basihyoid bone. It arises by a short tendon from the mandibular symphysis and, muscularly, from the inner surface of the mandible adjacent to the symphysis. It passes directly caudad, at first bordered on the lateral side by the m. genioglossus and in its further course by the m. mylohyoideus, which also covers much of its ventral surface. Throughout its length the muscle is in close contact with its fellow of the opposite side. It is inserted on the cranial border of the basihyoid bone.

Action: To draw the hyoid apparatus cranially.

Innervation: N. hypoglossus.

The **mm. jugulohyoideus** and **stylohyoideus** are described under the deep muscles of the face.

Muscles of Mastication

The **m. masseter** (Fig. 6–15) lies on the lateral surface of the ramus of the mandible ventral to the zygomatic arch. It projects somewhat beyond the ventral and caudal borders of the mandible. The muscle is covered by a strong, glistening aponeurosis, and tendinous intermuscular strands are interspersed throughout its depth. One can divide the muscle into three layers (superficial, middle, and deep), using the change of fiber direction as a guide to the separation between the layers.

The superficial layer, the strongest part, arises from the ventral border of the rostral half of the zygomatic arch. Its fibers pass caudoventrally and insert, partly, on the ventrolateral surface of the mandible. Some fibers project around the ventral and caudal borders of the mandible and insert on its ventromedial surface, as well as on a tendin-

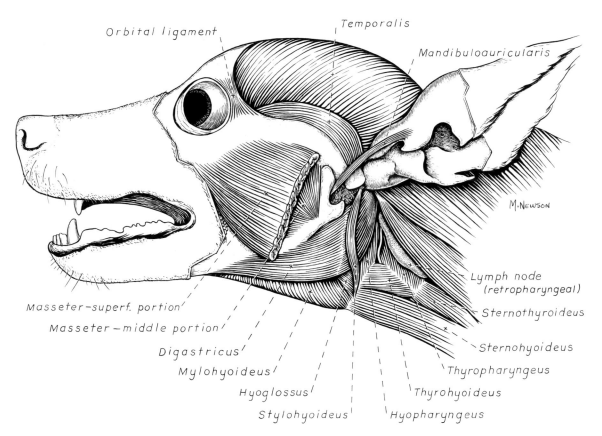

Figure 6–13.   Muscles of mastication, lateral aspect.

ous raphe which passes between the masseter and the m. pterygoideus medius. The tendinous raphe continues caudally from the angle of the jaw and attaches on the bone adjacent to the tympanic bulla. In specimens with well-developed masseters, this layer, at its ventral border, projects somewhat over the m. digastricus.

The middle layer, the weakest part, arises from the zygomatic arch, medial to the origin of the superficial layer and in part caudal to it. Most of its fibers pass ventrally to be inserted on the ventral margin of the masseteric fossa and the narrow area just ventral to the fossa. In some specimens a small bundle of fibers, which belong to this layer, run in a more rostral direction to be inserted on the rostroventral margin of the fossa.

The deep layer is impossible to isolate at its origin because many of its fibers intermingle with those of the temporalis. Some fibers, however, arise from the medial surface of the zygomatic arch. The majority of its fibers are directed caudoventrally and are inserted in the caudal part of the masseteric fossa and on the ridge adjacent to it. A few fibers pass down along the rostral margin of the temporal muscle to be inserted on the rostral ridge of the masseteric fossa.

Action: To raise the mandible in closing the mouth.

Innervation: N. massetericus of the ramus mandibularis, n. trigeminus.

The *m. temporalis* (Figs. 6–13, 6–14, 6–15) is the largest and strongest muscle of the head. It occupies the temporal fossa, from which it extends downward around the coronoid process of the mandible. During the course of its downward extension, it is related rostrally to the orbit and orbital fat, medially to the mm. pterygoidei and laterally to the m. masseter. Dorsolaterally it is covered by the caudoauricular muscles, the scutulum, and the ear. It arises largely from the parietal bone and to a lesser extent from the temporal, frontal, and occipital bones. The margins of the muscle at its origin are the orbital ligament and external frontal crest rostrally, the zygomatic arch laterally, the dorsal nuchal line caudally, and the external sagittal crest medially. Closely applied to the muscle, within these margins, is a strong, glistening fascia. In dolichocephalic dogs the temporal muscle meets its fellow of the opposite side and forms a mid-dorsal sulcus. In dogs with brachycephalic heads the temporal muscles usually do not meet on the midline, and the area is devoid of muscle, except for the caudoauricular muscles. From its large origin the muscle fibers curve forward and downward beneath the zygomatic arch to invest and insert on the coronoid process of the mandible, as far down as the ventral margin of the masseteric fossa. On the lateral side of the coronoid process the fibers are intermingled with fibers of the deep layer of the m. masseter; on the medial side the fibers lie in contact with the mm. pterygoidei. A bundle of muscle fibers arises from the dorsal nuchal line, near the base of the zygomatic process of the temporal bone, and sweeps forward dorsal and parallel to the zygomatic arch. It blends gradually into the main mass of the muscle.

Action: To raise the mandible in closing the mouth.

Innervation: N. temporalis of the ramus mandibularis, n. trigeminus.

The *m. pterygoideus lateralis* (Fig. 6–15) is a much smaller and shorter muscle than the m. pterygoideus medialis. It arises from the sphenoid bone in a small fossa, which lies ventral to the alar canal, round foramen, and orbital fissure. The ventral boundary of its origin is a bony ridge also on the sphenoid bone. This short muscle passes ventrolaterally and slightly caudally, to be inserted on the medial surface of the condyle of the mandible just ventral to its articular surface.

Action: To raise the mandible.

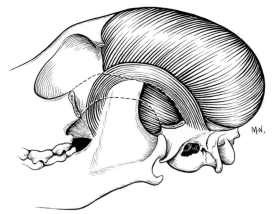

Figure 6–14. The m. temporalis, lateral aspect. (Zygomatic arch removed.)

Innervation: Nn. pterygoidei of the ramus mandibularis, n. trigeminus.

The *m. pterygoideus medialis* (Fig. 6–15) arises from the lateral surface of the pterygoid, palatine, and sphenoid bones. It passes caudolaterally to be inserted on the medial and caudal surfaces of the angular process of the mandible, and ventral to the insertion of the mm. temporalis and pterygoideus lateralis. Many fibers insert on a fibrous raphe that passes between the insertion of this muscle and the superficial layer of the masseter muscle. When viewed from the pharyngeal side, the medial pterygoid completely covers the lateral one.

The mandibular alveolar nerve passes across the lateral face of the m. pterygoideus medialis and the medial surface of the m. pterygoideus lateralis, thus separating the two muscles. The m. pterygoideus medialis extends to the caudal margin of the mandible and is inserted on the caudal margin and slightly on the caudomedial surface.

Action: To raise the mandible.

Innervation: Nn. pterygoidei of the ramus mandibularis, n. trigeminus.

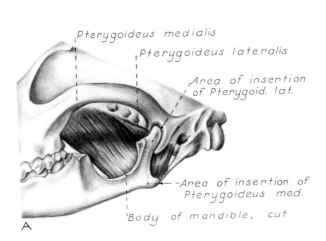

A

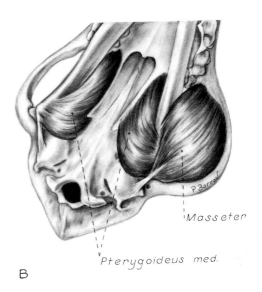

B

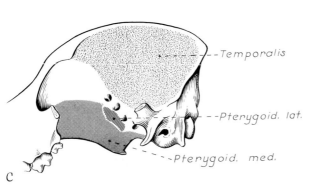

C

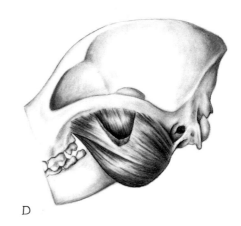

D

Figure 6–15. Muscles of mastication.
A. Mm. pterygoideus medialis and pterygoideus lateralis.
B. Mm. masseter and peterygoideus medialis.
C. Areas of origin of mm. temporalis, pterygoideus medialis, and pterygoideus lateralis.
D. M. masseter, cut to show the deep portion.

The *m. digastricus* is described with the deep muscles of the face.

Muscles of the Tongue

The *m. styloglossus* (Figs. 6–17, 6–18) extends from the stylohyoid bone to the tongue. It is composed of three muscle heads which insert in the tongue at different levels along its long axis.

The short head arises from the distal half of the caudal surface of the stylohyoid bone. It curves downward and forward across the lateral surface of the epihyoid bone. Immediately after crossing the epihyoid bone the fibers diverge and insert on the base of the tongue among the inserting fibers of the hyoglossal muscle.

The rostral head arises from the proximal half of the stylohyoid bone. These fibers curve downward and forward, pass over part of the inserting fibers of the short head, intermingle with fibers of the m. hyoglossus, and insert in the tongue, along its ventrolateral surface.

The long head arises just above and lateral to the origin of the fibers of the short head. These fibers immediately cross the stylohyoid bone, then curve downward and forward along the ventral border of the rostral head. They continue forward along the ventral midline of the tongue and across the lateral side of the genioglossus muscle, to their insertion on the ventral surface of the rostral half of the tongue, near the median plane.

Action: To draw the tongue backward when all three heads act together. Each muscle head depresses the tongue sector to which it is attached.

Innervation: N. hypoglossus.

The *m. hyoglossus* (Figs. 6–16, 6–17, 6–18) is located in the root of the tongue. It arises from the ventrolateral surface of the basihyoid and the adjoining end of the thyrohyoid bone. It runs forward dorsal to the m. mylohyoideus and lateral to the mm. geniohyoideus and genioglossus. At the base of the tongue it crosses the medial side of the m. styloglossus to be inserted in the root and caudal two-thirds of the tongue.

Action: To retract and depress the tongue.

Innervation: N. hypoglossus.

The *m. genioglossus* (Figs. 6–16, 6–17) is a thin, triangular muscle which lies in the intermandibular space, in and beneath the tongue. The apex of this triangular muscle corresponds to its origin on the medial surface of the mandible, just caudal to the origin of the geniohyoideus. The fibers run backward and upward in a sagittal plane. In their course the muscle fibers lie lateral to the m. geniohyoideus and dorsal to the m. mylohyoideus. The most rostral fibers run upward and forward, to be inserted on the midventral surface of the tip of the tongue. These fibers form the substance of the frenulum. The remaining fibers sweep upward and backward in a fanlike arrangement, to be inserted along the midventral surface of the tongue in close contact with the fibers of the corresponding muscles of the opposite side. A distinct bundle of fibers runs directly cau-

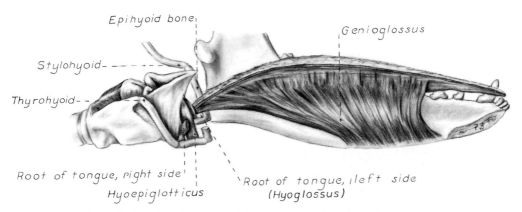

Figure 6–16.   The larynx, hyoid apparatus, and left half of the tongue.

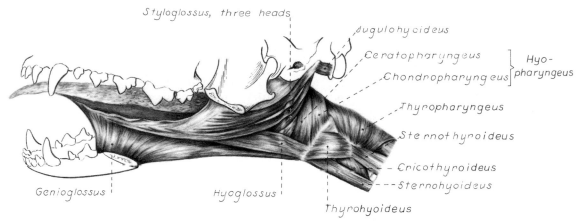

Figure 6-17. Muscles of the tongue and pharynx, lateral aspect.

dally, to be inserted on the basihyoid and ceratohyoid bones.

Action: To depress the tongue. The caudal fibers draw the tongue forward; the rostral fibers curl the tip of the tongue downward.

Innervation: N. hypoglossus.

The *m. propria linguae* consists of many muscular bundles which are located among the fascicles of insertion of the extrinsic muscles of the tongue. They are arranged bilaterally in four poorly delineated groups: (1) longitudinalis dorsalis, (2) longitudinalis ventralis, (3) transversus linguae, and (4) verticalis linguae. The dorsal longitudinal fibers lie directly under the dorsal mucosa of the organ and are well developed. The transverse and oblique fibers form a rather wide zone under the dorsal longitudinal bundles. A few long muscle strands lie ventral to the above-mentioned zone and compose the ventral longitudinal muscle. Thus the muscle bundles run in diverse directions.

Action: To proturde the tongue and bring about complicated intrinsic, local movements; to prevent the tongue from being bitten. The tongue functions in mastication and deglutition, as well as serving as the primary organ of taste. Bennett and Hutchinson (1946) have discussed the action of the tongue in the dog.

Innervation: N. hypoglossus.

For the complete structure of the tongue see Chapter 7, The Digestive Apparatus and Abdomen.

Muscles of the Pharynx

The *m. hyopharyngeus* (Figs. 6–11, 6–17, 6–19) is often divided into two parts: the m. chondropharyngeus and the m. ceratopharyngeus. The m. chondropharyngeus, the larger part, arises from the lateral surface of the thyrohyoid bone under cover of the hyoglossal muscle. The m. ceratopharyn-

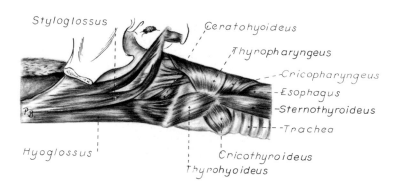

Figure 6-18. Muscles of the tongue and pharynx, deep dissection, lateral aspect.

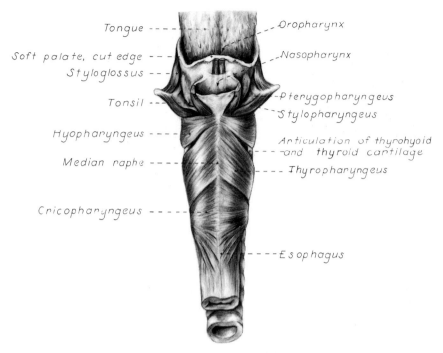

Tongue

Soft palate, cut edge

Styloglossus

Tonsil

Hyopharyngeus

Median raphe

Cricopharyngeus

Oropharynx

Nasopharynx

Pterygopharyngeus

Stylopharyngeus

Articulation of thyrohyoid and thyroid cartilage

Thyropharyngeus

Esophagus

Figure 6–19.   Muscles of the pharynx, dorsal aspect.

geus arises from the ceratohyoid bone. The muscle fibers of both parts form a muscle plate, the fibers of which pass upward over the larynx and pharynx to be inserted on the medial dorsal raphe of the pharynx, opposite the insertions of the muscles of the opposite side. Near their insertions the caudal fibers are overlaid by inserting fibers of the m. thyropharyngeus. The m. hyopharyngeus is the most rostral pharyngeal constrictor.

Action: To constrict the rostral part of the pharynx.

Innervation: Nn. glossopharyngeus et vagus.

The *m. thyropharyngeus* (Figs. 6–17, 6–19) lies on the larynx and pharynx just caudal to the hyopharyngeus muscle. It arises from the oblique line on the lamina of the thyroid cartilage, and goes upward and forward over the dorsal border of the thyroid lamina. The fibers spread out over the dorsal surface of the pharynx and insert on the median dorsal raphe of the pharynx, just caudal to the m. hyopharyngeus. Some of the most rostral fibers of insertion overlie fibers of the m. hyopharyngeus.

Action: To constrict the middle part of the pharynx.

Innervation: Nn. glossopharyngeus et vagus.

The *m. cricopharyngeus* (Figs. 6–18, 6–19) lies on the larynx and pharynx immediately caudal to the m. thyropharyngeus. It arises from the lateral surface of the cricoid cartilage and passes upward to be inserted on the median dorsal raphe. As the muscle fibers pass over the dorsal wall of the pharynx they blend, at their caudal margin, with muscle fibers of the esophagus.

Action: To constrict the caudal part of the pharynx.

Innervation: Nn. glossopharyngeus et vagus.

The *m. stylopharyngeus* (Figs. 6–20, 6–22) is a weak muscle that extends from the stylohyoid bone to the rostrodorsal wall of the pharynx. In most specimens the fibers arise from the caudal border of the proximal end of the stylohyoid bone; on some specimens, however, a few fibers arise on the epihyoid bone. From their origin the fibers run backward and inward beneath the constrictor muscles on the dorsolateral wall of the pharynx, where they are loosely arranged and intermingle with fibers of the m. palatopharyngeus.

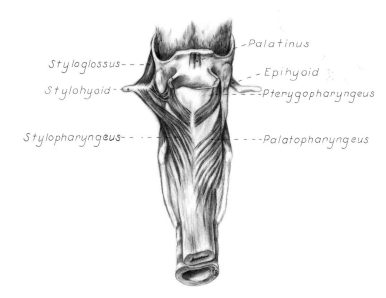

Styloglossus - - -
Stylohyoid -
Stylopharyngeus - - - -

- *Palatinus*
- *Epihyoid*
- - - *Pterygopharyngeus*
- - - -*Palatopharyngeus*

Figure 6–20. Muscles of the pharynx, deep dissection, dorsal aspect.

Action: To dilate, elevate, and draw the pharynx forward.

Innervation: Nn. glossopharyngeus et vagus.

The *m. palatopharyngeus* (Figs. 6–20, 6–21) is a poorly developed muscle, medial to the m. tensor veli palatini, whose fibers are loosely associated as they encircle the pharynx. Dyce (1957) divides the muscle into a dorsal and ventral portion. Most of the fibers arise from the soft palate and sweep obliquely upward and backward over the pharynx to the mid-dorsal line. Some fibers of the mm. pterygopharyngeus and stylopharyngeus blend with the m. palatopharyngeus on the dorsal wall of the pharynx. A few fibers run forward from their palatine origin and are dispersed in the soft palate, nearly as far forward as the hamulus of the pterygoid bone.

Action: To constrict the pharynx and draw it forward and upward.

Innervation: Nn. glossopharyngeus et vagus.

The *m. pterygopharyngeus* (Figs. 6–20, 6–21) arises from the hamulus of the ptery-

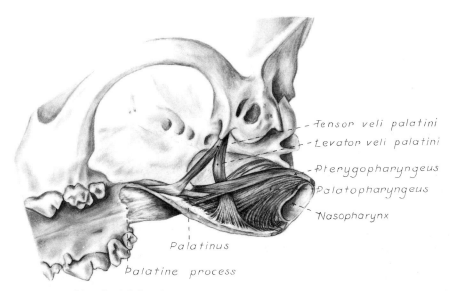

- *Tensor veli palatini*
- *Levator veli palatini*
- *Pterygopharyngeus*
- *Palatopharyngeus*
- *Nasopharynx*

Palatinus

Palatine process

Figure 6–21. Muscles of the pharynx and palate, deep dissection, ventrolateral aspect.

goid bone, passes backward lateral to the m. levator veli palatini, and continues upward over the pharynx to be inserted on the mid-dorsal raphe. Its fibers are intermixed with fibers of the m. palatopharyngeus and the m. stylopharyngeus as they radiate toward their insertions.

Action: To constrict the pharynx and draw it forward.

Innervation: Nn. glossopharyngeus et vagus.

Muscles of the Soft Palate

The *m. tensor veli palatini* (Fig. 6–21) is a very small muscle that arises from the muscular process at the rostral margin of the tympanic bulla. From its origin it passes downward over the wall of the pharynx to the hamulus of the pterygoid bone. At the hamulus the muscular fibers become tendinous and pass over a trochlear ridge on the hamulus. In their distal course these tendious fibers radiate forward and are dispersed in the palate.

Action: To stretch the palate between the pterygoid bones.

Innervation: Ramus mandibularis, n. trigeminus.

The *m. levator veli palatini* (Figs. 6–21, 6–22) is slightly larger than the m. tensor veli palatini. It arises from the muscular process adjacent to the tympanic bulla and passes downward and backward on the wall of the pharynx. In its distal course it passes be-tween the m. palatopharyngeus and the m. pterygopharyngeus and radiates to its insertion on the caudal half of the soft palate lateral to the m. palatinus.

Action: To raise the caudal part of the soft palate.

The *m. palatinus* (m. uvulae) (Fig. 6–21) is a small, straight muscle that runs longitudinally through the soft palate. It arises from the palatine process of the palatine bone and passes with its fellow to the caudal free border of the soft palate.

Action: To shorten the palate and curl the posterior border downward.

Muscles of the Larynx

The larynx, which has evolved from primitive gill arch supports, serves as a protective sphincter mechanism, in addition to subserving the function of sound production. The intrinsic muscles of the larynx are innervated by branches of the vagus nerve. The m. cricothyroideus is innervated by the ramus externus of the n. laryngeus cranialis. All other intrinsic muscles receive their motor supply via the n. laryngeus caudalis, the terminal portion of the n. recurrens. Laryngeal muscle innervation in the dog has been investigated by Vogel (1952). Pressman and Kelemen (1955) have reviewed the anatomy and physiology of the larynx in a variety of animals. Piérard (1963) studied the comparative anatomy of the larynx in the dog and

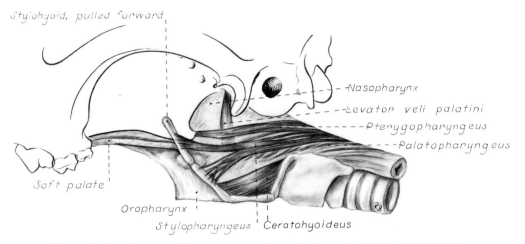

Figure 6–22. Muscles of the pharynx and palate, deep dissection, lateral aspect.

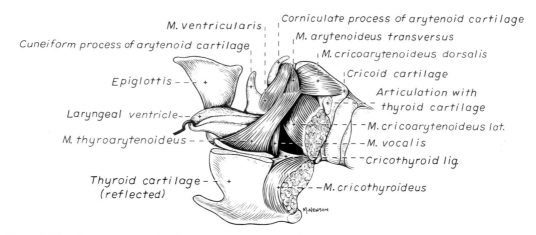

Figure 6–23. Laryngeal muscles, lateral aspect. (The thyroid cartilage is cut left of midline and reflected.)

other carnivores. Duckworth (1912) considered the plica vocalis and the tendency for subdivision of the thyroarytenoideus muscle mass in the dog and other animals.

The *m. cricothyroideus* (Figs. 6–18, 6–23) is a thick muscle on the lateral surface of the larynx between the thyroid lamina and the cricoid cartilage. From its attachment on the lateral surface of the cricoid cartilage (beneath the cricothyroid articulation), it runs upward and forward to attach to the caudal margin and medial surface of the thyroid cartilage. Some cranial fibers may attach ventrally close to the origin of the m. vocalis.

Action: To pivot the cricoid cartilage on its thyroid articulation, thus tensing the vocal cords.

The *m. cricoarytenoideus dorsalis* (Figs. 6–23, 6–25) arises from the entire length of the dorsolateral surface of the cricoid cartilage. The fibers run craniolaterally and converge at their insertion on the muscular process of the arytenoid cartilage. A few of the most lateral fiber bundles blend with the m. thyroarytenoideus.

Action: To open the glottis.

The *m. cricoarytenoideus lateralis* (Fig. 6–24) arises from the lateral and cranial surface of the cricoid cartilage. Its fibers pass upward and slightly forward to insert on the muscular process of the arytenoid cartilage between the m. cricoarytenoideus dorsalis above and the m. vocalis below.

Action: To pivot the arytenoid cartilage inward and close the rima glottis.

The *m. thyroarytenoideus* (Fig. 6–23) is the parent muscle mass which has given rise

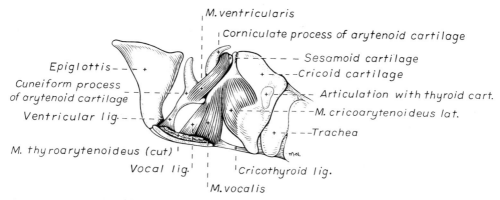

Figure 6–24. Laryngeal muscles, lateral aspect. (The thyroid cartilage is cut left of midline and removed; the mm. thyroarytenoideus, arytenoideus transversus, and cricoarytenoideus dorsalis have also been removed.)

to the m. ventricularis and the m. vocalis. It originates along the internal midline of the thyroid cartilage and passes caudodorsally to insert on the arytenoid cartilage at the raphe which represents the origin of the m. arytenoideus transversus. Dorsally the m. thyroarytenoideus (m. thyroarytenoideus externus of some authors) sends a few fibers to the m. ventricularis rostrally and to the m. cricoarytenoideus dorsalis caudally. The major middle portion of the m. thyroarytenoideus blends with the aponeurosis of the m. arytenoideus transversus superficially and attaches to the muscular process of the arytenoid cartilage deeply.

Action: To relax the vocal cord and constrict the glottis.

The *m. vocalis* (Fig. 6–24) is a medial division of the original thyroarytenoid muscle mass. It is also known as the m. thyroarytenoideus aboralis (Nickel, Schummer, and Seiferle 1954) or the thyroarytenoideus internus. The vocalis originates on the internal midline of the thyroid cartilage medial and partly caudal to the m. thyroarytenoideus. It inserts on the vocal process of the arytenoid cartilage, its greatest bulk being on the lateral side. Attached along the cranial border of the m. vocalis is the vocal ligament which can be distinguished grossly by its lighter color and finer texture.

Action: To draw the artenoid cartilage downward, thus relaxing the vocal cord.

The *m. ventricularis* (Figs. 6–23, 6–24, 6–25) is a cranial division of the thyroarytenoid muscle mass which has shifted its origin in the dog from the thyroid cartilage to the cuneiform process of the arytenoid cartilage. It is also known as the thyroarytenoideus oralis (Nickel, Schummer, and Seiferle 1954). The ventricularis lies medial to the laryngeal ventricle and possibly aids in dilating the ventricle. From its ventral origin on the cuneiform process, the ventricularis passes dorsally and slightly caudally to insert on the dorsal surface of the interarytenoid cartilage, where it meets its fellow of the opposite side. Occasionally an unpaired cartilage is present on the dorsal midline, above the intcrarytenoid cartilage, onto which the bulk of the fibers may insert. The m. ventricularis receives some connecting fibers from the cranial dorsal surface of the thyroarytenoideus.

Action: To constrict the glottis and dilate the laryngeal saccule.

The *m. arytenoideus transversus* (Figs. 6–23, 6–25) originates broadly on the muscular process of the arytenoid cartilage at the line of insertion of the thyroarytenoideus. It inserts on the lateral expanded ends and dorsal surface of the interarytenoid cartilage, meets its fellow fibers from the opposite side, and blends with the more dorsally located m. ventricularis, which spans the midline.

Action: To constrict the glottis and adduct the vocal folds.

The *m. hyoepiglotticus* (Fig. 6–16), a small, spindle-shaped muscle, arises from the medial surface of the ceratohyoid bone. It passes medially to the midline, then turns dorsally and passes to the ventral midline of the epiglottis to be inserted. The fibers of

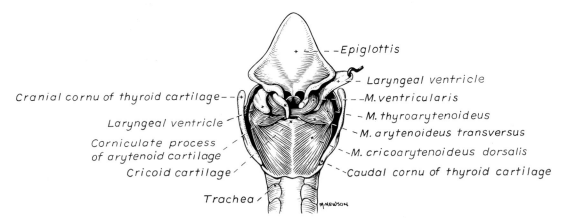

Figure 6–25. Laryngeal muscles, dorsal aspect. (The right corniculate cartilage has been cut, and the right laryngeal saccule is reflected.)

fellow muscles blend into a common tendon of insertion, which fades into the ventral surface of the epiglottis.

Action: To draw the epiglottis downward.

Fasciae of the Head

The **superficial fascia of the head** lies directly beneath the skin; for the most part it is easily displaceable, but in the muzzle it fuses with the skin. It contains the cutaneous muscles of the head, portions of the platysma, and the m. sphincter colli profundus. It covers the entire head like a mask and continues on the neck like a cylinder. It is divided into the pars temporalis, pars nasofrontalis, pars buccalis, pars parotideomasseterica, and pars submandibularis. In many places the special nerves and vessels for the skin pass through the superficial fascia of the head.

The **superficial temporal fascia** (*fascia temporalis superficialis*) conceals the muscles of the scutular group as well as the scutulum itself. Medially it goes into the superficial temporal fascia of the other side without attaching to the median system of cranial ridges. Rostrally it continues as the superficial nasofrontal fascia; more laterally and ventrally, however, and also displaceable with respect to the underlying tissue, it goes over the orbital ligament to the lids. Caudally it extends over the long levator of the pinna and becomes the superficial fascia which contains the platysma. Laterally it passes over to the pinna and, rostral to this, spreads over the zygomatic arch into the parotideomasseteric fascia.

The **superficial nasofrontal fascia** (*fascia nasofrontalis superficialis*) comes from the superficial temporal fascia and, containing the mm. frontalis and levator nasolabialis and their divisions, covers the nasofrontal region. It spreads out into the upper eyelid, the nose, and the upper lip; caudally, between the palpebral and labial commissures, it goes into the fascia of the cheek.

The **superficial buccal fascia** (*fascia buccalis superficialis*) comes from the superficial nasofrontal fascia. In addition to buccal portions of the platysma, it contains the pars palpebralis and pars intermedia (parts of the m. sphincter colli profundus) and covers the buccinator and the large facial vessels and nerves. The parotid duct is loosely surrounded as it lies behind the labial commissure. The fascia spreads out into the lips. Caudally it turns over the external surface of the m. masseter.

The **parotideomasseteric fascia** (*fascia parotideomasseterica*) is the continuation of the above-described portion of the superficial fascia covering the m. masseter and going to the external surface of the parotid gland and mandible; the branches of the facial nerve and the parotid duct are therefore surrounded by it. Dorsally this fascia goes over the zygomatic arch into the superficial temporal fascia and spreads out on the pinna of the ear. It contains the pars intermedia and pars auricularis of the m. sphincter colli profundus, and the platysma, with which it extends into the superficial cervical fascia caudally. Ventrally it goes into the submandibular fascia.

The **superficial submandibular fascia** (*fascia submandibularis superficialis*) is the intermandibular portion of the fascia of the head; it courses between the bodies of the mandible on either side as a continuation of the superficial buccal and masseteric fasciae, and covers the m. mylohyoideus and the body of the hyoid bone with its musculature. Caudally it runs into the region of the larynx and into the superficial cervical fascia.

The **deep fascia of the head** is found on all parts of the head. As the **deep temporal fascia** (*fascia temporalis profunda*) it is thick as it covers the temporal muscle and spreads out, enclosed by and attached to the external frontal crest, external sagittal crest, dorsal nuchal line, and the zygomatic process. If a part of the parietal bone is not covered by the temporal muscle, as frequently occurs in brachycephalic breeds, then this fascia fuses with the periosteum of the bone. Rostrally, the deep temporal fascia becomes the **deep nasofrontal fascia** (*fascia nasofrontalis profunda*), called the galea aponeurotica in man, and attaches to the orbital ligament. Ventrally the deep temporal fascia passes over the zygomatic arch and the masseter as the **deep masseteric fascia** (*fascia masseterica profunda*). It then spreads over the m. buccinator, extends into both lips, and passes over the mandible and larynx, as the **buccopharyngeal fascia** (*fascia buccopharyngea*). From the m. masseter, caudally, the buccopharyngeal fascia passes around the parotid gland, forming the fascia parotidea, crosses the digastricus, and goes beneath the mandibular gland and thence into the deep cervical fascia. Everywhere, on the head, the deep fascia lies beneath the large superficial vessels.

MUSCLES OF THE TRUNK

The trunk muscles are divided topographically into the muscles of the cervical, thoracic, and lumbar vertebrae; muscles of the lateral and ventral thoracic wall, including the diaphragm; muscles of the abdomen; and muscles of the tail.

The special muscles of the trunk are partly covered by those passing from the trunk to the limbs. This applies especially to those of the neck and the thoracic wall. In the lumbosacral region, the axial muscles continue into the dorsal caudal muscles, and the muscles of the pelvic limb overlap those of the trunk.

Muscles of the Cervical, Thoracic, and Lumbar Vertebrae

The muscles of the vertebrae, as far caudally as the sacrum, represent the trunk muscles in the narrow sense. They are grouped, aside from the cutaneous musculature, into five layers, which lie beside and one above another. Of these the two superficial layers and part of the third layer are discussed with the muscles of the thoracic limb.

The muscles of the first layer are: mm. trapezius and cleidocephalicus; the second layer: mm. latissimus dorsi and rhomboideus; the third layer: mm. serratus ventralis, serratus dorsalis, and splenius; the fourth layer: mm. iliocostalis, longissimus thoracis, longissimus cervicis, longissimus capitis, longissimus atlantis, spinalis et semispinalis dorsi et cervicis, and semispinalis capitis; and the fifth layer: mm. multifidus, interspinales, intertransversarii, and the dorsal muscles on the atlanto-occipital and axioatlantal joints — the mm. obliquus capitis caudalis, obliquus capitis cranialis, and rectus capitis dorsalis, consisting of three parts.

The *m. serratus dorsalis* (Figs. 6–26, 6–27, 6–33) is a wide, flat muscle partially covering the m. longissimus and the m. iliocostalis.

Lying under the mm. rhomboideus, serratus ventralis, and latissimus dorsi, it arises by a broad aponeurosis from the tendinous raphe of the neck and from the thoracic spines, and inserts on the proximal portions of the ribs. The muscle is completely divided into two portions (see the special investigations of Maximenko 1929, 1930).

The *m. serratus dorsalis cranialis*, also known as the inspiratory part, lies on the dorsal surface of the thorax, where its aponeurosis covers the m. splenius and its fleshy part covers the mm. longissimus dorsi and iliocostalis from ribs 2 to 10. The muscle arises by a broad aponeurosis from the superficial leaf of the thoracolumbar fascia and, by means of this, from the tendious raphe of the neck as well as from the spines of the first six to eight thoracic vertebrae. This aponeurosis fuses caudally with that of the mm. latissimus dorsi and serratus dorsalis caudalis. The fleshy portion of the muscle begins at about the dorsal border of the m. latissimus dorsi; it ends immediately lateral to the m. iliocostalis, with distinct serrations on the cranial borders and the lateral surfaces of ribs 2 to 10. The fibers of the muscle, as well as those of its aponeurosis, are directed caudoventrally.

Action: To lift the ribs for inspiration.

Innervation: Nn. intercostales (branches from the branch to the m. intercostalis externus).

The narrower *m. serratus dorsalis caudalis*, or expiratory part, consists of three rather distinctly isolated portions. These arise by a broad aponeurosis from the thoracolumbar fascia from which the m. obliquus externus abdominis and m. obliquus internus abdominis also arise. After extending cranioventrally, they end on the caudal border of the eleventh, twelfth, thirteenth, and, occasionally, also the tenth rib.

Action: To draw the last three or four ribs caudally for expiration.

Innervation: Branches from the nn. intercostales (from the trunk or the ramus medialis of the thoracic nerves).

The *m. splenius* (Fig. 6–27) is a flat, fleshy, triangular muscle with the caudal end as the apex and the cranial end as the base of the triangle. It lies on the dorsolateral portion of the neck, extending from the third thoracic vertebra to the skull. Its fibers run in a cranioventrad direction and cover the mm. semi-

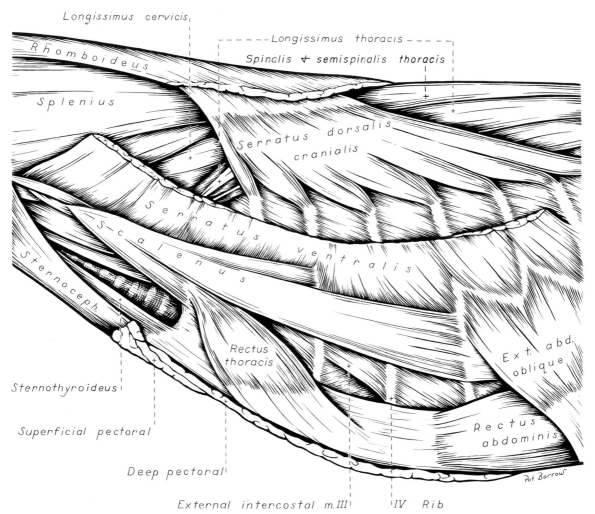

Figure 6–26. Muscles of neck and thorax, lateral aspect.

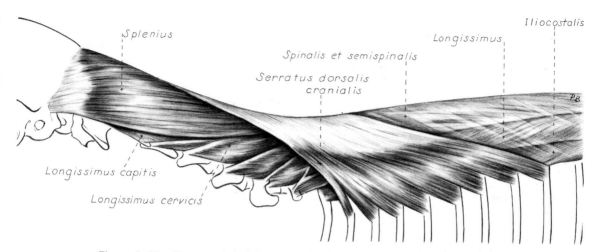

Figure 6-27. Topography of the mm. sphenius and serratus dorsalis cranialis.

spinalis capitis, longissimus capitis, and the terminal part of the m. spinalis et semispinalis dorsi. It arises by fleshy fibers from the end of the first and sometimes of the second thoracic spine, and from about 1 cm. of the ligamentum nuchae immediately in front of the first thoracic spine. A third point of origin is from the median dorsal raphe of the neck as far cranial as the first cervical vertebra. This tendinous raphe runs from the first thoracic spine, where it fuses with the ligamentum nuchae, cranially to the occiput. The final origin of the m. splenius is by an aponeurosis from the cranial border of the thoracolumbar fascia, which extends caudally to the fifth or sixth thoracic spine. At the cranial border of the atlas the m. splenius is enclosed in a coarse aponeurosis which inserts on the dorsal nuchal line of the occipital bone and the mastoid part of the temporal bone. The m. splenius may occasionally send a strong serration to the transverse process of the axis. At the lateral border of the atlas the dorsal surface of the m. longissimus capitis attaches firmly to the m. splenius and, by means of a strong tendon so formed, inserts along with the m. splenius on the mastoid part of the temporal bone.

Action: To extend and raise the head and neck. In unilateral action to draw the head and neck laterally. It also functions in fixation of the first thoracic vertebra.

Innervation: Nn. cervicales.

Epaxial Spinal Musculature

The dorsal trunk musculature, associated with the vertebral column and ribs, may be divided into three longitudinal muscle masses, each comprising many overlapping fascicles. The muscles act as extensors of the vertebral column and also produce lateral movements of the trunk when acting only on one side.

On the lateral aspect is the *m. iliocostalis* system; intermediately, the *m. longissimus* system; and deep medially, the *m. transversopinalis* system. Various fusions of these three primary segmental muscle masses result in patterns which differ according to the species. The specializations seen in the epaxial musculature, associated with the leaping-gallop type of locomotion, involve the development of long fascicles instead of short metameric ones, the shifting of muscle insertions onto the neural spines for better leverage, and the fusion of the caudal portions of the m. iliocostalis with the m. longissimus to form the *m. erector spinae*, also called the *m. sacrospinalis*. Slijper (1964) described the functional anatomy of the epaxial spinal musculature in a wide variety of mammals.

ILIOCOSTALIS SYSTEM

The *m. iliocostalis* (Fig. 6–28) consists of a series of fascicles lateral to the other epaxial muscles. The caudal members of the series arise on the ilium and constitute a lumbar portion, whereas the cranial fascicles extend to the seventh cervical vertebra and constitute the thoracic portion.

The *m. iliocostalis lumborum* is a strong muscle mass which arises from the pelvic surface of the wing of the ilium, the iliac crest, and from an intermuscular septum located between the m. iliocostalis and m. longissimus. This septum is attached to the ilium and the deep surface of the lumbodorsal fascia. As the fibers of the muscle run cranioventrad, strong lateral fascicles from the ends of all the lumbar transverse processes join them. The cranial end of the lumbar portion runs toward the ribs and is distinctly separated from the m. longissimus. With increasingly weaker fleshy serrations, the m. iliocostalis lumborum attaches to the thirteenth, twelfth, eleventh, and tenth ribs, and occasionally, by a long delicate tendon, to the ninth rib also.

The *m. iliocostalis thoracis* is a long, narrow muscle mass extending from the ribs, except the first and last, to the transverse process of the seventh cervical vertebra. Its origin lies medially under the cranial segments of the m. iliocostalis lumborum. It lies lateral to the m. longissimus and reaches its greatest size between the fifth and third ribs. It is composed of individual portions which originate on the cranial borders of the vertebral end of the ribs; they extend craniolaterally and, after passing over one rib, form a common muscle belly. From this belly, terminal serrations arise which, by means of long tendons (stronger cranially), end on the costal angles of the ribs and most cranially on the transverse process of the seventh cervical vertebra.

Action: Fixation of the vertebral column or lateral movement when only one side contracts; aids in expiration by pulling the ribs caudally.

Innervation: Dorsal branches of the nn. thoracales and lumbales.

LONGISSIMUS SYSTEM

The *m. longissimus* (Figs. 6–28, 6–29, 6–40A) is the medial portion of the m. erector spinae. Lying medial to the m. iliocostalis, its overlapping fascicles extend from the ilium to the head. The m. longissimus consists of thoracolumbar, cervical, and capital regional divisions. The m. sacrococaudalis lateralis can be regarded as the caudal continuation of the m. longissimus on the tail; this muscle is discussed with the tail muscles.

The *m. longissimus thoracis et lumborum* is the strongest muscle of the trunk in the lumbar and thoracic regions. Lateral to the spinous processes of the lumbar and thoracic vertebrae (which are covered by deeper muscles), and dorsal to the lumbar transverse processes and the ribs, it runs from the iliac crest to the last cervical vertebra. In the lumbar region, it is intimately fused with the m. iliocostalis lumborum to form the m. erector spinae (m. sacrospinalis). In the thoracic region the m. spinalis et semispinalis dorsi et cervicis arises from its strong aponeurotic covering. The thoracolumbar division reaches its greatest development in the cranial part of the lumbar region; in the thoracic region it gradually narrows, whereas the m. iliocostalis gets larger.

The *m. longissimus lumborum* (Eisler 1912) is covered by an exceptionally dense aponeurosis which is separated from the thoracolumbar fascia by fat. It arises caudally from the iliac crest and ventral surface of the ilium, and medially from the spinous processes and supraspinous ligament. Its fibers run craniolaterally. The aponeurosis is divided into many strong tendinous strands between which narrower intermediate portions extend. Cranially, it is dissipated at the fifth rib. From the eleventh to the seventh rib, it serves as an origin superficially for the m. spinalis et semispinalis thoracis et cervicis. In the lumbar region the m. longissimus sends off seven medially directed fascicles from the ilium and the intermuscular septum. These fascicles cover the roots of the lumbar transverse processes, and end on the accessory processes of the sixth to first lumbar vertebra. The weak, most caudal portion runs to a fleshy insertion on the arch of the seventh lumbar vertebrae and to the intervertebral disc of the lumbosacral joint. There are also independent, more dorsally placed, medial tendons going to the cranial articular processes of the seventh, sixth, and fifth lumbar vertebrae.

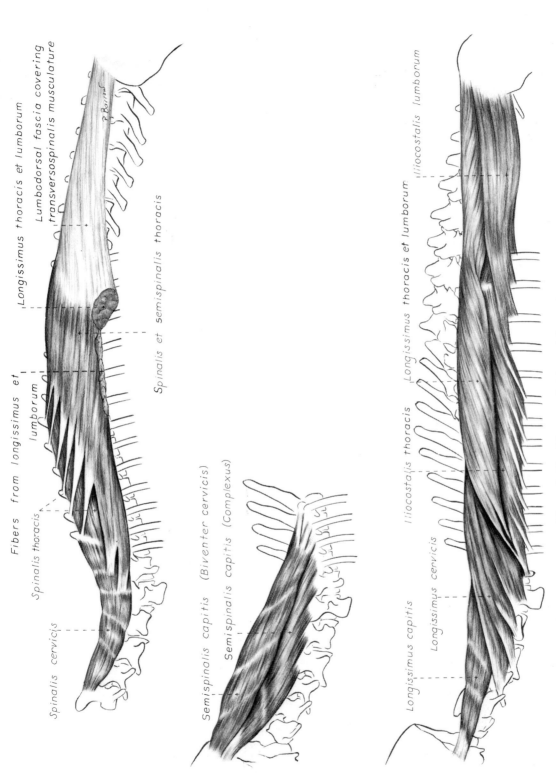

Longissimus thoracis et lumborum

Lumbodorsal fascia covering transversospinalis musculature

P. Barrow

Spinalis et semispinalis thoracis

Fibers from longissimus et lumborum

Spinalis thoracis

Spinalis cervicis

Semispinalis capitis (Biventer cervicis)

Semispinalis capitis (Complexus)

Iliocostalis lumborum

Longissimus thoracis et lumborum

Iliocostalis thoracis

Longissimus capitis

Longissimus cervicis

Figure 6–28. The superficial epaxial muscles.

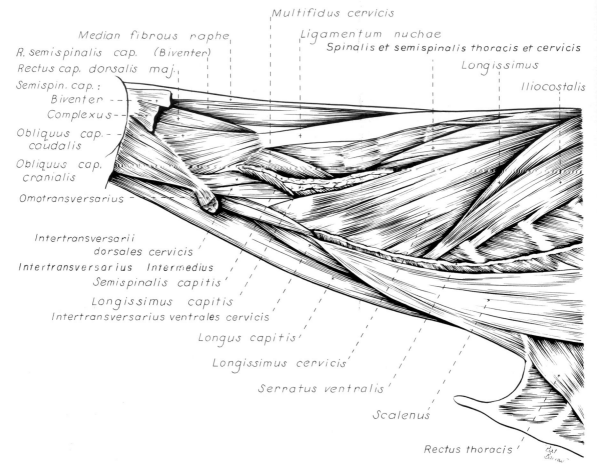

Multifidus cervicis

Median fibrous raphe,

R. semispinalis cap. (Biventer)

Rectus cap. dorsalis maj.

Semispin. cap.:
Biventer
Complexus

Obliquus cap. caudalis

Obliquus cap. cranialis

Omotransversarius

Ligamentum nuchae

Spinalis et semispinalis thoracis et cervicis

Longissimus

Iliocostalis

Intertransversarii dorsales cervicis

Intertransversarius Intermedius

Semispinalis capitis

Longissimus capitis

Intertransversarius ventrales cervicis

Longus capitis

Longissimus cervicis

Serratus ventralis

Scalenus

Rectus thoracis

Figure 6–29. Muscles of neck and head, deep dissection, lateral aspect.

The *m. longissimus thoracis* has serrations which run to the caudal borders of the ribs by means of broad tendinous leaves. Each tendinous leaf separates into a medial and a lateral terminal tendon, the edges thicken, and between them pass dorsal branches of the thoracic nerves. The medial tendons of these ventral serrations end on the accessory processes of the thirteenth to sixth thoracic vertebrae. Since accessory processes are lacking from the fifth to first thoracic vertebrae, the medial tendons insert on the caudal ends of the transverse processes. The lateral tendons of the m. longissimus thoracis insert on the thirteenth to sixth ribs, where they attach medial to the attachment of the m. iliocostalis on the edge of a flat groove adjacent to the costal tubercle. Cranial to the sixth rib the muscle becomes so narrow that

its tendons appear undivided. The terminal tendons end on the costal tubercles of the fifth to first ribs immediately lateral to the costal tubercular joint. Occasionally, further divisions of the terminal tendon insert on the transverse processes of the sixth and fifth cervical vertebrae, where they fuse with serrations of the m. longissimus cervicis.

Action: Extension of the vertebral column. Raising of the cranial portion of the body from the pelvis, sacrum, and loin; in conjunction with other muscles, fixation of the vertebral column; deflection of the back by fixation of the cervicothoracic junction; sudden raising of the caudal portion of the body, which is initiated by means of the rear extremities.

Innervation: Dorsal branches of the thoracic and lumbar nerves.

The m. longissimus in the lumbar and thoracic regions gives rise to serrations from its deep medial part. These follow the fiber direction of the m. longissimus, but, in contrast to it, they pass over only a small number of vertebrae. These are described under the system of the mm. intertransversarii.

The *m. longissimus cervicis* (Figs. 6–27, 6–28, 6–29) is a continuation of the m. longissimus thoracis, lying in the angle between the cervical and thoracic vertebrae. It is triangular in form, and in large dogs is 1 to 1.5 cm. thick. The muscle complex is composed of four serrations which are incompletely separable; each consists of a long lateral bundle and several short medial bundles; they are so arranged that a caudal serration partly covers its cranial neighbor.

Action: To extend the neck; in unilateral action to raise the neck obliquely and turn it to one side.

Innervation: Dorsal branches of cervical and thoracic nerves.

The *m. longissimus capitis* (Fig. 6–28) is a strong muscle 3.5 to 4.5 cm. wide and 5 to 7 mm. thick in large dogs; it lies medial to the mm. longissimus cervicis and splenius. It covers the m. semispinalis capitis along its ventral border and extends from the first three thoracic vertebrae to the temporal bone. It arises by separate bundles from the transverse processes of the third to first thoracic vertebrae in combination with corresponding serrations of the m. semispinalis capitis and on the caudal articular processes of the seventh to third or fourth cervical vertebrae. The muscle narrows gradually and is divided by one or two tendinous intersections. It runs over the dorsal surface of the atlas and, by means of a strong tendon, inserts on the mastoid part of the temporal bone. At the level of the atlas, it unites firmly with the m. splenius.

According to Bogorodsky (1930), in 20 per cent of specimens there is a deep portion, the *m. longissimus atlantis,* whose fibers come from the articular processes of the seventh to fourth cervical vertebrae and end on the edge of the wing of the atlas.

Action: Extension of the atlanto-occipital joint. The atlantal portion in unilateral action rotates the atlantoaxial joint, whereas in bilateral action it fixes the atlantoaxial joint.

Innervation: Dorsal branches of the cervical nerves.

TRANSVERSOSPINALIS SYSTEM

The most medial and deepest epaxial muscle mass consists of a number of different systems of fascicles which join one vertebrae. The nomenclature employed by various authors varies considerably (Plattner 1922, Winckler 1939, and Slijper 1946).

This portion of the trunk musculature is divided, according to Stimpel (1934), into: (1) the independent m. spinalis thoracis et cervicis; (2) the m. transverospinalis, composed of the m. semispinalis (thoracis et capitis), the m. multifidus (thoracis et cervicis), the m. sacrocaudalis dorsalis medialis, and the mm. rotatores (longi et breves), and (3) the mm. interspinales (thoracis et cervicis). The m. sacrocaudalis dorsalis medialis is considered to be the direct caudal continuation of the m. multifidus; it will be described with the other caudal muscles. The mm. spinalis thoracis et cervicis and semispinalis thoracis have intimate relationships with each other, so that they are considered together under the name m. spinalis et semispinalis thoracis et cervicis.

The *m. spinalis et semispinalis thoracis et cervicis* (Figs. 6–28, 6–29) of the dog cannot be considered as a pure m. spinalis, even though its segments run mainly between spinous processes, since it receives strands from the mamillary processes of some vertebrae. For this reason it is designated by a compound name, despite the fact that it is predominantly spinous in nature. As a strong, partly unsegmented, longitudinal muscle which consists largely of incompletely isolated segments, it lies lateral to the spinous processes of the thoracic vertebrae and dorsomedial to the m. longissimus thoracis. It runs on the cervical vertebrae to the spinous process of the axis ventral to the ligamentum nuchae. It arises from the tendinous leaf on the dorsal surface of the m. longissimus thoracis and, by means of this, directly from spinous processes of the first few lumbar vertebrae; it also arises from the spinous processes of the sixth to first thoracic and last cervical vertebrae and, finally, from the mamillary processes of the first two lumbar and last thoracic vertebrae. The large tendinous leaf on the outer surface of the m. longissimus, however, is attached to the iliac crest and the spinous processes of the second sacral to the last thoracic vertebrae. Considered superficially as a muscle mass, the spinalis et semispinalis thoracis et cervicis in the dog extends from the spinous process of the eleventh thoracic vertebra to the spinous process of the axis. This combined muscle is clearly separated into lateral and medial parts. The lateral part is the m. spinalis et semispinalis thoracis; the medial part is the m. spinalis cervicis.

The *m. spinalis et semispinalis thoracis* (Fig. 6–28) is the lateral part of the compound muscle which arises from the fascia of the m. longissimus. From the spinous processes of the eleventh to the seventh thoracic vertebra, this muscle with its nearly horizontal fibers is unsegmented; from the ninth thoracic vertebra forward it sends off gradually ascending bundles to the spinous processes of the thoracic vertebrae. The muscle divides into eight separate bundles, which insert on the spinous processes of the sixth thoracic to the sixth cervical vertebra by means of superficial tendons which become progressively more distinct cranially. The segment to the sixth cervical vertebra almost completely covers the segment to the seventh cervical vertebra. Each ends by well-developed tendinous leaves on the spinous processes. Each also gives off a wide, leaflike portion to the neighboring caudal, tendinous leaves of the m. spinalis cervicis. In the cranial thoracic region variations may occur, in that intermediate bundles with their own tendons may appear (Stimpel 1934).

The *m. spinalis cervicis* (Fig. 6–28) is the medial, flat muscular strand bearing four tendinous inscriptions. It arises from the tendon of the most cranial thoracic segment of the m. semispinalis thoracis and from the cranial border of the first thoracic spine, but it also receives a few bundles from the spinous process of the seventh cervical vertebra. Separated from the muscle of the opposite side only by the median ligamentous septum, it runs cranially ventral to the ligamentum nuchae. It inserts on the spinous processes of the fifth to second cervical vertebrae; it is covered in part by portions of the m. multifidus.

Action: To fix the thoracic vertebral column and to raise the neck.

Innervation: Medial branch of the dorsal branches of the cervical and thoracic nerves.

The *m. semispinalis capitis* (Figs. 6–28, 6–29) is the strong continuation to the head of the m. spinalis et semispinalis thoracis et cervicis. The capital portion of the semispinalis strand covers the cranial end of the m. spinalis et semispinalis thoracis et cervicis laterally; its broad origin is covered by the mm. longissimus and splenius. The muscle lies rather deep as it extends from the first five thoracic and the last cervical vertebra to the occiput. It surrounds each half of the ligamentum nuchae laterally and dorsally, meeting its fellow of the opposite side. The two muscles are separated only by the nuchal ligament and the median fibrous raphe. It is divided into the dorsally located m. biventer cervicis and the ventrally placed m. complexus, which can be separated as far as their insertions, despite the intimate connections between them.

The *m. biventer cervicis* (Fig. 6–28) arises, by three strong serrations medial to the m. longissimus cervicis and capitis, from the transverse processes of the fourth, third, and second thoracic vertebrae. Fascial strands also come from the lateral surfaces of the spinous processes underneath the m. semispinalis dorsi; other fibers are added to the dorsal border from the fascia spinotransversalis at the level of the shoulder. The m. biventer cervicis is firmly connected with the median fibrous raphe of the neck. It appears to be divided, by four (rarely five) very oblique tendinous inscriptions, into separate portions having longitudinal fibers. It inserts on a distinct, oval, rough area ventrolateral to the external occipital protuberance on the caudal surface of the skull.

The *m. complexus* (Fig. 6–28) arises from the caudal articular processes of the first thoracic to the third cervical vertebra in common with the m. longissimus capitis (laterally) and the m. multifidus cervicis (medially). The caudal segments are more fleshy; the one arising on the first thoracic vertebra has a tendinous covering medially which is also related to one of the portions of the m. multifidus. Fibers also arise in the fascia of the m. obliquus capitis caudalis somewhat cranial to the caudal border of the atlas. The fibers run craniomedially to end laterally on the dorsal nuchal line by means of a tendon coming from a strong superficial fibrous covering.

Action: To raise the head and neck; in unilateral action to flex the head and neck laterally.

Innervation: Dorsal branches of the nn. cervicales.

The *m. multifidus* (Figs. 6–30, 6–40A) is a muscle composed of numerous individual portions which overlap in segments and extend from the sacrum to the second cervical vertebra; it is augmented at both joints of the head, as well as in the tail, by more or less modified muscles — at the head by the mm. obliquus capitis caudalis and cranialis, in the tail by the m. sacrocaudalis dorsalis medialis. The m. multifidus as a segmental muscle extends from mamillary, transverse, or articular processes of caudally lying vertebrae to spinous processes of cranially lying ones. As a rule, two vertebrae are passed over by each bundle. The m. multifidus, aside from the oblique capital muscles, is divided into four portions: the pars lumborum, pars thoracis, pars cervicis, and the pars coccygeae, which is described with the caudal muscles as the m. sacrocaudalis dorsalis medialis.

The *m. multifidus lumborum* is a strong, seemingly homogeneous muscle which runs from the sacrum to the spinous process of the eighth or ninth thoracic vertebra. It is divided into 11 individual, flat portions which are united with each other. They originate from the three articular processes of the sacrum (including the mamillary process of the first caudal vertebra) and from the mamillary processes of the seventh lumbar to the twelfth thoracic vertebra. After the several parts pass over two segments, they end laterally on the ends of the spinous processes of the sixth lumbar to the ninth (occasionally eighth) thoracic vertebra immediately beneath the supraspinous ligament.

The *m. multifidus thoracis* lies more ventrally on the vertical column, and its segments are more vertical than those of the lumbar part. It arises by nine distinctly isolated portions on the mamillary and transverse processes of the eleventh to the third thoracic vertebra and inserts on the spinous processes of the eighth thoracic to the seventh cervical vertebra.

The *m. multifidus cervicis* is covered by the m. semispinalis capitis. It appears under the ventrolateral border of the m. spinalis et semispinalis thoracis et cervicis, where it extends from the articular process of the second thoracic vertebra to the spinous process of the axis. It consists of six incompletely separable individual portions which themselves are again partially divided into lateral principal, medial accessory, and deep accessory parts, according to Stimpel (1934); collectively they arise essentially from the articular processes.

Action: As a whole, the m. multifidus, along with the other dorsal back muscles, fixes the vertebral column, especially in bilateral action.

Innervation: Medial branches of the rami dorsales in the lumbar, thoracic, and cervical regions.

From the medial surface of the m. multifidus certain deep muscles have become extensively differentiated; these are the mm. rotatores longi and breves. In addition, there are the mm. interspinales lumborum, thoracis, and cervicis between the spinous processes, and the mm. intertransversarii caudal, lumborum, thoracis, and cervicis, which in general run between the transverse processes. The intertransverse muscles of the tail are described with the caudal muscles.

The *mm. rotatores* (Fig. 6–30) are developed as eight long and nine short rotators; in the dog they are confined strictly to the cranial thoracic region, where the pairs of articular processes are tangentially placed, thus allowing rotary movements.

The *mm. rotatores longi* extend between the transverse and spinous processes of two alternate vertebrae; the most caudal extends from the transverse process of the tenth to the spinous process (basal to the insertion of the corresponding segment of the multifidus) of the eighth thoracic vertebra, and the most cranial extends between corresponding points of the third and the first thoracic vertebra. These segments are more vertical than those of the m. multifidus, along the caudal border of which they appear.

The *mm. rotatores breves* pass between vertebrae. They are situated more deeply than are the long rotators. The most caudal belly runs between the transverse process of the tenth and the spinous process of the ninth thoracic vertebra, the most cranial

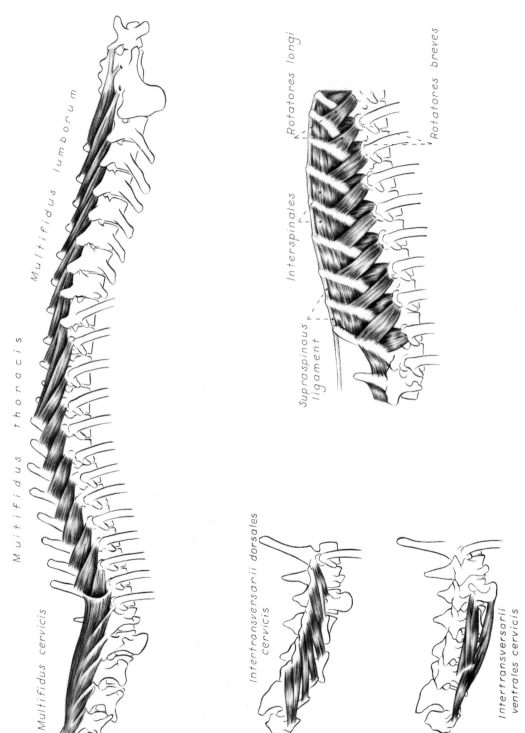

Figure 6–30. Deep epaxial muscles.

belly between similar points on the second and the first thoracic vertebra. Often this portion is surrounded extensively by tendinous tissue (Krüger 1929).

Action: Rotation of the greater cranial portion of the thoracic vertebral column about the longitudinal axis in unilateral action; otherwise, fixation.

Innervation: Medial branches of the rami dorsales of the thoracic nerves.

The *mm. interspinales* (Fig. 6–30) are distinctly separable into lumbar, thoracic, and cervical portions; the lumbar portion is covered by the m. multifidus. In the thoracic region, after removal of the mm. semispinalis and longissimus, the mm. interspinales are visible at the ends of the spinous processes. They run between contiguous edges of spinous processes and overlap these edges somewhat. They also extend between the spinous processes of the first thoracic to the fifth cervical vertebra.

Action: Fixation of the vertebral column.

Innervation: Medial branches of the rami dorsales of the spinal nerves.

The *mm. intertransversarii* (Fig. 6–30) are deep segments split off from the longissimus system. They are separable into caudal, lumbothoracic, and cervical parts, and, as delicate muscle bundles, they pass over one or two, or, at most, three vertebrae.

The *mm. intertransversarii caudae* are discussed with the muscles of the tail.

Only weak *mm. intertransverarii lumborum et thoracis* are formed on the trunk. These separate parts run between the mammillary processes of the seventh lumbar to the thirteenth or twelfth thoracic and the accessory processes of the fifth lumbar to the ninth thoracic vertebrae, and between the transverse processes of the twelfth to the eighth and those of the eighth to the fourth thoracic vertebrae.

The *mm. intertransversarii cervicis* are larger. They are divided into three separate muscle strands: a dorsal one, running between articular and transverse processes of the cervical vertebrae, and, corresponding to the intertransverse muscle of the lumbar and thoracic regions, an intermediate strand, and a ventral strand.

The *mm. intertransversarii dorsales cervicis* lie between the lines of insertion of the mm. longissimus cervicis, longissimus capitis, and semispinalis capitis. As a segmental

muscle strand it extends from the eminence on the cranial articular process of the first thoracic vertebra to the wing of the atlas. Its individual bundles run craniolaterally in the form of five indistinctly separated bellies from the first thoracic and the seventh to fourth cervical vertebrae to the transverse processes of the sixth to second cervical vertebrae. The cranial portion of the muscle extends from the eminence of the third and second cervical vertebrae to the caudal border of the wing of the atlas.

The *mm. intertransversarii intermedii cervicis* form a strand which is composed of five or six distinctly separable, delicate parts which extend only between transverse processes; they lie ventral to the insertion of the m. serratus ventralis cervicis and dorsal to the m. scalenus, and are partly covered by these two muscles. The segments course between the terminal tubercles of the ends of the transverse processes from the first thoracic to the second cervical vertebra. On the sixth cervical vertebra it is on the transverse process proper, and, from the fifth cervical vertebra forward, it is on the caudal branch of the transverse process and the border of the wing of the atlas; the most cranial portion runs under the dorsal m. intertransversarius of the axis. The deep fibers pass from segment to segment; the superficial ones pass over one segment.

The *mm. intertransversarii ventrales cervicis* run cranially from the m. scalenus and form a homogeneous longitudinal strand. This is found ventral to the m. scalenus and dorsal to the m. longus colli; it extends from the ventral border of the winglike transverse process of the sixth cervical vertebra to insert by three separate terminal segments on the caudal branch of the transverse process of the fourth, third, and second cervical vertebrae. This strand is covered by the m. scalenus caudally and by the intermediate portion of the mm. intertransversarii cranially.

At its cranial end the cervical vertebral column serves special functions. There is a corresponding special development of the first two cervical vertebrae, as well as of their joints. The specialized musculature dorsal and ventral to the atlas and axis is adapted to these special functions. There are three portions of the **m. rectus capitis** which runs between regions on the spine of the axis, the atlas, and the occiput, and which can be compared to the m. interspinalis; these are the mm. rectus capitis dorsalis major, intermedius, and minor. There are also two oblique muscles, the mm. obliquus capitis caudalis and cranialis, which can be considered modifications of the m. multifidus or derivatives of the m. intertransversarius.

The *m. rectus capitis dorsalis major* (Fig. 6–29) is covered by the m. semispinalis capitis as it runs between the spine of the axis and the squama of the occiput. It arises cranial to the attachment of the ligamentum nuchae on the caudal end of the spine of the axis, and it ends medially on the occiput. The dorsal portion of the m. obliquus capitis cranialis, which lies on the border of the wing of the atlas, also inserts on the ventrolateral part of the occiput.

The *m. rectus capitis dorsalis intermedius* is a strong, almost triangular muscle. Covered by the major part, it arises cranially on the axis and with diverging fibers runs over the atlas to the occiput, where it inserts on the ventral nuchal line. It is obviously only the deeper part of the m. rectus capitis dorsalis major, although it fuses in part with the small extensor of the head.

The *m. rectus capitis dorsalis minor* is a short, flat muscle, lying between the atlas and the occiput on the capsule of the atlanto-occipital joint immediately next to its fellow of the opposite side. It arises on the cranial edge of the dorsal arch of the atlas and inserts above the foramen magnum near the ventral nuchal line, where it fuses with the intermediate extensor of the head.

Action: All three portions extend the atlanto-occipital joint.

Innervation: Ramus dorsalis of n. cervicalis 1.

The *m. obliquus capitis caudalis* (Fig. 6–29) is a strong, flat muscle lying under the mm. semispinalis capitis and splenius; it covers the atlas and axis dorsally. It arises along the entire spinous process and the caudal articular process of the axis and runs obliquely craniolaterally over the capsule of the atlantoaxial joint to insert on the border of the wing of the atlas near the alar notch.

Action: Unilateral: rotation of the atlas and thus the head on the axis; bilateral: fixation of the atlantoaxial joint.

Innervation: Rami dorsales of the nn. cervicales 1 and 2.

The *m. obliquus capitis cranialis* (Fig. 6–29) extends obliquely craniolaterally over the atlanto-occipital joint; it lies under the m. splenius and is divided into two portions. The principal part arises on the lateroventral surface and lateral border of the wing of the atlas. Inclined dorsomedially, it runs over the jugular process and inserts on the mastoid part of the temporal bone and from there upward on the dorsal nuchal line. The accessory portion is a superficial flat belly extending to the atlantal end of the m. obliquus capitis caudalis; it takes its origin on the tip of the wing of the atlas and, provided with tendinous leaves, inserts between the principal portion and the m. rectus capitis dorsalis major on the dorsal nuchal line.

Action: Extension of the atlanto-occipital joint.

Innervation: Ramus dorsalis or n. cervicalis 1.

Muscles of the Ventral Neck Region

The muscles of the ventral neck region are divided into two groups. The first group is closely related to the trachea and esophagus and includes the large, superficially located mm. brachiocephalicus and sternocephalicus and the small, deeply located mm. sternohyoideus, sternothyroideus, and scalenus. The second group includes muscles that lie on the ventral surfaces of the cervical vertebrae: the mm. longus capitis, longus colli, rectus capitis ventralis, and rectus capitis lateralis.

The *m. brachiocephalicus* and *m. sternocephalicus* are described with the muscles of the thoracic limb.

The *m. sternohyoideus* is closely allied with the m. sternothyroideus throughout its course. It is described with the muscles of the hyoid apparatus.

The *m. sternothyroideus* (Figs. 6–11, 6–26, 6–48) lies deep to the m. sternohyoideus and has a similar tendinous intersection which divides the muscle into cranial and caudal portions. The m. sternothyroideus arises from the first costal cartilage, and passes up the neck covered by the m. sternocephalicus. Although weaker than the m. sternohyoideus, it covers more of the lateral surface of the trachea. It inserts on the lateral surface of the thyroid lamina.

Action: To draw the hyoid apparatus, larynx, and tongue caudally.

Innervation: Ramus ventralis of n. cervicalis 1.

The *m. scalenus* (Figs. 6–26, 6–32) bridges the space between the first three ribs and the cervical vertebrae. The muscle is divided into a superficial and a deep portion.

The *m. scalenus primae costae* arises as three portions on the cranial border of the first rib. The two deep segments, which are not clearly separated, run to the transverse process of the seventh cervical vertebra and to the winglike process of the sixth cervical vertebra. In its course it is covered proximally by the supracostal part of the m. scalenus, and distally by a superficial portion (also arising on the first rib). The superficial segment, on the other hand, extends from the first rib to the transverse processes of the fifth, fourth, and third cervical vertebrae.

The *m. scalenus supracostalis* forms the principal part of the superficial layer. In the form of two or three flat, distinctly separated portions it arises from the outer surfaces of the ribs and inserts on the cervical vertebrae. In so doing it covers parts of the m. scalenus primae costae, the caudal portions of the m. intertransversarius intermedius, and the first three or four serrations of the m. serratus ventralis. This part of the m. scalenus attaches to the transverse processes (caudal branches) of the fifth and fourth (and also, occasionally the third) cervical vertebrae. The dorsal portion arises underneath the corresponding segment of the serratus on the third rib and the middle portion on the fourth; both may arise in common on the fourth rib. The ventral portion is the longest; its origin from the eighth or ninth rib, by means of a long tendinous leaf, is covered by the m. obliquus abdominis externus.

Action: To draw the neck downward. In unilateral action. to bend the neck sideward. When the neck is fixed, the supracostal part can act in inspiration.

Innervation: Rami ventrales of the nn. cervicales and thoracales.

The *m. longus capitis* (Figs. 6–29, 6–31) is a long, flat muscle which lies on the lateral and ventral sides of the cervical vertebrae lateral to the m. longus colli. It arises from the caudal branches of the transverse processes of the sixth to the second cervical vertebra, and extends cranially to the axis,

where it receives a strong, tendinous leaf laterally. After crossing the atlanto-occipital joint, it inserts (tendinous laterally, muscular medially) on the muscular tubercle of the basioccipital, between the tympanic bullae.

Action: To flex the atlanto-occipital joint and to draw the neck downward.

Innervation: Rami ventrales of the nn. cervicales.

The *m. longus colli* (Fig. 6–31) is a long muscle composed of separate bundles; it lies adjacent to its fellow on the bodies of the first six thoracic and all of the cervical vertebrae, and thus is divided into thoracic and cervical

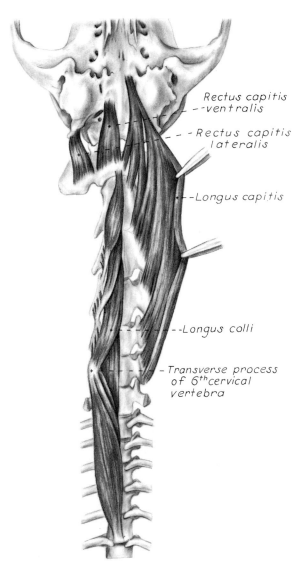

Rectus capitis ventralis

Rectus capitis lateralis

Longus capitis

Longus colli

Transverse process of 6ᵗʰ cervical vertebra

Figure 6–31. Ventral muscles of the vertebral column.

portions. On the neck the bilateral muscle is enclosed by the right and the left m. longus capitis. The thoracic portion consists of three incompletely separated parts which arise on the convex ventral surfaces of the first six thoracic vertebrae. These portions, complicated in their make-up, are provided with tendinous coverings. Diverging cranially from those of the opposite side, the fibers of these three portions become partly tendinous laterally; the medial fibers insert immediately beside this tendon on the ventral border of the wing of the sixth cervical vertebra, as well as on the transverse process of the seventh cervical vertebra. The continuation of the cervical portion consists of four separate bundles. These bundles arise on the ventral border of the transverse process of the sixth to the third cervical vertebra and end on the ventral spine of the next preceding vertebra. The caudal V-shaped segment formed by the muscles of both sides has lying in its angle the cranial part of the thoracic portion. The cranial segment ends on the ventral tubercle of the atlas.

Action: To draw the neck downward.

Innervation: Rami ventrales of the nn. cervicales.

The *m. rectus capitis ventralis* (Fig. 6–31) is a short, strong muscle which lies dorsal to the end of the m. longus capitis. It extends from the ventral arch of the atlas to the basioccipital bone. As it crosses the atlanto-occipital joint, it converges somewhat with its fellow of the opposite side.

Action: Flexion of the atlanto-occipital joint.

Innervation: Ramus ventralis of the n. cervicalis1.

The *m. rectus capitis lateralis* (Fig. 6–31) is a small muscle which lies lateral to the m. rectus capitis ventralis (separated from it by the ventral branch of the first cervical nerve). It originates on the ventral surface of the caudal half of the wing of the atlas (lateral to the n. rectus capitis ventralis); it passes sagittally toward the cranium over the atlanto-occipital joint, and inserts on the base of the jugular process of the occiput. This muscle can be considered a special portion of the m. intransversarius ventralis.

Action: Flexion of the atlanto-occipital joint.

Innervation: Ramus ventralis of n. cervicalis 1.

Fasciae of the Neck

The superficial and deep cervical fasciae are the direct continuations of the superficial and deep fasciae of the head.

The **superficial fascia of the neck** (*fascia colli superficialis*) is cylindrical in form, as it clothes the whole neck. It is delicate, lies directly under the skin, and is easily displaced. It originates from the superficial temporal, parotideomasseteric and intermandibular fasciae, and it continues caudally into the superficial omobrachial fascia and ventrally into the superficial trunk fascia of the sternal region. It contains the m. sphincter colli superficialis and the platysma; with these it covers the mm. trapezius, omotransversarius, cleidocephalicus, and sternocephalicus, and it bridges the external jugular vein which lies in the jugular groove. The bilateral portions of the fascia meet dorsally and ventrally. At the dorsal midline there is no special attachment to the underlying portions (median raphe) so that on the neck, just as on the back and loins, this fascia can be lifted in a big fold with the skin. In many places the smaller cutaneous vessels and nerves pass through the superficial fascia.

The **deep fascia of the neck** (*fascia colli profunda*) is a strong binder which extends under the mm. sternocephalicus, cleidomastoideus, omotransversarius, and cleidocervicalis; it covers the mm. sternohyoideus and sternothyroideus superficially and surrounds the trachea, thyroid gland, larynx, and esophagus, but passes over the large cervical vessels and nerves (nn. vagosympatheticus and recurrens, a. carotis communis, and v. jugularis internus) to the superficial surface of the mm. scalenus and longus colli. From these it goes to the superficial surface of the mm. serratus ventralis and rhomboideus, to end in the median raphe of the neck. Cranially, it continues as the external pharyngeal fascia and passes under the mandibular and parotid glands and finally ends on the hyoid bone and the base of the head. Under the salivary glands the deep cervical fascia is united with the deep fascia of the m. masseter. The dorsal part of the deep cervical fascia continues over the m. rhomboideus to the superficial surface of the scapula with its muscles, and thus runs into the deep fascia of the shoulder. On the surfaces of the cervical parts of the mm. serratus ventralis and scalenus, the deep cervical fascia continues to the thoracic parts of these muscles medial to the shoulder to become the deep thoracic fascia. Ventral to the m. scalenus it attaches to the first rib and the manubrium sterni. The deep fascia sends various divisions between the layers of muscles of the neck. One of these is a strong leaf which passes beneath the cervical parts of the mm. serratus ventralis, trapezius, and rhomboideus but external to the m. splenius.

One part of the deep cervical fascia ventral to the vertebral column is the **prevertebral fascia** (*fascia prevertebralis*), on which the superficial surface of the mm. longi colli lies. Cranially, it is attached to the base of the head; caudally, it continues with the mm. longi colli over the first six thoracic vertebrae into the thorax to unite with the **endothoracic fascia** (*fascia endothoracica*).

Finally, from the deep leaf of the cervical fascia, there detaches a delicate **proper fascia of the trachea** (*fascia tracheae propria*), which surrounds the trachea. Much loose connective tissue is accumulated around both the trachea and esophagus, to provide them with a high degree of displaceability. The carotid sheath is a special loose condensation of fascia in which the common carotid artery, internal jugular vein, lymphatics (tracheal duct), and the vagosympathetic trunk are located. It is located dorsolateral to the trachea and is the fascial union of the lateral and prevertebral parts of the deep cervical fascia.

Muscles of the Lateral and Ventral Thoracic Wall

The spaces between the ribs are filled by the mm. intercostales, which appear in a double layer, internal and external, and cross each other. Where these muscles occur between the costal cartilages, they are specifically called mm. intercartilaginei. Each m. intercostalis externus gives rise to a m. levator costarum proximally. The fibers which make up its almost spindle-shaped belly do not come from the following rib, but come rather from the transverse process of the corresponding thoracic vertebra. Cranially, on the thorax, the m. rectus thoracis covers

the superficial (ventral) ends of the first ribs; the m. transversus thoracis crosses the cartilages of the sternal ribs and the sternum deeply. The mm. retractor costae and subcostalis are special muscles of the last rib.

The **mm. intercostales externi** (Fig. 6–34) form the stronger outer layer of the intercostal spaces; they are 4 or 5 mm. thick in large dogs, but become weaker in the region of the floating ribs. They extend from the mm. levatores costarum, which are indistinctly set apart, to the costochondral junctions; they may also extend into the spaces between the costal cartilages as the mm. intercartilaginei. The fibers of the external intercostal muscle arise on the caudal border of each rib, and run caudoventrally to the cranial border of the next rib.

Action: Inspiration. For both external and internal intercostal muscles, the more proximal attachment is to be looked upon as the fixed point, since, for each arched rib, the points farther from the tubercle are relatively easier to move than those closer to the tubercle. Thus the m. intercostalis externus acts to increase the transverse diameter of the thorax.

Innervation: Muscular branches of the nn. intercostales 1 to 12.

The **mm. intercartilaginei externi** are unseparated continuations of the mm. intercostales externi into the interchondral spaces. They are lacking in the first two or three spaces because the external intercostals end proximal to or at the costochondral junctions. Distal to the ends of the external intercostals the internal intercostals make their appearance. With each successive segment, the external interchondral muscles extend farther distally, so that the ninth and the tenth interchondral spaces are completely filled, although occasional defects in the muscle are found. Although they are rather strongly developed in the false or asternal interchondral spaces, the muscle is completely absent in the twelfth interchondral space.

Action and Innervation: Same as for the mm. intercostales externi.

The **mm. levatores costarum** (Fig. 6–34) are present as 12 special formations of the external intercostal muscles. They are flat, spindle-shaped muscles covered by the mm. longissimus thoracis and iliocostalis; they are fleshy at their origins on the transverse processes of the first to twelfth thoracic vertebrae. After running caudoventrally to the angle of the rib next caudad, they end on the cranial borders of the second to thirteenth ribs. They then pass over into the mm. intercostales externi.

Action: Inspiration; the fixed point is the transverse process of the vertebra.

Innervation: Delicate little trunks of the nn. intercostales 1 to 12, given off shortly before their division into lateral and medial branches.

The *mm. intercostales interni* (Figs. 6–34, 6–36) form the weaker internal layer of the intercostal musculature; this layer is 2 or 3 mm. thick in large dogs. The internal intercostals extend from the vertebral column, where they leave free only a small triangular space adjacent to the vertebrae, to the distal ends of the ribs; they are only very slightly separated from the interchondral muscles. The fibers course from the cranial border of one rib to the caudal border of the rib next cranial to it. In this cranioventral course the fibers attain angles of inclination which, at the vertebral column, decrease from 78 to 71 degrees, and, at the sternum, from 68 to 54 degrees. Thus, they are steeper than the mm. intercostales externa, which they cross.

Action: Draw the ribs together.

Innervation: Nn. intercostales.

The *mm. subcostales* are located beneath the internal intercostal muscles at the vertebral ends of the ribs. They may bridge several ribs.

The *mm. intercartilaginei interni* are continuations of the mm. intercostales interni. They fill the interchondral spaces and are 4 or 5 mm. thick. After removal of the m. rectus abdominis, they appear at the sites where the external interchondral muscles are lacking.

Action and Innervation: Same as for the mm. intercostales interni.

The *m. rectus thoracis* was formerly known as the *m. transversus costarum.* It is a flat, almost rectangular muscle which runs

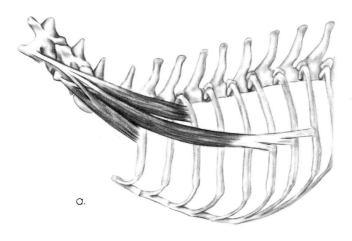

a.

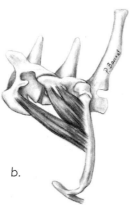

b.

Figure 6–32.   The scalenus muscles.
A. Superficial.
B. Deep.

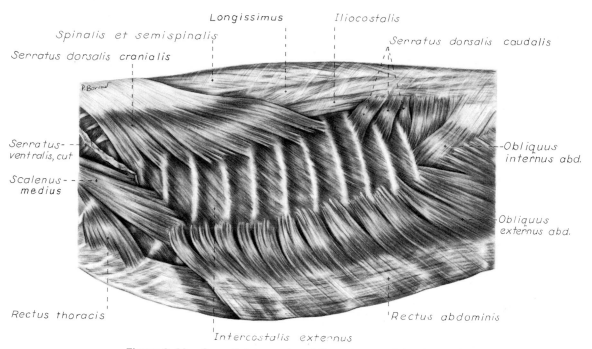

Figure 6–33. Superficial muscles of thoracic cage, lateral aspect.

caudoventrally from its origin on the first rib, opposite the most ventral portion of the m. scalenus (Fig. 6–32), to its insertion over the ventral ends of ribs 2, 3, and 4. Its aponeuro-

sis of insertion obliquely crosses the aponeurosis of the m. rectus abdominis and blends with the deep fascia of the trunk.

This muscle may correspond with the rare-

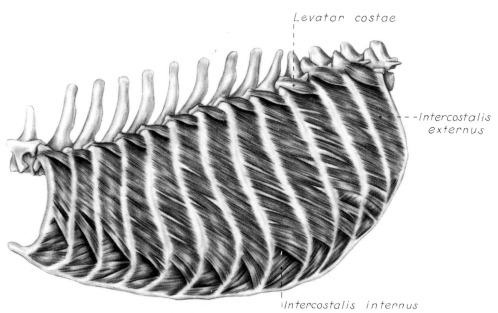

Figure 6–34. Deep muscles of thoracic cage, lateral aspect.

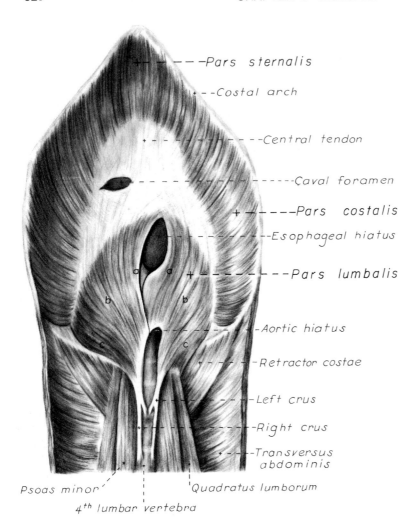

- - -Pars sternalis

- -Costal arch

- - -Central tendon

- - - - Caval foramen

- - - -Pars costalis

-Esophageal hiatus

- - -Pars lumbalis

-Aortic hiatus

-Retractor costae

- -Left crus

-Right crus

-Transversus abdominis

Psoas minor'

4th lumbar vertebra

'Quadratus lumborum

Figure 6–35. Diaphragm, abdominal surface. A = medial, B = intermediate, and C = lateral portions of pars lumbalis.

ly occurring m. sternalis of man. Artemenko (1929) regards it as a division of the m. obliquus externus abdominis. Sisson and Grossman (1953) preferred the term m. rectus thoracis because of its similar position to the m. rectus abdominis. According to Zimmermann (1927), relating it to the rectus abdominis is fallacious.

Action: Inspiration.

Innervation: Lateral branch of the nn. intercostales.

The *m. retractor costa* (Fig. 6–35) is a thin muscle lying under the tendon of origin of the m. transversus abdominis. It bridges the space between the transverse processes of the first three or four lumbar vertebrae and the last rib (Iwakin 1928). Seen from the interior, this thin muscle, the fibers of which cross those of the m. transversus abdominis, lies directly under the peritoneum. Over its

cranial border lies the arcus lumbocostalis of the diaphragm. Farther distally the caudal fiber bundles extend upon the last rib and partly encroach upon the peritoneal surface of the pars costalis of the diaphragm. The m. retractor costae belongs to the system of the m. intercostalis internus and is innervated by the last thoracic nerve (Kolesnikow 1928).

The *m. transversus thoracis* (Fig. 6–36) is a flat, fleshy muscle, lying on the inner surfaces of the sternum and sternal costal cartilages. It forms a continuous triangular leaf which covers the second to eighth costal cartilages. Only exceptionally do slips from its lateral border reach the sternum. A delicate, special bundle may be given off to the first costal cartilage. Its fibers arise by a narrow aponeurosis, on the lateral internal surface of the sternum, from the second sterne-

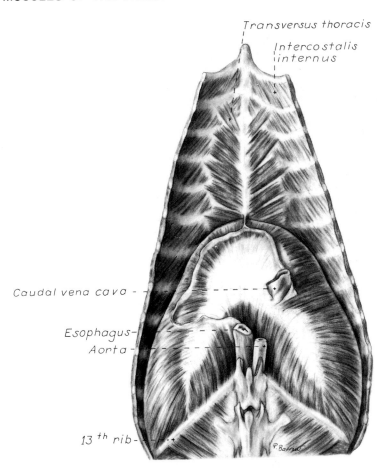

Transversus thoracis

Intercostalis internus

Caudal vena cava -

Esophagus-
Aorta -

13 *th rib* -

Figure 6–36. Diaphragm, thoracic surface.

Diaphragm

The **diaphragm** (*diaphragma*) (Figs. 6–35, 6–36) is a musculotendinous plate between the thoracic and abdominal cavities; it projects forward into the thoracic cavity like a dome. On the thoracic side, it is separated from the pleura by the endothoracic fascia; on the abdominal side, it is separated from the peritoneum by the transversalis fascia. Peripherally, this wall which separates the body cavities attaches to the ventral surfaces of the lumbar vertebrae, the ribs, and the sternum. The fibers of the diaphragm arise on these skeletal parts and radiate toward the tendinous center.

The **central tendon** (*centrum tendineum*) bra to the caudal end of the xiphoid process. They end with indistinct segmentations on the second to seventh costal cartilages, somewhat ventral to the costal symphyses.

of the diaphragm, in the dog, is relatively small. It consists of a triangular central area with dorsal extensions on each side. From the cranial aspect this tendinous area appears to be displaced somewhat ventrally. The two-layered disposition of the tendon fibers is easily followed. To the right, at the base of the body of the tendon, there is a concentric arrangement of strong fibers about the foramen venae cavae which courses slightly cranioventrally. On the columns of the tendon, fibers run in an arch from the crural musculature directly to those of the costal parts. Special strong reinforcements extend lengthwise along the borders. Fibers from the muscle surrounding the esophagus radiate on the body of the tendon to the sternal and ventral parts of the costal diaphragmatic musculature. Transverse fibers course from one side to the other as a reinforcing apparatus. Peculiar whorls are formed near the bases of both columns. Into the dorsal border of the foramen venae cavae, muscle fibers from the costal portion often radiate (Pancrazi 1928).

The muscular part of the diaphragm surrounds the central tendon on all sides, and its fibers stream into the latter in a radial direction. It is divided into the pars lumbalis, a pars costalis on each side, and the pars stenalis.

The *pars lumbalis* of the diaphragmatic musculature is formed by the right and left diaphragmatic crura. At the aortic hiatus (hiatus aorticus) they enclose the aorta, the azygos and hemiazygos veins, and the lumbar cistern of the thoracic duct. Although at first glance they appear to be symmetrical, they are not symmetrical in their construction or in the strength of their fibers. The right crus is considerably larger than the left. Each crus arises by a long bifurcate tendon, one part of which is longer and stronger and comes from the cranial edge of the body of the fourth lumbar vertebra. The shorter and somewhat weaker part of the tendon comes from the body of the third lumbar vertebra. Both portions of the tendon of each side unite (the right is considerably stronger) to form an almost sagittal tendon which appears medial to the m. psoas minor. The bilateral tendons press closely against the aorta, and, from their lateral surfaces in particular, they give rise to more and more muscle fibers. This results in a flat, fan-shaped muscle which bears a medial tendon of origin. The muscle parallels the dorsal thoracic wall. A tendinous strand descends on each side of the aorta, immediately anterior to the celiac artery, to form the aortic ring. Seen from the abdominal cavity, each crus of the diaphragm is a triangular muscle plate whose borders give rise to the tendinous portions; as a whole, this plate of muscle radiates forward toward the concavity of the diaphragmatic tendon. The muscle fiber arrangement is somewhat different in the two crura, although each is further differentiated into a lateral, intermediate, and medial portion.

The **lateral portion of the lumbar right crus** originates mainly from the tendon of origin coming from the third lumbar vertebra. It extends ventral to the psoas muscles in an almost transverse arch — the *arcus lumbocostalis*. The pleura and peritoneum encroach directly upon one another dorsal to the arch. After crossing the lumbar musculature, the fiber bundles of the lateral crus run toward those of the pars costalis with which they coalesce into a narrow tendinous band, the extension of the end of the column of the tendinous center. In the wedge between these portions is a triangular area which is free of muscle — the *trigonum lumbocostale;* only fascial coverings of the diaphragmatic musculature radiate into it. This portion of the peripheral diaphragmatic attachment crosses the m. retractor costae and the last rib ventrally. On each side the splanchnic nerves and the sympathetic trunk cross dorsal to the lumbocostal arch.

The **lateral portion of the left lumbar crus** arises in a similar way from its corresponding tendon; however, it has another special lateral division which radiates into the lumbocostal arch from the ventral border of the psoas. Thus on the left side the *trigonum lumbocostale* is muscular; therefore the relationships of the left lateral lumbar crus correspond entirely to those in man. The course of the fibers into the tendinous center is the same as on the right side.

The **intermediate portion of the right lumbar crus** derives its fibers from the principal part of the tendon of origin and from the right column of the tendinous aortic ring. On the left side, the fibers of this part come from the left column of the ring along its entire length. These fiber masses, only indistinctly separable from those of the crus mediale, radiate into the medial borders of the bilateral columns of the central tendon.

The musculature of the **medial portion of the lumbar crus** is the thickest (5 or 6 mm.) and originates asymmetrically on the two sides. On the right side its fibers originate from the terminal portion of the right column of the aortic ring; on the left side the fibers originate from the apical portion of the ring underneath the aorta itself. With blunt edges facing each other, the two parts extend ventrally until they reach the dorsal border of the body of the central tendon. The thick borders of the medial portions are fused by means of fibrous tissue. Distally they separate for the transmission of the esophagus with its vessels and the two vagal trunks, thus forming the **esophageal hiatus (*hiatus esophageus*)**. Ventral to this, they fuse by partial crossing of the fiber bundles. There is an evident asymmetry in the development of the right and left diaphragmatic crura: from the tendon of origin of the right crus come all three parts of the right crus plus the medial portion of the left crus. From the tendon of the left diaphragmatic crus come the intermediate and lateral portions.

The generally homogeneous *pars costalis* on each side consists of fibers radiating from the costal wall to the tendinous center. This muscle arises by indistinct serrations from the medial proximal part of the thirteenth rib, distal part of the twelfth rib, costochondral junction or symphysis of the eleventh rib, as well as the whole length of the tenth and ninth, and at the bend on the eighth costal cartilage. In the caudal part of the line of origin the serrations encroach distally on those of the m. transversus abdominis. In the region of the tenth, ninth, and eighth costal cartilages (often only the eighth alone) openings may be found which allow the passage of the first three cranial serrations of the m. transversus abdominis. The serrations of the diaphragm reach beyond those of the m. transversus abdominis and insert cranial and caudal to them on the corresponding costal cartilages. Interspersed with many radial, fatty strands, the bundles of the costal part run centrally into the lateral borders of the columns and body of the central tendon.

The *pars sternalis* of the diaphragm is an unpaired medial part unseparated from the bilateral costal portions. Its fibers arise on the base of the xiphoid cartilage, the adjacent transversalis fascia, and the eight costal cartilages. They extend dorsally to the apex of the body of the central tendon.

The diaphragm projects far into the thoracic cavity, and its costal part lies on the internal surface of the last few ribs. A capillary space between the diaphragm and the ribs, the *recessus phrenicocostalis*, is thus formed. This decreases on inspiration but increases in size on expiration. During active flattening of the summit of the diaphragm, the inflated lung pushes into the opened space, and upon cessation of the diaphragmatic action it is again pushed out of the space. Even during the most exteme inspiration the space is not entirely filled by the lung. Similar relationships exist in the region dorsal to the diaphragmatic crura over which (along the vertebra covered by the psoas muscles) a bilateral *recessus phreni-columbalis* extends backward to the middle of the lumbar vertebrae. In the midplane the diaphragm forms an arch bulging into the thoracic cavity; this arch extends freely downward from the first few lumbar vertebrae, passing cranioventrally over more than half of the height of the thoracic cavity; near the sternum it turns in a caudoventral direction. The summit of the diaphragm comes to lie at the junction of the middle and ventral thirds of the muscle. On expiration the diaphragm undergoes an excursion of at least 1½ thoracic segments at each respiration.

The muscle of the diaphragm is covered on the convex thoracic side by the fascia endothoracica and the pleura, on the concave abdominal side by a continuation of the fascia transversalis and the peritoneum. Both the fascia and serosa are so thin in the dog that over the tendinous portion they can only be seen microscopically.

The convex thoracic side of the diaphragm lies against the surface of the lungs, from which it is separated by a capillary space. At about the midplane of the thorax the mediastinum descends from the thoracic vertebrae; the two pleural leaves on either side of the mediastinum separate on the diaphragm to become its pleural covering. The attachment of the mediastinum is median only from the dorsal portion of the diaphragm to the esophagus. Ventral to the esophagus the mediastinum makes a strong deflection to the left, to return to the midplane just above the sternum. Here the pleura connecting to the postcava branches off in a convex arch.

In the dorsal part of the mediastinum the aorta, the azygos and hemiazygos veins, and the thoracic duct extend to the hiatus aorticus. The esophagus passes to the hiatus esophageus with the dorsal and ventral vagal trunks. On the right side the esophagus is covered by pleura, which comes from the mediastinum. In the ventral part of the mediastinum the left phrenic nerve lies in its own mediastinal fold, and the phrenicopericardial ligament runs to the diaphragm near the midline. The caudal vena cava and the right phrenic nerve reach the diaphragm in the plica vena cava. The stomach and liver attach by ligaments to the concave peritoneal surface of the diaphragm.

Action: Retraction of the diaphragmatic summit and thus inspiration.

Innervation: Nn. phrenici (from the ventral branches of the fifth, sixth, and seventh cervical nerves).

Muscles of the Abdominal Wall

From without inward the abdominal muscles are: the obliquus externus abdominis, the obliquus internus abdominis, the rectus abdominis, and the transversus abdominis. The m. rectus abdominis extends in the ventral abdominal wall on each side of the linea alba from the external surface of the thorax to the pecten ossis pubis. The mm. obliqui and the transversus are in the lateral abdominal wall. In general these muscles arise from the outer surface of the ribs, the lumbar region, or the tuber coxae to pass in the lateral wall to the ventral abdominal wall or to the pelvis. In the ventral wall the tendons of the two oblique muscles cross the rectus muscle superficially, while the tendon of the transverse muscle crosses deeply. In this way the "sheath of the rectus" (Fig. 6–39) is formed. The abdominal muscles are covered superficially by the large cutaneous muscle of the trunk (m. cutaneous trunci).

The oblique muscles, the fibers of which cross each other at about right angles, form the oblique girdle of the abdomen. The straight and transverse muscles, which also cross each other at right angles, form the straight girdle of the abdomen.

The *m. obliquus externus abdominis* (Figs. 6–33, 6–37, 6–39) is an expansive sheet covering the ventral half of the lateral thoracic wall and the lateral part of the abdominal wall. According to its origin, the muscle is divided into two parts. The *pars costalis* arises by indistinct serrations in a caudally rising line from the middle parts of the fourth or fifth to the twelfth rib, and the adjacent deep trunk fascia, which covers the external intercostal muscles. It is partly covered by the ventral edge of the m. latissimus dorsi at its origin. The unserrated *pars lumbalis* arises from the last rib and, in common with the pars costalis of the obliquus internus abdominis, from the principal lamina of the thoracolumbar fascia.

The cranial serrations of the muscle extend between the terminal serrations of the m. serratus ventralis and cover the terminal tendon of the longest part of the m. scalenus. The caudal serrations are higher on the cos-

tal wall than the cranial ones; thus the line of origin of the lumbar portion meets the lateral border of the m. iliocostalis.

The fibers of the external oblique muscle run caudoventrally, the caudal part being more horizontal than the cranial. In the ventral abdominal wall, 6 to 8 cm. from the midline in large dogs, it forms a wide aponeurosis. This can be differentiated into an abdominal and a pelvic tendon, separated by means of the **superficial** (**subcutaneous** or **external**) **inguinal ring** (*anulus inguinalis superficialis*).

The **abdominal aponeurosis,** or **tendon,** is by far the largest part of the aponeurosis of the m. obliquus externus abdominis; it is the part which arises from the pars costalis of the muscle. This flat tendon extends over the m. rectus abdominis to the linea alba, where it unites with that of the opposite side. It extends caudally also to attach to the pecten ossis pubis. The deep trunk fascia closely adheres to the aponeurosis, obscuring the direction of its fibers. The aponeurosis fuses deeply near the midline with the aponeurosis of the m. obliquus internus abdominis and with it forms the external leaf of the sheath of the rectus abdominis. It lies closely upon the superficial surface of the m. rectus abdominis, where it is intimately connected with the tendinous inscriptions of the rectus. Cranial to the pecten ossis pubis the abdominal tendon is separated from the pelvic tendon by the superficial inguinal ring, the medial crus of which it forms. The delicate fibers cover the caudal 2 to 3.5 cm. of the m. rectus abdominis. At the level of the superficial inguinal ring, the external leaf of the rectus sheath is rather sharply defined by a fibrous band. Strong fibers come from the opposite side and, after crossing the midline, stream upward in the direction of the tendon of origin of the m. pectineus. This tough strand becomes fixed, and with it the outer leaf of the rectus sheath, m. pectineus, and prepubic tendon pass to the iliopectineal eminence. This, in man, is called the ligamentum reflexum. At the same place, the tendinous fibers of the lateral crus of the superficial inguinal ring attach to form the strong caudal commissure.

The **pelvic aponeurosis,** or **tendon,** arises essentially from the lumbar part of the muscle; however, the serration arising from the twelfth rib also takes part. Like the abdomi-

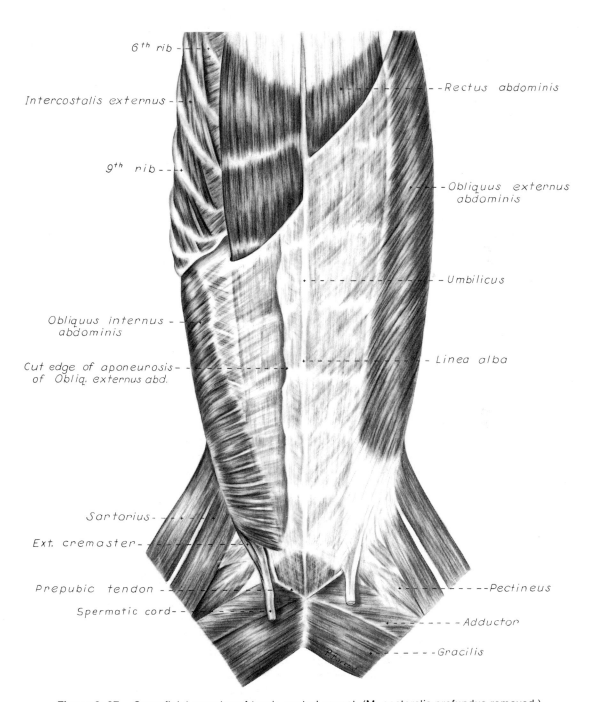

Figure 6–37. Superficial muscles of trunk, ventral aspect. (M. pectoralis profundus removed.)

nal tendon, it is intimately fused with the deep trunk fascia and ventrally is not separated from the abdominal tendon. It extends down into the niche between the abdominal wall and the femur. It crosses the contents of the inguinal canal and forms its lateral wall. The free dorsal border of the pelvic tendon courses along the femoral vessels which come through the vascular lacuna to enter the femoral triangle; ventrally the pelvic tendon is separated from the abdominal tendon by the sharp-edged sagittal slit, the superficial inguinal ring. The lateral crus of the pelvic tendon meets the medial crus of the abdominal tendon to form the cranial and caudal commissures of the superficial inguinal ring. Rarely, a small heterotopic bone may be present in the aponeurosis caudal to the medial crus where the m. pectineus arises (Baumeier 1908).

The **inguinal ligament** (*arcus inguinalis*) comes from the ilium and runs over the lateral surface of the m. iliopsoas to blend with the lateral part of the prepubic tendon. It serves as the origin for the principal part of the pars inguinalis of the m. obliquus internus abdominis.

The **prepubic tendon** (*tendo prepubicus*) is a tough tendinous mass which extends from the iliopectineal eminence and the tendon of origin of the m. pectineus to the same structures of the opposite side. It is firmly attached to the median pubic tubercle situated on the external surface of the symphysis caudal to the free edge. The paired portions of the tendon have a slightly caudomedial course. The m. rectus abdominis radiates into the prepubic tendon. On either side, the m. adductor longus arises close by.

A m. obliquus abdominis externus profundus is not rare. It is a plate of muscle consisting of two, exceptionally three, deep serrations beneath the principal muscle which gives rise to the pelvic tendon (Baum and Zietzschmann 1936).

Action: Along with other abdominal muscles, compression of the abdominal viscera. This action, known as abdominal press, aids in such vital functions as expiration, urination, defecation, and parturition. Flexion of the vertebral column when fellow muscles contract. Lateral bending of the vertebral column.

Innervation: Lateral ventral branches of the last eight or nine nn. thoracales and the lateral branches of the nn. iliohypogastricus and ilioinguinalis.

The **m. obliquus internus abdominis** (Figs. 6–33, 6–37, 6–39) is a flat muscle lying medial to m. obliquus externus abdominis in the lateral abdominal wall, where it is almost completely covered by the external oblique. Its fibers arise from the principal lamina of the thoracolumbar fascia caudal to the last rib, in common with the lumbar portion of the m. obliquus externus abdominis. It originates mainly from the tuber coxae. Some fibers arise also from the fascia covering the m. iliopsoas and the inguinal ligament. Its fibers in general run cranioventrally and thereby cross those of the external oblique muscle at approximately a right angle. The portion of the muscle arising from the lumbar region is divided according to its terminal insertion into a costal and an abdominal part; the portion coming from the inguinal ligament represents the inguinal part.

The strong cranial part, *pars costalis* proximally, is often separated from the middle part by a distinct fissure (containing vessels of the abdominal wall); its fleshy ending is on the thirteenth rib and on the cartilage of the twelfth rib.

The middle portion, *pars abdominalis*, gives rise to a broad aponeurosis at the lateral border of the m. rectus abdominis. This line of transition is often irregular; it extends from the bend of the twelfth costal cartilage to the iliopectineal eminence. This, together with the abdominal tendon of the m. obliquus externus abdominis extends over the outer surface of the rectus as part of the superficial leaf of the rectus sheath. It ends on the linea alba. A narrow cranial lamina of the aponeurosis is split off from the principal portion and runs over the inner surface of the m. rectus abdominis to aid in forming the deep leaf of the rectus sheath.

According to Kassianenko (1928), this part of the muscle becomes amplified at its cranial border (in 30 per cent of dogs) by one to three slender muscle bundles (pars costoabdominalis), which arise from the medial surface of the thirteenth, twelfth, and eleventh costal angles; their tendons are related to that portion of the tendon of the internal abdominal oblique muscle which helps make up the deep leaf of the rectus sheath.

The caudal portion of the m. obliquus internus abdominis, *pars inguinalis*, is rather distinctly separated from the middle portion by a fissure (containing vessels of the abdominal wall). This is the part of the muscle which comes from the tuber coxae, by means of a short aponeurosis, and from the inguinal ligament. The m. obliquus internus abdominis extends beyond the caudal border of the m. obliquus externus abdominis. It sends its outermost fibers in the region of the inguinal canal caudoventrally toward the linea alba. Arising from the caudal free border of

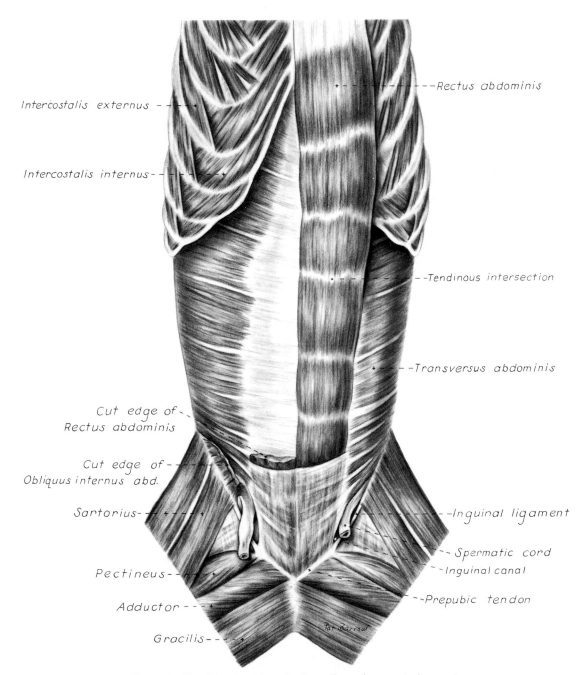

Figure 6–38.   Muscles of trunk, deep dissection, ventral aspect.

the internal oblique muscle are fibers which form the cremaster muscle in the male (Figs. 9–6, 9–7).

Action: Compression and support of the abdominal viscera.

Innervation: Medial branches of the last few nn. thoracales and the nn. iliohypogastricus and ilioinguinalis.

The *m. transversus abdominis* (Figs. 6–35, 6–38, 6–39) is the deepest abdominal muscle and, like the oblique muscles, it is developed into an extensive leaf which reaches a thickness of 2 to 4 mm. in large dogs. It lies in the lateral and ventral abdominal wall on the internal surface of the m. obliquus internus abdominis and adjacent costal cartilages; it arises from the eighth costal cartilage, last lumbar transverse process, and the tuber coxae. It is divided into a lumbar and a costal part.

The *pars lumbalis* arises by broad, short tendons from the transverse processes of all the lumbar vertebrae and the deepest division of the thoracolumbar fascia. This fascia completely surrounds the m. iliocostalis. Out of the dorsal parts of this fascial leaf arises the m. obliquus internus abdominis, which radiates into the lateral abdominal wall.

The *pars costalis,* not divided from the lumbar part, arises muscularly on the medial sides of the thirteenth and twelfth ribs and the eleventh to eighth costal cartilages in such a way that its line of origin crosses that of the diaphragm. From one to three serrations have pleural coverings. The entire muscle extends ventrally and slightly caudally from the internal surface of the thorax, 3 or 4 cm. cranial to the origin of the m. obliquus internus abdominis. The medial branches of the ventral divisions of the last few thoracic and the first few lumbar nerves run over the superficial surface of the m. transversus abdominis. The muscle is marked by these into several (usually six) "segments" which occur in the part behind the last rib, the remainder appearing medial to the costal arch. The muscle extends on the inner surface of the m. rectus abdominis and beyond its dorsal border before giving rise to its end aponeurosis. This forms a laterally convex line, the summit of which lies at the region of the umbilicus, 5 cm. from the midline; toward the xiphoid cartilage it lies only 1.5 cm. from the midline. The end aponeurosis forms most of the inner leaf of the sheath of the rectus abdominis. It unites inseparably at the linea

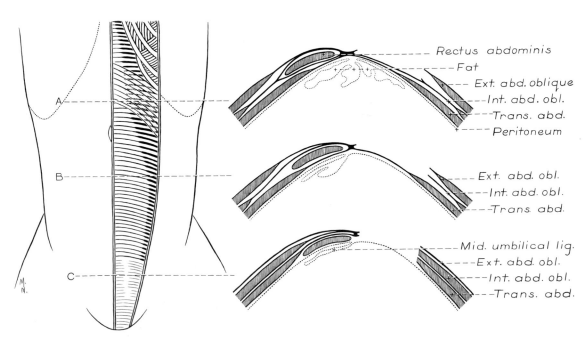

Figure 6–39. The sheath of m. rectus abdominis with cross sections at three levels.

alba with the external leaf. In the costal region, it unites only with part of the end aponeurosis of the m. obliquus internus abdominis.

The cranial part of the muscle, by the development of incomplete fissures, encroaches directly upon the m. transversus thoracis, and the end aponeurosis covers the outer surface of the free end of the xiphoid cartilage. The caudal part becomes aponeurotic in the region of the last two to four "segments" near the lateral edge of the m. rectus abdominis. This does not cross the deep surface of the rectus abdominis muscle; instead it traverses the outer surface to take part in the formation of the external leaf of the rectus sheath. It fuses toward the pelvis with a tendinous strand of the rectus. On the deep surface of the pelvic end of the rectus there is no aponeurotic covering; there is only a thin continuation of the fascia transversalis and peritoneum.

Action and Innervation: Same as for the internal abdominal oblique.

The *fascia transversalis* covers the inner surfaces of the mm. transversi abdominis. It runs between the iliac fascia on the ventral, lateral border of the m. iliopsoas and the ventral midline of the abdomen; during its course it covers the pelvic part of the m. rectus abdominis, which is free of the end aponeurosis of the transversus abdominis. Farther cranially, it fuses with the deep leaf of the rectus sheath. In the lateral abdominal wall it runs cranially to the diaphragm and continues on the abdominal surface of the latter, which it completely covers. The fascia transversalis may contain much fat. The fascia contains strong reinforcements of coarse elastic fibers which run in anastomosing strands from behind forward, thus crossing the course of the fibers of the m. transversus abdominis. These infiltrations come from the entire length of the m. iliopsoas, and are especially strong ventrally. Beneath the point of separation of the m. cremaster externus from the caudal border of the m. obliquus internus abdominis, the elastic masses are the thickest; toward the ribs they become correspondingly thinner. The peritoneum and the transversalis fascia evert to form the processus vaginalis just lateral to the caudal border of the m. rectus abdominis. The m. cremaster externus is located on its lateral and caudal sides.

The **inguinal ligament** (*arcus inguinalis* [*lig. inguinale*]) (Fig. 6–38) is closely related to the fascia transversalis and, like it, contains much elastic tissue. In comparison with that in other domestic animals it is, in the dog, a relatively incomplete structure and independent of the external oblique. It is a strong band extending in the iliac fascia from the tuber coxae obliquely over the m. iliopsoas, marking the caudal border of origin of the fascia transversalis. Together with this fascia it extends ventrolaterally along the m. iliopsoas to radiate into the transversalis fascia which covers the vaginal process. The main part of the inguinal ligament continues distally between the internal inguinal and femoral rings to attach to the lateral border of the prepubic tendon. By taking this course it forms the caudal border of the internal inguinal ring. At its ilial end, it gives origin to part of the m. obliquus internus abdominis. By fusing with the deep trunk, iliac, pelvic, and transversalis fasciae, it acts as a binder in closing the potential space which might exist between the pelvic and abdominal walls.

Budras and Wünsche (1972) substitute a concept of an inguinal arch for the inguinal ligament. Their inguinal arch is composed of lateral, middle, and medial parts, of which the lateral and middle parts often form an inconstant inguinal tract which joins the caudal border of the aponeurosis of the external abdominal oblique and continues on to join the tendon of origin of the pectineus at the prepubic tendon.

The **m. rectus abdominis** (Figs. 6–37, 6–38, 6–39) is a long, rather wide, flat muscle which extends, one on each side of the linea alba on the thoracic and abdominal walls between the external and internal leaves of the rectus sheath, and runs from the first costal cartilage to the pecten ossis pubis. Cranially, in large dogs, it is 7 to 8 cm. broad; caudally, it gradually narrows to 3.5 to 4 cm. Its thickness is 5 to 7 mm., decreasing toward the lateral border. The fibers of the muscle course longitudinally. It arises by a broad, flat tendon from the sternum and the first costal cartilage and rib, where it is covered by the terminal tendon of the m. rectus thoracis. It also has a fleshy origin by means of a special serration from the sternal portion of the ninth costal cartilage. As it passes over the ventral abdominal wall, it lies in a nearly horizontal position, with the medial border facing the linea alba. Occasionally the terminal portion of the muscle is wide enough to help in the formation of the medial wall of the inguinal canal and to appear at the level of the superficial inguinal ring. United by the linea alba and covered externally by a strong tendinous covering, the two recti end on the pecten ossis pubis, from one iliopectineal eminence to the other. At its insertion each muscle unites with the tendon of origin of the m. pectineus and the prepubic tendon. A conical, paired segment of superficial fibers, however, continues farther and ends on the tubercle on the ventral surface of the pelvic symphysis. This crosses the thickened border of the external leaf of the rectus sheath. This long muscle is divided into segments by three to six (usually five) transverse, zigzag, tendinous intersections (intersectiones tendineal). Their distinctness varies. Their number does not correspond with the number of entering nerves. Intimately attached to the tendinous intersections are fibers of the external leaf of the rectus sheath. The fibers of the internal leaf of the sheath are not as firmly attached. The first intersection is at the level of the seventh costal cartilage; the last segment is usually the longest; all other relations vary (Strauss 1927).

Action: All functions which are dependent upon abdominal press, such as expiration, urination, defecation, and parturition; support of the abdominal viscera; to bring the pelvis forward; flexion of the back.

Innervation: Medial branches of the ventral branches of the nn. thoracales and medial branches of the nn. iliohypogastricus and ilioinguinalis.

The **sheath of the rectus abdominis** (Figs. 6–37, 6–39) covers both surfaces of the rectus abdominis muscle. It is formed primarily by the aponeuroses of the other abdominal muscles. The external leaf of the rectus sheath consists of the wide and long aponeuroses of the m. obliquus externus abdominis, most of the aponeurosis of the m. obliquus internus abdominis, and, near its caudal end, a portion of the aponeurosis of the m. transversus abdominis. The internal leaf of the rectus sheath is formed by the end aponeurosis of the m. transversus abdominis, the fascia transversalis, and cranially, by an internal

leaf of the aponeurosis of the m. obliquus internus abdominis. At its pelvic end, the m. rectus abdominis lacks an internal aponeurotic covering, being covered here by only a thin continuation of the transversalis fascia and peritoneum.

The **inguinal canal** (*canalis inguinalis*) (Fig. 6–38), in both sexes, is a connective tissue-filled fissure between the abdominal muscles and their aponeuroses. In the male the inguinal canal serves as the passageway for the processus vaginalis and the descent of the testis; in the female it contains the vaginal process, the round ligament, and much fat. It is relatively short. It begins at the deep inguinal ring, which is formed by (1) the ventral end of the inguinal ligament, (2) the caudal border of the m. obliquus internus abdominis, and (3) the lateral border of the m. rectus abdominis. The deep inguinal ring is covered externally by the aponeurosis of the m. obliquus externus abdominis. The path of the canal is determined by the processus vaginalis. Since the latter pushes over the caudal border of the m. obliquus internus abdominis for a short distance, the medial wall of the inguinal canal is formed by the superficial surface of this muscle. The superficial surfaces of the aponeuroses of the mm. transversus abdominis and rectus abdominis also aid in forming the medial wall. The lateral wall is formed solely by the aponeurosis of the external oblique. The canal is open to the outside because the abdominal and pelvic parts of the aponeurosis of the m. obliquus externus abdominis separate and then come together in the slitlike, superficial inguinal ring (anulus inguinalis superficialis). In this way the pelvic tendon forms the lateral border, *crus laterale*, and the abdominal tendon forms the medial border, *crus mediale*, of the more or less sagittal slit. Where the borders meet, the cranial and caudal angles, or commissures, are formed. The caudal commissure is strong, as it is backed by the tendon of origin of the m. pectineus. The cranial commissure is much weaker, as the parallel strands of collagenous tissue which form the abdominal and pelvic parts of the aponeurosis of the external oblique are held together mainly by the deep trunk fascia. Unlike in man, there is a minimum of cross-over of fibers of the aponeurosis at the cranial angle.

The *linea alba* (Fig. 6–37) is a midventral

strip of collagenous tissue which extends from the xiphoid process to the symphysis pelvis. It serves for the main insertion of the abdominal transverse and external and internal oblique muscles. The medial borders of the right and left rectus muscles lie closely against its lateral borders. At the level of a transverse plane through the last ribs, the linea alba contains a scar, the umbilicus, (anulus umbilicalis), a remnant of the umbilical ring and cord. The linea alba, just caudal to the xiphoid process, is a little over 1 cm. wide and less than 1 mm. thick. It gradually narrows and thickens caudally. Caudal to the umbilicus it appears as a line, being less than 1 mm. wide but considerably thicker. It blends with the prepubic tendon and attaches to the cranial edge of the pelvic symphysis.

The *m. cutaneus trunci* (Figs. 6–47, 6–54), according to Langworthy (1924), is a derivative of the m. pectoralis profundus. As a thin leaf it covers almost the entire dorsal, lateral, and ventral walls of the thorax and abdomen. It begins caudally in the gluteal region and, running forward and downward, covers the dorsal and lateral surfaces of the abdomen and thorax. It ends in the axilla and on the caudal border of the deep pectoral. It lies in the superficial trunk fascia and is not attached to the vertebral spines. It is principally a longitudinal muscle; its origin is in the superficial gluteal fascia. The dorsal borders of the muscle on each side run parallel along the spines of the lumbar and thoracic vertebrae. Only in the region behind the scapula, where the muscle begins to extend downward on the thorax, do the fibers arise from the midline and meet those of the opposite side. Since this part of the muscle is also not attached to the spines of the vertebrae, it is free over the vertebral column to be included in raised folds of the skin. Its ventral border crosses, in the fold of the flank, to the lateral and ventral abdominal wall. The course of the fibers is slightly ventrocranial. Its craniodorsal border covers the m. trapezius, a portion of the m. infraspinatus, and the m. latissimus dorsi, and ends by means of the muscular axillary arch in the medial brachial fascia. The principal part of the muscle, however, with its loose fiber bundles, passes to the superficial surface of the m. pectoralis profundus adjacent to its free edge, where it ends in the super-

ficial thoracic fascia. The fibers of the ventral border coming from the flank reach each other in the midventral line caudal to the sternum. The gap thus existing between the muscles of the opposite sides is filled in by a division of the abdominal cutaneous muscle, right and left muscles spread out to the prepuce as the m. preputialis in the male, and to the mammary glands as the m. supramammaricus in the female.

The *m. preputialis* consists of longitudinal muscle strands filling the space between the opposite abdominal cutaneous muscles in the region of the xiphoid cartilage in the male. Toward the umbilicus a pair of muscular strands arise from the m. preputialis. They radiate into the prepuce in such a way that they come together archlike in the prepuce ventral to the glans. In so doing they are firmly united with each other and with the external preputial leaf.

The *m. supramammaricus* of the bitch is homologous with the m. preputialis of the male. In contrast to the muscle in the male, this muscle is more delicate and narrower, and is paired from its beginning. From the region caudal to the xiphoid cartilage they extend caudally in loose bundles, over the mammary gland complex, to the pubic region. Cranial to the paired inguinal mammary glands, each blends with the ipsilateral m. cutaneous trunci.

Action: The m. cutaneus trunci shakes the skin to remove foreign bodies and increase heat production. It also tenses the skin when required. The preputial muscle draws the prepuce over the glans after erection. The supramammary muscle aids in support of the mammary glands and perhaps in milk ejection.

Innervation: Cutaneous branches of the ventral branches of nn. cervicalis 8 and thoracalis 1 (Langworthy 1924).

MUSCLES OF THE TAIL

The caudal vertebrae are largely enclosed in muscles. The mm. sacrocaudalis dorsalis lateralis and medialis, dorsal in location, are extensors or levators of the tail. The mm. sacrocaudalis ventralis lateralis and medialis, ventral in location, are flexors or depressors of the tail. The mm. coccygeus, levator ani, and the intertransversari caudae, lateral

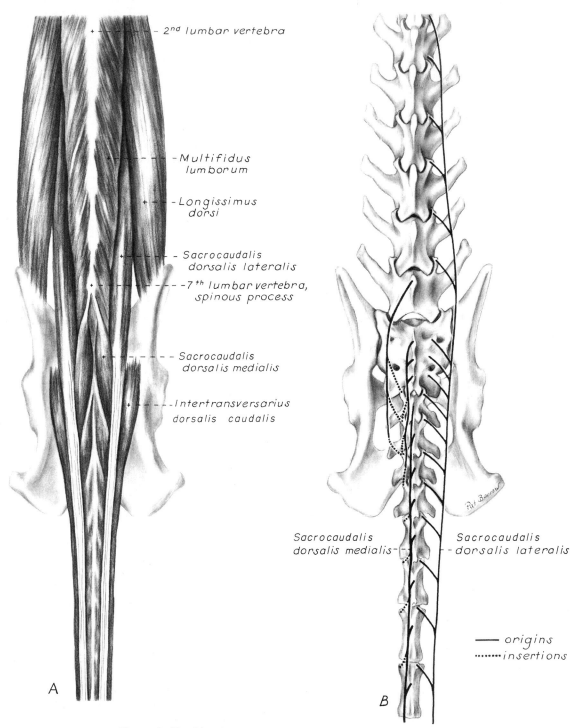

2nd lumbar vertebra

Multifidus lumborum

Longissimus dorsi

Sacrocaudalis dorsalis lateralis

7th lumbar vertebra, spinous process

Sacrocaudalis dorsalis medialis

Intertransversarius dorsalis caudalis

Sacrocaudalis dorsalis medialis

Sacrocaudalis dorsalis lateralis

——— origins
·········· insertions

A

B

Figure 6–40.   Muscles of lumbocaudal region.
A. Epaxial muscles, dorsal aspect.
B. Diagram of sacrocaudal muscles, dorsal aspect.

in location, are the lateral flexors of the tail. The dorsal muscles are direct continuations of the epaxial musculature of the trunk. The caudal muscles lie on the lumbar vertebrae, sacrum, and caudal vertebrae, and insert on the caudal vertebrae, exclusively. They have fleshy endings as well as tendinous ones of variable length. The most caudal tendons go to the last caudal vertebrae. Proximally the muscles, as well as the vertebral bodies, are larger. The caudal muscles of the dog resemble those of the cat (Schumacher 1910).

The **m. sacrocaudalis dorsalis lateralis**, or long levator of the tail (Fig. 6–40), is a flat, segmental muscle strand becoming stronger toward its dorsal border, 2.5 to 3 cm. high in the lumbosacral region in large dogs and 1 to 1.5 cm. thick at its free edge. It may be regarded as a continuation of the m. longissimus on the tail. In the caudal part of the lumbar region it lies between the m. longissimus, laterally, and the mm. multifidus lumborum and sacrocaudalis dorsalis medialis, medially. It is covered by the thick caudal dorsalis medialis. It has a fleshy origin from the aponeurosis of the m. longissimus and a tendinous origin from the mammillary processes of the first to sixth lumbar vertebra, the articular processes of the sacrum, and the mammillary processes of at least the first eight caudal vertebrae. It is indistinctly divided into long individual parts which partly cover one another. From this muscular belly which extends from the second sacral to the fourteenth caudal vertebra (when 20 caudal segments are present), there appear 16 thin, long tendons. These are arranged into a flat bundle by the accumulation of successive tendons, They lie embedded in the thick, deep caudal fascia. The first tendon ends on the mammillary process of the fifth caudal vertebra, the next ends on the sixth, and so on, to the last one. Cranial to their terminations a few take on a little tendon of the underlying segment of the m. sacrocaudalis dorsalis medialis

Action: Extension or lifting of the tail, possibly also to move it to one side.

Innervation: Branches of the plexus caudalis dorsalis.

The **m. sacrocaudalis dorsalis medialis**, or short levator of the tail (Fig. 6–40), is the direct continuation on the tail of the m. mul-tifidus and, like the latter, it is composed of relatively short, individual segments. It lies next to the median plane on the sacrum and caudal vertebrae and extends from the seventh lumbar to the last caudal vertebra. The individual segments can be isolated at the root of the tail. They are composed of deep, short muscle masses and a strong, superficial, long part which possesses a little tendon that spans four or five vertebrae. These individual muscles run between the spines of cranial vertebrae and the dorsolaterally located tubercles, as well as on the mammillary processes on the cranial ends of more caudal vertebrae. Toward the tip of the tail the muscle segments become shorter, smaller, and more homogeneous. They arise from the small processes which are dorsolateral to the caudal edge of the rodlike caudal vertebra. They pass over only one segment and end on dorsolateral humps which correspond to the mammillary processes of the lumbar vertebrae. The superficial tendons end in common with the long tendons of the m. sacrocaudalis dorsalis lateralis. Muscle fibers also accompany the tendons.

Action: Extension of the tail, possibly also lateral flexion.

Innervation: Branches of the plexus caudalis dorsalis.

The **m. sacrocaudalis ventralis lateralis**, or long depressor of the tail (Fig. 6–41), is strong; in large dogs, at the sacrum, it is 2.25 cm. high and 0.75 cm. thick. It consists of numerous long, individual parts which are arranged like those of the long levator and which end by means of long tendons from the sixth to the last segment. The first segment comes from the ventral surface of the body of the last lumbar vertebra and from the sacrum; the remainder arise from the ventral surfaces and the roots of the transverse processes of the caudal vertebrae. From the segmented bellies of the third and successive segments caudally, the individual long tendons arise and are embedded in the thick, deep caudal fascia. The first of these is attached to the ventrolateral tubercle (processus hemalis) of the proximal end of the sixth caudal vertebra, the second on the corresponding elevation of the seventh, and so on to the last caudal vertebra. Before inserting, each of these tendons acquires the little ten-

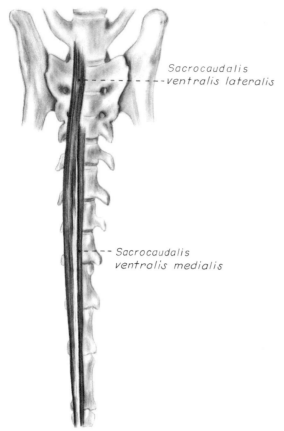

Sacrocaudalis
ventralis lateralis

Sacrocaudalis
ventralis medialis

Figure 6–41. Sacrocaudal muscles, ventral aspect.

don of the segment of the short depressor which has been crossed by the segment of the long depressor.

Action: Flexion of the tail and, occasionally, lateral movement.

Innervation: Branches of the plexus caudalis ventralis.

The *m. sacrocaudalis ventralis medialis,* or short depressor of the tail (Fig. 6–41), consists of segmental, short individual parts extending from the last sacral vertebra throughout the length of the tail. It lies against the ventral surface of the vertebrae and, with the muscle of the opposite side, forms a deep furrow (for the a. coccygea). At the pelvic outlet the bundles are very strong and the segmentation is indistinct. Soon, however, independent segments are separated out. The fibers of each of these segments arise essentially from the ventral surface of one vertebra. Superficially, a small, flat tendon is then formed. This unites with

the tendon of the long depressor which lies immediately lateral to it, and this common tendon then passes over the following segment to end on the hemal process of the next following vertebra.

Action and Innervation: Same as for the m. sacrocaudalis ventralis lateralis (*supra*).

The *m. intertransversarius dorsalis caudalis* (Figs. 6–40A; 6–42B) lies between the sacrum and the middle of the tail. In general, it consists of short, individual parts, of which only the first is well developed. This portion arises on the long, dorsal sacroiliac ligament, on the lateral part of the third sacral vertebra, and forms a large, round muscle belly which ends on the transverse process of the fifth or sixth caudal vertebra by means of a long tendon. In its course it receives supplementary fibers from the transverse processes of the first few caudal vertebrae. These deep elements gradually become independent muscles which extend from one transverse process to that of the following vertebra. They lie on the dorsal surfaces of the transverse processes or their rudiments, where they are partly covered by the long tendons of the levators. These muscle segments become so small in the caudal half of the tail that they can hardly be isolated. Superficial parts of the first large segment give rise to two or three long, flat tendons which extend to the thick caudal fascia and to the rudiment of the transverse process of the sixth or seventh or even the eighth caudal vertebra. This is the m. abductor caudae dorsalis, or m. coccygeus accesorius.

Action: With the m. intertransversarius ventralis caudalis, lateral flexion of the tail.

Innervation: Branches of the plexus caudalis ventralis.

The *m. intertransversarius ventralis caudalis* (Fig. 6–42B), situated ventral to the transverse processes, begins at the third caudal vertebra. It forms a round belly, composed of segments, and, at the base of the tail, is smaller than the dorsal muscle; however, it has a more constant size and is well segmented, and thus is easily traced to the end of the tail. Ventrally the muscle is covered by the long tendons of the long depressor of the tail. From the third to the fifth caudal vertebra the ventral and dorsal mm. intertransversarii are separated by the m.

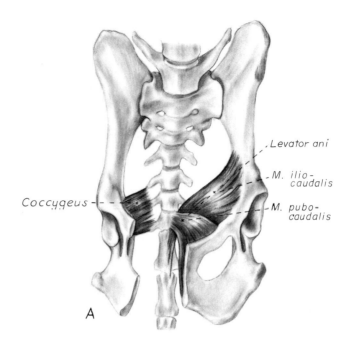

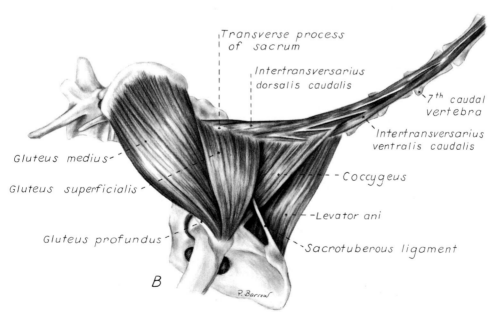

Figure 6–42.   Muscles of the pelvis.
A. Mm. levator ani and coccygeus, ventral aspect.
B. Caudal and gluteal muscles, lateral aspect.

coccygeus; otherwise they are separated by a strong intermuscular septum of the caudal fascia.

Action and Innervation: Same as for the m. intertransversarius dorsalis caudalis (*supra*).

The **pelvic diaphragm** (*diaphragma pelvis*) in quadrupedal mammals is the vertical closure of the pelvic cavity through which the rectum passes. The two muscles of the pelvic diaphragm are the m. coccygeus and the m. levator ani.

The **m. coccygeus**, formerly called the m. coccygeus lateralis (Figs. 6–42, 6–73, 6–88), is a strong muscle arising by means of a narrow tendon on the ischiatic spine cranial to the internal obturator muscle. It crosses the sacrotuberous ligament medially and, spreading like a fan, extends to the lateral surface of the tail. There it ends, between the mm. intertransversarii, on the transverse processes of the second to fifth caudal vertebrae. It is partially covered by the caudal branch of the m. gluteus superficialis.

Action: Bilateral: to press the tail against the anus and genital parts and, in conjunction with the depressors, to draw the tail between the rear legs. Unilateral: lateral flexion.

Innervation: Ventral branches of the third sacral nerve.

The **m. levator ani**, formerly known as the m. coccygeus medialis or the m. ilioischiopubococcygeus (Figs. 6–42, 6–73), lies cranial and medial to the coccygeus. It is a broad triangular muscle originating on the medial edge of the shaft of the ilium, on the inner surface of the ramus of the pubis, and on the entire pelvic symphysis. Bilaterally, the muscles spread out and radiate upward toward the root of the tail. In so doing, they surround a large median, fatty mass, as well as the genitalia and the rectum. Caudally, each encroaches upon the inner surface of the m. obturator internus. After decreasing in size, the muscle then appears at the caudal edge of the m. coccygeus, passes into the caudal fascia, and ends on the hemal process of the seventh caudal vertebra by means of a strong tendon immediately next to the tendon of its fellow of the opposite side. This muscle can be divided into a m. iliocaudalis and a m. pubocaudalis, between which the n. obturatorius passes. The fibers of both parts enter the tendon at an angle. The deep surface of the muscle is firmly covered by the pelvic fascia, which is also connected with the m. sphincter ani externus. Pettit (1962) has summarized many cases of perineal hernia in the dog and described their surgical repair in regard to the muscles of the pelvic diaphragm.

Action: Bilateral: to press the tail against the anus and genital parts; unilateral: to bring the tail cranially and laterally. The mm. levatores ani, in combination with the levators of the tail, cause the sharp angulation between the sixth and sev-

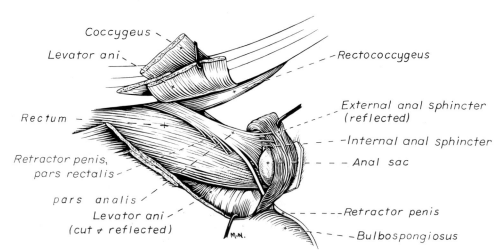

Figure 6–43.   Muscles of the anal region, lateral aspect.

enth caudal vertebrae which is characteristic for defecation; compression of the rectum.

Innervation: Ventral branches of the third (last) sacral and the first caudal nerve.

The *m. rectococcygeus* (Fig. 6-43) is a paired smooth muscle composed of fibers from the external longitudinal musculature of the rectum. The fibers sweep caudodorsally from the sides of the rectum and pass through the fascial arch formed by the attachment of the external anal sphincter to the fascia of the tail. Right and left portions of the muscle fuse beneath the third caudal vertebra. The median muscle thus formed lies between the ventral sacrocaudal muscles and passes caudally to insert on the fifth and sixth caudal vertebrae. The attachment of the rectococcygeus muscle on the tail serves to anchor the rectum and provide for caudal traction in defecation. Extension of the tail during defecation aids in evacuating the rectum because of the attachments of the mm. rectococcygeus, coccygeus, and levator ani. The mm. coccygeus and levator ani cross the rectum laterally and tend to compress it; the m. rectococcygeus, by shortening the rectum, aids in evacuation of the fecal column.

Action: To aid in defecation.

Innervation: Autonomic fibers from pelvic plexus.

The *m. sphincter ani internus* is the caudal thickened portion of the circular coat of the anal canal. It is composed of smooth muscle fibers and is smaller than the striated external anal sphincter. Between the two sphincter muscles, on either side, lies the anal sac. The duct from the anal sac crosses the caudal border of the internal sphincter muscle.

The *m. ani externus* (Fig. 6-43), composed of striated muscle fibers, surrounds the anus, covers the internal sphincter except caudally, and is largely subcutaneous. The cranial border of the external sphincter is united by fascia to the caudal border of the levator ani. Dorsally the external sphincter attaches mainly to the caudal fascia at the level of the third caudal vertebra. This attachment is such that a cranially directed concave fascial arch is formed, through which the rectococcygeus muscle passes. About half of the fibers of the external sphincter encircle the anus ventrally. The remaining superficial ventral fibers end on the urethral muscle and the bulbocavernosus muscle of the male. In the female comparable fibers blend with the constrictor vulvae.

The *m. retractor penis* (Fig. 6-43) is a band of muscle which arises ventrally on each side of the sacrum or first two caudal vertebrae. It was called the caudo-anal or caudo-cavernosus muscle in the cat by Straus-Durckheim (1845) and in the dog by Langley and Anderson (1895). (It was illustrated as the coccygeoanalis muscle in the dog by Miller, Christensen, and Evans 1964.) This muscle is now referred to as the m. retractor penis or clitoridis with a pars analis and a pars rectalis. At its origin on the vertebrae there is a considerable decussation of fibers ventral to the rectococcygeus muscle. Each band passes ventrocaudally across the lateral surface of the rectum, to which it contributes some fibers. It becomes wider distally as it passes behind the anal sac and into the sphincters. The bulk of its fibers appear to end near the duct of the anal sac, although some fibers insert in the external sphincter. Occasionally, a rudiment of a ventral anal loop may be present. A ventral portion of the muscle band, in combination with some fibers from the external sphincter, continues distally as the *retractor penis muscle*. Superficially the m. retractor penis is covered by the levator ani, with which there may be some fiber interchange.

FASCIAE OF THE TRUNK AND TAIL

On the trunk, as on other parts of the body, there is a superficial and a deep fascia, known collectively as the external fascia of the trunk. It covers the muscles and bones of the thorax and abdomen. In addition, there is an internal fascia of the trunk, which serves a special function in the formation of the body cavities.

The **internal fascia of the trunk** lies on the deep surfaces of the muscles of the body wall and on the superficial surfaces of the serous coverings of the cavities. In the thoracic cavity, it is the *fascia endothoracica;* in the abdominal cavity, the *fascia transversalis*. The latter covers the m. transversus abdominis on its deep surface and fuses ventrally with its aponeurosis. Cranially the fascia transversalis covers the diaphragm as a thin membrane. The internal trunk fascia is reinforced by yellow elastic tissue wherever it covers a movable or expansible structure, such as the diaphragm. The *fascia iliaca* covers the deep lumbar muscles and is connected with the last few lumbar vertebral bodies and with the ilium. The *fascia pelvina* clothes the pelvic cavity; it lies deeply on the bones and gluteal muscles concerned, and it continues on the pelvic surface of the muscles of the pelvic diaphragm. In obese dogs it contains much fat.

The **superficial external fascia of the trunk** (*fascia trunci superficialis*) is relatively thick; it covers the thorax and abdomen in a manner similar to that on other parts of the body. It extends cranially, dorsally, and laterally to the shoulder and neighboring parts of the brachium. The ventral part uses the sternal region to gain the neck and also sends connections to the superficial fascia of the medial surface of the thoracic limb. Caudally, direct continuations are found in the superficial gluteal fascia and, by means of the flank, on the cranial crural portions to the lateral and medial crural fasciae, and finally, in the pubic region, to correspondingly superficial fascial parts. There are no attach-

ments with the dorsal ends of the thoracic and lumbar vertebrae. Thus, as on the neck, the fascia can be picked up with folds in the skin. There are ventral fascial leaves to the prepuce and to the mammary glands. The superficial trunk fascia covers the mm. trapezius and latissimus dorsi, as well as parts of the pectoral muscles, omotransversarius, deltoideus, and triceps. In relation to the underlying structures, all parts of the superficial fascia are extremely displaceable; only on the shoulder is this mobility limited. Wherever there is great mobility, in well-nourished animals large quantities of subfascial fat are deposited.

The **deep external fascia of the trunk** (*fascia trunci profunda*) is a strong, shining, tendinous membrane; it begins on the ends of the spinous processes of the thoracic and lumbar vertebrae, from the supraspinous ligament. It passes over the epaxial musculature to the lateral thoracic and abdominal wall, to fuse with the fascia of the opposite side at the linea alba. In the sternal region it passes under the pectoral musculature on the sternum and costal cartilages. Caudally it is attached to the ilium.

The principal leaf of the thoracolumbar

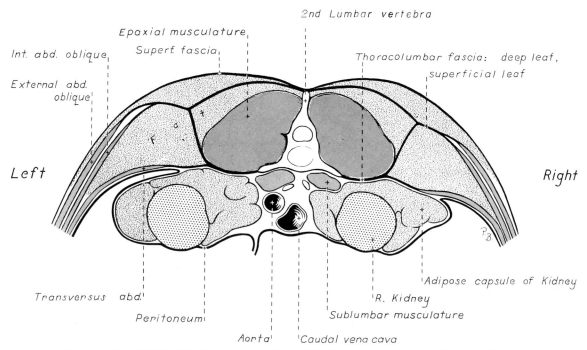

Figure 6–44. Schematic cross section through lumbar region, showing the fascial layers.

fascia is again divided into two superimposed leaves, and in well-nourished dogs is covered with large amounts of fat (Fig. 6–44). It passes under the mm. latissimus thoracis, trapezius, and rhomboideus to the medial surface of the base of the shoulder; here, it becomes the strong, eventually bilaminate fascia spinotransversalis. Especially in the lumbar region, a strong deeper leaf is separated from the thoracolumbar fascia; this is the aponeurosis of the mm. longissimus and iliocostalis. From its deep surface arise numerous fibers for both muscles. Moreover, it sends an intermuscular septum between the two muscles; after completely surrounding the lumbar part of the laterally projecting m. iliocostalis, it terminates on the free ends of the lumbar transverse process where the pars lumbalis of the m. transversus abdominis originates. Cranially this fascia gradually disappears; on the thorax it attaches to the ribs lateral to the m. iliocostalis. The principal lamina of the thoracolumbar fascia gives rise laterally to the m. serratus dorsalis cranialis. It becomes extremely strong as the superficial leaf of the fascia spinotransversalis. It passes over the m. splenius, where it encroaches upon the fascia colli profunda by means of a distinct border. In the lumbar region, the deeper layer of the principal lamina of the thoracolumbar fascia is the stronger leaf. At about the level of the lateral border of the epaxial musculature, it gives rise to portions of the mm. obliquus externus abdominis and obliquus internus abdominis. As the relatively delicate fascia trunci profunda passes to the lateral wall of the abdomen and thorax, it continues on the superficial surface of the m. obliquus externus abdominis and on the thoracic serrations of the m. serratus ventralis, with which it intimately fuses. Over the m. serratus ventralis thoracis and under the shoulder, the deep thoracic fascia is connected with the deep cervical fascia, these two fasciae meeting on the superficial surface of the mm. serratus ventralis cervicis and scalenus. In the abdominal region and in the caudal thoracic region the deep fascia of the trunk descends over the external surface of its pelvic and abdominal tendons and is more or less firmly united with them. Insofar as it has relationships with the abdominal tendon of the rectus muscle, the deep abdominal fascia

also takes part in the formation of the superficial leaf of the rectus sheath. Caudal to the lumbar region, the deep fascia of the trunk becomes the deep gluteal fascia, and from the lateral abdominal wall it becomes the crural fascia. In the region of the superficial inguinal ring, the deep trunk fascia has special significance. At the commissures of the crura it unites the diverging collagenous strands (cranial commissure) and tends to prevent enlargement of the ring during herniation. At the caudal commissure it blends with the tendon of origin of the m. pectineus. From the crura, especially the medial one, the deep fascia extends on the processus vaginalis and its contents, there being known as the external spermatic fascia.

Deep beneath the shoulder, the deep fascia of the trunk is represented by the *fascia spinotransversalis*, which is somewhat independent from the fascia thoracolumbalis; like its principal leaf, it consists of a superficial and a deep part, the superficial leaf being by far the stronger. Here under the shoulder it lies medial to the mm. rhomboideus, serratus ventralis, and latissimus thoracis, and lateral to the mm. semispinalis, longissimus, and iliocostalis. The fascia spinotransversalis is that portion of the deep trunk fascia which arises from the supraspinous ligament of the first 8 to 10 thoracic vertebrae. The two leaves are fused for 0.5 to 1 cm. from their origin.

The stronger superficial leaf of the fascia spinotransversalis is the aponeurosis of origin of the m. serratus dorsalis cranialis. Cranially, it has a distinct border; caudally, it becomes weaker and blends with the principal leaf of the thoracolumbar fascia. From this the m. serratus dorsalis caudalis arises. The cranial segment of the superficial fascial leaf seems to lie transversely over the origin of the m. splenius; however, it also sends a narrow tendinous strand ventrally over the mm. longissimus and iliocostalis to end on the transverse processes of the first thoracic and last cervical vertebrae.

The more delicate, deep leaf of the fascia spinotransversalis extends from a transverse plane through the third intercostal space to the last few thoracic vertebrae, where it goes into the deep layer of the thoracolumbar fascia. Its fibers extend transversely over the mm. semispinalis and longissimus to end with the lateral tendons of the m. longissimus on the ribs. Cranially the m. splenius arises from this leaf.

The **superficial and deep fasciae of the tail** (*fasciae caudae*) arise from the corresponding leaves of the gluteal fascia. The superficial is very insignificant; the thick, deep leaf provides thick connective tissue masses for special ensheathment of the long tendons of the mm. sacrocaudalis dorsalis lateralis and sacrocaudalis ventralis lateralis.

MUSCLES OF THE THORACIC LIMB

EXTRINSIC MUSCLES

The extrinsic muscles of the thoracic limb originate on the neck and thorax and extend to the shoulder or brachium as far distally as the elbow joint. They include a superficial layer of muscles lying directly upon the fascia of the shoulder and brachium, and a second, deeper layer, being in part medial and in part lateral to the shoulder and brachium. According to the points of attachment, the extrinsic muscles can be divided into those from the trunk to the shoulder and those from the trunk to the brachium. The *m. cucullaris*, a gill levator in fishes, is considered to represent a combination of several shoulder muscles in mammals whose primary derivatives are the *m. trapezius* and the *m. sternocleidomastoideus*. Donat (1972) has

described the various components of the cucullaris in domestic animals.

The **m. trapezius** is a broad, thin, triangular muscle (Figs. 6–45, 6–47). It lies under the skin and the cervical cutaneous muscle in the neck, and crosses the interscapular region of the shoulder. It arises from the median fibrous raphe of the neck and the supraspinous ligament of the thorax. Its origin extends from the third cervical vertebra to the ninth thoracic vertebra. The insertion is on the spine of the scapula. It is divided into a cervical and a thoracic portion by a tendinous band extending dorsally from the spine of the scapula.

The fibers of the comparatively narrow *pars cervicalis* arise on the mid-dorsal raphe of the neck. They run obliquely caudoventrally to the spine of the scapula, and end on the free edge of the spine. Only a small distal portion of the spine remains free for the attachment of the m. omotransversarius. This muscle cannot be separated from the ventral border of the trapezius near the spine.

The *pars thoracica* arises from the supraspinous ligament and the dorsal spine of the third to the eighth or ninth thoracic vertebra, and by an aponeurosis which blends with the lumbodorsal fascia. Its fibers are directed cranioventrally and end on the proximal third of the spine of the scapula.

The fibrous band which divides the m.

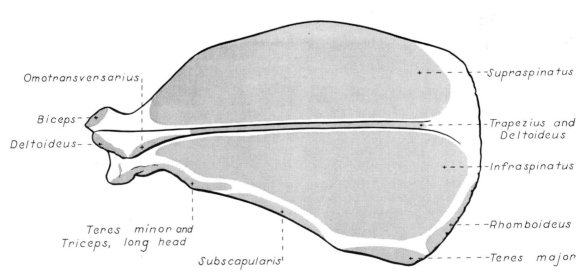

Figure 6–45. Left scapula, showing areas of muscle attachment, lateral aspect.

trapezius varies considerably. Sometimes it is lacking; sometimes it is broad and includes the dorsal border of the middle part of the entire muscle; sometimes it is interrupted. When it is present, it serves as a common attachment for the two parts of the m. trapezius.

Action: To elevate the limb and draw it forward.

Innervation: Dorsal branch of the n. accessorius.

The *m. omotransversarius* (Figs. 6–45, 6–47) lies at the side of the cervical vertebrae as a flat, narrow muscle. It arises on the distal portion of the scapular spine, as far as the acromion, and from that part of the omobrachial fascia which covers the acromial part of the m. deltoideus. It soon separates from the m. trapezius cervicis, dips under the m. cleidocervicalis, and proceeds over the mm. scalenus and the intertransversarius cervicalis, which cover the transverse processes of the cervical vertebrae dorsally, to the caudal end of the wing of the atlas. In large dogs it is at first as much as 4 cm. wide and 2 to 4 mm. thick; cranially it becomes narrower and thicker. Its ventral border is limited by the transverse processes of the cervical vertebrae.

Action: To draw the limb forward.

Innervation: N. accessorius.

The *m. rhomboideus* (Figs. 6–26, 6–45,

6–46), covered by the trapezius, fans out on the neck and on the scapular region of the back between the median line of the neck and the base of the scapula. It is in part flat and in part thick, and is divided into three parts.

The cervical part, *m. rhomboideus cervicis,* lies dorsolateral on the neck from the second or third cervical vertebra to the third thoracic vertebra. It arises on the tendinous raphe of the neck and the ends of the spinous processes of the first three thoracic vertebrae, and ends on the rough medial surface and on the edge of the base of the scapula, including the scapular cartilage close to the cervical angle. Near the scapula, in large dogs, it becomes as much as 1.5 cm. thick. From the cervical part, cranial to the fourth cervical vertebra, the *m. rhomboideus capitis,* is given off as a straplike muscle to the occiput. The thoracic portion, *m. rhomboideus thoracis,* arises on the spinous processes of the fourth to the sixth or seventh thoracic vertebra and inserts on the medial and partly on the lateral edge of the base of the scapula. This portion of the m. rhomboideus is covered by the m. lattisimus dorsi. The two portions are never clearly separated from each other and are often intimately bound together.

Action: To elevate the limb, pull the limb and shoulder forward or backward; to

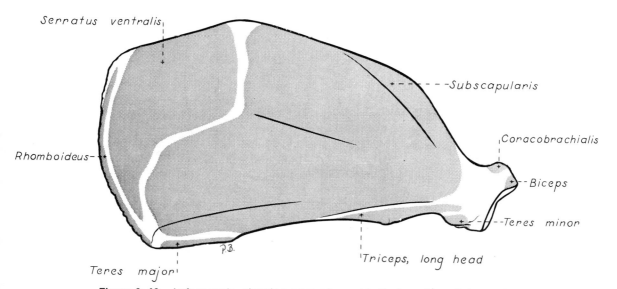

Figure 6–46. Left scapula, showing areas of muscle attachment, medial aspect.

draw the scapula against the trunk (in common with all the extrinsic muscles).

Innervation: Rami ventrales of nn. cervicales et thoracales.

The **m. serratus ventralis** (Figs. 6–26, 6–46) covers the caudal half of the lateral surface of the neck and the cranial half of the lateral thoracic wall; it is a very strong, fan-shaped muscle. It arises on the facies serrata of the scapula, its fibers diverging to form an angle of about 150 degrees. It ends on the transverse processes of the last five cervical vertebrae as the **m. serratus ventralis cervicis**, and on the first seven or eight ribs, somewhat ventral to their middle, as the **m. serratus ventralis thoracalis**. In large dogs the muscle is 1.5 to 2 cm. thick near the scapula. The terminal serrated edge of the cervical portion is not sharply defined; the individual slips insert between the m. longissimus cervicis and the m. intertransversarius. The thoracic portion has well-defined serrations which are covered in part by the m. scalenus. Its three or four caudal serrations interdigitate with those of the m. obliquus externus abdominis.

Action: Support of the trunk, to carry the trunk forward and backward; inspiration; to carry the shoulder forward and backward with respect to the limb.

Innervation: Cervical portion; rami ventrales of nn. cervicales; thoracic portion: n. thoracalis longus.

The **m. sternocephalicus** (Figs. 6–26, 6–47) in the dog can be more or less clearly

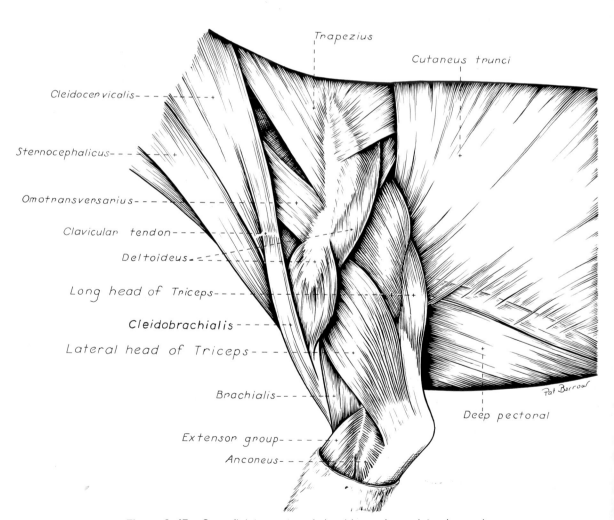

Figure 6–47. Superficial muscles of shoulder and arm, lateral aspect.

separated into mastoid and occipital parts. In large dogs this flat muscle is 2.5 to 3.5 cm. wide at the sternum and 10 to 14 mm. thick. It arises as a unit on the manubrium sterni and, covered only by skin and the m. cutaneous colli, runs to the mastoid part of the temporal bone and to the dorsal nuchal line of the occipital bone. At their origin the muscles of the two sides are intimately joined, but they separate at or before the middle of the neck, and each crosses under the external jugular vein of its own side, and encroaches closely upon the ventral edge of the ipsilateral m. cleidocervicalis.

The ventral portion, the *m. sternomastoideus*, separates as a strong, elliptical bundle and, united with the m. cleidomastoideus in a strong tendon, goes to the mastoid part of the temporal bone; the broader, thinner, dorsal segment, the *m. sterno-occipitalis*, attaches to the dorsal nuchal line as far as the midline of the neck by means of a thin aponeurosis. Because of the divergence of the two sternocephalic muscles, there is a space ventral to the trachea in which the bilateral mm. sternohyoideus and sternothyroideus appear. Here in the deep cervical fascia, additional fibers for the m. sternomastoideus may arise.

Action: To draw the head and neck to one side.

Innervation: Ventral branches of cervical nerves and branches from accessory nerve.

The *m. brachiocephalicus* (Figs. 6–49, 6–50, 6–54), lying on the neck under the m. sphincter colli superficialis and platysma as a long, flat muscle, extends between the brachium and the head and neck. Cranial to the shoulder the muscle is transversed by a clavicular remnant, a transverse, often arched, fibrous intersection or plate, the **clavicular intersection** (*intersectio clavicularis*). The vestigial clavicle is connected with the medial end of the clavicular tendon and lies under the muscle.

In man, in whom the clavicle is completely developed, the m. cleidomastoideus extends from the medial end of the clavicle to the head; from the lateral end of the clavicle the pars clavicularis of the m. deltoideus, which is closely joined with the two other portions of the deltoid muscle, extends to the arm. In the phylogenetic scheme, when the clavicle is reduced and shortened, the

origins of these muscles come closer together until they fuse. The muscle now extends from the arm to the head—the m. brachiocephalicus. The *m. cleidocephalicus*, which extends up the neck from the clavicular tendon, is further divided into two portions. In the dog it divides into a superficial m. cleidocervicalis (pars cervicalis), which broadens and attaches to the dorsal part of the neck, and a deep m. cleidomastoideus (pars mastoidea), which extends to the mastoid part of the temporal bone. However, the m. cleidobrachialis (pars brachialis of brachiocephalicus), which runs from the clavicular tendon to the humerus, corresponds to the pars clavicularis of the m. deltoideus of man.

The *m. cleidobrachialis*, 5 to 6 cm. broad and 5 to 8 mm. thick in large dogs, arises from a narrow part of the distal end of the humeral crest. It appears between the m. brachialis and m. biceps and, covering the shoulder joint cranially and somewhat laterally, ends on the clavicular tendon.

Action: Bilateral fixation of the neck.

Innervation: Ramus from the brachial plexus.

The *m. cleidocervicalis* is in an equally superficial position, and appears as a cranial extension of the m. cleidobrachialis from the clavicular tendon to the back of the neck. Nevertheless, there is no connection between fibers proximal and distal to the tendinous plate. Moreover, in the fascia of the triangle bounded by the mm. cleidocephalicus, pectoralis superficialis, and sternocephalicus there are scattered bundles coming from the medial edge of the m. cleidocephalicus. The m. cleidocervicalis gradually becomes broader and thinner as it goes to its aponeurosis of insertion on the fibrous raphe of the cranial half of the neck.

The *m. cleidomastoideus* is the deep cranial continuation of the m. cleidocephalicus anterior to the clavicular tendon. It is covered by the m. cleidocervicalis and m. sterno-occipitalis. It reaches a width of 2.5 to 3 cm. and a thickness of 7 to 10 mm., and is often split into two round bundles throughout its length. By means of a strong tendon it ends on the mastoid part of the temporal bone with the m. sternomastoideus, dorsal to which it lies.

Action: To draw the limb forward, and, acting bilaterally, to fix the neck.

Innervation: M. cleidocephalicus: n. accessorius, rami ventrales of the nn. cervicales; m. cleidobrachialis: n. axillaris.

The *m. latissimus dorsi* (Figs. 6–50, 6–51, 6–54), is a flat, almost triangular muscle which lies caudal to the muscles of the shoulder and brachium on the dorsal half of the lateral thoracic wall. It begins as a wide, tendinous leaf from the superficial leaf of the lumbodorsal fascia and thus from the spinous processes of the lumbar vertebrae and the last seven or eight thoracic vertebrae; and it arises muscularly from the last two or three ribs. Its fibers converge toward the shoulder. The cranial border of the muscle

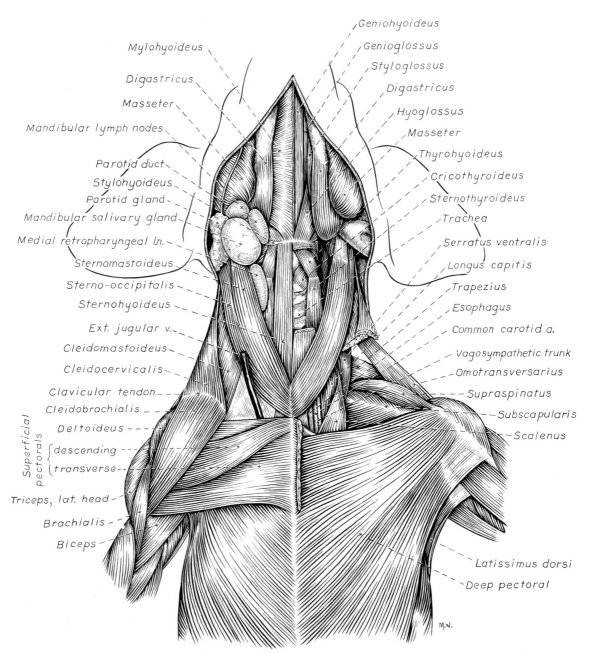

Figure 6–48. Superficial muscles of neck and thorax, ventral aspect.

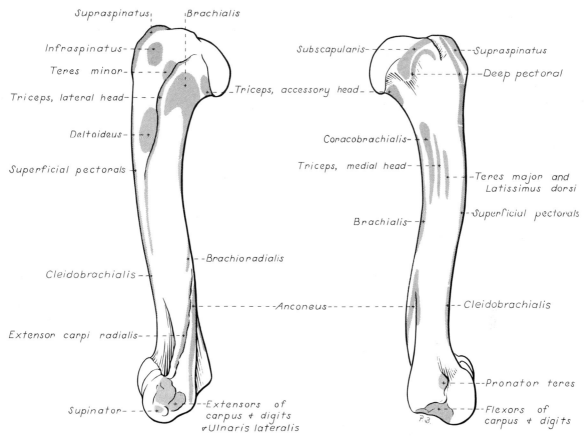

Figure 6–49. Left humerus, showing areas of muscle attachment, lateral aspect.

Figure 6–50. Left humerus, showing areas of muscle attachment, medial aspect.

lies under the m. trapezius thoracis, where it covers the caudal angle of the scapula. The apical end of the muscle encroaches upon the dorsal edge of the deep pectoral and with it goes under the shoulder and arm musculature, ending in an aponeurosis medially on the m. triceps; this aponeurosis partly blends with the tendon of the m. teres major to end on the teres tubercle and partly goes with the deep pectoral muscle to the medial fascia of the brachium. Laterally a tip of the m. cutaneus trunci joins it; this is near the origin of the m. tensor fasciae antebrachii. Since the ventral border of the m. latissimus dorsi gives off a bundle over the biceps to the m. pectoralis profundus and with it inserts aponeurotically on the crest of the major tubercle, the dog, like the cat, has a "muscular axillary arch" (Heiderich 1906 and Langworthy 1924).

Action: To draw the trunk forward and possi-

bly laterally; depress the vertebral column; support the limb, draw the limb against the trunk, draw the free limb backward during flexion of the shoulder joint.

Innervation: Nn. pectorales caudales, and n. thoracodorsalis.

The **mm. pectorales superficiales** consist of descending and transverse pectoral muscles formerly considered as one muscle (Figs. 6–48, 6–49, 6–50, 6–54). They lie under the skin on the cranioventral part of the thorax between the cranial end of the sternum and the humerus. Both arise paramedially on the cranial end of the sternum, run laterally and distally, and cover the m. biceps brachii. Then, with the m. cleidobrachialis, they pass between the mm. biceps brachii and brachialis and end, except for a small distal part, on the entire crest of the major tubercle of the humerus. Three divisions of the muscle are discernible because there are two slips of the more cranial **descending pectoral.**

Action: To support the limb, draw the limb inward, draw the limb forward or backward according to its position, and draw the trunk sideward.

Innervation: Nn. pectorales craniales and also branches from nn. cervicales 7 and 8 (Langworthy 1924).

The *m. pectoralis profundus,* or ascending pectoral (Figs. 6–48, 6–50, 6–54), is a broad muscle lying ventrally on the thorax; it can be divided into a major portion and a minor superficial, lateral portion. It extends between the sternum and the humerus and corresponds to the pars humeralis of the same muscle of some other animals. It arises from the first to the last sternebra and, with a superficial marginal portion as the pars abdominalis, from the deep fascia of the trunk in the region of the xiphoid cartilage. Its fibers run cranially and laterally toward the brachium. It covers the sternum and the cartilages of the sternal ribs from which it is separated by the aponeurosis of the mm.

rectus abdominis and rectus thoracis. After going underneath the superficial pectoral, the major part of the muscle largely inserts, partly muscularly and partly tendinously, on the minor tubercle of the humerus. An aponeurosis goes over the m. biceps brachii to the major tubercle. The superficial part, which originates from the abdominal fascia, and which is crossed laterally by the terminal fibers of the m. cutaneus trunci, goes to the middle of the humerus. There the m. latissimus dorsi and the m. cutaneus trunci attach to it. It then radiates into the medial fascia of the brachium.

The muscle in large dogs is 2 to 2.5 cm. thick in its cranial part; caudally it is thinner. The m. pectoralis superficialis covers it cranially.

Action: To pull the trunk up on the advanced limb; extend the shoulder joint; draw the limb backward. According to Slijper (1946), the m. pectoralis profundus, along with the m. serratus ventralis, plays an important role in supporting the trunk, since its humeral insertion is

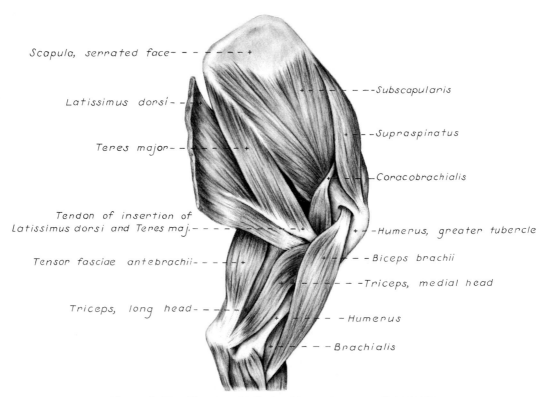

Figure 6–51.   Muscles of left shoulder and arm, medial aspect.

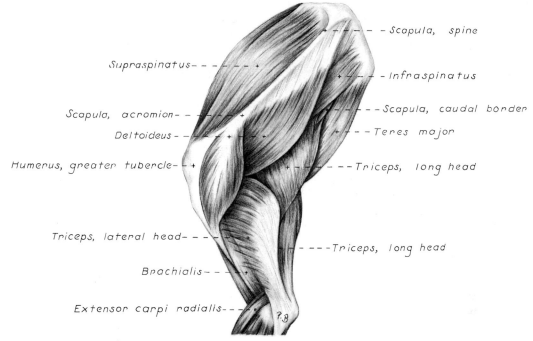

Supraspinatus- - - - - - -

Scapula, acromion- - -

Deltoideus - - - -

Humerus, greater tubercle- - -

Triceps, lateral head- - -

Brachialis - - -

Extensor carpi radialis - - -

- - - Scapula, spine

- - - Infraspinatus

- - Scapula, caudal border

- - Teres major

- - - Triceps, long head

- - - - Triceps, long head

Figure 6–52. Muscles of left shoulder and arm, lateral aspect.

considerably dorsal to its sternal origin.

Innervation: Nn. pectorales caudales, and also branches from nn. cervicalis 8 and thoracicus 1.

The lateral shoulder muscles. The lateral shoulder muscles, mm. supraspinatus and infraspinatus, occupy the scapular fossae. Superficially, the m. deltoideus and the m. teres minor traverse the flexor angle of the shoulder joint laterally.

The *m. supraspinatus* (Figs. 6–45 and 6–49 to 6–52) is covered by the mm. trapezius cervicis and omotransversarius. It fills the supraspinous fossa and curves over the lateral edge of the neck of the scapula. It arises from the entire surface of the supraspinous fossa, including the spine of the scapula, and from the edge of the neck of the scapula by numerous tendons from which the subscapularis also partly originates. Distally the strong muscular belly curves far around the neck of the scapula so that it also appears on the medial surface of the shoulder. The entire muscle ends with a short, extremely strong tendon on the free edge of the major tubercle of the humerus. In the distal third of the muscle, a strong tendinous fold develops which extends imto the terminal tendon.

The end of the muscle appears to be pennate. The caudal half of the muscle is covered by a glistening tendinous sheet from the spine of the scapula.

Action: Extension of the shoulder joint and forward advancement of the limb.

Innervation: N. suprascapularis.

The *m. infraspinatus* (Figs. 6–45, 6–49, 6–52) is covered largely by the m. deltoideus. It lies in the infraspinous fossa and extends caudally somewhat beyond the fossa. It arises from the fossa, the scapular spine, and the caudal border of the scapula, and finally from the tendinous sheet which covers it (shoulder aponeurosis and tendon of origin of the m. deltoideus). At the shoulder joint the fleshy muscle becomes a strong tendon which crosses the caudal part of the major tubercle. The infraspinous bursa is found here. The muscle ends distal to the tubercle. This tendon originates from the middle of the muscle so that it is circumpennate in form. Proximal to the infraspinous bursa, which is about 1 cm. in diameter in large dogs, there is constantly found a second, smaller one.

Action: The muscle is the outward rotator and abductor of the humerus and a flexor or extensor of the shoulder joint, depending on the position of the joint when the muscle contracts. Its tendon functions as a lateral collateral ligament of the shoulder joint.

Innervation: N. suprascapularis.

The *m. teres minor* (Figs. 6–45,6–46,6–49, 6–53) lies distocaudally on the scapula on the flexor side of the shoulder joint, where it is covered by the m. deltoideus and the m. infraspinatus. It arises by an aponeurosis which lies on the long head of the m. triceps, from the distal third of the caudal edge of the scapula, and primarily from the infraglenoid tubercle. It inserts by a short, strong tendon on a special eminence of the humeral crest above the deltoid tuberosity. It is covered on both sides by a tendinous sheet.

Action: Flexion of the shoulder joint.

Innervation: N. axillaris.

The *m. deltoideus* (Figs. 6–45, 6–47, 6–49, 6–52) is composed of two portions lying side by side. It lies superficially directly under the shoulder fascia between the scapular spine and the proximal half of the humerus and is covered to a great extent by an opalescent aponeurosis, from which it arises. This aponeurosis blends with the m. infraspinatus and comes from the scapular spine. Distal to the shoulder joint it becomes a tendinous sheet which slips under the acromial part, medially. This arises at the acromion; its oval, flat belly, which in large dogs is 1.25 to 1.5 cm. thick, crosses the lateral side of the shoulder joint, unites with the tendinous sheet of the scapular part, and ends partly in tendon and partly in muscle on the deltoid tuberosity. Over half of the acromial part is covered by an aponeurotic sheet composed of radiating fibers from which two distinct tendinous processes penetrate into the body of the muscle. The medial surface of both portions has an aponeurosis, which is weak distally as it attaches to the deltoid tuberosity. Between the acromial part and the tendon of the m. infraspinatus there is occasionally found a synovial bursa.

Action: Flexion of the shoulder joint, lifting of the humerus.

Innervation: N. axillaris.

The medial shoulder muscles. The medial shoulder muscles fill the subscapular fossa — m. subscapularis, or cross the flexor angle of the shoulder joint medially — the m. teres major.

The broad, flat *m. subscapularis* (Figs. 6–45, 6–46, 6–50, 6–51) lies in the subscapular fossa and overhangs the caudal edge of the scapula. It is covered by a shiny, tendinous sheet which sends four to six tendinous bands that divide the muscle into broad pennate portions. Three or four of these portions have separate, tendinous coverings on their free medial side. In the interior of the muscle there are tendinous bands which parallel the surface of the muscle. Correspondingly, the muscle has an exceedingly complicated system of fasciculi which run in many different directions. The m. subscapularis arises in the subscapular fossa, especially from the muscular lines on the caudal edge of the scapula and on the curved boundary line between the facies serrata and subscapular fossa. The muscle becomes narrower and is partly tendinous as it passes over the shoulder joint medially. It inserts by means of a short, very strong tendon on the minor tubercle of the humerus. The tendon unites intimately with the joint capsule.

Action: Primarily to adduct and extend the shoulder joint and to draw the humerus forward; during flexion of the joint, it aids in maintaining flexion. Its tendon functions as a medial collateral ligament.

Innervation: N. subscapularis.

The *m. teres major* (Figs. 6–45,6–46,6–50, 6–51) is a fleshy, slender muscle lying caudal to the m. subscapularis. It, as well as the m. subscapularis, arises at the caudal angle and the adjacent caudal edge of the scapula. Distally it crosses the mm. triceps and coracobrachialis as it diverges from the m. subscapularis. It inserts on the teres tubercle by a short, flat tendon, which blends with that of the m. latissimus dorsi. The lateral surface of the muscle bears a tendinous sheet which is strong distally; into this blends a similar tendinous sheet from the m. latissimus dorsi.

Action: Flexion of the shoulder joint, to draw the humerus backward.

Innervation: Branch of the n. axillaris.

THE BRACHIAL MUSCLES

The muscles of the brachium completely surround the humerus except for a small portion, mediodistally, which is left bare. Cranially are the extensors of the shoulder or flexors of the elbow joint — the mm. biceps

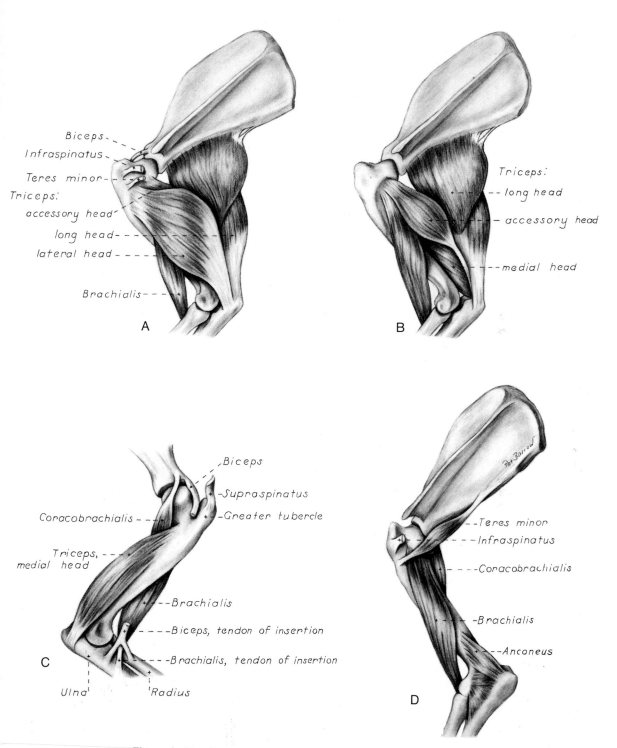

Figure 6–53. Deep muscles of the brachium.
A. Lateral aspect.
B. Lateral aspect. (Lateral head of triceps removed.)
C. Medial aspect.
D. Caudolateral aspect.

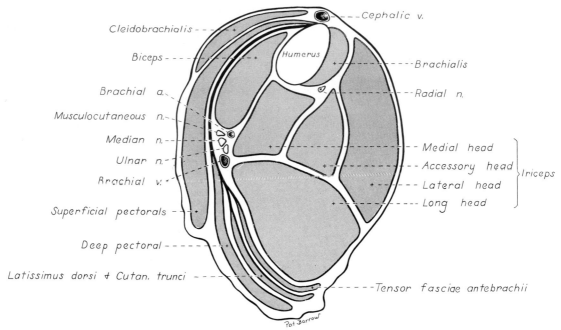

Figure 6–54. Schematic plan of cross section through the middle of the arm.

brachii, coracobrachialis, brachialis; caudally, the extensors of the elbow — the mm. triceps, anconeus, and tensor fasciae antebrachii. The shoulder is so attached to the lateral thoracic wall that the wall is covered as far as the third intercostal space. Accessibility of the heart for clinical examination would be diminished if the limb could not be drawn forward.

The cranial brachial muscles. The cranial brachial muscles include the m. biceps brachii, m. brachialis, and m. coracobrachialis.

The **m. biceps brachii** (Figs. 6–45, 6–51, 6–53, 6–54, 6–55) is a very homogeneous muscle. It begins on the tuber scapulae by means of a long tendon of origin which crosses the shoulder joint in a sharp curve to gain the cranial surface of the humerus through the intertuberal groove. Cranially, it invaginates the joint capsule deeply, and is held in place by a transverse band between the tubercles. The joint capsule reflects around the tendon as its synovial sheath. Distal to the trochlea the tendon becomes a strong, spindle-shaped muscle, which in large dogs is 3 to 4 cm. thick in the middle, and which extends from the medial to the cranial surface of the humerus. In the region

of the elbow joint the tendon of insertion splits into two parts. The stronger of the two inserts on the ulnar tuberosity and the weaker one inserts on the radial tuberosity. The terminal tendon of the m. brachialis inserts between the two parts of the tendon of insertion of the m. biceps brachii. Beginning at the tendon of origin, the muscle is covered by two extensive fibrous sheets which cover three-fourths to four-fifths of its length. The narrower one is applied to the side of the muscle next to the bone; the other is broader and covers the cranial and medial surfaces. Pushed into the interior of the muscle is a strong tendinous fold which, externally, is manifested by a groove. The fold does not reach the proximal tendon of origin; it makes the m. biceps brachii in the dog double pennate. The fibers of the m. biceps brachii run obliquely from both fibrous coverings to the interior fibrous fold, so that their length is less than one-fifth that of the entire muscle. The m. biceps brachii in the dog is not composed of long fibers, as is the case in man; rather, it shows the first step toward the acquisition of a passive tendinous apparatus (Krüger, 1929), which in the quadrupeds is necessary for the fixation of the shoulder joint when standing. Distally from

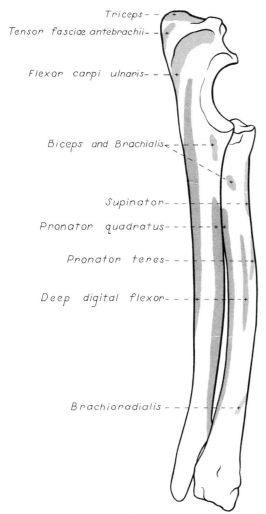

Triceps - -
Tensor fasciae antebrachii- -
Flexor carpi ulnaris- - -
Biceps and Brachialis -
Supinator- -
Pronator quadratus- - -
Pronator teres- -
Deep digital flexor- - -
Brachioradialis - -

Figure 6–55. Left radius and ulna, showing areas of muscle attachment, medial aspect.

the interior fold, there extends a tendinous strand in the groove between the m. extensor carpi radialis and m. pronator teres; it crosses these muscles and spreads out in the antebrachial fascia. It corresponds to the lacertus fibrosus of man.

Action: Flexion of the elbow joint.

Innervation: N. musculocutaneus.

The *m. brachialis* (Figs. 6–49, 6–53, 6–54) arises muscularly from the proximal part of the caudal surface of the humerus or proximal part of the musculospiral groove. It extends laterally as far as the humeral crest, and medially as far as the medial surface. It winds from the caudolateral to the cranial surface of the humerus in its course distally.

At the distal third of the humerus it becomes narrower, goes over the flexor surface of the elbow joint, lateral to the m. biceps brachii, and ends partly fleshy on that part of the tendon of the m. biceps brachii which goes to the radial tuberosity. The remainder becomes the tendon of insertion which goes to the ulnar tuberosity between the two tendons of the m. biceps brachii. The muscle is mostly covered by the m. triceps. Medially it is covered by a closely adherent fascial leaf which extends distally to the m. extensor carpi radialis.

Action: Flexion of the elbow joint.

Innervation: N. musculocutaneus, without the participation of the n. axillaris (Reimers 1925).

The *m. coracobrachialis* (Figs. 6–46, 6–50, 6–51), short and rather thick, arises on the coracoid process of the scapula by a long, narrow tendon which is surrounded by a synovial sheath (vagina synovialis m. coracobrachialis). The tendon extends obliquely caudodistally over the medial side of the shoulder joint and thus lies in a groove close to the tendon of the m. subscapularis. The muscle runs between the medial and accessory heads of the m. triceps brachii, ending on the crest of the minor tubercle, as well as caudal to the crest between the medial head of the m. triceps brachii and the m. brachialis. From its insertion a delicate tendinous leaf extends proximally over almost the entire muscle belly.

Action: Extension and adduction of the shoulder joint.

Innervation: N. musculocutaneus.

The caudal brachial muscles. The muscles which fill in the triangular space between the scapula, humerus, and olecranon form a mighty muscular mass. They are the extensors of the elbow joint. The principal part of this musculature is formed by the m. triceps brachii. The other extensors of the elbow joint in the dog are the m. anconeus and the m. tensor fasciae antebrachii.

The *m. triceps brachii* (Figs. 6–45 to 6–56) consists of four heads: caput longum, laterale, mediale, and accessorium, with a common tendon to the olecranon. Where this tendon crosses the grooves and prominences of the proximal end of the ulna, a synovial bursa is interposed.

The *caput longum of the m. triceps* forms a triangular muscle belly whose base lies on

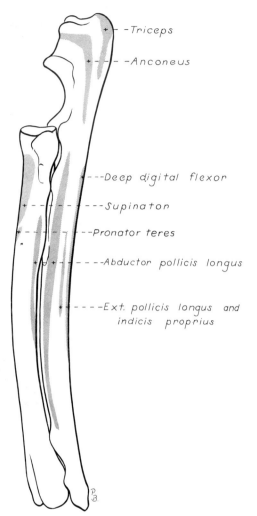

- -*Triceps*

- -*Anconeus*

---*Deep digital flexor*

---*Supinator*

--*Pronator teres*

- - - -*Abductor pollicis longus*

- - - -*Ext. pollicis longus and indicis proprius*

Figure 6–56. Left radius and ulna, showing areas of muscle attachment, lateral aspect.

the caudal edge of the scapula and the apex on the olecranon. The muscle arises partly fleshy and partly tendinously on the distolateral two-thirds of the caudal edge of the scapula and chiefly by tendon on the infraglenoid tuberosity. Its fibers, which are covered laterally by a rather weak and somewhat extensive fascia, converge toward the olecranon and end in a short, thick, round tendon. This tendon is attached to the caudal part of the olecranon, but under the lateral head, it is supplemented by a fascial sheet which is strong distally and which radiates between the long head and lateral head in a proximal direction. This fascia also embraces the cranial edge of the muscle. The

interior of the muscle reveals a weak, tendinous strand which is parallel to the surface. Between the terminal tendon and the cranial, grooved portion of the olecranon, there is a synovial bursa (bursa subtendinea olecrani), which may be over 1 cm. wide; here there is abundant fat. The muscle is interspersed with several tendinous bands and manifests distinct subdivisions. Near the scapula, the mm. deltoideus and teres minor are found laterally and the m. teres major lies medially.

The *caput laterale of the m. triceps* is a strong, almost rectangular muscle lying between the long head and the humerus. This muscle, which blends with the accessory head and which lies on the m. brachialis, arises on the humeral crest by an aponeurosis, which in small dogs is about 1 cm. wide. After emerging from the caudal border of the m. teres major, its fibers run toward the olecranon and terminate in a broad, short tendon which blends partially with the tendon of the long head and partially with the deep leaf of the antebrachial fascia.

The *caput mediale of the m. triceps* is a spindle-shaped muscle which arises tendinously on the crest of the minor tubercle between the point of insertion of the teres major and that of the m. coracobrachialis. A strong, tendinous fascia extends over the proximal two-thirds of the muscle. It attaches medially and independently on the olecranon. In addition, the tendon blends with that of the long head and continues into the antebrachial fascia. The bursa subtendinea olecrani is underneath the tendon.

The *caput accessorium of the m. triceps,* irregularly rectangular in cross section, lies on the caudal side of the humerus between the other heads of the m. triceps brachii and the m. brachialis. It arises from the proximal caudal part of the neck of the humerus, and becomes tendinous at the distal third of the humerus. The tendon is elliptical in cross section and blends with that of the long and lateral heads and thus inserts on the olecranon. The common tendon lies over the subtendinous bursa.

Action: Extend the elbow joint.
Innervation: N. radialis.

The short, strong *m. anconeus* (Figs. 6–47, 6–49, 6–50, 6–53, 6–56) lies on the caudal side of the distal half of the humerus between the epicondyles. It arises on the lateral epicondylar crest, the lateral epicondyle, and, since it almost completely fills the olecranon fossa, part of the medial epicondyle also. It ends on the lateral surface of the proximal end of the ulna and is mostly covered by the m. triceps brachii. It covers the proximal surface of the elbow joint capsule and one of its outpocketings.

Action: The m. anconeus, with the m. triceps brachii, extends the elbow joint, and helps tense the antebrachial fascia.

Innervation: N. radialis.

The *m. tensor fasciae antebrachii* (Figs. 6–51, 6–54, 6–55) is a flat, broad, straplike muscle which, in large dogs, is only 2 mm. thick; it lies on the caudal half of the medial surface and on the caudal edge of the long head of the m. triceps brachii. It arises above the "axillary arch" from the thickened epimysium of the lateral surface of the m. latissimus dorsi. It ends, in common with the m. triceps brachii, in a tendon on the olecranon, and independently in the antebrachial fascia. Occasionally one finds a synovial bursa between the muscle and the medial surface of the olecranon.

Action: It supports the action of the m. triceps brachii and is the chief tensor of the antebrachial fascia.

Innervation: N. radialis.

THE ANTEBRACHIAL MUSCLES

The muscles of the forearm embrace the bones in such a way that the distal two-thirds of the medial side of the antebrachial skeleton (especially the radius) is uncovered. The extensors of the carpus and digits lie dorsally and laterally. The carpal and digital joints of the thoracic limb have equivalent angles; that is, their extensor surfaces are directed dorsally or cranially. On the palmar side are the flexors of the joints. The mm. pronator teres and supinator serve to turn the forepaw about the long axis; these are found in the flexor angle of the elbow joint. Because most of the muscles appear on the palmar side, the antebrachium of the dog appears to be compressed laterally. Since the muscle bellies are located proximally and the slender tendons distally, the extremity tapers toward the paw.

The dorsolateral antebrachial muscles. The dorsolateral group of antebrachial muscles are represented chiefly by the extensors

of the carpal and digital joints. These are the mm. extensor carpi radialis, extensor digitorum communis, extensor digitorum lateralis, extensor carpi ulnaris, extensor pollicis longus et indicis proprius, and abductor pollicis longus. To these are added the mm. brachioradialis and the supinator in the flexor angle of the elbow joint. The majority of these muscles arise directly or indirectly from the lateral (extensor) epicondyle of the humerus.

The *m. brachioradialis* (Figs. 6–49, 6–57, 6–59), much reduced and occasionally lacking, is a long, narrow muscle in the flexor angle of the elbow joint. This muscle has also been called the m. supinator longus. Wakuri and Kano (1966) found the muscle present in 35 of 90 dogs examined. It is cranial in position between the superficial and the deep antebrachial fascia, and is intimately bound to the superficial leaf of the latter fascia. It

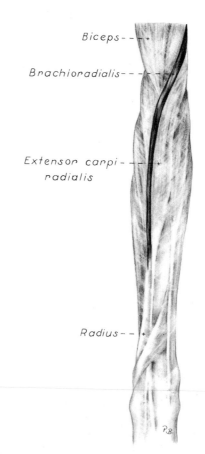

Figure 6–57. Forearm with antebrachial fascia, cranial aspect.

arises on the proximal end of the lateral condyloid crest of the humerus directly above the m. extensor carpi radialis. It extends cranially at first beside the m. extensor carpi radialis, then turns more medially and extends distally in the groove between the m. extensor carpi radialis and the radius. Between the third and the distal fourth of the bone it ends on the periosteum by a thin aponeurosis.

Action: Rotation of the radius dorsolaterally.

Innervation: N. radialis.

The *m. extensor carpi radialis* (Figs. 6–49, 6–58 to 6–61, 6–68) is a long, strong, fleshy muscle lying on the cranial surface of the radius medial to the m. extensor digitorum communis. It is the first muscle encountered after the free surface of the radius, when one palpates from the medial to the dorsal surface. The m. extensor carpi radialis arises on the lateral epicondylar crest of the humerus, united with the m. extensor digitorum communis for a short distance by an intermuscular septum. It forms a muscle belly which fades distally and splits into two flat tendons at the distal third of the radius. This muscle in the dog reminds one of the relations prevailing in man: an incomplete division into a weaker, more superficial, medial m. extensor carpi radialis longus, and a stronger, deeper, more lateral m. extensor carpi radialis brevis. The deep muscle is limited on its deep surface by a fascial leaf which extends from the

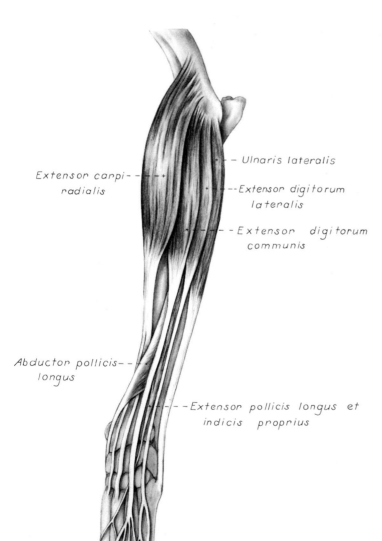

Extensor carpi-
radialis

Ulnaris lateralis

Extensor digitorum
lateralis

Extensor digitorum
communis

Abductor pollicis-
longus

Extensor pollicis longus et
indicis proprius

Figure 6–58. Superficial antebrachial muscles, craniolateral aspect.

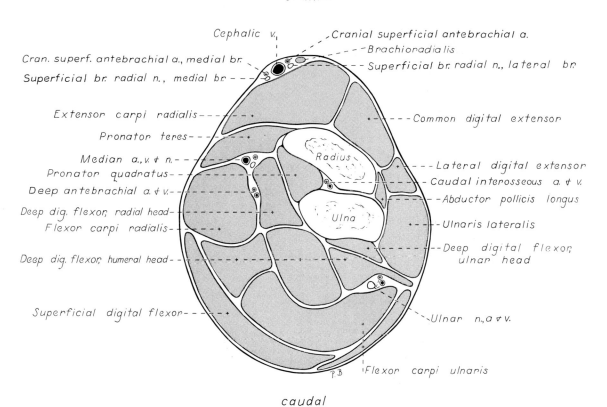

cranial

Cephalic v.

Cranial superficial antebrachial a.

Brachioradialis

Cran. superf. antebrachial a., medial br.

Superficial br. radial n., lateral br.

Superficial br. radial n., medial br.

Extensor carpi radialis

Common digital extensor

Pronator teres

Radius

Median a., v. & n.

Lateral digital extensor

Pronator quadratus

Caudal interosseous a. & v.

Deep antebrachial a. & v.

Abductor pollicis longus

Deep dig. flexor, radial head

Ulna

Flexor carpi radialis

Ulnaris lateralis

Deep dig. flexor, humeral head

Deep digital flexor, ulnar head

Superficial digital flexor

Ulnar n., a. & v.

P.B.

Flexor carpi ulnaris

caudal

Figure 6–59. Schematic plan of cross section of the forearm between the proximal and middle thirds.

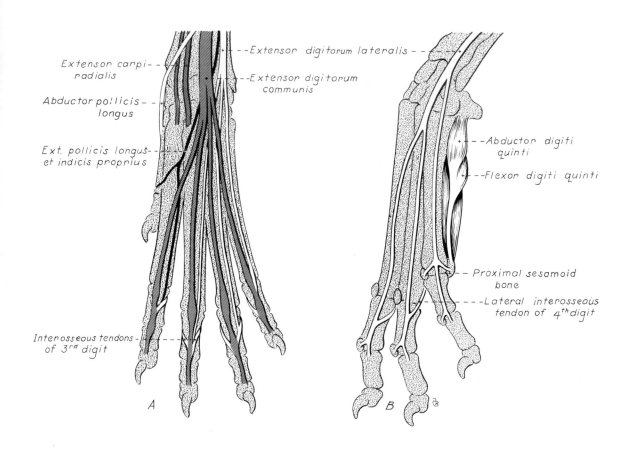

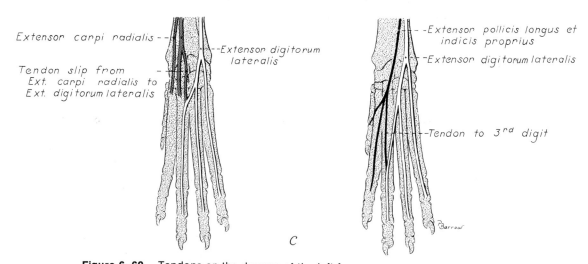

Figure 6–60. Tendons on the dorsum of the left forepaw.
A. Insertion of the common digital extensor.
B. Lateral aspect, tendons of the common digital extensor removed.
C. Two common variations.

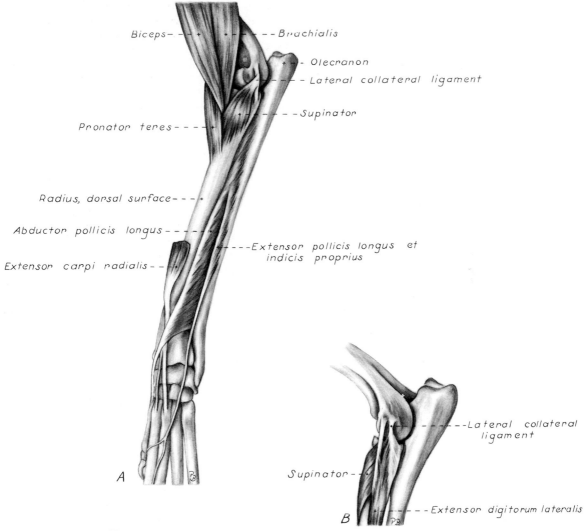

Biceps

Brachialis

Olecranon

Lateral collateral ligament

Supinator

Pronator teres

Radius, dorsal surface

Abductor pollicis longus

Extensor pollicis longus et indicis proprius

Extensor carpi radialis

A

Lateral collateral ligament

Supinator

Extensor digitorum lateralis

B

Figure 6–61. Antebrachial muscles.
A. Deep antebrachial muscles, craniolateral aspect.
B. Origins of supinator and extensor digitorum lateralis.

lateral epicondyle to its terminal tendon. Both tendons are closely approximated as they extend distally along the radius. By way of the middle sulcus of the radius, they gain the extensor surface of the carpus, where they lie in a groove formed by the extensor retinaculum. They are often surrounded by a synovial sheath. The tendons separate; one inserts on a small tuberosity on metacarpal II (m. extensor carpi radialis longus) and the other on metacarpal III (m. extensor carpi radialis brevis). From the aponeurosis covering the medial surface of the m. brachialis arises a fascial leaf which extends over the

proximal medial surface of the belly of the m. extensor carpi radialis, as does a fascial leaf from the m. biceps brachii. The most proximal part of the muscle lies on the joint capsule, which forms a bursa-like pocket at this point. In about half of all specimens the tendons are completely or almost completely surrounded by a common tendon sheath which extends from the beginning of the tendon to the proximal end of the metacarpus. A synovial bursa may exist at the proximal row of carpal bones under both tendons or only under the lateral tendon. A second bursa is occasionally found under the lateral tendon at the distal row of carpal bones. In other specimens, in place of the synovial sheath, one finds loosely meshed tissue.

Action: Extension of the carpal joint and flexion of the elbow joint.

Innervation: N. radialis.

The *m. extensor digitorum communis* (Figs. 6–49, 6–58 to 6–60, 6–66 to 6–68) lies on the craniolateral surface of the radius between the m. extensor carpi radialis and the m. extensor digitorum lateralis. It arises on the lateral epicondyle somewhat in front of and above the attachment of the ulnar collateral ligament of the elbow joint, and with a smaller portion from the antebrachial fascia. At its origin it is fused deeply with the m. extensor carpi radialis by a common aponeurosis which separates into two parts distally, one for each muscle. After the appearance of a corresponding number of superficial tendinous bands, the slender belly divides into four bellies and tendons distally; at first these lie so close together that the whole tendon appears to be undivided. At the same time the deep muscle fibers extend over to the medial tendon. The compound tendon, enclosed in a common synovial sheath, extends distally on the m. abductor pollicis longus and passes through the lateral distal sulcus of the radius, where it is covered by a strong, indistinct transverse ligament. After the tendon crosses the extensor surface of the carpal joint, the individual tendons separate from each other and pass on the extensor surface of the corresponding metacarpal bones and phalanges to the distal phalanges of digits II to V, inclusive. Here each tendon broadens into a caplike structure and ends on the dorsal portion of the edge of the horn of the distal phalanx, covered by the crura of the dorsal elastic ligaments. The m. extensor digitorum communis is composed of digital extensors II, III, IV, and V. Each tendon, at the distal end of the proximal phalanx, receives bilaterally thin check ligaments which cross obliquely from the palmar mm. interossei. The tendons of the lateral digital extensor unite with the tendons of the common digital extensor on digits III, IV, and V. Thus, all extensor tendons are deeply embedded in the dorsal fibrous tissue of the digits.

Under the origin of the m. extensor digitorum communis there extends an outpouching of the elbow joint capsule. The separa-

tion of the terminal portion of the muscle is usually described as distinct, although an undivided muscle is simulated. On the other hand, the tendons may fuse in part with one another; this is especially true for the tendons of digits IV and V. The muscle branch for digit II is the longest and becomes tendinous at the middle of the antebrachium. The three remaining muscle branches reach only to the middle third of the antebrachium. The synovial sheath which surrounds the tendon bundle of this muscle and that of the m. extensor pollicis longus et indicis proprius begins shortly after the muscle has become tendinous (in large dogs 3 to 4 cm. above the carpus). It reaches at least to the middle of the carpus, often to the proximal end of the metacarpus. Its fibrosa fuses with the periosteum of the radius and with the joint capsule of the carpus, that is, with the dorsal carpal ligament. Its mesotendon, which appears at its medial border, first covers the tendon of the m. extensor pollicis longus et indicis proprius and then the four tendons of the m. extensor digitorum communis. At the metacarpophalangeal joint the tendon glides on the sesamoid element which is embedded in the joint capsule; this sesamoid has an ossified nucleus, whereas those at the proximal interphalangeal joints remain cartilaginous.

Action: Extension of the joints of the four principal digits.

Innervation: N. radialis.

The *m. extensor digitorum lateralis* (Figs. 6–49, 6–58 to 6–60, 6–67, 6–68) is somewhat similar to the common extensor in strength; in the antebrachium it lies laterally on the radius between the m. extensor digitorum communis and the m. extensor carpi ulnaris. It covers the m. abductor pollis longus. The muscle has two bellies. It arises on the cranial edge of the ulnar collateral ligament of the elbow joint, and on the head and lateral tuberosity of the radius. A band from the ulnar collateral ligament runs under the muscle, then separates into two branches in its distal half, each half going into a tendon. The tendon adjacent to the common digital extensor is the weaker one and comes from a slender, distal fascial sheet; the other tendon arises from a considerably stronger, distal fascial leaf which lies next to the m. extensor carpi ulnaris. The tendons lie close together and usually are enclosed in a common synovial sheath. They pass through the groove between the distal ends of the radius and ulna, over the dorsolateral border of the carpus to the metacarpus, and then diverge from each other. The tendon of the stronger caudal belly extends from metacarpal V to the proximal phalanx of digit V, unites with the corresponding tendon of the m. extensor digitorum communis and ends with it on the distal phalanx as well as on the dorsal surface of the proximal ends of the proximal and middle phalanges. The tendon of the weaker belly divides at the carpus into two branches which extend obliquely under the tendons of the m. extensor digitorum communis medially to the third and fourth metacarpophalangeal joints; on the proximal phalanx of digits III and IV they unite with the corresponding tendons of the common digital extensor; often they also unite with one or both of the check ligaments which come from the m. interossei. They end principally on the distal phalanges of digits III and IV. The tendons of the lateral digital extensor are only about one-third the width of those of the common digital extensor.

In about one-half of all specimens, both tendons are enclosed in a common synovial sheath, which in large dogs begins 2.5 to 3 cm. above the carpus and often reaches the metacarpus. In others there is no distinct synovial sheath, but rather there is a space under both tendons bounded by the fascia. In exceptional specimens the tendon for digit III is independent and arises from the fascia distal to the carpus (Ziegler 1929).

Action: Extension of the joints of digits III, IV, and V.

Innervation: N. radialis.

The strong *m. extensor carpi ulnaris* (m. ulnaris lateralis) (Figs. 6–58, 6–59, 6–63, 6–67, 6–68) lies on the caudolateral side of the ulna, and is directly under the fascia. It arises on the lateral or extensor epicondyle of the humerus behind the ulnar collateral ligament of the elbow joint by a long, relatively strong tendon. At the middle of the antebrachium the strong distal tendinous band of the lateral surface, which eventually goes into the strong terminal tendon, takes on the delicate distal tendinous leaf of the medial surface. On the medial surface of the terminal tendon, fibers of the deeper muscle mass radiate into the broad tendon as far as the carpus; the tendon passes laterally over

the carpus, being held in place by connective tissue without a sulcus in which to glide. It ends laterally on the proximal end of metacarpal V. From the accessory carpal bone, two fiber bundles arise from the antebrachial fascia and cross each other to blend with the tendon of the m. extensor carpi ulnaris at the carpus. A tendinous fold, which parallels the surface, is concealed in the interior of the muscle. Under the tendon of origin of the muscle in older dogs, there is constantly a synovial bursa, 1 to 2 cm. in diameter; a second bursa is found occasionally, between the tendon and the distal end of the ulna.

Action: Extension of the carpal joint with weak lateral rotation. Sisson and Grossman (1953) said it functioned in abduction of the paw.

Innervation: N. radialis.

The *m. supinator* (Figs. 6–49, 6–55, 6–61) is broad and flat and is almost completely covered by a delicate fascia. It lies laterally in the flexor surface of the elbow, covered by the m. extensor carpi radialis and the digital extensors. It lies directly on the joint capsule and radius. It arises by a short, strong tendon on the ulnar collateral ligament of the elbow joint, and on the lateral epicondyle. It extends obliquely distomedially and, covering the proximal fourth of the radius, ends on its dorsal surface as far as the medial border. The muscle pushes a short distance under the border of the m. pronator teres.

Action: Rotation of the paw so that the palmar surface faces medially.

Innervation: N. radialis.

The *m. extensor pollicis longus et indicis proprius* (Figs. 6–56, 6–61, 6–68) is an exceedingly small, slender, flat muscle on the lateral side of the antebrachium, where it is covered by the m. extensor carpi ulnaris, and the lateral and common digital extensors. After arising from about the middle third of the dorsolateral border of the ulna, adjacent to the m. abductor pollicis longus, it runs distally, parallel to the ulna. Its fibers run obliquely distomedially. The muscle gradually crosses under the extensors, so that its extremely delicate tendon appears medial to the common extensor tendons at the carpal joint, where the two tendons are surrounded by a common synovial sheath. On the dorsal surface of metacarpal III the tendon divides into two parts. The medial portion expands

over metacarpal II to the distal end of metacarpal I, where it is buried in the fascia. The
lateral portion, after passing under the tendon of the m. extensor digitorum communis
to digit II, unites with it at the metacarpophalangeal joint. Rarely another weak tendon goes to digit III, and occasionally the
muscle splits into two bellies.

Action: Extension of digits I and II and adduction of digit I.

Innervation: N. radialis.

The *m. abductor pollicis longus* (Figs.
6–56, 6–59, 6–61, 6–62, 6–69), almost completely covered by the digital extensors, lies
in the lateral groove between the radius and
the ulna. It arises on the lateral surface of the
radius and ulna, and the interosseous membrane. Its fibers, which are directed obliquely medially and distally, blend into a narrow
but strong tendinous band which proceeds

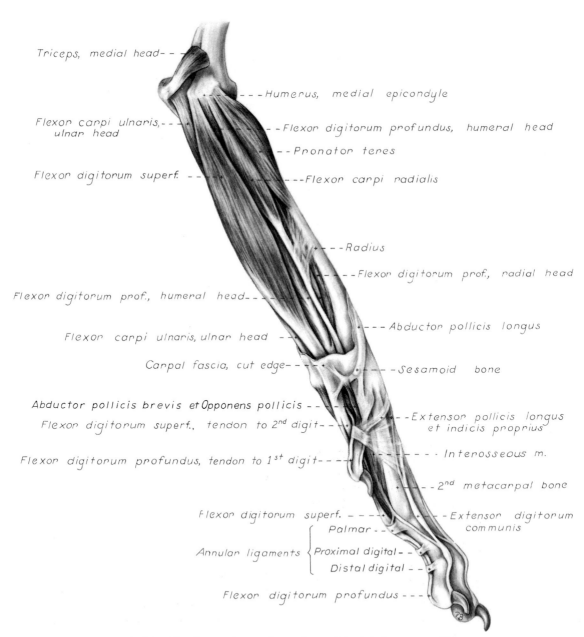

Figure 6–62. Muscles on the medial surface of the left forearm and forepaw.

along the dorsomedial border of the muscle and which becomes the terminal tendon toward the carpus, after it has bridged the gap between the tendons of the m. extensor digitorum communis and the m. extensor carpi radialis. Its tendon crosses the tendon of the m. extensor carpi radialis, passes into the medial sulcus of the radius, and crosses the medial border of the carpus under the short collateral ligament. Finally, the tendon inserts medially on the proximal end of metacarpal I, where a sesamoid bone is embedded in it.

Where the tendon goes over that of the m. extensor carpi radialis, there is usually a bursa or a short synovial sheath.

Action: Abduction and extension of the first digit; medial deviation of the forepaw.

Innervation: N. radialis.

The caudal antebrachial muscles. The caudal group of antebrachial muscles consists of the flexors of the carpus and digits: mm. flexor carpi radialis, flexor carpi ulnaris, flexor digitorum superficialis, and flexor digitorum profundus. To these are added the small mm. pronator teres and pronator quadratus, which do not extend beyond the antebrachium. Most of these muscles come from the medial or flexor epicondyle. They form the caudal part of the antebrachium.

When viewed medially, the muscles appear in the following order, beginning cranially: mm. pronator teres (only the proximal third of the forearm), flexor carpi radialis, flexor digitorum profundus (only the distal half of the forearm), flexor digitorum superficialis, and caput ulnare of the m. flexor carpi ulnaris (only in a very insignificant and proximal segment of the antebrachium). Seen from the palmar aspect, the caput ulnare of the m. flexor carpi ulnaris is medial to the m. flexor digitorum superficialis. However, at the distal half of the antebrachium, the caput humerale of the m. flexor carpi ulnaris pushes between these two. In the dog, the caput humerale is deep; in other animals, when viewed from the medial aspect, it is covered by the m. flexor digitorum superficialis. The m. extensor carpi ulnaris is next to the m. flexor carpi ulnaris on the lateral surface of the forearm.

The *m. pronator teres* (Figs. 6–50, 6–55, 6–59, 6–61, 6–62), round in cross section, crosses the medial surface of the elbow joint. It lies under the skin and fascia largely on the proximal third of the radius. It arises from the medial epicondyle in front of the m. flexor carpi radialis. The body of the muscle extends obliquely craniodistally and, upon forming a strong tendinous band, ends distal to the m. supinator on the medial border of the radius as far as its middle. Its internal surface is provided with a strong, proximal tendinous band.

Action: It rotates the forearm so that the dorsal surface tends to become medial. It may function only as a flexor of the elbow joint (Zimmermann 1928).

Innervation: N. medianus.

The *m. flexor carpi radialis* (Figs. 6–59, 6–62, 6–67, 6–69) lies in the medial part of the antebrachium directly under the skin and antebrachial fascia, where it covers the m. flexor digitorum profundus. It arises on the medial epicondyle behind the radial collateral ligament of the elbow joint between the m. pronator teres and the m. flexor digitorum profundus. It extends distally between the m. pronator teres and the m. flexor digitorum superficialis and, forming a short, thick fusiform belly, merges into a flat tendon near the middle of the radius. It receives a delicate supporting fascia from the radius throughout its entire length. At the flexor surface of the carpus it runs through the flexor retinaculum, where it is enclosed in a synovial sheath. At the metacarpus, it splits into two strong tendons which end on the palmar side of metacarpals II and III, very close to the proximal articular surface. The end of the muscle bears a lateral tendinous band and a delicate medial one. A projection from the joint capsule extends under the muscle at its origin.

Action: Flexion of the carpal joint.

Innervation: N. medianus.

The strong, flat *m. flexor digitorum superficialis* (Figs. 6–59, 6–62, 6–63, 6–66, 6–67, 6–69) in dogs, in contrast to that in other animals, lies directly beneath the skin and antebrachial fascia in the caudomedial part of the antebrachium. It covers the m. flexor digitorum profundus and the humeral head of the m. flexor carpi ulnaris. It arises by a short but strong tendon on the medial or flexor epicondyle cranial to the humeral head of the m. flexor carpi ulnaris and somewhat proximal to the flexor digitorum profundus. The fleshy muscle belly reaches far distally and becomes tendinous only a short distance above the carpus. Its tendon is strong, elliptical in cross section, about 1 cm. wide and 0.5 cm. thick. This tendon runs over the flexor surface of the carpus medial to the accessory carpal bone, but is not enclosed in the carpal synovial sheath, as it crosses the flexor retinaculum superficially. By means of this thick ligament it is separat- ed from the deep flexor tendon. In the proximal third of the metacarpus the tendon splits into four parts, which diverge to the second to fifth metacarpophalangeal joints. Lying in a palmar relationship to the corresponding terminal tendons of the deep flexor tendon, each extends over the respective proximal phalanx and ends on the proximal

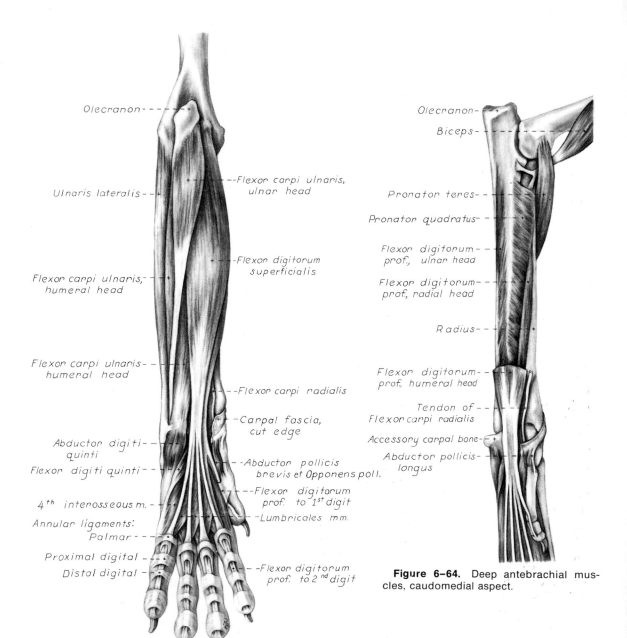

Olecranon

Ulnaris lateralis

Flexor carpi ulnaris, humeral head

Flexor carpi ulnaris, humeral head

Abductor digiti quinti

Flexor digiti quinti

4th interosseous m.

Annular ligaments:
Palmar

Proximal digital

Distal digital

Flexor carpi ulnaris, ulnar head

Flexor digitorum superficialis

Flexor carpi radialis

Carpal fascia, cut edge

Abductor pollicis brevis et Opponens poll.

Flexor digitorum prof. to 1st digit

Lumbricales mm.

Flexor digitorum prof. to 2nd digit

Figure 6-63. Antebrachial muscles, palmar aspect.

Olecranon

Biceps

Pronator teres

Pronator quadratus

Flexor digitorum prof., ulnar head

Flexor digitorum prof., radial head

Radius

Flexor digitorum prof. humeral head

Tendon of Flexor carpi radialis

Accessory carpal bone

Abductor pollicis longus

Figure 6-64. Deep antebrachial muscles, caudomedial aspect.

border of the palmar surface of the middle phalanx after being "perforated" by one of the deep flexor tendons passing to a distal phalanx. Each branch of the superficial tendon, at the metacarpophalangeal joint, forms a tubelike enclosure around the deep flexor tendon called a *manica flexoria*. The proximal edge of this sleeve projects a short distance proximally beyond the articular surfaces of the sesamoid bones. Distal to the sesamoids, the tube is so split for the passage of the deep tendon that the superficial tendon appears to be in two branches when it is viewed from the palmar side. The deep part of the sleeve, however, accompanies the enclosed tendon farther and attaches to the palmar, proximal edge of the middle phalanx throughout its breadth. The four terminal tendons of the muscle are as a rule of equal strength (in large dogs, about 5 mm. wide and 0.5 mm. thick). The branch to digit V is much weaker. At the metacarpophalangeal joint and at the proximal and middle digital joints, the branches of the superficial and deep digital flexor tendons are bridged by the three well-defined proximal, middle, and distal transverse ligaments.

Beneath the origin of the superficial digital flexor there is a synovial bursa 2 to 2.5 cm. long in large dogs. This communicates with a second bursa beneath the origin of the caput humerale of the m. flexor carpi ulnaris. Each of the four terminal tendons has a long digital synovial sheath. This sheath is described more fully below, in the discussion of the m. flexor digitorum profundus.

Action: Flexion of the proximal and middle digital joints of the four principal digits and thereby of the whole forepaw.

Innervation: N. medianus.

The *m. flexor carpi ulnaris* (Figs. 6–55, 6–59, 6–63, 6–65, 6–69) consists of two bellies, which converge into a tendon ending on the accessory carpal bone. The muscle lies caudolaterally on the antebrachium, with its weaker ulnar head most superficial and lateral to (and, in part, upon) the m. flexor digitorum superficialis. The much stronger humeral head is in the second layer of the palmar musculature beneath the m. flexor digitorum superficialis and upon the m. flexor digitorum profundus.

The rather flat **ulnar head** (*caput ulnare*), which is straplike in small dogs, arises medially on the palmar border of the proximal end of the ulna and is covered at its origin by the terminal tendon of the medial head of the m. triceps brachii. Above the middle of the antebrachium the ulnar head becomes a flat tendon which extends distally, lateral and palmar to the m. flexor digitorum superficialis covering the humeral head. Toward the accessory carpal bone the tendon gradually dips beneath the terminal tendon of the humeral head and ends independently on the accessory carpal bone. The strong antebrachial fascia fuses with it throughout its length.

The much stronger **humeral head** (*caput humerale*) arises on the medial or flexor epicondyle of the humerus by a short, strong tendon which is a close neighbor of that of the m. flexor digitorum superficialis. It ends by an equally short, strong tendon on the accessory carpal bone. This strong, flat muscle, as much as 3 cm. wide and 1 cm. thick, in the dog, in contrast to that in other mammals, is almost completely covered by the m. flexor digitorum superficialis. Only the lateral edge of its distal half and its terminal tendon encroach upon the antebrachial fascia. Both surfaces of its body are covered by a tendinous sheet. The palmar sheet is almost entirely a distal one; the dorsal tendinous sheet is equally extensive proximally and distally, and is provided with a narrow but strong tendinous sulcus which, near the middle, appears to be displaced somewhat medially. Thus the muscle has a complicated fiber structure.

Beneath the origin of this muscle is found a synovial bursa which communicates with the one beneath the origin of the m. flexor digitorum superficialis. A second bursa, beneath the terminal tendon, in large dogs extends proximally 1 to 1.5 cm. from the accessory carpal bone.

Action: Flexion of the forepaw with abduction.

Innervation: N. ulnaris.

The *m. flexor digitorum profundus* (Figs. 6–55, 6–59, 6–62, 6–65, 6–66, 6–69) consists of three heads and, generally speaking, forms the deepest layer of the caudal musculature of the forearm. It is covered by the mm. flexor carpi radialis, flexor digitorum superficialis, and flexor carpi ulnaris. Its bellies, along with the m. pronator quadratus, lie directly on the caudal surface of the radius and ulna. It consists of humeral, radial, and ulnar heads, which represent completely separate muscles whose tendons fuse to form the strong, deep digital flexor tendon. The homologization of these three muscles is still in dispute (Kajava 1922).

The **humeral head** (*caput humerale*) of the m. flexor digitorum profundus, as the strongest division of the entire muscle, consists of three bellies which are difficult to isolate. It is provided with tendinous sheets and bands and thus has a very complex fiber arrangement. The three bellies arise by a common short, strong tendon on the medial epicondyle of the humerus, immediately caudal to the tendon of origin of the m. flexor carpi radialis and covered by that of the m. flexor digitorum superficialis, and the humeral head of the m. flexor carpi ulnaris. The strongly constructed body of the muscle lies on the caudomedial side of the antebrachium in such a way that it appears partly enclosed between the radial head medially and deeply, and the ulnar head laterally. Near the carpus, or at the border between the carpus and antebrachium, the tendons of the humeral head fuse into a flat, but strong, main tendon which is grooved on its palmar surface. Just distal to the groove the weaker tendons of the radial and ulnar heads converge to form the deep flexor tendon.

The **radial head** (*caput radiale*) of the m. flexor digitorum profundus lies on the caudomedial surface of the radius, along the mm. pronator quadratus, pronator teres, flexor carpi radialis, and the humeral head of the flexor digitorum profundus. It arises, as the weakest division of the entire muscle, from the medial border and for a small distance also on the caudal surface of the proximal three-fifths of the radius. Near the carpus it forms a thin, flat tendon, which runs from a slender tendinous sheet on the proximal, caudal border of the muscle. This joins the strong tendon of the humeral head at the proximal border of the carpus. After being united for a short distance with the principal tendon, it again splits away to insert on digit I.

According to Kajava (1922), the muscle does not correspond to the m. flexor pollicis longus of man but is only a division of the m. flexor digitorum profundus. Exceptionally, the branch of the deep tendon going to the first digit is independent in the dog; in such a

case the tendon proceeds from the palmar carpal fascia (Ziegler 1931).

The **ulnar head** (*caput ulnare*) of the m. flexor digitorum profundus is stronger than the radial head. It is a flat muscle at the caudal side of the ulna, located among the m. extensor carpi ulnaris, flexor carpi ulnaris, and the humeral head of the m. flexor digitorum profundus. It is covered superficially by both the m. flexor and the m. extensor carpi ulnaris. It arises on the caudal border of the ulna from the distal portion of the medial ridge of the olecranon to the distal fourth of the ulna. Its fibers run obliquely distally and caudally to a strong, broad tendinous sheet which accompanies the palmar edge of the muscle almost from the level of the elbow joint. At the distal fourth of the antebrachium the muscle ends in a tendon which is soon united with the common tendon of the m. flexor digitorum profundus.

The deep flexor tendon crosses the flexor surface of the carpus in a groove which is converted into the carpal canal by the flexor retinaculum. The tendon is very wide and, because of great thickening of its edges, forms a palmar groove. In the proximal portion of the metacarpus, from its medial border, the deep flexor tendon gives off the round, weak tendon to digit I. Shortly thereafter the principal tendon divides into four branches, for digits II to V, which run distally, covered by the corresponding grooved branches of the m. flexor digitorum superficialis. At the level of the sesamoids of the metacarpophalangeal joints of digits II to V they pass through the tubular sheaths (manica flexoria) formed by the branches of the superficial digital flexor tendons. They emerge from the palmar sheaths, extend over the flexor surface of the distal digital joints, and end on the tuberosities of the distal phalanges of digits II to V.

The two digital flexor tendons on each of the four main digits are held in place by three transverse (annular) ligaments. The proximal one, lig. metacarpeum transversum superficialae (lig. anulare palmare), or palmar annular ligament, lies at the metacarpophalangeal joint and runs between the collateral borders of the sesmoid bones. The middle one, proximal digital annular ligament, lies on the proximal phalanx, at about its middle, and the distal one, distal digital annular ligament, lies immediately distal to the proximal interphalangeal joint. The distal one is lacking in digit I.

Synovial apparatus. Beneath the origin of the m. flexor digitorum profundus is a synovial bursa. The terminal tendon of the muscle, as it passes through the carpal canal, is partly or wholly surrounded by a synovial sheath. This extends from the distal end of the radius to the metacarpus. It may be replaced by a sac with rough, weak walls, without synovia. In the digits the digital synovial sheaths are formed around the individual branches of the deep flexor tendon. The branch going to the first digit has its own sheath, which extends from the middle of metacarpal I to the tuberosity of the distal phalanx. The principal tendons going to digits II to V, however, bear synovial sheaths which are also common for the superficial flexor tendons. These extend from the ends of the metacarpal bones to the tuberosities of the distal phalanges. These common synovial sheaths begin, in large dogs, 1 to 1.5 cm. proximal to the metacarpophalangeal articulations immediately proximal to the sesamoid bones in the region of the proximal ends of the rings formed by the superficial flexor tendons. At their origin the sheaths enclose only the deep flexor tendons. Somewhat farther distally, they also enclose the superficial flexor tendons. Here their fibrosa fuses with the proximal transverse ligaments. To each collateral border of the deeply situated and flattened superficial flexor tendons, there extends a mesotendon. Distally the synovial sheaths also enclose the middle transverse ligaments of the flexors so that they receive mesotendons extending to their proximal edges. Farther distally the deep branches alone are surrounded by synovial sheaths. The distal transverse ligaments, however, do not push into the synovial spaces, but fuse with the palmar and lateral walls of the synovial sheaths.

Action: The m. flexor digitorum profundus is the flexor of the forepaw (carpus and digital joints).

Innervation: The radial head as well as the deep and medial portions of the humeral head, n. medianus; the lateral portion of the humeral head and the ulnar head, n. medianus and n. ulnaris (Agduhr 1915).

The *m. pronator quadratus* (Figs. 6–55, 6–59, 6–64) fills in the space between the radius and the ulna medially. It is rhomboidal in outline, and covers the intcrosseous membrane, a portion of the medial surface of the ulna, and the caudal surface of the radius, except for the proximal and distal ends. It is covered by the m. flexor digitorum profundus. Its fibers run from the ulna obliquely, distally and medially, to insert on the radius.

Action: Turn the forepaw inward.

Innervation: N. medianus.

MUSCLES OF THE FOREPAW

In the forepaw are the tendons of the antebrachial muscles which insert on the metacarpal bones and phalanges, as well as the special muscles which are confined to the palmar surface of the forepaw. Digits I to V are thus covered by a large number of muscles. A portion of these special muscles lies between the large flexor tendons; another portion lies between these and the skeleton, either lying directly upon the four large metacarpal bones or occurring as special muscles of digits I, II, and V.

The muscles lying between the flexor tendons. These muscles include the m. interflexorus, m. flexor digitorum brevis, and the mm. lumbricales.

The weak *m. interflexorius* (Fig. 6–65) is the longest of the group. It arises at the level of the distal fourth of the antebrachium from the palmar tendinous sheet of the lateral superficial belly of the humeral head of the m. flexor digitorum profundus. As a slender, rounded muscle belly, it runs to the carpal joint lying between the digital flexors. It crosses under the flexor retinaculum with the deep flexor tendon, and its thin tendon splits into two (or threee) branches at the middle of the metacarpus. These accompany the branches of the m. flexor digitorum superficialis for digits III and IV, and occasionally digit II, and fuse with them.

Gurlt (1859) homologizes this muscle with the m. palmaris longus of man, whereas Ellenberger and Baum (1943) designate it as the m. palmaris longus accessorius. The name m. interflexorius is derived from Agduhr (1915) and Pitzorno (1905). According to Kajava (1923), the muscle could be designated m. interflexorius profundosublimis.

Action: Flexion of the forepaw.

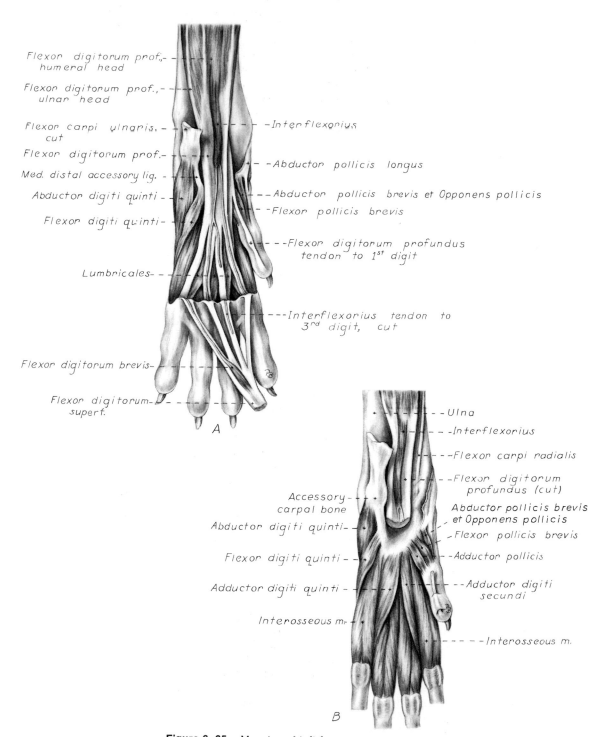

Figure 6–65.   Muscles of left forepaw.
A. Superficial muscles, palmar aspect.
B. Deep muscles, palmar aspect.

Innervation: N. medianus.

The *m. flexor digitorum brevis* (Fig. 6–65), the weakest of this group, is a delicate, only slightly fleshy muscle, which arises distal to the carpus from the palmarolateral surface of the superficial flexor tendon branch for digit V. It goes into a flat tendon, which occasionally takes the place of the whole muscle. It ends on the transverse ligament of the metacarpophalangeal joint.

Gurlt (1859) compares this with the m. palmaris brevis of man; according to Kajava (1923), however, one would have to regard this as a muscle mass in the carpal pad. Ellenberger and Baum (1943) describe this muscle as the m. palmaris brevis accessorius.

The *mm. lumbricales* (Figs. 6–63, 6–66) are three small muscles which are associated with the tendons of the flexor digitorum profundus. The first muscle arises from the contiguous sides of the flexor digitorum profundus tendons to the second and third digits; the second from the tendons to the third and fourth digits; and the third from the tendons to the fourth and fifth digits. They pass obliquely distally and laterally and end in thin tendons which are inserted on the proximal medial surfaces of the first phalanges of the third, fourth, and fifth digits. The tendon of the first muscle inserts on the third digit, the second on the fourth, and the third on the fifth (Leahy 1949).

Action: Flexion of the metacarpophalangeal joints.

Innervation: Deep branch of the n. ulnaris.

The muscles lying on the palmar side of the metacarpal bones. These include the interosseous muscles.

The fleshy *mm. interossei* (Figs. 6–65, 6–66, 6–69) are four in number. They lie on the palmar side of the four large metacarpal bones at the depth of the tendon branches of the flexor digitorum profundus. They are relatively strong and border on one another. They arise from the proximal ends of metacarpals II, III, IV and V, and from the joint capsule, and cover the entire palmar surfaces of these metacarpal bones. After coursing a short distance, each muscle divides into two branches, which attach by tendons to the proximal end of the first phalanx. A sesamoid is embedded in each tendon. A portion of

each tendon extends over the collateral borders of the joint and runs distally on the dorsal surface of the proximal phalanx to unite with the common extensor tendon.

Morphologically each of these muscles results from the fusion of two muscles. According to Kajava (1923), they represent the mm. flexores breves profundi, which are placed dorsal to the ramus palmaris profundus of the n. ulnaris and are innervated by it. Each of the four muscles is collaterally covered by a considerable tendinous fascia which extends far distally. According to Forster (1916), each muscle is invaginated by a tendinous sheet which comes from the diverging portions and which extends proximally, causing the duplicity of each muscle to be much more marked.

Action: Flexion of the metacarpophalangeal joints

Innervation: Deep branch of the n. ulnaris.

The special muscles of digit I. The rudimentary first, or medial, digit has three special muscles: an outward rotator, a flexor, and an inward rotator.

The *m. abductor pollicis brevis et opponens pollicis* (Figs. 6–62, 6–65, 6–69) arises from a tendinous band which comes from the synovial sheath of the superficial flexor tendon and goes to the sesamoid of the tendon of the m. abductor pollicis longus located on the medial side of the carpus. It ends in the ligamentous tissue at the metacarpopnalangeal joint of digit I. According to Kajava (1923), the muscle represents the radial head of the m. flexor pollicis brevis profundus.

Action: Flexion of digit I.

Innervation: Deep branch of the n. ulnaris.

The *m. flexor pollicis brevis* (Figs. 6–65, 6–69) is larger than the special abductor just described and lies between it and the m. adductor pollicis. It arises on the flexor retinaculum, runs obliquely to digit I, and ends on the sesamoid bone or on the proximal phalanx.

Action: Flexion of digit I.

Innervation: Deep branch of the n. ulnaris.

The *m. adductor pollicis* (Figs. 6–65, 6–69) is the strongest muscle of digit I. It arises as a small, fleshy muscle body between the special flexor and the m. interosseus of digit II

Text continued on page 372

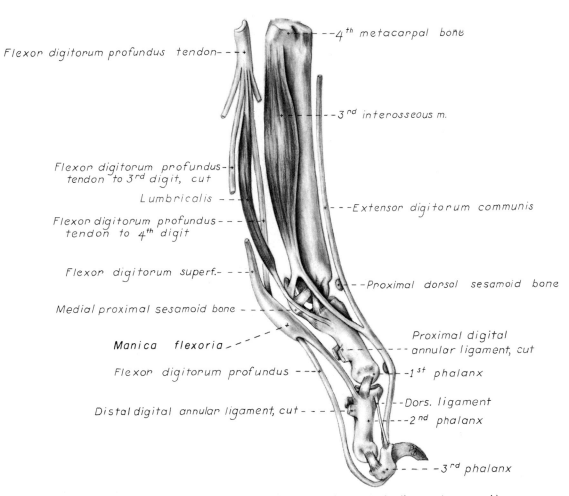

Flexor digitorum profundus tendon-- -

--4th metacarpal bone

--3rd interosseous m.

Flexor digitorum profundus-- tendon to 3rd digit, cut

Lumbricalis - - -

---Extensor digitorum communis

Flexor digitorum profundus- - - tendon to 4th digit

Flexor digitorum superf.- - - -

--Proximal dorsal sesamoid bone

Medial proximal sesamoid bone - - - - -

Proximal digital annular ligament, cut

Manica flexoria - - - -

Flexor digitorum profundus - - -

-1st phalanx

Distal digital annular ligament, cut - - - -

---Dors. ligament

2nd phalanx

- - - -3rd phalanx

Figure 6-66. The fourth digit, medial aspect. (Proximal annular ligament removed.)

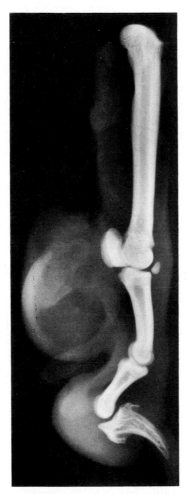

Figure 6–66. Radiograph. Lateral radiograph of a digit.

cranial

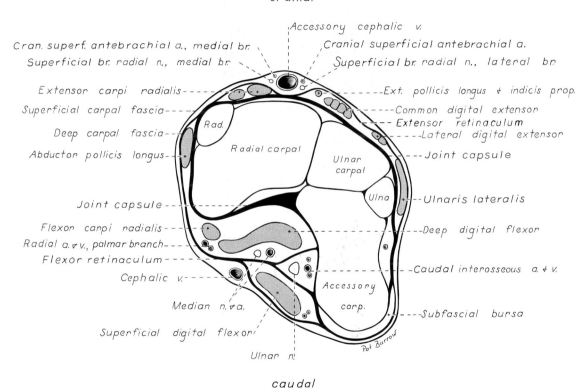

Accessory cephalic v.

Cran. superf. antebrachial a., medial br.

Cranial superficial antebrachial a.

Superficial br. radial n., medial br.

Superficial br. radial n., lateral br.

Extensor carpi radialis

Ext. pollicis longus + indicis prop.

Superficial carpal fascia

Common digital extensor

Extensor retinaculum

Deep carpal fascia

Lateral digital extensor

Abductor pollicis longus

Joint capsule

Rad.

Radial carpal

Ulnar carpal

Joint capsule

Ulna

Ulnaris lateralis

Flexor carpi radialis

Deep digital flexor

Radial a. + v., palmar branch

Flexor retinaculum

Caudal interosseous a. + v.

Cephalic v.

Accessory carp.

Subfascial bursa

Median n. + a.

Superficial digital flexor

Ulnar n.

Pat Barrow

caudal

Figure 6–67. Schematic cross section of forepaw through accessory carpal bone.

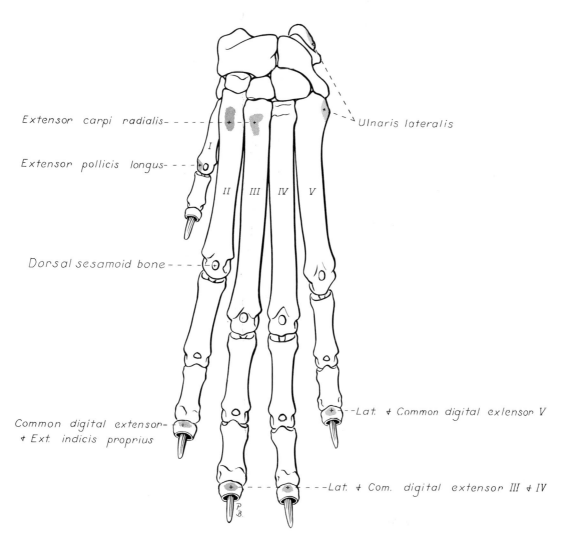

Figure 6–68. Left forepaw with muscle attachments, dorsal aspect.

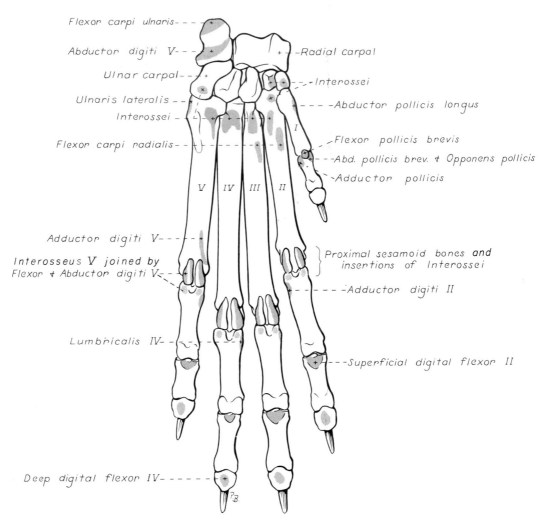

Figure 6–69. Left forepaw with muscle attachments, palmar aspect.

on the flexor retinaculum and ends on the lateral surface of the proximal phalanx of digit I.

The special muscles of digit V. The fifth digit, like the first, also has three special muscles: an adductor, a flexor, and an abductor.

The *m. adductor digiti quinti* (Figs. 6–65, 6–69) arises from the flexor retinaculum and extends as a slender belly obliquely in a lateral direction to end on the medial surface of metacarpal V and the proximal phalanx of digit V. Its proximal portion lies on the mm. interossei 3 and 4, and its distal portion lies between mm. interossei 4 and 5.
Action: Adduction of digit V.

Innervation: Deep branch of the n. ulnaris.

The *m. flexor digiti quinti* (Figs. 6–65, 6–69) arises on the ligament from the accessory carpal bone to metacarpal IV, runs obliquely over the m. interosseus of the fifth digit laterally, and by a thin tendon joins that of the m. abductor digiti quinti.
Action: Flexion of the fifth toe.
Innervation: Deep branch of the n. ulnaris.

The strong *m. abductor digiti quinti* (Figs. 6–65, 6–69) arises on the accessory carpal bone. It ends by means of a tendon which unites with the m. flexor digiti quinti on the lateral sesamoid and frequently by a thin tendon on the proximal phalanx of the fifth digit. It lies directly under the skin on the flexor retinaculum. On its deep surface the fusiform body bears a delicate, proximal tendinous leaf.

Action: Abduction of digit V.

Innervation: Deep branch of the n. ulnaris.

The special muscle of digit II. The second digit has only one special muscle, an adductor.

The *m. adductor digiti secundi* (Figs. 6–65, 6–69) arises on the flexor retinaculum between the m. interosseus 2 and the m. adductor digiti quinti, runs distally (between the mm. interossei 2 and 3), and ends by means of a tendon on the proximal end of the proximal phalanx of digit II.

Action: Adduction of digit II.

Innervation: Deep branch of the n. ulnaris.

Forster (1916) includes the adductors of the first, second, and fifth digits under the name mm. contrahentes digitorum, which in the dog are represented separately but which, when better developed, constitute a "contrahentes leaf." They lie palmarly on the ramus profundus of the n. ulnaris and are innervated by it.

Fasciae of the Thoracic Limb

The superficial and deep fasciae of the neck and thorax extend laterally over the shoulder and there form the superficial and deep fasciae of the shoulder and brachium. The whole limb is covered by this double system of connective tissue, the parts of which take their name from the portion of the limb that they cover.

The **superficial fascia** on the shoulder and brachium covers these portions of the extremity as a bridge laterally from the superficial neck and superficial trunk fasciae. In the distal brachial region, however, it is completely closed into a cylinder, since that portion of the superficial fascia in the region of the sternum goes over to the medial brachial surface. Cranially and laterally this covers the mm. brachiocephalicus, deltoideus, and triceps brachii, on which it unites strongly with the deep leaf of the brachial fascia. It covers the m. pectoralis superficialis medially. The v. cephalica is covered laterally by the superficial fascia and, as it crosses the flexor surface of the elbow joint, it lifts the fascia into a fold. Beyond the elbow joint the closely applied superficial fascia is extremely delicate and less easily movable with respect to the deep tissue. On practical grounds one also differentiates the special superficial fascia of the forearm, carpus, metacarpus, and digits, under which the cutaneous vessels and cutaneous nerves extend over long distances before they actually enter the skin.

The **deep fascia** of the lateral surface of the shoulder and brachium is called the *fascia omobrachialis lateralis*. From the deep fascia of the neck it runs along the m. rhomboideus; the superficial leaf of the deep fascia of the back runs over the mm. latissimus dorsi and trapezius to the lateral surface of the shoulder. Here it firmly attaches to the spine of the scapula after it has covered the mm. infraspinatus, supraspinatus, deltoideus, and triceps brachii, or as much of these muscles as make their appearance laterally. It extends distally as far as the elbow joint, attaching, between the mm. deltoideus and brachialis on one side and the m. cleidobrachialis on the other, to the crest of the major tubercle of the humerus. Distally, where the m. brachialis is free in the triangle between the mm. triceps brachii and brachiocephalicus, the fascia is especially strong; here it extends under the cephalic vein and surrounds the muscle loosely with intermuscular septa under the m. brachiocephalicus. Farther medially it covers the m. biceps brachii. The deep fascia is thinner and more firmly attached to the long and lateral heads of the m. triceps brachii than it is elsewhere. The lateral omobrachial fascia, including the origin of the m. tensor fasciae antebrachii, passes over the m. cleidobrachialis into the *fascia omobrachialis medialis*, which arises from the inner surface of the mm. subscapularis and teres major, and passes distally over the m. triceps brachii and the medial surface of the humerus to merge into the deep antebrachial fascia at the elbow joint. The *fascia antebrachii* covers the muscles of the forearm as a closely applied tube which is thickest medially. Between the extensor and flexor muscles it sends septa to the periosteum of the radius and ulna; these septa enclose the individual muscles, as well as small groups of muscles.

The fascia is intimately united with the extensors, more loosely covers the flexors, and is firmly fused to all free portions of the bones of the antebrachium. In the distal antebrachial region it is intimately joined with the connective tissue found between the tendons. In the groove between the tendons of the mm. extensor and flexor carpi ulnaris proximal to the accessory carpal bone, the fascia invaginates more deeply. At the carpus it becomes the **fascia of the forepaw** (*fascia manus*), which, as the dorsal and palmar deep fascia, ensheathes all tendons and superficial muscles of the forepaw distally and attaches to all projecting parts of bones. Even the cushions of all of the pads are closely united with the deep fascia. From the cushion of the carpal pad an especially strong band of the fascia extends directly laterally and a little proximally toward the distal end of the ulna; on the medial side another such band, the *flexor retinaculum*, goes to the medial border of the carpus, bridging over the flexor tendons. On the dorsal surface of the carpus are found transverse supporting fibers. Finally, on the dorsal surface of the metacarpus, there is formed a triangular, fibrous leaf which extends from metacarpal I obliquely laterally and distally to the branches of the common digital extensor tendon for digits II, III, and IV. Palmarly, on the metacarpophalangeal joints, thickened portions of the deep fascia (transverse ligaments) attach primarily to the sesamoid bones.

MUSCLES OF THE PELVIC LIMB

MUSCLES OF THE LOIN, HIP, AND THIGH

The pelvis and thigh are covered on all sides by muscles which are for the most part common to both of these body regions, so that the two groups cannot be sharply differentiated. The so-called hip muscles act mostly on the hip joint, but a few act also on the sacroiliac joint. The muscles of the thigh act primarily on the knee joint. The loin and hip muscles are divided into three groups. The **sublumbar** or inner loin muscles lie on the ventral surfaces of the lumbar vertebrae and ilium: mm. psoas minor, iliopsoas, and quadratus lumborum. The **rump** muscles lie on the lateral side of the pelvis: mm. gluteus superficialis, medius, and profundus, piriformis, and tensor fasciae latae. The **inner pelvic** muscles are in part inside the pelvis: mm. obturator internus, gemelli, and quadratus femoris.

The muscles of the thigh are designated as to their cranial, caudal, and medial positions. To the caudal group belong the mm. biceps femoris, semitendinosus, and semimembranosus (sometimes called the hamstring muscles). They form a strong, fleshy mass on the caudal side of the femur. The m. biceps femoris is lateral, the m. semitendinosus caudal, and the m. semimembranosus medial. To the cranial group belongs the m. quadriceps femoris, which forms the fleshy foundation of the cranial half of the thigh. A small m. capsularis and the m. sartorius cranialis are associated with it. The medial group includes the mm. gracilis, pectineus, adductores, obturator externus, and sartorius caudalis.

The sublumbar muscles. The sublumbar muscles arise on the ventral surfaces of the caudal thoracic and lumbar vertebrae and insert on the os coxae and femur; they lie on one another in several layers.

The *m. psoas minor* (Figs. 6–70 to 6–73) runs toward the pelvis ventromedially as it lies between the iliac fascia and peritoneum ventrally and the mm. iliopsoas and quadratus lumborum dorsally. Its muscular belly is weaker than that of the m. iliopsoas. It arises from the tendinous fascia of the m. quadratus lumborum at the level of the last thoracic vertebrae and on the ventral surface of the last thoracic vertebra and the first four or five lumbar vertebrae; it is separated from its fellow of the opposite side by an interval gradually increasing caudally so that the bodies of the lumbar vertebrae become visible. The strong, flat tendon fused with the iliac fascia comes out of a shining, tendinous leaf at the fifth lumbar vertebrae; it runs to the arcuate line and inserts on this line as far as the iliopectineal eminence.

Action: To steepen the pelvis, or to flex the lumbar part of the vertebral column.

Innervation: Lateral branches of the rami ventrales of lumbar nerves 1 to 4 or 5.

The *m. iliopsoas* (Figs. 6–71, 6–75) represents a fusion of the *m. psoas major* and the *m. iliacus*. It lies ventral to the m. qudratus lumborum and dorsal to the m. psoas minor, which it covers laterally and also medially

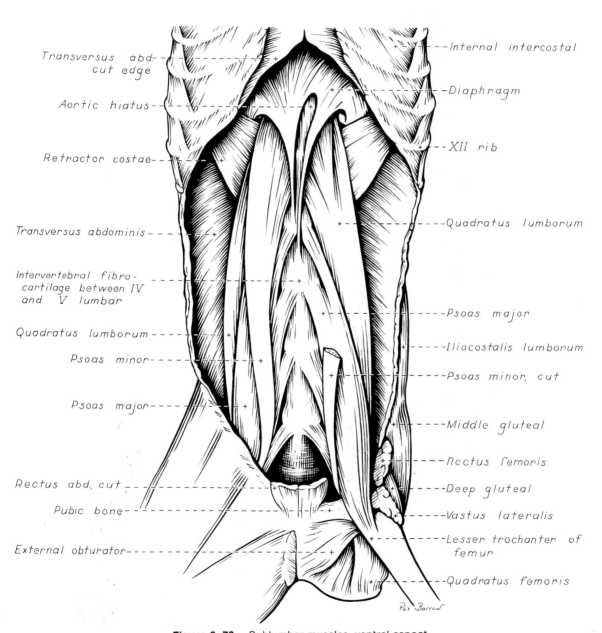

Transversus abd-
cut edge

Aortic hiatus

Retractor costae

Transversus abdominis

Intervertebral fibro-
cartilage between IV
and V lumbar

Quadratus lumborum

Psoas minor

Psoas major

Rectus abd., cut

Pubic bone

External obturator

Internal intercostal

Diaphragm

XII rib

Quadratus lumborum

Psoas major

Iliocostalis lumborum

Psoas minor, cut

Middle gluteal

Rectus femoris

Deep gluteal

Vastus lateralis

Lesser trochanter of
femur

Quadratus femoris

Pat Barrow

Figure 6–70. Sublumbar muscles, ventral aspect.

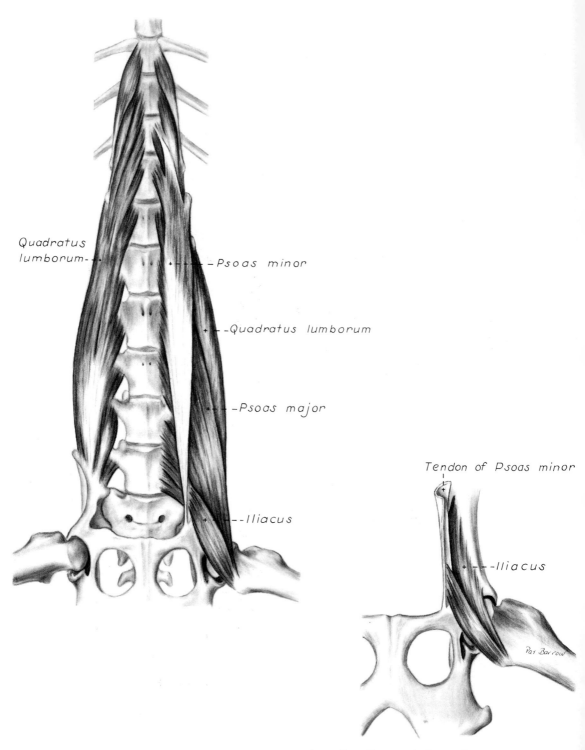

Figure 6-71. Sublumbar muscles, deep dissection, ventral aspect.

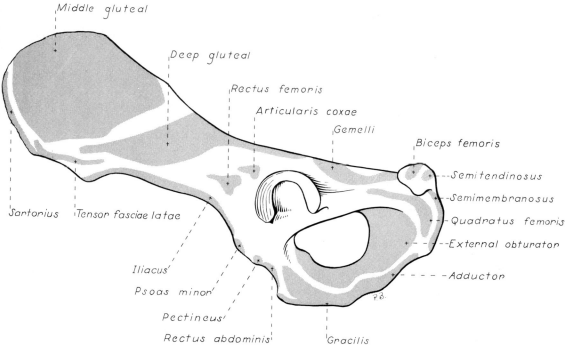

Figure 6–72.   Left os coxae, showing areas of muscle attachment, lateral aspect.

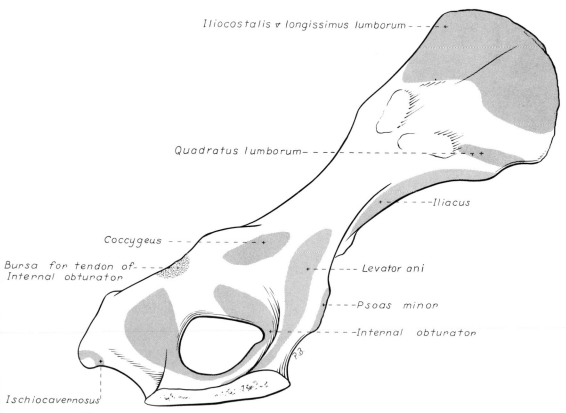

Figure 6–73.   Left os coxae, showing areas of muscle attachment, medial aspect.

caudal to its point of transition into tendon. It is narrow and tendinous at its origin on the transverse processes of lumbar vertebrae 2 and 3, where it lies lateral to the m. quadratus lumborum. It also attaches by means of the ventral aponeurosis of this muscle on lumbar vertebrae 3 and 4, and, finally, on the ventral and lateral surfaces of lumbar vertebrae 4 to 7. After this portion of the m. iliopsoas passes over the ilium as the m. psoas major it receives the m. iliacus from the smooth ventral surface of the ilium between the arcuate line and the lateral border of the ilium. The two muscle masses (m. psoas major and m. iliacus) composing the m. iliopsoas can be easily isolated. The m. iliopsoas attaches to the trochanter minor of the femur.

Action: To draw the pelvic limb forward by flexion of the hip joint; when the femur is fixed in position, flexion and fixation of the vertebral column; when the leg is extended backward, it draws the trunk backward.

Innervation: Branches of the rami ventrales of the lumbar nerves.

The *m. quadratus lumborum* (Figs. 6–70, 6–71, 6–73) is the most dorsal of the sublumbar muscles. It lies directly ventral to the bodies of the last three thoracic and all the lumbar vertebrae, as well as under the proximal portions of the last two ribs and the transverse processes of the lumbar vertebrae. It is covered ventrally from the first lumbar vertebra by the m. psoas minor and also, from the fourth lumbar vertebrae, by the m. psoas major. It has a thoracic and a lumbar portion. The thoracic portion of this muscle, which is rather strong in the dog, consists of incompletely isolated bundles which become tendinous; these bundles extend more or less distinctly caudolaterally from the bodies of the last three thoracic vertebrae to the transverse processes of the lumbar vertebrae as far as the seventh; it is covered by tendinous leaves dorsally and ventrally. It ends on the medial surface of the wing of the ilium between the articular surface and the caudal ventral iliac spine. The lateral portion of this muscle overhangs the transverse processes of the lumbar vertebrae, so that it also comes to lie on the ventral surface of the tendon of origin of the m. transversus abdominis.

Action: Fixation of the lumbar vertebral column.

Innervation: Rami of the ventral branches of the lumbar nerves.

The rump muscles. The rump muscles extend between the ilium and the thigh; they are arranged in several layers.

The *m. tensor fasciae latae* (Figs. 6–72, 6–74) is a triangular muscle which attaches proximally to the tuber coxae. With three more or less distinct slips, it radiates distally and caudally over the m. quadriceps femoris. It arises (partly superficially and partly deeply) from the aponeurosis of the m. gluteus medius and deeply from the tuber coxae. Laterally it covers the origin of the caudal belly of the m. sartorius. Distally it continues into the deep leaf of the fascia lata on a horizontal line running cranially from the trochanter major. Over this deep aponeurotic leaf runs the m. biceps femoris and a superficial leaf of this same fascia. Since this fascia runs over the m. quadriceps femoris to the patella, this muscle extends the stifle.

Action: Tension of the fascia lata, and thus flexion of the hip joint; extension of the stifle joint.

Innervation: N. gluteus cranialis.

The *m. gluteus superficialis* (Figs. 6–74, 6–75, 6–77), the most superficial gluteal muscle, is a rather small, flat, almost rectangular muscle. It extends between the sacrum and first caudal vertebra proximally and the trochanter major distally. The gluteal fascia covers this muscle loosely; it fuses with the muscle more intimately only at the proximal two-thirds of its cranial portion. The muscle arises from the gluteal fascia and thereby from the tuber sacrale of the ilium; neighboring portions come from the caudal fascia. The thick caudal portions come from the lateral part of the sacrum, from the first caudal vertebra, and from more than half of the proximal part of the sacrotuberous ligament. Its fibers converge distally and laterally and become tendinous; the tendon runs over the trochanter major and inserts on the weak trochanter tertius. This tendon fuses with the aponeurosis of the m. tensor fasciae latae. The m. gluteus superficialis covers portions of the mm. gluteus medius and piriformis, and also the sacrotuberous ligament. In large dogs it is 5 to 7 cm. wide and more than 1 cm. thick caudally. A thin, deep portion was seen

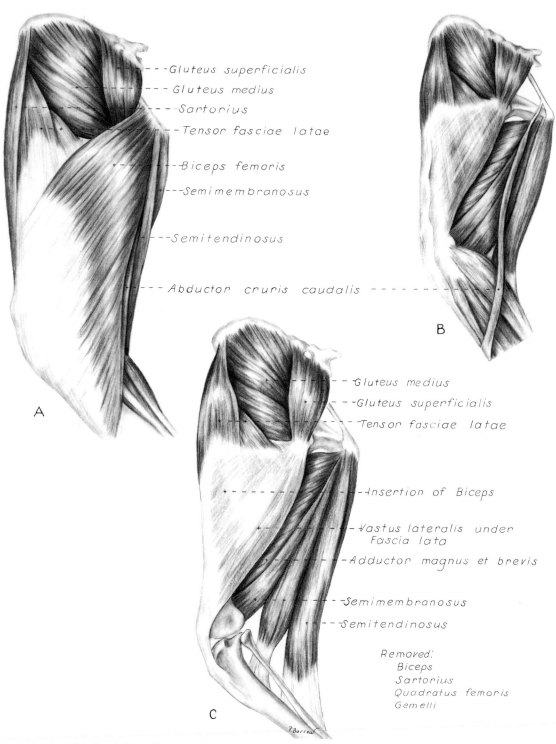

- - -Gluteus superficialis
- - - Gluteus medius
- - -Sartorius
- - -Tensor fasciae latae

- -Biceps femoris
- - -Semimembranosus

- - -Semitendinosus

- - -Abductor cruris caudalis - - - - - - - - - -

A

B

- - -Gluteus medius
- - -Gluteus superficialis
- - -Tensor fasciae latae

-Insertion of Biceps

-Vastus lateralis under
Fascia lata

-Adductor magnus et brevis

- -Semimembranosus

- - -Semitendinosus

Removed:
Biceps
Sartorius
Quadratus femoris
Gemelli

C

P.Barrow

Figure 6–74. Muscles of thigh.
A. Superficial muscles, lateral aspect.
B. Superficial muscles, lateral aspect. (Biceps femoris removed.)
C. Deep muscles, lateral aspect.

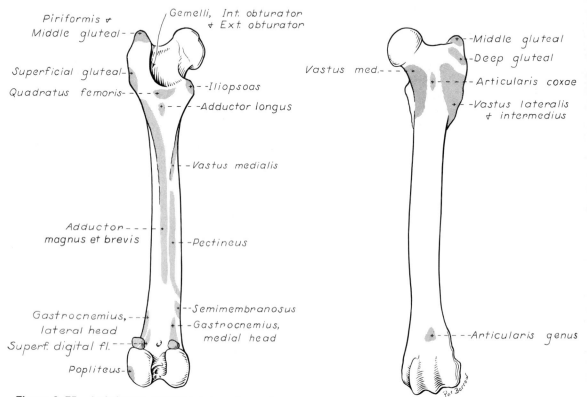

Figure 6–75. Left femur, showing areas of muscle attachment, caudal aspect.

Figure 6–76. Left femur, showing areas of muscle attachment, cranial aspect.

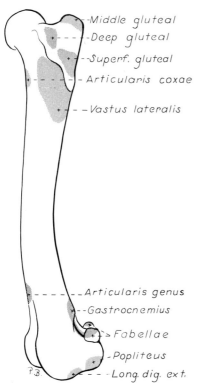

Figure 6-77. Left femur, showing areas of muscle attachment, lateral aspect.

by Ziegler (1934), lying under the caudal border of the muscle and having a special origin on the sacrotuberous ligament. Beneath its terminal tendon on the trochanter major, in about one-third of the specimens, there is a synovial bursa approximately 1 cm. wide.

Action: Extension of the hip joint.

Innervation: N. gluteus caudalis.

The strong *m. gluteus medius* (Figs. 6-72, 6-74, 6-75, 6-78) lies on the gluteal surface of the ilium, from which it takes its principal origin. It also arises from the iliac crest and from both angles (tuberosities) of the ilium. Some fibers also come from the dorsal sacro-iliac ligament and from the deep surface of the gluteal fascia, which is fused with the muscle caudal to the iliac crest. A large part of the muscle lies under the gluteal fascia and skin, and only caudally is it covered by the superficial gluteal muscle. In a craniodistal direction it extends over the m. gluteus profundus and ends by a short, strong tendon on the free end of the trochanter major. The muscle is 2.5 to 3.5 cm thick and 7 to 9 cm. wide. From the caudal border of the m. gluteus medius there is split off a narrow but strong, deep belly; this does not, however,

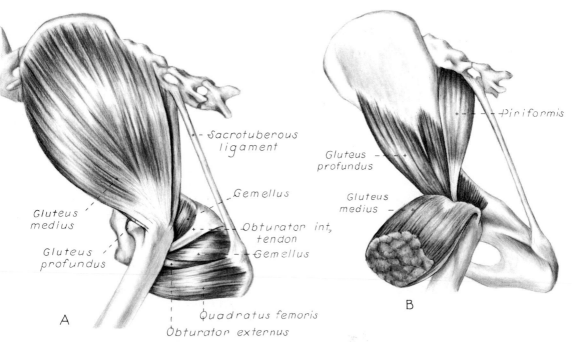

Figure 6-78. Muscles of the gluteal region.
A. Superficial muscles.
B. Deep dissection.

correspond to the m. gluteus accessorius of other animals. It arises on the transverse processes of the last sacral and first caudal vertebrae and on the sacrotuberous ligament, and ends on the trochanter major by a narrow, strong tendon which sinks into the principal tendon.

Action: Extension of the hip joint.

Innervation: N. gluteus cranialis.

The *m. piriformis* (Figs 6–78, 6–88) lies caudal and medial to the m. gluteus medius and is completely covered by the m. gluteus superficialis. The m. piriformis arises on the lateral surface of the third sacral and first caudal vertebrae. Its tendon of insertion joins that of the m. gluteus medius on the major trochanter. In the previous edition of this book the m. piriformis was confused with a deep slip of the m. superficialis, and what was called the caudal portion of the middle gluteal was in fact the piriformis. For an illustrated consideraton of this muscle and the gluteal region, see Henning (1965).

Action: Extension of the hip joint.

Innervation: N. gluteus caudalis.

The strong, fan-shaped *m. gluteus profundus* (Figs. 6–72, 6–76, 6–80, 6–88) is the deepest of the gluteal muscles. It is completely covered by the mm. gluteus medius and piriformis; at the same time it extends, for a considerable distance, caudal to the deep portion of the gluteus medius. It takes its origin laterally from the shaft of the ilium near the ischiatic spine. Its fibers converge over the hip joint distolaterally and form a short, strong tendon which ends cranially on the trochanter major distal to the insertion of the m. gluteus medius. Underneath the tendon of insertion, there is often a small synovial bursa.

Action: Extension of the hip joint, with some abduction.

Innervation: N. gluteus cranialis.

The inner pelvic muscles. The so-called "small pelvic association" includes a group of short muscles which lie caudal to the m. gluteus profundus and the hip joint. They extend from the inner and outer surfaces of the ischium to the femur. These are the mm. obturator internus, the gemelli, and the quadratus femoris.

The *m. obturator internus* (Figs. 6–73, 6–75, 6–80) is the strongest of this group. As a fan-shaped muscle it covers the obturator

foramen internally. It arises medial to the foramen on the pelvic surfaces of the rami of the pubis and ischium, and from the ischiatic arch. Its fibers converge laterally and, extending under the sacrotuberous ligament, are drawn over the lesser sciatic notch in a distinct, gliding surface directly behind the ischiatic spine. At the same time, the muscle turns almost at right angles, distolaterally, and forms a strong, flat tendon which is embedded deeply between the edges of the broader mm. gemelli, which lie beneath. The overhanging portions of the mm. gemelli, bearing a shining distal tendinous sheet, run from the edges of the lesser sciatic notch into the obturator tendon, so that the triple tendinous apparatus ends undivided in the trochanteric fossa. Caudally the tendon of the m. obturator externus accompanies it. Where the muscle glides over the ramus of the ischium, there is a very thin-walled synovial bursa, 1.5 to 2 cm. wide, which surrounds the edges of the tendon and may extend out on the underlying mm. gemelli. A second bursa, 2 to 3 mm. wide, lies under the mm. obturator and gemelli tendons in the trochanteric fossa between the trochanter major and the joint capsule.

Action: Outward rotation of the hip joint.

Innervation: N. ischiadicus.

The *mm. gemelli* (Figs. 6–72, 6–75, 6–80) represent a muscle which has been formed by the fusion of two parts and which lies between the terminal portions of the mm. obturator internus and obturator externus caudal to the m. gluteus profundus behind the hip joint. It overhangs the tendon of the m. obturator internus cranially and caudally. The mm. gemelli arise together on the outer surface of the ramus of the ischium in the arch under the gliding surface for the m. obturator internus and are covered superficially by a shining fascia which forms the floor of the groove for the obturator tendon. Altogether this tendon apparatus ends undivided in the trochanteric fossa.

Action: Outward rotation of the hip joint.

Innervation: N. ischiadicus.

The short, fleshy *m. quadratus femoris* (Figs. 6–72, 6–75, 6–78) arises on the ventral surface of the ischium medial to the lateral angle of the ischial tuberosity. It is surrounded by the origins of the mm. adductor magnus et brevis, semimembranosus, semitendinosus, biceps femoris, and obturator externus. Medial to the m. biceps femoris, it extends in an almost sagittal direction cranially; it bends slightly laterally and runs distally to reach the distal portion of the trochanteric fossa between the mm. obturator externus and adductor magnus et brevis. It ends immediately above the trochanter tertius.

Action: Extension and outward rotation of the hip joint.

Innervation: N. ischiadicus.

The caudal muscles of the thigh. The caudal thigh muscles are grouped about the ischial tuberosity, and some of them run to the lateral side of the thigh — the mm. biceps femoris and abductor cruris caudalis; others run to the medial side of the thigh — the mm. semitendinosus and semimembranosus.

The *m. biceps femoris* (Figs. 6–72, 6–74, 6–84, 6–85, 6–89) is a large, long muscle lying in the lateral part of the buttock and thigh and extending from the region of the ischial tuberosity to the middle of the tibia. It arises by two unequal heads, a cranial, superficial one, and a caudal, much smaller, deep one. The principal superficial head arises from the ventrocaudal end of the sacrotuberous ligament and partly from the lateral angle of the ischial tuberosity, where it is firmly united with the m. semitendinosus in an intermuscular leaf. The fibers of this muscle run nearly parallel. New fiber bundles, which arise from a strong, tendinous leaf on the deep surface of the muscle, are added distally. A portion of these laterally located fibers can be differentiated from the main mass of the cranial head as the middle branch of the m. biceps femoris. The deep head, covered by the superficial portion, arises under the medial tendinous origin of the latter from the ventral side of the lateral angle of the ischiatic tuberosity by a long, strong tendon. The slender muscle, at the middle third of the femur, passes over the lateral surface of the thigh at the caudal edge of the principal portion; here it broadens to unite with the middle head as the caudal branch of the m. biceps femoris, at the level of the popliteal space. The *m. abductor cruris caudalis* runs along its free edge. In the region of the femorotibial joint the entire m. biceps femoris, with slightly diverging fibers, runs on the lateral surface of the vastus lateralis and the gastrocnemius group,

and, as an aponeurosis, radiates into the fascia lata and the fascia cruris. By means of this aponeurosis the cranial head, by fusing with the covering of the m. quadriceps femoris, attaches to the patella and the patellar ligament and by this to the tibial tuberosity. The middle and caudal branches radiate into the fascia lata and crural fascia as these enclose all of the cranial and lateral muscles of the thigh and crus to insert on the cranial border of the tibia (tibial crest). The deep surface of the muscle is covered by a distinct perimysium. The covering appears strengthened distally as a distinct tendon. Where the muscle passes over the deep surface of the m. abductor cruris caudalis (which accompanies the caudal border of the m. biceps femoris), the tendon reaches a width of 5 mm. It lies under the abductor and, by means of the crural fascia, runs distally along the m. gastrocnemius. It curves in front of the main part of the calcanean tendon to end on the medial tuberosity of the tuber calcanei after it has united with a thick, tendinous strand coming from the mm. semitendinosus and gracilis. The **calcanean tendon** *(tendo calcaneus communis)*, also known as **Achilles' tendon,** consists of all those structures attaching to the point of the heel, or the tuber calcanei of the calcaneus. The tendons of the mm. flexor digitorum superficialis and gastrocnemius are its main components, although the mm. biceps femoris, semitendinosus, and gracilis also contribute to its formation. From the proximal end of the strand which aids in forming the calcanean tendon, fibers pass to the medial lip of the femur. Heavy, connective tissue fibers of the interfascicular septa of the m. biceps femoris enter the tendon and, without the aid of muscle fiber attachment, fasten it intimately so that a single functional unit results.

Action: Extension of the hip, stifle, and tarsal joints. The caudal part of the muscle raises the crus when the extremity is not bearing weight; this part, therefore, flexes the stifle at these times.

Innervation: N. ischiadicus. Cranial part: ramus distalis of the n. gluteus caudalis. Middle and caudal parts: ramus muscularis proximalis of the n. tibialis (Skoda 1908, Ziegler 1934, Nickel, Schummer, and Sieferle 1954). This double innervation supports the theory that the m.

biceps femoris of the dog is a m. gluteobiceps; its cranial belly is to be regarded as split off from the m. gluteus superficialis.

The straplike *m. abductor cruris caudalis* (Figs. 6–74, 6–88, 6–89), 10 mm. wide and 1 mm. thick, lies under the caudal edge of the m. biceps femoris. It arises by a long, flat tendon, which lies on the aponeurosis of the deep surface of the m. biceps femoris on the ventrocaudal edge of the sacrotuberous ligament near the ischial tuberosity. It extends under both heads of the m. biceps femoris on the lateral surface of the mm. quadratus femoris, adductor, and semimembranosus. At the level of the popliteal space, it appears superficially between the mm. biceps femoris and semitendinosus and, keeping close to the edge of the m. biceps femoris, crosses the shank in an arc and goes into the crural fascia, where it often extends beyond the end of the m. biceps femoris to the digital extensors. Because of its different innervation, this muscle cannot be looked upon as a division of the m. biceps femoris.

Action: With the caudal branch of the m. biceps femoris, it abducts the limb.

Innervation: N. ischiadicus.

The *m. semitendinosus* (Figs. 6–72, 6–74, 6–84, 6–89) is 2.5 to 3.5 cm. thick, and has four edges in cross section. It lies in the caudal part of the thigh between the cranial and lateral parts of the m. biceps femoris and the medial and cranial parts of the m. semimembranosus. It extends in an arc between the ischial tuberosity and the proximal segment of the shank, where it forms a large part of the caudal contour of the thigh. Its distal end diverges from the m. biceps femoris to the medial side of the m. gastrocnemius. The m. semitendinosus arises on the caudal and ventrolateral parts of the lateral angle of the ischiatic tuberosity between the mm. biceps femoris and semimembranosus. It extends distally at the caudal edge of the m. biceps femoris and diverges from it at the popliteal space to the medial side of the shank, where it follows the m. semimembranosus. By means of a strong, flat tendon it passes under the aponeurosis of the m. gracilis. Both of these aponeuroses represent portions of the crural fascia and, as such, attach to the medial surface of the tibia in front of the flexor muscles. Separate strands of both tendons, however, extend in the fascia in an arc cranially and upward toward the rough surface on the distal end of the cranial border of the tibia. From the caudal edge of the tendon of the m. semitendinosus (close to its origin) a tendon is separated which unites with a distinct, strong strand from the tendon of the m. gracilis. The conjoined tendons extend on the medial surface of the m. gastrocnemius medialis to the calcanean tendon and thus to the tuber calcanei. It is related to the corresponding tendon of the m. biceps femoris. The strands are attached to each other caudal to the main part of the calcanean tendon and the digital flexor tendons by the fascial bridge, which distally becomes progressively thicker.

Between the proximal and the middle third of the semitendinosus, and visible on the free surface, there is a delicate, but complete, tendinous inscription. This intercalation is probably the remains of the tendon of the m. caudofemoralis, which in lower mammals divides the semitendinosus transversely. This muscle has disappeared but its tendon remains as the inscription; the proximal portion of the muscle is to be regarded as a division of the m. gluteus superficialis. The two portions of the m. semitendinosus are provided with individual nerves.

Action: Extension of the hip, stifle, and tarsal joints; flexion of the stifle joint in the free non-weight-bearing limb.

Innervation: Ramus muscularis proximalis of the n. tibialis (special branches for the part found proximal to and the part found distal to the tendinous inscription).

The *m. semimembranosus* (Figs. 6–72, 6–75, 6–82, 6–87, 6–89) is entirely fleshy, has parallel fibers, and is oval to triangular in cross section. It crosses the m. semitendinosus medially and lies between the mm. biceps femoris and semitendinosus laterally, and the mm. adductor and gracilis medially. It arises caudal and medial to the m. semitendinosus on the lower surface of the rough portion of the tuber ischiadicum. It soon splits into two equally strong bellies which extend in a slight arch to the medial side of the stifle joint. The cranial belly, which is 3 to 3.5 cm. thick, joins the m. adductor magnus et brevis and ends in a short, flat tendon on the aponeurosis of origin of the m. gastrocnemius. It attaches also to the distal end of the medial lip of the femur, which runs distally on the shaft as a rough line to the medial condyle. The caudal belly partly covers the cranial belly from the lateral side. It goes into a somewhat longer, narrower tendon, which extends more distally than the m. semitendinosus over the m. gastrocnemius medialis to end under the tibial collateral ligament of the femorotibial joint on the margin of the medial condyle of the tibia.

Action: Extension of the hip and stifle joints with the stifle adducted.

Innervation: Ramus muscularis proximalis of the n. tibialis.

The cranial muscles of the thigh. The cranial thigh muscles extend between the pelvis and the femur proximally and the patella and tibial tuberosity distally. The four subdivisions of the m. quadriceps femoris form the bulk of the group. To this is added the insignificant muscle of the joint capsule of the hip, the m. capsularis, and the m. articularis genus proximal to the stifle joint.

The strong *m. quadriceps femoris* (Figs. 6–74 to 6–77, 6–81, 6–82, 6–84, 6–89) covers the femur cranially, laterally, and medially. Distally it forms a tendon which includes the patella within it and ends on the tibial tuberosity as the straight ligament of the patella; it fuses with the fascia lata and thereby with the aponeurosis of the mm. biceps femoris and sartorius. The quadriceps muscle consists of the rectus femoris (cranial), vastus lateralis (lateral), vastus medialis (medial), and sometimes a fourth belly, the vastus intermedius, directly underneath the rectus femoris and covering the cranial surface of the femur.

The *m. rectus femoris*, 2 to 3 cm. thick in large dogs, is round in cross section proximally and is laterally compressed distally; it is enclosed between the vasti in such a way that it overhangs them somewhat cranially. It arises by a short, strong tendon from the iliopubic eminence of the ilium, and appears between the m. sartorius and the m. tensor fasciae latae. Covered cranially and medially by the cranial belly of the m. tensor fasciae latae, it extends between the vastus lateralis and vastus medialis to the patella, which is included in its strong tendon as a sesamoid bone. This tendon continues distally as the straight ligament of the patella, over the stifle joint, to insert on the tibial tuberosity. Proximal to and at the sides of the patella,

islands of cartilage are found buried in the tendon. (For details, see the description of the stifle joint in Chapter 5.) The tendons of the vastus lateralis, vastus medialis, and the m. tensor fasciae latae fuse intimately with the patellar portion of the rectus tendon. Each collateral surface of the rectus bears a strong, tendinous leaf; the lateral surface takes on tendinous strands from the vastus lateralis.

The *m. vastus medialis*, 4 to 5 cm. wide and up to 2.5 cm. thick, arises in the proximal fifth of the femur on the intertrochanteric crest cranially. It arises medially on the line for the vastus medialis and, somewhat farther distally, on the proximal protion of the medial lip. It is laterally compressed and, with portions of the vastus intermedius, covers the distal portion of the rectus medially. Laterally and cranially it bears strong distal tendinous leaves; medially it bears a strong proximal one. Between these two tendinous systems the muscle fibers extend rather obliquely. Only above the patella do its tendinous elements fuse with those of the rectus femoris. The muscle ends on the patella, covered by the cranial belly of the m. tensor fasciae latae, and its principal portions encroach upon the mm. sartorius and pectineus. The vastus intermedius sends many fibers into the tendinous leaf.

The stronger *m. vastus lateralis* arises on the craniolateral part of the proximal fifth of the femur on the transverse line and on the line of the vastus lateralis to the lateral lip. It covers the rectus femoris laterally to some extent. Laterally, at the origin, it bears a strong tendinous leaf from which the majority of the muscle fibers extend obliquely medially and forward. The upper fibers go to the rectus femoris and radiate into its lateral aponeurotic covering as far as the terminal tendon. The vastus lateralis is inseparably united with the rectus femoris and its terminal tendon except proximally. The caudal fibers of the muscle are more longitudinal in direction; they extend from the lateral lip rather directly to the stifle joint, where their tendinous portion unites intimately with the joint capsule. There are firm connections with the fascia lata and with a strong aponeurotic leaf which pushes between the vastus lateralis and the m. biceps femoris to attach to the bone.

The *m. vastus intermedius* is the weakest

portion of the quadriceps. It arises with the vastus lateralis, which covers it, and also from the lateral part of the proximal fourth of the femur. Lying under the rectus, directly on the femur, its terminal tendinous leaf radiates into the m. vastus medialis.

Bursae. Under the tendon of origin of the m. rectus femoris, a small, synovial bursa is occasionally found. There is almost constantly a bursa (0.5 to 1.5 cm. in diameter) between the distal third of the muscle and the femur. A small (0.5 cm.) bursa is usually present under the terminal tendon of the vastus medialis and the vastus lateralis. In addition, the patellar joint capsule has a considerable proximal out-pocketing under the tendon of the quadriceps.

Action of the m. quadriceps: Extension of the stifle joint, tension of the fascia cruris.

Innervation: N. femoralis.

The *m. articularis coxae* (Figs. 6–72, 6–76, 6–79, 6–83) is a small, spindle-shaped muscle 2 to 4 cm. long in the region of the hip joint. It is placed laterally and caudally to the rectus femoris, and is covered laterally by the cranial border of the m. gluteus profundus. It arises with the rectus femoris on the iliopubic eminence and extends cranially and laterally over the capsule of the hip joint to the neck of the femur, where it attaches to the common ridge between the lateral and the medial vastus muscles.

Action: Flexion of the hip joint.

Innervation: N. femoralis.

The *m. articularis genus* (Fig. 6–76) is a small short muscle which arises from the cranial surface of the femur just proximal to the trochlea for the patella. It inserts on the tibial tuberosity. It is separated from the main portion of the m. quadriceps femoris by a delicate intermuscular septum. It is closely related to the proximal pouch of the stifle joint capsule as it courses distally.

Action: Extension of the stifle joint and possibly tension of the proximal pouch of the stifle joint capsule.

Innervation: N. femoralis.

The medial muscles of the thigh. The adductors are a strong group of muscles on the medial side of the thigh. Proximally, the long

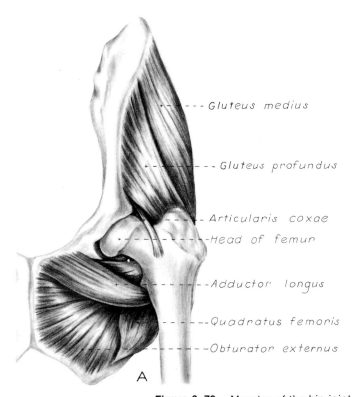
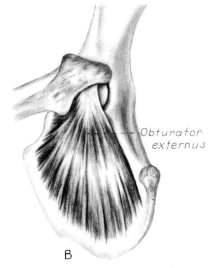

A
- - - - - Gluteus medius

- - - - - Gluteus profundus

- - Articularis coxae

-Head of femur

- - - - -Adductor longus

- - - - -Quadratus femoris

- - - - -Obturator externus

B
—Obturator externus

Figure 6–79. Muscles of the hip joint.
A. Ventral aspect.
B. Obturator externus, lateral aspect.

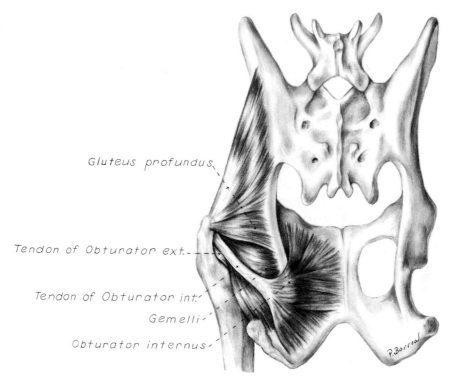

Gluteus profundus

Tendon of Obturator ext.

Tendon of Obturator int.

Gemelli

Obturator internus

P.Barr

Figure 6–80.   Muscles of the hip joint, dorsal aspect.

belly of the m. tensor fasciae latae lies on the medial side; distally, the m. semitendinosus takes part in the caudal border of the medial surface of the thigh. The adductors are arranged in superficial and deep layers. The superficial group includes the mm. sartorius and gracilis; the deep group is represented by the mm. pectineus, adductor, and obturator externus. Between the m. sartorius and the m. gracilis a broad gap exists proximally. In its depth the mm. adductor and pectineus can be seen caudally; the rectus femoris and vastus medialis cranially. Superficially, this gap is closed by the medial femoral fascia.

The *m. sartorius* (Figs. 6–72, 6–81, 6–84, 6–88, 6–89) is a long, flat muscle which extends in two, peculiar, straplike strands, each 3 to 4 cm. wide on the cranial contour of the thigh from the region of the tuber coxae to the medial surface of the stifle joint.

The **cranial part** arises on the iliac crest and the cranial ventral iliac spine, as well as from the lumbodorsal fascia. Along with the caudal belly it bounds the m. quadriceps femoris cranially and medially, and with the m. tensor fasciae latae it also bounds the

muscle laterally. The cranial belly of the m. sartorius is visible proximally on the lateral aspect of the thigh in front of the m. tensor fasciae latae. From the cranial border it slowly turns to the medial surface of the thigh, to pass into the medial femoral fascia immediately above the patella. There is a firm union with the tendon of the rectus femoris and the vastus medialis.

The **caudal part** lies close beside the cranial belly, entirely on the medial surface of the thigh. It arises on the bony ridge between the two ventral spines of the ilium, between the m. iliopsoas and the lumbar portion of the m. iliocostalis on the one side, and the mm. tensor fasciae latae and gluteus medius on the other. It runs over the medial surfaces of the vastus medialis and stifle joint, and forms an aponeurosis which blends with that of the m. gracilis. Radiating into the crural fascia, it ends on the cranial border of the tibia.

Action: To flex the hip; to advance and adduct the free thigh.

Innervation: N. femoralis (cranial branch) for both bellies.

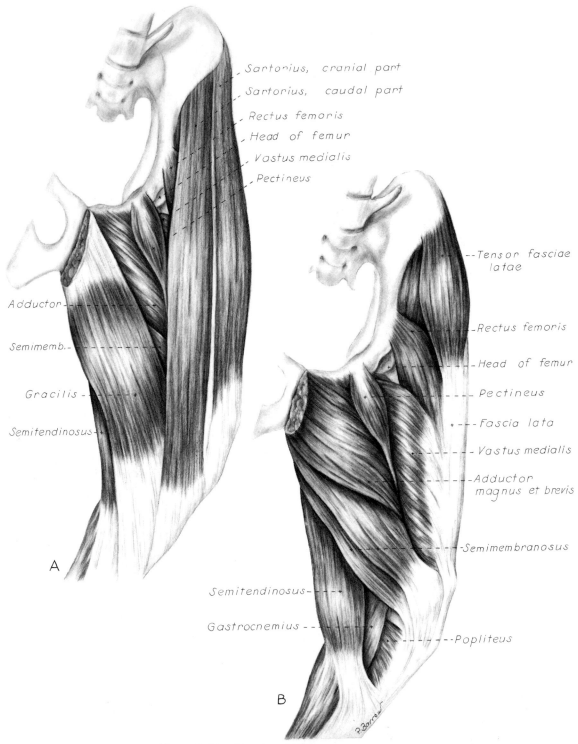

Sartorius, cranial part
Sartorius, caudal part
Rectus femoris
Head of femur
Vastus medialis
Pectineus

Adductor
Semimemb.
Gracilis
Semitendinosus

A

Tensor fasciae latae
Rectus femoris
Head of femur
Pectineus
Fascia lata
Vastus medialis
Adductor magnus et brevis
Semimembranosus

Semitendinosus
Gastrocnemius
Popliteus

B

Figure 6–81. Muscles of thigh.
A. Superficial muscles, medial aspect.
B. Deep muscles, medial aspect.

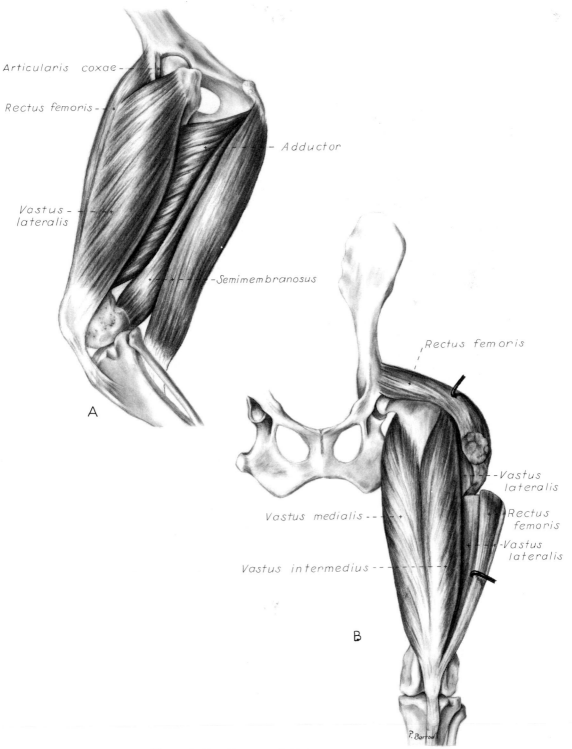

Articularis coxae

Rectus femoris

Adductor

Vastus lateralis

Semimembranosus

A

Rectus femoris

Vastus lateralis

Vastus medialis

Rectus femoris

Vastus lateralis

Vastus intermedius

B

P. Barrow

Figure 6–82. Muscles of thigh.
A. Deep muscles, lateral aspect.
B. Deep muscles, cranial aspect.

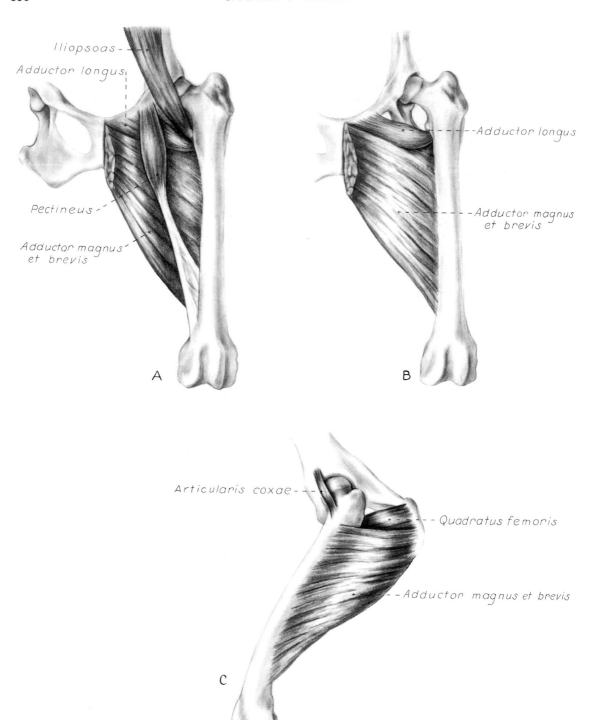

Figure 6–83. Muscles of thigh.
A. Mm. adductor magnus et brevis, adductor longus, pectineus, and iliopsoas, cranial aspect.
B. Mm. adductor magnus et brevis and adductor longus, cranial aspect.
C. Mm. adductor magnus et brevis, quadratus femoris, and articularis coxae, lateral aspect.

The *m. gracilis* (Figs. 6–72, 6–81, 6–84, 6–89), in the dog, forms an extensive, broad muscular sheet which is found in the superficial layer of the caudal portion of the inner surface of the thigh. Caudally the muscle becomes rather thick, and here it can be seen to a small extent from the lateral aspect. Its aponeurosis covers the medial surface of the m. adductor magnus et brevis. The muscle arises from the pelvic symphysis via a symphysial tendon (tendo symphysialis) which also serves as the origin for the m. adductor. The line of origin for the m. gracilis presents a distally convex arc which passes from the pecten ossis pubis to the ischial arch. Tendinous fibers extend from one side of this median, unpaired tendinous plate to the other. The m. gracilis passes over the m. adductor magnus et brevis as well as both bellies of the m. semimembranosus. At approximately the edge of this latter muscle it becomes a flat tendon which pushes under the m. sartorius, passes over the popliteal space with the m. gastrocnemius medialis and the end of the m. semitendinosus, to end along the entire length of the cranial border of the tibia. This end aponeurosis also spreads out into the crural fascia, and from its caudal border it sends a well-developed reinforcing band to the calcanean tendon of the semitendinosus. According to Frandson and Davis (1955), the caudal part of the gracilis, the part which attaches to the tuber calcis after joining the calcanean tendon of the semitendinosus, may rupture in racing dogs. Their findings suggest that the caudal part of the gracilis is an important extensor of the tarsus in the racing greyhound.

Action: Adduction of the thigh, extension of the hip joint.

Innervation: N. obturatorius.

The *m. pectineus* (Figs. 6–72, 6–75, 6–81, 6–83, 6–89) belongs to the deeper group of adductors, although in front of the m. gracilis it encroaches in part upon the medial fascia of the thigh. It is closely applied to the m. adductor magnus et brevis cranially, but is separated from the m. sartorius by the femoral fascia and vessels. Its spindle-shaped body is circular in cross section and may reach a thickness of 2 or 3 cm. It arises tendinously on the prepubic tendon and from the abdominal muscles which sink into this tendon; it also has a fleshy origin on the iliopectineal eminence. Distally the muscle

becomes flatter after it has passed under the m. sartorius. It fills the narrow space between the vastus medialis and the adductor in such a way that its wide and long tendon coming from tendinous sheets on two surfaces of the muscle passes over to the caudal surface of the femur after turning obliquely. The medial border of the tendon is thickened and ends on the raised ridge of the medial femoral condyle in common with the cranial belly of the m. semimembranosus. The thinner, principal portion of the tendon goes into the periosteum on the popliteal surface of the femur medial to the insertion of the m. adductor magnus et brevis.

Action: Adduction of the thigh, outward rotation of the leg.

Innervation: N. obturatorius.

The *mm. adductores* (Figs. 6–72, 6–75, 6–81, 6–83, 6–89) form the caudal part of the deep group of muscles next to the m. pectineus. In the dog they are represented by two separable muscles: the small fusiform m. adductor longus, and the much stronger m. adductor magnus et brevis. Both extend distally and laterally from the pelvic symphysis to the femur. See Budras (1972) for an illustrated review of the homology of the mm. adductores and the m. pectineus in domestic mammals, including the dog.

The *m. adductor longus* encroaches upon the proximal part of the m. pectineus. It is completely covered by the other part of the adductor. From its origin on the pubic tubercle, it extends cranial to the m. obturator externus and inserts on the lateral lip of the femur near the third trochanter. Its tendon covers the end of the m. quadratus femoris.

The *m. adductor magnus et brevis* arises along the entire pelvic symphysis and on the neighboring parts of the ischiatic arch, but primarily from the subpubic tendon and pelvic symphysis. The strong muscle belly of the large portion of the adductor extends obliquely under the m. gracilis and accompanies the distal portion of the m. pectineus. Under the m. sartorius it presses so closely against the vastus medialis that the m. pectineus with its tendon is compressed to a thin leaf. Here the m. adductor magnus et brevis receives a glistening tendinous leaf which unites firmly with the tendon of the m. pectineus. The muscle ends over the whole length of the lateral lip, from the trochanter tertius to a place near the origin of the m. gastrocnemius laterale. Its tendon, with that of the m. pectineus, blends with the periosteum of the popliteal surface of the femur. Laterally the muscle is covered by the m. biceps femoris. It extends between the m. quadratus femoris and the m. semimembranosus to the caudal surface of the femur. From the broad muscle body of the adductor magnus et brevis a less distinct portion can regularly be separated.

Action: Adduction and extension of the hip joint.

Innervation: N. obturatorius.

The *m. obturator externus* (Figs. 6–72, 6–75, 6–79) is a fan-shaped muscle which covers the obturator foramen externally. It arises on the ventral surface of the pelvis adjacent to the pubic symphysis. It is separated from the ischial symphysis by the mm. adductores. On some of its pennate parts, the muscle body bears delicate, tendinous strands from the origin of the muscle. The fibers converging toward the trochanteric fossa appear under the ramus of the ischium and form a strong tendon. The m. quadratus femoris, which takes a more sagittal course, is crossed on its deep surface in such a way that a small part of the obturator externus appears between the mm. quadratus femoris and gemelli laterally. It ends between these in the trochanteric fossa.

Action: To rotate the limb outward.

Innervation: N. obturatorius.

THE FEMORAL TRIANGLE AND ASSOCIATED STRUCTURES

The opening caudal to the abdominal wall for the passage of the iliopsoas muscle and its contained femoral nerve is known as the **muscular lacuna** (*lacuna muscularis*). It is bounded laterally and caudally by the os coxae, medially by the rectus abdominis, and cranially by the **iliopectineal arch** (*arcus iliopectineus*). This arch, composed of blended iliac and transversalis fascia, separates the muscular lacuna from the vascular lacuna.

The **vascular lacuna** (*lacuna vasorum*) lies craniomedial to the muscular lacuna, separated from it by the iliopectineal arch. It is bounded cranially by the caudal muscular border of the internal abdominal oblique and the inguinal ligament and medially by the rectus abdominis. It is separated from the superficial inguinal ring, which lies only a few millimeters craniolateral to it, by the inguinal ligament. It contains the femoral artery and vein and the saphenous nerve. The transversalis fascia which surrounds the femoral vessels as they pass through the vascular lacuna forms the **femoral sheath.** The funnel-shaped femoral sheath of man is divided into three compartments: a lateral one for the artery, an intermediate one for the vein, and a medial one partly occupied by lymph vessels and fat. It is into this medial compartment, known as the **femoral canal** in man, that a hernia may occur. The base of the femoral canal, in the vascular lacuna, is the **femoral ring.** It is closed by a condensation of extraperitoneal connective tissue and covered internally by parietal peritoneum. The femoral canal in the dog is considered by some authors to include the sheath and its contents (lymph vessels, femoral artery and vein, and saphenous nerve).

The **femoral triangle** (*trigonum femorale*) is the shallow triangular space through which the femoral vessels run to and from the leg. It is bounded cranially by the thin caudal belly of the m. sartorius and caudally by the m. pectineus. Its floor or lateral boundary is formed proximally for a short distance by the m. iliopsoas. The m. pectineus and m. vastus medialis complete this boundary. Superficially, the femoral artery and vein and the lymphatics are covered by the medial femoral fascia and skin. Because of the superficial position of the artery in the femoral triangle it is a favorable site for taking the pulse of the dog.

MUSCLES OF THE CRUS (TRUE LEG, SHANK, OR GASKIN)

On the crus the muscles lie on the cranial, lateral, and caudal surfaces of the crural skeleton, whereas the medial surface is essentially left free. Flexor and extensor groups are not separated on the crus as they are on the antebrachium. Cranially and laterally are found extensors of the digital joints and flexors of the tarsus; caudally lie flexors of the digital joints and extensors of the tarsus. These functional muscle groups are mixed on the crus because the tarsal joint is set at an angle opposite to that of the digital joints. The tarsal joint has its flexor surface dorsally, whereas each of the digital joints has its extensor surface dorsally; therefore, the muscles lying over the dorsal surface must be flexors of the tarsus and extensors of the digital joints.

The craniolateral muscles of the crus. The flexors of the tarsal joint which lie on the craniolateral side of the crus are the mm. tibialis cranialis, peroneus (fibularis) longus, extensor digitorum longus, extensor digitorum lateralis, and extensor hallucis longus. They are separated topographically into two groups according to the predominating muscles, not according to their function. One is the long digital extensor group; the other is the peroneal group. The former includes the mm. tibialis cranialis, extensor digitorum longus, and extensor hallucis longus, which lie for the most part on the cranial aspect of the tibia. The fibular group includes the mm. peroneus longus, peroneus brevis, and extensor digitorum lateralis, which lie on the lateral part of the crural skeleton.

The *m. tibialis cranialis* (Figs. 6–85, 6–88, 6–90, 6–91, 6–96) is a superficial, strong, somewhat flattened muscle lying on the cranial surface of the crural skeleton. It arises medial to the sulcus muscularis on the cranial portion of the articular margin of the tibia and on the laterally arched edge of the cranial border of the tibia. From its origin the muscle, which is about 3 cm. wide, passes over the craniomedial surface of the crus, and near its distal third becomes a thin, flat tendon. This tendon extends obliquely over the tarsus to the rudiment of metatarsal I, which is very often fused with the first tarsal bone and to the proximal end of metatarsal II. The threadlike tendon of the m. extensor hallucis longus encroaches closely upon the lateral edge of this tendon throughout its extent as far as the metatarsus, both turning toward the medial edge of the tarsus. It becomes increasingly removed from the long digital extensor, as this passes to the metatarsus lateral to the axis of the foot. On the distal part of the tibia all three tendons are bridged over by the broad proximal extensor retinaculum which extends obliquely upward in a medial direction. The long digital extensor tendon passes under the distal extensor retinaculum as it crosses the tarsus. This is a collagenous loop coming from the fibular tarsal bone. The end of the tendon often shows variations which frequently are associated with variations in the m. extensor hal-

lucis longus (Grau 1932). The fascia is thickened into a strand on the medial side of the muscle which reaches from the cranial border of the tibia to the proximal extensor retinaculum, and farther, a strand continues to the proximal end of metatarsal III. In its entirety, this strand may, perhaps, be comparable to the m. peroneus tertius of the horse. The terminal tendons of the mm. tibialis cranialis and extensor hallucis longus are, according to Walter (1908), surrounded by a synovial sheath between the proximal extensor retinaculum and the middle of the tarsus.

Action: Rotation of the hindpaw in a lateral direction, extension of digit II.

Innervation: N. peroneus.

The *m. extensor digitorum longus* (Figs. 6–77, 6–90, 6–91, 6–94), spindle-shaped, and at most 2 to 2.5 cm. thick, lies in the group of digital extensors on the tibia between the m. tibialis cranialis and the m. peroneus longus. Proximally it is covered by the m. peroneus longus; distally it lies free medial and caudal to the m. tibialis cranialis. It arises in the extensor fossa on the lateral epicondyle of the femur and passes through the sulcus muscularis of the tibia. The muscle belly bears a strong distal, cranial, tendinous leaf toward which all fibers converge; near the tarsus it goes into its terminal tendon. This tendon runs along the tendon of the m. tibialis cranialis and that of the m. extensor hallucis longus, laterally. At the distal fourth of the crus it is held in place by the proximal extensor retinaculum. Toward the tarsus it diverges from the other tendons and at the bend of the tarsus it is fastened by a second weaker, transverse ligament. At the same time the extensor tendon, from the very beginning, is split into four branches; these extend distally along the dorsal surfaces of the metatarsal bones and digits II, III, IV, and V. Each ends on a distal phalanx after it has received the collateral branches of the m. interosseus just proximal to the proximal digital joint. As in the pectoral limb, the dorsal elastic ligaments attach with them on the distal phalanx. A broadened portion of the tendon of the m. extensor digitorum lateralis unites with the branch for the fifth digit on the proximal phalanx. At the level of the proximal digital joint there is embedded in each branch a small, dorsal sesamoid element; the corresponding sesamoid of the metatarsophalangeal joint is located in the joint capsule. The tendon of origin of the long digital extensor is underlaid on its deep surface by a pouch (3 to 4 cm. long) of the meniscotibial portion of the stifle joint capsule. This is the so-called capsular synovial bursa. Since this bursa extends slightly around the caudal border of the tendon, it is also spoken of as a synovial sheath. The bundle of the four tendons of the long digital extensor is surrounded by a synovial sheath which extends from the beginning of the tendon at the proximal extensor retinaculum to the distal border of the tarsus, where the branches diverge. Accordingly, the sheath passes under both transverse ligaments; the mesotendon is sent out from the lateral surface.

Action: Extension of the digits; flexion of the tarsus.

Innervation: N. peroneus.

The *m. extensor hallucis longus* (Figs. 6–84, 6–90, 6–91, 6–94) is a delicate muscle band, elliptical in cross section, and covered by the mm. extensor digitorum longus and peroneus longus. It lies directly on the tibia. It arises on the cranial border of the fibula between the proximal and the middle third, and on the interosseous membrane cranial to the m. peroneus brevis. The muscle extends obliquely distomedially under the long digital extensor, on the m. peroneus brevis and then on the tibia. It is accompanied by the a. and v. tibialis cranialis. Near the distal fourth of the crus it goes into a fine tendinous strand which continues between the long extensor tendon, laterally, and the tendon of the m. tibialis cranialis, medially. The tendinous strand of the m. extensor hallucis longus passes over the dorsal surfaces of the tarsus and metatarsal II to the metatarsophalangeal joint of the second digit, where it is lost in the fascia beside the corresponding branch of the long extensor tendon. It occasionally broadens into the aponeurosis of the metatarsus, and, exceptionally, it ends on the rudiment of the hallux (first digit). If the first

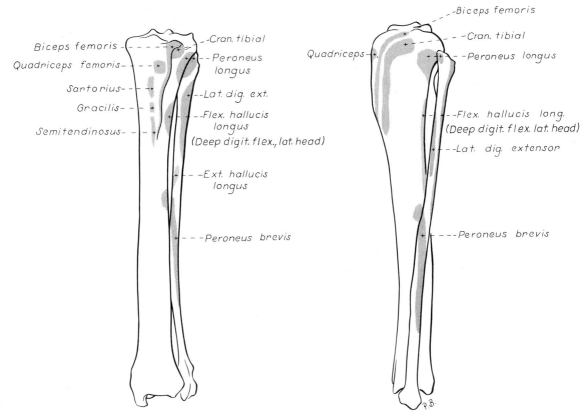

Figure 6–84. Left tibia and fibula, showing areas of muscle attachment, cranial aspect.

Figure 6–85. Left tibia and fibula, showing areas of muscle attachment, lateral aspect.

digit is present, the muscle may give off a tendinous branch to it. In about one-third of all specimens the muscle shows variations (see Grau 1932), which are often associated with variations in the m. tibialis cranialis. Often this muscle is stronger in smaller breeds than it is in larger ones. Occasionally the muscle is fused completely with the long digital extensor.

Action: Extension of digit II; extension of digit I also if it attaches to it.

Innervation: N. peroneus.

The short-bellied *m. peroneus (fibularis) longus* (Figs. 6–85, 6–90 to 6–93, 6–96) is the principal representative and the most superficial muscle of the fibular group. It lies in the proximal half on the lateral surface of the crus, between the m. tibialis cranialis and the m. flexor hallucis longus. It covers the proximal portions of the m. extensor digitorum lateralis and the m. peroneus brevis. In large dogs it may be 2 cm. thick. It arises on the lateral condyle of the tibia, the fibular collateral ligament of the femorotibial joint, and the proximal end of the fibula. Near the middle of the tibia it becomes an elliptical tendon which is enclosed in a tough fascial mass with the tendons of the mm. extensor digitorum lateralis and peroneus brevis. From the lateral surface of the tarsus it runs over the sulcus of the lateral malleolus. Running distally, cranial to the fibular collateral ligament, it crosses the tendon of the m. extensor digitorum lateralis and the m. peroneus brevis superficially; then, making use of the groove in the fourth tarsal, it passes in a sharp curve underneath the tendinous m. abductor digiti quinti to the plantar surface of the metatarsus, which it crosses transversely. It ends on the proximal end of metatarsals I, II, and V, and on the rudiment of the first digit opposite the insertion of the m. tibialis cranialis. The synovial sheath of the m. peroneus longus begins 3 to 4 cm. above the lateral malleolus and is provided with a mesotendon on its deep surface. It reaches approximately the end of the tarsus and may be divided. Beneath the plantar end of the tendon there is a synovial bursa which communicates with the joint capsule between the third and fourth tarsals.

Action: Rotation of the hindpaw so that the plantar surface faces laterally.

Innervation: N. peroneus.

The *m. extensor digitorum lateralis* (Figs. 6–84, 6–90 to 6–94) is the weakest muscle (1 cm. wide; 2 to 3 cm. thick) in the fibular group. It lies between the m. peroneus longus and the m. flexor hallucis longus on the m. peroneus brevis and fibula. It arises on the proximal third of the fibula. Superficially it has a strong distal, tendinous covering which, at the middle of the crus, becomes a thin tendon. This tendon lies between the cranially located tendon of the m. peroneus longus and the caudally located tendon of the m. peroneus brevis, as these cross the proximal lateral surface of the tarsus. As a rule the tendon of the lateral digital extensor lies in front of that of the m. peroneus brevis. Thus, the caudal groove of the lateral malleolus is crossed in common by these two tendons. Both the long fibular collateral ligament of the tarsal joint and the tendon of the m. peroneus longus are crossed obliquely on their deep sides by the tendon of the lateral digital extensor. It unites with the long digital extensor and the m. interosseus of the proximal phalanx of the fifth digit. The tendon of the lateral digital extensor is enclosed in a synovial sheath, having a medial mesotendon in common with that of the m. peroneus brevis. In large dogs this begins 1.5 to 2.5 cm. above the lateral malleolus, reaches almost to the middle of the tarsus, and almost always communicates with the capsule of the talocrural joint.

Action: Extension and abduction of digit V.

Innervation: N. peroneus.

The *m. peroneus (fibularis) brevis* (Figs. 6–85, 6–91, 6–93, 6–96) is the deepest muscle of the fibular group. It first appears distally under the lateral digital extensor between the m. peroneus longus and the m. flexor hallucis longus. It arises from the lateral surface of the distal two-thirds of the fibula and tibia, almost as far as the lateral malleolus. Here the muscle becomes completely tendinous, although the relatively strong tendon begins far proximad as a narrow, strong leaf. After crossing under the long fibular collateral ligament of the tarsus and the tendon of the m. peroneus longus, it runs farther between the m. peroneus longus and the tendon of the lateral digital extensor, to attach to the proximal end of metatarsal V. The tendon of the peroneus brevis is included in a common synovial sheath with the tendon of the m. extensor digitorum lateralis.

An insignificant bursa lies under the tendon at its insertion.

Action: Flexion of the tarsal joint.

Innervation: N. peroneus.

The caudal muscles of the crus. On the caudal side of the crus lie the extensor of the tarsal joint — the m. gastrocnemius; the flexors of the digital joints — the mm. flexor digitorum superficialis and profundus; and the flexor of the knee joint — the m. popliteus. The m. gastrocnemius is superficial and largely encloses the m. flexor digitorum superficialis. It covers the deep, digital flexor group consisting of the mm. flexor digitorum profundus, tibialis caudalis, and popliteus.

The *m. triceps surae* is a term used for the combined gastrocnemii and soleus when present.

The *m. gastrocnemius* (Figs. 6–75, 6–88, 6–90 to 6–93, 6–96) is divided into a lateral and a medial head covered by strong tendinous leaves and infiltrated by tendinous strands. The **lateral head** (*caput laterale*) arises by a large tendon on the lateral supracondylar tuberosity of the femur. The **medial head** (*caput mediale*) arises on the medial supracondylar tuberosity.

Each tendon of origin has a prominent sesamoid bone, the lateral and the medial fabella, respectively; these are cartilage covered on the side toward the femur, and they articulate with the corresponding condyle on a small, flat area. The two heads of the m. gastrocnemius almost completely enclose the m. flexor digitorum superficialis; they fuse with each other distally, forming a flat (in large dogs 5 to 6 cm. wide) muscle, the duplicity of which is accentuated distally by a middle tendinous plate. After crossing the superficial flexor tendon laterally, the tendon of the m. gastrocnemius attains its deep surface and ends on the tuber calcenei. Proximally the muscle is covered laterally by the m. biceps femoris and medially by the mm. semitendinosus, semimembranosus, and gracilis. Distally it lies under the fascia and skin. Beneath the m. gastrocnemius, directly on the tibia and fibula, are located the m. popliteus and the heads of the m. flexor digitorum profundus.

The **common calcanean tendon** (*tendo calcaneus communis*) is the aggregate of those structures which attach to the tuber calcanei. The tendon of the m. gastrocnemius, the main component, is crossed medially by that of the superficial digital flexor, which first lies cranial to the m. gastrocnemius but attains its caudal surface at the tuber calcanei. Joining these two tendons are those of the mm. biceps femoris, semitendinosus, and gracilis.

Action: Extension of the tarsal joint, flexion of the stifle joint.

Innervation: N. tibialis.

The *m. flexor digitorum superficialis* (Figs. 6–75, 6–90 to 6–93), infiltrated by tendinous strands, lies on the mm. flexor digitorum profundus and popliteus. It is enclosed to a great extent by the heads of the m. gastrocnemius and is compressed to an angular body which flattens distally and becomes broader. A small portion of the proximal segment and a narrow portion of the distal border appear beyond the m. gastrocnemius and thus encroach upon the fascia. The muscle is multipennate, because of the numerous tendinous folds which course through it. Proximally, it is firmly united with the lateral head of the m. gastrocnemius. It arises with the gastrocnemius on the lateral supracondylar tuberosity of the femur and on the lateral sesamoid. At the middle of the tibia the tendon of the m. flexor digitorum superficialis winds medially around the tendon of the m. gastrocnemius to gain its caudal surface. On the tuber calcanei it broadens like a cap and inserts collaterally on the tuber calcanei, united with the crural fascia and the calcanean tendon of the mm. biceps femoris and semitendinosus. The tendon divides twice, at the distal row of tarsal bones, forming four branches. These extend distally over metatarsals II, III, IV, and V. At the metatarsophalangeal joints, the branches are enclosed, in common with the corresponding branches of the deep flexor tendon, in a transverse ligament. Here they form the cylinders (manica flexoria), as on the thoracic limb, for the passage of the tendons of the deep digital flexor. The synovial apparatus and the transverse ligaments correspond with those of the thoracic limb. Underneath the tendon on the tuber calcanei, and extending proximally and distally from the tuber, there is an extensive synovial bursa, the *bursa calcanei.* Its proximal portion lies between the superficial digital flexor tendon and the tendon of the m. gastrocnemius; its distal portion lies between the superficial flexor tendon and the plantar ligament.

The portion of the muscle on the crus, as far as the tuber calcanei, corresponds to the m. plantaris of man, and its pedal portion corresponds to the m. flexor digitorum brevis of man.

Action: Flexion of the digits; extension and fixation of the tarsus, flexion of the stifle joint.

Innervation: N. tibialis.

The strong *m. flexor digitorum profundus* lies on the caudal surface of the tibia, covered by the m. gastrocnemius and the superficial digital flexor. It consists of the large, laterally located *m. flexor digiti I (hallucis) longus*, or **medial head of the deep digital flexor,** located between the fibular group and the gastrocnemius group, and the weaker, medially located m. flexor digitorum longus. The latter muscle, short and spindle-shaped, lies on the medial side, behind the tibia, between the m. popliteus and the m. flexor hallucis longus. In hoofed animals the m. tibialis caudalis, whose tendon joins that of the m. flexor hallucis longus, constitutes a third head of the deep digital flexor muscle. The caudal tibial muscle is independent of the two heads of the deep digital flexor in carnivores.

The *m. flexor digiti I (hallucis) longus,* or **lateral head of the deep digital flexor** (Figs. 6–86, 6–90 to 6–93, 6–95), arises from the caudal surface of the proximal three-fifths of the fibula and the proximal caudolateral border of the tibia. It also arises from the interosseous membrane. It has many tendinous strands, resulting in formation of the multipennate muscle. Caudally it is covered by a strong tendinous leaf from which the principal tendon arises and runs distally. At the level of the middle row of tarsal bones it fuses with the weaker tendon of the m. flexor digitorum longus to form the deep flexor tendon. Becoming wider and flatter, it divides into four branches at the middle of the metatarsus; these behave as do the tendons of the m. flexor digitorum profundus of the thoracic limb. In the region of the extensor surface of the tarsus, and proximal to it, a

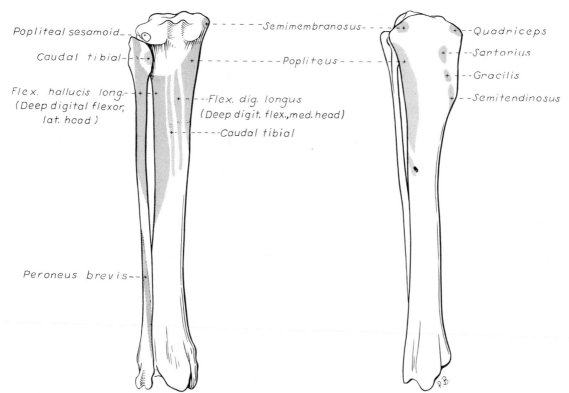

Figure 6–86. Left tibia and fibula, showing areas of muscle attachment, caudal aspect.

Figure 6–87. Left tibia and fibula, showing areas of muscle attachment, medial aspect.

synovial sheath is formed, which communicates with the capsule of the talocrural joint; its mesotendon passes to the caudal surface of the tendon. The tendon of the lateral head of the deep digital flexor is bound in the groove over the sustentaculum tali of the calcaneus by the flexor retinaculum.

The weak *m. flexor digitorum longus* (Figs. 6–86, 6–90, 6–92, 6–95) is a short, flat muscle lying medial to the m. flexor digiti I longus and lateral to the m. popliteus. At its origin it is narrow; it arises on the head of the fibula, the popliteal line, and the fascial leaf separating it from the m. flexor digiti I longus. Above the middle of the tibia it becomes a fine tendon which runs along the caudomedial border of the tibia with the even finer tendon of the m. tibialis caudalis. Both tendons pass through the groove of the medial malleolus, with the tendon of the tibialis caudalis lying cranial to that of the m. flexor digitorum longus. In this position they accompany the tibial collateral ligament of the tarsus to the level of the tibial carpal, where the tendon of the m. flexor digitorum longus unites with the tendon of the m. flexor digiti I longus to form the deep flexor tendon. The synovial sheath of the muscle begins 1 to 1.5 cm. above the medial malleolus and extends on the tendon to its union with the main tendon. A variable mesotendon passes to the tendon, laterally.

Action: Flexion of the digits, extension of the tarsus.

Innervation: N. tibialis.

The *m. tibialis caudalis* (Figs. 6–86, 6–90, 6–92), deep and medially placed, is completely separated in the dog from the heads of the flexor digitorum profundus, in contrast to the condition in the hoofed animals. As an insignificant spindle-shaped muscle, it lies between the m. popliteus and the m. flexor digiti I longus. It is covered by the m. flexor digitorum longus and lies directly on the caudal surface of the tibia. It arises on the medial part of the proximal end of the fibula and soon forms a very delicate tendon which extends distally in front of the somewhat stronger tendon of the m. flexor digitorum longus. It ends, as in man, on the medial ligamentous masses of the tarsus.

Exceptionally, the muscle may be lacking.

Action: Extension of the tarsus; outward rotation of the foot.

Innervation: N. tibialis.

The *m. popliteus* (Figs. 6–77, 6–86, 6–87, 6–90, 6–92) is a strong, triangular muscle lying in the popliteal space. It covers the joint capsule and the medial half of the proximal third of the tibia. It is covered caudally by the mm. gastrocnemius and flexor digitorum superficialis; it encroaches medially upon the m. semitendinosus and fascia. The popliteus arises on the plantar surface of the lateral condyle of the femur by a long tendon which contains a sesamoid bone. This tendon invaginates the joint capsule and crosses under the fibular collateral ligament of the femorotibial joint. The triangular muscle extends obliquely medially over the popliteal space to the medial border of the tibia and ends on its proximal third.

Action: Flexion of the stifle joint; inward rotation of the leg.

Innervation: N. tibialis.

MUSCLES OF THE HINDPAW

The muscles of the dorsal surface of the hindpaw. Only one muscle, the m. extensor digitorum brevis, is found on the dorsal surface of the hindpaw.

The weak *m. extensor digitorum brevis* (Figs. 6–94, 6–96) is a flat muscle, 2.5 to 3 cm. wide, on the dorsum of the foot; it lies on the distal row of tarsal bones and on the metatarsal bones. It is covered by the tendons of the m. extensor digitorum longus, the fascia, and skin. It consists of three heads, of which the middle is the longest. The heads arise from the distal part of the fibular tarsal bone and on the ligamentous masses on the flexor surface of the tarsus. Of the three terminal tendons, the lateral one goes to digit IV, the intermediate one to digits III and IV, and the medial one to digits II and III. Only the branch going to digit II radiates directly into the corresponding branch of the long digital extensor. All others unite on the proximal phalanx with the tendinous branches of the mm. interossei. The latter go to the dorsum of the paw and thereby unite secondarily with the digital extensor tendons.

Action: Extension of the digits.

Innervation: N. peroneus.

The muscles of the plantar surface of the hindpaw. The muscles of the plantar surface of the hindpaw, the *mm. interossei*, adductor digiti II, adductor digiti V, and lumbricales,

Text continued on page 407

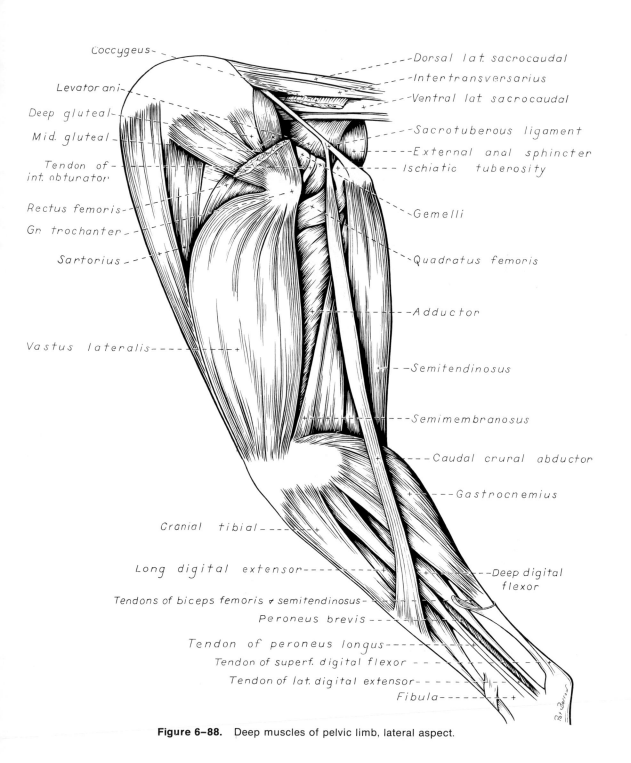

Figure 6–88. Deep muscles of pelvic limb, lateral aspect.

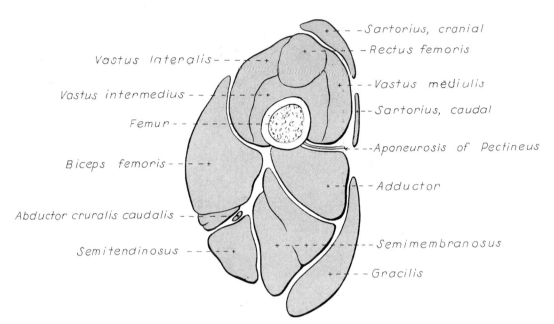

Vastus lateralis

Vastus intermedius

Femur

Biceps femoris

Abductor cruralis caudalis

Semitendinosus

Sartorius, cranial

Rectus femoris

Vastus medialis

Sartorius, caudal

Aponeurosis of Pectineus

Adductor

Semimembranosus

Gracilis

Figure 6–89.   Schematic cross section through left thigh.

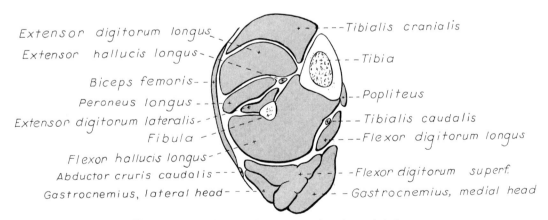

Extensor digitorum longus

Extensor hallucis longus

Biceps femoris

Peroneus longus

Extensor digitorum lateralis

Fibula

Flexor hallucis longus

Abductor cruris caudalis

Gastrocnemius, lateral head

Tibialis cranialis

Tibia

Popliteus

Tibialis caudalis

Flexor digitorum longus

Flexor digitorum superf.

Gastrocnemius, medial head

Figure 6–90.   Schematic cross section through left crus.

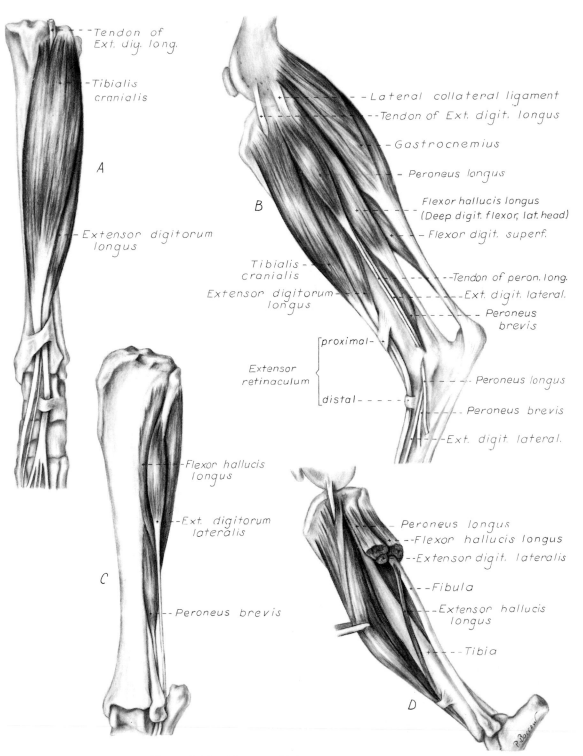

Figure 6–91. Muscles of left crus.
A. Superficial muscles, cranial aspect.
B. Superficial muscles, lateral aspect.
C. Deep muscles, cranial lateral aspect.
D. Deep muscles, lateral aspect.

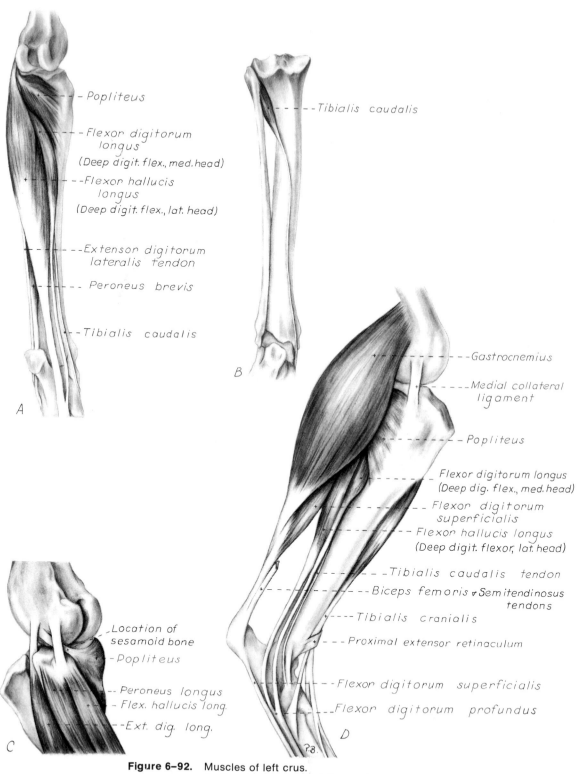

Figure 6-92.   Muscles of left crus.
A. Deep muscles, caudal aspect.
B. Tibialis caudalis, caudal aspect.
C. Muscles of crus at stifle joint, lateral aspect.
D. Muscles of crus, medial aspect.

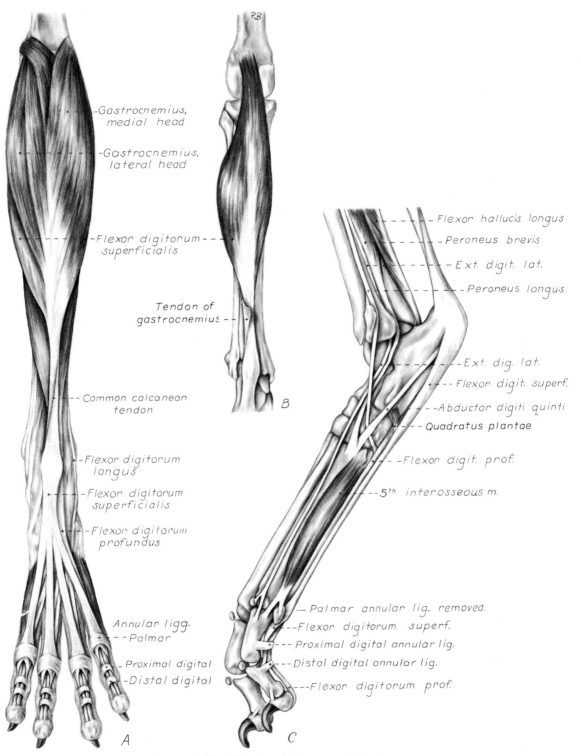

P.B.

Gastrocnemius,
medial head

Gastrocnemius,
lateral head

Flexor digitorum
superficialis

Tendon of
gastrocnemius

Common calcanean
tendon

Flexor digitorum
longus

Flexor digitorum
superficialis

Flexor digitorum
profundus

Annular ligg.
Palmar

Proximal digital
Distal digital

A

B

Flexor hallucis longus

Peroneus brevis

Ext. digit. lat.

Peroneus longus

Ext. dig. lat.

Flexor digit. superf.

Abductor digiti quinti

Quadratus plantae

Flexor digit. prof.

5th interosseous m.

Palmar annular lig. removed.

Flexor digitorum superf.

Proximal digital annular lig.

Distal digital annular lig.

Flexor digitorum prof.

C

Figure 6–93. Muscles of left crus and hindpaw.
A. Superficial muscles, plantar aspect.
B. Deep muscles of crus, plantar aspect.
C. Deep muscles, lateral aspect.

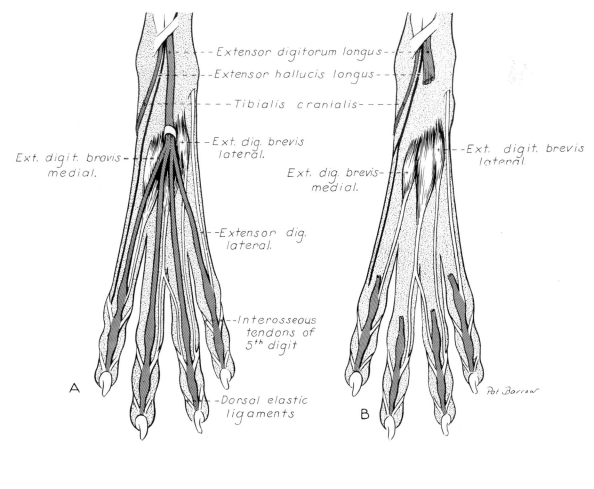

Extensor digitorum longus

Extensor hallucis longus

Tibialis cranialis

Ext. dig. brevis lateral.

Ext. digit. brevis medial.

Ext. digit. brevis medial.

Ext. digit. brevis lateral.

Extensor dig. lateral.

Interosseous tendons of 5th digit

Dorsal elastic ligaments

A

B

Pat Barrow

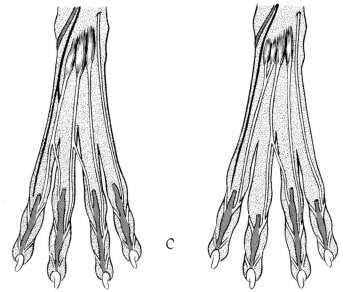

C

Figure 6–94. Extensor muscles of left hindpaw.
   A. Schematic plan of superficial extensor muscles, dorsal aspect.
   B. Schematic plan of deep extensor muscles, dorsal aspect.
   C. Two variations of the extensor digitorum brevis.

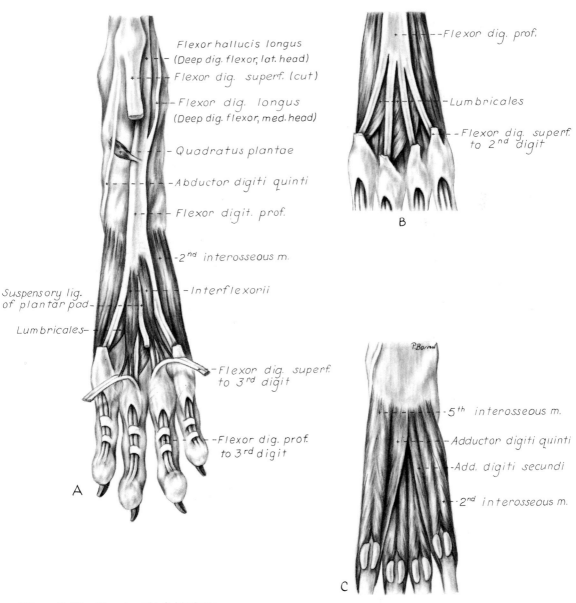

Figure 6–95.   Muscles of left hindpaw.
    A. Superficial muscles, plantar aspect.
    B. Deep muscles, plantar aspect.
    C. Deep muscles, plantar aspect. (Flexor digitorum profundus and lumbricales removed.)

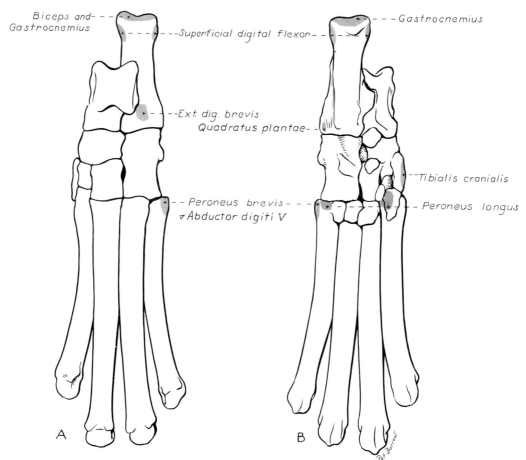

Figure 6–96. Left tarsal and metatarsal bones, showing areas of muscle attachment.
A. Dorsal aspect.
B. Plantar aspect.

behave as do the corresponding ones in the thoracic limb. The mm. abductor digiti V and interflexorii, which are united with the suspensory ligament of the metatarsal pad, are modified. When the first digit (dewclaw) is lacking, its muscles are also lacking, and there is no special flexor of the fifth digit as there is in the forepaw of the dog and in man.

These muscles belong to the region of innervation of the n. tibialis.

The *mm. lumbricales* (Fig. 6–95) lie between the four branches of the deep flexor tendon and are covered on their plantar surfaces by the mm. interflexorii and the suspensory ligament of the metatarsal pad. They are similar to those of the pectoral limb.

The *mm. interflexorii* (Fig. 6–95) are two flat, relatively strong muscles which are shorter than the corresponding unpaired muscle of the pectoral limb. They are located between the deep and superficial flexor tendons. They arise, with the suspensory ligament of the metatarsal pad, from the plantar surface of the tendon of the m. flexor digitorum profundus proximal to the middle of the tarsus; they end about 2 cm. proximal to the metatarsophalangeal joints on the tendons of the m. flexor digitorum profundus of digits III and IV, which partly cover them. The flat ligament for the metatarsal pad appears between the mm. interflexorii.

The *mm. adductores digiti II et V* (Fig. 6–95) lie on the plantar surfaces of the mm. interossei.

The *m. quadratus plantae* (Figs. 6–95, 6–96) is insignificant; it is the tarsal head of the m. flexor digitorum profundus. It arises on the lateral tuberosity of the fibular tarsal bone and the fibular collateral ligament of the tarsus. Directed mediodistally as a delicate muscle, it passes dorsal to the superficial flexor tendon to unite with the deep one. This is near the point where the tendon of the m. flexor digitorum longus joins the main tendon.

The *m. abductor digiti V* (Figs. 6–93, 6–95) is a very small, partially tendinous muscle. It arises from the tuber calcanei, under the tendon of the m. flexor digitorum superficialis. It courses distally superficial to the m. peroneus longus to insert with the m. peroneus brevis on the head of the fifth metatarsal bone.

The special muscles of digit I. The first digit of the dog is usually absent. When it is developed, it is known as the "dewclaw."

If the first digit is completely developed on the hindpaw of the dog, it receives special branches from an extensor and a flexor. Its own muscle, the fleshy *m. flexor digiti I brevis*, extends from the first tarsal bone and from the proximal end of metatarsal I by a short tendon to the head of the proximal phalanx or to the sesamoid apparatus found there.

Fasciae of the Pelvic Limb

The fasciae of the pelvic limb are classified as superficial and deep. These can be further divided in certain places, but the layers cannot always be separated from each other. In general, the deep one is the stronger.

The superficial fascia of the trunk continues dorsally on the lumbosacral region as the **superficial gluteal fascia** and passes over to the tail as the **superficial coccygeal fascia.** In obese specimens it is mainly separated from the deep fascia by a thick layer of fat. This is found primarily between the base of the tail and the ischial tuberosity. Blood vessels and nerves pass through it on their way to the skin. From the lateral abdominal wall the superficial trunk fascia continues on the lateral surface of the thigh; here, as the **superficial lateral fascia of the thigh,** as well as in the gluteal region, it conceals the origins of the m. cutaneus trunci. The superficial fascia passes over to the thigh almost as far as the patella; in the entire region of the m. biceps femoris there is an intimate union with the deep fascia. Otherwise the superficial fascia encloses the distal portion of the femur like a cylinder; a leaf of the superficial fascia pushes out over the m. sartorius and the femoral canal, as well as over the m. gracilis, to unite firmly with the deep leaf. The superficial fascia also envelops the crus, tarsus, metatarsus, and digits, as it does on the thoracic limb. Cutaneous vessels and nerves can be seen through the fascia over long distances before they themselves pass to the skin. The v. saphena medialis (magna) on the medial side, and the v. saphena lateralis (parva) on the lateral side, are located in the superficial fascia.

The **deep fascia of the pelvic limb** will be discussed more thoroughly. This fascia surrounds all portions of the extremity, with its bones, muscles, and tendons, like a tube. In the gluteal region is the **gluteal fascia.** This comes from the lumbodorsal fascia over the crest of the ilium; it continues as the deep coccygeal fascia on the tail. It is rather thick and covers the m. gluteus medius, which partly takes its origin from it; it is less firmly attached to the m. gluteus superficialis. Over these muscles and the m. tensor fasciae latae it radiates into the strong **lateral femoral fascia, or fascia lata,** on the lateral surface of the thigh. Be cause the end aponeurosis of the m. tensor fasciae latae dips into its deep surface, it is two-leaved over a considerable distance. The m. biceps femoris is covered firmly, but parts of the mm. semitendinosus and semimembranosus are covered more loosely; the fascia passes over both the caudal and cranial contours of the thigh, to its medial surface. The lateral femoral fascia and the **medial femoral fascia** thus join to form a cylinder on the thigh. The fascia lata passes from the m. biceps femoris cranially and covers the individual portions of the m. quadriceps femoris toward the medial surface as far as the vastus medialis. The medial fascia of the thigh reaches the medial surfaces of the mm. adductor and semimembranosus after going beneath the branches of the m. sartorius and after bridging the femoral canal. Into this fascia the m. gracilis sinks proximally; however, this muscle, like the caudal belly of the m. sartorius in particular, is also covered superficially by a thin leaf of the deep medial femoral fascia, which unites with the deeper lamella over its caudal edge. Distally, on the medial surface of the thigh, the a. and v. saphena and the n. saphenus are included between the two leaves. The fascia lata is attached to the lateral lip of the femur by an intermuscular septum between the m. biceps femoris and the vastus lateralis. The medial fascia is attached to the medial lip of

the femur by a septum behind the vastus medialis. Toward the stifle both portions of the fascia (lateral and medial) attach to the patella and to the corresponding condyles of the femur. The **fascia of the stifle joint** (*fascia genus*) is demarcated by its thickness and appears to be intimately united with the straight patellar ligament. Distally, it becomes the **crural fascia** (*fascia cruris*). The fascia of the shank is two-leaved. The superficial leaf is essentially the continuation of the lateral and medial femoral fasciae; it partially fuses with the superficial and deep fascia of the shank. It is completely lost on the metatarsus after passing over the tarsus. The deep leaf covers the muscles of the shank and the free-lying surfaces of the crural skeleton; laterally the fibers of the caudal branch of the m. biceps femoris and the m. abductor cruris caudalis radiate into it. Where the two leaves are united with each other, rigid fibrous strands are present, especially in the region of the calcanean tendon of the mm. biceps femoris and semitendinosus. Two other distinct strands exist on the medial side of the crus; one extends from the distal terminal tendon of the m. semimembranosus over the m. popliteus to the tibial crest, and the other extends along the caudal border of the m. tibialis cranialis to its terminal tendon. This tendinous strand is not homologous with the m. peroneus tertius of other animals. Moreover, the deep leaf forms special sheaths about the mm. tibialis cranialis, extensor digitorum longus, peroneus longus, extensor digitorum lateralis, flexor digitorum profundus, and popliteus, and, with the superficial flexor tendon, about the gastrocnemius group. On the individual portions of the foot, the relationships of the **fascia of the foot** (*fascia pedis*) are essentially the same as those of the fascia of the forepaw. There are, however, special relationships to the calcaneal tuber, the tendons, the malleoli, and the collateral ligaments of the tarsal joints.

BIBLIOGRAPHY

Acheson, G. H. 1938. The topographical anatomy of the smooth muscle of the cat's nictitating membrane. Anat. Rec. 71:297–311.

Agduhr, E. 1915. Anatomische, statische und experimentelle Untersuchungen über N. medianus and N. ulnaris, bes. deren motorisches Innervationsgebiet im Vorderarm von Haustieren, nebst einigen Bemerkungen über die Muskulatur desselben Gebietes und über N. musculocutaneus. Anat. Hefte 52:497–647.

Artemenko, B. A. 1929. Über die morphologische Bedeutung des M. transversus costarum. Anat. Anz. 68:248–255.

Barker, D., C. C. Hunt, A. K. McIntyre. 1974. Muscle receptors, 310 pp., *in* Handbook of Sensory Physiology. Vol. III/2. Berlin, Springer Verlag.

Basmajian, J. V. 1974. Muscles Alive; Their Functions Revealed by Electromyography. 3rd Ed. Baltimore, Williams & Wilkins.

Baum, H., and O. Zietzschmann. 1936. Handbuch der Anatomie des Hundes. Band I: Skelett- und Muskelsystem. Berlin, Paul Parey.

Baumeier, M. 1908. Zur vergleichenden Anatomie und Morphologie des Musculus obliquus abdominis ext. und der Fascia flava. Dissertation, Bern.

Bennett, G. A., and R. C. Hutchinson. 1946. Experimental studies on the movements of the mammalian tongue; protrusion mechanism of the tongue (dog). Anat. Rec. 94:57–83.

Bijvoet, W. F. 1908. Zur vergleichenden Morphologie des Musculus digastricus mandibulae bei den Säugetieren. Ztschr. Morph. u. Anthropol. 11:249–316.

Bloom, W., and D. W. Fawcett, 1962. Textbook of Histology. 8th Ed. Philadelphia, W. B. Saunders Co.

Boas, J. E. V., and S. Paulli. 1908. Über den allgemeinen Plan der Gesichtsmuskulatur der Säugetiere. Anat. Anz. 33:497–512.

Bogorodsky, B. W. 1930. Der laterale Strang der Dorsalmuskulatur bei den fleischfressenden Tieren. Anat. Anz. 69:82–121.

Bourne, G. H. 1972–73. The Structure and Function of Muscle: Vol. I, Structure, part I (1972); Vol. II, Structure, part 2 (1973); Vol. III, Physiology and Biochemistry (1973); Vol. IV, Pharmacology and Disease (1973). 2nd Ed. New York, Academic Press.

Bradley, O. C., and T. Grahame. 1959. Topographical Anatomy of the Dog. 6th Ed. New York, Macmillan Company.

Budras, K-D. 1972. Zur Homologisierung der Mm. adductores und des M. pectineus der Haussäugetiere. Zbl. Vet. Med. C. 1:73–91.

Chaine, J. 1914. Le digastrique abaisseur de la mandibule des mammifères. J. Anat. 50:248–319, 393–417, 529–703.

Chauveau, A. 1891. The Comparative Anatomy of the Domesticated Animals. Rev. by S. Arloing. 2nd English Ed. translated and edited by G. Fleming. New York, Appleton Co.

Donat, K. 1972. Der M. cucullaris und seine Abkömmlinge (M. trapezius und M. sternocleidomastoideus) bei den Haussäugetieren. Anat. Anz. 131:286–297.

Duckworth, W. L. H. 1912. On some points in the anatomy of the plica vocalis. J. Anat. 47:80–115.

Dyce, K. M. 1957. The muscles of the pharynx and palate of the dog. Anat. Rec. 127:497–508.

Edgeworth, F. H. 1935. The cranial muscles of vertebrates. Cambridge, University Press.

Eisler, P. 1912. Die Muskeln des Stammes. Jena, G. Fischer.

Ellenberger, W., and H. Baum. 1943. Handbuch der vergleichenden Anatomie der Haustiere. 18th Ed. Berlin, Springer.

Evans, H. E. 1959. Hyoid muscle anomalies in the dog (*Canis familiaris*). Anat. Rec. 133:145–162.

Field, E. J. 1960. Muscle regeneration and repair, *in* Structure and Function of Muscle, edited by G. H. Bourne. New York, Academic Press. Vol. III, pp. 139–170.

Forster, A. 1916. Die Mm. contrahentes und interossei manus in der Säugetierreihe und beim Menschen. Arch. Anat. Phys. (1916):101–378.

Frandson, R. D., and R. W. Davis. 1955. "Dropped muscle" in the racing greyhound. J. Am. Vet. Med. Assoc. 126:468–469.

Getty, R. 1975. Sisson and Grossman's Anatomy of the Domestic Animal. 5th Ed. Philadelphia, W. B. Saunders Co.

Gilbert, P. W. 1947. The origin and development of the extrinsic ocular muscles in the domestic cat. J. Morph. 81:151–194.

Grassé, P-P. 1971. Traité de Zoologie. Vol. 3, Mammifères. Musculature et Arthrologie. Paris, Masson et Cie.

Grau, H. 1932. Über einige Muskelvarietäten bei Haustieren besonders über Varietäten des M. extensor hallucis longus und des M. tibialis anterior beim Hunde. Anat. Anz. 74:218–227.

Gurlt, E. F. 1859. Handbuch der vergleichenden Anatomie der Haussäugetiere. 4th Ed. Stuttgart, Ebner and Seukert.

Ham, A. W., and T. S. Leeson. 1961. Histology. 4th Ed. Philadelphia, J. B. Lippincott.

Heiderich, F. 1906. Die Faszien und Aponeurosen der Achselhöhle. Anat. Hefte 30:517–557.

Henning, P. 1965. Der M. piriformis und die Nn. clunium Medii des Hundes. Zbl. Vet. Med. 12(3): 263–275.

Hess, A. 1954. Reactions of mammalian fetal tissues to injury: III. Skeletal muscle. Anat. Rec. 120:583–598.

Huber, E. 1922, 1923. Über das Muskelgebiet des N. facialis beim Hund, nebst allgemeinen Betrachtungen über die Fascialismuskulatur. Morph. Jahrb. 52:1–110, 354–414.

Huber, G. C. 1916. On the form and arrangement of fasciculi of striated voluntary muscle fibers. Anat. Rec. 11:149–168.

Huntington, G. S. 1903. Present problems of myological research and the significance and classification of muscular variations. Am. J. Anat. 2:157–175.

Iwakin, A. A. 1928. Zur Frage über die Homologie der ventralen Lumbalmuskulatur: I. Über den M. rectractor costae ultimae. Morph. Jahrb. 59:179–195.

Kajava, Y. 1922. Über Homologisierung einiger Muskeln der Hand unserer Haussäugetiere. Verhdlg. Anat. Ges. 55:136–153.

Kajava, Y. 1923. Die volare Handmuskulatur. Act. Soc. Med. Fennic. Duodecim 4:1–184.

Kassianenko, W. 1928. Zur vergleichenden Anatomie der Mm. intercartilaginei bei den Säugetieren. Ztschr. Anat. u. Entw. 85:166–177.

Kolesnikow, W. 1928. Zur Morphologie des M. iliocostalis. Ztschr. Anat. u. Entw. 88:397–404.

Krüger, W. 1929. Über den Bau des M. biceps brachii in seinen Beziehungen zur Funktion beim Menschen und bei einigen Haussäugetieren. Baum Festschrift, Hannover, pp. 139–147.

Langley, J. N., and H. K. Anderson. 1895. The innervation of the pelvic and adjoining viscera: III. The external generative organs. J. Physiol. 19:85–121.

Langworthy, O. R. 1924. The panniculus carnosus in cat and dog and its genetic relation to the pectoral musculature. J. Mammal. 5:49–63.

Leahy, J. R. 1949. Muscles of the Head, Neck, Shoulder and Forelimb of the Dog. Thesis, Cornell University, Ithaca, New York.

Le Gros, Clark, W. E. 1946. An experimental study of the regeneration of mammalian striped muscle. J. Anat. *80*:24–36.

Lockhart, R. D., and W. Brandt. 1938. Length of striated muscle fibers. J. Anat. 72:470.

Matthews, P. B. C. 1972. Mammalian muscle receptors and their central actions. Baltimore, Williams & Wilkins.

Maximenko, A. 1929, 1930. Material zum Studium der Mm. serrati dorsales der Säugetiere. Ztschr. Anat. u. Entw. *89*:156–170, 92:151–177.

Miller, M. E. 1958. Guide to the Dissection of the Dog, 3rd Ed. Ithaca, N. Y. Pub. by author.

Miller, M. E., G. C. Christensen, and H. E. Evans. 1964. Anatomy of the Dog. Philadelphia, W. B. Saunders Co.

Nickel, R., A. Schummer, and E. Seiferle. 1954. Lehrbuch der Anatomie der Haustiere. Band 1: Bewegungsapparat, Berlin, Paul Parey.

Pancrazi, G. 1928. Intorno al "foramen venae cavae" del diaframma dei mammiferi. Atti Soc. Nat. Modena 7:191–192.

Pettit, G. D. 1962. Perineal hernia in the dog. Cornell Veterinarian 52:261–279.

Piérard, J. 1963. Comparative Anatomy of the Carnivore Larynx. Thesis, Cornell University, Ithaca, New York.

Pitzorno, M. 1905. Musculi accessorii ad flexorum perforatum. Studi Sassarei Anno 4. Schwalbes Jahresbericht III:207.

Plattner, F. 1922. Über die ventral innervierte und die genuine Rückenmuskulatur bei drei Anthropomorphen. Morph. Jahrb. 52:241.

Pressman, J. J., and G. Kelemen. 1955. Physiology of the larynx. Physiol. Rev. 35:506–554.

Reighard, J., and H. S. Jennings. 1935. Anatomy of the Cat. 3rd Ed., by R. Elliott. New York, Henry Holt and Company.

Reimers, H. 1925. Die Innervation des M. brachialis der Haussäugetiere. Anat. Anz. 59:289–301.

Romer, A. S., and T. S. Parsons. 1977. The Vertebrate Body. 5th Ed. Philadelphia, W. B. Saunders Co.

Rouviere, H. 1906. Étude sur le development phylogenetique de certain muscles sushyoidiens. J. Anat. *42*:487–540.

Schumacher, S. von. 1910. Die Segmentale Innervation des Säugetierschwanzes als Beispiel für das Vorkommen einer "kollateralen" Innervation. Anat. Hefte *120*(Bd. 40):47–94.

Sisson, S., and J. D. Grossman. 1953. Anatomy of the Domestic Animals. 4th Ed. Philadelphia. W. B. Saunders Co.

Skoda, C. 1908. Eine beim Pferde vorkommende scheinbare Homologie des M. abductor cruris posterior der Carnivoren. Anat. Anz. 32:216–221.

Slijper, E. J. 1946. Comparative biologic-anatomical investigations of the vertebral column and spinal musculature of mammals. Kon. Ned. Akad. Wet., Verh. (Tweede Sectie) 42(5):1–128.

Stimpel, J. 1934. Die Morphologie des medialen Muskelstranges der Stammzone bei den Haustieren. Morph. Jahrb. 74:337–363.

Straus-Durckheim, H. 1845. Anatomie descriptive et comparative du Chat, type des mammiferes en général et des Carnivores en particulier. Vol. II. Syndesmologie et la Myologie. Paris.

Strauss, L. H. 1927. Der Rectus abdominis. Bruns Beiträge z. klin. Chir. *141*:684–698.

Van Harreveld, A. 1947. On the force and size of motor units in the rabbit's sartorius muscle. Am. J. Physiol. *151*:96–106.

Vogel, P. 1952. The innervation of the larynx of man and the dog. Am. J. Anat. 90:427–440.

Wakuri, H., and Y. Kano. 1966. Anatomical studies on the brachioradial muscle in dogs. Acta Anat. Nipponica *41*:222–231.

Walls, G. L. 1942. The Vertebrate Eye and Its Adaptive Radiation. Cranbrook Inst. of Science Bull. 19. Bloomfield Hills, Mich., The Cranbrook Press.

Walter, C. 1908. Die Sehnenscheiden und Schleimbeutel der Gliedmassen des Hundes. Dissertation, Dresden-Leipzig.

Winckler, G. 1939. Contribution à l'étude de la morphogénèse du muscle spinalis dorsi. Arch. Anat. Hist. Embr. 27:99.

Ziegler, H. 1929. Muskelvarietäten bei Haustieren. Ztschr. Anat. u. Entw. *91*:442–451.

Ziegler, H. 1931. Die Innervationsverhältnisse der Beckenmuskeln bei Haustieren im Vergleich mit denjenigen beim Menschen. Morph. Jahrb. 68:1–45.

Ziegler, H. 1934. Weitere Untersuchungen, über den M. glutaeobiceps von Hund (Canis familiaris) and Katze (Felis catus dom.). Morph. Jahrb. 73:385–391.

Zimmermann, A. 1927. Über die Quermuskeln der Rippen. Allattani Közlemenyek *24*:53–60.

Zimmermann, A. 1928. Zur vergleichenden Anatomie des M. pronator teres. Verhdlg. Anat. Ges. *66*:281–282.

Chapter 7

THE DIGESTIVE APPARATUS AND ABDOMEN

The **digestive apparatus** *(apparatus digestorius)*, consists of the oral cavity, pharynx, alimentary canal, and accessory organs. The accessory organs include the teeth, tongue, salivary glands, liver, gall bladder, pancreas, and anal sac.

The wall of the digestive tube is richly supplied with secretory epithelium and intrinsic glands. It is lined throughout by a mucous membrane which is continuous with the surface integument at the mouth and anus.

THE MOUTH AND ASSOCIATED STRUCTURES

MOUTH

The **mouth** *(os)*, in a restricted sense, includes only the opening between the lips. In a broad sense it designates the **oral cavity** *(cavum oris)*, in which are located the teeth and tongue. The oral cavity is divided into the vestibule and the oral cavity proper.

The **vestibule of the mouth** *(vestibulum oris)* is the space external to the teeth and gums and internal to the lips and cheeks. It opens to the outside anteriorly by means of the U-shaped slit, the **oral fissure** *(rima oris)*, between the lips. This opening is in essentially a frontal plane through the rostral margins of the orbits. When the mouth is closed, the vestibule communicates with the oral cavity proper by means of the interdental spaces, which vary greatly in size. A space on either side, caudal to the last cheek tooth and nearly 1 cm. long in large dogs, also establishes free communication between the two parts of the oral cavity.

The parotid and zygomatic salivary ducts open into the dorsocaudal part of the vestibule. The parotid duct opens through the cheek on the small **parotid papilla** *(papilla parotidea)*, located opposite the caudal part of the upper fourth premolar tooth, approximately 5 mm. from the fornix of the vestibule, which is formed by a reflection of the mucosa from the cheek to the gum. The main duct of the zygomatic gland opens lateral to the caudal part of the upper first molar tooth on a small papilla near the vestibular fornix (Fig. 7–1). A small ridge connects the main zygomatic and parotid duct openings. Usually one to four small accessory ducts from the zygomatic gland open caudal to the main duct. The submucous glands are few and are confined to the lower lip and the adjacent part of the cheek. The secretion of these glands is discharged through about 10 openings located opposite the four lower premolar teeth near the fornix of the vestibule.

The **oral cavity proper** *(cavum oris proprium)* is bounded dorsally by the hard palate and a small part of the adjacent soft palate; laterally and rostrally the dental arches and teeth form its boundary; the tongue and the reflected mucosa under it form the floor of this cavity. When the mouth is closed the

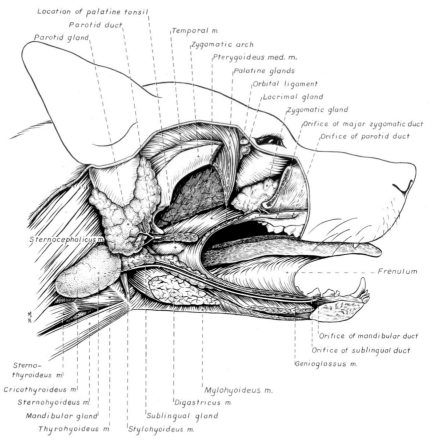

Labels on figure:
Location of palatine tonsil
Parotid duct
Parotid gland
Temporal m.
Zygomatic arch
Pterygoideus med. m.
Palatine glands
Orbital ligament
Lacrimal gland
Zygomatic gland
Orifice of major zygomatic duct
Orifice of parotid duct
Sternocephalicus m.
Frenulum
Orifice of mandibular duct
Orifice of sublingual duct
Genioglossus m.
Sterno-thyroideus m.
Cricothyroideus m.
Sternohyoideus m.
Mandibular gland
Thyrohyoideus m.
Stylohyoideus m.
Sublingual gland
Digastricus m.
Mylohyoideus m.

Figure 7–1. Salivary glands. (The right half of the mandible is removed.)

tongue nearly fills the oral cavity proper. The sublingual and mandibular ducts open under the body of the tongue, on the inconspicuous **sublingual caruncle** *(caruncula sublingualis).*

Extending caudally from the caruncle to about a transverse plane through the lower shearing teeth is a low ridge of mucosa about 2 mm. wide and 1 mm. high. This is the **sublingual fold** *(plica sublingualis).* It lies close to the body of the mandible and is formed by the underlying mandibular and sublingual ducts and a variable number of lobules of the polystomatic portion of the sublingual gland.

Just caudal to the upper central incisor teeth is the **incisive papilla** *(papilla incisiva),* a rounded eminence which extends caudally to blend with the first transverse ridge formed of the mucosa covering the hard palate. On each side of this papilla an **incisive duct** *(ductus incisivus)* opens. This duct has formerly been known as the incisive canal and the nasopalatine duct. This duct leaves the oral cavity by a slitlike opening and extends posterodorsally for 1 or 2

cm. through the palatine fissure to open into the floor of the nasal fossa. Before opening into the nasal fossa, the duct communicates with the cavity of the vomeronasal organ.

The oral cavity is continuous caudally with the isthmus of the fauces and with the oral pharynx.

LIPS

The **lips** *(labia oris)* form the rostral and most of the lateral boundaries of the vestibule. Upper and lower lips *(labia superior et inferior)* meet at the angles of the mouth *(angulus oris),* forming the commissures of the lip. The lips bound the oral fissure, the external opening of the oral cavity. Their margins are narrow and devoid of hair except the rostral two-thirds of the upper lip on either side. Toward the angle on either side, the caudal third progressively increases in width, to form a rounded border measuring as much as 1 cm. The margin of the lower lip as far caudally as a level through the canine teeth is devoid of hair over a zone about 5 mm. wide. Caudal to the

412

canine teeth this smooth zone increases to 1 cm. in width, and the narrow border becomes serrated by the formation of about 15 conical papillae several millimeters high.

No definite frenulum, or median mucosal fold, attaches the lower lip to the gum, and the median mucosal fold of the upper lip is poorly developed, being thick but narrow. The mucosa of the lower lip is firmly attached to the gum on either side in the space between the canine and the first premolar tooth (interdental space). A deep, straight, narrow cleft, the *philtrum,* marks the union of the two halves of the upper lip, anteriorly. The hair of both lips slopes caudoventrally. It is thinner and shorter in front, longer and thicker farther back. A few of these are **tactile hairs** *(pili tactiles).* On the upper lip and adjacent dorsal part of the muzzle the tactile hairs are imperfectly arranged in four rows. The wide orbicularis oris muscle and the insertions of several other facial muscles form the media of the lips.

CHEEKS

The **cheeks** *(buccae)* form the caudal portion of the lateral walls of the vestibular cavity. The cheeks are small in the dog because of the large mouth opening. The cavity of the cheeks runs medial to the masseter muscles and extends as far caudally as the attachment of the buccinator muscles on the mandible and maxilla at their alveolar borders opposite the last two cheek teeth of each jaw and the intervening anterior border of the basal half of the coronoid process. In rodents and most herbivores the cheeks are large and serve as storage space, especially during mastication and transportation. This function is of minor importance to the dog.

The morphology of the cheeks is closely related to that of the lips, with which they are continuous. The lips and the cheeks consist of three basic layers. The external layer is the hairy integument; the middle layer consists of muscle and fibroelastic tissue; and the inner layer consists of the mucosa. Projecting caudolaterally from the caudal part of the skin of the cheek are usually two coarse tactile hairs, 3 to 5 cm. long. The middle layer of the cheek consists primarily of the buccinator muscle, although lateral to

this are the m. zygomaticus and certain fasciculi of the cutaneous fascia. The dorsal buccal glands in carnivores are consolidated to form the zygomatic gland, a large mixed salivary gland located in the ventral part of the orbit. The ventral buccal glands consist of a few small, solitary glands located in the submucosa, rostral to the masseter muscle and medial to the fibers of the ventral part of the buccinator. The mucosa of the lips and cheeks is thinly cornified stratified squamous epithelium which, in some breeds, is partly or wholly pigmented. Although the lips are not used as prehensile organs, the various muscles in them provide for movement and expression of emotions such as anger or fear.

PALATE

The **palate** *(palatum)* (Fig. 7-2) is a partly bony, partly membranous partition separating the respiratory and digestive passages of the head. The nasal fossae and nasal pharynx lie above it; the oral cavity and oral pharynx lie below it. The bony hard palate is in front, and the membranous soft palate behind.

The **hard palate** *(palatum durum)* is formed by processes of the palatine, maxillary, and incisive bones on each side. The mucosa on the nasal side consists of pseudostratified ciliated columnar epithelium; that on the oral side consists of stratified squamous epithelium which is cornified. The hard palate is nearly flat. Laterally and rostrally it inclines slightly ventrally, and it is continuous with the alveolar processes of the upper jaw. Six to ten ridges and depressions cross it transversely on the oral side. Not all of these ridges are complete. In extremely brachycephalic heads they become nearly straight. Close inspection of the ridges reveals small blunt eminences.

The **soft palate** *(palatum molle* or *velum palatinum)* continues caudally from the hard palate at an irregularly transverse level, passing just caudal to the last upper molar teeth in mesaticephalic heads. In extremely brachycephalic heads the junction of the hard and soft palates is over 1 cm. caudal to this transverse level. The soft palate is particularly long in the dog. In preserved heads the epiglottis is usually seen to lie above the thick caudal border of the soft palate. Occa-

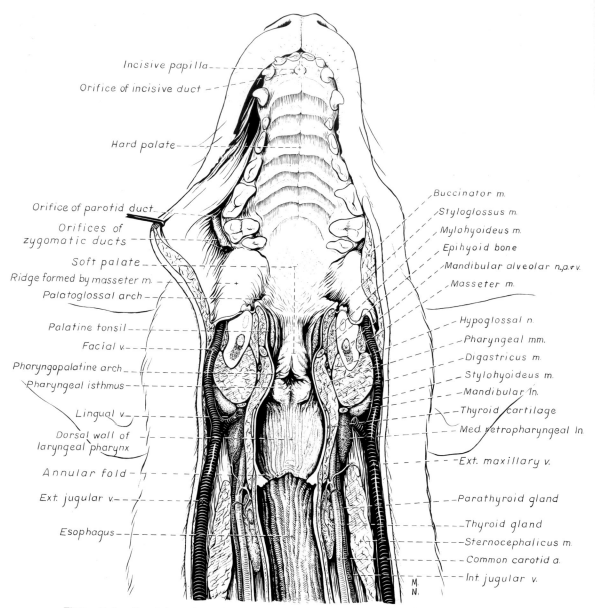

Incisive papilla

Orifice of incisive duct

Hard palate

Orifice of parotid duct

Orifices of zygomatic ducts

Soft palate

Ridge formed by masseter m.

Palatoglossal arch

Palatine tonsil

Facial v.

Pharyngopalatine arch

Pharyngeal isthmus

Lingual v.

Dorsal wall of laryngeal pharynx

Annular fold

Ext. jugular v.

Esophagus

Buccinator m.

Styloglossus m.

Mylohyoideus m.

Epihyoid bone

Mandibular alveolar n.,a.+v.

Masseter m.

Hypoglossal n.

Pharyngeal mm.

Digastricus m.

Stylohyoideus m.

Mandibular ln.

Thyroid cartilage

Med. retropharyngeal ln.

Ext. maxillary v.

Parathyroid gland

Thyroid gland

Sternocephalicus m.

Common carotid a.

Int. jugular v.

M. N.

Figure 7–2. Frontal section of head and neck through the digestive tube, ventral aspect.

sionally it may lie beneath the palate or even be recessed in the caudal margin of the soft palate. In brachycephalic breeds the soft palate may be so long as to interfere with the passage of air into the larynx. In the average dog's head the soft palate is 6 cm. long, 3 cm. wide, and 5 mm. thick where it is continuous with the hard palate. The soft palate gradually thickens, so that at the junction of its middle and caudal thirds it is about 1 cm. thick, after which it becomes

slightly thinner and ends in a concave border. From its ventrolateral part a thin elliptical fold extends laterally to form the ventral wall of the tonsillar sinus.

On each side the caudal border of the soft palate is continued to the dorsolateral wall of the **palatopharyngeal arch** *(arcus palatopharyngeus),* or caudal pillar of the soft palate. The **palatopharyngeal muscle** *(m. palatopharyngeus)* and the mucosa which covers it form this pillar. There is no palatoglossal arch or anterior pillar of the soft palate in the dog comparable to the structure of that name in man, since the dog lacks the palatoglossus muscle which forms the basis of the human arch. When the tongue is forcibly withdrawn from the mouth and moved to one side, a fold is developed on the opposite side, running from the body of the tongue to the initial part of the soft palate. Although this fold is not formed by muscle as in man, it seems feasible to regard it as the **palatoglossal fold,** which distinguishes it from the palatoglossal arch of man. That portion of the soft palate caudal to a transverse plane through the caudal borders of the pterygoid bones is known as the **palatine veil** *(velum palatinum).* No uvula is present.

Structure of the Soft Palate. The soft palate, from the ventral to the dorsal surface, consists of the following layered structures.

Stratified squamous epithelium, a continuation of that of the hard palate, covers the ventral surface of the soft palate. Unless stretched, it is thrown into many fine longitudinal folds and a few larger transverse ones. The mucosal folds are evidence of the mobility and slight elasticity of the soft palate. The stratified squamous epithelium does not end at the posterior border but curves around the border and runs forward a few millimeters on the respiratory surface.

The **palatine glands** *(glandulae palatinae)* form the thickest stratum of the organ. They are mixed glands opening on the surface of the soft palate by about 200 openings per square centimeter anteriorly; posteriorly the number decreases, but the openings increase in size. Near the caudal margin there are only about 10 openings per square centimeter. The mantle of the palatine glands in the rostral half of the soft palate is about 4 mm. thick, the glandular layer gradually becoming thicker and then thinner before it ends in the concave caudal border. The stra-

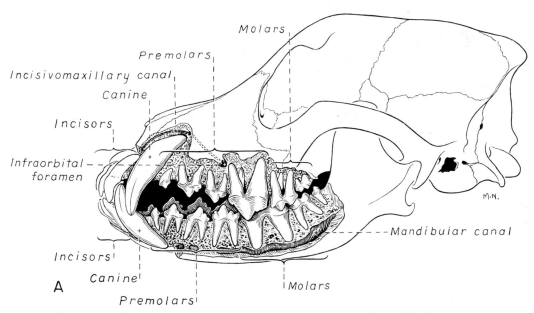

Figure 7–3. Jaws and teeth of an adult dog. Lateral view of jaws, sculptured to show tooth roots.

tum formed by the palatine glands is bounded on each side by the large medial pterygoid and the styloglossal muscle.

The muscles of the soft palate consist of the paired **palatine muscles** (*mm. palatini*) and the end ramifications of the paired **tensor and levator veli palatini muscles** (Fig. 7–4) which are nearly equal in size. These muscles are described with the muscles of the head. The right and left intrinsic palatine muscles lie close together on each side of the median plane, ventral to the palatine aponeurosis. The end ramifications of the paired extrinsic muscles, the levator and tensor veli palatini, blend with the palatine aponeurosis. The right and left **pterygopharyngeal muscles** pass lateral to the caudal part of the soft palate from origins on the pterygoid bones. Right and left **palatopharyngeal muscles** arise near the median plane from the palatine aponeurosis. They sweep laterally and upward in the pharynx, forming the bases for the palatopharyngeal arches.

Vessels of the Palate. The right and left **major palatine arteries** (*aa. palatinae majores*) are the main arteries to the hard palate. The main arteries to the soft palate are the **minor palatine arteries** (*aa. palatinae minores*), aided by the **ascending pharyngeal** (*a. pharyngea ascendens*) and the major palatine. They are dwarfed by the **palatine plexus** of veins of the soft palate, which lies mainly lateral to the two slender palatine muscles. The main part of the palatine plexus is located in the deep part of the glandular mantle of the soft palate. It arises from the smaller, less developed venous plexus of the hard palate, which is located in the submucosa covering the bony palate. The lymphatics go to the **medial retropharyngeal lymph nodes.**

Nerves of the Palate. The major palatine branch of the maxillary division of the **trigeminal nerve** (*n. trigeminus*) courses through the palatine canal and supplies sensory fibers to the oral side of the hard palate. The nasal side is supplied by the posterior nasal or sphenopalatine branch of the same nerve. The main sensory supply to the soft palate is the minor palatine branch of the **maxillary nerve** (*n. maxillaris*), which follows the minor palatine artery to enter the anterior end of the soft palate. Branches from the **glossopharyngeal** (*n. glosso-*

pharyngeus) and the **vagus nerve** (*n. vagus*) also enter the soft palate and supply the pterygopharyngeal and palatopharyngeal muscles. The vagi contribute most to the supply of these muscles; the glossopharyngeal nerves supply the lateral walls of the oral pharynx and, to a lesser extent, the soft palate. They are sensory also to the caudal part of the tongue.

Aponeurosis of the Palate. The **palatine aponeurosis** (*raphe palati*) consists essentially of the fanned-out terminal tendons of the right and left tensor veli palatini muscles. It is located between the ventral margins of the right and left perpendicular parts of the palatine bones and the pterygoid bones, and attaches rostrally to the caudal margin of the hard palate. It is thin but strong.

Lying directly dorsal to the palatine aponeurosis are the small mixed dorsal **palatine glands.** The epithelium which covers them, and on which their numerous ducts open, is of the pseudostratified ciliated columnar type. It is continued forward on the dorsum of the hard palate to line the nasal pharynx and the nasopharyngeal duct. Caudally, before reaching the caudal border of the soft palate, it is continued by stratified squamous epithelium which is the epithelium of the whole ventral surface of the palate.

TEETH

The **teeth** (*dentes*) (Fig. 7–3) are highly specialized structures which serve as weapons of offense and defense, as well as for the procuring, cutting, and crushing of food. Each tooth is divided into three parts. The **crown** (*corona dentis*) is the exposed part, the part which protrudes above the gums and is covered by the shiny white enamel. All the crowns of the teeth of the dog except the canines end in **tubercles** (*tubercula dentis*). The **neck** (*collum dentis*) is a slight constriction of the tooth located at the gum line, where the enamel ends. The **root** (*radix dentis*) is the portion below the gum, and for the most part it is embedded in the alveolus. Its pointed end is the **apex of the root** (*apex radicis dentis*). Many teeth have more than one root. Once teeth are fully erupted in the dog they cease growing. Because of the increased use and greater dependence on the teeth which comes about

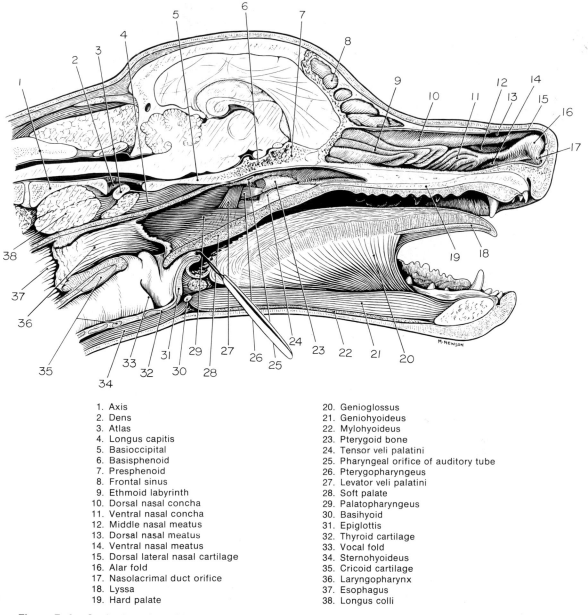

Figure 7–4. Sagittal section of head. (From Evans, H. E., and deLahunta, A., 1979, *Miller's Guide to the Dissection of the Dog,* Philadelphia, W. B. Saunders Co.)

1. Axis
2. Dens
3. Atlas
4. Longus capitis
5. Basioccipital
6. Basisphenoid
7. Presphenoid
8. Frontal sinus
9. Ethmoid labyrinth
10. Dorsal nasal concha
11. Ventral nasal concha
12. Middle nasal meatus
13. Dorsal nasal meatus
14. Ventral nasal meatus
15. Dorsal lateral nasal cartilage
16. Alar fold
17. Nasolacrimal duct orifice
18. Lyssa
19. Hard palate
20. Genioglossus
21. Geniohyoideus
22. Mylohyoideus
23. Pterygoid bone
24. Tensor veli palatini
25. Pharyngeal orifice of auditory tube
26. Pterygopharyngeus
27. Levator veli palatini
28. Soft palate
29. Palatopharyngeus
30. Basihyoid
31. Epiglottis
32. Thyroid cartilage
33. Vocal fold
34. Sternohyoideus
35. Cricoid cartilage
36. Laryngopharynx
37. Esophagus
38. Longus colli

through growth, two sets of teeth are necessary. The first set is fully erupted and functional early in the second month after birth. These teeth, known as **deciduous teeth** *(dentis decidui),* serve the animal during its most active puppyhood. Upon approaching maturity, when the jaws have become longer and larger and the small deciduous teeth are no longer adequate, they are shed and replaced by the **permanent teeth** *(dentes permanentes),* which last throughout adult life. These are larger than the deciduous teeth. The jaws continue to grow, so that additional permanent teeth are added in back of the permanent premolar teeth. These added teeth are the true molars. There are two on each side of the upper jaw and three on each side of the lower jaw.

Observations on the development of the deciduous teeth in the fetal dog have been reported by Hörmandinger (1958), Satrapa-Binder (1959) Williams (1961), and Williams and Evans (1978). Postnatal calcification and eruption of the deciduous and permanent teeth have been studied by Meyer (1942), Höppner (1956), Arnall (1961), and Kremenak (1967).

The teeth are arranged as **upper** and **lower dental arches** *(arcus dentalis superior et inferior).* The lower arch is narrower and usually shorter than the upper (Fig. 7–5A).

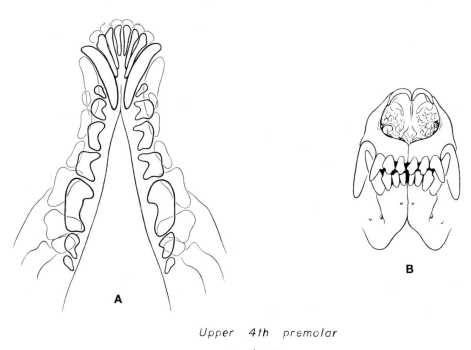

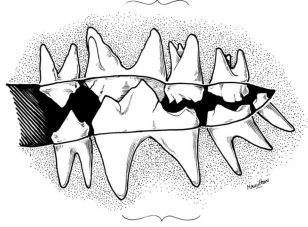

Upper 4th premolar

Lower 1st molar

C

Figure 7–5.   A. Superposition of upper and lower dental arches. (Upper teeth in light outline close lateral to the lowers.)

B. Bite of the incisor and canine teeth.

C. Bite of the shearing teeth. Medial view, right jaws.

The teeth which compose the dental arches are anchored in sockets, or alveoli, of the upper and lower jaws. They are so placed that many of the teeth fail to meet or else overlap or protrude between adjacent teeth lying opposite to them. The function of these nonocclusal or imperfectly occluding teeth is to grasp, puncture, or shear. Stockard (1941) refers to the area between the canine and carnassial teeth as the premolar carrying space. In dogs the food is bolted rather than masticated, and a complete occlusal surface is not necessary.

Tooth Surfaces. The surface of the tooth which faces the lip or cheek is the **vestibular surface** (*facies vestibularis*), formerly also called labial or buccal surface, and the surface which faces the tongue is the **lingual surface** (*facies lingualis*). The surface adjacent to the next tooth in the dental arch is the **contact surface** (*facies contactus*). For all teeth the contact surfaces are mesial and distal. The mesial surface of the first incisor is on the median plane between the first incisors. The opposite sides of each are their distal surfaces. The surface which faces the opposite dental arch is the masticatory or **occlusal surface** (*facies occlusalis*).

Tooth Groupings. The upper teeth are attached in the alveoli of the incisive (premaxillary) and maxillary bones. Those with roots embedded in the incisive bones are the **incisor teeth** (*dentes incisivi*). The upper incisors usually are placed slightly in front of the lower incisors. The incisor tooth nearest the midplane on each side in each jaw is incisor I, or central incisor. The second is incisor II, or intermediate incisor, and the third is incisor III, or corner incisor. They are long, slender teeth, arched slightly forward and laterally compressed. They increase in size from the central to the corner incisor. The central tubercles of the upper corner incisors are largest and slightly hooked posteriorly, whereas those of incisors I and II are irregularly trident transversely. Of the three projections or tubercles, the central one is largest and extends farthest distally; the small medial projections are farther from the gums than the lateral ones. A V-shaped ridge of enamel, with its apex near the gum, connects the side tubercles on the lingual surface. Tuber-

cles wear off early in adult life or remain to old age, depending on the chewing habits of the dog.

The **canine teeth** (*dentes canini*) are separated from the corner incisors by an interdental space of about 3 mm. on the upper jaw and by less than this distance on the lower jaw in mesaticephalic heads. They are by far the longest teeth of the dog, having large roots which are nearly two times as long as their crowns. The canine teeth of the upper jaw are about the same length as the lower ones, but are more massive. They are transversely compressed. When the mouth is closed, the crown of the lower canine tooth occupies the interval between the upper corner incisor and the canine tooth (Fig. 7–5B). The roots produce arciform alveolar juga which lie peripheral to the roots of the first premolar teeth. The canine teeth are pointed at each end and have a middle portion which may be slightly wider within the alveolus than at the gum line. This necessitates a flap operation of the maxilla over the root of the tooth to facilitate its proper removal.

All teeth caudal to the canines are often referred to as the cheek teeth (St. Clair and Jones 1957). They are divided into premolars and molars. In the permanent dentition there are four **premolar teeth** (*dentes premolares*) on each side in each jaw. The first premolar tooth in the dog is not replaced, and in this respect it is similar to the molars. It erupts between the fourth and fifth postnatal month and usually remains throughout life. Some authors (Cornevin and Lesbre 1894) consider the first premolar to belong to the deciduous dentition, whereas others (Tims 1895, Williams and Evans 1978) prefer to regard it as a member of the permanent dentition. Tims (1902) mentions other animals in which premolar I erupts late in the growth period and is not replaced.

The placement of the premolar teeth may be altered by changing the shape of the head through selective breeding, but regardless of head shape the teeth remain relatively constant in form and size (Stockard 1941). The first premolar is the smallest, and has a single peglike root. The last, or fourth, premolar is the largest; in the upper jaw it has three roots, in the lower, only two. The middle two premolar teeth are similar in both jaws. Each has two roots. The crowns of the first three premolars are similar to those of the corresponding teeth on the other side of the same jaw and of those in the opposite jaw. The first premolar tooth has a single tubercle which usually hooks slightly posteriorly. The ledge of enamel at the gum line is relatively straight medially and convex laterally. The second and third premolars have similar crowns. On each tooth the mesial border slopes diagonally backward to end in a strong, pyramid-shaped point which inclines slightly backward. The distal border of this tubercle is steep, ending in a relatively level, transversely rounded border. Adjacent to the base of the main tubercle on this border is a second smaller tubercle 1 or 2 mm. high. The distal margins of the crowns of the two middle premolars are prominent. They may form small tubercles.

The upper fourth premolar teeth are the largest cutting teeth of the upper jaw. They are the **carnassial** or **sectorial teeth** (dentes sectoria), which are sometimes referred to as the shearing teeth (Fig. 7–5C). Each has three stout diverging, conical roots. The two roots on the labial side form prominent alveolar juga that reach as far dorsally as a frontal plane through the infraorbital canals. The caudal root of each upper sectorial tooth is triangular and somewhat transversely flattened, becoming wide at the neck. The lingual root, a fourth shorter than the labial, is flattened in an oblique plane and slightly twisted. A small tubercle is located on the short crown which lies opposite its most medial part. According to Annis (1974), the upper fourth premolar is the tooth most commonly involved with root abscesses. Such an abscess of one of the lateral roots manifests itself usually by the formation of a persistent fistula which discharges on the face rostroventral to the eye. A permanent cure involves extraction of the affected tooth. To accomplish this, the tooth may be split so that the roots can be removed separately. The fourth premolar tooth in the lower jaw is similar in form to the two teeth in front of it but is slightly larger. It is not the carnassial tooth of the lower jaw.

The **molar teeth** (dentes molares) have no deciduous predecessors. There are two on each side of the upper jaw and three on each side of the lower jaw. In each jaw the first are the largest and the last are the smallest. The masticatory surfaces of the upper molars are multituberculate and are at two levels. The mesial parts are at a higher level than the distal. The mesial masticatory surfaces of the upper molar teeth are irregularly flattened and make normal contact with the last two molars and about the distal third of the first molar of the lower jaw. The lower molars, although multituberculate, contact only the mesial parts of the upper molar teeth.

Each upper molar tooth has three slightly diverging roots. The lingual root of each of these teeth is more massive than either of the two vestibular roots, although it is shorter. It is slightly compressed rostrocaudally and is so shaped that the greatest compression force is transmitted through it to the compact bone which lies adjacent to the neck of the tooth.

The lower molar teeth are peculiar in that the first is about twice as large as the other two together; it is the **sectorial** or **carnassial tooth** of the lower jaw. The distal third of each lower first molar tooth is adapted for crushing and grinding; the mesial portion is sharp and pointed, and suited for shearing action. The shearing portion of each lower first molar possesses the strongest tubercles of any teeth of the lower jaw and is roughly quadrilateral in form. This part of the crown is longer than the inner surface of the distal part of the crown of the upper fourth premolar tooth, with which it forms a scissors. A small fossa is formed in the hard palate, opposite its most salient tubercle, to receive the tubercle. For a discussion of the terminology used for tooth cusps and crests, see Szalay (1969) and Every (1972, 1974). Szalay uses a modified Cope-Osborn terminology for mammalian tritubercular cusps and crests. Every proposes a new term for each.

Dental Formulae. Since the teeth are grouped according to position and form, it is possible to express their arrangement as a dental formula. The abbreviation representing the particular teeth (I, incisor, C, canine; PM, premolar; M, molar) is followed by the number of such teeth on one side of the upper and the lower jaw.

The formula for the **deciduous dentition** of the dog is:

$$I \frac{3}{3} \ C \frac{1}{1} \ PM \frac{3}{3} = 28$$

The **permanent dentition** is represented as:

$$I \frac{3}{3} \ C \frac{1}{1} \ PM \frac{4}{4} \ M \frac{2}{3} = 42$$

Eruption of Teeth. The deciduous second, third, and fourth premolar teeth are replaced by the permanent premolars. The permanent teeth develop beneath or to one side of the deciduous teeth, which they are destined to dislodge. The permanent incisors erupt slightly posterior to the deciduous incisors. The central and intermediate deciduous incisors and the canine teeth of both jaws have usually erupted by the end of the first month. The corner incisors erupt during the fifth or sixth week; the deciduous premolars, between the fourth and eighth weeks. The time of eruption of the permanent teeth is rather closely correlated with the life span of the breed. Variations are numerous, but the life span of a Great Dane or St. Bernard is about 8 years; that of a setter, pointer, or hound, 13 years; and that of the toy breeds, 17 years. The larger the breed, the shorter the life span. The teeth erupt earliest in the large breeds. Table 7–1 gives the normal range of time for the eruption of each of the permanent teeth, which erupt at about the same time, in each jaw. Eruption of the permanent canine teeth usually precedes the shedding of the corre-

sponding deciduous ones; for two or three weeks, or even longer, the permanent canine is visible rostromedial to the corresponding deciduous tooth.

Structure of Teeth. The dense, pearly-white outer layer of the crown is **enamel** (*enamelum*). It is the hardest substance in the body, and gives sparks when struck with steel. Only about 5 per cent of enamel is organic matter. It is thickest on the wearing surfaces of the teeth, and its hardness gradually increases as the dog gets older. Glock et al. (1942) found that it is possible to remove enamel from the teeth of puppies up to 17 weeks of age by scraping with a scalpel.

Dentin (*dentinum*), similar to bone in chemical composition, forms the bulk of the tooth, enclosing the pulp cavity internally and underlying the enamel externally. It is known as "ivory," and is yellowish-white in color. It contains a few nerves, but through the processes of the cells (odontoblasts) that form it a conductive system is established which gives sensitivity to touch, cold, and other stimuli. Gradual erosion of the dentin, as occurs from wear in the aged, is not accompanied by pain, as the conducting fibers recede or calcify in advance of the wearing surface. Injured dentin is repaired under favorable conditions and is referred to as secondary dentin.

The **cementum** in the dog is a thin covering found only on the roots. Grossly, the cementum cannot be differentiated from the dentin which it covers.

The **pulp** (*pulpa dentis*), the only soft tissue contained in a tooth, is composed of sensory nerves, arteries, veins, lymphatic capillaries, and a rather primitive connective tissue which holds those structures together. The pulp is contained in the **pulp cavity** (*cavum dentis*). A small **apical foramen** (*foramen apicis dentis*), at the end of each root, allows free passage of the vessels and nerves in and out of the tooth through the **root canal** (*canalis radicis dentis*).

Dental Anomalies. Deviations from the normal number or placement of the teeth are the most common dental anomalies reported in dogs. Such anomalies seldom occur in the dolichocephalic breeds, but are encountered very commonly in the brachycephalic dogs. As the muzzle is shortened, deviation in placement of the teeth results. The size of the teeth does not decrease proportionately

TABLE 7–1. ERUPTION OF PERMANENT TEETH

Group	Tooth	Eruption Period
Incisors	Central	2 to 5 months
	Intermediate	2 to 5 months
	Corner	Most breeds 4 to 5 months
Canine		5 to 6 months
Premolars	First	4 to 5 months
	Second	6 months
	Third	6 months
	Fourth	4 to 5 months
Molars	First	5 to 6 months
	Second	6 to 7 months
	Third	6 to 7 months

with a reduction in the length of jaw (Stockard 1941). In one Boston terrier only three lower cheek teeth were present on one side of the lower jaw, and four on the other. There was no evidence of filled-in alveoli, and the dental arch was full. A few relatively large teeth, properly placed, are probably more efficient than many teeth placed at distorted angles.

Supernumerary teeth are occasionally found in all breeds. Extra incisors and premolars are most common. Usually, when an increase in the number of premolars occurs, two single-rooted premolars are located behind the canine. An extra incisor is commonly in series with the other incisors, although an extra incisor located on the palate has been recorded (Colyer 1936). Supernumerary teeth are usually unilateral, and occur in the upper jaw more frequently than in the lower. The number of teeth is seldom reduced, except in brachycephalic breeds. The greatest reduction involves the cheek teeth of the lower jaw. In old specimens lost teeth with a resultant obliteration of the alveoli must not be mistaken for a congenital decrease in number.

The third upper premolar is the first tooth to rotate as the muzzle is shortened by selective breeding. Later, all upper premolars may be rotated. The second and third premolars may rotate either clockwise or counterclockwise, although rotation which brings their anterior roots nearer the median plane is the more common. Little additional room is gained by rotation of the first premolar, since its width nearly equals its length. The long fourth upper premolar, however, usually rotates so that its mesial roots lie nearer the median plane than in the normal specimen. Sometimes the second and third premolars lie parallel to each other transversely or sagittally. The molar teeth seem to be little affected by rotation.

All anomalous placements of teeth are more common in the upper than in the lower dental arch. Wood and Wood (1933) describe a dog which had three permanent molars on one side of the upper jaw, and cite nine other accounts of this rare anomaly. These authors give four explanations for extra teeth in general: mechanical splitting of the tooth germ, retention of a deciduous tooth, mutation producing a new tooth, and reversion to an ancestral condition.

Deviation from the normal number of roots or fusion of the roots or of whole teeth is rarely found. Colyer (1936) and Meyer (1942) describe several different kinds of dental anomalies in the German shepherd. Graves (1948) cites the case of an English sheepdog which grew a third set of teeth after extraction of 11 of the permanent teeth when the dog was nine years old. Dechambre (1912) gives a brief description of an English toy terrier which had a total absence of all teeth. Zontine (1975) discusses radiographic techniques for studying the teeth of dogs.

GUMS

The **gums** *(gingivae)* are composed of dense fibrous tissue covered by smooth, heavily vascularized mucosae. The lamina propria is thick and strong. It extends around the necks of the teeth and down into the alveoli to be continuous with the alveolar periosteum. The gums are continuous externally with the mucosa of the vestibule, and internally with that of the floor of the oral cavity proper or of the hard palate. In those breeds with pigmented oral mucosa the gums are likewise pigmented.

THE TONGUE

By GREGORY A. CHIBUZO

Introduction

The **tongue** *(lingua)*, composed primarily of skeletal muscle, appears embryologically as a mesodermal swelling on the floor of the stomodeum and rostral foregut, overlying the first, second, and third branchial arches. There are two **lateral swellings** *(lateral lingual primordia)*, one **median tongue bud** *(tuberculum impar)* and one **median proximal swelling** *(the copula)* immediately caudal to the tuberculum impar (Fig. 7–6). These muscular swellings arise partly as a result of cellular migration from three occipital myotomes. Migrating cells form a condensation in the vicinity of the developing root fibers of the hypoglossal nerve, which innervates the hypoglossal musculature.

This hypoglossal muscle mass contributes most of the intrinsic and extrinsic lingual musculature by a ventrorostral migration into the floor of the first, second, and third branchial arches.

Part of the lingual musculature seems to develop in situ from the mesodermal swelling on the floor of the stomodeum. This view is supported by recent experimental results in the dog tongue which indicates that some facial motor neurons assist in the innervation of lingual musculature. Although the number of facial motor neurons involved is minimal when compared to the overwhelming contribution from the hypoglossal nerve, the finding supports a dual origin of the lingual musculature: in situ and occipital myotomal origins. The dual origin of lingual

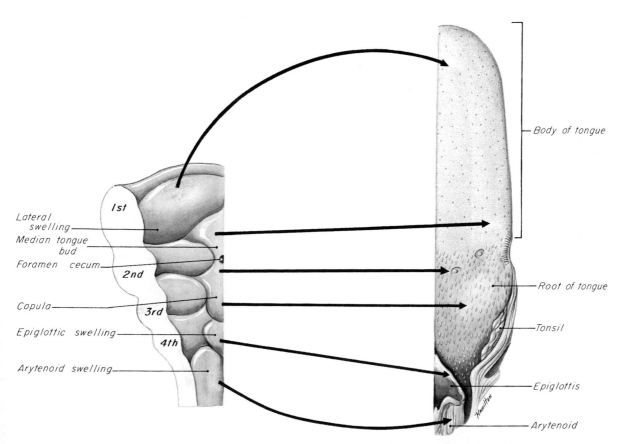

Figure 7–6.   Parts of the tongue and their origin from structures on the floor of the pharynx.

musculature has for a long time been a subject of much debate (Kingsbury 1926; Bates 1948; Langman 1975).

The tongue has a **body** *(corpus linguae)* and a **root** *(radix linguae)* separated by the **vallate papillae** *(papillae vallatae)* arranged in the form of a V. The rostral two-thirds of the tongue (body) is covered by the mucous membrane derived from ectodermal epithelium of the stomodeum. This portion of lingual epithelium receives sensory innervation from the lingual branch of the fifth cranial (trigeminal) nerve for pain, temperature, and touch sensations and from the chorda tympani, a branch of the seventh cranial (facial) nerve, for taste (gustatory). There is no visible foramen cecum or sulcus terminalis in the dog after birth. The mucous membrane, covering the root of the tongue, is derived from the endoderm of the foregut and receives its sensory innervations from the ninth cranial (glossopharyngeal) nerve for all sensations, including taste. Developmentally, these cranial nerves are associated with structures on sequential branchial arches: hence the trigeminal nerve (V) is the nerve of the first (mandibular) arch, the facial (VII) is the nerve of the second (hyoid) arch, the glossopharyngeal (IX) is the nerve of the third arch, and the laryngeal branch of the vagus nerve (X) is the nerve of the fourth arch. The hypoglossal nerve (XII) is the nerve of the hypoglossal musculature contributed from the occipital myotomes. Since these cranial nerves also supply different parts of the tongue, they reflect the embryological origin of the tongue and other lingual structures that they innervate.

The tongue of an adult dog (Fig. 7–7) is an elongated, mobile, muscular organ covered by cornified stratified squamous epithelium. It extends from its attachment on the basihyoid bone to its free tip at the mandibular symphysis. The dorsal surface of the tongue *(dorsum linguae)*, unlike the ventral surface *(facies ventralis linguae)*, is very rough owing to the presence of five types of cornified lingual papillae: filiform, fungiform, vallate, foliate, and conical papillae. The dorsum linguae is divided into two lateral halves by a **median groove** *(sulcus medianus linguae)* which extends from the tip of the tongue to the level of the caudal pair of the vallate papillae.

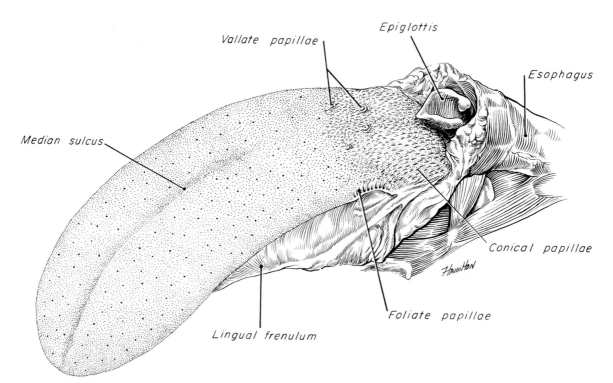

Figure 7–7. The tongue, dorsal aspect: large black dots represent fungiform papillae; finer dots represent filiform papillae.

The **margin of the tongue** *(margo linguae)* separates the dorsal and ventral surfaces of the tongue. The two margins meet rostrally in the formation of the **tip of the tongue** *(apex linguae)*, the thinnest and narrowest end of this muscular organ. Caudal to the apex, the tongue widens and gradually increases in thickness as more intrinsic and extrinsic lingual muscles are incorporated into the body of the tongue. The body is long and slender and lies between the apex and the root of the tongue. The root is the caudal one-third which bears conical, vallate, and foliate papillae (Fig. 7–7).

Filiform and fungiform papillae are located together on the ectodermally derived mucous membrane of the oral portion of the tongue, whereas the foliate, vallate, and conical papillae are located on the entodermally derived mucous membrane of the caudal one-third, or pharyngeal part, of the tongue. These two zones are demarcated in the fetus by the terminal sulcus, which is absent in the dog's tongue at birth. In the fetus it forms the rostral border of the V-shaped arrangement of the vallate papillae.

The **lingual mucosa** *(tunica mucosa linguae)* is thick and heavily cornified on the dorsal surface of the tongue but thin and less cornified on the ventral surface. It is composed of stratified squamous epithelium. This epithelium forms papillae on the dorsum of the tongue. On the ventral surface of the tongue it forms an unpaired, median mucosal fold, the **lingual frenulum** *(frenulum linguae)*, which primarily connects the body of the tongue to the floor of the mouth. A rounded fold of mucosa protruding on each side of the lingual frenulum on the ventral surface of the tongue is the **fimbriated plica** *(plica fimbriata)*. Its lateral border parallels that of the tongue. The fimbriated plica starts close to the rostral end of the frenulum but gradually disappears after a caudal extent of about 2 cm. Usually the sublingual vein, which can be used for venipuncture, is located in the submucosa between the lingual frenulum and the lateral border of the fimbriated plica.

Filiform Papillae

Filiform papillae *(papillae filiformes)* (Fig. 7–8A) are the smallest in size and the most numerous of all lingual papillae. Like the **fungiform papillae,** they are located on the dorsum of the rostral two-thirds of the tongue. Rarely do they extend caudal to the level of the vallate papillae. Filiform papillae are not threadlike, as the name implies, but rather are a complex of primary, secondary, and tertiary serrations. Each complex has a broad base with a dermal core. The dermal core forms the axis of the centrally placed single primary filiform papilla. Each secondary filiform papilla of each complex arises from the basal end of the primary and, like the latter, has a dermal core branching directly off the base of the primary basal core. There are generally four to six secondary filiform papillae around the base of the centrally placed primary filiform papilla. Generally, tertiary filiform papillae arise from the secondary. They are smaller than the secondary papillae and lack dermal cores. They are, therefore, totally made up of epithelial tissue.

These three subtypes of filiform papillae are inclined so that their tips point caudally. The surfaces of these papillae are well cornified to aid in licking and to protect deeper structures from injury. The cornification is greatest at the apex of the primary and secondary filiform papillae and least at their bases, which are bathed in a film of saliva and less exposed to wear. The tertiary filiform papillae, on the other hand, are completely cornified.

It is through the dermal cores (which are absent in the tertiary papillae) that the blood vessels and nerves reach these papillae. Only the lingual nerve, a branch of the trigeminal, innervates the filiform papillae, unlike the fungiform papillae, which receive a dual innervation from both the lingual and the chorda tympani, a branch of facial nerves (Miller 1974, Cheal and Oakley 1977).

Generally, 8 to 10 filiform papilla complexes are spaced around a single fungiform papilla (Fig. 7–8B). This arrangement creates a depression containing a fungiform papilla central to a circular arrangement of 8 to 10 taller filiform papilla complexes. The significance of this arrangement will be explained later.

Fungiform Papillae

Fungiform papillae *(papillae fungiformes)* are mushroom-shaped papillae

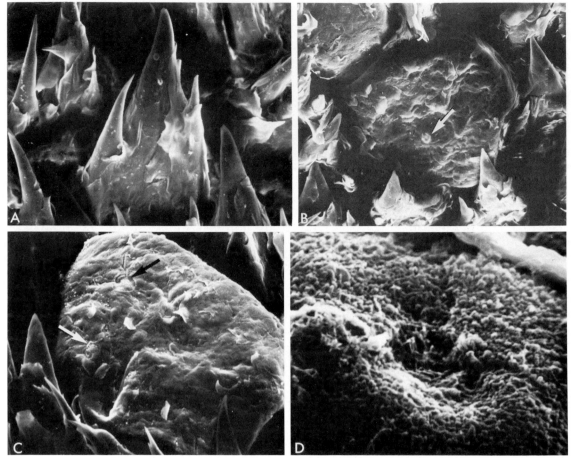

Figure 7–8. Scanning electron micrographs of filiform and fungiform papillae of the tongue.
A. Oblique view of filiform papillae on middle third of tongue.
B. Dorsal view of a fungiform papilla with a taste pore (*arrow*) surrounded by filiform papillae (250×).
C. Oblique view of a fungiform papilla showing taste pores (*arrows*).
D. Enlarged view of a taste pore from (B) (1390 ×) showing microvilli.

(Figs. 7–8B and C) on the rostral two-thirds of the tongue, among the rougher, more numerous filiform papillae. Occasionally, some are found behind the vallate papillae. They are the second most numerous lingual papillae and are most concentrated at the tip and sides of the tongue. The fungiform papillae in these locations are smaller in size than those elsewhere on the tongue. The median sulcus is devoid of fungiform papillae.

Fungiform papillae are shorter, broader, and less numerous than filiform papillae. Though their surfaces are covered by cornified stratified squamous epithelium, the cornification is much thinner than that in the filiform papillae. Each fungiform papilla has a proximal narrow base and an expanded distal end. The base may bear spinous cornified projections on only one side. Since the fungiform papillae are very thinly cornified but very well vascularized, the blood they contain gives them a dark pink appearance in the living state, making them prominent among the taller filiform papillae.

A scanning electron microscopic study of fungiform papillae (Fig. 7–8C) shows that some of them contain taste pores on their expanded surfaces. As many as eight taste pores have been seen on some papillae, but some contain no taste pores at all. These taste pores (Fig. 7–8D) open in the form of a rounded crater, fringed by a thin shelf of cornified epithelium. The rim of each fungiform taste pore is often slightly elevated above the surrounding surface of the papilla. At the center of this rim is a depression which contains numerous microvilli (Fig. 7–8D) which stand out distinctly at high magnifications. It is through these pores that taste fluids reach the taste buds. The depression created by the tall filiform papillae surrounding each short fungiform papilla tends to provide an environment for an adherent pool of taste fluid around the fungiform papillae and the associated taste pores.

A sagittal section of a fungiform papilla reveals a centrally located primary dermal core. Secondary dermal cores branch out from the primary core. Some of these secondary dermal cores border the taste buds which are located on the dorsal surface of the tongue and contact the oral cavity through the microvilli in the taste pores. Each taste bud — some fungiform papillae do not contain any — passes through the total thickness of epithelium and rests on a secondary dermal papilla. It is through the secondary dermal cores that nerves reach the taste buds and the adjacent epithelium. The fungiform papillae receive innervation from two sources: the chorda tympani to the taste buds and the lingual nerve to the epithelium of the papillae (Miller 1974). The terminal ends of taste nerves are non-myelinated. The dermal cores also offer passage for the blood vessels.

Vallate Papillae

In the dog, the **vallate papillae** *(papillae vallatae)* (Fig. 7–7) are located on the caudal third of the dorsum of the tongue. They mark the boundary between the filiform papillae of the oral part and the conical papillae of the pharyngeal part of the tongue. Like the foliate papillae, they are entodermal in origin. Usually there are three to six vallate papillae in the dog, most commonly four. Whenever an even number (four or six) are present, they are arranged in the form of a V evenly located on both sides of the sulcus medianus linguae. The point of the V is always directed caudally. When there is an odd number (three or five), they are asymmetrically arranged: one on one side of sulcus medianus linguae with two on the other; or two on one

side with three on the other. When asymmetrical, the V arrangement is absent because the dog, unlike man, does not have a median vallate papilla.

Each papilla may be simple or complex. The simple types are more prevalent in the even-numbered arrangement, whereas the complex types are seen more often in the odd-numbered arrangement. On the average, the diameter of vallate papillae is 1.5 to 2.5 mm. A simple vallate papilla (Fig. 7–9) has a deep moat around it. This moat, in the dog, is deficient on one side, and hence the old term "circumvallate" is not applicable. The dorsal surface of each papilla contains a central depression through which projects a secondary papilla (Fig. 7–10A and B), the significance of which is not known. This arrangement creates a complete central moat around the secondary papilla. In scanning micrographs (Fig. 7–11C) the top of some of the secondary papilla flares, displaying a rim around a central convexity. While this central elevation does not show a pit, it obviously is very uneven, being peppered with numerous tiny projections. A simple vallate papilla, therefore, has two moats. The central moat, which is complete, surrounds the secondary papilla; the outer moat, which is incomplete, surrounds the vallate papilla. The wall of the outer moat is formed by modified conical papillae arranged side by side in a closely packed manner around the vallate papilla.

While the vallate papillae and, occasionally, the walls of their moats contain taste buds, the secondary papillae contain none. The connective tissue core (the dermal papilla) forms secondary dermal papillae only on the upper half of the papilla. Most taste buds are located ventral to the level of the secondary dermal papillae on the epithelium of the basal half of the papillae. A tangential section that cuts only the epithelium of the vallate papilla contains 6 to 12 vertical rows of taste buds (Fig. 7–10C). Each row circles the papilla with taste buds. The number of taste buds per vallate papilla varies greatly. It is through the dermal papillae that blood vessels and nerves, which are terminal branches of the glossopharyngeal nerve, reach the taste buds.

Complex vallate papillae (Fig. 7–11A) occur occasionally, and, when present,

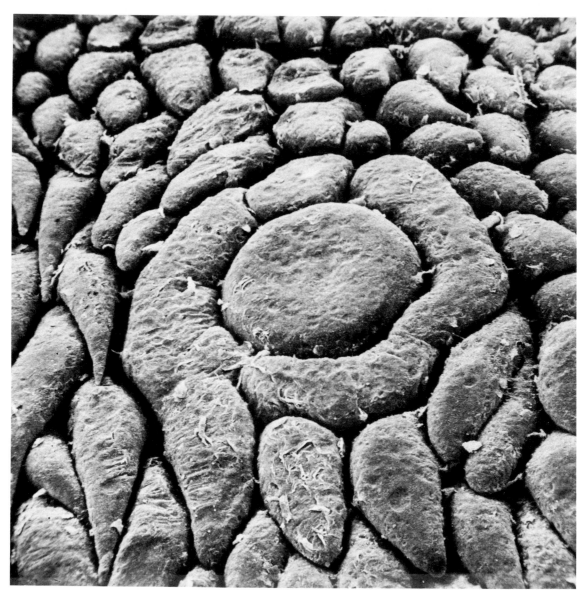

Figure 7–9. Vallate papilla of a puppy. (From Beidler, 1969, Olfaction and Taste III, New York, Rockefeller Univ. Press.)

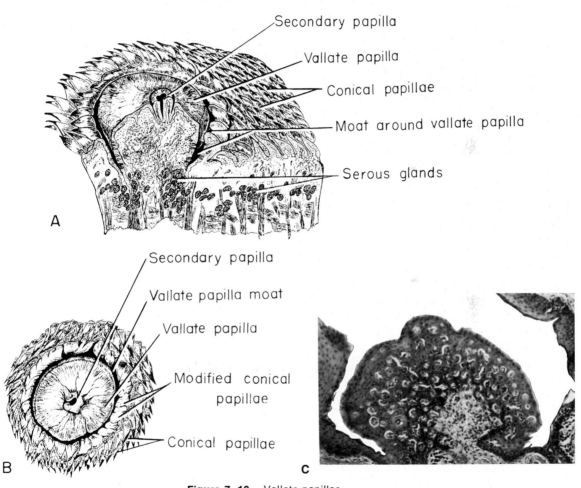

Secondary papilla
Vallate papilla
Conical papillae
Moat around vallate papilla
Serous glands

A

Secondary papilla
Vallate papilla moat
Vallate papilla
Modified conical papillae
Conical papillae

B   C

Figure 7–10.   Vallate papillae.
A. Longitudinal section.
B. Dorsal view.
C. Tangential section.

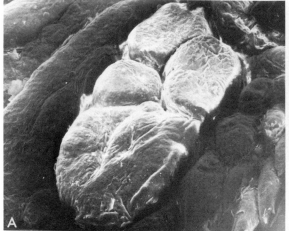
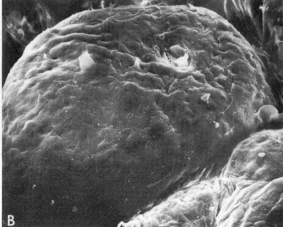
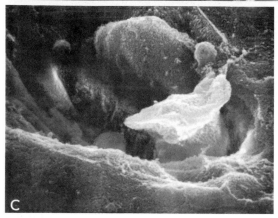

Figure 7–11. Complex vallate papilla (Scanning electron micrograph).

        A. Dorsal view (80 ×).

        B. Dorsal view of left middle component of (A) (400 ×).

        C. Dorsal view to show central moat and secondary papilla of (B) (3200 ×).

usually only one occurs per dog and generally on the side of the tongue with the least number of vallate papillae. Like the simple one, a complex vallate papilla has an outer moat. However, there are two to four individual simple vallate papillae inside this common outer moat. Each papilla of this complex resembles the simple type in every manner, including taste bud distribution. A central moat surrounds the secondary taste bud–free papilla of each member of the component vallate papilla (Figs. 7–11B and C).

    Serous gustatory glands *(glandulae gustatoriae),* or von Ebner's glands, exist at the base of the vallate papillae. They empty their secretions into the bottom of the moat. Associated with the ducts of these glands are lymphatic aggregates.

Foliate Papillae

    There are two groups of **foliate papillae** *(papillae foliatae)* in the dog. They are cov-

ered by cornified stratified squamous epithelium. Each group (Fig. 7–12A) is located on the dorsolateral aspect of the caudal third of the tongue immediately rostral to the palatoglossal fold. Such a group contains 8 to 13 papillae, alternating with 9 to 14 crypts (Fig. 7–12A and B) that parallel the papillae and separate them from one another in a leaflike arrangement. The long axes of these papillae (except the most central one) run obliquely from the side of the tongue toward the dorsum of the tongue in a radiating manner. The papillae, rostral and caudal to the central ones, spread out dorsally. The narrower ventrolateral end of this arrangement converges in an epithelial elevation. The taste fluids that drain out of the crypts are temporarily retarded by this elevation (Fig. 7–12A). The crypts are deepest and longest in the center of a group of foliate papillae and shallowest and shortest at the margins. The central crypts extend medially as diverticula beneath the surface epithelium. Their walls contain several taste buds whose pores

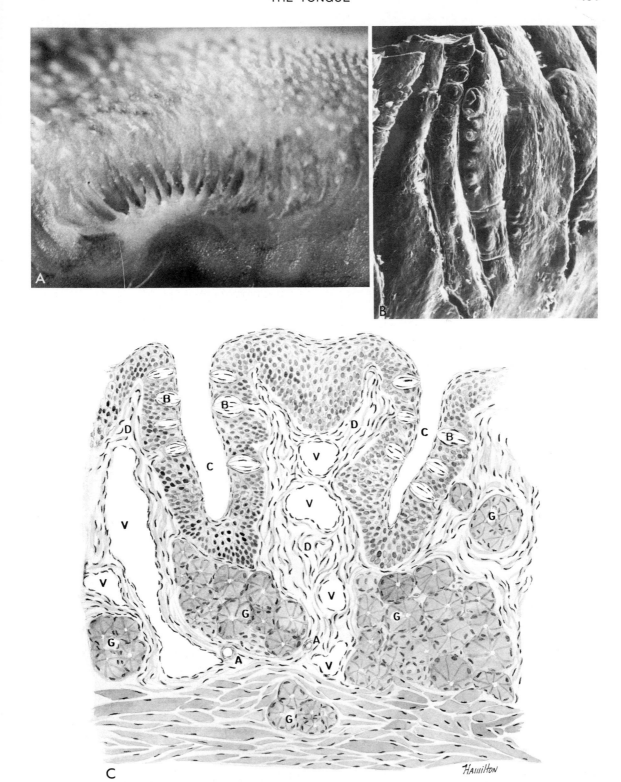

Figure 7–12. A. Foliate papilla, dorsolateral view.
B. Foliate papilla, oblique view (scanning electron micrograph 85 ×).
C. Section of foliate papilla showing taste buds, B; crypts, C; dermal papillae, D; serous glands, G; veins, V., and artery, A.

empty into the diverticula. Similarly, the papillae at the center of a foliate group are broader and longer than those located closer to the margins. Each group of foliate papillae is bordered dorsally, rostrally, and caudally by conical papillae. The lateral boundary is devoid of conical papillae. The epithelium in this area is very thinly cornified and contains the elevation cited previously. The papillae at the caudal and rostral parts of a group are small in size and very irregular in shape. They may resemble the adjacent conical papillae except that they are larger and lie between crypts.

An individual foliate papilla has a central primary dermal core. The secondary dermal papillae arise from the primary. The taste buds are located on the secondary dermal papillae and extend the total thickness of the epithelium (Fig. 7–12C). The majority of the taste buds are found on the sides of the central papillae. They occupy the upper half of the lateral walls of these papillae and are not found on the lower half. The taste pores open into the crypts. Unlike the vallate papillae, foliate papillae may contain taste pores that open on their dorsal surfaces. Irrespective of location in the diverticula, in the crypts, or on the dorsum of a papilla, foliate taste pores are in the form of circular craters with an almost uniform diameter. Scanning electron microscopy of these pores reveals a large number of microvilli.

The ducts of subepithelial serous glands, known as the gustatory salivary glands, (glands of von Ebner) are located at the bases of the papillae and open into the gustatory crypts. Also associated with the bases of some of the papillae are aggregates of lymphatic tissue. The foliate papillae are innervated by the glossopharyngeal nerve.

Conical Papillae

The **conical papillae** (*papillae conicae*) are found on the dorsum of the caudal one-third of the tongue. Each of them stands on a wide circular base and narrows to a thin, hard point at its apex (Fig. 7–13A). The caudally directed apex is more heavily cornified than the base. The area immediately rostral to the rostral pair of vallate papillae is a transitional zone containing a mixture of filiform and conical papillae. The conical papillae in the transitional zone are smaller in size than those caudal to it. The largest conical papillae are found caudal to the palatoglossal fold, but they are less crowded here. Scanning electron micrographs reveal that the unevenness of their surfaces is due to flakes of surface cornification about to fall out. Some conical papillae are modified to form the wall of the outer moat of the vallate papillae. In scanning electron micrographs it is not uncommon to see small conical papillae arising between the bases of full-sized conical papillae (Fig. 7–13B). The dermal core of the

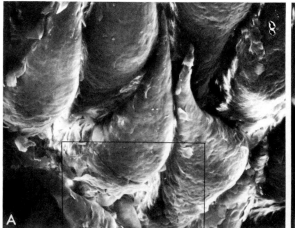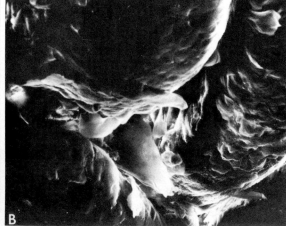

Figure 7–13.   A. Conical papillae (scanning electron micrograph 160 ×).
B. Conical papillae. Enlargement of area indicated in (A) to show a small conical papilla (350 ×).

conical papillae contains secondary dermal papillae. Conical papillae, like filiform papillae, are mechanical and tactile rather than gustatory in function. They are innervated by the glossopharyngeal nerve.

Marginal Papillae

The **marginal papillae** *(papillae marginales)* function in suckling and are present in newborn dogs. They disappear as the pups change from liquid to solid diets. They help in preventing milk from spilling over the tongue and aid in sealing the lips around the nipple for suction. These papillae (Fig. 7–14) are distributed along the margins of the rostral half of the pup's tongue (Habermehl 1952). They are least developed at the margin of the tip of the tongue, best developed at the margins of the tongue in the premolar carrying space, and absent at the margins of the caudal half of the tongue. The premolar carrying space is the area of lack of occlusal contact which extends from the canine teeth to the third premolars. Loss of liquid diet (milk) through this space is prevented by the dorsomedial folding of both the margins of the rostral half of the tongue and the associated marginal papillae during suckling. The marginal papillae adjacent to this space are each from 2.0 mm. to 4.0 mm. long and are arranged in a single compact row so as to obliterate spaces between successive papillae. These papillae are threadlike and narrower at their apices. Those located at the margin of the tip of the tongue are very small, and some may be less than 0.75 mm. long. The marginal papillae are mechanical and tactile rather than gustatory in function.

Taste Buds

A **taste bud** *(caliculus gustatorius)* (Fig. 7–15) is a pear-shaped group of epithelial cells located in the gustatory papillae: fungiform, vallate, and foliate papillae. The component cells of a taste bud lack tonofilaments, which are characteristic of the adjacent, fully differentiated epithelial cells that constitute the perigemmal cells. Lack of these tonofilaments confers a lighter appearance on taste bud cells. Beidler and Smallman (1965), using H[3] thymidine label and light microscopic autoradiography on the fungiform papillae of the rat, has conclusively demonstrated that cells enter the buds at the rate of one cell every 10 hours, with an average renewal time of 11 days. Taste bud cells are broad basally and narrow apically where they communicate with the oral cavity through an apical taste pore. Although blood capillaries lie very close to its base, the taste bud is separated from them and the underlying tissue by a basement membrane, so that blood capillaries are not in intimate contact with taste bud cells (Murray 1973).

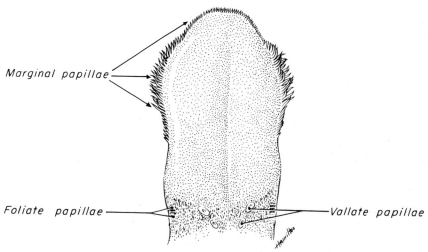

Figure 7–14. Tongue of a puppy, dorsal view.

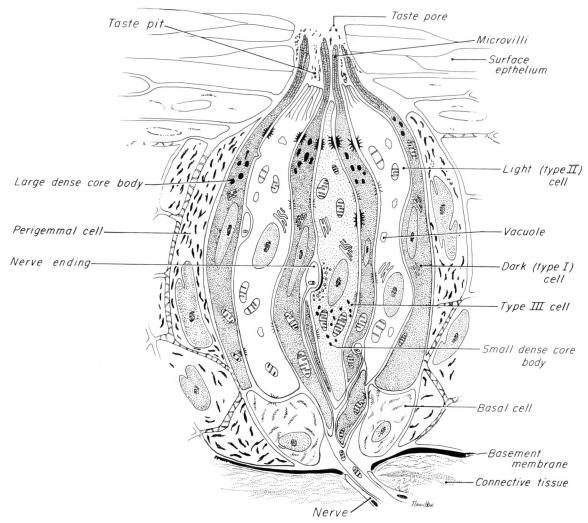

Figure 7–15. Schematic longitudinal section of a taste bud. Electron-microscopic composite.

The taste bud cells are closely applied to one another, although occasional spaces of some magnitude are present. Baradi (1965) suggested that these spaces might be important in taste function.

There are four types of cells in a taste bud (Murray and Murray 1967, Murray 1973): the dark, or type I, cell; the light, or type II, cell; the type III cell; and the basal cell. The nuclei of all four types are located below the upper third of the taste bud. Apically, three of the cell types are joined to each other by tight junctions (*zonula occludens*), but at deeper levels desmosomal connections (*macula adherens*) exist. Those cells that are dark and have apical, dense core particles above their nuclei are known as the dark, or type I, cells. Such cells are narrow and have extensive protoplasmic extensions that wrap around axons entering a taste bud. They also wrap around individual type II and type III cells, thus separating these cells from each other and making it impossible for them to contact each other.

Type I cells are the most numerous of all the cell types. Axons leave the wrapping but do not establish any synaptic contact with either type I or type II cells. Those cells that do not have dense core particles above or below their nuclei but have pale and vacuolated cytoplasm are designated as light, or type II, cells. Both types I and II cells have microvilli, but the microvilli of type II cells are shorter, broader, and more irregular. The mitochondria of the light cells (type II) are emptier than those of the dark cells (type I). Type I cells are considered supportive, whereas type II cells are considered degenerative.

Type III cells are more like type II cells except that they contain small, dark core particles which are located below the nucleus, unlike those of type I cells, which are larger and located above the level of the nucleus. Type III cells, which are in the minority, are the only cell type of the taste bud that has been shown to be in synaptic contact with nerve endings (Murray 1973). Since synaptic vesicles have been demonstrated only with type III cells, these cells are considered to be the ones involved in the transmission of stimulus to the sensory nerves that serve the taste buds. These synaptic vesicles, however, have not been revealed in the dog. Type III cells are located in the center of taste buds. Their terminal ends have not been shown to reach the taste pit but end bluntly just below it. The basal cells are located at the base of the taste bud.

It has been well documented that vallate and foliate taste buds disappear after glos-

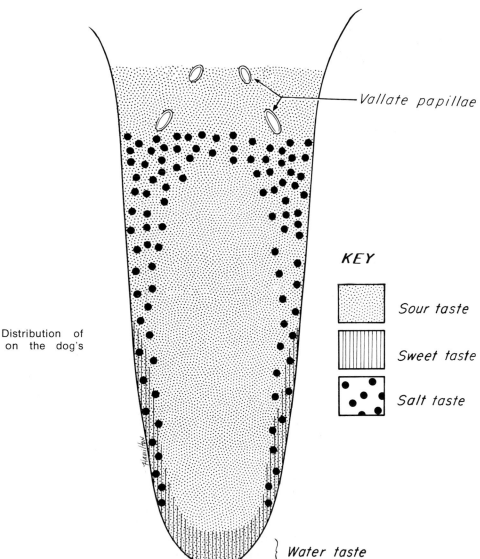

Figure 7–16. Distribution of taste modalities on the dog's tongue.

KEY

Sour taste

Sweet taste

Salt taste

Vallate papillae

} Water taste

sopharyngeal neurectomy and reappear following their reinnervation. Fungiform taste buds also disappear following transection of the chorda tympani in the dog (Olmsted 1921, Olmsted and Pinger 1936), in the rat (Guth 1957), and in the rabbit (El-Eishi and State 1974, and State and Dessouky 1977). This phenomenon indicates that the growth and maintenance of taste buds depend upon intact sensory nerves (State 1977) and that a humoral substance carried by the nerves is responsible for the maintenance of the taste bud (Olmsted 1920, Torrey 1934, State and Dessouky 1977). Andersson et al. (1950) mapped the qualities of taste sensitivity on the dorsum of the dog's tongue and compared them with those of the cat. They found that the cat lacks sweet-sensitive fibers, whereas in the dog an area restricted to rostral margins and tip of the tongue is sensitive to sweetness. The sensitivity to sweet solution decreases caudally and is minimal in the area just rostral to the vallate papillae (Fig. 7–16). The area of salt-taste sensitivity lies within the distribution of the chorda tympani (VII) and does not extend to the vallate area supplied by the glossopharyngeal nerve (IX). Sour-taste sensitivity is almost evenly distributed over the dorsum of the dog's tongue. The fungiform area either in part or as a whole is sensitive to sweetness, sourness, and saltiness. However, while sour taste extends over the fungiform and vallate areas, sweet-taste sensitivity is restricted to the tip and margins of the dog's tongue. This indicates that the areas innervated by the glossopharyngeal nerve are insensitive to sweet and salt tastes, while the areas innervated by the chorda tympani may be sensitive to sweetness, sourness, and saltiness. The tip of the dog's tongue is also sensitive to water taste (Liljestrand and Zotterman 1954).

The Tongue Muscles

The complex but precise movements of the dog's tongue are important for lapping water, food prehension, mastication, swallowing, and prevention of accidental biting of the tongue. These movements are dependent on the coordinated actions of the **extrinsic and intrinsic muscles** of the tongue *(musculi linguae)*. The extrinsic muscles are the m. styloglossus, the m. hyoglossus, and the m. genioglossus. The intrinsic muscle of the tongue (m. lingualis proprius) contains superficial longitudinal (fibrae longitudinales superficiales), deep longitudinal (f. long. profundae), perpendicular (f. perpendiculares), and transverse (f. transversae) muscle fibers.

The *m. styloglossus* is the most lateral of the extrinsic lingual muscles seen at the caudal third of the tongue. The medial border of the middle part of this muscle (Fig. 7–17) parallels the part of the hypoglossal nerve seen at this level. It is narrow proximally but wide and thin rostrally. Its fibers curve downward and forward to insert in the tongue (Fig. 7–18). The styloglossus muscle has three heads which insert at different levels along the axis of the tongue. In a caudorostral sequence these are the short, the long, and the rostral heads of the m. styloglossus. The short head arises from the distal half of the caudal surface of the stylohyoid bone and curves downward and forward over the epihyoid bone. Rostral to the epihyoid bone, its fibers spread out to insert on the proximal part of the tongue with the inserting fibers of the m. hyoglossus. The long head also curves downward and forward. It arises from the stylohyoid bone immediately dorsolateral to the origin of the short head. Rostrally, its fibers are lateral to those of the rostral head. These fibers curve ventrally and rostrally along the ventral surface of the tongue and cross the lateral surface of the genioglossus muscle to insert on the ventral surface of the rostral half of the tongue adjacent to its midline.

The rostral head of the m. styloglossus arises from the rostrodorsal surface of the proximal half of the stylohyoid bone. Its fibers pass dorsal to the insertion of the short head to insert along the lateral surface of the caudal part of the tongue. The fibers of insertion of the rostral head of the styloglossus mingle with those of the m. hyoglossus.

Action: Acting as a unit, the three muscle heads draw the tongue caudally. Acting independently, each head depresses the section of the tongue to which it is attached.

Innervation: N. hypoglossus.

The *m. hyoglossus* is located in the root of the tongue (Fig. 7–18). It arises from the ventrolateral surface of the basihyoid and

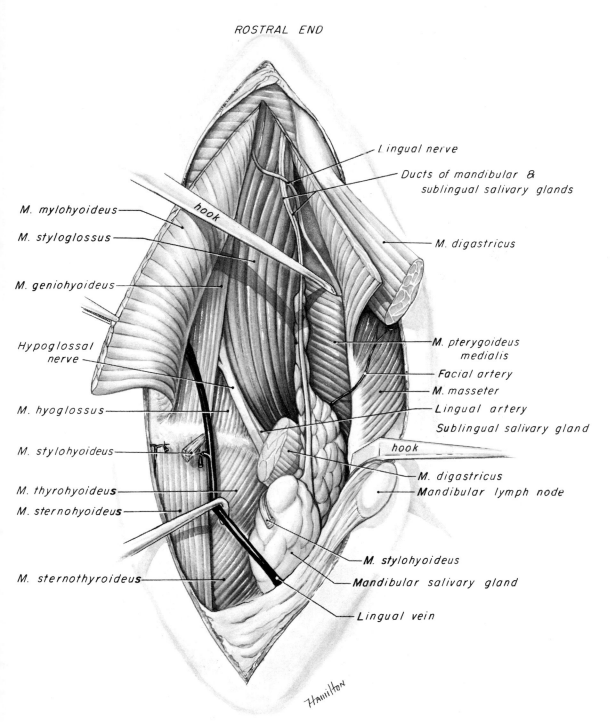

ROSTRAL END

Lingual nerve

Ducts of mandibular & sublingual salivary glands

M. mylohyoideus

M. styloglossus

hook

M. digastricus

M. geniohyoideus

Hypoglossal nerve

M. pterygoideus medialis

Facial artery

M. masseter

M. hyoglossus

Lingual artery

Sublingual salivary gland

M. stylohyoideus

hook

M. digastricus

M. thyrohyoideus

Mandibular lymph node

M. sternohyoideus

M. stylohyoideus

Mandibular salivary gland

M. sternothyroideus

Lingual vein

Hamilton

Figure 7–17. Structures of the intermandibular space and hyoid region.

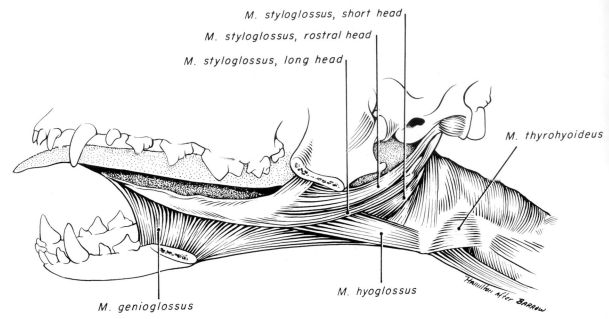

M. styloglossus, short head

M. styloglossus, rostral head

M. styloglossus, long head

M. thyrohyoideus

M. hyoglossus

M. genioglossus

Figure 7–18. Muscles of the tongue, lateral aspect.

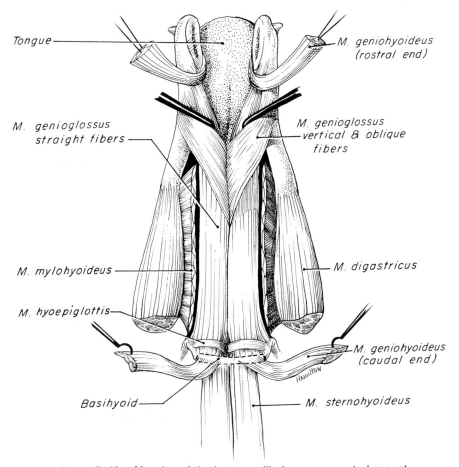

Tongue

M. geniohyoideus (rostral end)

M. genioglossus straight fibers

M. genioglossus vertical & oblique fibers

M. mylohyoideus

M. digastricus

M. hyoepiglottis

M. geniohyoideus (caudal end)

Basihyoid

M. sternohyoideus

Figure 7–19. Muscles of the intermandibular space, ventral aspect.

the adjacent end of the thyrohyoid bone. Rostrally, it is dorsal to the m. mylohyoideus but lateral to mm. geniohyoideus and genioglossus. At the base of the tongue, it crosses the medial side of the m. styloglossus and then inserts in the root and caudal two-thirds of the tongue.

Action: It retracts and depresses the tongue.

Innervation: N. hypoglossus.

The *m. genioglossus* (Fig. 7–19) is a fan-shaped muscle lying dorsal to the geniohyoideus. Its narrow end originates from the medial surface of the mandible adjacent to the mandibular symphysis just caudal to the origin of the m. geniohyoideus. It is composed of three muscle bundles best seen on its medial surface. Each bundle has a distinct mandibular origin near the mandibular symphysis. The bundles of the m. genioglossus are composed of the vertical, the oblique, and the straight fibers (Fig. 7–20).

The **vertical bundle** is located at the rostral part of m. genioglossus and inserts on the rostral half of the tongue just caudal to the lyssa. None of these fibers curves rostrally to the tip of the tongue. The vertical bundle arises from the ventromedial surface of the rostral end of the mandible caudal to the mandibular symphysis. It depresses the rostral half of the tongue just caudal to the lyssa.

The **oblique bundle** lies behind the vertical bundle. Some of its rostral fibers intermingle with the caudal fibers of the vertical bundle. It is narrower, longer, and more oblique than the vertical bundle. The fibers run ventrodorsal from its small ventral site of origin on the ventromedial aspect of the mandible, adjacent but caudal to the origin of the vertical bundle. It inserts on the caudal one-half of the tongue and, by its oblique orientation, helps to protrude the tongue.

The **straight bundle of the m. genioglossus** lies lateral to both the vertical and the oblique bundles. Its orientation parallels the body of the mandible. It originates with the m. geniohyoideus from the caudal border of the mandibular symphysis. However, its fibers of origin are medial to those of the

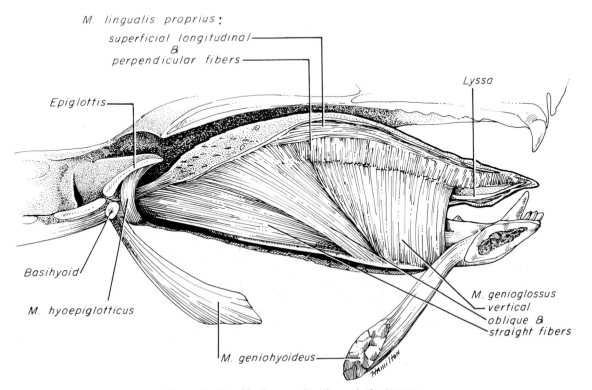

Figure 7–20. Median section through the tongue.

geniohyoideus. It has three insertion sites: the caudal one-third of the tongue, the basihyoid bone, and the ceratohyoid bone. Its principal action is to protrude the tongue. Caudally, one straight bundle is separated from its contralateral fellow by the m. hyo-epiglotticus (Fig. 7–20), which is located on the ventromedial surface of the basihyoid insertion of the straight bundle of genioglossus. Each m. hyoepiglotticus continues ventrally from the rostral side of the epiglottis to form a loop around the ventral surface of the insertion of the ipsilateral m. genioglossus on the basihyoid bone.

Action: As a unit, the genioglossus depresses and protrudes the tongue.
Innervation: N. hypoglossus.

The **intrinsic muscle of the tongue** (m. lingualis proprius) forms the core of the tongue and is arranged bilaterally in four groups.

The **superficial longitudinal fibers** (fibrae longitudinales superficiales) are located immediately under the dorsal lingual mucosa. They are best developed in the caudal half of the tongue (Fig. 7–21), where the fibers are organized into a compact mass on either side of the dorsal part of the septum of the tongue.

More rostrally (Fig. 7–22), the superficial longitudinal fibers are well spaced in distinct small groups arranged symmetrically so that the largest bundles are next to the septum of the tongue, while the smallest bundles are situated next to the edge of the tongue.

The **deep longitudinal fibers** (fibrae longitudinales profundae) are located in the ventral half of the tongue. They are less numerous and less organized than the superficial longitudinal fibers. They often intermingle with the inserting fibers of the extrinsic lingual muscles and the **perpendicular** (fibrae perpendiculares) and **transverse** (fibrae transversae) **fibers** of the intrinsic muscles. Rostrally, however, the deep longitudinal fibers organize into a compact mass. Here they also surround the lyssa (Figs. 7–20 and 7–22).

The transverse and the perpendicular fibers form an intricate network between the superficial and deep longitudinal muscle fibers. They occupy a wide area in the center of the tongue.

Action: As a unit, the intrinsic muscles protrude the tongue, bring about complicated intricate local move-

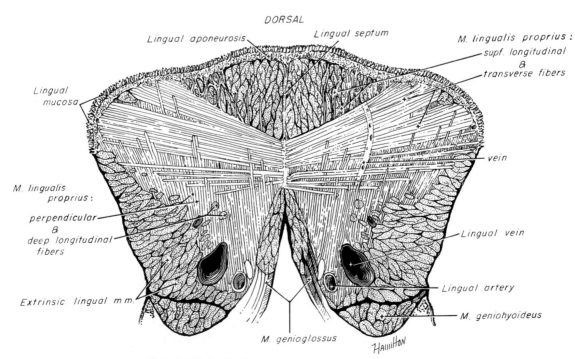

Figure 7–21. Transverse section through root of tongue.

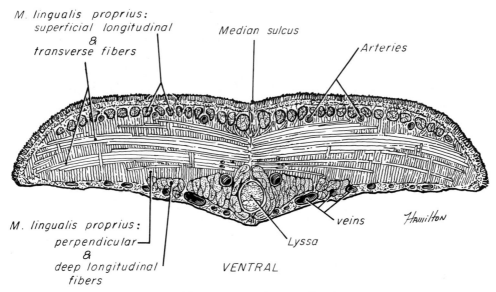

Figure 7-22. Transverse section near tip of tongue.

ments, and prevent the tongue from being bitten.

Innervation: N. hypoglossus.

According to Abd-el-Malek (1938), the superficial longitudinal portion of the m. lingualis proprius is the only contralateral deviator of the tongue. The deep longitudinal portion of the m. lingualis proprius, the styloglossus muscle, and the rostral fibers of hyoglossus and genioglossus (Abd-el-Malek 1938, Bennet and Hutchinson 1946), besides being homolateral deviators, are ventroflexors of the tip of the tongue. The tongue takes part in all stages of mastication (Abd-el-Malek 1955). During the "preparation" stage, the tongue becomes troughlike. During the "throwing" stage, it moves the food between the teeth. During the "guarding" stage, the tongue action prevents the food from falling back from the teeth. During the "sorting out" stage, the well-ground food particles are selected for molding into boluses, which are formed during the "bolus formation" stage prior to swallowing.

The Lyssa

The lyssa (Fig. 7–23) is a rodlike body, about 4 cm. long in a medium-sized dog, which lies on the median plane in the free end of the tongue. It extends from almost the tip of the tongue to the level of the rostral part of the vertical fibers of the genioglossus muscle (Figs. 7–20 and 7–22). Its rostral end is superficial beneath the mucosa, whereas its caudal end is buried among striated muscle fibers. It is encapsulated by a dense sheath of connective tissue. Adipose tissue, striated muscle, and, occasionally, islands of cartilage fill the capsule (Trautmann and Fiebiger 1957, Dellmann and Brown 1976). Whereas adipose tissue predominates in its ventral half, striated muscle occupies most of its dorsal half, especially in the middle part of the lyssa. It receives blood vessels and branches of the hypoglossal nerve. Bennett (1944) demonstrated that the lyssa was elongated when the tongue muscles were stimulated via the hypoglossal nerve and suggested that the lyssa may act as a stretch receptor.

The Glands of the Tongue

The tongue is rich in **salivary glands** (*glandulae linguales*) (Fig. 7–12C) of both serous and mucous types (Gómez 1961, Rakhawy 1975). Glands that are associated with the base of vallate or foliate papillae are exclusively serous. They are the **gustatory** (von Ebner's) **glands** (*glandulae gustatoriae*), whose ducts open into the gustatory furrow

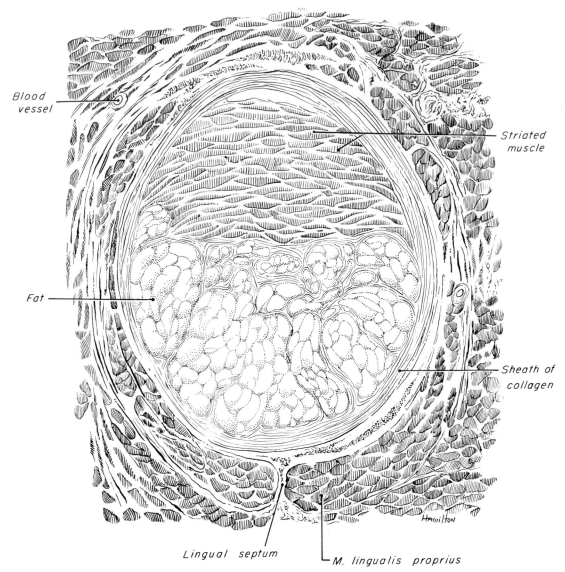

Blood
vessel

Striated
muscle

Fat

Sheath of
collagen

Lingual septum

M. lingualis proprius

Figure 7–23. Transverse section of the lyssa.

or moat. Exclusively serous glands are also present among the bundles of the intrinsic muscles of the tongue, especially in the caudal one-third of the tongue. Mucoserous glands are abundantly distributed in the lingual submucosa at the caudal one-third of the tongue. The lateral margins of the tongue also contain seromucous glands in the submucosa. Small ganglionic cell bodies in groups or in isolation are present near the acini and among the acinar cells (Gómez 1961, Rakhawy, 1972). These are postganglionic parasympathetic cell bodies that are innervated by preganglionic axons whose cell bodies are located in the salivatory nucleus (Chibuzo et al. 1979a). These axons reach the glands of the tongue through the chorda tympani and glossopharyngeal nerves, which are considered to be secretory motor nerves (Chibuzo et al. 1979a, b).

Blood Supply

The arterial supply to the tongue is primarily through the paired lingual arteries (Fig.

7–21) which enter the tongue deep to the hypoglossal nerve (Fig. 7–17). The **lingual artery** *(arteria lingualis)* crosses the medial surface of the m. hypoglossus at the root of the tongue and gives off muscular branches to the intrinsic and extrinsic muscles of the tongue as it runs to the tip of the tongue. The mm. genioglossus and geniohyoideus also receive some collateral arterial blood supply through the **sublingual artery** *(a. sublingualis)*. According to Nikolov and Schumacher (1973), primary arteries or muscular arterial branches of the lingual artery break up into secondary arteries, which form a capillary rete in the muscles. All lingual tributary veins (Fig. 7–22) are valveless (Nikolov and Schumacher 1973) and do not have parietal musculature except at their openings into the **lingual vein** (Figs. 7–21 and 7–17). Brown (1937) and Pritchard and Daniel (1953) demonstrated the presence of many arteriovenous anastomoses in the dog's tongue. Their predominantly superficial location in the tunica propria covering the dorsal surface of the tongue is indicative of their thermoregulatory function (Brown 1937).

Nerve Supply

The tongue is innervated by the lingual (V), chorda tympani (VII), glossopharyngeal (IX), and hypoglossal (XII) nerves (Figs. 7–24 and 7–17).

The **lingual nerve** *(n. lingualis)*, a branch of the trigeminal nerve, contains the general somatic afferent fibers that convey exteroceptive (tactile, pain, and thermal) impulses from the rostral two-thirds of the lingual mucosa. Their cell bodies of origin are situated in the trigeminal ganglion, from which their central axons enter the brain stem and terminate in the principal sensory trigeminal nucleus.

The *chorda tympani*, a branch of the facial nerve, supplies both special visceral afferent and general visceral efferent fibers to the tongue. The special visceral afferent component constitutes the gustatory fibers that innervate the fungiform taste buds distributed to mucosa of the rostral two-thirds of the tongue. The cell bodies of origin of these neurons are located in the geniculate ganglion, and their central axons enter the brain stem and terminate around neurons dorsal to

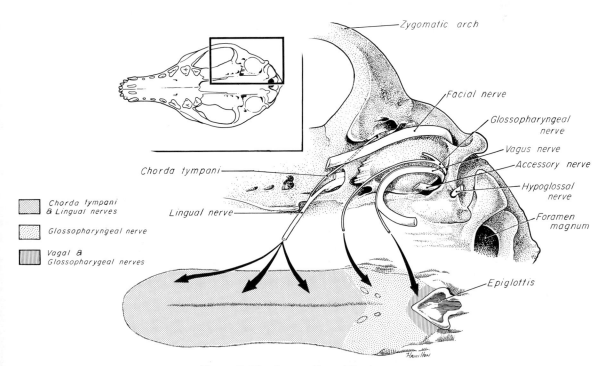

Figure 7–24. Innervation of the tongue.

the fasciculus solitarius. The general visceral efferent fibers innervate the glands of the tongue. Their parasympathetic postganglionic cell bodies are within or close to these glands in the tongue substance (Gómez 1961, Rakhawy, 1972). Their preganglioic cell bodies are located in the rostral part of the salivatory nucleus suprajacent to the facial nucleus (Chibuzo et al. 1979a). These general visceral efferent fibers are secretory and are vasodilators to the glands of the tongue.

The **glossopharyngeal nerve** (*n. glossopharyngeus*) carries special visceral afferent, general visceral afferent, and general visceral efferent fibers to the caudal one-third of the tongue. The special visceral afferent fibers innervate the vallate and foliate taste buds and therefore carry gustatory impulses from the caudal one-third of the tongue. The cell bodies of origin of these fibers are located in the glossopharyngeal ganglion. Their central axons enter the fasciculus solitarius, located in the brain stem. The general visceral afferent fibers carry exteroceptive (tactile, pain, and thermal) impulses from the mucosa of the caudal one-third of the tongue. The cell bodies of origin of this component are also situated in the glossopharyngeal ganglion, from which their central axons enter the brain stem and become part of the fasciculus solitarius. The general visceral efferent component of the glossopharyngeal nerve sends secretory and vasodilator fibers to the lingual glands located in the caudal part of the tongue. Their parasympathetic postganglionic cell bodies are located in the tongue substance close to or among these glands. However, the preganglionic cell bodies are in the salivatory nucleus dorsal to the facial nucleus (Chibuzo et al. 1979a). The dog does not have separate proximal and distal glossopharyngeal ganglia. They apparently have fused. What was formerly called the petrosal ganglion is now known as the *ganglion distale n. glossopharyngei* (N.A.V. 1973).

The **hypoglossal nerve** (*n. hypoglossus*) (Fig. 7–17) contains the general somatic motor fibers that innervate the mm. genioglossus, hyoglossus, styloglossus, and lingualis proprius. It has also been demonstrated that this nerve in the dog (Sobusiak et al. 1967) contains ganglionic cell bodies throughout its peripheral course. It has been suggested that these sensory-like neurons are sensory for deep sensation and are proba- (Chibuzo 1979) that some sensory meurons (Pearson 1945, Adatia and Gehring 1971). Recently, it has also been demonstrated (Chibuzo 1979) that some sensory neurons from the upper cervical ganglia reach the muscles of the dog's tongue via the hypoglossal nerve.

BIBLIOGRAPHY

Abd-el-Malek, S. 1938. A contribution to the study of the movements of the tongue in animals with special reference to the cat. J. Anat. 73:15–30.

————. 1955. The part played by the tongue in mastication and deglutition. J. Anat. 89:250–255.

Adatia, A. K., and E. N. Gehring. 1971. Proprioceptive innervation of the tongue. J. Anat. *110*:215–220.

Andersson, B., S. Landgren, L. Olsson, and Y. Zotterman, 1950. The sweet taste fibers of the dog. Acta Physiol. Scand. *21*:105–119.

Baradi, A. F. 1965. Intragemmal spaces in taste buds. Z. Zellforsch. Microsk. Anat. 65:313–318.

Bates, M. N. 1948. The early development of hypoglossal musculature in the cat. Am. J. Anat. 83:329–355.

Beidler, L. M., and R. L. Smallman. 1965. Renewal of cells within taste buds. J. Cell Biol. 27:263–272.

Bennett, G. A. 1944. The lyssa of the dog. Anat. Rec. 88:422.

Bennett, G. A., and R. C. Hutchinson. 1946. Experimental studies on the movements of the mammalian tongue; the protrusion mechanism of the tongue (dog). Anat. Rec. 94:57–83.

Brown, M. E. 1937. The occurrence of arterio-venous anastomoses in the tongue of the dog. Anat. Rec. 69:287–292.

Cheal, M., and B. Oakley. 1977. Regeneration of fungiform taste buds; temporal and special characteristics. J. Comp. Neurol. 172:609–626.

Chibuzo, G. A. 1979. Locations of primary motor and sensory cell bodies that innervate the dog's tongue. Ph.D. Thesis. Cornell Univ., Ithaca, N. Y.

Chibuzo, G. A., J. F. Cummings, and H. E. Evans. 1979a. Experimental investigation of the salivatory centers in the dog; evidence for trigeminal innervation. Anat. Rec. 193:162.

————. 1979b. Surgical procedures for exposure of the chorda tympani in dogs: a ventral approach. Cornell Vet. (In press).

Dellmann, H., and E. M. Brown. 1976. Textbook of Veterinary Histology. Philadelphia, Lea & Febiger.

El-Eishi, H. I., and F. A. State. 1974. The role of the nerve in the formation and maintenance of taste buds. Acta Anat. 89:599–609.

Gómez, H. 1961. The innervation of lingual salivary glands. Anat. Rec. *139*:69–76.

Guth, L. 1957. The effect of glossopharyngeal nerve

transection on the circumvallate papillae of the rat. Anat. Rec. *128*:715–731.

Habermehl, K. H. 1952. Über besondere Randpapillen an der Zunge neugeborener Säugetiere. Z. Anat. Entw. Gesch. *116*:355–372.

Kingsbury, B. F. 1926. Branchiomerism and the theory of head segmentation. J. Morph. Physiol. *42*:83–109.

Langman, J. 1975. Medical Embryology. Baltimore, The Williams & Wilkins Co.

Liljestrand, G., and Y. Zotterman. 1954. The water taste in mammals. Acta Physiol. Scand. *32*:291–303.

Miller, I. J. 1974. Branched chorda tympani neurons and interactions among taste receptors. J. Comp. Neurol. *158*:155–166.

Murray, R. G. 1973. The ultrastructure of taste buds, *In*: The Ultrastructure of Sensory Organs, edited by I. Friedman. London, North-Holland Publishing Co.

Murray, R. G., and A. Murray. 1967. Fine structure of taste buds of rabbit foliate papillae. J. Ultrastruct. Res. *19*:327–353.

Nikolov, Von Sp. D., and G. H. Schumacher. 1973. Zur Frage der Blutgefässversorgung der Zunge; mikrovaskularisation der Zungenmuskeln des Hundes. Dtsch. Stomatol. *23*:337–343.

Olmsted, J. M. D. 1920. The nerve as a formative influence in the development of taste buds. J. Comp. Neurol. *31*:465–468.

————. 1921. Effect of cutting the lingual nerve of the dog. J. Comp. Neurol. *33*:149–154.

Olmsted, J. M. D., and R. R. Pinger. 1936. Regeneration of taste buds after suture of lingual and hypoglossal nerves. Am. J. Physiol. *116*:225–227.

Pearson, A. A. 1945. Further observations on the intramedullary sensory type neurons along the hypoglossal nerve. J. Comp. Neurol. *82*:93–100.

Prichard, M. M. L., and P. M. Daniel. 1953. Arteriovenous anastomosis in the tongue of the dog. J. Anat. *87*:66–74.

Rakhawy, M. T. 1972. Phosphatases in the nervous tissue; the nature of the ganglionic nerve cells in the tongue. Acta Anat. *83*:356–366.

————. 1975. Phosphatases in the lingual glands of man and dog. Acta. Anat. *92*:607–614.

Sobusiak, T., R. Zimny, A. Obrebosski, and R. Skornicki. 1967. Ganglionic cells of the hypoglossal nerve in the dog. Folio Morphol. *26*:298–306.

State, F. A. 1977. Circumvallate papilla of dog following suture of the hypoglossal and glossopharyngeal nerves. Acta Anat. *98*:413–419.

State, F. A., and H. I. Dessouky. 1977. Effect of length of the distal stump of transected nerve upon the rate of degeneration of taste buds. Acta Anat. *98*:353–360.

Torrey, T. W. 1934. The relation of taste buds to their nerve fibers. J. Comp. Neurol. *59*:203–220.

Trautmann, A., and J. Fiebiger. 1957. Fundamentals of the Histology of Domestic Animals, translated and revised from 8th and 9th German editions, 1949, by R. E. Habel and E. L. Biberstein. Ithaca, N. Y., Comstock Publishing Assoc.

Salivary Glands

The salivary glands, broadly speaking, are all of those glands which pour their secretions into the oral cavity. These are the parotid, mandibular, sublingual, and zygomatic glands. Radiographic examination of the injected duct system of the salivary glands (sialography) is a practical means of discovering and treating salivary mucoceles (Harvey 1970, Glen 1972).

PAROTID GLAND

The **parotid gland** *(glandula parotis)* (Fig. 7–25) lies at the junction of the head and neck overlying the basal portion of the auricular cartilage. Its outline is V-shaped as viewed from the surface with the apex directed ventrally. The parotid gland is bounded caudally by the sternomastoideus and cleidocervicalis muscles, and rostrally by the masseter muscle and the temporomandibular joint. The gland weighs about 7 gm. and has an overall length of approximately 6 cm. It is thickest ventrally, measuring about 1.5 cm. The gland is a dark flesh color, with coarse lobulations visible through its thin capsule. It is divided into superficial and deep portions.

The **superficial portion** *(pars superficialis)* of the parotid gland consists of the two limbs of the V and the dorsally concave, thin-edged portion which connects the two limbs.

The **deep portion** *(pars profunda)* of the parotid gland is wedge-shaped and lies ventral to the cartilaginous external ear canal. It is dorsal to the cranial pole of the mandibular gland and extends medially toward the tympanic bulla and wall of the nasal pharynx.

Of the **three angles** of the parotid gland, one points dorsorostrally, one dorsocaudally, and the third ventrally. The **borders** are dorsal, rostral and caudal. The **superficial surface** is nearly flat transversely and only slightly convex longitudinally. It is crossed

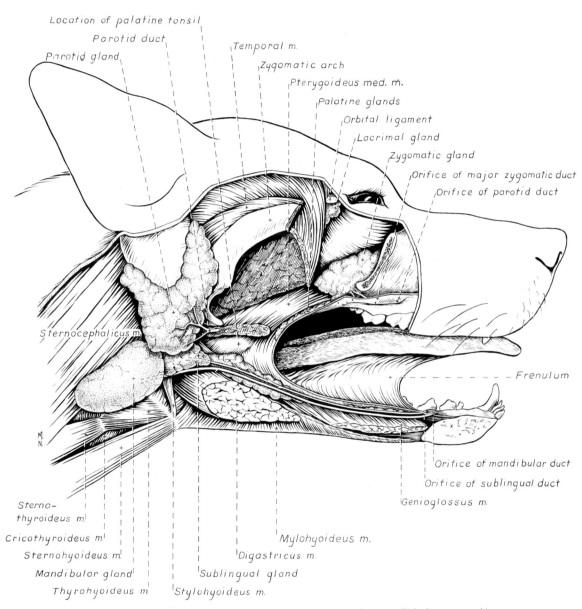

Location of palatine tonsil
Parotid duct
Parotid gland
Temporal m.
Zygomatic arch
Pterygoideus med. m.
Palatine glands
Orbital ligament
Lacrimal gland
Zygomatic gland
Orifice of major zygomatic duct
Orifice of parotid duct
Sternocephalicus m.
Frenulum
Orifice of mandibular duct
Orifice of sublingual duct
Genioglossus m.
Sterno-thyroideus m.
Cricothyroideus m.
Sternohyoideus m.
Mandibular gland
Thyrohyoideus m.
Stylohyoideus m.
Sublingual gland
Digastricus m.
Mylohyoideus m.

Figure 7–25.   Salivary glands. (The right half of the mandible is removed.)

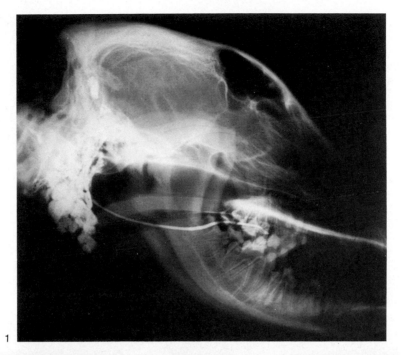

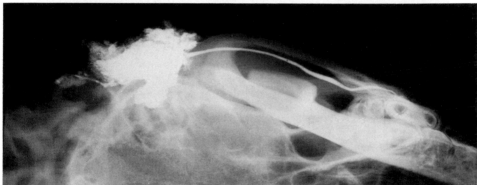

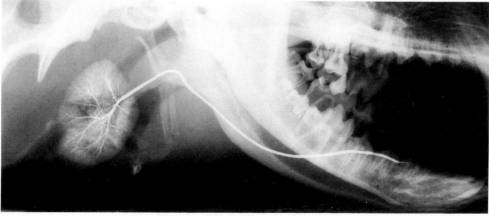

Figure 7–25. *Continued.* Radiograph 1. Lateral radiograph of the parotid gland and its duct.
Radiograph 2. Dorsoventral radiograph of the parotid gland and its duct.
Radiograph 3. Lateral radiograph of the mandibular gland and its duct.
Radiographs continued on the following page

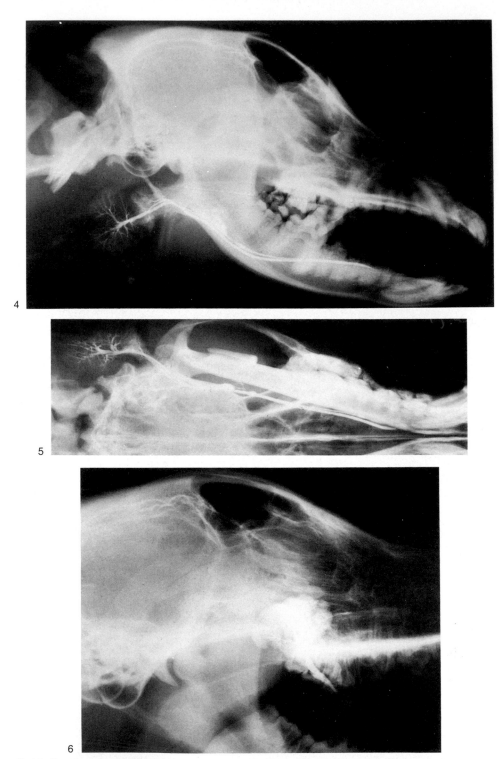

Figure 7–25. *Continued.*   Radiograph 4. Lateral radiograph of the mandibular and sublingual glands and their ducts. The sublingual gland lies rostral to and in contact with the mandibular gland.
Radiograph 5. Dorsoventral radiograph of the mandibular and sublingual glands and their ducts. The sublingual gland lies rostral to and in contact with the mandibular gland.
Radiograph 6. Lateral radiograph of the zygomatic gland and its ducts.

vertically by the straplike m. parotidoauricularis (also known as the m. depressor auriculae), which, in turn, is covered ventrally by fascicles of the platysma. Near the ventral angle the gland is usually tunneled by the internal maxillary vein. From under its rostral border emerge the palpebral, auriculotemporal, and the dorsal and ventral buccal nerves. The parotid lymph node usually lies for the most part under the rostral border near the ventral angle. The rostral auricular artery and vein and the transverse facial artery run under or along the rostral border. The caudal border is circled by branches of the intermediate auricular blood vessels. Some of these structures may run through the gland.

The dorsal border is not related to any large nerves or vessels, but its position is important because of its close proximity to the external ear canal. Blakeley (1957) gives data on the success of the operation to extend the intertragic incisure to a level of the horizontal part of the external ear canal. This operation is highly successful in correcting chronic otitis externa. In such an operation the dorsal border of the parotid gland, unless it was retracted, would be partly incised. The deep part of the parotid gland is related to the facial nerve and its terminal branches as they emerge caudal to the osseous external acoustic meatus. The deep portion of the gland is also related to the maxillary and superficial temporal arteries. The internal maxillary vein is related to the parotid gland at a more ventral and superficial level than are the nerves and arteries.

The **parotid duct** (*ductus parotideus*) is about 1.5 mm. in diameter and 6 cm. long. It is formed by two or three converging radicles which leave the ventral third of the rostral border of the gland and unite with each other on the masseter muscle several millimeters from the gland. The duct is rather closely united to the lateral surface of the masseter muscle by superficial fascia as it runs straight forward to the cheek, parallel, or nearly parallel, to the fibers of the masseter. Testoni, Lohse, and Hyde (1977) described the terminal part of the parotid duct as having two right angle curves, one medially and the other ventrally, before passing through the buccal mucosa. It opens into the buccal cavity at the rostral end of a

blunt ridge of mucosa by a small papilla. This is located opposite the caudal margin of the fourth upper premolar, or shearing tooth. The ridge of mucosa, on the rostral end of which the duct opens, runs caudally to the end of the dental arch lying in or near the fornix formed by the attachment of the cheek to the upper jaw.

Accessory parotid glands (*glandulae parotis accessoria*) are usually present on one or both sides. They range in size from single lobules to small oval glandular masses over 1 cm. long. They usually lie above the parotid duct and may be placed at any level along it. Their small ducts empty into the main parotid duct.

The fascia covering the parotid gland is thin and of the areolar type. At the borders of the gland it blends with the superficial fascia of the head, ear, and neck. Similar fascia lines the space in which the gland lies. The fascia separating the lobules is abundant and loose.

The **parotid artery**, a branch of the external carotid, is wholly distributed to the parotid gland and usually is its main blood supply. The caudal auricular, masseter, transverse facial, and rostral auricular arteries all send twigs to the parotid gland. The veins which drain the parotid are radicles of the **superficial temporal and great auricular veins**. The lymphatics from the parotid gland drain into the **parotid and medial retropharyngeal lymph nodes**. The parotid gland receives parasympathetic nerve fibers through the **auriculotemporal branch of the trigeminal nerve**. These fibers originally come from the glossopharyngeal nerve and run to the auriculotemporal in the short tympanic nerve. The sympathetic fibers reach the gland from the **external carotid plexus of nerves** by branches which follow mainly the parotid artery to the gland. Histologically, the parotid salivary gland is a serous, compound tubuloalveolar gland.

MANDIBULAR GLAND

The **mandibular gland** (*glandula mandibularis*) (Fig. 7–25) is an ovoid body lying largely between the external and internal maxillary veins just caudal to the angle of the jaw. It weighs approximately 8 gm., being a little heavier than the parotid gland.

Its lobules, which are of about the same dimensions as the parotid lobules, are fitted together much more compactly, with less connective tissue separating them. The whole organ is a light buff color and is not sharply separated from the smaller monostomatic portion of the sublingual gland. This portion of the sublingual gland butts against the ventral part of the cranial pole of the mandibular gland, on which it leaves a nearly flat, oblique impression.

From a superficial gross appearance this portion of the sublingual gland constitutes the pointed cranial pole of the salivary mass in this location, since both of the glands are contained within the same heavy fibrous capsule. The common capsule is a specialization of the buccopharyngeal fascia and does not send trabeculae into the glands. It is derived primarily from the deep cervical fascia. The mandibular gland has cranial and caudal poles and superficial and deep surfaces. Its cranial pole is truncate and is related to the major portion of the sublingual gland; the caudal pole forms an even arc vertically as it unites the superficial and deep surfaces at an acute angle. The superficial surface is slightly rounded in all planes and is grooved dorsally for the internal maxillary vein. Its dorsocranial part is variably overlapped by the parotid gland; ventrally it is related to the mandibular lymph node, or nodes, lying dorsal to the external maxillary vein. The deep surface is further divided into subsurfaces by the following structures on which it lies: the muscle and terminal tendon of the sternomastoideus, dorsocaudally; the medial retropharyngeal lymph node and larynx, medially; and the digastric and ribbon-like stylohyoid muscles, cranially.

The mandibular gland is a mixed gland. Stormont (1928) gives a brief review of the literature concerning the histology of the salivary glands of carnivores.

The **mandibular duct** *(ductus mandibularis)* leaves the medial surface of the gland near the ventromedial part of the impression formed by the sublingual gland. As the initial part of the duct runs rostromedially it lies in relation to the medial surface of the sublingual gland. In association with this gland the duct lies between the masseter muscle and mandible laterally, and the digastric muscle medially. In this location the duct runs obliquely anteromedially. Upon reaching the lateral part of the pharyngeal mucosa, it arches forward and, with lobules of the rostral part of the sublingual gland, runs in the intermuscular septum between the medially located m. styloglossus and the laterally located m. mylohyoideus. Throughout the remaining part of its course the mandibular duct is closely related to the sublingual duct, and is described with it following the description of the sublingual gland.

The largest artery supplying the mandibular gland, the glandular branch of the **facial artery**, enters the gland where the mandibular duct leaves it. Entering the dorsal part of the deep surface of the gland are one or two small branches from the **caudal auricular artery**. The chief vein draining the gland leaves its deep surface and terminates usually in the **lingual vein** as this vessel enters the external maxillary. A second vein leaves the caudal part of the gland and terminates in the **facial, maxillary,** or **lingual vein**. Its parasympathetic fibers come from the **facial nerve**. These fibers first run in the chorda tympani to the mandibular branch of the trigeminal nerve and continue in the lingual branch of the latter to the mandibular ganglion, where they synapse with postganglionic neurons. Secretory fibers from these cells run with the mandibular duct to the gland. Sympathetic fibers reach the gland by means of the **perivascular plexus around the glandular artery**. The lymphatics drain into the **medial retropharyngeal lymph node**.

SUBLINGUAL GLAND

The **sublingual gland** *(glandula sublingualis)* (Fig. 7–25) is the smallest of the four pairs of major salivary glands. It weighs about 1 gm. and consists of an aggregation of two or more lobulated masses. The nearly flat, truncated base, or posterior surface, of the largest division of the gland is closely related to the blunt anterior end of the mandibular gland. Both glands are enclosed in the same fibrous capsule, the sublingual gland being distinguishable by its slightly darker color. The tapered extremity of the sublingual gland extends rostromedially between the caudomedial border of the masseter muscle, rostrally. It curves rostrally upon reaching the lateral surface of the styloglossus, so that it lies medial to the body of the mandible.

Usually, separated from the rostral end of this portion of the gland, is an ovoid cluster of lobules which may reach a length of 3 cm. and a width of 1 cm. These lobules lie directly under the mucosa which reflects from the jaw to the tongue. Their secretion is poured into the main sublingual duct through four to six short excretory ducts. The sublingual duct lies ventral to the gland. Since these portions of the gland empty into the major sublingual duct, they are known as the *pars monostomatica* of the sublingual gland. The *pars polystomatica* consists of that portion of the sublingual gland which discharges its secretion directly into the oral cavity without its passing through the main sublingual duct. The gland consists of 6 to 12 small, usually isolated lobules of salivary tissue which lie under the mucosa on each side of the body of the tongue rostral to the lingual branch of the trigeminal nerve. They are so small that no definite sublingual fold is formed by them. They open into the ventral fornix of the vestibule by several ducts.

The **major sublingual duct** *(ductus sublingualis major)* is closely related to the mandibular duct throughout its course in the intermandibular space, being located dorsal to it. On reaching the caudal margin of the mylohyoid muscle which is located in a transverse plane less than 1 cm. rostral to the angle of the jaw, the two ducts pass medial and dorsal to this muscle. Rostral to the lingual branch of the trigeminal nerve, which crosses the lateral surfaces of the ducts at a transverse level just caudal to the orbital openings, the ducts lie between the genioglossal and mylohyoid muscles. The ducts open on a small **sublingual papilla** *(caruncula sublingualis)*, which is located lateral to the rostral end of the frenulum at its attachment caudal to the symphysis of the mandibles. According to Michel's (1956) observations (30 dogs), in two-thirds of the specimens the ducts open individually; in one-third they have a common opening. When the openings are separate, the mandibular opening lies rostral to the sublingual. Michel describes his technique for cannulating these ducts.

The glandular branch of the **facial artery** supplies the monostomatic portion, and the **sublingual artery**, a branch of the lingual, supplies the small polystomatic part of the sublingual gland. These arteries are accompanied by satellite veins which drain the gland. Lymphatic drainage is into the **medial retropharyngeal lymph node.**

ZYGOMATIC GLAND

The **zygomatic gland** *(glandula zygomatica)* (Figs. 7–1 and 20–16), also known as the orbital gland, weighs about 3 gm. and is located ventral to the zygomatic arch. Found only in the dog and cat, among the domestic mammals, it represents a caudal condensation of the largely unilobulated dorsal buccal glands of other mammals. It is globular to pyramidal in shape, with its base directed upward and backward and lying against the ventral part of the periorbita. It is surrounded by soft fat outside a feebly developed capsule. Because of the soft tissue adjacent to it, its lobules are much more distinct than are those of the mandibular gland, which it resembles in color. From its blunt apex, which lies lateral to that part of the maxilla containing the roots of the last upper molar tooth, it sends one major duct and two to four minor ducts to the caudal part of the buccal cavity. The major duct opens about 1 cm. caudal to the parotid papilla on the ridge of mucosa that extends to a plane through the caudal surface of the last upper cheek tooth. The smaller, minor ducts open on this ridge, caudal to the opening of the major duct.

Usually, the first branch of the **infraorbital artery,** as it enters the infraorbital canal, supplies the zygomatic gland. The main vein which leaves it enters the **deep facial vein,** which grooves its lateral surface.

PHARYNX

The **pharynx** *(cavum pharyngis)* (Fig. 7–26) is a passage which is, in part, common to both the respiratory and the digestive system. It extends approximately from a transverse plane through the head at the level of the orbital openings to a similar plane through the second cervical vertebra. Because of the long soft palate and the slight

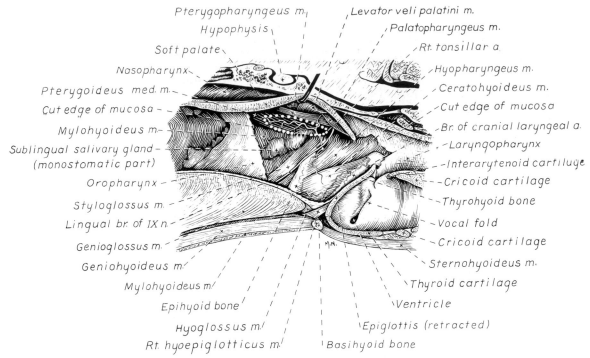

Figure 7–26. Median section through the pharynx. The location of the right palatine tonsil is indicated by a dotted circle. The buccopharyngeal fascia deep to the tonsil was removed and the soft palate elevated.

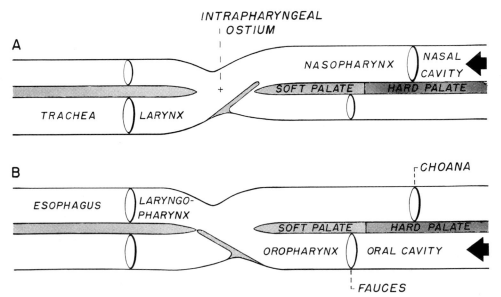

Figure 7–27. Diagram of the pharyngeal chiasma.
A. During respiration.
B. During deglutition.

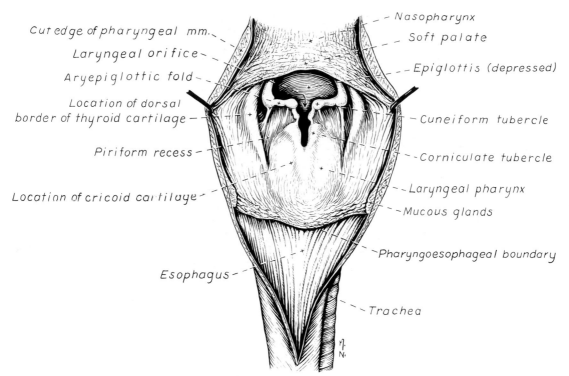

Figure 7–28. The pharynx opened mid-dorsally.

cervicocephalic angle in the dog, the parts of the pharynx are not entirely comparable with man. The pharynx of the dog consists of nasal, oral, and laryngeal parts (Fig. 7–27).

The **nasopharynx** *(pars nasalis pharyngis)* (Fig. 7–28) is the respiratory portion above the soft palate and extends from the choanae of the nasal cavity to the intrapharyngeal opening of the pharynx.

The **intrapharyngeal opening** *(ostium intrapharyngeum)* has formerly been called the pharyngeal isthmus or pharyngeal chiasma. It is the opening of the nasopharynx over the edge of the soft palate into the oral and laryngeal parts of the pharynx. The region is bounded caudally by the epiglottis and laterally by the mucosa covering the hyopharyngeus and ceratohyoideus muscles. Other structures associated with the nasopharynx are the pharyngeal tonsil and the openings of the auditory tubes, which are described with the respiratory system in Chapter 8.

Caudal to the ostium intrapharyngeum the digestive tube is continued dorsal to the larynx as the laryngeal part of the pharynx,

and the respiratory tube is continued ventrally as the larynx, which in turn is continuous with the trachea.

The **isthmus of the fauces** *(isthmus faucium)* is the short connection between the oral cavity and the oral part of the pharynx. It is bounded on each side by the palatoglossal arch, ventrally by the tongue, and dorsally by the soft palate.

The **oropharynx** *(pars oralis pharyngis)* extends from the isthmus of the fauces to the intrapharyngeal opening dorsally and to the larynx ventrally. It is bounded dorsally by the soft palate, ventrally by the root of the tongue, and laterally by the tonsillar fossa, with its contained palatine tonsil. The lateral wall of the oral part of the pharynx, which provides the site for the palatine tonsil, is called the *fauces.*

Unlike in man, the caudal portion of the pars oralis in the dog belongs to the pars nasalis owing to the extended soft palate.

The **palatine tonsil** *(tonsilla palatina)* is a long, relatively thin lymph node, located in the lateral wall of the oral part of the pharynx (fauces), just caudal to the palatoglossal arch. It is hidden from casual observation because of its position in the **tonsillar fossa** *(fossa tonsillaris)*. The medial wall of the sinus is formed by the thin, falciform **tonsillar fold** *(plica semilunaris)*, which is a fold from the ventral surface of the lateral portion of the soft palate.

The palatine tonsil is divided into a protruding fusiform portion which composes the major portion of the organ, and a usually smaller, deeper, minor portion which lies under the mucosa, forming the rostral part of the lateral wall of the tonsillar fossa. The deep portion of the gland may be formed as a result of tonsillitis, since it is usually absent in young, healthy specimens. It develops in the submucosa directly on the thick buccopharyngeal fascia. Lateral to this fascia lies the lingual branch of the glossopharyngeal nerve, and the styloglossal and medial pterygoid muscles.

The palatine tonsil has a long, narrow hilus which may be thickened in its middle so that rostral and caudal fossae are formed which separate the main portion of the tonsil from the lateral wall of the tonsillar sinus. When the major portion of the organ is forcibly pulled out of the tonsillar fossa, the deeper, minor portion is also drawn downward in a fold of mucosa formed by the traction, so that complete extirpation of the organ is made possible. The average dimensions of the major portion are: length 2.5 cm., width 0.5 cm., and thickness 0.4 cm. The pointed ends of the fusiform organ are not free but are firmly attached to the dorsolateral parts of the wall of the fossa.

The palatine tonsil has no afferent lymphatics. Its efferent vessels drain into the **medial retropharyngeal lymph node**. It receives its nutrition mainly from the **tonsillar artery**, which is derived from the lingual. The tonsillar artery enters the middle or widest portion of the tonsil by about three branches. The caudal pole of the tonsil may receive twigs from the hyoid branches of the lingual. The small veins from the tonsil enter the **palatine plexus of veins**. Besides the palatine tonsils, other lymphoid tissue lies in the base of the tongue and in the nasal part of the pharynx. These are referred to as the lingual and pharyngeal tonsils. The **lingual tonsil** *(tonsilla lingualis)* is so diffuse it cannot be seen in gross. The **pharyngeal tonsil** *(tonsilla pharyngea)*, or **adenoid**, is described with the nasal part of the pharynx in Chapter 8.

The **laryngopharynx** *(pars laryngea pharyngis)* (Fig. 7–28) is that portion of the pharynx which lies dorsal to the larynx. It extends from the ostium pharyngeum and the nasal part of the pharynx in front to the beginning of the esophagus behind. Its caudal limit, therefore, reaches to a transverse plane which passes through the caudal border of the cricoid cartilage and the middle of the axis. The pharyngoesophageal junction in the dog is distinctly marked by an annular ridge of tissue known as the *limen pharyngoesophageum*, formerly called the annular fold.

Although the function of the laryngeal part of the pharynx is both respiratory and alimentary, its chief importance is in deglutition. The bolus of food ingested is conveyed to it by the plunger-like movement of the base of the tongue. Six pairs of extrinsic muscles control the shape and size of the nasal and laryngeal parts of the pharynx. Three pairs of these muscles are constrictors, two pairs are shorteners, and the muscles of one pair act as a dilator. Dyce (1957) uses the name laryngopharyngeus to designate the combined cricopharyngeal and thyropharyngeal muscles, which are the two caudal pairs of constrictors. These two muscles are largely fused to each other, and they are also blended caudally with the spiral muscular coat of the esophagus. The muscles of the pharynx are described in Chapter 6.

The laryngopharynx receives its nutrition through the paired **pharyngeal branches of the cranial thyroid** and the **ascending pharyngeal arteries.** The soft palate, the dorsal wall of the oral part of the pharynx, receives most of its nutrition from the paired **minor palatine arteries.**

ALIMENTARY CANAL

ESOPHAGUS

The esophagus (Fig. 7–29), the first part of the alimentary canal, is the connecting tube between the laryngeal part of the pharynx and the stomach. In medium-sized dogs it is about 30 cm. long and 2 cm. in diameter when it is collapsed. Since this passage traverses most of the neck and all of the thorax, and ends upon entering the abdomen, it is divided into cervical, thoracic, and abdominal portions. It begins opposite the middle of the axis, dorsally, and the caudal border of the cricoid cartilage, ventrally. A plicated ridge of mucosa, the *limen pharyngoesophageum,* most prominent ventrally, is the internal demarcation between the pharynx and esophagus. The esophagus ends at the cardia of the stomach. The dorsal part of

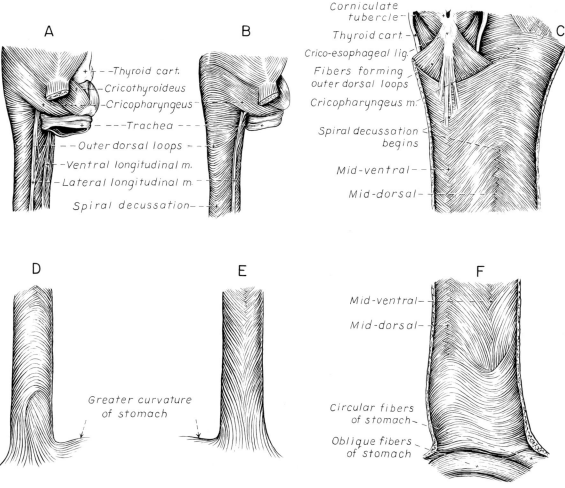

Figure 7–29. Musculature of esophagus.
A. Outer layer, cranial end, lateral ventral aspect.
B. Outer layer, cranial end, lateral dorsal aspect.
C. Inner layer, cranial end. (Esophagus opened on left side.)
D. Outer layer, caudal end, ventral aspect.
E. Outer layer, caudal end, dorsal aspect.
F. Inner layer, caudal end. (Esophagus opened on right side.)

the esophagus, as it terminates, lies ventral to the last thoracic vertebra, and the ventral part lies slightly cranial to this level. The site of termination of the esophagus may vary a vertebral segment cranially or caudally.

The **cervical portion** (*pars cervicalis*) of the esophagus is related mainly to the left longus colli and longus capitis muscles dorsally, and to the trachea ventrally and to the right. At its origin it starts to incline to the left, so that at the thoracic inlet it usually lies left lateral to the trachea. Its position here varies; in some specimens it is left dorsal and in others it is left ventral to the trachea. On the left side, the left common carotid artery, vagosympathetic nerve trunk, internal jugular vein, and tracheal duct run in the angle between the esophagus and the longus capitis muscle. The corresponding structures on the right side are located lateral to the trachea.

The **thoracic portion** (*pars thoracica*) of the esophagus extends from the thoracic inlet to the esophageal hiatus of the diaphragm. At first, it usually lies to the left of the trachea between the widely separated leaves of the dorsal part of the precardial mediastinum. It obliquely crosses the left face of the trachea to gain its dorsal surface as the trachea bifurcates into the bronchi ventral to the fifth and sixth thoracic vertebrae. In reaching this level it crosses the right face of the aortic arch and lies ventral to the right and left longus colli muscles. It is separated from these muscles here, as well as in the neck, by the prevertebral fascia. Caudal to a transverse plane through the termination of the trachea the esophagus lies nearly in the median plane as it passes between the two pleural sacs.

The aorta obliquely crosses the left side of the esophagus between the fifth and ninth thoracic vertebrae, and thereafter diverges from it in a progressive manner caudally so that at the diaphragm the two structures are separated by about 3 cm. The dorsal branches of both right and left vagal nerves run dorsocaudally across the sides of the esophagus and unite with each other on the dorsum of the esophagus, 2 to 4 cm. cranial to the dorsal part of the esophageal hiatus. The dorsal vagal trunk, so formed, continues to and passes through the dorsal part of the esophageal hiatus. The right and left ventral branches of the vagi unite immediately caudal to the root of the lungs to form the ventral vagal trunk. This trunk at first lies in contact with the esophagus and then arches ventrally in the ventral mediastinum before it passes through the esophageal hiatus with the esophagus.

The **abdominal portion** (*pars abdominalis*) of the esophagus is its wedge-shaped terminal part. Dorsally, the esophagus immediately joins the stomach. Ventrally, it notches the thin, dorsal border of the caudate lobe of the liver.

The esophagus is not uniform in either the thickness of its wall or the diameter of its lumen. In the cervical portion its wall averages about 4 mm. in thickness, and in the thoracic portion, 2.5 mm.; the wall is thickest in the abdominal portion, measuring about 6 mm. where it joins the stomach. As determined by the size of the approximately 10 primary longitudinal folds of its mucosa, the whole tube is capable of great dilatation. The least distensible parts occur at both its beginning and its end, and as it passes through the thoracic inlet.

COATS OF THE ESOPHAGUS

The esophagus has four coats: fibrous, muscular, submucous, and mucous. In the cervical region, the **fibrous coat,** or **adventitia** (*tunica adventitia*), blends with the deep cervical fascia (prevertebral fascia) dorsally and on the left, and with the fascia which forms the carotid sheath on the right. The adventitia of the thoracic and abdominal portions of the esophagus blends with the endothoracic and the transversalis fascia, respectively. It is largely covered by pleura in the thorax and with peritoneum in the abdomen. Where the esophagus is not covered by serosa, its adventitia blends with that proper fascia of the organs with which it comes in contact.

The **muscular coat** (*tunica muscularis*) (Fig. 7–29) consists essentially of two oblique layers of striated muscle fibers. The external muscular coat arises on the ventral side of the esophagus from the medial dorsal crest of the cricoid and the corniculate portions of the arytenoid cartilages by means of the **cricoesophageal tendon** (*tendo cricoesophageus*). This tendon is a distinct thin band of collagenous tissue about 0.5 cm. wide and 1 cm. long. It tapers to a point caudally as muscle fibers leave each side of it. It, more than any other structure, serves as the fixed point of cranial attachment of the esophagus (Sauer 1951).

The first few muscle fibers to arise arch sharply outward and upward on each side and meet dorsally as transverse fibers. They are not distinct from the caudal fibers of the cranially lying cricopharyngeal muscles. Helm (1907) appropriately called this initial portion of the external esophageal muscle coat the **cricoesophageal muscle.** The fibers from each side do not decussate mid-dorsally but blend with each other without demarcation. Although the first few fibers are nearly transverse in direction, the subsequent fibers become increasingly more oblique so that in fusing with their fellows they form progressively narrower loops or ellipses caudally.

The main musculature of the esophagus caudal to the cricoesophageal fibers is in the form of spiral fibers. These start about 5 cm. from the beginning of the esophagus and continue to within 5 to 10 cm. of the cardia of the stomach. The superficial fibers on one side become the deep ones on the other side. These apparently continuous oblique bundles spiral around the esophagus in such a way that they cross each other at nearly right angles in making up the two main muscular coats of the organ. The lines of decussations are dorsal and ventral in positon. The line of ventral decussation seems to stop about 10 cm. from the cardia, whereas the dorsal decussations end about 5 cm. from it. The decussations do not end abruptly, as there is a gradual shifting of the direction of the fibers of the inner and outer muscle coats.

The fibers of the inner coat become more transverse in direction, while those of the outer coat become more longitudinal, especially dorsally as they approach the cardia. About 3 cm. from the cardia, ventrally, many of the oblique fibers of the inner coat, instead of becoming more transverse, become nearly longitudinal and pass to the outside. They continue on the visceral wall of the stomach as its outer longitudinal layer. The longitudinal fibers of the dorsal surface of the esophagus (*m. esophageus longitudinalis dorsalis*) continue on the dorsal wall of the stomach. The nearly transverse, inner muscular fibers of the esophagus partly blend with the circular and oblique fibers of the stomach. The division between the striated musculature of the esophagus and the smooth musculature of the stomach cannot be determined by gross examination.

The cervical portion of the esophagus possesses, in addition to the two oblique coats, several poorly developed groups of longitudinal fibers. The right and left lateral longitudinal bands are best developed. They arise underneath the cricopharyngeal muscles from the fascia adjacent to the lateral borders of the dorsal cricoarytenoid muscles. Some fibers contributing to these bands come from the cricoesophageal tendon. The muscles are 1 to 2 mm. wide. They usually fade away on the caudal portion of the cervical part of the esophagus, but they may extend to the midthoracic part. Inner and outer ventral longitudinal esophageal muscle fibers can usually also be recognized. The inner ones arise from the cricoesophageal tendon and, after running about 2 cm., become dispersed on the main inner muscular coat. They lie on the dorsal surface of the ventral decussations which they partly cover. The outer longitudinal muscle fibers lie ventral to the ventral decussations. They are so feeble that they cannot always be dissected in gross with certainty. In some specimens extremely delicate longitudinal muscle bundles have been seen lying between the inner and outer muscular coats in the cranial part of the cervical portion of the esophagus.

The **submucous coat** (*tela submucosa*) loosely connects the mucous and the muscular coats. It allows the relatively inelastic mucous coat to be thrown into heavy longitudinal folds when the esophagus is contracted. It contains blood vessels, nerves, and mucous glands. Trautmann and Fiebiger (1957) state that the **esophageal glands** (*gl. esophageae*) form a continuous stratum that extends to the vicinity of the stomach. The **muscular layer of the musosa** (*lamina muscularis mucosae*), according to these authors, is present only in the caudal half of the esophagus and forms a continuous layer only near the stomach.

The **mucous coat** (*tunica mucosa*) is composed of a superficially cornified stratified squamous epithelium which contains the openings, at about 1 mm. intervals, of the ducts of the esophageal glands. In the collapsed esophagus it forms large and numerous longitudinal rugae or folds. Cardiac glands exist in the distal part of the esophagus.

VESSELS OF THE ESOPHAGUS

The arteries to the cervical portion of the esophagus are primarily twigs from the cranial and caudal **thyroid arteries.** The glandular mantle underlying the limen pharyngoesophageum is richly supplied by branches from the cranial thyroid artery. In the region of the thoracic inlet on the left side, a long but small descending branch from the left caudal thyroid artery anastomoses with an ascending branch from the bronchoesophageal. From this small anastomotic trunk, branches go to the esophagus. The esophageal portion of the **bronchoesophageal artery** is the main source of blood to the cranial two-thirds of the thoracic portion of the esophagus. The remaining part is supplied by esophageal branches of the aorta and/or **dorsal intercostal arteries,** and the terminal portion is supplied by the esophageal branch of the **left gastric artery.**

The veins which drain the esophagus are essentially satellites of the arteries which supply it. Those veins from the thoracic portion empty largely into the **azygos vein,** with the vein which accompanies the esophageal branch of the left gastric artery being a tributary of the portal system. Adjacent veins, like the arteries, anastomose with each other on the esophagus. According to Baum (1918), lymph vessels from the esophagus drain into the **medial retropharyngeal, deep cervical, cranial mediastinal, bronchial, portal, splenic,** and **gastric lymph nodes.**

NERVES OF THE ESOPHAGUS

The cricopharyngeal muscle and the cervical portion of the esophagus are sup-

plied with motor fibers from the small **pharyngoesophageal nerve,** which leaves the vagus just above the distal ganglion of the vagus (nodose of Hwang, Grossman, and Ivy 1948). These investigators also found that a variable portion of the cervical part of the esophagus contracted when the peripheral ends of the **vagus** or the **recurrent laryngeal nerves** were stimulated at the base of the neck. Most nerve twigs to the esophagus are too small to be seen in gross. There is probably an overlapping in the amount of the cervical part supplied by the pharyngoesophageal and recurrent laryngeal nerves. It is further evident from physiological experiments that these two nerves are reciprocal in the amount of the cervical portion of the esophagus that they supply.

Hwang, Grossman, and Ivy also found that a small branch from the pharyngeal branch of the vagus passed under the pharyngeal muscles to reach the esophagus. Stimulation of this branch produced contraction of the cranial half of the cervical portion. It is probable that the recurrent laryngeal nerves carry the afferent (Chauveau 1886) and some motor fibers to the cervical portion of the esophagus as well as providing both the motor and sensory nerve supply to the thoracic part as far caudally as the heart. The dorsal and ventral branches of the vagi and the vagal trunks they form supply the esophagus caudal to the heart.

Watson (1974) described three major regions of the esophagus in the dog as characterized by their innervation (Fig. 7–30): a cervical region supplied by paired pharyngoesophageal and paired pararecurrent laryngeal nerves; a cranial thoracic region supplied mainly by the left pararecurrent laryngeal nerve; and a caudal thoracic and abdominal region which is supplied by the vagal trunks. For most of its course in the neck the recurrent laryngeal nerve was paralleled by a slightly smaller nerve lying more dorsally and giving branches to both the trachea and the esophagus. This latter nerve was called the pararecurrent laryngeal nerve by Lemere (1932), and, as Watson points out, it is always present. (Lemere claimed that the pararecurrent nerves are motor, secretomotor, and sensory to the lower two-thirds of the cervical esophagus.)

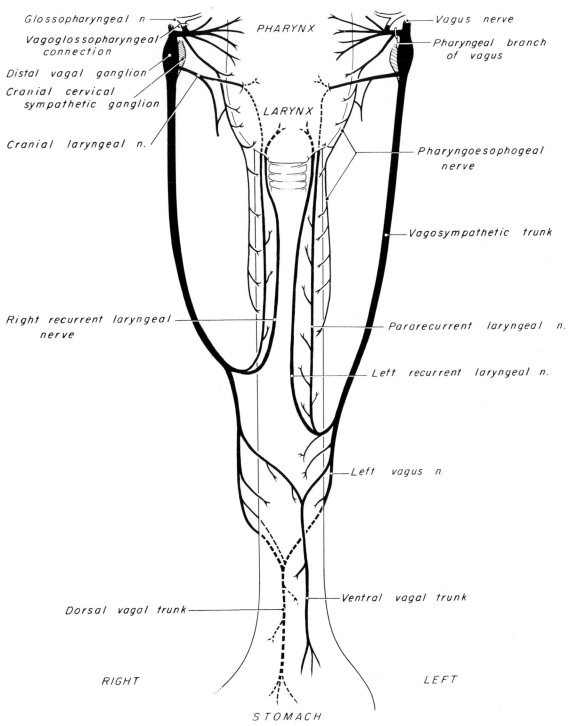

Figure 7–30. Innervation of the esophagus. (After Watson, Thesis, Massey Univ. 1974).

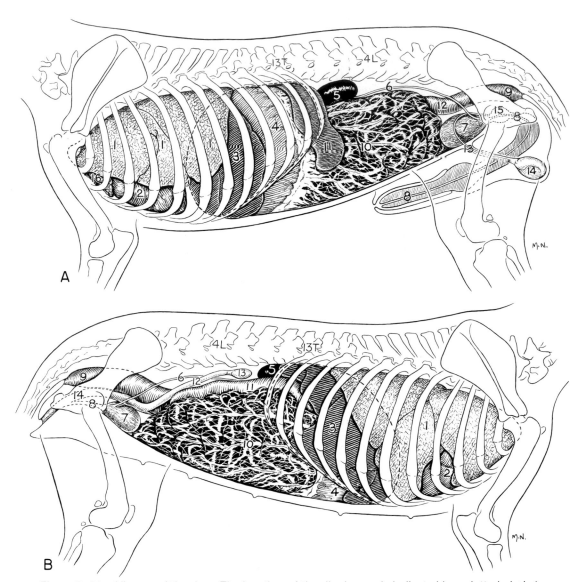

Figure 7–31.   Viscera of the dog. (The location of the diaphragm is indicated by a dotted circle.)
A. Viscera of male dog, left lateral aspect.   B. Viscera of female dog, right lateral aspect.

1. Left lung	1. Right lung
2. Heart	2. Heart
3. Liver	3. Liver
4. Stomach	4. Stomach
5. Left kidney	5. Right kidney
6. Ureter	6. Ureter
7. Bladder	7. Bladder
8. Urethra	8. Urethra
9. Rectum	9. Rectum
10. Greater omentum covering small intestine	10. Greater omentum covering small intestine
11. Spleen	11. Cecum
12. Descending colon	12. Descending duodenum
13. Ductus deferens	13. Right uterine horn
14. Left testis	14. Right ovary
15. Prostate	15. Vagina
16. Thymus	

ABDOMEN

The **abdomen** (abdominal region, or *regio abdominis*) (Figs. 7–31 to 7–33) is that part of the trunk which extends from the diaphragm to the pelvis. It contains the largest cavity in the body, the **abdominal cavity** *(cavum abdominis)*. Caudally, the abdominal cavity is continuous with the pelvic cavity, the division between the two being a plane through the pelvic inlet, or brim of the pelvis. The abdominal cavity is a muscle- and bone-bounded cavity. It is lined internally by the transversalis fascia, which in turn is covered in most places by the peritoneum.

The peritoneum forms the potential **peritoneal cavity** *(cavum peritonei)*, which is contained largely, but not exclusively, within the abdominal cavity. In both sexes the pelvic cavity contains the pelvic portion of the peritoneal cavity. Vaginal processes, always well developed in the male and usually present in the female, also exist as extra-abdominal extensions of the peritoneal cavity. The abdominal cavity is truncate and cone-shaped, with a concave cranial end or base. It contains the abdominal viscera, which include primarily the flexuous alimentary canal and its two associated, indispensable glands, the liver and the pancreas. Also included are the spleen, female reproductive tract, many nerve plexuses, vessels, and lymph nodes. It is bounded cranially by the diaphragm; dorsally by the lumbar vertebrae, the sublumbar muscles, and the crura of the diaphragm; bilaterally by the diaphragm, the two oblique and transverse abdominal muscles, and a small portion of the shaft of

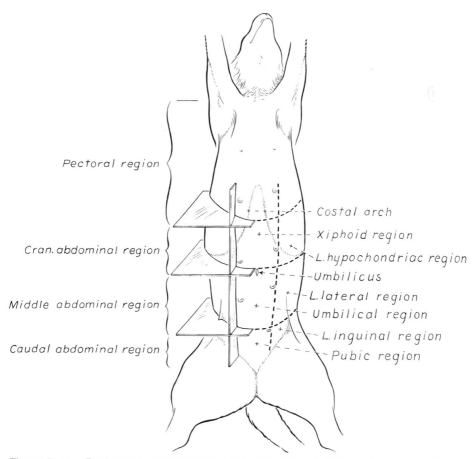

Figure 7–32.   Regions of the abdomen as determined by sagittal and transverse planes.

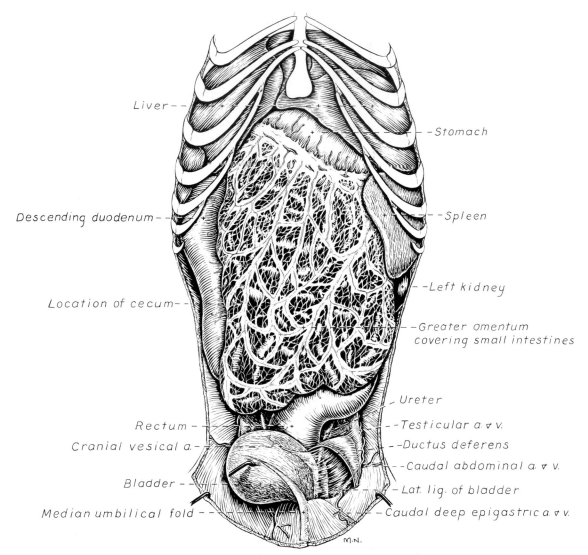

Liver

Stomach

Descending duodenum

Spleen

Left kidney

Location of cecum

Greater omentum covering small intestines

Ureter

Rectum

Testicular a. & v.

Cranial vesical a.

Ductus deferens

Caudal abdominal a. & v.

Bladder

Lat. lig. of bladder

Median umbilical fold

Caudal deep epigastric a. & v.

Figure 7–33. Abdominal viscera of male dog, ventral aspect.

the ilium on each side. Ventrally, the right and left rectus abdominis muscles and their sheaths form the abdominal wall.

There are three unpaired apertures in the diaphragm: the **esophageal hiatus**, for passage of the esophagus, vagal nerve trunks, and esophageal vessels; **the caval foramen**, for passage of the caudal vena cava; and the **aortic hiatus**, for passage of the aorta, thoracic duct, and the azygos and hemiazygos veins. Paired slitlike openings existing dorsal to the diaphragm are formed ventrally by the dorsal edge of the diaphragm, and dorsally by the psoas

muscles. At these sites the pleura and peritoneum are separated only by the fused endothoracic and transversalis fasciae. The sympathetic trunk and splanchnic nerves pass dorsal to the lumbocostal arch on each side.

Caudally, the abdominal cavity communicates freely with the pelvic cavity at the pelvic inlet. In the fetus, there is a relatively large opening, the umbilical aperture, located midventrally, which serves for the passage of the umbilical blood vessels, the small vitelline duct, and the stalk of the allantois. After these structures are disrupted at birth, this opening rapidly closes, forming a faint scar, the umbilicus, on the midventral line. There is a passage on each side in the caudoventral part of the abdominal wall, called the inguinal canal, for the passage of the vaginal process and the spermatic cord in the male and the round ligament in the female. In both sexes the external pudendal vessels and the genital nerve pass through the caudal part of the inguinal canal. Another pair of abdominal openings in the caudal part of the abdominal wall are the right and left vascular lacunae. The femoral artery, vein, lymphatics, and saphenous nerve, surrounded by transversalis fascia, pass through each vascular lacuna.

The **pelvic cavity** (*cavum pelvis*) begins at the pelvic inlet as a continuation of the abdominal cavity, and in a broad sense is regarded as a division of it. It is bounded dorsally by the sacrum and first vertebra; bilaterally, it is bounded in front of the ilia and behind these by the coccygeus, levator ani and middle gluteal muscles. It ends at the pelvic outlet, which is bounded by the first caudal vertebra dorsally, the sacrotuberous ligaments and middle gluteal muscles bilaterally, and the ischial arch ventrally. The floor is formed in part by the coccygeal muscles but mainly by the pubes and ischii, which lie largely peripheral to these thin muscular sheets. The pelvic cavity contains the rectum and the urethra in both sexes, the vagina and part of the vestibule in the female, and a part or all of the prostate in the male.

The abdominal and pelvic cavities are lined by fascia throughout. In most places this fascia attaches to muscles or bones peripherally and blends with the subserous areolar tissue centrally. The fascia was previously named according to the region or parts it covered, so that parts of it were known by such names as diaphragmatic, transversalis, iliac, internal spermatic, and pelvic fascia. The term **transversalis fascia** (*fascia transversalis*) is used to include all of these fascial divisions. The subserous areolar tissue forms the medium whereby the peritoneum is united with the transversalis fascia. In obese specimens the large fat deposits around the kidneys, in the falciform ligament, in the pelvis,

and around the vaginal rings are located in the subserous fascia.

REGIONS OF ABDOMEN

For convenience of description the abdomen is divided into three primary regions by two transverse and two sagittal planes. Each of the regions has two or more subdivisions (Fig. 7–32).

The more cranial of the two transverse planes passes through the most caudal border of the costal arch. This plane therefore passes through the caudal portion of the second lumbar vertebra. The more caudal of the two transverse planes crosses the body at a level through the ventral cranial iliac spines, or the most cranial parts of the wings of the ilia. This plane bisects the caudal part of the sixth lumbar vertebra. By means of these two planes the abdomen is divided into three segments, which are named, from before backward, the cranial, middle, and caudal abdominal regions. Of these three segments, the cranial abdominal region is by far the largest, as it extends far forward in the thoracic cage, where it is limited cranially by the diaphragm. The caudal abdominal region is smallest, as it ends caudally at the pelvic inlet.

The two imaginary sagittal planes, which further divide the abdomen into smaller regions, pass on each side midway between the ventral cranial iliac spine and the median sagittal plane. The three parts of the cranial abdominal region are the **right and left hypochondriac regions** (*regio hypochondriaca dextra et sinistra*) and the median **xiphoid or epigastric region** (*regio xiphoidea*). The middle abdominal region includes the unpaired, median ventral area of the abdomen, known as the **umbilical region** (*regio umbilicalis*), and the **right and left lateral regions** (*regio lateralis dextra et sinistra*). The lateral regions include the **flank** (*regio plicae lateris*), and each contains an expansive **paralumbar fossa** (*fossa paralumbalis*), which forms a ventral arc and a dorsal, straight base located ventrolateral to the transverse processes of the lumbar vertebrae and the iliocostalis muscle. The three parts of the caudal abdominal segment are the **right and left inguinal re-**

gions *(regio inguinalis dextra et sinistra)* and the median unpaired **pubic region** *(regio pubica).* In addition to the abdominal regions already named, there are preputial and mammary regions.

RELATIONS OF ABDOMINAL ORGANS

The greater omentum, or epiploon (Fig. 7–33), is a fat-streaked, lacy, double reflection of peritoneum which covers most of the abdominal contents ventrally and on the sides. It lies principally between the parietal peritoneum and the intestinal mass. The greatest caudal extension of the liver occupies the right hypochondriac region.

The stomach, when empty, does not contact the abdominal wall, but when moderately filled it lies against the xiphoid and left hypochondriac portions of this wall caudal to the liver. The completely filled stomach, especially in pups, lies largely in contact with the xiphoid and umbilical regions, ventrally, and the right and left lateral portions of the abdominal wall, and reaches caudally to a transverse plane just caudal to the umbilicus.

The left kidney contacts the dorsal part of the left lateral abdominal wall. The spleen, separated from the greater curvature of the stomach by an outward fold of the gastrosplenic ligament, contacts an oblique zone of the abdominal wall from the left kidney to the midventral line or even beyond this. A small portion of the right kidney immediately caudal to the last rib and a large portion of the descending part of the duodenum lie directly in contact with the dorsal part of the right sublumbar region. The urinary bladder, nestled in the greater omentum but not covered ventrally by it, is the only visceral organ which lies in contact with the abdominal wall in the pubic region.

All abdominal organs vary normally in size and position. The stomach, uterus, urinary bladder, and spleen vary more than the other organs because of their ability to undergo marked changes in size and shape. The gravid uterus alters the position of all the other movable abdominal organs more markedly than do any of the others. It always occupies the most ventral position in the abdomen because it contains no gas and is therefore the heaviest freely movable abdominal organ. In advanced pregnancy it nearly completely fills the ventral half of the abdominal cavity. It is impossible to assign a constant position for any one of the abdominal organs, especially those which are attached by peritoneal folds. Roentgenograms and hardened anatomical preparations clearly reveal the normal variations which exist from time to time and in different specimens.

The Peritoneum

The peritoneum is a **serous membrane** *(tunica serosa)* which, like other serous membranes, is made up of a surface mesothelium composed of squamous cells, and a connective tissue ground work, or stroma. The connective tissue stroma is composed of yellow elastic and white fibrous tissue. A serous membrane is histologically largely connective tissue, whether it is found in the thoracic, pericardial or abdominal cavity.

Peritoneum, like other serous membranes, is united with the transversalis fascia by areolar tissue, known as the **tela subserosa.** In certain locations, in well-nourished animals, this stratum contains considerable fat. Peripherally, it is coextensive with the stroma of the peritoneum, and in most places it is grossly inseparable from the transversalis fascia, which serves to attach the peritoneum to the underlying muscle and bone.

The lining of the abdominal cavity and its coextensive pelvic and scrotal cavities, as well as the covering of and the reflections from the organs of the abdomen, is peritoneum. It is distributed as though in the embryo all the organs developed in the walls of the abdominal, pelvic, or scrotal cavities and, as they grew, pushed the peritoneum before them as elastic veils. Some structures, such as the liver, kidneys and gonads, do develop in this way. The abdominal part of the alimentary canal arises very early in the embryo between the two layers of peritoneum, which form a midsagittal partition separating the abdominal part of the celom into right and left parts. Most of the ventral part of this fold becomes obliterated shortly after it forms. The parts which are left go to the liver, the stomach, and the beginning of the duodenum. Because most of the ventral part of the mesentery (that ventral to the digestive tube before it has rotated) becomes obliterated, the definitive appearance of the tube suggests that it migrated ventrally from the dorsal abdominal wall; actually it develops between the two peritoneal layers.

The peritoneum serves to reduce friction between parts. A small amount of viscous fluid is produced for this purpose. Whenever two peritoneal surfaces in contact fail to move for an appreciable time, the mesothelium of the apposed surfaces is absorbed and the connective tissue re-enforcement, the stroma, ceases to be differentiated from that of adjacent parts; thus, peritoneal sheets become obliterated, and viscera adhere to neighboring structures. From a relatively simple disposition in the embryo, the arrangement of the peritoneum in an adult becomes a complicated maze of mesenteries and ligaments.

The peritoneum may be divided into:

The **parietal peritoneum** *(peritoneum parietale),* which covers in large part the inner surface of the walls of the abdominal, pelvic, and scrotal cavities.

The **visceral peritoneum** *(peritoneum viscerale),* which covers the organs of the abdominal, pelvic, and scrotal cavities, wholly or in part.

The **connecting peritoneum** consists of double sheets of peritoneum extending between organs or connecting them to the parietal peritoneum. These peritoneal folds are referred to as mesenteries, omenta, or ligaments. A **mesentery** *(mesenterium),* in a restricted sense, passes from the abdominal wall to the intestine. It is wide and contains many vessels. In a broader sense, a mesentery is any wide serous fold which attaches organs to a wall and serves as a route by which the nerves and vessels reach the organs. A **ligament** passes from a wall to an organ, or from an organ to an organ, and is usually narrow and contains few vessels.

The cavities enclosed by serous membrane are closed cavities, except for the peritoneal cavity in most female animals. In the bitch there is an opening at the abdominal end of each uterine tube and thus, through the genital tract, to the outside. No organs or tissues are located in the peritoneal cavity (except at the time of ovulation, when an egg ruptures from the ovary). In life this is an almost non-existent cavity containing only enough lubricating fluid to moisten the apposed peritoneal surfaces, both between different organs and between the organs and the parietal peritoneum.

Organs which lie against the walls of the abdominal or pelvic cavities and which are covered only on one surface by peritoneum are said to be retroperitoneal. Most of these organs are small and are embedded in fat. Organs which project freely into the abdominal, pelvic, and scrotal cavities and receive a nearly complete covering of peritoneum are termed intraperitoneal.

The **common dorsal mesentery** *(mesenterium dorsale commune)* is the peritoneal fold which leaves the dorsal abdominal wall and reflects, directly or indirectly, around most of the freely movable organs of the abdominal cavity. It can be demonstrated by grasping the abdominal part of the digestive tube along with the pancreas and spleen and moving them ventrally. In thin dogs it can be seen to leave the aorta, providing a route by which the celiac, cranial, and caudal mesenteric arteries, autonomic nerves, lymphatics, and radicles of the portal vein pass to or from the intestine and other organs. It is the retained part of the serous partition which divided the celom dorsal to the alimentary canal into right and left halves. At an early stage in development the alimentary canal is relatively straight, thus allowing the common dorsal mesentery to exist as a simple median partition, the *mesenterium dorsale primativum.* When the parts of the digestive tube differentiate, those parts of the common dorsal mesentery going to the several parts receive specific names, as follows: to the stomach, *mesogastrium;* to the duodenum, *mesoduodenum;* to the jejunoileum, *mesojejunoileum;* to the colon, *mesocolon;* and the rectum, *mesorectum.* In fat dogs, the apposed surfaces of the two layers of peritoneum of the common dorsal mesentery are separated by fat which is deposited along the arteries, and because of this, neither the abdominal aorta nor the caudal part of the caudal vena cava is visible.

A transverse section of the abdomen just

cranial to the tuber coxae reveals the disposition of the peritoneum in its simplest form. Tracing the peritoneum to the right from the inner surface of the left rectus abdominis, one finds it is closely united by transversalis fascia to the inner sheath of that muscle. At the linea alba it is thrown into an extremely delicate plica about 3 cm. high, the **middle umbilical fold** *(plica umbilicalis medialis)*. In the fetus, this fold contained the stalk of the allantois, but in the adult no trace of this fetal structure exists.

After covering the right rectus sheath, the peritoneum runs dorsally on the lateral abdominal wall. Dorsolaterally, it sends a thin, narrow plica ventromedially, which surrounds the external spermatic vessels in the male. It covers the variable amount of fat occupying the groove lateral and ventral to the sublumbar muscles and continues medially in the male to approximately the midplane. In its course across the sublumbar muscles, it loosely covers the ureter as the latter crosses the external iliac vessels. Near the midplane it covers the right sympathetic trunk, and opposite the body of the seventh lumbar vertebra it leaves this bone and runs ventrally to the descending colon.

After nearly completely encircling the colon, the peritoneum runs back to the vertebra so that, between the dorsal abdominal wall and the colon, it forms a fold, consisting of two layers of peritoneum. Between these layers is located a small amount of transversalis fascia, the caudal mesenteric vessels, and the lumbar colonic nerves. This fold and its contents constitute the **descending mesocolon** *(mesocolon descendens)*. The peritoneum, in running from the descending mesocolon across the left abdominal wall to the starting point ventrally, covers the same structures on the left side as it does on the right. In the female the **broad ligament of the uterus** *(lig. latum uteri)* leaves each side of the sublumbar region as a peritoneal fold to surround the uterine horn. At the level under consideration, a lateral fold leaves the broad ligament to surround the round ligament of the uterus.

PELVIC PERITONEAL EXCAVATIONS

The peritoneum of the pelvis will now be traced by following it in a sagittal section just lateral to the median plane. By starting ventrally at a transverse plane passing through the tuber coxae and advancing caudally, one finds the peritoneum runs on the sheath of the rectus abdominis and passes over the ventral part of the pelvic inlet into the pelvic cavity. Within the first 2 cm. after entering the pelvic cavity the peritoneum reflects dorsally on the neck of the bladder of the female or on the prostate gland of the male, forming the shallow **pubovesical pouch** *(excavatio pubovesicalis)*. After reflecting around the cranial surface of the bladder, the peritoneum leaves the dorsal surface of its neck in the female or the prostate gland in the male and, forming an acute angle in the male caudal to the farthest caudal extension of the pubovesical excavation, reflects cranially on the ventral surface of the rectum to form a **rectovesical pouch.**

The caudal extension of the rectovesical pouch is not as great in the female as in the male, and it differs also in that it is divided into ventral and dorsal parts by the presence of the vagina and uterus and the right and left broad ligaments, which attach these organs to the pelvic walls laterally. In this way, the **vesicogenital pouch** *(excavatio vesicogenitalis)* is formed ventrally and the **rectogenital pouch** *(excavatio rectogenitalis)* is formed dorsally. The caudal angles of the reflections of the peritoneum from one pelvic organ to the next dorsal to it are located progressively farther caudally in the male when they are traced in parasagittal planes from the pubis to the tail. A line drawn from a point about 1 cm. caudal to the cranial border of the pubis to the transverse process of the third caudal vertebra indicates their most caudal extensions in the male. In the female, the rectogenital pouch extends farther caudally than the vesicogenital, but neither extends as far caudally as does the undivided rectovesical pouch, which takes their place in the male.

To understand the definitive positions of the connecting peritoneum to the alimentary tract and the liver, it is helpful to know about the primary rotations the tube makes and the differential growth that certain parts undergo. Early in development, the dorsal part of the stomach grows faster than the ventral part, with the result that it rotates on its axis to the left. Simultaneously with this rotation the stomach comes to lie in nearly a transverse position with the outlet located on the right while the inlet remains in nearly the median plane. Distal to the stomach the anlagen of the liver and pancreas appear as diverticula of the mucosa of the duodenum. The liver grows ventrally between the peritoneal layers, forming the ventral mesentery, and the pancreas grows into the dorsal mesentery. The cecum develops as a diverticulum of the lateral wall of the proximal part of the ascending colon.

The abdominal part of the alimentary canal grows many times the length of the abdominal cavity. Therefore, the parts with long mesenteries, mainly the jejunum and to a lesser extent the ileum, become greatly folded and occupy no constant location in the abdominal cavity. During development a part of the tube, in the form of a U-shaped loop, is accommodated in a normal umbilical hernia which lasts but a short time. The cranial mesenteric artery lies in the mesentery which connects the two limbs of the loop. As the loop forms, it rotates about three-fourths of a circle in a counterclockwise direction, as viewed ventrally, around the axis of the artery. Because of this rotation, the proximal part of the large intestine hooks around the cranial mesenteric artery from the left, and the small intestinal mass moves caudal to the axis of the artery by passing to the right of it.

The various derivatives of the common dorsal mesentery and the ventral mesentery will now be described as they run between the abdominal wall and the various parts of the alimentary canal.

The peritoneal folds which, in the adult, leave approximately the greater and lesser curvatures of the stomach are known as the greater and lesser omentum, respectively. The greater omentum, the most specialized serous fold in the body, is derived from the dorsal mesogastrium. The lesser omentum, small and relatively simple, is that portion of the ventral mesogastrium which extends between the liver and the lesser curvature of the stomach and the initial part of the duodenum.

GREATER OMENTUM

The **greater omentum** (*omentum majus*), or epiploon (Figs. 7–33, 7–34, 7–35), usually reaches as a lacy apron from the stomach to the urinary bladder, covering the intestinal coils ventrally and on the sides. Zietzschmann (1939) divided the greater omentum into a large bursal portion and a smaller splenic and veil portion. Each portion is composed of a double peritoneal sheet in which there are streaks of fat located around the arteries which run through it. The omentum forms one of the major fat storehouses in obese specimens. Between the streaks of fat the peritoneum is transparent because of its thinness.

In the postmortem specimen the omentum appears as a fenestrated membrane which can be manually separated into a superficial and deep leaf.

The **bursal portion** of the greater omentum attaches cranioventrally to most of the greater curvature of the stomach. Always closely related to the ventral abdominal wall, it extends caudally as far as the urinary bladder. Regardless of the fullness of the bladder, the bursal portion passes dorsally between that organ and the intestinal coils, and then reflects upon itself and returns to the region dorsal to the stomach, intimately covering the jejunal coils in its course. Because of this folding of the greater omentum, the intestinal coils are covered by two layers of the greater omentum.

The more superficial, ventral layer (*paries superficialis*) was formerly called the parietal part, and the deeper, dorsal layer (*paries profundus*) was known as the visceral part. Between these two layers is a potential cavity, the lesser peritoneal cavity, or the omental bursa. Except for the large constant opening, the epiploic foramen, the omental bursa is a closed sac. The epiploic foramen is bounded ventrally by the peritoneum covering the portal vein and dorsally by that covering the caudal vena cava. The greater omentum is considerably longer and wider than the abdomen; to accommodate this bulk, it is folded in around the margin of the intestinal coils. These infoldings usually occur as follows: caudally, cranial to the bladder; on the right, medial to the descending portion of the duodenum; and on the left, ventral to the sublumbar muscles, left kidney, and the fat which surrounds these structures. Owing to movements of the intestinal tract, however, the infoldings of the bursa undergo frequent changes.

The **splenic portion** is also called the **gastrosplenic ligament** (*lig. gastrolienale*). Al-

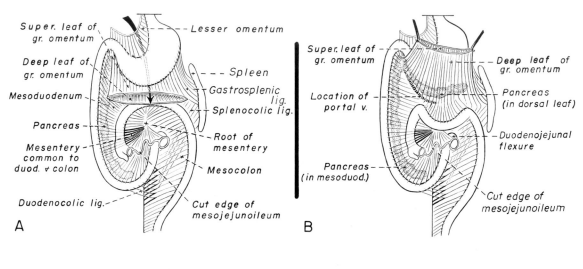

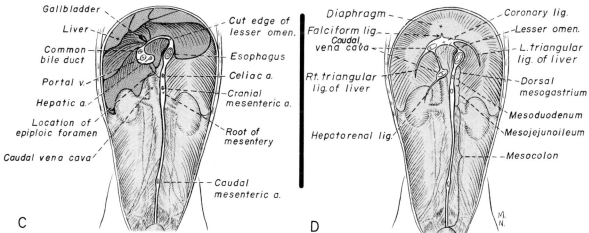

Figure 7–34. Peritoneum.
A. Plan of visceral and connecting peritoneum, ventral aspect. The greater omentum is transected caudal to the stomach. Arrow in omental bursa.
B. Plan of peritoneum with greater omentum reflected cranially. The transverse colon is displaced caudally.
C. Plan of the dorsal reflections of the connecting and parietal peritoneum. The stomach and intestines removed.
D. Plan of the dorsal reflections of the connecting and parietal peritoneum. All abdominal viscera removed.

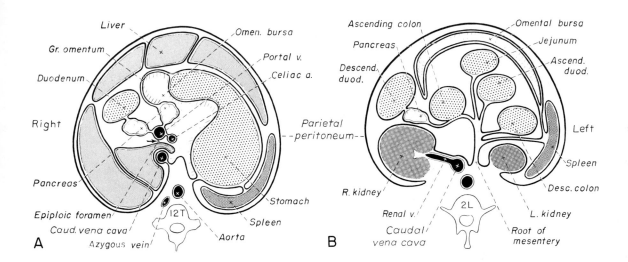

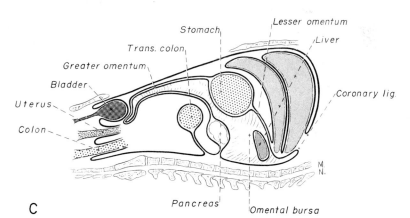

Figure 7–35. Peritoneal schema as viewed with the dog in a supine position.
A. Transverse section through the epiploic foramen.
B. Transverse section through the root of the mesentery.
C. Sagittal section.

though Zietzschmann (1939) considers this portion as extending beyond the ends and hilus of the spleen, in this description the term will be limited to that portion of the greater omentum which extends from the diaphragm, fundus, and greater curvature of the stomach to a line which parallels the hilus of the spleen. The spleen is not located between the two peritoneal layers which form the gastrosplenic ligament but is interposed in an outpocketing of the superficial peritoneal layer which forms the parietal sheet. When taut, the gastrosplenic ligament is about 5 cm. wide ventrally. The large splenic vessels and nerves which approach the spleen near the middle of its hilus give origin to the left gastroepiploic vessels which course through the ligament to the greater curvature and fundus of the stomach.

The **veil portion** is the smallest of the three portions of the greater omentum and the only portion that does not form a part of the omental bursa. It contains between its peritoneal leaves, cranially, the left extremity and caudal margin of the left lobe of the pancreas. Its borders form approximately a rectangle, each lateral margin measuring about 20 cm. in length. The cranial margin is about 10 cm. long and the caudal one 7 cm. The right or mesocolic margin blends with the left peritoneal leaf of the left mesocolon opposite the attachment of the duodenocolic ligament which blends with the right peritoneal leaf of the left mesocolon. Its left margin is free and usually contains a fine but strong filament. Its cranial margin blends with the dorsal peritoneal leaf of the visceral layer of the bursa portion. A relatively fat-free side plica leaves the left lateral peritoneal leaf of the main portion of the omental veil. Cranially, this side fold attaches to the visceral surface of the spleen, usually at the beginning of the distal fourth, where it extends at a right angle to the hilus and runs toward the caudal border of the organ.

The greater omentum has many functions, yet when it is largely removed, it does not regenerate. Webb and Simer (1940) removed the greater omentum in three dogs and found that the health of the dogs was not impaired in any way. The greater omentum is used frequently by the surgeon because of the important part it plays aiding in the revascularization of tissues which have had their normal blood supply impaired. The mobility of the omentum is facilitated by peristalsis. Higgins and Bain (1930) found that there are two systems of lymphatic drainage from the abdomen, one associated with the gastrointestinal tract which passes through the mesenteries to the cisterna, and the other associated with the omentum and the diaphragm which passes through the ventral portion of the mediastinum to the cervical lymph ducts. The omental lymphatics function more to hold and isolate foreign material in the peritoneal cavity than they do to transport such material to larger channels.

LESSER OMENTUM

The **lesser omentum** *(omentum minus)* is the largest derivative of the ventral mesentery, but it is not nearly as voluminous or as complex as is the greater omentum, which it resembles in structure, although it contains less fat. It loosely spans the distance from the lesser curvature of the stomach and the initial part of the duodenum to the porta of the liver. Between the liver and the cardia of the stomach it attaches to the margin of the esophageal hiatus of the diaphragm. It becomes continuous with the mesoduodenum on the right. The course of the bile duct from the liver to the duodenum may be regarded as the division between them. The papillary process of the liver is loosely enveloped by the lesser omentum as it projects into the vestibule of the omental bursa (Fig. 7–35A). The portion of the lesser omentum which goes to the duodenum from the liver is known as the **hepatoduodenal ligament** *(lig. hepatoduodenale),* and that portion which passes from the liver to the stomach is the **hepatogastric ligament** *(lig. hepatogastricum).*

The **omental bursa** *(bursa omentalis)* (Fig. 7–35), or lesser peritoneal cavity, is collapsed in life. The organs bounding it cranially are the visceral wall of the stomach, the caudate lobe of the liver, and the left lobe of the pancreas, which is located largely in the visceral wall of the greater omentum. The omental bursa has but one large, constantly present opening into the greater peritoneal cavity, which is called the **epiploic foramen** *(foramen epiploicum).* It is a narrow passage, about 3 cm. long, which lies to the right of the median plane, medial to the caudate process of the liver. It is bounded dorsally by the caudal vena cava and ventrally by the portal vein. The foramen leads into the **vestibule** *(vestibulum bursae omentalis)* or antechamber of the omental bursa, from which three recesses radiate.

The **dorsal recess** *(recessus dorsalis omentalis)* is that portion of the omental bursa into which projects the papillary process of the liver. It is bounded ventrally by the lesser omentum, dorsally and cranially by the liver, and caudally partly by the lesser curvature of the stomach. It freely communicates with the caudal recess over the dorsal wall of the stomach.

The **caudal recess** *(recessus caudalis omentalis)* is the main cavity of the omental bursa. It is enclosed by the bursa portion of the greater omentum and extends caudally and laterally from the stomach to the urinary bladder. The two laminae of the greater omentum do not fuse with each other caudally, thus reducing the size of the omental bursa, nor does any portion of it attach to the colon as it does in some other species (e.g., man and horse).

The omental bursa is closed caudally from the superficial wall reflecting upon itself to form the deep wall. Cranially, the closure of the omental bursa is more involved. The dorsal wall of the stomach and the liver largely fill the "mouth" of the bursa. At the cardia, the inner peritoneal layer of the greater omentum becomes continuous with the similar layer of the lesser omentum as the two layers are connected by the visceral peritoneum covering the dorsum of the cardia. On the right side the inner peritoneal layers of the two omenta converge on the medial surface of the cranial part of the duodenum, thus closing the bursa at this site. At other places around its periphery the inner peritoneal layers of the deep and superficial sheets become continuous. The other peritoneal layers of the two walls are also continuous except on the left, where the veil portion of the greater omentum is formed.

The **splenic recess** *(recessus lienalis)* is largely non-existent unless the abdominal cavity is opened and the spleen displaced. In such instances, it is a recess of the omental bursa opposite the hilus of the spleen.

Three folds invaginate the inner peritoneal sheet of the omental bursa or the whole bursa wall. These are formed by the three branches of the celiac artery and the nerve plexuses which surround them. The **right gastropancreatic fold** *(plica gastropancreatica dextra)* is a low peritoneal fold which contains the common hepatic artery. It extends obliquely across the medial face of the portal vein just within the vestibule from the epiploic foramen. The largest fold is formed by an upward displacement of the bursa wall produced by the splenic artery and nerves and their continuation opposite the hilus of the spleen as the left gastroepiploic artery and nerve plexus. The **left gastropancreatic fold** *(plica gastropancreatica sinistra)* is formed by the left gastric artery and nerve plexus. It is short and extends from the celiac artery to the left extremity of the lesser curvature of the stomach. Surrounded by fat, it continues on the dorsal wall of the body of the stomach just caudal to the attachment of the lesser omentum.

The various peritoneal folds which attach the liver and digestive tube to the abdominal wall or to other viscera are treated in the descriptions of the several organs.

STOMACH

The **stomach** *(ventriculus* [gaster]) (Figs. 7–36, 7–37), the largest dilatation of the alimentary canal, is a musculoglandular organ interposed between the esophagus and the small intestine. An axis through it is shaped somewhat like the letter **C** rotated 90 degrees in a counterclockwise direction. It varies greatly in size. It stores and partly mixes the food, while its intrinsic glands intermittently add enzymes, mucus, and hydrochloric acid. The stomach lies largely in a transverse position, more to the left of the median plane than to the right of it. It forms an extensive concavity in the caudal surface of the liver, and when it is empty it is located completely cranial to the thoracic outlet. The inlet of the stomach is called the **cardia,** and the outlet the **pylorus.** The major divisions of the stomach are the cardiac portion, fundus, body, and pyloric portion. It possesses visceral and parietal surfaces, and greater and lesser curvatures.

The **visceral surface** *(facies visceralis)* presents a convex outer surface which faces mainly dorsally, but also caudodextrally. It lies in contact with the left lobe or limb of the

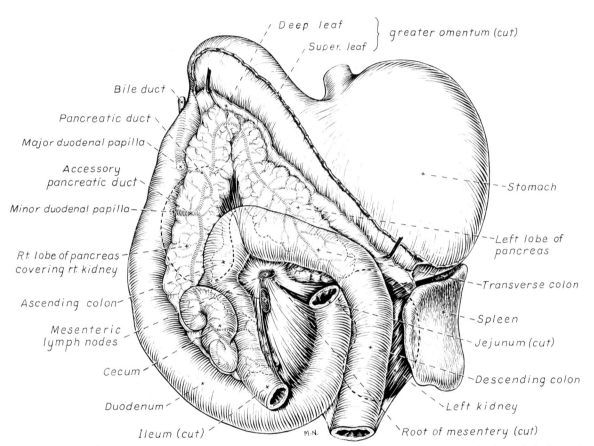

Figure 7–36.   Abdominal viscera, ventral aspect. The pancreas in situ; the position of the kidneys is indicated by a dotted line.

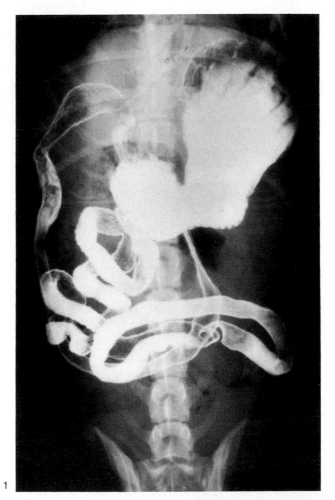

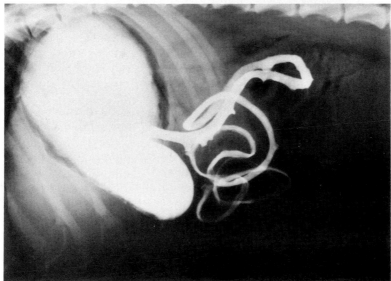

Figure 7–36. *Continued.*   Radiograph 1. Dorsoventral radiograph of the stomach and small intestine eight minutes after the administration of contrast medium.
Radiograph 2. Lateral radiograph of the stomach and duodenum.

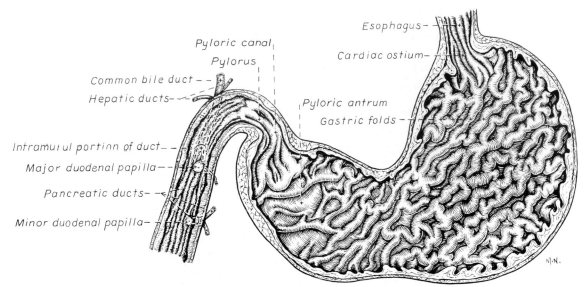

Figure 7–37.   Longitudinal section of stomach and proximal portion of duodenum.

pancreas and is separated from the intestinal mass and left kidney by the visceral (or dorsal) leaf of the greater omentum. The **parietal surface** *(facies parietalis)* faces to the left and cranially as well as ventrally. In the contracted state the stomach lies in contact with the liver, in which it produces an extensive gastric impression. The dilated stomach extends beyond the liver chiefly to the left and ventrally.

CURVATURES OF STOMACH

The **greater curvature** *(curvatura ventriculi major)* is convex, and extends from the cardia to the pylorus. It is approximately 30 cm. long in a stomach moderately filled. The superficial leaf of the greater omentum attaches to the greater curvature except on the left, where its line of attachment runs obliquely across the dorsal wall of the stomach to form, with the lesser omentum at the cardia, a closure of the omental bursa at this site.

The **lesser curvature** *(curvatura ventriculi minor)* also runs from the cardia to the pylorus, and is the shortest distance between these two parts. It does not form an even concavity but is in the form of a 50 to 70 degree angle, the **angular notch** *(incisura angularis).* Lying within this angle is the

papillary process of the liver. The pyloric antrum and pylorus lie to the right of the papillary process, and the body of the stomach lies to the left of it. The caudal edge of the lesser omentum attaches to it.

REGIONS OF STOMACH

The **cardiac portion** *(pars cardiaca)* of the stomach is that portion which blends with the esophagus. The four coats of both organs blend with each other. The greatest differences occur in the muscular and mucous layers as these are traced from the esophagus to the stomach. The opening into the stomach is the *ostium cardiacum.*

The **fundus of the stomach** *(fundus ventriculi)* is the rather large blind outpocketing located to the left and dorsal to the cardia. The esophagus joins the stomach in such a way that on the right its surface continues with that of the lesser curvature of the stomach without definite demarcation. On the left the **cardiac notch** *(incisura cardiaca)* is formed between the cardia and the bulging fundic part. The **body of the stomach** *(corpus ventriculi)* is the large middle portion of the organ. It extends from the fundus on the left to the pyloric portion on the right. If two transverse planes are projected through the stomach at right angles to its curved long

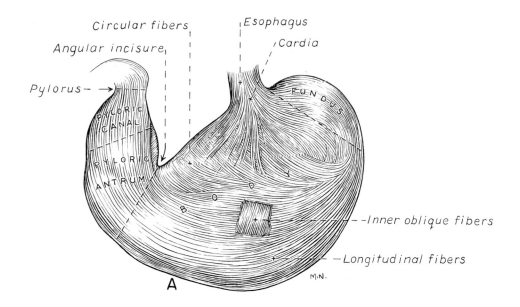

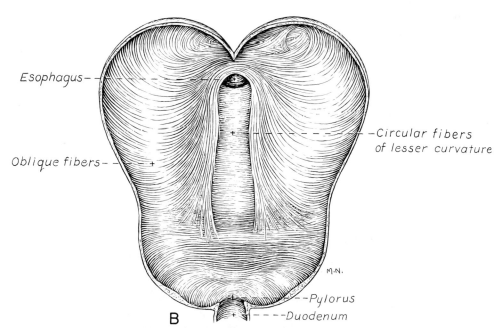

Figure 7–38. Musculature of stomach.
A. Outer layer, ventral aspect. Window cut to show inner oblique fibers.
B. Inner layer, the stomach opened along the greater curvature.

axis, one through the caudal face of the cardia above and the other through the angular notch below, these planes will limit the beginning and end of the body of the stomach. The shortest path which ingesta can take in passing from the cardia to the pyloric portion of the stomach is known as the **gastric groove** *(sulcus ventriculi)*. This path follows the lesser curvature of the stomach.

The **pyloric part** *(pars pylorica)* (Figs. 7–37, 7–38) is approximately the distal third of the stomach as measured along the lesser curvature. It is always somewhat sacculated as it unites the body of the organ to the duodenum. The pyloric part is irregularly funnel-shaped toward the pylorus, which is directed cranially. The initial two-thirds is thin-walled and expanded to form the **pyloric antrum** *(antrum pyloricum)*. The distal third is contracted and bent so that the greater curvature (caudal side) is three or four times longer than the opposite side. It is largely surrounded by a heavy, double sphincter which forms the narrowest part of the cavity of the stomach. The passage through the pyloric part is known as the **pyloric canal** *(canalis pyloricus)*.

The muscular sphincter at the entrance to the duodenum is the *pylorus* and its lumen, the *ostium pyloricum*.

SHAPE, POSITION, AND CAPACITY

The empty, contracted stomach is not only contained within the caudal part of the thoracic cage, but it is also nestled within the caudal concavity of the liver, being completely separated from the abdominal wall. When the stomach increases in size as the result of filling, the fundus enlarges first and pushes caudodorsally. It tends to displace the liver ventrally as the stomach comes in contact with the left lateral abdominal wall and diaphragm. The body of the stomach is the second part to fill and expand. It is the largest division of the organ, as well as the part capable of the greatest dilation. During filling, it migrates caudoventrally and makes extensive contact with the abdominal wall. It is particularly distensible in puppies, and when maximally expanded it may extend from a transverse plane through the eighth thoracic vertebra to a plane caudal to the umbilicus. This necessitates an expansion of the abdomen and a crowding of the intestinal

mass and spleen caudally and slightly dorsally. According to Grey (1918) the normal stomach possesses a striking capacity to adjust its size to the volume of its contents with only minimal changes in intragastric pressure.

Secord (1941) states that food remains in the dog's stomach from 10 to 16 hours. Zietzschmann (1938) states that the pylorus varies least in position as the stomach fills. It is always more cranial and ventral than the cardia. The pyloric portion is the last part to expand, the antrum expanding more than the canal. The pyloric portion functions chiefly as an ejection mechanism by which the partly digested stomach contents, the *chyme*, is forced through the pyloric canal and squirted into the duodenum. The empty or the partly filled stomach is shaped like a C, with its convex surface facing caudoventrally and to the left.

The capacity of the stomach varies from 0.5 to 8 liters. Greater ranges in relative size are present in puppies than in adult dogs. Ellenberger and Baum (1943) cite Neumayer, who gives the capacity as 100 to 250 cc, per kilogram of body weight. The average empty stomach of a 15-kg. dog weighs about 100 gm.

COATS OF STOMACH

The **serous coat** *(tunica serosa)* completely covers the stomach except for an extremely narrow line on the distal half of the greater curvature which continues obliquely across the dorsal surface of the stomach to the cardia. A second similar line on the stomach which is not covered by peritoneum extends from the cardia along the lesser curvature to the duodenum. The smooth peritoneal sheets of serous membrane which intimately cover the dorsal and ventral walls of the stomach fuse just distal to these lines to form the greater and lesser omenta. In obese specimens a small irregular strip of fat and the nerves and vessels serving the organ widen the line along which the omenta arise. The two peritoneal layers which form these folds separate at the cardia to extend forward on the abdominal portion of the esophagus. At the duodenum the peritoneal leaves are similarly forced apart by the duodenum. The peritoneal leaves of the lesser omentum become continuous with those of the mesoduodenum over the common bile duct. The serosa of the stomach is extremely thin and elastic. It adheres closely to the stomach musculature by a scanty amount of subserous tissue.

The **muscular coat** *(tunica muscularis)* of the stomach (Fig. 7–38) consists essentially of an outer longitudinal and an inner circular layer of smooth muscle fibers. To these layers are added oblique fibers over the body of the stomach.

The outer **longitudinal layer** *(stratum longitudinale)* is continuous with the essentially outer longitudinal layers of both the duodenum and the esophagus. The longitudinal fibers on and adjacent to the lesser curvature, as traced from the esophagus, spray out and end before reaching the angular incisure. Those fibers on the dorsal and ventral walls of the body of the stomach end before reaching the middle of the body, whereas the longitudinal fibers on the greater curvature continue uninterruptedly from the esophagus to the duodenum. On both the dorsal and ventral walls of the stomach at about the junction of the fundus and body there is a kind of muscular whorl formed by the bundles of fibers and the longitudinal layer changing position and direction. At and adjacent to the angular incisure there are no longitudinal fibers, so that the circular fibers of the inner coat become superficial. As the longitudinal fibers cover the pyloric portion they are particularly heavy on the sides between the curvatures, but no definite pyloric ligaments are formed, as in man.

The inner **circular layer** *(stratum circulare)* of the stomach is more complete and specialized than is the longitudinal layer. At the cardia the circular layer is thickened to form the feeble **cardiac sphincter** *(m. sphincter cardiae)*. This sphincter is augmented on the greater curvature by the acquisition of a condensation of the transversely running inner oblique fibers. At the approximate junction of the fundus and the body on both the dorsal and ventral walls, the circular coat enters into the formation of a muscular fiber interchange with the longitudinal coat. The circular coat is not covered by the longitudinal coat in and adjacent to the angular incisure so that in this region it receives a peritoneal covering, superficially. Deeply, throughout the length of the lesser curvature from the cardia to the pyloric antrum, there is a muscular trough, about 2 cm. wide, which is formed on the sides by the parallel longitudinal parts of the oblique fibers and deeply by the circular fibers.

Surrounding the pyloric canal, the inner circular layer is well developed, as Torgersen (1942) has pointed out. The musculature is thickest as it crosses the greater curvature. The pylorus, which opens into the duodenum, is also surrounded by a circular muscle termed the **pyloric sphincter** *(m. sphincter pylori)*.

The **oblique fibers** *(fibrae obliquae)* are adjacent to the submucosa. They appear to arise from a heavy transverse stratum which is arched across the dorsal (greater curvature) boundary of the cardiac orifice. These fibers run distally and outward toward the pylorus and the greater curvature in each wall. The oblique fibers thus are spread like a fan; in a moderately distended, medium-sized stomach, those bundles nearest the lesser curvature are essentially parallel. The fibers next peripheral to the parallel ones fan out and end on the inner surface of the circular fibers of the distal part of the body of the stomach; the most proximal oblique fibers become almost transverse proximally and blend with those of the circular layer in augmenting the size of the dorsal part of the feeble cardiac sphincter.

The **submucous coat** *(tela submucosa)* consists of a strong but thin elastic layer of areolar tissue which more firmly attaches to the mucosa than to the muscularis. It contains the finer branches of the gastric vessels and nerves. In the contracted organ it is thrown into folds which occupy the centers of the relatively inelastic plicae of the mucous coat.

The **mucous coat** *(tunica mucosa)* consists of a columnar surface epithelium, a glandular lamina propria, and a lamina muscularis mucosae consisting of muscular fibers which may be irregularly interwoven or stratified (Trautmann and Fiebiger 1957). In the contracted empty or even a moderately distended organ, the mucosa and much of the underlying submucosa are thrown into folds, the *plicae gastricae*. These folds are largely longitudinal in direction and very tortuous except adjacent to the lesser curvature, where the folds are less crowded and are relatively straight. In a strongly contracted stomach the mucosal folds, which may be 1 cm. high, lie closely adjacent to each other. The normal color of the mucosa in the body and fundus of a fresh stomach is pink to grayish red. In the pyloric region it is lighter in color. The color varies with the amount of contained blood, as well as the freshness of the material.

Under magnification the mucosa is seen to possess about 40 raised areas for every square centimeter. These are the *areae gastricae*. In the pyloric region these areas are elongated in a longitudinal direction, but elsewhere they are polygonal and rendered distinct by small surrounding furrows. Each of the small gastric areas on their surfaces and sides is stippled with numerous minute openings, the *foveolae gastricae*. The foveolae are longest, about 0.68 mm. in the pyloric region. They gradually shorten toward the cardia and disappear, so that none exist in the cardiac gland region which is adjacent to the esophagus (Mall 1896). According to Mall, there are 1,000,000 foveolae in the stomach of the dog, and each of these has about 16 gastric glands opening into it. There may be folds, *plicae villosae*, between the gland openings.

GLANDS OF STOMACH

The glands of the stomach are known as the **gastric glands** *(glandulae gastricae)*. They are branched tubular glands with necks and bodies which reach nearly to the lamina muscularis mucosae. According to Trautmann and Fiebiger (1957), in older carnivores a double-layered lamina subglandularis intervenes between the lamina muscularis mucosae and the blind ends of the glands. According to these authors, the lamina propria contains the gastric glands, and in certain areas, exclusive of the cardiac gland zone, the glands are divided into groups by heavier strands of supporting tissue which contain muscle fibers from the lamina muscularis mucosae. Three types of gastric glands are recognized in the dog. These are the cardiac glands, the gastric glands proper, and the pyloric glands.

The **cardiac glands** *(glandulae cardiacae)*, according to Haane (1905), are found in a narrow zone around the cardia. Cardiac glands are also scattered along the lesser curvature, according to Ellenberger (1911).

The **gastric glands proper** *(glandulae gastricae [propriae])*, or fundic glands, occupy about two-thirds of the gastric mucosa. This includes the left extremity, or fundus, and the body of the stomach. It is exclusive of the pyloric part and the cardiac gland region.

The **pyloric glands** *(glandulae pyloricae)* are found in the pyloric part of the stomach. Between the pyloric and gastric glands proper, according to Bloom and Fawcett (1975), there exists in the dog a zone of intermediate glands which reaches a width of 1 to 1.8 cm. Harvey (1906) states that when the stomach is flattened out the intermediate zone is 2 to 3 cm. wide. The difference between the various gland zones is in the type of cells they contain, and therefore in the nature of the secretion they produce. The mucosal regions of the stomach are not coextensive with the gross divisions of the organ with the same or comparable names. There is considerable intermixing of the glands of each gland area with those of adjacent areas. The intermediate gland zone is formed in this manner. The cardiac glands of the stomach are similar to the cardiac glands in the distal portion of the esophagus. The pyloric glands imperceptibly blend with those of the duodenum, and as they do they come to lie in the submucosa. Both the type of cells in the gastric glands and their morphology and location have influenced the naming of the mucosal areas of the stomach.

According to Bloom and Fawcett (1975), four types of glandular cells are found in the stomach mucosa. These are: (1), chief, or zymogenic, cells, which contain granules that are believed to contain pepsinogen, the precursor of the chief gastric enzyme, pepsin; (2) parietal cells, which are spherical or pyramidal cells lying next to the basement membrane, and are considered the source, probably indirectly, of the hydrochloric acid of the gastric juice; (3) mucous neck cells, which are located in the necks of the gastric glands, filling the spaces between the parietal cells, and which produce mucus; (4) argentaffin cells, which are moderately abundant in the proper gastric glands and less frequent in the pyloric glands, and which are numerous in the first part of the small intestine. The glands of the cardiac, intermediate, and pyloric regions function mainly to produce mucus; the proper gastric glands produce hydrochloric acid indirectly and the enzyme pepsin. As in most other areas of the alimentary canal, lymph nodules are scattered throughout the mucosa of the stomach. Some of these extend through the lamina muscularis mucosae into the submucosa (Bensley 1902).

VESSELS OF STOMACH

The main arteries to the stomach are the **left** and **right gastric arteries,** which run along the lesser curvature, and the **left** and **right gastroepiploic arteries,** which run along the greater curvature. The larger left gastric artery anastomoses with the right gastric at the beginning of the pyloric antrum. The epiploic vessels anastomose with each other on the greater curvature of the body of the stomach. In addition to these arteries, two or more long branches leave the terminal part of the splenic artery and supply a portion of the fundus of the stomach. The arterial branches which actually enter the musculature of the organ along the greater curvature run greater distances under the serosa and are more nearly vertical to the parent trunks than the comparable branches along the lesser curvature. The veins from the stomach are satellites of the arteries supplying the organ. The **left gastric** and **left gastroepiploic veins** are tributaries of the gastrosplenic vein. The **right gastric** and **right gastroepiploic veins** are tributaries of the gastroduodenal vein. The blood from the stomach enters the liver through the portal vein. The lymphatics from the stomach all eventually drain into the **hepatic lymph nodes.** Baum (1918) states that most of these vessels drain into the left hepatic node after first having passed through the splenic and gastric nodes. A few lymphatics from the stomach drain into the right hepatic node after first having passed through the duodenal node.

NERVES OF STOMACH

The stomach is supplied by parasympathetic fibers from the **vagi** and by sympathetic fibers from the **celiac plexus.** The ventral vagal trunk, after passing through the esophageal hiatus, immediately sends two to four small branches to the pylorus and liver. Other branches go to the lesser curvature of the stomach. The dorsal vagal trunk also sends branches to the lesser curvature and to the ventral wall of the stomach and continues across the celiac plexus to reach and follow branches of the celiac and cranial mesenteric arteries. The sympathetic fibers to the stomach reach it by traveling on the numerous gastric branches of the celiac artery. They come from the **celiacomesenteric plexus.**

SMALL INTESTINE

The **small intestine** *(intestinum tenue)* extends from the pylorus of the stomach to the ileocolic orifice leading into the large intestine. It is the longest portion of the alimentary canal, having an average length, in the living animal, of 3.5 times the length of the body. After death and the cessation of peristaltic contractions, the intestine increases in length owing to the loss of muscular tonus. Williams (1935), Alvarez (1940), and Nickel, Schummer, Seiferle, and Sack (1973) cite intestinal measurements in the dog. The small intestine consists of three main parts, the relatively fixed and short proximal loop, or **duodenum;** the freely movable, long, distal portion, the **jejunum;** and the very short terminal part, the **ileum.**

DUODENUM

The duodenum (Fig. 7–36) is the first and most fixed part of the small intestine. It is about 25 cm., or 10 inches, long. It begins in the upper half of the right hypochondriac region opposite the ninth intercostal space. It runs mainly caudally to a transverse level through the tuber coxae, makes a U-shaped turn, and runs obliquely craniosinistrally to be continued by the jejunum to the left of the root of the mesentery. This loop formed by the duodenum is more marked in puppies than in adults (Mitchell 1905). Both the pancreatic and the bile ducts open into the duodenum. The acid chyme which enters it from the stomach is mixed with the alkaline secretions from the liver, pancreas, and small intestinal glands. Because of the high nutritive content of the material ingested, most free-living intestinal parasites are found in the duodenum. For descriptive purposes the duodenum is divided into four portions. These are the cranial, descending, caudal, and ascending portions.

The **cranial portion** *(pars cranialis)* of the duodenum arises from the pylorus in such a way that its right wall is considerably longer than its left wall. This configuration brings about the acute **cranial duodenal flexure** *(flexura duodeni cranialis)* formed by the cranial portion of the duodenum. The pyloric part of the stomach runs cranially, whereas the cranial portion of the duodenum runs essentially caudally. The cranial portion of the duodenum is also known as the duodenal cap or bulb. It lies opposite the ninth and tenth ribs and the intervening intercostal space. Ventrally, it is separated from the stomach by the greater omentum. Dorsally and laterally it lies in contact with the liver, and medially it is in contact with the pancreas.

The **descending portion** *(pars descendens)* of the duodenum, about 15 cm. long, runs caudally from the cranial portion nearly to the pelvic inlet. Its lateral surface lies in contact with the parietal peritoneum of the upper flank, and with the right lateral and right medial lobes of the liver, cranially. Dorsally, the right lobe or limb of the pancreas lies in contact with it. Medially, it is related primarily to the cecum caudally and the ascending colon cranially, but it is separated from these by the infolded greater omentum.

The **transverse portion** *(pars transversa [pars caudalis])* of the duodenum is also known as the caudal flexure of the duodenum *(flexura duodeni caudalis).* It connects the descending and ascending portions from right to left and is about 5 cm. long. It lies in a frontal (horizontal) plane. The transverse portion usually lies ventral to the body and right transverse process of the sixth lumbar vertebra. A full bladder and colon would force it cranially. The initial segment of this portion, at the right, lies in contact with the parietal peritoneum of the sublumbar region and the uterine horn ventral to the deep circumflex iliac vessels. It is continued by a rounded angle which is somewhat less than a U-turn as the ascending portion of the duodenum diverges slightly from the descending portion. The transverse portion of the duodenum is related to the terminal portion of the ileum and to the jejunum ventrally.

As the **ascending portion** *(pars ascendens)* of the duodenum runs obliquely forward and to the left from the caudal flexure, it crosses obliquely the dorsally lying ureters, sympathetic trunks, caudal vena cava, aorta, and lumbar lymphatic trunks. Ventrally, the ascending portion is related to the coils of the jejunum. On the left it approaches the descending colon, and makes a sweeping curve ventrally to form the **duodenojejunal flexure** *(flexura duodenojejunalis).* At this flexure the jejunum continues the duodenum ventrally, caudally, and to the left, and enters into the formation of numerous coils and kinks (festoons), which constitute most of the intestinal mass.

Attachments and Peritoneal Relations. The first two parts or right portion of the duodenum is located in the free border of the *mesoduodenum.* The ventral peritoneal layer of the mesoduodenum, near the stomach, is continuous with the ventral layer of the lesser omentum over the bile duct. The dorsal peritoneal layer is coextensive with the dorsal layer of the lesser omentum as traced over the portal vein and through the epiploic foramen. Upon leaving the right portions of the duodenum, the two peritoneal leaves which form the mesoduodenum as traced to the left extend directly to the right lobe of the pancreas or merge for a short distance before covering this portion of the gland. The peritoneal leaves merge again upon leaving the left side of the right pancreatic lobe and pass to and cover the ascending part of the duodenum. Upon leaving the ascending duodenum, the mesoduodenum becomes continuous with the right peritoneal leaf of the descending mesocolon. Caudally, the two peritoneal layers of the mesoduodenum form a triangular fold with a free caudal border which runs from the mesocolon in the region of the pelvic inlet obliquely craniodextrally to the caudal duodenal flexure. This triangular attachment of the initial part of the ascending duodenum is called the **duodenocolic ligament** *(lig. duodenocolicum).*

The **duodenal fossa** *(fossa duodenalis)* (Fig. 7–39) is comparable to the inferior

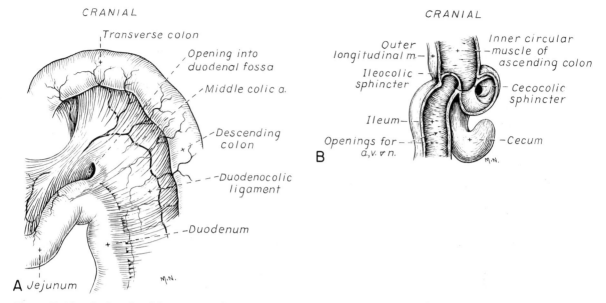

Figure 7-39.   A. Duodenal fossa, ventral aspect.
B. Dissection of inner circular and outer longitudinal muscle layers of ileum, cecum, and colon, dorsal aspect.

duodenal fossa of man. It may be absent, or, when present, it may admit the end of the little finger. It varies in length up to 4 cm. Its opening faces cranially and is usually situated about 2 cm. caudal to the duodenojejunal flexure. The fossa runs caudally along the attached border of the ascending duodenum. When maximally developed, it reaches a transverse level through the most cranial extension of the duodenocolic ligament. This simple peritoneal pocket is bounded by the duodenum ventrally and on the right, and by the left leaf of the descending mesocolon dorsally and on the left.

JEJUNUM AND ILEUM

The jejunum and ileum compose the bulk of the small intestine. The jejunum begins at the left of or caudal to the root of the mesentery at the duodenojejunal flexure, and the ileum ends by opening into the initial portion of the ascending colon as the **ileal papilla** *(papilla ilealis)* with an **ileal orifice** *(ostium ileale)* and an associated **sphincteric muscle** *(m. sphincter ilei).* The orifice lies usually between the descending

and ascending portions of the duodenum. In contrast to the relatively fixed duodenum, the jejunoileum is the most mobile and free part of the entire alimentary canal. It is suspended by the long mesentery from the cranial part of the sublumbar region and is therefore also known as the mesenteric portion of the small intestine *(pars intestinum tenue mesenteriale).* No definite gross, microscopic, or developmental manifestations mark the division between jejunum and ileum. This division was made by early investigators from the gross appearance of this portion of the bowel. According to Field and Harrison (1947) Galen applied the term "jejunum" to the middle portion of the small intestine because it is usually empty or appears emptier than the rest. The term "ileum" is applied to the relatively short, contracted terminal portion of the small intestine in domestic animals. Although there are distinctive differences, particularly in the mucosa, between a typical portion of the jejunum and that of the ileum, there are no sufficiently marked gross differences in the character of the walls to differentiate the jejunum from the ileum. The ileum of the dog is approximately the last 15 cm. of the small intestine.

POSITION OF SMALL INTESTINE

The *jejunum* is located ventrocaudal to the empty stomach. It is separated from the ventral and lateral abdominal wall only by the deep and superficial sheets of the greater omentum. It is related dorsally to the large intestine, duodenum, pancreas, kidneys, caudal vena cava, aorta, sympathetic nerve trunks, and lumbar lymphatics. The jejunum rarely extends into the pelvic cavity caudally because the urinary bladder and the rectum largely fill the pelvic inlet. The spleen, through the greater omentum, is related to the craniosinistral part of the jejunum. The ileum is the terminal part of the small intestine. In man, the distal three-fifths of the jejunoileum is regarded as ileum, the proximal two-fifths as jejunum. Most veterinary anatomists regard only the short, terminal, usually contracted part of the small intestine as ileum. In the dog, unlike man, the ileum contains fewer aggregated lymph follicles than the more proximal part of the small intestine, including the duodenum (Titkemeyer and Calhoun 1955). Its terminal part usually crosses the ventral surface of the descending colon. Its termination of the ileocolic valve is located within the duodenal loop or ventral to the ascending part of the duodenum. The ileum is readily identified by the ileocecal fold and its contained antimesenteric ileal vessels.

MESENTERY OF SMALL INTESTINE

The **ileocecal fold** (*plica ileocecalis*) is a narrow but usually long, variably developed plica of peritoneum which continues proximally on the ileum from the area of adhesion of the cecum to the ileum. It is rarely over 1 cm. wide at its origin, and when it is double at this site it is considerably narrower. It varies in length from 2 to 30 cm. A streak of fat located in its ileal attachment surrounds the antimesenteric ileal vessels. Its distal end is reduced to a low ridge of peritoneum formed by the underlying vessels and fat. Frequently the vessels can be traced proximally on the ileum beyond the ending of the fold.

The **mesentery** (*mesenterium*) (Fig. 7–34) is also known as the great or proper mesentery, or the mesojejunoileum, to differentiate it from the various other portions which are derived from the common dorsal mesentery. In the adult it is continuous in front with the mesogastrium (deep sheet of the greater omentum) and behind with the descending mesocolon. Embryonically, the great mesentery is continuous with the mesoduodenum in front and with the ascending mesocolon behind. Through rotation, torsion, and differential growth of the various portions of the alimentary canal the definitive continuations of the great mesentery have been altered. The great mesentery is in the form of a large fan hanging from the cranial part of the sublumbar region. In its free distal border is located the convoluted jejunum and ileum. It is about 20 cm. wide and 5 mm. thick. Its length or greatest dimension varies greatly, depending on where the measurement is taken. It is only about 1.5 cm. long at its parietal attachment to the aorta and diaphragmatic crura opposite the second lumbar vertebra. This portion of the great mesentery is known as the **root of the mesentery** (*radix mesenterii*). It is the thickest portion because it includes the cranial mesenteric artery, intestinal lymphatics, and the large mesenteric plexus of nerves which surround the artery. The free or intestinal border of the great mesentery is its longest part. It extends from the duodenojejunal flexure, proximally, to the ileocolic junction, distally. It is as long as the jejunoileum, or about 250 cm. in life. It is greatly folded or ruffled as it follows the turns of the intestine. The peripheral part of the mesentery is much thinner than the root. Even in obese specimens the great mesentery does not contain a great amount of fat, and it is deposited most abundantly only along the larger vessels, so that large translucent areas are present between the vascular branches and arcades.

COATS OF SMALL INTESTINE

The small intestine, like the other parts of the alimentary tract, is composed of mucous, submucous, muscular, and serous tunics.

The **mucous coat** (*tunica mucosa*), throughout the small intestine of the dog, presents a free surface which is velvety owing to the presence of innumerable **intestinal villi** *(villi intestinales)*. The single-layered surface cells are of two types. One type consists of the columnar cells which function in absorption, and the other type consists of the goblet, mucus-producing cells. The deeper part of the mucosa is occupied largely by the **intestinal glands** *(glandulae intestinales)* and diffuse lymphoid tissue and single follicles. In about 22 areas throughout the small intestine of the dog the lymphoid follicles are grouped together to form the **aggregated follicles** *(noduli lymphatici aggregati)*. Titkemeyer and Calhoun (1955) found the aggregated follicles to be circumscribed elevations measuring about 2 cm. by 1.5 cm. They are more numerous in the proximal portion of the small intestine than in the ileum, many being found in the duodenum. In the distended bowel they are visible through the serosa and are more numerous in the side walls of the intestines than in that part of the wall opposite the mesentery. Titkemeyer and Calhoun also described a special connective tissue layer about 30 μm thick between the intestinal glands and the lamina muscularis mucosae. This layer, found only in the dog, is apparently similar to the layer in a comparable location in the stomach. The *lamina muscularis mucosae*, according to the previously mentioned investigators, was found to be three times thicker in the dog than it was on an average in the other domestic animals studied. It is definitely divided into inner circular and outer longitudinal layers. In the dog, the **duodenal glands** *(glandulae duodenales)* differ from the intestinal glands in that they closely resemble the pyloric glands of the stomach. They are located only around a narrow zone of the duodenum adjacent to the pylorus. The main portion of each gland is located in the submucosa, and a portion may extend into the lamina propria of the mucosa. Warren (1939) estimates that the ratio of mucosal area to serosal area for the whole small intestine of the dog is 8.5 to 1. The villi account for most of this large surface area, since the dog has no circular mucous folds.

The **submucous coat** *(tela submucosa)* resembles that of the stomach and large intestine. It loosely binds together the mucous

and muscular layers. The smaller blood vessels, lymphatics, and the submucous nerve plexus are located in it. Trautmann and Fiebiger (1957) state that the aggregated follicles are located mainly in the submucosa, with only a small portion of the lamina propria of the mucosa.

The **muscular coat** *(tunica muscularis)* consists of a relatively thin outer longitudinal layer *(stratum longitudinale)* and a thicker inner circular layer *(stratum circulare)*.

The **serous coat** *(tunica serosa)* of the small intestine is composed of the peritoneum, which completely covers the duodenum except along the lines where it is attached, including the duodenocolic ligament, and a small elongated area where it leaves the pancreas to reflect around the duodenum. The jejunum and ileum are also truly intraperitoneal organs. The only parts not covered by peritoneum are along the lines of the mesenteric attachment and on the antimesenteric side of the terminal portion of the ileum where the cecum is loosely fused to the ileum and where this attachment is continued by the ileocecal fold.

VESSELS OF SMALL INTESTINE

The large middle portion of the small intestine, the jejunum, is supplied by 12 to 15 **jejunal arteries**, which are branches of the cranial mesenteric. The duodenal branches from both the **cranial** and **caudal pancreaticoduodenal arteries** supply the duodenum, the most proximal jejunal artery anastomosing with the most distal duodenal branch of the caudal pancreaticoduodenal artery. The ileum is supplied on its mesenteric side by an ascending ramus for the **accessory cecal artery**, and on its antimesenteric side it is supplied by the ileal branches of the **ileocecal artery**. The main ileal branch runs in the areolar tissue, connecting the cecum to the ileum. Upon leaving the adhered area between the two viscera, it continues in the ileocecal fold along the whole ileum as the *ramus antimesenterialis* and anastomoses with the most distal jejunal artery in the musculature of the small intestine. From the terminal arcades which lie closely adjacent to the intestine the short, irregular *vasa recti*, upon reaching the intestine, run variable distances on the mesenteric half of the musculature before perforating it to supply the submucosa and the mucosa (Noer 1943). According to Morton (1929), the duodenum has a much richer blood supply than does the ileum, and it produces 5 to 10 times more fluid. The lymph vessels from the jejunum and ileum drain primarily into the **right** and **left mesenteric lymph nodes.** Some lymph from the duodenum is carried to the **hepatic lymph nodes** and to the **duodenal lymph node**, when it is present. Lymphatics from the ileum also drain into the **colic lymph nodes.** The nerve fibers to the mesenteric portion of the small intestine come to it from the **vagus** and **splanchnic nerves** by way of the **celiac** and **cranial mesenteric plexuses.**

LARGE INTESTINE

The **large intestine** (*intestinum crassum*) is short and unspecialized. The large intestine of the dog and cat resembles that of man more than it does that of the other domestic animals. Neither haustra nor tenia exist, and no vermiform process or sigmoid colon is present. In general, it is a simple tube, only slightly larger in diameter than the small intestine. Its most important function is the dehydration of its fecal contents. The large intestine is divided into cecum, colon, rectum, and anal canal. It begins at ileal sphincter and ends at the anus.

CECUM

The **cecum** (Figs. 7–39, 7–40) is usually described as the first part of the large intestine, but this is not true in the dog because

the ileum, the terminal part of the small intestine, communicates only with the colon, and the cecum exists as a diverticulum of the proximal portion of the colon. The openings of the ileum and cecum into the colon are closely associated. The cecum is extremely variable in size and form. In the live animal it is about 5 cm. long and 2 cm. in diameter at its colic end. It irregularly tapers to the rather blunt **apex**, which is less than 1 cm. in diameter and usually points caudoventrally or is located transversely. The large middle portion of the organ may be referred to as the **body**. When it is detached and straightened, the length of the cecum is over twice what it is when the cecum is attached. The only communication of the cecum is with the beginning of the ascending colon by means of the **cecocolic orifice** (*ostium cecocolicum*). This opening lies approximately 1 cm. from the ileocolic orifice. The **cecocolic sphincter** (*m. sphincter cecocolicum*) is a specialization of the inner circular muscular coat which guards the cecocolic orifice. It is about 0.5 cm. in diameter when partly constricted. The cecum is attached to the terminal portion of the ileum by fascia and peritoneum throughout most of its length. At the apical end of the body beyond its attachment to the ileum extends the single or double **ileocecal fold** (*plica ileocecalis*). When single, this fold is triangular, with its free caudal border measuring 0.5 to 1 cm. in width and only slightly more in length. It does not leave the apex of the cecum but the concavity of the terminal flexure. A low peritoneal ridge containing fat and the antimesenteric ileal vessels continues beyond the fold

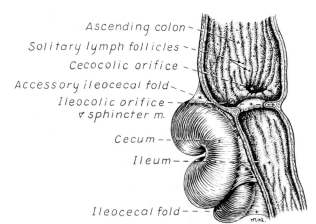

Ascending colon
Solitary lymph follicles
Cecocolic orifice
Accessory ileocecal fold
Ileocolic orifice
v sphincter m.
Cecum
Ileum
Ileocecal fold

Figure 7–40. Longitudinal section through ileocolic orifice, ventral aspect.

as far as 30 cm. A smaller fold, devoid of visible vessels, extends from the base of the cecum to the ascending colon. It is the accessory ileocecal fold (Kadletz 1929). It may also be double, enclosing a small peritoneal fossa. The peritoneal folds affect the definitive form of the cecum. The twisting of the cecum is less marked in puppies than in adults (Mitchell 1905). Bradley and Grahame (1948) imply that the variations of its flexures are formed by its not having a wide mesentery. As the cecum grows out from the colon, it is apparently restrained from growing in a straight line by its attachment to the ileum. The flexures develop quite irregularly. In its simplest form the cecum is sigmoid in shape, but more often it is in the form of an irregular corkscrew with a large U-shaped kink extending to the left from its ileal attachment. The cecum is located to the right of the median plane, usually within the duodenal loop. It lies dorsal to and occasionally partly surrounded by the coils of the jejunum, ventral to the right transverse processes of the second to fourth lumbar vertebrae. In rare instances it contacts the right lateral abdominal wall.

COLON

The **colon** (Fig. 7–36) is divided into ascending, transverse, and descending portions and their connecting flexures. The colon lies in the dorsal part of the abdominal cavity and is shaped like a shepherd's crook or question mark. The hooked part of the colon lies cranial and to the right of the root of the mesentery. The cranial part of the crook is the transverse colon, the short right portion is the ascending, or right, colon. The flexure which unites these two parts is known as the **right colic flexure** (*flexura coli dextra*). The transverse colon is continued by the descending colon at the flexure located to the left of the root of the mesentery. This bend is called the **left colic flexure** (*flexura coli sinistra*). The colon measures about 2 cm. in diameter throughout its length and is approximately 25 cm. long in a beagle-type preserved specimen. Because of its shorter mesentery, the colon does not vary as much in position or in length as does the small intestine.

The **ascending colon** (*colon ascendens*) begins at the ileocolic sphincter, runs cran-ially, and ends at the right-angled right colic flexure. It usually is about 5 cm. long, but this varies greatly. In rare instances the ascending colon is lacking, or it may lie cranial or even to the left of the mesenteric root. In these specimens apparently the whole gut failed to rotate around the cranial mesenteric artery as it usually does. Typically, the ascending colon is related to the mesoduodenum and the right limb of the pancreas dorsally, where it lies ventral to the right kidney. The descending part of the duodenum bounds it on the right. The small intestinal mass lies adjacent to the ascending colon ventrally and on the left. Cranially the ascending colon lies in contact with the stomach unless this viscus is greatly distended, in which case the stomach displaces the small intestinal mass caudally and lies ventral to the relatively fixed ascending colon.

The **transverse colon** (*colon transversum*) forms an arc which runs from right to left cranial to the cranial mesenteric artery and the dorsal part of the mesojejunoileum or root of the mesentery. Like the ascending colon, with which it is continous at the right colic flexure, it may fail to form or be placed to the left of the mesenteric root owing to the failure of the gut to rotate sufficiently. It is related cranioventrally to the stomach and craniodorsally to the left limb of the pancreas. Ventrally and caudally it lies in contact with the coils of the small intestine. It is about 7 cm. long.

The **descending colon** (*colon descendens*) is the longest segment of the colon. It extends from the left colic flexure to a transverse plane passing through the pelvic inlet, where it is continued by the rectum without demarcation. It is about 12 cm. long and usually quite straight. It follows the curvature of the left lateral abdominal wall, usually covered by the greater omentum, from the dorsal part of the left costal arch to a point ventral to the promontory of the sacrum. It lies closely applied dorsally to the iliopsoas muscle, but at its beginning it lies in contact with the left lateral or occasionally with the ventral surface of the left kidney. The left ureter lies dorsal to its medial border initially, but farther caudally this tube obliquely crosses the dorsal surface of the colon and curves around the caudal part of its lateral border before emptying into the bladder. Usually the ascending portion of the duodenum lies adjacent to its medial or mesenteric border and the spleen crosses it laterally; elsewhere it is bounded by the small intestinal mass. The uterus and the bladder lie ventral to its terminal part. The body of the uterus always lies in contact with it, and the bladder, if it is distended sufficiently to extend cranial to the uterine body, lies in contact with it at a more cranial level. In the male an enlarged prostate gland replaces the uterus as a ventral boundary.

Mesentery of Colon

The **mesocolon** (Fig. 7–34) is divided into the same parts as the colon which it suspends. No part of the colon of the dog is retroperitoneal, nor does any part of the greater omentum attach to it. As the proximal part of the colon hooks around the cranial side of the root of the mesentery, the mesocolon of this part attaches to it. The two peritoneal sheets which largely compose the mesocolon do not run uninterruptedly at all places to their central attachments. Therefore, the various parts will be described separately. The **ascending mesocolon** (*mesocolon ascendens*) is the mesentery of the ascending colon. It is shortest at its origin, where the ileum, colon, and cecum come together. In some specimens these parts are directly attached by areolar tissue to the left mesenteric lymph node, but usually a short mesentery exists. Cranially, this mesentery is the beginning of the ascending mesocolon; caudally, it is the end of the great mesentery or the mesoileum. The medial peritoneal sheet of the ascending mesocolon attains a width of 2 to 3 cm. at the right colic flexure. Its length varies with the length of the ascending colon. It is continuous centrally with the great mesentery, which covers the right mesenteric lymph node, cranial mesenteric vessels, and nerves. A small colic fossa is usually formed by a thin, circular plica of peritoneum that bridges between the ascending and transverse colon at the right colic flexure. The left sheet of the ascending mesocolon is interrupted by the mesoduodenum attaching to it. It becomes coextensive with the right peritoneal leaf of the root of the mesentery by continuing on the large cranial mesenteric vein.

The **transverse mesocolon** (*mesocolon transversum*) runs directly to the mesenteric root. It is about 3 cm. wide at the right colic flexure, but as it crosses the median plane, it increases in width to about 5 cm. at the left colic flexure. The length of the transverse mesocolon, determined by the length of the transverse colon, is usually about 7 cm. It is continous at the right colic flexure with the ascending mesocolon and at the left colic flexure with the descending mesocolon. The **descending mesocolon** (*mesocolon descendens*) is continuous without demarcation with the mesorectum at the pelvic inlet. The width of the descending mesocolon is greatest at the left colic flexure, where it measures about 5 cm. Its width decreases caudally, so that is is only about 2 cm. wide as it is continued in the pelvis by the mesorectum. Its parietal attachment is about 12 cm. long, and its visceral attachment is approximately twice this length. The dorsal part of the descending mesocolon represents the common dorsal mesentery and attaches to the aorta approximately in the median plane. Both its right and left peritoneal leaves are interrupted in a frontal plane about 1 cm. from their aortic attachments. The secondary fold which blends with the right peritoneal sheet is the duodenocolic ligament. It extends as far caudally as the pelvic inlet before it is effaced. On the left side, a loose, fat-streaked plica, a fold from the greater omentum and spleen, blends with the left peritoneal sheet of the descending mesocolon. It attaches at the same dorsoventral level as does the duodenocolic ligament on the right side and fades completely a few centimeters cranial to the pelvic inlet. In addition to carrying the caudal mesenteric artery and the large cranially coursing caudal mesenteric vein and fat, one to several left colic lymph nodes are located around the terminal branches of the caudal mesenteric artery.

RECTUM

The **rectum** (Fig. 7–41) begins at the pelvic inlet, where it is continuous cranially with the descending colon; it ends ventral to the second or third caudal vertebra, at the beginning of the anal canal. It is straight, about 5 cm. long, and 3 cm. in diameter. Dorsally, the rectum is attached to the ventral surface of the sacrum by the thin, 1 cm.

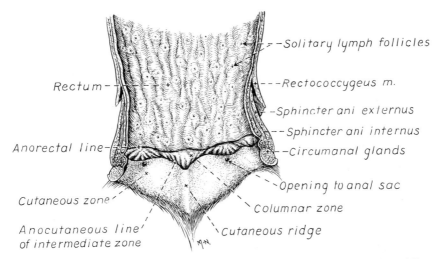

Rectum—

Anorectal line—

Cutaneous zone'

Anocutaneous line'
of intermediate zone

– –Solitary lymph follicles

– – –Rectococcygeus m.

–Sphincter ani externus

– –Sphincter ani internus

– –Circumanal glands

`Opening to anal sac

`Columnar zone

`Cutaneous ridge

Figure 7–41. The rectum and anal canal, opened to the left of the mid-dorsal line.

wide *mesorectum*. Caudally, the mesorectum becomes narrow and ends at a point usually opposite the second coccygeal vertebra. The peritoneal sheets of the mesorectum are continued on the sides of the pelvis as the parietal peritoneum; caudally the visceral peritoneum from the rectum reflects forward at an acute angle in a frontal plane to become coextensive with the parietal peritoneum, which is derived from the lateral portion of the mesorectum. In this manner, a **pararectal fossa** (*fossa pararectalis*) is formed on each side, right and left of the terminal portion of the rectum. Ventrally, the peritoneum blends with that of the rectovesical excavation of the male or with the rectouterine excavation of the female. The rectum is bounded dorsally by the right and left ventral sacrocaudal muscles. Laterally it is bounded primarily by the levator ani muscle. The visceral branch of the internal iliac artery obliquely crosses the lateral side of the initial portion of the rectum. After the bifurcation of this vessel, the prostatic (male) or vaginal (female) artery continues the direction of the parent vessel, and the internal pudendal runs parallel to the rectum for about 2 cm. before passing lateral to the levator ani muscle. In a craniocaudal sequence the rectum is crossed laterally first by the obturator, then by the ischiatic, pelvic, and anal nerves. The pelvic plexus lies lateral to the middle portion of the rectum. The hypogastric nerve enters it from in

front, and the pelvic nerve or nerves enter it from above. Ventrally, the rectum is bounded by the vagina in the female, and by the urethra in the male. When the prostate gland is small, it lies within the pelvis or at the pelvic brim, and bounds the rectum at that place. When it is large, the prostate lies largely cranial to the pelvic inlet. The most prominent feature of the rectal mucosa is the presence of approximately 100 **solitary lymph nodules** (*noduli lymphatici solitarii*). These nodules are each about 3 mm. in diameter and 1 mm. high. The free surface of each is umbilicated, forming a crater, or rectal pit.

ANAL CANAL

The **anal canal** (*canalis analis*) (Fig. 7–41) is the terminal, specialized portion of the alimentary canal. It is about 1 cm. long and extends from the termination of the rectum to the anus. The anal canal lies ventral to the fourth caudal vertebra and is surrounded by both the smooth and the striated anal sphincter muscle. The mucosa of the anal canal is divided into cutaneous, intermediate, and columnar zones.

The **cutaneous zone** (*zona cutanea*), the most caudal of the three zones of the anal canal, is divided into external and internal portions (Martin 1923). The **anus**, the terminal opening of the alimentary canal, may be located in a plane separating the two portions of the cutaneous zone. Thus the outer cutaneous zone is not properly a zone of the anal canal, because it lies outside the canal. It is feasible, however, to describe the two zones together, as the division between them varies with the movement of the tail and the degree of fullness of the rectum. Except during defecation, the anus is closed. In an animal's normal position, with the tail hanging, the anus is indicated by a transverse groove; however, if the tail is slightly raised, the boundaries of the anus form an irregular isosceles triangle. The dorsal or longer border is not straight in old male dogs but is ventrally arched owing to the large mass of gland tissue located under it. The shorter ventrolateral borders converge in forming a V with the apex pointing ventrally. A low ridge, about 2 mm. wide and 8 mm. long, continues from the apex of the V toward the pudenda. The inner portion of the cutaneous zone is about 4 mm. wide, and in life its surface is moist. The duct from the anal sac opens on this zone, about 2 mm. from its cranial limit, and in the depths of the lateral angle of the anus. The outer cutaneous zone may be defined as the relatively hairless zone peripheral to the anus. It varies greatly in width, particularly in adult male dogs, owing to the varied development of the underlying circumanal glands. Because these glands probably grow throughout life in the male (Parks 1950) and their full development tends to result in a loss of hair, the outer cutaneous zone may attain a width of 4 cm. in large, old male dogs. The cutaneous zone of the dog is studded by small elevations, on the crests of which open the ducts of the circumanal glands.

The **intermediate zone** (*zona intermedia*) is usually less than 1 mm. wide, and is in the form of an irregular, sharp-edged scalloped fold, which is divided into four arcs. Known as the **anocutaneous line** (*linea anocutanea*), it completely encircles the anal canal. Its mucosa, like the surface of the cutaneous zone, is also stratified squamous epithelium.

The **columnar zone** (*zona columnaris*) is so named because it contains longitudinal or oblique ridges, or **anal columns** (*columnae anales*), which run forward from the anocutaneous line for about 7 mm. Caudally, the adjacent anal columns are united by the fold which forms the anocutaneous line. In this way, a large number of pockets are formed, called **anal sinuses** (*sinus anales*). The anal sinuses are contained within the four arches of the anocutaneous line, thus producing its scalloped appearance. The smaller of the four arches are dorsal and ventral; the larger ones are located laterally. The columns are not uniform in either length or direction. Some disappear after running only a few milimeters cranially; most end in a line which encircles the anal canal, known as the **anorectal line** (*linea anorectalis*).

The **anal sacs** (*sinus paranalis*) (Fig. 7–42), one on each side of the anal canal, are approximately spherical sacs which are located between the inner smooth and the outer striated sphincter muscle of the anus. The anal sacs vary in size from a pea to a marble, the average diameter being a little less than 1 cm. The excretory duct of each sac is about 5 mm. long and 2 mm. in diameter, and opens near the cranial end of the furrow between the dorsal and lateral parts of the inner cutaneous zone of the anus adjacent to the intermediate zone. In about 10 per cent of dogs the opening of the anal sac is located in the broad depression formed by the lateral arch of the anocutaneous line on each side. The anal sacs are of considerable clinical importance. They frequently become enlarged, owing to accumulated secretion, or they may become abscessed and painful, causing constipation. Infrequently they rupture to the outside, lateral to the anus, producing anal fistulas. For a discussion of the anal sac and its secretions, see Grau (1935), Ewer (1973), and Doty and Dunbar (1974).

Glands of Anus.   Three gland areas are located in relation to the anus. These comprise the circumanal glands, the anal glands, and the glands of the anal sacs.

The **circumanal glands** (*glandulae circumanales*) are located around the anus in a subcutaneous zone which may reach a radius of 4 cm. and a thickness of 8 mm. Circumanal gland elements are also found in the walls of the anal sac ducts (Parks 1950), and may extend peripherally a short distance under the skin, which contains abundant hair. Parks found that the circumanal gland is a bipartite structure consisting of a superficial sebaceous portion and a deep

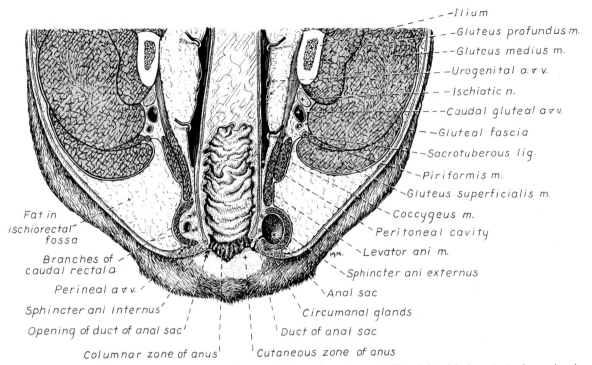

- - - -Ilium
- - -Gluteus profundus m.
- - -Gluteus medius m.
- - Urogenital a. v.
- - Ischiatic n.
- - - Caudal gluteal a v v.
- - Gluteal fascia
- - Sacrotuberous lig.
- Piriformis m.
Gluteus superficialis m.
Coccygeus m.
Peritoneal cavity
Levator ani m.
Sphincter ani externus
Anal sac
Circumanal glands
Duct of anal sac
Cutaneous zone of anus

Fat in
ischiorectal
fossa
Branches of
caudal rectal a.
Perineal a v v.
Sphincter ani internus
Opening of duct of anal sac
Columnar zone of anus

Figure 7–42. Section through anus in horizontal plane, dorsal aspect. The right side is cut at a lower level, through the duct of the anal sac.

non-sebaceous part. Since the non-sebaceous cell is capable of transforming and in certain cases does transform into a sebaceous cell, Parks concluded that the circumanal gland is potentially a sebaceous gland.

Scattered among the circumanal glands are apocrine glands, and sweat glands are located in a zone 2 to 4 mm. wide directly peripheral to the anal orifice. Probably because the circumanal glands continue to grow throughout life in the unaltered male, adenomas of this region are common in old male dogs.

The **anal glands** (*glandulae anales*), according to Trautmann and Fiebiger (1957), are tubuloalveolar glands which open to the outside in the intermediate zone. Their secretion is fatty in the dog. Bradley and Grahame (1948) state that the microscopic anal glands are laterally located, cranial to the circumanal glands. Ellenberger and Baum (1943) describe these as a band of grape-shaped glands, 5 mm. wide.

The **glands of the anal sac** (*glandulae sacci paranales*) lie in the wall of the sac and open into it. They are composed of large, coiled, apocrine, sudoriparous tubules. Similar tubules lie in the wall of the duct, which also contains sebaceous acini. The anal sac, therefore, is a reservoir for the secretion of the glands of its wall, the secretion being a foul-smelling, serous to pasty liquid. According to Montagna and Parks (1948), both the anal sac and its duct are lined by cornified, stratified squamous epithelium, subjacent to which lies a thick mantle of glandular tissue embedded in a connective tissue stroma rich in diffuse lymphatic tissue.

Special Muscles of Rectum and Anal Canal

The **internal anal sphincter** (*m. sphincter ani internus*) (Fig. 7–41) is the caudal thickened part of the circular coat of the anal canal. It is composed of smooth, and therefore involuntary, muscle. Its size, particularly its width, is much less than that of the external anal sphincter. It is lined by submucosa on its inner surface, and is separated from the external anal sphincter by a small amount of fascia. On its lateral external surface on each side lies the anal sac, which is interposed largely between the two sphincter muscles. The duct from the anal sac crosses the caudal border of the internal sphincter, which extends slightly farther caudally than does the external anal sphincter.

The **external anal sphincter** (*m. sphincter ani externus*) averages about 1.5 cm. in width and 0.5 to 1.5 mm. in thickness. It is more than 2 cm. wide dorsally, and usually about 1 cm. wide ventrally. On the sides, because of the underlying anal sacs, it forms variably developed bulges. It is largely a circular band of striated, and therefore voluntary, muscle, and is the chief guardian of the lumen of the anal canal. It lies largely subcutaneously on the sides; ventrally, its fibers decussate somewhat and may spread out and end on the urethral muscle and the origin of the bulbospongiosus muscle of the male. In the female, the comparable fibers blend with the constrictor vulvae. About half of the deeper fibers of this sphincter from each side continue across to the other side, ventrally; in both sexes the fibers of the superficial half blend with the muscles of the external genitalia. On the sides the cranial border of the external anal sphincter is united by fascia to the caudal borders of the levator ani muscles. Dorsally, the muscle becomes wider and attaches mainly to the fascia of the tail opposite the third caudal vertebra. This attachment is made in such a way that a cranially facing, concave fascial arch is formed through which pass the smooth, paired rectococcygeal muscles, the extremities of the arch being continued by small tendons to the caudal fascia.

A band of smooth muscle fibers about 3 mm. wide and less than 1 mm. thick arises on each side from the ventral surface of the sacrum or first caudal vertebra and sweeps caudoventrally. As it obliquely crosses the rectum, it contributes some fibers to it and then fans out between the anal sac and the internal anal sphincter. This band of muscle fibers was called the coccygeoanal muscle by Straus-Durckheim (1845) and Langley and Anderson (1895) but is known now as the **pars analis of the retractor penis muscle** (see Fig. 6–43). The major portion of this muscle appears to end near the duct of the anal sac, with some fibers inserting in the external anal sphincter. The minor ventrola-teral portion of the pars analis, in combination with fibers from the external anal sphincter, continues distally as the mixed (smooth and striated) **retractor penis muscle** of the male.

The **rectococcygeal muscle** (*m. rectococcygeus*) (Chapter 9) is a paired, smooth muscle at its origin, composed of a condensation of many of the outer longitudinal fibers from each side of the rectum. These fibers sweep caudodorsally from the sides of the rectum and pass dorsally through the fascial arch formed by the attachment of the external anal sphincter to the fascia of the tail. Right and left portions lie closely together at this site, ventral to the third caudal vertebrae, and fuse. Caudally, the muscle lies on the bodies of successive caudal vertebrae, in the goove formed by the apposed ventral sacrocaudal muscles, and runs caudally to attach on the bodies of the fifth and sixth caudal vertebrae. The attachment of the rectococcygeus on the tail serves as an anchorage whereby the muscle can stabilize the anal canal and rectum and prevent their being pulled cranially by a peristaltic wave, or, by its contraction, it can move the anal canal and rectum caudally during defecation. The movement of the tail during the act of defecation has a direct influence in evacuating the rectum not only through the action of the rectococcygeus but also by the action of the coccygeus and levator ani muscles. The coccygeus muscles cross the rectum laterally and tend to compress the tube while the rectococcygeus, by shortening, aids the circular muscle coat in moving the fecal column to the outside.

Vessels and Nerves of Anal Canal

The mucosa of the anal canal and the sphincter muscles which surround it receive their blood mainly from the **right** and **left caudal rectal arteries** which anastomose in the anal sphincters. The long **cranial rectal artery** may extend far enough caudally to furnish some blood to the anal canal. Thus the blood from the anal canal returns to the heart both by the portal system, through the **cranial rectal, caudal mesenteric, portal,** and **hepatic veins** and **caudal vena cava,** and by the systemic system, through the **caudal rectal** and **perineal veins,** which are tributaries of the internal pudendal or the perineal and caudal gluteal veins. The internal pudendal and caudal gluteal veins unite to form the internal iliac vein. The internal and external iliac veins unite to form the common iliacs, which in turn unite to form the caudal vena cava. The venous blood is returned from the anal canal through the veins which are the satellites of the arteries of supply. The skin of the perineum and the underlying circum-anal glands are served largely by the perineal vessels, usually from both the internal pudendal and caudal gluteal arteries. The lymph vessels from the anal canal drain into the **sacral lymph nodes**, if present, and the **internal iliac lymph nodes**. The external anal sphincter muscle, being striated and voluntary, is supplied by the anal branch of the **pudendal nerve**. The involuntary, smooth internal anal sphincter and recto-coccygeus are supplied by autonomic fibers from the **pelvic plexuses**. The parasympathetic portion comes to it through the **pelvic nerves**, branches from usually the first, second, and third sacral nerves, and the sympathetic portion is derived from the **hypogastric nerves** which arise from the caudal mesenteric ganglion.

COATS OF LARGE INTESTINE

All except the terminal part of the large intestine has the usual four coats as found in the small intestine. These are the mucous, submucous, muscular, and serous tunics, or coats.

The **mucous coat** (*tunica mucosa*) of the large intestine differs from that of the small intestine in that there are no aggregated lymph nodules or intestinal villi. Solitary lymph nodules, however, are numerous, and can be counted from the outside in the dilated gut. Although present throughout the large intestine, they are most numerous in the rectum. The mucosa of the large intestine, except in the anal canal, contains no folds which cannot be effaced by distention. Many folds which appear similar to those of the contracted stomach are present in the contracted large intestine. Depending on the type of contraction, these plicae are either longitudinal or circular. The **intestinal glands** (*glandulae intestinales*) of the large intestine are longer, straighter, and richer in goblet cells than are those of the small intes-

tine (Trautmann and Fiebiger 1957). They are lined by a columnar epithelium which is continuous with the columnar epithelium of the mucosal surface of the lumen of the gut. The *lamina muscularis mucosae* consists of muscle fibers which are poorly arranged in two strata.

The **submucous coat** (*tela submucosa*) does not differ appreciably from that of the small intestine. Many of the solitary lymph nodules are located partly within it. This tunic contains the submucous nerve plexuses and many vessels in the meshes of loose connective tissue.

The **muscular coat** (*tunica muscularis*) is uniform in thickness. The *stratum longitudinale* is not concentrated in muscular bands, as it is in man, horse, and pig, and no haustra are present. The fibers forming the longitudinal stratum sweep dorsocaudally from the sides of the rectum and, opposite the first or second caudal vertebra, leave the dorsum of the rectum to form the smooth *m. rectococcygeus*, which passes dorsal to the external anal sphincter and attaches to the bodies of the fifth and sixth caudal vertebra. The *stratum circulare* of the large intestine resembles that of the small intestine, except that it is heavier. Its caudal portion forms the **internal anal sphincter muscle**.

The **serous coat** (*tunica serosa*) resembles that of the small intestine. It covers the colon, cecum, and much of the rectum, as the visceral peritoneum. The anal canal and the caudal portion of the rectum are retroperitoneal. See the description of the pararectal and rectovesical or rectouterine excavations, in the discussion of the peritoneum, above.

PERINEUM

The perineum is the region of the pelvic outlet. On the surface of the body of the dog it is limited by the tail above, the scrotum or beginning of the vulva below, and by the skin which covers the paired superficial gluteal and internal obturator muscles and the tubera ischiorum on the sides. Deeply, the perineum is bounded by the third caudal vertebra above, the sacrotuberous ligaments on the sides, and by the arch of the ischium below. The digestive system terminates in the anus. The urogenital system in both sexes continues distal to the perineal region to occupy the pudendal region. The bones, muscles, blood vessels, nerves, and glands of the pelvic outlet are described in the appropriate chapters, so that at this place only the ischiorectal fossa, and the perineal and, to a lesser extent, the pelvic fasciae will be described. The perineal region holds much interest for the surgeon because of the frequency of occurrence of adenomas and perineal hernias in old male dogs.

The **ischiorectal fossa** (*fossa ischiorectalis*) is the deep, wedge-shaped depression located lateral to the terminal pelvic portions of the digestive and urogenital tubes. The lateral boundary of the fossa is the levator ani and coccygeus muscles which obliquely cross the lateral surface of these tubes as the muscles pass essentially sagittally from the os coxae to the caudal vertebrae. The most caudal part of the medial boundary is the external anal sphincter dorsally, and the constrictor vulvae of the female and the retractor penis of the male ventrally. The lower lateral and ventral boundary is formed by the internal obturator muscle, the dorsal and upper lateral boundary by the superficial gluteal muscle. Cranially and ventrally, the fossa forms a narrow fornix where the medially located coccygeus muscles arise adjacent to the origin of the laterally located internal obturator muscle. This angle is located opposite the bodies of the ilium and ischium and along the entire pelvic symphysis. The ischiorectal fossa in well-nourished dogs is filled with fat which is bounded peripherally by the skin.

The **perineal fascia** (*fascia perinei*) is a term used by veterinary anatomists to include those parts of the adjacent fasciae from the tail, rump, and thigh which converge at the anus, enclosing the pelvic outlet. It is divided into superficial and deep strata.

The **superficial perineal fascia** (*fascia perinei superficialis*) forms the feeble matrix in which the fat of the ischiorectal fossa is elaborated. It is a single-layered, loose fascia, but it is abundantly developed. The numerous small perineal vessels and nerves which stream caudally over the ventral part of the pelvic outlet lie in the superficial perineal fascia. It is continuous with the superficial fascia of the rump, thigh, and tail.

The **deep perineal fascia** (*fascia perinei profunda*) covers the dorsomedial surface of the internal obturator muscle. It firmly attaches to the dorsal subcutaneous portion of the tuber ischii ventrally, and the adjacent portion of the obliquely running sacrotuberous ligament caudolaterally. Craniolaterally it becomes continuous with the deep gluteal fascia of the superficial gluteal muscle.

In the male the deep perineal fascia is tightly applied to the dorsal surfaces of the right and left ischiocavernosus muscles and to the unpaired bulbospongiosus muscle which lies between them. In both sexes the deep perineal fascia, but no muscle fibers, extends cranially between the digestive and urogenital tubes at the pelvic outlet. Developed in the caudal part of this frontally located fascial partition is the retractor penis muscle of the male and the constrictor vulvae muscle of the female. Just cranial to the muscles which pass from the external anal sphincter to the pudenda in both sexes, collagenous fibers directly unite the complex musculature between the anal canal and the vagina or the bulb of the penis. This median fascial union, if sufficiently heavy or differentiated as a fibromuscular node, is spoken of as the "perineal body" (*centrum tendineum perinei*).

LIVER

The **liver** (*hepar*) (Figs. 7–43, 7–44) is the largest gland in the body. It is both exocrine and endocrine in function. The bile, which is its exocrine product, is largely stored in the gall bladder before being poured into the descending portion of the duodenum. Its endocrine substances released into the blood stream function in the intermediary metabolism of fats, sugars, and some nitrogenous products.

PHYSICAL CHARACTERISTICS

The average weight of the liver in 91 dogs was 450 gm. The average weight of these dogs was 13.3 kg. In this sampling of adult mongrel dogs of both sexes the weight of the liver averaged 3.38 per cent of the body weight. Thus, the liver weighed approximately a pound in those dogs whose average

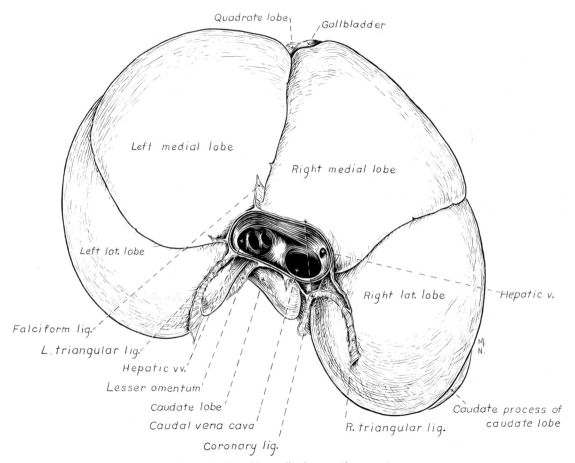

Figure 7–43. Liver, diaphragmatic aspect.

weight was about 27 pounds. The liver is relatively much heavier in the puppy than in the aged dog. The fresh liver is a deep red color, firm in consistency, but friable. In a 30-pound, hound-type dog, its dorsoventral dimension is 14 cm., its width 12 cm., and its thickness 6 cm.

SURFACES, BORDERS, AND RELATIONS

The **diaphragmatic surface** *(facies diaphragmatica)*, or parietal surface, of the liver is strongly convex in all directions, as it lies mainly in contact with the diaphragm. It is so strongly convex that more of this surface faces bilaterally and dorsally than faces cranially and ventrally. Therefore, this surface can be logically subdivided into a **right part**, a **left part**, a **cranial part**, a **dorsal part**, and a **ventral part**. The right part is largest, as it extends from the cranial part, which lies opposite the sixth intercostal space to the last, or twelfth, intercostal space; on the left side the caudal border usually lies opposite the tenth intercostal space. The cranial part is marked by a shallow, broad, indistinct **cardiac impression** *(impressio cardiaca)* which is located largely to the left of the median plane. The dorsal part is deeply notched, approximately in the median plane, by the caudal vena cava and the esophagus. The caudal vena cava lies to the right of the esophagus as they groove the liver.

The **visceral surface** *(facies visceralis)* of the liver is irregularly concave and faces mainly caudoventrally and to the left. It lies in contact with the stomach, duodenum, pancreas, and right kidney. All but the pancreas produce impressions on the organ hardened in situ. The liver is completely invested with peritoneum except at the hilus and where the gall bladder is fused to it. The centrally located **papillary process** *(processus papillaris)* of the caudate lobe, which protrudes caudally from its dorsal part, is the most prominent feature of the visceral surface.

The **gastric impression** *(impressio gastrica)* occupies the whole left half of the visceral surface when the stomach is moderately full. The pyloric part, owing to its relatively fixed position, produces an oblique impression across the middle portion of the gland, where it lies in contact with the caudal face of the gall bladder. When the stomach is empty, the coils of small intestine contact the visceral surface of the liver through the greater omentum but leave no impressions on it.

The **duodenal impression** *(impressio duodenalis)* begins at the junction of the right and quadrate lobes as the most cranial indentation of the visceral surface. This impression at first runs to the right, then it makes a sweeping arch caudoventrally and finally runs caudodorsally, essentially paralleling and lying dorsal to the right ventral border of the organ. The right limb of the pancreas lies dorsomedial to the cranial and descending portions of the duodenum in contact with the liver, but leaves no impression on it.

The **renal impression** *(impressio renalis)* is a deep, nearly hemispherical fossa formed by the cranial pole of the right kidney projecting into the most caudodorsal portion of the liver. Ventrally, the liver covers more than half of the kidney. Because of the close proximity of the caudal vena cava to the right adrenal gland, this gland does not leave an impression on the liver.

The **porta of the liver** *(porta hepatis)* is the hilus of the organ. The hepatic vessels and nerves and the bile duct communicate with the gland through the porta. The nerves and arteries enter the porta dorsally, the biliary duct leaves ventrally, and the portal vein enters between the two. It is located on the dorsal third of the visceral surface, ventrodextral to the attachment of the papillary process. From the porta, the deep fissures which subdivide the organ diverge toward the lateral and ventral surfaces, so that the liver, opposite and dorsal to the porta, is solid.

Dorsal, ventral, right and left lateral borders *(margo dorsalis, ventralis, lateralis dexter et lateralis sinister)* are recognized. They are sharp-edged and continuous around the periphery of the organ except dorsally, where this circumferential margin is effaced by the deep, broad notch which contains in its depths the caudal vena cava and esophagus. In addition to the main clefts which subdivide the liver into lobes, there are a few short fissures which cut into the borders of the organ. In the dog, the **fissure for the round ligament** *(fissura lig. teretis)* is the caudoventral portion of the interlobar fissure between the quadrate lobe on the right and the left lateral lobe on the left. In the puppy, the round ligament (umbilical vein of the fetus) runs from the umbilicus to the porta of the liver in or just caudal to this cleft.

LOBES AND PROCESSES

The liver is divided into four lobes and four sublobes, as well as two processes, by deeply running fissures.

The **left hepatic lobe** *(lobus hepatis sinister)* is that portion of the liver which lies entirely, or almost entirely, to the left of the median plane. This lobe forms from a third to nearly a half of the total liver mass. Its parenchyma is usually completely divided into two sublobes, as follows: The **left lateral hepatic lobe** *(lobus hepatis sinister lateralis)* begins dorsally under the left crus of the diaphragm, where it is about 3 cm. wide. Traced ventrally, it crosses under the left portion of the tendinous center and then under the left portion of the muscular periphery of the diaphragm. Its diaphragmatic surface gradually becomes wider, until it reaches a width of 4.5 to 5 cm. in its middle, after which it gradually becomes narrower and ends in a point dorsal to the last sternebra. The lateral border may protrude as much as 2 cm. caudal to the ventral portion of the costal arch, but in some specimens it is completely contained within the rib wall. The dorsal portion partially caps the body of the stomach. The visceral surface of the left lateral lobe is concave peripherally as it lies on the fundus and body of the stomach. Centrally, it is partly covered by the papillary process of the caudate lobe. This central portion of the lobe is slightly convex, forming the **omental tuber** *(tuber omentale)*. It lies adjacent to the lesser omentum covering the papillary process and is formed by the moldable hepatic tissue protruding toward the lesser curvature of the stomach.

The **left medial hepatic lobe** *(lobus hepatis sinister medialis)* varies from being nearly triangular to oval in outline as seen from the diaphragmatic surface. The fissure which separates it from the left lateral lobe begins from 1.5 to 3 cm. from the most caudoventral portion of the organ. It exists as a deep, curved cleft, which usually completely separates the two portions of the left lobe of the liver. It extends to the porta. Dorsally, in some specimens, the two portions are joined by a narrow but deep bridge of liver

tissue; otherwise, the two portions are joined together only by the intrahepatic vessels and nerves. The two portions of the left lobe are separated from the quadrate and right lobes by a deep fissure, nearly midsagittal in location, which extends to the porta and nearly to the esophageal notch.

The **quadrate lobe** *(lobus quadratus)* is a deep wedge of liver tissue which lies essentially in the median plane, where it is interposed in the fissure which separates the right medial and the left lobes, being fused to a certain extent to the former. Its diaphragmatic surface is fusiform, and it extends neither to the ventral border nor to the notch for the esophagus and caudal vena cava. The middle of its right surface is smoothly excavated by the left half of the **fossa for the gall bladder** *(fossa vesicae felleae)*. In nearly half of the specimens examined, the quadrate lobe did not reach the visceral surface of the liver.

The **right hepatic lobe** *(lobus hepatis dexter)* is smaller than the left hepatic lobe and lies completely to the right of the median plane. It lies between transverse planes passed through the upper portions of the sixth and the tenth intercostal spaces. Like the left hepatic lobe it is divided into medial and lateral sublobes.

The **right medial hepatic lobe** *(lobus hepatis dexter medialis)* is fused to the medially lying quadrate lobe. The degree of fusion varies; in some specimens only the dorsal portions of these always closely adjacent lobes are fused; in others the fusion extends nearly to the fossa for the gall bladder, leaving only a short fissure extending dorsally from the fossa to separate the two lobes. The right medial lobe is always longer than the right lateral lobe and is the portion which extends caudally beyond the ventral portion of the costal arch if any portion of the right lobe protrudes beyond it. Its diaphragmatic surface is in the form of a curved triangle, with its base dorsomedial and its apex ventromedial in location. It is also triangular in cross section as it is wedge-shaped, possessing a concave, slightly fissured medial border which extends to the visceral surface of the organ. The right half of the **fossa for the gall bladder** is located on its medial face opposite the comparable excavation on the quadrate lobe.

The **right lateral hepatic lobe** *(lobus hepa-*

tis dexter lateralis) is shaped roughly like a laterally compressed hemisphere with a slightly concave base. Cranially it is overlapped by the right medial lobe; caudally it overlaps the caudate process of the caudate lobe and is usually fused to it, lateral to the postcava. Its most ventral extension lies opposite the distal portion of the middle third of the caudal border of the right medial lobe.

The **caudate lobe** *(lobus caudatus)* is composed of the caudate and papillary processes and the isthmus of liver tissue which connects them. This isthmus is compressed between the postcava dorsally and the portal vein ventrally. It is a bridge of hepatic tissue which is about 1.5 cm. long, 1 cm. wide, and 0.5 cm. thick.

The **papillary process** *(processus papillaris)* is pyramidal to tongue-shaped, and is usually partly subdivided by one or two fissures. The more constant fissure separates the frenular part from the body of the process. This process is loosely enveloped by the lesser omentum and lies in the lesser curvature of the stomach. It projects by an acute angle to the left and forward from its attachment to the caudate lobe.

The **caudate process** *(processus caudatus)* forms the most caudal portion of the liver as it extends to a plane through the twelfth intercostal space or last rib on the right side. Its caudolateral portion is deeply recessed by the cranial half of the right kidney. The outline of its diaphragmatic surface forms nearly an equilateral triangle as it lies mainly ventral to the right kidney. The parenchyma of the caudate lobe is usually partly fused to the right lateral lobe, but occasionally the two portions are completely separated, or other variations may exist.

PERITONEAL ATTACHMENTS AND FIXATION

The liver is almost completely enveloped by peritoneum, which forms its **serous coat** *(tunica serosa)*. The serous coat is fused to the underlying **fibrous capsule** *(capsula fibrosa perivascularis)*, a thin but strong layer, composed mainly of collagenous tissue, which closely invests the surfaces of the liver and sends interlobular trabeculae into the gland substance. At the porta the fibrous coat becomes heavier and is continued into the interior of the liver in association with the vascular and nervous structures which serve the gland. The only parietal attachment of the liver is to the diaphragm by means of continuations of its serous and fibrous coats in the form of the coronary ligament and several small folds which radiate from it. These folds are the two right triangular, the usually single left triangular, and the falciform ligaments. The hepatorenal ligament and the lesser omentum also attach to the liver.

The **coronary ligament of the liver** *(lig. coronarium hepatis)* is not a true peritoneal ligament, since the two sheets of peritoneum which form it are not in the form of a fold but are irregularly separated. The term refers to the line of peritoneum which reflects around a triangular "bare area" of the liver, about 2 cm. long on each side, and is continued on the dorsal surface of the caudal vena cava and the tributaries which enter it from the diaphragm. The coronary ligament is irregular in outline. It reflects the close embryonic relationship between the diaphragm and liver. Its stellate border gives rise to the three or more triangular ligaments and is coextensive with the dorsal part of the falciform ligament.

The **right triangular ligament** *(lig. triangulare dextrum)* is a plica of peritoneum which extends between the diaphragm and the dorsal part of the right lateral lobe. Its free lateral border is 1 to 5 cm. wide. As it passes medially, it becomes progressively narrower until its two formative peritoneal layers become continuous with the right peritoneal leaf of the coronary ligament as this bounds the bare area of the liver on the right. It is usually longer than it is wide. A second, smaller right triangular ligament regularly goes from the diaphragm to the diaphragmatic surface of the right medial lobe. Other smaller but similar plicae are present.

The **left triangular ligament** *(lig. triangulare sinistrum)*, like the comparable right ligament, may also be double or triple. If there are two the caudal member is larger than the cranial and contains the **fibrous appendix of the liver** *(appendix fibrosa hepatis)*, when this is present. This fibrous appendix is a narrow, thin tapering band of atrophic hepatic tissue located in or near the free border of the ligament. It is present in only a small number of adult specimens. When a second triangular ligament is present on the left side, it runs from the left medial lobe to the diaphragm.

The **falciform ligament of the liver** *(lig. falciforme hepatis)* (Fig. 7–43) is a remnant of the ventral mesentery which extends between the liver and the diaphragm and ventral body wall caudally to the umbilicus. The middle portion of the falciform ligament in the dog usually becomes wholly or partly obliterated before birth so that the umbilical vein, which in early fetal life is located in its free border, usually has no peritoneal attachment immediately before birth. The proximal end of the falciform ligament — that part extending from the umbilicus to the diaphragm — remains as a fat-filled irregular fold which may weigh several pounds in obese specimens. The distal portion of the falciform ligament may disappear completely, but usually it remains as a thin, avascular fold which extends from the dorsal end of the fissure between the right and left lobes to the coronary ligament. When present, the left peritoneal sheet of the falciform ligament becomes coextensive with the left portion of the coronary ligament, and the right peritoneal sheet becomes coextensive with the ventral portion.

The **hepatorenal ligament** *(lig. hepatorenale)* is a delicate peritoneal fold which extends from the medial portion of the renal fossa to the ventral surface of the right kidney lateral to the fat which fills its hilus. It is not constant. The **lesser omentum** *(omentum minus)* is a thin, lacy, fat-streaked, loose peritoneal fold which is that remnant of the ventral mesentery which extends from the liver to the lesser curvature of the stomach and cranial part of the duodenum.

STRUCTURE

The free surface of the liver is firmly covered by the thin peritoneum superficially and the equally thin fibrous capsule which sends septa into the gland. On close observation the surface of the liver presents a fine mottled appearance. The delicate dappling is due to the contrast in color between the dark, small, polygonal units of liver parenchyma and the lighter connective tissue surrounding them. These units, called **hepatic lobules** *(lobuli hepatis)*, are the smallest grossly visible functional divisions of the organ. Each lobule is about 1 mm. in diameter and is composed of curved sheets of cells which enclose numerous, blood-filled cavities known as the liver sinusoids. According to Elias (1949), the sinusoids of the dog are intermediate in form between the saccular and tubular types which are found in some other mammals. The sheets or plates of cells which form their walls are one cell thick and contain openings which allow free passage of the intersinusoidal blood.

BLOOD AND LYMPH VESSELS

In the centers of the lobules there are typically the single **central veins** *(venae centrales)*. These constitute the beginning of the efferent, or outgoing, venous system of the liver. Adjacent central veins fuse to form the **interlobular veins** *(venae interlobulares)*. The interlobular veins unite with each other to form finally the **hepatic veins** *(vv. hepaticae)*, which empty into the postcava. Arey (1941) and others have shown that in the dog, but not in the cat, there are spiral and circular muscle fibers in the walls of the central and (sublobular) interlobular veins. The sphincter action produced by these muscle fibers restricts venous drainage, producing precisely the effects of experimental shock. Prinzmetal et al. (1948) have shown that glass beads measuring 50 to 180 μm. did not pass from an afferent vessel to the efferent hepatic vessel draining the area supplied by the injected afferent vessel. The hepatic veins convey to the caudal vena cava all the blood which the liver receives through the portal vein and proper hepatic arteries.

The **portal vein** *(v. portae)* (Fig. 7–44) brings the functional blood to the liver from the stomach, intestines, pancreas, and spleen. About four-fifths of the blood entering the liver reaches it by the portal vein (Markowitz et al., 1949). The **proper hepatic arteries** *(aa. hepaticae propriae)* furnish the liver with the blood which nourishes its cells — the nutritional supply. The parenchymal cells are bathed by mixed blood from the portal vein and proper hepatic arteries so that they receive nutrition from both. The proper hepatic arteries supply primarily the liver framework, including its capsules and the walls of the blood vessels, the intrahepatic biliary duct system, and the nerves.

Although only about one-fifth of the blood

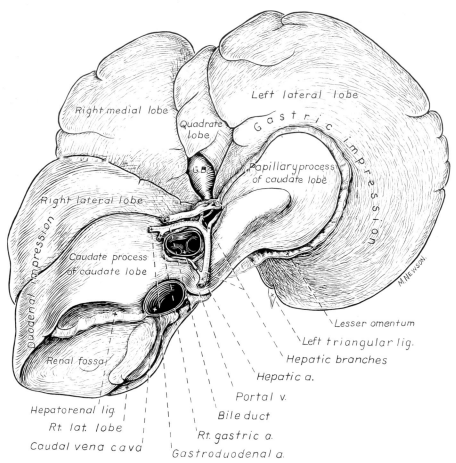

Right medial lobe

Left lateral lobe

Quadrate lobe

Gastric impression

Papillary process of caudate lobe

Right lateral lobe

G.B.

Caudate process of caudate lobe

Duodenal impression

M.NEWSON.

Lesser omentum

Left triangular lig.

Hepatic branches

Renal fossa

Hepatic a.

Portal v.

Hepatorenal lig.

Bile duct

Rt. lat. lobe

Rt. gastric a.

Caudal vena cava

Gastroduodenal a.

Figure 7–44.   Liver, visceral aspect.

coming to the liver reaches it through the proper hepatic arteries, their occlusion usually results in death if such occlusion is not accompanied by massive doses of penicillin. Markowitz et al. (1949) found that, without antibiotic treatment following ligation of the arteries at the portal fissure, gangrene results. In research on the arterial blood supply to the liver, it should not be overlooked that there are never less than two, and in some specimens there are as many as five, proper hepatic arteries which leave the common hepatic artery, and most of these arteries branch before they enter the lobes of the liver. See the treatise of Payer et al. (1956) on the surgical anatomy of the arteries to the liver of the dog.

In the fetal pup there is a shunt from the umbilical vein to the hepatic venous system,

known as the *ductus venosus*. The ductus venosus becomes fibrotic after birth and is known as the *ligamentum venosum*. In the stillborn pup it is several millimeters long and about 2 mm. wide. It extends obliquely from left to right in the porta hepatis, where it lies ventral to the attachment of the papillary process. The lymph vessels from the liver freely anastomose with those of the gall bladder (McCarrell et al., 1941). They drain into the **hepatic** and **splenic lymph nodes.**

NERVES

The liver is supplied by both afferent and efferent fibers through the **vagi** and by sympathetic fibers from the **celiac plexus.** The vagal fibers reach the abdomen by passing through the diaphragm with the esophagus as the dorsal and ventral vagal (esophageal) nerve trunks. Chiu (1943) has shown that in a representative dog two branches leave the ventral vagal trunk and one leaves the dorsal at the level of the cardia. They pass obliquely to the right in the lesser omentum toward the porta and supply the liver parenchyma and biliary system. McCrea (1924) describes the abdominal distribution of the vagi in the rabbit, cat, and dog. Possibly the liver also receives vagal fibers through the portion of the dorsal vagal trunk which joins the celiac plexus. Chiu (1943) also mentions the possibility of a coronary nerve reaching the liver in the dog. The sympathetic fibers reach the liver through the **splanchnic nerves, celiac ganglia,** and **celiac plexus,** and continue on the common and proper hepatic arteries as the plexuses of these arteries. Alexander (1940) states that in some specimens the biliary system receives afferent fibers from the phrenic nerves. He also confirmed that the hepatic artery receives only sympathetic fibers.

BILE PASSAGES AND GALL BLADDER

The bile, produced by the sheets of liver cells surrounded by the blood sinuses, is discharged into the minute bile canaliculi, or bile capillaries, which lie between these cells. The canaliculi unite to form the plexiform **interlobular ducts** *(ductuli interlobulares),* which lie in the interstitial tissue between the lobules. Finally the interlobular ducts of various sizes unite to form the lobar or bile ducts *(ductuli biliferi),* which are variable in number and termination. The extrahepatic bile passages consist of the **hepatic ducts** *(ductuli hepaticae)* from the liver, the **cystic duct** *(ductus cysticus)* to the gall bladder, and the **bile duct** *(ductus choledochus)* to the duodenum. The right and left hepatic ducts may be single or double. One

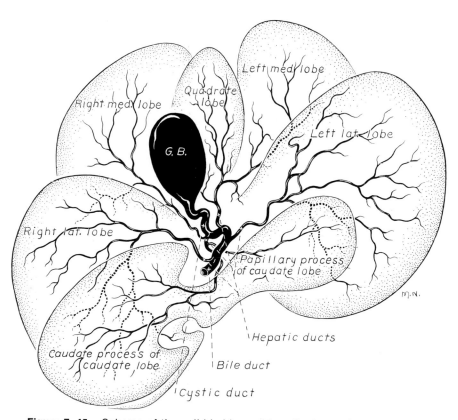

Figure 7–45. Scheme of the gall bladder and hepatic ducts, visceral aspect.

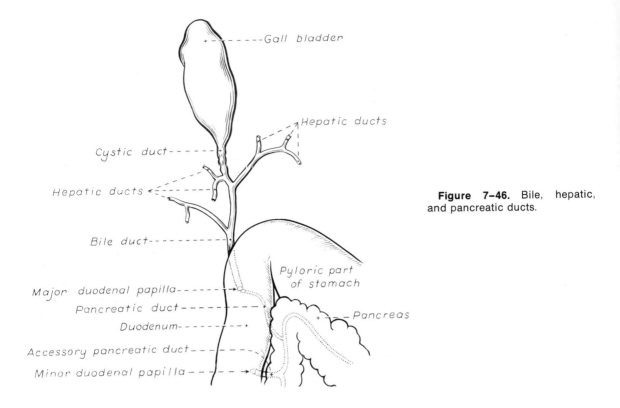

Figure 7–46. Bile, hepatic, and pancreatic ducts.

of the many possible patterns of hepatic duct termination is illustrated by Figure 7–45.

The **gall bladder** *(vesica fellea)* (Fig. 7–45) stores and concentrates the bile; the function of its mucoid secretion, like that of mucus generally, is for lubrication and protection (Ivy, 1934). Although the vagal nerve fibers are motor to its musculature, Winkelstein and Aschner (1924) state that the gall bladder displays variations in tonicity but seems to posssess little contractile power. Intra-abdominal pressure chiefly due to the in-spiratory phase of respiration effects a large variation in pressure within the gall bladder. The gall bladder is a pear-shaped vesicle which lies between the quadrate lobe me-dially and the right medial lobe laterally. When distended, it extends through the thickness of the liver to its diaphragmatic surface and contacts the diaphragm. Its ca-pacity in a beagle-sized dog is 15 ml. (Mann, Brimhall, and Foster, 1920). It is about 5 cm. long and 1.5 cm. in its greatest width in such specimens. The blind, rounded, cranial end of the gall bladder is known as the **fundus** *(fundus vesicae felleae)*; the large middle portion, as the **body** *(corpus vesicae felleae);*

and its slender, tapering caudodorsally directed extremity, as the **neck** *(collum vesi-cae felleae).*

The **cystic duct** *(ductus cysticus)* (Fig. 7–46) in a topographical sense may be re-garded as the beginning of the biliary duct system. It extends from the neck of the gall bladder to the site of its junction with the first tributary from the liver. From this level distally to the duodenum the main excretory channel which receives bile from the hepa-tic ducts is known as the **bile duct** *(ductus choledochus).* Higgins (1926) has recorded double cystic and bile ducts in the dog. In the dog, the lobar ducts do not unite to form the hepatic duct, as they do in man, but enter the main trunk of the excretory tree which, as stated above, may be regarded as beginning with the cystic duct. The distal or caudal portion of the bile duct enters the dorsal or mesenteric wall of the duodenum. This portion of the bile duct may be known as the free portion, in contrast tp the intra-mural portion, which extends obliquely through the duodenal wall. The free portion is about 5 cm. long and 2.5 mm. in diameter as it courses through the lesser omentum.

The intramural portion of the duct and its mode of emptying into the lumen of the duodenum have been studied by many investigators.

Eichhorn and Boyden (1955) have analyzed the structure of the choledochoduodenal junction in the dog. They illustrate and describe from wax reconstructions and maceration specimens the intramural portion of the bile duct and its musculature in both the fetus and the adult. In their review of the literature they call attention to the early work by Oddi (1887), which included studies on the dog. The bile duct has an intramural length of 1.5 to 2 cm. It terminates on a small hillock located at the end of a low longitudinal ridge representing its intramural course. The bile duct opens in the center of a small rosette upon the hillock, and to one side is the slitlike opening of the ventral pancreatic duct. The site of this combined opening of the bile duct and the pancreatic duct is spoken of as the major duodenal papilla. About 3 cm. distal to this opening lies a second low hillock, upon which the accessory pancreatic duct from the dorsal pancreas opens.

Eichhorn and Boyden have verified the existence of a double layer of smooth muscle around the intramural portion of the bile duct. The outer layer is formed by the *tunica muscularis* of the duodenum. The inner layer, or *musculus proprius*, begins in the infundibular portion of the bile duct and extends in the submucosa to the termination of the duct as the sole investing muscle. As such, the musculus proprius forms a variable ring of muscle which surrounds the terminations of the bile and pancreatic ducts to form the *m. sphincter ampullae hepatopancreaticae*. Ensheathing the remaining intramural portion of the bile duct the musculus proprius constitutes the *m. sphincter ductus choledochi*. Eichhorn and Boyden point out that in the dog the downgrowth of a septum of the tunica muscularis on the mucosal side of the duct creates a muscular funnel through which the bile duct must pass. This feature makes the discharge of bile dependent to a large degree upon the activity of the duodenum.

Apparently, great variation exists in the amount of musculature which is present at the termination of the common bile duct. Halpert (1932) found proper muscle present in only 1 of 25 dogs he examined. Casas (1958) states that in the dog the sphincter of Oddi consists of three layers of muscle.

PANCREAS

The pancreas (Figs. 7–36, 7–46) is yellowish gray when preserved and pinkish gray in life. It is a rather coarsely lobulated, elongate gland. The lobules, as Revell (1902) has pointed out, produce a nodular surface with irregularly crenated margins. The pancreas is located in the dorsal part of both the epigastric and the mesogastric abdominal segment, caudal to the liver. Like the liver, the pancreas has both an exocrine and an endocrine function. Its exocrine secretion, the pancreatic juice, the most important of the digestive secretions, is conveyed to the descending portion of the duodenum by one or several ducts, usually two. It is a clear alkaline secretion containing three principal enzymes, one of which reduces proteins, one fats, and the third carbohydrates. Insulin, a protein hormone, is the endocrine secretion produced by the islet cells. This hormone keeps the sugar content of the blood at a constant level, and in its absence a fatal sugar diabetes sets in.

The weight of the pancreas averaged 31.3 gm. in 76 dogs, with an average weight of 13.8 kg., the pancreas thus accounting for an average of 0.227 per cent of total body weight in this group of adult mongrel dogs of both sexes. The pancreas, therefore, weighs about 1 ounce in a dog whose weight is approximately 30 pounds. These data compare favorably with Mintzlaff's (1909) findings in 30 dogs, although his specimens were, on an average, larger. The average total length of the pancreas in a 30-pound dog is approximately 25 cm., or 10 inches.

The pancreas, when hardened in situ, is in the form of a V which lies in a frontal plane with the apex pointing forward. The gland is basically divided into a thin, slender right lobe, and a shorter, thicker, and wider left lobe. The two lobes are united at the body, formerly called the pancreatic angle, which lies caudomedial to the pylorus.

LOBES AND RELATIONS

The **right lobe** of the pancreas (*lobus pancreatis dexter*) lies in the mesoduodenum near or in contact with the dorsal portion of the right flank. It extends from a transverse plane through the middle of the ninth intercostal spaces to one through the fourth lumbar vertebra. The right lobe varies in width from 1 to 3 cm. and in thickness up to 1 cm. Its length is approximately 15 cm., or 6 inches, in a beagle-type dog. The right lobe is positioned in the mesoduodenum in such a way that its round, flat, caudal extremity lies in the concavity of the duodenal loop. By traction the gland can be separated for a distance of about 3 cm. from the various parts of the duodenum which form the loop, since the mesoduodenum at this place is loose.

As the right lobe runs obliquely cranially toward the pylorus it becomes narrow and flattened dorsoventrally, so that dorsal and ventral surfaces are formed. Upon contacting the initial part of the descending duodenum, it becomes molded to this organ. The caudal part of the right lobe of the pancreas is related to the sublumbar fat containing the ureter and to the ventral surfaces of the right kidney and the caudate process of the liver. The right lobe of the pancreas is related ventrally to the ileum and cecum caudally, and to the ascending colon cranially. Loops of the jejunum contact those portions of its ventral surface which are not already in contact with more fixed viscera. In some specimens the right lobe of the pancreas and the adjacent descending part of the duodenum have gravitated lateral and even ventral to the jejunal coils.

The **body** of the pancreas (*corpus pancreatis*), unites the two lobes of the pancreas in an angle of about 45 degrees, which is open sinistrocaudally. Cranially, it lies closely applied to the caudosinistral portion of the pyloric region, which forms a large concave impression on the cranial portion of the body. Caudal to this impression, the pancreas is about 1 cm. thick and 3 cm. wide. The portal vein crosses the dorsal portion of the body. As the gastropancreatic artery and gastroduodenal vein disappear into the pancreas at this place, they are crossed on their right side by the bile duct, which lies adjacent to the duodenum.

The **left lobe** of the pancreas (*lobus pancreatis sinister)* lies in the deep leaf of the greater omentum. It begins at the body and runs caudosinistrally. It is about two-thirds as long and half again as wide as the right

lobe, measuring 10 cm. or 4 inches, in length, and 4 cm., or 1.6 inches, in width. Its **dorsal surface** (*facies dorsalis*), on the right, is related to the caudate process of the liver and then, in succession on the left, to the portal vein, caudal vena cava, and aorta. It ends in the left part of the sublumbar region in close relation to the cranial pole of the left kidney and the middle portion of the spleen. A full stomach alters these relations. The **ventral surface** (*facies ventralis*) of the left lobe of the pancreas is related ventrocaudally to the transverse colon and ventrocranially to the dorsal wall of the stomach.

An **accessory pancreas** (*pancreas accessorium)* is occasionally found in the dog. Baldyreff (1929) cites cases in which the aberrant gland was located in the wall of the gall bladder and in the caudal part of the great mesentery. Pancreatic bladders have been described by various authors as occurring in the cat, but none have been recorded in the dog (Boyden 1925).

DUCTS OF PANCREAS

The pancreas nearly always has two excretory ducts (Figs. 7–36, 7–37), in conformity with the dual origin of the gland, one anlage arising dorsally from the duodenum and the other ventrally at the termination of the bile duct (Revell 1902). These two ducts usually intercommunicate within the gland, since the parenchyma of the whole gland is elaborated around them. In the adult, the two portions of the gland are fused without any demarcation to indicate their dual origin. Revell, however, points out that, when the two ducts do not communicate within the gland, the pancreatic duct drains the right lobe and the dorsal duct drains the left lobe. Although this is the basic pattern by which the pancreatic ducts form in the domesticated mammals, great variations exist among the different species and within the same species.

The largest excretory duct of the pancreas in the dog is the **accessory pancreatic duct** (*ductus pancreaticus accessorius*), which opens into the duodenum as the minor duodenal papilla. The **pancreatic duct** (*ductus pancreaticus)* is the smaller duct in the dog and may occasionally be absent. The latter is associated with the opening of the common bile duct and usually enters the duodenum on the major duodenal papilla alongside the common bile duct. From its formation at the union of the ducts from the two lobes in the dog to the site where it perforates the intestinal wall, the accessory pancreatic duct is about 3 to 4 mm. long and 2 mm. wide. The union of the two lobar ducts to form the main duct may occur at any level up to the intestinal wall, or rarely, the two ducts may open separately (Revell 1902). According to Bottin (1934), the two pancreatic ducts open separately into the duodenum in about 75 per cent of dog specimens, and they always communicate with each other in the gland.

In the 50 dogs of all ages and breeds and of both sexes which Nielsen and Bishop (1954) studied by the use of radiopaque mediums, the duct system of the canine pancreas could be divided into five main types. In type 1 (46 per cent) a single main duct, formed by a tributary from each lobe uniting in a Y junction, entered from the duodenum at the (minor) ventral duodenal papilla. In this group there was also an additional duct, arising from the left lobar duct, which frequently followed a most indirect and tortuous route and entered the duodenum at or near the major duodenal papilla. Type 2 (22 per cent) was similar to type 1, except that the small dorsal duct arose from the right lobar duct instead of the left, and crossed over the duct of the left lobe before entering the duodenum. In type 3 (16 per cent), each lobe had its own excretory duct. The ducts crossed, the one from the right lobe emptying into the dorsal duodenal papilla with the bile duct and that from the left lobe emptying on the ventral duodenal papilla. A fine and often tortuous shunt connected the two ducts. In type 4 (8 per cent), the ducts from the two lobes anastomosed in the Y formation, but there was no other duct leading into the duodenum or into the bile duct. In two of the four specimens there was a small anastomosis within the pancreas between the ducts of the two lobes. In type 5 (8 per cent), there were three orifices into the duodenum from the ducts from the pancreas, and in one specimen there were two additional small ducts, one emerging on either side of the minor duodenal papilla.

Other variations of the ducts of the canine pancreas exist, as Revell (1902) and Mintzlaff (1909) have shown. The main duct from each lobe occupies the approximate center of the lobe and is joined at right angles by tributaries from the adjacent parenchyma. Because the gland is ribbon-like, the small ducts from the adjacent parenchyma enter largely on opposite sides, the openings being spaced at 0.5- to 1.5-cm. intervals.

The opening of the pancreatic duct is closely associated with that of the bile duct. In two out of three specimens Eichhorn and Boyden (1955) found the slitlike orifice of the pancreatic duct located distal to that of the bile duct; others have described this opening as proximal to that of the bile duct. The accessory pancreatic duct usually opens into the duodenum 28 mm. from the opening of the bile duct into the duodenum, or approximately 8 cm. from the pyloric sphincter (Nielsen and Bishop 1954). Its entry into the duodenum resembles that of the bile duct in that a ridge of mucosa is formed, with a slight elevation at its distal end on which the opening is located.

The accessory pancreatic duct, like the pancreatic duct, but unlike the bile duct, runs through the duodenal wall rather directly. The opening through the mesenteric wall of the proximal portion of the descending duodenum is frequently located to the left of the cranial pancreaticoduodenal vessels, whereas the bile and pancreatic ducts open to the right of these vessels.

Eichhorn and Boyden (1955) described and illustrated the musculature of the pancreatic and bile ducts of the dog (sphincter of Oddi). Kyosola and Rechardt (1974) described its innervation.

BLOOD AND LYMPH VESSELS

The main vessels to the right lobe of the pancreas are the pancreatic branches of the **cranial** and **caudal pancreaticoduodenal arteries** which anastomose in the gland. The left extremity of the left lobe of the pancreas is primarily supplied by the pancreatic branch of the **splenic artery.** It also receives small branches from the common **hepatic artery,** as this vessel may groove the dorsal surface of the organ, and the left limb regularly receives, near the pancreatic angle, one or two branches from the **gastroduodenal artery.** Small pancreatic branches directly from the **celiac artery** may supply a small portion of the left limb of the pancreas near its free end.

The **caudal pancreaticoduodenal vein**, a satellite of the artery of the same name, is the principal vein from the right pancreatic lobe. It is the last tributary to enter the cranial mesenteric vein and, unlike the intestinal veins which empty into it, it enters the larger vessel from the cranial side. The left lobe of the pancreas is drained primarily by two veins which terminate in the last 2 cm. of the splenic vein. The satellite of the small branch of the cranial pancreaticoduodenal artery which supplies the left lobe near the pancreatic angle drains this part of the gland.

The lymphatics from the pancreas drain into the **duodenal lymph node**, if present, and into the **hepatic, splenic,** and **mesenteric lymph nodes.**

NERVES

Most sympathetic fibers come from the **celiac plexus** and reach the organ by following the pancreatic branches of the cranial pancreaticoduodenal and celiac arteries. It is probable that the caudal part of the right lobe receives sympathetic fibers from the **cranial mesenteric plexus** which follow the caudal pancreaticoduodenal artery and its pancreatic branches. McCrea (1924) states that, in the dog, vagal (parasympathetic) fibers reach the pancreas as fine twigs which run with the splenic branch of the celiac artery and with the cranial mesenteric artery, presumably along the caudal pancreaticoduodenal branch. These findings accord with those of Richins (1945) in the cat.

BIBLIOGRAPHY

Alexander, W. F. 1940. The innervation of the biliary system. J. Comp. Neurol. 72:357–370.

Alvarez, W. C. 1948. An Introduction to Gastro-Enterology. 3rd Ed. New York, Paul B. Hoeber, Inc.

Annis, J. R. 1974. Dental surgery, in Canine Surgery, edited by J. Archibald. Evanston, Ill., American Veterinary Publications, Inc., pp. 313–328.

Arey, L. B. 1941. Throttling veins in the livers of certain animals. Anat. Rec. 81:21–33.

————. 1974. Developmental Anatomy. 7th Ed. Philadelphia, W. B. Saunders Co.

Arnall, L. 1961. Some aspects of dental development in the dog; calcification of crown and root of the deciduous dentitions. J. Small Anim. Pract. 1:169–173.

Baldyreff, E. B. 1929. Report of an accessory pancreas on the ileum of a dog. Anat. Rec. 43:47–51.

Baum, H. 1918. Das Lymphgefässystem des Hundes. Berlin, Hirschwald.

Bennett, G. A. 1944. The lyssa of the dog. (Abstr.) Anat. Rec. 88:422.

Bennett, G. A., and R. C. Hutchinson. 1946. Experimental studies on the movements of the mammalian tongue; the protrusion mechanism of the tongue (dog). Anat. Rec. 94:57–83.

Bennett, G. A., and A. J. Ramsay. 1941. Experimental studies on the movements of the mammalian tongue; movements of the split tongue (dog). Anat. Rec. 79:39–51.

Bensley, R. R. 1902. The cardiac glands of mammals. Am. J. Anat. 2:105–156.

Blakeley, C. L. 1957. Otorrhea and Surgical Drainage. Pp. 309–320, in Canine Surgery, edited by K. Mayer, J. V. Lacroix, and H. P. Hoskins, 4th Ed. Evanston, Ill., American Veterinary Publications, Inc.

Bloom, W., and D. W. Fawcett. 1975. A Textbook of Histology, 10th Ed. Philadelphia, W. B. Saunders Co.

Bottin, J. 1934. Contribution à l'Étude de l'Anatomie des Canaux Excreteurs du Pancréas chez le Chien. C. R. Soc. biol. (Paris) 117:825–827.

Boyden, E. A. 1925. The problem of the pancreatic bladder. Am. J. Anat. 36:151–183.

Bradley, O. C., and T. Grahame. 1948. Topographical Anatomy of the Dog. 5th Ed. London, Oliver & Boyd.

Brown, M.E. 1937. The occurrence of arteriovenous anastomoses in the tongue of the dog. Anat. Rec. 69:287–292.

Casas, A. P. 1958. Contribution à l'étude du sphincter d'Oddi chez Canis familiaris. Acta anat. (Basel) 34:130–153.

Chauveau, A. 1886. The Comparative Anatomy of the Domestic Animals. New York, D. Appleton & Company.

Chiu, S. L. 1943. The superficial hepatic branches of the vagi and their distribution to the extrahepatic biliary tract in certain mammals. Anat. Rec. 86:149–155.

Colyer, F. 1936. Variations and Diseases of the Teeth of Animals. London, J. Bale Sons and Danielson, Ltd.

Cornevin, C. E., and F. X. Lesbre. 1894. Traité de l'Age des Animaux Domestiques. Paris, J. B. Bailliere et Fils.

Dechambre, M. 1912. Absence totale des dentes chez un chien. Rec. Méd. vét. 89:67–68.

Doty, R. L., and I. Dunbar. 1974. Color, odor, consistency and secretion rate of anal sac secretions from male, female, and early-androgenized female beagles. Am. J. Vet. Res. 35:729–731.

Dyce, K. M. 1957. The muscles of the pharynx and palate of the dog. Anat. Rec. 127:497–508.

Eichhorn, E. P., Jr., and E. A. Boyden. 1955. The choledochoduodenal junction in the dog — a restudy of Oddi's sphincter. Am. J. Anat. 97:431–451.

Elias, H. 1949. A re-examination of the structure of the mammalian liver; parenchymal architecture. Am. J. Anat. 84:311–333.

Ellenberger, W. 1911. Handbuch der vergleichenden mikroskopischen Anatomie der Haustiere. Vol. 3. Berlin, Paul Parey.

Ellenberger, W., and H. Baum. 1943. Handbuch der vergleichenden Anatomie der Haustiere. 18th Ed. Berlin, Springer.

Every, R. G. 1972. A New Terminology for Mammalian Teeth Founded on the Phenomenon of Theogosis. Christchurch, N.Z., Pegasus Press.

Every, R. G. 1974. Thegosis in prosimians, in Prosimian Biology, edited by R. D. Martin, G. A. Doyle, and A. C. Walkers. London, Duckworth.

Ewer, R. 1973. The Carnivores. Ithaca, N.Y., Cornell Univ. Press.

Field, E. J., and R. J. Harrison. 1947. Anatomical Terms, Their Origin and Derivation. Cambridge, Heffer.

Getty, R. 1975. Sisson and Grossman's Anatomy of the Domestic Animals. 5th Ed. Philadelphia, W. B. Saunders Co.

Glen, J. B. 1972. Canine salivary mucoceles: the results of sialographic examination and surgical treatment of fifty cases. J. Small Anim. Pract. 13:515–526.

Glock, G. E., H. Mellanby, M. Mellanby, M. M. Murray, and J. Thewlis. 1942. A study of the development of dental enamel in dogs. J. Dent. Res. 21:183–199.

Grau, H. 1935. Der After von Hund und Katze unter biologischen und praktischen Gesichtspunkten. Tierärztl. Rundschau 41:351–354.

Graves, E. F. 1948. An unusual dentition in a dog. J. Am. Vet. Med. Assoc. 113:40.

Grey, E. G. 1918. Observations on the postural activity of the stomach. Am. J. Physiol. 45:272–285.

Haane, G. 1905. Über die Cardiadrüsen und die Cardiadrüsenzone des Magens der Haussäugetiere. Arch. Anat. Physiol. 11–32.

Halpert, B. 1932. The choledocho-duodenal junction — a morphological study in the dog. Anat. Rec. 53:83–102.

Harvey, B. C. H. 1906. A study of the gastric glands of the dog and of the changes they undergo after gastroenterostomy and occlusion of the pylorus. Am. J. Anat. 6:207–239.

Harvey, C. E. 1970. Sialography in the dog. J. Am. Vet. Radiol. Soc. 10:18–27.

Helm, R. 1907. Vergleichende anatomische und histologische Untersuchungen über den Oesophagus der Haussäugetiere. Inaug. Diss., Zürich.

Higgins, G. M. 1926. An anomalous cystic duct in the dog. Anat. Rec. 33:35–41.

Higgins, G. M., and C. G. Bain. 1930. The absorption and transference of particulate material by the greater omentum. Surg. Gynec. Obstet. 50:851–860.

Höppner, N. 1956. Röntgenologische Untersuchungen über Gebiss und Zahnentwicklung beim Hunde von Geburt bis zum Ende des Zahnwechsels. Vet. Diss., Free Univ., Berlin.

Hörmandinger, J. 1958. Untersuchungen über die Zahnentwicklung im Oberkiefer von Hundefeten. Vet. Diss., Wien.

Hwang, K., M. I. Grossman, and A. C. Ivy. 1948. Nervous control of the cervical portion of the esophagus. Am. J. Physiol. 154:343–357.

Ivy, A. C. 1934. The physiology of the gall bladder. Physiol. Rev. 14:1–102.

Jemerin, E. E., and F. Hollander. 1938. Gastric vagi in the dog. Proc. Soc. exp. Biol (N.Y.) 38:139–146.

Kadletz, M. 1929. Über eine Blinddarmvarietät beim Hund, nebst Bemerkungen über die Lage, Gestalt und Entwicklungsgeschichte des Hundeblinddarmes. Morph. Jb. 60:469–479.

Kremenak. C. R. Jr. 1967. Dental exfoliation and eruption chronology in Beagles. J. Dent. Res. 46:686–693.

————. 1969. Dental eruption chronology in dogs: deciduous tooth gingival emergence. J. Dent. Res. 48:1177–1184.

Kyösola, K., and L. Rechardt. 1974. The antomy and innervation of the sphincter of Oddi in the dog and cat. Am. J. Anat. 140:497–533.

Lemere, F. 1932. Innervation of the larynx. I. Innervation of laryngeal muscles. Am. J. Anat. 51:417–437.

Mall, F. 1896. The vessels and walls of the dog's stomach. Johns Hopkins Hosp. Rep. 1:1–36.

Mann, F. C., S. D. Brimhall, and J. P. Foster. 1920. The extrahepatic biliary tract in common domestic and laboratory animals. Anat. Rec. 18:47–66.

Markowitz, J., A. Rappaport, and A. C. Scott. 1949. The function of the hepatic artery in the dog. Am. J. Dig. Dis. 16:344–348.

Martin, P. 1923. Lehrbuch der Anatomie der Haustiere. Vol. 4. Stuttgart, Schickhardt and Ebner.

McCarrell, J. D., S. Thayer, and C. K. Drinker. 1941. The lymph drainage of the gallbladder together with the composition of liver lymph. Am. J. Physiol. 133:79–81.

McCrea, E. D. 1924. The abdominal distribution of the vagus. J. Anat. 59:18–40.

Meyer, Leo. 1942. Das Gebiss des deutschen Schäferhundes mit besonderer Berücksichtigung der Zahnalterbestimmung und der Zahnanomalie. Vet. Diss., Zürich.

Michel, G. 1956. Beitrag zur Topographie der Ausführungsgänge der Gl. mandibularis und der Gl. sublingualis major des Hundes. Berl. Münch. tierärztl. Wschr. 69:132–134.

Mintzlaff, M. 1909. Leber, Milz, Magen, Pankreas des Hundes. Diss., Leipzig.

Mitchell, P. C. 1905. Intestinal tract of mammals. Trans. zool. Soc. 17:437–536.

Montagna, W., and H. F. Parks. 1948. A histochemical study of the glands of the anal sac of the dog. Anat. Rec. 100:297–318.

Morton, J. 1929. The differences between high and low intestinal obstruction in the dog; an anatomic and physiologic explanation. Arch. Surg. 18:1119–1139.

Nickel, R., A. Schummer, E. Seiferle, and W. Sack. 1973. The Viscera of the Domestic Animals. 2nd Ed. Translated and revised by W. O. Sack. New York, Springer-Verlag.

Nielsen, S. W., and E. J. Bishop. 1954. The duct system of the canine pancreas. Am. J. Vet. Res. 15:266–271.

Noer, R. 1943. The blood vessels of the jejunum and ileum; a comparative study of man and certain laboratory animals. Am. J. Anat. 73:293–334.

Oddi, R. 1887. D'une disposition a sphincter speciale de l'ouverture du canal choledoque. Arch. Ital. Biol. 8:317–332.

Parks, H. F. 1950. Morphological and cytochemical observations on the circumanal glands of dogs. Thesis, Cornell Univ., Ithaca, N.Y.

Payer, V. J., J. Riedel, J. Minar, and R. Moravec. 1956. Der extrahepatale Abschnitt der Leberarterie des Hundes vom Gesichtspunkt der chirurgischen Anatomie. Anat. Anz. *103*:246–257.

Prinzmetal, M. E., E. M. Ornitz, Jr., B. Simkin, and H. C. Bergman. 1948. Arterio-venous anastomoses in liver, spleen and lungs. Am. J. Physiol. *152*:48–52.

Revell, D. G. 1902. The pancreatic ducts of the dog. Am. J. Anat. *1*:443–457.

Richins, C. A. 1945. The innervation of the pancreas. J. Comp. Neurol. *83*:223–236.

Robinson, B. 1895. The peritoneum of the dog. Amer. Practit. *20*:368–376.

Satrapa-Binder, N. 1959. Ein Beitrag zur Zahnentwicklung im Unterkiefer von Hundefeten. Vet. Diss., Wien.

Sauer, M. E. 1951. The cricoesophageal tendon. Anat. Rec. *109*:691–699.

Schilling, I. A., F. W. McKee, and W. Welt. 1950. Experimental hepatic-portal arteriovenous anastomoses. Surg. Gynec. Obstet. *90*:473–480.

Schutz, C. B. 1930. The mechanism controlling migration of the omentum. Surg. Gynec. Obstet. *50*:541–544.

Secord, A. C. 1941. Small animal dentistry. J. Amer. Vet. Med. Assoc. *98*:470–476.

Seiferle, E., and L. Meyer. 1942. Das Normalgebiss des deutschen Schäferhundes in den verschiedenen Altersstufen. Vjschr. naturforsch. Gesellsch. (Zürich) *87*:205–252.

St. Clair, L. E., and N. D. Jones. 1957. Observations on the cheek teeth of the dog. J. Am. Vet. Med. Assoc. *130*:275–279.

Stockard, C. R. 1941. The Genetic and Endocrinic Basis for Differences in Form and Behavior. Amer. Anat. Memoirs 19. The Wistar Inst. Anat. Biol., Philadelphia.

Stormont, D. L. 1928. The Salivary Glands. Pp. 89–135 *in* Special Cytology, edited by E. V. Cowdry, Ed. 1, vol. I. New York, Paul B. Hoeber.

Szalay, F. S. 1969. Mixodectidae, Microsyopidae, and the insectivore-primate transition. Bull. Am. Museum Nat. Hist. *140*:193–330.

Testoni, F. J., C. L. Lohse, and R. J. Hyde. 1977. Anatomy and cannulation of the parotid duct in the dog. J. Am. Vet. Med. Assoc. *170*:831–834.

Tims, H. W. 1895. Notes on the dentition of the dog. Anat. Anz. *11*:537–547.

Tims, H. W. 1902. On the succession and homologies of the molar and premolar teeth in mammalia. J. Anat. Physiol. *3*:321–343.

Titkemeyer, C. W., and M. L. Calhoun. 1955. A comparative study of the structure of the small intestines of domestic animals. Am. J. Vet. Res. *16*:152–157.

Torgersen, J. 1942. The muscular build and movements of the stomach and duodenal bulb. Acta Radiol. (Stockh.) Suppl. *45*:1–191.

Trautmann, A., and J. Fiebiger. 1957. Fundamentals of the Histology of Domestic Animals. Translated and revised from the 8th and 9th German editions, 1949, by R. E. Habel and F. L. Biberstein. Ithaca, N.Y., Comstock Publishing Assoc.

Warren, R. 1939. Serosal and mucosal dimensions at different levels of the dog's small intestine. Anat. Rec. *75*:427–437.

Watson, A. G. 1974. Some aspects of the vagal innervation of the canine esophagus, an anatomical study. Masters Thesis. Massey Univ., N.Z.

Webb, R. L., and P. H. Simer. 1940. Regeneration of the greater omentum. Anat. Rec. *76*:449–454.

Williams, R. C. 1961. Observations on the Chronology of Deciduous Dental Development in the Dog. Thesis, Cornell Univ., Ithaca, N.Y.

Williams, R. C., and H. E. Evans. 1978. Prenatal dental development in the dog, *Canis familiaris:* chronology of tooth germ formation and calcification of deciduous teeth. Zbl. Vet. Med. C. Anat. Histol. Embryol. 7:152–163.

Williams, T. 1935. The anatomy of the digestive system of the dog. Vet. Med. *30*:442–444.

Winkelstein, A., and P. W. Aschner, 1924. The pressure factors in the biliary duct system of the dog. Am. J. Med. Sci. *168*:812–819.

Wood, A. E., and H. E. Wood. 1933. The genetic and phylogenetic significance of the presence of a third upper molar in the modern dog. Am. Midland Nat. *14*:36–48.

Zietzschmann, O. 1938. Lage und Form des Hundemagens. Berl. Münch. tierärztl. Wschr. *10*:138–141, and Vet. Rec. *50*:984–985.

—————. 1939. Das Mesogastrium dorsale des Hundes mit einer schematischen Darstellung seiner Blätter. Morph. Jb. *83*:325–358.

Zontine, W. J. 1975. Canine dental radiology: radiographic technique, development, and anatomy of the teeth. J. Am. Vet. Radiol. Soc. *16*:75–83.

Chapter 8

THE RESPIRATORY APPARATUS

The respiratory apparatus or system consists of the lungs and the air passageways which lead to the sites of gaseous exchange within the lungs. Various structures associated with these passageways modify or regulate the flow of air, serve as olfactory receptors, facilitate water and heat exchange, and make phonation possible. The nasal cavity and conchae (formerly called turbinates) warm and moisten the air, and remove foreign material from it. The pharynx serves as a passageway for both the respiratory and the digestive system. The larynx guards the entrance to the trachea, functions in vocalization, and regulates both the inspiration and the expiration of air. The trachea is a non-collapsible tube, lined by ciliated epithelium. It divides into the principal bronchi, and the air passageways continue in the two lungs as lobar bronchi, segmental bronchi, bronchioles, alveolar ducts, alveolar sacs, and alveoli. The terminal divisions are located in the elastic, well-vascularized lungs, which passively expand and collapse in response to changes in intrathoracic pressure, created by action of the muscles of the diaphragm and thoracic wall.

NOSE AND NASAL PORTION OF THE PHARYNX

Nose

The **nose** *(nasus)*, in a broad sense, refers to the **external nose** *(nasus externus)* and its associated **nasal cartilages** *(cartilagines nasi)*, as well as to the internal nose, or **nasal cavity** *(cavum nasi)*. In terms relating to nasal structures or diseases the root *rhin-*, from the Greek *rhinos* for nose, is frequently employed. The facial portion of the respiratory system and the rostral portions of the upper and lower jaws collectively constitute what is called the **muzzle.** In dolichocephalic breeds the muzzle is long and may account for half of the total length of the skull. In brachycephalic breeds, the shortened muzzle often is the cause of respiratory difficulties.

EXTERNAL NOSE

The external nose consists of a fixed bony case and a movable cartilaginous framework. The cartilaginous portion is movable or distensible by virtue of several skeletal muscles associated with the muzzle. The short hair on the skin of the nose is directed caudally on the mid-dorsal surface and gradually slopes in a caudoventral direction laterally, where it is continued on the lips. The apical portion of the nose is flattened and devoid of hair. It is called the **nasal plane** *(planum nasale)* and includes the **nostrils** *(nares)*, which are separated from each other by a groove, or *philtrum.* The integument of the nasal plane presents epithelial elevations or papillary ridges which result in patterns characteristic for each individual. For this reason nose prints may be used as a means of identification in the dog, similar to the way fingerprints are used in man (Horning et al. 1926).

The lateral walls of the bony portion of the nose are formed by the incisive bones and maxillae, whereas the roof is formed by the paired nasal bones. The concave rostral ends of the nasal bones, dorsally, and the incisive bones, laterally and ventrally, bound the largest opening into the skull. This opening, called the **bony nasal aperture** *(apertura nasi ossea),* formerly called the piriform aperture, is wider ventrally than dorsally and lies in an oblique plane. In life this opening is bounded rostrally by

the nasal cartilages. The aggregate of these cartilages, with their ligaments and covering skin, comprises the **movable portion of the nose** *(pars mobilis nasi).* The movable part of the nose ends in a truncated **apex** *(apex nasi).*

CARTILAGES OF THE NOSE

The mobile part of the external nose has a framework composed entirely of the nasal cartilages (Fig. 8–1). These include the unpaired septal cartilage, the paired dorsal parietal and ventral parietal cartilages, and

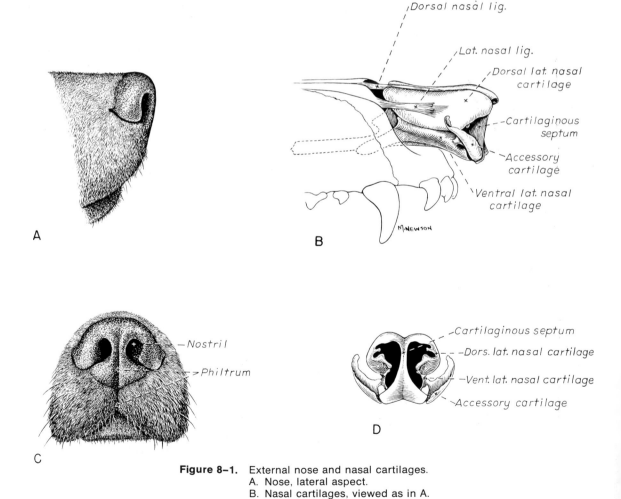

Figure 8–1. External nose and nasal cartilages.
  A. Nose, lateral aspect.
  B. Nasal cartilages, viewed as in A.
  C. Nose, rostral aspect of planum nasale.
  D. Nasal cartilages, as viewed in C.

the paired accessory cartilages. Related to the ventral part of the septal cartilage is the vomeronasal cartilage.

The **septal cartilage of the nose** *(cartilago septi nasi)* is a perpendicular median plate which separates most of the nasal cavity into right and left nasal fossae. It is a rostral continuation of the perpendicular plate of the ethmoid bone which fails to ossify. In the region of the aperture of the osseous skull, it is lacking over a distance of about 1 cm. so that the nasal septum in this region is formed of the **membranous nasal septum** *(pars membranacea septi nasi),* which connects the cartilaginous immovable caudal part with the mobile rostral part.

The caudal part of the cartilaginous nasal septum is thicker ventrally, where it lies in the septal groove of the vomer, than it is dorsally, where it blends with the thin, conjoined ventral processes of the nasal bones. It presents a prominent caudal process which occupies the space between the osseous nasal septum dorsally and the groove in the vomer ventrally.

The rostral part of the cartilaginous nasal septum continues a median course forward from the membranous portion of the nasal septum (Fig. 8–2). It lies between the right and left nasal vestibule. The rostral border of this portion of the septum is divided into

right and left laminae. The cleft between the two leaves is deeper and much wider ventrally than it is dorsally. It forms a depressed triangular area ventrally. The dorsal portion of each lamina is rolled laterally to form the dorsolateral nasal cartilage. Arising from the ventral portion of the rostral part of the septal cartilage is the small ventrolateral nasal cartilage, which turns upward and inward toward the dorsolateral nasal cartilage (Fig. 8–1 D). The accessory cartilage is united by collagenous tissue to the ventrolateral nasal cartilage.

The **dorsolateral nasal cartilage** *(cartilago nasi lateralis dorsalis)* is the most expansive of the cartilages in the mobile part of the external nose. On each side it is a continuation of half of the dorsal portion of the septal cartilage. From this dorsal origin it is rolled into a tube by curving outward, downward, and inward. It is about 2.5 cm. long. Its widest portion is its rostral half, which in large heads attains a width of 1 cm. Caudally, it joins the dorsal part of the bony aperture, to which it is attached by fibrous tissue along the concave border of the nasal bone. The free rolled-in border of the dorsolateral nasal cartilage is greatly thickened rostrally and contains a plexus of blood vessels which form a meshwork in the collagenous tissue directly caudal to the nostril. It be-

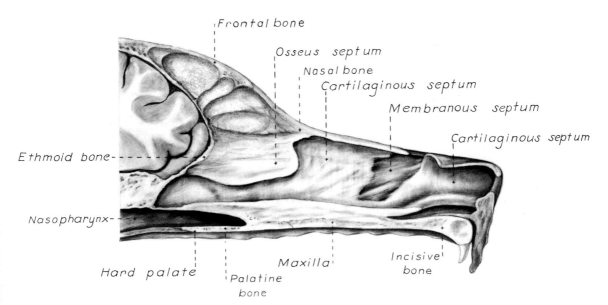

Figure 8–2. Sagittal section, showing nasal septum.

comes much thinner caudally. At a transverse plane through the bony aperture it blends with the rostral extremity of the ventral nasal concha (formerly called maxilloturbinate). The free medial border of the dorsolateral cartilage curves ventral to the thicker free lateral border of the ventrolateral *cartilage* to a transverse plane through the caudal angle of the midlateral slit of the nostril.

The **ventrolateral nasal cartilage** *(cartilago nasi lateralis ventralis)* is a continuation of the rostral portion of the lateral half of the septal cartilage. Caudally, its origin moves obliquely dorsad on the lateral surface of the ventral part of the septal cartilage. It is slightly shorter and about one-fourth as wide as the dorsolateral cartilage. As it rolls dorsally, it is neither of uniform thickness nor of uniform curvature. Rostrally, it runs forward into the apex of the nose. It ends caudal to the lateral leaf of the septal cartilage adjacent to the articulation of this cartilage with the accessory cartilage. Caudally, it assumes a sigmoid shape, in cross section, being bent in such a way that its free border is added to the free border of the dorsolateral cartilage in forming the cartilaginous basis of the fold which continues the ventral nasal concha to the vestibule.

The **accessory cartilage** *(cartilago acces-*

soria) is a laterally convex leaf which articulates with the ventrolateral angle of the wide ventrally divided portion of the septal cartilage and extends dorsoposteriorly to the lateral surface of the expanded portion of the dorsolateral nasal cartilage. For a considerable portion of its length, it lies directly under the integument that covers the ventral surface of the midlateral slit in the nostril.

A second small accessory cartilage is occasionally located directly dorsal to the septal cartilage in the groove formed by the origins of the right and left dorsolateral cartilages. Its position is only a few millimeters rostral to the internasal suture.

The mobile part of the nose is moved and the shape of the nostrils is altered by the action of intrinsic muscles and of the nasal part of the levator labii maxillaris and of the levator nasolabialis muscles which are inserted on these cartilages.

VOMERONASAL ORGAN

The paired **vomeronasal organ** *(organum vomeronasale)* (Figs. 8–3; 8–6) long known as Jacobson's organ, is located in the rostral base of the nasal septum as a tubular pocket of olfactory epithelium partially enclosed by a scroll of cartilage *(cartilago vomerona-*

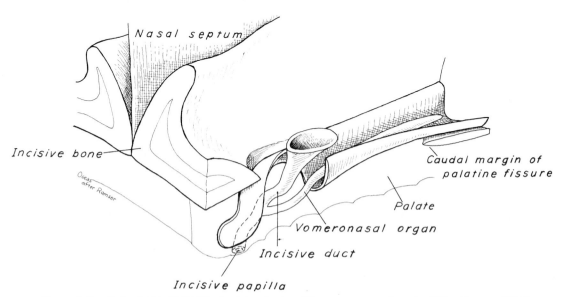

Figure 8–3.   Schematic view of the incisive duct and the vomeronasal organ. (After Ramser 1935).

salis). Each vomeronasal organ opens rostrally into an incisive duct which connects the nasal and oral cavities.

The olfactory nature of this organ in the mammal was long suspected, but its specific function is as yet unknown, although there is evidence that it plays a role in sexual behavior and kin recognition (Estes 1972). It has been investigated by Read (1908) and Ramser (1935) in the cat and dog; McCotter (1912) in the opossum and other mammals; Mann (1961) in the bat; Kratzing (1971) in the sheep; and Estes (1972) in several ungulates. The role of olfaction for social communication in mammals has been reviewed by Eisenberg and Kleiman (1972).

In connection with the probable sexual role of the vomeronasal organ is the behavioral "olfactory reflex," "lip-curl," or "Flehmen" reaction (described by Schneider 1930–1935) in many mammals, which perhaps provides mechanical assistance in clearing the passageways to the vomeronasal organ and in aspirating odorants so they may contact the ciliated epithelium of the organ to reach receptor sites.

The **incisive duct** *(ductus incisivus),* formerly called the nasopalatine duct or Stensen's canal, passes through the palatine fissure and connects the nasal and oral cavities. The oral orifice of each duct lies lateral to the incisive papilla, behind the central incisor teeth. The paired ducts pass dorsocaudally to open into each nasal fossa. The incisive duct is not enclosed in a bony canal, as in man.

LIGAMENTS OF THE NOSE

Three ligaments, composed of one paired and one unpaired, attach the mobile part of the nose to the dorsal portion of the osseous muzzle (Fig. 8–1 B). The **dorsal nasal ligament** *(lig. nasale dorsale)* is a single band of collagenous tissue which runs from the dorsal accessory cartilage to the dorsum of the nasal bones. The **lateral nasal ligament** *(lig. nasale laterale),* one on either side, is a collagenous band which runs from the midlateral surface of the dorsal parietal cartilage to the border of the bony nasal aperture directly dorsal to the end of the nasomaxillary suture. The ligaments of the nose are

best developed in old dogs of the working breeds.

THE NASAL CAVITY

The **nasal cavity** *(cavum nasi)* is the facial portion of the respiratory passageway. It extends from the nostrils to the choanae, being divided into right and left halves by the nasal septum. The **septum** consists of a bony portion *(septum nasi osseum),* a cartilaginous portion *(cartilago septi nasi),* and a membranous portion. Each half of the nasal cavity has a respiratory and an olfactory region.

Each half of the nasal cavity begins at the nostril with the nasal vestibule and ends with the nasopharyngeal meatus and choana. The nasal fossa is divided into four principal air channels and several smaller ones (Fig. 8–5). During development the growth of laminae from the lateral and dorsal walls of the nasal cavity results in the formation of conchae which largely fill the cavity and restrict the flow of air. The air passages thus created between the conchae are called the nasal meatuses.

The **nostril** *(naris),* the opening into the nasal vestibule, is a curved opening which is much wider dorsomedially than it is ventrolaterally. It possesses more than usual importance, because in some brachycephalic dogs the opening is too restricted and interferes with respiration. Leonard (1956) devised an operation whereby the transverse diameter of the nostril may be increased.

The **alar fold** *(plica alaris),* which is an extension of the ventral nasal concha (or maxilloturbinate) terminates within the vestibule by a bulbous enlargement that fuses to the wing of the nostril. The **wing of the nostril** *(ala nasi)* is the thickened dorsolateral portion of the nostril. The wing of the nostril contains much of the dorsolateral and accessory nasal cartilages. It is the most mobile portion of the nostril, because it receives the terminal fibers of the nasal portions of the mm. levator labii maxillaris and levator nasolabialis.

The **nasal vestibule** *(vestibulum nasi)* is not an empty antechamber, as it is in man, but rather it is largely obliterated by the large bulbous end of the alar fold which

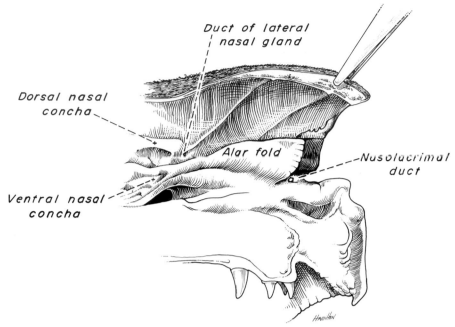

Figure 8–4. The nasal vestibule opened to show the entrance of the duct of the lateral nasal gland and the nasolacrimal duct.

extends into it. Because the end of the alar fold is fused to the inner surface of the wing of the nostril, it acts to divert the incoming air. Upon entering the vestibule through a nostril, air is diverted medially and ventrally into the largest meatus of the nose. The **nasolacrimal duct** *(ductus nasolacrimalis),* which conducts the lacrimal secretion from the eye, opens into the vestibule by an orifice located at the external end of the attached margin of the alar fold. The small duct of the lateral nasal gland opens into the dorsal vestibule dorsocaudal to the opening of the nasolacrimal duct (Fig. 8–4).

Nasal Conchae

The **nasal conchae** *(conchae nasales)* are cartilaginous or slightly ossified scrolls covered with nasal mucosa which occupy the major portion of each half of the nasal cavity. They include the dorsal, ventral, and ethmoidal conchae, which were formerly called the dorsal nasoturbinate, the maxilloturbinate, and the ethmoturbinate, respectively.

The **dorsal nasal concha** *(concha nasalis dorsalis)* (Figs. 8–4, 8–5) consists of an elongated, slightly curled scroll of the first endo-

turbinate which is attached to the crista ethmoidalis of the ethmoid and nasal bones. It is not homologous to the superior concha of man. A mucosal fold continues the concha into the vestibule of the nose.

The **ventral nasal concha** *(concha nasalis ventralis)* (Fig. 8–5), formerly called the maxilloturbinate, is a tightly folded series of scrolls which occupies the rostral part of the nasal cavity and is attached to the *crista conchalis* on the medial surface of the maxilla. It extends from the level of the first to the third premolar teeth.

The **ethmoidal conchae** *(conchae ethmoidales)* (Fig. 8–5) are part of the ethmoidal labyrinth *(labyrinthus ethmoidalis)* that fills the caudal part of the nasal cavity. Developmentally, they are outgrowths of the ethmoid bone covered with nasal mucosa. The numerous delicate, bony scrolls are known as ethmoturbinates and are further subdivided into ectoturbinates and endoturbinates. All are attached to the external lamina and cribriform plate of the ethmoid. The six ectoturbinates are small and lie on the dorsal aspect of the labyrinth, whereas the four endoturbinates lie ventrally and fill the caudal portion of the cavity. The first endoturbinate extends rostrally and forms the

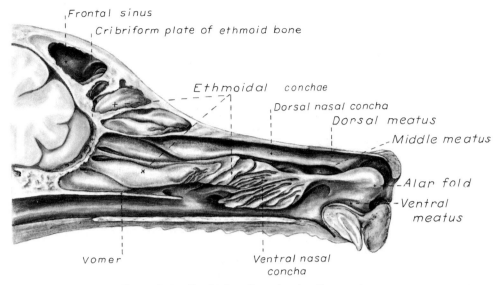

Figure 8–5.   Sagittal section, showing the conchae.

bony scroll for the concha nasalis dorsalis, while the second endoturbinate forms the concha nasalis medialis (Graeger 1958). One or more scrolls of the ethmoturbinates usually extend into the frontal sinus.

Olfactory nerves ramify in the mucosa of the ethmoturbinate scroll that extends into the frontal sinus, according to Read (1908), and they also reach into the mucosa of the sinus opposite the cribriform plate. The epithelium on the olfactory part of the sinus has a brown color. The mucosa of the nasal cavity is richly supplied with nerves and blood vessels (Figs. 8–6, 8–7). Olfactory nerves supply about half of the ethmoturbinates, the caudal half of the nasal septum, and a good portion of the roof and lateral walls of the nasal cavity.

After the inhaled air leaves the nasal vestibule it traverses the longitudinal **nasal meatuses** to reach the nasal part of the pharynx.

Nasal Meatuses

The **dorsal nasal meatus** *(meatus nasi dorsalis)* is a passage through the dorsal part of each nasal cavity. It lies between the dorsal nasal concha and the ventral surface of the nasal bone. Laterally it is limited by the nasoturbinate crest, from which the concha arises. Medially the dorsal nasal meatus be-

comes confluent with the common nasal meatus.

The **middle nasal meatus** *(meatus nasi medius)* lies between the dorsal nasal concha dorsally and the dorsal part of the numerous scrolls composing the ventral nasal concha. Throughout the long middle portion of the middle nasal meatus its width is approximately 1 mm. At its rostral end it presents a dilatation, and caudally its lateral portion is divided into several parts.

The **atrium of the middle meatus** *(atrium meatus medii)* is an ellipsoidal dilatation which connects the nasal vestibule with the middle nasal meatus. The atrium is formed ventrally by the narrow handle of the club-shaped mucosal alar fold which runs forward and upward from the ventral nasal concha. Dorsally it is bounded by the relatively straight rostral portion of the dorsal nasal concha. In large mesaticephalic heads it is approximately 5 mm. deep, 5 mm. wide, and 2 cm. long.

Laterally, the caudal part of the middle nasal meatus is divided by the scrolls of the second endoturbinate from the ethmoid bone into several air passages which lie between these scrolls.

The **ventral nasal meatus** *(meatus nasi ventralis)* is located between the ventral nasal concha and the dorsal surface of the hard palate. It is narrow rostrally as it leaves the nasal vestibule. It gradually wides caudally and attains a width of 1 cm. at the large nasomaxillary opening, where it continues ventral to the transverse lamina or floor plate of the ethmoid bone as the nasopharyngeal meatus. This wide portion of the nasal cavity is located in a transverse plane through the caudal portions of the fourth upper premolar teeth, where the middle, ventral, and common nasal meatuses converge.

The **common nasal meatus** (*meatus nasi communis*) is a longitudinal narrow space on either side of the nasal septum. Laterally, it is bounded by the nasal bone and ventral nasal concha. Above, below, and between these bones it is coextensive with the dorsal, middle, and ventral nasal meatuses, respectively.

The **nasopharyngeal meatus** (*meatus nasopharyngeus*) extends on either side from the caudal dilated portion of the ventral nasal meatus to the choana. It is a short passage with a much longer lateral than medial wall. It is bounded laterally by the maxillary and palatine bones, dorsally by the transverse lamina of the ethmoid bone, ventrally by the palatine bone, and medially by the vomer.

The *choanae* are the openings of the two nasopharyngeal meatuses into the nasal portion of the pharynx. They are oval in shape and oblique in position.

PARANASAL SINUSES

The paranasal sinuses, which are also connected with the respiratory passageways, are described with the skeletal system. They include a maxillary recess, a frontal sinus, and a sphenoidal sinus. The maxillary recess is not called a sinus because it is not enclosed in the maxilla. However, the opening to the recess, in life, is narrowed considerably by the mucosa-covered orbital lamina of the ethmoid bone and constitutes a functional cavity. The frontal sinus is divided into rostral, medial, and lateral compartments. The sphenoidal sinus of the dog is only a potential cavity because it is filled by an endoturbinate scroll.

NASAL MUCOSA

The various types of epithelia which line the nasal cavity and coat its associated structures are spoken of collectively as the nasal mucosa. The transition from the more peripheral respiratory type of epithelium to the deeper-lying olfactory epithelium is not abrupt. The receptors for the sense of smell

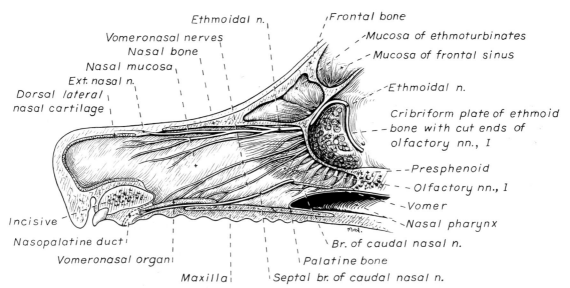

Figure 8-6. Sagittal section, showing distribution of nerves on the septal mucosa.

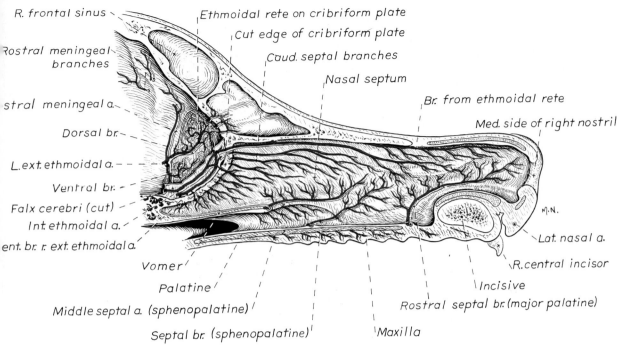

Figure 8–7.   Sagittal section, showing arteries of the nasal septum.

are located primarily on the ethmoturbinates, which lie in the caudomedial and caudodorsal parts of the nasal cavity. The olfactory epithelium and sensory nerves, so beautifully demonstrated by various techniques in several mammals by Bojsen-Møller (1964, 1967, 1975), are developed from the olfactory placode and include olfactory nerves, vomeronasal nerves, and the nervus terminalis. A richly vascularized and innervated mucosa (Figs. 8–6, 8–7) is present in the nasal cavity, vomeronasal organ, and in some mammals on a small area of the septum called the septal olfactory organ.

Under normal conditions of inspiration, the respiratory and olfactory currents of air are associated. When a dog deliberately wants to sample the environment, the nostrils are dilated, and with a forced inspiration the dog sniffs the air. This act provides a greater volume of inspired air, which takes a more dorsal course around the ethmoturbinates, where the olfactory receptors are most numerous.

GLANDS OF THE NOSE

The **lateral nasal gland** (*glandula nasalis lateralis*) (Fig. 8–8) is a serous gland which

is located in the mucosa of the maxillary recess near the opening of this recess into the nasal cavity.

The lateral nasal gland was first described in the dog by Steno in 1662 and more recently in other mammals by Bojsen-Møller (1964). The functional significance of the lateral nasal gland as part of the thermoregulatory system in the dog was postulated by Schmidt-Nielsen et al. (1970), based on airflow studies. This was confirmed by Blatt et al. (1972), who measured the amount of secretion correlated with heat-load and panting.

The lateral nasal gland is thickest at the level of the fourth upper premolar, where its ducts unite and pass rostrally to form one major duct which opens on the lateral wall of the vestibule (Fig. 8–8). The opening of the nasolacrimal duct into the ventral vestibule is visible externally, whereas the opening of the duct of the lateral nasal gland into the dorsal vestibule is hidden from view by the alar fold (Fig. 8–8). The lateral nasal duct can be entered by a cannula with a diameter of 0.038 to 0.048 mm. after surgical reflection of the dorsal wall of the vestibule. The opening lies dorsal and caudal to that of the nasolacrimal duct, where air flow is

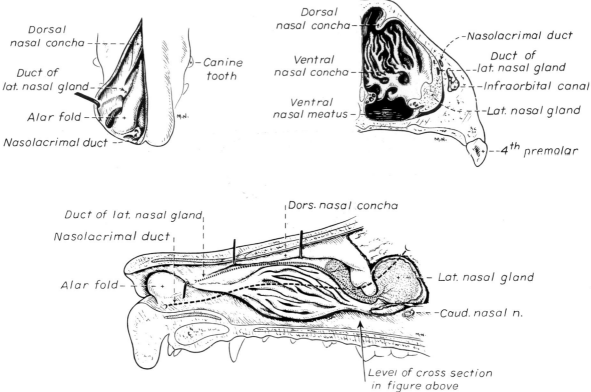

Figure 8–8. Topography of the lateral nasal gland and its duct in the dog. (From Evans, Anat. Rec. 1977.)

most rapid because of the narrowness of the passageway.

The lamina propria of the mucosa of the respiratory part also contains serous, mucous, and mixed tubuloalveolar glands. These glands are also present in the mucosa of the nasal vestibule. Goblet cells are present throughout the respiratory region, and olfactory glands which contain yellow pigment granules are located in the olfactory epithelium.

The **nasolacrimal duct** (*ductus nasolacrimalis*) carries the serous secretion from the conjunctival sac to the nasal vestibule. A characterisic of a healthy dog is a moist nose, which is maintained in part by the combined secretions of the lacrimal and lateral nasal glands. An accessory opening of the nasolacrimal duct may occasionally be found at the level of the canine tooth, lateral to the ventral nasal conchae (Nickel. Schummer, Seiferle, and Sacks 1973).

FUNCTIONAL CONSIDERATIONS

The rostral end of the nose in the dog is stiffened by cartilages and moved by the levator nasolabialis and levator labii superioris muscles. Dilation of the nostril alters the conformation of the nasal vestibule and changes the flow pattern of inspired air. This is most noticeable when the dog is presented with a stimulating scent, which, for analysis, must reach the deeper ethmoturbinals. There may also be a relationship between changes in air-flow pattern and the functioning of the vomeronasal organ.

Much evaporative cooling in the dog is accomplished by heat exchange from the lungs and respiratory passageways. Schmidt-Nielsen et al. (1970) suggested that the lateral nasal gland supplies the water necessary for heat dissipation during panting. This function of the lateral nasal gland is in a sense analogous to that of the sweat glands of man. Hyperventilation to get rid of excess heat can result in drying and hypertrophy of the respiratory mucosa unless moisture is provided. The laryngeal and nasal glands of mammals provide the necessary fluid. Blatt et al. (1972) cannulated the duct of the lateral nasal gland, subjected the dogs to various heat-loads, and collected the secretions. It was found that a rise from 25° to 40° C. results in an increased flow about 40 times as great.

The nasal cavity of the dog is not an empty antechamber, as in man, but rather a restricted residual cavity encircling the bulbous end of the alar fold, which extends into it. The alar fold is an extension of the ventral nasal concha which attaches to the wing of the nostril. Air entering the nostril is diverted dorsally, medially, and ventrally around the obstructing alar fold, with a resultant increase in velocity and evaporative effect.

Perhaps the alar fold or ridges in the vestibule act as a swell body to produce cyclic alternating changes in the air passageway in order to protect the nasal mucosa from constant desiccation. In humans there is a cyclic distention of the cavernous tissues in the conchae and septum which results in an alternating air flow through right and left nasal chambers. Bojsen-Møller and

Fahrenkrug (1971) found a similar nasal cycle in rabbits and rats.

Nasal Portion of Pharynx

The **nasal portion of the pharynx** *(pars nasalis pharyngis)*, also called the **nasal pharynx** or **nasopharynx,** extends from the choanae to the intrapharyngeal ostium. The **intrapharyngeal ostium** *(ostium intrapharyngeum)*, formerly called the pharyngeal isthmus, is formed cranial to the larynx by the crossing of the digestive passageway (oral pharynx) and the respiratory passageway (nasal pharynx) (Fig. 8–9). The rostral part of the nasal pharynx is bounded by the hard palate ventrally, the vomer dorsally, and the palatine bones bilaterally. Although the middle and caudal portions of the pharynx are bounded dorsally by the base of the skull and the muscles which attach to it, its ventral boundary is the mobile, long soft palate (Fig. 8–10). At each act of swallowing, the cavity of the caudal part of the nasal pharynx is obliterated by the pressure of the material swallowed and the root of the tongue forcing the soft palate dorsally.

On each lateral wall of the nasal pharynx, above the middle of the soft palate, is an oblique slitlike opening, about 5 mm. long,

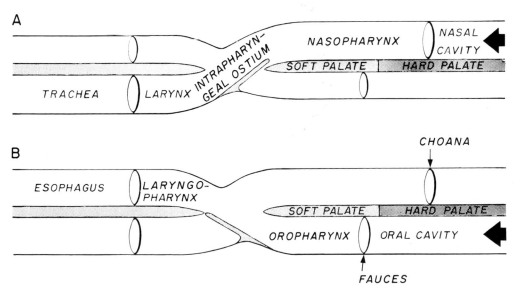

Figure 8–9. Diagram showing relation of portions of pharynx to esophagus and trachea.
   A. During normal respiration.
   B. During deglutition.

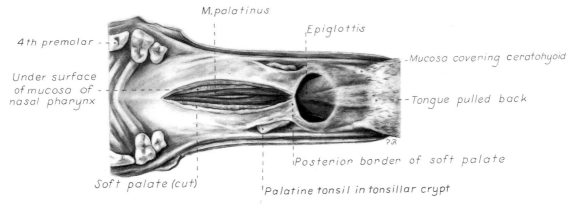

Figure 8-10.   Soft palate and epiglottis, ventral aspect.

which is the **pharyngeal opening of the auditory tube** *(ostium pharyngeum tubale auditivae)*. The opening is located directly caudal to the caudal border of the pterygoid bone and faces cranioventrally.

The **auditory tube** *(tuba auditiva)*, or eustachian tube, extends between the cavity of the middle ear and the cavity of the nasal pharynx. It serves to equalize the atmospheric pressure on the two sides of the tympanic membrane.

LARYNX

The larynx is a musculocartilaginous organ guarding the entrance to the trachea, which serves as an air passageway, aids vocalization, and prevents the inspiration of foreign material. The valvular function of the larynx, by means of the epiglottis, is vital, since it is across its inlet that all substances swallowed must pass in their course from the oral pharynx through the laryngeal pharynx to the esophagus. Negus (1949) has described and illustrated the comparative anatomy of the larynx from fish through mammals and Piérard (1965) studied the dog and other carnivores. The larynx is located directly caudal to the root of the tongue and the soft palate, ventral to the atlas. It is about 6 cm. long in a medium-sized dog, nearly half of this length being occupied by the epiglottic cartilage, which lies in front of the laryngeal opening. The intrinsic muscles of the larynx control the size of the laryngeal inlet, the size and

shape of the glottis, and the positions of the laryngeal cartilages.

CARTILAGES OF THE LARYNX

The **laryngeal cartilages** *(cartilagines laryngis)* (Figs. 8–11 to 8–18) are the epiglottic, thyroid, cricoid, arytenoid, sesamoid, and interarytenoid cartilages. Only the arytenoid cartilage is paired.

The **epiglottic cartilage** *(cartilago epiglottica)* forms the basis of the epiglottis. In outline, the rostral margin of the cartilage forms a thin, dorsally concave triangle with its **apex** pointing forward. The **epiglottis** resembles a sharp-pointed spade. Its dorsocaudal surface, known as the **aboral surface** *(facies aboralis)*, is concave. The opposite surface, called the **oral surface** *(facies oralis)* because it faces the oral pharynx, is convex. The oral surface is attached to the middle of the body of the hyoid bone by the short, stout hyoepiglottic muscle. On either side of the median mucous fold which covers the muscle is a deep pocket of mucosa, called the *vallecula,* which may attain a depth of 1.5 cm. Each vallecula is limited laterally by a small fold of stratified squamous epithelium running from the oral surface of the epiglottis near its caudolateral angle to the lateral wall of the laryngeal part of the pharynx. The **stalk** of the epiglottis *(petiolus epiglottidis)* is in the form of a thickened handle of fibrous tissue which unites the midcaudal portion of the epiglottis and the dorsal rostral surface of the thyroid cartilage. The normal position of the epiglottis allows the apex to rest above the soft palate.

The **thyroid cartilage** *(cartilago thyroidea)* is the largest cartilage of the larynx. It forms the middle portion of the laryngeal skeleton and is open dorsally. It consists of **right and left laminae** *(lamina dextra et sinistra)*, which are united ventrally to form a short but deep trough. An inconspicuous **oblique line** *(linea obliqua)* serves primarily for the insertion of the sternothyroid muscle. Each lamina is expanded dorsally to form transversely thin processes, the **rostral and caudal cornua** *(cornu rostralis et caudalis)*. Separating the rostral and caudal thyroid cornua from the thyroid laminae on each side are the **rostral and caudal thyroid notches** *(incisura thyroidea rostralis et caudalis)*, respectively. The cranial laryngeal nerve and the laryngeal artery pass through the rostral thyroid notch. Where the two laminae fuse ventrally, a slight **laryngeal prominence** *(prominentia laryngea)* is formed. The laryngeal prominence, known as the "Adam's apple" in man, is not visible externally in the dog, but it can be palpated.

The caudal border of the thyroid cartilage possesses the median deep **caudal thyroid notch** *(incisura thyroidea caudalis)*, whereas the cranial border is slightly convex from side to side. The caudal border of the thyroid cartilage is united to the ventral arch of the cricoid cartilage by the **cricothyroid ligament** *(ligamentum cricothyroideum)*. The rostral border is joined to the thyrohyoid bones by the **thyrohyoid membrane** *(membrana thyrohyoidea)*.

The **cricoid cartilage** *(cartilago cricoidea)* is the only cartilage of the larynx which forms a complete ring. The dorsal portion is approximately five times wider than the ventral portion. The expanded dorsal part is the **lamina of the cricoid cartilage** *(lamina cartilaginis cricoideae)*. It possesses a **median crest** *(crista mediana)* for muscle attachment. Occasionally a pair of vascular foramina are located in the lamina, one on each side of the cranial portion of the crest. The **arch of the cricoid cartilage** *(arcus cartilaginis cricoideae)* extends ventrally from the lamina and completes the enclosure of the caudal part of the cavity of the larynx. It is bilaterally concave in a transverse direction. The cricoid cartilage possesses two

pairs of articular facets. An indistinct pair of facets, for articulation with the apices of the caudal cornua of the thyroid cartilage, are located at the junction of the lamina and the arch about 1 mm. from the caudal border. A more prominent pair of facets, for articulating with the arytenoid cartilages, are located on the rostral border of the lamina on each side of the median crest. Both pairs of facets are enclosed in articular capsules and form synovial joints with the cartilages with which they articulate. The sides of the ventral arch are gradually reduced in width ventrally. Midventrally, in a medium-sized dog, the narrowest part of the arch is only 5 mm. long; in such a dog the mid-dorsal lamina would be approximately 2 cm. long.

The **arytenoid cartilage** *(cartilago arytenoidea)* is an irregular cartilage, one on either side, which articulates with the rostrodorsal border of the cricoid cartilage. When the laryngeal cartilages are viewed laterally, the arytenoid is largely hidden from view by the thyroid lamina.

The morphology of the arytenoid cartilage varies greatly in different species of mammals, so that what may appear as a process of the arytenoid in one species may be a separate cartilage in another. In the dog, the arytenoid cartilage embodies the corniculate cartilage and the cuneiform cartilage of other mammals. As a result, this compound cartilage may be described as possessing a corniculate process, a muscular process, a vocal process, and a cuneiform process.

The **articular surface** *(facies articularis)* is a slightly oval, concave facet on the caudal border of the arytenoid, which faces caudomedially and receives the cricoid articular facet, to form the cricoarytenoid articulation *(articulatio cricoarytenoidea)*.

The **muscular process** *(process muscularis)* is a relatively thick, rounded process which is located directly lateral to the articular surface. The m. cricoarytenoideus dorsalis inserts on this process.

The **corniculate process** *(processus corniculatus)* is the longer and more caudal of the two dorsal processes which form the dorsal margin of the laryngeal inlet. In man and several other mammals it exists as a separate cartilage *(cartilago corniculata)*, sometimes spoken of as Santorini's cartilage.

The **vocal process** *(processus vocalis)* is a
Text continued on page 524

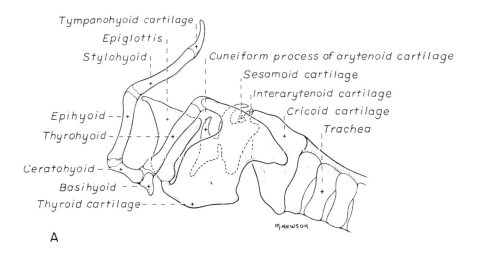

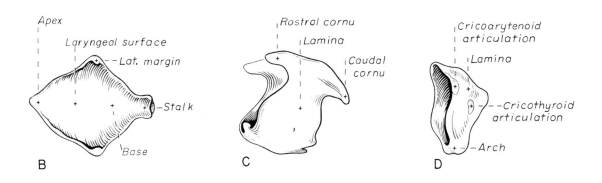

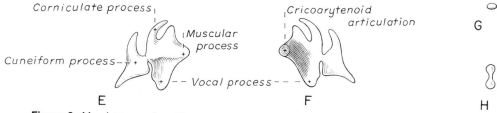

Figure 8-11.   Laryngeal cartilages.
   A. Scheme of laryngeal cartilages and hyoid apparatus, lateral aspect.
   B. Epiglottis, dorsal aspect.
   C. Thyroid cartilage, lateral aspect.
   D. Cricoid cartilage, lateral aspect.
   E. Arytenoid cartilage, lateral aspect.
   F. Arytenoid cartilage, medial aspect.
   G. Interarytenoid cartilage.
   H. Sesamoid cartilage, dorsal aspect.

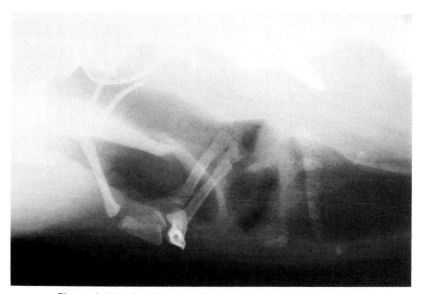

Figure 8–11. Radiograph. The hyoid apparatus and larynx.

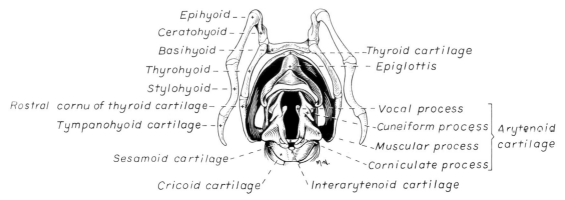

Figure 8–12. Laryngeal cartilages and hyoid apparatus, dorsal aspect.

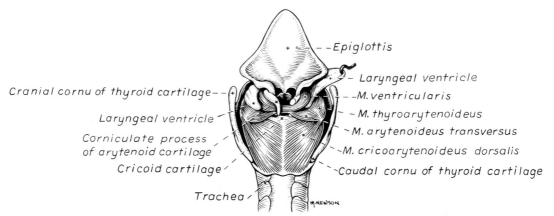

Figure 8–13. Laryngeal muscles, dorsal aspect. (The right corniculate cartilage has been cut and the right laryngeal ventricle reflected.)

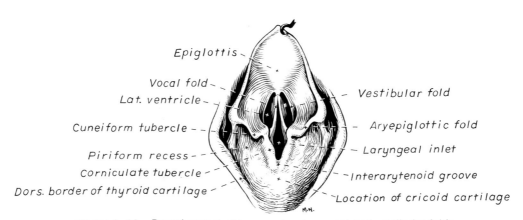

Figure 8–14. Dorsal aspect of larynx, showing vocal and vestibular folds.

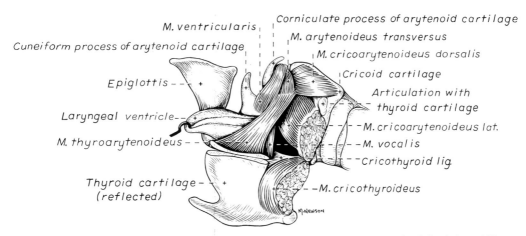

Figure 8–15. Laryngeal muscles, lateral aspect. (The thyroid cartilage is cut to the left of the midline and re-flected.)

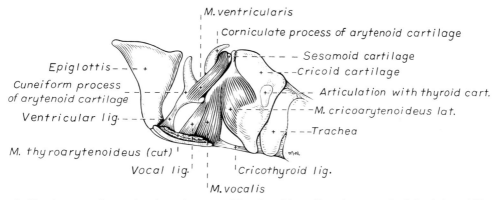

Figure 8–16. Laryngeal muscles, lateral aspect. (The thyroid cartilage is cut to the left of the midline and removed. The mm. thyroarytenoideus, arytenoideus transversus, and cricoarytenoideus dorsalis have been removed.)

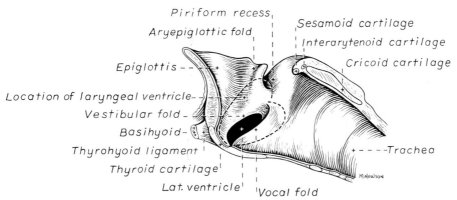

Figure 8–17. Midsagittal section of the larynx. (The dotted lines show the extent of the laryngeal ventricle and the lateral ventricle.)

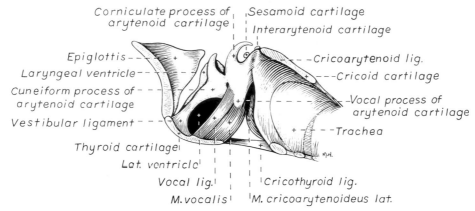

Figure 8–18. Midsagittal section of the larynx. (The mucosa has been removed to expose muscles and ligaments.)

caudal ventral projection of the arytenoid cartilage. It is about 3 mm. thick and 5 mm. long at its base. The vocal ligament and the m. vocalis, from the thyroid cartilage, attach to the vocal process (Fig. 8–18).

The **cuneiform process** (*processus cuneiformis*) is the most rostral portion of the arytenoid cartilage. It is connected by a narrow neck (Fig. 8–11) to the main portion of the arytenoid cartilage, and is considered by some authors to be a separate cartilage. It exists as an inconstant separate cartilage in man (cartilago cuneiformis), sometimes spoken of as Wrisberg's cartilage. In the horse it is fused to the epiglottic cartilage rather than to the arytenoid, and in the ox it is lacking. In the dog, the cuneiform process is roughly triangular in shape. The ventral portion lies in the aryepiglottic fold, and the dorsal portion aids in forming the laryngeal inlet. Attached to the cuneiform process are the ventricular ligament and the m. ventricularis (Fig. 8–16). Duckworth (1912) suggested that the cuneiform cartilage arose in mammals from the lateral margin of the epiglottis.

The **sesamoid cartilage** (*cartilago sesamoidea*) (Fig. 8–12) is an oval or dumbbell-shaped nodule located cranial to the cricoid lamina and between the arytenoid cartilages. It is occasionally paired, in which case an intersesamoid ligament or fibrous union joins the two. Primarily, the sesamoid cartilage appears to be intercalated in the transverse arytenoid muscle. There is frequently a small contact surface with the dorsal portion of each arytenoid cartilage.

The **interarytenoid cartilage** (*cartilago interarytenoidea*) is small, flat, and easily overlooked. It lies cranial to the cricoid lamina and caudodorsal to the transverse arytenoid muscle and sesamoid cartilage. In this superficial position the interarytenoid cartilage is embedded in connective tissue which attaches the arytenoid cartilages and the cricoesophageal tendon to the cricoid lamina.

CAVITY OF LARYNX AND LARYNGEAL MUCOSA

The **cavity of the larynx** (*cavum laryngis*) is divided into three transverse segments: the vestibule, or antechamber, which opens in the pharynx by the aditus laryngis; a middle, narrow portion, called the glottis; and the infraglottic cavity, which lies caudal to the glottis.

The **laryngeal opening** (*aditus laryngis*) lies directly caudal to the intrapharyngeal ostium. Air entering or leaving the larynx can travel either by way of the nasal part, or by way of the oral part of the pharynx. In lolling (rapid breathing with the tongue hanging out), most of the air passes through the mouth and oral pharynx; in slow, shallow breathing it passes through the nasal cavity and nasal pharynx. The margin of the laryngeal opening forms an imperfect triangle, with the base located caudally. The margin of the epiglottis forms its lateral boundaries and apex. The caudal boundary is formed by the right and left aryepiglottic folds.

Each **aryepiglottic fold** *(plica aryepiglottica)* (Fig. 8–14) runs from the dorsal portion of the arytenoid cartilage and the closely associated corniculate process to the caudolateral angle of the epiglottic cartilage. Two prominent tubercles which are separated by a deep notch are present in the fold. The more dorsocaudal tubercle is formed by the underlying corniculate process and is called the **corniculate tubercle.** It is a rounded process approximately 1 cm. long by 0.5 cm. wide. It and the opposite tubercle form the most dorsal and the least expandable part of the laryngeal opening. Ventral to these tubercles the dorsal parts of the arytenoid cartilages lie close together to form the **interarytenoid groove** *(incisura interarytenoidea).* This groove communicates with the middle rostral portion of the laryngeal pharynx. The **cuneiform tubercle** *(tuberculum cuneiforme)* forms a relatively heavy cone-shaped projection which is united in a sagittal plane with the corniculate tubercle; it is also united in a transverse plane to each lateral angle of the epiglottic cartilage by means of a loose plica of mucosa. The channel lying external to each aryepiglottic fold is called the **piriform recess** *(recessus piriformis).* When food or fluid is forced past the closed laryngeal opening in deglutition, it largely occupies these recesses, which constitute the ventrolateral portions of the laryngeal pharynx.

The **laryngeal vestibule** *(vestibulum laryngis)* extends from the laryngeal opening to the ventricular folds. It is a funnel-shaped cavity which opens freely dorsorostrally. It is bounded ventrally by the mucosa covering the large, dorsally concave epiglottis. The cranial part of the wall on either side is also composed of the epiglottis. Dorsorostral to the ventricular folds the flattened cuneiform processes form its wall. The **rim of the vestibule** *(rima vestibuli)* is the caudal opening of the vestibule. It is bounded bilaterally by the vestibular folds and the mucosa covering the cuneiform processes. The ventral boundary is formed by the mucosa covering the thyroid cartilage directly caudal to the thyroepiglottic ligament.

The **vestibular** (ventricular) **fold** *(plica vestibularis)* (Fig. 8–17) is a short, wide plica of mucosa, containing a few elastic fibers, which runs from the expanded ventral margin of the cuneiform cartilage to the cranial dorsal surface of the thyroid cartilage. It is less than one-half the dorsoventral diameter of the larynx, and its width approximates that of the vocal fold which lies caudal to it.

The **glottis** consists of the paired arytenoid cartilages dorsally and the paired vocal folds ventrally which form a narrow passageway into the larynx, which is called the *rima glottidis.* The portion of the rima glottidis between the vocal folds is the **intermembranous part** *(pars intermembranacea)*; that between the medial surfaces of the arytenoid cartilages is the **intercartilaginous part** *(pars intercartilaginea).* The rima glottidis is the most important part of the larynx, from the standpoint of veterinary medicine, because it is the narrowest part of the laryngeal passageway, and it contains the vocal folds which are necessary for vocalization — barking, baying, whining, and growling.

The **vocal fold** *(plica vocalis)* (Fig. 8–17), on either side, extends from the vocal process of the arytenoid cartilage to the dorsocaudal part of the trough of the thyroid cartilage. It is approximately 13 mm. long and 6 mm. wide in a medium-sized dog. It is separated from the vestibular fold by the slitlike opening of the lateral ventricle. The **vocal ligament** *(lig. vocale)* is a strap of elastic fibers which is enclosed in the vocal fold. It forms the supporting framework for the cranial border of the vocal fold and is covered by mucous membrane. It measures 1 to 2 mm. in maximum thickness and has a thin cranial border. Caudally, it is continuous with the m. vocalis, which is a portion of the thyroarytenoideus muscle mass.

The **ventricle of the larynx** *(ventriculus laryngis)* includes the slight dorsoventral depression between the vestibular and vocal folds, as well as the saccule which lies medial to the thyroid cartilage and lateral to the vocal and vestibular folds. The mucosa of the laryngeal ventricle is continuous with that of the rima glottidis and rima vestibuli through an opening which maintains a constant width of about 1.5 mm. and a length equal to that of the rostral border of the vocal fold which forms the caudal border of the opening. In the production of sound the vocal and vestibular folds can vibrate into the cavity of the glottis.

The **infraglottic cavity** *(cavum infraglotti-cum)* of the larynx extends from the rima glottidis to the cavity of the trachea. It constitutes a third subdivision of the larynx. The infraglottic cavity is wide dorsally, corresponding to the lamina of the cricoid cartilage, and narrow ventrally. Because of this disparity in the lengths of the dorsal and the ventral wall of the infraglottic cavity, the cavity of the larynx is in nearly a frontal plane, whereas that of the trachea is in an oblique caudoventral plane.

Pressman and Kelemen (1955), in their review of laryngeal physiology, consider the comparative functional anatomy responsible for the sphincter action of the larynx. Closure or constriction of the larynx may be accomplished at three levels: at the inlet (aryepiglottic folds) by muscular action during deglutition; at the vestibular (ventricular) folds by passive action for creating intrathoracic or intra-abdominal pressure; and at the vocal folds by muscular action during phonation. All three levels may be involved in normal closure, or only one level may be.

INNERVATION OF THE LARYNX

The nerves of the larynx in the dog (Fig. 8–19) have been reported upon by Franzmann (1907), Lemere (1932a, b and 1933), Vogel (1952), and Bowden and Scheuer (1961).

The **cranial laryngeal nerve** leaves the vagus at the level of the nodose ganglion. It divides into an external branch, which supplies the m. cricothyroideus, and an internal branch, which receives fibers from the mucosa of the larynx cranial to the vocal folds. The internal branch usually anastomoses *(ramus anastomoticus)* with the caudal laryngeal nerve. A nerve from the pharyngeal plexus (pharyngeal ramus of the vagus or medial laryngeal nerve) may also supply the m. cricothyroideus. The distribution of sensory fibers in the dog and man is similar, according to Pressman and Kelemen (1955).

The **caudal laryngeal nerve** is the motor supply to all of the intrinsic muscles of the larynx except the m. cricothyroideus. It is the terminal segment of the recurrent laryngeal nerve, which in turn originates in the thorax by leaving the vagus. Réthi (1951)

reports that cutting the intracranial portion of the accessory nerve results in degeneration of the vast majority of the fibers in the recurrent nerve. Cutting the intracranial portion of the vagus leaves the motor fibers of the recurrent nerve intact, whereas the motor component of the cranial laryngeal nerve shows complete degeneration. Section of the recurrent trunk or of the cranial laryngeal trunk is followed by degeneration similar to that which follows intracranial nerve severance. Several investigators — (Lemere (1932a), Piérard (1963), and Watson (1974) — describe an inconstant pararecurrent nerve which leaves the recurrent near its origin in the thoracic cavity, parallels the recurrent nerve along the trachea, and has variable anastomoses along its course. It is continuous cranially with the ramus anastomoticus of the cranial laryngeal nerve.

TRACHEA

The trachea runs from a transverse plane through the middle of the body of the axis to a similar plane through the fibrocartilage between the fourth and fifth thoracic vertebrae. It is a relatively non-collapsible tube which extends from the cricoid cartilage of the larynx to its **bifurcation** *(bifurcatio tracheae)* dorsal to the cranial part of the base of the heart. The crest of the partition at the side where the trachea divides into the two principal bronchi is called the **tracheal carina** *(carina tracheae)*. Approximately 35 C-shaped hyaline **tracheal cartilages** *(cartilagines tracheales)* form the skeleton of the trachea. The space left by failure of these cartilages to meet dorsally is bridged by fibers of the smooth, transversely running **tracheal muscle** *(musculus trachealis)* and connective tissue. The rings so formed are united in a longitudinal direction by bands of fibroelastic tissue, called the **annular ligaments of the trachea** *(ligg. anularia [trachealia])*. They are about 1 mm. wide, compared with the 4 mm. width of each tracheal cartilage. The annular ligaments allow considerable intrinsic movement of the trachea without breakage or collapse of the tube. Macklin (1922) reconsidered the network of longitudinal elastic fibers in the tunica propria of the

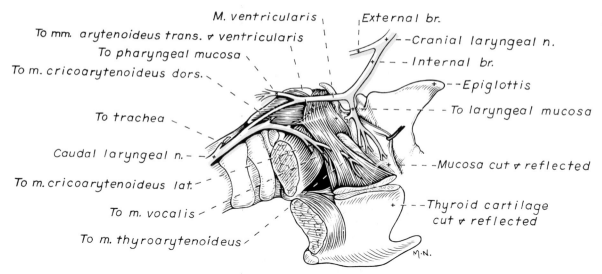

M. ventricularis

To mm. arytenoideus trans. ↋ ventricularis

To pharyngeal mucosa

To m. cricoarytenoideus dors.

External br.

Cranial laryngeal n.

Internal br.

Epiglottis

To trachea

To laryngeal mucosa

Caudal laryngeal n.

To m. cricoarytenoideus lat.

To m. vocalis

To m. thyroarytenoideus

Mucosa cut ↋ reflected

Thyroid cartilage cut ↋ reflected

M. N.

Figure 8–19.   Distribution of the laryngeal nerves. Lateral aspect.

trachea and bronchial tree. This elastic membrane beneath the epithelium is thick in the trachea and larger bronchi but thin in the terminal respiratory passageways. It provides a recoil mechanism for the lung.

BRONCHIAL TREE

The bronchial tree (Figs. 8–20, 8–21) begins at the bifurcation of the trachea by the formation of a right and a left **principal bronchus** *(bronchus principalis [dexter et sinister])*. Each principal bronchus divides into **lobar bronchi** *(bronchi lobares)*, which are also known as secondary bronchi. These supply the various lobes of the lung, and are named according to the lobe supplied. Within the lobe of the lung the lobar bronchi divide into **segmental bronchi** *(bronchi segmentales)*, which are sometimes referred to as tertiary bronchi. The segmental bronchi and the lung tissue which they ventilate are known as **bronchopulmonary segments** *(segmenta bronchopulmonalia)*. Adjacent bronchopulmonary segments normally communicate with each other in the dog. Various injection and reconstruction techniques have been employed to delineate these segments in the dog (Angulo et al. 1958, Tucker and Krementz 1957, Boyden and Tompsett 1961, and Kilpper and Stidd 1973).

The segmental bronchi usually branch dichotomously (Miller 1937) into small bronchi. This process of branching continues until the respiratory bronchioles are formed. The bronchi are cylindrical tubes which are kept by flattened, overlapping, curved cartilages. The cartilaginous elements end when the diameter of the terminal bronchiole is reduced to 1 mm. or less. In addition to the cartilages, the bronchioles have spiral bands of smooth muscle in their walls which continue peripherally on the respiratory (alveolar) bronchioles.

The **respiratory bronchioles** *(bronchioli respiratorii)* give rise to **alveolar ducts** *(ductuli alveolares)*, **alveolar sacs** *(sacculi alveolares)*, and **pulmonary alveoli** *(alveoli pulmonis)*. The respiratory portion of the bronchial tree in the dog, including the arteries, veins, and lymphatics, was modeled and described by Miller (1900, 1937). Boyden and Tompsett (1961), in their excellent paper on the postnatal growth of the lung in the dog, point out the substantial reduction postnatally in the number of non-respiratory branches of both the axial bronchus and the peripheral bronchioles. At birth, many non-respiratory peripheral bronchioles lined by cuboidal epithelium are converted into respiratory units. Concomitantly, new alveoli and alveolar sacs arise along the terminal bronchioles. Boyden and Tompsett conclude that the number of non-respiratory

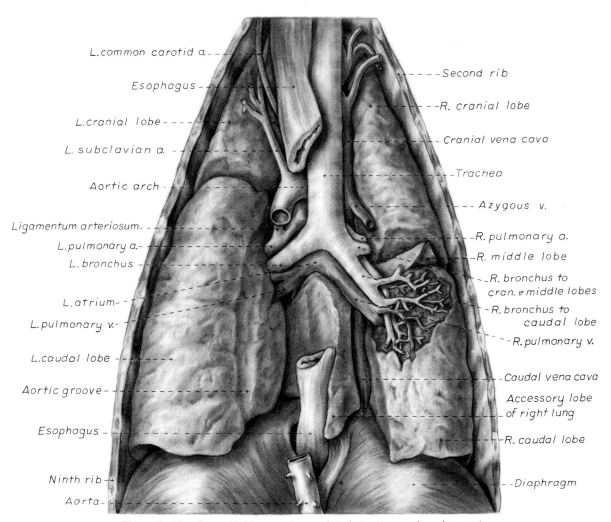

Figure 8–20. Bronchial tree and associated structures, dorsal aspect.

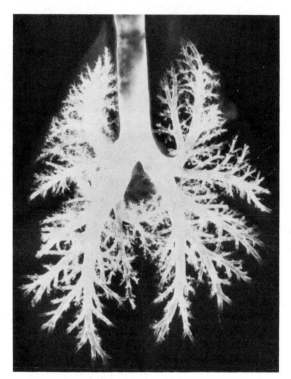

Figure 8–20. Radiograph. The bronchial tree.

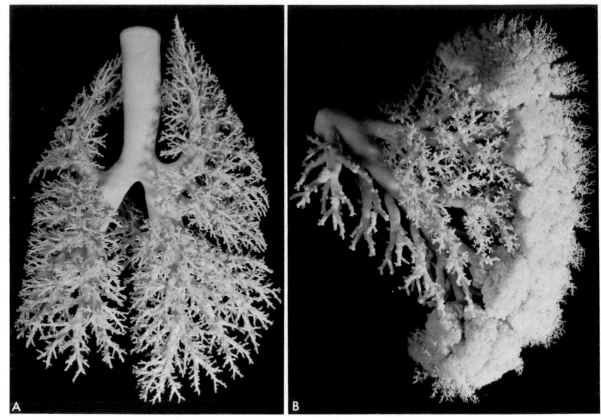

Figure 8-21. A. Silicone cast of the bronchial tree of a Beagle pruned through the terminal bronchioles.
B. Silicone cast of the bronchial tree, differentially pruned.
(From Kilpper and Stidd 1973.)

branches on the axial stem (dorsocaudal bronchus of the intermediate lobe) is reduced from 38 rami before birth to 32 at maturity.

Loosli (1937) studied the microscopic structure of adult alveoli in several species of mammals, including the dog. He described and illustrated interalveolar communications in both normal and pathologic lungs, and pointed out the significance of these pores in the normal mechanism of respiration and in the spread of infection.

THORACIC CAVITY AND PLEURAE

To obtain a clear understanding of the lungs and how they function passively in the mechanical act of breathing, it is necessary first to understand the morphology of the thoracic cavity and its lining membrane, the pleurae.

THORACIC CAVITY

The **thoracic cavity** *(cavum thoracis)* (Fig. 8-22), in a narrow sense, is bounded by the subserous endothoracic fascia. In a wider sense, its walls are formed by the muscles, bones, and ligaments of the thoracic wall.

The **endothoracic fascia** *(fascia endothoracica)* is the areolar tissue which attaches the costal and diaphragmatic pleurae to the underlying muscles, ligaments, and bone. The endothoracic fascia is scanty where it closely attaches the costal pleura to the ribs. Dorsally and ventrally it dips into the mediastinal space and becomes the connective tissue that invests the organs and other structures which lie in the mediastinum.

Cranially it passes through the thoracic inlet and is continued into the neck. It blends with the deep cervical fascia, particularly with the prevertebral portion of this fascia.

The thoracic wall is formed bilaterally by the ribs and the intercostal muscles, and dorsally by the bodies of the thoracic vertebrae and the intervening fibrocartilages. Cranial to the sixth thoracic vertebra the right and left longus colli muscles cover the thoracic vertebral bodies and lie directly under the pleura and endothoracic fascia. Ventrally, the narrow sternum and the paired flat transversus thoracis muscles contribute to the thoracic wall. Caudally, the base of the thoracic cavity is formed by the dome-shaped, obliquely placed, musculotendinous diaphragm.

The shape of the thoracic cavity of the dog differs greatly from that in man. Its walls are laterally compressed, with the result that in dogs of usual proportions its average dorsoventral dimension is greater than either the average lateral or craniocaudal measurement. Cranially, the thoracic cavity opens to the exterior at the thoracic inlet. The external contour of the thorax differs from the internal limits of the thoracic cavity. In cross section the thorax is roughly oval in shape, wider above than below, and long dorsoventrally. A cross section of the thoracic cavity cranial to the diaphragm is heart-shaped, with the apex located ventrally. The base, located dorsally, is widened to accommodate the epaxial muscles, thoracic vertebral bodies, aorta, and smaller associated structures. The thorax is a laterally compressed cone with a base (diaphragm) which is convex cranially. The greatest cranial encroachment of the diaphragm is to a transverse plane through the sixth intercostal spaces and about 5 cm. dorsal to the sternum. In addition to being convex, the diaphragm is oblique in position; the most cranial dorsal attachment is approximately eight vertebral segments caudal to its most cranial ventral attachment. From an origin on the inner surfaces of the ribs and costal cartilages, the diaphragm bilaterally extends almost directly cranially, forming a slitlike space or recess. The recess is formed by the diaphragm centrally, and the ribs and intercostal structures peripherally. In normal respiration the diaphragm undergoes a craniocaudal excursion of about 1½ vertebral segments, yet even in forced inspiration the margin of the lungs never completely invades the recesses so created. In a beagle-type dog the diaphragm protrudes forward from its costal attachment for a maximum distance of 6 cm. For the details of the intrinsic structure of the diaphragm, see Figures 6–35 and 6–36. The thoracic cavity contains the lungs, heart, thymus gland, and lymph nodes. The structures which partly or completely traverse the thoracic cavity are the aorta, and cranial vena cava and caudal vena cava, azygos and hemiazygos veins, thoracic duct and smaller lymph vessels, esophagus, and vagal, phrenic, and sympathetic nerves.

The **thoracic inlet** (*apertura thoracis cranialis*) is the roughly oval opening into the cranial part of the thoracic cavity. It is bounded bilaterally by the first pair of ribs and their cranially extending costal cartilages. In a beagle-type dog its dorsoventral dimension is about 4 cm. Its greatest width is about one-fourth less than its dorsoventral dimension. The aperture is wider dorsally than ventrally. The first thoracic vertebra and the paired longus colli muscles bound the thoracic inlet dorsally; the manubrium of the sternum bounds it ventrally. Traversing the aperture are the trachea, esophagus, vagosympathetic nerve trunks, recurrent laryngeal nerve, phrenic nerve, first two thoracic nerves, and several vessels. The apices of the pleural cavities lie in the thoracic inlet.

MEDIASTINUM

The mediastinum is the space between the right and left pleural sacs which encloses the thymus, heart, aorta, trachea, esophagus, the vagus nerves, and other nerves and vessels. It is divided by the heart into three transverse divisions. The portion of the mediastinum lying in front of the heart is known as the **cranial mediastinum** (*mediastinum craniale*). The portion containing the heart is called the **middle mediastinum** (*mediastinum medium*). The **caudal mediastinum** (*mediastinum caudale*) is that part of the mediastinum lying caudal to the heart. In man the mediastinum contains a considerable amount of collagenous tissue and is sufficiently strong so that one lung can be collapsed independently of the

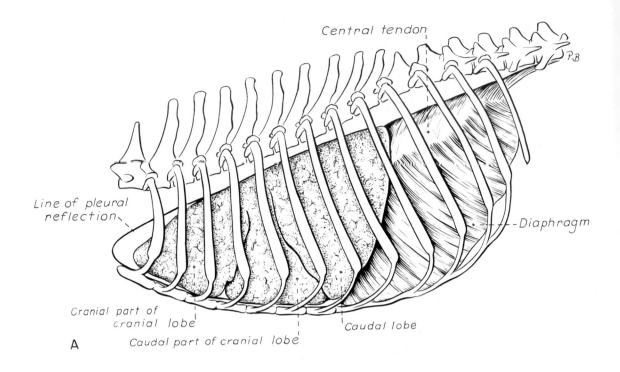

A

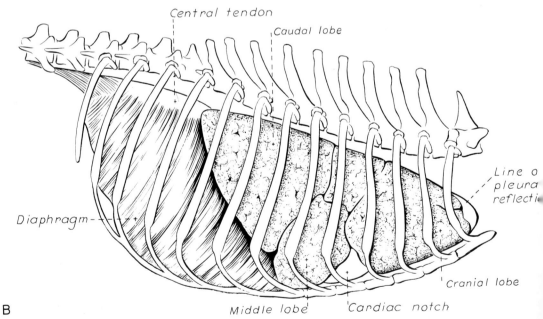

B

Figure 8–22. Thoracic cage and lungs (lungs hardened in situ).
A. Left side.
B. Right side.

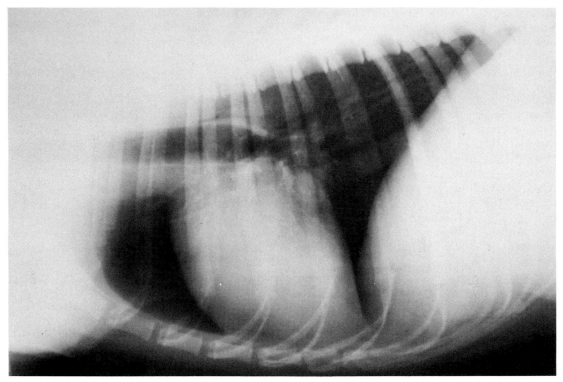

Figure 8–22. Radiograph. Lateral radiograph of the thorax.

other. In the dog the tissue in the mediastinum is extremely scanty, but the pleura which covers it is not fenestrated. In all mammals the mediastinum contains the esophagus, trachea, heart, and the nerves and vessels which enter or leave the heart. Only the caudal vena cava, certain lymph vessels, and the right phrenic nerve have separate folds as they partly or completely traverse the thoracic cavity. The thymus, trachea, and thoracic duct are located primarily in the cranial portion.

For purposes of description, the mediastinum may also be divided into dorsal and ventral portions by a frontal plane passing through the roots of the lungs. In the dog the ventral surface of the heart is attached to the sternum by the folds of pleura which extend from the thoracic inlet to the diaphragm. The minute space located between these portions of the pleurae may be called the **infracardiac bursa** *(cavum mediastini serosum [bursa infracardiaca]).* The cranial portion of the ventral mediastinum is occupied by the thymus in growing dogs. It also contains the paired internal thoracic arteries and veins.

PLEURAE

The pleurae are the serous membranes which cover the lungs, line the walls of the thoracic cavity, and cover the structures in the mediastinum, or in places form the mediastinum. The pleurae form two complete sacs, one on either side, which are known as the pleural cavities.

Each **pleural cavity** *(cavum pleurae)* in life is essentially only a potential cavity, because it contains only a capillary film of fluid which moistens the flat mesothelial cells paving its surface. Except for this capillary fluid, the visceral pleura of the lungs, or pulmonary pleura, lies in contact with the wall or parietal pleura. Only when gas (air) or fluid collects between the pulmonary and parietal pleurae and prevents a lung from expanding does it exist as a real cavity. The right pleural cavity is larger than the left because of displacement of the postcardial mediastinal wall to the left side. The pleural cavities do not communicate with each other, although their medial walls and the tissue between them are poorly developed.

For purposes of description, the pleura is designated as the parietal and the pulmonary pleura.

The **parietal pleura** *(pleura parietalis)* forms the walls of the pleural cavities. It is further designated as **costal, mediastinal, and diaphragmatic.**

The **costal pleura** *(pleura costalis)* is that portion of the pleura which attaches to the inner surfaces of the lateral walls of the thoracic cavity. The true costal part is firmly adherent to the inner surfaces of the ribs and is thin. The costal pleura, which covers the inner surfaces of the intercostal muscles, the aorta, and related structures, is thicker and less firmly attached to them. The pleura and the underlying endothoracic fascia possess considerable elasticity.

The **mediastinal pleura** *(pleura mediastinalis)* may be divided into four parts to facilitate description. These are the ventral, cranial (precardial), middle (pericardial), and caudal (postcardial) mediastinal pleurae.

The **ventral mediastinal pleura** is composed of the ventral portions of the three other parts of the mediastinal pleura. It is much more delicate and expansive than the comparable part in man, which is restricted to the area ventral to the heart. In old dogs it extends as a loose midventral fold from the thoracic inlet to the diaphragm. Ventrally, it attaches to the sternum. The pleural reflections which form the ventral mediastinal pleura continue from the dorsum of the sternum on the transversus thoracis muscles to form the right and the left **costomediastinal recess** *(recessus costomediastinalis).* The ventral borders of the various lobes of lungs cranial to the diaphragm lie in these recesses. The cranial part of the ventral mediastinal pleura continues dorsally as the precardial mediastinal pleura. In young dogs it covers the ventral portion of the thymus gland, and in all dogs an elevation of the pleura is produced bilaterally by the paired internal thoracic vessels before these vessels disappear ventral to the transversus thoracis muscles. The cranial portion of the ventral mediastinal pleura averages slightly over 1 cm. in width in beagle-sized dogs. Because

of its loose attachments, it may be increased two or three times this size without rupturing. It extends from the third to the sixth sternebra ventrally. Dorsally, the mediastinal pleural sheets separate and enclose the fibrous pericardium as the middle mediastinal pleura. The ventral mediastinal pleura widens caudal to the heart as it forms a sagittally triangular sheet. Ventrally, it attaches to the dorsum of the sternum caudal to the sixth sternal segment. Dorsally, it attaches to the diaphragm.

The **cranial mediastinal pleura** covers most of the thymus gland, which is large in young dogs. In old dogs fatty remnants of the gland and vessels persist. Coursing between the pleural sheets in a caudoventral direction are the right and left internal thoracic vessels. Dorsally the pleural leaves separate to cover the precava, brachiocephalic artery, thoracic duct, and the phrenic, vagal, recurrent laryngeal, cardiosympathetic, and cardiovagal nerves. The sternal and the cranial mediastinal lymph nodes also lie in the cranial mediastinal space. Above the trachea and esophagus the pleural leaves clothe the sides of the longus colli muscles and reflect on the thoracic walls as the costal pleurae.

The **middle mediastinal pleura** leaves the sternum as a delicate membrane which is a continuation of the costal pleura. This portion of the ventral mediastinal pleura may be 4 cm. wide cranially, 1.5 cm. wide where the heart lies closest to the sternum, and approximately 2 cm. wide caudally. The right and left middle mediastinal pleural sheets, upon reaching the fibrous pericardium, diverge from each other to enclose the heart. The middle mediastinal pleura may enclose a portion of the thymus gland when this organ is well developed. The right leaf of the middle mediastinal pleura, upon leaving the fibrous pericardium, covers the right phrenic nerve, cranial vena cava, trachea, and longus colli muscle before bending ventrally as the costal pleura of the right thoracic wall. The left leaf of the middle mediastinal pleura, upon leaving the fibrous pericardium dorsally, covers the left phrenic and vagal nerves, the left subclavian artery, and the left longus colli muscle before it reflects on the lateral thoracic wall as the costal pleura. The large central portion of the middle pleura, after passing over the phrenic nerves, continues dorsally on the pericardium, and at the hilus of the lung of either side the right and left pleural sheets reflect on the roots of the respective lungs and become the pulmonary pleurae.

The **caudal mediastinal pleura** lies caudal to a transverse plane passing through the apex of the heart. For descriptive purposes it will be divided into sagittal and oblique parts. The sagittal part is a continuation of the middle mediastinal pleura. It is a triangular fold which bridges the space between the heart cranially, the sternum ventrally, and the diaphragm caudally. It contains the sternopericardiac ligament. The left pleural layer of the fold for the caudal vena cava is elaborated from its right pleural leaf. As a fold from the left pleural leaf, the oblique portion runs to the ventral part of the left costodiaphragmatic recess. The cranial pleural sheet of this fold reflects acutely forward to cover the diaphragm, and the caudal leaf continues on the lateral thoracic wall as the costal pleura. The diaphragmatic attachment of the oblique portion irregularly attaches to the diaphragm at an increasing dorsocranial level. The line of attachment of this portion of the mediastinal pleura follows the left convex portion of the diaphragm parallel to the muscle bundles which compose its muscular periphery.

The **diaphragmatic pleura** (*pleura diaphragmatica*) is the pleural covering of the diaphragm. It is heavier at the muscular periphery of this dome-shaped partition, and more loosely attached, than it is on the tendinous center. Where the diaphragm is attached to the lateral thoracic wall the diaphragmatic pleura reflects acutely forward to form the costal pleura. In life a capillary space is present on each side between the diaphragm and the caudal lateral wall of the thorax. These spaces are called the **right** and **left costodiaphragmatic recesses** (*recessus costodiaphragmatici*). After death the diaphragm and its pleura lie in contact with the costal pleura throughout a circular zone which is about 4 cm. in width. The right and left costodiaphragmatic recesses extend cranioventrally on each side of the mediastinal pleura as the right and left **costomediastinal recesses** (*recessus costomediastinales*). These pleural spaces are formed between the ventral mediastinum and the lateral thoracic walls. Although in cross section they form acute angles, they are large enough to

receive the ventral borders of the lobes of the lungs during inspiration.

The apical portion of each pleural sac extends through the thoracic inlet into the base of the neck, forming a pleural pocket known as the **pleural cupula** (*cupula pleurae*). The left cupula is the larger and extends further cranial to the first rib than does the right. The right pleural cupula is wide dorsoventrally but extends only about half as far forward as does the left.

The **pulmonary pleura** (*pleura pulmonalis*) tightly adheres to the surfaces of the lungs and follows all of their irregularities. It is the visceral portion of the pleura. Its greatest intrapulmonary extensions correspond to the adjacent free surfaces of the lobes of the lungs. In all interlobar fissures, except that between the cranial and caudal parts of the cranial lobe of the left lung, the pleura extends to the bronchus of the lungs. Between these two parts of the left cranial lobe the lung parenchyma does not extend to the bronchus, and therefore the pleura does not extend as deeply in this fissure as it does in others. In some specimens the borders of the lungs are notched or fissured by clefts of varying depth. These clefts occur more often on the cranial lobes.

The **pulmonary ligament** (*lig. pulmonale*) is a triangular fold of pleura on each side which leaves the respective lung caudal to the hilus. It is continuous with the pleura covering the root or hilus of the lung, and contains no visible structures. On the left side it extends about 3 cm. caudodorsally from the large pulmonary vein leaving the caudal lobe of the left lung, and ends in a falciform border which stretches from the medial surface of the lung to the left sheet of mediastinal pleura directly ventral to the aorta. The right pulmonary ligament is about the same size and shape as the left. It leaves the acute dorsal border of the right caudal lobe, extends over the right portion of the intermediate, or accessory, lobe, and becomes the right sheet of mediastinal pleura which covers the esophagus and aorta. It ends in a falciform border about 1 cm. cranial to the diaphragm. The pulmonary ligaments are relatively avascular. The pleurae which form them are continuous with the caudal portions of the middle mediastinal pleurae which reflect on the roots of the lungs. Small connecting pleural plicae occasionally extend between adjacent lobes of a lung at its hilus.

The *plica venae cavae* is a thin, loose fold of pleura which surrounds the caudal vena cava. It occupies the triangular space bounded by the caudal vena cava dorsally, the pericardium cranially, and the diaphragm caudally. Because of its delicate nature the right portion of the intermediate lobe of the right lung is visible through it. The space is known as the **mediastinal recess** *(recessus mediastini)*. The right phrenic nerve runs from the base of the heart to the diaphragm in a separate plica, only a few millimeters wide, which leaves the right pleural leaf of the main plica immediately ventral to the caudal vena cava. The plica venae cavae leaves the ventral third of the diaphragm about 1 cm. peripheral to its tendinous center. Ventrally, it blends with the sagittal portion of the caudal mediastinal pleura. The plica venae cavae on the right and the oblique portion of the caudal mediastinal pleura on the left form a pocket between the heart and diaphragm in which is located the intermediate lobe of the right lung. The sagittal portion of the postcardial mediastinal pleura is located ventral to this space where it is formed by the left pleural layer of the plica venae cavae on the right and right pleural layer of the oblique portion of the caudal mediastinum on the left. Where these layers come together the pericardiacodiaphragmatic ligament runs from near the apex of the pericardium to the diaphragm.

Histologically, the pleura is more delicate in the dog than it is in other domestic animals. It contains smooth muscle fibers, and under the epithelium is a dense network of elastic fibers which separate the true serosa from the collagenous subserosa. The subserosa also contains elastic fibers, mainly on its deep surface, where they communicate with those of the lobules of the lung (Trautmann and Fiebiger 1957). The surface of the pleura is covered by a pavement layer of flat mesothelial cells. In life the pleura is covered by a capillary film of fluid so that the friction between the pulmonary pleura and the parietal pleura or between adjacent layers of the pulmonary pleura is reduced to a minimum. Elastic fibers are also present in the parietal pleura, so that both pulmonary and parietal pleura are capable of stretching. The pulmonary pleura is more tightly adherent to the lung parenchyma than is the parietal pleura to the thoracic wall.

LUNGS

The **lung** *(pulmo)* is the organ in which oxygen from the atmosphere and carbon dioxide from the blood are exchanged. The lungs serve a passive function in the mechanical act of respiration. The diaphragm, when it contracts, enlarges the pleural cavity by moving caudally. When the intercostal muscles contract and draw the ribs cranially, the size of the thoracic cavity is also increased, and thus air is drawn into the lungs because of the negative pressure which is produced. Aiding in expulsion of the air from the lungs are the abdominal muscles, which contract and force the abdominal viscera against the caudal surface of the diaphragm.

The two lungs *(pulmo sinister et dexter)* possess many features in common. Each has a slightly concave **base** *(basis pulmonis)*, which lies adjacent to the diaphragm, and an **apex** *(apex pulmonis)*, which lies in the thoracic inlet. The apex of the left lung is more pointed and extends farther forward than the apex of the right lung (Bourdelle and Bressou 1927). The curved lateral surface of each lung is called the **costal surface** *(facies costalis)*, and the flattened surface which faces the opposite lung is called the **medial surface** *(facies medialis)*. Because the vertebral bodies protrude ventrally from the dorsal wall of the thorax and intervene between the two lungs, this portion of the medial wall of each lung is known as the **vertebral part** *(pars vertebralis)*. The remaining ventral portion of each medial wall faces the mediastinum and is known as the **mediastinal part** *(pars mediastinalis)*. The medial surface of each lung is deeply indented by the heart over an area between the third and sixth ribs. This is called the **cardiac impression** *(impressio cardiaca)*. On each side the pleura covering the pericardial sac is displaced sufficiently by the underlying heart so that its ventral portion lies in contact with the costal pleura. The **cardiac notch of the right lung** *(incisura cardiaca pulmonis dextri)* is V-shaped, with the apex located dorsally. The right cardiac notch is formed by the ventrally diverging borders of the cranial and middle lobes. Its dorsal apex lies opposite the beginning of the distal fourth of the fourth rib. On the left side usually no obvious cardiac notch is formed.

Each lung has a **diaphragmatic surface** (*facies diaphragmatica*), which is concave, because it lies against the convex surface of the diaphragm. The diaphragmatic surface of the right lung is about one-third larger than that of the left lung, this larger area being caused by the intermediate lobe of the right lung extending ventrally and to the left, ending in a process at the apex of the heart.

The margin along the vertebral part of the lung is the **dorsal margin** (*margo dorsalis* [*obtusus*]) and extends from the apex to the base of the lung. The costal surface of each lung is continuous with the medial surface at an acute angle ventrally, lying in the costomediastinal recess. This margin, extending from the apex to the base of the lung, is called the **ventral margin of the lung** (*margo ventralis*). Caudally, the ventral margin of the lung is continuous with the peripheral margin of the base of the lung, or the **basal margin** (*margo basalis*). The combined ventral and basal margins constitute the **acute margin** (*margo acutus*).

The area of each lung which receives the bronchi and furnishes passages for the pulmonary and bronchial vessels and nerves is known as the **hilus of the lung** (*hilus pulmonis*).

The **root of the lung** (*radix pulmonis*) consists of the aggregate of those structures which enter the organ at the hilus. Each lung is divided down to its root by the **oblique fissure** (*fissura obliqua*). The left lung is divided into two nearly equal masses by this fissure. The oblique fissure of the right lung is located about 1 cm. caudal to the oblique fissure of the left lung, but at a comparable angle. Each lung is further divided by deep fissures into three or four lobes. The surfaces of adjacent lobes which lie in contact with each other are called the **interlobar surfaces** (*facies interlobares*).

SHAPE OF LOBES AND POSITION OF INTERLOBAR FISSURES

Left Lung. The **cranial part of the cranial lobe of the left lung** (*pulmo sinistra, lobus cranialis, pars cranialis*) was formerly called the apical lobe (Fig. 8–23). It is transversely compressed between the heart and the lat-

eral thoracic wall. Its caudal margin may lie medial or lateral to the caudally lying caudal portion of the cranial lobe. In a 22-pound dog it is 10 cm. long, 3 cm. wide, and 1 cm. thick. It extends from the dorsal part of the fifth rib to and through the thoracic inlet, where its apex lies not only cranial to a transverse plane through the first ribs but also largely to the right of the median plane. It is the only lobe in the lung of the dog which regularly fuses with an adjacent lobe. In the left lung the parenchyma of the cranial and caudal parts of the cranial lobe are fused over a transverse distance of 2.5 cm. from the vertebral border to the fissure between the lobes.

The **caudal part of the cranial lobe of the left lung** (*pulmo sinister, lobus cranialis, pars caudalis*) was formerly called the cardiac lobe. It presents a thin dorsocranially convex border which overlies the caudal thickened portion of the cranial part of the cranial lobe, or these features and positions of the adjacent lobes are reversed. The ventral margin of the caudal part of the cranial lobe of the left lung lies nearly in a frontal plane 1 cm. from the midventral line. The left lung does not possess a cardiac notch.

The **caudal lobe of the left lung** (*pulmo sinister, lobus caudalis*) is pyramidal in shape and is completely separated from the cranially lying caudal part of the cranial lobe by the oblique fissure, which extends from the costal surface to the root of the lung. When the lungs are moderately distended, the fissure begins at the vertebral end of the sixth rib and ends near the costochondral junction of the seventh rib. This lobe of the lung was formerly called the diaphragmatic lobe.

Right Lung. The right lung (Fig. 8–24) is divided into cranial, middle, accessory, and caudal lobes. The right oblique fissure separates the cranially lying cranial and middle lobes from the caudally lying accessory and caudal lobes.

The **cranial lobe of the right lung** (*pulmo dexter, lobus cranialis*) extends from the dorsal part of the oblique fissure cranially and ventrally to the right of the median plane. This lobe was formerly called the apical lobe. It does not end in a definite apex, as does the left lung, but rather its most cranial and ventral part has a gentle curved convex border which goes from an acute, cranial, and dorsal margin to a slightly convex sur-

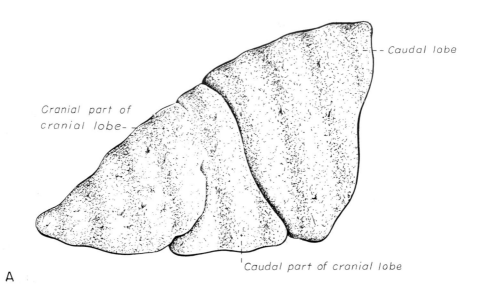

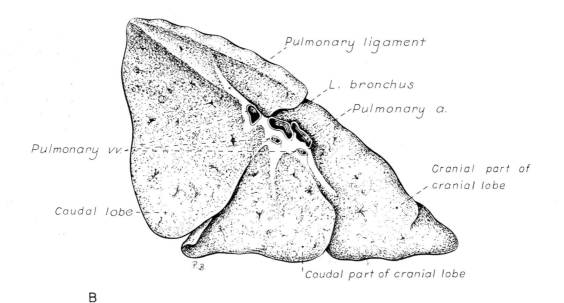

Figure 8–23. Left lung.
A. Lateral aspect.
B. Medial aspect.

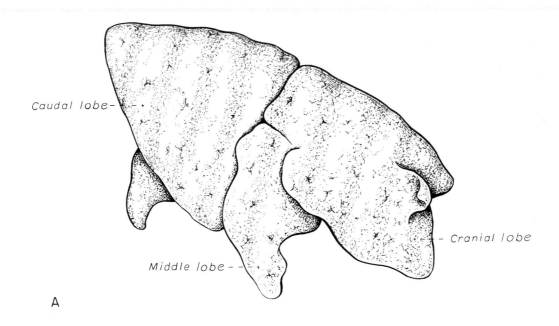

A

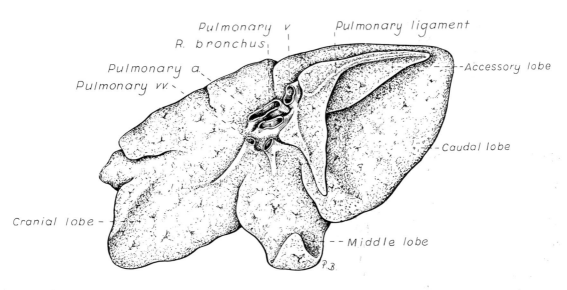

B

Figure 8–24. Right lung.
A. Lateral aspect.
B. Medial aspect.

face cranial to the heart. This portion of the cranial lobe extends across the median plane to the left side, whereas its most cranioventral portion lies in contact with the caudal portion of the apex of the left lung which extends across the midline to the right side. The caudoventral margin of the cranial lobe of the right lung is separated from the craniodorsal portion of the middle lobe by the curved **horizontal fissure** *(fissura horizontalis)*.

The **middle lobe of the right lung** *(pulmo dexter, lobus medius)* begins at the horizontal fissure, where its costal surface is broad and tapers to a narrow, pyramid-shaped ventral extremity which lies caudal or caudosinistral to the apex of the heart. Its medial surface is deeply excavated by the heart, resulting in the cardiac impression. The cranioventral one-fourth of its border diverges sharply from the rounded transversely located caudal portion of the cranial lobe, thus exposing a portion of the sternocostal surface of the heart to the thoracic wall. This notch in the right lung is the **cardiac notch** *(incisura cardiaca)*.

The **accessory lobe of the right lung** *(pulmo dexter, lobus accessorius)* is the most irregular of all of the lobes of the lungs. It is also called the intermediate lobe. Caudally it is molded against the diaphragm; cranially it lies in contact with the diaphragmatic surface of the heart and the adjacent portions of both caudal lobes. The accessory lobe possesses a thickened middle portion and three processes — a dorsal, a ventral, and a right lateral. The dorsal process is a sharp-pointed pyramid-shaped eminence which extends caudally in contact with the caudoventral face of the dorsomedial portion of the caudal lobe of the right lung. Its free caudal apex does not reach as far caudally as do the caudal lobes. The ventral process of the accessory lobe runs almost directly ventrally to the dorsal surface of the sixth sternebra. It is wedged in the space between the diaphragm and the diaphragmatic surface of the heart. Separating the dorsal and the right lateral lobe is the notch through which the caudal vena cava and the right phrenic nerve pass.

The **caudal lobe of the right lung** *(pulmo dexter, lobus caudalis)* is similar in shape and comparable in location to the left lobe, except that it lies to the right of the median

plane. It is smaller than the left caudal lobe, and does not extend as far ventrally as does the left caudal lobe. Furthermore, its diaphragmatic surface is irregularly excavated in its central part by the accessory lobe. Around its periphery it is concave in all directions, in conformity with the convex surface of the diaphragm against which it lies.

RELATIONS OF LUNGS TO OTHER ORGANS

The heart produces large impressions on the medial surfaces of each lung. The **cardiac impression of the right lung** *(impressio cardiaca pulmonis dextri)* is a deep excavation of the medial or mediastinal surfaces of the right cranial lobe cranially, the middle lobe laterally, and the accessory lobe caudally. Ventrally and on the right the cranial and middle lobes fail to cover the heart, so that the pericardial pleura lies in contact with the costal pleura over an area which is V-shaped, with the apex located dorsally. This notch in the right lung is called the **cardiac notch of the right lung** *(incisura cardiaca pulmonis dextri)*. The cardiac notch exposes about 5 square cm. of the sternocostal surface of the heart to the thoracic wall. The ventral margin of the left cranial lobe is arciform; it is located, approximately in a frontal plane, about 1 cm. from the median plane, ventrally. The **notch for the caudal vena cava** *(incisura venae cavae caudalis)* (see Fig. 11–2) is located between the dorsal and the right lateral process of the accessory lobe. Passing through this notch in close association with the caudal vena cava is the right phrenic nerve as it runs to the diaphragm. Both structures are surrounded by the plica venae cavae, which attaches to the diaphragm. The medial surfaces of the cranial lobes of the lungs lie in contact with the thoracic portion of the thymus gland when the thymus is present. When it is fully developed, the thymus gland ends caudally opposite the left fifth costal cartilage. It therefore enters into the formation of the cardiac impression of the caudal part of the left cranial lobe.

PULMONARY VESSELS

The pulmonary arteries carry non-aerated blood from the right ventricle of the heart to the lungs for gaseous exchange. The pulmonary veins return aerated blood from the lungs to the left atrium of the heart. McLaughlin, Tyler, and Canada (161) have shown that the spatial and functional arrangements of the pulmonary vessels differ greatly among various species. In the dog, the pulmonary artery, in addition to supplying the distal portion of the respiratory bronchiole, alveolar duct, and alveoli, continues on to supply the thin pleura.

The **pulmonary trunk** (*truncus pulmonalis*) is the stem artery arising from the fibrous pulmonary ring which extends into the media of the pulmonary trunk peripherally and centrally serves for the attachment of muscle fibers from the conus arteriosus. The pulmonary trunk bifurcates into the left and right pulmonary arteries, which ramify in the left and right lungs. The **left pulmonary artery** (*a. pulmonalis sinistra*) curves dorsally cranial to the vein from the cranial part of the cranial lobe which crosses the main bronchus to the lobe. Just prior to this crossing, a pulmonary arterial trunk arises and bifurcates. The larger terminal branch runs cranially as the main vessel to the cranial part of the left cranial lobe. The branch to the caudal part of the left cranial lobe lies cranial to the bronchus and caudal to the large vein.

The portion of the pulmonary artery which ramifies in the left caudal lobe is larger than all the other branches of the left pulmonary artery combined. Before entering the lobe, it passes dorsal to the bronchus, which bifurcates to ventilate the cranial and caudal parts of the left cranial lobe. Upon entering the lobe, it branches irregularly to vascularize the whole lobe. The veins from all of the lobes compose the most ventral part of the root of the lung.

The **right pulmonary artery** (*a. pulmonalis dextra*) is shorter than the left. It runs caudolaterally ventral to the left lobar bronchi and dorsal to the large left lobar veins. The artery divides unequally into a small branch which runs to the right cranial lobe and a large branch which courses caudally into the right caudal lobe. Near the origin of the large artery to the caudal lobe the relatively small right middle lobar artery runs laterally and enters the dorsal third of the lobe. It is related to the dorsal surface of its satellite vein and lies dorsocranial to the right middle lobar bronchus. It may arise from the right cranial lobar artery. The pulmonary lobar artery to the accessory lobe of the right lung enters the thickened middle portion of the lobe and trifurcates — a branch supplying each of the three processes of the lobe. This lobar artery lies ventral to the bronchus to this lobe and dorsal to its satellite vein.

The **pulmonary veins** (*vv. pulmonales*) (see Fig. 11–3) are variable in number. Except for that from a limited area around the hilus, all of the blood distributed to the bronchial tree is returned by the pulmonary veins. There is one main vein from each lobe, although there may be two veins which drain the right cranial lobe. The latter veins anastomose to form a larger vein which immediately receives the vein from the right middle lobe so that blood from all of these areas is returned to the right lateral part of the left atrium by a single large vein 1 cm. in diameter. The blood from the right caudal and accessory lobes is drained by a single trunk which lies to the right of a similar trunk from the left caudal lobe. The pulmonary lobar veins from the left lung usually open individually into the dorsum of the left atrium. The pulmonary veins lie most ventrally. The pulmonary arteries circle dorsocaudally above the veins, and the lobar bronchi are insinuated between the arteries and the veins.

BRONCHIAL VESSELS

The small **bronchial arteries** are variable in origin, although in the majority of dogs the parent trunk is the bronchoesophageal artery (see Fig. 11–50), which arises from the right fifth intercostal artery close to its origin from the aorta. The bronchoesophageal artery crosses the left face of the esophagus and contributes an esophageal branch before entering the root of the lung as the bronchial artery. In its course, the bronchial artery supplies the tracheobronchial lymph nodes, peribronchial connective tissue, and the bronchial mucous membrane. At the level of the respiratory bronchiole the bronchial artery terminates in a capillary bed which is continuous with that of the pulmonary artery. Miller (1937) and, more recently, McLaughlin, Tyler, and Canada (1961) were unable to demonstrate normally occurring bronchial artery–pulmonary artery anasto-moses in the dog.

Notkovitch (1957) found single bronchial arteries on each side in 75 per cent of his specimens, and double bronchial arteries on each side in 10 per cent. In all cases they arose from the first to the fourth right inter-costal artery. Berry, Brailsford, and Daly (1931) found the right and left bronchial arte-ries arose from the right sixth intercostal artery by a common trunk which also sup-plied twigs to the esophagus. They also de-scribed small bronchial vessels which sup-plied the hilus of the lung and which arose from the pericardiacophrenic or internal tho-racic arteries.

True **bronchial veins** are found only at the hilus of the lung. They empty into the azygos vein or intercostal vein at the level of the seventh thoracic vertebra.

PULMONARY LYMPHATICS

The afferent lymph vessels from the lobes of each lung run to the **tracheobronchial lymph nodes** of the respective side and to the middle tracheobronchial lymph node. From these locations lymph is drained via a chain of cranial mediastinal lymph nodes. Correll and Langston (1958) state that a crossing of the lymphatic vessels draining the lower lobes of the lung occurs but is uncommon. Reference is made to Chapter 13 and to Fig-ure 13–8, for further information on the pul-monary lymphatics.

BIBLIOGRAPHY

Allison, A. C. 1953. The morphology of the olfactory system in the vertebrates. Biol. Rev. 28:195–244.

Angulo, A. W., V. P. Kownacki, and E. C. Hessert, Jr. 1958. Additional evidence of collateral ventilation between adjacent bronchopulmonary segments. Anat. Rec. 130:207–211.

Berry, J. L., J. F. Brailsford, and I. B. Daly. 1931. The bronchial vascular system in the dog. Proc. Roy. Soc. B 109:214–228.

Blatt, C. M., C. R. Taylor, M. B. Habal. 1972. Thermal panting in dogs: the lateral nasal gland, a source of water for evaporative cooling. Science 177:804–805.

Bojsen-Møller, F. 1964. Topography of the nasal glands in rats and some other mammals. Anat. Rec. 150:11–24.

————. 1967. Topography and development of anterior nasal glands in pigs. J. Anat. 101:321–331.

————. 1975. Demonstration of terminalis, olfactory, trigeminal, and perivascular nerves in the rat nasal septum. J. Comp. Neurol. 159:245–256.

Bojsen-Møller, F., and J. Fahrenkrug. 1971. Nasal swell bodies and cyclic changes in the air passage of the rat and rabbit nose. J. Anat. 110:25–37.

Bourdelle, E., and C. Bressou. 1927. Le cul de sac anterieur de la caveté pleurale chez les carnivores en particulier chez le chien et chez le chat. Rec. Méd. vét. 103:457–466.

Bowden, R. E. M., and J. L. Scheuer. 1961. Comparative studies of the nerve supply of the larynx in eutherian mammals. Proc. zool. Soc. London 136:325–330.

Boyden, E. A., and D. H. Tompsett. 1961. The postnatal growth of the lung in the dog. Acta anat. (Basel) 47:185–215.

Correll, N. O., Jr., and H. T. Langston. 1958. Pulmonary lymphatic drainage in the dog. Surg. Gynecol. Obstet. 107:284–286.

Duckworth, W. L. H. 1912. On some points in the anatomy of the plica vocalis. J. Anat. Physiol. 47:80–115.

Eisenberg, J. F., and D. G. Kleiman. 1972. Olfactory communication in mammals. Ann. Rev. Ecol. Systemat. 3:1–32.

Estes, R. 1972. The role of the vomeronasal organ in mammalian reproduction. Mammalia 36:315–341.

Evans, H. E. 1977. The lateral nasal gland and its duct in the dog. Anat. Rec. 187:574–575.

Franzmann, A. F. 1907. Beiträge zur vergleichenden Anatomie und Histologie des Kehlkopfes der Säuge-tiere mit besonderer Berücksichtigung der Haus-saügetiere. Bonn, C. Georgi.

Graeger, K. 1958. Die Nasenhöhle und die Nasenne-benhöhlen beim Hund unter besonderer Berucksich-tigung der Siebbeinmuscheln. Dtsch. Tierärztl. Wschr. 65:425–429, 468–472.

Horning, J. G., A. J. McKee, H. E. Keller, and K. K. Smith. 1926. Nose printing your cat and dog patients. Vet. Med. 21:432–453.

Kilpper, R. W., and P. J. Stidd. 1973. A wet-lung tech-nique for obtaining Silastic rubber casts of the respiratory airways. Anat. Rec. 176:279–287.

Kratzing, J. 1971. The structure of the vomeronasal organ in the sheep. J. Anat. 108:247–260.

Lemere, F. 1932a. Innervation of the larynx. I. Innerva-tion of laryngeal muscles. Am. J. Anat. 51:417–437.

————. 1932b. Innervation of the larynx; II. Ramus anastomoticus and ganglion cells of the supe-rior laryngeal nerve. Anat. Rec. 54:389–407.

————. 1933. Innervation of the larynx; III. Experimental paralysis of the laryngeal nerves. Arch. Otolaryngol. 18:413–424.

Leonard, H. C. 1956. Surgical relief for stenotic nares in a dog. J. Amer. Vet. Med. Assoc. 128:530.

Loosli, C. G. 1937. Interalveolar communications in normal and in pathologic mammalian lungs. Arch. Path. 24:743–776.

Macklin, C. C. 1922. A note on the elastic membrane of the bronchial tree of mammals with an interpretation of its functional significance. Anat. Rec. 24:119–135.

Mann, G. 1961. Bulbus olfactorius accessorius in chiroptera. J. Comp. Neurol. *116*:135–144.

McCotter, R. E. 1912. The connection of the vomeronasal nerves with the accessory olfactory bulb in the opossum and other animals. Anat. Rec. 6:299–317.

McLaughlin, R. F., W. S. Tyler, and R. O. Canada. 1961. A study of the subgross pulmonary anatomy in various mammals. Am. J. Anat. *108*:149–165.

Miller, W. S. 1900. Das Lungenläppchen, seine Blut- und Lymphgefässe. Arch. Anat. Physiol., Anat. Abtheilung. 197–228.

— — — — — . 1937. The Lung. Springfield, Ill., Charles C Thomas.

Negus, V. E. 1949. The Comparative Anatomy and Physiology of the Larynx, London, W. Heinemann Ltd.

— — — — — . 1958. The Comparative Anatomy and Physiology of the Nose and Paranasal Sinuses. Edinburgh, Livingstone.

Nickel, R., A. Schummer, E. Seiferle, and W. Sack. 1973. The Viscera of the Domestic Animals. 2nd Ed. Translated and revised by W. O. Sack. New York, Springer-Verlag.

Notkovitch, H. 1957. Anatomy of the bronchial arteries of the dog. J. Thor. Surg. *33*:242–253.

Piérard, J. 1963. Comparative Anatomy of the Carnivore Larynx. Thesis, Cornell Univ.

— — — — — . 1965. Anatomie comparée du larynx du chien et d'autres carnivores. Can. Vet. J. 6:11–15.

Pressman, J. L., and G. Keleman. 1955. Physiology of the larynx. Physiol. Rev. *35*:506–554.

Ramser, R. 1935. Zur Anatomie des Jakobsonschen Organs beim Hunde. Dissertation, Friedrich-Wilhelms-Univ., Berlin.

Read, E. A. 1908. A contribution to the knowledge of the olfactory apparatus in dog, cat, and man. Am. J. Anat. *8*:17–47.

Réthi, A. 1951. Histological analysis of the experimentally degenerated vagus nerve. Acta morph. Acad. Sci. Hung. Tome I, Fasc. 2:221–230.

Schmidt-Nielsen, K., W. L. Bretz, and C. R. Taylor. 1970. Panting in dogs: unidirectional air flow over evaporative surfaces. Science *169*:1102–1104.

Schneider, K. M. 1930. Das Flehmen. Zeit. f. Gesamte Tiergartnerei Leipzig *3*:183–198.

Sussdorf, M. V. 1911. Der Respirationsapparat, *in* Handbuch der vergleichenden mikroskopischen Anatomie der Haustiere, by W. Ellenberger. Berlin, Paul Parey.

Trautmann, A., and J. Fiebiger. 1957. Fundamentals of the Histology of Domestic Animals (translated and revised from the 8th and 9th German editions, 1949, by R. E. Habel and E. Biberstein). Ithaca, N. Y., Comstock Publishing Assoc.

Tucker, J. L., Jr., and E. T. Krementz. 1957. Anatomical corrosion specimens; I. Heart-lung models prepared from dogs. Anat. Rec. *127*:655–665; II. Bronchopulmonary anatomy in the dog, *ibid.*, pp. 667–676.

Vogel, P. H. 1952. The innervation of the larynx of man and the dog. Am. J. Anat. *90*:427–447.

Watson, A. G. 1974. Some Aspects of the Vagal Innervation of the Canine Esophagus, An Anatomical Study. Masters Thesis, Massey Univ., N Z.

Chapter 9

THE UROGENITAL APPARATUS

By GEORGE C. CHRISTENSEN

The urogenital apparatus (*apparatus urogenitalis*) is made up of the urinary organs (*organa uropoëtica*) and the reproductive organs (*organa genitalia masculina et feminina*).

THE URINARY ORGANS

The **urinary organs** (*organa uropoëtica*) include the **kidneys** (*renes*), the **ureters,** the **bladder** (*vesica urinaria*), and the **urethra** (*urethra masculina* or *urethra feminina*).

THE KIDNEYS

The **kidney** (*ren*), which is *nephros* in Greek (Figs. 9–1, 9–2), is a reddish-brown, paired, bean-shaped structure lying against the sublumbar muscles on either side of the vertebral column. The kidneys maintain the ionic balance of the blood by excreting urine as waste after reabsorbing about 99 per cent of the filtrate.

Each kidney has a cranial and a caudal pole, a medial and a lateral border, and a dorsal and ventral surface. The cranial and caudal extremities are joined by a convex lateral border and a straight medial border.

The medial border is indented by an oval opening, the hilus, which opens into a space, the renal sinus. The hilus transmits the ureter, renal artery and vein, lymph vessels, and nerves. Of these structures, the renal artery is the most dorsal, and the renal vein the most ventral. The nerves and lymphatics lie in close relationship to the vein.

Both kidneys are retroperitoneal, located in the sublumbar region, one on either side of the aorta and caudal vena cava. The dorsal surface of each kidney is less convex than the ventral surface. The dorsal surface is in partial contact with the areolar tissue underlying the sublumbar muscles, whereas the entire ventral surface is covered by peritoneum.

The cranial pole of each kidney is covered with peritoneum on both the dorsal and the ventral surface; only the ventral sur-

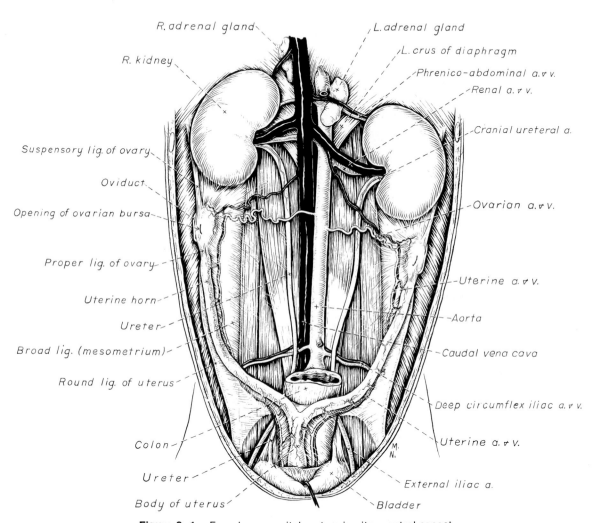

Figure 9–1. Female urogenital system in situ, ventral aspect.
See Figure 9–1 Radiograph on following page

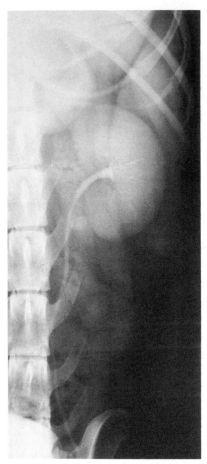

Figure 9–1 Radiograph. Dorsolateral radiograph of the left kidney and ureter.

in length, 4 to 5 cm. in width, and 3 to 4 cm. in thickness. The weight of the freshly excised kidney averages 25 to 35 gm. Finco et al. (1971) made kidney measurements on radiographs of 27 normal male dogs prior to direct measurement at necropsy in order to establish a basis for estimating kidney size radiographically. Kidney weight and volume were highly correlated; kidney length was best correlated with kidney weight. Several tables of normal values and variations are presented.

INVESTMENT AND FIXATION

A strong, thin **fibrous capsule** (*capsula fibrosa*) covers the surface of the kidney. The capsule dips inward at the hilus to line the walls of the sinus and to form the adventitia of the renal pelvis. It also invests the renal vessels and nerves before they pass into the hilus. The fibrous capsule of normal kidneys is easily removable, except in the renal sinus, where it is adherent to blood vessels and to the **pelvic recess** (*recessus pelvis*). Fat of the **adipose capsule** (*capsula adiposa*), in which the kidney is partially embedded, extends through the hilus into the sinus.

POSITION AND RELATIONS

The craniolateral surface of the left kidney is in contact with the dorsal end of the medial surface of the spleen, the greater omentum, and the greater curvature of the stomach. Cranially, it is in contact with the pancreas and left adrenal gland. Dorsally, the kidney, with its adipose capsule, is related to the quadratus lumborum, transversus abdominis, and psoas muscles, as well as to the deep layer of the lumbodorsal fascia underlying the retroperitoneal or pararenal fat. Caudally, the left kidney of the female is in contact with the descending colon and the mesovarium. The peritoneum on the ventral surface of the kidney blends with the peritoneum suspending the ovary. In the male, the renal peritoneum is reflected onto the dorsal body wall as parietal peritoneum. Medially, the left kidney of the male is related to the descending colon, mesocolon, and ascending duodenum. The descending colon is also related to the ventral surface of the kidney. The medial edge of

face of the caudal pole is covered. The kidneys thus lie in an oblique direction, tilted cranioventrally. The right kidney is more firmly attached to the dorsal wall than is the left, and has a correspondingly larger retroperitoneal area.

Both kidneys are invested with a fibrous capsule, are embedded in adipose tissue, and are held in position by subperitoneal connective tissue. The kidneys are not rigidly fixed; they may move during respiration or be displaced by a full stomach. In some lean animals, it is possible to examine the kidneys by deep abdominal palpation. The right kidney lies more cranially than the left, and is in contact with the liver. Grandage (1975) has considered some effects of posture on the radiographic appearance of the kidney.

The kidney in the dog averages 6 to 9 cm.

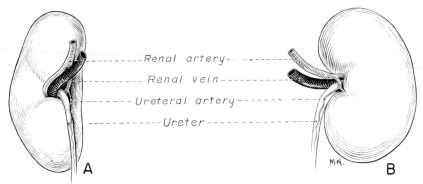

Figure 9-2. Right kidney, and vessels of hilus.
A. Medial aspect.
B. Dorsal aspect.

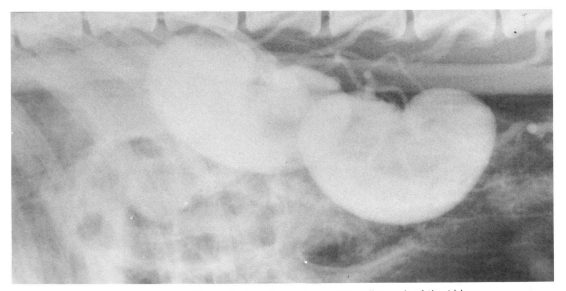

Figure 9-2 Radiograph. Lateral radiograph of the kidneys.

the left kidney is located approximately 1 cm. from the mid-dorsal line in an average-sized dog; the cranial pole, or extremity, lies about 5 cm. caudal to the upper third of the last rib.

The right kidney has its cranial pole embedded in the fossa of the caudate lobe of the liver. This extremity is located at the level of the thirteenth rib. It may be a few centimeters craniad or caudad, depending on the degree of gastric or, in the female, uterine distention. It may be in contact with the diaphragm and retractor costae muscle. The right adrenal gland is also related to its cranial pole. Medially, the right kidney is in

close proximity to the caudal vena cava and, ventrally, it is in contact with the right limb of the pancreas and the ascending colon.

The **renal sinus** (*sinus renalis*) is the cavity of the kidney. Its opening on the medial border of the kidney is called the **renal hilus** (*hilus renalis*). The sinus contains the renal pelvis, a variable amount of adipose tissue, and branches of the renal artery, vein, lymphatics, and nerves. After they pass through the sinus, the vessels and nerves enter the parenchyma of the kidney.

The **renal pelvis** (*pelvis renalis*) (Fig. 9-3) is a funnel-shaped, saclike structure that collects urine from the collecting ducts of the kidney. Urine collects in the pelvis before it passes into the ureter. Since the kidney of the dog is unipyramidal, there are no calyces connected to the renal pelvis. However, the pelvis of the kidney is elongated in a craniocaudal direction, and is curved to conform with the lateral border of the kidney. It extends into the renal parenchyma

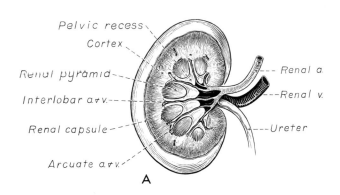

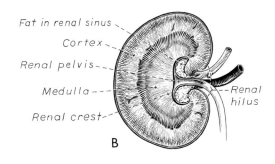

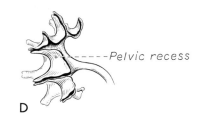

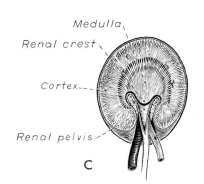

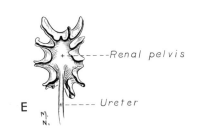

Figure 9–3. Details of structure of left kidney.
A. Dorsal aspect, dissected in frontal plane.
B. Dorsal aspect, midfrontal plane.
C. Cross section.
D. Cast of renal pelvis, dorsal aspect.
E. Cast of renal pelvis, medial aspect.

both dorsally and ventrally by means of curved diverticula. There are generally five or six diverticula curving outward and toward the median plane from each border of the pelvis. The cranial and caudal extremities of the renal pelvis are designated terminal recesses, and the funnel-shaped portion of the pelvis is termed the middle recess.

STRUCTURE

The parenchyma of the kidney is made up of an internal **medulla** (*medulla renis*) and an external **cortex** (*cortex renis*) (Fig. 9–3). When the kidney is cut longitudinally, the medulla appears striated. The portion of the medulla closest to the renal sinus is comparatively light in color and projects into the renal pelvis. This portion, the **renal crest** (*crista renalis*), is thickest longitudinally. Transverse, curved secondary ridges project from each side of the common papilla. These are in contact with the diverticula (*recessus pelvis*) of the renal pelvis. The pyramid, from which the renal crest projects, is smooth in section and is marked with fine striae that converge from base to apex. These striations are formed by the renal tubules. The renal crest is marked on its intrapelvic surface by a variable number of foramina, which are the openings of the papillary ducts. These ducts carry urine into the renal pelvis. The branches of the renal artery and vein diverge between the diverticula of the renal pelvis to reach the cortex.

The sectioned peripheral portion of the renal parenchyma, the cortex, is granular in appearance owing to the presence of renal corpuscles and convoluted tubules. When the kidney is cut in a frontal plane, numerous cut ends of arcuate arteries and veins are apparent at the corticomedullary junction. The thickness of the renal cortex is approximately the same as the transverse diameter of the renal medulla. The peripheral surface of the cortex is covered by the fibrous capsule.

RENAL TUBULES

The **nephron** (Fig. 9–4), composed of a **renal corpuscle** (*corpuscula renis*) and a portion of **straight tubule** (*tubulus renalis rectus*), is the unit for urine production. The double-layered **glomerular capsule** (Bowman's capsule) begins the renal tubule. It is invaginated by a spherical confluence of blood capillaries, the glomerulus. The glomerulus and capsule together form the renal (malpighian) corpuscle. Renal corpuscles are present in the renal cortex, but not in the medulla. From each glomerula capsule (*capsula glomeruli*), the proximal convoluted tubule (coiled and twisted in appearance) descends toward the medullary portion of the kidney through the pars convoluta (peripheral region of the cortex) to the pars radiata, where it straightens out as the descending limb of Henle's loop. This narrow tube extends into the medullary pyramid. After progressing downward for a variable distance, it makes a U-turn and again runs peripherally, increases in diameter, and becomes the ascending limb of Henle's loop. It now reaches the pars convoluta as the distal convoluted tubule.

After further twisting and coiling, it extends into the pars radiata, becomes smaller in diameter, and opens into a straight collecting tubule along with other distal convoluted tubules. The collecting tubule runs through the medulla, united with other collecting tubules, and opens onto the renal crest as a **papillary duct** (*ductus papillaris*). Moffat(1975) notes that there is little or no correlation between concentrating power and the length of Henle's loop or the relative thickness of the tubule wall in various species of mammals. For comparative information on the mammalian kidney, see Sperber (1944). Studies of renal morphology and function also appear in Vimtrup (1928), Smith (1943, 1951), Sperber (1944), Moffat (1975), Rouiller and Muller (1969–71), and Brenner and Rector (1976).

VESSELS AND NERVES

The kidney is a highly vascular organ. Briefly, blood enters the renal artery from the aorta, goes through end arteries, interlobar vessels, arcuate vessels, interlobular arteries, and finally to glomeruli via afferent arterioles. Efferent arterioles leave the glomeruli and course directly into the outer layer of the medulla, giving rise to long cap-

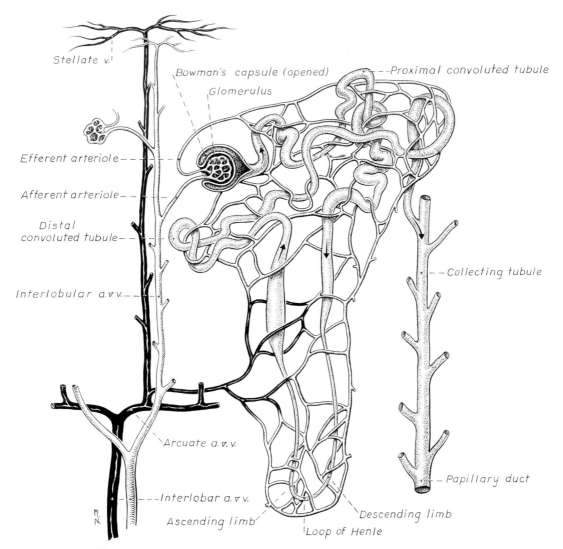

Figure 9–4. Schema of vessels and tubules of structural unit of kidney.

illary nets which extend to the apical end of the pyramid, or they branch directly into intertubular capillary networks.

The **renal artery** (*arteria renis*) bifurcates into dorsal and ventral branches. The site of bifurcation is extremely variable (Christensen 1952). Variations in the renal artery are common, ranging from a single vessel to one with numerous branches or to completely doubled renal arteries. The two primary branches of the renal artery, end branches, divide into two to four **interlobar arteries** (*aa. interlobares renis*). These branch into arcuate arteries at the corticomedullary junction. The **arcuate arteries**

(*aa. arcuatae*) radiate toward the periphery of the cortex, where they redivide into numerous **interlobular arteries** (*aa. interlobulares*). **Afferent arterioles** (*vasa afferentes*) leave these to supply the glomeruli and thence the efferent arterioles. The mean glomerular diameter in a 35-pound dog is 170 μm. According to Rytand (1938), there are 408,100 glomeruli in one canine kidney, with a total glomerular volume of 1247 cu. mm. Finco and Duncan (1972) found a correlation between kidney and nephron size and the body size of the dog.

Venous drainage of the kidney stems from the numerous **stellate veins** (*venulae stella-*

tae) in the fibrous capsule. These connect with veins of the adipose capsule and empty into **interlobular** (*venae interlobulares*), **arcuate** (*vv. arcuatae*), and **interlobar** (*vv. interlobares*) and, finally, into the main trunk of the **renal vein** (*vena renis*), which joins the caudal vena cava. Venous arcuate vessels, unlike their arterial counterparts, unite to form elaborate arches. Arcuate veins span the medulla to join the dorsal and ventral parts of the kidney.

The reported existence of two independent circulatory pathways in the kidney, emanating from the renal artery (Trueta et al. 1947), was not confirmed by Moyer et al. (1950), Schlegel and Moses (1950), or Christensen (1952). Daniel et al. (1952) stress the view that blood passing through either the cortex or the medulla in normal animals passes through glomeruli. White (1940) found that, under conditions normally encountered in the dog, glomeruli are open all of the time. However, the possibility that severe hemorrhage or sudden increments of circulating epinephrine may bring on glomerular intermittence is not excluded. Renal vascularization has also been studied by Huber (1907), MacCallum (1926, 1939), Morison (1926), Fitzgerald (1940), and Fuller and Huelke (1973).

Capsular and parenchymal lymphatics are connected to interlobular plexuses which pass into trunks that leave the kidney at the hilus (Trautmann and Fiebiger 1952). They terminate in the lumbar lymph nodes. According to Peirce (1944), lymphatics in the kidney accompany the interlobular, arcuate, and interlobar vessels, surrounding them in an irregular network. The periarterial rete is thicker than the perivenous network. Cortical and perirenal lymphatics anastomose.

Renal nerves, consisting of myelinated and unmyelinated fibers derived from sympathetic and parasympathetic (vagus) nerves, form dense plexuses around the renal blood vessels. They also innervate the renal tubules and the musculature of the renal pelvis.

ANOMALIES

Malformations are fairly common. Congenital renal cysts and polycystic kidneys may occur in the dog. Isolated renal cysts are more commonly found. Uremia may result from bilaterally cystic kidneys.

Other congenital anomalies of the kidney include hypoplasia (very small kidney), aplasia (failure of one kidney to develop), and hypogenesia (bilateral rudimentary development) (Pearson and Gibbs 1971; Höfliger 1971).

Fetal lobations may persist in the adult dog. Variations in size, shape, and position are not infrequent. Variations in the arrangement of the renal arterial supply are frequently observed (Christensen 1952), as are venous anomalies.

THE URETERS

The **ureters** are fibromuscular, slightly flattened tubes that carry urine from the kidneys to the bladder. The diameter of a single ureter measures 0.6 to 0.9 cm. when it is distended. The length of the ureter depends on the size of the animal, averaging between 12 and 16 cm. in a 35-pound dog. The right ureter is slightly longer than the left, because of the more cranial position of the right kidney; it is also longer in the male than in the female.

The ureter begins at the renal pelvis, which receives urine from the renal crest. Running caudoventrally and mesially toward the urinary bladder, it is bounded dorsally by the psoas muscles and ventrally by the peritoneum (Fig. 9–1). The ureters lie dorsal to the internal spermatic vessels in the male and to the utero-ovarian artery and vein in the female. The right ureter lies in close association with the caudal vena cava and is 1 to 2 cm. lateral to the aorta. The ureters pass ventral to the deep circumflex iliac and external iliac arteries and veins. In the male, the ureter crosses dorsal to the ductus deferens, 2 cm. from the junction of the ductus deferens with the neck of the bladder. It enters between the two layers of peritoneum forming the lateral ligament of the bladder and reaches the dorsolateral surface of the bladder just caudal to its neck. In the female, it reaches the lateral ligament of the bladder after being associated with the broad ligament of the uterus. The ureters enter the bladder obliquely and open by means of two slitlike orifices.

STRUCTURE

The muscular wall of the ureter is divided into three thin layers: an outer longitudinal, a middle circular, and an inner longitudinal. Only longitudinal fibers are present at the junctions of the ureters with the bladder. The **ureteral mucosa** (*tunica mucosa*) is made up of transitional epithelium.

VESSELS AND NERVES

The blood supply to the ureter is derived from the **renal artery** (cranial ureteral artery) and the **prostatic** or **vaginal artery** (caudal ureteral artery). Cranial and caudal ureteral arteries anastomose on the ureter. The ureteral arteries have venous counterparts. The autonomic nerves to the ureter come from the celiac and pelvic plexuses.

ANOMALIES

Congenital anomalies include duplication, ectopic openings into the vagina, ureteral atresia, or dilatation of the renal pelvis.

The latter condition may result from an obstructed ureter. Obstruction may be caused by calculi, tumors, scars, ligatures, or developmental anomalies.

THE URINARY BLADDER

The **urinary bladder** (*vesica urinaria*) (Fig. 9–5) is a hollow, musculomembranous organ that varies in form, size, and position, depending on the amount of urine it contains. It receives urine from the kidneys, through the ureters, and stores it until it is disposed of through the urethra. The bladder in a 25-pound dog is capable of holding 100 to 120 ml. of urine without being overly distended. When relaxed, the bladder in a 25-pound dog measures 17.5 cm. in diameter and 18 cm. in length. When contracted, it measures 2 cm. in diameter and 3.2 cm. in length. The bladder may arbitrarily be divided into a **neck** (*cervix vesicae*) connecting with the urethra, and a **body** (*corpus vesicae*).

The ventral surface of the bladder is separated from the abdominal wall, just cranial to the pubis, by visceral and parietal layers of peritoneum. The greater omentum fre-

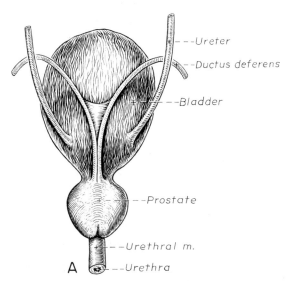

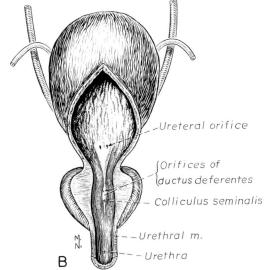

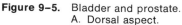

Figure 9–5.   Bladder and prostate.
A. Dorsal aspect.
B. Ventral aspect, partially opened on midline.

quently occupies the space between peritoneal layers. Dorsally, the bladder is in contact with the small intestine (jejunum and, frequently, ileum) and with the descending colon (in the female), cranial to the divergence of the uterine horns from the body of the uterus. In the male the deferent ducts lie dorsal to the neck of the bladder, whereas in the female the cervix and body of the uterus are in contact with the dorsal surface of the bladder (see Fig. 9–25). When empty, the bladder lies entirely, or almost entirely, within the pelvic cavity. The space on each side of the bladder is occupied by the ureters and small intestine.

STRUCTURE

There are three layers of muscle in the wall of the urinary bladder, similar to the arrangement of muscle fibers in the ureter: outer and inner longitudinal layers, and a relatively thick middle circular layer. The muscle fibers all take on an oblique appearance at the urethral-bladder junction. The mucosa of the urinary bladder, like that of the ureter and renal pelvis, is made up of transitional epithelium. It is irregularly folded when the bladder is empty. The mucosal folds disappear during vesical distention. A loose tela submucosa lies between the mucosa and the muscular layer.

Internally, a triangular area near the neck of the bladder is termed the **trigone** of the bladder (*trigonum vesicae*). The apex of the trigone is at the urethral orifice, and the base is indicated by a line connecting the ureteral openings. This area is free from the characteristic mucosal folds, but poorly developed ridges, converging toward the urethral crest, denote the boundaries of the trigone.

FIXATION

The reflections of the peritoneum from the lateral and ventral surfaces of the urinary bladder to the lateral walls of the pelvis and to the ventral abdominal wall are known as ligaments of the bladder. These are made up of double layers of peritoneum separated by intercalated blood vessels, nerves, lymphatics, and adipose tissue, as well as by the ureters, deferent ducts, and vestiges of embryonic structures. The largest peritoneal fold, the **median ligament of the bladder** (*lig. vesicae medianum*) is reflected from the ventral surface of the bladder to the symphysis pelvis and the midventral line of the abdominal wall as far cranially as the umbilicus. It is median in position and triangular in shape. The **middle ligament of the bladder** in the fetus contains the urachus (stalk of the embryonic allantois). This normally disappears shortly after birth, leaving only the peritoneal fold. Even in the adult, a vestigial fibrous urachus may sometimes be found in the free edge of the ligament. In an average-sized dog the umbilical ligament is widest caudally (6 cm.) and narrows down to form an acute angle with the abdominal wall at the umbilicus. Caudally, the ligament ends approximately at the level of the vaginovestibular junction in the female, and at the level of a transverse plane through the middle of the prostate gland in the male.

The **lateral ligaments of the bladder** (*ligg. vesicae laterales*) connect the lateral surfaces of the bladder to the lateral pelvic walls. They are also triangular in shape. The lateral ligaments of the bladder in the fetus extend cranially to the umbilicus, and each contains an umbilical artery (round ligament of the bladder) and a ureter. Before birth, the bilateral umbilical arteries (branches of the internal iliac arteries) carry blood from the fetus to the placenta and are components of the umbilical cord. When the cord is severed at birth, the arteries retract and become fibrous cords between the bladder and the umbilicus. The narrowed lumen of each vessel remains patent between the internal iliac artery and the bladder, where the relatively minute cranial vesical artery leaves the umbilical artery to vascularize the vertex and body of the bladder. The lateral ligaments of the bladder, in the female, blend laterally with the **broad ligament of the uterus** (*mesometrium*) as well as with the lateral pelvic wall. The ureter and round ligament of the bladder cross at nearly right angles to each other at the junction of the broad and lateral ligaments. The ureter is the more mesial of the two structures. In the male, the ureter and ductus deferens cross each other a few centimeters from the entrance of the ureters into the bladder (Fig. 9–5). The ductus deferens is located in the deferential or genital fold of peritoneum, which arises from the peritoneal fold containing the internal spermatic vessels and nerves. The peritoneal pocket between the rectum and the bladder is the rectovesical pouch. It is partially subdivided by the genital fold into the rectogenital and vesicogenital pouches. In the female, the rectogenital and the vesicogenital pouch are completely separate. A small pubovesical pouch, between the bladder and pubis, is present in the male. In the female, the peritoneum is reflected directly from the bladder to the ventral abdominal wall at the level of the pubic brim, so that a pubovesical pouch normally does not exist.

VESSELS AND NERVES

The urinary bladder receives its blood supply through the **cranial vesical artery,** a branch of the umbilical artery, and through the **caudal vesical artery,** a branch of the urogenital artery. The latter is the termination of the visceral branch of the internal iliac artery.

The plexus of veins on the urinary bladder drains primarily into the internal pudendal veins.

The lymphatics of the bladder drain into the **hypogastric** and **lumbar lymph nodes.**

The urinary bladder receives its innervation from three sources: somatic, sympathetic, and parasympathetic. The **pudendal nerve,** derived from sacral nerves 1, 2, and 3, supplies its somatic innervation. It is distributed to the external sphincter of the bladder and to the striated musculature of the urethra. Volitional impulses of urination pass through this nerve. Sympathetic inner-

vation to the bladder is by means of the **hypogastric nerve**; parasympathetic impulses are carried by the **pelvic nerve.** Stimulation of the pelvic nerves results in contraction of the bladder and relaxation of the sphincter, with subsequent urination. Petras and Cummings (1978) describe the location of the cells of origin for the sympathetic and parasympathetic innervation of the urinary bladder and urethra in the dog. Their study, utilizing horseradish peroxidase injected into the bladder and urethra and counterstained with cresyl violet, demonstrates the presence of both sympathetic and parasympathetic intramural ganglia and axons. Thus, there is a direct preganglionic sympathetic pathway to the urinary bladder and urethra in addition to the normal postganglionic sympathetic innervation.

According to Jacobson (1945), section of the hypogastric nerves and excision of the lumbar sympathetic ganglia in the dog result in a temporary decrease in the capacity of the bladder and a slight increase in its tone. This emphasizes the fact that the principal innervation of the bladder is by way of the sacral or pelvic nerves (parasympathetic).

Anomalies of the urinary bladder include diverticula, strictures of the neck of the bladder, urachal cysts, and patent urachus. An enlarged prostate may be the indirect cause of dilatation of the bladder.

THE URETHRA

Both the male and the female urethra are discussed with the genital organs.

THE REPRODUCTIVE ORGANS
The Male Genital Organs

The male genital organs (*organa genitalia masculina*) (Figs. 9–6, 9–7) consist of the scrotum, the two testes (located within the scrotum), the epididymides, the deferent ducts, the prostate (an accessory gland), the penis, and the urethra (the common passageway of urine and semen).

THE SCROTUM

The **scrotum** is a membranous pouch divided by a median septum into two cavities, each of which is occupied by a testis, an epididymis, and the distal part of the spermatic cord. The median partition is

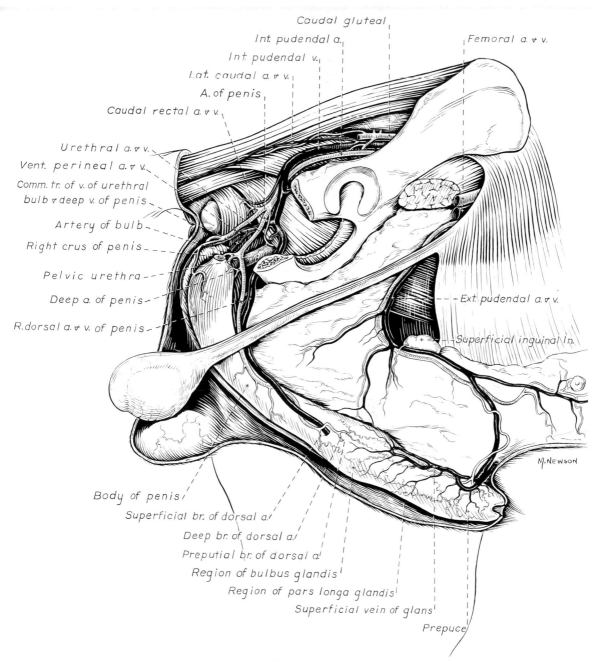

Figure 9–6. Topographic relations of the penis and other pelvic structures. (The right ischium is removed.) (From Christensen 1954.)

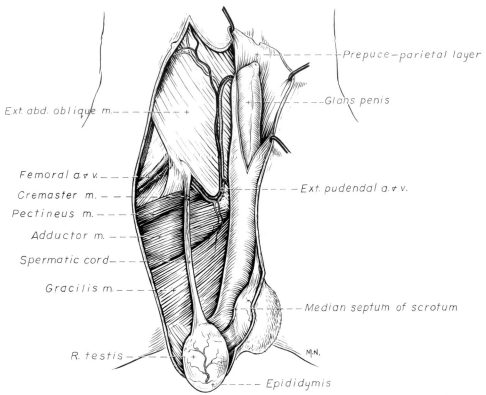

Ext. abd. oblique m.

Femoral a. & v.

Cremaster m.

Pectineus m.

Adductor m.

Spermatic cord

Gracilis m.

R. testis

Prepuce—parietal layer

Glans penis

Ext. pudendal a. & v.

Median septum of scrotum

Epididymis

Figure 9–7.   Male genitalia, ventral aspect.

made up of all the layers of the scrotum except the skin. In the dog, the scrotum is located approximately two-thirds of the distance from the preputial opening to the anus. It lies between the thighs and has a spherical shape, indented in an oblique craniocaudal direction by an indistinct raphe scroti. The left testis is usually farther caudad than the right, allowing the surfaces of the testes to glide on each other more easily and with less pressure.

STRUCTURE

The scrotal integument is pigmented and covered with fine scattered hairs. Sebaceous and tubular (sudoriparous) glands are well developed. Deep to the outer integument of the scrotum is the dartos, a poorly developed layer of smooth muscle mixed with collagenous and elastic fibers. The *dartos* forms a common lining for both halves of the scrotum and also helps to form the scrotal septum (Fig. 9–8C), but is frequently incomplete in the septum. Dorsally, the tissue forming the septum

blends with the abdominal fascia. Contraction of the tunica dartos causes the integument of the scrotum to retract and draw the testes close to the body.

Extending into each scrotal sac is an evaginated pouch of peritoneum, the **vaginal process** (*processus vaginalis peritonei*), covered by fascias of the abdominal wall. The vaginal process and fascia wrap the descended testis and spermatic cord in such a way as to result in a double-walled vaginal tunic (see Figs. 9–10, 9–11). The outer wall, or parietal (superficial, common) vaginal tunic, is separated by a space, the cavum vaginale, from the visceral (deep or proper) vaginal tunic. The cavum vaginale, or cavity of the vaginal process, is continuous with the peritoneal cavity in domestic animals (unlike in the human) at the vaginal ring.

The development of the vaginal process in the male and female is similar. As the evaginating peritoneum passes through the inguinal canal, it is invested by the **transversalis fascia (internal spermatic fascia)** as it enters the deep inguinal ring and by

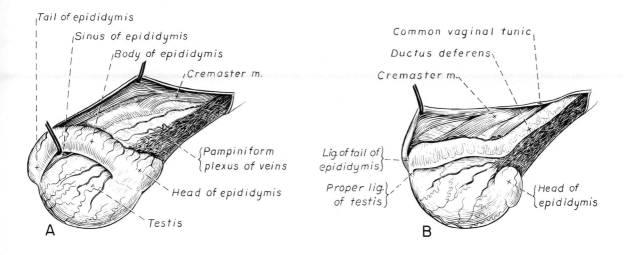

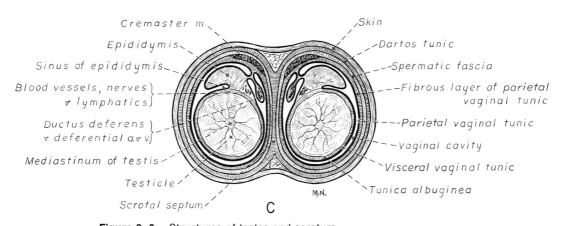

Figure 9–8. Structures of testes and scrotum.
A. Right testis, lateral aspect.
B. Left testis, medial aspect.
C. Schematic cross section through scrotum and testes.

the **internal abdominal fascia (cremasteric fascia)** and **external abdominal fascia (external spermatic fascia)** as it exits from the superficial inguinal ring. The combined fascias form the **spermatic fascia,** which covers the spermatic cord.

The cremaster muscle arises from the caudal free border of the internal abdominal oblique (or occasionally from the transversus abdominis) and inserts upon the parietal vaginal tunic at the level of the testis. The action of the muscle is to pull the testis closer to the body in conjunction with the dartos of the scrotum.

The scrotum, because of its thin, hairless skin, its lack of subcutaneous fat, and its ability to contract toward the body, functions as a temperature regulator for the testis and epididymis. Recent evidence indicates that the epididymis, as the site of sperm storage, is the most heat-sensitive region of the male reproductive tract.

VESSELS AND NERVES

The principal blood vessel to the scrotum is the **external pudendal artery.** The cremasteric artery arises from the femoral or deep femoral artery. The scrotal arteries run along the cranioventral surface of the testis, superficial to the common vaginal tunic. The perineal branches of the external pudendal artery supply the scrotum in part. The draining veins follow the same course in reverse.

The genital ramus, a branch of the genitofemoral nerve (from the third and fourth lumbar nerves), innervates the skin of the scrotum, the inguinal region, and the prepuce. The perineal nerve, a branch of the pudendal (from sacral nerves 1, 2, and 3), helps supply this region.

THE TESTIS

The testes, or male gonads (Fig. 9–8), are located within the scrotum. Each testis is oval in shape and thicker dorsoventrally than from side to side. The length of the testis in a 25-pound dog averages 2.8 to 3.1 cm.; the width (dorsoventral diameter), 2 to 2.2 cm.; and the thickness, 1.8 to 2 cm. The fresh organ weighs 7.8 to 8.2 gm.

In normal position, the testis of the dog is situated obliquely, with the long axis running dorsocaudally. The epididymis is adherent to the dorsolateral surface of the organ, with its tail located at the caudal extremity of the testis, and its head at the cranial end.

STRUCTURE

The surface of the testis is covered by the visceral vaginal tunic, a serous membrane continuous with the parietal peritoneum of the abdominal cavity. Deep to the vaginal tunic is the tunica albuginea, a dense, white fibrous capsule. At the epididymal attachment to the dorsomedial border of the testis, the tunica albuginea joins the mediastinum testis by means of connective tissue lamellae (septula testis), which converge centrally. The mediastinum testis is a cord of connective tissue running lengthwise through the middle of the testis. The lobuli testis (wedge-shaped portions of testicular parenchyma) are bounded by the septula. The lobuli contain the seminiferous tubules (tubuli seminiferi contorti), a large collection of twisted canals. Spermatozoa are formed from the epithelial lining of the tubules, which contains spermatogenic cells and sustentacular (Sertoli) cells. Straight tubules (tubuli seminiferi recti) are formed by the union of the seminiferous tubules of a lobule. These congregate in the rete testis. The mediastinum testis contains a network of confluent spaces and ducts called the rete testis. Testicular blood vessels and lymphatics enter and leave through the mediastinum. The lobuli testis also contain glandular cells (of Leydig), which are thought to produce testosterone, an internal secretion (Pollock 1942, Hooker 1944).

ATTACHMENTS

The caudal ligament of the epididymis, from the caudal extremity of the testis, connects the testis and epididymis to the tunica vaginalis parietalis. The ligament of the tail of the epididymis, a vestige of the embryonic gubernaculum testis, runs between the caudal reflections of the vaginal tunics, not entering the vaginal cavity. Indirectly, the testis is stabilized by the spermatic cord and its reflected vaginal tunics.

VESSELS AND NERVES

The testicular artery and the artery of the ductus deferens supply the testis and epididymis. The testicular artery (homologue of the ovarian artery of the female) arises from the ventral surface of the aorta at the level of a transverse plane through the fourth lumbar vertebra. The right artery originates cranial to the left, corresponding to the embryonic positions of the testes. The artery of the ductus deferens, a branch of the prostatic artery from the visceral branch of the internal iliac, follows the ductus deferens from the vaginal ring to its origin. It sends branches to the epididymis and anastomoses with the testicular artery. The testicular vein follows the arterial pattern but forms an extensive pampiniform plexus in the spermatic cord, surrounding the internal spermatic artery, lymphatics, and nerves. The right testicular vein empties into the caudal vena cava at the level of the origin of its arterial counterpart. The left drains into the left renal vein. Harrison (1949) has made a detailed comparative study of the vascularization of the mammalian testis.

The testicular and epididymal lymphatics anastomose into a variable number of trunks which drain into the lumbar lymph nodes.

The nerve supply to the testis is derived from the sympathetic division of the autonomic nervous system (Kunz 1919a). The **testicular (internal spermatic) nerves** accompany the spermatic arteries distally and enter the gland with either the blood vessels or the efferent ducts. Indirectly, they are derived from the fourth, fifth, and sixth lumbar ganglia of the sympathetic trunk. The blood vessels and smooth muscle fibers in the testis receive a sympathetic nerve supply, but the seminal epithelium and the interstitial secretory tissue do not. Elimination of the sympathetic nerve supply to the testis is followed by degeneration of the seminal epithelium and hypertrophy of the interstitial secretory tissue (Kuntz 1919b). The degenerative changes are considered to be the result of paralysis of the blood vessels in the spermatic cord and testis.

ANOMALIES

Cryptorchidism is the most important congenital anomaly of the testis. This condition is comparatively frequent, and is believed to be hereditary in some instances. In a cryptorchid animal one or both testes are retained either in the abdominal cavity (in the region of the inguinal canal) or between the superficial inguinal ring and the scrotum. Sterile, cryptorchid dogs usually possess normal sexual desire. More rare than cryptorchidism is anorchidism, bilateral absence of the testes, or the presence of only one testis. According to Runnells (1954), testicular tumors of dogs have been reported to cause anatomical alterations, such as atrophy of the opposite testis, metaplasia of the epithelium of the prostatic urethra, enlargement and alopecia of the prepuce, and enlargement of the prostate gland. Male pseudohermaphroditism and true hermaphroditism have been reported in the dog (Lee and Allam 1952, Brodey et al. 1954).

THE EPIDIDYMIDES

The **epididymides** (Fig. 9-8) are the organs where spermatozoa are stored before ejaculation. The epididymis is comparatively large in the dog and consists of an elongated structure composed of a long convoluted tube the coils of which are joined by collagenous connective tissue. It lies along the dorsolateral border of the testis. The **head** *(caput)* of the epididymis begins on the medial surface of the testis but immediately twists around the cranial extremity to attain the lateral side. It is slightly larger than the remainder of the epididymis. It continues as the **body** *(corpus)*, which runs along the dorsolateral surface of the testis, and then as the **tail** *(cauda epididymidis)*, which is attached to the caudal extremity of the testis by the proper ligament of the testis. It is continued craniodorsally up the spermatic cord as the ductus deferens. The curved epididymis has its concave surface in juxtaposition with the testis. Its medial edge is attached to the testis by the visceral vaginal tunic. This dips in between the lateral edge of the epididymis and the testis, under the body of the epididymis, forming a potential space, the *sinus epididymis (bursa testicularis)*. The sinus is limited cranially and caudally by the epididymal head and tail, which adhere tightly to the testis.

STRUCTURE

In addition to being a storage place for spermatozoa, the epididymis slowly transports spermatozoa to the ductus deferens. Circular smooth muscle fibers aid seminiferous tubule secretion in moving the germ cells. The length of the epididymis and the slowness of spermatozoa movement are important in allowing the spermatozoa to complete their maturation process.

THE DUCTUS DEFERENS

The **deferent duct** *(ductus deferens)* (Figs. 9-9, 9-11) is the continuation of the duct of the epididymis. Beginning at the tail of the epididymis, it runs along the dorsomedial border of the testis, ascends in the spermatic cord, and enters the ab-

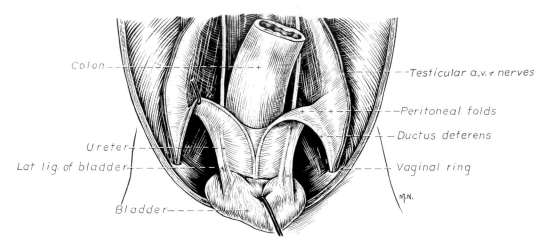

Colon

Testicular a.,v. & nerves

Peritoneal folds

Ductus deferens

Ureter

Lat. lig. of bladder

Vaginal ring

Bladder

M.N.

Figure 9–9.   Urogenital ligaments of the male, ventral aspect.

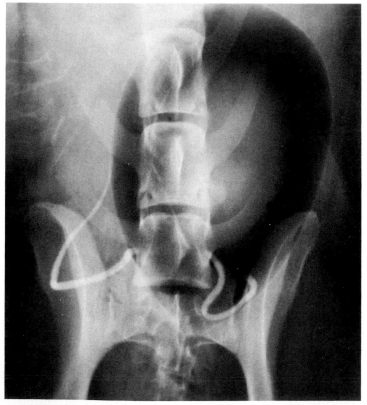

Figure 9–9 Radiograph. Dorsolateral radiograph of the bladder and ureters.

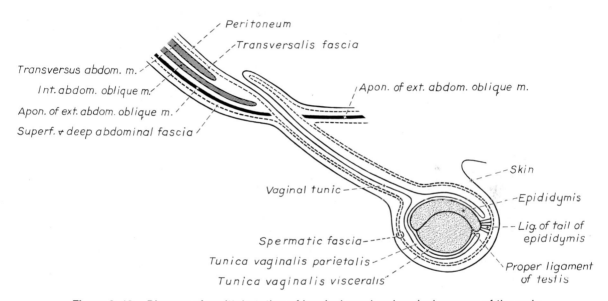

Figure 9–10. Diagram of sagittal section of inguinal canal and vaginal process of the male.

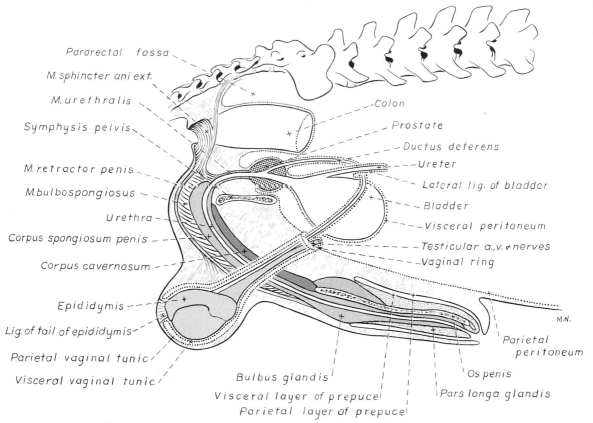

Pararectal fossa
M. sphincter ani ext.
M. urethralis
Symphysis pelvis
M. retractor penis
M. bulbospongiosus
Urethra
Corpus spongiosum penis
Corpus cavernosum
Epididymis
Lig. of tail of epididymis
Parietal vaginal tunic
Visceral vaginal tunic

Colon
Prostate
Ductus deferens
Ureter
Lateral lig. of bladder
Bladder
Visceral peritoneum
Testicular a., v. & nerves
Vaginal ring

M.N.

Parietal peritoneum

Bulbus glandis
Visceral layer of prepuce
Parietal layer of prepuce

Os penis
Pars longa glandis

Figure 9–11. Diagram of peritoneal reflections and the male genitalia.

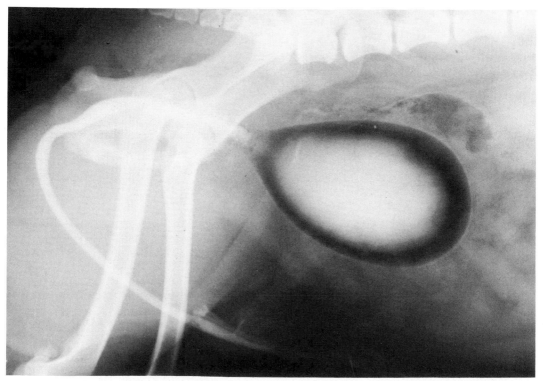

Figure 9–11 Radiograph. Lateral radiograph of the bladder and urethra in the male.

dominal cavity through the inguinal canal. Running in a fold of peritoneum, it crosses ventral to the ureter at the lateral ligament of the bladder and penetrates the prostate to open into the urethra, lateral to the colliculus seminalis.

In a 25-pound dog, the ductus deferens averages 17 to 18 cm. in length and 1.6 to 3 cm. in diameter. The epididymal end of the duct is slightly tortuous, but it straightens out in its course along the medial surface of the testis. It is attached to the testis, along with the artery and vein of the ductus deferens, by a special fold of the proper vaginal tunic called the *mesoductus deferens*. In the abdominal cavity the deferential fold of peritoneum leaves the vaginal ring, to which it is attached at one edge, and blends with the peritoneal fold containing the testicular vessels, nerves, and lymphatics. The ductus deferens lies 3.4 cm. from the body wall when the deferential fold of peritoneum is stretched out. The right and left deferent ducts come into close apposition approximately 2 cm. before they penetrate the prostate gland. For about 1.5 cm. before they contact each other, the ducts are joined by a peritoneal trigone. Peritoneum covering the pelvic portion of the ductus deferentes is reflected ventrally onto the prostate, bladder, and ureters. Dorsally, it is reflected over the prostate and then onto the ventral surface of the rectum.

STRUCTURE

The layers of the ductus deferens are the *tunicae adventitia, muscularis*, and *mucosa*. Three layers of smooth muscle are generally recognized; an outer and an inner longitudinal, and a middle circular. The mucosa is made up of simple or pseudostratified columnar epithelium. The terminal portion of the ductus deferens and the adjacent area of the urethra contain branched tubular glands. A distinct ampulla of the deferent duct is not obvious in the dog.

VESSELS AND NERVES

The **artery of the ductus deferens** is a branch of the prostatic artery, which, in turn, arises from the visceral branch of the internal iliac. It accompanies the ductus deferens to the epididymis, which it also supplies with blood. The artery of the ductus deferens anastomoses with the testicular artery in the spermatic cord. The **vein of the ductus deferens** runs in the spermatic cord with the deferent duct, and empties into the internal iliac vein. The lymphatics drain into the **hypogastric** and **medial iliac lymph nodes.** Nerves to the ductus deferens are autonomic, arising from the pelvic plexus. The **hypogastric nerve** (sympathetic) supplies the pelvic part of the ductus deferens. Parasympathetic fibers are thought to be distributed only to the epididymis and musculature of the ductus deferens.

THE SPERMATIC CORDS

Each **spermatic cord** *(funiculus spermaticus)* is composed of the ductus deferens and the testicular vessels and nerves, along with their serous membrane coverings, the *mesoductus deferens* and the *mesorchium*. These structures pass through the inguinal canal during the descent of the testis. The spermatic cord begins at the deep inguinal ring, the point at which its component parts converge to leave the abdominal cavity. The constituents of the spermatic cord are: the ductus deferens, which arises from the tail of the epididymis, leaves the vaginal ring, runs caudomedially in the deferential fold of peritoneum, and enters the prostate gland before opening into the prostatic urethra; the small artery of the ductus deferens, which arises from the prostatic artery; and the vein of the ductus deferens, which drains into the internal iliac vein. The testicular artery, originating from the ventral surface of the aorta, is also in the cord. The left testicular artery arises approximately 4 cm. cranial to the origin of the caudal mesenteric artery, and the right arises 4.6 cm. cranial to the same artery. The testicular artery runs laterally and caudally, crossing the ventral surface of the ureter, at which point it is joined by the testicular vein and nerve. The peritoneal fold enclosing the testicular structure is attached to the abdominal wall in a line slightly lateral to the junction of the transversus abdominis and psoas muscles. If the fold of peritoneum is stretched out, its edge is 4 cm. from the body attachment at the widest point. The left testicular vein empties into the left renal vein and the right into the caudal vena cava. The plexus of the testicular nerves arises from the area of the sympathetic trunk between the third and the sixth ganglion. The testicular lymph vessels pass to the lumbar lymph nodes.

The components of the spermatic cord are joined together by loose connective tissue and peritoneum, which is designated as the **vaginal tunic.** At the abdominal inguinal ring, the parietal peritoneum dips into the inguinal canal as the **parietal vaginal tunic** *(tunica vaginalis parietalis).* The peritoneal ring formed by the vaginal process passing through the deep inguinal ring is termed the **vaginal ring.** The vaginal ring averages 0.5 to 1 cm. in diameter in the male. There is usually an irregular mass of fat at the vaginal ring, covered by peritoneum. It overlaps the cranial border of the ring and probably acts like a valve to decrease the possibility of intestinal or omental herniation. The fat mass may be in two separate parts.

The ductus deferens, with its vessels, is enveloped by one fold of peritoneum at the vaginal ring, and the testicular vessels and nerves are covered by another. The double layer of peritoneum uniting these two folds to each other and to the edge of the vaginal ring is termed the **mesorchium.** It may be compared to the mesentery, which attaches the intestines to the abdominal wall.

Along the path of the spermatic cord from the deep inguinal ring to the testis, the relationships of the vaginal tunics and the enclosed structures remain constant. The tunics also reflect over the testis, joining along the dorsomedial border of the organ. A small circumscribed area on the tail of the epididymis is free of tunic, allowing the **ligament of the tail of the epididymis** (embryonic *gubernaculum testis*) to attach the epididymis to the spermatic fascia.

The **inguinal canal** is a sagittal slit between the abdominal muscles. The **inguinal ring** is located approximately 1 cm. craniomedial to the femoral ring. The femoral ring affords passage for the femoral

vessels. The inguinal canal is bounded medially by the the rectus abdominis muscle, cranially by the internal oblique muscle, and both laterally and caudally by the aponeurosis of the external abdominal oblique muscle. The spermatic cord dips in between medial the lateral branches of the caudal deep epigastric arteries which lie along the borders of the internal ring. The **superficial ring,** located 2 to 4 cm. lateral to the linea alba, is merely a slit in the aponeurosis of the external abdominal oblique muscle. The cranial wall of the inguinal canal is made up of the transversus abdominis and internal abdominal oblique muscles, as well as the aponeurosis of the external abdominal oblique muscle. Only the latter forms the caudal wall of the canal.

As the spermatic cord and testis pass through the inguinal canal, transversalis fascia (underlying parietal peritoneum) is reflected onto them, and is here known as **internal spermatic fascia.** The combined superficial and deep abdominal fascia, from the outer surface of the external abdominal oblique muscle, is reflected onto the spermatic cord as it emerges from the inguinal canal. It then lies superficial to the internal spermatic fascia and is known as the **external spermatic fascia.** The cremaster muscle, a caudal fasciculus of the internal abdominal oblique muscle, passes down between the internal and the external spermatic fascia.

Scrotal and inguinal hernias occur in male dogs, scrotal hernia being the more common of the two. A fold of omentum or a loop of intestine protrudes into the vaginal process and lies in the vaginal cavity (between the visceral and the parietal vaginal tunic), forcing the testis and scrotal wall apart. The hernia may be bilateral or unilateral. Scrotal hernia is characterized by an elongated enlargement of the scrotum and a dilatation of the inguinal canal. For a description of surface palpation of the superficial inguinal ring, see McCarthy (1976). Inguinal hernia (intestines or omentum pushing into the inguinal canal peripheral to the vaginal process) is recognizable as a soft, fluctuating enlargement to one side of the penis.

A malformed, undeveloped scrotum (infantile scrotum) is the result of undescended testes. The condition may be unilateral or bilateral, depending on whether one or both testes are retained or absent.

THE PROSTATE GLAND

The **prostate gland (prostata)** (Fig. 9–5) is a musculoglandular mass that completely encompasses the proximal portion of the male urethra at the neck of the bladder. It is the only accessory genital gland present in the male dog. The prostate develops from a series of symmetrical buds of the pelvic urethra which appear about the sixth week of gestation (Price 1963). The size and weight of the prostate varies, depending upon the age, breed, and body weight of the dog (O'Shea 1962). In most dogs, progressive enlargement occurs with age (Schlotthauer and Bollman 1936). The latter authors found that in all dogs in which there was more than 0.7 gm. of prostate per kilogram of body weight, the prostate was abnormal on histological examination. They suggested 0.7 gm. prostate per kilogram of body weight as the upper limit of normality for the dog. O'Shea (1962) divides prostatic growth into three phases: normal growth in the young adult, hyperplasia during the middle of adult life, and senile involution.

The prostate is bounded dorsally by the rectum and ventrally by the symphysis pubis and ventral abdominal wall. Its craniocaudal position is age-dependent, as discussed by Gordon (1961). The prostate lies entirely within the abdominal cavity until the urachal remnant breaks down at about two months of age. From that time until sexual maturity the gland is confined to the pelvic cavity. With sexual maturity it increases in size and extends cranially. By 4 years of age over half of the gland is abdominal, and by 10 years of age the entire gland is in the abdomen. The degree of bladder distention was not found to alter these relationships to any significant degree. The prostate is androgen-dependent, and castration at any age results in a marked reduction in size (Hansel and McEntee 1977).

The dorsal surface of the prostate is separated from the ventral surface of the rectum by the two layers of the fold of peritoneum which bounds the rectogenital space (Fig. 9–11). The ventral surface of the prostate is retroperitoneal; the ventral sheet of the lateral ligament of the bladder is not continued onto the prostate (Gordon 1960). A layer of fat usually covers the ventral surface. In mature dogs, the caudal third of the dorsal surface of the prostate is attached to the rectum by a fibrous band (Gordon 1960).

The prostate is semioval in transverse section; the dorsal surface is flattened. A mid-dorsal sulcus is usually palpable per rectum. The prostatic part of the urethra passes through the gland somewhat dorsal to its center. A prominent median septum divides the gland into right and left lobes. Each lobe is further divided into lobules by capsular trabeculae. The lobules consist of numerous compound tubuloalveolar glands lined by columnar epithelium. Ducts from these glands enter the urethra throughout its circumference. The **capsule of the prostate** (*capsula prostatae*) is comparatively thick. Smooth muscle fibers are found throughout the capsule, and muscle fibers from the wall of the urinary bladder extend onto the dorsal surface of the capsule.

The two deferent ducts enter the craniodorsal surface of the prostate. They lie adjacent to each other, one on either side of the median plane. They run caudoventrally through the dorsal part of the gland to open into the urethra by two slits on each side of the colliculus seminalis.

The function of the prostate is not entirely understood. Prostatic secretion contains citrate, lactate, cholesterol, and a number of enzymes. It is believed to be essential to provide an optimum environment for sperm survival and motility. Dog semen is unique in its absence of reducing sugars supplied by the accessory glands in other species; the source of readily metabolizable energy for the spermatozoa is unknown (Hansel and McEntee 1977).

VESSELS AND NERVES

The vascular and nervous supply of the prostate have been reported by Gordon (1960) and Hodson (1968). The **prostatic artery** (*a. prostatica*) arises from the internal pudendal at the level of the second or third sacral vertebrae, although it may arise from the umbilical (Hodson 1968) near its origin. The prostatic gives rise to the **artery of the ductus deferens** (*a. ductus deferentis*), which is homologous with the uterine artery in the female. The artery of the ductus deferens gives rise to the **caudal vesicular artery** *(a. vesicalis caudalis)*, which is larger than the parent vessel, before turning to follow the ductus toward the testis, where it anastomoses with terminal branches of the testicular artery. The caudal vesicular artery gives branches to the ureter *(ramus uretericus)* and urethra *(ramus urethralis)* and then ramifies on the surface of the bladder, anastomosing with the contralateral caudal vesicular and cranial vesicular arteries. When the cranial vesicular artery is not present, the caudal vesicular supplies the entire bladder.

The prostatic artery continues caudoventrally and gives off the small **middle rectal artery** *(a. rectalis media)* before ramifying on the surface of the prostate. These branches penetrate the capsule on the dorsolateral surface of the gland to become subcapsular arteries (Hodson 1968). Radial tributaries pass along the capsular septae toward the urethra to supply the glandular tissue. Cavernous tissue, continuous with the corpus spongiosum, surrounds the prostatic urethra. It is supplied by the artery of the bulb of the penis.

Anastomoses occur between the prostatic vessels and the urethral artery and the cranial and caudal rectal arteries, which complicate prostatectomy. The venous network of the gland drains by way of the **prostatic** and **urethral veins** into the internal iliac vein. The prostatic lymph vessels empty into the **iliac lymph nodes**.

The nerve supply to the prostate is closely allied to the vasculature. The **hypogastric nerve**, which supplies sympathetic innervation to the prostate, follows the artery of the deferent duct. The **pelvic nerve**, which may be single or double (Gordon 1960), accompanies the prostatic artery as far as the lateral surface of the rectum. Here it forms the **pelvic plexus**, together with branches of the hypogastric nerve. The middle portion of the pelvic plexus forms the **prostatic plexus**, from which fine nerve fibers innervate the gland itself. Parasympathetic stimulation increases the rate of glandular secretion.

ANOMALIES AND VARIATIONS

Prostatic hypertrophy is common in older male dogs. Excessive enlargement leads to urinary and rectal obstruction, resulting in clinical signs of tenesmus, stranguria and dysuria (O'Shea 1962). Perineal hernia often accompanies prostatic hypertrophy (Greiner and Betts 1975).

Carcinoma of the prostate occurs much more commonly in the dog than in any other domestic animal, although it occurs less frequently than in man (Leav and Ling 1968). Prostatic hypertrophy is also peculiar to man and dog, and thus the dog is an important model for prostatic cancer research.

Cysts of the prostate are common in dogs three years of age and older. Occasionally these reach enormous proportions and require surgical drainage. Prostatic abscesses and, less frequently, prostatic calculi are also encountered.

The classic surgical technique for canine prostatectomy was described by Archibald and Cawley (1956). Numerous modifications have since appeared. The entire prostatic urethra is resected and the pelvic urethra anastomosed to the neck of the bladder. Any operative technique in this area must preserve the caudal vesicular artery, which supplies the bladder, and the pelvic nerves, which are necessary for the voluntary control of micturition.

THE PENIS

The male copulatory organ, the **penis**, is composed of three principal divisions: the **root** *(radix penis)*, the **body** *(corpus penis)*, and the **distal portion** *(glans penis)*. The latter is subdivided into a bulbus glandis and a pars longa glandis. The glans penis is enveloped by epithelium and, in non-erection, the glans is entirely withdrawn into the prepuce. The prepuce is attached to the ventral abdominal wall except for its distal open end, which is free. The penis has two primary surfaces, a **dorsal** *(dorsum penis)* and a **ventral** or **urethral surface** *(facies urethralis)*.

Measurements of the non-erect penis in over 150 mature dogs of assorted breeds show that its length ranges from 6.5 to 24 cm., with an average of 17.9 cm.

The root and body, which are directly continuous, are made up of the *corpora cavernosa penis*, the ventrally located *corpus spongiosum penis*, containing the penile urethra, and the enlarged proximal end of the *os penis*. The corpora contain enlarged venous spaces. The root is attached to the tuber ischii, and the parts of the root to the left and the right are termed the crura. Each crus is made up of the proximal part of the corpus cavernosum and the ischiocavernosus muscle covering it. The body of the penis begins at the blending of the crura. The corpus cavernosum arises, one on each side, from the ischial tuberosity, and each runs distally in the dorsolateral part of the body of the penis as far as the os penis. A fibrous median septum completely separates the right and the left cavernous body in the dog. Superficially, each corpus cavernosum is enveloped by a thick layer of collagenous and elastic fibers, the tunica albuginea.

The extremes and averages of the dorsoventral diameter of the crus and corpus cavernosum penis of one side (from 150 specimens) are as follows: of the crus — 2.8 to 7.4 mm., average 5.2 mm.; of the middle of the corpus cavernosum penis — 0.7 to 2.4 mm., average 1.5 mm.; of the apex — 0.7 to 2.5 mm., average 1.6 mm. Similar figures for the side-to-side measurements are: crus — 2.8 to 5.7 mm., average 4.2 mm.; of the middle of the corpus cavernosum — 1.8 to 5.4 mm., average 3.4 mm.; apex — 1.5 to 4.9 mm., average 3.2 mm.

The **corpus spongiosum** (formerly called corpus cavernosum urethrae), (Figs. 9–12, 9–13) is located in a groove on the urethral side of the penis and surrounds the penile urethra throughout its course. The **bulb of the penis** (formerly called the urethral bulb) is a bilobed *expansion* of this erectile body, located between the crura at the ischial arch (Fig. 9–12). The corpus spongiosum narrows in diameter from its orgin just within the pelvic cavity until it dips into the glans penis. There it gives off numerous shunts that supply the bulbus glandis with blood. It then continues in the pars longa glandis to the external urethral orifice. The cavernous spaces are generally much larger than those of the corpus cavernosum. From one to four valves are located in each of the venous connections between the bulbus glandis and the corpus spongiosum, preventing blood from leaving the bulbus glandis by this route. At the distal quarter of the glans the corpus spongiosum diverges ventrally from its groove in the os penis.

The corpus spongiosum in the non-erect penis varies greatly in diameter, depending on the size of the dog. Average dorsoventral diameters are as follows: urethral bulb — 11.7 mm.; middle of corpus cavernosum urethrae — 4.1 mm.; corpus cavernosum urethrae just proximal to the bulbus glandis — 4.6 mm. Average side-to-side diameters are: bulb — unilateral, 8.5 mm., bilateral, 18 mm.; middle of corpus cavernosum urethrae — 5.1 mm.; caudal to the bulbus glandis — 3.4 mm.

In contrast to the glans of man and the stallion, the glans of the dog is not a simple cap-shaped expansion of the corpus spongiosum. The glans of the dog is a bipartite structure, consisting of a *bulbus glandis* and a *pars longa glandis*. The distal three-fourths of the glans is made up primarily of the pars longa glandis. The distal half of the bulbus glandis is overlapped by the caudal third of the pars longa glandis.

The bulbus glandis, a cavernous expansion of the corpus spongiosum, surrounds the proximal part of the os penis. Its thickest part and area of greatest potential expansion is located on the dorsal surface of the penile bone. The bulbus glandis contains large venous sinuses bounded by trabeculae rich in elastic tissue. The pars longa glandis and bulbus glandis are separated from each other by a connective tissue septum. This layer continues distally as a sheath for the corpus spongiosum and the os penis. Although similar to the bulbus glandis in structure, the pars longa glandis is not capable of comparable expansion. However, the pars longa glandis deforms into a flattened corona when the penis is intromitted in the vagina, and the ligament from the distal cartilage of the os restricts the forward movement of the central area of the pars longa glandis and thus indents it. (Hart and Kitchell 1965). This latter action aids in containing the ejaculate in the fornix of the vagina and in the region of the cervix.

The bulbus glandis averages 2.4 cm. in length, compared with 5.3 cm. for the pars longa glandis. In diameter, the bulbus glandis averages 2.3 cm. dorsoventrally and 2 cm. from side to side. Comparable diameters for the pars longa glandis are: dorsoventrally — 1.4 cm. distal to the bulbus and 1.7 cm. in middle; side to side — 1.2 cm. distal to the bulbus and 1.5 cm. in middle.

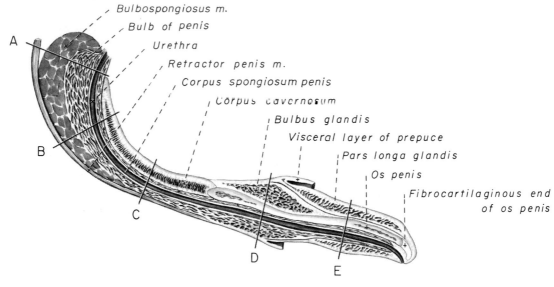

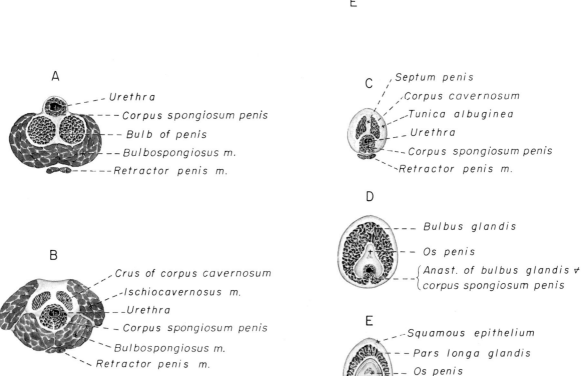

Figure 9–12.   Internal morphology of the penis. Upper drawing, a parasagittal section. A to E, cross sections at five levels indicated by letters on upper drawing. (From Christensen 1954.)

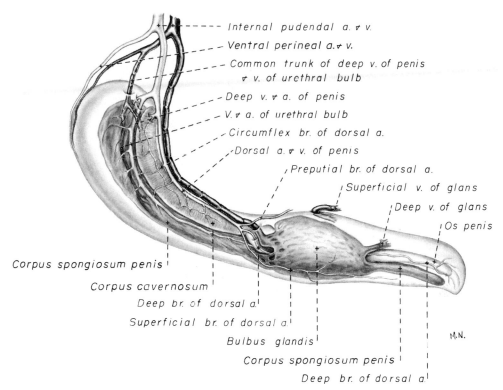

- - - Internal pudendal a. & v.
- - Ventral perineal a. & v.
- Common trunk of deep v. of penis
  & v. of urethral bulb
- Deep v. & a. of penis
- V. & a. of urethral bulb
- Circumflex br. of dorsal a.
- Dorsal a. & v. of penis
- Preputial br. of dorsal a.
- Superficial v. of glans
- Deep v. of glans
- Os penis

Corpus spongiosum penis
Corpus cavernosum
Deep br. of dorsal a.
Superficial br. of dorsal a.
Bulbus glandis
Corpus spongiosum penis
Deep br. of dorsal a.

M.N.

Figure 9–13. Semidiagrammatic view of penis. The pars longa glandis and the muscles of the root are illustrated as if transparent. The vessels of only one side are shown. (From Christensen 1954.)

The *os penis* or *baculum* (Fig. 9–12), is long and tapers in a proximodistal direction. The proximal end, located in the body of the penis just caudal to the bulbus glandis, is comparatively broad. It is thicker dorsoventrally than from side to side. The distal end of the bone is small in diameter and is extended by a slightly curved fibrocartilaginous projection. The proximal two-thirds of the bone is indented ventrally by a distinct groove. The corpus spongiosum and the penile urethra occupy the groove until they emerge from the urethral groove toward the external urethral orifice. The os penis holds

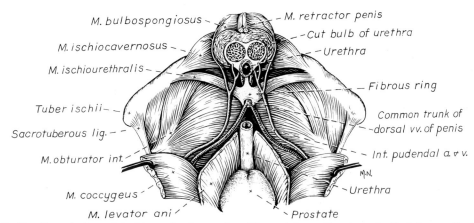

M. bulbospongiosus
M. ischiocavernosus
M. ischiourethralis
Tuber ischii
Sacrotuberous lig.
M. obturator int.
M. coccygeus
M. levator ani

M. retractor penis
Cut bulb of urethra
Urethra
Fibrous ring
Common trunk of dorsal vv. of penis
Int. pudendal a. & v.
Urethra
Prostate

M.N.

Figure 9–14. Dorsal view of ischiourethral muscles. (The pelvic urethra is cut off at the urethral bulb and removed.)

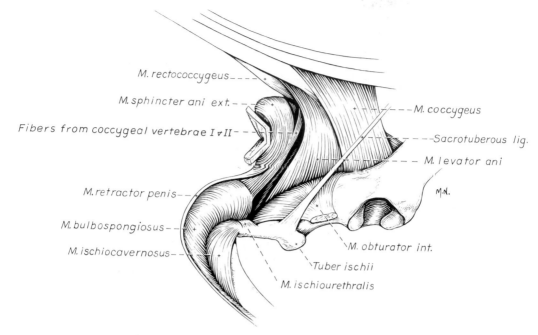

M. rectococcygeus

M. sphincter ani ext.

Fibers from coccygeal vertebrae I & II

M. retractor penis

M. bulbospongiosus

M. ischiocavernosus

M. coccygeus

Sacrotuberous lig.

M. levator ani

M.N.

M. obturator int.

Tuber ischii

M. ischiourethralis

Figure 9–15.   Root of penis with superficial muscles, lateral aspect.

the organ relatively stiff when it is not in erection. Didier (1946) has made a comprehensive study of the os penis.

There are four paired extrinsic penile muscles in the dog (Figs. 9–14, 9–15, 9–16).

The **retractor penis muscles**, composed principally of smooth muscle fibers (Fisher 1917), arise indirectly from the first and second caudal vertebrae, by fascial slips, and blend with the anal sphincters and levator

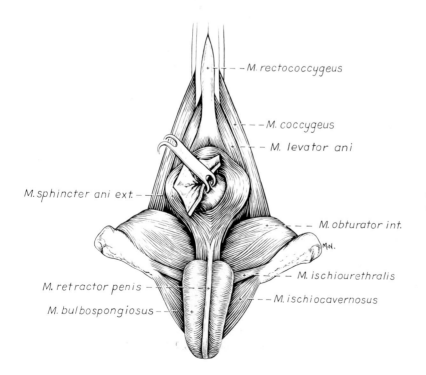

M. rectococcygeus

M. coccygeus

M. levator ani

M. sphincter ani ext.

M. obturator int.

M.N.

M. ischiourethralis

M. retractor penis

M. ischiocavernosus

M. bulbospongiosus

Figure 9–16.   Superficial muscles of male perineum, caudal aspect.

ani muscles. Each runs ventrally along the peripheral border of the anal sphincter to the urethral surface of the penis, and inserts in the penis at the fornix of the prepuce. Bands of muscle fibers leave the retractor penis at the level of the caudal edge of the scrotum and disperse in the septum scroti. The short, broad, paired **ischiocavernosus muscles** cover the crura of the penis. Each originates from the ischial tuberosity and has a broad insertion upon the corpus cavernosum. The **bulbospongiosus muscles,** consisting mainly of transverse fibers, cover the superficial surface of the bulb of the penis. Each arises from the external anal sphincter and fuses with the retractor penis muscle at the proximal third of the body of the penis. Each **ischiourethralis muscle** (m. compressor venae dorsalis penis of Houston 1830) originates from the dorsal surface of the ischial tuberosity and inserts into a fibrous ring which encircles the common trunk of the right and left dorsal veins of the penis (Fig. 9–14). A strong, short ligament attaches the fibrous ring to the area adjacent to the caudal edge of the symphysis pelvis. There is also a ligament which attaches the fibrous ring to the concave surface of the penis at the level of the ischial arch. The two ligaments represent the superficial layer of the urogenital diaphragm of man.

STRUCTURE

The corpus spongiosum contains numerous sinuses, separated by connective tissue trabeculae, in which pass the blood vessels and nerves. Smooth muscle, fibrous connective tissue, and adipose tissue are found in the intersinusoidal lamellae of the corpus cavernosum. The glans penis is covered by stratified squamous epithelium. The bulbous glandis morphologically resembles and is continuous with the corpus spongiosum. The pars longa glandis, also similar to the corpus spongiosum structurally, has a dense network of arteries and veins in its trabeculae. Profuse vascular anastomoses are located under the epithelial surface of the entire glans penis.

VESSELS AND NERVES

The principal source of blood to the penis is the **internal pudendal artery**, a ramification of the visceral branch of the internal iliac. This is augmented by the **external pudendal artery**, which anastomoses with the preputial branch of the dorsal artery of the penis (Fig. 9–6). After giving off the vaginal, urethral, and caudal rectal arteries, the internal pudendal gives rise to the **perineal artery**, which supplies the superficial part of the penile root and the perineum (Fig. 9–17). The artery of the penis is that portion of the internal pudendal between the perineal artery and the three principal vessels of the penis: artery of the bulb, deep artery of the penis, dorsal artery of the penis. Typically, the artery of the bulb arises from that of the penis proximal to the deep artery of the penis. At this point, the artery of the penis is continued by the dorsal artery (Fig. 9–13).

The paired **arteries of the bulb** diverge into two or three branches which divide again before entering the corpus spongiosum. These supply the spaces and tissue of the corpus spongiosum, the penile urethra, and the pars longa glandis (Christensen 1954). The principal trunk is partially coiled in the non-erect state. Its branches anastomose with the end branches of the dorsal artery of the penis as well as with branches of the deep artery.

The **deep artery of the penis** gives off two to five branches and passes through the tunica albuginea to enter the corpus cavernosum. In this cavernous body, the artery again divides into clumps of spiral or looped vessels, the **helicine arteries**, which open directly into the cavernous spaces. According to Vaerst (1938), Kiss (1921), and von Ebner (1900), helicine arteries retain their spiral shape in the non-erect penis, owing to contracted myoepithelium.

The **dorsal artery of the penis** runs diagonally distally to the bulbus glandis, anastomosing with the deep artery and artery of the bulb. Proximal to the glans penis, the dorsal artery trifurcates into a preputial branch, a deep branch, and a superficial branch. The preputial branch runs dorsodistally over the bulbus glandis, supplying the dorsal surface of the pars longa glandis as well as anastomosing with the external pudendal artery in the parietal wall of the prepuce. The superficial branch runs ventrodistally deep to the epithelium of the glans, extending almost to the tip of the glans. The deep branch of the dorsal artery dips into

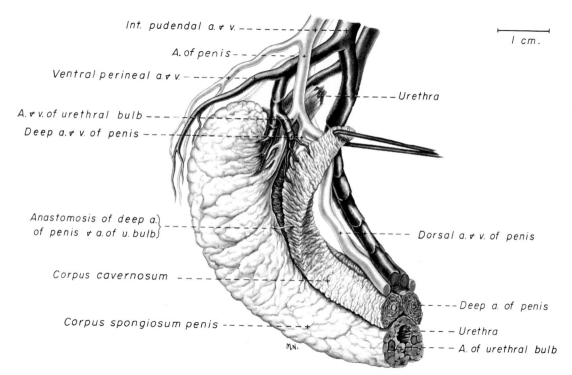

Figure 9–17.   Corrosion preparation of proximal half of the penis. (From Christensen 1954.)

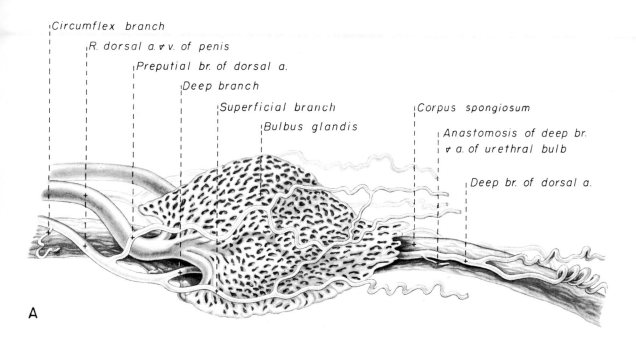

A

Circumflex branch
R. dorsal a. ʋ v. of penis
Preputial br. of dorsal a.
Deep branch
Superficial branch
Bulbus glandis
Corpus spongiosum
Anastomosis of deep br. ʋ a. of urethral bulb
Deep br. of dorsal a.

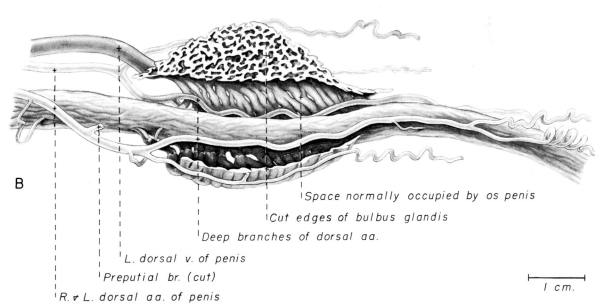

B

Space normally occupied by os penis
Cut edges of bulbus glandis
Deep branches of dorsal aa.
L. dorsal v. of penis
Preputial br. (cut)
R. ʋ L. dorsal aa. of penis

1 cm.

Figure 9–18. Drawings of corrosion specimen of bulbus glandis and part of corpus cavernosum urethrae. (The os penis has been removed.) (From Christensen 1954.)

    A. Superficial view showing distribution of branches of dorsal artery of the penis.

    B. The near half of the bulbus glandis is cut away, showing the route of the deep branches of the dorsal arteries.

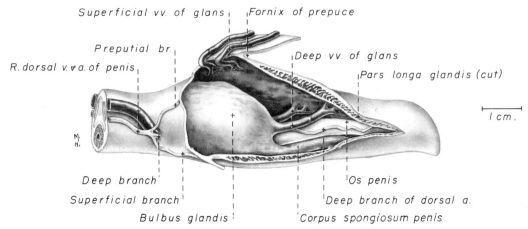

Figure 9–19. Internal morphology of the glans penis. (The pars longa glandis has been slit and partially reflected.) (From Christensen 1954.)

the tunica albuginea and reaches the dorsolateral surface of the os penis, deep to the bulbus glandis (Fig. 9–18). It passes into the pars longa glandis, terminating near the penile tip. Distal to the bulbus glandis, a large anastomotic branch is given off to the corpus spongiosum. The three branches of the dorsal artery of the penis and the external pudendal artery supply blood to the pars longa glandis.

The **internal and external pudendal veins** drain blood from the penis. The internal pudendal joins the internal iliac vein, and the external pudendal drains into the external iliac. The iliac veins on each side unite to form the common iliac vein, the two common iliac veins then converging into the caudal vena cava.

The intrinsic penile veins partially parallel the arteries at the root of the penis (Fig. 9–17). The dorsal veins of the penis are united at the ischial arch for a short distance before they diverge into the right and left internal pudendal veins. Unlike the corresponding arteries, the dorsal vein of the penis and the vein of the bulb unite in a common trunk before entering the internal pudendal vein. The perineal vein empties into this common trunk.

The **dorsal veins of the penis** arise from either side of the bulbus glandis and run along the dorsolateral surface of the penile body as far as the ischial arch. The superficial vein of the glans runs from the dorsal surface of the pars longa glandis to the external pudendal vein (Fig. 9–19).

The **dorsal vein of the penis** (deep dorsal vein of Deysach, 1939), in its course between the bulbus glandis and the common venous trunk, has distinct semilunar valves regularly spaced along its entire length (Fig. 9–20). The **vein of the bulb** drains the cavernous spaces of the proximal half of the corpus spongiosum, arising at the junction of its proximal and middle thirds. Typically, two valves are located in the vein of the bulb between its emergence from the cavernous body and its junction with the deep vein of the penis. The common trunk of the deep vein and the vein of the bulb receives the perineal vein. One to five valves are present between the origin of the common trunk and its junction with the perineal vein.

The **superficial vein of the glans** arises from the deep surface of the pars longa glandis, which it helps to drain, and runs dorsoproximally to the fornix of the prepuce, where it bends acutely and drains into the external pudendal vein by two or more connections (Figs. 9–6, 9–19). One or more valves are present in each branch of the vein.

The **deep vein of the glans** drains blood from the pars longa glandis into the bulbus glandis. It arises, on each side of the midline, from the middle of the deep surface of the pars longa glandis and runs proximally along the dorsolateral surface of the os penis to enter the bulbus glandis. A semilunar valve prevents blood in the bulbus glandis from going to the pars longa glandis.

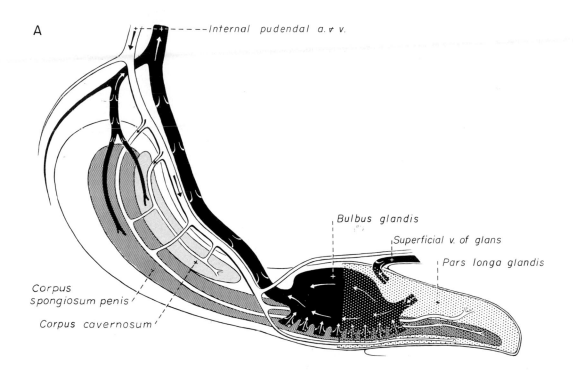

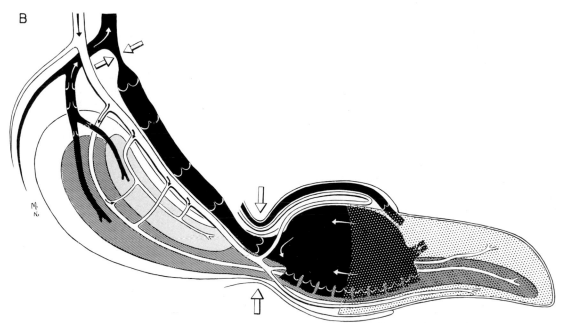

Figure 9-20. Diagrams of circulatory pathways of the penis. (From Christensen 1954.)
A. In non-erection.
B. In erection.

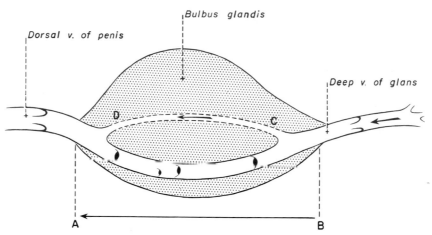

Figure 9–21. Diagram of venous pathways in the bulbus glandis connecting the deep vein of the glans and the dorsal vein of the penis. The ventral shunt (B to A) is the principal route when the penis is relaxed; the dorsal shunt (C to D) is utilized during erection. (From Christensen 1954.)

There is a double connection, on each side of the os penis, between the deep vein of the glans and the dorsal vein of the penis (Fig. 9–21). A ventral shunt, through irregularly located openings, receives blood from the venous spaces of the bulbus glandis and the corpus spongiosum. The openings are directed proximally and are bordered distally by a lip of endothelium which diverts blood toward the dorsal vein of the penis. There is also a dorsal shunt through the bulbus glandis, which is narrower in diameter and less clearly defined than the ventral shunt. Numerous branches go from the dorsal shunt into the spaces of the bulbus. Blood going through the dorsal shunt disseminates into cavernous spaces of the bulbus, whereas blood in the ventral shunt is directed into the dorsal vein of the penis without detouring through the venous sinuses of the bulbar erectile tissue.

Nerves emanating from the pelvic and sacral plexuses supply the penis. These are the paired **pudendal nerve** and the paired **pelvic nerve** (n. erigens of Eckhard, 1863). The **pelvic plexus** lies on the pelvic wall dorsal to the prostate gland, lateral to the rectum. It receives sympathetic fibers through the hypogastric nerve, which runs caudally from the caudal mesenteric plexus. The pelvic plexus receives parasympathetic fibers through the ventral branches of the first, second, and sometimes the third sacral nerves. Infrequently, only fibers from the second sacral nerve go to the plexus.

Fibers leave the pelvic plexus and go to the bladder, prostate, pelvic urethra, rectum, and the penis. Sensory (afferent) fibers are present in the pelvic nerve, as well as efferent parasympathetic fibers, according to Gruber (1963). The **hypogastric nerves** (sympathetic) are responsible for ejaculation and prostatic secretion.

The pudendal nerve (mixed) arises from all three sacral nerves, enters the sacral plexus, and then gives off perineal and rectal branches before continuing as the **dorsal nerve of the penis.** Perineal branches supply the skin around the anus and scrotum. Rectal branches go to the external anal sphincter, ischiocavernosus, bulbospongiosus, ischiourethralis, urethralis, and the retractor penis muscles. The smooth muscle fibers of the retractor penis also receive sympathetic and parasympathetic impulses (Oppenheimer 1938). The dorsal nerves of the penis pass cranially, on the dorsolateral surface of the organ, to the glans penis (sensory nerves of the glans).

Lymph vessels from the penis drain into the **superficial inguinal lymph nodes.**

MECHANISM OF ERECTION

Erection in the dog results from the filling of the spaces of the cavernous bodies with blood. Stimulation of the pelvic nerve causes increased penile blood pressure, partial inhibition of venous drainage (Fig. 9–20), and dilatation of the arteries in the penis (Christensen 1954).The phenomenon of delayed erection in the dog is due to slow engorgement of the bulbus glandis and the pars longa glandis.

Initially, the muscles in the helicine branches of the deep arteries and the arteries of the urethral bulb relax, allowing more blood to enter the cavernous bodies. Arterial blood pressure in the penis rises, but venous outlets are still sufficient to accommodate the increased inflow of blood. The corpus spongiosum receives the greater share of the blood through the artery of the bulb of the penis. Venous blood continues to flow into the bulbus glandis and into the vein of the bulb. Arterial blood is shunted from the artery of the bulb into the corpus cavernosum via anastomotic branches. The deep vein of the penis is not of sufficient diameter to drain the increased amount of arterial blood emptying into the cavernous spaces from the helicine arteries. Internal pressure against the tunica albuginea causes a stiffening of the corpus cavernosum. The intrinsic veins tend to be compressed.

In the second stage, after intromission has occurred, partial venous occlusion becomes a factor in erection of the glans penis. The increased arterial blood supply is largely directed into the dorsal artery of the penis. Blood in the artery of the bulb is shunted to the dorsal artery through circumflex branches. Impulses in the pudendal nerve stimulate contraction of the extrinsic penile muscles: the ischiocavernosus, ischiourethralis, and bulbospongiosus. The ischiourethralis contracts enough to lessen the free flow of venous blood through the common trunk of the dorsal veins. When the paired ischiourethralis muscles contract, the fibrous ring (encircling the common venous trunk) is pulled caudally and laterally, narrowing the lumen of the ring and partially squeezing the vein. The fibrous ring is anchored to the symphysis pelvis by a strong, short ligament. Muscular contraction does not affect the arteries, which lie outside the encircling fibrous ring. The ischiocavernosus and bulbospongiosus muscles, upon contracting, slow down egress of

blood through the deep vein, vein of the bulb of the penis, and the common trunk of these veins. Before intromission, the superficial veins of the glans, as well as the dorsal veins of the penis, allow free flow of venous blood away from the glans. With constriction of the veins leaving the two cavernous bodies, venous blood is further directed into the bulbus glandis via shunts from the corpus spongiosum. More arterial blood enters the capillaries of the pars longa glandis through the branches of the dorsal artery of the penis and the artery of the bulb. Constriction of the sphincter muscle of the female vestibule causes increased occlusion of the dorsal veins of the penis and also prevents blood from leaving the pars longa glandis through the superficial vein of the glans (Fig. 9–20B). Because of their relatively thick muscular walls and greater intrinsic pressure, the arteries of the penis are not occluded by extrinsic muscular contraction.

Since the superficial veins of the glans no longer permit outflow of blood, all blood in the pars longa is directed toward the deep veins of the glans. The spaces of the bulbus glandis receive venous blood from two sources: deep veins of the glans and the corpus spongiosum. When the dorsal veins of the penis permit free exit of blood from the bulbus, the blood traverses the ventral shunt in the bulbus without expanding the erectile tissue. With the dorsal veins partially occluded, and the inflow of blood greatly increased, blood entering the bulbus through the deep vein of the glans is forced into the dorsal bulbar shunt (Fig. 9–21). Because of the abundance of openings in the dorsal shunt, excess venous blood is permitted to enter and engorge the spaces of the bulbus glandis. The caliber of the dorsal veins of the penis is not sufficient for them to accommodate this increased amount of blood. Valves in the deep veins of the glans and in the venous connections with the corpus spongiosum prevent blood from leaving the bulbus by any route except through the now inefficient dorsal penile veins.

The engorgement of the bulbus glandis and pars longa glandis is facilitated by relaxation of the smooth muscles of the intersinusoidal trabeculae and by stretching of the elastic fibers in the trabeculae.

The branches of the dorsal artery of the penis lose their helicine-like appearance as the pars longa glandis is distended. The deep branch of the dorsal artery uncoils and straightens out during the height of erection. As the glans penis becomes engorged and firm, less arterial blood can be accommodated, and it is shunted from the dorsal artery to the corpus cavernosum, increasing its stiffness (Fig. 9–20B).

After ejaculation, the extrinsic penile muscles relax and arterial blood pressure drops to normal. Venous pressure declines as the erectile bodies shrink. The bulbus glandis decreases in diameter sooner than the pars longa glandis, owing to its greater venous drainage. Elastic recoil of the intersinusoidal trabeculae helps force blood out of the glans.

For further physiological discussion of the mechanism of erection, refer to: Eckhard (1863), Francois-Franck (1895), Langley (1896), Hart and Kitchell (1965), Nitschke (1966), and Dorr and Brody (1967). Hart (1972) recorded tonic contractions of the ischiourethralis muscle, which inhibits venous return in the dorsal vein, and rhythmical contractions of the bulbospongiosus, which forces blood distally. Purohit and Beckett (1976) found high arterial pressures in the corpora cavernosa that were correlated with contractions of the ischiocavernous muscles. This indicates a more active role for the corpora cavernosa than was previously believed. Grandage (1972) divides coitus into a first stage (mounted) and a second stage of longer duration, when the male dismounts and faces in the opposite direction during the "tie" or "lock." During the second stage there is a 180° bend in the middle of the body of the penis which occludes the emissary veins and prevents detumescence. Using radiographic techniques, Grandage was able to determine that the penis displaced the cervix to the level of the sacral promentory. The glans and about 3 cm. of the body enter the vagina. During this paradox of flexible rigidity, the dorsal surface of the penis remains dorsal, since it is a bend of the corpus penis and not a twist. Although the sperm-rich fraction of the ejaculate is passed within 80 seconds of first-stage coitus, another 30 ml. of seminal fluid is produced during second-stage coitus, which probably aids passage into the uterus

and possibly stimulates peristaltic contractions of the uterine tubes. Study of psychosexuality in dogs has been made by Gantt (1938, 1949).

ANOMALIES AND VARIATIONS

Congenitally, the penis may be short or bent. Short retractor penis muscles may inhibit erection. Anomalies may be corrected by surgical treatment, but if the condition is congenital it is not advisable to use the animal for breeding purposes.

The flaccid penis varies greatly in length and diameter, depending on individual variations as well as on breed and size of the animal. Physiological changes caused by temperature, urination, and sexual excitement also contribute to marked variations of size.

THE MALE URETHRA

The urethra is the canal that conveys urine from the bladder to the exterior of the body. The male **urethra** (*urethra masculina*) carries both urine and seminal secretions to the distal end of the pars longa glandis. It varies from approximately 10 to 35 cm. in length, depending on the size of the dog. In a 25-pound mature dog, it averages 25 cm. in length. It is divisible into three portions: the prostatic, the membranous, and the cavernous, or penile.

The **prostatic portion** (*pars prostatica*) passes through the prostate gland (Fig. 9–5). It extends from the urinary bladder to the caudal edge of the prostate, where the membranous urethra begins. The walls of the relaxed prostatic urethra are made up of a variable number of longitudinal mucosal folds. When distended, all the folds, except the dorsally located **urethral crest** (*crista urethralis*), are obliterated. The **seminal hillock** (*colliculus seminalis*) is an oval enlargement, located at the center of the urethral crest, which protrudes into the lumen of the urethra. The center of the colliculus seminalis contains a minute opening into a tiny tube (*uterus masculinus*), which runs craniodorsally into the prostate. The uterus masculinus (**prostatic utricle, utriculus prostaticus**) is a homologue of the caudal portion of the vagina (fused müllerian ducts) in the female. The deferent ducts open on each side of the colliculus seminalis on slightly different transverse planes. The openings are usually not visible macroscopically unless fluid is forced from the ducts into the urethra. Also, numerous prostatic ducts open into the urethra adjacent to and surrounding the urethral crest. Instead of being round or oval, a cross section through the middle of the prostatic urethra appears U-shaped.

The **membranous portion** of the urethra *(pars membranacea)* is located between the prostate gland and the point where the urethra dips into the bulb of the penis. Since the dog does not possess a urogenital diaphragm, this part of the urethra is located dorsal to the symphysis pelvis only as far caudally as the ischial arch. When the urinary bladder is contracted, the length of the membranous urethra averages 4 to 5 cm. in a 25-pound dog.

The **cavernous, or penile, portion** of the urethra begins at the entrance of the membranous urethra into the bulb of the corpus spongiosum and extends to the external opening in the distal end of the pars longa glandis. There is no fossa navicularis along its course through the glans penis, such as is present in the horse.

The prostatic and membranous portions of the urethra are lined by transitional epithelium. The mucosa of the cavernous urethra, also folded when relaxed, is transitional except near the external urethral opening, where it changes to stratified squamous epithelium similar to that covering the glans (Trautmann and Fiebiger 1952). The crypts between the mucosal folds are called the lacunae of Morgagni. The stratum vasculare of the membranous urethra varies slightly in its distribution. Typically, it extends from the bulb caudally to a point one-half to one-third of the distance to the prostate gland. This layer is composed of highly vascular erectile tissue which is continuous in the penis with the corpus spongiosum. Peripheral to the vascular layer is the glandular layer (urethral glands) and the muscular layer that extends from the prostate, which it overlaps slightly, to the point where the membranous urethra ends at the urethral bulb. The urethral muscle is composed of an inner layer of smooth muscle fibers, primarily longitudinal in direction, and an outer layer of transversely running striated musculature which is separated dorsally by a thin, longitudinal fibrous raphe.

VESSELS AND NERVES

The prostatic urethra is supplied with blood through the **prostatic artery.** The membranous urethra is supplied by small **urethral arteries** which branch off the internal pudendal, urethral, or prostatic arteries (Fig. 9–6). The cavernous urethra is supplied in the same manner as the corpus spongiosum, through the **artery of the urethral bulb.** The **urethral veins** are satellites of the arteries, draining into the internal pudendal vein. The smooth muscles of the urethra are innervated by the autonomic nerves derived from the **pelvic plexus.**

VARIATIONS

The length and diameter of the urethra vary within wide limits. When the penis is flaccid, the urethral mucosa is folded longitudinally and the lumen is obliterated. During urination or ejaculation, the urethral walls are distended. Only that part of the cavernous urethra which passes in the ventral groove of the os penis is limited in its expansion. Fracture of the os (Stead 1972) can obstruct the urethra. The urethra may infrequently open on the ventral surface of the penis (hypospadias) or on the dorsal surface (epispadias). Hypospadias is considered to result from failure of the urethral groove to close normally, whereas epispadias is caused by an embryonic displacement of cells forming the cloacal membrane, resulting in a reversal of the cloacal membrane location.

Woodburne and Lapides (1972) studied the size and shape of the dog ureter during peristaltic enlargement. They found it to enlarge 17 times (up to 2 mm.²) during diuresis by thinning of the muscle coats. The collapsed ureter has a stellate lumen with the epithelial surfaces in contact.

THE PREPUCE

The **prepuce** *(preputium)* (Fig. 9–6, 9–11) is a complete tubular sheath of integument which, when the penis is not erect, contains and covers the pars longa glandis and part of the bulbus glandis. It is firmly attached to and continuous with the skin of the ventral abdominal wall.

STRUCTURE

The prepuce is composed of two layers of integument dorsally, except for the cranial 1 to 3 cm., where it is free of the abdominal wall and there are three layers. Ventrally and laterally there are three layers of integument. The outer layer is skin. The inner layers, parietal and visceral, are made up of stratified squamous epithelium which is smooth and thin, and stippled with lymph nodules and nodes. These are more numerous on the parietal layer and along the fornix of the preputial cavity than on the visceral layer. The **parietal, or middle, layer** is a continuation of the outer skin layer onto the wall of the preputial cavity. It extends to the fornix, which is located in a transverse plane through the middle of the bulbus glandis. In erection, the parietal layer is reflected from the preputial orifice directly onto the proximal half of the bulbus glandis and the body of the penis. The **visceral, or inner, layer** extends from the preputial fornix to the external urethral orifice, where it is continuous internally with the cavernous urethra. Each **preputial muscle** (*protractor preputii*) is a divergent strip of cutaneus trunci muscle which extends from the area of the xiphoid cartilage to the dorsal wall of the prepuce. There it inserts in apposition with the preputial muscle of the opposite side. Two functions are attributed to the preputial muscles — to prevent the cranial free end of the prepuce from hanging loosely in non-erection and to pull the prepuce back over the glans penis after erection. During erection, the preputial muscles are relaxed.

VESSELS AND NERVES

The parietal layer of the prepuce is supplied by an intricate, anastomosing network of arteries located between the outer and the middle layer of the prepuce. The arteries are branches of the **external pudendal artery** and the preputial branch of the **dorsal artery of the penis** (Fig. 9–6). The inner, visceral layer of the prepuce, which invests the glans penis, is supplied by the three anastomosing branches of the dorsal artery of the penis, the external pudendal artery, and to a lesser degree by the **artery of the bulb of the penis.**

The veins from the parietal layer drain into the **external pudendal vein.** The visceral layer is drained by the **superficial vein of the glans** into the external pudendal vein and, deeply, by the **dorsal vein of the penis** via the **deep vein of the glans** and the *bulbus glandis.*

Preputial lymph vessels go to the **superficial inguinal lymph nodes.**

Sensory (afferent) innervation of the visceral layer of the prepuce, covering the glans, is through the **dorsal nerve of the penis** to the **pudendal.** Superficial branches of the **ilioinguinal** and **iliohypogastric nerves** innervate the parietal layer.

ANOMALIES AND VARIATIONS

The two most common anomalies of the prepuce are phimosis and paraphimosis. Phimosis is the existence of an abnormally small preputial opening, which prevents protrusion of the penis. Paraphimosis is the inversion of the preputial opening forming a constrictive ring after penile protrusion, which prevents the return of the glans penis into the preputial cavity. Persistence of the preputial frenulum may also result in the inability to protrude the penis.

Female Genital Organs

The female genital organs consist of ovaries, uterine tubes (oviducts), uterus, vagina, and vulva. The ovaries produce the female sex cells (oocytes), which are transported by the uterine tubes to the uterus.

There implantation takes place, if fertilization has occurred, and the embryo is developed. The vagina is a canal leading from the uterus to the external genitalia, the vulva.

THE BROAD LIGAMENTS

The ovaries, oviducts, and uterus are attached to the dorsolateral walls of the abdominal cavity and to the lateral walls of the pelvic cavity by paired double folds of peritoneum called the right and left **broad ligaments** (*plicae latae uteri*) (Fig. 9–1, 9–22). Each broad ligament contains an ovary, uterine tube, and uterine horn. It also contains vessels and nerves to the genitalia and finger-like streaks of fat. It does not support or suspend the genitalia in the body cavities, but rather unites its components.

The broad ligament is attached dorsally along or near the junction of the psoas and transversus abdominis muscles. Cranially, it is attached by means of the suspensory ligament of the ovary to the junction of the middle and distal thirds of the last rib. The ligament is reflected off the vagina onto the rectum dorsally, ventrally onto the urethra and bladder, and in a curved line laterally onto the wall of the pelvic cavity as far as the internal inguinal ring. The ligament is broadest at the level of the ovary, tapering from there to its cranial and caudal extremities. In a 25-pound dog, the distance

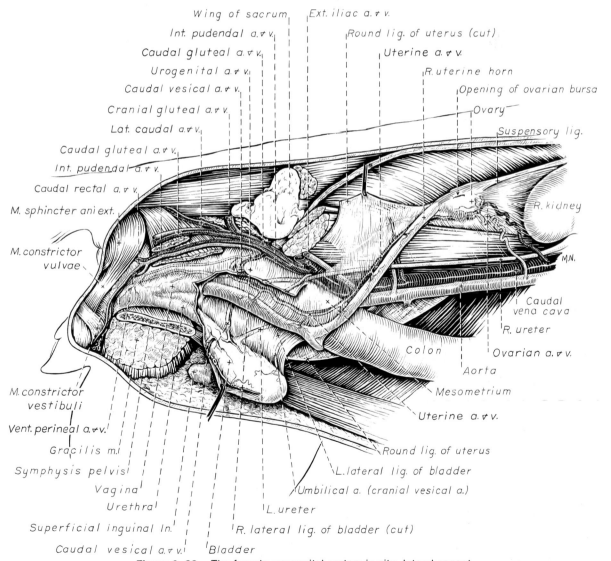

Figure 9–22. The female urogenital system in situ, lateral aspect.

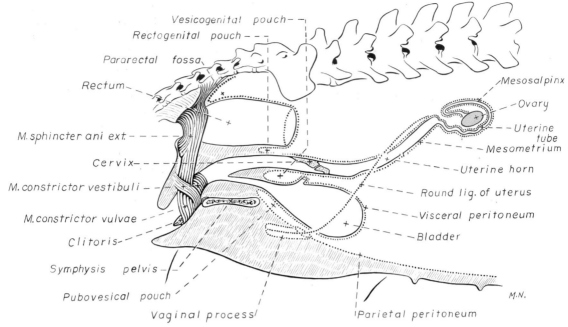

Figure 9–23. Diagram of peritoneal reflections and the female genitalia.

spanned by the broad ligament at the ovarian level varies from 6 to 9 cm. A peritoneal fold arises from the lateral surface of the broad ligament and extends from the ovary to or through the inguinal canal. It contains the round ligament of the uterus in its free border (Fig. 9–23). The peritoneal extension of the broad ligament through the inguinal canal into the subcutaneous region of the vulva, known as the **vaginal process**, corresponds to the peritoneal outpocketing (vaginal tunics) into the scrotum of the male. The vaginal process of the female varies from none at all, when the peritoneum does not even dip into the internal inguinal ring (no vaginal ring), to a long process of peritoneum containing the round ligament of the uterus and fat. It extends into the subcutaneous tissue of the labia. The round ligament of the uterus and the ovarian ligaments contained in the broad ligament are described more fully in the discussions of the uterus and ovaries.

Morphologically, the broad ligament is divided into three regions: mesovarium, mesosalpinx, and mesometrium. The *mesovarium* is that part of the broad ligament which attaches the ovary to the dorsolateral region of the abdominal wall. The cranial

boundary of the broad ligament, where the suspensory ligament of the ovary attaches to or near to the thirteenth rib, marks its beginning, and it ends at a transverse plane just caudal to the ovary. It contains the utero-ovarian vessels (Fig. 9–22).

The *mesosalpinx,* another double fold of peritoneum, extends laterally from the dorsal peritoneal layer of the mesovarium. It curves around the dorsal and ventrolateral borders of the ovary to attach to the medial surface of the broad ligament just dorsal to the ovary. It encloses the ovary within a small peritoneal cavity, the ovarian bursa. The ovarian bursa is variable in size, depending on the age and size of the animal. It averages 2 cm. in length in a 25-pound dog. The amount of fat between the peritoneal layers of the mesosalpinx depends on the condition of the animal. The bursa is completely closed except for a narrow slit in its ventromedial surface, connecting it with the peritoneal cavity (Figs. 9–22, 9–24). In a 25-pound dog, the opening averages 0.8 cm. in length. Topographically, it is located caudoventral to the kidney. Several structures come close together at the opening of the bursa. These are the proper and suspensory ligaments of the ovary, and the fimbriae arising from the ovarian end of the uterine tube. The entire uterine tube lies between the peritoneal layers of the mesosalpinx.

The *mesometrium* begins at the cranial edge of the uterine horn, where it is continuous with the mesovarium, and extends caudally to a point where the peritoneum of the broad ligament reflects onto the bladder and the colon. It leaves the uterine horn, the body of the uterus, the cervix, and the cranial part of the vagina to attach along the abdominal and pelvic walls. The cranial and

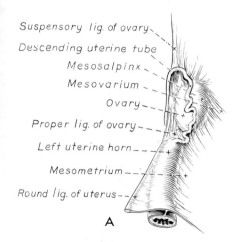

Suspensory lig. of ovary
Descending uterine tube
Mesosalpinx
Mesovarium
Ovary
Proper lig. of ovary
Left uterine horn
Mesometrium
Round lig. of uterus

A

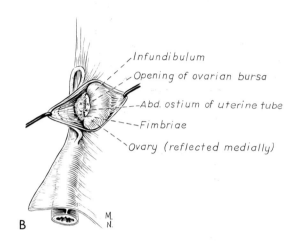

Infundibulum
Opening of ovarian bursa
Abd. ostium of uterine tube
Fimbriae
Ovary (reflected medially)

B

M. N.

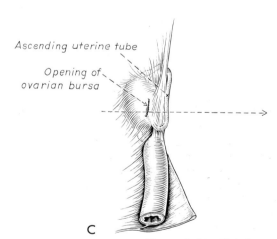

Ascending uterine tube
Opening of ovarian bursa

C

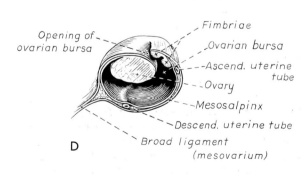

Opening of ovarian bursa
Fimbriae
Ovarian bursa
Ascend. uterine tube
Ovary
Mesosalpinx
Descend. uterine tube
Broad ligament (mesovarium)

D

Figure 9–24. Relations of left ovary and ovarian bursa.
A. Lateral aspect.
B. Lateral aspect, ovarian bursa opened.
C. Medial aspect.
D. Section through ovary and ovarian bursa.

caudal uterine arteries and veins run between the peritoneal layers of the mesometrium.

THE OVARIES

The **ovaries** *(ovaria)*, or female gonads, are paired oval organs, located in the abdominal cavity caudal to the kidneys. They are the sites of oocyte maturation, as well as the source of certain hormones. In a 25-pound dog, an ovary averages 1.5 cm. in length, 0.7 cm. in width, 0.5 cm. in thickness, and 0.3 gm. in weight. In its normal position, an ovary may be described as having cranial and caudal poles, medial and lateral borders, and dorsal and ventral surfaces. The ovary is smooth in appearance before estrus, which occurs for the first time between 6 and 9 months of age. In multiparous bitches its surface is rough and nodular.

In a sexually mature, 25-pound dog, the left ovary is located approximately 12 cm. caudal to the middle of the thirteenth rib and 1 to 3 cm. caudal to the corresponding kidney. Typically, it lies between the abdominal wall and the left colon. The right ovary is located approximately 10 cm. caudal to the last rib of the right side. The ventral border and medial surface of the ovary are in contact with the mesovarium. In a young animal it lies ventral to the adipose capsule of the right kidney and dorsal to the descending duodenum. In animals that have undergone numerous pregnancies, both right and left ovaries migrate caudally and ventrally.

LIGAMENTS

In addition to the mesovarium (cranial portion of the broad ligament), the ovary has two other ligamentous attachments. The **suspensory ligament of the ovary** *(plica suspensoria ovarii)* is attached cranially to the middle and ventral thirds of the last one or two ribs. Caudally, it attaches to the ventral aspect of the ovary and mesosalpinx, lying between the opening of the ovarian bursa and the ascending uterine tube (Fig. 9–24). The suspensory ligament lies between the two layers of peritoneum and forms the cranial portion of the free border of

the broad ligament. It is continued caudally by the **proper ligament of the ovary** (**ovarian ligament,** *chorda utero-ovarica).* This in turn attaches to the cranial end of the uterine horn. There it is continuous with the round ligament of the uterus, which extends caudally toward the vaginal process. Both the proper and the suspensory ovarian ligament are composed of connective tissue mixed with smooth muscle fibers.

STRUCTURE

The ovary is divided into a medulla and a cortex. The *medulla (zona vasculosa)* contains blood vessels, nerves, lymphatics, smooth muscle fibers, and connective tissue fibers. The *cortex* consists of a connective tissue stroma which contains a large number of follicles. For a review of ovarian structures in several mammals, including the dog, see Mossman and Duke (1973).

The connective tissue condenses to form the *tunica albuginea* around the ovary. The tunica albuginea is covered by peritoneum referred to as the **superficial epithelium of the ovary.** (The latter term replaces the unsuitable but much used name, germinal epithelium.) Follicles are present deep to the tunica albuginea. A **follicle** consists of an oocyte and its surrounding granulosa cells, enclosed in a basement membrane which separates the follicle from the ovarian stroma.

With each estrus a number of follicles mature. The granulosa cells at first form a single cuboidal layer around the oocyte, which constitutes a **primary follicle** *(folliculus ovarici primarii).* With further maturation, several layers of granulosa cells are formed around the oocyte. Eventually, a cavity, filled with follicular fluid, forms within the granulosa cell mass. Such a follicle is designated as a **vesicular follicle** *(folliculus ovarii vesiculosi),* formerly termed a graafian follicle. At one end of the cavity there is a hillock, *cumulus oophorus,* which contains the maturing ovum. In intimate contact with the ovum is a clear membrane, the *zona pellucida.* This is surrounded by a layer of radially arranged granulosa cells, the *corona radiata.* As the **follicular fluid** *(liquor folliculi)* increases, the follicle migrates to the periphery of the ovary. When the follicle is under considerable tension by pressure of the fluid, it ruptures, and the ovum is released into the **ovarian bursa**, which is actually an outpocketing of the peritoneal cavity (Fig. 9–24). Strassmann (1941) describes a "theca cone" in the ovary of the dog, human, and other mammals. It is formed by eccentric growth of the follicle and the one-sided proliferation of the theca interna toward the ovarian surface. It has the special function of bringing the follicle up to that part of the ovarian surface from which the ovum can escape to the peritoneal cavity. Not all follicles which begin to mature proceed to ovulation. The vast majority of follicles degenerate at all stages of development throughout the life of the bitch, as they do in all mammals (Perry 1972).

After ovulation, relatively slight hemorrhage occurs, filling the follicular cavity. As this is resorbed, the *corpus luteum* (yellow body) is formed from the granulosa and theca interna cells. If fertilization does not take place, the corpus luteum gradually degenerates into a connective tissue scar, the *corpus albicans*. If the ovum is fertilized, the corpus luteum remains fully developed throughout pregnancy. After parturition, it regresses. Involution of the corpus luteum again allows vesicular follicles to mature. Estrus in the bitch usually occurs twice a year, in the spring and again in the fall. For a general account of reproduction in mammals, several comprehensive reviews are available. Among them are Austin and Short (1972), Segal et al. (1973), and Cole and Cupps (1977).

The cyclic changes of the ovary and the hormonal control of estrus are discussed by Perry (1972) and Concannon, Hansel, and Visek (1975). The latter authors have measured plasma estrogen, luteinizing hormone (LH), and progesterone levels in cycling bitches. Changes in sexual behavior correlated with hormone levels have been reported by Concannon, Hansel, and McEntee (1977).

Raps (1948) made a study of the developmental changes in the dog ovary, from two days after birth to the sixth postnatal month. His data show that primordial ovocytes surrounded by follicular epithelial cells are present at 4 days of age; that at 15 days true primary follicle formation with granulosa cell development occurs; and that antrum formation is not observable until 6 months of age. A laparoscopic study of the ovary of a cycling bitch was made by Wildt et al. (1977).

VESSELS AND NERVES

The ovary is supplied with blood through the **ovarian artery**. Homologous to the testicular artery of the male, the ovarian artery arises from the aorta approximately one-third to one-half the distance from the renal arteries to the deep circumflex iliac arteries. Usually the right ovarian artery arises slightly cranial to the left (Fig. 9–1). The degree of uterine development determines the tortuosity and size as well as the position of the artery. In a nulliparous animal, the artery extends laterally almost at right angles from the aorta, whereas in late pregnancy it is drawn cranioventrally, along with the ovary, by the enlarged, heavy uterus. In addition to supplying the ovary, the ovarian artery supplies branches to the adipose and fibrous capsules of the kidney. In addition, small tortuous branches supply the uterine tube and uterus. Caudally, the artery anastomoses with the **uterine artery** (a branch of the urogenital artery). Through this anastomotic connection, the uterine artery may be considered as a supplementary source of arterial blood to the ovary. The arteries to the ovary supply the parenchyma of the medulla and cortex as well as the thecae of the follicles. Capillary loops become extensive during follicular enlargement but recede or disappear during corpus luteum regression.

The right and left **ovarian veins** have different terminations. The right vein drains into the caudal vena cava, whereas the left enters the left renal vein. Similar to the corresponding arteries, the **uterine vein** and ovarian vein anastomose between the peritoneal layers of the broad ligament. The ovarian vein receives a tributary which comes from the medial edge of the suspensory ligament of the ovary and the lateral surface of the kidney. In some instances, the vein will also anastomose with the deep circumflex iliac vein. The arteries and veins of the ovaries in the dog have been studied by Del Campo and Ginther (1974).

The lymphatics drain into the **lumbar lymph nodes**. Polano (1903) has demonstrated the ovarian lymphatics of the dog.

The nerve supply to the ovaries is via the sympathetic division of the autonomic nervous system. The nerves reach the ovaries by way of the **renal** and **aortic plexuses**, which receive nerve fibers from the fourth, fifth, and sixth lumbar ganglia of the sympathetic trunk (via the caudal mesenteric plexus). They accompany the ovarian artery to the ovary. The ovarian blood vessels receive an abundant sympathetic nerve supply, but, according to Kuntz (1919a), the ovarian follicles and interstitial secretory tissue are devoid of sympathetic innervation.

ANOMALIES AND VARIATIONS

The ovaries may be hypoplastic, displaced, or completely missing. Rarely, an ovary may descend through the inguinal canal, in the manner of a testis, to rest in the vulvar region. Follicular cysts rarely occur. According to McEntee and Zepp (1953), ovarian tumors are relatively infrequent in the dog.

Physiologically, the ovary changes in size and shape during pregnancy (development of a large corpus luteum) and with advanced age (contracture of the corpora albicantia). Pregnancy will cause the ovary to shift cranioventrally in position.

UTERINE TUBE

The **uterine tube** or oviduct (*tuba uterina*) transports the ocytes to the uterus. Each uterine tube is located between the peritoneal layers of the mesosalpinx and connects the peritoneal cavity with the uterine cavity. The uterine tube averages 4 to 7 cm. in length and 1 to 3 mm. in diameter. The ovarian extremity of the tube, the infundibulum, is located near the edge of the opening into the **ovarian bursa** (Fig. 9–24). The *infundibulum* is the funnel-shaped origin of the uterine tube which narrows into a minute opening, the abdominal ostium. The edges of the infundibulum are fringed by numerous diverging, finger-like processes, the *fimbriae*. Fimbriae are usually visible projecting out of the opening of

the ovarian bursa. They mark the junction of peritoneum (mesosalpinx) with the mucous membrane lining the oviduct. The fimbriae are most visible from within the ovarian bursa. They are very vascular. From the time of ovulation until envelopment by the fimbriae, oocytes are actually in the peritoneal cavity. At the time of ovulation, the fimbriae are swollen and capable of movement so as to engulf an ovulated oocyte and block the entrance to the ovarian bursa. Follicular fluid and tubal mucous also play a role in drawing the oocyte into the uterine tube. Almost all oocytes ovulated reach the uterine horns.

From the abdominal ostium, the uterine tube at first runs craniolaterally between the ventral layers of the mesosalpinx. Approximately halfway between the caudal extremity of the ovary and the cranial tip of the uterine horn, it bends sharply cranially and runs along the free edge of the suspensory ligament of the ovary. Approximately 0.5 cm. cranial to the gonad, it swings onto the dorsal aspect of the suspensory ligament and, still between peritoneal layers of the mesosalpinx, runs in a tortuous manner caudomesially toward the ovary. At the middle of the ovary it curves caudolaterally toward the apex of the uterine horn, where it terminates. The opening of the uterine tube into the horn of the uterus is called the uterine ostium.

Oocytes are moved down the uterine tube toward the uterus principally by peristaltic movements rather than by action of cilia. Fertilization, the union of ovum and sperm, normally takes place in the infundibulum.

STRUCTURE

The uterine tube is covered by a *tunica serosa* which is composed of the peritoneum making up the mesosalpinx. The muscular layer of the duct is composed primarily of circular bundles of fibers, but a variable number of longitundinal and oblique fibers are also present. The muscular layer reaches its greatest development near its union with the circular muscles of the uterine horn, where it forms the tubouterine junction. The innermost layer (*tunica mucosa*) is made up of partially ciliated simple columnar epithelium, the motion of the cilia being directed toward the uterine horn. The mucosa is folded slightly (*plicae tubariae*) near the uterine ostium.

VESSELS AND NERVES

The uterine tube is supplied by the **ovarian** and **uterine arteries**. The two vessels anastomose near the cranial extremity of the uterine horn. The veins are satellites of the arteries. The lymphatics follow the ovarian lymph ducts to the **lumbar lymph nodes**. Anderson (1927), Sampson (1937), and Ramsey (1946) have worked out the detailed anatomy of the lymphatics of the uterine tube in various species of mammals. For the most part, the numerous lymphatic channels in the mucosal folds of the infundibulum and fimbriae, as well as those of the tube, drain into the lymph vessels of the mesosalpinx.

Nerves to the uterine tube are derived primarily from the thoracolumbar (sympathetic) division of the autonomic nervous system and pass through the **aortic** and **renal plexuses**. Fibers from the **pelvic plexus** (parasympathetic) also innervate the uterine tube (Mitchell 1938).

ANOMALIES AND VARIATIONS

Associated with the uterine tube in the mesosalpinx are the epoophoron and paroophoron. The epoophoron is homologous with the embryonic male appendix epididymis and efferent duct, whereas the paroophoron is homologous with the paradidymis (tubules) of the testis. In the female, these structures are rudimentary and represent the embryonic mesonephros and mesonephric (wolffian) duct. Hydatids of Morgagni (hydatides terminales) in the adult female are pedunculated cysts, located on the fimbriae, which represent part of the epoophoron. The paroophoron may persist in the mesosalpinx and occasionally give rise to cysts. Congenital stenosis of the oviduct is not infrequent.

THE UTERUS

The uterus is a hollow muscular organ which serves as the habitation for the developing young. It gives attachment to the fer-tilized ovum and functions as a source of fetal nourishment. It also serves as the route by which the sperm may reach the oocyte in the uterine tube.

The uterus consists of a neck or cervix (cervix uteri), a body (corpus uteri), and two horns (cornua uteri). It is a tubular, Y-shaped organ which communicates with the oviducts cranially and the vagina caudally (Fig. 9–25). Its size varies considerably, depending on age, size of animal, number of previous pregnancies, and whether the animal is currently pregnant. In nulliparous mature female dogs of 25 pounds weight, the uterine horns average 10 to 14 cm. long and 0.5 to 1 cm. in diameter. They diverge from the body of the uterus at a point 4 to 5 cm. cranial to the symphysis pelvis. The uterine body is 1.4 to 3 cm. long and 0.8 to 1 cm. in diameter. The **cervix** (*cervix uteri*) averages 1.5 to 2 cm. long. A small caudal portion of the cervix may protrude 0.5 to 1 cm. into the vagina. The diameter of this intravaginal cervix is approximately 0.8 cm. The gravid uterus in the latter third of pregnancy (Fig. 2–15C) may occupy any portion of the abdominal cavity, and the uterine horns frequently change in their orientation to one another, although they are somewhat limited by the suspensory and round ligaments. During uterine distention, the horns flex near the middle of their dorsal surfaces, bending the uterus cranially and ventrally upon itself. Both the ovary, with its suspensory ligament, and the vagina are drawn slightly cranioventrally.

The **uterine horns** (*cornua uteri*) are musculomembranous tubes, slightly flattened dorsoventrally, which unite at the body of the uterus. The horns are located entirely within the abdomen. The cranial tip of each horn is connected to the ovary by the proper ligament and indirectly by the broad ligament. The oviduct opens into the uterine horn at or near its tip. From a lateral view, the horns have an **S**-shaped appearance, with the ovarian ends being most dorsad. Typically, the right horn is slightly longer than the left.

The **body** of the uterus (*corpus uteri*) is usually located in both the pelvic and the abdominal cavity. Generally, the largest portion is in the abdomen, and in multiparous bitches the entire uterine body may be located cranial to the brim of the pelvis.

The body extends from the point of convergence of the uterine horns to the cervix (Fig. 9–25). There are three openings into the uterine body: one from each uterine horn and one from the internal orifice of the cervix. An internal musculomembranous projection extends 1 cm. into the body of the uterus, separating the horns. This partition is not discernible externally. The canal of the cervix is directed caudoventrally from uterus to vagina. The cervix lies diagonally across the uterovaginal junction. Its ventral border attaches to the uterine wall cranial to its dorsal attachment. Consequently, the **internal orifice** of the cervical canal (*ostium uteri internum*) is facing almost directly dorsally, whereas the **external orifice** (*ostium uteri externum*) is directed toward the vaginal floor. The external uterine orifice opens on a hillock projecting into the vagina on a dorsal median fold (Pineda et al. 1973). The cervical canal averages 0.5 to 1 cm. in length and is closed during pregnancy by a mucous plug.

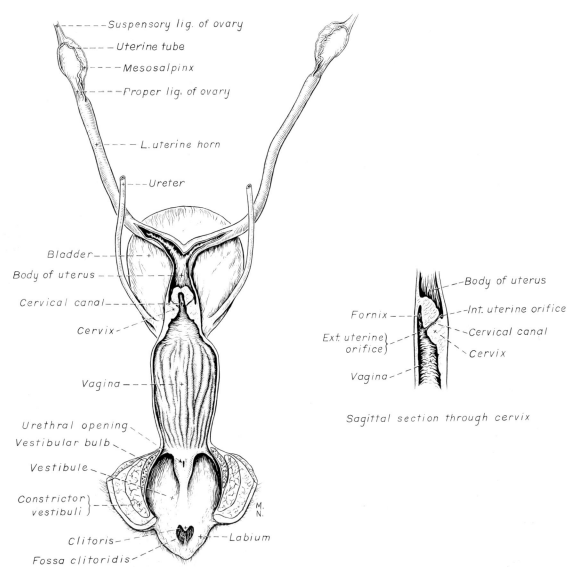

Figure 9–25. Dorsal view of female genitalia, partially opened on midline. Smaller view shows sagittal section through cervix.

RELATIONS

Dorsally, the uterus is in contact with the descending colon, the psoas muscles, the transverse abdominal muscle, and the ureters. Ventrally, it contacts the urinary bladder, the greater omentum, the jejunum, the ileum, and the descending duodenum (Fig. 9–22). The parietal and visceral peritoneum investing these structures is, of course, interposed between the above organs and structures as well as between the neighboring blood vessels, nerves, and lymphatics. The **rectouterine space**, a potential peritoneal cavity between the rectum and the uterus, is continuous with the pararectal fossa. The **vesicouterine space**, between the urinary bladder and the uterus (with its attached broad ligament), is separated from the paravesical fossa by double peritoneal folds, the lateral ligaments of the bladder.

LIGAMENTS

The **broad ligaments**, containing some fat and unstriped muscle, attach the uterus and ovaries to the body wall. The *mesometrium* is that part of the broad ligament which attaches the uterus to the dorsolateral body wall. The *mesovarium* and *mesosalpinx* have been discussed in the descriptions of the ovary and of the uterine tube. The mesometrium begins on a transverse plane through the cranial end of the uterine horn and extends caudally as far as the cranial end of the vagina. It is attached peripherally to the lateral pelvic wall. The medial surfaces of the uterine horns are connected to each other for approximately 1 cm. by a triangular-shaped, double layer of peritoneum. The mesometrium and the lateral ligament of the bladder fuse at their attachments to the pelvic wall.

The **round ligament of the uterus** *(ligamentum teres uteri)* is attached to the cranial tip of the ipsilateral uterine horn and is a caudal continuation of the proper ligament of the ovary that has shifted its origin caudally. These ligaments consist largely of smooth muscle, allowing for stretching during pregnancy. The round ligament runs in the free edge of the peritoneal fold given off from the lateral surface of the mesometrium. It extends caudally, ventrally, and mesially,

toward the deep inguinal ring. In most bitches the round ligament, with its investment of peritoneum, passes through the inguinal canal and terminates subcutaneously in or near the vulva. Zietschmann (1928) found the vaginal process to be absent in some dogs. The vaginal process is accompanied in its course through the inguinal canal by the external spermatic nerve, the external pudendal artery and vein, and fat. The fascial layers enveloping the vaginal process are the same as those described with the spermatic cord and vaginal tunics of the male.

STRUCTURE

The uterus is made up of three tunics: serosa, muscularis, and mucosa. The *tunica serosa* is the layer of peritoneum which covers the entire uterus. It is continuous with the mesometrium of the broad ligament.

The *tunica muscularis (myometrium)* consists of a thin, longitudinal outer layer and a thick, circular inner layer of involuntary muscle. Within the circular layer, close to its junction with the longitudinal layer, is a vascular layer containing blood vessels, nerves, and circular and oblique muscle fibers. The circular layer is especially thick in the region of the cervix.

In describing the process of labor in the bitch, Rudolph and Ivy (1930) discuss the action of uterine musculature in detail. Briefly, the fetus is advanced by a strong circular contraction that progresses like a cylindrical band, and by a longitudinal shortening. The "retreat" of the fetus is prevented by a persistent longitudinal contraction. In the body of the uterus, transverse circular contraction, with some longitudinal shortening, moves the fetus into the vagina. Contraction of the abdominal muscles, as well as of the vaginal musculature, causes the final expulsion of the fetus. Reynolds (1937) describes uterine motility, during estrus, as being a series of simple myometrial contraction waves, since intermediation of an intrinsic innervation is not essential to it. Verma and Chibuzo (1974) found that sectioning the hypogastric and pelvic nerves did not alter the frequency or the amplitude of uterine contractions. For an account of birth, see Naaktgeboren and

Slijper (1970); and for endocrine parameters of pregnancy and parturition, Concannon et al. (1977).

The *tunica mucosa* (*endometrium*) is the thickest of the three uterine tunics. Facing the lumen of the uterus is a layer of low columnar epithelium whose cells are only temporarily ciliated (Trautmann and Fiebiger 1952). Simple branched tubular glands are present in the lamina propria. Opening into the uterine cavity, these glands are generally very long and are separated by shorter, inconstant glands or crypts. The long glands in the bitch show relatively little branching or coiling, in contrast with those of the mare or cow. They generally traverse the entire thickness of the endometrium. Grossly, the mucosal surface of the uterus is reddish in color and may either be smooth or contain low longitudinal ridges which obliterate the uterine cavity in the non-pregnant state. The cervical canal does not contain the relatively high mucosal folds observed in other domestic animals, but is closed in pregnancy by a mucous plug and muscular contraction.

VESSELS AND NERVES

The uterus is supplied with arterial blood via the **ovarian** and **uterine arteries**. See Del Campo and Ginther (1974) for the details of vascularization as seen in cleared tissues. The origin of the ovarian arteries from the aorta has been discussed with the description of the ovaries and of the uterine tubes. The ovarian artery anastomoses with the uterine artery, one of the principal branches of the vaginal artery. The artery enters the mesometrium at the level of the cervix (Figs. 9–1, 9–22). Upon entering the broad ligament, the artery lies relatively close to the body of the uterus. It diverges from the uterine horn until it approaches the cranial extremity of the horn, where it anastomoses with the ovarian artery. The uterine artery ramifies in the wall of the uterus and in the mesometrium. Branches supply both sides of the uterine horn.

The **uterine** and **ovarian veins** follow a course similar to that of the arteries, except at their terminations. The right ovarian vein empties into the caudal vena cava at the level of the right ovary, whereas the left enters the left renal vein. Both ovarian veins are very tortuous in their course between the peritoneal layers of the broad ligament.

Studying the physiological aspects of uterine circulation during pregnancy, Reynolds (1949) found two distinct phases in the adjustment of the uterine vessels to the shape and size of the conceptus. First, there is progressive stretching of the blood vessels and, secondly, the blood vessels during the latter part of gestation separate from one another without increase of length. Burwell and his coworkers (1938) found that blood pressure in the femoral and uterine veins was elevated during pregnancy in the dog. However, pressure in the uterine vein is higher than that in the femoral vein. The uterine veins drain into the caudal vena cava. Two of the principal functions served by rhythmic uterine contractions during estrus, according to Fagin and Reynolds (1936), may be production of an increased volume flow of blood through enlarged, hyperemic vessels and removal of any edematous fluid.

The lymphatics from the uterus pass to the **hypogastric** and **lumbar lymph nodes**. The vagina possesses a dense plexus of lymphatics in its tunica propria.

The uterus receives sympathetic and visceral afferent fibers through the **hypogastric plexus** and parasympathetic and visceral afferent fibers via the **pelvic nerves**.

THE VAGINA

The **vagina** is a musculomembranous, highly dilatable canal, extending from the uterus to the vulva. The term vulva is used to include the vestibule, clitoris, and labia (see discussion, *infra*). Cranially, the vagina is limited by the fornix and the intravaginal cervix (Fig. 9–25). The cervix may protrude 0.5 to 1 cm. into the vagina, and is 0.8 cm. in diameter. The fornix is the slitlike space cranioventral to the intravaginal part of the cervix. The length of the dorsal vaginal wall is less than that of the ventral wall because of the oblique situation of the cervix. Caudally, the vagina ends just cranial to the urethral opening. It is demarcated from the vestibule by a transverse mucosal ridge that extends 1 cm. dorsally on each side of the midventral line. No definite hymen is located at this point in the bitch, although its vestige may sometimes be found at the vaginovestibular junction. In a 25-pound dog, the vagina averages 10 to 14 cm. long and 1.5 cm. in diameter. Both the length and diameter of the vagina increase considerably during pregnancy and during parturition. The longitudinal folds (*rugae*) of the vaginal mucosa are high, allowing for great expansion in diameter (Fig. 9–25). Smaller transverse folds connecting the longitudinal folds permit craniocaudal stretching of the vagina.

RELATIONS

The cranial portion of the vagina is covered dorsally by peritoneum which reflects onto the colon, forming the **rectouterine pouch** (Fig. 9–23). Ventrally, this portion of the vagina has a peritoneal covering which reflects onto the bladder, forming the **vesicouterine pouch**. Laterally, the dorsal and ventral peritoneal coverings of the vagina fuse and become part of the broad ligaments. The caudal half of the vagina is retroperitoneal, being connected dorsally to the rectum and ventrally to the urethra by means of loose connective tissue. Laterally, the caudal part of the vagina is related to the vaginal blood vessels and nerves and to the ureters. The right and left ureters, with their peritoneal coverings, cross the lateral surface of the uterovaginal junction. The vagina is also related laterally to the caudal muscles. The portion of the vagina which is located retroperitoneally depends to a large extent on the fullness of the bladder and rectum.

STRUCTURE

The vaginal walls are made up of an inner mucosal layer, a middle smooth muscle layer, and an external coat of connective tissue and peritoneum (cranially). The *tunica mucosa* is non-glandular, stratified squamous epithelium. The epithelium changes in appearance during the various stages of the estrus cycle. It is cornified during the heat period and, according to Trautmann and Fiebiger (1952), intraepithelial glands have been found during this stage of the sexual cycle. The *tunica muscularis* is composed of a very thin inner layer of longi-

tudinal muscle, a thick circular layer, and a thin outer longitudinal layer. The inner longitudinal and circular layers encircle the external uterine orifice. The outer longitudinal layer blends with the muscular layer of the body of the uterus. The submucous tissue contains a rich plexus of blood vessels. The ducts of Gartner, vestigial remains of the caudal portion of the wolffian duct, are usually absent.

VESSELS AND NERVES

Arterial blood is supplied to the vagina via the **vaginal artery**, a branch of the internal pudendal. In addition to its vaginal distribution, the artery also supplies branches to the urethra and vestibule. Its urethral branches anastomose with the caudal vesical artery, and its vestibular branches (cranial vestibular) anastomose with caudal vestibular branches from the terminal part of the urogenital. The **vaginal veins** are satellites of the arteries and drain into the internal pudendal veins. The lymphatics drain into the **internal iliac lymph nodes**. The vagina is innervated by sympathetic and parasympathetic nerves from the **pelvic plexus** and by sensory afferent fibers via the **pudendal nerve**.

ANOMALIES AND VARIATIONS

Normal physiological changes account for considerable variation in the size and shape of the vagina. During pregnancy, the epithelium proliferates and muscular fibers hypertrophy. Cornification of vaginal epithelium is manifest in estrus. Stenosis of the vagina may occur as a congenital anomaly. Failure of the müllerian ducts to unite completely in fetal life may result in vaginal as well as uterine duplication.

THE VULVA AND FEMALE URETHRA

The Vulva

The **vulva** (*pudendum femininum*), or **external genitalia** (*partes genitales externae*), consists of three parts: vestibule, clitoris, and labia.

The **vestibule** (*vestibulum vaginae*) is the space connecting the vagina with the external genital opening (Fig. 9–25). It develops from the embryonic urogenital sinus, the common opening for genital and urinary tracts. The space is variable in size, depending on the size of the animal and whether or not she is pregnant. In a non-pregnant, mature 25-pound dog, the external vulvar opening or cleft is approximately 3 cm. long. The distance from the ventral commissure of the vulva to the urethral opening is 5 cm., and the diameter of the vaginovestibular junction is 1.5 to 2 cm.

The **urethral tubercle** is a ridgelike projection on the ventral floor of the vestibule, near the vaginovestibular junction. It contains the **external urethral orifice** *(ostium urethrae externum)*. The tubercle, widest cranially, narrows caudally to an apex located at a point approximately half the distance from the urethral opening to the clitoris. A shallow fossa or depression is present on each side of the tubercle. The mucosa of the vestibule is not covered with distinct ridges, as is the mucosa of the vagina, but is relatively smooth and red.

The *labia (labia pudendi)*, or lips, form the external boundary of the vulva. Homologous with the scrotum of the male, the labia fuse above and below the vulvar cleft to form dorsal and ventral vulvar commissures. The labia are soft and pliable, being composed of fibrous and elastic connective tissue, smooth muscle fibers, and an abundance of fat. The vaginal processes, containing the round ligaments of the uterus, end in the subcutaneous connective tissue of the labia (Fig. 9–23). These are not present in all animals (see description of uterine ligaments). The distance between the dorsal commissure and the anus is 8 to 9 cm. The dorsal commissure lies at or slightly below a frontal plane passing through the symphysis pelvis. The ventral portions of the labia, with their uniting commissure, form a pointed projection extending downward and backward from the body.

The *clitoris*, the homologue of the male penis, is composed of **paired roots** *(crura clitoridis)*, a **body** *(corpus clitoridis)*, and a *glans (glans clitoridis)*. The roots and body are homologues of the male corpora cavernosa penis, and the glans clitoridis (possessing erectile tissue) is homologous with the glans penis, although it is not bipartite in structure. The body of the clitoris in the dog is composed primarily of fatty rather than erectile tissue. It is covered by a thick tunica albuginea. The clitoris of the dog does not normally contain any structures comparable to the os penis, the corpus spongiosum, or the urethra of the male. Instead, in the bitch, there are elongate masses of erectile tissue, the **vestibular bulbs** *(bulbi vestibuli)*, lying deep to the vestibular mucosa and united to each other by an isthmus. They lie in close proximity to the corpus clitoridis, corresponding to the bulb of the urethra in the male. The vestibular bulbs are each supplied by a terminal branch of the internal pudendal artery, homologous to the artery of the urethral bulb in the male. The glans clitoridis, erectile in structure, is very small and projects into the fossa clitoridis. The fossa is partially folded over the glans clitoridis dorsally. This fold corresponds to the male prepuce. The free part of the clitoris (glans) is about 0.6 cm. long and 0.2 cm. in diameter in an average-sized dog; the distance from the ventral commissure of the vulva to the glans clitoridis is 2 to 3 cm., and that from the ventral commissure to the fundus of the fossa of the clitoris is 3 to 4 cm. The opening of the fossa is approximately 1 cm. in diameter.

An *os clitoridis* can develop in response to an altered hormone balance and may be over 2.5 cm long. Grandage and Robertson (1971) reported an os clitoridis in a normal Welsh corgi bitch that subsequently mated and whelped. The bone was seen radiographically to be 13 mm. long, laterally compressed, and pointed at its apex.

Nitschke (1970) has described the structure of the clitoris and vagina in the dog.

STRUCTURE

Recognition of the location of the external urethral orifice is important for the purpose of catheterization. Not uncommonly, the clitoris or the fossa clitoridis is mistaken for the urethral opening, and the catheterization attempt is unsuccessful. The urethra opens on a tubercle 4 to 5 cm. cranial to the ventral commissure of the vulva.

The mucosal surface of the vulva is covered by stratified squamous epithelium. A variable number of lymph nodules may cause prominences to appear on the mucosa. Small minor vestibular glands, lobular in structure, open ventrally on each side of the median ridge connected to the urethral tubercle. The glands are located deep to the constrictor vestibuli muscles. The body of the clitoris consists of fat, elastic connective tissue, and a peripheral tunica albuginea. The glans clitoridis, made up of erectile tissue, contains numerous sensory nerve endings. The vestibular bulbs are also composed of cavernous tissue. The labia, covered with stratified squamous epithelium, are rich in sebaceous and tubular glands and also contain fat, elastic tissue, and smooth muscle fibers.

MUSCLES

In addition to the usual unstriped muscle fibers, similar to those of the vagina, the vulva possesses two striated circular muscles (Figs. 9–26, 9–27). The most cranial of the two is the strong **constrictor vestibule muscle.** It is incomplete on the dorsal surface of the vestibule, but fuses along its caudal border to the external sphincter of the anus. Its fibers run diagonally in a cranioventral direction, encircling the urethra, vestibule, and caudal portion of the vagina

before it joins its fellow of the opposite side. It constricts the vestibule.

Immediately caudal to the constrictor vestibuli muscle is the relatively weak, thin **constrictor vulvae muscle** (the muscle of the labia). This muscle is continuous dorsally with the external anal sphincter, arising from the caudal fascia ventral to the first and second caudal vertebrae, and encircles the vulva and vestibule about 1 cm. caudal to the point where the urethra disappears into the genital tract. The constrictor vulvae muscle blends with the vestibular constrictor to a slight degree. The vulvar constrictors fuse together below the vulva, cranial to the ventral commissure. They lift the labia dorsally prior to intromission of the penis, allowing it to enter the vagina more easily. The vestibular and vulvar constrictors together are homologous with the bulbospongiosus muscles of the male. They lie superficial to the vestibular bulbs. Their counterparts in the male are peripheral to the bulb of the penis. The **ischiourethralis muscles** arise from the caudomedial surface of the tuber ischii, on each side, and insert upon the poorly developed central tendon of the perineum.

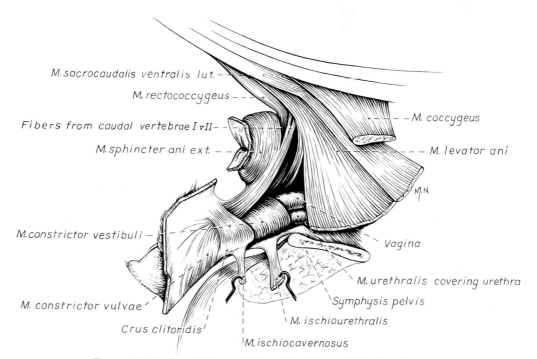

Figure 9–26. Constrictor muscles of female genitalia, lateral aspect.

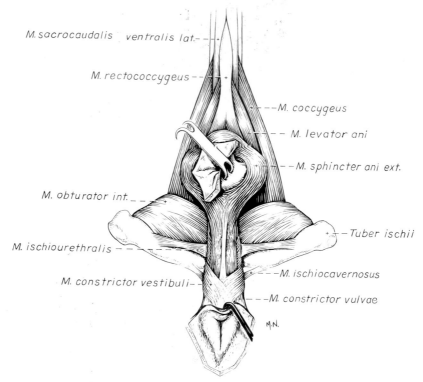

M. sacrocaudalis ventralis lat.

M. rectococcygeus

M. coccygeus

M. levator ani

M. sphincter ani ext.

M. obturator int.

Tuber ischii

M. ischiourethralis

M. ischiocavernosus

M. constrictor vestibuli

M. constrictor vulvae

M.N.

Figure 9–27.   Constrictor muscles of female genitalia, caudal aspect.

The **ischiocavernosus muscles** are small in the female. They arise bilaterally from the caudal edge of the ischium and attach to the crura clitoridis. This is similar to the manner in which they insert upon the corpora cavernosa in the male.

The Female Urethra

The urethra of the bitch *(urethra femin-ina)* (Fig. 9–22) corresponds to that portion of the male urethra which lies cranial to the prostatic utricle. It is about 0.5 cm. in diameter and 7 to 10 cm. long. It originates from the urinary bladder at or near the cranial edge of the symphysis pelvis. It extends caudodorsally to enter the genital tract approximately 0.5 cm. caudal to the vaginovestibular junction. Its dorsal wall is in close apposition to the ventral wall of the vagina. Structurally, the female urethra resembles that of the male. It is linked by folded mucous membrane, allowing the urethral lumen to expand considerably when under pressure. The mucosa is non-glandular, and the submucosa is highly vascular. Lymph nodules are also present. The musculature of the female urethra consists of outer and inner longitudinal and middle circular layers of unstriped muscle. The smooth muscles become less conspicuous near the entrance of the urethra into the vestibule. At the external urethral orifice, voluntary muscle encircles all but the dorsal surface of the urethra, which is in close contact with the vestibule. These circular fibers form a strong sphincter at the external orifice.

VESSELS AND NERVES

The external genitalia and urethra of the female are supplied with blood through the **vaginal** and the **external and internal pudendal arteries.** The external pudendal artery sends branches to the labia (cranial labial artery), corresponding to the scrotal branches in the male. The vaginal artery supplies the vulva by means of the cranial vestibular branches, from the vaginal ramus, and the caudal vestibular branches from the termination of the urogenital artery. The clitoris is supplied by branches of the internal pudendal artery, corresponding to the dorsal and deep penile arteries. The vestibular bulb is also supplied by the internal pudendal artery (homologous to the artery of the bulb in the male).

The **bilateral veins** from the clitoris (**dorsal veins of clitoris**) join each other at the ischial arch and then separate again (after a distance of 1 or 2 cm.) into **internal pudendal veins** which drain into the internal iliac veins. The vestibular bulb is drained by a separate tributary of the internal pudendal. Valves are apparent in most of the veins, including the common trunk of the dorsal veins of the clitoris. The dorsal arteries and veins of the clitoris are not actually dorsal to the clitoris, in the manner of the comparable penile vessels. They curve ventrally around the ischial arch and run caudoventrally along that surface of the clitoris which corresponds to the dorsum of the penis. Lymphatic drainage compares with that of the external genitalia of the male.

The sensory afferent nerves to the external genitalia are derived from the **pudendal** and the **genital nerves.** Both nerves innervate the labia. The glans clitoridis receives its sensory nerves from the pudendal (**dorsal nerve of the clitoris**). Motor impulses to the urethral muscle and to the vestibular and vulvar constrictors also pass through the pudendal nerve. Autonomic innervation to the external genitalia and urethra in the female is through the **hypogastric** and **pelvic nerves.** It includes principally sympathetic fibers which innervate the musculature of the blood vessels.

ANOMALIES AND VARIATIONS

Incomplete degeneration of the wolffian (mesonephric) ducts may be the cause of congenital anomalies. During pregnancy, the mucosa of the vulva will exhibit changes similar to those occurring in the vagina. Normal physiological changes during the estrus cycle (variations of size and shape) are also apparent.

EMBRYOLOGY OF THE UROGENITAL SYSTEM

The development of the mammalian urogenital system is briefly summarized here. For specific details, one should refer to textbooks of embryology: Arey (1974), Willis (1962), Langman (1975), and Moore (1977). The homology of the structures in the male and female is shown in Table 9–1.

The embryonic intermediate cell mass divides during early fetal life into a urinary, or nephric, region and a genital region. Segmental tubules (the pronephros) develop in the cranial portion of the nephric region (mesomere). The tubules, one pair per segment, grow caudally and form a longitudinal duct (pronephric duct), which runs to the primitive cloaca. The funnel-shaped connection of each segment with the celom (body cavity) is called the nephrostome. Renal arteries push out from the aorta to form glomeruli, one per segment. Caudal to the pronephros, other segmental tubules unite with the pronephric duct to form the mesonephros, or wolffian body. As this occurs, the pronephros degenerates, and the pronephric duct becomes what is called the mesonephric, or wolffian, duct. The mesonephric tubules are more complex than their predecessors. Although the pronephros is functionless in mammals, the mesonephros is thought to serve as a temporary organ of excretion. Unlike the pronephros, the mesonephric glomerulus is internal, and the nephrostome does not act as a mouth for the tubule. Later, a bud develops from the mesonephric duct near its entrance into the cloaca. This bud gives rise to the renal pelvis, ureter, and the collecting tubules of the definitive, or metanephric, kidney. The secretory tubules and Bowman's capsules of the metanephros develop from the caudal portion of the nephric ridge.

The three types of kidneys in vertebrates appear successively in craniocaudal sequence both ontogenetically and phylogenetically. The pronephros, or more commonly the mesonephros, is the functional kidney of anamniotes (fish and amphibians), whereas the metanephros is the functional

TABLE 9–1 HOMOLOGIES OF GENITAL ORGANS IN MALE AND FEMALE MAMMALS

Male	Female
Testis	Ovary
Mesorchium	Mesovarium
Appendix testis	Abdominal ostium of uterine tube
Gubernaculum	Round ligament of uterus, proper ligament of ovary
Prostatic urethra (cranial to utricle)	Urethra
Prostatic urethra (caudal to utricle) and membranous urethra	Vestibule
Penis	Clitoris
Glans penis	Glans clitoridis
Corpus spongiosum penis	Vestibular bulbs
Corpus cavernosum	Corpus cavernosum clitoridis
Scrotum	Labia
Scrotal raphe	Dorsal commissure of labia
Prepuce	Fold of fossa clitoridis

kidney of adult amniotes (reptiles, birds, and mammals).

The gonads develop from the ventromedial portion of the intermediate cell mass (genital ridge). The sex of the embryo is genetically determined at fertilization, but is not distinguishable morphologically until after day 30 of gestation.

The developing male testes join the cranial portion of the mesonephric tubules, which become the epididymis. The middle tubules form the ductus deferens, and the caudal ones persist as the vestigial paradidymis and ductuli aberrantes. The mesonephric tubules do not join the ovaries in the female, persisting as the vestigial epoophoron and paroophoron. The mesonephric ducts also degenerate during the course of embryonic development. In rare instances, Gartner's ducts (remnants of the caudal parts of the mesonephric ducts) open near the vaginovestibular junction in the adult.

The ovaries and testes migrate caudally during fetal development. In addition, the male gonads leave the body cavity and descend into the scrotum, moving from their dorsolateral positions in the abdomen toward the inguinal canals, and migrating between the abdominal wall and the parietal peritoneum. During early fetal development, when the testes are still high in the abdomen, parietal peritoneum begins to push into the inguinal canals. These peritoneal outpocketings are the primordia of the vaginal processes. Each vaginal process evaginates through its inguinal canal into the scrotum. During the latter part of embryonic life, each testis descends through the inguinal canal into the scrotum, passing between the vaginal process and the scrotal wall. The testis always remains outside the peritoneal cavity, but both the testis and gubernaculum are enveloped by peritoneum before descent begins. The shortening of the gubernaculum, both actual and relative, is thought to draw the testis into the scrotum. Some workers attribute the testicular descent to a process of normal herniation through a canal previously dilated by the gubernaculum (Wensing 1968–1973). The peritoneal layers within the scrotum form the tunics for the testis and spermatic cord. Unlike in the human, there is no disappearance of tunics in the area between the abdominal cavity and the testis. The vaginal process contains a cavity which is continuous, throughout life, between the peritoneal cavity of the abdomen and the vaginal cavity of the scrotum. The ductus deferens and the spermatic vessels and nerves are pulled into the scrotum by the testis and epididymis, causing the ductus deferens to loop over the ureter.

In the female, the ovaries remain in the abdominal cavity, suspended by the broad ligaments. The vaginal processes of the bitch, containing the round ligaments of the uterus, migrate toward the labia in a manner similar to the migration of the vaginal processes of the male toward the scrotum. The female counterpart of the male gubernaculum fails to unite in the same way because of the persistence of the fused müllerian ducts.

Paired müllerian ducts also develop in both male and female (Fig. 9–28). In the bitch, the müllerian ducts open cranially into the peritoneal cavity as the abdominal openings of the uterine tubes. Caudally, they unite, in part, to form the uterine tubes and the bicornuate uterus and, completely, to form the vagina. In the male, the müllerian ducts degenerate, except for their cranial and caudal ends. Cranially, they remain

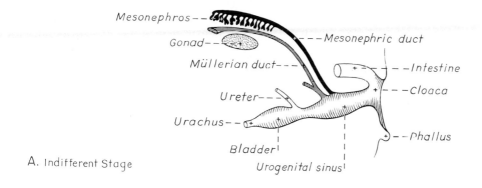

A. Indifferent Stage

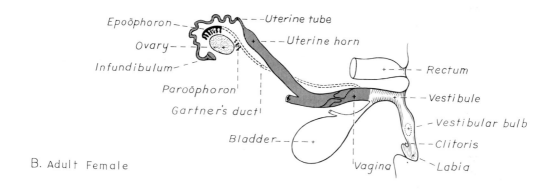

B. Adult Female

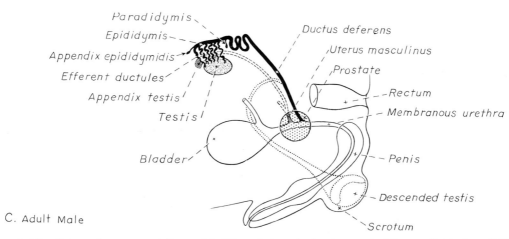

C. Adult Male

Figure 9–28. Schematic representations of indifferent stage in development of the genital system of the dog, and the genital system in the adult female and the adult male dog.

as the vestigial appendix testis. Caudally, the ducts unite and persist as the uterus masculinus (prostatic utricle), opening on the colliculus seminalis. The broad ligament of the embryo encloses the wolffian and müllerian ducts in its free edge.

The terminal blind end of the primitive hindgut (cloaca) divides into the rectum dorsally and the urogenital sinus ventrally. The latter, receiving the paired wolffian and müllerian ducts, develops into the urinary bladder, the urethra, and the vestibule of the vagina in the female. The wolffian duct in the male becomes the ureter and ductus deferens. The ureter opens into the bladder, and the ductus deferens opens into that part of the urethra which connects the bladder

with the urogenital sinus. The prostate gland, endodermal in origin, arises from the urethral epithelium. The prostatic tubules originate as solid epithelial projections from the prostatic urethra.

During early fetal life the external genitalia are in an indifferent or indeterminate condition (See Chapter 2). Rounded labioscrotal swellings bound the phallic eminence. The ventral surface of the phallus contains a urethral groove. The phallus develops into the penis in the male and the clitoris in the female. In the female, the clitoris arises from the indeterminate phallus, the urethral groove forms the vestibule, and the genital tubercles become the labia of the vulva.

BIBLIOGRAPHY

Allam, M. W. 1952. The Vagina. Pp. 462–469 *in* Canine Surgery, edited by J. V. Lacroix and H. P. Hoskins. 3rd Ed. Evanston, Ill., American Veterinary Publications, Inc.

Anderson, D. H. 1927. Lymphatics of the fallopian tube of the sow. Contr. Embryol. Carneg. Instn *19*:135–147.

Archibald, J. 1957. The canine prostate. N. Am. Vet. *38*:253–255.

Archibald, J., and A. J. Cawley, 1956. Canine prostatectomy. J. Am. Vet. Med. Assoc. *128*:173–177.

Arey, L. B. 1974. Developmental Anatomy. 7th Ed. Philadelphia, W. B. Saunders Co.

Austin, C. R., and R. V. Short. 1972. Reproduction in Mammals. Cambridge, Cambridge Univ. Press.

Baum, H. 1918. Das Lymphgefässsystem des Hundes. Berlin, Hirschwald.

Bensley, R. R., and R. D. Bensley. 1930. The structure of the renal corpuscle. Anat. Rec. *47*:147–175.

Blakely, C. L. 1952a. The Vulva. Pp. 470–478 *in* Canine Surgery, edited by J. V. Lacroix and H. P. Hoskins. 3rd Ed. Evanston, Ill., American Veterinary Publications, Inc.

– – – –. 1952b. The Kidneys. Pp. 501–503 *in* Canine Surgery, edited by J. V. Lacroix and H. P. Hoskins. 3rd Ed. Evanston, Ill., American Veterinary Publications, Inc.

Brenner, B. M., and F. C. Rector. 1976. The Kidney. 2 Vols. Philadelphia, W. B. Saunders Co.

Brodey, R. S., J. E. Martin, and D. G. Lee. 1954. Male pseudohermaphroditism in a toy terrier. J. Am. Vet. Med. Assoc. *125*:368–370.

Burwell, C. S., W. D. Strayhorn, D. Flickinger, M. B. Corlette, E. P. Bowerman, and J. A. Kennedy. 1938. Circulation during pregnancy. Arch. Intern. Med. *62*:979–1003.

Christensen, G. C. 1952. Circulation of blood through the canine kidney. Am. J. Vet. Res. *13*:236–245.

– – – – –. 1954. Angioarchitecture of the canine penis and the process of erection. Am. J. Anat. *95*:227–262.

Christensen, K., E. Lewis, and A. Kuntz. 1951. Innervation of the renal blood vessels in the cat. J. Comp. Neurol. *95*:373–385.

Cole, H. H., and P. T. Cupps. 1977. Reproduction in Domestic Animals. 3rd Ed. New York, Academic Press.

Concannon, P., W. Hansel, and W. J. Visek. 1975. The ovarian cycle of the bitch: plasma estrogen, LH, and progesterone. Biol. Reprod. *13*:112–121.

Concannon, P., W. Hansel, and K. McEntee. 1977. Changes in LH, progesterone and sexual behavior associated with preovulatory luteinization in the bitch. Biol. Reprod. *17*:604–613.

Concannon, P. W., M. E. Powers, W. Holder, and W. Hansel., 1977. Pregnancy and parturition in the bitch. Biol. Reprod. *16*:517–526.

Daniel, P. M., C. N. Peabody, and M. M. L. Prichard. 1952. Cortical ischemia of the kidney with maintained blood flow through the medulla. Q. J. Exp. Physiol. 37:11–18.

Del Campo, C. H., and O. J. Ginther. 1974. Arteries and veins of uterus and ovaries in dogs and cats. Am. J. Vet. Res. *35*:409–415.

Deysach, L. J. 1939. The comparative morphology of the erectile tissue of the penis with especial emphasis on the probable mechanism of erection. Am. J. Anat. *64*:111–131.

Didier, R. 1946. Étude systématique de l'os pénien des mammifères. Mammalia. *10*:78–91.

Dorr, L. D., and M. J. Brody, 1967. Hemodynamic mechanisms of erection in the canine penis. Am. J. Physiol. *213*:1526–1531.

Dukes, H. H. 1947. The Physiology of Domestic Animals. 6th Ed. Ithaca, N.Y., Comstock Publishing Co.

von Ebner, V. 1900. Über klappenartige Vorrichtungen in den Arterien der Schwellkörper. Anat. Anz., Ergänzungsh, *18*:70–81.

Eckhard, C. 1863. Untersuchungen über die Erektion des Penis beim Hunde. Beitr. Anat. Physiol. 3:123–166.

Ellenberger, W., and H. Baum. 1943. Handbuch der vergleichenden Anatomie der Haustiere. 18th Ed. Berlin. Springer.

Evans, H. M., and H. H. Cole. 1931. An Introduction to the Study of the Oestrus Cycle in the Dog. Memoirs Univ. of Calif. Vol. 9, No. 2.

Fagin, J., and S. R. M. Reynolds. 1936. The endometrial vascular bed in relation to rhythmic uterine contractility, with a consideration of the functions of the intermittent contractions of oestrus. Am. J. Physiol. 117:86–91.

Finco, D. R., and J. R. Duncan. 1972. Relationship of glomerular number and diameter of body size of the dog. Am. J. Vet. Res. 33:2447–2450.

Finco, D. R., S. K. Kneller, and R. B. Barrett. 1971. Radiologic estimation of kidney size of the dog. J. Am. Vet. Med. Assoc. 159:995–1002.

Fisher, H. G. 1917. Histological structure of the retractor penis muscle of the dog. Anat. Rec. 13:69–75.

Fitzgerald, T. C. 1940. The renal circulation of domestic animals. Am. J. Res. 1:89–95.

Francois-Franck, C. A. 1895. Recherches sur l'innervation vasomotrice du penis; topographie des nerfs constricteurs et dilatateurs. Arch. Physiol. Norm. Path. 19:122–138, 744–816.

Fuller, P. M., and D. F. Heulke. 1973. Kidney vascular supply in the rat, cat, and dog. Acta Anat. 84:516–522.

Gantt, W. H. 1938. Extension of a conflict based upon food to other physiological systems and its reciprocal relations with sexual functions. Am. J. Physiol. 123:73–74.

– – – –. 1949. Psychosexuality in Animals. Pp. 33–51 in Psychosexual Development in Health and Disease, edited by P. H. Hoch and Joseph Zubin. New York, Grune & Stratton.

Gersh, I. 1937. The correlation of structure and function in the developing mesonephros and metanephros. Contr. Embryol. Carneg. Instn 26:33–58.

Gordon, N. 1960. Surgical anatomy of the bladder, prostate gland and urethra in the male dog. J. Am. Vet. Med. Assoc. 136:215–221.

– – – –. 1961. The position of the canine prostate gland. Am. J. Vet. Res. 22:142–146.

Grandage, J. 1972. The erect dog penis: A paradox of flexible rigidity. Vet. Rec. 91:141–147.

– – – –. 1975. Some effects of posture on the radiographic appearance of the kidneys of the dog. J. Am. Vet. Med. Assoc. 166:165–166.

Grandage, J., and B. Robertson. 1971. An os clitoridis in a bitch. Aust. Vet. J. 47:346.

Greiner, T. P., and C. W. Betts. 1975. Diseases of the prostate gland. Pp. 1274–1306 in Vol. 2 of Textbook of Veterinary Internal Medicine, edited by S. J. Ettinger. Philadelphia, W. B. Saunders Co.

Gruber, C. M. 1933. The autonomic innervation of the genito-urinary system. Physiol. Rev. 13:497–609.

Hansel, W., and K. McEntee. 1977. Male reproductive processes. Pp. 801–824 in Dukes' Physiology of Domestic Animals, edited by M. Swensen. Ithaca, N.Y., Cornell Univ. Press.

Harrison, R. G. 1949. The comparative anatomy of the blood supply of the mammalian testis. Proc. zool. Soc. London 119:325–344.

Hart, B. L. 1972. The action of extrinsic penile muscles during copulation in the male dog. Anat. Rec. 173:1–5.

Hart, B. L., and R. L. Kitchell. 1965. External morphology of the erect glans of the dog. Anat. Rec. 152:193–198.

Hisaw, F. L. 1947. Development of the graafian follicle and ovluation. Physiol. Rev. 27:95–119.

Hodson, N. 1968. On the intrinsic blood supply to the prostate and pelvic urethra in the dog. Res. Vet. Sci. 9:274–280.

Höfliger, H. 1971. Zur Kenntnis der kongenitalen unilateralen Nierenagenesie bei Haustieren. Schweiz. Arch. Tierheilke 113:330–337.

Hooker, C. W. 1944. The postnatal history and function of the interstitial cells of the testis of the bull. Am. J. Anat. 74:1–37.

Houston, J. 1830. Compressor venae dorsalis penis. Dublin Hosp. Rep. and Comm. in Med. and Surg. 5:459–493.

Huber, G. C. 1907. The arteriolae rectae of the mammalian kidney, Amer. J. Anat. 6:391–406.

Huggins, C., and P. V. Moulder. 1944. Studies on the mammary tumors of dogs. J. Exp. Med. 80:441–454.

Jacobson, C. E. 1945. Neurogenic vesical dysfunction. J. Urol. (Baltimore) 53:670–695.

Johnson, A. D., W. R. Gomes, and N. L. Vandemark. 1970. The testis. 3 Vols. Vol. 1, Development, Anatomy, Physiology. New York, Academic Press.

Kiss, F. 1921. Anatomisch-histologische Untersuchungen über die Erektion. Z. ges. Anat. 61:455–521.

Kunde, M. M., F. E. D'Amour, A. J. Carlson, and R. G. Gustavson. 1930. Studies on metabolism; the effect of estrin injections on the metabolism, uterine endometrium, lactation, mating, and maternal instincts in the adult dog. Am. J. Physiol. 95:630–640.

Kuntz, A. 1919a. The innervation of the gonads in the dog. Anat. Rec. 17:203–219.

– – – – –. 1919b. Experimental degeneration in the testis of the dog. Anat. Rec. 17:221–234.

Langley, J. N. 1896. The innervation of the pelvic and adjoining viscera. J. Physiol. 20:372–406.

Langman, J. 1975. Medical Embryology. 3rd Ed. Baltimore. The Williams & Wilkins Co.

Latimer, H. B. 1951. The growth of the kidneys and the bladder in the fetal dog. Anat. Rec. 109:1–12.

Leav, I., and G. V. Ling. 1968. Adenocarcinoma of the canine prostate. Cancer 22:1329–1345.

Lee, D. G. and M. W. Allam. 1952. True unilateral hermaphroditism in a dog. Univ. Penn. Bull., Vet. Ext. Quart. No. 128, pp. 142–147.

MacCallum, D. B. 1926. The arterial blood supply of the mammalian kidney. Am. J. Anat. 38:153–175.

– – – – –. 1939. The bearing of degenerating glomeruli on the problem of the vascular supply of the mammalian kidney. Am. J. Anat. 65:69–103.

Markee, J. E. 1953. The Urogenital System. Pp. 1493–1569 in Morris' Human Anatomy, edited by J. P. Schaeffer. 11th Ed. New York, The Blakiston Co.

Marshall, E. K., Jr. 1934. The comparative physiology of the kidney in relation to theories of renal secretion. Physiol. Rev. 14:133–159.

Marshall, F. H. A., and E. T. Halnan. 1917. On the post-oestrous changes occurring in the generative organs and mammary glands of the non-pregnant dog. Proc. Roy. Soc. B 89:546–559.

McCarthy, P. H. 1976. The anatomy of the superficial inguinal ring and its contained and adjacent structures in the live Greyhound — a study by palpation. J. Small Anim. Pract. 17:507–518.

McEntee, K., and C. P. Zepp, Jr. 1953. A Study of Canine and Bovine Ovarian Tumors. (Abst.) Ann. Rep. New York State Veterinary College, Cornell Univ.

Mitchell, G. A. G. 1938. The innervation of the ovary, uterine tube, testis and epididymis. J. Anat. (Lond.) 72:508–517.

Moffat, D. B. 1975. The Mammalian Kidney. London, Cambridge Univ. Press, 263 pp.

Moore, C. R. 1942. The physiology of the testis and application of male sex hormone. J. Urol. (Baltimore) 47:31–44.

Moore, K. L. 1977. The Developing Human: Clinically Oriented Embryology. Philadelphia, W. B. Saunders Co., 411 pp.

Morison, D. M. 1926. A study of the renal circulation, with special reference to its finer distribution. Am. J. Anat. 37:53–93.

Mossman, H. E. 1938. The homology of the vesicular ovarian follicles of the mammalian ovary with the coelom. Anat. Rec. 70:643–656.

Mossman, H. W., and K. L. Duke. 1973. Comparative Morphology of the Mammalian Ovary. Madison, Univ. Wisconsin Press, 461 pp.

Moyer, J. H., H. Conn, K. Markley, and C. F. Schmidt. 1950. An attempt to demonstrate vascular by-passes in the kidney. Am. J. Physiol. 161:250–258.

Naaktgeboren, C., and E. J. Slijper. 1970. Biologie der Geburt. Berlin, Paul Parey, 225 pp.

Nitschke, T. 1966. Der m. compressor venae dorsalis Penis s. Clitoridis des Hundes. Anat. Anz. 118:193–208.

— — — —. 1970. Diaphragma pelvis, Clitoris und Vestibulum vaginae der Hündin. Anat. Anz. 127:76–125.

Oppenheimer, M. J. 1938. Autonomic control of the retractor penis in the cat. Am. J. Physiol. 122:745–752.

O'Shea, J. D. 1962. Studies on the canine prostate gland. 1. Factors influencing its size and weight. J. Comp. Pathol. 72:321–331.

Pearson, H., and C. Gibbs. 1971. Urinary tract abnormalities in the dog. J. Small Anim. Pract. 12:67–84.

Peirce, E. C. 1944. Renal lymphatics. Anat. Rec. 90:315–335.

Perry. J. S. 1972. The Ovarian Cycle of Mammals. New York, Hafner, 219 pp.

Petras, J. M., and J. F. Cummings. 1978. Sympathetic and parasympathetic innervation of the urinary bladder and urethra. Brain Res. 153:363–369.

Pineda, M. H., R. A. Kainer, and L. C. Faulkner. 1973. Dorsal median postcervical fold in the canine vagina. Am. J. Vet. Res. 34:1487–1491.

Polano, O. 1903. Beiträge zur Anatomie der lymphbahnen im menschlichen Eierstock. Mschr. Geburstch. Gynäk. 17:281–295, 466–496.

Pollock, W. F. 1942. Histochemical studies of the interstitial cells of the testis. Anat. Rec. 84:23–30.

Price, D. 1963. Comparative aspects of development and structure in the prostate, in Biology of the Prostate and Related Tissues, edited by Nat. Cancer Inst. Monograph 12:1–27.

Purohit, R. C., and S. D. Beckett. 1976. Penile pressures and muscle activity associated with erection and ejaculation in the dog. Am. J. Physiol. 231:1343–1348.

Ramsey, A. J. 1946. Lymphatic vessels of the fallopian tube. Anat. Rec. 94:524.

Raps, G. 1948. The development of the dog ovary from birth to six months of age. J. Am. Vet. Med. Assoc. 9:61–64.

Reynolds, S. R. M. 1937. The nature of uterine contractility; a survey of recent trends. Physiol. Rev. 17:304–334.

— — — —. 1949. Adaptation of uterine blood vessels and accommodation of the products of conception. Contr. Embryol. Carneg. Instn 33:1–19.

Rouiller, C., and A. F. Muller. 1969–1971. The Kidney. 4 Vols. New York, Academic Press.

Rudolph, L., and A. C. Ivy. 1930. Physiology of the uterus in labor: experimental study of the dog and rabbit. Am. J. Obstet. Gynecol. 19:317–335.

Runnells, R. A. 1954. Animal Pathology. 5th Ed. Ames, Iowa, Iowa State College Press.

Rytand, D. A. 1938. The number and size of mammalian glomeruli as related to kidney and to body weight with methods for their enumeration and measurement. Am. J. Anat. 62:507–520.

Sampson, J. A. 1937. The lymphatics of the mucosa of the fimbriae of the fallopian tube. Am. J. Obstet. Gynecol. 33:911–930.

Schlegel, J. U., and J. B. Moses. 1950. A method for visualization of kidney blood vessels applied to studies of the crush syndrome. Proc. Soc. Exp. Biol. (N.Y.) 74:832–837.

Schlotthauer, C. F. 1937. Diseases of the prostate gland in the dog. J. Am. Vet. Med. Assoc. 43:176–187.

Schlotthauer, C. F., and J. L. Bollman. 1936. The prostate gland of the dog. Cornell Vet. 26:342–349.

Segal, S. J., R. Crozier, P. A. Corfman, and P. G. Condliffe. 1973. The Regulation of Mammalian Reproduction. Springfield, Ill., Charles C Thomas.

Shannon, J. A. 1939. Renal tubular excretion. Physiol. Rev. 19:63–93.

Sisson, S., and J. D. Grossman. 1953. Anatomy of the Domestic Animals. 4th Ed. Philadelphia, W. B. Saunders Co.

Smith, H. W. 1943. Lectures on the Kidney. Porter Lectures, Series 9. Lawrence, Kansas, University of Kansas.

— — — —. 1951. The Kidney. New York, Oxford University Press.

Speert, H. 1948. The normal and experimental development of the mammary gland of the Rhesus monkey, with some pathological correlations. Contr. Embryol. Carneg. Instn 32:9–65.

Sperber, I. 1944. Studies on the mammalian kidney. Zool. Bidr. Upps. 22:249–432.

Stead, A. C. 1972. Fracture of the os penis in the dog — two case reports. J. Small Anim. Pract. 13:19–22.

Stephenson, H. C. 1939. Urinary calculi of small animals. J. Amer. Vet. Med. Assoc. 95:309–315.

Strassmann, E. O. 1941. The theca cone and its tropism toward the ovarian surface, a typical feature of growing human and mammalian follicles. Am. J. Obstet. Gynecol. 41:363–378.

Trautmann, A., and J. Fiebiger. 1952. Fundamentals of the Histology of Domestic Animals. (Translated and revised from 8th and 9th German editions, 1949, by R. E. Habel and E. L. Biberstein.) Ithaca, New York, Comstock Publishing Assoc.

Trueta, J., A. E. Barclay, P. Daniel, K. J. Franklin, and

M. Prichard. 1947. Studies of the Renal Circulation. Springfield, Ill., Charles C Thomas.

Vaerst, L. 1938. Über die Blutversorgung des Hunde-penis. Morph. Jb. *81*:307–352.

Verma, O. P., and G. A. Chibuzo. 1974. Hormonal in-fluences on motility of canine uterine horns. Am. J. Vet. Res. *35*:23–26.

Vimtrup, B. J. 1928. On the number, shape, structure, and surface area of the glomeruli in the kidneys of man and mammals. Am. J. Anat. *41*:123–151.

Wensing, C. J. G. 1968–1973. Testicular Descent in Some Domestic Mammals. I, II, and III. Amsterdam, Koninkl. Nederl. Akademie Van Wetenschappen, Proc. Ser. C 71, 1968; 76, 1973.

White, H. L. 1940. Observations indicating absence of glomerular intermittence in normal dogs and rabbits. Am. J. Physiol. *128*:159–168.

Wildt, D. E., C. J., Levin and S. W. J. Seager. 1977. Laparoscopic exposure and sequential observation of the ovary of the cycling bitch. Anat. Rec. *189*:443–450.

Willis, R. A. 1962. The Borderland of Embryology and Pathology. 2nd Ed. London, Butterworth, 641 pp.

Woodburne, R. T., and J. Lapides. 1972. The ureteral lumen during peristalsis. Am. J. Anat. *133*:255–258.

Zietzschmann, O. 1928. Über den Processus vaginalis der Hündin. Dtsch. tieräztl. Wschr. *36*:20–22.

Zuckerman, S. 1977. The Ovary. 2nd Ed. 3 Vols, New York, Academic Press.

Chapter 10

THE ENDOCRINE SYSTEM

By RONALD L. HULLINGER

GENERAL FEATURES OF THE ENDOCRINE GLANDS

The endocrine system differs from other body systems in that the component organs are not in direct continuity. They are positioned throughout the body in widely separated locations. Hormone synthesis is the one principal function shared by all components of the system. The various organs of the endocrine system are linked by the circulatory system: blood vessels, lymphatic vessels, and tissue fluids. There are some morphological similarities in these organs: a general sparsity of stromal connective tissue; an extensive blood vascular and lymphatic supply; epithelioid cell types which compose parenchymal racemi, follicles, sheets, or acini; and an absence of secretory ducts. The specific structural features may vary in amount and kind from one endocrine gland to another. The term endocrine system, therefore, refers to a functional relationship of various cells, tissues, and organs and not to a structurally contiguous set of component organs. The name endocrine has a derivational meaning taken from the Greek words *endon*, meaning "within," and *krinein*, meaning "to separate" — thus, separating internally. The study of the structure and functioning of this system of units which secrete internally is termed endocrinology.

Endocrine glands exercise a major control over the organism. They supplement and augment the function of the nervous system in response to stimuli from both the internal and the external environments. These stimuli directly or indirectly affect the specific metabolism of epithelioid cells of the endocrine tissues and organs and some nerve tissues, causing them to release into the intercellular space relatively small quantities of substances termed **hormones.** Hormones are the secretory products of parenchymal cells found singly or, more often, in aggregates as endocrine tissues and in endocrine organs. The hormones released by these cells may diffuse directly to other loci via the interstitial fluids or pass into blood or lymphatic vessels to be distributed to more distant receptors. Receptors, which may be cells, tissues, or entire organs, are modulated in their activity by the hormone. Sites so affected are referred to as target areas, and the hormones are considered *trophic* substances.

The cellular components of the endocrine glands produce these trophic substances which diffuse into the tissue fluids of the organism. Tissues and glands organized from these cells are characteristically without ducts and thus have been called the *glandulae sine ductibus*, the ductless glands. The tissue fluids immediately surrounding the secretory cells become the transport media for the secretions of the glandular parenchyma. Nearly all cells, and therefore nearly all tissues and organs of the body, yield, in varying quantities, humoral or fluid secretions as products and by-products of their internal metabolism. Current findings suggest that most of these materials do not function as hormones. However, some of these metabolites diffuse into the organism and, as hormones, affect specific target sites. Some excite, whereas others suppress the activity of these target organs, tissues, or cells.

Components of the endocrine system are classified according to their major functional activity. Primary endocrine organs are the *hypophysis* (**pituitary gland**), **thyroid gland, parathyroid gland, pineal gland** *(epiphysis),* and **adrenal gland.** Their major function is the production of hormones. The **testis, ovary, pancreas,** and **placenta** have both major endocrine and exocrine functions. Other organs, such as the **kidney, liver,** and **thymus,** have an endocrine function which is secondary to their other functions. Still other tissues and cells — the gastric and intestinal mucosa and cells of the general connective tissues — produce known hormonal substances. The effects of hormones can be demonstrated physiologically, but in many cases the specific cell for the production of the hormones has yet to be determined.

The size of the endocrines singly or as a group is unimpressive, as is the extent of their distribution. But these secretory centers utilize information from the exteroceptive and interoceptive portions of the nervous system. By chemoreceptive and chemostimulative means these ductless glands exert structural and functional control over the organism.

The endocrine parenchyma may develop in the embryo and fetus from cells derived from any one of the three germ layers. The adrenal gland, however, develops from only two. In such instances groups of cells from varied origins may retain discrete and autonomous functions. Prior to birth, the endocrine structures of the fetus are under direct influence of the maternal endocrines. Although the endocrines are significantly developed at birth, the major differentiation of structural and functional relationships is affected postnatally. Once established, this system regulates normal

growth, propels the organism to sexual maturity, controls metabolism and homeostasis, and precipitates senescent change.

A normally functioning endocrine control system is essential to the integration of the functioning body systems. A malfunctioning endocrine system causes dramatic changes in body form, function, and behavior. Diabetes insipidus, hypophyseal adenoma, diabetes mellitus, hyperthyroidism, interstitial cell tumor, and hypercorticoidism are a few of the major endocrine disturbances affecting the dog. Bloom (1959) presents a general discussion of endocrine diseases affecting the dog.

The endocrine organs vary in structure and function among the breeds (Stockard 1941), among individuals, seasonally, and, at a cellular level, diurnally. The morphology of the endocrines can change rapidly in response to variations in normal physiological activity. In many of the endocrine tissues there is a storage of precursors or products of cellular synthesis leading to the formation of the active hormone. The adrenal cortex, interstitial endocrine cells of the testis, and endocrine cells of the corpus luteum store neutral fats and cholesterol in large quantities. The thyroid produces great amounts of thyroglobulin and stores such materials in the follicular lumina. In light of this "reserve capacity" and the normal variation in cell numbers or percentages within the same individual, abnormal changes leading to or resulting from disease of the endocrine system are very difficult to determine by morphological means. Such findings suggestive of hyper- or hypofunctioning must be correlated with a case history, clinical signs, and assays for hormone production. Hormonal assay is accomplished primarily by direct measurement of blood levels or levels of urinary excretion of hormone metabolites. Most tests are only recently developed, are not yet reliable, and have not provided sufficient data upon which to base normal values. Thus, most clinical diagnoses of endocrine disease are made from indirect information relative to levels of secretory product.

The dog continues to serve as a model for endocrine research and, as such, has provided much of the knowledge relating to the functions of this system. The term hormone was first applied by Bayliss and Starling

(1902) in their observations of the mechanisms of pancreatic secretion in the dog. In 1922, when Banting and Best made medical history by discovering the effects of pancreatic extract in reducing blood sugar, the dog was the experimental animal. The dog was also the test animal used for the demonstration of secretin (Ivy and Oldberg 1928, Kosaka and Lim 1930).

Michaelson (1970) presented a review of the general anatomical features of the endocrine glands of the dog, along with an account of specific functional and clinical parameters. Much of the structural and functional detail relative to the endocrine system of the dog has been inferred from work with other mammals, including man, and specific information about the morphology of the dog endocrines is still widely scattered in the literature. Stockard (1941) presented detailed information on the morphology of the endocrines in many breeds of dogs. Venzke (1976) discusses the macroscopic anatomy of selected endocrine organs of the dog. Compendia such as those edited by Harris and Donovan (1966) for the pituitary gland, Pitt-Rivers and Trotter (1964) for the thyroid gland, and Jones (1957) for the adrenal cortex contain excellent comparative information, including extensive references to the dog. Endocrine morphology varies considerably with the breed and with the individual, depending upon age, nutrition, environment, and health. Most endocrine research in dogs is conducted using the medium-sized, mesaticephalic breeds.

HYPOPHYSIS

The *hypophysis,* or **pituitary gland,** is a reddish appendage attached at the ventral midline to the diencephalon (Fig. 10–1). The Greek term *hypophysis* conveys this positional meaning, *hypo* meaning under and *physis* meaning growth — thus, the growth on the undersurface of the brain. The term "pituitary" is derived from a historical interpretation by Vesalius concerning the function of this gland as the source of nasal exudate, *pituita* or phlegm (Field and Harrison 1957).

The size of the hypophysis varies greatly among breeds of dogs and among individuals of the same breed (Stockard 1941, White and Foust 1944, Hewitt 1950, Latimer 1941 and 1965, Hanström 1966). In the adult of the mesaticephalic breeds, the size of the unpreserved gland is approximately 1 cm.

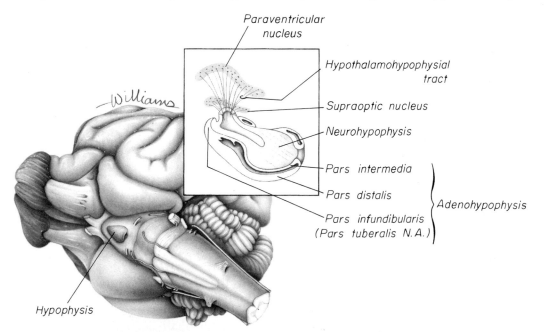

Figure 10–1. Hypophysis attached to ventral midline of brain, left caudoventrolateral view. Inset is a schematic representation of the midsagittally sectioned hypophysis and the extension into the neurohypophysis of nerve fibers from cell bodies found in the supraoptic and paraventricular nuclei of the hypothalamus.

in length, 0.7 cm. in width, and 0.5 cm. in depth. Its weight is about .06 gm. in the male. The hypophysis of the larger dog shows an absolute increase in size but a relative decrease in proportion to body weight. When other factors such as breed and nutrition are constant, the hypophysis of the female is somewhat larger than that of the male. It is also larger in the gravid female than in the non-gravid female. (Latimer 1941, White and Foust 1944).

Although small, this organ plays a major regulatory role in the entire endocrine system. The close structural positioning of the glandular and nervous segments of this gland is symbolic of its function in interrelating the nervous and endocrine systems. So extensive are the influences of the hypophysis upon cells, tissues, and organs that it is often referred to as the "master gland" of the body. The early work of Putnam et al. (1929), which reported the effects of administering pituitary extracts to young dogs, demonstrated the widespread changes in body tissues due to injections of extracts from the "master gland" and indicated the significance of this endocrine gland. The works of Crowe et al. (1910) and Dandy and Reichert (1925) established that the hypophysis was essential for the maintenance of life.

Macroscopic Features

The hypophysis occupies a bony recess in the basisphenoid (os basisphenoidale) (Fig. 4–16). The recess is a shallow, oval depression, the fossa hypophysialis. The rostral and caudal margins of the fossa are formed by the tuberculum sellae and the dorsum sella, respectively. When the fossa is viewed dorsally, the rostral and caudal clinoid processes accentuate the boundaries of the fossa and form a bony complex termed the sella turcica. In the dog the fossa is quite shallow and is lined by the external, or endosteal, layer of dura mater. The inner, or meningeal, layer of the dura forms the diaphragma sellae. The latter does not pass directly into the fossa with the external dural layer but extends partially over the dorsal aspect of the fossa to provide an incomplete septum. The primary attachments of this septum, or diaphragm, are by way of

the clinoid processes. A large oval foramen is present in the center of the diaphragm, which loosely encircles the stalklike connection of the hypothalamus to the hypophysis as it passes into the fossa. This thin meningeal layer then continues around the main portion of the gland as a delicate capsule, thus eliminating any subdural space. Schwartz (1936) demonstrated that the subarachnoid space does not invest the hypophysis.

The space created by the separation between the inner and outer dural layers contains prominent cavernous and intercavernous sinuses. Large cavernous sinuses bound the hypophysis laterally and are connected by intercavernous sinuses. The larger intercavernous sinus passes just caudal to the hypophysis, whereas the smaller intercavernous sinus is present in only some individuals and, when present, passes rostral to the hypophysis. The rostral portion of the middle meningeal artery and the anastomotic ramus of the external ophthalmic artery lie within each of the cavernous sinuses. The internal carotid artery also courses through each of the cavernous sinuses on the lateral margins of the hypophysis from the dorsum sella rostral to the region of the optic chiasm. The oculomotor, trochlear, and abducent nerves and the ophthalmic branch of the trigeminal nerve pass in close proximity to the hypophysis. The interpeduncular cistern surrounds the attachment of the hypophysis to the hypothalamus, and within the cistern lies the circulus arteriosus. A bony wall separates the hypophyseal fossa from the sphenoid sinus, which lies rostroventral to it. The positional relations of these structures to the hypophysis present a degree of surgical risk and, as reported by Harrison (1964) and emphasized by Farrow (1969), may account for many of the signs accompanying hypophyseal disease.

Mesoscopic Features

As noted by Hanström (1966), much of the literature demonstrates individual and breed differences in the morphology of the hypophysis. The hypophysis of the adult (Figs. 10–1, 10–2) has grossly visible rounded protuberances, the adenohypophysis and

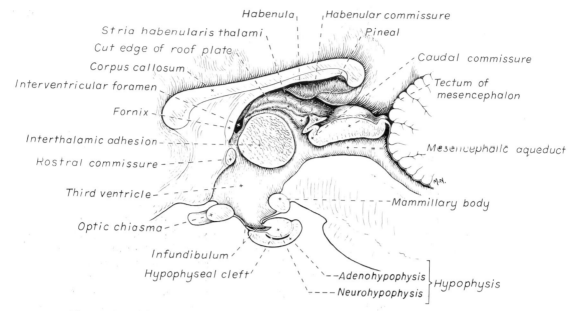

Figure 10–2. The diencephalon, median section. (From Evans and deLahunta 1971.)

the *neurohypophysis*. The adenohypophysis, being composed of glandular parenchyma and having an extensive blood supply, appears reddened and friable in comparison to the pallor and the fibrous texture of the neurohypophysis. A median section of the hypophysis and the hypothalamus, when examined with slight magnification, reveals further subdivisions of the organ (Fig. 10–1). The gland is suspended from the midline of the hypothalamus by a cylindrical stalk. This stalk is an extension of the **median eminence** from the *tuber cinereum* of the hypothalamus. It is the proximal portion of the neurohypophysis, the *pars proximalis neurohypophysis (infundibulum)*. In most dogs the third ventricle continues as an invagination, the *recessus neurohypophysis*, into the infundibulum, but it rarely passes into the more distal portion of the neurohypophysis. The pars proximalis neurohypophysis is continuous with the distal enlargement, the *pars distalis neurohypophysis (lobus nervosus, N.A.)*, which is the major portion of the neurohypophysis.

The principal axis of the gland is in a horizontal plane. The largest portion of the adenohypophysis lies ventrorostral to the pars distalis neurohypophysis, but it invests nearly all of the surface of this part of the neurohypophysis and forms three major subdivisions. The adenohypophysis contains a large, compressed vesicle which, when seen in slices of the gland, appears as a cleft. It is termed the **hypophyseal cleft** (*cavum hypophysis*) and is a remnant of development. The inner portion of the adenohypophysis is in direct contact with the pars distalis neurohypophysis and is termed the *pars intermedia adenohypophysis* owing to its location between the two major parts of the hypophysis. The largest portion of the adenohypophysis remains separated from the pars intermedia by the hypophyseal cleft and forms the distal portion of the adenohypophysis, *pars distalis adenohypophysis*. The adenohypophysis also extends as a cuff or collar around the pars proximalis neurohypophysis to envelop part of the median eminence. This is the *pars infundibularis adenohypophysis*.

There are some differences in terminology for the parts of the hypophysis between Nomina Anatomica (N.A.) and Nomina Anatomica Veterinaria (N.A.V.). Nomina Histologica (N.H.) agrees with Nomina Anatomica, but Nomina Histologica Veterinaria (N.H.V.) does not entirely parallel Nomina Anatomica Veterinaria.

For clarification, the listing below of 1978 N.A.V. terms shows the variants in N.A. (= N.H.) and N.H.V.

Hypophysis (Glandula pituitaria)
Adenohypophysis (Lobus anterior)
 Pars infundibularis adenohypophysis
  = Pars tuberalis (N.A.)
  = Pars proximalis adenohypophysis (N.H.V.)
 Pars intermedia adenohypophysis
  = Pars intermedia (N.A.)
 Pars distalis adenohypophysis
  = Pars distalis (N.A.)
Neurohypophysis (Lobus posterior)
 Pars proximalis neurohypophysis (Infundibulum)
  = Infundibulum (N.A.)
 Pars distalis neurohypophysis
  = Lobus nervosus (N.A.)

Developmental Anatomy

By the 7-mm. stage of development in the dog, a small portion of oral ectoderm lining the dorsum of the stomodeum contacts the ventral surface of the neural tube. This contact or adhesion between oral and neural ectoderm is maintained while differential growth and the resultant proliferation of mesoderm continues in the head region. The neural ectoderm retains its relative position, and the portion of oral ectoderm is drawn away from the stomodeum, first as a cul-de-sac and then as a closed vesicle separated from the developing oral cavity. With continued differentiation of the neural ectoderm, a small projection or evagination develops at the midline of the ventral surface of the mesencephalon at the point of adhesion to the oral ectoderm. This structure, the *sacculus infundibuli,* becomes surrounded by the collapsing vesicle of oral ectoderm. The adjoining mesenchyme develops the stroma and vascularization for this parenchymal primordium of the hypophysis (Herring 1908b, Kingsbury and Roemer 1940, Stockard 1941, Latimer 1965).

This duality of origin results in an organ with segregated structure and function. The neural ectodermal portion forms the neurohypophysis with pars proximalis and pars distalis. The vesicle of oral ectoderm invests the neurohypophyseal primordium on all of its surfaces except for a small area at the distal, caudal extremity. A portion of the vesicle extends to envelop the pars proximalis neurohypophysis. The surfaces of the vesicle contacting the pars distalis neurohypophysis develop into the pars intermedia adenohypophysis. The remaining portion of the vesicle, which does not directly contact the neurohypophysis, develops into the major glandular portions, the pars proximalis adenohypophysis and the pars distalis adenohypophysis.

Microscopic Features

The general microscopic features of this gland are those of an endocrine gland and a segment of nerve tissue. The stroma is formed by a capsule of delicate pial connective tissues which forms around the neurohypophysis during development and remains as a boundary between it and the adenohypophysis subdivisions. The adenohypophysis is enveloped by a delicate investment of collagenous connective tissue of the arachnoid which binds the adenohypophysis to the adjoining inner layer of dura mater. At a point representing the original connection to the oral ectoderm on the midventral surface of the adenohypophysis, the adenohypophysis is attached to the inner dural layer, which is fused to the outer layer. This attachment is easily separated and usually accounts for some separation artifact in preparations of the pars distalis adenohypophysis. When a small remnant of development called the *parahypophysis* is present, it is attached at this location (Kingsbury and Roemer 1940). The stroma which forms a capsule for both the neurohypophysis and the adenohypophysis is originally quite delicate and increases in amount only slightly with advancing age. The blood vessels are invested by limited amounts of adventitial connective tissues, and the parenchymal cells of the adenohypophysis are supported by a delicate interstitium of reticular fibers.

The pars distalis adenohypophysis has a parenchyma of epithelioid cell types arranged in small, interconnected clusters, *racemi endocrinocyti,* which are bounded and permeated by numerous sinusoids. The aggregates of cells are quite varied in size and shape and are arranged three-dimensionally as anastomosing lattices. The close proximity of nearly all parenchymal cells to a sinusoid is maintained throughout. By the use of specific staining techniques, especially immunohistochemistry, cellular biochemistry and structure have been related to the specific hormonal secretions of these cells. With routine staining there is a separation of three distinctive cell populations: *endocrinocytus acidophilus, endocrinocytus basophilus,* and *endocrinocytus chromophobus.* The frequency of occurrence of each cell type varies according to the age, sex, breed, and physiological state of the individual. Francis and Mulligan (1949) reported that in the adult male dog the acidophilic cells outnumbered the basophilic cells approximately five to one and that the chromophobic cells occurred with three times the frequency of the basophilic cells. Stockard (1941) reported that the basophilic cells were outnumbered 30 to 1 by the acidophilic cells. White and Foust (1944) reported no significant sex differences and a ratio of 11 to 1, acidophilic to basophilic cells.

The acidophilic endocrine cells have an affinity for acidic dyes, staining well with the acid dyes of the bichromic and trichromic procedures. The dye is taken up by the granules of the cytoplasm, which are stored secretory products. These cells are smaller than the basophilic cells, measuring approximately 15μm. in diameter. The acidophilic cells are located adjacent to sinusoids, in most cases being displaced only by basophilic cells from that site. Their distribution within the pars distalis adenohypophysis is generally uniform, with only a slight increase in their numbers near the inner aspect. During pregnancy the proportion of acidophilic cells increases and remains elevated, constituting approximately 65 per cent of cells, until the cessation of lactation, when their numbers return to prepregnancy levels. These cells are of two types, one which produces a somatotrophin protein and another which produces a lactotrophin protein.

The basophilic endocrine cells possess cytoplasmic granules which have a moderate affinity for the basic component of routine laboratory stains. The granules are composed of glycoprotein and react positively when stained with dyes specific for that component. The cells are larger than the acidophilic cells, measuring approximately 20μm. in diameter, and are more elongated. Like the acidophilic cells, these cells are generally observed along a sinusoid. They occur in greater numbers at the periphery of the pars distalis adenohypophysis. Stockard (1941) has reported that basophilic cells proliferate during proestrus. Thyrotrophin and the gonadotrophins are the hormones produced by specific basophilic cell types.

The chromophobic cells do not stain well with routine dyes. They occur in moderate numbers near the central region of a racemus. Their nuclei are easily visible, but their cytoplasmic boundaries are difficult to determine. There is equivocal evidence that some of these cells, which lack marked staining affinity, produce adrenocorticotrophin (Goldberg and Chaikoff 1952, Mikami 1956, Purves 1966). Ricci and Russolo (1973) present immunocytological evidence that suggests this hormone is produced by the chromophils.

Many investigators have described a fourth cell type in the dog. Most believe it to be a functional variant (Wolfe and Cleveland 1932, Hartman et al. 1946, Smith et al. 1953, Purves and Griesback 1957). Kagayama (1965) and Gale (1972) term this cell a stellate or **follicular cell.** Goldberg and Chaikoff (1952), Carlon (1967), and Gale (1972) describe six cell types based upon specific histochemistry or structural features.

The pars proximalis adenohypophysis is composed of epithelioid cells packed between sinusoids in anastomosing walls or muralia. Occasionally, small follicles are seen. The cells are uniform in size, being low columnar, and in staining characteristics, being largely chromophobic. There are often a few cells typical of the chromophobic cells of the pars distalis adenohypophysis mixed among cells of the pars proximalis adenohypophysis near that segment. Their cytoplasm is finely granular and reacts slightly positive with dyes for glycogen. When follicles occur, they are filled with a homogeneous glycoprotein and lined by a simple columnar epithelium. The tissues are also continuous with those of the pars intermedia adenohypophysis. A specific function for this part of the adenohypophysis has not been determined.

The parenchyma of the pars intermedia adenohypophysis is a complexly folded pseudostratified columnar epithelium. It envelops nearly all of the .pars distalis neurohypophysis and is separated from it by only a delicate band of vascularized, pial connective tissue. The remnant of the vesicular space, the hypophyseal cleft, separates it from the pars distalis adenohypophysis. The pars intermedia adenohypophysis blends with the pars distalis adenohypophysis as it reflects distally to become continuous with that part. Where it is continuous with the pars proximalis adenohypophysis, the folded epithelium projects into the lumen of the hypophyseal cleft. These folds are more prominent in the brachycephalic breeds. Infrequently, there are villus-like projections extending for short distances into the neurohypophyseal tissues. The parenchyma is normally devoid of sinusoids but has access to numerous pial capillaries at the neurohypophyseal surface. Often these vessels can be seen projecting into the connective tissue septa of the folds. The epithelioid cells stain only faintly. Melanocyte-stimulating hormone is secreted by these cells.

Cells of the pars distalis neurohypophysis and the pars proximalis neurohypophysis are chiefly glial cells which are termed *glio-cyticentrales (pituicytes)*. These cells support nerve fibers in an extensive neuropil. These nerve fibers are extensions of cell bodies found in the nuclei of the hypothalamus and compose the hypothalamohypophyseal tract (Fig. 10–1). From these processes are released neurohumors produced in the hypothalamus and termed vasopressin and oxytocin.

Vascularization

The blood supply and venous drainage of all the hypophyseal regions are structurally interrelated. The functional interdependence of the hypophyseal subdivisions is facilitated by their common vascularization.

The arterial supply of the hypophysis arises from two major sources, the **internal cartoid arteries** and the **caudal communicating arteries** (Fig. 10–3). Several branches passing directly to the region of the gland arise from the rostral and caudal intercarotid arteries and the caudal communicating arteries. The number of vessels from these sources which converge upon the hypophysis is extensive and was described by Dandy and Goetsch (1910) as appearing "like spokes to the hub of a wheel." The **rostral intercarotid artery** provides a variable number of branches, 4 to 10, to the region of the pars proximalis neurohypoph-

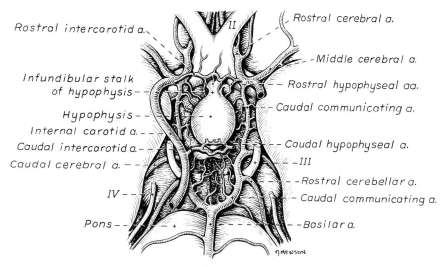

Figure 10–3. The vascularization of the hypophysis, ventral aspect.

ysis (Basir 1932). An equally extensive group of vessels arises from the **rostral communicating arteries.** In addition, a vessel arises from each of the internal carotids and proceeds toward the pars proximalis neurohypophysis. Less prominent vessels may also originate from the caudal aspect of the arterial circle. All of these vessels pass centripetally toward the pars proximalis neurohypophysis, where they join in forming a plexus, the **mantle plexus** (Green 1951). The plexus is incorporated in the meningeal investment of the hypophysis.

From this mantle plexus many arterioles enter the median eminence and provide capillaries to the **primary blood capillary network** *(rete hemocapillare primarium).* On the rostral and lateral surfaces of the median eminence major vessels arise from the mantle plexus or as direct branches of the intercarotid vessels which are termed the **rostral hypophyseal arteries.** Several of these also provide capillaries to the median eminence and the initial portion of the pars proximalis neurohypophysis and join in the primary blood capillary network (Akmayev 1971a). These capillaries receive neurohumoral substances called releasing factors, which are subsequently carried from this region of the hypothalamus, called the **median eminence gland** by Reichlin (1974), to the hypophysis via a portal blood vascular system (Green 1966, Campbell 1970). Akmayev (1971b) reported that those nuclei of the median eminence which have the greatest vascularity have the smallest caliber capillaries. A major portion of the venous drainage from this capillary network returns to the surface of the pars proximalis neurohypophysis and is collected by veins which run parallel to its outer surface. These veins supply the sinusoids of the adenohypophysis. The capillaries (sinusoids) of the adenohypophysis form the **secondary blood capillary network** *(rete hemocapillare secundus).* The veins which connect these two capillary networks are the **hypophyseal portal vessels** *(venulae portis hypophysis),* meaning the gateway vessels to the gland. This portal circulation in the dog is less obvious than in some other species owing to the relatively short pars proximalis neurohypophysis in the dog (Green 1966).

From the caudal rim of the mantle plexus a small number of arterioles pass directly to the neurohypophysis at its distal extremity. In this region the adenohypophysis does not complete its investment of the pars distalis neurohypophysis. These arterioles, termed the **caudal hypophyseal arteries,** enter the parenchyma of the pars distalis neurohypophysis and distribute as capillaries. A few small branches may pass to the pars distalis adenohypophysis.

The pars proximalis adenohypophysis receives its blood supply from the hypophyseal portal vessels and the sinusoids of the pars distalis adenohypophysis; the latter also receives its supply from the portal vessels. The pars intermedia adenohypophysis lacks an intraepithelial vascular network. Instead, the epithelial cells rest on a basement membrane in close proximity to capillaries, the *rete intermedius,* in the stromal tissues coursing between the pars intermedia adenohypophysis and the pars distalis neurohypophysis.

The neurohypophysis receives a blood vascular supply from the rostral hypophyseal arteries supplying the median eminence and the pars proximalis neurohypophysis. The pars distalis neurohypophysis also receives some small number of caudal hypophyseal arteries. Unlike in the adenohypophysis, within the neural tissue there are profiles of arterioles and some muscular arteries.

The regulation of the adenohypophysis by the hypothalamus is made possible by the architecture of the portal system. Structural and functional data suggest that blood entering the hypophysis from these multiple sources will eventually percolate through the sinusoids of the pars distalis adenohypophysis. From these sinusoids venous drainage is by way of vessels which exit from the gland parenchyma and empty into the cavernous and intercavernous sinuses (Dandy and Goetsch 1910, Morato 1939, Green 1951, Green 1966).

Innervation

The nerve supply to the hypophysis passes primarily to the neurohypophysis. The parenchyma of the neurohypophysis is primarily composed of axonic processes which extend from cell bodies in the hypothalamic regions (Dandy 1913). The majority of these cell bodies are located in the supraoptic and paraventricular nuclei and, to a lesser degree, in other hypothalamic regions (Green 1951). The processes extend into the pars distalis neurohypophysis. This collection of fibers is termed the **hypothalamohypophyseal tract.** These are unmyelinated nerve fibers which convey and release a neurosecretory product. Sympathetic fibers arising from the cranial cervical ganglion pass by way of the tunica externa of the carotid artery and its branches to the vessels of the hypophysis. From this pathway along the carotid distribution, sympathetic plexus fibers that pass to the hypophysis have been described (Dandy 1913, Truscott 1944, Green 1951). A parasympathetic innervation has not been described. The results of ablation experiments suggest that hormonal output from the adenohypophysis is not under direct autonomic control. Yamada et al. (1956) and

Green (1966) review the literature on innervation to portions of the adenohypophysis.

THYROID GLAND

The thyroid tissue in most dogs forms paired structures, although each gland, *glandula thyroidea*, is referred to as a lobe. It is an elongate, dark red mass attached to the outer surface of the proximal portion of the trachea (Figs. 10–4, 10–5). They are positioned laterally and somewhat ventrally on the trachea, spanning the initial five to eight tracheal rings on their respective side. The size of the thyroid is variable, depending upon the breed and individual (Marine 1907, Stockard 1941). In the adult of the medium-sized breeds, the fresh gland is approximately 5 cm. in length, 1.5 cm. in width, and 0.5 cm. in thickness, with the dorsal margin of the gland being somewhat

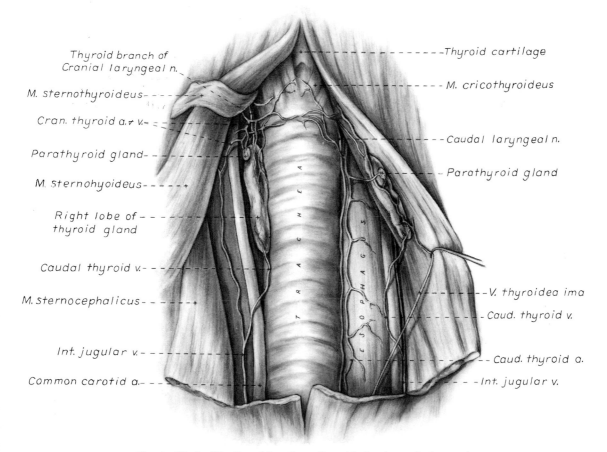

Figure 10–4. The thyroid and parathyroid glands, ventral aspect.

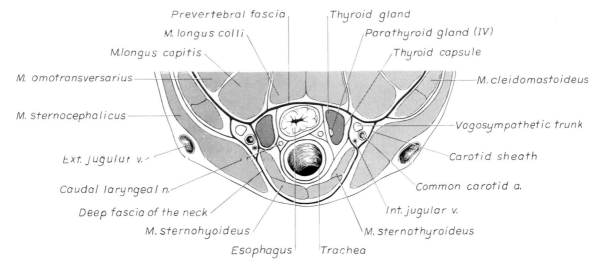

Figure 10–5. Schematic transverse section of ventral neck passing through thyroid glands.

thicker than its ventral counterpart. The thyroid is the largest of those ductless glands which perform only an endocrine function. By virtue of its size and position, the thyroid is unique among the exclusively endocrine organs in that it can be palpated during a physical examination, especially when enlarged. The weight of the fresh gland of the adult is quite variable. Data comparing the thyroid mass to body weight revealed a ratio of approximately 0.1 gm. thyroid per kg. body weight (Mulligan and Francis 1951, Gilmore et al. 1940).

The thyroids exert a major control on the metabolic processes of the body and affect most body systems. The thyroid hormone is synthesized by the glandular parenchyma, stored intercellularly, and released from the gland into the circulation. Foster et al. (1964a and b) demonstrated the presence of a second hormone produced by the thyroid called thyrocalcitonin, or calcitonin, which lowers blood calcium by stimulating calcium uptake by the skeleton.

Macroscopic Features

Each gland is embedded, in the deep cervical fascia and is closely adherent to the trachea (Fig. 10–5). The sternocephalicus and sternohyoideus muscles pass immediately lateral to the convex surface of each gland, and the sternothyroideus covers each thyroid on its ventral surface.

The cranial pole of the right thyroid lobe lies at the level of the caudal border of the cricoid cartilage of the larynx. Caudally, it extends to the region of the fifth tracheal ring. It is bounded dorsolaterally by the common carotid artery, the internal jugular vein, the tracheal duct, and the vagosympathetic trunk. Fibers of the recurrent laryngeal nerve pass dorsally in close association with the right thyroid lobe. The stroma of the thyroid is continuous with and tightly bound within the proper fascia of the trachea and blends with the fascia of the carotid sheath.

The left thyroid lobe is generally indistinguishable from its counterpart in size and shape. It does differ in positional relationships, being further caudal, extending from the third through the eighth tracheal rings. Its dorsolateral boundary is formed by the esophagus. The caudal laryngeal nerve passes dorsal to the gland, and the trachea provides its medial boundary. The components of the carotid sheath on the left side are displaced by the esophagus and therefore are not in contact with the gland.

Mesoscopic Features

The thyroid gland of the adult has an organization of tissues based primarily upon follicles. The stromal connective tissues are best developed at the dorsal aspect, where many of the main vessels enter and leave the gland. From this dorsal median mass of stroma extend the septa which incompletely subdivide the gland into lobules. The larger parenchymal follicles can be resolved with only slight magnification.

In some dogs there is a narrow connection between the two lobes. When present, this bridge, *isthmus glandularis*, composed of glandular parenchyma, passes horizontally as a band on the ventral surface of the trachea to connect the caudal aspects of each lobe. Such an isthmus is more frequently observed in the brachycephalic breeds, but can also be found in individuals of most other breeds.

The thyroid glands are intimately related structurally to the parathyroid glands. Each thyroid normally is related to a pair of these endocrine glands. One parathyroid is usually found near the cranial dorsolateral margin of the gland, but the positional relationship of the thyroid and the cranial parathyroid varies. This parathyroid *(glandula parathyroidea externa [III])* in some individuals is a satellite only; in others it is indented into the surface of the thyroid and enveloped by the thyroid fascia, and in others it is embedded within the thyroid parenchyma. The second parathyroid *(glandula parathyroidea interna [IV])* is more often not "para" in position, but is embedded within the thyroid at a depth which is variable, generally in the caudal portion of the gland (Fig. 10–5).

Developmental Anatomy

The parenchyma of the thyroid develops from structures of the pharyngeal entoderm. Beginning at the 4-mm. stage, the midventral surface of the pharynx gives rise to a tissue thickening, the **thyroid diverticulum** (Godwin 1936). Further growth in size occurs predominantly at the distal end of the diverticulum. Its connection to the pharynx narrows to become a stalk before separating completely. If a connection persists, it may appear as a duct, the **thyroglossal duct,** with functional glandular epithelium and numerous cysts along its course.

After separating from its attachment to the pharynx, the median thyroid diverticulum forms a bilobed mass connected by a broad isthmus, oriented transversely on the ventral aspect of the trachea. As the primordium progresses caudally, it passes near the developing pharyngeal pouches. A ventral component of the fourth pharyngeal pouch, which does not form parathyroid tissue, separates and fuses (12.5 to 17 mm.) with the developing median thyroid mass to form the lateral thyroid mass (Godwin 1937a, Boyd 1964). These components continue their development as a single unit, and any subsequent recognition of the contributing components is impossible. The isthmus narrows and in many cases remains only as a fibrous band connecting the caudal aspect of each gland. The orientation of each gland eventually parallels the trachea. The stroma of the glands is developed from the surrounding mesoderm.

An outstanding feature of thyroid development in the dog is the frequent occurrence of accessory thyroid tissue *(glandula thyroidea accessoria)*. Islets of the rapidly proliferating cells of the thyroid primordia separate from the main mass and become incorporated in the developing structures of the branchium and thorax. Functional accessory thyroid tissue is frequently found along the trachea, at the thoracic inlet, within the mediastinum, and along the thoracic portion of the descending aorta. Swarts and Thompson (1911) reported accessory thyroid tissue in the pericardial sac of 24 of 30 dogs examined. Halsted (1896), French (1901), Godwin (1936), and Kameda (1972) reported on the occurrence of numerous accessory thyroid tissues in the dog, and Smithcors (1964) reported some accessory tissues in all embryos and nearly one-half of adults examined.

As the main mass of developing thyroid parenchyma extends first ventrocaudally and then laterally, it incorporates cells which in the adult will function to produce calcitonin (thyrocalcitonin). The origin of these cells has long been considered pharyngeal entoderm of the fourth branchial pouch. These cell groups of the developing neck, **ultimobranchial bodies,** fuse with and become dispersed within the developing thyroid parenchyma (Godwin 1937a). As thyroid development proceeds, these cells occupy positions satellite to and between thyroid follicles (Nonidez 1932a and b) and differentiate to form the **parafollicular endocrine cells** *(endocrinocyti parafolliculares)* and the **interfollicular endocrine cells** *(endocrinocyti interfolliculares),* respectively. Evidence presented by Pearse and Polak (1971) suggests that the ultimobranchial bodies are invaded by cells of the neural crest prior to their migration into the developing thyroid. They suggest a neural crest rather than pharyngeal entoderm as the tissue of origin for the parafollicular cells. Some interfollicular tissue forms **interfollicular islets** *(insulae interfolliculares)*, which are believed to be stem cells for the formation of new thyroid follicles.

The developing cords of the thyroid analogue, which are solid initially, branch and separate to form isolated groupings which differentiate into small follicles prior to birth.

Microscopic Features

The thyroid gland is sharply delineated from adjacent tissues and organs but does not possess a markedly fibrous capsule. Its delicate *capsula thyroideae* limits the gland and is continuous with the surrounding fasciae. Septa and trabeculae, conveying the arteries, veins, and lymphatic vessels, subdivide the gland and separate much of the gland into lobules, *lobulae thyroideae*. From this stroma the fine reticular fibers pass to form the interstitium which supports the follicles and conveys the extensive capillary network.

The follicles, *folliculae thyroideae*, are the principal parenchymal units. These spheres vary in size from 50 to 900 μm. in diameter. Sectional geometry makes the determination of maximum and minimum diameters difficult (Wissig 1964). Venzke (1940) reported that for puppies under three months of age the follicular diameter varies from 30 to 160 μm. The follicles are filled with a homogeneous mass of glycoprotein termed *colloid*. Each follicle is lined by a simple epithelium, the cells of which vary in shape (from high columnar to squamous) from one follicle to another, depending upon the functional status of the follicle. These cells form the thyroglobulin and the hormones tetraiodothyronine (thyroxin) and triiodothyronine. The follicular epithelial cell, termed the *endocrinocytus follicularis*, is the most numerous of the parenchymal cells.

Another parenchymal cell, the *endocrinocytus parafollicularis* (also "C" cell), named for its close position to the follicle, is larger and lighter staining than the follicular endocrine cell (Vicari 1937, Kameda 1971, 1973). Roediger (1973) distinguished between "intrafollicular" and "parafollicular" cells. The former are within the lining epithelium of the follicle; the latter are adjacent to it. As demonstrated by Kalina and Pearse (1971), these cells secrete calcitonin (thyrocalcitonin). The parafollicular endocrine cells are commonly seen bordering the follicle and displacing follicular epithelial cells. They may also occur singly or in groupings between follicles as *endocrinocyti interfolliculares*. Hedhammar et al. (1974) reported that these cells decrease markedly with aging. Teitelbaum et al. (1970) have demonstrated follicle profiles of parenchyma composed entirely of the "C" cells.

In the interfollicular region there are cells which have morphological and staining features similar to those of the follicular cells. A grouping of these interfollicular cells is an *insula interfollicularis*, which appears to be a pool of reserve cells, capable of forming new follicles (Zechel 1931).

Vascularization

The vascular supply of the thyroid is quite extensive, equal to or exceeding that of any other similar parenchymal volume in the body. The principal supply arises from two vessels: the **cranial and caudal thyroid arteries** *(arteria thyroidea cranialis, arteria thyroidea caudalis)*. The cranial thyroid artery commonly arises as the first major branch from the common carotid. This artery terminates in branches supplying the larynx and associated structures and supplies a major vessel to the thyroid. It approaches the cranial pole of the thyroid and passes parallel to the dorsal surface of the gland to anastomose with the caudal thyroid artery. The caudal thyroid artery most commonly arises from the brachiocephalic artery and runs along the lateral surface of the trachea on the left side to join with the respective cranial thyroid artery.

From the vessel formed by the anastomosis of the thyroid arteries dorsal to each gland arise vessels in various numbers and patterns which approach the dorsal surface. Some smaller vessels may pass to the ventral aspect of the gland. These branches, both dorsal and ventral, bifurcate before entering the gland and supply the lateral and medial surfaces. From the branching of the cranial thyroid artery a small vessel continues directly to the cranial parathyroid.

At the surface of the gland and within the delicate capsule these vessels anastomose freely across the surface to form a blood vascular network (*rete arteriosum*) (Major 1909). Passing via septa and trabeculae into the gland, the distributing arteries possess marked valvelike tunica media tissues, arterial cushions (Modell 1933). These vessels subdivide to the capillary level. Each follicle is enveloped by an extensive, fenestrated sinusoidal network, and the tissues between follicles are equally well supplied with these channels (*rete hemocapillare perifolliculare*) (Fujita and Murakami 1974).

Modell (1933) described arteriovenous anastomoses deep within the gland and noted large and small varieties which he hypothesized shunted blood past proportionate amounts of parenchyma, indirectly regulating secretion from the gland. Venous output is via a system of venules and veins which generally parallel the arterial system. Numerous valves have been reported in the veins of the thyroid (Modell 1933). Venous flow from the thyroid is primarily from the **cranial and caudal thyroid veins**, which exit from the respective poles of the gland. At the caudal margin of the larynx the cranial thyroid vein joins the internal jugular vein, which, as it passes to the caudal neck region, receives the caudal thyroid vein. Smithcors (1964) reported that in many cases an unpaired vessel near the midline of the trachea receives a large tributary from the middle segment of the left thyroid and passes to enter the brachiocephalic vein.

Satellite to the follicles and from within the interfollicular region arise numerous lymphatic capillaries (*plexus lymphocapillaris perifollicularis*). These beginnings as dilated cul-de-sacs have been reported by Rienhoff (1938), who called them bursella. He observed that each follicle was only partially enveloped by these endothelial capillaries. The lymphatic vessels do not relate as closely to the follicles as do the blood sinusoids. They merge and flow as larger vessels toward the surface of the gland. In the septa these lymphatic vessels become quite large and run satellite to the blood vessels. Valves

are numerous in these thin-walled channels (Baber 1877). At the surface beneath the capsule they join a lymphatic plexus. Rienhoff (1938) demonstrated large lymphatic trunks draining the cranial aspect of the gland toward the cranial deep cervical lymph node and suggested that in 10 per cent of cases a collateral circulation of lymph may occur via lymphatics communicating between glands at a point corresponding to the location of the isthmus. Lymphatics draining the caudal aspect of the gland pass to the caudal deep cervical lymph nodes. Mahorner et al. (1927) reported direct connections of these draining lymphatics with veins of the caudal aspect of the neck. In the majority of cases the efferents are ultimately drained on the right side by the right lymphatic duct and on the left by the left tracheal duct.

Innervation

The innervation to the thyroid is via the **thyroid nerve**, formed of fibers from the cranial cervical ganglion and cranial laryngeal nerve. These fibers run in close association with the cranial thyroid artery (Nonidez 1931, Ross and Moorhouse 1938). Sympathetic postganglionic fibers are distributed as a plexus in the adventitia of the interfollicular blood vessels (Mikhail 1971). Ganglionic cells in the interfollicular space suggest the presence of a preganglionic supply. It has been proposed that both types of fibers might indirectly regulate secretion by controlling the rate of blood flow to major portions or to the entire gland (Nonidez 1935, Cunliffe 1961). Other fibers may pass to the gland from the caudal cervical ganglion via the perivascular plexuses. Terminal endings have not been demonstrated in association with the follicles. Prolonged stimulation of the thyroid nerve or denervation via transplantation of the thyroid does not produce secretory or histological change in the parenchyma (Mason et al. 1930, Ross and Moorhouse 1938).

PARATHYROID GLAND

The **parathyroid glands** (*glandulae parathyroidea*) which are closely related to the thyroid glands, are smaller than most endocrine organs. The parathyroid tissue is generally circumscribed, occurring as small ellipsoid discs measuring 2 to 5 mm. in diameter and 0.5 to 1 mm. in width. Mulligan and Francis (1951) were unable to demonstrate a correlation between body weight and parathyroid weight. As a result of having small, compact parenchymal cells which do not store significant amounts of precursors or products, the glands have a purplish coloration as fresh tissue.

As pointed out in the preceding discussion of the thyroid, the parathyroids usually occur as four structurally independent glands in close association with the thyroid glands — one applied to the surface and one embedded within each thyroid. The Greek term *parathyroid*, meaning "applied to the shield," holds that meaning only for the so-called "parathyroid III," which is normally found on the cranial dorsolateral surface of the respective thyroid. The remaining gland is embedded within the thyroid at various depths, most often within the caudal portion of each gland. Those that are beside the thyroid can be visualized readily. Those deeply embedded within the thyroid can be seen upon fresh dissection.

Like the hypophysis and adrenal gland, this composite of thyroid and parathyroid brings together tissues of different functions and perhaps different origins. Mulligan and Francis (1951) reported frequent variations in the relationship of thyroid and parathyroids and in the number and distribution of parathyroids. The parathyroids produce parathyroid hormone, which acts to mobilize body stores of calcium.

Mesoscopic Features

The position of parathyroid bodies is generally in close relation to the thyroid. Consequently, extirpation of the thyroid without regard for the preservation of the parathyroid and its support tissues results in death to the dog, owing to the precipitous loss of blood calcium. The external parathyroid gland is most frequently located at the cranial dorsolateral edge of the thyroid, occupying a shallow indentation in this gland. The profile and curvature of the thyroid gland is generally unaltered by the presence of the parathyroid. The internal parathyroid is fre-

quently found embedded within the caudal portion of the thyroid (Vicari 1937) but has been reported with greatest frequency on the dorsolateral surface (Godwin 1937b).

Developmental Anatomy

The four most regularly occurring parathyroids are also named parathyroid III and IV, indicating the external and internal glands respectively. This number designation refers to their pharyngeal pouch point of origin. At approximately the 7-mm. stage in the dog embryo there is a proliferation of cells on the dorsocranial aspect of the third pharyngeal pouch. This budding of pharyngeal entoderm gives rise to a short projection of cells which invades the branchial mesenchyme. From a similar location on the fourth pharyngeal pouch there develops a second evagination of entoderm, also known as the ultimobranchial body. At first the growth is that of a solid cord of cells followed by a hollowing out to form a duct, as in the case of an exocrine duct development. With differential growth of the branchial region, these connections are lost, and the resulting islets of entoderm associate with the developing thyroid entoderm. Those cells from pouch III, excluding the lateral thyroid primordium, associate with the cranial pole of the thyroid, and those from pouch IV become associated with and frequently embedded within the caudal portion of the developing thyroid (Godwin 1937b).

Microscopic Anatomy

The parenchyma of the parathyroid gland is arranged as an anastomosing reticulum of epithelioid cell types. In two dimensions these cells are arranged in coils or glomeruli. The cells have little cytoplasm and thus are compactly arranged. Numerous sinusoids separate the parenchyma into racemi and provide for secretory release. (Godwin 1937a, Vicari 1938, Bensley 1947).

A delicate stroma of collagenous connective tissues envelops the gland as a capsule (*capsula parathyroideae*), and from this project trabeculae carrying vascular and nervous tissues into the gland. The parenchymal cells (*endocrinocyti parathyroidea*) are supported by fine reticular fibers which are barely visible in a routine preparation. Bergdahl and Boquist (1973) describe light, dark, and oxyphilic parenchymal cells in the glands of both young and adult dogs.

Accessory parathyroid tissue (*glandula parathyroidea accessoria*) was reported by Marine (1914). He observed that following parathyroidectomy 5 to 6 per cent of the dogs did not show signs of tetany, and he attributed his observations to the presence of accessory parathyroid tissue. In similar studies, Reed et al. (1928) concluded that the incidence of accessory parathyroid tissue was only about 3 per cent. Routine serial sections of the thyroid frequently reveal accessory parathyroids (Godwin 1937a). Such accessory tissues may also migrate caudally into the thorax with the thymus (Godwin 1937b, Mulligan and Francis 1951).

Vascularization and Innervation

The blood supply to the parathyroids is directly related to that of the thyroid. The external parathyroid receives its vascular supply by one or more small branches which are direct continuations of the cranial thyroid artery. The internal parathyroid receives its supply from vessels surrounding the thyroid parenchyma. In the peripheral stroma there are small arteriolar networks which supply the sinusoids, which in turn perfuse the parenchyma.

Venous and lymphatic drainage of these small glands is accomplished by the corresponding structures in the thyroid. The innervation to the parathyroid is likewise related to that of the thyroid. Mikhail (1971) described nerve fibers ending in close association with the parenchymal cells and terminal ganglion cell bodies in the substance of the gland. Yeghiayan et al. (1972) report a well-developed adrenergic innervation to the blood vessels of the parathyroid.

PINEAL GLAND

The **pineal gland** (*glandula pinealis*), or *epiphysis cerebri* of Nomina Histologica, is an unpaired, cream-colored, wedge-shaped, small excrescence on the dorsal midline surface of the diencephalon (Figs. 10–2, 14–12).

Mesoscopic Features

The pineal gland forms the caudal boundary point of the roof of the third ventricle. The polyp-like growth extends into the potential space between the cerebellum caudally and the approximation of the two cerebral hemispheres dorsally and laterally. Its size in the dog, approximately 3 mm. by 1.5 mm. by 1 mm., requires that one use care in order to preserve this organ in these relationships (Venzke and Gilmore 1940, Zach 1960, Oksche 1965).

The structural and functional relationships of this gland and its role in the endocrine system of mammals are currently subjects of active investigation. Its antigonadotrophic influences have been established by observation of precocious sexual development subsequent to tumor formation in the pineal, which reduces the functional parenchyma, and following pinealectomy. Research into the function of the pineal in recent years has explored the role of the pineal as a photoreceptor-photointegrator of the endocrine system and its relationship to the sympathetic nervous system (Wurtman and Cardinali 1974).

Developmental Anatomy

Except for a small amount of stroma derived from mesoderm, the pineal is developed from neural ectoderm. In the developing pineal, the ependymal layer proliferates, forming the expanded mantle layer. This focal proliferation just dorsal to the posterior commissure forms the parenchyma of the pineal. The pial investment is at first an encapsulation, but with continued growth the parenchyma becomes folded and the gland is subdivided by septa. Primary support for the parenchyma is provided by the glial cell population. The pineal is quite small at the time of birth and is difficult to observe. Postnatal growth and differentiation continues progressively until sexual maturation, when the growth subsides and the gland begins a slow regression. With age there may be an increase in the size and number of intercellular concretions termed **brain sand** (*acervulus cerebri*). They are seen infrequently in the dog.

Microscopic Anatomy

The parenchyma of the pineal is composed of *endocrinocyti pinealis*, which stain well with routine dyes and are embedded in an extensive neuropil. These cells outnumber the support cells, or *gliocyti centrales*. The pia mater provides an outer limit of the gland, and only moderate amounts of connective tissues are found in a delicate capsule, septa, and trabeculae. Through these stromal structures course the blood vessels and the accompanying postganglionic sympathetic fibers.

Vascularization and Innervation

The blood vascular supply arises from the pial circulation, providing to the septa and trabeculae a generous number of arterioles. These vessels penetrate into the parenchyma and form sinusoids which are fenestrated as in the hypophysis. A lymphatic drainage for this organ has not yet been described.

The innervation of the pineal is of primary interest. The efferent fibers are of postganglionic sympathetic variety, having their origin at the **cranial cervical ganglion**. These fibers enter the adventitia of the vessels supplying the head and follow those branches to the pineal. Hartmann (1957) and Zach (1960) described fibers of cerebral origin which enter the pineal from the habenular commissure but, for the most part, return to the habenular commissure without innervating endocrine cells of the pineal. An occasional neuron cell body can be seen within the parenchyma. Ariëns-Kappers (1965) has presented evidence for a contact termination of autonomic fibers upon the pinealocytes.

ADRENAL GLAND

The **adrenal gland** *(glandula suprarenalis)* is composed of two structurally and functionally different tissues which have unique developmental histories. Each adrenal gland is composed of an outer cortex and an inner medulla. The adrenal gland is located near the craniomedial border of the kidney. In man the topographic relationship of the adrenal gland in the standing position led to the use of the term "suprarenal gland" for humans and other primates. The fourth edition of Nomina Anatomica has reinstated *glandula adrenalis* as an alternate term.

The adrenal cortex is a major steroid-producing organ, the secretions of which are indispensable for regulating mineral balance via normal kidney function and most useful for the regulation of carbohydrate metabolism. The adrenal medulla has a modulating effect upon the immediate response of the nervous system to stress. The cortex and medulla combine to influence markedly the organism's response to both acute and chronic stress.

Macroscopic Features

The left adrenal gland is the larger of the two glands (Fig. 10–6). Its cranial portion is somewhat flattened dorsoventrally and oval in outline, and its caudal projection is cylindrical. The right adrenal gland has an acute angular bend with its vortex projecting cranially. Positioned near the kidney, its longer segment projects caudally along the caudal vena cava, and the shorter segment projects toward the cranial pole of the right kidney. The left adrenal gland, lying beneath the lateral process of the second lumbar vertebra, is not as cranial in position as the right, which is beneath the lateral process of the last thoracic vertebra. Baker (1936) reported that adult males of mixed breeds have adrenal glandular tissue weighing approximately 1.14 gm. The female in diestrus has glandular tissue weighing approximately 1.24 gm. The left and right glands generally differ in weight, but this difference has not been found to be statistically significant.

The left adrenal gland is retroperitoneally positioned near the craniomedial border of the left kidney. This adrenal gland is firmly bound in the loose collagenous connective tissue of the fascia. Thus, it is more structurally related by position to the abdominal aorta than to the left kidney. Its dorsal border is applied closely to the body of the psoas minor muscle and the lateral process of second lumbar vertebra. Medially, it is bounded by the abdominal aorta at a position just caudal to the origin of the cranial mesen-

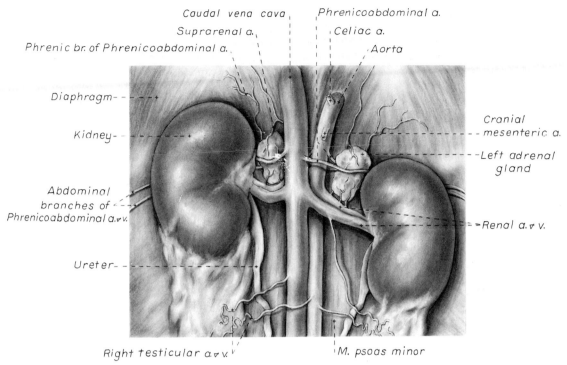

Figure 10–6. The adrenal glands, ventral aspect.

teric artery and adjacent to the origin of the phrenicoabdominal artery. This latter vessel courses over its dorsal surface at the midpoint of the gland. The caudal border of the left adrenal gland is formed by the renal artery and vein. Its ventral surface is bisected by the phrenicoabdominal vein and is covered to varying degrees by the spleen. Laterally, its boundary is formed by the kidney.

The right adrenal gland is also retroperitoneal in position, but it is near the hilus of the corresponding kidney. Its firm connective tissue attachments bring it into close proximity with the caudal vena cava as its immediate medial boundary. Often, the capsule of the right adrenal gland is continuous with the tunica externa of the caudal vena cava. The psoas minor and the crus of the diaphragm form its dorsal border. The right phrenicoabdominal artery crosses its dorsal surface, and the mass of the right kidney covers this adrenal gland on its ventrolateral surface. As a result of these organ relationships, this gland assumes a triangular or wedge shape in transectional profile. Its ventral surface is bisected by the phrenicoab-

dominal vein, and the cranial two-thirds of this adrenal gland is covered by the caudal extension of the right lateral hepatic lobe of the liver. Both glands lie in a generous bed of retroperitoneal fat.

The adrenal cortex usually completely invests the adrenal medulla. The medulla comes closest to the outer surface at the hilus of each gland. Subtle and easily overlooked, the hilus is located near the midpoint of the medial surface and serves as the exit point for the adrenal vein or veins. In some instances the medulla may extend to the capsule, especially near the hilus.

As with the hypophysis, the major subdivisions of the parenchyma can be resolved with the unaided eye. In the fresh preparation the cortex is white or faintly yellow owing to the large amount of lipid storage in the cortical parenchyma; the medulla is dark brown or black.

Developmental Anatomy

The adrenal cortex and adrenal medulla have different developmental origins. The cortex is the first to develop, originating from mesenchymal cells of the celomic mesoderm. The initial mass of these mesodermal cells proliferates near the genital ridge and condenses to form an elongate, spherical group of cells termed the **fetal cortex**. Almost immediately, a second migration of mesenchyme begins, and it eventually envelops the fetal cortex and differentiates to form the permanent, or **adult, cortex** while the fetal cortex regresses. The medullary parenchyma arises from neural crest cells, which migrate from their point of origin into the developing mesodermal mass, penetrate into this mesenchyme, and assume a central position characteristic of the adrenal medulla in the adult. The adrenal medulla is by function a sympathetic ganglion, and it develops by a process similar to that of the ganglia of the sympathetic trunk. Numerous ganglionic cells can be seen in the medulla at birth, but their number decreases with age. This migration of ectodermal cells of the neural crest is not completed until after birth, and often, even in the adult, islets of medullary tissue can be found within the cortical parenchy-

ma, within the capsule, or as satellite structures of the adrenal glands (Saleh et al. 1974).

The **adrenal capsule**, *capsula adrenalis*, develops as a condensation of mesenchyme at the periphery of the cortex. The outer portion, *pars fibrosa*, becomes a fibrous supportive stroma. The inner portion of the capsule at birth and in the young remains quite cellular and during that period is termed the *pars cellulosa*. Until the development of the outer cortical zone, the *zona arcuata*, this inner cellular layer of the adrenal capsule may serve as the stem cell population for the generation of additional adrenal cortical parenchyma.

Masses of neural crest cells develop elsewhere in the abdomen. These so-called *paraganglia* and the adrenal medulla react histochemically to reduce the salts of chromium and other heavy metals and, because of that property, are called chromaffin cells. The structure and function of the paraganglia closely parallel that of the adrenal medulla.

Adrenal cortical tissue also occurs ran-

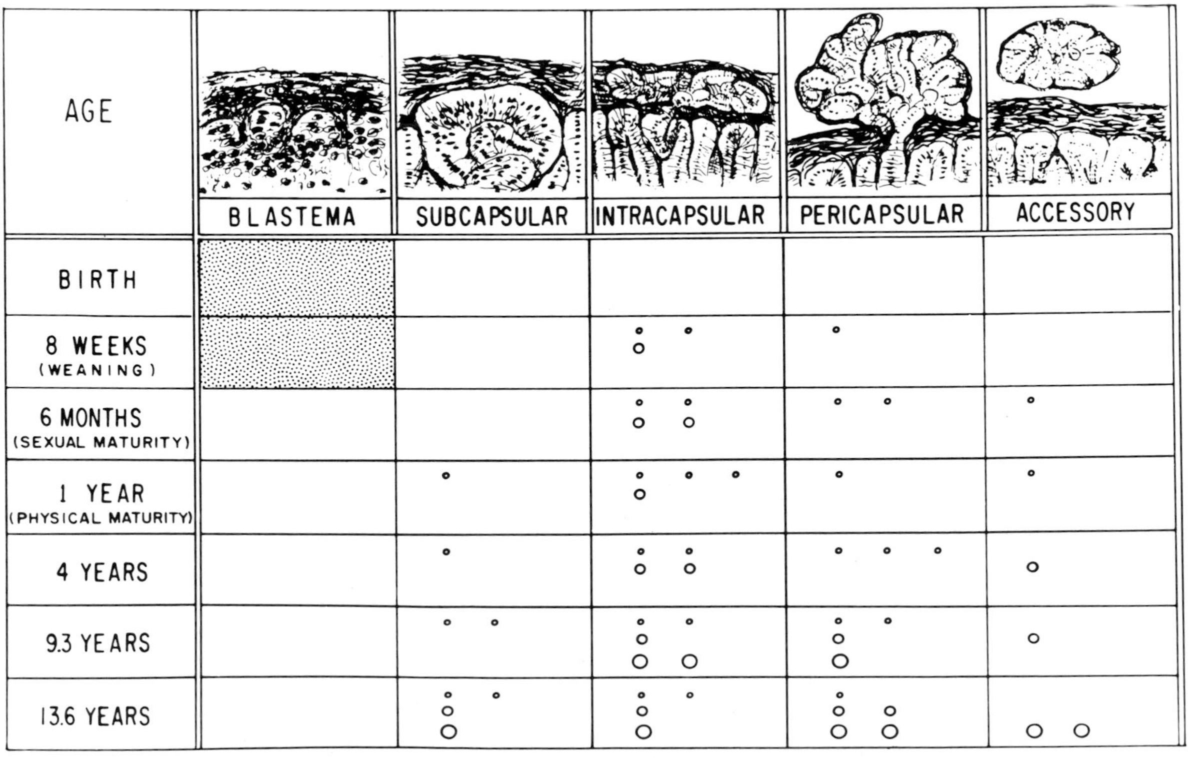

AGE	BLASTEMA	SUBCAPSULAR	INTRACAPSULAR	PERICAPSULAR	ACCESSORY
BIRTH					
8 WEEKS (WEANING)			∘ ∘ ∘	∘	
6 MONTHS (SEXUAL MATURITY)			∘ ∘ ∘ ∘	∘ ∘	∘
1 YEAR (PHYSICAL MATURITY)		∘	∘ ∘ ∘ ∘	∘	∘
4 YEARS		∘	∘ ∘ ∘ ∘	∘ ∘ ∘	∘
9.3 YEARS		∘ ∘	∘ ∘ ∘ ∘ ∘	∘ ∘ ∘ ∘	∘
13.6 YEARS		∘ ∘ ∘ ⊙	∘ ∘ ∘	∘ ∘ ∘	∘ ∘

Figure 10–7. Chart plotting the frequency of cortical tissue nodules (number of spheres), size of nodules (size of spheres), and type of nodule against age. (From Hullinger, R. L., Zbl. Vet. Med. C. Anat. Histol. Embryol. 7:1–27, 1978.)

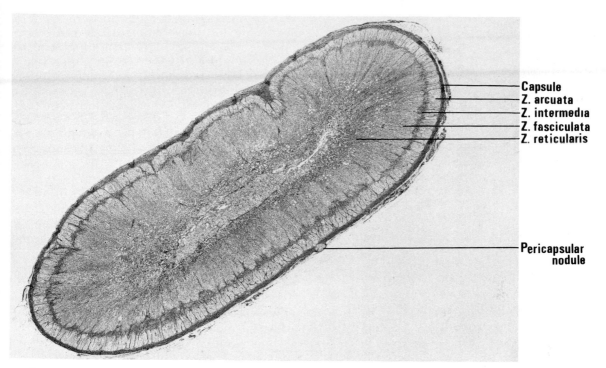

Figure 10–8. Mesoscopic features of the left adrenal gland of the normal, mature dog. Two-year-old female Beagle, transverse section. Mallory's triple staining. Magnification 15×.

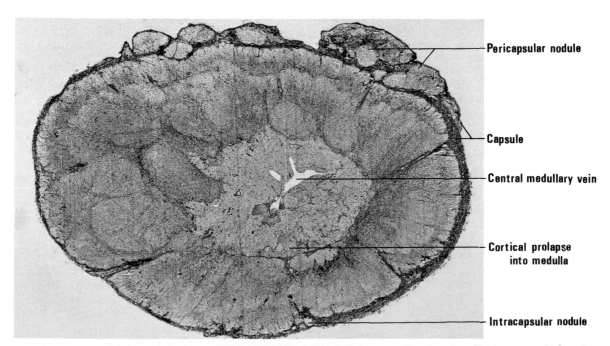

Figure 10–9. Mesoscopic features of the left adrenal gland of the normal, aging dog. Twelve-year-old female Corgi, transverse section. Mallory's triple staining. Magnification 15×.

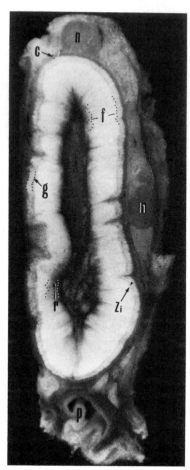

Figure 10–10. Mesoscopic features of the right adrenal gland of a dog of 1.0 year. Female Beagle, transverse section, 5 mm. thick, unstained, magnified 25 ×. Structures indicated include capsula fibrosa (c), zona glomerulosa (g), zona fasciculata (f), zona reticularis (r), zona intermedia (Zi), sympathetic ganglia (n), posterior vena cava (p). (From Hullinger, R. L., Zbl. Vet. Med. C. Anat. Histol. Embryol. 7:1–27.)

domly as accessory aggregates, satellite to or incorporated within various abdominal organs. Those aggregates associated with the gland occur as nodules, owing to compensatory hyperplasia and increase in size and number with advancing age (Figs. 10–7, 10–9).

Mesoscopic Features

The mesoscopic organization of the adrenal gland is based upon concentric lamellae or shells of cortical parenchyma which envelop a central medulla (Figs. 10–8, 10–10). The medulla, or heart of the gland, is separat-

ed from the cortex by a delicate network of reticular and loose collagenous connective tissues, the *septum corticomedullae*. The innermost cortical layer is applied to all surfaces of the undulating medullary contour. The parenchyma of this inner zone is disposed in a relatively random and loose network and is termed the *zona reticularis*. In most dogs this zone will compose the innermost 25 per cent of the cortex. Upon examination with the unaided eye, the zone reticularis appears as the darker zone of the cortex. This appearance is due to the relatively greater numbers and size of the sinusoids in the parenchyma of this zone and a relatively lesser amount of lipid storage in the cytoplasm of these cells.

The next cortical zone, moving from inner to outer cortex, is the largest. It typically composes 50 per cent or more of the cortex. This *zona fasciculata* is named such because of its appearance in two dimensions. It is actually composed of anastomosing plates or muralia of cells which radiate toward the periphery. This zone appears yellow to white in the fresh specimen. The cells of the outer one-third of this zone store more lipid and are somewhat larger than those of the inner two-thirds. These regions are called *pars externa* and *pars interna*, respectively.

A narrow *zona intermedia corticalis* is at the outer surface of the zona fasciculata. This small, dark-appearing region composes less than 5 per cent of the total cortex and in the dog is believed to function as a blastemic region for replacement cells of the adult cortex (Hullinger 1978, Hullinger and Getty 1971).

The outermost cortical zone is the *zona arcuata*. This zone, constituting approximately 25 per cent of the adrenal cortex, is composed of cells arranged in arches and nestled into a stromal template provided by the inner surface of the capsule.

Microscopic Features

A cursory view of the microscopic features of this endocrine organ reveals zonal homogeneity of component cell types (*endocrinocyti corticales* and *endocrinocyti medullares*) and the extensive vascularity (*rete hemocapillare sinusoideum*).

The stroma of the adrenal gland, *stroma glandulae adrenalis*, varies considerably in amount and kind. It ranges from the thick fibrous capsule found in the aged adult to the fine reticular support fibers of the parenchymal cells in the medulla. When compared to other endocrine organs, the capsule of the adrenal gland is especially prominent. In the adult it is composed, for the most part, of dense, irregular collagenous connective tissue with some scattered elastic fibers. Smooth muscle fibers found in some other species have not been demonstrated in the dog's adrenal capsule (Bloodworth and Powers 1968). On the capsular surface the fibrous portion is continuous with the fascia and periadrenal adipose tissue. Its inner aspect is extended as septa and a few trabeculae into the cortical parenchyma. These septa become confluent and, as thin walls or sheets, envelop the parenchyma of the zona arcuata. At the inner margins of the zona arcuata these stromal elements send delicate fibrous branches laterally to compartmentalize the outer cortical zone. From this connective tissue boundary which has formed within the zona intermedia trabeculae and numerous fine reticular fibers project into the inner aspects of the adrenal cortex. Coursing through the parenchyma of the zonae fasciculata and reticularis, these fibers provide support for the functional cells and the network of sinusoids. Most fibers terminate at the corticomedullary boundary and become a part of the moderate connective tissue investment of the medulla. In comparison to the cortex, the medulla contains only moderate amounts of connective tissue as reticular fibers for support of the parenchyma and vascular channels. The stroma increases in amount with age.

The zona arcuata is composed of vermiform whorls of columnar epithelioid cells which mimic to some degree the arrangement of columnar cells on the villi of the small intestine. The structural differentiation of this zone occurs for the most part after birth. The cellular groupings are initially in coils or glomeruli, as in the *zona glomerulosa* of other species, but by six weeks of age most dogs have adrenal cortices with a distinct zona arcuata morphology. These columnar epithelial cells are oriented as a simple or a pseudostratified epithelium, as if they composed an epithelium on a free sur-

face. Thus, it is likely that two surfaces or poles of these cells may bound a loose collagenous connective tissue rich in sinusoids. The nucleus is centrally located, and neutral lipids and cholesterol precursors for steroidogenesis are stored in great quantities at both poles of the cell. These cells produce the mineralocorticoids essential for the maintenance of electrolyte balance and of life.

Cells of the zona intermedia appear between the parenchyma of the zona arcuata and zona fasciculata (Figs. 10–8, 10–10). These are observed as small, polygonal cells with only small amounts of cytoplasm. There is little cytoplasmic lipid stored in these cells, and the nuclei suggest that the cells are still undifferentiated. The cells appear to be compressed tightly into the network of collagenous and reticular fibers found here at the junction of the two major cortical zones which bound it. Its location and cytological appearance have led others to refer to this zone as the zone of compression, zone of transition, and intermediate zone. The cells of this zone are not prominent in most other species. Hullinger (1968) and Hullinger and Getty (1971) hypothesized that these were a blastemic population of cells developing as the zona arcuata forms and that this cell group is, in effect, a displaced inner cellular layer from the capsule. The cells of this zone are continuous with those of the zona arcuata and zona fasciculata, and there is not a sharp morphological boundary separating these zones.

The cells of the zona fasciculata are polygonal in shape and are arranged in plates, forming a *murus complexus*, as shown by Elias and Pauly (1956), which radiate from the zona reticularis. These plates or walls compose anastomosing networks of labyrinths which are separated by an equally complex system of sinusoids and delicate reticular connective tissue. The functional parenchymal units of this zone communicate via the intercellular space with a sinusoid on multiple surfaces. Histochemical staining affinities reveal that these cells are rich in neutral fats and cholesterol, in accordance with their functional status. This cell population produces glucocorticoids.

The zona reticularis is composed of cells similar in appearance to those of the zona fasciculata. Their cytoplasm contains a lesser amount of lipid, which in some cells is stored as large droplets. The plates of this zone are direct continuations of the zona fasciculata but differ in that they are quite randomly arranged; the overall arrangement is much looser, and the intervening sinusoids are larger. These cells produce the sex steroids.

Sinusoids of the cortex form a continuous network throughout the cortex and provide a venous drainage passing centripetally to the corticomedullary boundary. These sinusoids penetrate the moderate amount of stroma at the corticomedullary capsule to become confluent with the sinusoids of the medulla. The number and specific distribution of these channels has been investigated in an attempt to relate this to the apparent functional specificity of the cortical zones. The medullary sinusoids anastomose to form larger venous sinuses.

Endocrine cells of the medullary parenchyma are larger than those of the cortex and are polygonal in shape. Most stain darkly, and a heterogeneity of affinities for dye suggests a functional difference. On occasion, ganglionic cells are dispersed among the parenchyma. The function of the medullary cells is the production of the catacholamines epinephrine and norepinephrine, which are termed *endocrinocytus lucidus* and *endocrinocytus densus*, respectively.

Vascularization

As with the hypophysis, the blood supply and venous drainage of the adrenal cortex and medulla is structurally and perhaps functionally interrelated.

The arterial supply of the adrenal gland arises from several major vessels (Flint 1900, Ljubomudrov 1939). The branches from these vessels passing to the adrenal gland are numerous and of small caliber. They include branches from the arteria phrenicoabdominalis, the rami adrenales craniales; branches from the aorta abdominalis, the arteria adrenales mediae; abdominal branches of the arteriae lumbale; and branches from the arteria renalis, the rami adrenales caudales. These provide 20 to 30 contributing arterioles which approach the gland from all surfaces, enter the fibrous portion of the capsule, and anastomose to

form a network *(rete arteriosum capsulare)*. Numerous vessels plunge from the capsular network into the cortex. Some of these — according to Flint (1900), about 50 — pass as larger arterioles and small muscular arteries in the trabeculae and septa and descend directly to the corticomedullary boundary. Here they supply oxygen-rich blood to the medullary parenchyma. The supply to the sinusoids of the cortex is via small arterioles from the capsular network which pass in the smaller trabeculae and septa which separate the arches of the zona arcuata. These vessels pass to the zona intermedia and form a second network in the connective tissues there *(rete arteriosum subcapsulare)*. From this network arise the sinusoids of the zona arcuata and the inner zones of the cortex *(rete hemocapillare sinusoideum)*.

The sinusoids of the cortex and medulla become confluent, as outlined above, and join to form the large medullary sinuses and a *plexus venosus medullae*. These sinuses pass toward the hilus of the adrenal gland and are drained via the adrenal vein. Smithcors (1964) describes a venous tree in the medulla which is independent of the medullary sinuses and which joins with the sinuses to form the adrenal vein. Arteriovenous anastomoses have been observed in the connective tissues around the adrenal gland (Brondi and Castorina 1953). The adrenal veins of each gland terminate differently because of their position relative to the caudal vena cava. The right adrenal vein joins directly with the caudal vena cava, but the left adrenal vein enters the left renal vein.

According to Verhofstad and Lensen (1973), lymphatic vessels in the adrenal gland form extensive plexuses in the capsule, cortex, corticomedullary boundary, and medulla. There is also a well-developed lymphatic plexus surrounding the central vein of the medulla.

Innervation

The innervation of the adrenal cortex has been difficult to demonstrate (Wilkinson 1961). Saleh et al. (1974) reported multipolar nerve cells in all regions of the adrenal cortex. They propose a hypothalamic control of cortical secretion by nervous as well as humoral means. The parenchyma of the adrenal medulla is actually a sympathetic ganglion specialized for neurohumoral release. Fibers passing to the medulla travel through the cortex, accompanying the medullary arteries in the major cortical trabeculae and septa. These fibers are, for the most part, preganglionic sympathetic fibers which can be traced from the splanchnic supply through the celiac, splanchnic, and adrenal ganglia.

PARS ENDOCRINA PANCREATIS

The endocrine pancreas of the dog has served as the classic model for the exploration of insulin-deficiency diabetes by Banting and Best (1922).

The endocrine pancreas is a composite of many thousands of small endocrine racemi (*insulae pancreaticae*) scattered randomly among the exocrine pancreatic acini. Each racemus is composed of several endocrinocyti, the normal functioning of which is essential for the proper regulation of blood sugar levels. The endocrine pancreas is essential for maintaining life.

Developmental and Mesoscopic Anatomy

The pancreatic islets develop from the epithelial outgrowths of foregut entoderm. The exocrine acini and endocrine islets share the same initial development. Subsequently, some endodermal cells become isolated and lose contact with the branching cords of cells which eventually become the exocrine ducts of the pancreas. The cells of these isolated clusters undergo hypertrophy and differentiate to form the typical heterogeneous cell population of the endocrine islet.

The islets are distributed randomly among the exocrine acini, but there are differences in the cellular components of the endocrine islets. The pancreas develops from a dorsal and ventral pancreatic anlage. These two differing sites of origin may play a determining role in the cytogenesis of the alpha endocrine cell (*endocrinocytus alpha*); Bencosme and Liepa (1955) reported that the ventral pancreas in the dog is devoid of alpha endocrine cells.

The endocrine islets vary in size from approximately 1500 to 20,000 μm.^{2} and are composed of from 10 to 120 epithelioid cells (Acosta et al. 1969). The number and size of the islets increase progressively from the right lobe to the left lobe of the pancreas. Perfusion of the pancreas with neutral red allows one to observe without magnification the number and distribution of the larger endocrine islets. The volume of islet tissue in the dog differs significantly between individuals without apparent relation to the volume of pancreas or body weight (Acosta et al. 1969).

Microscopic Features

The endocrine islet is a highly vascularized epithelioid cell parenchyma with small amounts of collagenous and reticular connective tissue fibers separating it from the adjoining exocrine acini. The term islet is used rather loosely in describing endocrine cell groups. Acosta et al. (1969) considered 10 endocrine cells as the minimum number composing an islet and suggested that it was also necessary to establish the existence of a separate set of blood capillaries in order to consider such cell groups as pancreatic endocrine islets. Bordi et al. (1972) reported that all of the intrainsular cell types can be found singly in extrainsular locations. Hellman et al. (1962) described the islets of the uncinate process as smaller and possessing a different cellular composition from those in other portions of the pancreas.

The endocrine cells in small numbers or in islets can be detected by routine staining procedures, whereby they stain faintly in contrast to the exocrine parenchymal cells. The principal cells of the islet are the *endocrinocytus alpha* and the *endocrinocytus beta*. The cells are closely positioned and somewhat randomly oriented in each racemus. With routine staining, the alpha cells show a marked cytoplasmic granulation, occur in fewer numbers (approximately 20 per cent of the total), and tend to be positioned more peripherally in the islets. The beta cells have only a delicate granulation, compose nearly 75 per cent of the islet cells, and assume a more central position in the islets. A delta cell (Kobayashi and Fujita 1969) and "F" cell have also been described as additional endocrine cell types of the endocrine islets in the dog (Munger et al. 1965). These investigators also reported that the beta and delta cells were found in islets of all regions of the pancreas. In the uncinate process, the alpha cells were lacking, and in their place were "F" cells, called alpha$_1$ cells by Hellman et al. (1962). Munger et al. (1965) suggested that alpha cells might be further limited to only the tail portion of the left lobe of the pancreas. This segregation of the alpha cell population was first described by Bencosme et al. (1955) and has provided an excellent model for the study of alpha cell function.

Kobayashi and Fujita (1969) emphasized the large number of nerve fibers in contact with the islet cells. These authors described a complex forming between the endocrine cells of the islet and the nerve fibers. The endocrine cells normally respond to altered blood sugar levels; the alpha cell reacts to a lowered level by releasing glucagon, and the beta cell to an elevated level by releasing insulin.

ENTEROENDOCRINE CELLS

Intraepithelial endocrine cells occur singly in the epithelial lining of the tunica mucosa of the digestive tract. The basal surface of these *enteroendocrine* cells is in contact with the epithelial basement membrane. These cells are positioned quite closely to the lumen of the digestive tract, and in the pyloric gland region of the stomach and the duodenal segment of the small intestine, some of these cells have an apical projection reaching the lumen. The distribution of organelles and inclusions in these cells, coupled with experimental evidence, suggests that they are releasing their secretions basally (Kobayashi and Fujita 1974). Such a morphology indicates that these endocrine cells may be directly monitoring the external environment.

Pearse (1966) has proposed that all enteroendocrine cells have a common ancestry of neural crest cells, but Andrew (1974) and Pictet et al. (1976) have shown that the precursors of the enteroendocrine cells are present in the gut before the migration of the neural crest cells.

Cells originally described in the pancreatic endocrine islets have now been ob-served in the gastric and intestinal mucosal epithelium. Solcia et al. (1970), Forssmann (1970), and Polak et al. (1971) reported alpha endocrine cells in the gastric wall and attributed to them the release of enteroglucagon. Fujita and Kobayashi (1974) considered the delta cell of the gastric and intestinal mucosa to be the gastrin-producing cell, but Sasagawa et al. (1974) concluded that the delta cell produced secretin. Inage (1974) concluded that the "F" cells of the pancreas (Munger et al. 1965) and a structurally similar enterochromaffin cell (Bencosme and Liepa 1955) were actually different cell types.

ENDOCRINE TISSUES OF THE OVARY

In the ovary the gametogenic exocrine function is initiated and regulated by the ovarian endocrine functions. As a sequel to follicular atresia, especially of the secondary and tertiary follicles, thecal luteal cells of the internal thecal layer remain in the dense cellular connective tissue stroma of the ovary as an interstitial parenchymal cell population. These cells in other species have been called the **interstitial gland** tissue (Guraya 1973).

Guraya and Greenwald (1964) described clusters and single epithelioid cells widely distributed in the stroma and varying in number and distribution between individuals. These investigators reported that some of these cells arose from invaginations of surface epithelial cells and that others developed from single cells forming from thecal and stromal tissues of normal and atretic follicles. Stott (1974) suggested that the interstitial gland tissue developed from an isolation of granulosal cells from large atretic follicles. Guraya (1973) outlined the development of the interstitial gland tissue from the hypertrophy of the theca interna and adjoining stromal tissues of medium-sized and maturing atretic follicles.

These generally sparsely distributed cells are believed to produce androgens and small amounts of estrogens and progestins.

With the continued maturation of those follicles destined to rupture and discharge oocytes from the ovary, there is the formation of an **internal thecal** (*theca interna*)

layer adjacent to the follicle. This inner investment or case is composed of epithelioid cells which have differentiated from the mesenchymally derived stromal support tissues of the ovary. The endocrine cells of the theca (*endocrinocyti thecales*) are somewhat elongated or spindle-shaped and are enveloped by a network of fine reticular fibers. Numerous blood capillaries form a network which permeates this tissue. These cells produce estrogens.

After ovulation and the formation of a *corpus hemorragicum*, cells of the internal theca and of the follicular epithelium accumulate large amounts of lipid and hypertrophy to form the *corpus luteum*. Those cells arising from the internal theca are called **theca lutein endocrine cells** and those arising from the granulosa cells are called **follicular lutein endocrine cells.** The capillary network of the theca interna vascularizes the corpus luteum. Balboni (1973) noted that in endocrine tissues of the ovary the blood capillaries have an attenuated and fenestrated wall, as in the sinusoids of the major endocrine organs. The corpus luteum produces progestrone. It is a transient endocrine tissue, recurrent with each estrous cycle (most breeds having two per year, one in the spring and one in the fall) and retained in full functional maturity during the gestation period.

PLACENTAL ENDOCRINE TISSUES

It has been inferred from the work done with other species that the placenta of the dog has an endocrine as well as an exocrine function. Wynn and Björkman (1968) and Anderson (1969) provide structural evidence that suggests such a function for carnivores. During the latter two-thirds of gestation, there is a relatively undifferentiated *cytotrophoblast* which forms as a simple cuboidal basal layer beneath the outer *syncytiotrophoblast* layer of the fetal epithelium. Both layers are well vascularized with fetal blood capillaries. Extensive decidual cells were reported by Anderson (1969). It is yet to be determined which cells, if any, produce the hormones of pregnancy. Sokolowski et al. (1973) described the changes in various endocrine organs during the estrous cycle and pregnancy.

ENDOCRINE TISSUES OF THE TESTIS

The testis, like the ovary, has both an exocrine and an endocrine function; the gametogenic or holocrine (cytogenic) function results in the production of spermatozoa and of testosterone.

With the onset of puberty, in the *interstitium testis* between the developing seminiferous tubules there is a differentiation of islets of epithelioid cells, the *endocrinocyti interstitiales*. These cells synthesize large amounts of the male steroid hormones, androgens, the primary active component of which is testosterone (Setoguti et al. 1974). The cells, which, according to Kothari et al. (1972) normally compose about 15 per cent of the testicular volume, are supported by a delicate connective tissue stroma and supplied by a rich network of blood and lymphatic capillaries. The distribution of these vessels to the parenchyma has not been reported in great detail for the dog, but the extent of their distribution might be inferred from observations of several species of domesticated and common laboratory mammals. When comparing the distribution of lymphatics among the endocrine cells of the testis, Fawcett et al. (1973) described an elaborate, varied relationship among the species. On the basis of their work, one would expect to find conspicuous and centrally located lymph capillaries in the dog.

Structural features and histopathological evidence also suggest an endocrine function for the *epitheliocytus sustentans* (supporting or nurse cell) of the seminiferous epithelium.

ENDOCRINE CELLS OF THE KIDNEY

At the vascular pole of the renal glomeruli, the smooth muscle cells of the tunica media of both the afferent and the efferent arterioles are epithelioid in morphology and contain secretory granules. Each of these is an *endocrinocytus myoideus*, or **juxtaglomerular cell** of the **juxtaglomerular complex,** which also includes epithelial cells of the macula densa of the distal tubular portion of the nephron and granulated cells of the extraglomerular mesangium.

The granules of the juxtaglomerular cells in the dog have proven difficult to demonstrate histochemically. These granules have been shown in other species to contain the hormone renin. Renin is released from these cells in response to altered blood volume, blood pressure, and ionic concentration. Renin acts on a blood protein to bring about the formation of angiotensin II. In addition to other functions, angiotensin II increases the release of mineralocorticoids from the zona glomerulosa of the adrenal cortex, which in turn act upon the distal tubule of the kidney nephron to bring about greater resorption of sodium.

The smaller laboratory animals have been used as the model for this important endocrine function. As more is understood of its action and as more sophisticated measurement techniques are developed, the dog may provide a reliable model for investigating this system. For related reading, consult Davis et al. (1961), Rojo Ortega et al. (1970), and Granger et al. (1971).

BIBLIOGRAPHY

Acosta, J. M., J. C. Buceta, J. E. Pons, L. E. P. de Figari, O. A. M. R. Galli, E. E. Weinschelbaum, and J. C. Soloaga. 1969. Distribution and volume of the islets of Langerhans in the canine pancreas. Acta Physiol. Lat. Am. *19*:175–180.

Akmayev, I. G. 1971a. Morphological aspects of the hypothalamic-hypophyseal system. II. Functional morphology of pituitary microcirculation. Z. Zellforsch. *116*:178–194.

——————. 1971b. Morphological aspects of the hypothalamic-hypophyseal system. III. Vascularity of hypothalamus, with special reference to its quantitative aspects. Z. Zellforsch. *116*:195–204.

Anderson, J. W. 1969. Ultrastructure of the placenta and fetal membranes of the dog. I. The placental labyrinth. Anat. Rec. *165*:15–36.

Andrew, A. 1974. Further evidence that enterochromaffin cells are not delivered from the neural crest. J. Embryol. Exp. Morphol. *31*:589–598.

Ariëns-Kappers, J. 1965. Survey of the innervation of the epiphysis cerebri and the accessory pineal organs of vertebrates. Pp. 87–153 *in* Progress in Brain Research, Vol. 10, edited by J. Ariëns-Kappers and J. P. Schade. New York, American Elsevier Publishing Co.

Baber, E. C. 1877. On the lymphatics and parenchyma of the thyroid gland of the dog. Q. J. Micr. Sc. *17*:204–212.

Baker, D. D. 1936. Studies on the suprarenal glands of dogs. I. Comparison of the weights of suprarenal glands of mature and immature male and female dogs. Am. J. Anat. *60*:231–252.

Balboni, G. C. 1973. The problem of the ovarian stroma. Arch. Ital. Anat. Embriol. *78*:37–58.

Banting, F. G., and C. H. Best. 1922. The internal secretion of the pancreas. J. Lab. Clin. Med. *7*:251–266.

Basir, M. A. 1932. The vascular supply of the pituitary body in the dog. J. Anat. *66*:387–397.

Bayliss, W. M., and E. H. Starling. 1902. The mechanism of pancreatic secretions. J. Physiol. *28*:325–353.

Bencosme, S. A., and E. Liepa. 1955. Regional differences of the pancreatic islet. Endocrinol. *57*:588–593.

Bencosme, S. A., E. Liepa, and S. S. Lazarus. 1955. Glucagon content of pancreatic tissue devoid of alpha cells. Proc. Soc. Exp. Biol. Med. *90*:387–392.

Bensley, S. H. 1947. The normal mode of secretion of the parathyroid gland of the dog. Anat. Rec. *98*:361–381.

Bergdahl, L., and L. Boquist. 1973. Parathyroid morphology in normal dogs. Pathol. Eur. *8*:95–103.

Bloodworth, J. M., Jr., and K. L. Powers, 1968. The ultrastructure of the normal dog adrenal. J. Anat. *102*:457–476.

Bloom, F. 1959. The endocrine glands. Chapter 12, pp. 419–454 *in* Canine Medicine, edited by H. D. Hoskins, J. V. LaCroix, and Karl Mayer. 2nd Ed. Santa Barbara, Calif., American Veterinary Publications, Inc.

Bordi, C., R. Togni, A. Costa, and A. Bertani. 1972. Extrainsular endocrine cells of the dog pancreas. A light microscopic study. Endokrinologie *60*:39–50.

Boyd, J. D. 1964. Development of the human thyroid gland. Chapter 2, pp. 9–31 *in* The Thyroid Gland, Vol. I, edited by R. Pitt-Rivers and W. R. Trotter. Washington, Butterworths.

Brondi, C., and S. Castorina. 1953. Il circdo arterioso del currene nel cane. Minerva Chir. *8*:380–383.

Bunting, F. G., and C. H. Best. 1922. Internal secretion of pancreas. J. Lab. Clin. Med. *7*:251–266.

Campbell, H. J. 1970. Control of the anterior pituitary gland by hypothalamic releasing-factors. Chapter 10, pp. 152–172 *in* The Scientific Basis of Medicine Annual Reviews, British Postgraduate Federation. London, Athlone Press.

Carlon, N. 1967. Cytologic du lobe antérieur de l'hypophyse du chien. Z. Zellforsch. *78*:76–91.

Crowe, S. J., H. Cushing, and J. Homans. 1910. Experimental hypophysectomy. Johns Hopkins Hosp. Bull. *21*:127–169.

Cunliffe, W. J. 1961. The innervation of the thyroid gland. Acta Anat. *46*:135–141.

Dandy, W. E. 1913. The nerve supply to the pituitary body. Am. J. Anat. *15*:333–343.

Dandy, W. E., and E. Goetsch. 1910. The blood supply of the pituitary body. Am. J. Anat. *11*:137–150.

Dandy, W. E., and F. L. Reichert. 1925. Studies on experimental hypophysectomy; effect on the maintenance of life. Johns Hopkins Hosp. Bull. *37*:1–13.

Davis, J. O., C. R. Ayers, and C. C. J. Carpenter. 1961. Renal origin of an aldosterone-stimulating hormone in dogs with thoracic caval constriction and in sodium-depleted dogs. J. Clin. Invest. *40*:1466–1474.

Elias, H., and J. E. Pauly. 1956. The structure of the human adrenal cortex. Endocrinology *58*:714–789.

Evans, H. E., and A. deLahunta. 1974. Miller's Guide to the Dissection of the Dog. Philadelphia, W. B. Saunders Co.

Farrow, B. R. H. 1969. Chromophobe adenoma of the pituitary in a dog. Vet. Rec. *84*:609–610.

Fawcett, D. W., W. B. Neaves, and M. N. Flores. 1973. Comparative observations on intertubular lymphatics and the organization of the interstitial tissue of the mammalian testis. Biol. Reprod. *9*:500–532.

Field, F. J., and R. J. Harrison. 1957. Anatomical Terms: Their Origin and Derivation. 2nd Ed. Cambridge, W. Heffer & Sons.

Flint, J. M. 1900. The blood-vessels, angiogenesis, organogenesis, reticulum and histology of the adrenal. Johns Hopkins Hosp. Rept. *39*:153–230.

Forssmann, W. G. 1970. Ultrastructure of hormone-producing cells of the upper gastrointestinal tract. Pp. 21–70 *in* Origin, Chemistry, Physiology and Pathophysiology of the Gastrointestinal Hormones, edited by W. Creutzfeldt. New York, F. K. Schattauer Verlag.

Foster, G. V., A. Baghdiantz, M. A. Kumar, E. Slack, H. A. Soliman, and I. MacIntyre. 1964a. Thyroid origin of calcitonin. Nature *202*:1303–1305.

Foster, G. V., I. MacIntyre, and A. G. E. Pearse. 1964b. Calcitonin production and the mitochondrion-rich cells of the dog thyroid. Nature *203*:1029–1030.

Francis, K. C., and R. M. Mulligan. 1949. The weight of the pituitary gland of the male dog in relation to body weight and age, with a differential cell count of the anterior lobe. J. Morphol. *85*:141–161.

French, C. 1901. The thyroid gland and thyroid glandules of the dog. J. Comp. Med. Vet. Arch. *22*:1–14.

Fujita, T., and S. Kobayashi. 1974. The cells and hormones of the GEP endocrine system. Chapter 1, pp. 1–16 *in* Gastro-Entero-Pancreatic Endocrine System: A Cell-Biological Approach, edited by T. Fujita. Baltimore, The Williams & Wilkins Co.

Fujita, H., and T. Murakami. 1974. Scanning electron microscopy on the distribution of the minute blood vessels in the thyroid gland of the dog, rat and rhesus monkey. Arch. Histol. Jap. *36*:181–188.

Gale, T. F. 1972. An electron microscopic study of the pars distalis of the dog adenohypophysis. Z. Anat. Entwickl-Gesch. *137*:188–199.

Gilmore, J. W., W. G. Venzke, and H. L. Foust. 1940. Growth changes in body organs. Part II. Growth changes in the thyroid of the normal dog. Am. J. Vet. Res. *1*:66–72.

Godwin, M. C. 1936. The early development of the thyroid gland in the dog with especial reference to the origin and position of accessory thyroid tissue within the thoracic cavity. Anat. Rec. *66*:233–251.

— — — — —.1937a. Complex IV in the dog with special emphasis on the relation of the ultimobranchial body to interfollicular cells in the postnatal thyroid gland. Am. J. Anat. *60*:299–330.

— — — — —. 1937b. The development of the parathyroids in the dog with emphasis upon the origin of accessory glands. Anat. Rec. *68*:305–325.

Goldberg, R. C., and I. L. Chaikoff. 1952. On the occurrence of six cell types in the dog anterior pituitary. Anat. Rec. *112*:265–274.

Granger, P., J. M. Rojo-Ortega, S. C. Pérez, R. Boucher, and J. Genest. 1971. The renin-angiotensin system in newborn dogs. Can. J. Physiol. Pharmacol. *49*:134–138.

Green, J. D. 1951. The comparative anatomy of the hypophysis, with special reference to its blood supply and innervation. Am. J. Anat. *88*:225–311.

— — — — —. 1966. The comparative anatomy of the portal vascular system and of the innervation of the hypophysis. Vol. I., Chapter 3, pp. 127–146 *in* The Pituitary Gland, edited by G. W. Harris and B. T. Donovan. Los Angeles, Univ. Calif. Press.

Guraya, S. S. 1973. Interstitial gland tissue of mammalian ovary. Acta Endocrinol. (Suppl.) *171*: 1–27.

Guraya, S. S., and G. S. Greenwald. 1964. A comparative histochemical study of interstitial tissue and follicular atresia in the mammalian ovary. Anat. Rec. *149*:411–434.

Halsted, W. S. 1896. An experimental study of the thyroid gland of dogs, with especial consideration of hypertrophy of this gland. Johns Hopkins Hosp. Rep. *1*:373–409.

Hanström, B. 1966. Gross anatomy of the hypophysis in mammals. Vol. I., pp. 1–57 *in* the Pituitary Gland, edited by G. W. Harris and B. T. Donovan. Los Angeles, Univ. Calif. Press.

Harris, G. W., and B. T. Donovan (eds.). 1966. The Pituitary Gland. 3 Vols. Los Angeles, Univ. Calif. Press.

Harrison, R. G. 1964. The ductless glands. Pp. 529–551 *in* Cunningham's Textbook of Anatomy, edited by G. J. Romanes. 11th Ed. New York, Oxford Univ. Press.

Hartman, J. F., W. R. Fain, and J. M. Wolfe. 1946. A cytological study of the anterior hypophysis of the dog with particular reference to the presence of a fourth cell type. Anat. Rec. *95*:11–27.

Hartmann, F. 1957. Über die Innervation der Epiphysis cerebri einiger Säugetiere. Z. Zellforsch. *46*:416–429.

Hedhammar, A., F. Wu, L. Krook, H. Schryver, A. deLahunta, J. P. Whalen, R. A. Kallfelz, E. A. Nunez, H. F. Hintz, B. E. Sheffy, and G. D. Ryan. 1974. Overnutrition and skeletal disease: an experimental study in growing Great Dane dogs. Cornell Vet. *64* (Suppl. 5): 1–160.

Hellman, B., A. Wallgren, and C. Hellerström. 1962. Two types of islet alpha cells in different parts of the pancreas of the dog. Nature *194*:1201–1202.

Herring, P. T. 1908a. The histological appearances of the mammalian pituitary body. Q. J. Exp. Physiol. *1*:121–159.

— — — — —. 1908b. The development of the mammalian pituitary and its morphological significance. Q. J. Exp. Physiol. *1*:161–185.

Hewitt, W. F., Jr. 1950. Age and sex differences in weight of pituitary glands in the dog. Proc. Soc. Exp. Biol. Med. *74*:781–782.

Hullinger, R. L. 1968. A Histocytological Study of Age Changes in the Canine Adrenal Gland Studied by Light and Electron Microscopy. Unpublished Ph.D. Thesis, Ames, Iowa, Iowa State University.

— — — — —. 1978. Adrenal cortex of the dog (Canis familiaris): I. Histomorphologic changes during growth, maturity, and aging. Zbl. Vet. Med. C. Anat., Histol. Embryol. 7:1–27.

Hullinger, R. L., and R. Getty. 1971. The genesis and maintenance of the canine adrenal cortex from birth to one year of age. XIX Congreso Mundial de Medicina Veterinaria y Zootecniz 2:563.

Inage, T. 1974. Fluorescence histochemical and electron microscopic observations on the enterochromaffin cells in the dog pancreas. Chapter IX, pp. 96–100 *in* Gastro-Entero-Pancreatic Endocrine System: A Cell-Biological Approach, edited by T. Fujita. Baltimore, The Williams & Wilkins Co.

Ivy, A. C., and E. Oldberg. 1928. A hormone mechanism for gall-bladder contraction and evacuation. Am. J. Physiol. 86:599–613.

Jones, I. C. 1957. The Adrenal Cortex. New York, Cambridge Univ. Press.

Kagayama, M. 1965. Follicular cells in the pars distalis of the dog pituitary gland. An electron microscopic study. Endocrinology 77:1053–1060.

Kalina, M., and A. G. E. Pearse. 1971. Ultrastructural localization of calcitonin in C-cells of dog thyroid; an immunocytochemical study. Histochemie 26:1–8.

Kameda, Y. 1971. The occurrence of a special parafollicular cell complex in and beside the dog thyroid gland. Arch. Histol. Jap. 33:115–132.

— — — — —. 1972. The accessory thyroid glands of the dog around the intrapericardial aorta. Arch. Histol. Jap. 34:375–391.

— — — — —. 1973. Electron microscopic studies on the parafollicular cells and parafollicular cell complexes in the dog. Arch. Histol. Jap. 36:89–105.

Kingsbury, B. F., and F. J. Roemer. 1940. The development of the hypophysis of the dog. Am. J. Anat. 66:449–469.

Kobayashi, S., and T. Fujita. 1969. Fine structure of mammalian and avian pancreatic islets with special reference to D cells and nervous elements. Z. Zellforsch. 100:340–363.

— — — — —. 1974. Emiocytotic granule release in the basal-granulated cells of the dog induced by intraluminal application of adequate stimuli. Chapter 4, pp. 49–58 in Gastro-Entero-Pancreatic Endocrine System: A Cell-Biological Approach, edited by T. Fujita. Baltimore, The Williams & Wilkins Co.

Kosaka, T., and R. K. S. Lim. 1930. Demonstration of the humoral agent in fat inhibition of gastric secretion. Proc. Soc. Exp. Biol. Med. 27:890–891.

Kothari, L. K., D. K. Srivastava, P. Mishra, and M. K. Patni. 1972. Total Leydig cell volume and its estimation in dogs and in models of testis. Anat. Rec. 174:259–264.

Latimer, H. B. 1941. The weight of the hypophysis in the dog. Growth 5:293–300.

— — — — —. 1965. Changes in relative organ weights in the fetal dog. Anat. Rec. 153:421–428.

Ljubomudrov, A. P. 1939. The blood supply of the suprarenal glands in the dog. Arkhiv Anat. Gistol. Embriol. 20:220–224. (English Summary, 381–382).

Mahorner, H. R., H. D. Caylor, C. F. Schlotthauer, and J. de J. Pemberton. 1927. Observations on the lymphatic connections of the thyroid gland in man. Anat. Rec. 36:341–347.

Major, R. H. 1909. Studies on the vascular system of the thyroid gland. Am. J. Anat. 9:475–492.

Marine, D. 1907. On the occurrence and physiological nature of glandular hyperplasia of the thyroid (dog and sheep) together with remarks on important clinical problems. Johns Hopkins Hosp. Bull. 18:359–364.

— — — — —. 1914. Observations on tetany in dogs. J. Exp. Med. 19:89–105.

Mason, J. B., J. Markowitz, and F. C. Mann. 1930. A plethysmographic study of the thyroid gland of the dog. Am. J. Physiol. 94:125–134.

Michaelson, S. M. 1970. Endocrine system. Chapter 17, pp. 412–499 in The Beagle as an Experimental Dog, edited by A. C. Andersen. Ames, Iowa, Iowa State University Press.

Mikami, S. 1956. Cytological changes in the anterior

pituitary of the dog after adrenalectomy. J. Fac. Agric. Iwate Univ. 3:62–68.

Mikhail, Y. 1971. Intrinsic nerve supply of the thyroid and parathyroid glands. Acta Anat. 80:152–159.

Modell, W. 1933. Observations on the structure of the blood vessels within the thyroid gland of the dog. Anat. Rec. 55:251–269.

Morato, M. J. X. 1939. The blood supply of the hypophysis. Anat. Rec. 74:297–320.

Mulligan, R. M., and K. C. Francis. 1951. Weights of thyroid and parathyroid glands of normal male dogs. Anat. Rec. 110:139–143.

Munger, B. L., F. Caramia, and P. E. Lacy. 1965. The ultrastructural basis for the identification of cell types in the pancreatic islets. II. Rabbit, dog and opossum. Z. Zellforsch. 67:776–798.

Nonidez, F. J. 1931. Innervation of the thyroid gland. II. Origin and course of the thyroid nerves in the dog. Am. J. Anat. 48:299–329.

— — — — —. 1932a. The origin of the "parafollicular" cell, a second epithelial component of the thyroid of the dog. Am. J. Anat. 49:479–495.

— — — — —. 1932b. Further observations on the parafollicular cells of the mammalian thyroid. Anat. Rec. 53:339–347.

— — — — —. 1935. Innervation of the thyroid gland. III. Distribution and termination of nerve fibers in dogs. Am. J. Anat. 57:135–170.

Oksche, A. 1965. Survey of the development and comparative morphology of the Pineal Organ. Pp. 3–30 in Progress in Brain Research, Vol. 10, edited by J. Ariëns-Kappers and J. Schade. New York, American Elsevier Publishing Co.

Pearse, A. G. E., and J. M. Polak. 1971. Cytochemical evidence for the neural crest origin of mammalian ultimobranchial C cells. Histochemie 27:96–102.

Pearse, A. G. E. 1966. Common cytochemical properties of cells producing polypeptide hormones, with particular reference to calcitonin and the thyroid C cells. Vet. Rec. 79:587–590.

Pictet, R. L., L. B. Rall, P. Phelps, and W. J. Rutter. 1976. The neural crest and the origin of the insulin-producing and other gastrointestinal hormone-producing cells. Science. 191:191–192.

Pitt-Rivers, R., and W. R. Trotter. 1964. The Thyroid Gland. Washington, Butterworths.

Polak, J. M., S. Bloom, I. Coulling, and A. G. E. Pearse. 1971. Immunofluorescent localization of enteroglucagon cells in the gastrointestinal tract of the dog. Gut 12:311–318.

Purves, H. D. 1966. Cytology of the adenohypophysis. Vol. I, pp. 147–232 in The Pituitary Gland, edited by G. W. Harris and B. T. Donovan. Los Angeles, Univ. Calif. Press.

Purves, H. D., and W. E. Griesbach. 1957. A study on the cytology of the adenohypophysis of the dog. J. Endocrinol. 14:361–370.

Putnam, T. J., E. B. Benedict, and H. M. Teel. 1929. Studies in acromegaly VIII. Experimental canine acromegaly produced by injection of anterior lobe pituitary extract. Arch. Surg. 18:1708–1736.

Reed, C. I., R. W. Lackey, and J. I. Payte. 1928. Observations on parathyroidectomized dogs, with particular attention to the regional incidence of tetany and to the blood mineral changes in this condition. Am. J. Physiol. 84:176–188.

Reichlin, S. 1974. Neuroendocrinology. Chapter 12,

pp. 776–831 *in* Textbook of Endocrinology, edited by R. H. Williams. 5th Ed. Philadelphia, W. B. Saunders Co.

Ricci, V., and M. Russolo. 1973. Immunocytological observations on the localization of ACTH in the hypophysis of the dog. Acta Anat. 84:10–18.

Rienhoff, W. F. 1938. The lymphatic vessels of the thyroid gland in the dog and in man. Arch. Surg. 23:783–804.

Roediger, W. E. W. 1973. A comparative study of the normal human neonatal and the canine thyroid C cell. J. Anat. 115:255–276.

Rojo-Ortega, J. M., P. Granger, R. Boucher, and J. Genest. 1970. Studies on the distribution of the JGI in the renal cortex of dogs and beavers. Nephron 7:61–66.

Ross, W. D., and U. H. K. Moorhouse. 1938. The thyroid nerve in the dog and its function. Q. J. Exp. Physiol. 27:209–214.

Saleh, A. M., N. N. Y. Nawar, and I. Kamal. 1974. A study on the adrenal ganglion and adrenal gland of the dog. Anat. Rec. 89:345–351.

Sasagawa, T., S. Kobayashi, and T. Fujita. 1974. Electron microscope studies on the endocrine cells of the human gut and pancreas. Chapter II, pp. 17–38 *in* Gastro-Entero-Pancreatic Endocrine System: A Cell-Biological Approach, edited by T. Fujita. Baltimore, The Williams & Wilkins Co.

Schwartz, H. G. 1936. The meningeal relations of the hypophysis cerebri. Anat. Rec. 67:35–51.

Setoguti, T., E. Haruhiko, and T. Shimizu. 1974. Electron microscopic studies on the testicular interstitial cells. Arch. Histol. Jap. 37:97–108.

Smith, E. M., M. L. Calhoun, and E. P. Reineke. 1953. The histology of the anterior pituitary, thyroid and adrenal of thyroid-stimulated purebred English Bulldogs. Anat. Rec. 117:221–240.

Smithcors, J. F. 1964. The Endocrine System. Chapter 16, pp. 807–836 *in* Anatomy of the Dog, by M. E. Miller, G. C. Christensen, and H. E. Evans. Philadelphia, W. B. Saunders Co.

Sokolowski, J. H., R. G. Zimbelman, and L. S. Goyings. 1973. Canine reproduction: reproductive organs and related structures of the nonparous, parous, and postpartum bitch. Am. J. Vet. Res. 34:1001–1013.

Solcia, E., G. Vassallo, and C. Capella. 1970. Cytology and cytochemistry of hormone producing cells of the upper gastrointestinal tract. Pp. 3–29 *in* Origin, Chemistry, Physiology and Pathophysiology of the Gastrointestinal Hormones, edited by W. Creutzfeldt. New York, F. K. Schattauer Verlag.

Stockard, C. R. 1941. The Genetic and Endocrine Basis for Differences in Form and Behavior. Am. J. Anat. Memoir 19. Philadelphia, Wistar Institute of Anatomy and Biology.

Stott, G. G. 1974. Granulosal cell islands in the canine ovary: histogenesis, histomorphologic features, and fate. Am. J. Vet. Res. 35:1351–1355.

Swarts, J. L., and R. L. Thompson. 1911. Accessory thyroid tissue within the pericardium of the dog. J. Med. Res. 29:299–308.

Teitelbaum, S. L., K. E., Moore, and W. Shieber. 1970. C cell follicles in the dog thyroid: demonstrated by in vivo perfusion. Anat. Rec. 168:69–78.

Truscott, B. L. 1944. The nerve supply to the pituitary of the rat. J. Comp. Neurol. 80:235–255.

Venzke, W. G. 1940. Histology of the thyroid glands of dogs 16 weeks of age. Proc. Iowa Acad. Sci. 46:439–441.

— — — — . 1976. Carnivore endocrinology. Chapter 54, pp. 1590–1593 *in* Sisson and Grossman's Anatomy of the Domestic Animals, edited by R. Getty. Philadelphia, W. B. Saunders Co.

Venzke, W. G., and J. W. Gilmore. 1940. Histological observations on the epiphysis cerebri. Proc. Iowa Acad. Sci. 47:409–413.

Verhofstad, A. A. J., and W. F. J. Lensen. 1973. On the occurrence of lymphatic vessels in the adrenal gland of the white rat. Acta Anat. 84:475–483.

Vicari, E. M. 1937. Observations on the nature of the parafollicular cells in the thyroid gland of the dog. Anat. Rec. 68:281–285.

— — — — — . 1938. Variations in structure of the parathyroid glands of dogs. Anat. Rec. 70 (Suppl. 3):80–81.

White, J. B., and H. L. Foust. 1944. Growth changes in body organs III. Growth changes in the pituitary of the normal dog. Am. J. Vet. Res. 5:173–178.

Wilkinson, I. M. S. 1961. The intrinsic innervation of the suprarenal gland. Acta Anat. 46:127–134.

Wissig, S. L. 1964. Morphology and cytology. Chapter 3, pp. 32–70 *in* The Thyroid Gland, Vol. I, edited by R. Pitt-Rivers and W. R. Trotter. Washington, Butterworths.

Wolf, J. M., and R. Cleveland. 1932. Cell types found in the anterior hypophysis of the dog (abstr.). Anat. Rec. 52:43–44.

Wurtman, R. J., and D. P. Cardinali. 1974. The pineal organ. Chapter 13, pp. 832–840 *in* Textbook of Endocrinology, edited by R. H. Williams. 5th Ed. Philadelphia, W. B. Saunders Co.

Wynn, R. M., and N. Björkman. 1968. Ultrastructure of the feline placental membrane. Am. J. Obstet. Gynecol. 91:533–549.

Yamada, H., S. Ozama, and R. Endo. 1956. Histological studies on the mammalian pituitary gland with special reference to the innervation. Bull. Tokyo Med. Dent. Univ. 3:55–65.

Yeghiayan, E., J. M. Rojo-Ortega, and J. Genest. 1972. Parathyroid vessel innervation: an ultrastructural study. J. Anat. 112:137–142.

Zach, B. 1960. Topographie und mikroskopisch-anatomischer Feinbau der Epiphysis cerebri von Hund und Katze. Zentralbl. Veterinaermed. 7:273–303.

Zechel, G. 1931. Follicular destruction in the normal thyroid of the dog. Surg. Gynecol. Obstet. 52:228–232.

Chapter 11

THE HEART AND ARTERIES

PERICARDIUM AND HEART

PERICARDIUM

The **pericardium,** or heart sac, is the fibroserous envelope of the heart. It can be divided into an outer fibrous and an inner serous part. The serous part consists of a parietal and a visceral layer.

The **fibrous pericardium** *(pericardium fibrosum)* (Fig. 11–1) is a thin, tough sac which contains the heart, serous pericardium, and a small amount of fluid. The greater part of its outer surface is covered by the pericardial mediastinal pleura. In young dogs the thymus is in contact with a variable portion of its cranial surface. The inner surface of the fibrous pericardium is intimately lined by the parietal layer of the serous pericardium. The base of the fibrous pericardium is continued on the great arteries and veins which leave and enter the heart. It blends with the adventitia of these vessels. Its apex is continued to the ventral part of the muscular periphery of the diaphragm in the form of a dorsoventrally flattened band of yellow elastic fascicles, the **sternopericardiac ligament** *(lig. sternopericardiaca).* This is nearly 1 cm. wide, less than 1 mm. thick, and about 5 mm. long.

The **serous pericardium** *(pericardium serosum)* (Fig. 11–1) forms a closed cavity into which about one-half of its wall is invaginated by the heart to form its visceral layer, the smooth, outer covering of the heart. The uninvaginated part forms the parietal layer as it covers the inner surface of the fibrous pericardium. The **pericardial cavity** *(cavum pericardii)* is located between the two layers of the serous pericardium. It is the smallest of the three great body cavities and, unlike the other two, contains 0.3 to 1 ml. of a clear, light yellow fluid, the *liquor pericardii.*

The **parietal layer** *(lamina parietalis)* of the serous pericardium is so firmly fused to the fibrous pericardium that no separation is possible. It is composed of interlacing collagenous fibers which are paved on the inside by mesothelium but on the outside are undistinguishable from the tissue of the fibrous pericardium.

The **visceral layer** *(lamina visceralis)* of the serous pericardium, or *epicardium,* is attached firmly to the heart muscle, except primarily along the grooves where fat and the coronary vessels or their branches intervene. Its smooth mesothelial surface is underlaid by a stroma which contains elastic fibers. The division between the parietal and visceral layers of the serous pericardium is marked by an undulating line which follows the highest part of the pericardial cavity around and across the base of the heart. The aorta and pulmonary trunk, the two great arteries leaving the heart, are united by a tube of epicardium and areolar tissue at their origins so that, caudal to them, a part of the pericardial

632

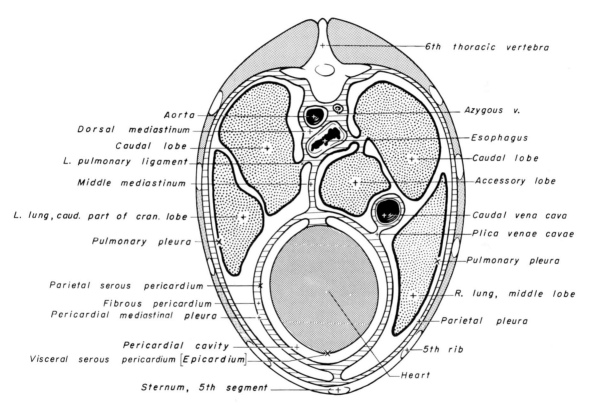

Aorta

Dorsal mediastinum

Caudal lobe

L. pulmonary ligament

Middle mediastinum

L. lung, caud. part of cran. lobe

Pulmonary pleura

Parietal serous pericardium

Fibrous pericardium

Pericardial mediastinal pleura

Pericardial cavity

Visceral serous pericardium [Epicardium]

Sternum, 5th segment

6th thoracic vertebra

Azygous v.

Esophagus

Caudal lobe

Accessory lobe

Caudal vena cava

Plica venae cavae

Pulmonary pleura

R. lung, middle lobe

Parietal pleura

5th rib

Heart

Figure 11–1. Schematic transverse section of the thorax.

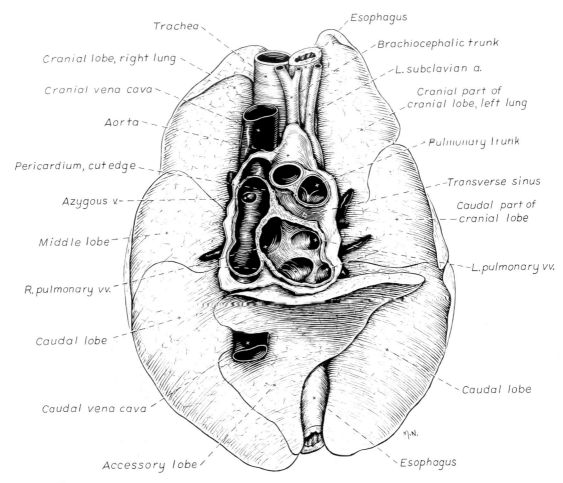

Figure 11-2.   Dorsal wall of the pericardial sac, with adjacent structures, ventral aspect.

cavity curves transversely across the base of the heart. This is the **transverse sinus of the pericardium** (*sinus transversus pericardii*) (Fig. 11-2), a **U**-shaped passage between the right and left sides of the pericardial cavity.

HEART

The **heart** (*cor*) (Figs. 11-3, 11-4, 11-5) is the muscular pump of the cardiovascular system. The musculature and conducting system of the heart are spoken of collectively as the **myocardium.** (See Langer and Brady 1974). For current references on the heart, blood vessels, and circulation, see Chevalier (1976), Cliff (1976), and Abramson (1976). The heart is cone-shaped and obliquely placed in the thorax so that its

base (*basis cordis*) faces dorsocranially and its apex (*apex cordis*) is directed ventrocaudally. A small, roughly triangular area of its dorsocaudal surface adjacent to the apex is related to the diaphragm. The large remaining part of its circumference is largely covered by the lungs and faces the sternum and ribs. The heart is divided internally by a transversely curved, longitudinal septum, lying in an oblique plane, into a cranioventral (right) and a caudodorsal (left) part. These in turn are partly divided transversely into the blood-receiving chambers, the atria, and the pumping chambers, the ventricles. The internal partitions are indicated on the surface of the organ by grooves. The caudodorsal part of the heart, consisting of the **left atrium** (*atrium sinistrum*) and the **left ventricle** (*ventriculus sinister*), receives blood from

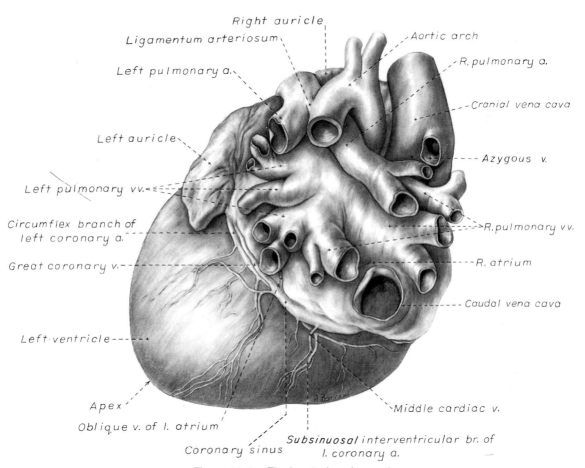

Figure 11–3. The heart, dorsal aspect.

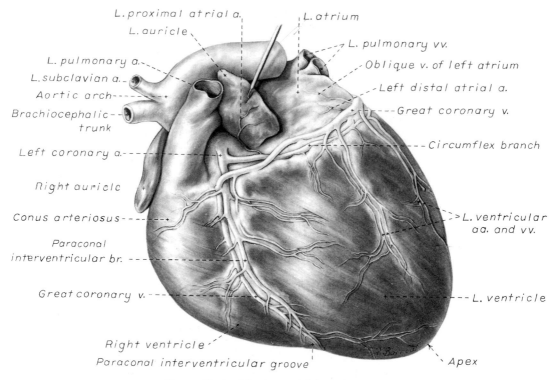

Figure 11–4. The heart, left lateral aspect.

L. proximal atrial a.
L. auricle
L. atrium
L. pulmonary vv.
Oblique v. of left atrium
Left distal atrial a.
Great coronary v.
Circumflex branch
L. ventricular aa. and vv.
L. ventricle
Apex
L. pulmonary a.
L. subclavian a.
Aortic arch
Brachiocephalic trunk
Left coronary a.
Right auricle
Conus arteriosus
Paraconal interventricular br.
Great coronary v.
Right ventricle
Paraconal interventricular groove

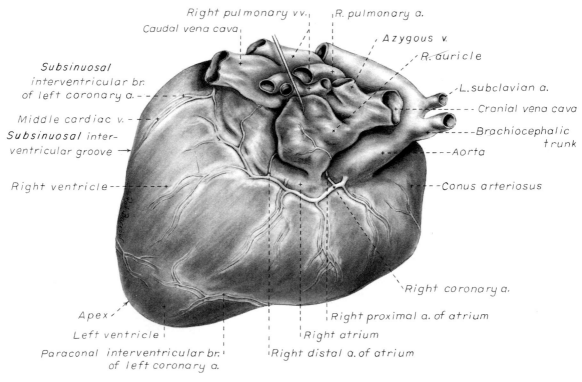

Figure 11–5. The heart, right lateral aspect.

Right pulmonary vv.
Caudal vena cava
R. pulmonary a.
Azygous v.
R. auricle
L. subclavian a.
Cranial vena cava
Brachiocephalic trunk
Aorta
Conus arteriosus
Right coronary a.
Right proximal a. of atrium
Right atrium
Right distal a. of atrium
Subsinuosal interventricular br. of left coronary a.
Middle cardiac v.
Subsinuosal interventricular groove
Right ventricle
Apex
Left ventricle
Paraconal interventricular br. of left coronary a.

the lungs and pumps it to all parts of the body (systemic circulation). The cranioventral part of the heart, consisting of the **right atrium** *(atrium dextrum* and **right ventricle** *(ventriculus dexter),* receives blood from all parts of the body and pumps it to the lungs (pulmonary circulation).

ORIENTATION

The heart forms a part of the mediastinum, the partition which separates the two pleural cavities. It is the largest organ in this tissue-filled septum located between the walls of the mediastinal pleurae. Covered by its pericardium, the heart extends from the third rib to the caudal border of the sixth rib. Roentgenograms show that variations in position occur among the breeds, strains, and individuals, and in the same animal according to age, condition, and the presence of pathological processes. A longitudinal axis through the heart tips forward about 45 degrees from a vertical plane. The base, therefore, faces dorsocranially and the bulk of it lies above a frontal plane dividing the thorax into dorsal and ventral halves. The apex points caudoventrally where it lies slightly to the left and caudal to a transverse plane through the most cranial part of the diaphragm. It touches the cranial surface of a transverse plane through the thorax which bisects the caudal part of the seventh sternebra. Normally the lungs cover most of the surface of the heart. On the right side the cardiac notch of the lung allows a variable area of the heart, covered by its fibrous pericardium and three layers of serosa, to contact the lateral thoracic wall. The cardiac notch of the right lung is **V**-shaped, with the apex directed dorsally. This allows a greater exposure of the heart on the right side than on the left, where the ventral border of the lung is usually not notched. The lung is thin adjacent to the lateral surfaces of the heart so that, in spite of the lung covering it, the beat of the organ can be easily heard and felt through the thoracic wall. The conformation of the dog has most to do with the ease with which the heart beat can be felt and heard. The heart beat is most pronounced in thin, narrow-chested, athletic-type animals.

SIZE AND WEIGHT

The dog heart is frequently used in studies of cardiac hypertrophy, and several surveys have been made of the normal values. Herrmann (1925) includes data on 200 dogs and cites heart weight to body weight ratios averaging 8.10 grams per kilogram of body weight for males and 7.92 for females. Northrup, Van Liere, and Stickney (1957) analyzed Herrmann's data and found that the sex differences were not significant. However, when they studied the heart weight to body weight ratios of an additional 346 adult dogs and 135 pups, they found that females have significantly smaller ratios than males. Small adults were found to have higher ratios than large adults, although pups had significantly smaller ratios than those in any adult category. The average ratio for 169 adult males was 7.74, and for 177 females it was 7.56. The range for 346 adult dogs was 4.53 to 11.13 grams per kilogram. House and Ederstrom (1968) found that heart weight to body weight ratios increased during postnatal development from 7.17 grams per kilogram in the newborn to 8.87 grams per kilogram in adults. Latimer (1961) has presented data on the ratios between the weights of the walls of the right and left ventricles in 46 dogs. The right ventricular wall accounted for 22 per cent of the total heart weight, the left ventricular wall for 39 per cent.

In the average young dog the heart weight is approximately 1 per cent of the body weight.

SURFACE TOPOGRAPHY

The **coronary groove** *(sulcus coronarius)* marks, on the surface of the heart, the separation of the atria and ventricles. It contains much fat, which surrounds the coronary vessels. The coronary groove encircles the heart except cranioventrally, where the dorsal part of the right ventricle (pulmonary conus) intervenes.

The **interventricular grooves** are indistinct surface markings of the separation of the right and left ventricles. Since they neither indent the substance of the heart nor contain as much fat as does the coronary groove, vessels are visible in them; however, most vessels stream toward the apex on the surface of the ventricles, rather than following the interventricular grooves. Obliquely traversing the cranioventral surface of the heart is the **paraconal interventricular groove** *(sulcus interventricularis paraconalis),* formerly known as the left, ventral, or cranial longitudinal sulcus or groove. It begins on the caudodorsal side of the pulmonary conus, where it is covered by the auricular portion of the left atrium; it ends before reaching the apex of the heart. The **subsinuosal interventricular groove** *(sulcus interventricularis subsinosus),* formerly known as the right, dorsal, or caudal longitudinal sulcus, is a short, straight, shallow furrow which marks, on the dorsocaudal surface of the heart, the approximate position of the interventricular septum.

Two surfaces of the heart are recognized in Nomina Anatomica Veterinaria as *facies auricularis,* the former left side of the heart of quadrupeds, and *facies atriales,* the opposite side. The facies auricularis of N.A.V. corresponds more or less to the *facies sternocostalis* of Nomina Anatomica for man. In man, the surface which faces the diaphragm is the **diaphragmatic surface** *(facies diaphragmatica).* This lies in an oblique plane and is flattened. The other surface, the **sternocostal surface** *(facies sternocostalis),* is rounded and, as the name implies, faces toward the sternum and ribs. This constitutes about two-thirds of the surface area of the cone-shaped organ. A line drawn on the left side from the root of the pulmonary trunk to the apex would follow an ill-defined border, known as the **left border** (margo sinister), located at the junction of the sternocostal and diaphragmatic surfaces on the left side. A similar poorly defined border on the right side is known as the **right margin** *(margo dexter).* The **base** of the heart *(basis cordis)* is the hilus of the organ. It is the craniodorsal portion and receives the great veins and emits the great arteries. The atria also enter into its formation. The **apex** of the heart *(apex cordis)* is formed by the looping of a swirl of muscle fibers at the apex of the left ventricle. It forms the most ventrocaudal part of the organ.

ATRIA

The **right atrium** *(atrium dextrum)* (Fig. 11–6) receives the blood from the systemic veins and most of the blood from the heart itself. It lies dorsocranial to the right ventricle. It is divided into a main part, the *sinus venarum cavarum,* and a blind part which projects forward and downward, the **right auricle** *(auricula dextra).* The main openings of the right atrium are four in number. The **coronary sinus** *(sinus coronarius)* is the smallest of these and enters the atrium from the left. Dorsal to it is the large **caudal vena cava** *(vena cava caudalis),* which enters the heart from behind. The caudal vena cava returns blood from the abdominal viscera, part of the abdominal wall, and the pelvic limbs. The **cranial vena cava** *(vena cava cranialis)* is about the same size as the caudal vena cava. It enters the heart from above and in front. In the dog the azygos vein usually enters the cranial vena cava, although occasionally it enters the atrium directly. The azygos vein drains blood back to the heart from part of the lumbar region and the caudal three-fourths of the thoracic wall. The cranial vena cava returns blood to the heart from the head, neck, thoracic limbs, the ventral thoracic wall, and the adjacent part of the abdominal wall. The **right atrioventricular orifice** *(ostium atrioventriculare dextrum)* is the large opening from the right atrium into the right ventricle. This opening and the valve which guards it are described with the right ventricle.

Other features of the right atrium are the tuberculum intervenosum, crista terminalis, fossa ovalis, limbus fossae ovalis, and the mm. pectinati.

On the dorsal wall of the right atrium is a transverse ridge of tissue placed between the two caval openings. This is the **intervenous tubercle** *(tuberculum intervenosum).* It diverts the converging inflowing blood from the two caval veins into the right ventricle. Just caudal to the intervenous tubercle on the medial wall of the atrium is a slitlike depression, the *fossa ovalis,* which varies from one to sev-

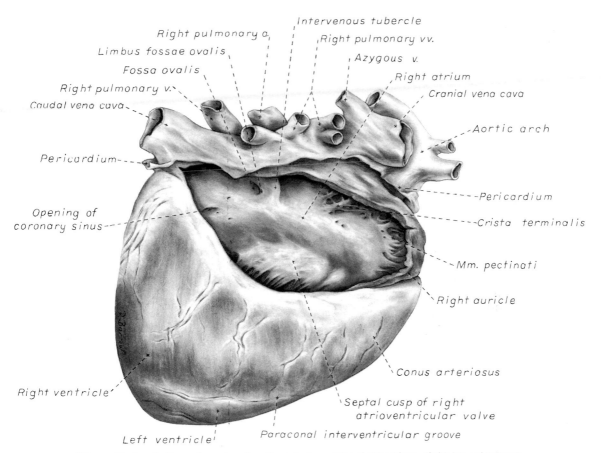

Figure 11–6. A dissection showing the interior of the right atrium, right lateral aspect.

eral millimeters in depth. The crescent-shaped ridge of muscle which projects from the caudal side of the intervenous tubercle and deepens the fossa is the *limbus fossae ovalis*. In the fetus there is an opening at the site of the fossa, the *foramen ovale*, which allows blood to pass from the right to the left atrium. The foramen usually closes during the first few postnatal weeks. Even though a small anatomical opening frequently persists, it is closed physiologically because the obliquity of the foramen enables the greater pressure in the left atrium to close the passage.

The **right auricle** *(auricula dextra)* is the ear-shaped pouch of the right atrium which extends cranioventrally. The internal surface of the wall of the right auricle is strengthened by freely branching, interlacing muscular bands, the *mm. pectinati*. These are also found on the lateral wall of the atrium proper. Most of the pectinate muscles radiate from a semilunar crest, a thick ridge of tissue, which is placed between the entrance of the cranial vena cava and the atrioventricular opening. This is the *crista terminalis*. It is also the dorsal separation of the sinus venarum cavarum and the auricle. On the external surface of a dilated heart a poorly defined groove, the *sulcus terminalis,* lies opposite the crista terminalis at the ventral part of the junction of the cranial vena cava with the atrium. Small veins empty into the right atrium through openings in the pits between the mm. pectinati. Everywhere, the internal surface of the heart is lined with a thin glistening membrane, the *endocardium.*

The **left atrium** *(atrium sinistrum)* forms the left, dorsocaudal part of the base of the heart. The mm. pectinati are confined to its **left auricle** *(auricula sinistra),* which is similar to the right auricle in shape and structure. It lies caudal to the pulmonary conus and pulmonary trunk. Its apex covers the proximal end of the paraconal interventricular groove. The left auricle lies along the right auricle, caudal to it, for nearly its entire length, the two being separated by the pulmonary trunk. Elsewhere on the surface of the heart there is no distinct indication of the separation of the two atria. Internally, the atria are separated by the **interatrial septum** (*septum interatriale*). Five or six openings (*ostia venarum pulmonalium*) mark the entrance of the pulmonary veins into the atrium. The two or three veins from the right lung cross over the right atrium and open into the craniodorsal part of the chamber; the three veins from the left lung usually empty into its caudodorsal part. The veins from the diaphragmatic lobes are larger than the others and are most caudal in position. Frequently a thin concave flap of

tissue is present on the cranial part of the septal wall. This is the **valve of the foramen ovale** (*valvula foraminis ovalis*).

FIBROUS BASE

The fibrous base of the heart, or "cardiac skeleton" (Fig. 11–7B), is the fibrous tissue, containing some cartilage, which separates the thin atrial musculature from the much thicker muscle of the ventricles. Only the special neuromuscular tissue, the atrioventricular bundle, extends continuously through the fibrous base from the atria to the ventricles. Baird and Robb (1950) have reconstructed the principal parts of the conduction system of a puppy's heart. Their dissections and reconstruction show the close spatial relationship which exists between the primary conduction and supporting tissues of the dog's heart. The cardiac skeleton is in the form of two narrow **fibrous rings** (*anuli fi-*

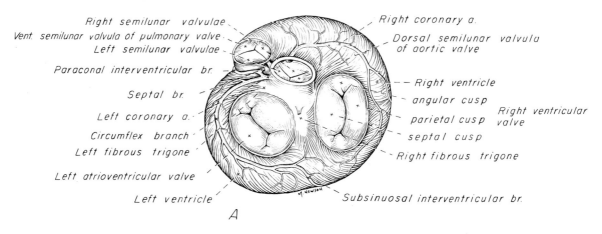

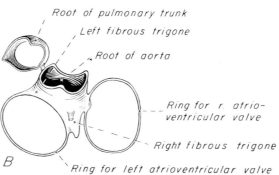

Figure 11–7. The base of the heart.
    A. Atrioventricular, aortic, and pulmonary valves, craniodorsal aspect.
    B. The fibrous base of the heart, craniodorsal aspect.

brosi), one surrounding each atrioventricular orifice, and two scalloped cuffs, or rings, one surrounding each arterial orifice.

Of the two arterial rings, the **aortic fibrous ring** *(anulus fibrosus aorticus)* is the better developed. Peripherally the collagenous fibers which form it give way to the yellow elastic fibers composing the tunica media of the wall of the ascending aorta. There are three points on its aortic margin, each of which conforms to the attachments of two adjacent semilunar valvulae of the aortic valve. The projections between the attachments of the dorsal and right and the dorsal and left semilunar valvulae are best developed and are composed of hyaline cartilage. Between the points, the margin of the aortic fibrous ring is semilunar so that its scalloped circumference is in the form of three arcs. To the right of the aortic fibrous ring the fibrous base of the heart is best developed. Here, in the interval between the right and left atrioventricular ostia, it consists of the thickened parts of the aortic and left atrioventricular rings as they lie adjacent to each other. Because of its triangular shape, it is called the **right fibrous trigone** *(trigonum fibrosum dextrum).* The **left fibrous trigone** *(trigonum fibrosum sinistrum)* is smaller than the right and lies in the triangle between the aortic and left atrioventricular ostia to the left of the right fibrous trigone. Both trigones contain cartilage and are united through the medium of the opposed aortic and left atrioventricular fibrous rings. In the ox and horse the trigone ossifies in part.

The **pulmonary fibrous ring** *(anulus fibrosus pulmonalis)* is similar to the aortic ring. It serves for the attachment of the pulmonary semilunar valvulae, and distally it blends with the media of the pulmonary trunk. It is much weaker than the aortic ring. The muscle fibers of the conus arteriosus attach to its distal surface. Between the apposed surfaces of the pulmonary and aortic anuli there is a short mass of collagenous tissue, the **ligament of the conus,** which unites these two structures.

The **atrioventricular fibrous rings** *(anuli fibrosi atrioventriculares)* (Fig. 11–7B) are thin rings of collagenous tissue to which the muscle fibers of the atrial and ventricular walls attach. From their endocardial edges emanate the atrioventricular valves. Although the proximal parts of the ventricular musculature attach to their distal surfaces, the bulk of the fibers of this muscle at its origin lie peripheral to the rings as the muscle tissue bulges proximally after arising from them. These fibrous rings are so delicate they are best seen in longitudinal sections of the heart cut through the atrioventricular junctions.

VENTRICLES

The ventricles form the bulk of the heart. Together they form a conical mass, the apex of which is also the apex of the left ventricle. The right ventricle is ventral and cranial to the left and arciform in shape (Fig. 11–8).

The **right ventricle** *(ventriculus dexter)* (Fig. 11–9) receives the systemic blood from the right atrium and pumps it to the lungs. Its outline is in the form of a curved triangle, as it is molded on the surface of the caudodorsally lying conical-shaped left ventricle. It is crescentic in cross section, and its long axis extends from the subsinuosal interventricular groove to the pulmonary trunk.

The large opening through which blood enters the right ventricle is the **right atrioventricular ostium** *(ostium atrioventriculare dextrum)* between the **cusps** of the right atrioventricular valve. Blood leaves the chamber through the **pulmonary os-**

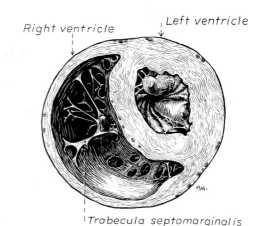

Right ventricle   Left ventricle

Trabecula septomarginalis

Figure 11–8. Cross section through the ventricles, craniodorsal aspect.

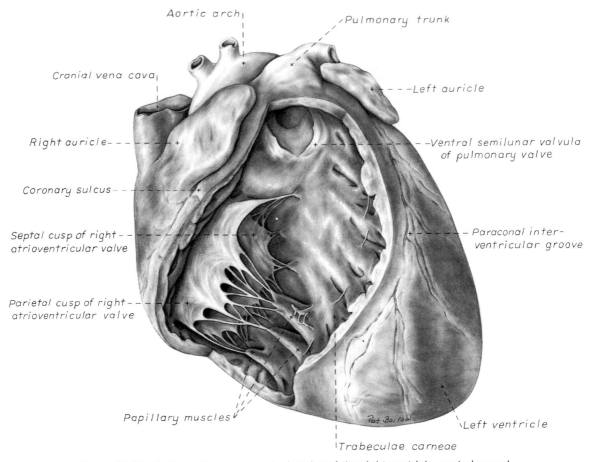

Aortic arch

Pulmonary trunk

Cranial vena cava

Left auricle

Right auricle

Ventral semilunar valvula
of pulmonary valve

Coronary sulcus

Septal cusp of right
atrioventricular valve

Paraconal inter-
ventricular groove

Parietal cusp of right
atrioventricular valve

Papillary muscles

Left ventricle

Trabeculae carneae

Figure 11–9.   A dissection showing the interior of the right ventricle, ventral aspect.

tium (*ostium trunci pulmonalis*) and is received by the pulmonary trunk. The *conus arteriosus* is the funnel-shaped part of the right ventricle which lies between planes going through the two openings of the ventricle. It is embraced by the right auricle externally. The **supraventricular crest** (*crista supraventricularis*) is a blunt, obliquely placed ridge of muscle between the origin of the conus arteriosus and the atrioventricular opening.

Both ventricles contain muscular ridges and projections with small, usually deep and oblong depressions between. The conical-shaped muscular projections which give rise to the chordae tendineae are called **papillary muscles** (*musculi papillares*). They are the most conspicuous features of the ventricular walls. There are usually three main papillary muscles in the right ventricle, although great variation

exists. Typically they arise from the apical third of the septal wall, 1 to 3 cm. from its junction with the outer wall. Their branched chordae tendineae, which prevent eversion of the valve, go to the free border and adjacent ventricular surface of the ventral cusp of the right atrioventricular valve. Commonly the papillary muscle which lies farthest to the right is larger than the others, and, when it is, its apex is usually bifid. The papillary muscle which lies farthest to the left may be bifid also. In such specimens the middle muscle is absent. A single compound papillary muscle may replace three main ones. Other variations are common. Near the angle formed by the septal and the outer wall, and caudodorsal to the most sinistral large papillary muscle, is a small, blunt to conical papillary muscle which gives rise to chordae tendineae going to the most caudodorsal part of the ventral cusp. Truex and Warshaw (1942), in the 12 dogs they studied, found only three large papillary muscles from the septum and a small constant papillary muscle of the conus. The numerous chordae tendineae which go to the septal or dorsal cusp arise from the septal wall directly or from muscular ridges or papillae located peripheral to the septal or dorsal cusp, when this valvula lies against the septum.

The *trabeculae carneae* are myocardial ridges which project mainly from the outer wall of the ventricle. Since they are endocardial-covered portions of the deepest muscular layer of the right ventricle, their long axes parallel the directions of these fibers. They run from the base and converge toward the apex. Those on the septal wall are largely adjacent to the ventral interventricular sulcus, and they largely parallel the axis of the heart. Some of the fossae coalesce near the ventral interventricular groove so that muscular columns or pillars are formed.

The *chordae tendineae* are fibromuscular cords which arise from the apices of the papillary muscles or directly from smaller, blunter elevations on the wall. The cords branch, at their valvular extremity, as they approach the free border of the cusps. The smaller cords blend with the tunica media of these cusps at the points of their free borders, and the larger strands fan out on the ventricular surfaces of the cusps.

The *trabecula septomarginalis* (Fig. 11–8), formerly called moderator band, is a branched or single muscular strand which extends across the lumen of the right ventricle. It usually leaves the septal wall of the right ventricle near or from the base of the largest papillary muscle and runs to the peripheral wall. The extremity of the trabecula usually branches repeatedly as it blends with the muscular ridges of the outer right ventricular wall. It serves the morphological role of conducting Purkinje fibers from the right branch of the atrioventricular bundle across the lumen of the cavity. Instead of the usual single band, there may be two or more anastomosing strands forming a loose plexus.

The **left ventricle** (*ventriculus sinister*) (Fig. 11–10) is conical in shape, with its apex forming the apex of the heart. It receives the oxygenated blood from the lungs by way of the left atrium and pumps it to most parts of the body through the aorta. The **left atrioventricular orifice** (*ostium atrioventriculare sinistrum*) is the large opening between the left atrium and the left ventricle closed by the left atrioventricular valve. The **aortic orifice** (*ostium aortae*), located near the center of the base of the heart, is the opening from the left ventricle into the ascending aorta. The left ventricle is characterized by its thick wall, conical shape, and two large papillary muscles. Its wall is three or four times thicker than that of the right ventricle. Truex and Warshaw (1942), in 12 specimens, found the mean thickness of the left and the right ventricle to be 13 and 4.2 mm., respectively.

The two large **papillary muscles** (*musculi papillares*) of the left ventricle come from its outer wall. They are heavy, smooth rolls of myocardium which have compound apices and give rise to the stout chordae tendineae of this chamber. The **dorsal papillary muscle** (*musculus papillaris dorsalis*) lies near the subsinuosal interventricular groove; the **ventral papillary muscle** (*musculus papillaris ventralis*) is closer to the paraconal interventricular groove. Adjacent to the ventral papillary muscle, near its attachment, is a fine network of muscular strands which come from the septal wall. From findings in similar strands from other species (Truex and Warshaw 1942), it is probable that these contain fascicles of Purkinje fibers derived from the left branch of the atrioventricular bundle en route to the ventral papillary muscle and general ventricular musculature. The strand nearest the atrioventricular opening is larger than the others and extends obliquely to the ventral papillary muscle from the septal wall. It may be the only one present. Near the apex of the ventricle some of the fine threads which compose this network extend under the endocardium which covers the crests of the muscular ridges. Other threads bridge the sulci between the larger trabeculae carneae in their courses to the ventral papillary muscle. The chordae tendineae from the ventral papillary muscle go to both the ventral and the dorsal part of the

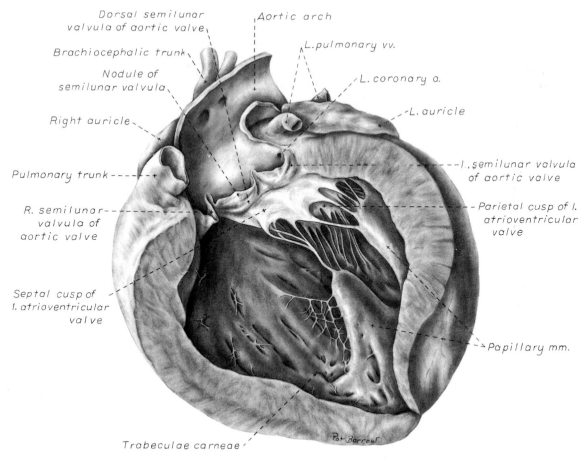

Figure 11-10. A dissection showing the interior of the left ventricle, left lateral aspect.

left atrioventricular valve. Those which go to the ventral or septal part are shorter and arise closer to the ventricular wall than those which go to the dorsal or outer part of the valve. The dorsal papillary muscle is separated from the ventral muscle by a deep cleft. The origin, course, and termination of the chordae tendineae from this muscle are similar to those of the chordae tendineae from the ventral muscle. Both papillary muscles extend to within a few millimeters of the apex of the ventricle.

The **interventricular septum** (*septum interventriculare*) consists of an inconspicuous, small, proximally located membranous part and a large, thick muscular part. The **membranous part** (*pars membranacea*) of the interventricular septum is the thinnest part and is the last part of the septum to form embryonically. This is brought about in man (Odgers 1938) by the fusion of the atrioventricular cushions and not by a downward growth of the aortic (pulmonoaortic) septum to meet the interventricular septum. In man, this closure is completed by the end of the second month of gestation. In the dog the membranous part can be seen by transmitted light under the septal cusp of the right atrioventricular valve adjacent to the origin of the aorta. When the foramen fails to close, a subaortic defect or interventricular foramen is left. The **muscular part** (*pars muscularis*), which constitutes the bulk of the interventricular septum, is formed by the myocardium of the combined walls of the two ventricles as they lie adjacent to each other.

MYOCARDIUM

The myocardium, or heart muscle of the atria, is not divided into distinct layers, except at the interatrial septum. Here the deep fibers of the two chambers fold in and lie adjacent to each other, to form this partition, whereas the superficial fibers are common to both atria. Muscular fibers encircle the ostia of the systemic and the pulmonary veins as they empty into the atria. Between the pectinate muscles the musculature is so thin that the heart wall is translucent at these places. The fixed point of the atrial musculature, like that of the ventricles, except for the atrioventricular bundle, is the fibrous base of the heart.

The musculature of the ventricles in the dog was described by Thomas (1957). There are superficial, middle, and deep layers. All muscle fasciculi of the superficial layer arise from the fibrous base and return to it. The superficial layer is common to both ventricles. When the heart is viewed from the base, these bundles run toward the apex, showing a clockwise twist. At the apex of the heart they turn in and run toward the base in such a manner that they cross, at right angles, the superficial fibers running down. They form the papillary muscles of the left ventricle. The superficial fibers penetrate the middle layer of the right ventricle to become its deep layer, after which they are disposed as on the left side. The middle layer forms the bulk of the ventricular walls. These are spiral or circular muscle masses which interdigitate between the two chambers. They primarily decrease the size of the lumen of the ventricles, and the superficial and deep layers shorten and twist the organ. The apex of the heart is quite thin, being formed by muscle fasciculi of the superifical layer of the left ventricle as they swirl in figure-of-eight fashion to form the apex of the chamber.

ATRIOVENTRICULAR VALVES

The **atrioventricular valves** (*valvae atrioventriculares*) (Figs. 11–7, 11–9, 11–10) are irregular, serrated cusps which are located in the atrioventricular ostia. They are the intake valves to the ventricles. They prevent blood from returning to the atria during the systolic phase of the heart beat. Peripherally, they attach to the fibrous rings which separate the musculature of the atria from that of the ventricles. When the ventricles contract, the valves are kept

from being pushed into the atria by the chordae tendineae. The chordae tendineae attach to the ventricular surfaces of the valvulae. The strongest cords can be followed as ridges under the endocardium to their attached borders, whereas the weakest cords go to the points of the serrations and disappear. The medium-sized cords go as far as the middle part of the ventricular sufaces of the valvulae before blending with the stratum proprium. Although some blood vessels have been described in the valves adjacent to their attached borders, the bulk of the nutrition of the valves is derived from the free blood in the heart. Smith (1971) found a nerve network in the atrioventricular and semilunar valves of the dog which concentrated near the points of attachment of the chordae tendineae. Nerves composing the atrioventricular valve plexus were derived from the atrial subendocardial network. In general, the tricuspid valve was better innervated than the mitral valve.

The **right atrioventricular valve** (*valva atrioventricularis dextra*) in man is also known as the **tricuspid valve.** The valve in the dog consists basically of two cusps. The cusp of the right atrioventricular valve which attaches to the fibrous ring adjacent to the septum is called the **septal** or **dorsal cusp** (*cuspis septalis*), known in the N.A. as cuspis anterior. The cusp which attaches to the fibrous ring adjacent to the ventral wall is the **parietal** or **ventral cusp** (*cuspis parietalis*), known in the N.A. as cuspis posterior. The extremities of the dorsal and ventral cusps become narrower and merge or, in some specimens, small secondary cusps are formed at these sites.

According to Detweiler (1955), tricuspid valve lesions cause murmurs which are usually heard most distinctly at the fourth right intercostal space at the level of the costochondral junction, the so-called "tricuspid area."

The **left atrioventricular valve** (*valva atrioventricularis sinistra*), also known as the **bicuspid** or **mitral valve,** is basically similar to the right atrioventricular valve in form and structure, but it is made on a heavier scale. The chordae tendineae as well as the papillary muscles from which they arise are several times larger in the left ventricle than they are in the right.

This stronger construction of the left intake valve, as compared with the right, is necessary, as the blood leaving the left ventricle through the aorta is under about four times more pressure than that which leaves the right ventricle through the pulmonary trunk. The division of the left atrioventricular valve into cusps is indistinct. The part which arises from the fibrous ring adjacent to the septum is wider than that coming from the remainder of the ring so that when the valve is open and viewed from the atrial side it is tricuspid in appearance.

According to Detweiler (1955), murmurs associated with disease of the mitral valve are generally most intense at the fifth left intercostal space above the middle of the lower third of the thorax, the so-called "mitral area."

AORTIC AND PULMONARY VALVES

The **aortic valve** (*valva aortae*) (Figs. 11–7, 11–10) consists of **right, left,** and **septal semilunar cusps,** or valvulae (*valvulae semilunares dextra et sinistra et septalis*). These consist of a fibrous tissue stroma covered on each surface by endothelium. Opposite their free borders they are attached to the aortic fibrous ring. In the middle of the free borders of the semilunar cusps are **nodules** (*noduli valvularum aortae*). Extending from each nodule toward the periphery of the valvulae are the **lunulae** (*lunulae valvularum semilunarium*). These represent the areas of contact with the adjacent cusps when the valve is closed. The nodules close the space that would otherwise be left open by the coming together of the three contiguous arcs. On the vessel side, or peripheral to each of the semilunar cusps, the wall of the aorta is dilated to form the three **aortic sinuses** (*sinus aortae*). These, like the cusps, are right, left, and dorsal in position, with the right and left coronary arteries leaving the right and left sinuses. The widening of the base of the ascending aorta, formed by the aortic sinuses, is the **aortic bulb** (*bulbus aortae*). At the end of systole, or heart contraction, the pressure in the aorta is greater than that in the left ventricle, and a back pressure is developed which closes the valve and dilates

the sinus. The coronary arteries leaving the sinus are affected by the changes in pressure.

The **valve of the pulmonary trunk** (*valva trunci pulmonalis*) (Figs. 11–7, 11–9) lies cranial to the aortic valve and is similar to it in construction. Since the blood pressure developed in the pulmonary trunk which it guards is not as great as that in the aorta, the development of the cusps and related structures is not as great. All parts present in the aortic valve are represented in the pulmonary. The valve of the pulmonary trunk consists of **right, left** and **intermediate semilunar cusps,** or valvulae (*valvulae semilunares dextra et sinistra et intermedia*). The intermediate cusp corresponds to the *valvula semilunaris anterior* of the N.A.

According to Detweiler (1955), the "aortic area," the area in which sounds associated with lesions of the aortic valve may be heard, is located at the fourth left intercostal space slightly below a line drawn through the point of the shoulder. Sounds associated with functioning of the pulmonary trunk valve can also best be heard at this place. Furthermore, such sounds are audible on each side and both farther forward and more ventrally than are the sounds of the aortic valve.

CONDUCTION SYSTEM

No part of the conducting system of a preserved dog's heart can be adequately identified in gross dissection without special preparation. The conducting system consists of three parts, which are closely integrated physiologically: (1) the sinoatrial node, (2) the atrioventricular node, and (3) the atrioventricular bundle. Baerg and Bassett (1963) described a method to stain the conduction tissue with paladium iodide.

The **sinoatrial node** (*nodus sinuatrialis*) appears to be the center which initiates the heart beat and also regulates the interval between beats. It is located in the terminal crest at the confluence of the cranial vena cava, sinus venarum cavarum, and auricular orifice. When it is destroyed experimentally, the heart beat slows or stops (Dukes 1955). The sinoatrial node is composed of Purkinje fibers little modified from those which compose the atrioventricular conduction system. Nonidez (1943) has shown that both the sinoatrial and the atrioventricular node are supplied by postganglionic parasympathetic nerve terminals.

The **atrioventricular bundle** *(fasciculus atrioventricularis)* begins as a mass of Purkinje fibers known as the **atrioventricular node** *(nodus atrioventricularis)*. This is about 1.5 mm. in diameter in the dog, according to Baird and Robb (1950), and shows little histological differentiation from the bundle. Nonidez (1943) demonstrated that the atrioventricular node recieved a richer parasympathetic supply than did the sinoatrial node. The atrioventricular node begins in the septal wall of the atrium about 5 mm. cranioventral to the opening of the coronary sinus and craniodorsal to the septal cusp of the right atrioventricular valve. From this apparently blind beginning, the atrioventricular bundle runs forward and downward through the fibrous base of the heart. As it does so it divides into right and left branches. These lie closely under the endocardium of the septal wall of the right and left ventricles. The right septal branch crosses the cavity of the right ventricle in the septomarginal trabecula of this chamber. It arborizes in the outer ventricular wall of the right ventricle, where it ends. The left septal branch is more diffuse than the right one and partly traverses the cavity of the left ventricle to the outer wall of this chamber in bundles which are smaller and more branched than those on the right side. The ramification of the conduction system of the heart is more complicated than the previous description indicates. Baird and Robb (1950) found that there were transitions from the atrioventricular node to the atrial muscle and that various parts of the atrioventricular bundle blended with the general ventricular musculature. Abramson and Margolin (1936) say that in the dog, as in other species, the myocardial branches, as they leave the subendothelial plexus, tend to pass perpendicularly to the endocardium in the left ventricular wall and obliquely in the right ventricular wall. They further state that the interventricular septum is traversed by myocardial Purkinje fibers which arise from the adjacent subendothelial Purkinje networks. No one has demonstrated a structural link between the sinoatrial and atrioventricular node or bundle.

Blood Vessels of the Heart

CORONARY ARTERIES

The right and left coronary arteries and their branches supply the muscle of the heart. They arise from the aortic bulb immediately distal to the aortic valve. Higginbotham (1966) investigated the question of variation in coronary artery patterns between closely related pedigreed dogs and mongrels. The conclusion was that there is as much, if not more, variation in purebred dogs.

The **right coronary artery** *(a. coronaria dextra)* (Figs. 11–5, 11–7A) is smaller than the left coronary artery and measures about 1.5 mm. in diameter and 5 cm. in length. It arises from the right sinus of the aorta and makes a sweeping curve to the right and ventrocranially, lying in the fat of the coronary groove. Its initial part is bounded by the pulmonary trunk and the conus arteriosus craniolaterally, and dorsally it is covered by the right auricle. The right coronary artery supplies the bulk of the outer wall of the right ventricle. As a result of ligation studies, Donald and Essex (1954) found that an average of 66 per cent of the free right ventricular wall normally receives its blood through the right coronary artery. According to Kazzaz and Shanklin (1950), the right coronary artery rarely extends as far as the borders of the right ventricle and in no case goes beyond them. It also sends branches to the right atrium, and small twigs to the initial parts of the aorta and pulmonary trunk. In 20 per cent of specimens, according to Moore (1930), an **accessory right coronary artery** *(a. coronaria dextra accessoria)* arises closely adjacent to the main right coronary artery from the right sinus of the aorta and runs about 2 cm. out on the conus arteriosus, where it largely becomes dissipated.

According to Pianetto (1939), there are 4

to 9 ventricular branches of the main right coronary artery. Most of these are small vessels which arborize on and in the right ventricular wall. They run at right angles to the long axis of the cavity, and none extend as far as the ventral interventricular groove. One of these, the **right marginal branch** *(ramus marginalis dextra),* is larger, longer, and more branched than the others. It supplies the middle part of the right lateral ventricular wall.

The atrial branches from the right coronary artery are variable in development. Kazzaz and Shanklin (1950), Meek et al. (1929), Moore (1930), and Pianetto (1939) all describe an atrial branch, larger than the others, which leaves the distal half of the artery, traverses the sulcus terminalis and, as it does so, supplies the sinoatrial node. This vessel anastomoses with one or more atrial branches which arise from the circumflex vessel. Usually one or two small atrial rami leave the right coronary artery both proximal and distal to the branch destined for the sinoatrial node. The distal branches supply the caudal part of the right atrium adjacent to the coronary sulcus. A terminal twig ends on the caudal vena cava. The first branch or two which leave the dorsal surface of the right coronary artery supply the right auricle.

The normal anastomoses between the right coronary artery and adjacent vessels are small. Donald and Essex (1954) found that, in those dogs which survived the ligation of the right coronary artery, four demonstrable anastomoses could be found after 30 days. These were as follows:

1. By means of the right (anterior) and left atrial (auricular) arteries.

2. Through the large, often tortuous, connection of the (left) circumflex and the right coronary artery.

3. Through small connections between the paraconal interventricular (anterior descending) and the ventricular (lateral) branches of the right coronary artery.

4. By means of several branches of the subsinuosal interventricular (descending branch of the left circumflex) and the last large ventricular (lateral) branch of the right coronary artery.

The **left coronary artery** *(a. coronaria sinistra)* (Figs. 11–4, 11–7, 11–11) is a short trunk about 5 mm. long and nearly as wide. It always terminates in the circumflex and paraconal (ventral) interventricular branches and, in many specimens, in a septal branch also.

The **circumflex branch** *(ramus circumflexus),* about 1.5 cm. in diameter, lies in the coronary groove as it extends to the left, then winds dorsocaudally and to the right across the caudodorsal surface of the heart. On reaching or approaching the subsinuosal (dorsal) interventricular groove, it turns toward the apex of the heart and is known as the subsinuosal interventricular branch. The combined lengths of the circumflex and subsinuosal interventricular branches is approximately 8 cm. The circumflex branch has ventricular and atrial branches.

According to Kazzaz and Shanklin (1950), there are 4 to 11 ventricular branches. Pianetto (1939) found two to six principal ventricular branches. Most of the hearts examined in the present study had five main ventricular branches leaving the circumflex branch, but great variation was found. When the left ventricular branches of the paraconal interventricular branch are well developed, they are a major source of supply to the outer wall of this chamber, and the ventricular branches of the circumflex are confined to a triangular area adjacent to the coronary and subsinuosal interventricular grooves. When the first or first few branches of the circumflex branch cross the superficial muscle fibers of the left ventricle obliquely, they are short; they are much longer when they parallel the superficial muscle fibers as well as the left ventricular branches of the paraconal interventricular vessel. Usually the longest branch leaving the circumflex lies adjacent and parallel to the subsinuosal interventricular branch. This is appropriately known as the **left marginal branch** *(ramus marginalis sinister),* as it follows the left border down the surface of the left ventricle.

The atrial branches of the circumflex branch of the left coronary artery, which supply the left atrium and auricle, are small and variable. The first branch arises deep to the great coronary vein and, extending dorsally and toward the center of the base of the heart, supplies the deep surface of the left atrium and the ventral

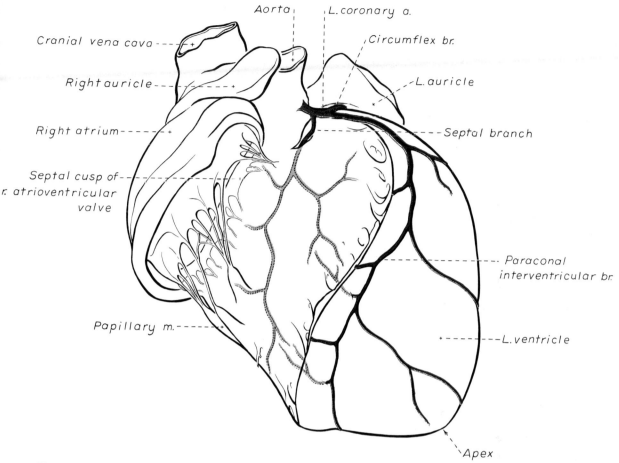

Figure 11–11. Branches of the left coronary artery, ventral aspect. The right ventricle has been removed.

part of the interatrial septum. It is the largest of the right atrial branches and, according to Meek et al. (1929), its terminal branches may partly encircle the termination of the cranial vena cava, where they anastomose with the right atrial vessel to the sinoauricular node. It may be the main source of supply to this node. At least two other small branches cross the lateral surface of the great coronary vein or the coronary sinus and supply the right atrium.

The **subsinuosal interventricular branch,** or caudal descending coronary artery (*ramus interventricularis subsinuosus*), is a continuation of the circumflex branch in or near the poorly defined subsinuosal interventricular groove. It is over 1 mm. wide and about 3.5 cm. long, and has left, right, and septal branches. It is shorter than the paraconal interventricular branch, as its

terminal branches usually end at about the junction of the middle and distal thirds of the ventricular mass. They may reach to the apex of the heart, or even extend beyond it. Usually three or four branches supply the adjacent musculature of the left ventricle, and usually five or six larger and longer branches arborize in the adjacent part of the right ventricle. The septal branches supply a narrow zone of the interventricular septum adjacent to the subsinuosal interventricular groove.

The **paraconal interventricular branch,** or cranial descending coronary artery, is about the same diameter as the circumflex branch, and averages 1.5 mm. in width. It is about 7 cm. long as it winds obliquely and distally from left to right across the sternocostal surface of the heart in the paraconal interventricular groove. It usually extends beyond the apex of the heart (Christensen 1962). It has left ventricular, right ventricular, and septal branches.

The **left ventricular branches** are long and large. Usually there are seven, which in general decrease in size toward the apex. As they largely parallel the superficial muscle fibers of the left ventricle they lie in grooves on its surface. Usually the branch which arises near the end of the proximal third of the parent vessel is longer than the others and supplies the apex of the heart. Most of the left ventricular branches are quite free of superficial collateral branches.

The **right ventricular branches** *(rami ventriculares dextri),* usually five in number, supply a strip, about 2 cm. wide, of the right outer ventricular wall adjacent to the paraconal interventricular groove. The first branch is prominent as it partly encircles the conus arteriosus adjacent to the origin of the pulmonary trunk. This is the conal branch. It anastomoses with the conal branch of the right coronary artery. Most of the remaining branches are short and form small anastomoses with the branches of the right coronary artery and the subsinuosal interventricular branch.

The **septal branch** *(ramus septalis)* (Fig. 11–11), as found by Donald and Essex (1954) in the 125 specimens they studied, arose as follows:

1. Paraconal (ventral) interventricular branch, 48 per cent.
2. As one of the three terminal branches of the left coronary artery, 27 per cent.
3. Left coronary artery, 19 per cent.
4. Aorta, 5 per cent.
5. Circumflex branch, 1 per cent.

Immediately after entering the interventricular septum, the septal branch usually runs obliquely toward the apex of the heart, giving off major and minor branches along its course. Initially it runs obliquely to the right ventricular side of the septum. In this course it may parallel the atrioventricular fibrous ring as it lies under the endocardium adjacent to the dorsal half of the septal cusp of the right atrioventricular valve. The first half of the artery lies closely under the endocardium of the right ventricle; the second half penetrates deeply into the septum. It supplies all the main papillary muscles of the right ventricle.

According to Donald and Essex (1954), it supplies 70 to 75 per cent of the interventricular septum. These authors and many others have found that the anastomoses between the septal artery and the adjacent arteries of the canine heart are not sufficiently large or numerous to permit retrograde filling of the septal artery with injected dye. The subsinuosal and paraconal interventricular branches supply the periphery of the interventricular septum. The paraconal vessel contributes much more blood than does the subsinuosal vessel (Christensen and Campeti 1959).

CARDIAC VEINS

The **cardiac veins** *(venae cordis),* although in many instances satellites of the arteries to the heart, do not take the names of the comparable vessels. Most of the blood to the heart is returned to the right atrium by a short, wide trunk, the coronary sinus. Some of the ventral cardiac veins and the smallest cardiac veins open into the cavities of the heart directly.

The **coronary sinus** *(sinus coronarius)* (Figs. 11–3, 11–6) is the dilated terminal end of the great coronary vein. It is about 2 cm. long and 5 to 8 mm. in diameter. It lies in the fat of the dorsodextral part of the coronary groove below the caudal vena cava and above the terminal part of the circumflex branch of the left coronary artery. It opens into the right atrium ventral to the termination of the caudal vena cava. It may be partly covered by a few muscle fibers derived from the left atrium. Frequently a small, inefficient semilunar valve is located at its termination.

The **great coronary vein** *(v. cordis magna)* (Figs. 11–3, 11–4) lies in the dorsal part of the coronary groove as it circles the diaphragmatic surface of the heart from the left. It arises near the apex of the heart and ascends toward the base in the paraconal interventricular groove, where it is usually paired. Along its course it collects numerous veins from the ventricles and small twigs from the left atrium. Most of those from the ventricles are paired, in that a vein lies on each side of the comparable artery. At its termination in the coronary sinus the diameter of the passage increases threefold, according to Meek et al. (1929). At this place two veins usually enter the great coronary vein or the coronary sinus. The larger branch, which may not be paired, ascends from near the apex of the heart and is known as the **dorsal vein of the left ventricle** (*v. dorsalis ventriculi sinistri*). It is the largest vein draining this chamber. Frequently a smaller vein, the **oblique vein of the left atrium** (*v. obliqua atrii sinistri*) can be seen entering the coronary sinus after emerging from under the pulmonary veins (Fig. 11–3). Its importance lies in the fact that, like the coronary sinus, it is a vestige of the embryonic left common cardinal vein. It may persist as a left cranial vena cava (Fig. 12–1), or it may be non-patent.

The **middle cardiac vein** (*v. cordis media*) (Fig. 11–5) ascends in the subsinuosal interventricular groove in company with or near the subsinuosal interventricular artery. It is a large paired vessel which collects tributaries from both ventricles and empties into the coronary sinus near its termination. Near the apex of the heart, where the subsinuosal and paraconal interventricular grooves merge, the middle and great coronary veins anastomose.

The **right cardiac veins** (*vv. cordis dex-trae*), homologous to the *vv. cordis anteriores* of N.A., consist of several rather long, narrow vessels which may not be paired as they ascend to the ventral part of the coronary groove from the right ventricle. They either enter the small cardiac vein or open into the right atrium directly.

The **small cardiac veins** (*vv. cordis minimae*) were formerly known as the veins of Thebesius. These are microscopic channels of venule size which open into every chamber of the heart and, within the myocardium, anastomose with both the coronary arteries and other coronary veins. It is possible that when gradually occluding lesions develop the blood may flow in a retrograde direction in these valveless veins, which hypertrophy, to provide nourishment for the affected part. Pina et al. (1975) studied the Thebesian veins of the heart in 48 dogs using a corrosion fluorescence method. They found five types of terminations: arborform, sinuous, brushlike, canaliculated, and stellate. Although these small cardiac veins are found in the walls of all chambers of the heart, they are most often found in the right ventricle (81 per cent of dogs) and right atrium (77 per cent of dogs).

PULMONARY ARTERIES AND VEINS

PULMONARY TRUNK

The **pulmonary trunk** (*truncus pulmonalis*) and its branches are the only arteries in the body which carry unaerated (venous) blood. The trunk arises from the pulmonary fibrous ring at the conus arteriosus and, after a course of about 4 cm., it divides into the right and left pulmonary arteries. At the conus it is flanked by the right and left auricles. Farther out on the trunk, but still within the fibrous pericardium, fat usually masks the true length and position of the intrapericardial part of the vessel. The ventral, lateral, and caudal surfaces of the proximal three-fourths of the vessel are covered by serous pericardium and fat. The distal fourth of the vessel serves for the attachment of the fibrous pericardium and can be examined without opening the pericardial cavity. The pulmonary trunk, along its entire medial surface, contacts the aorta. The two vessels form a slight spiral as they obliquely cross each other. The *ligamentum arteriosum* is a connective tissue remnant of the fetal ductus arteriosus which arises near the bifurcation of the pulmonary trunk and passes to the aorta. House and Ederstrom (1968) found that the ductus arteriosus was open in all puppies less than four days of age and did not close anatomically until they were seven or eight days of age. (For radiographic aspects of patent ductus, see Buchanan 1972).

The **right pulmonary artery** (*a. pulmonalis dextra*) is about 2 cm. long and 1 cm. in diameter. It leaves the pulmonary trunk at nearly a right angle and runs to the right, where at first it is in contact with the concavity of the arch of the aorta and later with the right bronchus. It lies obliquely across the base of the heart between the cranial and caudal venae cavae. Its first branch enters the right cranial lobe of the lung. About 1 cm. distal to the origin of this branch the vessel divides into numerous vessels which supply the cranial, middle, caudal, and accessory lobes of the right lung.

The **left pulmonary artery** (a. pulmonalis sinistra) is shorter and slightly smaller in diameter than the right artery. It is partly covered dorsally at its origin by the left bronchus. The artery then passes obliquely across the pulmonary vein coming from the cranial (formerly apical) lobe and divides unevenly into two or more branches. The smaller branch or branches enter the cranial part of the cranial lobe; the large one enters the bulk of the lung, where it subdivides and supplies the caudal part of the cranial (formerly cardiac) lobe and the caudal (formerly diaphragmatic) lobe of the lung. The relations of the pulmonary arteries within the lungs are described in Chapter 8, Respiratory System.

PULMONARY VEINS

The **pulmonary veins** (vv. pulmonales) are valveless and return aerated blood from the lungs to the left atrium. Unlike the pulmonary arteries, the pulmonary veins from each of the lobes of the lungs as a rule retain their separate identity to the heart. An exception to this is frequently found in the veins from the right caudal and accessory lobes, which fuse just before entering the atrium. Not uncommonly, the vein from the left caudal lobe joins the venous trunk from the right side, a single opening in the left atrium thus serving for the return of the blood from the right and left caudal and accessory lobes. Other variations are common; not infrequently, two veins may enter the heart directly from one of the lobes. Usually, however, several veins converge and empty into the heart by a single vessel which is from a few millimeters to about 1.5 cm. in length. In medium-sized dogs all pulmonary veins are over 5 mm. in diameter as they enter the heart. The pulmonary veins may be divided into the right pulmonary veins (vv. pulmonales dextrae) from the right lung, and the left pulmonary veins (vv. pulmonales sinistrae) from the left lung. The pulmonary veins within the lungs are described with the intrinsic structure of the lungs, in Chapter 8.

SYSTEMIC ARTERIES

AORTA

The aorta leaves the left ventricle near the center of the base of the heart. It is a thick-walled vessel through which all the systemic blood of the body passes. All of the large systemic arteries arise directly from it. For descriptive purposes, it may be divided into an ascending and a descending portion, separated by the aortic arch. The initial part attaches to the fibrous base of the heart, is largely located within the pericardium, and is known as the **ascending aorta** (aorta ascendens). It is about 2 cm. long before it makes a **U**-turn dorsocaudally and to the left as the **aortic arch** (arcus aortae). The remainder of the aorta, from the arch to its terminal iliac branches, is the **descending aorta** (aorta descendens). The descending aorta may be divided further into a **thoracic part** (aorta thoracica) and an **abdominal part** (aorta abdominalis).

The ascending aorta, at its origin, is slightly expanded to form the **bulb of the aorta** (bulbus aortae). Peripheral to each of the three semilunar cusps which forms the aortic valve are the **sinuses of the aorta** (sinus aortae) (see discussion under aortic valve, p. 646). The aggregate of the sinuses form the bulb of the aorta. The coronary arteries arise from the aortic bulb.

The normal development of the heart and aortic arches in several domestic animals is discussed by Krediet (1962), in a thesis on anomalies of the arterial trunks in the thorax.

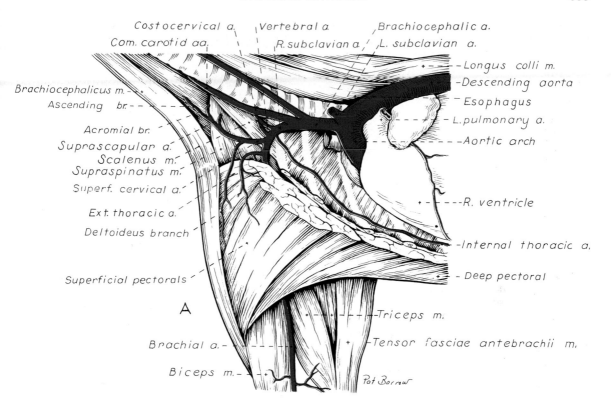

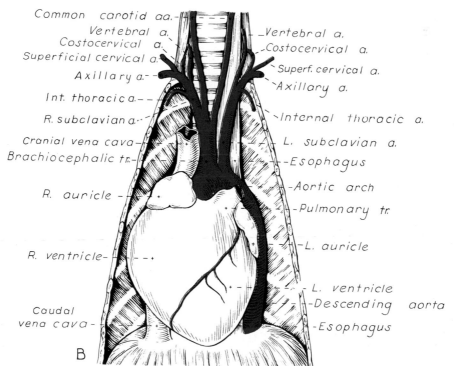

Figure 11–12. The aortic arch and great vessels.
A. Branches of the right subclavian artery, medial aspect.
B. The heart and great vessels, in situ, ventral aspect.

AORTIC ARCH

Arteries of the Head, Neck, and Thorax

The blood supply of the head and neck and the thoracic limbs leaves the aorta through two great vessels arising from the aortic arch, the brachiocephalic trunk and the left subclavian artery.

BRACHIOCEPHALIC TRUNK

The **brachiocephalic trunk** (*truncus brachiocephalicus*) (Fig. 11–12), the first large artery from the aortic arch, passes obliquely to the right and cranially across the ventral surface of the trachea. It is about 4 cm. long and 8 mm. in diameter. The left common carotid artery is the first branch to leave the brachiocephalic trunk. Frequently, a small branch leaves the brachiocephalic artery close to the heart to aid in the supply of the thymus and pericardium. The brachiocephalic artery terminates in the right common carotid and the right subclavian arteries. This termination is medial to the first rib or first intercostal space of the right side.

The **common carotid arteries** (*aa. carotides communes*) arise from the brachiocephalic artery about 1 cm. apart. Of 123 dogs examined, the right and left common carotids arose from the brachiocephalic artery by a common trunk in four specimens. The interval between the origins of the common carotids varies from 15 to less than 1 mm. When a **bicarotid trunk** (*truncus bicaroticus*) is formed, it usually arises from the brachiocephalic artery about 2 cm. distal to its origin from the aorta opposite the first intercostal space or second rib.

The **left common carotid** (*a. carotis communis sinistra*) usually arises opposite the vertebral end of the second rib and ventral to the trachea. Its relations are similar to those of the right vessel as it traverses the neck, except that it is on the left side and is loosely bound to the esophagus dorsomedially by the deep cervical fascia. Its branches and termination are similar to those of the right vessel, which is described here.

The **right common carotid** (*a. carotis*

communis dextra) diverges from the left and obliquely crosses the ventrolateral surface of the trachea as it runs toward the head. Throughout the neck it lies in the angle formed by the longus colli or longus capitis dorsally, the trachea ventromedially, and the brachiocephalicus and sternocephalicus laterally. At the thoracic inlet the vagosympathetic nerve trunk becomes associated with the dorsal surface of the artery and remains bound to it during its course through the neck. The internal jugular vein is also associated with the common carotid artery in the middle half of the neck. The fascia which binds these structures together and attaches them rather loosely to adjoining parts is the carotid sheath. It is a part of the poorly developed deep cervical fascia. The common carotid artery terminates at or near a transverse plane through the body of the hyoid bone by dividing into internal and external carotid arteries. The internal carotid, much smaller than the external one, leaves the medial side of the parent vessel and immediately runs through the deep structures of the head to the brain. The external carotid is the main supply to either half of the head. The branches of the common carotid arteries are the caudal thyroid occasionally, the cranial thyroid, and the terminal external and internal carotid vessels.

Clendenin and Conrad (1978) studied collateral vessel development following bilateral carotid artery occlusion. During the postoperative period of eight weeks, there was no indication of significant neurological deficit, hair loss, or skin changes in any of the dogs. Examination of the vascular casts revealed two consistent bilateral anastomotic connections: (1) internal carotid artery to ascending pharyngeal artery; and (2) internal carotid artery to internal maxillary artery. The pattern of anastomotic connections observed after bilateral ligation of the common carotids was different from that observed by the same authors after unilateral occlusion.

The **caudal thyroid artery** (*a. thyroidea caudalis*) (Fig. 11–13) is a small vessel which usually arises from the brachiocephalic between the origin of the common carotids. It is not constant in its origin; it may arise from the brachiocephalic, the

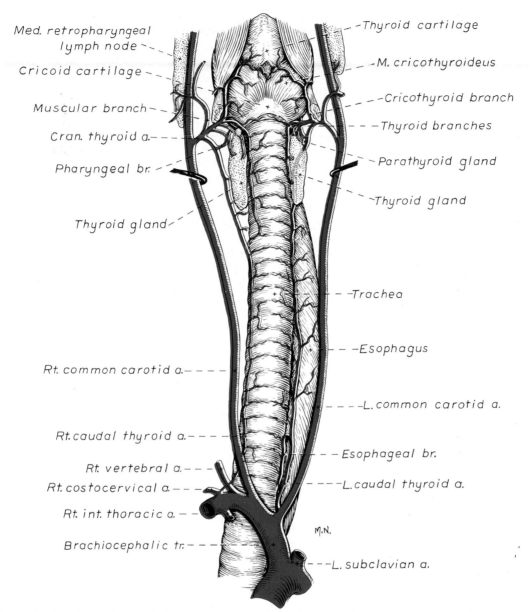

Med. retropharyngeal
lymph node

Cricoid cartilage

Muscular branch

Cran. thyroid a.

Pharyngeal br.

Thyroid gland

Rt. common carotid a.

Rt. caudal thyroid a.

Rt. vertebral a.

Rt. costocervical a.

Rt. int. thoracic a.

Brachiocephalic tr.

Thyroid cartilage

M. cricothyroideus

Cricothyroid branch

Thyroid branches

Parathyroid gland

Thyroid gland

Trachea

Esophagus

L. common carotid a.

Esophageal br.

L. caudal thyroid a.

M.N.

L. subclavian a.

Figure 11–13. The relation of the common carotid arteries to the larynx, trachea, and related structures, ventral aspect.

left subclavian, or the ascending cervical, a branch of the superficial cervical artery. It occasionally comes from the costocervical trunk on the right side. The most common origin is in the form of a short trunk from the brachiocephalic, giving rise to the right and left caudal thyroid arteries, which run cranially toward the respective lobes of the thyroid gland. They lie on the trachea and in contact with the respective borders of the esophagus. Branches from

the caudal thyroid arteries are freely supplied to the esophagus, trachea, caudal cervical ganglia, and nerves in the region of the thoracic inlet. They anastomose with the larger cranial thyroid arteries.

The **cranial thyroid artery** (*a. thyroidea cranialis*) (Fig. 11–13) is a short vessel which arises from the common carotid opposite the caudal part of the larynx. It is the largest and the only constantly present branch of the common carotid. It has the following branches: thyroid, pharyngeal, cricothyroid, and muscular. The branches vary in distribution and constancy. Frequently the thyroid branches and the pharyngeal and cricothyroid branches come directly from the common carotid as two separate trunks.

The **thyroid branches** (*rr. thyroidei*) are those which run in a caudal direction to the thyroid lobe. Their number and location vary; usually, however, several branches enter the dorsal and ventral borders of the lobe from its middle to the cranial pole and diverge as they ramify on its lateral and medial surfaces. Thus the blood supply to the thyroid lobe may be divided into dorsal and ventral groups of vessels. One ramus, usually from the dorsal group, and more often larger than the others, extends from the cranial pole of the thyroid lobe caudally past the dorsal border of the gland. This branch continues caudally in association with the recurrent laryngeal nerve and anastomoses with the caudal thyroid artery, giving off **esophageal** and **tracheal** branches along its course. When the caudal thyroid artery is well dveloped, the cranial vessel is reduced in a reciprocal ratio. In specimens with well-developed caudal thyroid arteries, most of the cervical parts of the trachea and esophagus are supplied by them.

The **pharyngeal branch** (*r. pharyngeus*) leaves the cranial side of the cranial thyroid artery, or it may leave in common with one of the thyroid branches. It is the smallest of the branches of the cranial thyroid artery. It runs obliquely dorsocranially and supplies twigs to the beginning of the esophagus; ventrally, a twig enters the larynx in company with the recurrent nerve; continuing forward, the pharyngeal branch supplies the constrictor muscles of the pharynx.

The **cricothyroid branch** (*r. cricothyroideus*) is a freely branching vessel which leaves the cranial thyroid artery and runs cranioventrally over the cricothyroid muscle. Twigs go to the sternohyoideus, sternothyroideus, thyrohyoideus, and cricothyroideus. End-twigs go through the cricothyroid membrane to the mucosa of the caudal compartment of the larynx, where they anastomose with the laryngeal artery. Right and left vessels anastomose on the cricothyroid membrane.

The **muscular branches** (*rr. musculares*) go dorsolaterally to both parts of the sternocephalicus and the mastoid part of the brachiocephalicus. Only small parts of these muscles are supplied by the muscular branches. The cranial thyroid artery also sends twigs to the capsule of the submandibular salivary gland and submandibular and medial retropharyngeal lymph nodes. A large muscular branch may go to the longus capitis and longus colli. This usually comes from the external carotid vessel near its origin, as described under the muscular branches of that vessel.

The **external carotid artery** (*a. carotis externa*) (Figs. 11–14, 11–15) is the main continuation of the common carotid to the head. It is about 4 cm. long and forms a sigmoid flexure as it winds its way under the cranial end of the hypoglossal nerve, submandibular salivary gland, and digastric muscle. It is bounded deeply by the muscles of the larynx and pharynx. The following branches leave the external carotid artery after its bifurcation from the internal carotid: occipital, cranial laryngeal, ascending pharyngeal, lingual, facial, caudal auricular, parotid, superficial temporal, and maxillary arteries.

The **occipital artery** (*a. occipitalis*) is most frequently the first branch of the external carotid. It may, however, arise in the angle formed by the splitting of the common carotid into the external and the internal carotid. In some dogs it arises an appreciable distance out on the external carotid, and therefore it will be regarded as a branch of this vessel. It is slightly smaller than the internal carotid, measuring about 1.5 mm. in diameter. In general, the vessel takes a tortuous course dorsally, its terminal branches anastomosing with those of its fellow. This occurs caudal to the external occipital protuberance in the dorsal straight muscles of the head. The vessel initially runs dorsocranially, being crossed laterally by the hypoglossal nerve and medially by the internal carotid artery. A larger structure, which bears important relations to the initial part of the vessel, is the medial retropharyngeal lymph node, lying caudal and partly over the vessel. The boundary, craniolaterally, is the digastric muscle, and, medially, the last four cranial nerves, although the accessory nerve may lie lateral to it. Usually the first

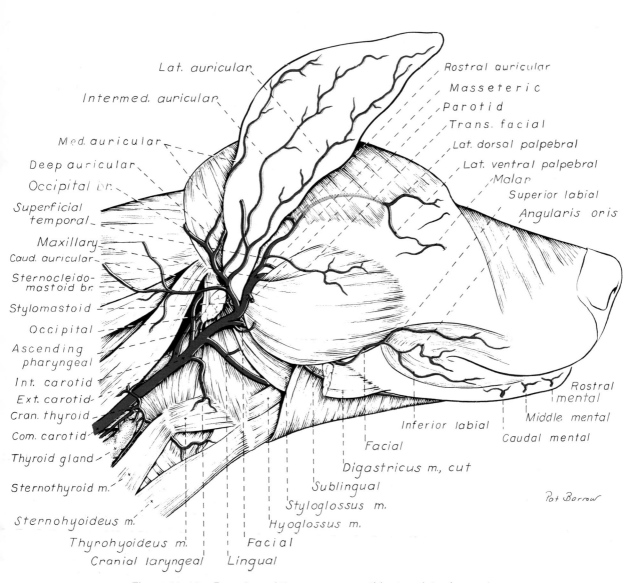

Figure 11–14. Branches of the common carotid artery, lateral aspect.

15 mm. of the occipital artery is free of branches. When it reaches the condyloid fossa, several branches arise. The vessel then forms an arc around the caudal surface of the jugular process and, lying between the two nuchal lines, runs to the median plane dorsally. In this location it is covered by the brachiocephalicus, splenius, obliquus capitis cranialis and the semispinalis capitis muscles. The branches of the occipital artery are: occasionally a ramus to the cranial pole of the medial retropharyngeal lymph node, a condyloid artery, a cervical ramus, a descending ramus, a caudal meningeal artery, and occipital branches.

A **condyloid artery** (*a. condylaris*) may arise from the cervical branch of the occipital or directly from the occipital. It enters the petro-occipital fissure and, in association with the accessory nerve, passes through the petrobasilar canal and finally becomes dissipated in the dura at the ventral end of the pyramid. It supplies twigs to the middle and inner ear. A branch goes to the digastricus before the vessel enters the fissure. Davis and Story (1943) described this vessel as the inferior (caudal) tympanic in the canids on which they worked.

The **cervical branch** (*r. cervicalis*) is the second branch of the occipital. It usually is a short trunk which arises medial to the jugular process and dorsolateral to the last four cranial nerves. One branch descends under the wing of the atlas, where it lies close to the bone and supplies the atlanto-occipital joint capsule. It also supplies parts of the rectus capitis ventralis, obliquus capitis cranialis, longus capitis, and rectus capitis lateralis muscles. Some branches end by forming a feeble ventral anastomosis with the vertebral artery. Others enter minute foramina on the ventral surface of the atlas in the region of the transverse foramen. The ascending branch runs medial to the last four cranial nerves as they leave the skull. It sends minute twigs to these nerves, as well as to the cranial cervical ganglion and the two ganglia on the glossopharyngeal and the vagus nerve. The main muscular branch continues cranially past these ganglia and ends primarily in the longus capitis and rectus capitis ventralis. Some twigs go to the mu-

cosa of the roof of the pharynx, where they anastomose with the ascending pharyngeal artery.

The **descending branch** (*r. descendens*) is the largest branch of the occipital. It is nearly as large as the continuation of the parent vessel. A branch enters the origin of the digastricus from the proximal end of the descending branch or from the occipital artery directly. The descending vessel takes origin about 3 mm. from the cervical branch and goes directly under the obliquus capitis cranialis to the alar notch of the atlas. Here it becomes associated with the ventral division of the first cervical nerve and the vertebral artery and vein. The vertebral artery is several times larger than the descending branch of the occipital. An anastomosis between the two vessels occurs. The descending branch continues dorsomedially in the obliquus capitis caudalis, the cranial part of which it supplies. It also supplies the cranial part of the dorsal straight muscles of the head and the semispinalis capitis.

The **caudal meningeal artery** (*a. meningea caudalis*) leaves the occipital as it lies between the nuchal lines of the occipital bone. This is about 1 cm. from the base of the jugular process. The vessel is about 0.7 mm. in diameter, and the extracranial part is 5 mm. long. It goes through the supramastoid foramen and ramifies in the dura of the occipital cranial fossa. Some branches supply the tentorium cerebri just dorsal to the pyramid.

From the digastricus to the mid-dorsal line, where the occipital artery anastomoses with its fellow, numerous branches arise. Most of these are dissipated in the muscles which attach to the caudal surface of the skull. Some, however, ramify in the temporal muscle and the caudoauricular musculature, where they anastomose with branches of the great auricular artery.

The **cranial laryngeal artery** (*a. laryngea cranialis*) (Fig. 11–14) is usually the second branch of the external carotid artery, arising ventrally nearly opposite the occipital artery. Sometimes this vessel, which is less than 1 mm. in diameter, arises from the common carotid at its termination. Near its origin it supplies one or two small twigs to the sternomastoid muscle. Another branch runs dorsally and supplies the

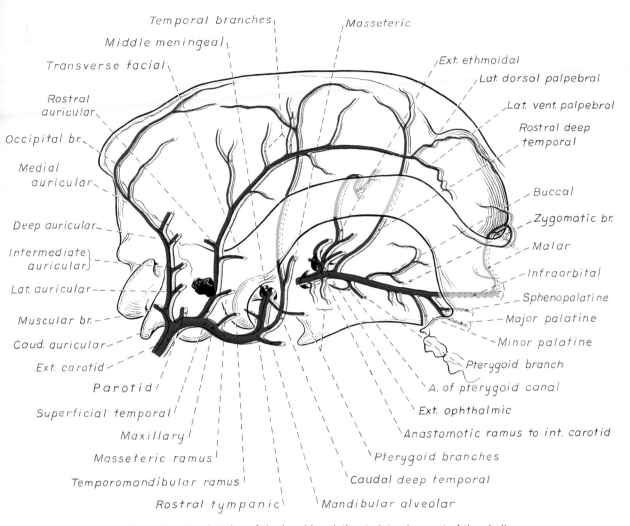

Figure 11–15. Arteries of the head in relation to lateral aspect of the skull.

dorsal portions of the cricopharyngeal, thyropharyngeal, and hyopharyngeal muscles. The main continuation of the vessel runs ventrally with the cranial laryngeal nerve over the surface of the larynx and disappears in the triangle formed by the mm. thyrohyoideus, thyropharyngeus, and hyopharyngeus. It perforates the thyrohyoid membrane and supplies most of the mucosa and intrinsic muscles of the larynx. Cranial branches have been found which supply the m. hyoglossus and in which they may anastomose with the lingual artery. Caudally and ventrally, twigs ramify in the thyrohyoideus, where an anastomosis occurs with the cricothyroid branch of the cranial thyroid artery.

A **muscular branch** (r. muscularis), slightly less than 1 mm. in diameter, frequently leaves the dorsal surface of the external carotid at its origin. Usually this origin is in the same transverse plane as the origin of the occipital artery. It runs directly to the ventral surface of the longus capitis, on which it arborizes. Its most caudal branches also supply the longus colli. In many specimens, the branch just described arises from the cranial thyroid artery or from the external carotid directly or by a short trunk with the ascending pharyngeal.

The **ascending pharyngeal artery** (a. pharyngea ascendens) (Fig. 11–16) is a small, freely branching vessel that arises from the external carotid in common with or close to the occipital artery. When the relatively large muscular branch to the longus capitis and longus colli arises here, instead of from the cranial thyroid, it also may be closely related to the ascending pharyngeal or arise in common with it. Thus from a medial origin the ascending pharyngeal runs dorsomedially on the pharyngeal mucosa medial to the tympanic bulla. The pharyngeal branch of the vagus is the only nerve to cross its medial surface. The artery extends as far forward as the pharyngeal opening of the auditory tube, where its terminal branch anastomoses with the loop of the internal carotid artery. Its branches are the palatine and pharyngeal.

The **palatine branches** (*rr. palatini*) are a few small branches which leave the initial part of the ascending pharyngeal. They run ventrally in the lateral wall of the pharynx to the soft palate, where they supply the extensive palatine glands, and the palatine mucosa and muscles. These branches anastomose with their fellows, as well as with the tonsillar branch of the lingual.

The **pharyngeal branches** (*rr. pharyngei*) are distributed to the musculature and mucosa of the cranial part of the pharynx as well as to the ventral axial muscles, mainly the longus capitis. The main ascending pharyngeal artery lies external to the pharyngeal musculature directly against the bulla. It sends many branches to the cranial part of the roof and sides of the pharynx. After giving origin to these, the artery continues to the external carotid foramen. Here an unusual vascular occurrence takes place. The internal carotid, after having traversed the carotid canal, forms a loop which fills the external carotid foramen. The ascending pharyngeal artery ends by anastomosing with the internal carotid loop. (In the adult domestic cat, where the internal carotid is not patent, the ascending pharyngeal artery extends through the external carotid foramen and contributes directly to the formation of the arterial circle at the base of the brain. This distal part of the ascending pharyngeal in the domestic cat is morphologically the internal carotid — Davis and Story 1943.)

The **lingual artery** (*a. lingualis*) (Figs. 11–14, 11–16, 11–18) is usually the largest collateral branch of the external carotid. It leaves the parent trunk just medial to the digastric muscle, runs cranioventrally in company with the hypoglossal nerve, and enters the tongue medial to the hyoglossal muscle, and lateral to the genioglossus. (The hypoglossal nerve does not accompany the vessel during its initial intramuscular course but runs lateral to the hypoglossal muscle. After a course of about 4 cm. in this location, the nerve usually crosses the medial surface of the lingual artery and then, in contact with its dorsal surface, extends to the tip of the tongue.) At the root of the tongue two or more branches are given off which can be traced to the hyoid and pharyngeal muscles and the palatine tonsil. These are (1) the **hyoid branches** (*rr. hyoidei*), and (2) the **tonsillar artery** (*a. tonsillaris*). The tonsillar artery leaves the dorsal surface of the lingual opposite the lateral surface of the ceratohyoid bone and runs dorsally rostral to it. In its course dorsocaudally, it perforates the styloglossal muscle to become related medially to the pharyngeal mucosa. The vessel then enters the tonsil near its middle by three or more minute branches. Twigs from the hyoid branches enter the caudal end of the tonsil and anastomose with the tonsillar artery from the lingual. The twigs from the caudally lying hyoid branches may be the major supply of the palatine tonsil. All other branches of the lingual are destined for the supply of the muscles of the tongue. The artery lies near the midline, near the ventral part of the organ. The lingual vein accompanies the artery only in its rostral third. The vein lies in a more ventral and superficial position than the artery in the remainder of the organ. According to Prichard and Daniel (1953), arteriovenous anastomoses occur in the tongue of the dog. The lingual and sublingual arteries anastomose in the tongue.

The **facial artery** (*a. facialis*), (Figs. 11–14, 11–16) is about 3 cm. long and 1.5 mm. in diameter. It arises near the angle of the jaw, 1 cm. from the lingual artery, and for the first centimeter is bounded medially by the styloglossal muscle; it then runs rostrally superficial to the stylohyoid muscle. The masseter muscle is related to it

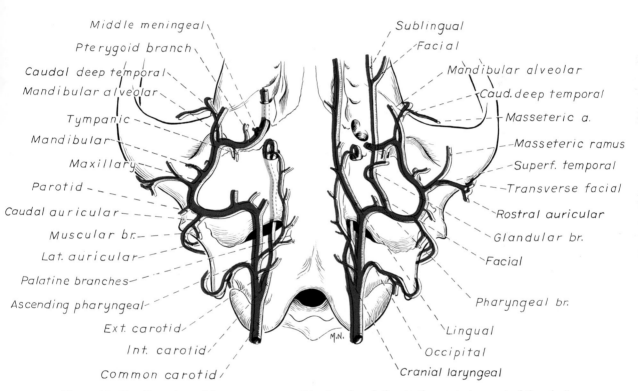

Middle meningeal
Pterygoid branch
Caudal deep temporal
Mandibular alveolar
Tympanic
Mandibular
Maxillary
Parotid
Caudal auricular
Muscular br.
Lat. auricular
Palatine branches
Ascending pharyngeal
Ext. carotid
Int. carotid
Common carotid

Sublingual
Facial
Mandibular alveolar
Caud. deep temporal
Masseteric a.
Masseteric ramus
Superf. temporal
Transverse facial
Rostral auricular
Glandular br.
Facial
Pharyngeal br.
Lingual
Occipital
Cranial laryngeal

M.N.

Figure 11–16. Branches of the common carotid artery in relation to the ventral aspect of the skull.

dorsally and laterally, and the digastricus lies ventral to it. The artery runs forward, but its deviation from the horizontal depends on the degree of closure of the jaws. The facial artery gives rise to a glandular branch and to muscular branches, before its first large collateral branch, the sublingual artery, arises.

The **glandular branch** (*r. glandularis*) is the largest but not necessarily the first branch to leave the initial part of the facial artery. It is the main supply to the mandibular and sublingual salivary glands.

The **muscular branches** (*rr. musculares*) are usually two small vessels which supply the adjacent parts of the digastricus, pterygoideus medialis and, occasionally, the styloglossus. The largest of these branches are smaller than the glandular branch and arise rostral to it. A twig may supply a small patch of mucosa in the region of the caudal pole of the palatine tonsil. The branch to the mucosa, when present, usually anastomoses with the ascending pharyngeal.

The **sublingual artery** (*a. sublingualis*) arises from the facial medial to the ventral part of the body of the mandible, in the depths of a deep cleft which is bounded laterally by the masseter, medially by the caudal part of the mylohyoideus, and ventrally by the digastricus. The sublingual artery parallels the medial surface of the mandible near its ventral border and is distributed largely to the mylohyoid and the rostral belly of the digastric muscle, although some branches perforate the mylohyoideus and supply the genioglossus and geniohyoideus. It is accompanied by a satellite vein and the mylohyoid nerve. It anastomoses with the lingual and ventral labial arteries. Near the middle of the body of the mandible the **submental artery** (*a. submentalis*) is given off. This runs to the ventral surface of the symphysis of the mandible, supplying this region and the incisor teeth. Other branches are distributed to the muscles and mucosa in the region of the frenulum of the tongue.

The **inferior labial artery** (*a. labialis inferior*) arises about 1 cm. from the ventral border of the mandible rostral to the masseter muscle. As it runs rostrally, it lies along the ventral border of the orbicularis oris. Some fibers of this muscle

may actually cover the artery during its course forward. At the caudal mental foramen it anastomoses with the caudal mental artery. The inferior labial artery sends twigs across the ventral border of the mandible which anastomose with the sublingual artery.

The **angular artery of the mouth** (*a. angularis oris*) arises from one to several centimeters peripheral to the origin of the inferior labial. It takes a rather tortuous course to the commissure of the lips, where usually one branch extends to the dorsal and the other to the ventral margin. The angularis oris supplies in part the buccinator, orbicularis oris, and the skin and mucosa of this region. It anastomoses with the superior and inferior labial arteries, and may anastomose with the caudal mental artery.

The **superior labial artery** (*a. labialis superior*) is the termination of the facial artery and ramifies on the cheek and nose. Small branches may extend dorsally to the orbit and anastomose with the terminal branches of the lateral inferior palpebral and malar arteries; others run forward in the orbicularis oris and anastomose with the lateral nasal artery. Fine twigs, following the buccal nerves caudally, anastomose with those of the masseter artery, which runs forward. It supplies mainly the orbicularis oris and levator nasolabialis.

The **caudal auricular artery** (*a. auricularis magna*) (Figs. 11–14, 11–15, 11–16) arises at the base of the annular cartilage from the dorsocaudal surface of the external carotid. It circles around the caudal half of the base of the ear. It is a medium-sized vessel which at first lies under the parotid salivary gland, then is located more dorsally under the caudoauricular group of muscles. For convenient exposure of the origin of the caudal auricular, it is necessary to remove the digastric muscle which lies caudoventral to it. Its branches may vary considerably in origin and disposition. They are: (1) stylomastoid, (2) glandular, (3) muscular, (4) lateral auricular, (5) intermediate auricular, (6) deep auricular, (7) medial auricular, and (8) occipital.

The **stylomastoid artery** (*a. stylomastoidea*) is the smallest branch of the caudal auricular. It leaves the caudoventral surface of the vessel and runs directly to the stylomastoid foramen in company with the facial nerve, which is supplies. Sometimes the vessel is double. It is located directly caudal to the external acoustic process.

The **glandular branches** (rr. glandulares) go to the parotid and mandibular salivary glands. These may arise not from the caudal auricular directly but from its sternocleidomastoideus or lateral auricular branches. The mandibular branches enter the dorsal surface of the gland; the parotid branches enter the deep surface of the parotid salivary gland.

The **sternocleidomastoideus branches** (rr. sternocleido-mastoidea) are one or two large vessels which supply the cranial parts of the brachiocephalicus and sternocephalicus. They also supply the skin, platysma, and subcutaneous fat, and finally anastomose with the superficial cervical artery. The muscular branch located at a deeper level runs under the sternocephalicus and brachiocephalicus and supplies the cranial end of the splenius muscles. Occasionally, branches supply the medial retropharyngeal lymph node. The dorsal deep part of the muscular branch anastomoses with the occipital artery.

The **lateral auricular branch** (r. auricularis lateralis) arises from the muscular branch to the splenius or from the intermediate auricular. It is a large artery which branches as it passes through or in contact with the caudal border of the parotid salivary gland. It extends distally on the caudal surface of the auricular cartilage, near its lateral border. It usually extends beyond the cutaneous pouch 1 cm. or more and terminates by anastomosing with the intermediate auricular branch.

The **intermediate auricular branch** (r. auricularis intermedia) is the largest artery to the ear. It arises about 1 cm. peripheral to the lateral auricular deeply under the caudoauricular muscles. During its initial course toward the apex of the ear, it sends branches to the caudoauricular muscles. After emerging, it sends many anastomosing branches both laterally and medially over the auricular cartilage. Many twigs pass through the foramina in the auricular cartilage to its concave surface.

The **deep auricular artery** (a. auricularis profunda) usually arises independently from the great auricular peripheral to the origin of the intermediate auricular. Occasionally, it arises from the intermediate or medial auricular. It is a small vessel which runs distally about 2 cm. and passes through the space between the tragus and the anthelix to supply part of the dermis of the cartilaginous external acoustic meatus.

The **medial auricular branch** (r. auricularis medialis), about the same size as the lateral auricular, arises about 1 cm. from the considerably larger intermediate auricular. It crosses the caudal part of the temporal muscle under cover of the auricular cartilage. At the scutiform cartilage, it becomes subcutaneous and continues along the rostral border of the cartilage to within 2 cm. of the apex of the ear. It anastomoses freely with the intermediate auricular branch and the rostral auricular branch of the superficial temporal. Branches also perforate the cartilage and extend around its margin to supply the fascia and dermis of the concave surface.

The **occipital branch** (r. occipitalis) is the main peripheral continuation of the caudal auricular after the auricular branches are given off. It enters the caudal part of the temporal muscle and at first nearly parallels the dorsal nuchal line. It supplies a large caudal part of the temporal muscle and finally anastomoses with the caudal deep temporal artery. Branches from this or a separate vessel from the caudal auricular also supply parts of the caudoauricular muscles and anastomose with the ascending cervical branch of the superficial cervical artery.

The **parotid artery** (a. parotis) is a small vessel which arises, 5 to 15 mm. distal to the origin of the caudal auricular, from the dorsal surface of the external carotid artery as this artery crosses the ventral end of the annular cartilage; it may arise from the caudal auricular artery. Thus, the vessel arises under the parotid gland and immediately enters it. It freely branches in the gland. Although the periphery of the parotid gland is supplied by parotid rami from adjacent vessels, such as the caudal auricular and superficial temporal, its main supply in the dog is the parotid artery. This vessel sends twigs to the facial nerve and the most dorsal mandibular lymph node, and may supply the skin.

The **superficial temporal artery** (a. tem-

poralis superficialis) (Figs. 11–14, 11–15) is the smaller of the terminal branches of the external carotid. Its diameter is approximately 1.5 mm., compared with 4 mm. for the maxillary, the other terminal branch. It arises in front of the base of the auricular cartilage and at first extends dorsally. As it crosses the zygomatic arch, it makes a sweeping curve rostrally and, about 1 cm. above it, dips under the heavy, deep temporal fascia. During part of its subsequent course toward the eye, it actually lies in the temporal muscle. Opposite the orbital ligament the superficial temporal perforates the deep temporal fascia and divides into its two terminal branches, which lie in the superfical fascia. The branches of the superficial temporal artery are the (1) masseteric, (2) transverse facial, (3) rostral auricular, (4) temporal, (5) superior and (6) inferior palpebral arteries.

The **masseteric branch** *(r. masseterica)* is a relatively large branch, usually over 1 mm. in diameter, which arises from the rostral side of the superficial temporal near its origin or from the maxillary artery directly. Hidden by the parotid salivary gland, it runs forward and enters the deep surface of the masseteric muscle, where it passes rostroventrally between the muscle and the masseteric fossa. Usually several other fine branches arise from the vessel and supply other structures. In about half of the specimens, they come off separately from the superficial temporal close to the masseteric branch. Some twigs enter the parotid salivary gland and parotid lymph node. Others run forward on the face with the dorsal and ventral buccal branches of the facial nerve and anastomose with arterial branches of the dorsal labial; still other branches supply the skin and occasionally the temporomandibular joint capsule.

The **transverse facial artery** *(a. transversa faciei)* is no larger than the nutrient branches which accompany the buccal nerves. It usually arises distal to the masseteric from the rostral border of the superficial temporal. It emerges from under the parotid salivary gland, usually after the artery has divided. One branch follows the zygomatic branch of the facial nerve, and the other runs parallel and ventral to the

zygomatic arch in company with the auriculotemporal branch of the trigeminal nerve.

The **rostral auricular branch** *(r. auricularis rostralis)* arises distal to the transverse facial on the opposite or caudal side of the superficial temporal. It is larger than the transverse facial, but less than 1 mm. in diameter. It runs between the upper rostral part of the parotid gland and the temporal muscle. It supplies both of these and finally ends in the rostrauricular muscles near the tragus.

The **temporal branches** *(rr. temporales)* arise from the distal half of the superficial temporal. These are variable in number, size, and origin. Usually two to five branches leave the dorsal surface of the vessel and are distributed to the substance of the temporal muscle. From the ventral surface of the vessel an average of two dissectable rami are present. These also supply the temporal muscle. Some branches run medial to the zygomatic arch and then ventral to it, to supply the masseter. The larger temporal branches anastomose with the deep temporal arteries of the maxillary.

The **lateral superior palpebral artery** *(a. palpebralis superior lateralis),* about 1 mm. in diameter, is about twice as large as the lateral inferior palpebral artery. It arises opposite the orbital ligament and, by a tortuous course at the junction of the upper eyelid and frontal bone, extends toward the medial canthus of the eye. It freely branches along its course, sending twigs to the various structures which form the upper eyelid. Twigs also supply the muscles, fascia, and the skin covering the temporal muscle and the subcutaneous part of the frontal bone. It forms an anastomosis with the small arteries that leave the dorsal part of the orbit and with those that perforate the skull through small foramina in the region of the frontonasomaxillary suture.

The **lateral inferior palpebral artery** *(a. palpebralis inferior lateralis)* sends branches to the caudal half of the lower eyelid. Several branches pass ventrally across the zygomatic arch and masseter. Here in the superficial fascia these branches anastomose with the transverse facial and malar arteries. The palpebral ar-

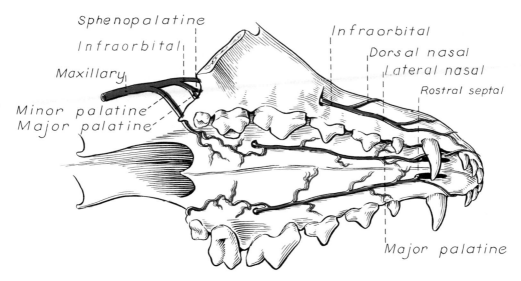

Sphenopalatine

Infraorbital

Maxillary

Minor palatine
Major palatine

Infraorbital

Dorsal nasal

Lateral nasal

Rostral septal

Major palatine

Figure 11–17. Terminal branches of the maxillary artery.

teries are the terminal branches of the superficial temporal artery.

The **maxillary artery** (*a. maxillaris*) (Figs. 11–15, 11–16, 11–17, 11–20, 11–22) gives off many branches which supply the deep structures of the head lying outside the braincase. It is the larger of the two terminal branches of the external carotid and, in a medium-sized dog, measures about 4 mm. in diameter. It is the main continuation of the external carotid. For convenience in describing its branches, it may be divided into three parts: the mandibular portion, the pterygoid portion, and the pterygopalatine portion. The mandibular portion extends to the alar canal. The pterygoid portion lies in the canal, and the pterygopalatine portion extends from the alar canal across the pterygopalatine fossa. No branches arise from the vessel as it passes through the alar canal.

The **first part, or mandibular portion, of the maxillary artery** includes that part of the artery from the point where the superficial temporal leaves the external carotid to the alar canal. It begins at the base of the ear, where the vessel reaches its most dorsal level and is covered by the parotid salivary gland. It continues the arch formed by the external carotid forward and downward to the caudal border of the mandible, where it is bounded laterally by the masseter muscle. On reaching the mandible, the artery changes its course

and runs medially, lying against the caudal part of the temporomandibular joint capsule as it does so. It actually follows the ventral border of the retroarticular process closely, and since this border is convex the artery also makes a ventral arch, lying, as it makes the arch, on the pterygoid muscles. Before entering the alar canal, the vessel is embraced by the mandibular division of the trigeminal nerve dorsally and the chorda tympani ventrally. The first part ends by making a bend forward and entering the alar canal. The following vessels leave the first part of the maxillary artery: mandibular branch, mandibular alveolar, caudal deep temporal, tympanic, pterygoid branch, and middle meningeal.

The **mandibular branch** (*r. mandibularis*) is the main supply to the caudal part of the temporomandibular joint capsule. Sometimes two or three branches are present, instead of one. The branch or branches leave the dorsal surface of the maxillary 5 to 15 mm. distal to the origin of the superficial temporal. When more than a single vessel is present, they are small and threadlike.

The **mandibular alveolar artery** (*a. alveolaris inferior*) (Fig. 11–19) measures slightly over 1 mm. in diameter and enters the mandibular canal after a course of about 1 cm. It arises from the ventral surface of the first part of the maxillary artery. Sometimes a trunk is formed from which the mandibular alveolar and caudal deep temporal arise in common. After entering the mandibular canal, the alveolar artery of the mandible closely follows the ventral border of the bone. It runs from the mandibular foramen to the middle mental foramen. During its course in the mandible, it sends many small twigs through the apical foramina to the roots of the teeth (Boling 1942) and others to the bone itself. The mandibular nerve is dorsolateral in the mandibular canal. The artery is in the middle, and the vein is ventromedial to the artery. Usually a considerable amount of fat surrounds these structures. Three vessels continue rostrally from the mandibular alveolar to supply the rostral part of the lower jaw. These are the caudal, middle, and rostral mental arteries.

The **caudal mental artery** (*a. mentalis caudalis),* with its satellite nerve and vein, leaves the caudal mental foramen and runs to the lower lip. It is much smaller than the middle mental artery, with which it anastomoses. It also anastomoses with the inferior labial artery.

The **middle mental artery** (*a. mentalis media)* is the largest of the three mental vessels and is the main blood supply to the rostral part of the lower jaw. It leaves the middle mental foramen, which is located in the ventral half of the mandible, ventral to the first two cheek teeth. With its accompanying vein and nerve, it supplies the skin, tactile hair follicles, and other soft structures. It forms an anastomosis with the rostral and caudal mental arteries. It is the main continuation of the alveolar artery of the mandible.

The **rostral mental artery** (*a. mentalis rostralis)* is the smallest of the three mental arteries. It leaves the mandibular alveolar less than 1 cm. caudal to the middle mental foramen and, with its satellite vein and nerve, runs in the delicate incisivomandibular canal, which closely follows the ventral border of the body of the mandible to the rostral mental foramen. It anastomoses with its fellow of the opposite side, as well as with the middle mental artery.

The **caudal deep temporal artery** (*a. temporalis profunda caudalis)* arises from the ventral surface of the maxillary just distal to or in common with the mandibular alveolar. It immediately crosses the lingual, mylohyoid, and alveolar branches of the trigeminal nerve, as well as the lateral pterygoid muscle. It enters the temporal muscle and extensively arborizes in it. It also sends rami which accompany the mylohyoid and lingual nerves. Most of the branches, however, are confined to that part of the temporal muscle lying medial to the coronoid process. It forms anastomoses with the rostral deep temporal, the occipital branches of the caudal auricular, and the temporal branches of the superficial temporal. One branch passes with the masseteric nerve through the mandibular notch to the masseter muscle. This is the **masseteric branch** (*r. massetericus).* It anastomoses with the masseteric branch of the superficial temporal, which is the main supply to the masseter, as it runs on its deep surface. The caudal deep temporal artery is accompanied by a satellite nerve and vein or veins.

The **tympanic artery** (*a. tympanica)* is a small, inconstant branch of the maxillary artery. It may arise from the caudal deep temporal artery. It usually leaves the maxillary medial to the temporomandibular joint and enters one of the small foramina located in a depression medial to the joint. It courses through the temporal bone into the middle ear. Davis and Story (1943) describe a similar vessel for the cat under the name of tympanica anterior.

Branches leave the ventral surface of the maxillary caudal to its entrance into the alar canal and arborize in the medial and lateral pterygoid muscles. Only small caudal portions of the pterygoid muscles are supplied by this source. Twigs also supply the origins of the tensor and levator veli palatini, the pterygopharyngeus, the palatopharyngeus, and the mucosa of the nasal pharynx.

The **middle meningeal artery** (*a. meningea media)* (Fig. 11–15) leaves the dorsal surface of the maxillary before this vessel enters the alar canal. It is about 1 mm. in diameter and runs through the oval foramen, which is closely adjacent to the maxillary artery. A notch, or still more rarely a foramen (foramen spinosum), is formed in the rostral wall of the oval foramen for the passage of the vessel. Within the cranial cavity the middle meningeal artery gives

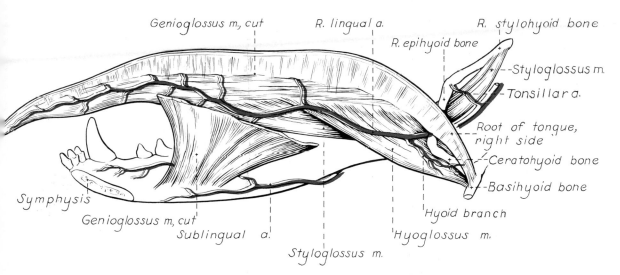

Figure 11–18.   Lingual and sublingual arteries, medial aspect.

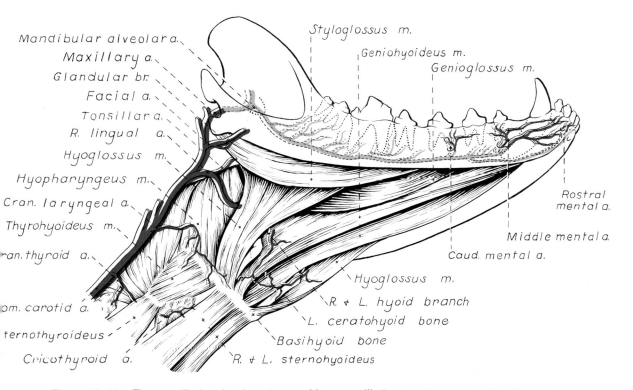

Figure 11–19.   The mandibular alveolar artery and intermandibular structures, ventrolateral aspect.

off the ramus anastomoticus, which runs medially and is about equal in size to the parent artery. The ramus anastomoticus joins the anastomotic artery of the external ophthalmic. After giving off the ramus anastomoticus, the middle meningeal artery follows the vascular groove on the cerebral surface of the skull. It runs in company with two satellite veins along the lateral border of the suture between the petrous and squamous parts of the temporal bone. It then passes almost directly dorsally across the middle part of the braincase and bifurcates into rostral and caudal branches. At its termination along the mid-dorsal line, it anastomoses with its fellow of the opposite side. The middle meningeal artery is the largest of the meningeal arteries. Its branches leave the parent vessel at right angles and run both rostrally and caudally. They supply the dura and adjacent portions of the skull.

The *ramus anastomoticus* enters the cavernous sinus and makes two to four loops before joining the anastomotic artery from the external ophthalmic lateral to the hypophysis.

The **second part, or pterygoid portion, of the maxillary artery** is about 1 cm. long, lies in the alar canal, and gives off no branches.

The **third part, or pterygopalatine portion,** lies on the lateral side of the lateral pterygoid muscle and crosses it obliquely. The following vessels leave the pterygopalatine portion of the maxillary artery: external ophthalmic, artery of the pterygoid canal, pterygoid branches, rostral deep temporal, buccal, minor palatine, and infraorbital arteries, and a trunk which gives rise to the major palatine and sphenopalatine arteries.

The **external ophthalmic artery** (*a. ophthalmica externa*) (Figs. 11–15, 11–22), whose origin was formerly called the orbital artery, gives rise to vessels supplying the orbit and anastomoses with vessels inside the skull. Tandler (1899) originated the term orbital artery, which Davis and Story (1943) and Jewell (1952) adopted in their works. Ellenberger and Baum (1943) called it the external ophthalmic artery in ruminants and in the horse, and this is now the accepted N.A.V. term for all domestic animals.

The external ophthalmic artery arises from the dorsal surface of the maxillary immedi-

ately after it leaves the alar canal. It is bounded medially by the maxillary nerve, laterally by the zygomatic and lacrimal nerves. The zygomatic and lacrimal nerves, as well as the terminal part of the external ophthalmic artery, are encased within the periorbita. The external ophthalmic artery typically gives rise to the external ethmoidal and an anastomotic ramus to the internal carotid.

The origin of the retractor bulbi muscle receives most of its blood from rami which leave the external ophthalmic artery. About 15 mm. before reaching the eyeball, this artery forms one or two flexures in the fat on the dorsal surface of the optic nerve. It runs to the medial side of the optic nerve, where it anastomoses with the smaller internal ophthalmic artery. From the union of the external and internal ophthalmic arteries, two to four **long posterior ciliary arteries** (*aa. ciliares posteriores longae*) arise and run to the eyeball. On reaching the sclera which surrounds the optic nerve, the vessels break up into several branches, **short posterior ciliary arteries** (*aa. ciliares posteriores breves*), which extend through the cribriform area and ramify in the choroid part of the vascular coat as the **choroid arteries** (*aa. choroideae*). Other branches do not perforate the sclera in the cribriform area but continue, closely applied to the sclera, toward the cornea, and are called the **episcleral arteries** (*aa. episclerales*). The retinal arterioles arise as branches of the short posterior ciliary arteries where the latter penetrate the sclera (see Chapter 20, The Eye). Retinal vessels emerge around the periphery of the optic papilla as nine or more arterioles which radiate into the retina. According to Catcott (1952), the venules which lie in the retina are three or four in number and converge to the center of the optic papilla. Great variation exists among dogs and between the eyes of the same dog.

The **anastomotic artery** (*a. anastomotica*) (Fig. 11–22) leaves the external ophthalmic or even the maxillary artery, according to Jewell (1952), close to the orbital fissure which it traverses. It sends minute twigs to the dura and to the nerves which pass through the orbital fissure. Sometimes the vessel is double throughout part or all of its course. It enters the cavernous sinus and receives the ramus anastomoticus from the middle meningeal artery. It continues cau-

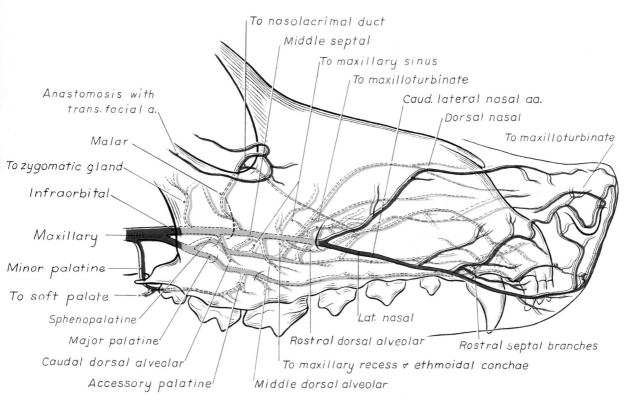

To nasolacrimal duct
Middle septal
To maxillary sinus
To maxilloturbinate
Caud. lateral nasal aa.
Dorsal nasal
To maxilloturbinate

Anastomosis with trans. facial a.
Malar
To zygomatic gland
Infraorbital
Maxillary
Minor palatine
To soft palate
Sphenopalatine
Major palatine
Caudal dorsal alveolar
Accessory palatine
Middle dorsal alveolar
To maxillary recess & ethmoidal conchae
Rostral dorsal alveolar
Lat. nasal
Rostral septal branches

Figure 11–20. Scheme of the terminal branches of the maxillary artery, lateral aspect.

dally as a single tortuous vessel and unites with the internal carotid at a transverse plane which passes through the dorsum sellae. Thus it is possible for blood to pass from the maxillary artery to the internal carotid by the external ophthalmic and anastomotic arteries.

The **external ethmoidal artery** *(a. ethmoidalis externa)* (Figs. 11–22, 11–24) is the rostral branch of the external ophthalmic artery. It is homologous with the anterior and posterior ethmoidal arteries of man (Davis and Story 1943). It makes an initial curve dorsally across the lateral surface of the ocular muscles, where it runs through the plexus formed by the ophthalmic vein in this location. It sometimes gives off branches to the dorsal oblique muscle and to the frontal bone. It then makes one or two more bends and enters the larger, more dorsally located ethmoidal foramen in company with its small satellite vein. The entrance of the artery into the ethmoidal foramen is unusual in that the vessel enters it from above and in front and not from the side from which the vessel approaches it. Also peculiar is the fact that a separate, smaller, and more ventral ethmoidal foramen conducts the ethmoidal nerve. Usually there are dorsal and ventral muscular branches which supply the muscles of the eyeball. These arteries arise independently from the rostral surface of the external ethmoidal artery, but occasionally they arise by means of a common trunk from the external ethmoid or directly from the orbital artery. The external ethmoidal artery, after passing through the ethmoidal foramen, reaches the dura. In the dura, between the cribriform plate and the olfactory bulb, it divides into a dorsal and a ventral branch. These branches anastomose rostrally and form an arterial circle on the lateral wall of the ethmoidal fossa. Many small branches leave this arterial circle and reunite, so that an ethmoidal rete is formed. The internal ethmoidal arteries, from the rostral cerebral arteries, run in the falx cerebri to the cribriform plate, where they anastomose and aid in forming the ethmoidal rete. Many branches pass through the cribriform plate from the rete to supply the mucosa of the ethmoturbinates and the nasal septum. Those to the nasal septum are the **caudal septal arteries** *(aa. septales caudales).* Another branch, the **rostral meningeal artery** *(a.*

meningea rostralis), runs dorsally in the dura at the caudal margin of the cribriform plate, passes through the inner table of the frontal bone, and enters the mucoperiosteum on the floor of the major compartment of the frontal sinus. After running caudally in the frontal sinus, it passes through the inner table of the frontal bone 5 to 10 mm. from the midsagittal plane and arborizes in the dura ventral to the caudal part of the frontal sinus. It forms a delicate anastomosis with the middle meningeal artery. There is considerable variation in the size and distribution of the branches of the external ethmoidal artery.

If a common trunk of origin is formed for the muscular branches, it is usually short. Each of its resultant branches dips between adjacent recti muscles to the fat which lies between these and the retractor bulbi. Many branches are dispersed to the recti and oblique muscles of the eye and to the eyeball itself. The zygomatic and lacrimal arteries, which are small, arise from the dorsal muscular branch.

The **ventral muscular branch** *(r. muscularis ventralis)* (Fig. 11–23) extends toward the globe of the eye between the ventral and lateral recti, although some branches pass to the medial side. The muscles it supplies are primarily the lateral and ventral recti and the ventral portions of the retractor bulbi. It also supplies the medial rectus, the gland of the third eyelid, and the conjunctiva of the lower eyelid near the fornix. One small arterial branch runs with a branch of the oculomotor nerve to the ventral oblique muscle. It anastomoses with the dorsal muscular branch and the ciliary arteries.

The **dorsal muscular branch** *(r. muscularis dorsalis)* of the external ethmoidal artery arises in common with, or 1 cm. from, the ventral muscular branch, which it exceeds slightly in size. It crosses the proximal third of the lateral rectus obliquely and passes between the lateral and dorsal recti toward the globe of the eye. In its course it sends branches to the lateral and dorsal recti, dorsal oblique, retractor bulbi, and levator palpebrae muscles. On reaching the globe of the eye it gives off one small branch which extends dorsally over the eyeball and ends in the bulbar conjunctiva under the upper eyelid. In its course it supplies the terminal part of the levator palpebrae muscle and a portion

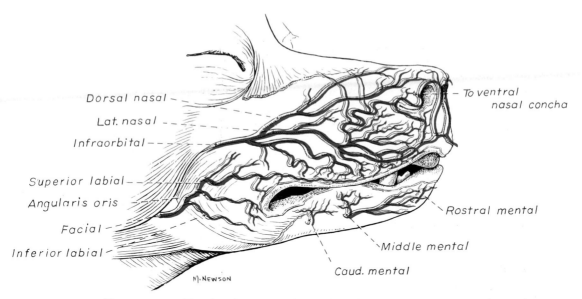

Figure 11–21.   Terminal branches of the infraorbital and facial arteries.

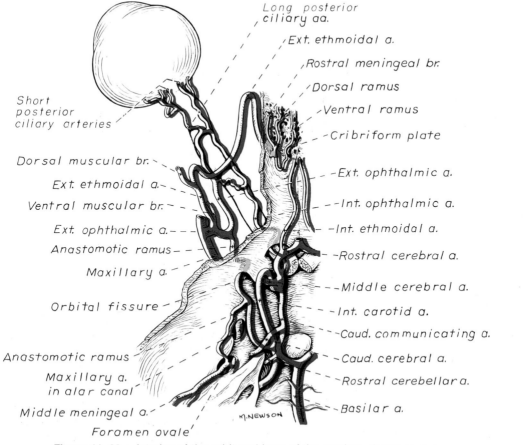

Figure 11–22.   Arteries of the orbit and base of the cranium, dorsal aspect.

of the lacrimal gland. This part is comparable to the supraorbital artery of man, whereas the dorsal muscular branch represents the superior muscular set of vessels in man. As the dorsal muscular branch crosses the lateral rectus, it divides into lacrimal and zygomatic branches, which follow the respective nerves. The arterial twigs supplying the muscles of the orbit are peculiar in that they run centrifugally. The main arteries run deeply in the muscular cone and issue their fine branches peripherally.

The **pterygoid branch** (*r. pterygoideus*) of the maxillary artery (Fig. 11–15) is a freely branching muscular twig to the medial and lateral pterygoids. It is about 0.5 mm. in diameter and arises opposite the origin of the rostral deep temporal artery. Its origin is about 2 mm. peripheral to the origin of the external ophthalmic artery, but it usually arises from the opposite side. It may arise from the medial side of the external ophthalmic artery. Several branches supply both the lateral and the medial pterygoid muscles. Occasionally, a twig can be traced through the muscle into the pterygoid canal. The pterygoid branch anastomoses with the muscular branch of the buccal artery, which supplies part of the medial pterygoid.

The **lacrimal artery** (*a. lacrimalis*), larger than the zygomatic artery, accompanies its satellite nerve and supplies the lacrimal gland. It passes deep to the orbital ligament and terminates in the conjunctiva and skin of the upper eyelid.

The threadlike **zygomatic artery** (*a. zygomatica*) follows the zygomatic nerve to the lacrimal gland, the skin, the conjunctiva near the lateral canthus of the eye, and the adjacent lower eyelid. The lacrimal and zygomatic areries may anastomose with each other. The lacrimal occasionally joins the superior palpebral; the zygomatic usually unites with the inferior palpebral.

The **rostral deep temporal artery** (*a. temporalis profunda rostralis*) (Fig. 11–15) is a vessel less than 1 mm. in diameter which arises close to the external ophthalmic artery. It may be double. From the dorsal surface of the maxillary it runs dorsally between the temporal muscle and the caudal part of the frontal bone. The small rostral deep temporal artery enters the temporal muscle near the middle of its rostral border and arborizes in the muscle. Accompanied by two satellite veins, it forms an anastomosis in the temporal muscle with the superficial and caudal deep temporal arteries.

The **buccal artery** (*a. buccalis*) (Fig. 11–15) arises from the ventrolateral surface of the maxillary about 1 cm. distal to the origin of the deep temporal. It is nearly 1 mm. in diameter as it leaves the maxillary at an acute angle and runs toward the cheek. Usually near its origin a small branch is given off to the medial pterygoid muscle. It soon becomes related to the buccinator nerve which accompanies it to the cheek. A tiny twig is given off to the ventral portion of the zygomatic gland, and larger twigs are distributed to the masseter, temporal, and buccinator muscles. The vessel finally terminates in the region of the soft palate and the pterygomandibular fold.

The **minor palatine artery** (*a. palatina minor*) (Figs. 11–17, 11–25) arises from the ventral surface of the maxillary or one of its terminal branches dorsal to the last upper cheek tooth. It is less than 0.5 mm. in diameter and passes ventrally through a notch in the caudal part of the maxilla. It is distributed to the adjacent soft and hard palate. The branch to the soft palate runs nearly the whole length of this part and lies close to the midplane. It supplies the palatine glands, musculature, and mucosa. Fine twigs anastomose with the ascending pharyngeal and the major palatine arteries. Occasionally a branch of the minor palatine artery sends a twig to the zygomatic gland.

The infraorbital artery and a common trunk which gives rise to the sphenopalatine and major palatine arteries are the terminal branches of the maxillary. They are nearly equal in size.

The **common trunk of the sphenopalatine and major palatine arteries** (Figs 11–17, 11–25) arises from the maxillary rostral to the origin of the minor palatine artery. It usually has a single, but sometimes a double, muscular ramus to the rostral portion of the medial pterygoid muscle. The muscular ramus may arise from the maxillary or from one of the terminal branches of the common trunk.

The **major palatine artery** (*a. palatina major*) arises from the common trunk as one of its terminal branches. The vessel, which is slightly more than 1 mm. in diameter, passes through the caudal palatine foramen and the palatine canal with a delicate vein and rela-

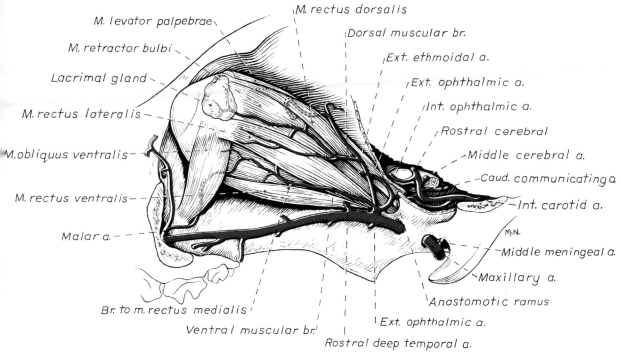

Figure 11–23. Arteries of the orbit and extrinsic ocular muscles, lateral aspect.

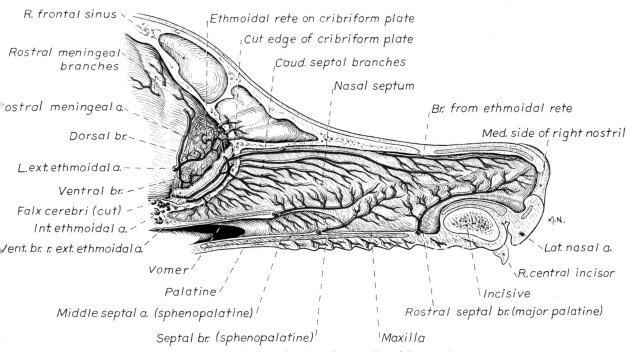

Figure 11–24. A sagittal section showing arteries of the nasal septum.

tively large satellite nerve. Within the palatine canal the nerve and artery divide so that two or more sets of major palatine arteries and nerves emerge on the hard palate. the main channel draining the area of the hard palate does not follow the major palatine artery through the palatine canal but runs caudally in the soft tissue of the hard palate as a spongy, poorly developed venous plexus. The plexus continues in the soft palate, where it lies dorsal to the palatine glands. It empties into the maxillary vein caudal to the temporomandibular joint, ventrolateral to the tymphanic bulla. Some of the nerve and artery branches pass through the accessory palatine foramina located caudal to the major palatine foramen. The arteries anastomose with each other, and the most caudal branch anastomoses with the minor palatine. The most rostral branch is the main continuation of the major palatine artery. The palatine groove on the surface of the hard palate, in which the vessels lie, is situated midway between the alveoli and the midline. Anastomoses between the right and left palatine vessels occur throughout their course. The major palatine artery and nerve usually leave the palatine groove midway between the palatine fissure and the major palatine foramen. They extend through the palatine venous plexus in their rostral course so that they lie closely under the oral mucosa. The groove rostral to the plane in which the artery and nerve leave it contains a portion of the palatine venous plexus. The major palatine artery supplies the mucosa of the oral surface of the hard palate, the periosteum, and the bone which forms the alveoli. A small branch passes through the palatine fissure and anastomoses with a branch of the sphenopalatine artery, which supplies the mucosa on the nasal side of the hard palate. A small artery extends rostrolaterally, passes through the interdental space between the canine and corner incisor teeth, and anastomoses with the lateral nasal artery. The **rostral septal branches** *(rr. septi rostrales)* (Fig. 11–24) from the major palatine artery run dorsomedially and supply that part of the septum caudal to the area supplied by septal branches of the lateral nasal and rostral to the area supplied by the middle septal artery from the sphenopalatine. By an extensive, fine arterial plexus they anastomose with

adjacent vessels. The major palatine artery continues forward, branching profusely, and in back of the incisor teeth turns toward the midline and anastomoses with its fellow. At the anastomosis a small vessel runs dorsally through the incisive foramen and joins with the right and left lateral nasals as these anastomose with each other at the midplane. This anastomotic branch is small as it passes dorsally through the interincisive suture.

The **sphenopalatine artery** *(a. sphenopalatine)* (Fig. 11–25), which is the other terminal branch of the common trunk, is over 2 mm. in diameter and leaves the pterygopalatine fossa by passing through the sphenopalatine foramen with its satellite nerve and vein. On reaching the nasopharyngeal duct, the artery runs rostroventrally under the mucoperiosteum and on the dorsal surface of the palatomaxillary suture to a point ventral to the opening into the maxillary recess. Here the sphenopalatine artery swings dorsorostrally for a few millimeters and divides into a dorsal and a ventral branch, and a branch which goes to the ventral nasal concha. All of the terminal branches of the sphenopalatine artery are collectively known as the **caudal lateral nasal arteries** *(aa. nasales caudales laterales).*

As the main sphenopalatine vessel makes its dorsal bend, the ventral vessel continues forward to supply the mucoperiosteum of the side and floor of the nasal fossa and the adjacent middle portion of the nasal septum. A small artery leaves the dorsal surface of this vessel and, curving dorsocaudally, runs toward the eye on the nasolacrimal duct. This twig supplies blood to the rostral part of the duct and anastomoses with the twig of the malar artery, which supplies its caudal part. Beyond the origin of the small artery to the nasolacrimal duct, the ventral vessel continues forward and slightly medially; its terminal twigs anastomose with a branch of the major palatine, which ascends through the palatine fissure.

The dorsal branch arises near the opening into the maxillary sinus aligned with a transverse plane passing between the third and fourth upper premolar teeth. This vessel runs dorsorostrally and bifurcates: one branch supplies the ventral part of the nasal concha and the mucoperiosteum lateral to it; the other branch goes to the dorsal part of

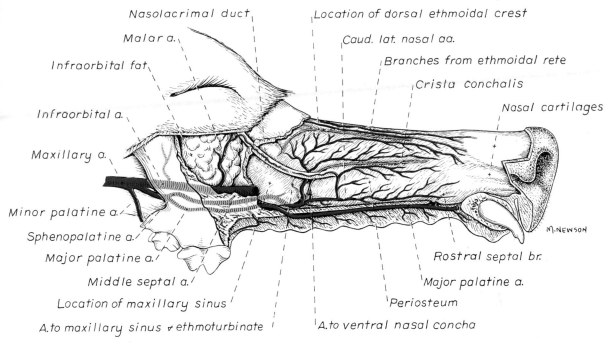

Nasolacrimal duct

Malar a.

Infraorbital fat

Infraorbital a.

Maxillary a.

Minor palatine a.

Sphenopalatine a.

Major palatine a.

Middle septal a.

Location of maxillary sinus

A. to maxillary sinus & ethmoturbinate

Location of dorsal ethmoidal crest

Caud. lat. nasal aa.

Branches from ethmoidal rete

Crista conchalis

Nasal cartilages

M. NEWSON

Rostral septal br.

Major palatine a.

Periosteum

A. to ventral nasal concha

Figure 11–25. A dissection showing arteries of the lateral nasal wall.

this bone and has an extensive anastomosis with a vessel which runs rostrally in endoturbinate I from the ethmoidal rete.

The branch which goes to the ventral nasal concha is short, medially inclined, and variable in origin. It may come from either the dorsal or the ventral branch previously described. It goes to the caudal part of the conchal (maxilloturbinate) crest and divides into five or six small arteries which arborize on the primary scrolls into which the bone is divided. It anastomoses rostrally with a branch of the lateral nasal artery, which curves around the dorsal part of the nostril and extends caudally on the ridge of tissue which is continuous with the conchal crest. In addition to the ventral, dorsal, and ventral conchal parts of the caudal lateral nasal arteries, smaller branches supply the mucosa and bone of the maxillary recess and the rostral parts of the ethmoturbinates and a large part of the middle of the nasal septum. The twigs to the maxillary recess arise from the caudal side of the dorsal branch of the caudal lateral nasal as this vessel runs in the mucoperiosteum which forms the rostroventral and rostrolateral parts of this cavity. A twig to the caudal part of the maxillary recess

may leave the sphenopalatine shortly after it enters the nasopharyngeal duct. Many other branches supply the mucoperiosteum of the floor and sides of the ventral nasal meatus. The **middle septal artery** (*a. septi media*) is the first branch of the sphenopalatine artery after it leaves the sphenopalatine foramen. It runs from the mucoperiosteum and the plate of bone separating the nasopharyngeal duct and the nasal fundus to the middle part of the nasal septum. Rostrally it anastomoses with the rostral septal branches from the major palatine, and caudally it anastomoses with the caudal septal branches from the ethmoidal rete. All the branches of the sphenopalatine and the middle septal arteries form voluminous arterial plexuses in the mucoperiosteum, which they supply. Numerous anastomoses also occur between adjacent vessels.

The **infraorbital artery** (*a. infraorbitalis*) (Figs. 11–17, 11–20, 11–21) is the main continuation of the maxillary across the medial pterygoid muscle. Accompanied by the maxillary division of the trigeminal nerve, it leaves the pterygopalatine fossa, gives off the caudal dorsal alveolar artery, and passes through the maxillary foramen to enter the infraorbital canal. It gives off a branch to the zygomatic gland, caudal dorsal alveolar, malar, middle dorsal alveolar, and rostral dorsal alveolar. It terminates by dividing into the lateral and dorsal nasal arteries. These terminal arteries arise either before or after the vessel has passed through the infraorbital foramen (Christensen and Toussaint 1957).

The **caudal dorsal alveolar artery** *(a. alveolaris dorsalis caudalis)* is a small vessel which may arise from the minor palatine or either of the terminal branches of the maxillary. It usually arises from the ventral surface of the infraorbital before the latter enters the infraorbital canal. The caudal dorsal alveolar divides and runs directly to the alveolar canals of the last two molar teeth. These are minute arterial branches accompanied by satellite nerves and veins.

The **malar artery** *(a. malaris)* arises from the dorsal surface of the infraorbital after this vessel enters the infraorbital fossa. Near its origin a small branch is given off, which supplies the ventral oblique muscle and passes along its deep surface to anastomose with the ventral muscular branch of the external ophthalmic or external ethmoidal artery. The main trunk runs to the medial canthus of the eye superficial to the periorbita. During its course it gives off a delicate branch which enters the nasal fossa in company with the nasolacrimal duct. The terminal twigs go mainly to the lower eyelid, where they anastomose with the inferior palpebral and the transverse facial arteries.

The **middle dorsal alveolar branches** *(rr. alveolares maxillares media)* leave the ventral surface of the infraorbital artery as it runs through the infraorbital canal. They run short distances with their satellite nerves and veins and enter the alveolar canals of the three roots of the shearing, or last premolar, tooth.

The **rostral dorsal alveolar arteries** *(aa. alveolares dorsales rostrales)* consist of one relatively large branch, which enters the incisive canal, and two or more smaller branches, which enter the maxilla rostral to the infraorbital foramen. The smaller twigs supply the alveoli and possibly the roots of the second and third premolar teeth. The main branch arches over the root of the canine tooth and terminates in the roots of the incisors. Thus this artery, which is accompa-

nied by its satellite vein and nerve, has an intraosseous course throughout its length. It supplies the first premolar, the canine, and incisor teeth of the same side.

The **lateral nasal artery** *(a. lateralis nasi)* (Figs. 11–17, 11–21) is the larger of the two terminal branches of the infraorbital. It measures slightly more than 1 mm. in diameter at its origin at the infraorbital foramen. It runs forward into the muzzle with many large infraorbital nerve branches and anastomoses with branches of the superior labial. It first crosses under the maxillonasolabialis muscle and then runs among the fibers of the orbicularis oris. The vessel branches profusely and supplies the upper lip and snout as well as the follicles of the vibrissae, or tactile hairs. It furnishes blood to the rostral part of the upper lip and the adjacent part of the nose. At the philtrum the vessel anastomoses with its fellow and sends a relatively large branch dorsally between the nostrils and another branch caudally in the mucosa of the parietal cartilage.

The **dorsal nasal artery** *(a. dorsalis nasi)* travels rostrodorsally across the lateral surface of the nose to its dorsal surface. It runs under and supplies the levator nasolabialis muscle, and then it continues to supply the structures of the dorsal surface of the rostral half of the muzzle. It anastomoses with its fellow of the opposite side, as well as with the lateral nasal, ventral nasal, and caudal lateral nasal arteries, and their middle septal branches.

The **internal carotid artery** *(a. carotis interna)* (Figs. 11–16, 11–22, 11–29) arises with the external carotid as the smaller of the two terminal branches of the common carotid. Other arteries, which arise in close association with this vessel, are the occipital and ascending pharyngeal. The termination of the common carotid is directly medial to the medial retropharyngeal lymph node, which is bound to the artery and adjacent structures by the fascia which forms the carotid sheath. At a still more lateral level is the sternocephalic muscle. The internal carotid artery at first runs dorsorostrally across the lateral surface of the pharynx. At its origin, from the dorsal surface of the parent artery, is a bulbous enlargement, the **carotid sinus** *(sinus caroticus)*, which is about 3 mm. in diameter and 4 mm. long (see Chapter 18, The Autonomic Nervous System). The vessel then narrows to approximately 1 mm. The internal carotid gives off no branches before entering the petrooccipital fissure. Just before entering this depression, it crosses the lateral surface of the cranial cervical sympathetic ganglion and the medial surface of the digastric muscle. The artery enters the caudal carotid foramen in the petrorostral fissure and traverses the carotid canal. On leaving the internal carotid foramen, which is the rostral opening of the carotid canal, it passes ventrally through the external carotid foramen, forms a loop, and re-enters the cranial cavity through the same foramen. Frequently, a small twig from the ascending pharyngeal artery anastomoses with the loop formed by the internal carotid. On re-entering the cranial cavity the internal carotid runs at first obliquely toward the dorsum sellae, then directly rostral to the optic chiasm. The vessel first perforates one layer of the dura mater and runs a short distance in the blood-filled cavernous sinus which separates the dura into two layers. It then perforates the second layer of dura and the arachnoid, and comes to lie in the subarachnoid space. On entering this space it trifurcates as the middle cerebral, rostral cerebral, and caudal communicating arteries. A small **rostral intercarotid artery** arises from the trifurcation. While in the cavernous sinus, the internal carotid artery forms an anastomosis with the anastomotic artery of the external ophthalmic.

The **caudal intercarotid artery** *(a. intercarotica caudalis)* (Fig. 11–26) is a small vessel which leaves the first part of the internal carotid as it enters the cavernous sinus. The vessel runs obliquely toward the midline and joins with its fellow, caudal to the hypophysis. It is closely applied to the dura of the cavernous and intercavernous sinuses. It gives off a branch which perforates the dura surrounding the hypophysis and supplies the caudal lobe. This is the **caudal hypophyseal artery** *(a. hypophyseos caudalis)*. Occasionally, the caudal intercarotid artery arises from the anastomotic artery.

The **caudal communicating artery** *(a. communicans caudalis)* (Figs. 11–27, 11–28) leaves the caudal surface of the internal carotid after it perforates the dura and arachnoid and enters the subarachnoid space. It forms the lateral and caudal thirds of the arterial circle. Caudally it anastomoses with the basilar artery. It is readily identified by the fact that the third cranial nerve crosses its dorsal surface.

The **arterial circle of the brain** *(circulus arteriosus cerebri)* (Figs. 11–27, 11–28) is an elongated arterial ring on the ventral surface of the brain, formed by the right and left internal carotids and the basilar artery. From the arterial circle, on each side, arise three vessels which supply the cerebrum. These are the rostral, middle, and caudal cerebral arteries (Fig. 11–33). The rostral cerebellar arteries from the circulus arteriosus cerebri and the caudal cerebellar arteries from the basilar artery supply the cerebellum; pontine and medullary branches of the basilar supply the pons and medulla oblongata. All

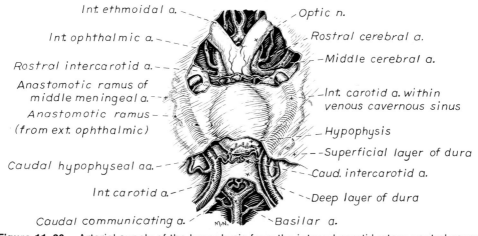

Figure 11–26. Arterial supply of the hypophysis from the internal carotid artery, ventral aspect.

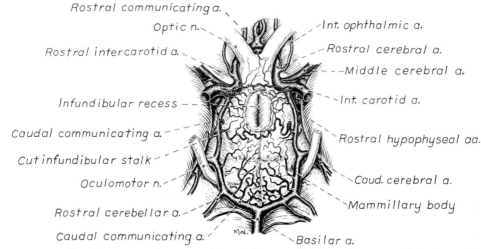

Figure 11–27. The circulus arteriosus cerebri and the superficial arterial supply of the hypothalamus.

of these vessels form anastomoses with adjacent vessels on the surface of the brain. A rich capillary network is found in the cortex, whereas the white matter of the brain has a less abundant supply. The arterial circle ensures the maintenance of constant blood pressure in the terminal arteries and provides alternate routes by which blood can reach the brain.

Several **rostral hypophyseal arteries** *(aa. hypophyseos rostrales)* leave the caudal communicating artery and run over the tuber cinereum to the stalk of the hypophysis. These, with their fellows, supply the major portion of the gland. The pars nervosa, however, is supplied by the caudal hypophyseal artery, a branch of the caudal intercarotid artery.

The **middle cerebral artery** *(a. cerebralis media)* (Figs. 11–30, 11–31) is the largest vessel which supplies the brain. It leaves the internal carotid as a terminal branch about 1 mm. from the origin of the caudal communicating. It lies at first on the rostral perforated substance, where it gives rise to the **choroid artery** *(a. choroidea).* This enters the lateral ventricle at the apex of the piriform gyrus, circles around the cerebral peduncle with the hippocampus, and supplies the vessels of the choroid plexus of the lateral ventricle. The middle cerebral then crosses in front of the piriform lobe and divides into at least two large branches, which supply the whole cortex of the lateral surface of the cerebral

hemisphere. The vessels follow the sulci in some places, and run over the gyri in others. Terminal **cortical branches** *(rami corticales),* in the form of minute twigs, enter the cortex and richly supply it. The **central branches** *(rami centrales)* leave the middle cerebral near its origin in the form of several branches which supply the basal nuclei and adjacent tracts. The middle cerebral artery anastomoses with the rostral and caudal cerebral arteries.

The **rostral cerebral artery** *(a. cerebralis rostralis)* (Fig. 11–32) arises lateral to the optic chiasm and runs dorsal to the optic nerve in rostromedial direction. On reaching the longitudinal fissure, it unites with its fellow. This side-to-side union of right and left rostral cerebral arteries is usually about 2 mm. long, after which the two vessels separate. In some specimens there is an arterial bridge rather than a broad union connecting the right and left vessels. When an arterial bridge is present, it is called the **rostral communicating artery** *(a. communicans rostralis).* The rostral cerebral artery runs dorsally to the genu of the corpus callosum, turns backward along the corpus callosum, and anastomoses with the caudal cerebral, which comes into the longitudinal fissure from behind. Numerous tortuous and freely branching vessels leave the dorsal and rostral surfaces of the rostral cerebral artery. These extend dorsally to the lateral sulcus, where they anastomose with the middle cerebral.

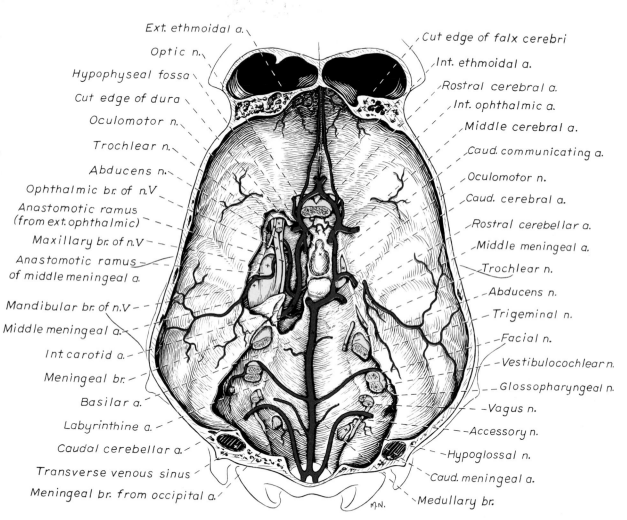

Figure 11–28. Dorsal aspect of the base of the skull showing arteries and nerves. The dura is partially removed on the left side, opening the cavernous sinus.

The area of their distribution to the dorsal surface of the hemisphere is largely rostral to the lateral sulcus. The rostral cerebral artery, like the middle vessel, not only supplies the cortex with its cortical branches, but also sends branches into the medullary substance. Farther ventrally, toward the olfactory bulbs, they are confined to the longitudinal fissure.

The **internal ophthalmic artery** (*a. ophthalmica interna*) (Fig. 11–22) is homologous with the ophthalmic of man. It is less than 0.5 mm. in diameter, and leaves the rostral cerebral artery. It follows the dorsal surface of the optic nerve through the optic canal, and may be double. As the artery travels rostrolaterally with the optic nerve, it passes from the dorsal surface to the medial surface of the nerve. At a location approximately 15 mm. caudal to the bulbus oculi, the internal ophthalmic artery anastomoses with the external ophthalmic artery. From the anastomosis of the two vessels three or four ciliary arteries and the central artery of the retina arise. Their distribution is discussed in connection with the description of the external ophthalmic artery.

The **internal ethmoidal artery** (*a. ethmoidalis interna*) (Fig. 11–24) is a small artery which arises from the ventral part of the rostral cerebral artery and runs toward the cribriform plate. It lies near the attached portion of the falx cerebri, where it parallels the medial olfactory stria. Upon reaching the most rostral portion of the ethmoidal fossa, it anastomoses with the ventral branch of the external ethmoidal artery, forming a rete. Most of the branches of the internal ethmoidal artery pass through the cribriform plate to supply the ethmoturbinates and the nasal septum. Branches on the nasal septum anastomose with the middle septal branch of the sphenopalatine as well as with the rostral septal branch of the major palatine arteries.

SUBCLAVIAN ARTERY

The **subclavian artery** (*a. subclavia*) (Figs. 11–35, 11–36) is the intrathoracic portion of the parent vessel to each pectoral limb. It arises on the left side from the arch of the aorta and on the right side as a terminal branch of the brachiocephalic trunk. It is continued at the cranial border of the first rib on each side by the axillary artery. The name

subclavian implies that the vessel lies under the clavicle. This is not the case in quadrupeds, because the thorax is laterally compressed so that the clavicle, even when it is well developed, lies ventrolateral to the thoracic inlet. The right subclavian artery arises medial to the first right intercostal space and is about 2 cm. long. The left subclavian artery arises medial to the left third intercostal space and is about 6 cm. long. Since the four arteries which arise from each subclavian have similar origins and distributions, only a single description of them will be given. All arise medial to the first rib or first intercostal space.

The **vertebral artery** (*a. vertebralis*) (Figs. 11–30, 11–34) is the first branch of the subclavian. It arises from the dorsal surface on the ventrolateral side of the trachea. As it ascends to the transverse foramen of the sixth cervical vertebra, it crosses the trachea obliquely on the right side (the trachea and esophagus on the left side), and the longus colli and the lateral surface of the transverse process of the sixth cervical vertebra on each side. Near the thoracic inlet, on each side, a small muscular branch supplies the peritracheal fascia and, running forward, terminates in the caudal part of the longus capitis. The vertebral artery, before entering the transverse canal at the sixth cervical vertebra and at every intervertebral space cranial to this, sends dorsal and ventral **muscular branches** (*rami musculares*) into the adjacent musculature. The dorsal branches are distributed to the scalenus, intertransversarius colli, serratus ventralis cervicalis, and omotransversarius. The ventral branch supplies mainly the longus capitis and longus colli, although some twigs go to the brachiocephalicus and sternocephalicus. Arising as a rule separately, but occasionally from a short common trunk, the dorsal and ventral branches follow the dorsal and ventral divisions of the corresponding spinal nerve. They are accompanied by satellite veins. According to Wisnant et al. (1956), four or five anastomoses occur between the muscular branches of each vertebral artery and the costocervical artery. These workers indicate that a secondary anastomosis exists between the vertebral and superficial cervical arteries. These combined anastomoses are large enough to sustain life in the majority of dogs when both the vertebral and common carotid

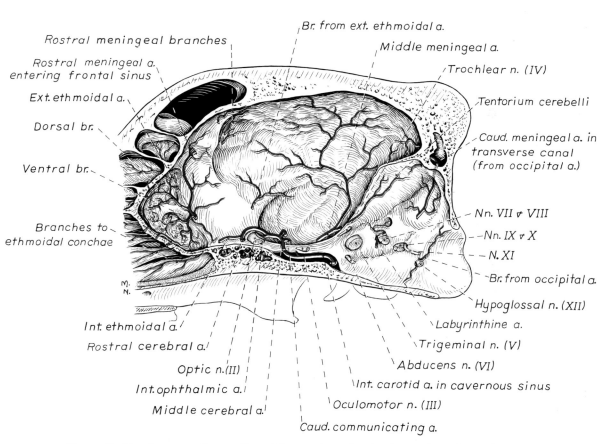

Rostral meningeal branches

Br. from ext. ethmoidal a.

Middle meningeal a.

Rostral meningeal a. entering frontal sinus

Trochlear n. (IV)

Ext. ethmoidal a.

Tentorium cerebelli

Dorsal br.

Caud. meningeal a. in transverse canal (from occipital a.)

Ventral br.

Nn. VII v VIII

Nn. IX v X

N. XI

Branches to ethmoidal conchae

Br. from occipital a.

Int. ethmoidal a.

Hypoglossal n. (XII)

Rostral cerebral a.

Labyrinthine a.

Optic n.(II)

Trigeminal n. (V)

Int. ophthalmic a.

Abducens n. (VI)

Middle cerebral a.

Int. carotid a. in cavernous sinus

Oculomotor n. (III)

Caud. communicating a.

Figure 11–29. A paramedian section of the cranium, showing internal arteries and nerves.

arteries are ligated bilaterally at the base of the neck.

From the medial surface of the vertebral artery, usually opposite the muscular branches, arise the first seven cervical **spinal branches** *(rami spinales).* These enter the spinal canal at each of the first seven cervical intervertebral foramina. Within the spinal canal, each divides into a small dorsal and a slightly larger ventral branch. The ventral branches are all united through the medium of the **ventral spinal artery** *(a. spinalis ventralis)* (Fig. 11–30). This is an unpaired vessel which lies at the ventral median fissure and extends along the length of the spinal cord. It sends segmental twigs into the ventral median fissure to the gray matter of the spinal cord. Other branches, lying in the pia, partially encircle the spinal cord and supply the ventral and lateral white matter. The dorsal branches of the spinal arteries follow the dorsal nerve root to the spinal cord, where they are dissipated without a continuous dorsolateral trunk being formed. The largest spinal ramus is usually the third cervical, but occasionally the fourth cervical spinal branch equals it in size. The vertebral artery divides unequally into a large dorsal and a small ventral branch (Fig. 11–34). The ventral branch anastomoses with the small cervical branch of the occipital artery ventral to the wing of the atlas. The larger dorsal branch, at the transverse foramen of the atlas, anastomoses with the descending branch of the occipital. The vertebral artery enters the spinal canal by passing through the lateral foramen of the atlas. (This portion of the vertebral artery was formerly called the cerebrospinal artery.) (Fig. 11–30). It perforates the dura and arachnoid and divides into cranial and caudal branches, which anastomose to form the basilar artery. The **basilar artery** *(a. basilaris)* runs along the ventral surface of the brain stem and is the largest source of blood to the brain via the circulus arteriosus cerebri (Anderson and Kubicek 1971).

The **costocervical trunk** *(truncus costocervicalis)* (Figs. 11–35, 11–36) arises from the subclavian artery 5 to 10 mm. peripheral to the origin of the vertebral artery. Since it courses dorsally and the vertebral courses cranially, the costocervical on the left side crosses first the lateral surface of the vertebral artery, then the esophagus; on the right it crosses the trachea. On either side the vessel crosses the longus capitis muscle and terminates as it enters the first intercostal space by dividing into the small, caudally running a. intercostalis suprema and the larger, dorsally running a. cervicalis profunda. On either side the vessel lies largely medial to the first rib and has only one collateral branch.

The **dorsal scapular artery** *(a. scapularis dorsalis)* (Fig. 11–36), formerly known as the transverse artery of the neck, arises from the cranial surface of the costocervical trunk, at an acute angle, at about the middle of the medial surface of the first rib. It runs mainly dorsally and leaves the thoracic cavity in front of the first rib. From its initial part it sends at least one large branch caudally into the thoracic part of the serratus ventralis and mainly two or three smaller branches cranially into the cervical part of the serratus ventralis. The dorsal scapular artery at the proximal end of the first rib inclines dorsocaudally, crosses the lateral surface of the first costotransverse joint, and arborizes extensively in the dorsal part of the thoracic portion of the serratus ventralis. It gives origin to the eighth cervical spinal branch as it passes the intervertebral foramen. In some specimens this branch arises from that part of the vessel which supplies the thoracic part of the serratus ventralis. During its course, it obliquely crosses the deeply lying deep cervical artery.

The **deep cervical artery** *(a. cervicalis profunda)* is the dorsocranially extending terminal branch of the costocervical trunk. It leaves the thorax at the proximal end of the first intercostal space. A medium-sized vessel usually leaves the parent artery here and, extending dorsally, arborizes mainly in the semispinalis capitis in the region of the withers. The main part of the deep cervical runs craniomedially to the median plane, supplying along its course the deep structures of the neck, particularly the semispinalis capitis, multifidus cervicis, longissimus capitis, spinalis et semispinalis thoracis et cervicis, and the terminal fasciculus of the thoracic portion of the longissimus. It anastomoses with the dorsal muscular branches of the vertebral artery and, in the cranial part of the neck, with the descending branch of the occipital artery. It gives origin to the first thoracic spinal branch.

The **thoracic vertebral artery** *(a. vertebra-*

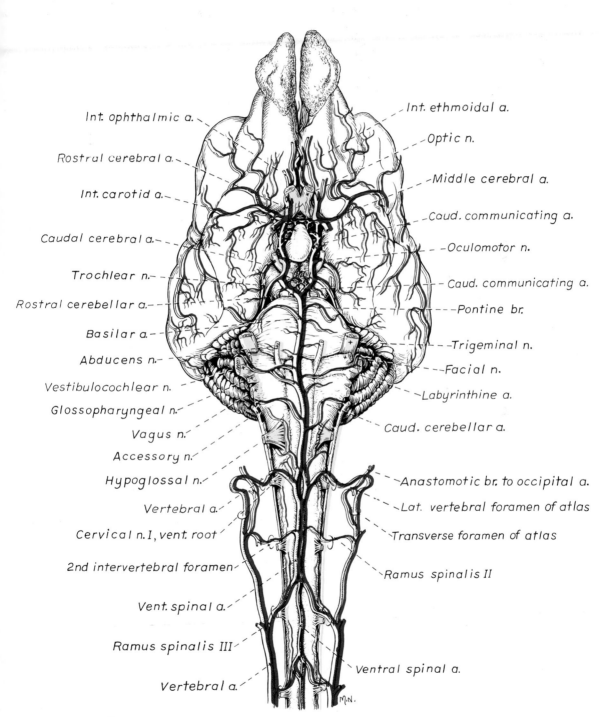

Figure 11–30. Arteries of the brain and cervical spinal cord, ventral aspect.

lis thoracica) leaves the costocervical trunk in the proximal end of the first intercostal space. It extends caudally to the third and occasionally the fourth intercostal space, where it anastomoses with the dorsal (aortic) intercostal artery of that space. It passes through the costotransverse foramen above the neck of the rib and therefore is not homologous to the supreme intercostal artery of other domestic animals. However, a supreme intercostal artery ventral to the rib is sometimes seen in the dog together with the thoracic vertebral artery. Usually the vessel is smallest in the third intercostal space, and therefore the anastomosis may be regarded to be at this site, but this is variable. Caudal to each of the ribs it crosses, it sends a small dorsal intercostal vessel ventrally, which anastomoses with the intercostal vessels from the first two or three intercostal spaces and with the ventral intercostal arteries of the internal thoracic in the third or fourth intercostal spaces, or both. The small second and third, and occasionally the fourth, thoracic spinal branches arise from the thoracic vertebral.

In addition to the three main branches of the costocervical trunk, smaller vessels exist. A constant branch, the **first intercostal artery** *(a. intercostalis prima)* leaves the costocervical at its origin and, extending under the pleura covering the first two or three intercostal spaces and the intervening ribs, supplies the principal intercostal vessels of these spaces. The dorsal portions of its intercostal branches anastomose with the smaller dorsal intercostal arteries from the thoracic vertebral. The ventral portions anastomose with the ventral intercostal arteries from the internal thoracic.

Small branches leave the costocervical trunk near its origin and supply the adjacent musculature. Usually a small ramus accompanies the common carotid artery a short distance up the neck. The costocervical trunk and its branches are accompanied by satellite veins.

The **superficial cervical artery** *(a. cervicalis superficialis)* (Figs. 11–36, 11–37) is more nearly homologous with the thyrocervical trunk of man than is any other artery. It arises from the cranial surface of the subclavian medial to the first rib and opposite the origin of the internal thoracic artery. It is a long, meandering artery which lies in the

angle between the shoulder and the neck. It lies dorsal to the pectoral, brachiocephalic, and omotransverse muscles, and ventral to the brachial plexus. It has five named branches in addition to several small muscular twigs to the muscles which lie adjacent to it.

The **deltoid branch** *(ramus deltoideus),* formerly called descending branch, arises from the superficial cervical artery about 3 cm. from its origin. Occasionally the deltoid branch arises from the internal thoracic artery instead of from the superficial cervical. It runs distolaterally on the brachium in the groove between the pectoral muscles, which bound it caudally, and the brachiocephalicus, which bounds it cranially and partly covers it. Ite ends in the distal third of the brachium in either the superficial pectoral brachiocephalicus, or biceps brachii. Peripheral to the origin of the descending branch, the superficial cervical sends one or more branches to the muscles which lie ventral to the trachea.

The **suprascapular artery** *(a. suprascapularis)* leaves the caudal side of the superficial cervical about 2 cm. distal to the origin of the deltoid branch. Accompanied by the suprascapular nerve, it goes through the triangular space bounded by the subscapularis, supraspinatus, and pectoralis profundus. On reaching the neck of the scapula, it divides into a large lateral and a small medial branch. The lateral branch passes under the supraspinatus to the lateral surface of the scapula, on which it ramifies. It supplies the supraspinatus and sends one large and several small nutrient arteries into the bone. Passing across the distal end of the spine of the scapula, it sends branches to the infraspinatus, teres minor, and the shoulder joint. Near the caudal border of the scapula, it anastomoses with the circumflex scapular artery. The medial branch of the suprascapular passes between the subscapular muscle and the medial surface of the neck of the scapula. It supplies both of these as well as a part of the shoulder joint. It ends in an anastomosis with the circumflex scapular artery.

The **ascending cervical artery** *(a. cervicalis ascendens)* leaves the superficial cervical peripheral to the origin of the suprascapular. Frequently the artery is double. The profusely branching ascending cervical is considerably smaller than the suprascapular.

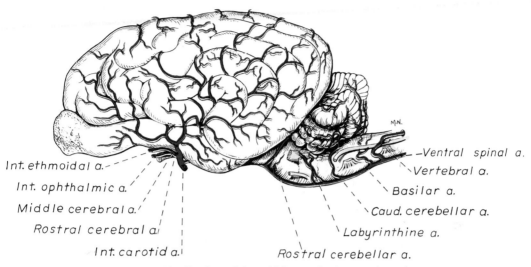

Int. ethmoidal a.

Int. ophthalmic a.

Middle cerebral a.

Rostral cerebral a.

Int. carotid a.

—Ventral spinal a.

Vertebral a.

Basilar a.

Caud. cerebellar a.

Labyrinthine a.

Rostral cerebellar a.

Figure 11–31. Distribution of the middle cerebral artery, lateral aspect.

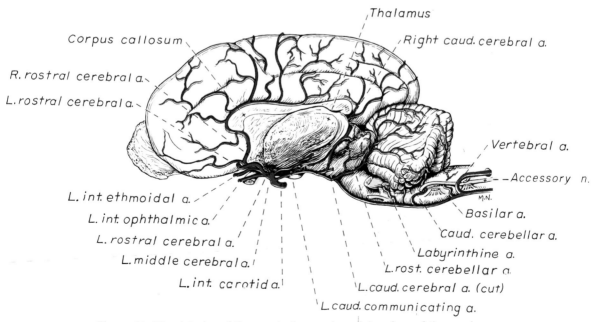

Thalamus

Corpus callosum

Right caud. cerebral a.

R. rostral cerebral a.

L. rostral cerebral a.

Vertebral a.

Accessory n.

L. int. ethmoidal a.

L. int. ophthalmic a.

L. rostral cerebral a.

L. middle cerebral a.

L. int. carotid a.

Basilar a.

Caud. cerebellar a.

Labyrinthine a.

L. rost. cerebellar a.

L. caud. cerebral a. (cut)

L. caud. communicating a.

Figure 11–32. Arteries of the cerebellum and medial surface of the cerebrum.

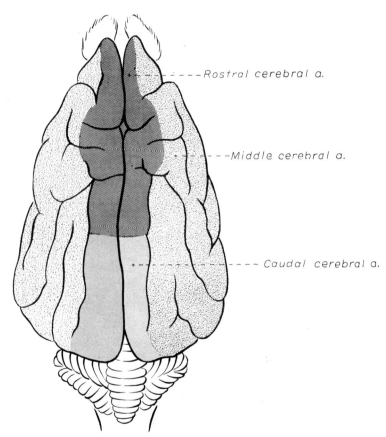

Figure 11–33. Areas supplied by the cerebral arteries, dorsal aspect.

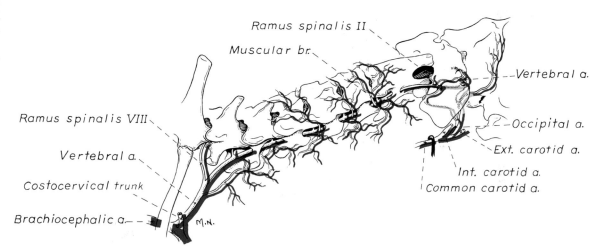

Figure 11–34.   The vertebral artery in relation to the cervical vertebrae, lateral aspect.

It courses cranially, medial to the brachiocephalicus and lateral to the scalenus. It supplies the sternocephalicus, the cervical portions of the brachiocephalicus, rhomboideus, omotransversarius, scalenus, and the prescapular lymph nodes. In the cranial half of the neck, its terminal branches are distributed chiefly to the omotransversarius and brachiocephalicus. Some branches anastomose with the cervical branch of the caudal auricular artery.

The **supraspinous artery** (*a. supraspinatus*) is given off the superficial cervical 1 cm. or less peripheral to the origin of the ascending cervical, and is usually larger than the latter. The supraspinous artery circles around the cranial border of the supraspinatus and goes into its lateral surface, being largely distributed by ramifying proximally through it. A few terminal branches, however, reach as far as the infraspinatus. In some specimens a small branch extends distally over the tendon of the supraspinatus and the major tubercle of the humerus, and anastomoses with the deltoid branch. Under the brachial part of the brachiocephalicus deep branches anastomose with the suprascapular artery.

The **prescapular branch** (*ramus prescapularis*) (Fig. 11–37) is the terminal part of the superficial cervical artery. It is a continuation of the parent artery after the supraspinous artery has been given off. It runs in the space between the shoulder and neck, caudal to the prescapular lymph nodes, which it supplies. It reaches the ventrocranial border of the cervical part of the trapezius. At this point it divides into an ascending and a descending branch.

The ascending branch usually becomes superficial, sending many branches dorsocranially into the superficial fascia and the cutaneous muscles which cover the cleidocervicalis. This branch varies greatly in development; when it is fully developed, it may anastomose with the cervical branch of the caudal auricular artery in the cranial third of the neck. It may remain relatively deep, supplying the superficial muscles of the neck.

The deltoid branch passes under the cervical part of the trapezius, to which it sends many branches. It terminates in the muscle near the cranial angle of the scapula. The branches of the superficial cervical are accompanied by satellite veins.

The **internal thoracic artery** (*a. thoracica interna*) (Figs. 11–36, 11–38) leaves the caudoventral surface of the subclavian opposite the origin of the superficial cervical. It runs caudoventrally in a narrow, lateral pleural plica from the precardial mediastinum to the craniomedial border of the transversus thoracis. Lying parallel to the sternum, it passes under the transversus thoracis and runs caudally above the sternal ends of the costal cartilages and the intervening interchondral spaces. It ends just inside the costal arch at the thoracic outlet by dividing into the small musculophrenic and the large cranial epigastric artery.

The **pericardiacophrenic artery** (*a. pericardiacophrenica*) (Fig. 11–39) is a small vessel which leaves the caudal side of the internal thoracic artery near its origin and runs with the phrenic nerve to the pericardium. In addition to supplying the precardial, intrathoracic part of the phrenic nerve, the left vessel may send one or more branches to the precardial mediastinum and the thymus, when this is well developed. At the level of the heart on both sides the pericardiacophrenic artery anastomoses with the branch from the musculophrenic artery which courses cranially on the nerve from the diaphragm. Twigs leave the vessel to supply the pericardium. A ventral bronchial branch may course to the root of the left lung.

The main **thymic branches** (*rami thymici*) usually leave the precardial parts of the internal thoracic artery as it passes through the thymus. Usually a single thymic branch supplies each lobe, but more than one vessel may be present.

Bronchial branches may leave the left pericardiacophrenic artery, or they may come from the internal thoracic arteries directly (Berry, Brailsford, and Daly 1931). They go to the roots of the lungs and furnish the minor blood supply to the bronchi, bronchial lymph nodes, and connective tissue. They are frequently absent.

The **mediastinal branches** (*rami mediastinales*) supply the ventral part of the mediastinum. Usually two to four branches run directly into the precardial mediastinum from the internal thoracic arteries. Those to the cardiac and postcardial mediastinum perforate the origin of the overlying transverse thoracic muscle and extend vertically to either the pericardium or the diaphragm.

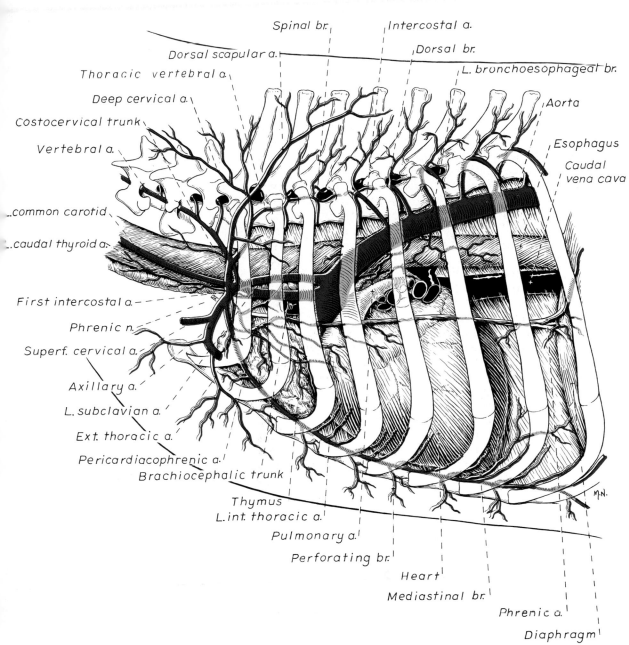

Figure 11-35. Arteries of the left thorax.

Coming from the various phrenic arteries are mediastinal branches which extend forward into the ventral part of the mediastinum. Other twigs from the phrenic arteries ramify in the plica venae cavae.

The **perforating branches** *(rami perforantes)* are straight, short, ventrally directed branches which leave the ventral surface of the internal thoracic artery. One is present in each interchondral space except the first and the last (eighth), and occasionally the seventh. These lie close to the lateral surfaces of the sternebrae and give off **sternal branches** *(rami sternales)* to them. The perforating branches also supply the internal intercostal and pectoral musculature adjacent to them. They are continued subcutaneously near the sternum as **ventral cutaneous branches** along with their satellite veins and comparable nerves. The twigs which supply the medial portions of the thoracic mammary glands are called **mammary branches** *(rami mammarii)*. These, present only when the glands are developed, come from the fourth, fifth, and sixth vessels.

The **ventral intercostal branches** *(rr. intercostales ventrales)* (Fig. 11–38) usually are double for each of the interchondral spaces, starting with the second and ending with the eighth. Starting with the caudal artery of the eighth space, all remaining arteries come from the musculophrenic artery. Those from the ventral surface of the internal thoracic arise singly, so that a small artery lies on each side of the costal cartilages, except the first, which has a delicate single artery. The artery caudal to the cartilage is slightly larger than the one cranial to it, and is accompanied by the intercostal nerve, in addition to the satellite vein. These double arteries anastomose with each other across the ribs. The ventral intercostal artery lying caudal to the rib anastomoses with the dorsal intercostal artery. The ventral intercostal artery lying cranial to the rib anastomoses with the collateral branch of the dorsal intercostal. These vessels lie, for the most part, in the endothoracic fascia closely under the pleura. Occasionally, some fibers from the internal intercostal muscles cover them. They supply the ventral part of the costal pleura, the adjacent intercostal musculature, and the costal cartilages.

The **musculophrenic artery** *(a. musculophrenica)* (Fig. 11–38) is the smaller, lateral, terminal branch of the internal thoracic. It arises under the caudal part of the transverse thoracic muscle opposite the eighth interchondral space close to the sternum. It runs caudodorsolaterally, in the angle formed by the diaphragm and the lateral thoracic wall, where it lies in a small amount of fat covered by the pleura. After it has traveled about one-fourth of the length of the costal arch, it perforates the diaphragm and comes to lie under the peritoneum. It ascends on the inner surface of the costal arch by following the margins of the interlocked digitations of attachments of the diaphragm and the transverse abdominal muscle. Along its course it sends the **ventral intercostal branches** *(rr. intercostales ventrales)* dorsally in the caudal part of the eighth interchondral space, and two each for spaces 9 and 10. The single terminal branch of the musculophrenic anastomoses with the eleventh dorsal intercostal artery caudal to the diaphragm. These ventral intercostal branches form feeble anastomoses with the ventral parts of the eighth, ninth, and tenth dorsal intercostal arteries. Numerous small branches leave the musculophrenic to supply the muscular periphery of the diaphragm. Some of these end in the postcardial ventral mediastinum and plica venae cavae. A small branch runs forward on both sides with the postcardial portion of the phrenic nerve and anastomoses with the small pericardiacophrenic artery. Fewer branches ramify in the adjacent abdominal wall. Both sets of branches anastomose with the phrenicoabdominal artery, and each is accompanied by a satellite vein.

The **cranial epigastric artery** *(a. epigastrica cranialis)* (Fig. 11–36) is the larger, medial terminal branch of the internal thoracic. It arises dorsal to the eighth interchondral space lateral to the sternum under the m. transversus thoracis. It perforates the diaphragm, and in the angle between the costal arch and the xiphoid process it gives rise to the cranial superficial epigastric artery. It continues to run on the deep surface of the rectus abdominis as the cranial epigastric artery. This was formerly called the cranial deep epigastric artery.

At a transverse level through the umbilicus, many of its branches enter the rectus abdominis and shortly thereafter anastomose with the cranially running terminal branches of the caudal epigastric artery. It is

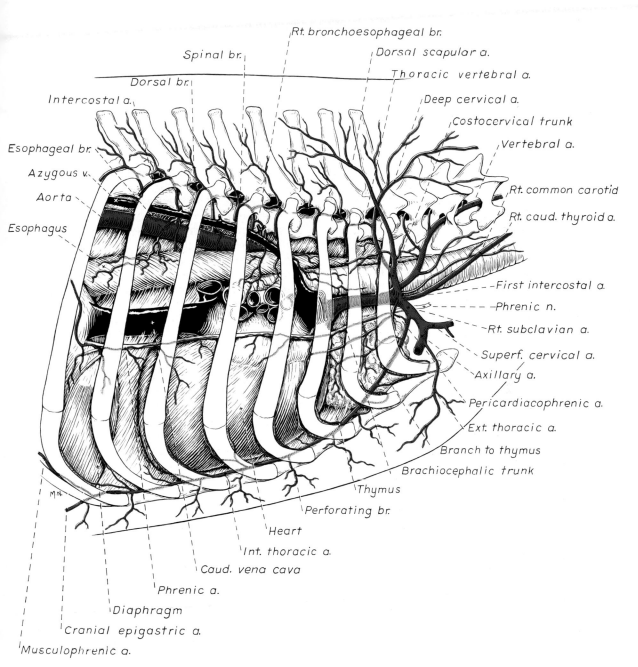

Figure 11–36. Arteries of the right thorax.

the primary blood supply to the middle portion of the rectus abdominis. It is accompanied by its laterally lying satellite vein.

The ventral abdominal wall has two arterial channels on each side of the median plane which connect the thoracic circulation with that of the pelvic limbs. One of these is superficial, and the other is deep. The deep epigastric vessels are always well developed, but the superficial ones reach their maximum size only during the height of lactation. The deep vessels anastomose feebly with each other across the linea alba. As it passes through the diaphragm, it may supply a sizeable branch.

The **superficial cranial epigastric artery** (*a. epigastrica cranialis superficialis*) in a lactating bitch may be larger than the deep artery. It runs through the rectus abdominis and its sheath and enters the subcutaneous tissue between the caudal thoracic and the cranial abdominal mammae. It sends most of its branches caudolaterally and is the chief supply to the cranial abdominal mamma. Caudal to this gland, several of its many branches anastomose with the end-branches of the cranially running superficial caudal epigastric artery.

Arteries of the Thoracic Limb

AXILLARY ARTERY

The **axillary artery** (*a. axillaris*) (Fig. 11–41) is a continuation of the subclavian artery and extends from the cranial border of the first rib to the distal border of the conjoined tendon of the teres major and latissimus dorsi muscles. At first it lies lateral to its satellite vein, then cranial to it. In general, the nerves from the brachial plexus lie lateral to the axillary vessels. (The musculocutaneous nerve is cranial, the radial is lateral, and the median-ulnar trunk is caudal. The axillary vein, at its termination, lies directly medial to the median-ulnar trunk.) The axillary artery has four primary branches: the external thoracic, lateral thoracic, subscapular, and cranial circumflex humeral arteries.

The **external thoracic artery** (*a. thoracica externa*) is usually the first branch of the axillary. It arises about 1 cm. lateral to the first rib and curves around the craniomedial border of the deep pectoral in company

with the nerve and two satellite veins, supplying the superficial pectoral muscle, to which it is distributed.

The **lateral thoracic artery** (*a. thoracica lateralis*) arises from the caudal surface of the axillary artery about 2 cm. from the cranial border of the first rib. Occasionally its origin is located distal to the large, caudodorsally running subscapular artery. It runs caudally in the axillary fat and crosses the lateral surface of the axillary lymph node, which it supplies. It also supplies an area of the latissimus dorsi ventral to the area supplied by the thoracodorsal artery. Other branches supply the deep pectoral and cutaneous trunci muscles. Its lateral **mammary branches** supply the dorsolateral portions of the cranial and caudal thoracic mammary glands when these are fully developed. The lateral thoracic artery is accompanied by a satellite vein and nerve.

The **subscapular artery** (*a. subscapularis*) (Figs. 11–40, 11–41, 11–42) may be larger than the continuation of the axillary in the true arm. This great size is explained by the fact that the vessel supplies a greater muscular mass in the shoulder and arm than is present in the remainder of the limb. It arises from the caudal surface of the axillary usually just peripheral to the origin of the lateral thoracic artery. It runs obliquely in a dorsocaudal direction between the subscapularis and teres major, and becomes subcutaneous near the caudal angle of the scapula. It has the following principal branches: thoracodorsal, caudal circumflex humeral, circumflex scapular, and collateral radial arteries.

The **thoracodorsal artery** (*a. thoracodorsalis*) is a large artery which leaves the caudal surface of the subscapular less than 1 cm. from its origin. It runs caudally, usually supplying a part of the teres major as it crosses the medial surface of the distal end of the muscle, and terminates in the latissimus dorsi and skin. A satellite vein and nerve accompany the artery.

The **caudal circumflex humeral artery** (*a. circumflexa humeri caudalis*) leaves the lateral surface of the subscapular artery at about the same level as the thoracodorsal and plunges immediately between the head of the humerus and the teres major. It is the principal source of blood to all four heads of the large triceps muscle. The **collateral radi-**

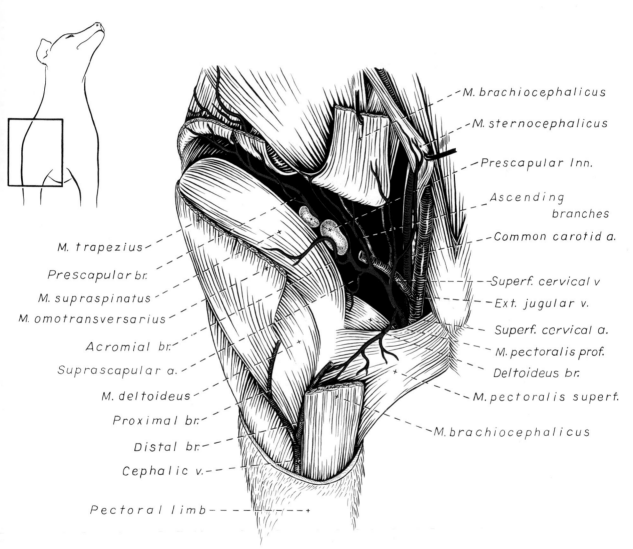

Figure 11–37. Branches of the omocervical artery.

al artery of the humerus leaves the distal surface of the caudal circumflex humeral about 1 cm. from its origin and takes a direct course distally, lateral to the terminal ends of the teres major and latissimus dorsi, and the medial head of the triceps. It lies medial to the accessory head and caudal to the brachial muscle. It may terminate, as the **nutrient artery of the humerus,** by entering the nutrient foramen near the middle of the caudal surface of the bone. In most specimens, as Miller (1952) has pointed out, the collateral radial artery, after sending the nutrient artery into the humerus, continues obliquely distocranially on the brachial muscle and anastomoses with the proximal collateral radial artery above the flexor angle of the elbow joint. During its course it supplies branches to the brachialis and heads of the triceps, which lie along its course. This vessel is accompanied by its small satellite vein and the radial nerve.

The main part of the caudal circumflex humeral artery arborizes extensively in the triceps as ascending and descending branches. The caudal part of the shoulder joint capsule, infraspinatus, teres minor, and coracobrachialis also receive twigs from this vessel. Some of the proximal branches anastomose with the small circumflex scapular artery, and some of the descending branches anastomose with the deep brachial artery. The anastomosis between the caudal and cranial circumflex humeral arteries in man is classic; in dogs, there is only a small union between the homologous arteries. The main stem of the caudal circumflex leaves the triceps mass and enters the deep face of the deltoideus. Other branches extend between the deltoideus laterally and the long and lateral heads of the triceps medially, to appear subcutaneously near the middle of the lateral surface of the brachium. Some of these branches extend proximally to anastomose with the descending branch of the superficial cervical; others run distally and anastomose with the superficial brachial artery. The caudal circumflex humeral artery is accompanied through the fleshy part of the brachium by its satellite vein and the axillary nerve.

The **circumflex scapular artery** (*a. circumflexa scapulae*) is a small vessel which leaves the cranial surface of the subscapular artery and, extending obliquely dorsocran-

ially between the subscapularis medially and the long head of the triceps laterally, reaches the caudal border of the scapula near its middle. Here it divides into a medial and a lateral branch. The medial branch ramifies in the periosteal part of the subscapularis, and the lateral branch arborizes in a similar manner in the infraspinatus. Minute twigs enter the bone from both medial and lateral parts.

Distal to the origin of the circumflex scapular artery, there are muscular branches to the proximal ends of the teres major, subscapularis, infraspinatus, deltoideus, and latissimus dorsi. A large patch of skin covering the region over and caudal to the caudal angle of the scapula is supplied by the terminal cutaneous branches. The subscapular artery and its branches are accompanied by satellite veins.

The **cranial circumflex humeral artery** (*a. circumflexa humeri cranialis*) usually arises from the medial surface of the axillary proximal to the origin of the subscapular. It may arise from the axillary, distal to the origin of the subscapular, or from the subscapular artery itself. It is a small vessel which curves around the neck of the humerus under the tendon of origin of the biceps brachii after crossing the insertion of the coracobrachialis. A relatively large branch supplies the proximal end of the biceps, and smaller twigs go to the coracobrachialis and to the conjoined teres major and latissimus dorsi. A branch extends proximally to supply the cranial part of the joint capsule. In the region of the major tubercle of the humerus, the two circumflex humeral vessels join with each other as well as with the suprascapular above and with the descending branch of the superficial cervical below. Occasionally the supraspinous also joins in this anastomosis.

BRACHIAL ARTERY

The **brachial artery** (*a. brachialis*) (Figs. 11–40, 11–41) is a continuation of the axillary. It begins at the distal border of the conjoined tendons of the teres major and latissimus dorsi and becomes the median artery in the antebrachium, where it continues into the forepaw as the main blood supply. The brachial artery in the brachium lies caudal to the musculocutaneous nerve and

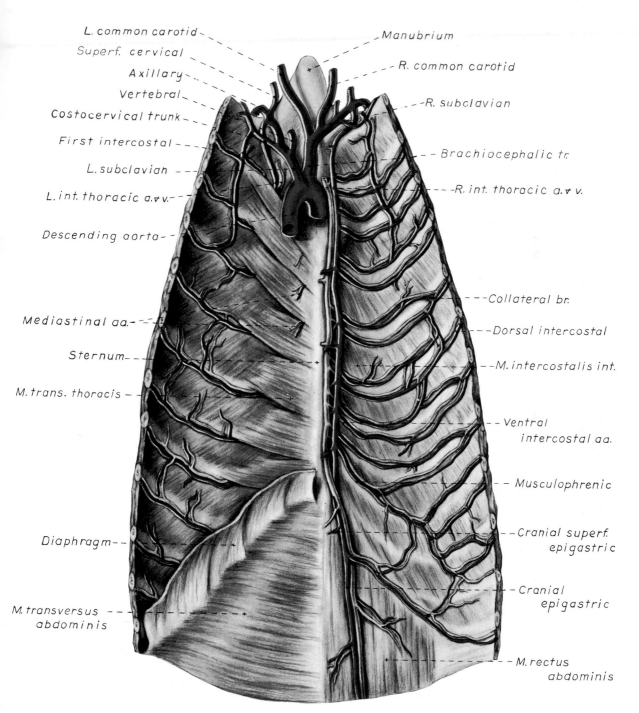

L. common carotid
Superf. cervical
Axillary
Vertebral
Costocervical trunk
First intercostal
L. subclavian
L. int. thoracic a. & v.
Descending aorta
Mediastinal aa.
Sternum
M. trans. thoracis
Diaphragm
M. transversus abdominis

Manubrium
R. common carotid
R. subclavian
Brachiocephalic tr.
R. int. thoracic a. & v.
Collateral br.
Dorsal intercostal
M. intercostalis int.
Ventral intercostal aa.
Musculophrenic
Cranial superf. epigastric
Cranial epigastric
M. rectus abdominis

Figure 11–38. The internal thoracic arteries, dorsal aspect, in relation to the sternum.

biceps muscle, medial to the medial head of the triceps and humerus, and cranial to the median nerve and axillary vein. The deep pectoral muscle and a nerve connecting the musculocutaneous and median nerves form the medial boundary of the brachial artery in the brachium. It crosses the distal half of the humerus obliquely and becomes the median artery as it passes under the pronator teres muscle. The collateral branches of the brachial artery as it lies in the arm are the deep brachial, bicipital, collateral ulnar, superficial brachial, transverse cubital, recurrent ulnar, and common interosseus.

The **deep brachial artery** (*a. brachialis profunda*) leaves the caudal side of the brachial in the proximal third of the arm. Occasionally the artery is double. The deep brachial enters the medial and long heads of the triceps; a smaller branch enters the medial head of the triceps, and a larger one, after a course of over 1 cm., is distributed to the long head. Within the triceps the relatively small deep brachial anastomoses with the caudal circumflex humeral artery proximally and with the collateral ulnar distally. It is accompanied by a satellite vein. The radial nerve enters the triceps cranial to the artery.

The **bicipital artery** (*a. bicipitalis*) is frequently called the muscular ramus to the biceps. It may be double and usually, when it is double, one branch arises from the proximal collateral radial artery. The bicipital artery, when it is single, usually arises from the medial surface of the brachial at the junction of the middle and distal thirds of the arm. The bicipital artery anastomoses with the muscular branch to the biceps from the cranial circumflex humeral artery. It is accompanied by a satellite vein. It runs distally and enters the distal end of the biceps brachii. It may arise from the proximal collateral radial artery.

The **collateral ulnar artery** (*a. collateralis ulnaris*) (Figs. 11–40, 11–43) arises from the caudal surface of the brachial in the distal third of the arm. Usually the first branch runs proximocaudally to enter the medial surface of the triceps. This may be paired with one branch leaving the brachial directly. Another branch runs distally toward the palmar side of the forearm, with the ulnar nerve below. Usually a strong branch arises

from this vessel proximal to the elbow joint and, running under the medial head of the triceps and anconeus, plunges into the olecranon fossa. It supplies primarily the fat and the pouch of the elbow joint capsule which are located here. The branch which continues distally arborizes in the proximal parts of the flexor muscles of the antebrachium. An anastomosis exists with a proximally extending branch from the accessory interosseous artery between the ulnar and humeral heads of the deep digital flexor. Slightly caudal to the branch which runs with the ulnar nerve is a superficial branch which runs distally across the medial surface of the elbow joint and continues in the subcutaneous tissue of the proximal half of the palmar surface of the antebrachium. It supplies the skin here and anastomoses with a proximally extending subcutaneous twig from the caudal interosseous artery. It is accompanied by a vein and the caudal (palmar) cutaneous antebrachial nerve.

The **superficial brachial artery** (*a. brachialis superficialis*) was formerly called the proximal collateral radial artery. It leaves the cranial surface of the brachial about 3 cm. proximal to the elbow joint and extends obliquely distocraniad to its flexor surface. After it crosses the tendon of the biceps, it gives off a cutaneous branch to the skin of the medial surface of the antebrachium. In the region of the cephalic vein, under which it runs, it gives off a medial branch to the digits and continues as the cranial superficial antebrachial artery.

The **medial branch** (*ramus medialis*) extends from the flexor surface of the elbow joint to the medial part of the forepaw. In its course down the antebrachium it lies on the extensor carpi radialis, where it is bounded laterally by the large antebrachial part of the cephalic vein and medially by the small medial branch of the superficial radial nerve. At the carpus it anastomoses with the dorsal branch of the radial artery before continuing to the dorsomedial part of the metacarpus. The resultant branches of this anastomosis run on the dorsal carpal ligament and there form the medial part of the poorly defined **dorsal rete of the carpus** (*rete carpi dorsale*). The lateral part of this rete is formed by the distal dorsal interosseous artery. The deep dorsal metacarpal arteries, which the dorsal

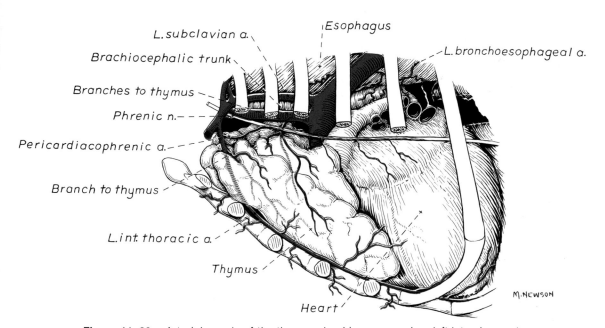

Figure 11-39. Arterial supply of the thymus gland in a young dog, left lateral aspect.

rete of the carpus contributes to the fore-paw, are described under "Arteries of the Forepaw."

The superficial brachial artery continues as the cranial superficial antebrachial artery. It runs transversely, laterally across the distal end of the biceps brachii and, upon emerging from under the cephalic vein, bends distally and accompanies this vein on its lateral side throughout the antebrachium. A small ascending branch, which arises as it turns distally, anastomoses with the descending branch of the superficial cervical artery. The cranial superficial antebrachial artery is somewhat larger than its medial branch, but, like it, is long and sparsely branched. Throughout its antebrachial course, it is flanked medially by the large cephalic vein of the antebrachium and laterally by the lateral branch of the superficial radial nerve. On the proximal part of the metacarpus the artery trifurcates. The three resultant arteries are described under "Arteries of the Forepaw."

The **transverse cubital artery** (*a. transversa cubiti*) (Fig. 11–44) was formerly called the distal collateral radial artery. It arises from the lateral surface of the brachial, about 1 cm. before that vessel runs under the pronator teres. The transverse cubital, which is about as large as the superficial brachial, runs under the distal end of the biceps brachii and brachialis. On reaching the extensor carpi radialis, it breaks up into many branches, most of which supply this muscle. Occasionally a branch runs proximally in company with the radial nerve and anastomoses with that part of the nutrient artery of the humerus which descends beyond the foramen. Besides the branches which go to the extensor carpi radialis, there are twigs which go to the supinator, common digital extensor, and brachialis. Anastomoses with the cranial interosseous and superficial brachial sometimes occur.

The **recurrent ulnar artery** (*a. recurrens ulnaris*) is a small vessel which leaves the caudal side of the median at the proximal end of the radius and extends from the caudal border of the pronator teres into the flexor group of muscles. The artery first runs under the flexor carpi radialis near its origin and contributes to its supply. It continues caudally through the humeral head of the deep digital flexor, which it supplies, and

terminates mainly in the superficial digital flexor. Two traceable anastomoses are present; one of these is with the collateral ulnar artery, as the recurrent ulnar runs proximally over the medial epicondyle of the humerus, and the other is with the palmar antebrachial on the deep surface of the superficial digital flexor. Davis (1941) describes two recurrent ulnar arteries in the dog.

The **common interosseous artery** (*a. interossea communis*) (Fig. 11–45) is the largest branch of the median artery. It is about 3 mm. in diameter and 1 cm. long, as it runs from the lateral surface of the brachial to the interosseous space at the proximal end of the pronator quadratus. This places the artery about 1 cm. distal to the elbow joint. Before entering the interosseous space, it sends one or more muscular twigs cranially into the pronator teres and gives off the large caudally running accessory interosseous artery. Within the interosseous space it terminates by dividing into the caudal interosseous and cranial interosseous arteries.

The **ulnar artery** (*a. ulnaris*) accompanies the ulnar nerve in the antebrachium (Davis 1941). It was formerly called the accessory interosseous in the dog. It can be exposed by separating the humeral from the ulnar head of the deep digital flexor. The large ulnar nerve, which lies closely applied to the lateral border of the deep digital flexor, largely covers the artery. On leaving the common interosseous artery, the ulnar artery courses obliquely distocaudally across the medial surface of the ulna and enters the deep digital flexor. A recurrent branch extends proximally between the radial and ulnar heads of the deep digital flexor and anastomoses with the collateral ulnar artery. The bulk of the artery continues distally, however, in association with the humeral head of the deep surface of the flexor carpi ulnaris. Near the carpus the ulnar and the antebrachial anastomose. The small trunk vessel which results passes distally into the carpal canal, where it usually anastomoses with the caudal interosseous artery. The ulnar supplies largely the ulnar and humeral heads of the deep digital flexor and the corresponding heads of the flexor carpi ulnaris.

The **caudal interosseous artery** (*a. interos-*

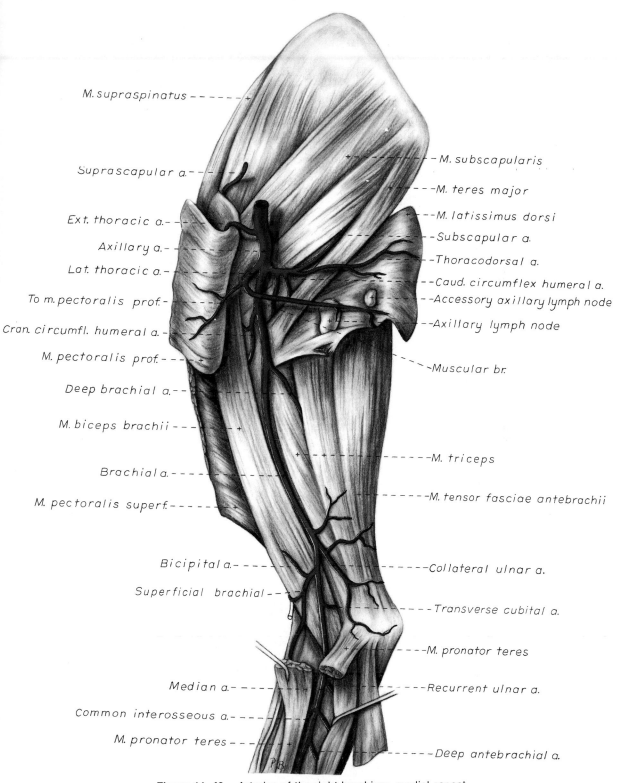

Figure 11–40. Arteries of the right brachium, medial aspect.

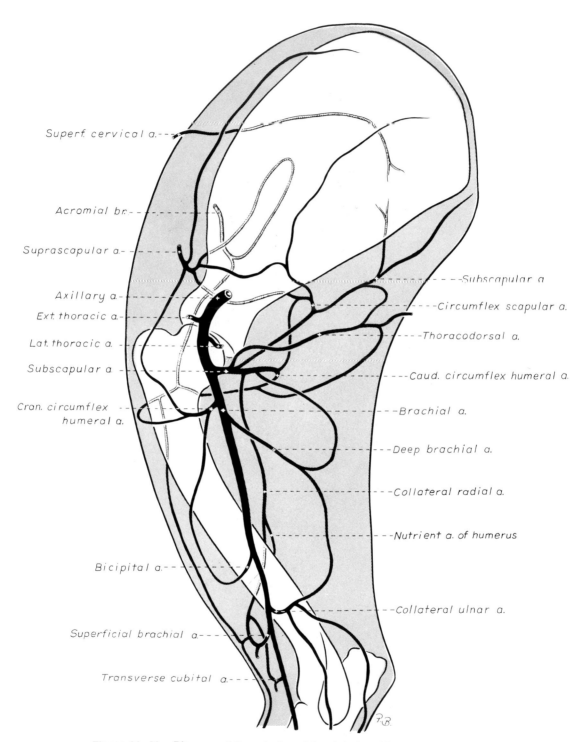

Figure 11–41.   Diagram of the arteries of the right brachium, medial aspect.

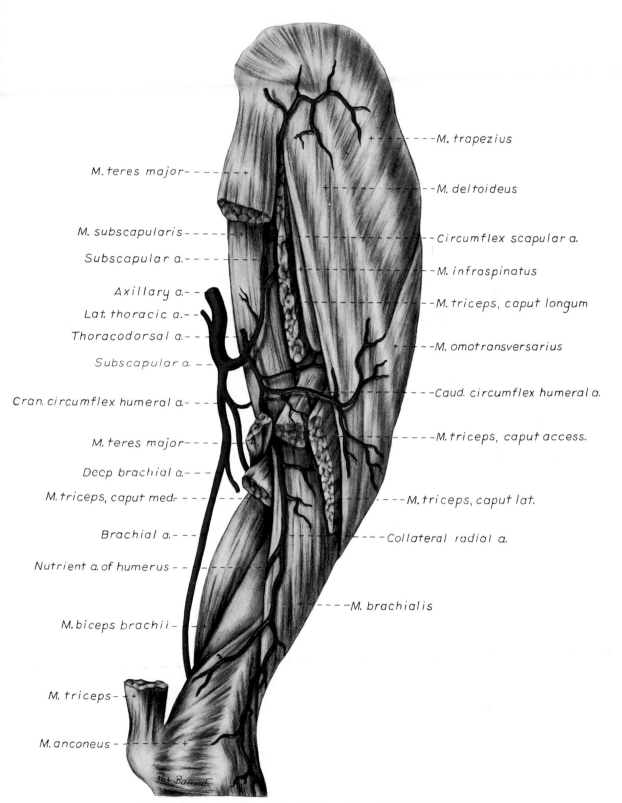

M. teres major

M. subscapularis

Subscapular a.

Axillary a.

Lat. thoracic a.

Thoracodorsal a.

Subscapular a.

Cran. circumflex humeral a.

M. teres major

Deep brachial a.

M. triceps, caput med.

Brachial a.

Nutrient a. of humerus

M. biceps brachii

M. triceps

M. anconeus

M. trapezius

M. deltoideus

Circumflex scapular a.

M. infraspinatus

M. triceps, caput longum

M. omotransversarius

Caud. circumflex humeral a.

M. triceps, caput access.

M. triceps, caput lat.

Collateral radial a.

M. brachialis

Figure 11–42. Arteries of the right brachium, caudolateral aspect.

sea caudalis) lies between the apposed surfaces of the radius and ulna. The pronator quadratus muscle lies on the palmarolateral side of the artery. In its course down the forearm, the artery supplies many small branches to adjacent structures. The pronator quadratus and the ulnar and radial heads of the deep digital flexor receive twigs on the caudal side of the forearm. The abductor pollicis longus, common digital extensor, lateral digital extensor, and extensor pollicis longus et indicis proprius receive branches on the dorsolateral side of the forearm. In the proximal half of the forearm, it forms a feeble anastomosis with the cranial interosseous artery. At the junction of the proximal and middle thirds of the radius the **nutrient artery of the radius** *(a. nutriciae radii)* extends distally into the bone. At a similar location on the ulna, the **nutrient artery of the ulna** *(a. nutriciae ulnae)* extends proximally into the ulna. At the base of the styloid process of the ulna the caudal interosseous artery bifurcates. One of these arteries, the **distal cranial interosseous ramus** *(r. interosseus)* (Fig. 11–45) leaves the dorsal side of the interosseous space proximal to the carpus. From under the abductor pollicis longus it runs to the dorsal carpal ligament and aids in the formation of the lateral part of the dorsal carpal rete. A recurrent branch extends proximally and anastomoses with the ulnar artery on the deep digital flexor.

The **cranial interosseous artery** *(a. interossea cranialis)* continues in the direction of the common interosseous after the caudal interosseous arises. It is a small vessel which emerges from the interosseous space, about 2 cm. distal to the lateral epicondyle of the humerus. It enters the deep surface of the proximal extremities of the extensor carpi ulnaris and the lateral and common digital extensors. Small twigs also go to the pronator quadratus, supinator, and the flexor muscle adjacent to the proximal third of the ulna. The caudolateral part of the elbow joint receives twigs from under the caudal border of the extensor carpi ulnaris. Anastomoses exist between the cranial interosseous artery and the collateral ulnar and caudal interosseous arteries.

The **median artery** *(a. mediana)* (Figs. 11–43, 11–44) is the largest artery of the forearm and was formerly considered to be only an antebrachial part of the brachial artery. It begins after the brachial artery gives rise to the common interosseous artery. The first branch of the median is the deep antebrachial artery, which supplies the muscles on the caudal side of the forearm, and the terminal branches of the median supply the digits. The main branches of the median artery are the **deep antebrachial, radial,** and **superficial palmar arch.**

The **deep antebrachial artery** *(a. antebrachialis profunda)* (Fig. 11–43) arises from the palmar surface of the median about 1 cm. distal to the origin of the common interosseous. It is a branched vessel, about 1 mm. in diameter, which runs distocaudally under the flexor carpi radialis into the deep digital flexor. It supplies the flexor carpi radialis, superficial and deep digital flexors, and the flexor carpi ulnaris. It anastomoses prominently with the recurrent ulnar under the superficial digital flexor and with the ulnar artery under the humeral head of the flexor carpi ulnaris in the distal fourth of the antebrachium. Frequently the small common trunk formed by this anastomosis joins the deep antebrachial in the carpal canal. It traverses the antebrachium in such a way that a series of branches leave the vessel, as it lies in the deep digital flexor, and terminate in the superficial digital flexor. These appear in a linear series of about eight vessels lying 1 to 2 cm. apart. The deep antebrachial artery is accompanied by a branch of the median nerve and a satellite vein.

The distal part of the median artery is the principal source of blood supply to the forepaw. It lies on the medial borders of the radial and humeral heads of the deep digital flexor under the antebrachial fascia and tendon of the flexor carpi radialis. It obliquely crosses the humeral head of the deep digital flexor, which it grooves. Usually a small twig to the medial surface of the carpus is its only branch in the antebrachium. As it passes through the carpal canal with the tendon of the deep digital flexor, it lies lateral to the median nerve. It emerges from the carpal canal in the palmar groove of the deep flexor tendon. A small branch to the carpal pad arises from the median just distal to the carpal canal. Lying between the superficial and deep flexor tendons in the proximal part of the metacarpus, the vessel anastomoses with a small branch of the cau-

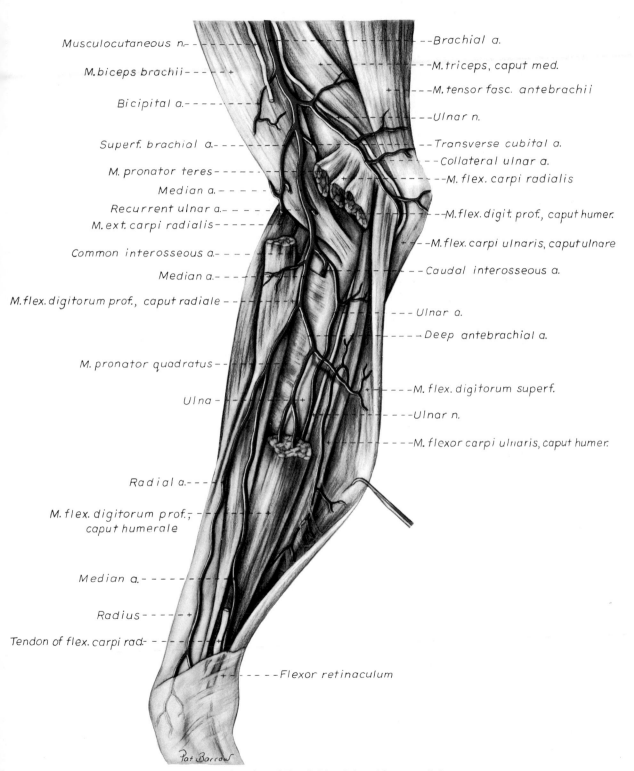

Musculocutaneous n.

M. biceps brachii

Bicipital a.

Superf. brachial a.

M. pronator teres

Median a.

Recurrent ulnar a.

M. ext. carpi radialis

Common interosseous a.

Median a.

M. flex. digitorum prof., caput radiale

M. pronator quadratus

Ulna

Radial a.

M. flex. digitorum prof.,
caput humerale

Median a.

Radius

Tendon of flex. carpi rad.

Flexor retinaculum

Brachial a.

M. triceps, caput med.

M. tensor fasc. antebrachii

Ulnar n.

Transverse cubital a.

Collateral ulnar a.

M. flex. carpi radialis

M. flex. digit. prof., caput humer.

M. flex. carpi ulnaris, caput ulnare

Caudal interosseous a.

Ulnar a.

Deep antebrachial a.

M. flex. digitorum superf.

Ulnar n.

M. flexor carpi ulnaris, caput humer.

Pat Barrow

Figure 11–43. Arteries of the right antebrachium, medial aspect.

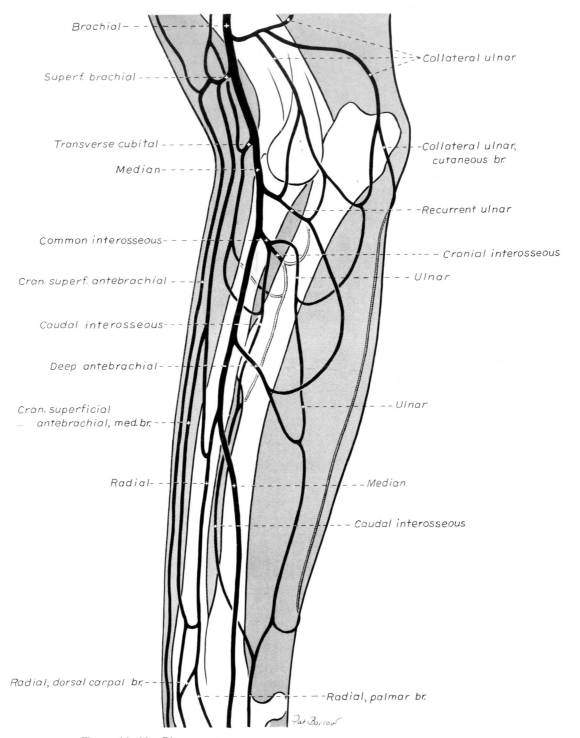

Figure 11-44. Diagram of the arteries of the right antebrachium, medial aspect.

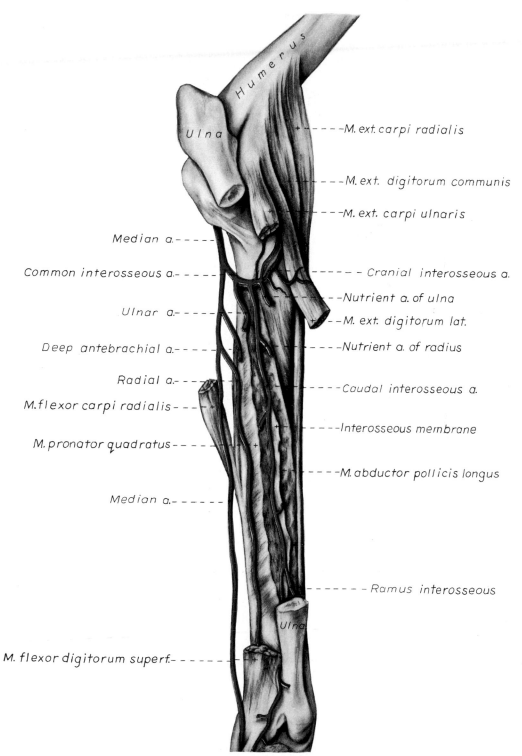

Figure 11–45. Arteries of the right antebrachium, caudolateral aspect. (The shaft of the ulna is removed.)

dal interosseous to form the **superficial palmar arch** (*arcus palmaris superficialis*). This arch is not apparent without close inspection, as the median artery dominates in the formation of the arch as well as in the supply of blood to the paw. Essentially the median artery terminates as three principal palmar metacarpal arteries.

The **radial artery** (*a. radialis*) (Fig. 11–43), from its origin just proximal to the middle of the forearm, runs distally under the aponeurotic origin of the flexor carpi radialis. It closely follows the caudomedial border of the radius in the forearm. At the carpus it divides into palmar and dorsal branches. The **dorsal branch** (*ramus carpeus dorsalis*) supplies the dorsal part of the carpal joint capsule. It contributes to the formation of the dorsal rete of the carpus by sending a branch to anastomose with the medial branch of the superficial brachial artery. The **palmar branch** (*ramus carpeus palmaris*) runs toward the first digit by passing in the superficial part of the transverse carpal ligament. It is a small vessel which anastomoses with the larger caudal interosseous artery in the interosseous musculature of the proximal part of the metacarpus. The **deep palmar arch** (*arcus palmaris profundus*) is formed by this anastomosis.

ARTERIES OF THE FOREPAW

The arteries of the forepaw may be divided into a dorsal and a palmar set, each of which is further divided into a superficial and a deep series. By convention the superficial arteries of the metapodium are named the **dorsal** or **palmar common digital arteries,** whereas the deep arteries are called the **dorsal** or **palmar metacarpal** or **metatarsal arteries.** Digital arteries that arise from the bifurcation of the dorsal and palmar common digital arteries are called **proper digital arteries.** All are small and of minor importance, except for the superficial series of the palmar set, which are the main source of blood supply to the digits and the footpads.

The **dorsal common digital arteries I, II, III, and IV** (*aa. digitales dorsales communes*) (Figs. 11–46, 11–48) are formed by the trifurcation of the cranial superficial antebrachial artery and by a direct continuation of the medial branch of the cranial superficial antebrachial artery into the metacarpus from the carpus. These small vessels first lie on, then between, the tendons of the common digital extensor as both diverge unequally. The distal portions of the main arteries sink into the distal portions of the intermetacarpal spaces and join the dorsal metacarpal arteries. The first artery anastomoses with the corresponding palmar common digital artery.

The **dorsal metacarpal arteries II, III, and IV** (*aa. metacarpeae dorsales*) arise from the distal part of the dorsal rete of the carpus. They are the smallest of all the metacarpal arteries, as they lie in the dorsal grooves between adjacent metacarpal bones. Proximally, they send branches to the deep palmar arch, and distally, communicating branches are sent to the palmar metacarpal arteries. They terminate by anastomosing with corresponding arteries of the superficial series to form the common dorsal digital arteries.

The **dorsal common digital arteries II, III, and IV** are each about 1 cm. long as they terminate opposite the metacarpophalangeal joints. Anastomotic twigs leave them to go to the corresponding palmar common digital arteries. The dorsal common digital arteries, upon reaching the skin which ensheathes the digits, divide into the **axial and abaxial dorsal proper digital arteries II, III, and IV** (*aa. digitales propriae dorsales axialis et abaxialis II, III, et IV*). These go to the dorsal parts of the contiguous sides of adjacent digits and, as branched cutaneous vessels, extend to the claws.

The palmar set of arteries of the forepaw, like the dorsal set, is divided into a superficial and a deep series of arteries, but, unlike the dorsal set, these vessels arise from the superficial and deep palmar arches. The median artery dominates in the formation of the superficial series so completely that the contribution of the caudal interosseous artery, which anastomoses with it to form the **superficial palmar arterial arch** (*arcus palmaris superficialis*), is minor. From this arch arise the relatively large palmar common digital arteries.

The **palmar common digital artery I** (*a. digitalis palmaris communis I*) (Fig. 11–47, 11–48) runs about 1 cm. before entering the space between the first digit and the metacarpus. It is the first branch from the meta-

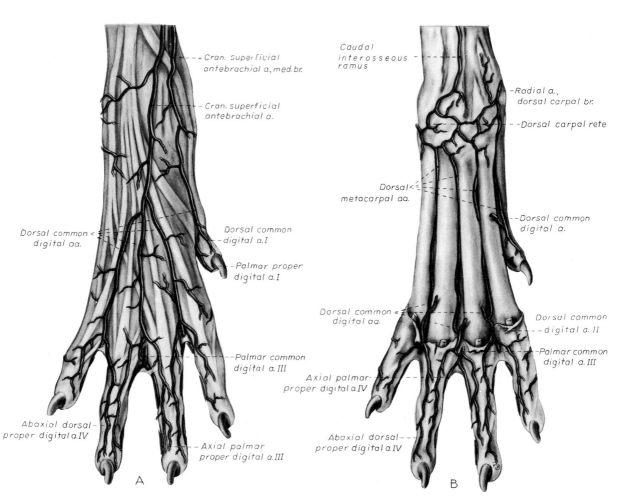

Figure 11–46. Arteries of the right forepaw.
A. Superficial arteries of the right forepaw, dorsal aspect.
B. Deep arteries of the right forepaw, dorsal aspect.

carpal part of the median artery or the superficial palmar arterial arch. It anastomoses with the common digital dorsal artery I. The main part of the median artery or the medial limb of the superficial palmar arterial arch gives rise to the **common digital palmar arteries II, III, and IV.** These run distally under the superficial flexor tendons, where, proximal to the large metacarpal footpad, each sends a branch into the proximal part of the pad. At the distal ends of the intermetacarpal spaces, the arteries anastomose with the comparable palmar metacarpal arteries. The palmar common digital arteries are accompanied by small sensory nerve branches from the median nerve. The satellite veins accompany the arteries only in the distal half of the metacarpus.

The **palmar metacarpal arteries II, III, and IV,** *(aa. metacarpeae palmares)* arise from the **deep palmar arch** *(arcus palmaris profundus)*, which is formed by the palmar branch of the radial artery anastomosing with the terminal part of the caudal interosseous artery. This arch lies distal to the flexor retinaculum, where it is covered by the interosseous muscles. The palmar metacarpal arteries run distally between these muscles, which they supply, and anastomose with the corresponding palmar common digital arteries. These unions take place more than 1 cm. proximal to the metacarpophalangeal joints. The deep vessels send the several nutrient arteries into the proximal third of the palmar surfaces of the four main metacarpal bones. They are accompanied by satellite veins and companion nerves.

The **palmar common digital arteries II, III, and IV** are continuations of the superficial vessels after they receive the palmar metacarpal arteries. Each of these three pairs of arteries sends a prominent branch into the center of each of the three parts into which the distal portion of the large metacarpal footpad is divided. After running about 1 cm., the palmar common digital arteries divide into the **axial abaxial and palmar proper digital arteries** *(aa. digitales propriae palmares axialis et abaxialis)*, which supply the adjacent palmar sides of contiguous digits. The proper vessels lying on the digits which are closest to the axis through the paw are the chief arteries to the digits. In fact, those that go to the palmar

surfaces of the opposite, or abaxial, sides are mainly small cutaneous branches. However, in the palmar vascular canals of the third phalanges, the abaxial and axial proper digital arteries anastomose. Branches given off from these terminal arches nourish the corium of the claws and the third phalanges.

THORACIC AORTA

The **thoracic aorta** *(aorta thoracica)* (Figs. 11–49, 11–50) continues from the aortic arch, opposite the fourth thoracic vertebra, and extends to the caudal border of the second lumbar vertebra. There it enters the abdominal cavity by passing between the obliquely placed crura of the diaphragm to become the abdominal aorta. It is accompanied by the azygos vein, thoracic duct, and the cisterna chyli as it passes through the aortic hiatus of the diaphragm. At its beginning, the bulk of the thoracic aorta lies to the left of the median plane, being displaced to this position by the horizontally running esophagus, which crosses it on the right. It lies in the mediastinal septum and inclines slightly to the right and dorsally as it runs to the diaphragm. Caudally it is separated from the bodies of the thoracic vertebrae by a small amount of fat. The thoracic duct lies in this fat as it follows the dorsal surface of the aorta forward to the midthoracic region.

The branches of the thoracic aorta may be divided into visceral and parietal. The visceral branches are the bronchial and esophageal. The parietal branches include the intercostal, last thoracic, and the first two lumbar arteries.

VISCERAL BRANCHES

The bronchial and intrathoracic esophageal branches vary in number and origin. The chief nutritional blood supply to the lungs are the right and left **bronchial branches** *(rami bronchiales)* of the **bronchoesophageal artery** *(a. bronchoesophagea)*. This vessel also sends ascending and descending **esophageal branches** *(rami esophagei)* to most of the intrathoracic portion of the esophagus. The bronchoesophageal artery usually arises from the right fifth intercostal artery close to the aorta. Variations

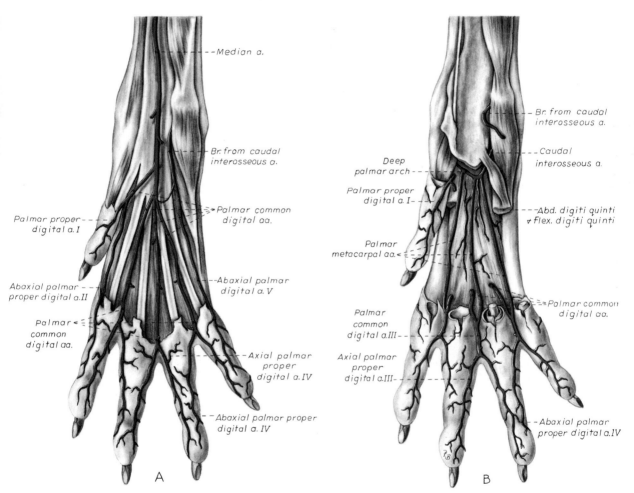

Figure 11–47. Arteries of the right forepaw.
 A. Superficial arteries of the right forepaw, palmar aspect.
 B. Deep arteries of the right forepaw, palmar aspect.

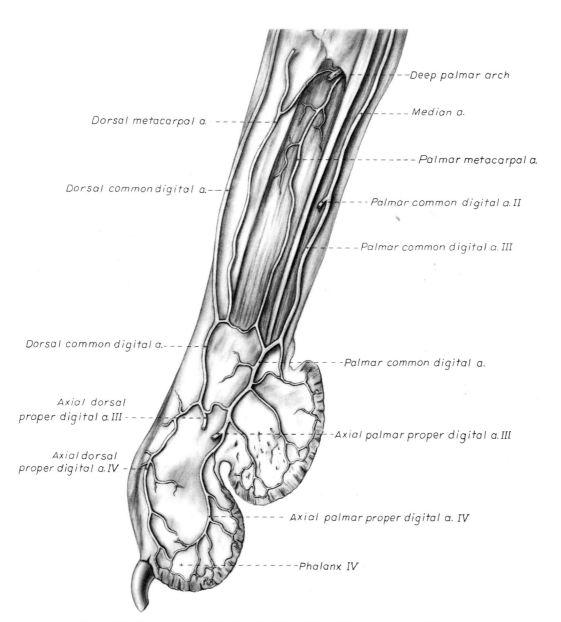

Dorsal metacarpal a.

Dorsal common digital a.

Dorsal common digital a.

Axial dorsal
proper digital a. III

Axial dorsal
proper digital a. IV

Deep palmar arch

Median a.

Palmar metacarpal a.

Palmar common digital a. II

Palmar common digital a. III

Palmar common digital a.

Axial palmar proper digital a. III

Axial palmar proper digital a. IV

Phalanx IV

Figure 11–48. Arteries of the fourth digit of the right forepaw, medial aspect.

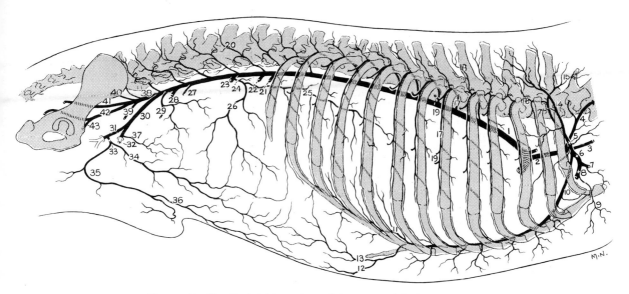

Figure 11–49. Parietal branches of the aorta, right lateral aspect.

1. Aorta
2. Brachiocephalic trunk
3. Right common carotid
4. Vertebral
5. Costocervical trunk
6. Right subclavian
7. Superficial cervical
8. Axillary
9. External thoracic
10. Right internal thoracic
11. Musculophrenic
12. Cranial superficial epigastric
13. Cranial epigastric
14. Dor. scapular.
15. Deep cervical

16. Supreme intercostal or thoracic vertebral
17. Intercostal
18. Dorsal branch of intercostal
19. Lateral cutaneous branches of intercostal
20. Lumbar
21. Celiac
22. Cranial mesenteric
23. Right renal
24. Phrenicoabdominal
25. Phrenic
26. Cranial abdominal
27. Right testicular or ovarian
28. Caudal mesenteric
29. Deep circumflex iliac
30. Right external iliac

31. Deep femoral
32. Femoral
33. Pudendoepigastric trunk
34. Caudal epigastric
35. External pudendal
36. Caudal superficial epigastric
37. Caudal abdominal
38. Right internal iliac
39. Right umbilical
40. Median sacral
41. Parietal branch of internal iliac
42. Visceral branch of internal iliac
43. Urogenital

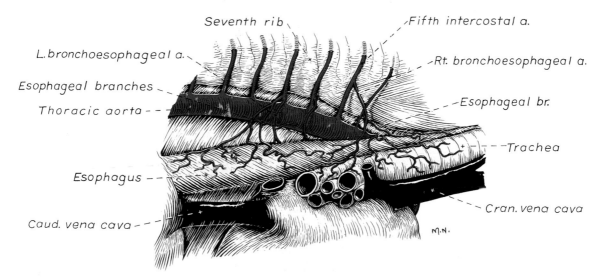

Figure 11–50. Bronchoesophageal arteries, right lateral aspect.

described by Berry et al. (1931) include: a branch from the right fifth or sixth intercostal artery to the right lung, direct branches from the first part of the descending aorta, and a single twig from the left sixth intercostal artery to the left lung. The esophageal branches come primarily from the bronchoesophageal artery in the form of ascending and descending vessels. The descending branches anastomose with several long esophageal branches that arise from two or more of the right intercostal arteries caudal to the origin of the bronchoesophageal. These anastomose with each other, and the last esophageal branch anastomoses with the esophageal branch of the left gastric artery, which ascends through the diaphragm on the esophagus. The ascending branches anastomose with the esophageal branch from the caudal thyroid. There is a small but definite arterial passage from the caudal thyroid to the left gastric artery in or on the wall of the esophagus.

Most bronchial and esophageal arteries have satellite veins. Usually the bronchoesophageal vein empties into the azygos at the level of the seventh thoracic vertebra. According to Berry et al. (1931) only a capillary anastomosis exists between the bronchial and pulmonary circulations.

Three or four small arteries leave the dorsal part of the thoracic aorta and run ventrally over its sides. Two or more of these are distributed to the dorsal part of the mediastinal septum as **mediastinal branches** (*rami mediastinales*). In the caudal part of the thorax one or more of these vessels may arise from the ventral surface of the aorta. Other branches arise from the bronchoesophageal, aorta, or intercostal vessels and run

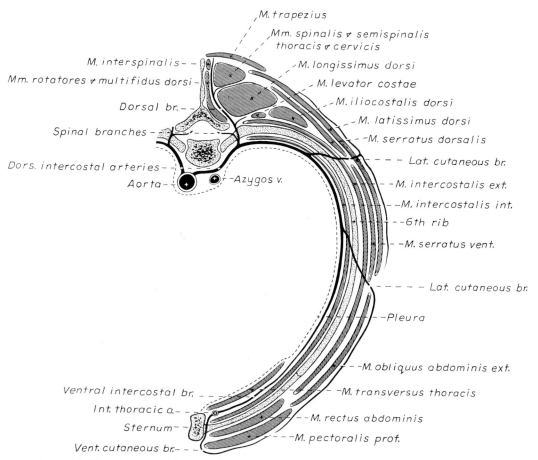

Figure 11–51. Scheme of the arteries of the thoracic wall in cross section, caudal aspect.

to the fibrous pericardium as the **pericardial branches** *(rami pericardiaci).*

PARIETAL BRANCHES

The **intercostal arteries** *(aa. intercostales)* (Figs. 11–51, 11–52) are 12 in number on each side. The first three or four are branches of the thoracic vertebral artery, and the last eight or nine are branches of the aorta. The artery which follows the caudal border of the last (thirteenth) rib is the dorsal costoabdominal artery. All of the aortic intercostal arteries arise from the dorsal surface of the aorta and are similar in distribution. The fifth right intercostal artery may serve as a common trunk for the fourth, occasionally the third, and, in rare instances, for the second intercostal artery, according to Reichert (1942). The first left aortic intercostal artery is rarely farther forward than the fourth. Right and left arteries arise close together, and occasionally the first pair come from a common trunk. In the midthoracic region the paired vessels may be separated by as much as 4 mm. at their origins,

and this spacing continues throughout the remainder of the thorax. Right and left vessels are not symmetrical in their origins; the right vessels usually arise caudal to the left. From the midthoracic region caudally the dorsal intercostal arteries run directly laterally across the bodies of the vertebrae. Each dorsal intercostal artery gives off a dorsal branch.

The **dorsal branches** *(rami dorsales)* arise from the intercostal arteries lateral to the bodies of the vertebrae. They pass directly dorsally in the medial portion of the m. longissimus, obliquely across the spines of the vertebrae, and end in the epaxial muscles or the skin as small **dorsal cutaneous branches.** Each has a **spinal branch** *(ramus spinalis),* which crosses the dorsal surface of the spinal nerve to reach its caudal border. It follows this border centrally to the nerve roots, after which it is distributed like the spinal branches of the vertebral artery. After the dorsal branches arise, the intercostal arteries extend laterally dorsal to the pleura, sympathetic trunks, azygos vein (on the right side), and hemiazygos vein (at the last

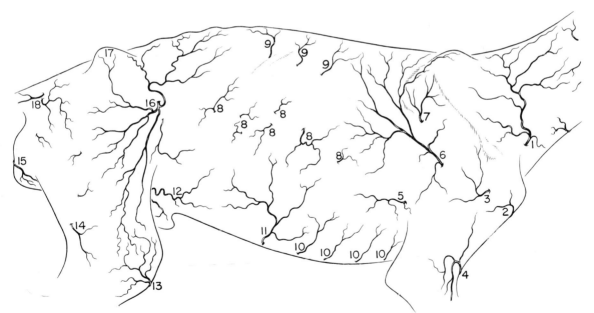

Figure 11–52.   Superficial arteries of the trunk.

1. Superficial cervical branch of omocervical
2. Cranial circumflex humeral
3. Caudal circumflex humeral
4. Proximal collateral radial
5. Lateral thoracic
6. Cutaneous branch of thoracodorsal
7. Cutaneous branch of subscapular
8. Distal lateral cutaneous branches of intercostals
9. Proximal lateral cutaneous branches of intercostals
10. Ventral cutaneous branches of internal thoracic
11. Cranial superficial epigastric
12. Caudal superficial epigastric
13. Medial genicular
14. Cutaneous branch of caudal femoral
15. Perineal
16. Deep circumflex iliac
17. Tuber coxae
18. Cutaneous branches of superficial lateral coccygeal

two or three intercostal spaces on the left side). The intercostal arteries supply segmental branches to the overlying epaxial musculature. Some of these branches become cutaneous along the lateral border of the iliocostalis. They supply the dorsal part of the latissimus dorsi and cutaneus trunci, which they perforate to reach a paramedian band of skin. On their way to the skin, they are accompanied by the dorsal lateral cutaneous nerve of their segment. Sometimes a single artery bifurcates, so that each of its branches accompanies intercostal nerves which arise from adjacent segments.

Hughes and Dransfield (1959) have found that in the regions of the body where the skin is covered by hair, there are superficial, middle, and deep arterial and venous plexuses. The deep plexus may be arranged in more than one layer.

The **lateral cutaneous branches** (*rami cutanei laterales*) form a second series of cutaneous branches (Marthen 1939). These start usually at the third segment and perforate the external intercostal muscle and the thoracic part of the serratus ventralis when it is present. The distal lateral cutaneous branches appear along the ventral border of the latissimus dorsi. They continue laterally through the cutaneus trunci and subcutaneous fascia to terminate in long ventral and short dorsal branches. In the cranial part of the series, the dorsal and ventral branches usually arise independently from the intercostal arteries. They are accompanied by satellite veins and the lateral cutaneous nerves. The nerves are constantly present, although the vessels may be missing in some of the segments. The comparable second nerve is the intercostobrachial nerve, which goes largely to the skin of the brachium.

The **collateral branches** (*rami collaterales*) arise from the intercostal arteries near the beginning of the ventral half of the thorax. They are the last of four to eight oblique branches which run ventrocranially across the medial surfaces of each of the last 9 or 10 ribs to reach the intercostal spaces cranial to them. They descend to anastomose with the ventral intercostal arteries, which ascend in front of the costal cartilages. They supply the ventral halves of the intercostal muscles which lie immediately cranial to the ribs. Usually the last three

dorsal intercostal arteries (intercostal arteries 10, 11, and 12) terminate medial to the costal arch and ventrocranial to the intercostal spaces of the same number. After circling the costal margin of the diaphragm, each artery gives a branch to the diaphragm and another to the musculature of the lateral abdominal wall. The phrenic components of these vessels anastomose with the phrenic branches of the phrenicoabdominal in the muscular periphery of the diaphragm, and the abdominal branches anastomose with the cranial abdominal artery. The end branches of the musculophrenic anastomose with the tenth or eleventh dorsal intercostal artery.

The **dorsal costoabdominal artery** (*a. costoabdominalis dorsalis*) is homologous with the subcostal artery of man. It is similar to the intercostal arteries in origin and course, but it is smaller. It runs dorsolaterally to become related to the caudal border of the last rib. It is covered at first by the pleura, then in succession by the psoas minor and the retractor costae muscle. It is accompanied by a cranially lying vein, but the ventral branch of the last thoracic nerve diverges caudally from the artery and passes ventral to the first lumbar transverse process. The last thoracic spinal artery runs outward and backward to enter the intervertebral foramen dorsal to the nerve. Branches leave the spinal artery and supply a segment of the epaxial musculature. The dorsal costoabdominal artery anastomoses with the cranial abdominal branch of the phrenicoabdominal artery via the ventral ramus of the costoabdominal artery.

The first two pairs of **lumbar arteries** (*aa. lumbales I et II*) arise as the last branches of the thoracic aorta because of the attachment of the crura of the diaphragm on the third and fourth lumbar vertebrae. They are distributed like other typical lumbar arteries. The first lumbar artery anastomoses with the dorsal costoabdominal and the second lumbar arteries and may effect a union with the cranial abdominal branch of the phrenicoabdominal.

ABDOMINAL AORTA

The **abdominal aorta** (*aorta abdominalis*) (Figs. 11–49, 11–53) is that portion of the

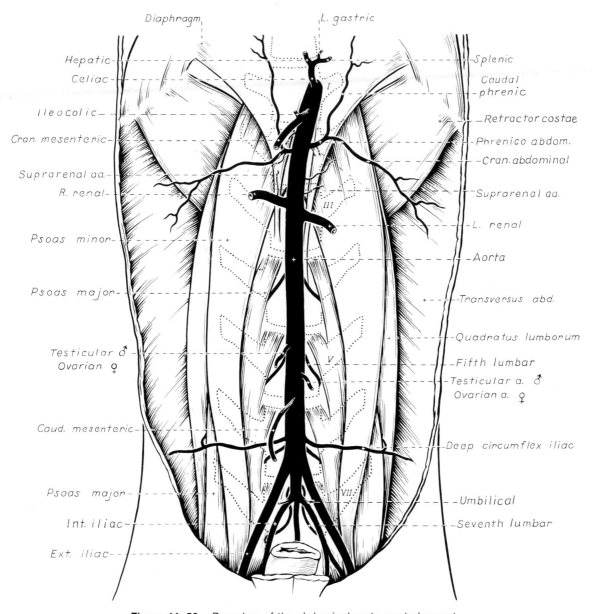

Figure 11-53. Branches of the abdominal aorta, ventral aspect.

descending aorta which lies in the abdomen. The diaphragm which separates the thoracic from the abdominal cavity is perforated so obliquely by the aorta that the first visceral branch of the abdominal aorta is at a more cranial plane than the origin of the second lumbar arteries. The abdominal aorta terminates opposite the seventh lumbar vertebra by bifurcating into right and left internal iliac and middle sacral arteries. Cranially, it lies in the median plane between the crura of the diaphragm; caudally, it is displaced slightly to the left by the caudal vena cava. It lies in the furrow formed by the right and left iliopsoas muscles. The branches of the abdominal aorta supply both visceral and parietal structures.

Unpaired Visceral Branches of Abdominal Aorta

The **celiac artery** (*a. celiaca*) (Figs. 11–54 to 11–56) arises from the ventral surface of the abdominal aorta as its first visceral branch. It is approximately 4 mm. in diameter and 2 cm. long. At its origin it is closely flanked on its lateral sides by the crura of the diaphragm. It is related to the stomach on the left and the liver and adrenal gland on the right. The left limb of the pancreas bounds it caudally. Although the vessel is large, its size is exaggerated by the celiac plexus of nerves which surrounds it. Its terminal branches are the hepatic, splenic, and left gastric arteries. The celiac artery usually trifurcates, although in some specimens the left gastric and splenic arteries arise by a short common trunk. Kennedy and Smith (1930) recorded a case in which the splenic arose from the cranial mesenteric. Small inconstant pancreatic and phrenic branches may arise from the celiac.

The **hepatic artery** (*a. hepatica*) (Fig. 11–57) runs cranioventrally and to the right, in a groove of the pancreas. Opposite the porta of the liver it sends three to five rather long branches into the hilus of the organ. These furnish the nutritional blood to the gland. When three branches are present, the first branch, the **right lateral branch** (*ramus dexter lateralis*), goes to the right portion of the liver, the caudate and right lateral lobes. The **right middle branch** (*ramus dexter medialis*) goes to the right medial lobe, dorsal part of the quadrate, and part of the left medial lobe; this vessel may be replaced by two or more arteries. The **left branch** (*ramus sinister*), shortest of the three, supplies the large left lateral lobe, the quadrate lobe, and part of the left medial lobe. From this branch arises the **cystic artery** (*a. cystica*). This vessel leaves the left branch, about 1 cm. before it enters the liver, and ramifies by two or more branches primarily on that surface of the gallbladder which is attached to the liver. After giving off its branches, the hepatic artery terminates as the small right gastric and the much larger gastroduodenal artery.

The **right gastric artery** (*a. gastrica dextra*) leaves the hepatic artery at nearly a right angle and, running in the lesser omentum at the pylorus, continues to the lesser curvature of the stomach. It frequently arises from one of the hepatic branches, in which case the hepatic artery becomes the gastroduodenal after the last hepatic branch. The right gastric artery sends branches to both the parietal and the visceral surface of the pylorus, pyloric antrum, and the lesser omentum. It anastomoses with the much larger left gastric artery close to the pylorus on the pyloric antrum as it runs toward the cardia in the lesser omentum.

The **gastroduodenal artery** (*a. gastroduodenalis*) runs across the first part of the left limb of the pancreas to the medial surface of the duodenum, where it terminates. It may issue delicate twigs to the pylorus, and constantly sends one or more larger branches into the left limb of the pancreas. It terminates as the right gastroepiploic and the cranial pancreaticoduodenal artery.

The **right gastroepiploic artery** (*a. gastroepiploica dextra*) leaves the pancreas from the medial surface of the duodenum and enters the greater omentum. It lies about 1 cm. from the greater curvature of the stomach as it runs in the greater omentum toward the cardia. It sends branches to the stomach at intervals of about 5 mm. Most of these **gastric branches** divide on reaching the greater curvature of the stomach into a branch which goes to the visceral surface. These anastomose in the musculature of the organ with gastric branches of the right and left gastric arteries which lie on the lesser curvature. Long, freely branching **epiploic branches** leave the opposite or omental side of both gastroepiploic arteries and ramify mainly in the parietal leaf of the greater omentum. Grävenstein (1938) demonstrated that there is a strong epiploic branch which leaves the right gastroepiploic artery near its origin and runs caudally in the parietal leaf of the greater omentum near its right border. A second vessel with a similar origin essentially parallels the first, but runs in the visceral layer of the greater omentum near its right border. The comparable vessels near the left border of the greater omentum are continuations of the splenic artery near the apex of the spleen. There are many other smaller epiploic branches which arise at short intervals from the gastroepiploic arteries as they lie in the omentum just peripheral to the greater curvature of the stom-

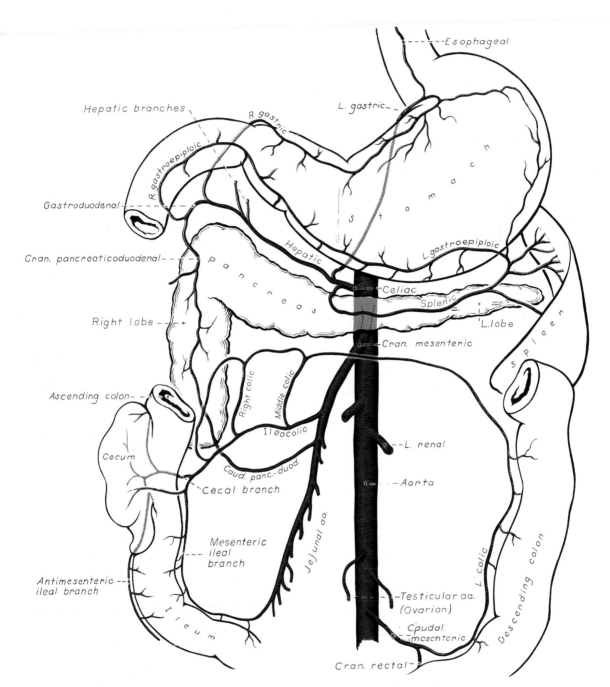

Figure 11–54. Diagram of the visceral branches of the aorta with their principal anastomoses, ventral aspect.

ach. These vessels anastomose freely with each other. The fat of the omentum is deposited around the epiploic vessels. The right and left gastroepiploic arteries anastomose with each other opposite the beginning of the pyloric antrum. The long left gastroepiploic artery on its way to the stomach supplies most of the epiploic vessels to the greater omentum.

The **cranial pancreaticoduodenal artery** (*a. pancreaticoduodenalis cranialis*) is the slightly larger terminal branch of the gastroduodenal. It continues the parent artery in the mesoduodenum and enters the pancreas at the junction of its right and left limbs. Shortly before entering the right limb of the pancreas, it sends a small branch to the left limb which, after extending a short distance, anastomoses with the pancreatic branch of the splenic, which is the chief supply to this part of the gland (Cadete 1973). The principal part of the cranial pancreaticoduodenal artery continues caudally in the right limb of the pancreas, sending **pancreatic branches** to the right limb of the pancreas and **duodenal branches** through the gland to the duodenum. In the caudal half of the right limb of the pancreas, it anastomoses with one or more pancreatic branches coming from the caudal pancreaticoduodenal. The main part of the artery usually leaves the caudal third of the right limb of the pancreas, traverses a part of the mesoduodenum, and comes to lie on the mesenteric border of the descending duodenum. Near the caudal flexure it anastomoses with the main duodenal branch of the caudal pancreaticoduodenal artery.

The **splenic artery** (*a. lienalis*) (Fig. 11–58) is that branch of the celiac which runs to the left and lies in a groove of the left limb of the pancreas near its free end. It is usually over 2 mm. in diameter and gives off 3 to 5 long primary branches as it courses in the greater omentum toward the apical or ventral third of the spleen. The first branch usually is the **pancreatic branch** (*ramus pancreaticus*), which leaves the right side of the vessel at its origin and enters the apex of the left limb of the pancreas. It is the main supply to this limb of the gland and anastomoses with the smaller pancreatic branch from the cranial pancreaticoduodenal artery. Usually a single vessel to the pancreas is present at this site, but occasionally this is

augmented by as many as three other small twigs. Arising from the cranial surface of the splenic artery are two long vessels which run toward the proximal half of the spleen. They do not enter the gland directly, but pass at right angles to the long hilus of the spleen, where they send **splenic branches** to it. They then continue in the gastrosplenic ligament to the greater curvature of the stomach, where they supply the **short gastric arteries** (*aa. gastricae breves*) to the fundus, which anastomose with the much larger gastric branches from the left gastric artery. The main part of the splenic artery approaches the hilus of the spleen near its middle, where it sends many branched splenic twigs into the distal half of the gland. This main trunk, now called the **left gastroepiploic artery** (*a. gastroepiploica sinistra*), continues in the gastrosplenic ligament to the greater curvature of the stomach. It runs near this curvature toward the pylorus, where it anastomoses with the small right gastroepiploic artery. Other sizeable branches leave it and run toward the cardia, where the terminal vessel anastomoses with an adjacent, more proximal branch of the splenic. The **gastric arteries** in this region run considerable distances in the subserosa before entering the muscularis. They anastomose with the gastric branches of the left gastric artery which come from the lesser curvature, and finally the terminal part of the vessel joins the small right gastroepiploic artery near the beginning of the pyloric antrum. All primary divisions of the splenic, as they run through the greater omentum and the gastrosplenic ligament, give off **omental (or epiploic) branches** to the omentum. These are disposed like the comparable branches from the right gastroepiploic. The vascular pattern in the visceral leaf of the greater omentum is similar to that in the parietal.

The **left gastric artery** (*a. gastrica sinistra*) arises from the cranial surface of the celiac as its smallest terminal branch. It may be double. When single, it may form a common trunk with the splenic. It runs cranially in the lesser omentum to the fundus of the stomach adjacent to the cardia. It sends long subserous branches to both surfaces of the organ, with the heavier branches going to the parietal surface. An anastomosis between right and left gastric arteries takes place in

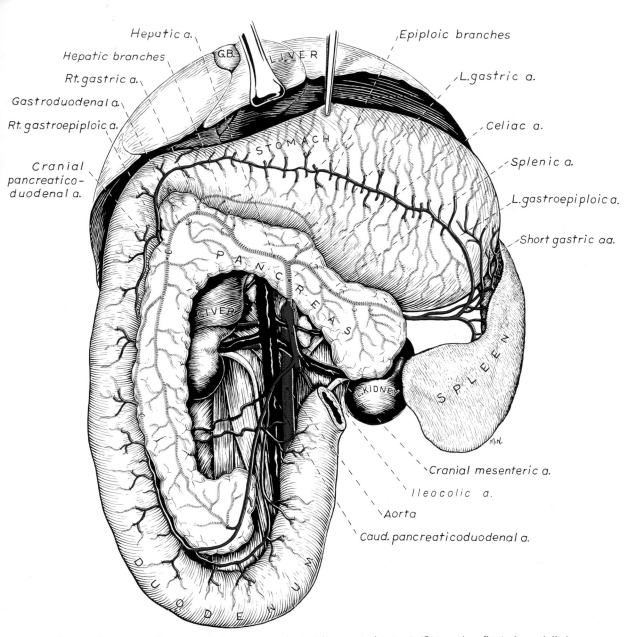

Figure 11–55. Celiac and cranial mesenteric arteries, ventral aspect. (Stomach reflected cranially.)

the lesser omentum. In addition to supplying the fundus, small **epiploic branches** go to the lesser omentum, and one or more **esophageal branches** *(rami esophagei)* run through the esophageal hiatus to supply the caudal part of the esophagus. These branches anastomose with the esophageal rami from the thoracic aorta.

The celiac artery is surrounded by the artery celiac plexus of nerves and the two celiac ganglia which are intimately connected with both the adrenal and the cranial mesenteric nerve plexuses.

Lymph vessels from the stomach, spleen, pancreas, and a portion of the duodenum run dorsally in association with the celiac trunk to the cisterna chyli. The hepatic, splenic, and left gastric arteries and their branches have accompanying lymphatics and satellite veins which are radicles of the portal vein.

The **cranial mesenteric artery** *(a. mesenterica cranialis)* (Figs. 11–59, 11–60) is the largest visceral branch of the aorta. This unpaired artery arises from the ventral surface of the abdominal aorta about 5 mm. caudal to the origin of the celiac artery opposite the first or second lumbar vertebra. It is about 5 mm. in diameter and is surrounded by the cranial mesenteric plexus of autonomic nerves. It passes ventrocaudally through the dorsal mesentery and acts as an axis around which the whole small and large intestine rotates during development. As it extends into the intestinal mass, it is loosely bounded by the duodenum on the right and caudally. This is followed distally by the colon, which hooks around the artery in such a way that the ascending colon lies to its right, the transverse colon in front, and the descending colon to the left. The cranial mesenteric ganglion and plexus are so intimately associated with the celiac ganglia and plexuses that it is usual to find a combined celiacomesenteric plexus and ganglion which obscures the origin of both vessels. Peripheral to the plexus are the long mesenteric lymph nodes and the portal vein. The first two branches of the cranial mesenteric artery arise from opposite sides of the artery, about 2 cm. from its origin. One of these, the ileocolic, runs cranially in the transverse mesocolon, while the other, the caudal pancreaticoduodenal, from a caudal origin, runs to the right and cranially. The remaining part of the artery gradually diminishes in size as some 14 jejunal

arteries arise from it. Ileal arteries terminate the cranial mesenteric.

The **ileocolic artery** *(a. ileocolica)* gives origin in succession to the middle colic, and right colic, and continues as the ileocolic.

The **middle colic artery** *(a. colica media)* arises from the first few millimeters of the left side of the ileocolic or from the cranial mesenteric directly. It runs cranially in the transverse mesocolon, where it usually makes one spiral turn. It bifurcates about 2 cm. from the left colic flexure. One branch runs distally in the descending mesocolon and, after supplying about half of the descending, or left colon, anastomoses with the left colic artery, a branch of the caudal mesenteric. The other branch swings in the opposite direction and forms an arcade with the smaller accessory middle colic artery.

An **accessory middle colic artery** may supply the proximal portion of the transverse colon. When present, it forms terminal arcades with the middle and right colic arteries, from which vasa recti arise which supply all the transverse colon and a part of the right colon also. This vessel may be absent or double.

The **right colic artery** *(a. colica dextra)* is a small vessel which arises as the last branch of the ileocolic. It runs in the right mesocolon toward the right colic flexure, giving off branches to the distal part of the right colon and adjacent part of the transverse colon. It forms a definite arcade with the accessory middle colic and a weaker anastomosis with the colic branch of the ileocolic.

The **ileocolic artery** *(a. ileocolica)* was formerly called the ileocecocolic artery because it supplies all three structures. Before the main artery disappears between the ileum and cecum, the **colic branch** *(ramus colicus)* arises and goes to the proximal part of the right colon. Within the wall of the colon the colic branch anastomoses with the right colic artery. The **mesenteric ileal branch** *(ramus ilei mesenterialis) (r. caecalis accessorius* of Thamm 1941) arises near the colic branch and runs to the ventral side of the ileocolic junction. It supplies small twigs to the junction, sends an ascending ramus proximally along the mesenteric border of the ileum to anastomose with the last intestinal (ileal) artery, and sends a branch distally to anastomose on the right colon with the

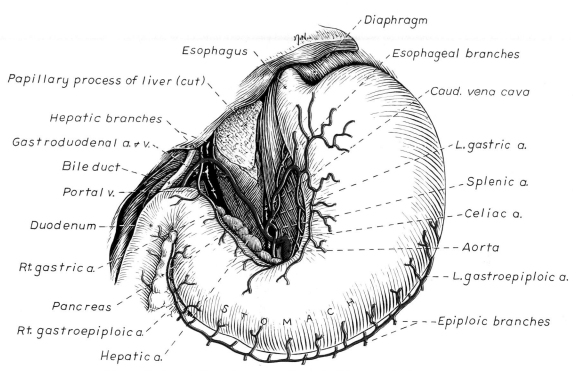

Diaphragm

Esophagus

Papillary process of liver (cut)

Hepatic branches

Gastroduodenal a. & v.

Bile duct

Portal v.

Duodenum

Rt. gastric a.

Pancreas

Rt. gastroepiploic a.

Hepatic a.

Esophageal branches

Caud. vena cava

L. gastric a.

Splenic a.

Celiac a.

Aorta

L. gastroepiploic a.

Epiploic branches

STOMACH

Figure 11-56. Celiac artery, ventral aspect. (Stomach displaced to left.)

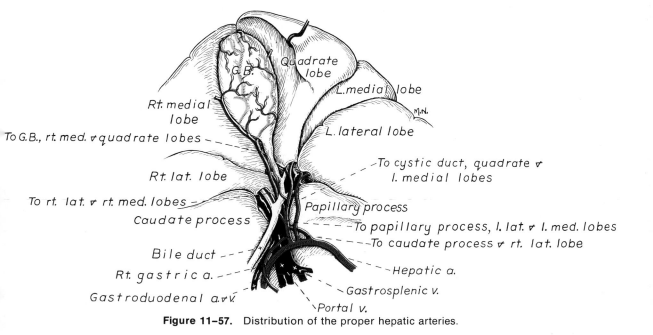

G.B.

Quadrate lobe

L. medial lobe

Rt. medial lobe

L. lateral lobe

To G.B., rt. med. & quadrate lobes

Rt. lat. lobe

To cystic duct, quadrate & l. medial lobes

To rt. lat. & rt. med. lobes

Papillary process

Caudate process

To papillary process, l. lat. & l. med. lobes

Bile duct

To caudate process & rt. lat. lobe

Rt. gastric a.

Hepatic a.

Gastroduodenal a. & v.

Gastrosplenic v.

Portal v.

Figure 11-57. Distribution of the proper hepatic arteries.

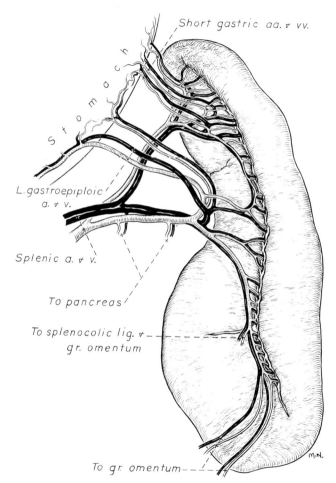

Figure 11–58. Blood supply of the spleen. (The cranial border is reflected laterally.)

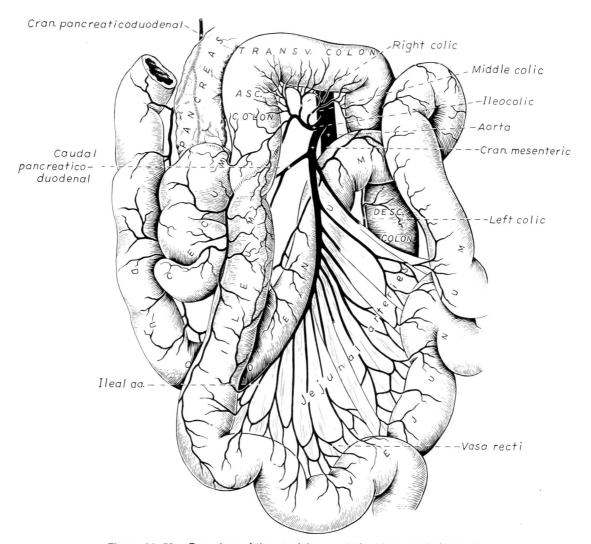

Figure 11–59. Branches of the cranial mesenteric artery, ventral aspect.

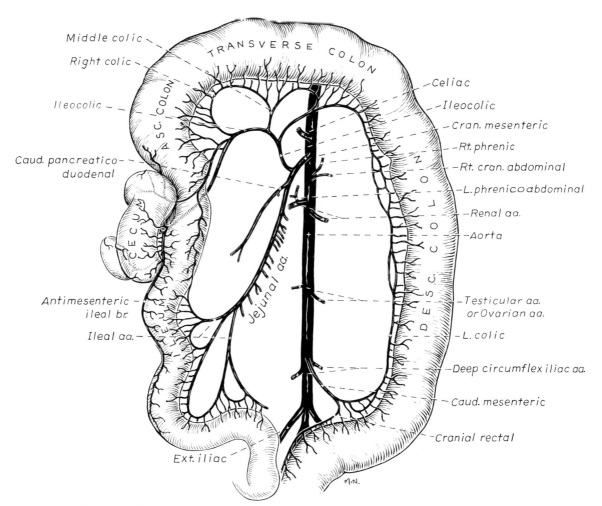

Figure 11–60.   Branches of the cranial and caudal mesenteric arteries, ventral aspect.

colic branch. Variations in the vascularity of this portion of the digestive tube are extremely common.

The **ileocecal artery** continues across the dorsal surface of the ileocolic junction. It leaves the subserosa and disappears in the areolar tissue uniting the cecum to the ileum. It sends **ileal branches** *(rami ileales)* to the ileum and **cecal branches** *(rami cecales)* to the cecum. The main part of the vessel, after emerging from between the ileum and cecum, continues proximally in the ileocecal fold as the *ramus ilei antimesenterialis*. This vessel extends along the ileum and anastomoses with the last jejunal artery in the musculature of the small intestine. The ileocolic artery and its branches are accompanied by plexuses of autonomic nerves which take the names of the vessels they follow, and all have satellite veins and lymphatics.

The **caudal pancreaticoduodenal artery** *(a. pancreaticoduodenalis caudalis)* (Fig. 11–55) arises from the cranial mesenteric opposite the origin of the ileocolic and runs to the right in the mesentery to the descending portion of the duodenum near the caudal flexure. On passing the tip of the right limb of the pancreas, it sends from one to three **pancreatic branches** into the gland, which anastomose in the caudal third of this limb with similar branches from the cranial pancreaticoduodenal. The **duodenal branch,** as it runs obliquely to the mesenteric border of the descending portion of the duodenum, supplies a small branch to the duodenum which, in the region of the caudal flexure, anastomoses with the first jejunal artery. The main branch anastomoses on the duodenum with the larger duodenal branch of the cranial pancreaticoduodenal artery.

The **jejunal arteries** *(aa. jejunales)* (Fig. 11–59) are 12 to 15 in number. Some of these are larger than others, and branch after they have traveled only a few millimeters. They arise from the caudal or convex side of the cranial mesenteric artery. The proximal 8 to 10 vessels are covered on each side by the two large jejunal lymph nodes. Usually the first five to seven arteries arise closely together; rarely are they separated by more than 3 mm. Following this cluster of vessels which go to the proximal part of the jejunum, the artery is free of branches for about 1 cm. The cranial mesenteric then gives rise to the branches that go to the distal part of the jejunum and finally to the two ileal arteries. The ileal arteries branch on approaching the intestine and join to form primary and secondary arcades, which lie directly adjacent to the intestinal wall. According to Noer (1943), tertiary arcades are rudimentary and, when present, consist of communications between the branchings of the secondary arcades. There is little variation of the vascular pattern throughout the jejunum, ileum, and colon.

The *vasa recti* leave the terminal arcades and go directly to the intestine. They are short and irregular. Numerous lateral twigs unite with similar adjacent vessels. Most bifurcate on entering the wall of the intestine; these branches circle the gut on opposite sides and typically anastomose with each other on the antimesenteric border. Eisberg (1924) states that the vasa recti pierce the muscular coats in the mesenteric quarters of the small intestine and the antimesenteric quarters of the large intestine. There is a well-defined arterial anastomosis along the mesenteric border in addition to those between the vasa recti. These anastomoses are small and few in the duodenum. They are moderate in number in the jejunum and occur in large numbers in the ileum and colon. Within the gut there are two arterial networks. The subserous one is well developed as compared with man and is derived from the mural network. The mural network lies largely in the submucosa and is formed by direct and plexiform anastomoses (Noer 1943).

Branches are supplied at irregular intervals from the proximal halves of the jejunal arteries to the jejunal lymph nodes, one of which lies on each side of the jejunal mesentery.

The cranial mesenteric artery terminates as ileal and jejunal branches. The ileal branch forms a terminal arcade with the last jejunal artery and issues vasa recti to the proximal part of the ileum and the distal part of the jejunum. The jejunal branch bifurcates. The proximal branch forms an arcade with the ileal branch, whereas the distal branch runs about 6 cm. along the mesenteric border of the ileum as the mesenteric ileal branch. The first jejunal artery anastomoses with the duodenal part of the caudal pancreaticoduodenal, and all jejunal and ileal arteries anastomose with each other. Thus a vascular channel is formed along the whole small intestine. The cranial mesenteric artery and its branches are accompanied by the cranial mesenteric plexus of nerves, the intestinal lymph trunk, and the cranial mesenteric vein.

The **caudal mesenteric artery** (*a. mesenterica caudalis*) (Fig. 11–60) arises from the ventral surface of the aorta about 4 cm. cranial to the termination of the aorta or opposite the caudal part of the fifth lumbar vertebra. It runs caudoventrally in the left mesocolon to the mesenteric border of the left colon, where it divides into the proximally running left colic and the distally running cranial rectal arteries. The caudal mesenteric artery is accompanied by the caudal mesenteric plexus, which begins about 1 cm. from the origin of the artery. No veins or lymph vessels accompany the main caudal mesenteric artery.

The **left colic artery** (*a. colica sinistra*) follows the mesenteric border of the left colon cranially and anastomoses with the middle colic at the junction of the proximal and distal halves of the left colon. No arcade is formed; numerous vasa recti run from it to the left colon. It is accompanied by the left colic vein and autonomic nerves.

The **cranial rectal artery** (*a. rectalis cranialis*), formerly known as the cranial hemorrhoidal, descends along the mesenteric border of the left colon and rectum. It supplies many branches to the rectum and may anastomose with the middle rectal and caudal rectal arteries.

Paired Visceral Branches of Abdominal Aorta

The **renal arteries** (*aa. renales*) (Fig. 11–53) arise asymmetrically from the lateral surfaces of the abdominal aorta. The right renal artery arises about 2 cm. cranial to the left renal artery in conformity to the more cranial position of the right kidney. This places it about 4 cm. caudal to the origin of the cranial mesenteric artery. In a large dog, it is 5 cm. long and 4 mm. in diameter. The left renal artery is about the same diameter as the right, but is only about 3 cm. long. The fact that the caudal vena cava lies to the right of the aorta accounts for the greater length of the right renal artery, which crosses its dorsal surface. Each renal artery supplies two or three branches to the caudal pole of the suprarenal gland and a small cranial ureteral branch to the ureter, although this vessel may arise from the aorta directly. About 1 cm. outside the renal hilus, each divides typically into a dorsal and a ventral branch. According to Christensen (1952), each of these further divides into two, four, or even seven interlobar arteries, or they may not divide at all before entering the hilus of the kidney. At the corticomedullary junction numerous relatively straight arcuate arteries continue the interlobar arteries in this zone. Only in mature or elderly specimens are the arcuate arteries arched toward the cortex. These interanastomose and give rise to the numerous interlobar arteries of the cortex. These in turn give rise to the afferent vessels which are continued as the glomerular arterioles. The renal arteries may be double in 20 per cent of dogs, particularly on the left side. Christensen (1952) found 29 double renal arteries in 117 specimens. Reis and Tepe (1956) examined 500 dogs and reported that 99.4 per cent had a single right renal artery, whereas 12.8 per cent had double left renal arteries, and in 0.4 per cent the left renal arteries were triple. The renal artery and its larger branches are accompanied by satellite veins and plexuses of autonomic nerves. Both the arteries and the veins have companion lymphatics, according to Peirce (1944). The left renal vein, like the right renal artery, is longer than its fellow and empties into the caudal vena cava more caudally than does the opposite vessel. According to Kazzaz and Shanklin (1951), the dog has a system of stellate veins on the surface of its kidney. Those on the lateral side drain into the interlobar veins; those on the medial side go directly into the renal vein. According to these authors, the cat has radially arranged veins on the surface of its kidney.

The **suprarenal (adrenal) arteries** (*aa. suprarenales*) in the dog usually arise from the phrenicoabdominal, the aorta, and the renal artery. Quite commonly, the cranial pole of the adrenal gland receives a branch from the cranial mesenteric artery and occasionally from the celiac also. Ljubomudrov (1939) states that there are 20 to 30 or more fine arteries that supply each gland, and that some of these may arise from the artery to the adipose capsule and from a lumbar artery, unilaterally, in addition to the sources already cited. Flint (1900) gives a classic description of the suprarenal gland and details its blood supply.

Flint's (1900) phrenic and accessory phrenic arteries represent the phrenic part of the phrenicoabdominal of modern terminology, whereas his lumbar artery is the abdominal part. When they reach the adrenal, Flint divides the vessels into three groups: capsular, cortical, and medullary. The capsular vessels are represented by a rather poorly defined system of arterioles which arise from all the arteries going to the gland. These form a plexus from which most of its branches, of arteriole size, enter the cortex. The arterioles divide and form capillaries which follow the reticular septa between the coiled columns of cortical cells which they nourish. The arteries of the medulla, according to Flint, number about 50. They come from the capsular plexus and run at right angles to it through the cortex to the corticomedullary junction, where most turn and run short distances in this area as they give off finer vessels which supply the medulla. While traversing the cortex, they neither divide nor supply any branches to it.

The arteries which supply the testes and ovaries arise from the ventral third of the circumference of the aorta in the midlumbar region or opposite the fibrocartilage between the fourth and fifth lumbar vertebrae. The right artery usually arises several millimeters cranial to the left, in conformity with the more cranial location of the right gonad.

The **testicular artery** (*a. testicularis*) (Fig. 11–54) is also called the internal spermatic artery. It is a small vessel, much longer and straighter than its homologue, the ovarian artery. It runs laterally or craniolaterally across the sublumbar muscles and the ventral surface of the ureter. Each makes a sweeping bend caudally to the deep inguinal ring, and lies in a special plica of peritoneum, which may be 3 to 4 cm. wide. At the deep inguinal ring it becomes a constituent of the spermatic cord and runs to the testis with its companion vein and nerve plexuses in the free border of the mesorchium. It is the only artery supplying the testis and epididymis.

The **ovarian artery** (*a. ovarica*), the homologue of the testicular artery, varies in size and tortuosity, depending on the age and past genital activity of the bitch. The ovarian artery (Preuss 1959) has also been called the utero-ovarian artery or the internal spermatic artery. The vessel arises from the ventral surface of the aorta at the midlumbar level. Like its homologue, it lies ventral to the caudal vena cava. Each gives a minute branch to the capsule of the kidney. Other twigs may go to the periaortic fat, peritoneum, and the adventitia of the caudal vena cava. Medial to the ovary the vessel bifurcates or trifurcates. The resultant branches are very tortuous, even in immature females, as they course in the broad ligament on their way to the ovary. Small branches from these supply the peritoneum and fat of the ovarian bursa, and the uterine tube. One or more of these branches continue caudally to the ovarian end of the uterine horn and anastomose with the uterine artery, which runs cranially in the broad ligament. The ovarian artery is accompanied by a plexus of sympathetic nerves, a satellite vein, and lymph vessels.

Parietal Branches of Abdominal Aorta

The paired **lumbar arteries** (*aa. lumbales*) are seven in number. The first two pairs arise from the thoracic aorta, and the last five pairs come from the dorsal surface of the abdominal aorta. Those in the cranial part of the series are 4 mm. apart at their origins, and usually the right arteries arise 3 to 6 mm. caudal to those on the left. The vessels are progressively closer at their origins as they are traced caudally, and the last lumbar arteries may arise from a common trunk. Each pair of vessels runs caudolaterally across the ventrolateral surface of the body of the lumbar vertebra of the same serial number. The first three or four pairs arise opposite the disc cranial to the vertebra of the comparable number; the last three or four pairs arise opposite the body of the comparable vertebra. For this reason the last several arteries are less oblique as they run on the side of the vertebral bodies, covered in their courses by the sublumbar muscles which they supply. Near their origins each vessel gives off a small, short nutrient branch which enters the body of the vertebra. The arteries parallel the dorsal branches of the spinal nerves as they enter the epaxial muscles caudal to the craniolaterally extending transverse processes. Here each artery divides into a **spinal branch** (ramus spinalis), which runs with the nerve into the spinal canal (see discussion of vertebral artery for description), and a **dorsal branch** (ramus dorsalis), which runs dorsocaudally in the longissimus past the lateral surface of the cranially inclined articular process of the vertebra behind it. The latter gives off many branches to the epaxial muscles in its course to the subcutaneous fat and skin near the midline. Small branches also leave the dorsal branch near the origin of the spinal branch and run laterally on the ventral surfaces of the transverse processes which they supply. The adjacent internal oblique and transversus abdominis muscles receive delicate terminal twigs from these. The seventh lumbar arteries differ from the others in that they may arise as a common trunk from the terminal part of the aorta or from the middle sacral artery. Each vessel as it courses laterally runs dorsal to the lumbosacral nerve plexus and divides into dorsal and caudal branches at the iliosacral joint. The dorsal branch enters the epaxial muscles by passing through the cranial angle formed by the ilium and the vertebral column. The caudal branch runs into the pelvis in company with the sympathetic trunk. It supplies this, and at the ganglion impar continues into the ventral sacrocaudal muscles. The caudal branch may also send a branch laterally to the pelvic surface of the wing of the ilium. The lumbar arteries are accompanied by satellite veins.

The **phrenicoabdominal artery** (a. phrenicoabdominalis) (Fig. 11–53) is the common trunk for the caudal phrenic artery and the cranial abdominal artery. It arises from the lateral surface of the aorta between the cranial mesenteric and renal arteries. The right artery occasionally arises from the corresponding renal vessel, and the caudal phrenic and abdominal parts may arise separately. The artery runs caudolaterally, parallel to the ribs, and crosses the ventral surface of the psoas muscles and the dorsal surface of the adrenal gland. (The corresponding vein grooves the ventral surface of the adrenal gland.) As it enters the fat lying lateral to the sublumbar muscles, it sends a small phrenic branch forward to a part of the diaphragm and a branch into the psoas muscle.

The main **caudal phrenic artery** (a. phrenica caudalis) arises within 1 cm. of the origin of the phrenicoabdominal and runs forward to ramify on the ventrocaudolateral surface of the crus of the diaphragm. In its subperitoneal course along the medial border of the dorsal extension of the tendinous center of the diaphragm, it usually sends two branches ventrolaterally, which redivide laterally as they cross the tendinous part of the diaphragm and enter its muscular periphery. The branches in general follow the course of the muscle fibers peripherally and anastomose within the muscle with the phrenic branches of the tenth, eleventh, and twelfth intercostal arteries.

The **cranial abdominal artery** (a. abdominalis cranialis) continues the direction of the parent artery into the abdominal wall after the phrenic vessels arise. It runs medial to the narrow aponeurosis of origin of the transversus abdominis, then perforates the muscle to ramify extensively between it and the internal abdominal oblique. Usually the vessel divides after perforating the transversus abdominis. The cranial branch runs toward the costal arch; the caudal branch diverges from it and supplies the middle zone of the lateral abdominal wall. The cranial abdominal artery anastomoses cranially with the phrenic vessels, ventrally with the cranial and caudal deep epigastric arteries, and caudally with the deep branch of the deep circumflex iliac artery. The cranial abdominal artery is accompanied by the cranial iliohypogastric nerve and a satellite vein.

The paired **deep circumflex artery** (a. circumflexa ilium profunda) (Fig. 11–53) arises from the lateral surface of the aorta about 1 cm. cranial to the origin of the external iliac artery, ventral to the sixth lumbar vertebra. The right artery usually arises a few millimeters cranial to the left, its initial part lying between the caudal vena cava dorsally and the cranial pole of the medial iliac lymph node ventrally. It runs laterally across the ventral surface of the sublumbar muscles and enters the abdominal wall ventral to the tuber coxae. At the lateral border of the psoas major, it usually sends a cranial twig to this muscle and to the overlying quadratus lumborum. It terminates as deep and superficial branches.

The **deep branch** leaves the parent vessel before this perforates the abdominal wall and, running forward and downward, extends lateral to the transversus abdominis. The main vessel sends branches dorsally, which accompany the lumbar nerves to the lateral surface of the transversus abdominis. The end branches of this vessel anastomose with the cranial and caudal abdominal arteries. It is the main supply of the caudodorsal fourth of the abdominal wall.

The **superficial branch** perforates the abdominal wall between the lumbar and inguinal portions of the internal oblique and the cutaneus trunci to reach subcutaneous structures. It is stellate in form as it subdivides into diverging branches which arborize ventrally and caudally. One or two strong branches spray out over the rump and loin to anastomose with their fellows at the mid-dorsal line. The caudal branches extend about halfway to the caudal border of the thigh before they anastomose with cutaneous twigs of the caudal femoral artery, which emerge from the substance of the hamstring muscles. The ventral branches are longer than the others as they run downward and backward over the thigh and stifle joint to the crus. The most cranial of the ventral branches ends in the fold of the flank. The cranial branches are smallest as they radiate in the subcutaneous fat of the caudal parts of the loin and abdomen.

The deep circumflex iliac artery is accompanied by a satellite vein which lies caudal to it. At the lateral border of the psoas major, it is joined by the lateral cutaneous femoral nerve.

Arteries of the Pelvic Limb

EXTERNAL ILIAC ARTERY

The **external iliac artery** (*a. iliaca externa*) (Fig. 11–53), is the largest parietal branch of the abdominal aorta. This paired vessel arises from the lateral surface of the aorta ventral to the disc between the sixth and seventh lumbar vertebrae. It runs caudoventrally and is related near its origin to the common iliac vein and the psoas minor muscle. Farther distally it lies on the iliopsoas muscle. The ureter and the ductus deferens of the male or the uterine horn of the female, lying in their peritoneal folds, cross the artery at nearly right angles medially. Small, slender branches usually leave the vessel to supply the adjacent fat or to run in the broad

ligament of the female. The small caudal abdominal artery may arise from it near the pubis, but its only constant branch is the deep femoral artery. It is continued outside the abdominal wall by the femoral artery. Its satellite vein lies caudolaterally.

The **caudal abdominal artery** (*a. abdominalis caudalis*) (Figs. 11–49, 11–65) is a small vessel which usually arises from the cranial surface of the external iliac just after the deep femoral arises, but it may arise from the deep femoral, the pudendoepigastric trunk, or the caudal deep abdominal artery (Marthen 1939). From the region of the vascular lacuna it runs forward on the deep surface of the internal oblique and divides typically into dorsal and ventral branches. On reaching the caudal border of the transversus abdominis, they arborize on the medial surface of the internal oblique. Branches perforate this muscle to reach the external oblique. The ventral branch lies about 2 cm. from, and parallel to, the lateral border of the rectus abdominis. It anastomoses with the caudal epigastric artery. The dorsal branch passes dorsally and anastomoses with the deep branch of the deep circumflex iliac artery. The **cremasteric artery** (*a. cremasterica*) is a small twig which runs toward the inguinal canal and supplies the fat lying around the abdominal inguinal ring. It enters the inguinal canal and supplies the cremaster muscle. According to Joranson et al. (1929) it, along with the artery of the ductus deferens, continues to the testis. Blood supplied through these vessels is not sufficient to prevent atrophy of the organ when the testicular artery is interrupted.

The **deep femoral artery** (*a. profunda femoris*) (Fig. 11–65) is about 3 cm. long, about half of it lying within the abdomen and half outside. It arises from the caudomedial surface of the external iliac at an angle of about 45 degrees, and runs obliquely distocaudally over the medial surface of the external iliac vein to leave the abdominal cavity by passing through the caudal part of the vascular lacuna. It sends a small branch over the pelvic surface of the pubis to the levator ani muscle, where it anastomoses with the obturator branch which ascends through the obturator foramen. Another small branch enters the laterally lying iliopsoas. Its principal intra-abdominal branch is the short pudendoepigastric trunk. After leaving the abdomen, the deep femoral becomes the **medial circumflex femoral** and passes between the quadriceps femoris and the medially lying pectineus. Before reaching the large adductor muscle, the artery gives rise to the caudally running obturator branch.

The **pudendoepigastric trunk** *(truncus pudendoepigastricus)* (Figs. 11–49, 11–65) is short and may extend to the abdominal inguinal ring before bifurcating terminally. Rarely, it is absent, in which case the terminal caudal epigastric and the external pudendal arteries arise separately from the deep femoral.

The **caudal epigastric artery** *(a. epigastrica caudalis)* (Fig. 11–49) runs cranially to lie on the dorsal surface of the rectus abdominis directly under the peritoneum. When it reaches the caudal border of the sheath of the rectus abdominis, it runs under this. It grooves the dorsal surface of the muscle as it parallels the linea alba in its branched, cranial course. At about the junction of the cranial and middle thirds of the abdominal part of the rectus abdominis, it anastomoses with the cranial epigastric artery in the substance of the muscle. It sends three or more branches laterally, which anastomose with the terminal abdominal branches of the deep circumflex iliac artery in the middle third of the abdomen. Small medial branches supply the relatively avascular linea alba. It is accompanied by a small satellite vein.

The **external pudendal artery** *(a. pudenda externa)* (Fig. 11–61) arises as the ventral terminal branch of the pudendoepigastric trunk. It leaves the abdominal cavity through the caudal part of the inguinal canal. After emerging through the superficial inguinal ring, it continues caudoventrally to the cranial border of the gracilis. It then arches cranially and becomes related to the dorsal surface of the superficial inguinal lymph node. Here in the fat-laden superficial abdominal fascia, it gives origin to the caudal superficial epigastric artery which continues along the mammary row. At this place the external pudendal starts the second arc of the sigmoid flexure it describes and, running ventral to the superficial inguinal lymph node, enters the fat ventral to the subpubic tendon. After emerging caudally from between the legs, it terminates in the vulva as the **ventral labial branch** *(ramus labialis ventralis)*. In the male, the comparable vessel may be extremely small or absent. Rarely is it large. It terminates as the **cranial scrotal branch** *(ramus scrotalis cranialis)*. When the cranial scrotal branch is absent, the external pudendal continues along the prepuce as the caudal superficial epigastric artery. The external pudendal artery is accompanied by a laterally lying satellite vein.

The **caudal superficial epigastric artery** *(a. epigastrica superficialis caudalis)* (Fig. 11–62) is small in the male as it runs forward to supply the prepuce, superficial inguinal lymph node, fascia, fat, and skin. It usually ends before it reaches the umbilicus. In the female it may be the largest artery of the abdominal wall. Its size is in direct proportion to the state of development of the mammary glands it supplies. From its origin, on the convex side of the second arc of the external pudendal artery, it runs forward deep to the inguinal mammary gland and medial to its nipple. It supplies many branches to this potentially pendulous gland. As it advances cranially, it enters the caudal abdominal mammary gland and divides into several branches, some of which become subcutaneous around the nipple. Many of these branches anastomose with like branches from the cranial superficial epigastric artery between the cranial and caudal mammary glands. The vessel takes on special significance because of the high incidence of mammary tumors in the dog. It is accompanied by a satellite vein and lymphatics which drain into the superficial inguinal lymph node.

The **medial circumflex femoral artery** *(a. circumflexa medialis)* (Figs. 11–63, 11–65) is the continuation of the deep femoral artery beyond the vascular lacuna. The medial circumflex femoral artery obliquely crosses the surface of the iliopsoas and vastus medialis, sending twigs into each, and aborizing extensively in the deep part of the adductor. On the caudal surface of the femur distal to the major trochanter, a nutrient artery enters the femur at the junction of the proximal and middle thirds.

The **obturator branch** *(ramus obturatorius)* (Figs. 11–63, 11–65) of the medial circumflex femoral artery, after running a few millimeters caudally, ascends through the cranial part of the obturator foramen, lateral to the obturator nerve. This vessel takes the place of the obturator artery, a

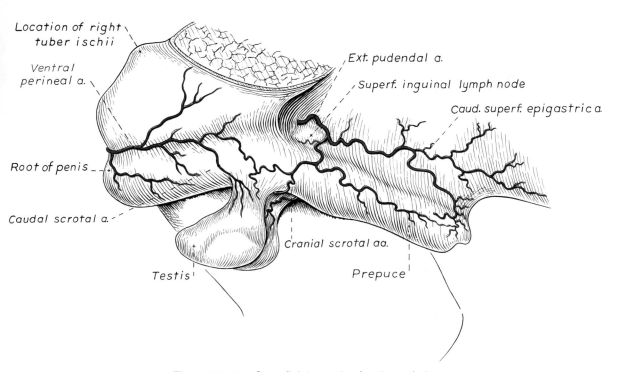

Figure 11–61. Superficial vessels of male genitals.

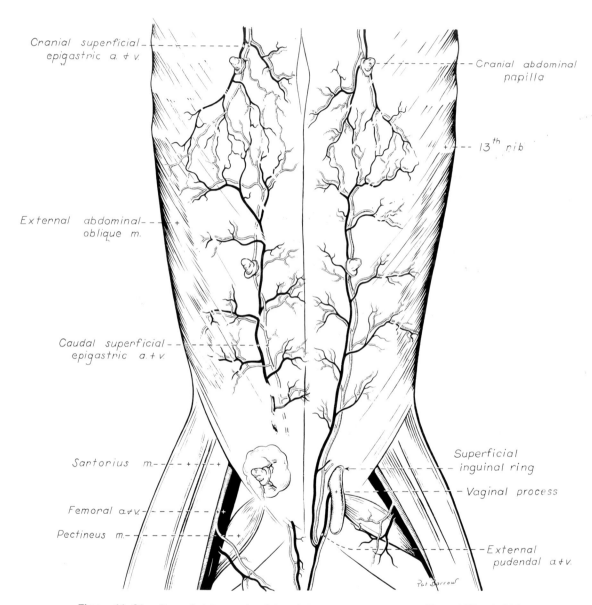

Figure 11–62. Superficial vessels of the abdomen, ventral aspect. (From Miller 1958.)

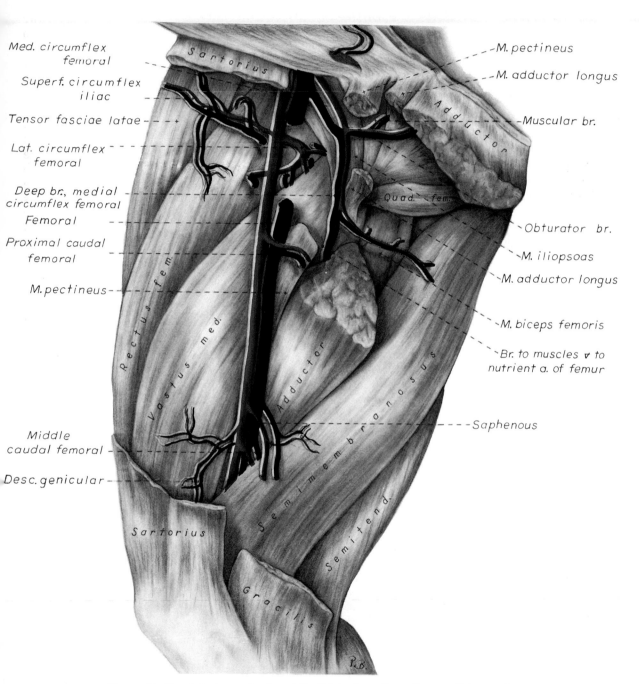

Figure 11–63. Arteries and veins of the thigh, deep dissection, medial aspect.

branch of the internal iliac as found in man and horse. It furnishes branches to the levator ani and coccygeus muscles and the external and internal obturator muscles. On the pelvic surface of the pubis it anastomoses with the small branch from the deep femoral which supplies the cranial portion of the levator ani muscle. Most of the intrapelvic portion of the obturator branch terminates in the internal obturator muscle. The main part of the obturator branch supplies the proximal end of the massive adductor muscle. About 5 mm. distal to the origin of the branch which ascends through the obturator foramen, a small branch extends laterally to enter the trochanteric fossa and supply portions of the muscles which insert there. Its chief importance concerns the branches which supply the caudal part of the hip joint capsule and the neck of the femur. It ends in the semimembranosus. Small branches also supply the quadratus femoris, pectineus, external obturator, and the semimembranosus. The main terminal part of the vessel, the transverse branch, *(r. transversus)* enters the semimembranosus about 3 cm. from the tuber ischii and anastomoses with the caudal gluteal artery. A deeper portion of the vessel *(r. profundus)* descends in the adductor near the femur and anastomoses with the lateral circumflex femoral artery.

FEMORAL ARTERY

The **femoral artery** *(a. femoralis)* (Figs. 11–63 to 11–65) continues the external iliac artery from the vascular lacuna through the thigh. It in turn is continued back of the stifle joint by the popliteal artery. Throughout the proximal half of the thigh, it lies cranial to its satellite vein and either caudal or medial to the saphenous nerve. Here, it lies superficially in the **femoral triangle** *(trigonum femorale),* which is the favored site for taking the pulse of the dog. The femoral vein is large and securely placed in the femoral triangle so that venipunctures can be made without compressing it. The femoral vessels are covered only by the thin skin and the deep medial femoral fascia. Upon leaving the femoral triangle at the middle of the thigh, the femoral artery and vein incline laterally along the medial border of insertion

of the adductor, where they are covered by the caudal belly of the semimembranosus. Upon reaching the popliteal surface of the femur, the femoral vessels are continued between the lateral and medial heads of the gastrocnemius as the popliteal vessels. The branches of the femoral artery in the order in which they arise are: superficial circumflex iliac, lateral circumflex femoral, muscular branches, proximal caudal femoral, middle caudal femoral, saphenous, descending genicular, and distal caudal femoral.

The **superficial circumflex iliac artery** *(a. circumflexa ilium superficialis)* is a small artery which runs dorsocranially over the medial surface of the rectus femoris to reach the septum between the caudal belly of the sartorius and the tensor fascia lata. It arises from the lateral surface of the femoral artery close to the lateral circumflex femoral artery, or it may arise from this vessel by a short common trunk. Its first branch usually is the principal vessel to the tensor fasciae latae. It also sends branches to the underlying rectus femoris. The main vessel then bifurcates into dorsal and ventral branches, under the thin, caudal belly of the sartorius. The dorsal branch supplies the proximal fourth of the cranial belly of this muscle and becomes subcutaneous at the cranial ventral iliac spine; the ventral branch runs distocranially and supplies the middle portion of the cranial belly of the sartorius. It has a satellite vein.

The **lateral circumflex femoral artery** *(a. circumflexa femoris lateralis),* formerly called the cranial femoral artery in this text, leaves the caudolateral surface of the femoral about 5 mm. from the abdominal wall. It immediately disappears between the rectus femoris and the vastus medialis. It is about 2 mm. in diameter at its origin, where it is crossed cranially by the saphenous nerve and caudally by the femoral vein. Usually its first branch is small and enters the iliopsoas near its insertion on the trochanter minor. Another small branch may extend lateral to the proximal part of the cranial border of the vastus lateralis and enter the caudal, deep part of the tensor fasciae latae. As the main artery loses contact with the iliopsoas to bend distally in the quadriceps, it sends a large branch caudally, which bifurcates upon reaching the neck of the femur. The

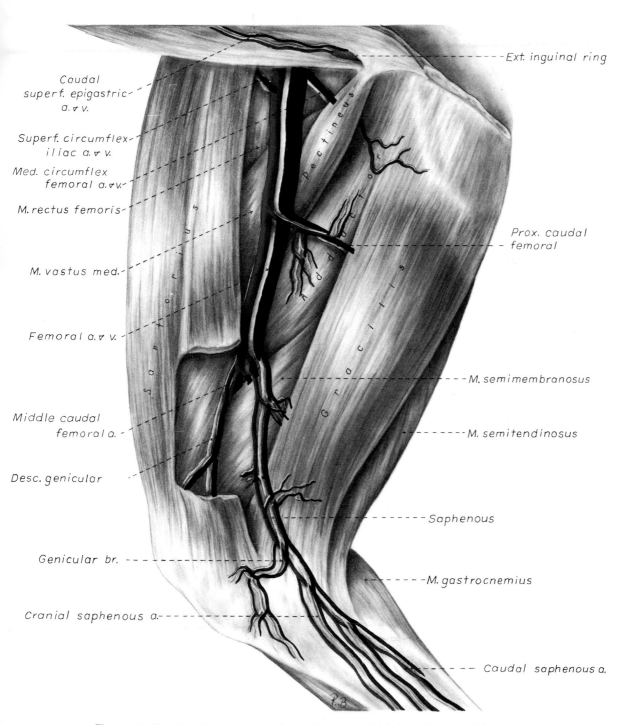

Caudal
superf. epigastric-
a. v v.

Superf. circumflex-
iliac a. v v.

Med. circumflex
femoral a. v v.-

M. rectus femoris-

M. vastus med.-

Femoral a. v v. -

Middle caudal
femoral a. -

Desc. genicular

Genicular br.

Cranial saphenous a.

Ext. inguinal ring

Pectineus

Adductor

Gracilis

Sartorius

Prox. caudal
- femoral

M. semimembranosus

M. semitendinosus

Saphenous

M. gastrocnemius

Caudal saphenous a.

Figure 11–64. Arteries and veins of the thigh, superficial dissection, medial aspect.

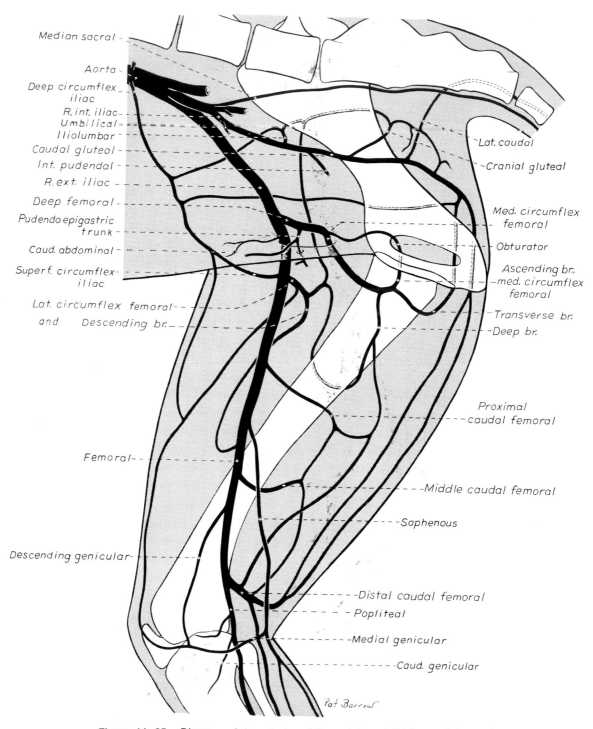

Figure 11–65.   Diagram of the arteries of the pelvis and thigh, medial aspect.

resultant **ascending branch** *(ramus ascendens)* sends many twigs to the cranial part of the hip joint capsule, then continues dorsally to supply the insertions of the deep and middle gluteal and the tensor fasciae latae muscles. As the **descending branch** *(ramus descendens)* bends distolaterally around the caudal surface of the rectus femoris, it sends branches to the vasti, rectus femoris, and tensor fasciae latae. It is the principal source of blood to the quadriceps. It usually forms a feeble anastomosis with the medial circumflex femoral artery caudal to the femur. It is accompanied by a satellite vein, which lies distal to it, and the large femoral nerve, which lies proximal to it. The lateral circumflex femoral of this description is comparable to the transverse ramus of man. Hermann (1940) and Ellenberger and Baum (1943) state that the lateral circumflex femoral may arise from the femoral separately. There is disagreement about the homology of this vessel in man and the domestic animals.

The **muscular branches** *(rami musculares)* of the femoral are variable in origin, number, size, and distribution. As the femoral artery passes distally from the point where the lateral circumflex femoral arises, it lies adjacent to the cranial border of the vastus medialis proximally and inclines laterally around its caudal border distally. Distal to the origin of the lateral circumflex femoral, a muscular branch arises from the lateral surface of the femoral, which bifurcates into cranial and caudal vessels. Two separate diverging vessels may arise directly from the femoral. The caudal vessel may end in the femoral sheath or terminate in the proximal end of the pectineus. The slender cranial branch goes to the deep surface of the caudal belly of the sartorius.

The **proximal caudal femoral artery** *(a. caudalis femoris proximalis)* leaves the caudal side of the femoral, extends distocaudally, and crosses the end of the pectineus and then the adductor. Upon reaching the cranial border of the gracilis, at the junction of its proximal and middle thirds, it runs under it to enter its deep surface. It sends one or more branches to the pectineus and the adductor. Usually three branches go to the adjacent muscles before the femoral sinks into the substance of the thigh. Often two of these

arise from the cranial side of the vessel and after short courses disappear into the vastus medialis. One or two smaller twigs quite regularly leave the caudal side of the femoral and disappear into the distal part of the adductor.

The **middle caudal femoral artery** *(a. caudalis femoris media)* arises from the femoral. It runs caudodistally and ramifies in the adductor and semimembranosus. It anastomoses with the distal caudal femoral artery and is accompanied by a satellite vein.

The **saphenous artery** *(a. saphena)* (Fig. 11–64), less than 1 mm. in diameter, arises from the medial surface of the femoral just before the femoral disappears under the semimembranosus. It runs distally across the medial surface of the semimembranosus, where it and its accompanying vein and nerve lie between the converging borders of the caudal belly of the sartorius in front and the gracilis behind. It supplies small twigs to these muscles. As it passes over the medial surface of the stifle, it sends a single or a paired **genicular branch** *(r. articularis genus)* to the skin and superficial fascia covering the medial side of the joint. A proximal branch runs dorsally and anastomoses with the superficial part of the deep circumflex iliac artery. Opposite or distal to the tibial condyle the saphenous terminates in a small dorsal and a larger plantar branch.

The **cranial branch** of the saphenous artery *(r. cranialis)* obliquely crosses the subcutaneous part of the tibia in its course distocranially. It then bends around the medial border of the cranial tibial muscle and continues distally in the superficial fascia covering this muscle to traverse the flexor surface of the tarsus. Opposite the tarsus or distal to it, the dorsal branch anastomoses with the superficial ramus of the cranial tibial. Two or three delicate rami leave the dorsal branch and supply the fascia, periosteum, skin, and cranial tibial muscle. In the proximal part of the metatarsus it terminates in the three dorsal common digital arteries, which are described under "Arteries of the Hindpaw."

The **caudal branch** of the saphenous artery *(r. caudalis)* is the direct continuation of the saphenous artery in the crus after the dorsal branch arises. It lies medial to the tibia opposite the cleft between the medial head of the gastrocnemius and the bone. It soon becomes related to the flexors of the digits, and with its small accompanying vein and tibial nerve crosses the medial surface of the tarsus to enter the metatarsus. It has one prominent branch in the crus.

The **descending genicular artery** (*a. genus descendens*) usually arises from the femoral distal to the origin of the saphenous, but it may arise in common with it. It runs distally between the vastus medialis and the semimembranosus to the medial surface of the stifle joint. In this course it sends two or more small, short twigs into the vastus medialis. It lies at a deeper level than does the genicular branch of the saphenous artery. Upon reaching the medial epicondyle, it divides into articular branches. End-branches supply the medial part of the femoropatellar and the medial division of the femorotibial joint capsules. The descending genicular is the main blood supply to the stifle joint. It is accompanied by a satellite vein.

The **distal caudal femoral artery** (*a. caudalis femoris distalis*) (Fig. 11–66) arises from the caudolateral surface of the femoral approximately 1 cm. proximal to its entrance into the gastrocnemius to become the popliteal artery. Usually it runs caudodistally on the gastrocnemius and sends its first and largest branch laterally into the distal end of the biceps femoris, but this branch may come from the femoral directly. Most of this vessel ascends in the muscle, but one or more twigs continue through the muscle to end as cutaneous branches to the caudolateral part of the thigh. Another twig enters the insertion of the adductor. The small artery which crosses the lateral surface of the femur to enter the vastus lateralis comes from either the femoral or the distal caudal femoral. Several branches run distally and are distributed to both heads of the gastrocnemius. About 2 cm. from its origin, the distal caudal femoral artery crosses the lateral surface of the tibial nerve and sends a large branch proximally into the semimembranosus. Coming from this muscular ramus, or arising independently distal to it, is a smaller branch to the semitendinosus. The terminal part of the vessel becomes cutaneous in the upper caudal part of the crural region. The large descending branches of the distal caudal femoral artery supply the gastrocnemius; the large ascending branches are the chief supply to the hamstring muscles. Their principal anastomoses (Fig. 11–65) are with the caudal gluteal artery, but there are also anastomoses with branches of the deep femoral and with the muscular branches of the femoral which reach the hamstring muscles. The distal caudal femoral artery is accompanied by the large proximally lying lateral saphenous vein. This receives the veins which accompany the branches of the artery. The main lymphatics from the limb distal to the stifle ascend over the gastrocnemius in relation to this artery.

POPLITEAL ARTERY

The **popliteal artery** (*a. poplitea*) (Fig. 11–66) is the continuation of the femoral through the popliteal fossa. It begins by entering the gastrocnemius and terminates in the interosseous space between the tibia and fibula distal to their heads. It divides into the small caudal tibial and the much larger cranial tibial artery. Passing between the two heads of the gastrocnemius, it crosses the medial surface of the superficial digital flexor and over the flexor surface of the stifle joint. It inclines laterally under the popliteus and, upon leaving it, perforates the origin of the flexor hallucis longus to reach the interosseous space. It is accompanied by a small satellite vein. It has small genicular and muscular branches.

The **caudal genicular arteries** (*aa. genus caudales*) are small branches which run to the caudal surface of the stifle joint. Opposite the femoral condyles, a medium-sized genicular branch arises from each of the medial and lateral surfaces of the popliteal artery and diverge from each other to reach the distal ends of the medial and lateral collateral ligaments. These vessels then arborize on the deep surfaces of the medial and lateral heads of the gastrocnemius. Two or three small articular branches leave the cranial surface of the popliteal and supply the caudal part of the femorotibial joint capsule and the cruciate ligaments. The greatest vascular supply to the stifle joint comes from the caudal side.

The **muscular branches** (*rami musculares*) are represented by two branches to the gastrocnemius, which also go to the collateral

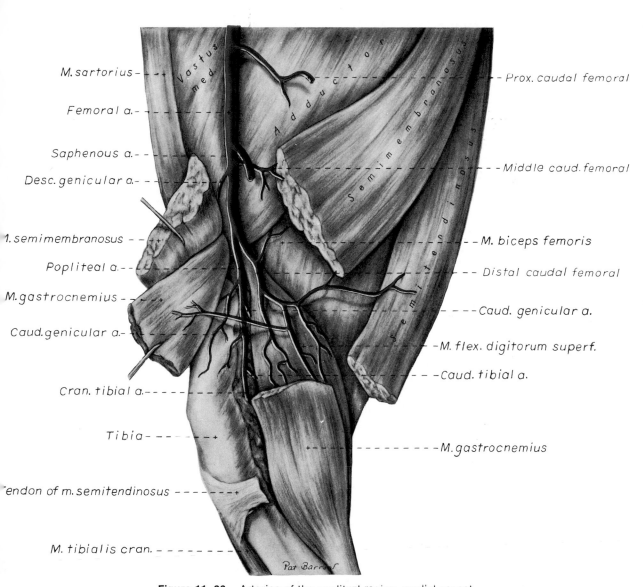

Figure 11-66. Arteries of the popliteal region, medial aspect.

ligaments, and by a branch which leaves the caudal surface of the popliteal artery and runs to the caudal surface of the popliteal muscle. Before the popliteal artery terminates, it gives a branch to the small, proximal tibiofibular joint capsule which continues beyond this to descend in the flexor hallucis longus.

The **caudal tibial artery** (*a. tibialis caudalis*), leaves the caudal surface of the popliteal at the interosseous space and plunges into the flexor hallucis longus. It runs distally in this muscle, giving off medial and lateral branches. It supplies the **nutrient artery of the tibia** (*a. nutriciae tibiae*) which enters the nutrient foramen on the caudal surface near the lateral border at the junction of the proximal and middle thirds of the tibia.

The **cranial tibial artery** (*a. tibialis cranialis*) (Figs. 11–66, 11–67) continues the popliteal artery between the tibia and fibula after the caudal tibial arises. It inclines laterally as it runs distally. It crosses under the peroneus longus to gain the deep surface of the long digital extensor. Here it is partly separated from the bone by the small, flat extensor hallucis longus which lies lateral to the artery. The superficially lying cranial tibial muscle and the lateral digital extensor muscle receive almost their entire blood supply from branches of the cranial tibial artery. The largest three or four muscular branches arise from the first 2 cm. of the artery. The first branch runs proximally and supplies the long digital extensor as it lies in the sulcus muscularis of the tibia, and smaller twigs are distributed to the stifle joint capsule. The second muscular branch runs under the long extensor and is distributed to the cranial tibial muscle. Small anastomoses exist between the cranial tibial and the distal caudal femoral and the cranial branch of the saphenous. The **superficial branch** (*ramus superficialis*) is very small in diameter, but long. It reaches the superficial crural fascia between the peroneus longus and the long digital extensor at the beginning of the distal third of the crus. Here, in association with the superficial peroneal nerve, it sends a branched twig to the lateral malleolus. The artery continues to the flexor surface of the tarsus, where it anastomoses with the cranial branch of the saphenous artery. If this does not occur, it continues into

the hindpaw as the dorsal metatarsal artery II.

ARTERIES OF THE HINDPAW

The cranial tibial artery is continued opposite the talocrural joint as the **dorsal pedal artery** (*a. dorsalis pedis*) (Fig. 11–70). Several small branches arise from it to supply adjacent structures. The **medial and lateral tarsal arteries** (*a. tarsea medialis et a. tarsea lateralis*) run deeply to the sides of the tarsus and end in the collateral ligaments. The **arcuate artery** (*a. arcuata*) (Fig. 11–67) is a small vessel which leaves the lateral side of the dorsal pedal artery at the proximal end of the metatarsus and runs transversely to disappear in the ligamentous tissue of the lateral and plantar sides of the proximal end of the metatarsus. It sends two or three branches proximally to be distributed to the deep structures of the flexor surface of the tarsus. Distally, it gives rise to the third and fourth dorsal metatarsal arteries.

The dorsal pedal artery becomes the **perforating metatarsal artery** (*a. metatarsea perforans*) as it passes from the dorsal to the plantar surface of the hindpaw, by passing between the proximal ends of the second and third metatarsal bones.

The arteries of the metatarsus and digits are similar to those of the metacarpus and digits. Forepaws and hindpaws are similar also in that the main arteries supplying them lie on their flexor sides. (In accordance with the N.A., the superficial arteries of the metapodium are designated aa. digitales communes, and the deep arteries are termed aa. metatarsea or metacarpea. Digital arteries that originate from the bifurcation of aa. digitales communes are called aa. digitales propriae.)

The **dorsal common digital arteries II, III, and IV** (*aa. digitales dorsales communes II, III, et IV*) (Fig. 11–70) arise subcutaneously, over the tendon of the long digital extensor, from a small trunk formed by the anastomosis of the cranial branch of the saphenous and the superficial branch of the cranial tibial. Typically, this bifurcates into medial and lateral branches at the proximal end of the metatarsus. The medial branch becomes the second dorsal common digital artery, and the lateral branch divides into the third and

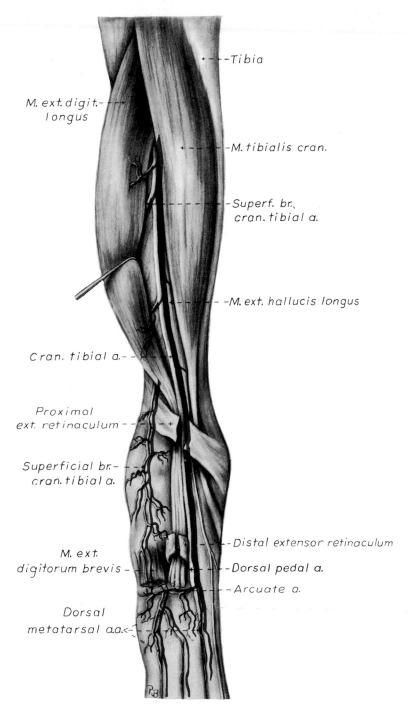

Tibia

M. ext. digit.-
longus

M. tibialis cran.

Superf. br.,
cran. tibial a.

M. ext. hallucis longus

Cran. tibial a.-

Proximal
ext. retinaculum

Superficial br.-
cran. tibial a.

M. ext.
digitorum brevis-

Distal extensor retinaculum

Dorsal pedal a.

Arcuate a.

Dorsal
metatarsal a.a.<

Figure 11–67. Branches of the cranial tibial artery, cranial aspect of right hind leg.

fourth. Occasionally, an arch formed by the anastomoses of the medial and lateral branches gives off the superficial set. The superficial branch of the tibial and the cranial branch of the saphenous may continue independently to form the axial and abaxial arteries of the dorsal superficial set. When a first digit or dewclaw is present (Fig. 11–70), its dorsal part is supplied by the abaxial and axial dorsal proper digital arteries (aa. digitalis dorsales propriae). These vessels arise from the **dorsal common digital artery I** (*a. metatarsea dorsalis I*), which comes from the cranial branch of the saphenous and the dorsal pedal artery. When the first digit is missing, as it usually is, the first dorsal common digital artery becomes dissipated on the medial part of the metatarsus. The dorsal common digital arteries II, III, and IV anastomose with the dorsal metatarsal arteries in the distal ends of the intermetatarsal spaces.

The **dorsal metatarsal arteries II, III, and IV** (*aa. metatarseae dorsales*) (Fig. 11–67) are similar to the comparable arteries of the forepaw. (No first dorsal metatarsal artery is present, even when a first digit exists.) The second dorsal metatarsal artery arises from the dorsal pedal artery as it enters the second intermetatarsal space. The third and fourth arteries arise from the arcuate artery either separately or from a common stem. The three dorsal metatarsal arteries anastomose with their superficial counterparts to continue the dorsal common digital arteries.

The dorsal common digital arteries II, III, and IV are short vessels which give off **axial and abaxial dorsal proper digital arteries II, III, and IV** (*aa. digitales dorsales propriae II, III, et IV*). Each dorsal common digital artery, or one of its proper branches, anastomoses by a short connection deeply in the interdigital cleft with the comparable plantar common digital artery.

As the caudal branch of the saphenous artery approaches the tarsus, it gives off several twigs, **tarsal branches** (*rami tarsici*), to the skin and fascia of the medial surface of the tarsus. Distal and medial to the tuber calcanei the **medial and lateral plantar arteries** (*a. plantaris medialis et a. plantaris lateralis*) arise. These are only slightly larger than the largest tarsal branches. The lateral plantar artery usually arises about 1 cm. proximal to the origin of the medial vessel. It

runs diagonally distolaterally on the plantar ligament under the superficial flexor tendon and, gaining the lateral border of the deep flexor tendon, runs under it. The medial plantar artery runs at first along the medial border of the deep flexor tendon, then under it, to anastomose in the interosseous musculature with the lateral plantar artery. The **plantar common digital artery I** (*a. digitalis plantaris communis*) leaves the medial side of the medial plantar before it turns under the deep flexor tendon. It runs on the plantar surface of the first digit as the **plantar digital artery I** (*a. digitalis plantaris I*). It anastomoses with the larger dorsal digital artery I. The short trunk formed by the anastomosis of the two plantar arteries in turn anastomoses with the perforating metatarsal artery to form the **deep plantar arch** (*arcus plantaris profundus*). Occasionally two anastomoses exist between these vessels. The perforating metatarsal limb is much the larger of the two parts which form the arch. The metatarsal part of the caudal branch of the saphenous artery gives rise to the plantar common digital arteries.

The **plantar common digital arteries II, III, and IV** (*aa. digitales plantares communes*) (Fig. 11–68) arise from the terminal part of the caudal branch of the saphenous as it crosses the superficial digital flexor tendons. In the clefts between the digits they anastomose with the larger plantar metatarsal arteries.

The **plantar metatarsal arteries II, III, and IV** (*aa. metatarseae plantares II, III, et IV*) (Fig. 11–69) arise from the plantar arch. Between their origins, branches arise which supply interosseous muscles. The plantar metatarsal arteries are the chief blood supply to the hindpaw distal to the tarsus. As they run distally, they lie between adjacent interosseous muscles. They anastomose with the corresponding members of the common digitals near the distal ends of the intermetatarsal spaces. One or two small anastomotic branches extend dorsally from about the last centimeter of each of these arteries and anastomose with either the dorsal metatarsal arteries or the dorsal common digital arteries.

The plantar common digital arteries II, III, and IV are continued by the anastomoses of the corresponding members of the two plantar series. These and their branches are

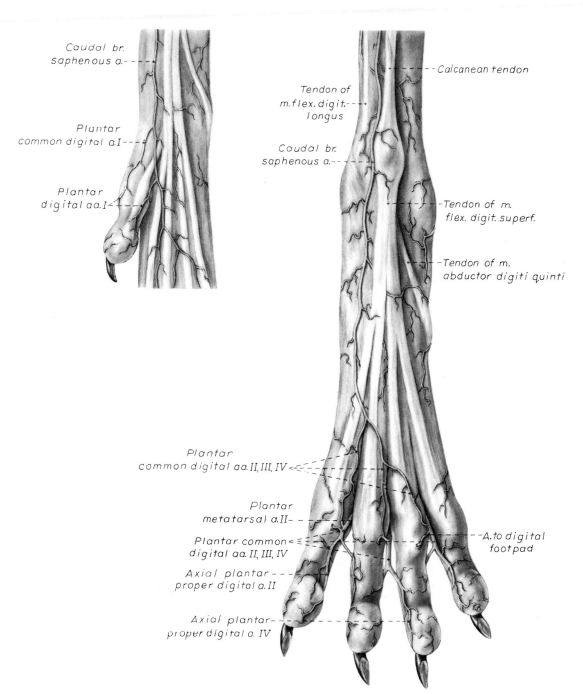

Caudal br.
saphenous a.--

Plantar
common digital a.I---

Plantar
digital aa.I--

Calcanean tendon

Tendon of
m.flex.digit.--
longus

Caudal br.
saphenous a.--

Tendon of m.
flex. digit. superf.

Tendon of m.
abductor digiti quinti

Plantar
common digital aa.II, III, IV--

Plantar
metatarsal a.II--

Plantar common--
digital aa. II, III, IV

Axial plantar---
proper digital a.II

Axial plantar--
proper digital a. IV

A.to digital
footpad

Figure 11–68. Superficial arteries of the right hindpaw, plantar aspect.

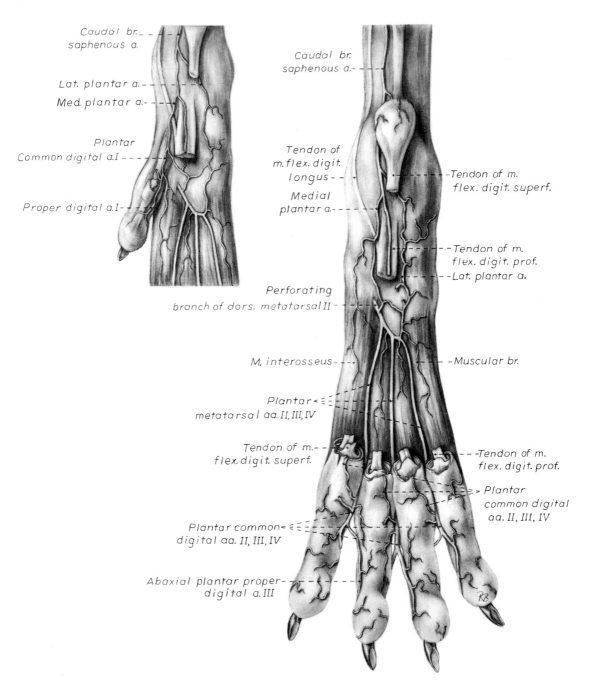

Figure 11–69.   Deep arteries of the right hindpaw, plantar aspect.

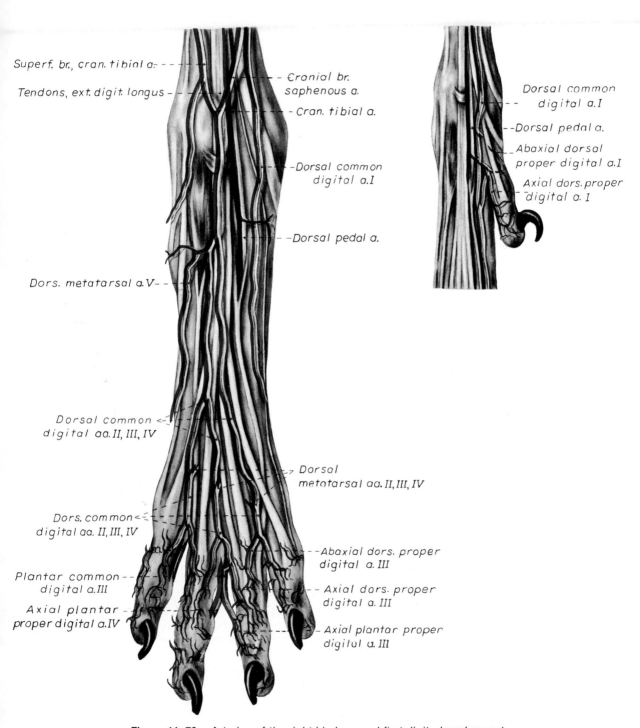

Superf. br., cran. tibial a.

Tendons, ext. digit. longus

Cranial br. saphenous a.

Cran. tibial a.

Dorsal common digital a.I

Dorsal pedal a.

Dors. metatarsal a.V

Dorsal common digital aa. II, III, IV

Dorsal metatarsal aa. II, III, IV

Dors. common digital aa. II, III, IV

Abaxial dors. proper digital a. III

Plantar common digital a.III

Axial dors. proper digital a. III

Axial plantar proper digital a.IV

Axial plantar proper digital a. III

Dorsal common digital a.I

Dorsal pedal a.

Abaxial dorsal proper digital a.I

Axial dors. proper digital a. I

Figure 11–70.   Arteries of the right hindpaw and first digit, dorsal aspect.

similar to the palmar common digital arteries. The superficial and deep dorsal and plantar arteries of the hindpaw are accompanied by corresponding nerves. With the exception of the plantar metatarsal arteries, all are accompanied by satellite veins.

Internal Iliac Artery

The **internal iliac artery** *(a. iliaca interna)* (Figs. 11–71, 11–72), with its fellow and the smaller, unpaired median sacral artery, terminates the aorta at the level of the seventh lumbar vertebra. It is about half the diameter of the external iliac, or 2.5 mm. in a medium-sized male or non-gravid female. The internal iliac artery terminates as the caudal gluteal artery and the internal pudendal.

The **umbilical artery** *(a. umbilicalis)* arises from the internal iliac about 0.5 cm. from its origin, but it may arise from the terminal portion of the aorta. In the fetus it carries blood to the placenta and is the main artery of the pelvis. In about half of the specimens, it sends one or more twigs to the cranial end of the bladder as the **cranial vesical arteries** *(aa. vesicales craniales)*. In the adult dog the lumen of the artery is usually obliterated distal to the origin of these arteries, forming the **lateral umbilical ligament** *(ligamentum umbilicale laterale).* Ellenberger and Baum (1943) and Sisson and Grossman (1953) call the urogenital artery, which supplies much of the pelvic parts of the urinary and genital systems, the umbilical artery. Since this artery never formed a part of the vascular path from the fetus to the placenta, the term is a misnomer; the artery is not homologous with the comparable vessel of man.

The **internal pudendal artery** *(a. pudenda interna)* (Figs. 11–71, 11–72) is the smaller, more ventral branch of the internal iliac artery. It lies in contact ventrolaterally with its satellite vein as it runs caudally on the terminal tendon of the psoas minor muscle. Upon reaching the origin of the levator ani, it gives rise to the prostatic artery in the male (or the vaginal artery in the female) and continues as the internal pudendal artery. The prostatic and vaginal arteries were formerly known as the urogenital or urethrogenital artery.

The **prostatic artery** *(a. prostatica)* of the male (Fig. 11–71) is homologous with the **vaginal artery** of the female. It lies in the pelvic fascia with the pelvic plexus of nerves. As the prostatic artery passes ventrally across the rectum, it gives origin to the **caudal vesicle artery** of the bladder, with its **ureteral** and **urethral branches,** and to the **artery of the ductus deferens.** Before reaching the prostate gland, the prostatic artery gives rise to the **middle rectal artery.**

The **caudal vesical artery** *(a. vesicalis caudalis)* is a branched vessel which contacts the bladder at its neck and ramifies over its caudolateral surface. Its medial branches anastomose with their fellows both dorsally and ventrally on the organ. Cranially, the caudal vesical artery may anastomose with the cranial vesical from the umbilical artery. If the cranial vesical is lacking, the paired caudal vesical artery supplies the whole organ. It is accompanied by a satellite vein and lymph vessels. Sometimes two caudal vesical arteries are present on each side.

In the male the small **artery of the ductus deferens** *(a. ductus deferentis)* (Fig. 11–71), is homologous with the uterine artery of the female. Upon reaching the ductus deferens at the point where it enters the dorsum of the prostate, it runs proximally toward the testis and anastomoses in the epididymis with the testicular artery. It is accompanied by a satellite vein, lymphatics, and a nerve component from the pelvic plexus.

When the prostate is hypertrophied to the extent that it is an abdominal organ, the prostatic artery and related structures are carried forward with the gland. Two or more branches pass over the lateral surface of the prostate, and others enter the parenchyma of the organ. Some of these extend through the gland to supply the prostatic portion of the urethra and the colliculus seminalis. The **urethral branches** *(rami urethrales)* are the several branches which arise from the caudal extension of the prostatic artery. The first branch sends twigs to the caudal part of the prostate, as well as to the urethra. The remaining branches aborize on the caudal part of the membranous urethra and its surrounding urethral muscle and, bending distally around the ischial arch, enter the root of the penis, where they anastomose with the artery of the bulb. The caudal portion of the prostatic and its branches are accompanied by satellite veins, lymphatics, and autonomic nerves.

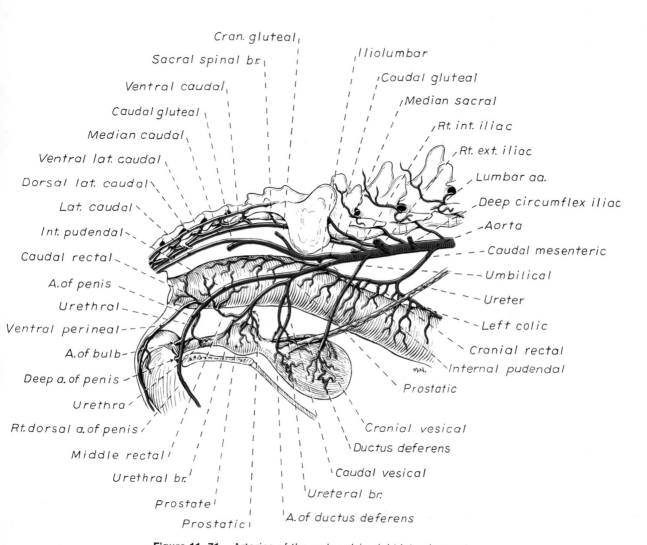

Figure 11–71. Arteries of the male pelvis, right lateral aspect.

The **middle rectal artery** *(a. rectalis media)* leaves the dorsal side of the prostatic artery in the male. It arborizes on the wall of the rectum and anastomoses with the cranial and caudal rectal arteries.

The **vaginal artery** *(a. vaginalis)* of the female (Fig. 11–72) courses ventrally across the rectum and divides into two primary branches. The more cranial branch gives rise to a **urethral artery**, a **uterine artery**, **vaginal branches**, and a **caudal vesicle artery** with **ureteral and urethral branches.** The caudal branch of the vaginal artery gives rise to a **middle rectal artery** and terminates by ramifying on the vagina.

The **uterine artery** *(a. uterina)* (Fig. 11–72) is the main artery to the uterus in all species and has also been called the middle uterine artery, caudal uterine artery (Ellenberger and Baum 1943), or cervicouterine artery (Barone and Pavaux 1962). It is variable in size, depending upon the reproductive state. The artery enters the broad ligament at the level of the cervix and passes cranially on the lateral side of the body of the uterus. Shortly after the uterine horns diverge, the uterine artery lies 1 to 4 cm. from the uterine horn in the broad ligament and sends branches onto the mesometrial surface of the uterus. Near the origin of the uterine artery, a branch arises which supplies the cranial end of the vagina. The uterine artery has a satellite vein and accompanying autonomic nerve plexuses and lymphatics.

The **caudal vesical artery** *(a. vesicalis caudalis)* contacts the neck of the bladder, as in the male, and ramifies over the caudolateral surface of the bladder. It has a **ureteral branch** ascending the ureter and a **urethral branch** coursing caudally over the neck of the bladder.

The **middle rectal artery** *(a. rectalis media)* leaves the dorsal surface of the vaginal artery and ramifies in the wall of the rectum, where it forms anastomoses with both cranial and caudal rectal arteries.

The terminal portion of the vaginal artery aborizes in the wall of the vagina.

The internal pudendal artery continues along the dorsal border of the ischiatic spine, where it lies lateral to the coccygeus muscle and medial to the gluteal and piriform muscles. Occasionally the caudal gluteal and lateral caudal arteries arise from the internal pudendal. The internal pudendal is free of branches until it reaches the ischiorectal fossa, where it gives off a trunk for the caudal rectal and the perineal arteries (Thamm 1941). After this trunk arises, opposite the anal sac or caudal to the caudal border of the levator ani, the internal pudendal continues as the artery of the penis or clitoris.

The **caudal rectal artery** *(a. rectalis caudalis)* may arise from the internal pudendal cranial to the perineal artery, but the two usually arise from a common trunk. The caudal rectal runs medially and divides into dorsal and ventral branches as it reaches the anal canal just cranioventral to the anal sac. In its course medially it gives off a lateral branch to this sac. Its dorsal part lies under the external anal sphincter and anastomoses with the opposite artery mid-dorsally. It sends branches to and through the external anal sphincter to supply the circumanal glands, which are usually hypertrophied in old male dogs. It anastomoses in a plexiform manner with the cranial rectal artery on the dorsal part of the anal canal. The ventral branch anastomoses with its fellow so that an arterial circle is formed around the anal opening. This is frequently plexiform in nature. In some specimens the caudal rectal artery arises as an unpaired vessel from the middle caudal artery opposite the sixth caudal vertebra.

The **ventral perineal artery** *(a. perinealis ventralis)* (Figs. 11–71, 11–72, 11–73) usually arises from the internal pudendal in common with the caudal rectal. It courses superficially and supplies the skin and fat at the pelvic outlet. Ventrally, its deeper branches may aid in forming the arterial ring around the anus, after which it leaves the pelvic outlet and runs distally to supply the caudal part of the scrotum as the **dorsal scrotal branch** *(ramus scrotalis dorsalis).* In the female it sends a long branched vessel distally, which ends in the vulva as the **dorsal labial branch** *(ramus labialis dorsalis).* The internal pudendal artery is accompanied by the pudendal nerve, and its branches have satellite veins and lymphatics.

The **artery of the penis** *(a. penis)* (Fig. 11–71) terminates the internal pudendal artery in the male (Christensen 1954). It begins where the ventral perineal artery leaves the internal pudendal or if this vessel should arise by a common trunk with the caudal rectal, then the origin of the trunk

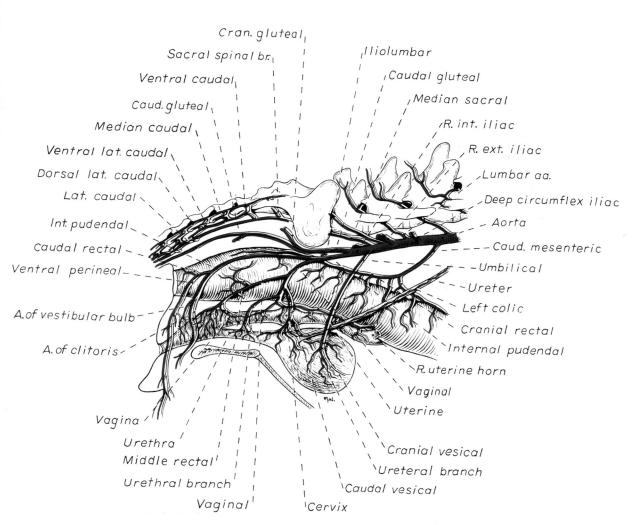

Figure 11–72. Arteries of the female pelvis, right lateral aspect.

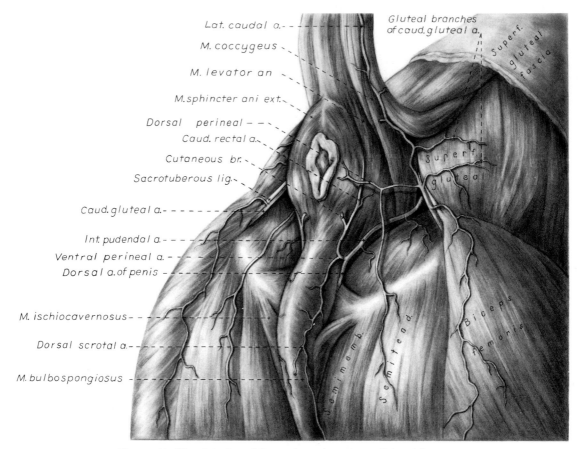

Figure 11–73. Arteries of the male perineum, caudolateral aspect.

marks the beginning of the artery of the penis. It is about 2 cm. long and devoid of collateral branches. It terminates by bifurcating or, more commonly, giving off in succession the artery of the bulb of the penis, deep artery of the penis, and the dorsal artery of the penis.

The **artery of the bulb of the penis** (*a. bulbi penis*) is a short artery which divides initially into two or three branches, which then redivide within the bulb. The bulb, corpus spongiosum, penile urethra, and pars longa glandis receive blood through the artery of the bulb. A homologous vessel, the **artery of the vestibular bulb,** is present in the female.

The **deep artery of the penis** (*a. profunda penis*) divides into two to five branches and passes through the tunica albuginea to supply the corpus cavernosum penis and os penis.

The **dorsal artery of the penis** (*a. dorsalis penis*) runs on the dorsal surface of the penis distally to the bulbus glandis. It anastomoses with the deep artery and the artery of the bulb. According to Christensen (1954), the artery trifurcates into deep, superficial, and preputial branches. For a discussion of the finer distribution of the arteries and veins of the penis and the mechanism of erection, refer to Chapter 9, The Urogenital Apparatus.

The **artery of the clitoris** (*a. clitoridis*) (Fig. 11–72) is homologous with the artery of the penis. It is a minute vessel which supplies the fat, erectile tissue, and integument which compose the clitoris. It is the terminal part of the visceral branch of the internal pudendal artery.

The **caudal gluteal artery** (*a. glutea caudalis*) is the larger, more terminal branch of the internal iliac. Its proximal portion was formerly called "the parietal branch of the internal iliac." The caudal gluteal artery lies

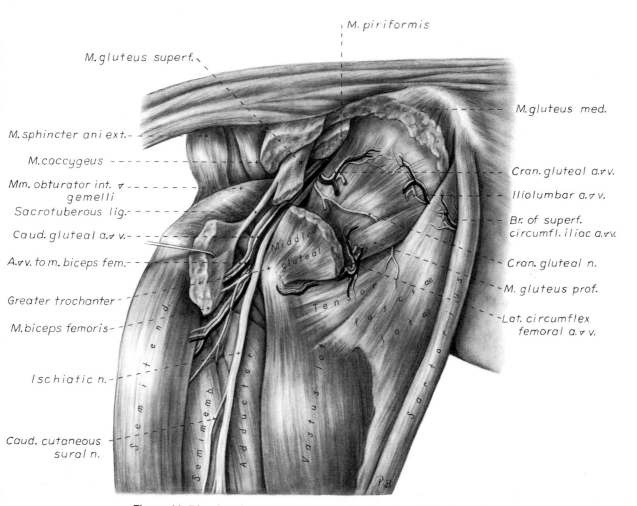

Figure 11-74. Arteries and veins of the gluteal region, lateral aspect.

ventral to the base of the sacrum caudomedial to the iliopsoas muscle and gives off an iliolumbar artery before continuing toward the ischiatic spine parallel to the internal pudendal artery. Before reaching the muscles of the hip, the caudal gluteal gives rise to a lateral caudal (coccygeal) artery to the tail and a dorsal perineal to the anus.

The **iliolumbar artery** (*a. iliolumbalis*) (Figs. 11–74, 11–75), less than 1 mm. in diameter, leaves the pelvic cavity between the iliopsoas muscle and the base of the sacrum. It arises from the caudal gluteal artery or from the internal iliac directly. A small twig leaves its caudal side to run with the lumbosacral plexus of nerves. As it crosses the border of the ilium, it sends the **nutrient artery of the ilium** (*a. nutricia ilia*) caudoventrally to enter the nutrient foramen. The main part of the vessel crosses the cranioventral border of the wing of the ilium at the caudal ventral iliac spine and plunges into the cranial muscles of the rump and thigh. Branches are supplied to the iliopsoas and the quadratus lumborum as they cross the artery. The principal branches go to the proximal parts of the middle gluteal, and other branches supply the abdominal wall. The iliolumbar anastomoses with the cranial gluteal, caudally, and with the superficial circumflex iliac, cranially. It is accompanied by a small satellite vein.

The **cranial gluteal artery** (*a. glutea cranialis*) (Fig. 11–74), about 1.5 mm. in diameter, runs dorsocaudally across the lateral surface of the ischiatic nerve and enters the overlying middle gluteal after passing over the cranial part of the greater ischiatic incisure near the caudal dorsal iliac spine. Upon reaching the deep surface of the middle gluteal, it divides. A small branch extends dorsally between the middle gluteal and the piriformis, both of which it supplies, and becomes superficial at the sacrocaudal junction. The main part of the vessel, with the cranial gluteal nerve and a satellite vein, runs craniolaterally between the deep and middle gluteals. It is the main supply to the middle gluteal and anastomoses with the iliolumbar artery, which enters the muscle cranially. It also anastomoses with the lateral circumflex femoral, which enters the middle gluteal after crossing the cranial surface of the hip joint.

From the ventral surface of the caudal gluteal artery, a long cutaneous branch arises which leaves the pelvic outlet at the ischiorectal fossa and courses toward the root of the penis in the fat of the pelvic outlet. It supplies the fat and skin which cover the upper part of the caudal surface of the thigh. It may anastomose with the perineal artery or may replace it functionally.

The **lateral caudal artery** (*a. caudalis* [*coccygea*] *lateralis*) (Fig. 11–75) leaves the caudal gluteal about 4 cm. caudal to the origin of the cranial gluteal. It leaves the pelvis by passing caudodorsally between the tail and the superficial gluteal muscle. A cutaneous branch arises at this level and, curving around the sacrotuberous ligament and superficial gluteal muscle, enters the superficial gluteal fascia. It supplies the skin and fascia of the rump as far forward as the crest of the ilium. Branches leave both the dorsal and the ventral surface of the vessel at irregular intervals and supply the skin and adjacent fascia of the tail. At about the sixth caudal vertebra, it inclines dorsally and, lying on the deep caudal fascia dorsal to the transverse processes, runs to the tip of the tail. In the distal two-thirds of the tail it sends many branches to the skin and muscles along its dorsolateral aspect. Other branches run across the discs and anastomose with the median caudal artery; shorter branches run deeply to anastomose with the deep lateral caudal arteries. Arteriovenous anastomoses exist in the caudal third of the tail. The large satellite vein lies ventral to the lateral caudal artery.

The continuation of the caudal gluteal artery distal to the origin of the lateral caudal artery passes toward the tuber ischii in relation to the sacrotuberous ligament caudolaterally, the ischiatic nerve cranioventrally, and its satellite vein medially. It gives a muscular branch to the superficial gluteal, and sends twigs to the piriformis, obturator internus, and gluteus medius. It anastomoses with the cranial gluteal artery in the gluteus medius. The main caudal gluteal artery passes over the ischiatic spine with the ischiatic nerve and satellite vein, and divides into several branches as it enters the biceps femoris. One branch enters the proximal end of the semitendinosus, and another enters the semimembranosus. The **satellite artery of the ischiatic nerve** (*a. comitans n. ischiadicus*), a branch of the caudal gluteal,

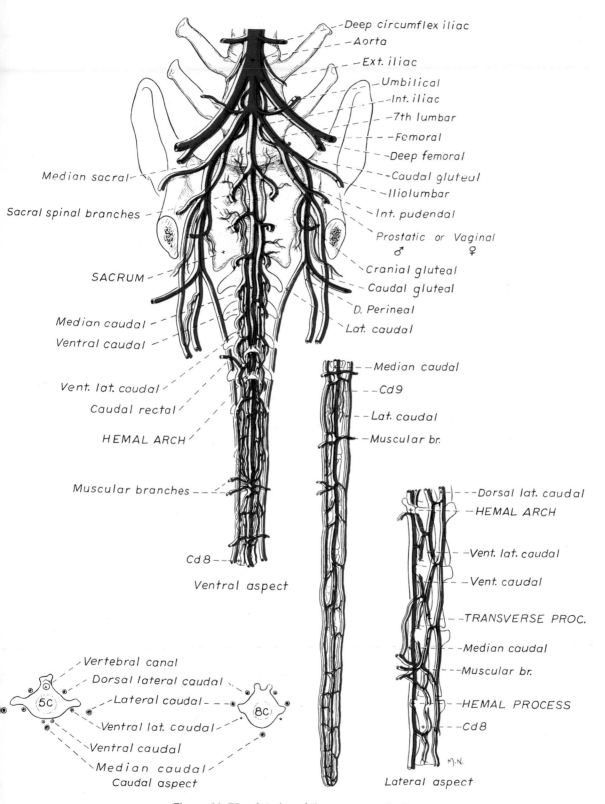

Deep circumflex iliac
Aorta
Ext. iliac
Umbilical
Int. iliac
7th lumbar
Femoral
Deep femoral
Caudal gluteal
Iliolumbar
Int. pudendal
Prostatic or Vaginal
♂   ♀
Cranial gluteal
Caudal gluteal
D. Perineal
Lat. caudal

Median sacral
Sacral spinal branches

SACRUM

Median caudal
Ventral caudal

Vent. lat. caudal
Caudal rectal
HEMAL ARCH

Muscular branches

Cd 8

Ventral aspect

Median caudal
Cd 9
Lat. caudal
Muscular br.

Dorsal lat. caudal
HEMAL ARCH

Vent. lat. caudal
Vent. caudal
TRANSVERSE PROC.
Median caudal
Muscular br.
HEMAL PROCESS
Cd 8

Vertebral canal
Dorsal lateral caudal
Lateral caudal
Ventral lat. caudal
Ventral caudal
Median caudal
Caudal aspect

5C   8C

Lateral aspect

Figure 11–75.   Arteries of the sacrum and tail.

joins the nerve caudal to the trochanter major and runs distally with it. The quadratus femoris, obturator internus, and gemelli may also receive branches from the proximal part of the artery. The caudal gluteal is the main supply to the biceps femoris and semitendinosus, although an extensive anastomosis with the distal caudal femoral and deep femoral arteries exists within these muscles.

The **dorsal perineal artery** *(a. perinealis dorsalis)* originates from the caudal gluteal artery shortly after the latter emerges from the ischiorectal fossa. The dorsal perineal artery passes medially to supply the external anal sphincter and ventrally to supply cutaneous areas of the perineum (Fig. 11–73).

MEDIAN SACRAL ARTERY

The **median sacral artery** *(a. sacralis mediana)* (Fig. 11–75) is the direct continuation of the aorta caudally, after the internal iliac arteries arise. It usually arises opposite the body of the seventh lumbar vertebra as an unpaired median vessel which is slightly less than 2 mm. in diameter. It crosses under the promontory of the sacrum with its dextrally lying satellite vein and enters the fat-filled furrow between the right and left medial ventral sacrocaudal muscles. In this region it usually gives off two pairs of **sacral spinal branches** *(rami spinales)*, which enter the ventral sacral foramina. Arising from these spinal branches are twigs to the adjacent ventral sacrocaudal muscles.

Variations of the arteries to the tail are numerous. Typically, there are seven longitudinal arterial trunks in the proximal third of the tail as follows: 1 (unpaired) median caudal artery, 2 (paired) lateral caudal arteries, 2 (paired) dorsal lateral caudal arteries, and 2 (paired) ventrolateral caudal arteries.

The **median caudal artery** *(a. caudalis [coccygea] mediana)* is a direct continuation caudally of the median sacral artery at the first caudal vertebra. It runs midventrally on the caudal vertebrae, where it lies between the right and left medial ventral sacrocaudal muscles. It passes through the fourth, fifth, and sixth (if present) hemal arches, and then between the successive hemal processes. Throughout most of its course, segmental

arteries which run caudolaterally arise opposite the bodies of the vertebrae. They pass ventral to the transverse processes of the corresponding vertebra and give fine twigs to the adjacent structures. At irregular intervals, usually the length of two or three segments, there are ventral branches which supply the skin. Other branches successively leave the medial caudal from alternate sides, starting at the eighth caudal vertebra, each anastomosing with the lateral caudal artery of the same side. Only the two lateral caudal arteries and the median caudal artery reach the tip of the tail, where they anastomose. They lose their segmental character in the last few segments and anastomose with each other in a plexiform manner. The right and left ventral caudal arteries may be the first branches to leave the median caudal.

The paired **ventral caudal artery** *(a. caudalis [coccygea] ventralis)* Fig. 11–75), if present, arises asymmetrically with its fellow as the first branches of the median caudal or the last branches of the median sacral. They may be important collateral branches that give rise to the segmental arteries which supply the vertebrae and surrounding soft tissue. Typically, they rejoin the median caudal beyond the pelvic outlet after passing through or around the hemal arches.

Delicate bilateral or unilateral longitudinal arterial connections may exist caudal to the pelvic outlet, extending over three to five segments. These closely resemble the ventral caudal arteries which are located cranial to them.

More constant are the paired **dorsal and ventral lateral caudal arteries** *(aa. caudales [coccygea] laterales dorsales et ventrales)*. Less than 1 mm. in diameter, these vessels are joined by the segmental arteries at each caudal vertebra. The segmental arteries, upon reaching the transverse processes, bifurcate into dorsal and ventral branches. The resultant ventral branches arch medially and anastomose with the next segmental arteries caudal to them. In this way a small segmental arterial channel is formed, which consists of a series of lateral arches lying ventral to the transverse processes. Collectively, this segmental arterial passage is known as the ventrolateral caudal artery. It ends at about the eighth caudal vertebra. The dorsal branches pass directly dorsally on the lateral surfaces of the discs and anastomose at right angles with the dorsolateral caudal artery. This artery begins as a continuation caudally of the last sacral segmental artery. This longitudinal arterial channel is straighter than the ventrolateral caudal artery but, like it, is joined by all the caudal segmental arteries of its side. It passes deep to the dorsal caudal musculature and lies ventral to the articular processes. Each sends branches at every three to six segments into the skin and more numerous and irregular branches into the dorsal tail musculature. As the muscles and vertebrae decrease in size toward the end of the tail, the dorsolateral caudal artery decreases in size also, so that it can no longer be followed caudal to the ninth caudal vertebra. The dorsal and ventral lateral caudal arteries are accompanied by the dorsal and ventral caudal nerve trunks. The only recognizable veins from the tail are the right and left lateral caudal veins.

Located along the tail may be small arteriovenous anastomoses called *corpora caudalia [coccygea]*.

BIBLIOGRAPHY

Abramson, D. I. 1976. Circulation in the Extremities. New York, Academic Press.

Abramson, D. I., and S. Margolin. 1936. A Purkinje conduction network in the myocardium of the mammalian ventricles. J. Anat. 70:250–259.

Anderson, W. D., and W. Kubicek. 1971. The vertebral-basilar system of dog in relation to man and other mammals. Am. J. Anat. 132:179–188.

Ashley, L. M. 1945. A determination of the diameter of ventricular myocardial fibers in man and other mammals. Am. J. Anat. 77:325–363.

Baerg, R. D., and D. L. Bassett. 1963. Permanent gross demonstration of the conduction tissue in the dog heart with palladium iodide. Anat. Rec. 146:313–317.

Baird, J. A., and J. S. Robb. 1950. Study, reconstruction and gross dissection of the atrioventricular conducting system of the dog heart. Anat. Rec. 108:747–763.

Barone, R., and C. Pavaux. 1962. Les vaisseaux sanguins du tractus génital chez les femelles domestiques. Bull. Soc. Sc. Vet. Lyon 64:33–51.

Berry, J. L., J. F. Brailsford, and I. de Burgh Daly. 1931. The bronchial vascular system in the dog. Proc. Roy. Soc. London, Series B 109:214–228.

Boling, L. R. 1942. Blood vessels to the dental pulp. Anat. Rec. 82:25–34.

Bradley, O. C., and T. Grahame. 1959. Topographical Anatomy of the Dog. 6th Ed. New York, Macmillan Co.

Buchanan, J. W. 1972. Radiographic aspects of patent ductus arteriosus in dogs before and after surgery. Acta Radiol. (Suppl.) 319:271–278.

Cadete, L. A. 1973. The arteries of the pancreas of the dog. An injection corrosion and microangiographic study. Am. J. Anat. 137:151–157.

Catcott, E. J. 1952. Ophthalmoscopy in canine practice. J. Am. Vet. Med. Assoc. 121:35–37.

Chevalier, P. A. 1976. The Heart and Circulation. Stroudsburg, Pa., Dowden, Hutchinger and Ross Inc.

Christensen, G. C. 1952. Circulation of blood through the canine kidney. Am. J. Vet. Res. 13:236–245.

————. 1954. Angioarchitecture of the canine penis and the process of erection. Am. J. Anat. 95:227–262.

————. 1962. The blood supply to the interventricular septum of the heart — a comparative study. Am. J. Vet. Res. 23:869–874.

Christensen, G. C., and F. L. Campeti. 1959. Anatomic and functional studies of the coronary circulation in the dog and pig. Am. J. Vet. Res. 20:18–26.

Christensen, G. C., and S. Toussaint. 1957. Vasculature of external nares and related areas in the dog. J. Am. Vet. Med. Assoc. 131:504–509.

Clendenin, M. A., and M. C. Conrad. 1979. Collateral vessel development following unilateral chronic carotid occlusion in the dogs. Am. J. Vet. Res. 40:84–88.

Cliff, W. J. 1976. Blood Vessels. London, Cambridge Univ. Press.

Davis, D. D. 1941. The arteries of the forearm in carnivores. Zool. Series Field Museum of Natural History 27:137–227.

Davis, D. D., and H. E. Story. 1943. The carotid circulation in the domestic cat. Zool. Series Field Museum of Natural History 28:3–47.

Dellbrügge, K. F. W. 1940. Die Arterien des weiblichen Geschlechtsapparates vom Hunde. Morph. Jahrb. 85:30–48.

Detweiler, D. K. 1955. Clinical aspects of canine cardiology. Univ. of Penn. Vet. Ext. Quart. No. 137:39–62.

Donald, D. E., and H. E. Essex. 1954. Pressure studies after inactivation of the major portion of the canine right ventricle. Am. J. Physiol. 176:155–161.

Dukes, H. H. 1955. The Physiology of Domestic Animals. 7th Ed. Ithaca, N.Y., Comstock Publishing Assoc.

Eisberg, H. B. 1924. Intestinal arteries. Anat. Rec. 28:227–242.

Ellenberger, W., and H. Baum. 1943. Handbuch der vergleichenden Anatomie der Haustiere. 18th Ed. Berlin, Springer.

Fitzgerald, T. C. 1940. The renal circulation of domestic animals. Am. J. Vet. Res. 1:89–95.

Flint, J. M. 1900. The blood-vessels, angiogenesis, organogenesis, reticulum, and histology, of the adrenal. Contribution to the Science of Medicine, pp. 153–228. Baltimore, The Johns Hopkins Press.

Grävenstein, H. 1938. Über die Arterien des grossen Netzes beim Hunde. Morph. Jahrb. 82:1–26.

Hermann, G. 1940. Über die Arterien der Hinterglied-masse des Hundes, insbesondere ihr topographisches Verhalten. Dissertation, Hannover. 31 pp.

Hermann, G. R. 1925. Experimental heart disease; I. Methods of dividing hearts, with sectional and proportional weights and ratios for two hundred normal dogs' hearts. Am. Heart J. *1*:213–231.

Higginbotham, F. H. 1966. Ventricular coronary arteries of Beagles. J. Atheroscler. Res. (Amst) 5:474–488.

House, E. W., and H. E. Ederstrom. 1968. Anatomical changes with age in the heart and ductus arteriosus in the dog after birth. Anat. Rec. *160*:289–295.

Hughes, H. V., and J. W. Dransfield. 1959. Blood supply to the skin of the dog. Brit. Vet. J. *115*:299–310.

Jewell, P. A. 1952. Anastomoses between internal and external carotid circulation in the dog. J. Anat. *86*:83–94.

Joranson, Y., V. E. Emmel, and H. J. Pilka. 1929. Factors controlling the arterial supply of the testis under experimental conditions. Anat. Rec. *41*:157–176.

Kazzaz, D., and W. M. Shanklin. 1950. The coronary vessels of the dog demonstrated by colored plastic (vinyl acetate) injection and corrosion. Anat. Rec. *107*:43–59.

————. 1951. Comparative anatomy of the superficial vessels of the mammalian kidney demonstrated by plastic (vinyl acetate) injection and corrosion. J. Anat. *85*:163–165.

Kennedy, H. N., and A. W. Smith. 1930. An abnormal celiac artery in the dog. Vet. Rec. (London) *10*:751.

Krediet, P. 1962. Anomalies of the arterial trunks in the thorax and their relation to normal development. Thesis, Utrecht. pp. 1–108.

Langer, G. A., and A. J. Brady. 1974. The Mammalian Myocardium. New York, John Wiley & Sons.

Latimer, H. 3. 1961. Weights of the ventricular walls of the heart in the adult dog. Univ. Kansas Science Bull. XLII: 3–11.

Ljubomudrov, A. P. 1939. The blood supply of the suprarenal glands in the dog. Arkhiv Anat. Grist. i Embryol. *20*:220–224 (Eng. Summary 381–382).

Marthen, G. 1939. Über die Arterien der Körperwand des Hundes. Morph. Jahrb. *84*:187–219.

Meek, W. J., M. Keenan, and H. J. Theisen. 1929. The auricular blood supply in the dog; I. General auricular supply with special reference to the sinoauricular node. Am. Heart J. *4*:591–599.

Miller, M. E. 1952. Guide to the Dissection of the Dog. 3rd Ed. Ithaca, N.Y. Pub. by author.

Moore, R. A. 1930. The coronary arteries of the dog. Am. Heart J. 5:743–749.

Noer, R. 1943. The blood vessels of the jejunum and ileum; A comparative study of man and certain laboratory animals. Am. J. Anat. 73:293–334.

Nonidez, J. F. 1943. The structure and innervation of the conductive system of the heart of the dog and rhesus monkey as seen with a silver impregnation technique. Am. Heart J. 26:577–597.

Northup, D. W., E. J. Van Liere, and J. C. Stickney.

1957. The effect of age, sex, and body size on the heart weight-body weight ratio in the dog. Anat. Rec. *128*:411–417.

Odgers, P. N. B. 1938. The development of the pars membranacea septi in the human heart. J. Anat. 72:247–259.

Peirce, E. C. 1944. Renal lymphatics. Anat. Rec. *90*:315–335.

Pianetto, M. B. 1939. The coronary arteries of the dog. Am. Heart J. *18*:403–410.

Pina, J. A. E., M. Correia, and J. G. O'Neill. 1975. Morphological study on the thebesian veins of the right cavities of the heart in the dog. Acta Anat. 92:310–320.

Preuss, F. 1942. Arterien und Venen des Hinterfusses vom Hund, vorzüglich ihre Topographie. Dissertation, Hannover. 24 pp.

————. 1959. Die A. vaginalis der Haussäugetiere. Tierärzt. Wchnschr. 72:403–416.

Prichard, M. M. L., and P. M. Daniel. 1953. Arteriovenous anastomoses in the tongue of the dog. J. Anat. 87:66–74.

Reichert, F. L. 1924. An experimental study of the anastomotic circulation in the dog. Bull. Johns Hopkins Hosp. 35:385–390.

Reis, R. H., and P. Tepe. 1956. Variations in the pattern of renal vessels and their relation to the type of posterior vena cava in the dog (*Canis familiaris*). Am. J. Anat. 99:1–15.

Sisson, S., and J. D. Grossman. 1953. Anatomy of the Domestic Animals. 4th Ed. Philadelphia, W. B. Saunders Company.

Smith, R. B. 1971. Intrinsic innervation of the atrioventricular and semilunar valves in various mammals. J. Anat. *108*:115–121.

Speed, J. G. 1943. The thoraco-acromial artery of the dog. Vet. J. 99:163–165.

Tandler, J. 1899. Zur vergleichenden Anatomie der Kopfarterien bei den Mammalia. Denkschr. Akad. Wiss. Wien Math.-naturwiss. Kl. 67:677–784.

Thamm, H. 1941. Die arterielle Blutversorgung des Magendarmkanals, seiner Anhangsdrüsen (Leber, Pankreas) und der Milz beim Hunde. Morph. Jahrb. 85:417–446.

Thomas, C. E. 1957. The muscular architecture of the ventricles of hog and dog hearts. Am. J. Anat. *101*:17–58.

Trautmann, A., and J. Fiebiger. 1952. Fundamentals of the Histology of Domestic Animals, translated and revised from 8th and 9th German editions, 1949, by R. E. Habel and E. L. Biberstein. Ithaca, N.Y., Comstock Publishing Assoc., Cornell Univ. Press.

Truex, R. C., and L. J. Warshaw. 1942. The incidence and size of the moderator band in man and mammals. Anat. Rec. *82*:361–372.

Whisnant, J. P., C. H. Millikan, K. G. Wakim, and G. P. Sayre. 1956. Collateral circulation of the brain of the dog following bilateral ligation of the carotid and vertebral arteries. Am. J. Physiol. *186*:275–277.

Chapter 12

VEINS

GENERAL CONSIDERATIONS

Veins follow the same general course as do the arteries, although variations in their number, size, and course are more frequent than in arteries. The accompanying veins are known as satellite veins, or *venae comitantes*, and often take the same name as the artery they accompany (Fig. 12–1). Although the smaller satellite veins are frequently double, the larger veins are single, as are most of the deep veins. All systemic veins have thin walls, and most have large lumina in comparison with the arteries. The pressure in the veins is low, and the blood flows much more slowly in them than in the arteries. Since there is generally no pulse in the venous system, the movement of blood depends primarily on pressure relations in the thorax and on muscular activity. The contraction of muscles results in compression of the veins, thus propelling the contained blood toward the heart. Negative pressure in the thorax during inspiration and the presence in most veins of semilunar valves which prevent backflow augment this effect of the skeletal and visceral muscles. Veins farthest from the heart contain the most valves. For an overview, see Shepherd and Vanhoutte (1975) as well as the older classic work by Franklin (1937).

The venous passages in the dura of the central nervous system are known as sinuses. In the extremities, the veins may be divided into superficial and deep sets. The veins of the superficial set are large and are clinically important because of the frequent necessity of making venipunctures for drawing blood or injecting liquids. In this location they act in cooling the blood as they communicate not only with the deep veins but also with extensive subcutaneous, interconnecting venous plexuses. When the animal is cooled, these plexuses and the larger superficial veins contract so that most of the blood from the extremities must be returned to the heart via the deep veins; this prevents heat loss. When the animal is warmed, and during work, in short-haired specimens the superficial veins and their connecting plexuses dilate and are plainly visible beneath the skin; this dilatation provides a means of heat dissipation.

CRANIAL VENA CAVA

The **cranial vena cava** (*v. cava cranialis*), formerly called precava (Fig. 12–2), is an unpaired vessel, 1.5 to 2 cm. in diameter and 8 to 12 cm. long. It lies ventral to the trachea, and is in contact with the esophagus on the left side. It is also in contact with the thymus when this gland is fully developed. It runs through the precardial mediastinum and is the most ventral of the several structures which course through the thoracic inlet. It is formed, at a level cranial to the thoracic inlet, by the convergence of the right and left brachiocephalic veins. These form an angle, open cranially, of about 90 degrees, so that each

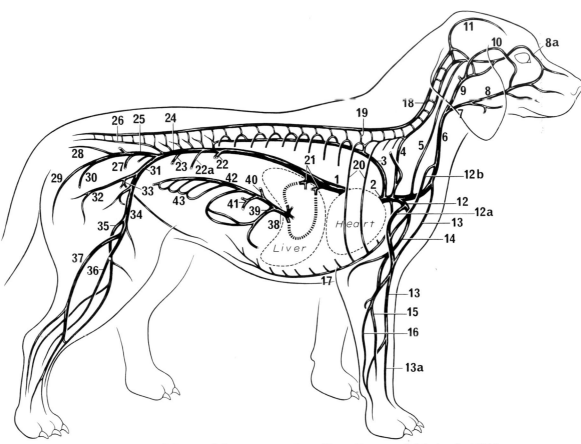

Figure 12–1. Schema of the venous system. (From Evans and deLahunta 1971.)

1. Caudal vena cava	21. Hepatic
2. Cranial vena cava	22. Renal
3. Azygos	22a. Testicular or ovarian
4. Vertebral	23. Deep circumflex iliac
5. Internal jugular	24. Common iliac
6. External jugular	25. Right internal iliac
7. Linguofacial	26. Median sacral
8. Facial	27. Prostatic or vaginal
8a. Angularis oculi	28. Lateral caudal
9. Maxillary	29. Caudal gluteal
10. Superficial temporal	30. Internal pudendal
11. Dorsal sagittal sinus	31. Right external iliac
12. Axillary	32. Deep femoral
12a. Axillobrachial	33. Pudendoepigastric trunk
12b. Omobrachial	34. Femoral
13. Cephalic	35. Medial saphenous
13a. Accessory cephalic	36. Cranial tibial
14. Brachial	37. Lateral saphenous
15. Radial	38. Portal
16. Ulnar	39. Gastroduodenal
17. Internal thoracic	40. Splenic
18. Right vertebral venous plexus	41. Caudal mesenteric
19. Intervertebral	42. Cranial mesenteric
20. Intercostal	43. Jejunal

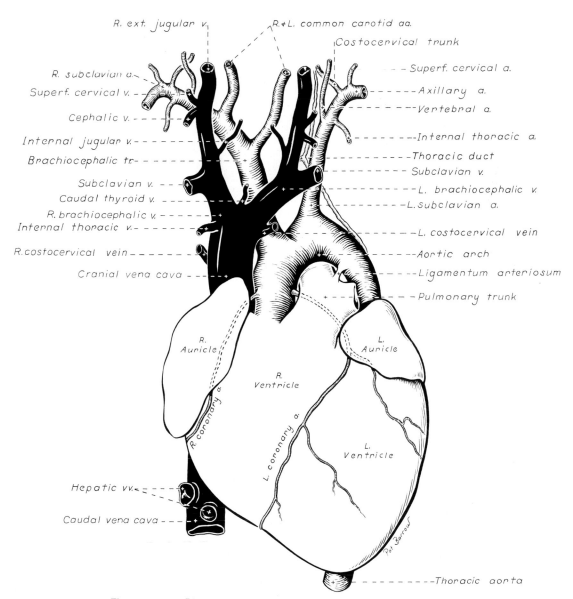

R. ext. jugular v.

R.+L. common carotid aa.

Costocervical trunk

R. subclavian a.

Superf. cervical v.

Cephalic v.

Internal jugular v.

Brachiocephalic tr.

Subclavian v.

Caudal thyroid v.

R. brachiocephalic v.

Internal thoracic v.

R. costocervical vein

Cranial vena cava

Superf. cervical a.

Axillary a.

Vertebral a.

Internal thoracic a.

Thoracic duct

Subclavian v.

L. brachiocephalic v.

L. subclavian a.

L. costocervical vein

Aortic arch

Ligamentum arteriosum

Pulmonary trunk

R. Auricle

L. Auricle

R. Ventricle

R. coronary a.

L. coronary a.

L. Ventricle

Hepatic vv.

Caudal vena cava

Pat Barrow

Thoracic aorta

Figure 12–2. Diagram of the heart and great vessels, ventral aspect.

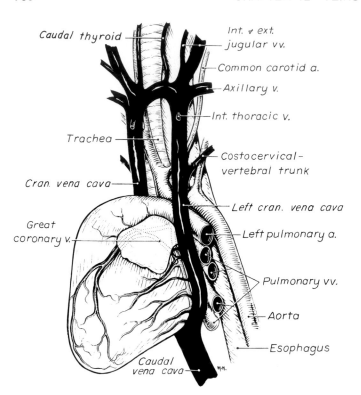

Caudal thyroid

Int. & ext. jugular vv.

Common carotid a.

Axillary v.

Int. thoracic v.

Trachea

Costocervical-vertebral trunk

Cran. vena cava

Left cran. vena cava

Great coronary v.

Left pulmonary a.

Pulmonary vv.

Aorta

Esophagus

Caudal vena cava

Figure 12–3. Persistent left cranial vena cava.

vein enters the cranial vena cava at an angle of approximately 45 degrees with the median plane. The cranial vena cava empties into the cranial part of the right atrium. Stoland and Latimer (1947) describe a dog with a persistent left cranial vena cava. The vein was about equal in size to the normal right vessel, and was continued caudally in the left part of the coronary groove to the coronary sinus. In reviewing the incidence of anomalous cranial venae cavae, they state that a total of 207 cases had been seen by various authors to that date. A persistent left cranial vena cava (Fig. 12–3) has also been reported by Schaller (1955), Buchanan (1963), Gomerčič (1967), Hutton (1969), and Buergelt and Wheaton (1970).

The **azygos vein** (*vena azygos*), with its tributaries, is discussed after the description of the other tributaries of the cranial vena cava.

The **costocervical vein** (*v. costocervicalis*) (Fig. 12–2) is medial to the proximal end of the first rib. The left vein runs lateral to the left subclavian artery and empties into the dorsolateral surface of the cranial part of the cranial vena cava. It may

terminate in the left brachiocephalic vein. The right costocervical vein, about 1 cm. caudal to the left at its termination, crosses the lateral surface of the trachea and enters the cranial vena cava ventral to the brachiocephalic artery and vagus nerve. The right costocervical vein is about 3 cm. long; that of the left side is about 4 cm. Both vessels are approximately 5 mm. in diameter.

In addition to the frequent union of the vertebral vein, there are four main tributaries to the costocervical vein in the dog. All are satellites of arteries, although they are less constant in position. They include the **dorsal scapular vein** (*v. scapularis dorsalis*), formerly called the transverse colli; the **first dorsal intercostal vein** (*v. intercostalis dorsalis I*); **the deep cervical vein** (*v. cervicalis profunda*), of which the thoracic vertebral is a tributary; and the **supreme intercostal vein** (*v. intercostalis suprema*).

Intervertebral veins (*v. intervertebrales*), pass through the intervertebral foramen and are connected to the vertebral plexus. The last or eighth intervertebral vein regularly bifurcates into a cranial communicating vein which joins the vertebral vein and a caudal communicating vein which empties into the deep cervical vein. According to Worthman (1956), the supreme intercostal vein receives the second and third thoracic intervertebral veins on each side. In almost half of his specimens, the fourth thoracic intervertebral vein on the left also emptied into the supreme intercostal vein of the same side.

The **vertebral vein** (*v. vertebralis*) begins by the confluence of the vein of the hypoglossal canal, if present, and the sigmoid sinus in the petro-occipital fissure. The small internal jugular vein also arises wholly or partly at this confluence. The vertebral vein is about 1.5 mm. in diameter as it runs caudally across the ventrolateral part of the atlanto-occipital joint, then under the wing of the atlas to the transverse foramen. It receives the first intervertebral vein which leaves the vertebral plexus, passes through the intervertebral foramen of the atlas and then through the alar notch, and joins the vertebral vein in the atlantal fossa. Worthman (1956) and Dräger (1937) call the initial portion of the vertebral the occipital vein, although only a segment of it parallels the occipital artery. The dog, however, may be regarded as not having a true occipital vein (v. occipitalis), because the dorsal nuchal area supplied by the occipital artery is drained by the occipital emissary vein and the ventral portion supplied by the occipital artery is drained by the vertebral vein.

After passing through the transverse foramen of the atlas from the atlantal fossa, the vertebral vein receives a small muscular tributary, and a slightly larger communication from the caudal portion of the first intervertebral vein and the large second intervertebral vein. The portion of the vertebral vein which passes through the transverse foramen of the axis is its smallest part. There is usually a small anastomotic vein between the second and third intervertebral veins. Most of the blood is conveyed caudally by way of the large vertebral venous plexuses. The remaining portion of the vertebral vein and its tributaries are satellites of the comparable portions of the vertebral artery and its branches.

The **internal thoracic vein** (*v. thoracica interna*) (Fig. 12–4) is unpaired at its ter-mination in the middle of the ventral surface of the cranial vena cava in about half of all specimens. In such instances it usually ranges from 1 to 4 cm. in length. In the specimens in which it is paired, the right vein usually enters the cranial vena cava, whereas the left enters the left brachiocephalic vein. The peripheral part of the internal thoracic vein and its branches are prominent satellites of the comparable parts of the internal thoracic artery. Tributaries of the internal thoracic vein parallel the arteries and include pericardicophrenic, thymic, mediastinal, intercostal, and musculophrenic veins. The cranial epigastric vein of the abdomen is an extension of the internal thoracic vein.

The **brachiocephalic vein** (*v. brachiocephalica*) (Fig. 12–4) merges with its fellow of the opposite side cranial to the thoracic inlet to form the cranial vena cava. Each is about 2 cm. long and 1 cm. in diameter, and is formed by the joining of the caudally coursing external jugular and the medially coursing subclavian vein. (Unlike the comparable artery, no part of the venous channel coming from the thoracic limb normally lies within the thorax. There is, therefore, no basis for naming a portion of the channel for venous return from the thoracic limb the subclavian vein.) The merging brachiocephalic veins lie ventral to the trachea and esophagus as the most ventral structures in this region. In addition to the external jugular and axillary veins, the thyroidea ima and, occasionally, the caudal thyroid (Fig. 12–4), and the internal jugular veins enter the brachiocephalic.

The **caudal thyroid vein** (*v. thyroidea caudalis*) is an unpaired vein, about 1 mm. in diameter, which arises primarily from the deep surfaces of the sternothyrohyoid muscles, but on one or both sides its most cranial tributary may arise in the thyroid lobe or lobes. It terminates usually in the cranial angle formed by the merging brachiocephalic veins.

The **internal jugular vein** (*v. jugularis interna*), about 1 mm. in diameter, is formed in the petro-occipital fissure by the confluence of the vertebral vein, sigmoid sinus, and, occasionally, the vein of the hypoglossal canal. It regularly receives a relatively large cranial communicating

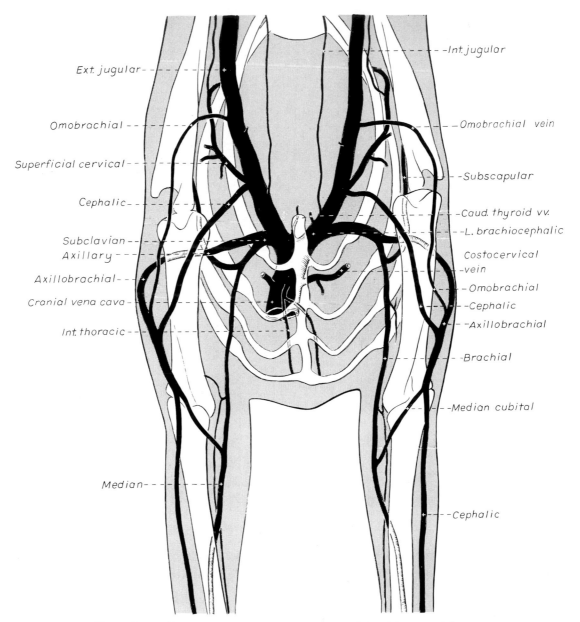

Ext. jugular

Omobrachial

Superficial cervical

Cephalic

Subclavian
Axillary

Axillobrachial

Cranial vena cava

Int. thoracic

Median

Int. jugular

Omobrachial vein

Subscapular

Caud. thyroid vv.

L. brachiocephalic

Costocervical
vein

Omobrachial

Cephalic

Axillobrachial

Brachial

Median cubital

Cephalic

Figure 12–4. Diagram of the veins of the neck and shoulders, cranial aspect.

vessel from the caudal end of the ventral petrosal sinus. Its initial portion may be double. The internal jugular lies at first in association with the internal carotid artery and then in the sheath of the common carotid. In the vicinity of the larynx, it receives an anastomotic twig from the laryngeal or pharyngeal tributary of the lingual. Caudal to the larynx it receives the **cranial thyroid vein** (*v. thyroidea cranialis*) from the cranial pole of the thyroid lobe. Opposite this tributary there is occasionally a second anastomotic connection with the external jugular vein. A twig from the medial retropharyngeal lymph node is also received here. Frequently, it receives the small **middle thyroid vein** (*v. thyroidea media*), which comes from the caudal pole

of one or both thyroid lobes. On either or both sides, this vein may terminate in the brachiocephalic vein rather than the internal jugular, in which case it is considered to be the caudal thyroid vein. Occasionally, there is an unpaired vein which drains the thyroid lobes and the deep surface of the sternothyroid muscles and terminates in the fork of the brachiocephalic veins. This was formerly known as the v. thyroidima. The internal jugular vein usually terminates in the caudal portion of the external jugular vein; rarely it terminates in the brachiocephalic vein.

The **external jugular vein** (*v. jugularis externa*) (Fig. 12–5), unlike that of man, is the main channel for return of venous blood from the head. It begins by the

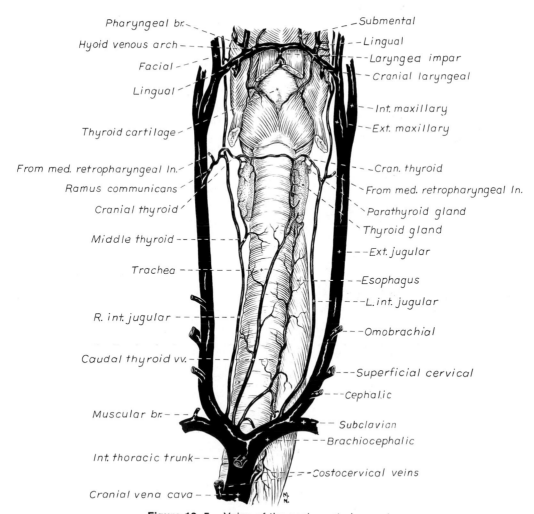

Figure 12–5. Veins of the neck, ventral aspect.

union of the linguofacial and maxillary veins, caudal to the mandibular gland or at a transverse plane through the cricoid cartilage and the axis. It is about 1 cm. in diameter and 12 cm. long. In the adult it contains a few nonfunctional valves which are irregular in their spacing. As the external jugular runs caudally in the superficial fascia, it crosses the lateral surface of the brachiocephalic muscle obliquely. It lies directly under the skin and is commonly used for venipuncture in dogs which are too small for the procedure to be feasible in the smaller veins of the extremities. At the cranial border of the shoulder, it receives the omobrachial and cephalic veins which ascend from the brachium. About 2 cm. caudal to the termination of the omobrachial, the external jugular receives the **superficial cervical** (*v. cervicalis superfi-*

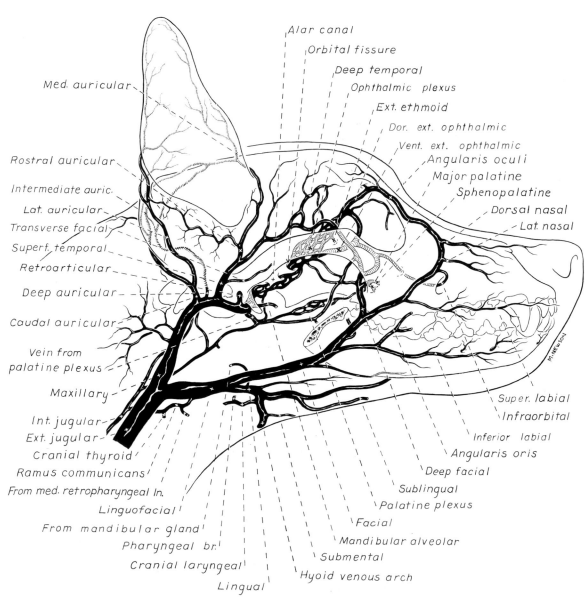

Figure 12–6. Superficial veins of the head, lateral aspect.

cialis), but this is variable. At its termination it usually receives the internal jugular vein on its medial side.

The **linguofacial vein** (*v. linguofacialis*) (Fig. 12–6) begins by the confluence of the lingual and facial veins ventral to the mandibular gland. It may have one or more tributaries from the capsule of the mandibular gland. It regularly receives the **glandular vein** (*v. glandularis*), which leaves the caudal pole of the mandibular gland. It contains a valve at its termination.

The **facial vein** (*v. facialis*) (Figs. 12–6, 12–7) begins on the dorsolateral surface of the muzzle, covered by the levator nasolabialis. It is formed by the confluence of the smaller **dorsal nasal vein** (*v. nasalis dorsalis*), which drains the dorsolateral surface of the nose, and the larger angular vein of the eye.

The **angular vein of the eye** (*v. angularis oculi*) is about 3 mm. in diameter and 2 cm. long. Blood may flow in either direction in it, as it lacks valves. It receives a tributary from the surface of the frontal bone. The angularis oculi vein may anastomose with the superficial temporal vein. It disappears from the surface of the face

by curving around the dorsomedial border of the orbital margin, to become the dorsal external ophthalmic vein. It usually receives an emissary vein from the superficial surfaces of the frontal and nasal bones.

The **dorsal external ophthalmic vein** (*v. ophthalmica externa dorsalis*) runs about 2 cm. caudally into the orbit and forms the **ophthalmic plexus** (*plexus ophthalmicus*). The plexus lies within the periorbita. It extends to the orbital fissure, and therefore lies in the caudal two-thirds of the orbital fossa. The plexus becomes consolidated at the orbital fissure and, after traversing it, joins the cavernous sinus. Dorsally, a small branch runs through the optic canal to join its fellow and the dorsal petrosal sinus of its side; ventrally the plexus is continued caudally outside the alar canal and is to be regarded as the beginning of the maxillary vein. A second connection between the ophthalmic system and the maxillary exists here in the form of a vein which runs through the alar canal. This vein also communicates with the cavernous sinus by an emissary vein which traverses the round foramen. The ventral external ophthalmic vein (see Fig. 20–33) joins the dorsal ophthalmic vein in the

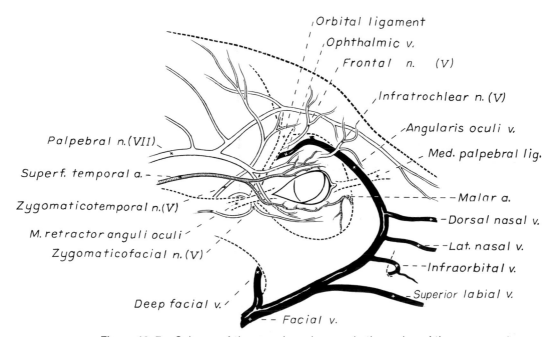

Figure 12–7.   Scheme of the vessels and nerves in the region of the eye.

ophthalmic plexus. Caudal to the eyeball an anastomotic branch unites the dorsal and ventral external ophthalmic veins. The ventral external ophthalmic vein turns ventrally, receives the ventral vorticose veins and a branch from the third eyelid before anastomosing with the deep facial vein.

Two emissary veins join the ophthalmic plexus. One of these is the **external ethmoidal vein** (*v. ethmoidalis externa*), a satellite of the like-named artery, which passes through the ethmoidal foramen. The other emissary vein is the **frontal diploic vein** (*v. diploica frontalis*), from the diploë of the frontal bone, which passes through a small foramen in the supraorbital process. The latter vein may join the ophthalmic vein before the plexus is formed.

The **lateral nasal vein** (*v. lateralis nasi*) is a satellite of the lateral nasal artery. It is a tributary which enters the facial vein.

The **infraorbital vein** (*v. infraorbitalis*) is about 1 mm. in diameter and 1 cm. long. It communicates with the ventral side of the facial vein, which lies dorsal to the infraorbital foramen. It has tributaries from the infraorbital nerve and adjacent musculature. Caudally, it unites with the sphenopalatine vein to form a short common trunk which anastomoses with the reflex vein in the rostral part of the pterygopalatine fossa.

The **malar vein** (*v. malaris*) is a small tributary which arises mainly in the skin of the lower eyelid and terminates in the dorsal surface of the facial vein.

The **superior labial vein** (*v. labialis superior*) runs caudally along the dorsal margin of attachment of the buccinator muscle and enters the facial vein lateral to the rostral end of the zygomatic arch. It drains blood from the upper lip and the dorsal part of the cheek.

The **angular vein of the mouth** (*v. angularis oris*) is a small tributary from the commissure of the lips which enters the facial vein caudal to the commissure.

The **deep facial vein** (*v. faciei profunda*) (Fig. 12–6) has no companion artery and is significant for its deep course and many anastomoses. A former name was vena reflexa (see Preuss 1954). It is 2 to 4 mm. in diameter at its terminal end, which lies in the fascia cranial to the masseter muscle, about 1.5 cm. ventral to the zygomatic arch. It arises in the ventral part of the orbital and the adjacent pterygopalatine fossa. The main portion of the vein arches dorsomedially and anastomoses with the ventral external ophthalmic vein on the floor of the orbit. (See Chapter 20.) An anastomosis with the superficial temporal vein frequently occurs as a small vein which obliquely crosses the lateral surface of the zygomatic arch. A small vein unites the deep facial with the maxillary by running across the lateral surfaces of the pterygoid muscles.

Tributaries which enter the deep facial vein may form a short venous trunk by the union of the sphenopalatine, infraorbital, and occasionally the major palatine. The major palatine is small, if it is present at all. These veins are satellites of the comparable arteries. The main venous drainage of the hard palate is by a poorly formed venous plexus which is continuous with the much more salient venous plexus of the soft palate. Therefore, the venous drainage of the hard and soft palates is not chiefly through satellites of the arteries supplying them, but by means of the veins of the palatine plexuses which drain into the right and left maxillary veins. Constantly, one or two veins leave the ventral part of the ocular muscles to enter the reflex vein. It also receives tributaries from the maxilla as it curves from the deep surface of the masseter before entering the facial vein.

The **inferior labial vein** (*v. labialis inferior*), a satellite of its artery, runs along the ventral border of attachment of the buccinator muscle. It receives at its termination a vein which arises in the intermandibular space. This vessel runs along the margin of insertion of the digastricus, then over the lateral surface of the mandible. It may terminate in the facial vein. Throughout the course of the facial vein small twigs from the skin and fascia enter it. Ellenberger and Baum (1943) illustrate a small anastomotic vein located between the facial and the superficial temporal veins, as well as small twigs coming from the mandibular lymph nodes.

The **lingual vein** (*v. lingualis*) (Fig. 12–6) is the ventral tributary which joins the facial to form the external maxillary. It begins in the tip of the tongue, and as it courses caudally it is augmented by numerous tributaries from this organ. It lies in areolar tissue in association with the lingual artery and hypoglossal nerve, lateral to the genioglossal and medial to the hypoglossal muscles. About 1 cm. rostral to the body of the hyoid bone it crosses the dorsal border of the hyoglossus and comes to lie dorsal to the mylohyoideus. As it runs out from under the caudal border of the mylohyoideus it is joined by the *v. sublingualis*. This vein begins in the rostral part of the lingual frenulum and runs caudally on the dorsal surface of the mylohyoideus. It lies directly under the thin mucosa between the lingual frenulum and the fimbriated fold which lies lateral to the frenulum. It is occasionally used for venipuncture, but this is ill advised, as the exceedingly loose tissue which surrounds the vessel allows considerable hemorrhage to occur. Rostrally, this vein is accompanied by its satellite artery and the lingual branch of the trigeminal nerve. It carries blood from the frenulum and the closely adjacent sublingual and mandibular ducts and the polystomatic part of the sublingual gland.

The **hyoid venous arch** (*arcus hyoideus*) (Fig. 12–5) is a constant large, unpaired vein, about 3 mm. in diameter and 3 to 4 cm. long, which lies ventral to the basihyoid bone. It usually connects right and left lingual veins about 1 cm. caudal to the termination of the sublingual veins. Petit (1929) illustrates this vessel as extending between the two sublingual veins. It may be double. It receives on each side of the midline the caudally running small *v. submentalis*, which usually begins as a single vessel in the midline between the fellow mylohyoid muscles. It receives delicate tributaries from both the mylohyoid and the geniohyoid muscles. Entering the caudal surface of the arcus hyoideus at the midline is the delicate unpaired *v. laryngea impar*. It anastomoses with the delicate v. laryngea cranialis and usually with end tributaries of the thyroid veins.

The **cranial laryngeal vein** (*v. laryngea cranialis*) (Fig. 12–5), a satellite of the like named artery, leaves the larynx ventral to the cranial corner of the thyroid cartilage

in company with the artery and cranial laryngeal nerve. It joins the lingual vein about 1 cm. from its termination. It may send a communicating branch to the internal jugular vein. The **pharyngeal vein** (*v. pharyngea*) is a variably formed tributary which usually arises in a small venous plexus located between the vagosympathetic nerve trunk and the internal carotid artery on the lateral wall of the pharynx. It usually sends a communicating branch to the internal jugular vein. It is a tributary of the cranial laryngeal vein which enters it just lateral to the larynx.

The **maxillary vein** (*v. maxillaris*) (Fig. 12–6) begins ventral to the alar canal by a continuation and later a consolidation of the extension of the ophthalmic plexus. The formation and anatomy of the venation in this location are complicated and variable. Usually a small vein lies in the alar canal and receives an emissary vein from the cavernous sinus through the round foramen. Rostrally the vein in the alar canal joins the ophthalmic plexus, whereas caudally it joins the maxillary vein. Here also the maxillary receives an emissary vein from the oval foramen and the *v. meningea media*, which is a satellite, usually double, of the corresponding artery. About 5 mm. caudal to the oval foramen, two more veins join the maxillary. One of these is small and comes from the pterygoid canal; the second is larger, passes through the external carotid foramen, and connects internally at the confluence of the ventral petrosal and cavernous sinuses. The maxillary vein winds laterally, caudal to the retroarticular process, and receives the vein of the palatine plexus.

The **venous palatine plexus** (*plexus venosus palatinus*) (Fig. 12–8) is a rather loose network of veins in the soft palate. The largest elements of this plexus are 0.5 mm. in diameter. Rostrally the plexus anastomoses with the sphenopalatine and deep facial veins.

The **temporomandibular articular vein** (*v. articularis temporomandibularis*) (Fig. 12–6) was formerly called the retroglenoid or retroarticular vein. It leaves the skull through the retroarticular foramen and more than doubles the size of the maxillary vein by joining it caudal to the tem-

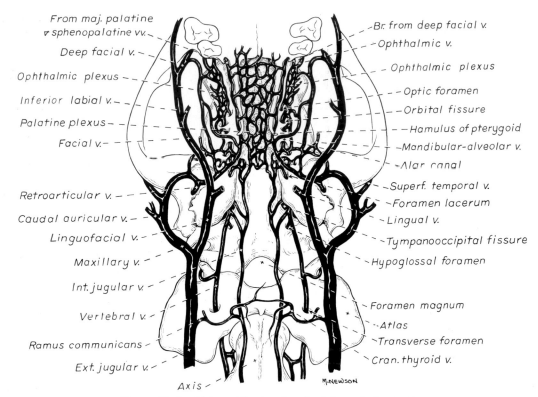

Figure 12–8. Veins of the head and neck, ventral aspect.

poromandibular joint. The intracranial formation of this vein is described with the veins of the central nervous system.

The **inferior** or **mandibular alveolar vein** (*v. alveolaris inferior*) is the satellite of the comparable artery. It leaves the mandibular foramen and at once receives a branch from the musculature medial to the mandible, mainly from the m. temporalis as the **deep temporal vein** (*v. temporalis profunda*). Entering the maxillary about 5 mm. caudal to the entry of the mandibular alveolar is the **masseteric vein** (*v. masseterica*). This small tributary comes from the upper caudal border of the masseter muscle and curves medial to the caudal border of the mandible before it terminates.

The **superficial temporal vein** (*v. temporalis superficialis*) (Fig. 12–6) is about 2.5 mm. in diameter as it terminates in the dorsal surface of the maxillary. The vein crosses ventral to the base of the ear, under cover of the parotid gland. The dorsal tributary arises dorsomedial to the orbit

by occasionally anastomosing with the small frontal vein, a tributary of the v. angularis oculi. The ventral tributary takes a dorsal course in the caudal part of the orbital fossa. It runs under the deep temporal fascia, crosses under the orbital ligament, and continues medial to the zygomatic arch to anastomose with a branch of the deep facial vein. This anastomotic channel is over 1 mm. in diameter. A small, third anastomotic channel between the v. temporalis superficialis and the veins of the orbit is formed as follows: a branch of the v. temporalis superficialis runs through the rostral portion of the temporal muscle to enter the orbital fossa and anastomoses with the ophthalmic plexus.

The superficial temporal vein, after its formation by the three anastomotic branches, runs under the deep temporal fascia caudally. It receives numerous tributaries from the temporal muscle in its course toward the base of the ear. The *v. auricularis rostralis* is a small vein which begins in the interauricular musculature and skin and runs transversely, under the rostroauricular muscles, rostral to the base of the ear. It receives twigs from the skin and auricular muscles and the base of the pinna itself. It terminates in the superficial temporal vein. The *v. transversa facei* is a small tributary which arises in the fascia ventral to the zygomatic arch. It empties into the superficial temporal about 1 cm. ventral to the termination of the much larger rostral auricular vein. According to Ellenberger and Baum (1943), the rostral auricular vein terminates in the caudal auricular vein 50 per cent of the time. As the superficial temporal grooves the rostral border of the parotid gland, it receives one or more *rami parotidei* from it.

The **caudal auricular vein** (*v. auricularis caudalis*) (Fig. 12–6) is formed by the lateral and intermediate auricular veins. The medial auricular vein drains into the rostral auricular vein, which joins the maxillary close to the caudal auricular vein. Between the rostral and caudal auricular veins the deep auricular enters the maxillary. The marginal veins, the **medial** and **lateral auricular**, anastomose with each other near the tip of the pinna on the caudal, or convex, side. Unlike the caudal auricular artery, the caudal auricular vein anastomoses with the rostral auricular by means of a venous circle which lies on the cervicoauricular and interauricular muscles. This venous circle receives tributaries from the adjacent muscles. The caudal auricular vein terminates in the caudal maxillary vein. A portion of the caudal auricular vein is bridged superficially by parotid gland tissue.

One or two veins enter the maxillary near its termination. These come from the dorsally lying skin and underlying brachiocephalic muscle. Occasionally one of these ends in the external jugular vein. A communicating vein may be located between the internal jugular and the maxillary vein.

AZYGOS SYSTEM OF VEINS

The **right azygos vein** (*v. azygos dextra*) (Fig. 12–9) is about 8 mm. wide at its junction with the cranial vena cava where the latter terminates in the right atrium oppo-

site the right third intercostal space. It begins on the median plane ventral to the body of the third lumbar vertebra by anastomosing with the single trunk formed by the merging of the right and left third lumbar intervertebral veins. Lying in the fat with the lumbar lymphatic trunk, it runs forward, flanked by the tendons of the crura of the diaphragm and the psoas major muscles. In the caudal third of the thorax it inclines slightly to the right, where it lies in the angle formed by the vertebral bodies and the aorta. Here, covered only by pleura, it ascends to the base of the heart, hooks around the root of the right lung, and empties into the termination of the cranial vena cava at a right angle. It has the following tributaries: lumbar, costoabdominal, dorsal intercostal, and bronchoesophageal veins.

The first two **lumbar**, the **costoabdominal**, and the **dorsal intercostal veins** (*vv. lumbales et costoabdominales dorsales et intercostales dorsales*), except the first three intercostals on the right side and the first three or four dorsal intercostals on the left side, are received by the azygos or the hemiazygos vein. The (third) fourth and fifth dorsal intercostal veins anastomose with each other so that a longitudinal venous trunk is formed which usually terminates in the proximal end of the sixth dorsal intercostal vein. This pattern is usually bilaterally symmetrical in its formation, but the left venous trunk is longer than the right because it crosses under the body of the fifth thoracic vertebra to reach the azygos vein.

The first two lumbar veins are smaller than the others. They are received by the initial part of the azygos vein as it extends forward from an anastomosis with the stem, which receives the right and left third lumbar veins. The initial part of the azygos vein may be double as it forms this union. The azygos vein carries most of the blood from the vertebral venous plexus via the intercostal and lumbar veins to the cranial vena cava (Bowsher 1954). The intervertebral veins of the thorax are single vessels which join the intercostal veins at the highest points of the several intercostal spaces. The lumbar intervertebral veins are double as they traverse the interverte-

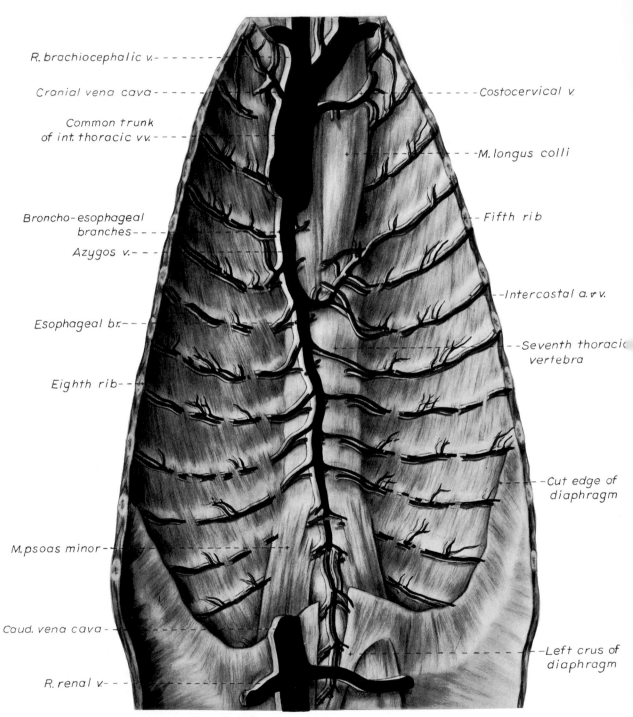

R. brachiocephalic v.

Cranial vena cava

Common trunk
of int. thoracic vv.

Broncho-esophageal
branches

Azygos v.

Esophageal br.

Eighth rib

M. psoas minor

Caud. vena cava

R. renal v.

Costocervical v.

M. longus colli

Fifth rib

Intercostal a. & v.

Seventh thoracic
vertebra

Cut edge of
diaphragm

Left crus of
diaphragm

Figure 12–9.   The azygos vein, ventral aspect.

bral foramina. They then quickly unite to form the several lumbar veins (Worthman 1956).

The **hemiazygos vein** (*v. hemiazygos*) is about 2 mm. in diameter and is extremely variable. It lies on the left side of the aorta and connects the caudal vena cava with the azygos vein. It runs from the left phrenicoabdominal vein near its termination in the caudal vena cava through the aortic hiatus and usually anastomoses cranially with the ninth or tenth left dorsal intercostal vein close to the vertebral bodies. The hemiazygos vein receives the left costoabdominal vein and the last two or three left dorsal intercostal veins. Occasionally there is an anastomosis between the left phrenicoabdominal and the left costabdominal vein. In such specimens the hemiazygos is absent.

The **esophageal** and **bronchoesophageal veins** (*vv. esophageae et bronchoesophageae*) are variable and small. They are satellites of the comparable arteries. The bronchoesophageal veins, larger and more constant than the others, terminate in the azygos vein, usually at the level of the seventh thoracic vertebra. The esophageal veins, with delicate mediastinal tributaries, terminate in the azygos vein caudal to the termination of the bronchoesophageal. There are usually two of these, 1 to 3 cm. apart, which cross the right face of the aorta to empty into the ventral surface of the azygos vein. Occasionally, there is an anastomosis between the azygos or hemiazygos vein and the deep circumflex iliac vein.

VEINS OF THE THORACIC LIMB

The veins of the thoracic limb may be divided into superficial and deep sets.

SUPERFICIAL VEINS OF THE THORACIC LIMB

The **cephalic vein** (*v. cephalica*) (Figs. 12–4, 12–10) is a branch of the external jugular. It is the only large superficial vein of the thoracic limb. It begins as a transverse vein, the superficial palmar venous arch, which crosses the palmar side of the distal third of the fourth and third metacar-

pal bones. Here it is well protected by the heavy metacarpal pad. It runs proximally from the superficial arch on the palmar side of the interosseous muscles directly caudal to the interosseous space between the second and third metacarpal bones. It passes superficial to the flexor retinaculum (formerly called the palmar carpal transverse ligament), parallel to the carpal canal. It usually receives three tributaries here. One of these comes from a band of skin extending to the elbow, whereas the other two are tributaries from the flexor muscles of the antebrachium. Upon gaining the cranial surface of the antebrachium, it is known as the **antebrachial part of the cephalic vein**. From the carpus it runs proximally until it reaches the cranial surface of the extensor carpi radialis, and then it follows this muscle to the flexor angle of the elbow joint. It is at this point that the vessel is commonly compressed to raise the vein. It is flanked on its medial and lateral sides by the superficial antebrachial artery and its medial branch as well as by the medial and lateral branches of the superficial radial nerve. It is 3 to 5 mm. in diameter and lies directly under the skin, loosely surrounded by the superficial fascia. Because of its size, location, and ease of compressibility, it is the favored site for venipuncture in the dog. The v. cephalica antebrachii is augmented by receiving the **accessory cephalic vein** (*v. cephalica accessoria*) at the beginning of the distal fourth of the antebrachium. This vein, about 2 mm. in diameter at its termination, begins on the dorsum of the metacarpus and passes proximally over the carpus and the dorsum of the distal portion of the antebrachium before joining the cephalic vein. It is joined by a tributary from the first digit and skin of the second metacarpal bone at the distal end of the antebrachium.

The **median cubital vein** (*v. mediana cubiti*) (Fig. 12–11) extends between the median vein at the flexor angle of the elbow joint and the cephalic vein of the arm. It is about 2 mm. in diameter and 2 cm. long. It crosses the distal end of the biceps obliquely as it runs proximolaterally to anastomose with the cephalic vein near the lateral border of the biceps.

The **brachial part of the cephalic vein** continues the antebrachial part from the

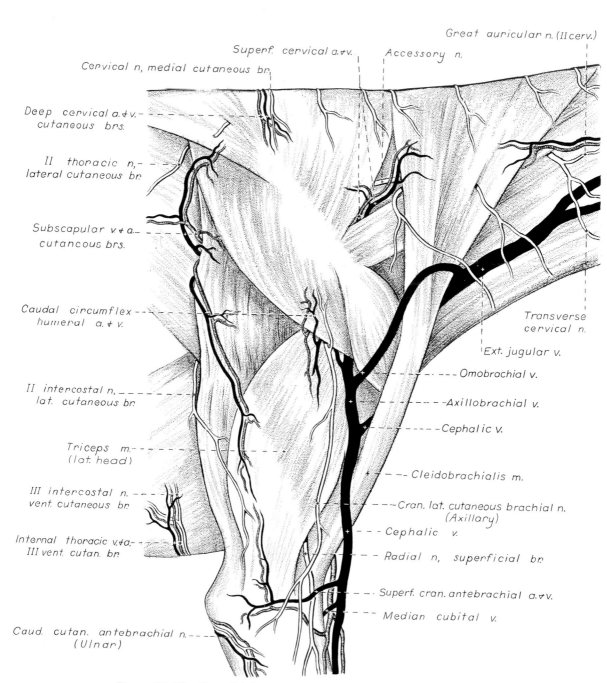

Figure 12–10. Superficial structures of the scapula and arm, lateral view.

flexor angle of the elbow joint. It runs proximally, crosses the cleidobrachialis muscle and joins the axillobrachial vein on the lateral surface of the triceps. The cephalic vein runs proximomedially under the brachiocephalicus at the junction of the middle and distal thirds of the brachium. It enters the external jugular between the omobrachial and the axillary veins and receives a tributary from the

major tubercle of the humerus and two or three more from the brachiocephalicus and pectoral musculature. This portion of the cephalic vein was formerly known as the distal communicating branch.

The **axillobrachial vein** (*v. axillobrachialis*) was formerly considered a continuation of the cephalic vein. It courses over the lateral head of the triceps and passes behind the humerus. Caudal to the

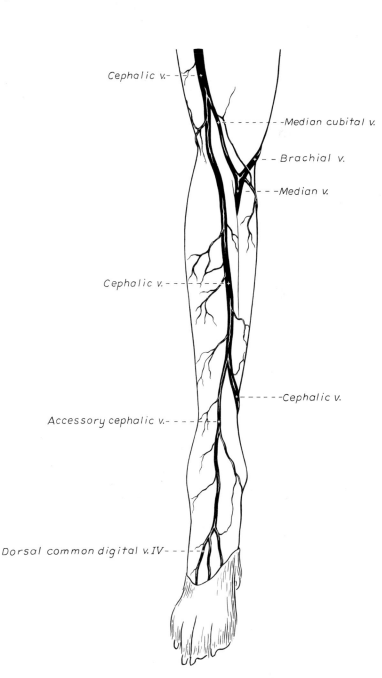

Figure 12–11. Veins of the right antebrachium, cranial aspect.

shoulder joint, the axillobrachial vein anastomoses with both the axillary and the subscapular vein. It receives large tributaries from the triceps in its course through it and terminates in the axillary vein.

The **omobrachial vein** (*v. omobrachialis*) was formerly called the proximal communicating vein of the cephalic. It leaves the axillobrachial about 2 cm. proximal to the cephalic and runs superficially at first on the deltoideus. It then arches forward and inward, crosses the brachiocephalicus, and enters the lateral surface of the external jugular about 3 cm. cranial to the end of the distal communicating branch. The omobrachial vein has no muscular tributaries and receives only small vessels from the skin and fascia.

DEEP VEINS OF THE THORACIC LIMB

The deep veins of the thoracic limb are represented at the proximal end of the carpus by the radial, ulnar, and the interosseous branch of the common interosseous vein. These three veins lie on the palmar surface of the antebrachium and anastomose distally not only with each other but also with the cephalic and accessory veins.

The **radial vein** (*v. radialis*) (Fig. 12–12) is very small, being only about one-third as large as its satellite artery. It arises from the **deep palmar venous arch** (*arcus venosus profundus palmaris*). It follows the mediocaudal border of the radius, where it is covered by the deep antebrachial fascia. It joins the small ulnar vein proximal to the origin of the terminal radial and ulnar branches of the brachial artery.

The **ulnar vein** (*v. ulnaris*) is about the same diameter as its companion artery. It leaves the supracarpal venous arch and runs proximally in the deep digital flexor. It receives tributaries from most of the flexor muscles lying in the antebrachium. It joins the small radial vein at about the junction of the proximal and middle thirds of the antebrachium to form the median vein.

The **median vein** (*v. mediana*) describes a sigmoid flexure before becoming the brachial vein in the arm. It is continued in the axilla as the axillary vein after it traverses most of the brachium. The median vein receives the **palmar antebrachial vein** (*v. antebrachialis palmaris*) about 1.5 cm. distal from the place where it collects the relatively large **common interosseous vein** (*v. interossea communis*), which enters its caudal side. The common interosseous

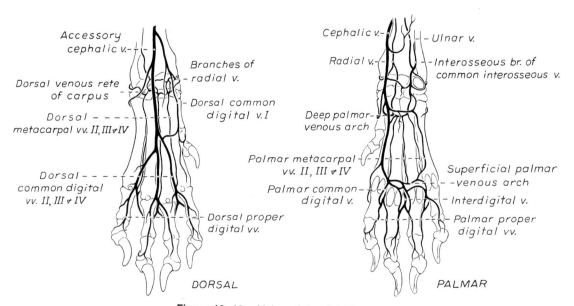

Figure 12–12.   Veins of the right forepaw.

vein has one tributary, the **interosseous branch** (*ramus interosseus*). This latter branch connects distally with the supracarpal, the deep palmar, and the proximal palmar venous arches. The median vein communicates with the cephalic via the median cubital vein. It then crosses the biceps in company with the superficial antebrachial artery and becomes the **brachial vein** (*v. brachialis*) caudal to the brachial artery. It receives in succession the **bicipital vein** (*v. bicipitalis*), the **proximal collateral ulnar vein** (*v. collateralis ulnaris proximalis*), and the **deep brachial vein** (*v. profunda brachii*), which may be double. These are satellites of their companion arteries.

The **axillary vein** (*v. axillaris*) (Fig. 12–13) is a continuation of the brachial vein. It receives the axillobrachial vein, cephalic vein, cranial circumflex humeral vein, lateral thoracic vein, and subscapular vein. The **subscapular vein** (*v. subscapsularis*) receives the following tributaries, which are satellites of the corresponding arteries: the **cutaneous branches** (*rami cutanei*), the **circumflex vein of the scapula** (*v. circum-*

flexa scapulae), the **thoracodorsal vein** (*v. thoracodorsalis*), and the **caudal circumflex vein of the humerus** (*v. circumflexa humeri caudalis*). Before terminating, the subscapular vein frequently bifurcates so that one branch may enter the subclavian or axillobrachial and the other branch may enter the beginning of the axillary or termination of the brachial. Usually three **pectoral branches** (*rami pectorales*) enter the medial side of the axillary at its termination, and another tributary comes from the nerves leaving the brachial plexus.

VEINS OF THE FOREPAW

The veins of the forepaw (manus) (Fig. 12–12), like the arteries, nerves, and lymphatics, are divided into a dorsal and a palmar set. These are not as completely divided into superficial and deep series as are the arteries in the metacarpus, and only single series exist dorsally and palmarly in the digits.

The single **dorsal proper digital veins II, III, IV, and V** (*vv. digiti dorsales propriae*)

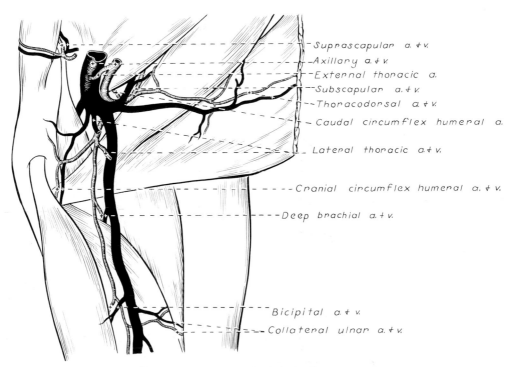

Figure 12–13. Vessels of the axillary region.

begin in the *arcus venosus digitales* formed by the anastomoses of the dorsal and palmar sets of proper digital veins. They occur on the medial aspects of digits II and III and on the lateral aspects of digits IV and V. These digital arches collect small tributaries from the digital pads and the corium of the claws. The dorsal proper digital veins run proximally on the dorsum of the digits and receive communicating branches from the palmar proper digital veins.

The **dorsal common digital veins I, II, III, and IV** (*vv. digitales dorsales communes*) continue proximally on the extensor tendons from their formation by the confluence of the dorsal digital veins and the palmar communicating branches. The lateral dorsal common digital vein IV runs proximally and joins III. The trunk formed by this union continues the axis of the fourth vessel and in turn is joined by the dorsal common digital vein II to form the **accessory cephalic vein** (*v. cephalica accessoria*). At the level of the first digit a small anastomosis usually occurs between the dorsal common digital vein I and the termination of vein II on the medial surface of the second metacarpal bone.

The **dorsal metacarpal veins I, II, III, and IV** (*vv. metacarpeae dorsales I, II, III, et IV*) are delicate veins which lie in the dorsal grooves between the main metacarpal bones. They anastomose with the corresponding dorsal common digital veins at the junction of the middle and distal thirds of the metacarpus and with the poorly formed dorsal rete of the carpus. The blood from the dorsal part of the first digit and the medial surface of the second metacarpal drains into the accessory cephalic vein some 4 cm. proximal to the carpus via the **dorsal common digital vein I.** This vein collects a tributary from the dorsal venous rete of the carpus.

The **dorsal venous rete of the carpus** (*rete carpi dorsale*) is a minute, poorly defined plexus of veins on the dorsal surface of the distal row of carpal bones. The dorsal metacarpal veins drain into it, and the accessory cephalic, dorsal common digital vein I, and the palmar set of veins carry blood from it.

The palmar set of veins of the forepaw begin usually as the single but occasionally double **palmar proper digital veins II, III, IV and V** (*vv. digitales palmares propriae*). These commence at the palmar extremities of the sagittally placed digital venous arches and run proximally on the palmar surfaces of the proximal interphalangeal joint and the adjacent phalanges. Usually the middle two veins divide on the first phalanges into medial and lateral branches. The apposed (axial) branches converge toward an axis through the paw and anastomose with the communicating veins from the dorsal set and with each other or extend singly to the superficial palmar venous arch and anastomose with it. The abaxial branches anastomose with the palmar proper digital veins nearest them. Thus, the medial branch anastomoses with the second palmar proper digital vein, and the lateral branch anastomoses with the fifth. In this way the **palmar common digital veins II, III, and IV** (*vv. digitales palmares communes*) are formed. These immediately anastomose with veins of the dorsal set and then continue to anastomose with the **superficial (distal) palmar venous arch** (*arcus palmaris superficialis*). This arch is formed by an anastomosis of the cephalic vein, medially, and the fourth palmar metacarpal vein, laterally. It lies deeply under the metacarpal pad, on the palmar surfaces of the metacarpophalangeal joints. Occasionally the arch is double, and other irregularities exist.

The **palmar metacarpal veins II, III, and IV** (*vv. metacarpeae palmares*) are small satellites of the deep palmar metacarpal arteries. They lie between the fleshy interosseous muscles and run from the superficial (distal) palmar venous arch proximally to anastomose with the deep (proximal) palmar venous arch.

The **deep palmar venous arch** (*arcus palmaris profundus*) lies under the origins of the palmar muscles and follows the distal border of the thick flexor retinaculum (palmar carpal ligament). Medially, it connects with the cephalic vein superficially and with the radial vein deeply; laterally, it anastomoses deeply with both the ulnar and the interosseous branches of the common interosseous vein. Usually a second venous arch exists here, connecting the cephalic vein with the interosseous branch across the superficial surface of the superficial flexor tendon. It lies subcutaneously, just distal to the carpal pad. The palmar skin and small carpal pad and the skin on the medial surface of the second metacarpal and digit are frequently drained by a single vein which terminates in the medial surface of the cephalic opposite the carpus. In some specimens a communicating vein connects the cephalic with the dorsal common digital vein I near the level of the metacarpophalangeal joint.

CAUDAL VENA CAVA

The **caudal vena cava** (*v. cava caudalis*), or postcava (Fig. 12–14), begins in contact with the ventral surface of the seventh lumbar vertebra by convergence of the common iliac veins. It is about 1 cm. in diameter in large dogs and lies in the furrow formed by the right and left psoas major and minor muscles. At its beginning the aorta lies to the left of it, since the aorta terminates ventral to the left common iliac vein. In its course cranially it gradually inclines ventrally until it reaches the medial part of the caudate lobe of the liver. It then inclines ventrally and to the right at a slightly sharper angle, and deeply grooves or tunnels the caudate lobe of the liver as it passes in it before reaching the diaphragm. It passes through the obliquely placed foramen venae cavae of the diaphragm, which is about 3 cm. to the right of the median plane. The intrathoracic portion of the caudal vena cava, about 4 cm. long, lies in a special pleural fold, the plica venae cavae, in company with the right phrenic nerve. Here both the nerve and vessel lie in a deep groove of the accessory lobe of the right lung before the vein terminates in the caudal part of the right atrium.

Reis and Tepe (1956), after examining 500 dogs for variations in the caudal vena cava and renal veins, list only two variants. One of these had been described previously by Kadletz (1928) and represented a circumaortic venous ring at the level of the renal veins. The other aberration was a persistence of the left supracardinal vein of the fetus. (The right supra-

cardinal forms the definitive caudal vena cava.) This was found in 2.6 per cent of the 500 dogs examined. Aberrations in the development of the caudal vena cava of the cat are common, compared with those in the dog. Reis and Tepe (1956) state that there are probably four longitudinal venous pathways in the lumbar region of the canine fetus. These are the right and left caudal (posterior) cardinal veins and the right and left supracardinal veins. Any one of these venous pathways other than the normal right supracardinal may persist, or any combination of them, to give any one of 15 different anomalous caudal vena cava types. The caudal vena cava in the dog is essentially devoid of muscle (Franklin 1937). It has the following tributaries, in addition to the formative common iliac veins: lumbar, deep circumflex iliac, right testicular or right ovarian, renal, phrenicoabdominal, hepatic, and phrenic veins.

The **lumbar veins I, II, III, IV, V, VI, and VII** (*vv. lumbales*) are satellites of the corresponding arteries. The first two lumbar veins are tributaries of the azygos vein on the right side and of the hemiazygos on the left. The members of the third pair of lumbar veins anastomose with each other directly ventral to the body of the third lumbar vertebra. From this small venous yoke a small median unpaired vein runs forward to become the azygos vein. In a similar manner a larger unpaired median vessel runs caudoventrally and enters the common trunk formed by the anastomoses of the right and left fourth lumbar veins. This trunk vein is the largest vessel entering the caudal vena cava from the lumbar vertebrae. Because of these venous anastomoses in the middle of the lumbar region, blood can flow forward to the heart either by the azygos vein or by the caudal vena cava. The members of the fifth and sixth pairs of lumbar veins anastomose with each other to form common trunks which enter the ventral surface of the caudal part of the caudal vena cava. The right and left seventh lumbar veins empty into the right and left common iliac veins, respectively.

The **deep circumflex iliac vein** (*v. circumflexa ilium profunda*) is a satellite of the corresponding artery and lies caudal to

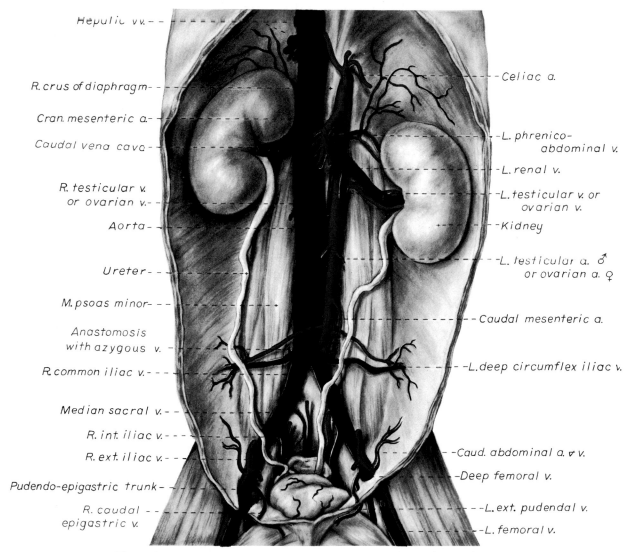

Hepatic vv.

R. crus of diaphragm

Cran. mesenteric a.

Caudal vena cava

R. testicular v. or ovarian v.

Aorta

Ureter

M. psoas minor

Anastomosis with azygous v.

R. common iliac v.

Median sacral v.

R. int. iliac v.

R. ext. iliac v.

Pudendo-epigastric trunk

R. caudal epigastric v.

Celiac a.

L. phrenico-abdominal v.

L. renal v.

L. testicular v. or ovarian v.

Kidney

L. testicular a. ♂ or ovarian a. ♀

Caudal mesenteric a.

L. deep circumflex iliac v.

Caud. abdominal a. & v.

Deep femoral v.

L. ext. pudendal v.

L. femoral v.

Figure 12–14.   The caudal vena cava and its main tributaries, ventral aspect.

it. The superficial and deep tributaries have the same anastomoses as do the comparable arteries.

The **right testicular vein** (*v. testicularis dextra*) enters the ventral surface of the caudal vena cava about 2 cm. caudal to the termination of the right renal vein. It collects tributaries from the testis and epididymis and becomes greatly coiled as it continues in the free border of the mesorchium. This coiled and flexuous arrangement of the testicular vein is known as the **pampiniform plexus** (*plexus pampiniformis*). The plexus is intertwined with the testicular lymphatics, artery, and nerve plexus as they form a funiculus which is located cranial to the ductus deferens and its blood vessels, as they all traverse the inguinal canal. The vein straightens on approaching the vaginal ring. Throughout its oblique intra-abdominal course to the caudal vena cava, it lies in a plica of peritoneum which may be 4 cm. wide near the inguinal canal. The testicular vein is joined by one or two small tributaries drom the adipose and true renal capsules a few centimeters before its termination. It is accompanied, except at its termination, by the testicular artery, nerves, and lymphatics.

The **right ovarian vein** (*v. ovarica dextra*) is shorter and less tortuous than the homologous right testicular vein. It begins by two or three tributaries from the right ovary and surrounding fat. These are tortuous and plexiform. The caudal or uterine tributary freely anastomoses in the broad ligament opposite the cranial end of the uterine horn with the larger uterine vein.

The **renal veins** (*vv. renales*) (Fig. 12–14) are about 8 mm. in diameter. The right renal vein is about 3 cm. long, whereas the left is 4 cm. long. Each begins at the hilus of the respective kidney by convergence of the two veins which arise near the poles of the kidney by collecting the interlobar veins. Christensen (1952) described the intrarenal morphology of the renal veins of the dog. The renal veins may take an oblique course forward to reach the caudal vena cava. The renal veins contain valves at their terminations. The left renal vein receives the left gonadal vein, the **left testicular vein** (*v. testicularis sinistra*) or the **left ovarian vein** (*v.*

ovarica sinistra). Except for the difference in termination, the left gonadal vein closely resembles its fellow on the right, which empties directly into the caudal vena cava. The left gonadal vein or the left renal vein receives the **left cranial ureteric vein** (*v. ureterica cranialis sinistra*) several centimeters before its termination, whereas the right renal vein usually receives the **right cranial ureteric vein** (*v. ureterica cranialis dextra*) near the hilus of the kidney. In the 500 dogs examined by Reis and Tepe (1956), double renal veins were found five times on the right side but never on the left.

The **phrenicoabdominal veins** (*vv. phrenicoabdominales* (Fig. 12–14) are about 4 mm. in diameter as each terminates in the lateral surface of the caudal vena cava about 1 cm. cranial to the renal vein of the same side. Each grooves the ventral surface of the corresponding suprarenal gland as the terminal 1 cm. of the vein passes under the gland. Here it receives the **adrenal veins** (*vv. suprarenales*) which, unlike the arteries, drain entirely into the phrenicoabdominal as it lies in the groove of the gland (Flint 1900). These veins are short and inconspicuous. The formative phrenic and cranial abdominal parts of the phrenicoabdominal vein are satellites of the arteries of the same names.

The **hepatic veins** (*vv. hepaticae*) (Fig. 12–14) are embedded, wholly or in part, in the liver parenchyma. They are therefore short and receive numerous tributaries. They terminate along the lateral and ventral surfaces of the last 4 cm. of the intraabdominal portion of the caudal vena cava. The largest hepatic vein serves the left lateral, left medial, quadrate, and part of the right medial lobe. It enters the left ventral part of the caudal vena cava at the foramen venae cavae and is the most cranially located of all the hepatic veins. From the right, usually two major hepatic veins enter the caudal vena cava, about 2 cm. apart. The cranial tributary comes from the right lateral and partly from the right medial lobe. The caudal tributary comes mostly from the caudate lobe. These vessels are about 3 mm. in diameter. There are a score or more of hepatic tributaries, ranging under 1 mm. in diameter, which drain into the caudal vena cava at various places as it courses through the liver.

The **phrenic veins** (*vv. phrenica*) are represented by a single tributary on each side, beginning in the ventral part of the muscular periphery of the diaphragm. At the lateral junction of the muscular and tendinous parts, each empties into the caudal vena cava as it passes through the foramen venae cavae. There is usually a small trunk vein which empties into the caudal vena cava on the thoracic side of the foramen venae cavae. This is formed by two or more tributaries which drain the extensive but flimsy plica venae cavae.

PORTAL VEIN

The **portal vein** (*v. portae*) (Fig. 12–15) with its tributaries from the viscera forms a portal system. It arises from a capillary bed and ends in the liver. It collects blood from the pancreas, spleen, and all of the gastrointestinal tract except the anal canal. It is about 1.2 cm. in diameter at the porta of the liver, where it terminates. It lies deeply buried among the abdominal viscera. It runs cranially from its formation in the root of the mesojejunum, dorsal to the junction of the right and left limbs of the pancreas. It continues forward from the pancreas in association with the hepatic artery and plexus of autonomic nerves to form the ventral boundary of the epiploic foramen. The portal vein is formed by the confluence of the cranial and caudal mesenteric and the splenic vein. Great variations exist in the patterns of formation of these veins. The cranial mesenteric vein is always the largest vessel.

The **cranial mesenteric vein** (*v. mesen-*

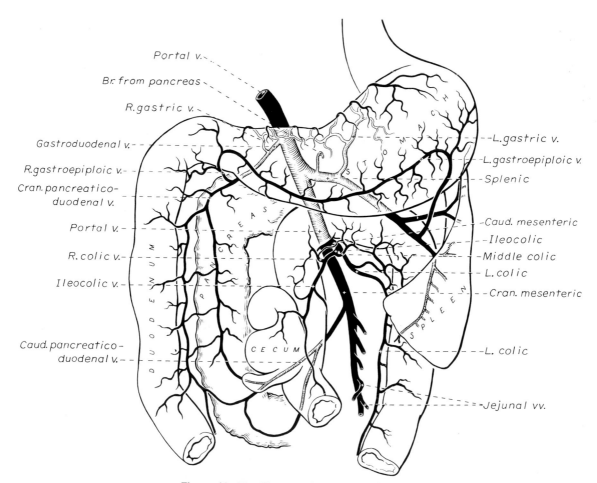

Figure 12–15.   The portal vein, ventral aspect.

terica cranialis) is 8 to 10 mm. in diameter at its termination. It collects approximately 12 **jejunal** and **ileal veins** (*vv. jejunales et ilei*), which are satellites of the corresponding arteries and, like them, are divided into a proximal and a distal series. There are only primary formative arcades, and these are largest in the middle of the series and smaller at each end. Some of the vasa recta are double as they flank the straight arteries. The most distal ileal vein anastomoses with the ileal branch of the ileocolic, whereas the most proximal jejunal vein forms an arcade with the **caudal pancreaticoduodenal vein** (*v. pancreaticoduodenalis caudalis*). This latter vessel is the last tributary to enter the cranial mesenteric and is like the other intestinal vein, except that it enters the cranial side of the cranial mesenteric.

The **caudal mesenteric vein** (*v. mesenterica caudalis*) is not a satellite of the like-named artery. It begins in the pelvic cavity as the **cranial rectal vein** (*v. rectalis cranialis*), which is a satellite of the cranial rectal artery. In the rectal plexus of veins at the pelvic outlet, it anastomoses with the caudal rectal vein, which drains blood into the caval system of veins. The cranial rectal vein continues forward from the pelvic inlet in the left mesocolon to the place where the cranial rectal artery becomes associated with it. Here it is continued as the **left colic vein** (*v. colica sinistra*) and collects approximately 25 left colic branches, which are satellites of the vasa recta of the left colic artery. Some of these may be double. The last tributary to enter the left colic vein is larger than the other, as it collects blood from the left colic flexure as well as from the adjacent middle and left parts of the colon. Its middle colic tributary anastomoses in an arcade with the middle colic vein. At the left colic flexure, the caudal mesenteric vein enters the mesojejunum, in which it crosses the left face of the cranial mesenteric artery and associated structures. It joins the cranial mesenteric vein to the right of the left limb of the pancreas to form the portal vein.

The **ileocolic vein** (*v. ileocolica*) enters the caudal mesenteric vein after receiving the cecal, right colic, and middle colic veins. The **right colic vein** (*v. colica dextra*) drains the distal part of the right colon, the adjacent right colic flexure, and the beginning of the transverse colon. The right colic anastomoses with the colic tributary of the ileocolic by the formation of a weak arcade. The **middle colic vein** (*v. colica media*) enters the ileocolic less than 1 cm. from its termination. It is formed by the vasa recta from the transverse colon and the proximal and distal anastomotic arcades, which are also partly formed by the right and left colic veins, respectively. Occasionally the right and middle colic veins form a common terminal trunk.

The **splenic vein** (*v. lienalis*) (Fig. 12–15) is one-half to two-thirds as large as the cranial mesenteric vein. It is about 5 mm. in diameter and 1.5 cm. long. It is formed by the confluence of the smaller caudally running left gastric vein and the larger splenic vein. The **left gastric vein** (*v. gastrica sinistra*) is formed by several veins which come from the lesser curvature of the stomach adjacent to the cardia. Like the corresponding artery, it anastomoses with the **right gastric vein**. The splenic vein receives tributaries from the long hilus of the spleen. The **pancreatic veins** (*vv. pancreaticae*) are represented by two tributaries from the left limb of the pancreas; they may terminate separately in the last 2 cm. of the splenic. The **left gastroepiploic vein** (*v. gastroepiploica sinistra*) is a satellite of the left gastroepiploic artery; it comes from the greater curvature of the stomach and collects many epiploic tributaries along its course.

The **gastroduodenal vein** (*v. gastroduodenalis*) (Fig. 12–15) is 3 to 4 mm. in diameter and empties into the portal vein about 1.5 cm. from the hilus of the liver. Its chief formative tributary is the **cranial pancreaticoduodendal vein** (*v. pancreaticoduodenalis cranialis*). Its pancreatic and duodenal tributaries begin as small anastomoses with the pancreatic and duodenal parts of the **caudal pancreaticoduodenal vein** (*v. pancreaticoduodenalis caudalis*) near the caudal flexure of the duodenum. At its termination it receives a small tributary from the left limb of the pancreas. The **right gastroepiploic vein** (*v. gastroepiploica dextra*) and the **right gastric vein** (*v. gastrica dextra*) are satellites of their companion arteries. They sometimes unite before blending with the larger cranial pancreaticoduodenal to form the gastroduodenal vein. They anastomose with the left gastroepiploic and left gastric veins, respectively.

The portal vein divides, upon entering the liver, into a small right branch, which is dispersed in the right lateral and the right medial lobe, and a large left branch, which breaks up in the remainder of the liver.

VEINS OF THE PELVIC LIMB

SUPERFICIAL VEINS OF THE PELVIC LIMB

The large superficial veins of the pelvic limb, exclusive of the hindpaw, are the lat-

eral saphenous and the superficial branch of the deep circumflex iliac on the lateral side and the medial saphenous, saphenous, and the proximal part of the femoral on the medial side. All of these veins, except the deep circumflex iliac, are commonly used for venipuncture.

The **lateral saphenous vein** (*v. saphena lateralis [parva]*) (Fig. 12–16) begins by collecting its dorsal branch from the flexor surface of the tarsus and its plantar branch from the lateral surface. The **cranial branch** (*ramus cranialis*) is 2 to 4 mm. in diameter as it inclines caudally in its proximal course. It obliquely crosses the lateral surface of the distal end of the tibia and is here frequently used for venipuncture. At the space be-

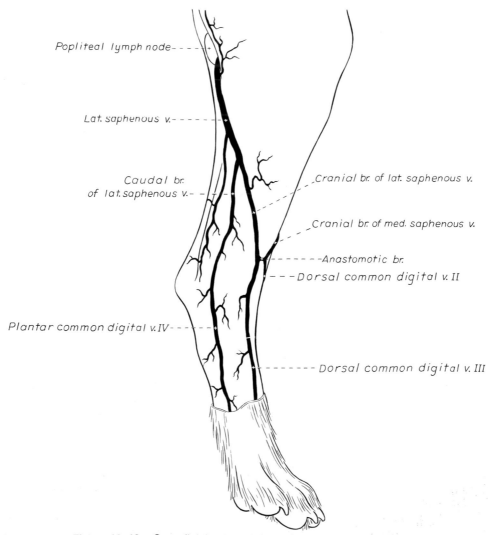

Figure 12–16. Superficial veins of the right hindleg, lateral aspect.

tween the deep caudal crural muscles and the beginning of the calcanean tendon, it receives the smaller **caudal branch** (*ramus caudalis*) and becomes the lateral saphenous vein. This crosses the beginning of the calcanean tendon and receives a small tributary from the skin covering the calcanean tuberosity. It continues its subcutaneous course proximally on the caudal surface

of the gastrocnemius to the popliteal lymph node. It runs deep to the node and follows the intermuscular septum between the m. biceps femoris and the m. semitendinosus to terminate in the femoral vein in the popliteal fossa.

The **superficial branch** of the **deep circumflex iliac vein** (*v. circumflexa ilium profunda*) drains the caudal half of the dorsal

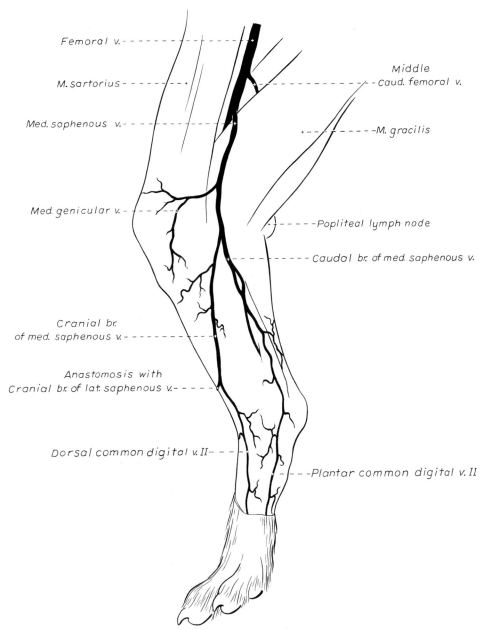

Figure 12–17. Superficial veins of the right hindleg, medial aspect.

two-thirds of the skin of the abdominal wall and the cranial half of the rump and proximal part of the thigh.

The **medial saphenous vein** (*v. saphena medialis* [*magna*]) (Fig. 12–17) begins by the confluence of its cranial and caudal branches medial to the stifle joint. The **cranial branch**, traced from the flexor surface of the tarsus where it anastomoses with the cranial branch of the lateral saphenous, run proximally across the m. tibialis cranialis and the tibia. It is about 1 mm. in diameter. The medial saphenous vein and its main tributaries are satellites of the saphenous artery and its main branches. A prominent **medial genicular vein** from the stifle joint joins the saphenous vein above the confluence of the cranial and caudal branches. In most specimens a tributary from the m. gracilis is received by the medial saphenous about 1 cm. before its termination in the femoral. It lies under the thin but strong medial femoral fascia between the caudal belly of the m. sartorius and the m. gracilis. It is firmly anchored by fascia between these muscles and is a common site for venipuncture. It joins the femoral vein at the apex of the femoral triangle.

The **femoral vein** (*v. femoralis*) (Figs. 12–14, 12–17) is accessible for venipuncture as it lies in the femoral triangle. This segment of the vessel is about 8 cm. long and 5 mm. in diameter. It lies caudal to the femoral artery, from which the pulse can easily be taken, and cranial to the small saphenous nerve. It is sufficiently large and well attached to permit injections without compression.

The cutaneous veins of the pelvic limb, exclusive of the hindpaw, are largely tributaries of the superficial veins. However, the skin of the caudal part of the rump, the region of the pelvic outlet, and the caudolateral part of the thigh are drained by deeply lying veins. A cutaneous area over the caudal part of the middle gluteal muscle drains into the cranial gluteal vein, whereas a cutaneous tributary which drains into the terminal part of the caudal lateral vein and perineal veins serves the skin of the pelvic outlet. A wide caudolateral zone of skin of the thigh is drained by three or more tributaries of a large vein which descends in the biceps femoris and empties into the terminal part of the lateral saphenous vein. Others

are formative tributaries of the caudal and deep femoral veins.

DEEP VEINS OF THE PELVIC LIMB

The deep veins of the pelvic limb, exclusive of the hindpaw, are largely satellites of the neighboring arteries. The **cranial tibial vein** (*v. tibialis cranialis*) lies medial to its companion artery, which it approaches in size. It begins on the flexor surface of the tarsus as a continuation of the dorsal common digital vein II with a contribution from the medial tarsal vein. It receives tributaries from the cranial crural group of muscles mainly at their proximal ends. It passes between the tibia and fibula, runs under the m. popliteus, and unites with the delicate **caudal tibial vein** (*v. tibialis caudalis*) to form the **popliteal vein** (*v. poplitea*). The popliteal vein (Fig. 12–19) traverses the popliteal notch of the tibia to enter the popliteal fossa. Here it receives a medial and a lateral tributary from the sides of the stifle joint. These enter the popliteal vein proximal to the joint. A cluster of veins enters this venous channel at the proximal end of the m. gastrocnemius, where the popliteal vein becomes the femoral vein. The largest of these is the lateral saphenous, which is three times larger than any other vein distal to the stifle.

The large **middle caudal femoral vein** (*v. caudalis femoris media*), 3 mm. in diameter, drains the biceps femoris. It begins as cutaneous twigs from the caudolateral surface of the thigh. Muscular tributaries ascend from the gastrocnemius. Others enter it from in front and medially from the quadriceps, adductors, and hamstring muscles. At the distal end of the femoral triangle, in the vicinity of entry of the medial saphenous into it, the femoral receives a tributary from the adductor and semimembranosus. In its course through the femoral triangle, the femoral vein receives a tributary which is the satellite of the most proximal muscular branch of the femoral artery. This vein arises from the medial surface of the proximal half of the m. gracilis and receives a large tributary from the proximal part of the adductor muscles. A small vein frequently arises in the region of the subcutaneous inguinal ring and enters the caudal part of the femoral. Entering the cranial side of the femoral vein, 1 or 2 cm. from the abdominal wall, is a long vein which begins in the distal part of the cranial belly of the m. sartorius. It runs proximally on the m. vastus medialis and, after receiving a large tributary from the m. rectus femoris, passes under the femoral artery and unites with the **superficial circumflex iliac vein** (*v. circumflexa ilium superficialis*).

The **lateral circumflex femoral vein** (*v. circumflexa femoris lateralis*), formerly known as the cranial femoral vein, is the largest and the last tributary of the femoral vein, which it enters laterally. As a satellite of the lateral circumflex femoral artery, it begins in the skin of the proximal, lateral surface of the thigh and the adjacent dorsally lying gluteal region. Its formative tributaries cross ventral to the middle gluteal muscle near its insertion and then unite and pass the cranial aspect of the neck of the femur and the caudal surface of the origin of the m. rectus femoris to enter the lateral surface of the femoral at the vascular lacuna. It may receive the superficial circumflex iliac vein at its termination, when this vein does not enter the femoral directly. The femoral vein lies ventromedial to the m. iliopsoas as it passes through the abdominal wall to become the **external iliac vein** (*v. iliaca externa*). This venous trunk is about 6 mm. in diameter and 3 cm. long. It arches dorsocranially across the medial surface of the m. iliopsoas and ventral to the promontory of the sacrum and unites with the internal iliac vein to form the **common iliac vein** (*v. iliaca communis*). It has but one large tributary, the **deep femoral vein** (*v. femoralis profunda*), which enters the caudal side of the external iliac on the abdominal side of the vascular lacuna. Its most caudal tributaries, like those of its accompanying artery, are cutaneous twigs from the skin of the upper caudal part of the thigh. These merge to form larger vessels as they enter the proximal part of the hamstring musculature. Within these heavy muscles, further blending occurs so that the single vein emerges from between the m. pectineus medially and the m. iliopsoas laterally. At its termination it receives the large **pudendoepigastric trunk** (*truncus venosus pudendoepigastricus*), which is disposed like the comparable artery and its branches. Two small veins enter the external iliac or deep femoral

vein. One enters caudally from the adipose tissue inside the pelvic inlet, and the other enters cranially from the m. rectus abdominis.

The **internal iliac vein** (*v. iliaca interna*) (Figs. 12–14, 12–19), unlike the internal iliac artery, is not divided into visceral and parietal parts. As a single vein it lies between these two arteries. Its tributaries are satellites of the branches of the two parts of the internal iliac artery. It is formed caudally by the merging of the internal pudendal and caudal gluteal veins.

The **caudal gluteal vein** (*v. glutea caudalis*) (Fig. 12–18) arises mainly in the m. biceps femoris, with smaller tributaries coming from the mm. semimembranosus and semitendinosus. It ascends through the lesser ischiatic foramen lateral to the internal obturator muscle, medial to its companion artery, and caudodorsal to the ischiatic nerve. At the pelvic outlet it collects a superficial tributary from the proximal caudal half of the thigh and another one from the internal obturator muscle. Lateral to the m. coccygeus a branch from the fat of the ischiorectal fossa and the first (with, occasionally, the second) sacral intervertebral vein enter the caudal gluteal vein.

The **internal pudendal vein** (*v. pudenda interna*) is formed at the root of the penis by the convergence of the *v. dorsalis penis* and *truncus venosus profundus penis et bulbi*. These are satellites of the comparable arteries. The trunk so formed usually receives the perineal vein. (For a complete description of the veins of the penis and male perineum the reader is referred to Chapter 9, Urogenital Apparatus.) The internal pudendal vein receives the third and sometimes the second sacral intervertebral vein. In the female the vein of the clitoris takes the place of the three veins from the penis.

The **vein of the clitoris** (*v. clitoridis*) begins in the vestibular bulb and is not sharply differentiated caudally from the vestibular plexus. The **vestibular plexus** (*plexus vestibuli*) is a closely knit venous plexus which completely surrounds the external vulvar opening. It extends about 1 cm. cranially from the dorsal commissure of the vulva on its dorsal wall. Ventrally, it widens to such an extent that it runs proximally on the urethra. This network, the **urethral venous plexus** (*plexus venosus*

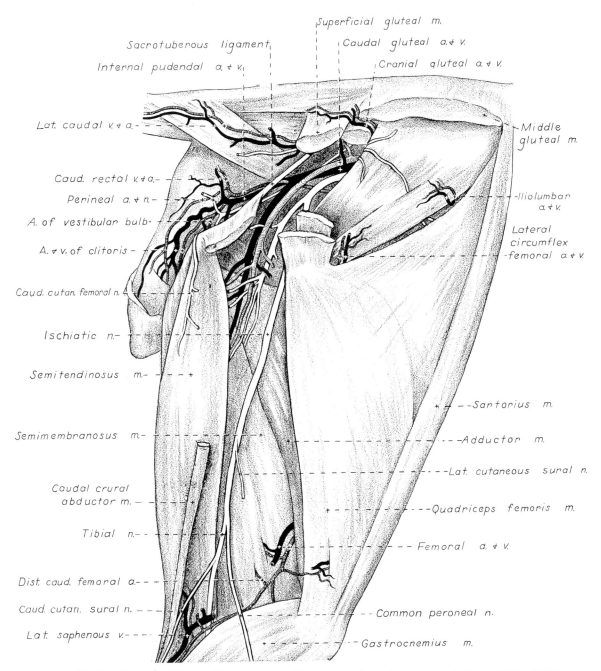

Figure 12–18. Deep structures of the gluteal and femoral regions, lateral aspect. (From Miller 1958.)

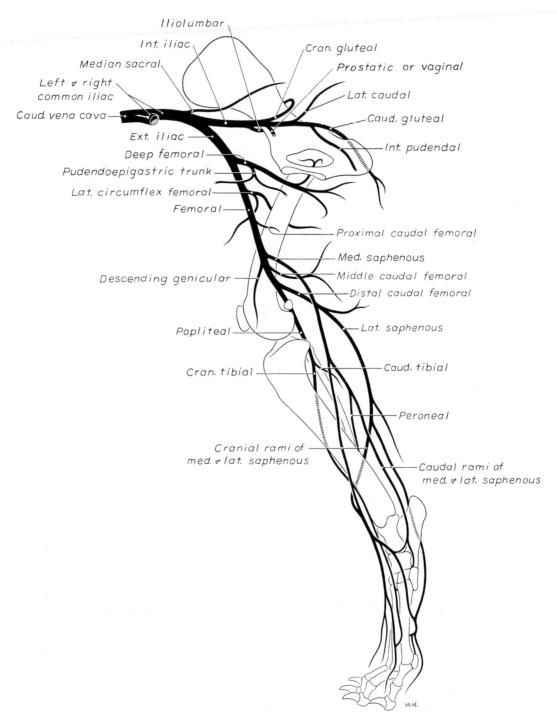

Figure 12-19. Veins of the pelvic limb, medial view.

urethralis), surrounds the female urethra and splays out on the neck of the bladder, between the heavy muscular coat and the mucosa. Both the vestibular bulbs and the plexus lie deep to the strong, partly divided constrictor vestibuli muscle. Between the vestibular bulbs ventrally, the vestibular venous plexus reflects the division which occurs in the constrictor vestibuli muscle in such a way that the plexus caudal to the division is thick and annular in nature, whereas that cranial to the division is thin and longitudinal. After leaving the vestibular bulb, the vestibular vein receives (opposite the ventral part of the external anal sphincter) the **caudal rectal vein** *(v. rectalis caudalis),* which comes directly from the external anal sphincter, anal sac, and the wall of the anal canal. It anastomoses with the cranial rectal vein via the rectal plexus of veins. Since the cranial rectal vein is indirectly a tributary of the portal and the caudal rectal vein is a like tributary of the caudal vena cava, the rectal plexus serves to unite the two systems. The **rectal venous plexus** *(plexus venosus rectalis)* is poorly developed. It lies on and in the musculature of the rectum just cranial to the anal canal. It is partly covered by the anal sacs when these are large. The **perineal vein** *(v. perinealis)* from the cutaneous anal structures ends in the termination of the vaginal vein or directly in the internal pudendal vein.

The **lateral caudal vein** *(v. caudalis coccygea lateralis)* is the main venous drainage from the tail. It is about 3 mm. in diameter at the ischiorectal fossa, through which it runs to enter the caudal gluteal vein. It receives segmental branches as it runs forward from the free end of the tail. At the pelvic outlet it receives a large tributary from the skin of the tail and the adjacent part of the rump. It contains valves at intervals of about 1 cm.

The **prostatic vein** *(v. prostatica)* of the male is homologous with the **vaginal vein** *(v. vaginalis)* of the female. They were formerly known as the urogenital vein.

The prostatic vein originates on the prostate gland and receives tributaries from the caudal vesicle of the bladder, including ureteral and urethral branches, and from the vein of the ductus deferens as well as from the middle rectal vein.

The vaginal vein originates in the wall of the vagina and receives the uterine vein, the caudal vesicle vein with a ureteral branch, and the middle rectal vein.

In the female the development of the vaginal vein is directly proportional to the genital activity of the bitch. In late pregnancy and immediately after parturition the uterine tributary is larger than all the others together. In the non-gravid bitch the cranial and caudal tributaries are similar to those in the male. The cranial tributary is formed by the caudally running **uterine vein** *(v. uterina)* meeting at a right angle with the **caudal vesical vein** *(v. vesicalis caudalis).* (The **caudal ureteral vein** [*v. ureterica caudalis*] ends in the proximal part of the caudal vesical.) The cranial and caudal ureteral veins anastomose. The caudal vesical vein anastomoses with the cranial vesical vein, if one is present. The uterine vein begins in an anastomosis with the uterine tributary of the ovarian vein. This anastomosis takes place opposite the caudal third of the uterine horn. The main vessel follows the companion artery and therefore lies as much as 3 to 4 cm. out in the broad ligament caudally. There is, however, a smaller venous channel which lies in the uterine wall opposite the attached border of the uterine horn and extends throughout its length.

This vein is connected, by about five communicating veins, with the main uterine vein, which lies in the broad ligament. From both sides of the uterine wall it receives many tortuous tributaries which anastomose with each other but do not form a definite uterine plexus. Near the ovarian end of the uterine horn, the converging uterine veins anastomose with each other and with the uterine extension of the ovarian vein. The vaginal vein arises in the submucosa of the wall of the vagina as a fine **vaginal venous plexus.** Unlike the comparable artery, it does not serve the urethra. Caudally, it anastomoses with the vestibular plexus of veins.

The prostatic and vaginal veins empty into the medial surface of the internal iliac opposite the **cranial gluteal vein** *(v. glutea cranialis).*

The cranial gluteal vein is a satellite of the like artery and therefore drains most of the proximal part of the middle gluteal muscle. It usually has a prominent cutaneous tributary which drains the skin over the proximal dorsal part of the rump and also receives the first and occasionally the second intervertebral vein.

The **iliolumbar vein** (*v. iliolumbalis*) is a satellite of the iliolumbar artery. It crosses the cranial border of the wing of the ilium and terminates in the lateral side of the internal iliac vein at its termination. It receives the seventh lumbar intervertebral vein (Worthman 1956).

The **common iliac vein** (*v. iliaca communis*) is about 8 mm. in diameter and 5 cm. long. It is formed by the confluence of the external and internal iliac veins on the tendon of the psoas minor about 2 cm. cranial to its insertion. From their origins the right and left common iliac veins converge and merge, usually ventral to the sixth lumbar vertebra, to form the caudal vena cava. Occasionally, the union of the common iliac veins is unusually far forward, but this anomaly is more common in the cat than in the dog, according to Darrach (1907) and Huntington and McClure (1920).

The **median sacral vein** (*v. sacralis media*) (Fig. 12–14) is the unpaired median vein which receives the middle caudal vein from the tail. It runs forward between the right and left ventral sacrocaudal muscles

and terminates, according to Worthman (1956), in both the right and the left common iliac vein, or in only one of them. It is a small vein, about 1 mm. in diameter, usually devoid of significant tributaries. In one specimen it was large and collected as single short tributaries the first two right and left sacral intervertebral veins and the last pair of lumbar intervertebral veins, after they had united in a common trunk.

VEINS OF THE HINDPAW

The veins of the hindpaw (pes) (Fig. 12–20) are divided into a dorsal and a plantar set. In the metatarsus these are further divided into dorsal and plantar common digital veins lying superficially, with dorsal and plantar metatarsal veins lying deeply. Anastomoses occur between the dorsal and plantar series at both the proximal and the distal ends of the intermetatarsal spaces. The veins of the hindpaw in the dog have been described by Preuss (1942).

The **digital venous arches** (*arci venosi digitales*) are anastomoses between the dorsal and plantar proper digital veins. One for each digit, they lie on the abaxial sides of

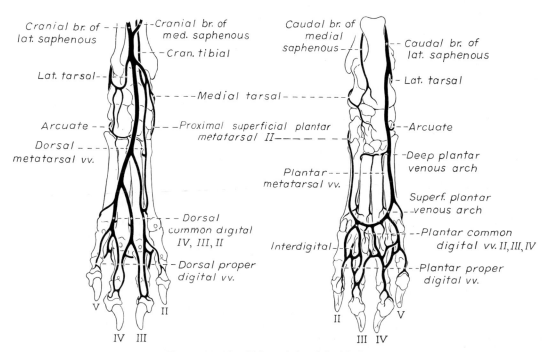

Figure 12–20. Veins of the right hindpaw.

the distal ends of the second phalanges and receive tributaries from the claws, digital pads, and terminal phalanges. The single or paired dorsal and plantar common digital veins arise proximally from the arches.

The **dorsal proper digital veins II, III, IV, and V** *(vv. digitales dorsales propriae)* arise from the axial and abaxial sides of the digits. They receive anastomoses from the plantar digital veins at the distal end of the metatarsus, in the respective digital clefts.

The **dorsal common digital vein III** *(v. digitalis dorsalis communis)* is formed by the axial dorsal proper digital veins of the third and fourth digits. The **dorsal common digital veins II and IV** are formed by the second and fifth dorsal digital veins, which anastomose with the abaxial dorsal branches of the third and fourth dorsal proper digital veins, respectively. The dorsal common digital veins anastomose with the plantar metatarsal veins near the distal ends of the metatarsal bones. The third and fourth dorsal common digital veins anastomose with each other at the middle of the metatarsus, and the resultant common trunk is joined near the proximal end of the metatarsus by the second dorsal common digital vein. The common venous trunk formed by the three dorsal common digital veins is the **cranial branch of the lateral saphenous vein** *(ramus cranialis v. saphenae lateralis)*, which is called the v. metatarsea dorsalis superficialis proximalis II by Preuss (1942). This vessel continues proximally on the long digital extensor tendon to the distal end of the crus. Opposite the talocrural joint it receives the **lateral tarsal vein** *(v. tarsea lateralis)*. The lateral tarsal vein also connects with the caudal branch of the lateral saphenous vein and receives tributaries from the joint capsule, skin, and arcuate vein.

The **dorsal metatarsal veins II, III, and IV** *(vv. metatarseae dorsales)* are small veins which lie in the grooves between adjacent metatarsal bones. They enter the arcuate vein proximally.

The **cranial branch of the medial saphenous vein** *(ramus cranialis v. saphenae medialis)* parallels the tendon of the m. tibialis cranialis as it runs proximally over the flexor surface of the tarsus. Upon leaving the tarsus, it receives the **medial tarsal vein** *(v. tarsea medialis)* from the plantar surface of

the tarsus. At the talocrural joint the cranial branch of the medial saphenous vein receives a long anastomotic branch from the second dorsal common digital vein. This anastomotic branch is the v. metatarsea dorsalis superficialis proximalis II of Preuss (1942). It receives an anastomotic branch from the deep plantar venous arch which passes between metatarsal bones II and III.

The **arcuate vein** *(v. metatarsea arcuata)* crosses the dorsal surfaces of the proximal ends of the metatarsal bones. It receives the threadlike dorsal metatarsal veins, and anastomoses with the caudal branch of the lateral saphenous vein and with the more proximal, lateral tarsal vein.

The cranial branches of the lateral and medial saphenous veins are either confluent or joined by a short anastomosis at the level of the distal end of the tibia. Figure 12–20 represents confluent saphenous veins, whereas Preuss (1942) illustrates a short anastomosis.

The plantar set of veins of the hindpaw begins as single or paired **plantar proper digital veins** *(vv. digitales plantares propriae)*. These originate as plantar continuations of the digital venous arches, and collect tributaries from the digital pads, skin, and terminal phalanges.

The **plantar common digital veins** *(vv. digitales plantares communes)* are short and drain into the **superficial plantar venous arch** *(arcus venosus plantaris superficialis)*. The arch is continued medially by the cranial branch of the medial saphenous vein and laterally by the caudal branch of the lateral saphenous vein. The plantar metatarsal portions of the medial and lateral saphenous veins are called the v. metatarsea plantaris superficialis proximalis II and the v. metatarsea plantaris superficialis proximalis IV, respectively, by Preuss (1942).

The **caudal branch of the lateral saphenous vein** *(ramus caudalis v. saphenae lateralis)* passes proximally in the superficial metatarsal fascia along the lateral surface of the metatarsus. At the proximal end of the metatarsus it receives the **deep plantar venous arch** *(arcus venosus plantaris profundus)*, formed by the plantar metatarsal veins.

The **caudal branch of the medial saphenous vein** *(ramus caudalis v. saphenae medialis)* is the smallest of the main saphenous branches. It begins from the larger medial tarsal vein, distal to the medial malleolus. In the crus it is related to the caudal branch of the saphenous artery and joins the larger cranial branch of the medial saphenous vein opposite the proximal end of the tibia.

The two or three **plantar metatarsal veins** (*vv. metatarseae plantares*) are small and lie in the intermetatarsal grooves or on the plantar surface of the third and fourth metatarsal bones. Near the middle of the metatarsus they anastomose with the dorsal common digital veins by passing between the respective metatarsal bones. Proximally, the plantar metatarsal veins enter the deep plantar venous arch.

The medial and lateral saphenous veins are the main vessels which return blood from the hindpaw. The lateral saphenous is appreciably larger than the medial saphenous. As its cranial tributary obliquely crosses the distal lateral surface of the tibia, it may be used for venipuncture.

VEINS OF THE CENTRAL NERVOUS SYSTEM

VENOUS SINUSES OF THE CRANIAL DURA MATER

Within the dura, usually between its periosteal and meningeal parts, and in certain places within large osseous canals, there are venous passages into which the veins of the brain and of its encasing bone drain. These passages, known as the **sinuses of the dura mater** (*sinus durae matris*), in the dog are not confined exclusively to the dura. By means of the these passages the blood is conveyed from the brain and skull to the paired maxillary, internal jugular, and vertebral veins, and to the vertebral venous plexuses. They lack a tunica media and tunica adventitia in their walls, and they do not have valves in their lumina. They are divided into dorsal and ventral sets, which freely intercommunicate. The dorsal set consists of the unpaired dorsal sagittal and straight sinuses and the paired transverse sinus. The ventral set consists of the double, unpaired intercavernous, and the paired cavernous, sigmoid, basilar, and dorsal and ventral petrosal sinuses. The cranial venous sinuses have been studied in a variety of vertebrates by Hofmann (1901), and in the dog by Zimmermann (1936) and Reinhard, Miller, and Evans (1962). A comparative developmental study of the cranial venous system in man and dog was made by Padget (1957).

The **dorsal sagittal sinus** (*sinus sagittalis dorsalis*) (Fig. 12–21) begins by the confluence of the right and left **rhinal veins** which come from the osseous nasal septum and its covering mucosa, and the olfactory bulbs and their dural coverings. From near the middle of the cribriform plate, where the sagittal sinus is formed, it runs caudally in the attached edge of the falx cerebri. It therefore lies directly ventral to the sagittal suture and the interparietal process of the occipital bone. The sagittal sinus collects the dorsal cerebral and most of the diploic veins and is usually joined by the straight sinus near its termination. It measures about 3 mm. in diameter at the foramen impar, which it traverses to form a junction with the right and left transverse sinuses.

The **straight sinus** (*sinus rectus*) (Fig. 12–21) usually drains into the caudal part of the sagittal sinus before it enters the foramen impar, but it may pursue an independent course through the accessory foramen impar and join the confluence of the sinuses within the occipital bone. Another frequent variation is for the sagittal sinus to bifurcate after the straight sinus has entered it. The straight sinus is about 1.5 mm. in diameter and 5 mm. long. It begins at the free caudal margin of the falx cerebri by the merging of the great cerebral vein, ventrally, and the vein of the corpus callosum, dorsally. The straight sinus lies in the caudal border of the falx cerebri as the right and left parts of the tentorium cerebelli join it. It lies rostrodorsal to the tentorium ossium.

The **transverse sinus** (*sinus transversus*) (Fig. 12–22) is paired. Each begins middorsally by receiving the sagittal and occasionally the straight sinus, and merges with its fellow to form the **confluence of the sinuses** (*confluens sinuum*). This triple or even quadruple merging of sinuses is located within the dorsal part of the occipital bone and may be asymmetrical. From the confluens sinuum the transverse sinus runs laterally in the transverse canal for about the proximal two-thirds of its length and

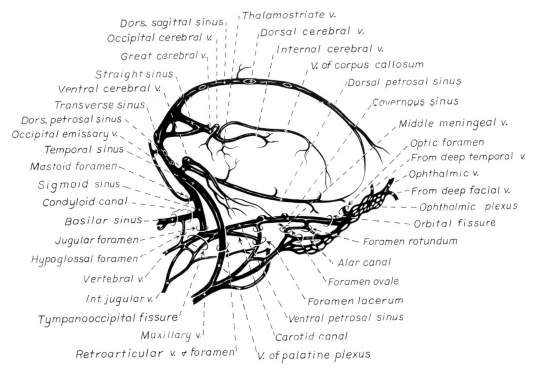

Figure 12–21. Diagram of the cranial venous sinuses, lateral aspect. (From Reinhard, Miller, and Evans 1962.)

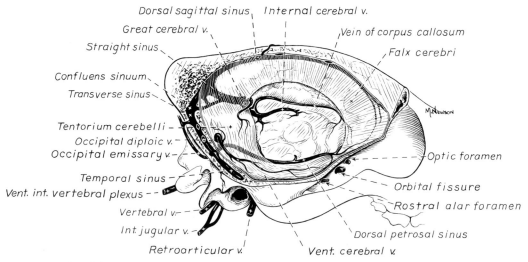

Figure 12–22. Cranial venous sinuses, lateral aspect. (Right cerebral hemisphere removed.) (From Reinhard, Miller, and Evans 1962.)

then continues in the transverse groove. It terminates at the distal end of the transverse groove by dividing into the temporal and sigmoid sinuses. The temporal sinus, larger than the sigmoid, continues in the direction of the transverse sinus through the temporal meatus, whereas the sigmoid sinus bends downward and backward on its way to the petro-occipital fissure.

Both the transverse and the sigmoid sinus have a connection with the **occipital emissary vein** *(v. emissaria occipitalis)*. This vein lies on the caudal surface of the skull and drains blood from the deep muscles on the cranial part of the neck. Right and left veins lie ventral to the ventral nuchal line as they form a prominent anastomosis with each other dorsally. The occipital emissary vein drains the area supplied by the dorsal part of the occipital artery. The smaller linkage with the transverse sinus occurs lateral to the confluence of the sinuses. The larger connection between this vein and the sinus system passes through the mastoid foramen as it joins the first bend of the sigmoid sinus. The rostral side of the proximal third of the transverse sinus receives an occipital diploic vein, and a second such vessel may enter the sinus as it leaves the transverse

canal. The short, caudally running dorsal petrosal sinus enters the transverse sinus near the ventral end of the transverse groove.

The **temporal sinus** *(sinus temporalis)* (Figs. 12–21, 12–22) of Zimmermann (1936) is the rostroventral continuation of the transverse sinus. It lies in the temporal meatus and therefore is placed between the petrous and squamous parts of the temporal bone. It receives no tributaries. At the retroarticular foramen it becomes the **retroarticular vein** *(v. retroarticularis)*, which, after a course of about 1 cm., empties into the maxillary vein as one of its largest tributaries. The retroarticular vein forms a shallow vertical groove on the caudal side of the retroarticular process. It arises from a small plexus of veins which lies between the cartilaginous external acoustic process and the ventral part of the squamous temporal bone. The plexus receives one or two tributaries from the musculature of the cranial part of the neck.

The **sigmoid sinus** *(sinus sigmoideus)* (Fig. 12–23) is the roughly S-shaped caudoventral continuation of the transverse sinus. It begins by forming an arc around the proximal end of the petrous temporal bone. The

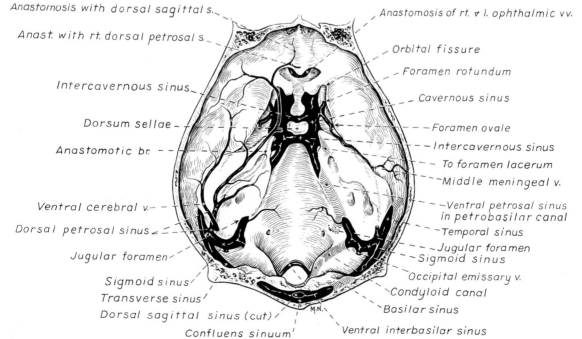

Figure 12–23. Cranial venous sinuses, dorsal aspect. (Calvaria removed.) (From Reinhard, Miller, and Evans 1962.)

first arc is continued by the second arc, which lies medial to the petro-occipital synchondrosis. The sinus terminates after traversing the jugular foramen by continuing as the internal jugular vein. At this junction the ventral petrosal sinus enters the venous channel from in front, and the vertebral vein leaves from behind. The sigmoid sinus receives one or two delicate meningeal veins from the medulla oblongata. Its largest connection is with the **condyloid vein** (*v. condyloidea*). From the junction of the two arcs forming the sigmoid sinus the condyloid vein passes through the condyloid canal and continues as the basilar sinus, which becomes the vertebral venous plexus. Pilcher (1930), by means of dye injections into the sagittal sinus, demonstrated that the sagittal, transverse, sigmoid, and vertebral plexus system is the main venous drainage from the brain. Occasionally, there is an osseous canal between the condyloid canal and the hypoglossal foramen. This canal conducts the **vein of the hypoglossal canal** (*v. canalis hypoglossi*), which extends from the condyloid vein to the dorsal surface of the initial part of the vertebral vein in the petro-occipital fissure.

The **dorsal petrosal sinus** (*sinus petrosus dorsalis*) (Fig. 12–21) is a caudal extension, beginning at the free border of the tentorium cerebelli opposite the distal end of the petrosal crest, of the basal vein of the cerebrum and the vein which runs with the trigeminal nerve (Zimmermann 1936). The sinus almost at once comes to lie on the lateral surface of the pyramid, which it grooves. About 6 mm. from its junction with the transverse sinus, it receives the ventral cerebral vein. This vein is 1 mm. in diameter and is as large as the sinus it enters. The confluence of the dorsal petrosal and transverse sinuses forms an acute angle ventrally, lateral to the temporal meatus.

The **ventral petrosal sinus** (*sinus petrosus ventralis*) (Fig. 12–23) extends between the caudal end of the cavernous sinus and the ventral end of the sigmoid sinus. It lies in the petrobasilar canal and is an intraosseous caudolateral extension of the cavernous sinus. A smaller venous channel lies in the laterally adjacent and parallel carotid canal. It connects the same parent sinuses as does the ventral petrosal; this venous channel contains the internal carotid artery.

The paired **cavernous sinus** (*sinus cavernosus*) (Fig. 12–23) plays a key role in the ventral venation of the brain. The right and left sinuses lie on the respective sides of the floor of the middle cranial fossa and extend from the orbital foramina to the petrobasilar canals. Rostrally, each communicates through the orbital foramen with the ophthalmic plexus of veins. Laterally, each gives off emissary veins which run through the round, oval, and the external carotid foramina to enter the internal maxillary vein. Caudally, each is continued by the ventral petrosal sinus and indirectly is connected with the vertebral venous plexus. Laterally, the middle meningeal vein, a satellite of the middle meningeal artery, enters the cavernous sinus opposite the round foramen. The two cavernous sinuses are connected medially, by means of the large but short rostral and caudal **intercavernous sinuses** (*sinus intercavernosi*), in front of and behind the stalk of the dorsum sellae. The expanded dorsal part of the dorsum sellae covers the middle portion of the usually larger rostral intercavernous sinus, which lies directly caudal to the hypophysis. The usually smaller caudal intercavernous sinus runs across the caudal surface of the base of the dorsum sellae and is only about 2 mm. long. It unites the right and left cavernous sinuses where they lie closest together. This sinus may be absent. A third delicate intercavernous connection may exist rostral to the hypophysis. The cavernous sinus contains, free in its lumen, the rostral portion of the middle meningeal artery and the anastomotic ramus of the external ophthalmic artery. The cavernous sinuses contain no trabeculae, but a few stabilizing threads attach to the arteries as they perforate the wall of the sinus (Zimmermann 1936).

The **basilar sinus** (*sinus basilaris*) (Fig. 12–21) is the venous link between the condyloid vein and the vertebral plexus (called the ventral occipital sinus in the first edition). It is about as wide as it is long, measuring approximately 5 mm. in each direction. It is transversely compressed and converges toward the opposite sinus as it lies on the medial surface of the lateral part of the occipital bone. The basilar sinus becomes the **internal vertebral venous plexus** as it leaves the foramen magnum to run across the medial surface of the lateral mass of the atlas. It is always connected to its fellow sinus ventrally by the transversely running, flat ventral interbasilar sinus, and there may be a dorsal connection also. The caudal part of the medulla lies between the right and the left basilar sinus.

The **ventral interbasilar sinus** *(sinus interbasilaris ventralis)* (Fig. 12–23) is a stout, flat transverse venous passage which connects the right and left basilar sinuses just inside the foramen magnum. It lies on the floor and adjacent sides of the occipital bone. Its rostral border is irregular, as it apparently sends small, finger-like processes between the two layers of the dura.

The **dorsal interbasilar sinus** *(sinus interbasilaris dorsalis)* may be absent. It is a dorsal transverse channel or plexus which unites the right and left basilar sinuses. It may be coextensive with the first dorsal internal vertebral venous plexus, which lies under the cranial lip of the arch of the atlas. These plexuses, which are lateral to the junction of the brain stem and spinal cord, might be entered in cisternal punctures at the atlanto-occipital joint.

VEINS OF THE BRAIN

The **veins of the brain** *(venae cerebri),* like the dural sinuses into which they drain, do not contain valves, and their walls contain no muscular coat. They empty into the sinuses in a direction usually opposite to the flow of blood in the sinuses. Those from the cerebral hemispheres may be divided into cortical and central veins. The cortical veins may be divided further into dorsal and ventral cerebral veins. The central veins, which drain into the great cerebral vein, are the corpus callosal, basal, internal cerebral, and thalamostriate veins. The cerebellar veins are divided into dorsal and ventral veins. The veins of the brain stem are the medullary and pontine veins.

The **dorsal cerebral veins** *(vv. cerebri dorsales)* (Fig. 12–21) are paired but are not bilaterally symmetrical. All enter the sagittal sinus. They drain the cortex of nearly the whole cerebrum. Although for the most part they lie in the sulci, they often run across the gyri. From one to four dorsal cerebral veins enter the rostral half of the sagittal sinus. These come from the cortex of the frontal lobe. Usually the largest and longest dorsal cerebral vein arises in and on the gyri

dorsal to the rhinal fissure. It runs over the coronal and caudal sigmoid gyri to reach the cruciate sulcus. It collects many tributaries from neighboring gyri and sulci along its course. The last tributary to enter it is the parietal diploic vein just before it enters the sagittal sinus from the cruciate sulcus. One or two small dorsal cerebral veins enter the caudal half of the dorsal sagittal sinus from each hemisphere.

The **ventral cerebral vein** *(v. cerebri ventralis)* (Figs. 12–22, 12–23) drains most of the cortex of the temporal lobe. One branch arises from the dorsolateral portion of the lobe, and the other branch comes from the piriform area. This branch forms a groove on the lateral surface of the pyramid, as it runs dorsocaudally to unite with the more dorsal branch before entering the dorsal petrosal sinus. Sometimes this union fails, and the branches terminate independently in the dorsal petrosal sinus.

The **great cerebral vein** *(v. cerebri magna)* (Fig. 12–21) is the sole channel for return of venous blood from all the deep or central ganglionic veins of the cerebrum. It begins by the confluence of the dorsally located vein of the corpus callosum and the ventrally located internal cerebral veins and the thalamostriate vein. It runs in the triangle formed by the cerebral hemispheres and the vermis of the cerebellum caudally. At its termination it receives a tributary from the occipital lobe. Bedford (1934) experimentally produced occlusions of the great cerebral vein in the dog and noted the rapid establishment of a collateral circulation.

The **vein of the corpus callosum** *(v. corporis callosi)* (Fig. 12–22), unpaired, begins by collecting a small tributary from the medial surface of each hemisphere, rostral to the corpus callosum. It runs caudally, dorsal to the corpus callosum, where it is connected to the free margin of the falx cerebri by arachnoid. During this course it receives minute tributaries from the medial surfaces of the hemispheres. At the splenium it receives the single **choroidal vein** *(v. choroidea),* which is a caudal continuation of the choroid plexus of the third ventricle.

The **thalamostriate vein** *(v. thalamostriata)* is another tributary which enters the caudal end of the vein of the corpus callosum. It arises in the thalamus and corpus striatum.

The **internal cerebral veins** *(vv. cerebri internae)* (Fig. 12–22) are present in the dog as a cluster of tributaries from the dorsal midbrain. Dorsal to the mesencephalon, right and left vessels anastomose with each other, as well as with the basal vein on the respective side.

The cerebellar veins, like the cerebral veins, lie in the pia and tend to follow the sulci. They are divided into dorsal and ventral sets.

The **dorsal cerebellar veins** *(vv. cerebelli dorsales)* are right and left vessels which drain into the right and left transverse sinuses by one or two stems adjacent to the confluence of the sinuses. They arise and lie one on either side in the fissure between the two lateral hemispheres and the central vermis. The numerous fine tributaries which enter them lie mostly in the fine sulci between the thin, closely packed gyri. Two or more venous threads run across the gyri of the vermis from origins in the caudal colliculi.

The **ventral cerebellar veins** *(vv. cerebelli ventrales)* are one or two minute vessels, on each side, which lie between the lateral hemispheres and the medulla. They collect veins from the brain stem primarily, but also from the cerebral hemispheres, and drain into the sigmoid or basilar sinuses.

The **medullary and pontine veins** lie on the ventral and lateral surfaces of the medulla oblongata and pons. Those from the pons are transverse, whereas the medullary veins lie lateral to the pyramids. They are about 0.2 mm. in diameter and lie about 4 mm. from the basilar artery. They are moderately sinuous and collect many fine twigs from each side of the ventral median fissure. The lateral tributaries may also come from a longitudinal vessel which crosses the olive. It receives fine tributaries from the cerebellar hemispheres. The medullary veins empty laterally into the basilar sinuses. The pontine veins, one on each side, arise midventrally and run laterally over the pons or between the pons and the trapezoid body to reach the transverse fissure. They drain into the sigmoid sinus.

VEINS OF THE DIPLOË

The **diploic veins** *(vv. diploicae)* (Fig. 12–24) are present only in those places where there is cancellous or spongy bone (diploë) uniting the two tables of the skull. There are thus no diploic veins in the lateral wall of the braincase, where the two tables are fused, or rostrally, where the two tables of the frontal bone are widely separated to form the frontal air sinuses. The dog has frontal, parietal, and occipital diploic veins.

The **frontal diploic vein** *(v. diploica frontalis)* drains into the angular vein of the eye from the diploë located in the triangle between the roof of the cranium and the caudal part of the frontal sinus. It runs rostrolaterally, and leaves the skull by the small frontal foramen to enter the angular vein of the eye ventral to the zygomatic (supraorbital) process. Caudally, it anastomoses with either the rostral part of the sagittal sinus or a large dorsal cerebral vein by a single or a double vessel.

The **parietal diploic vein** *(v. diploica parietalis)* arises in the diploë of the medial part of the parietal bone 1 to 2 cm. from the midline by anastomosing with the occipital diploic vein or veins. It terminates in the mid-dorsal part of the sagittal sinus or in the adjacent large dorsal cerebral vein. When single, it lies in the diploë of the cerebral juga opposite the ectolateral sulcus. When it is double, the two parts are connected by a dense rete.

The **occipital diploic vein** *(v. diploica occipitalis)* arises from an anastomosis with the parietal diploic vein and enters the transverse sinus about 2 cm. from the midline. It is frequently double or triple; when it is triple, the second vessel enters the sinus close to the midline, and the third enters the proximal part of the middle third. Zimmermann (1936) illustrates a small diploic vein in the interparietal process which enters the confluence of the sinuses.

MENINGEAL VEINS

The **meningeal veins** lie in and drain the dura mater. The **rostral meningeal vein** is a delicate vessel which begins in the dura covering the frontal lobe. It frequently leaves a faint shallow groove on the cerebral surface of the frontal bone. It perforates the inner table of the frontal bone and joins the frontal diploic vein in the region of the ethmoidal fossa. The **middle meningeal vein**

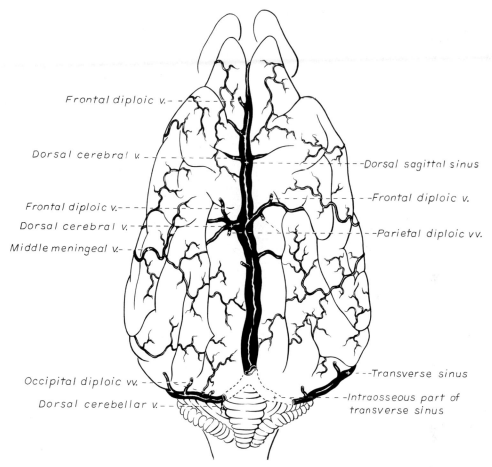

Figure 12–24. Veins of the brain, dorsal aspect.

(Fig. 12–24) is more extensive in its ramifications within the dura. After the major tributaries converge over the lateral surface of the brain, the vein courses ventrorostrally along the petrous temporal bone. Within the cranium, at the foramen ovale, it joins a venous sinus which connects the cavernous sinus with the internal maxillary vein.

VEINS OF THE SPINAL CORD AND VERTEBRAE

The vertebral vein system constitutes an alternate route for the return of blood from the body to the heart via anastomoses with rostral systemic veins and the azygos vein, which in effect bypasses the caval system. The vertebral plexuses are in direct communication with the cranial venous sinuses and, since no valves exist in either, blood may flow cranially or caudally, depending on pressure relations. Batson (1940, 1957), in discussing the concept of the vertebral vein system and its active physiological role, makes reference to anatomical and clinical observations in both man and animals. Dräger (1937), Worthman (1956), and Reinhard, Miller and Evans (1962) have described and illustrated the vertebral sinuses in the dog.

The **internal vertebral venous plexus** (*plexus vertebralis internus ventralis*) (Figs. 12–25, 12–26, 12–27) was formerly called the *sinus vertebrales*. It consists of left and right thin-walled, flattened, valveless vessels which extend from the skull to the caudal vertebrae. They lie on the floor of the vertebral canal in the epidural fat. As paired trunks coursing through the verte-

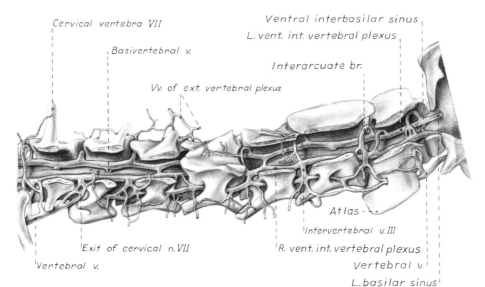

Figure 12–25.   Cervical vertebral veins, right lateral aspect. (From Reinhard, Miller, and Evans 1962.)

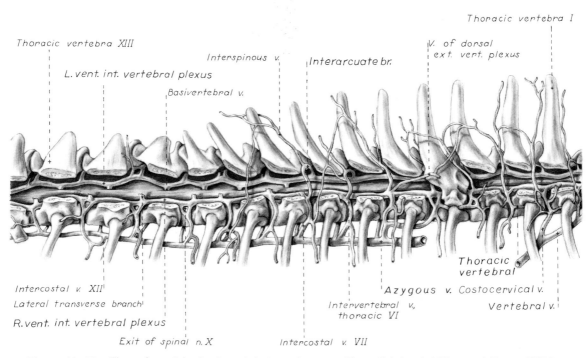

Figure 12–26.   Thoracic vertebral veins, right lateral aspect. (From Reinhard, Miller, and Evans 1962.)

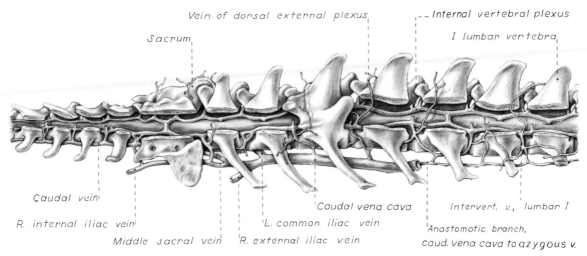

Figure 12–27. Lumbar, sacral, and caudal vertebral veins, right lateral aspect. (From Reinhard, Miller, and Evans 1962.)

bral canal, they diverge from each other at the intervertebral foramina and approach each other over the vertebral bodies. They are largest in the cervical region. Within the arch of the atlas, where they originate as continuations of the basilar sinuses, they may appear ampullated (Worthman 1956). Their diameter is reduced at the junction of the last cervical and first thoracic vertebrae, and remains constant from there to the level of the fourth or fifth lumbar vertebra. Caudal to this level the vertebral plexus vessels decrease in size, and they may fuse within the fourth to sixth caudal vertebrae or terminate as fine venules in the tail musculature. Along the course of the vertebral plexus, there are frequent anastomoses between the right and left channels. Some of these anastomoses are superficial, whereas others are beneath the dorsal longitudinal ligament or within the vertebral body.

Within the vertebral canal the vertebral plexus receives the **spinal veins** (*vv. spinales*), which follow the nerve roots to the intervertebral region on each side. The most prominent connections of the right and left internal vertebral plexuses are the interarcuate branches.

The **basivertebral veins** (*vv. basivertebrales*) are usually paired tributaries which arise within the vertebral bodies or from the soft tissues ventral to the vertebrae or from anastomoses with paravertebral veins. They ascend through osseous canals in the vertebral bodies and join the longitudinal verte-

bral plexuses. In the cervical region they begin in the ventral vertebral venous plexuses by an anastomosis with muscular tributaries of the vertebral veins within the longus colli muscle. In some cranial segments of the thoracic region no basivertebral veins are evident; more caudally, single basivertebral veins, at their beginnings, anastomose with intercostal veins. In the lumbar region the basivertebral veins are largest. They are usually paired and connect with the lumbar veins by means of the ventral venous plexuses. The sacral and caudal vertebrae usually have no basivertebral veins.

The **intervertebral veins** (*vv. intervertebrales*) are present at every intervertebral foramen, providing communication between the vertebral plexuses and extravertebral veins. The first few are single on each side, but most are double with one part lying in the caudal and the other in the cranial notch of the contiguous vertebrae. When they are double, the emerging roots of the spinal nerve lie between them, or may be ringed at the intervertebral foramen by dorsal and ventral anastomoses between the double veins. In this way, a venous cushion surrounds the nerve roots at their union. This may include the dorsal root ganglion. Even when the intervertebral veins are single, they may arise as paired veins. This arrangement is regularly found in the caudal few segments of both the cervical and the thoracic region. The intervertebral veins take the names and numbers of the intervertebral spaces through which they pass, except for the first two sacral intervertebral veins, which pass through the two ventral sacral foramina on each side. They have major extravertebral anastomoses as follows:

Cervical I through VIII, with the vertebral vein.

Thoracic I, II, and III, with the costocervical and thoracic vertebral vein.

Thoracic IV (left side), about 50 per cent with the thoracic vertebral vein.

Thoracic IV or V (left side) through thoracic IX or X, with the azygos vein.

Thoracic IX or X (left side) through thoracic XIII, with the hemiazygos vein.

Thoracic IV (right side) through lumbar III, with the azygos vein.

Lumbar IV (V) and V (VI), with the caudal vena cava.

Lumbar VI and VII, with the internal iliac or directly into common iliac or the caudal vena cava.

Lumbar VII with the internal iliac.

Sacral I (II), with cranial gluteal vein.

Sacral II (III), with internal pudendal vein.

Caudal I through IV, with the middle sacral and the internal iliac vein.

In the thoracic and lumbar regions the intervertebral veins empty into the intercostal vein before joining the larger channels.

The **interarcuate branches** (*rami interarcuales*) of the internal vertebral plexus were called arcuate veins by Ellenberger and Baum (1943) and Reinhard, Miller, and Evans (1962). Dräger (1937) and Worthman (1956) called them interarcuate veins. As the main components of the internal vertebral plexus, they are most prominent in the cervical and thoracic regions. They arise in the epaxial musculature as **interspinous veins** or veins of the dorsal external vertebral venous plexus. They approach the interarcuate spaces, pierce the ligamenta flava, and enter the spinal canal. Within the spinal canal the interarcuate branches of the right and left sides frequently join each other at the apex of the internal vertebral arch. Longitudinal anastomoses between successive interarcuate branches are also present. The first five pairs of cervical interarcuate branches, of which the third pair is the largest, enter the vertebral plexuses. Those from the fifth cervical to the fifth or sixth thoracic empty into the intervertebral veins. Arcuate veins are lacking between the ninth thoracic and the seventh lumbar, and in the caudal region.

The **dorsal external vertebral venous plexus** (*plexus venosus vertebralis externus dorsalis*) is formed by anastomoses between adjacent intervertebral and interspinous veins of the same and of the opposite side, being best developed in the cervical and cranial thoracic regions. Tributaries from superficial and deep epaxial veins also partipicate in the anastomoses.

The **ventral external vertebral venous plexus** (*plexus venosus vertebralis externus ventralis*) is not very extensive in the dog. Some ventral tributaries of the intervertebral veins are formed by anastomoses ventral to the vertebral bodies. In the cervical and lumbar regions several subvertebral tributaries join to form a long, median vessel which enters an intervertebral vein. Several radicles from the vertebral bodies join the ventral external vertebral plexus.

The **internal vertebral venous plexus** (*plexus vertebralis internus ventralis*), lying within the vertebral canal, is formed by both dorsal and ventral anastomosing tributaries from the spinal cord, and communicates with the external vertebral venous plexuses. The interarcuate branches are the most prominent components of the internal vertebral plexus, although they are often incomplete between the fifth and seventh cervical vertebrae, as well as between the ninth thoracic and seventh lumbar vertebrae. The internal vertebral venous plexus frequently surrounds the exit of the spinal nerve. The plexus is best developed in the first two cervical segments. At the atlantooccipital joint it is coextensive with the interbasilar and basilar sinuses.

BIBLIOGRAPHY

Audell, L., L. Jönsson, and B. Lannek. 1974. Congenital portacaval shunts in the dog. A description of three cases. Zentralbl. Veterinaermed. [A] *21*:797–805.

Batson, O. V. 1940. The function of the vertebral veins and their role in the spread of metastases. Ann. Surg. *112*:138–149.

—————. 1957. The vertebral vein system. Caldwell lecture, 1956. Am. J. Roentgenol. 78:195–212.

Bedford, T. H. B. 1934. The great vein of Galen and the syndrome of increased intracranial pressure. Brain 57:1–24.

Bowsher, D. 1954. A comparative study of the azygos venous system in man, monkey, dog, cat, rat and rabbit. J. Anat. 88:400–407.

Buchanan, J. W. 1963. Persistent left cranial vena cava in dogs: Angiocardiography, significance and coexisting anomalies. J. Am. Vet. Radiol. Soc. 4:1–8.

Buergelt, C. D., and L. D. Wheaton. 1970. Dextroaorta, atopic left subclavian artery, and persistent left cephalic vena cava in a dog. J. Am. Vet. Med. Assoc. 156:1026–1029.

Christensen, G. C. 1952. Circulation of blood through the canine kidney. Am. J. Vet. Res. 13:236–245.

Darrach, W. 1907. Variations in the postcava and its tributaries as observed in 605 examples in the domestic cat. Anat. Rec. 1:30–33.

Dräger, K. 1937. Über die Sinus columnae vertebralis des Hundes und ihre Verbindungen zu Venen du Nachbarschaft. Morph. Jahrb. 80:579–598.

Ellenberger, W., and H. Baum. 1943. Handbuch der vergleichenden Anatomie der Haustiere. 18th Ed., Berlin, Springer.

Evans, H. E., and A. deLahunta. 1971. Miller's Guide to the Dissection of the Dog. Philadelphia, W. B. Saunders Co.

Ewing, G. O., P. F. Suter, and C. S. Bailey. 1974. Hepatic insufficiency associated with congenital anomalies of the portal vein in dogs. J. Amer. Anim. Hosp. Assoc. 10:463–476.

Flint, J. M. 1900. The blood-vessels, angiogenesis, organogenesis, reticulum, and histology of the adrenal. Contributions to the Science of Medicine, pp. 153–228. Baltimore, The Johns Hopkins Press.

Franklin, K. J. 1937. A Monograph on Veins. Springfield, Ill., Charles C Thomas.

Gomerčič, H. 1967. Vena cardinalis sinistra persistens in a dog. Vet. Archiv. 37:307–334.

Hofmann, M. 1901. Zur vergleichenden Anatomie der Gehirnund Rückenmarksvenen der Vertebraten. Ztschr. Morph. u. Anthropol. 3:239–299.

Huntington, G. S., and C. F. W. McClure. 1920. The development of the veins in the domestic cat (Felis domestica) with especial reference (1) to the share taken by the supracardinal veins in the development of the postcava and azygos veins and (2) to the interpretation of the variant conditions of the postcava and its tributaries as found in the adult. Anat. Rec. 20:1–30.

Hutton, P. H. 1969. The presence of a left cranial vena cava in a dog. Br. Vet. J. 125:xxi–xxii.

Kadletz, M. 1928. Über eine Missbildung im Bereiche der Vena cava caudalis beim Hunde. Ztschr. Anat. u. Entw. 88:385–396.

Padget, D. H. 1957. The development of the cranial venous system in man from the viewpoint of comparative anatomy. Contr. to Embryol. 36:81–140, Carnegie Inst., Wash.

Petit, M. 1929. Les veins superficielles due chien. Rév. Vét. et J. Méd. Vét. 81:425–437.

Pilcher, C. 1930. A note on the occipito-vertebral sinus of the dog. Anat. Rec. 44:363–367.

Preuss, F. 1942. Arterien und Venen des Hinterfusses vom Hund, vorzüglich ihre Topographie. Dissertation, Hanover. 24 pp.

—————. 1954. Gibt es eine V. reflexa? Tierärztl. Umschau. 9:388–389.

Reinhard, K. R., M. E. Miller, and H. E. Evans. 1962. The craniovertebral veins and sinuses of the dog. Am. J. Anat. 111:67–87.

Reis, R. H., and P. Tepe. 1956. Variations in the pattern of renal vessels and their relation to the type of posterior vena cava in the dog (Canis familiaris). Am. J. Anat. 99:1–15.

Schaller, O. 1955. Persistent left vena cava cranialis in domestic mammals especially carnivores. Z. Anat. Entwicklungsgesch. 119:131–155.

Shepherd, J. T., and P. M. Vanhoutte. 1975. Veins and Their Control. Philadelphia, W. B. Saunders Co.

Stoland, O. O., and H. B. Latimer. 1947. A persistent left superior vena cava in the dog. Trans. Kansas Acad. Sci. 50 84–86.

Worthman, R. P. 1956. The longitudinal vertebral venous sinuses of the dog: I. Anatomy; II. Functional aspects. Am. J. Vet. Res. 17:341–363.

Zimmermann, G. 1936. Über die Dura mater encephali und die Sinus der Schädelhöhle des Hundes. Ztschr. Anat. u. Entw. 106:107–137.

Chapter 13

THE LYMPHATIC SYSTEM

GENERAL CONSIDERATIONS

The lymphatic system, consisting of a network of permeable capillaries, variously sized collecting ducts, a filtering mechanism in the form of lymph nodes, and conducting channels which enter the great veins of the heart, serves as an adjunct to the venous part of the vascular system. By means of veins and lymphatics, the blood and tissue fluid are returned from the capillary bed and tissue spaces to the general circulation. Lymph, like blood, is both fluid and corpuscular. It contains red blood cells, some nucleated cells, lymphocytes, and histiocytes. Mononuclear cells are mostly small lymphocytes. Lymph from each region of the body has a characteristic composition. Battezzati and Donini (1973) review the history, techniques, and current status of our understanding of the lymphatic system and include extensive literature citations. For a clear presentation of the cells and tissues of the immune system, see Weiss (1972). While the blood is traversing from the arterial to the venous side of the capillary bed, fluid and proteins escape from it into the tissue spaces.

According to Yoffey and Courtice (1956), as much as 50 per cent of the total circulating protein escapes from the blood vessels in the course of a day. This extravascular protein and tissue fluid, which is used in part for cell nutrition, readily enters the lymphatic capillaries, along with foreign particles, if any are present.

The clear, colorless fluid, known as lymph, is returned slowly to the heart via lymphatic ducts which empty mainly into the jugular or the cranial vena cava. Blalock et al. (1937) attempted complete blockage of the lymphatic return in 52 dogs and were apparently successful in three instances. Anastomoses exist between lymphatics and veins in several organs (Clark and Clark 1937), in addition to the main connections at the base of the neck. Zajac (1972) reported communications between the cisterna chyli and the caudal vena cava.

Lymphatic capillaries are simple, transparent endothelial tubes which arise, according to Sabin (1911), from two sets of paired sacs (jugular and iliac) and two unpaired sacs (retroperitoneal and cisterna chyli) by endothelial sprouting. These sacs do not persist as such in the adult, except as pathological structures. Yoffey and Courtice (1956) state that in mammals the sacs ultimately become primary lymph nodes, whereas the secondary nodes develop along the course of the lymph ducts.

The larger lymphatic collecting vessels are surrounded by smooth muscle and a fibrous adventitia. They occasionally exhibit intrinsic pulsations, although the flow of lymph depends mainly on the movement of adjacent muscles.

Lymph vessels are not necessarily present wherever there are veins. There are no lymph vessels in the brain and spinal cord, or in bone marrow. Although lymphatics are present in the fascial planes between muscles (Yoffey and Courtice 1956), there are none within skeletal muscle. In the spleen, lymphatics are observed only in the capsule and the thickest trabeculae, but not in the pulp. The mucous membranes and the skin are richly supplied with lymph vessels. The lymph capillaries in the villi of the intestine absorb and transport emulsified fat or chyle, and therefore appear milky. They are known as **lacteals**. The lymphatics from the liver (Drinker 1946) carry proteinized lymph to the thoracic duct, and hence to the blood, thus providing a route for stored or newly formed protein.

The large **lymphatic vessels** *(vasa lymphatica)* have thinner walls than do comparably sized veins, but contain more valves. When the flow of lymph is obstructed, the vessels become distended, and, owing to constrictions at the valves, they often resemble a string of beads in appearance. The blockage of lymphatics results in an accumulation of tissue fluid and consequent swelling, known as lymphedema. Lymphatic vessels, when cut, remain open longer than do comparable blood vessels, but they have remarkable regenerative capacities. Reichert (1926) found that the lymph vessels of the thigh of the dog regenerated rapidly after being transected. The cutaneous lymphatics began to regenerate after four days, and the deep lymphatics after eight days. Meyer (1906) found no regeneration after ligating and resecting 3 to 5 mm. of the large lymphatic trunks in the leg of the dog.

Since lymphatics contain numerous valves, they are difficult to inject in a retrograde direction. The valves usually possess two cusps, but they may consist of a single flap. They may best be observed by producing congestion after ligation, by injecting dye particles peripherally, or in edematous material. Prier, Schaffer, and Skelley (1962) have performed direct lymphangiography in the dog, using a radiopaque medium injected into metatarsal lymphatics. Clinical applications of lymphography include identification of lymph node lesions, evaluation of lymphatic blockage, and visualization of the progress of therapy on lymphatic lesions. (Fischer 1959, Fischer and Zimmerman 1959, Skelley et al. 1964, and Suter 1969).

Research on the anatomy and physiology of the lymphatic system has resulted in a considerable fund of knowledge, although many questions remain unanswered. Drinker (1942) commented on the observations of Aselli (in 1627) and of Pecquet (in 1651) on the lymph vessels and nodes in the dog. Aselli described and illustrated the mesenteric lacteals and mesenteric lymph node ("pancreas of Aselli"); Pecquet described the receptaculum chyli ("cistern of Pecquet") and the thoracic duct, including its termination. Rusznyák, Földi, and Szabo (1967) credit Rudbeck (in 1652) and Bartholinus (in 1653) for having recognized the lymphatic system as an entity. The English edition of the monograph on lymphatics and lymph circulation by Rusznyák, Földi, and Szabo (1967), revised and enlarged from the Hungarian, German, and Russian editions, contains an extensive bibliography (over 1700 citations). Frequent reference is also made therein to the morphology and experimental physiology of the lymphatic system in the dog. A classical topographical description of the lymph nodes and vessels in the dog was written by Hermann Baum in 1918. The illustrations used in this chapter have been reproduced from that work, *Das Lymphgefässytem des Hundes*, through the kind permission of Springer Verlag, Berlin. The terminology has been changed to conform to N.A.V., as explained by Grau and Barone (1970).

Chretien et al. (1967) described the distribution of lymph nodes in the dog and discussed the effects of surgical excision of the thymus, spleen, and all gross lymph nodes on the survival of slain allografts. They injected pontamine blue intravenously three or more days prior to dissection and found all lymphoid tissue was stained a deep blue. Excision of lymphoid tissue did not significantly alter the allograft response.

LARGE LYMPH VESSELS

The **thoracic duct** *(ductus thoracicus)* (Fig. 13–5) is the chief channel for return

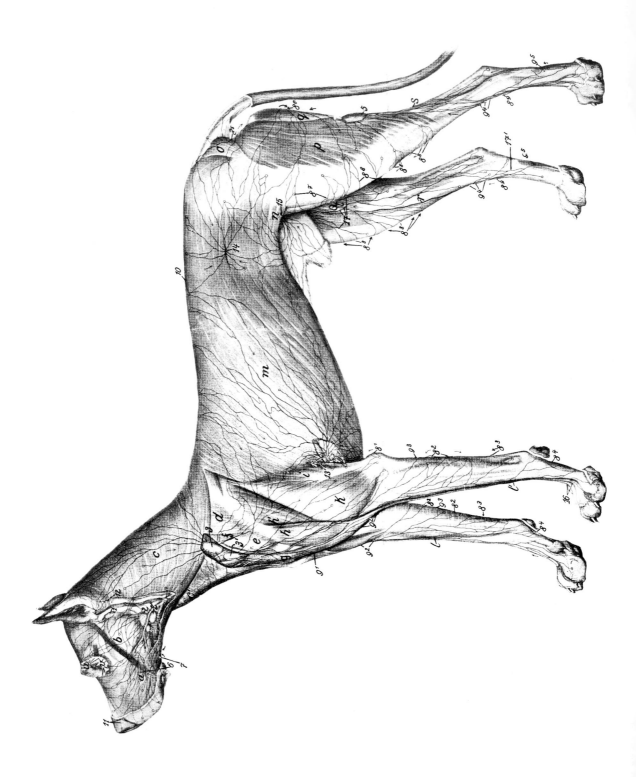

Figure 13–1. Superficial lymph vessels of the dog. (From Baum.)

1. Parotid lymph node
2. 2', 2". Mandibular lymph nodes
3. 3'. Superficial cervical lymph nodes
4. Accessory axillary lymph node
5. Popliteal lymph node
6. Lymph vessels of the gums on the buccal side of the maxillary teeth
7. Lymph vessels of the gums on the buccal side of the mandibular teeth
8' to 9⁹. Lymph vessels which course to the medial side of the limb (those numbered 8⁵ to 8⁸ go to the superficial inguinal lymph nodes)
9'. Lymph vessels of the skin of the cranial pectoral region
9² to 9⁵. Lymph vessels which course to the lateral side of the limb
10. Lymph vessel which crosses over the dorsal midline.
11. Lymph vessels of the muzzle
12. Lymph vessel which passes deep to the medial retropharyngeal lymph node
13. 13'. Lymph vessels which go to the axillary lymph node
14. Lymph vessels which go to the medial iliac lymph node
15. Lymph vessels which enter the superficial inguinal lymph nodes
16. Lymph vessels which pass from the palmar to the dorsal side of the forepaw
17. 17'. Lymph vessels which go to the superficial inguinal lymph nodes
   a. Cheek muscles
   b. Masseter muscle
   c. c'. Platysma and sphincter colli muscle

d. M. trapezius cervicalis
e. M. omotransversarius
f. M. supraspinatus
g. M. brachiocephalicus
h. h'. M. deltoideus
i. k. Long and lateral head of the M. triceps brachii
l. Antebrachial cephalic vein
l'. Accessory cephalic vein
m. M. cutaneus trunci
n. Fold of the flank
o. M. gluteus superficialis
p. M. biceps femoris
q. M. semitendinosus
r. Medial saphenous vein
s. Lateral saphenous vein
t. Medial femoral lymph node
u. v. Upper and lower eyelid

of the lymph of the body. Lymph from the right thoracic limb and shoulder and the right side of the neck and head is returned via the right lymphatic duct.

The thoracic duct begins in the sublumbar region, or between the crura of the diaphragm as a cranial continuation of the cisterna chyli. This dilated portion of the lymph channel receives the intestinal trunk, which runs dorsally from the abdominal viscera. It is formed by the lumbar lymph trunks, which are the forward continuations of the lymph vessels from the pelvic limbs, via the efferents from the iliac lymph nodes.

The exact origin of the thoracic duct is somewhat arbitrarily assigned, as the morphology of the cisterna chyli is so erratic. Typically, the cisterna chyli is an elongated sac, dilated in the middle and constricted at both ends, which lies dorsodextral to the aorta from the fourth to the first lumbar vertebra. The thoracic duct is considered to begin between the crura of the diaphragm where the cisterna attains its minimum width. It runs cranially from a point opposite the first lumbar vertebra on the right dorsal border of the thoracic aorta and the ventral border of the azygos vein to the sixth thoracic vertebra. Here it inclines to the left, running between the azygos vein and the aorta, and then obliquely crosses the ventral surface of the fifth thoracic vertebra to enter the precardial mediastinum. It runs cranioventrally in the precardial mediastinal septum, and in about one-half of the specimens it terminates singly at the junction of the left external jugular vein with the cranial vena cava (Baum 1918). In many specimens the cranial portion of the thoracic duct bifurcates or trifurcates. These branches are usually connected by cross branches so that a coarse, widespread plexus is formed. The termination of the end parts of the thoracic duct varies as to both the veins they empty into and the places on the vessels at which they empty. When the thoracic duct is single, it usually presents a terminal dilation immediately before it becomes constricted to perforate the venous wall. Similar ampulla-like dilations may be present on each of the branches when the thoracic duct terminates in multiple vessels. Freeman (1942), who examined the termination of the thoracic duct in 25 dogs, found that it had no branches in 3 and was divided in 5, and that it sent branches to the right side in 8, to the azygos vein in 5, and to both the right side and the azygos vein in 4. Suter and Green (1971) describe a case of chylothorax in a dog with an abnormal termination of the thoracic duct.

LYMPHOID TISSUE

Lymphoid tissue is present in all classes of vertebrates, but reaches its highest development in mammals, in which circumscribed organs, the lymph nodes, occur along the lymph vessels. The spleen, thymus, and bone marrow also contain lymphoid tissue, although bone marrow functions primarily in forming erythrocytes, and leukocytes rather than lymphocytes. Yoffey and Courtice (1956) estimate that the total amount of lymphoid tissue in the mammalian body is about 1 per cent. The proportional distribution and form of the lymphoid tissue in the various species of mammals vary greatly. The dog and cat have but one or two large nodes at each nodal station.

Lymph nodes are always located in the course of lymph vessels. Those vessels which enter the node are known as **afferent lymph vessels** *(vasa afferentia)*. They break up into many minute vessels before perforating the capsule of the node. Many lymphatics perforate the nodal capsule, along with the artery and vein serving the organ, and anastomose to form a single efferent vessel. The **efferent lymph vessels** *(vasa efferentia)* are those which leave the node at the hilus. Probably all lymph vessels pass through at least one node (Yoffey and Courtice 1956). Some pass through several nodes. The lymph vessels therefore form portal systems comparable to the venous portal system of the mammalian liver and the arterial portal system of the kidney in lower vertebrates.

Certain lymphoid organs have only efferent lymphatics. Examples of these are the tonsillar masses of the pharynx, and the solitary and aggregated nodules in the mucous membrane of the digestive system

(Ehrick 1929). The spleen, thymus, and bone marrow are interposed not in the lymphatic system but rather in the blood vascular system. Lymphoid tissue, wherever found, probably reaches its greatest development at sexual maturity. Endocrine and sex differences affect the lymphoid tissue, but investigators are not in agreement about these effects (Yoffey and Courtice 1956).

LYMPHOCENTRUM

A **lymphocentrum,** as defined in the N.A.V., is "a lymph node or group of lymph nodes that occurs constantly in the same region of the body and receives afferent vessels from approximately the same region in all species."

LYMPH NODES

The **lymph node** *(lymphonodulus)* is the structural and functional unit of the lymphatic system. It serves two important functions. It acts as a filter of the blood and as a germinal center for lymphocytes, the most numerous of the white blood cells. Lymph nodes are located in those places where they are afforded maximum protection yet produce minimal interference with the functioning of the skeletal, muscular, and blood vascular systems. They are thus found in the small fat storehouses at the flexor angles of joints, in the mediastinum and mesentery, and in the angles formed by the origin of many of the larger blood vessels. Each node consists of a capsule, containing elastic and smooth muscle fibers, and an internal framework consisting of septa and trabeculae. There is a convex surface, and a small flat or concave area, the **hilus,** which is usually not prominent. Internally, the node contains a poorly defined **cortex** and **medulla.** The structural unit of the lymph node is the lymph nodule. Each nodule contains light-colored central areas in which the lymphocytes are formed. Most nodules are located in the cortex, where they are partly surrounded by the lymph sinus. The lymph sinus is a lymph-circulating space located under the capsule and partly along the septa and larger trabeculae. The lymphoid tissue of the medulla is in the form of anastomosing cords of lymphocytes with few nodules.

LYMPH FOLLICLES

Solitary and aggregated follicles are found mainly in the wall of the digestive tube. They differ from nodes, in that lymph vessels arise in, rather than pass through, them; thus they contain only efferent lymphatics. The various tonsils and the aggregated follicles are described with the digestive system. The solitary and small aggregated follicles are particularly abundant in the cecum, rectum, anal canal, prepuce, and third eyelid. The lymphoid tissue of the third eyelid is located on its bulbar side. It occasionally becomes infected and hypertrophied, causing follicular conjunctivitis. It may protrude around its free border in the form of a reddened tumor (tumefied orbital gland). The lymph follicles of the prepuce are commonly infected in old male dogs, and an almost continuous purulent exudate is discharged from the sheath. The submucous lymph follicles of the large intestine can be seen clearly in the cadaver under proper lighting conditions if the intestine is inflated.

REGIONAL ANATOMY OF THE LYMPHATIC SYSTEM

The larger lymph vessels and the lymph nodes of the body will be described regionally, under the following headings: head and neck, thoracic limb, thorax, abdominal and pelvic walls, genital organs, abdominal viscera, and pelvic limb. The spleen and thymus are discussed separately later.

LYMPH NODES AND VESSELS OF THE HEAD AND NECK

The parotid lymph nodes at the rostral base of the ear and the mandibular lymph nodes at the angle of the mouth are constant lymph nodes of the head. The tonsillar ring of lymph tissue is described with the pharynx.

The **parotid lymph nodes** (*lymphonodi parotidei*) (Figs. 13–2, 13–3) is a bean-shaped node located under the rostrodorsal border of the parotid gland on the caudal parts of the zygomatic arch and adjacent masseter muscle. In a medium-sized dog it is about 1 cm. long, 0.5 cm. wide, and 0.3 cm. thick. Occasionally, a second or even a third node may be present (Baum 1918).

The afferent lymphatics to the parotid lymph node come from the cutaneous area of the caudal half of the dorsum of the muzzle and the side of the cranium, including the eyelids and associated glands, the external ear, the temporomandibular joint, and the parotid gland. Its two or three efferent ducts run between the digastric muscle and the parotid gland to the large retropharyngeal lymph node.

The **mandibular lymph nodes** (*lymphonodi mandibulares*) (Figs. 13–2, 13–3) form a group of two or three nodes, or rarely as many as five, which lie ventral to the angle of the jaw. A flattened three-sided node, with borders about 1 cm. long, lying above the external maxillary vein and a long, ovoid node lying below the external maxillary vein constitute what is probably the most common arrangement. The node lying ventral to the external maxillary vein usually is over 2 cm. long and about 1 cm. wide. It is flattened transversely. In more than one-third of the specimens examined by Baum (1918), two or more nodes replaced this single node on one or both sides. In only 16 out of 36 specimens was the grouping of the mandibular nodes on the two sides similar. The dorsal node may be double, but this condition is rare.

The afferent lymphatics to the mandibular lymph nodes come from all parts of the head not drained by the afferent lymphatics of the parotid node. There is overlapping in the areas of drainage, so that the eyelids and their glands, and the skin of the dorsum of the cranium and the temporomandibular joint drain into both nodal stations. The efferent lymphatics of the mandibular lymph nodes go primarily to the ipsilateral medial retropharyngeal lymph node. The nodes composing the group are connected with each other, as well as with the contralateral medial retropharyngeal lymph node. The 8 to 10 efferent lymphatics anastomose with each other and form a plexus as they pass over the pharynx. Before reaching the medial retropharyngeal node, they unite to form three to five small trunks which enter the ventrolateral surface of the node. According to Battezzati and Donini (1973), Bronzini in 1935 injected India ink into the cerebellomedullary cistern of a dog and observed dye in the nasal mucosa, pharynx, and submandibular lymph nodes. This communication between the subarachnoid spaces and the lymphatic vessels of the nasal mucosa and pharynx can be of considerable importance in disease processes.

The **medial retropharyngeal lymph node** *(lymphonodus retropharyngeus mediale)* (Figs. 13–2, 13–3) is the largest node found in the head and neck. It largely and sometimes completely takes the place of the cranial members of both the superficial and the deep series of the cervical and the pharyngeal nodes as found in man. It is an elongated, transversely compressed node, with a more pointed caudal end, and is

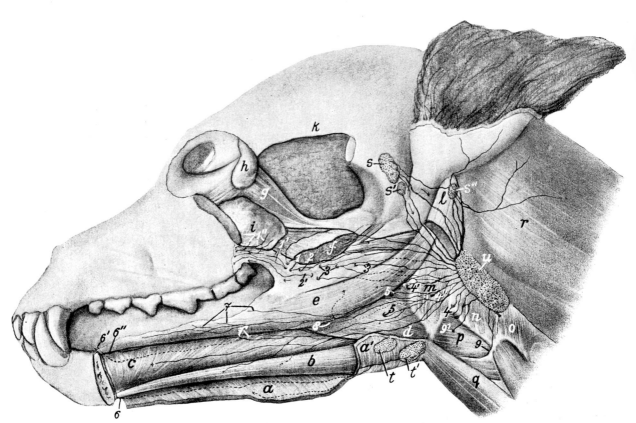

Figure 13–2. Lymph vessels of the tongue, the tongue muscles, the soft palate, and the larynx of the dog. (After Baum.)

a, a'. M. mylohyoideus
b. M. geniohyoideus
c. M. genioglossus
d. M. hyoglossus
e. M. styloglossus
f. M. pterygoideus (cut)
g. M. rectus lateralis
h. Lacrimal gland
i. Zygomatic gland
k. Temporal muscle
l. Posterior belly of M. digastricus
m. M. ceratopharyngeus
n. M. thyropharyngeus
o. M. cricopharyngeus

p. M. hypothyroideus
q. M. sternohyoideus
r. M. splenius
s,s'. Parotid lymph nodes
s". Lateral retropharyngeal lymph node
t, t'. Mandibular lymph nodes
u. Medial retropharyngeal lymph node

l, l'. Lymph vessels of the hard and soft palate
l". Lymph vessels of the zygomatic gland which join 1'
2. Lymph vessels of the soft palate
2'. Lymph vessel of fold of tonsillar sinus

3, 3'. Lymph vessels of the tonsil
4, 4', 4". Lymph vessels of the soft palate, the tonsil, the base of the tongue, and the mucous membrane of the upper pharyngeal cavity, which pass between the mucous membrane and muscles
5. Lymph vessels of the tongue which go through the M. hyoglossus
6, 6', 6". Lymph vessels of the tip of the tongue
7, 7', 8. Lymph vessels of the body of the tongue
9, 9'. Lymph vessels of the larynx

about 5 cm. long and nearly 2 cm. wide. It lies under the wing of the atlas in the triangle bounded by the m. digastricus cranially, the m. longus colli dorsally, and the pharynx and larynx ventrally. The mastoid parts of both the m. brachiocephalicus and m. sternocephalicus largely cover the node laterally, although its cranioventral part is related to the mandibular gland. Coursing along its medial surface is the terminal portion of the common carotid artery, as well as the hypoglossal, vagus, and sympathetic nerves, and the internal jugular vein.

In 10 out of 47 specimens Baum found two medial retropharyngeal lymph nodes present on one or both sides; in one-third of his specimens a lateral retropharyngeal lymph node was present. This node is less than 1 cm. in diameter and lies at the dorsal border of the mandibular gland in the fat caudal to the cartilaginous external acoustic meatus. It is completely or partially covered by the caudal part of the

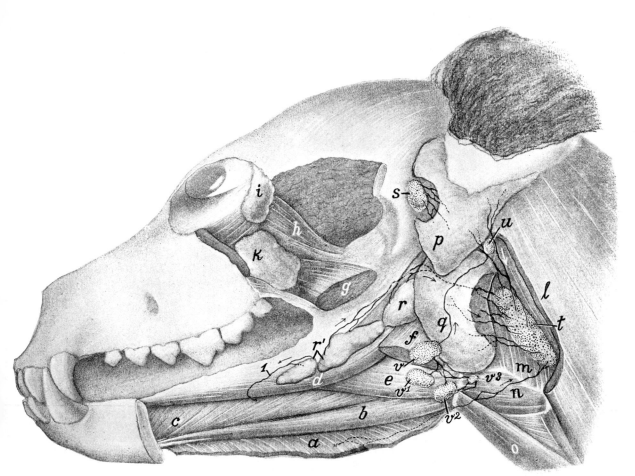

Figure 13–3. Lymph vessels and nodes associated with the salivary glands of the dog. (After Baum.)

a. M. mylohyoideus (reflected)
b. M. geniohyoideus
c. M. genioglossus
d. M. styloglossus
e. M. hyoglossus
f. Posterior belly of M. digastricus (cut)
g. M. pterygoideus
h. M. rectus lateralis
 i. Lacrimal gland

k. Zygomatic gland
 l. M. sternomastoideus and M. cleidomastoideus (each partially removed)
m. Pharyngeal musculature
n. M. hyothyroideus
o. M. sternohyoideus
p, q. Parotid gland and submaxillary gland (each partially removed)
r, r'. sublingual glands

s. Parotid lymph node
t. Medial retropharyngeal lymph node
u. Lateral retropharyngeal lymph node
v¹ to v³. Mandibular lymph nodes

1. Lymph vessel of the sublingual gland which goes to a mandibular lymph node (cut)

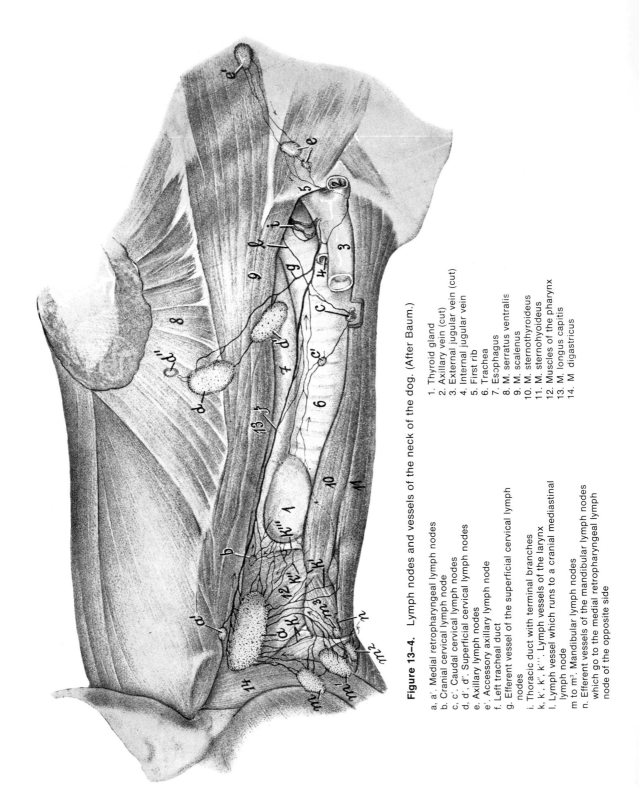

Figure 13–4. Lymph nodes and vessels of the neck of the dog. (After Baum.)

a, a'. Medial retropharyngeal lymph nodes
b. Cranial cervical lymph node
c, c'. Caudal cervical lymph nodes
d, d', d''. Superficial cervical lymph nodes
e. Axillary lymph nodes
e'. Accessory axillary lymph node
f. Left tracheal duct
g. Efferent vessel of the superficial cervical lymph
   nodes
i. Thoracic duct with terminal branches
k, k', k'', k'''. Lymph vessels of the larynx
l. Lymph vessel which runs to a cranial mediastinal
   lymph node
m to m'. Mandibular lymph nodes
n. Efferent vessels of the mandibular lymph nodes
   which go to the medial retropharyngeal lymph
   node of the opposite side

1. Thyroid gland
2. Axillary vein (cut)
3. External jugular vein (cut)
4. Internal jugular vein
5. First rib
6. Trachea
7. Esophagus
8. M. serratus ventralis
9. M. scalenus
10. M. sternothyroideus
11. M. sternohyoideus
12. Muscles of the pharynx
13. M. longus capitis
14. M digastricus

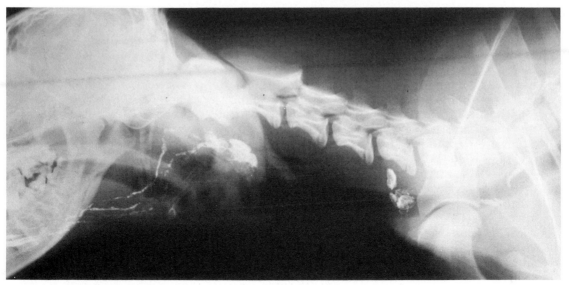

Figure 13–4   Radiograph. Lateral radiograph of the cervical region showing the retropharyngeal and superficial cervical lymph nodes.

parotid gland. The afferent lymphatics of these two small accessory nodes come from the structures lying adjacent to them. Their efferent vessels drain into the large medial retropharyngeal lymph node.

The afferent lymphatics of the medial retropharyngeal lymph node come from all the deep structures of the head which have lymphatics. Thus, the tongue, the walls of the oral, nasal, and pharyngeal passages, the salivary glands, and the deep parts of the external ear drain into this node. It also receives afferent lymphatics from the larynx, esophagus, and the non-cutaneous, non-mucous structures of the neck which have lymphatics. The efferent ducts from the parotid and mandibular lymph nodes drain into the medial retropharyngeal lymph node also. Yoffey and Drinker (1938) found, lying between the hamulus and the pharyngeal opening of the auditory tube, four or five lymph vessels which came from the floor and side walls of the nose. The deep lymphatics of the nasal cavity drain into the retropharyngeal lymph nodes and pass via cervical lymphatics to the region of the caval veins. Some of the tracheal lymphatics of the neck communicate with bronchial lymphatics and reach the bronchial lymph nodes. Thus it is possible for lymph from the nasal cavity to reach the lung area.

The **right and left tracheal trunks** *(truncus trachealis dexter et sinister),* 2 to 4 mm. wide, arise from the caudal pole of the ipsilateral medial retropharyngeal lymph node or its efferent ducts. Yoffey and Drinker (1938) state that the tracheal trunk (their cervical duct) is usually single as it leaves the caudal pole of the node, but it may be double or in the form of a plexus. As it runs down the neck, it lies in or adjacent to the lateral wall of the carotid sheath. These investigators found no additional nodes along the course of the tracheal trunk. The left tracheal trunk usually terminates in the thoracic duct. The right tracheal trunk usually terminates in the angle formed by the merging of the right external jugular and the right axillary vein to form the brachiocephalic vein, but the termination may be to one side of the angle. Baum (1918) states that the efferent lymph duct from the large superficial cervical nodes doubles the size of the right tracheal trunk as it enters this channel 2 or 3 cm. from the first rib in large dogs. The resultant vessel is known as the **right lymphatic duct** *(ductus lymphaticus dexter).* It is about 5 mm. wide and 1 cm. long; it empties into the right axillary vein or the angle formed by the merging of this vein with the right external jugular. Commonly, the right lymphatic duct bifurcates and then reunites, forming a circle before it terminates. Sometimes the forked condition persists, so that the right lymphatic duct enters the venous system, in front of the first rib, by two channels. Todd and Bernard (1973), using the Falck fluorescence technique, observed a network of sympathetic fibers on the surface of the cervical lymphatic ducts in dogs. The fibers came from sympathetic plexuses on blood vessels outside the carotid sheath (most likely via segmental communications with the vertebral nerve plexus).

The **superficial cervical lymph nodes** (*lymphonodi cervicales superficiales*) (Fig. 13–4) usually consist of two nodes, one lying dorsal to the other in the adipose tissue on the serratus ventralis and the scalenus in front of the supraspinatus. They are covered superficially by the thin cleidocervicalis, the omotransversarius and, at their dorsal end, by the trapezius. The omocervical artery and vein lie medial to the caudal parts of the nodes as they course in front of the shoulder in the groove between the shoulder and neck. The more ventral superficial cervical lymph node may encroach on the trachea on the right side, and the esophagus and trachea on the left side. Occasionally, a single node is present on each side, but more commonly three or more nodes replace the usual two. Most of the nodes are oval and somewhat flattened. They are collectively about 3 cm. long and less than 1 cm. thick.

The afferent lymphatics come mainly from the skin of the caudal part of the head, including the pharyngeal region, and a part of the pinna, the lateral surface of the neck, and the whole thoracic limb, except a variable region on the medial side of the brachium and antebrachium, the shoulder, and the cranial part of the thoracic wall. Mahorner et al. (1927), by making injections into the thyroid gland of dogs, revealed efferent connections with the cervical lymph nodes, the cervical lymphatic trunks, and the veins at the base of the neck. The efferent ducts connect members of the group when more than one node is present, and as one to three trunks they descend over the serratus ventralis and the scalenus to merge with the

tracheal trunk on the right side to form the right lymphatic duct. On the left side they empty into the thoracic duct. On either or each side they may empty into the external jugular vein directly.

The **deep cervical lymph nodes** (*lymphonodi cervicales profundi*) (Fig. 13–4) are located along the cervical portion of the trachea on each side. They are exceedingly small and vary in number. They are customarily divided into cranial, middle, and caudal nodes. In the dog one or more of these nodes are frequently absent. They range considerably in size; they may be barely visible, or several millimeters long. The smaller nodes are spherical to ovoid; the larger ones are usually elongated and parallel to the long axis of the trachea.

If a cranial deep cervical lymph node is present, it is located between the caudal end of the medial retropharyngeal lymph node and the thyroid gland. It lies either dorsomedial to the thyroid gland along the carotid sheath, or on the pharynx cranial to the thyroid. It was absent in 19 of Baum's 64 dissections. The middle deep cervical lymph node was present in only 4 of 50 specimens Baum examined, and in only 1 of these 4 was it present on both sides. It usually lies along the carotid sheath, but it may lie ventral to the trachea in the middle third of the neck. The caudal deep cervical lymph node was present in 17 of 56 specimens examined by Baum, but in only two specimens was it present bilaterally. In 11 of 17 specimens a single node was located on the ventral surface of the trachea, and in others there were two or more nodes lying on the ventral surface of the caudal third of the cervical part of the trachea.

The afferent lymph vessels to the deep cervical lymph nodes come from the larynx, thyroid gland, trachea, esophagus, and the last five or six cervical vertebrae. The efferent lymphatics of each cranially located node become a part of the afferent lymphatics of the node located next caudally. In this way the cranial deep cervical node receives lymphatics from the medial retropharyngeal node. Those from the caudal deep cervical lymph nodes empty into the right lymphatic duct, or into the thoracic duct on the left. On either side they may empty into the tracheal trunk or into a cranial mediastinal node.

LYMPH NODES AND VESSELS OF THE THORACIC LIMB

The **axillary lymph node** *(lymphonodus axillaris)* (Fig. 13–4) usually is the only lymph node of the thoracic limb. In Baum's (1918) series of 43 specimens, double nodes were found in 10, and in 6 of these the double nodes occurred on both sides. The main axillary lymph node is usually in the form of a disc about 2 cm. in diameter, although the diameter may range from 0.3 to 5 cm. It lies 2 to 5 cm. caudal to the shoulder joint in the angle formed by the diverging brachial and subscapular blood vessels. It is bounded laterally by the teres major, medially by the transversus thoracis, and ventrally by the dorsal border of the deep pectoral muscle. The **accessory axillary lymph node** *(lymphonodus axillaris accessorius),* when present (probably more often than Baum's figures indicate), lies caudal to the principal node in the fascia between the adjacent borders of the deep pectoral and latissimus dorsi muscles, caudal to the muscles of the brachium. It varies in size from less than 1 mm. to 1.5 cm. When it is large, the main node is correspondingly reduced in size. In 4 of 29 specimens Baum found a third node in close relation to the main axillary node. In only one of his specimens was it present on both sides.

The afferent vessels of the axillary lymph node or nodes come mainly from the thoracic wall and the deep structures of the thoracic limb. Those of the thorax extend beyond it so that vessels arise from the deep structures of the neck and from the abdomen. Both the thoracic and the cranial abdominal mammary glands of each side have lymphatics which drain into the axillary nodes. An anastomosis between the afferent lymphatics of the axillary nodes and those of the superficial inguinal nodes sometimes occurs between the cranial and caudal abdominal mammary glands. The axillary and accessory axillary lymph nodes are connected with each other by lymph vessels. The efferent lymph vessels from the axillary nodes of each side course cranially and unite with

each other to form one or more anastomosing larger trunks which lie on the transversus thoracis. The efferent trunks pass medial to the axillary vein, curve around the first rib, and empty by one or several branches on the left side into the thoracic duct, left tracheal trunk, left external jugular vein, or into all of these. On the right side the axillary efferent lymphatics empty into the right tracheal trunk, the right lymphatic duct, the right external jugular vein, or into all three of these.

LYMPH NODES AND VESSELS OF THE THORAX

The lymph nodes of the thorax may be divided into parietal and visceral groups. The parietal group includes the sternal and intercostal nodes; the visceral group includes the mediastinal and tracheobronchial nodes. The parietal nodes are smaller and less constant in number and location than are the visceral nodes.

The **sternal lymph node** *(nodus lymphaticus sternalis)* (Figs. 13–5, 13–6) is usually represented by a single node on each side. In others a single median node, which may be located either right or left of the median plane, serves both sides. The node may be lacking completely, or, in rare instances, a double node may be present on one side.

When one node is present on each side, it lies immediately cranial to the transversus thoracis muscle and medial to the second costal cartilage or second interchondral space cranioventral to the internal thoracic blood vessels. Typically, the node is ellipsoidal in shape and 2 mm. to 2 cm. in length. If a single node is present it may be dumbbell-shaped. When two nodes are present on one side, they may lie close together and be mistaken for a single node.

The afferent lymphatics of the sternal node on each side lie under the transversus thoracis in the fat lying between this muscle and the dorsal surfaces of the sternal ends of the costal cartilages. Occasionally, a single vessel is present. It arises in the abdominal wall, perforates the diaphragm near the middle of the costal arch, and, running under the pleura in the phrenicocostal sinus, extends forward and

Text continued on page 819

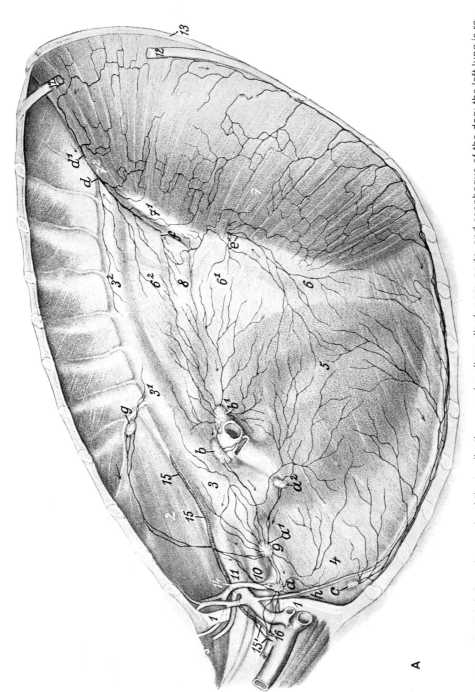

Figure 13–5. A. Lymph vessels of the mediastinum, pericardium, diaphragm, aorta, and esophagus of the dog; the left lung is removed, and the left wall of the thorax is almost completely removed, as is the m. transversus thoracis. (After Baum.)

1. First rib
2. M. longus colli
3. 3^1, 3^2. Aorta
4. Precardial mediastinum
5. Pericardium and cardial mediastinum
6. 6^1, 6^2. Postcardial mediastinum
7. 7^1, 7^2. Diaphragm
8. Esophagus
9. Cranial vena cava
10. Brachiocephalic artery
11. Left subclavian artery

12. Twelfth rib
13. Thirteenth rib
14. Costocervical vein
15. $15'$. Thoracic duct
16. Union of the axillary and jugular veins with the subclavian vein
a, a^1, a^2. Cranial mediastinal lymph nodes
b. Left tracheobronchial lymph node
b'. Middle tracheobronchial lymph node
c. Sternal lymph node

d, d'. Lymph vessels which run to the cranial lumbar aortic lymph nodes
e. Lymph vessels which enter the abdominal cavity through the diaphragm and empty into the splenic, gastric, left portal, or cranial lumbar aortic lymph nodes
f. Lymph vessel which enters the abdominal cavity with the esophagus
g. Intercostal lymph noce
h. An efferent vessel which goes to the right side and appears as number 10 on Figure 13–6

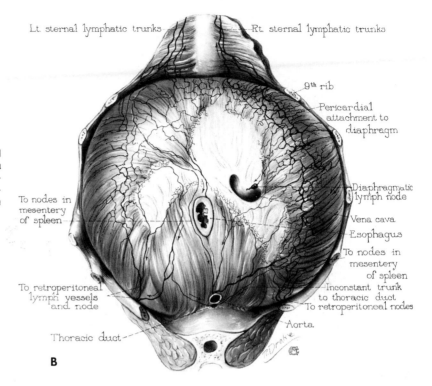

Figure 13–5. B. The pleural surface of the diaphragm in a dog after the intraperitoneal injection of a graphite preparation. (From Higgins and Graham 1929.)

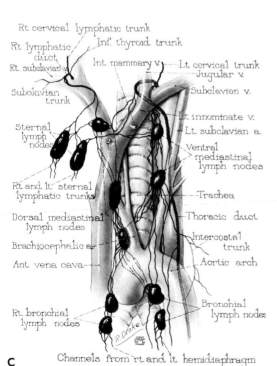

Figure 13–5. C. The arch of the aorta and mediastinum, showing lymph channels in a dog given an intraperitoneal injection of a graphite solution. (From Higgins and Graham 1929.)

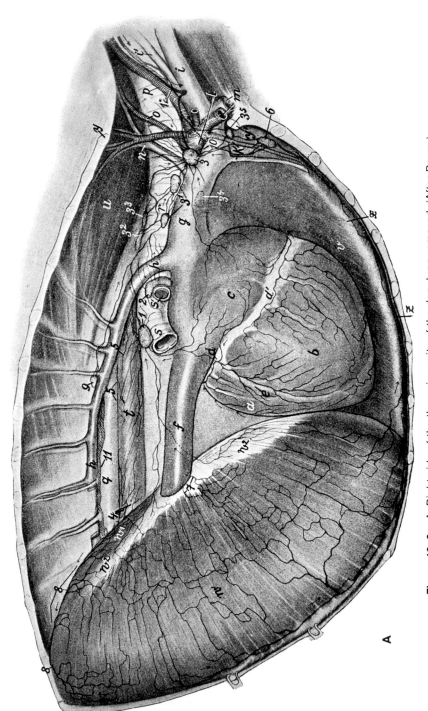

Figure 13-6. A. Right side of the thoracic cavity of the dog, lung removed. (After Baum.)

A

a. Left ventricle
b. Right ventricle
c. Right auricle
d, d'. Coronary sulcus
e. Right longitudinal sulcus
f. Caudal vena cava
g. Cranial vena cava
h. Azygos vein
i. External jugular vein
i'. Internal jugular vein
k. Internal thoracic artery and vein
l. Right subclavian artery
m. Right axillary artery and vein
n. Right costocervical vein
o. Right vertebral artery
p. Right common carotid artery

q. Aorta
r. Trachea
s. Right main bronchus
s'. Right eparterial bronchus
t. Esophagus
u. M. longus colli
v. Left M. transversus thoracis
w, w¹, w². Pars costalis, pars lumbalis, and tendinous
   parts of the diaphragm
x. Sternum
y, y'. Dorsal and ventral piece of the first rib
z. Right M. transversus thoracis (cut)

1, 2. Middle and right tracheobronchial lymph node
3 to 3⁵. Cranial mediastinal lymph nodes

4. Lymph vessels of the esophagus which enter the
   abdominal cavity
5. Lymph vessels of the esophagus which turn to
   the left and empty into the left tracheobronchial
   lymph node
6. Sternal lymph node
7. Lymph vessels which empty into the gastric,
   splenic, portal, or cranial lumbar lymph node
8. Lymph vessels which go to the cranial lumbar
   lymph node
9. Intercostal lymph node
10. Efferent vessel of a left cranial mediastinal lymph
    node
11. Thoracic duct
12. Right tracheal duct

Illustration continued on opposite page

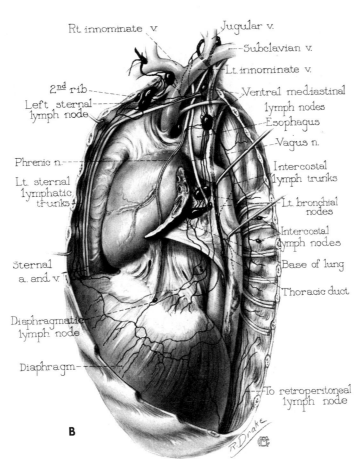

Figure 13–6. *Continued.*   B. The right half of the thorax after the intraperitoneal injection of a graphite preparation. (After Higgins and Graham 1929.)

Illustration continued on following page

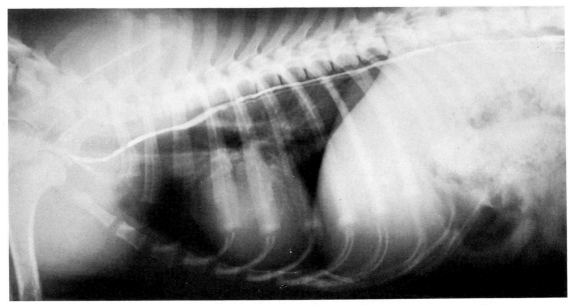

Figure 13–6   Radiograph 1. Lateral radiograph of the thorax showing the course of the thoracic duct.

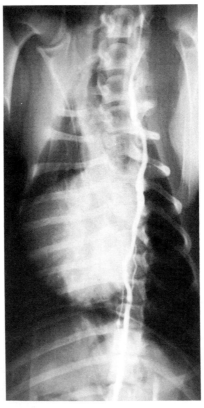

Figure 13–6   Radiograph 2. Dorsoventral radiograph of the thorax showing the thoracic duct inclining to the right at the level of the fifth thoracic vertebra.

downward to dip under the transversus thoracis. According to Baum (1918), it receives tributaries from the ribs, sternum, serous membranes, thymus, adjacent muscles, and mammary glands. Stalker and Schlotthauer (1936) state that lymphatics were not found to penetrate the thoracic or abdominal walls. Clinical observations indicate that the sternal lymph nodes receive no afferent vessels from the mammary glands. In the absence of the sternal lymph nodes, the afferent vessels which would otherwise drain into them drain into the mediastinal nodes. The one to three efferent lymphatics from the sternal node on each side run in front of the internal thoracic blood vessels, where they form a plexus. The right vessels terminate in the right lymphatic duct, the left in the thoracic duct. Many variations exist.

Higgins and Graham (1929) introduced 40 to 50 ml. of a graphite preparation into the peritoneal cavity of dogs to trace the lymphatic drainage routes. Within 10 minutes to one hour the lymphatic channels on the surface of the diaphragm were completely delineated. Muscular activity hastened the absorptive process. The major lymphatic drainage from the peritoneal cavity appeared to be via the diaphragm to the **sternal lymphatics.** A group of channels course cranially from the diaphragm (Fig. 13–5A) onto the sternum parallel to the internal thoracic artery and vein. Occasionally, plexuses are formed. These sternal lymphatic trunks continue into a series of two to four sternal lymph nodes which lie near the ventral ends of the first three ribs (Fig. 13–6). In the region of the sternal lymph nodes, numerous collecting ducts unite the lymphatics of the right and left sides. In general, the drainage of the right sternal lymphatics is to the right lymphatic duct, and drainage from the left lymphatics is toward the thoracic duct (Fig. 13–5B). Thus it appears that the thoracic duct plays a relatively insignificant part in the drainage of the peritoneal cavity.

The **intercostal lymph node** (*lymphonodus intercostalis*) (Figs. 13–5 to 13–7) was found in only 14 of 54 specimens by Baum (1918), and in only 2 of these was it present on both sides. This small spherical node lies in the vertebral end of either the

fifth or the sixth intercostal space under the sympathetic trunk, caudal to the intercostal artery. According to Baum it receives a portion of those lymph vessels which pass into the thoracic cavity through the last six to eight intercostal spaces. It probably also receives lymphatics from the ribs, vertebrae, pleura, and aorta. These vessels lie on the thoracic vertebrae and longus colli muscle. The efferent vessels go to the mediastinal nodes.

The **mediastinal lymph nodes** (*lymphonodi mediastinales*) (Figs. 13–5 to 13–9) vary in number and shape. Most of them are associated with the large vessels of the heart which run through the dorsal part of the precardial mediastinum. Unlike most other animals, the mediastinal lymph nodes in the dog are confined to the precardial mediastinum or surface of the heart. Although they lie in the precardial mediastinal septum, most nodes are not visible through the pleura from both sides, even in emaciated specimens. For this reason right and left nodes are described. In young animals some of these nodes are partly embedded in the thymus.

On the left side the mediastinal lymph nodes vary in number from one to six and in length from less than 1 mm. to 3 cm. Most of these nodes are oblong, and they lie along the cranial vena cava and the brachiocephalic, left subclavian, and costocervical arteries. When several nodes are present on the left side, one of these is located opposite the first intercostal space either cranial or caudal to the left costocervical vein. If only a single node exists, it is always located cranial to the costocervical vein. Small kernel-like nodes lying between the dorsally lying left subclavian and the ventrally lying brachiocephalic artery, or in the left groove between the trachea and esophagus, are not apparent unless they are exposed by removal of the overlying pleura and fat.

On the right side the mediastinal lymph nodes usually number two or three, with a maximum of six. The disc-shaped node lying between the right costocervical vein and the cranial vena cava is most constant. It may be double. In large dogs it measures over 1 cm. in diameter and may partly cover both veins between which it lies. Additional nodes are frequently found

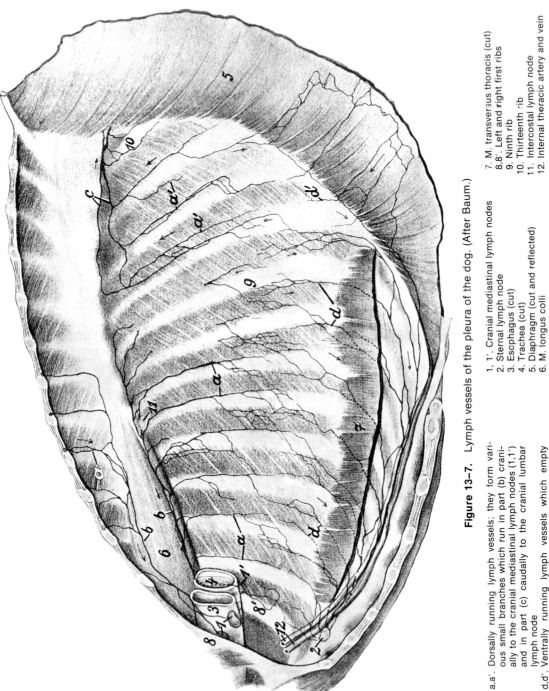

Figure 13-7. Lymph vessels of the pleura of the dog. (After Baum.)

a,a'. Dorsally running lymph vessels; they form various small branches which run in part (b) cranially to the cranial mediastinal lymph nodes (1,1') and in part (c) caudally to the cranial lumbar lymph node

d,d'. Ventrally running lymph vessels which empty into the sternal lymph node (2). Some of them (d') first run on the diaphragm

1, 1'. Cranial mediastinal lymph nodes
2. Sternal lymph node
3. Esophagus (cut)
4. Trachea (cut)
5. Diaphragm (cut and reflected)
6. M. longus colli

7. M. transversus thoracis (cut)
8,8'. Left and right first ribs
9. Ninth rib
10. Thirteenth rib
11. Intercostal lymph node
12. Internal thoracic artery and vein

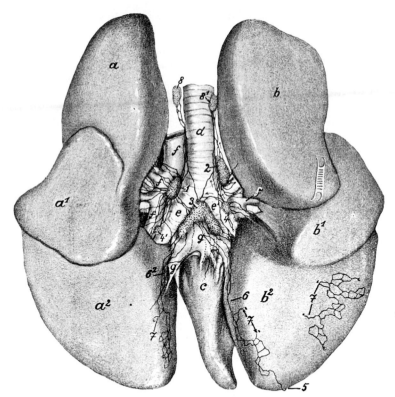

Figure 13–8. Lymph nodes and vessels of the lungs and bronchi of the dog.

a, a¹, a². Cranial and caudal parts of cranial lobe and caudal lobe of the left lung
b, b¹, b². Cranial, middle, and caudal lobes of the right lung
c. Accessory lobe
d. Trachea
e, e'. Left and right main bronchus
f. Pulmonary artery and its branches

g. Pulmonary veins
1, 2, 3. Left, right, and middle tracheo-bronchial lymph nodes
4, 4'. Pulmonary lymph nodes
5. Subserosal lymph vessels which pass around the acute margin to the dia-phragmatic surface and there pass deeply

6, 6². Subserosal lymph vessels which course along the attachment of the pulmonary ligament
7. Subserosal lymph vessels which pass deeply
8, 8'. A left and a right mediastinal lymph node

Figure 13–9. Left side of the heart of the dog with injected lymph vessels. (After Baum.)

a. Right ventricle
b. Left ventricle
c. Right auricle
d. Left auricle
e. Pulmonary artery (cut)
f. Pulmonary veins (cut)
g. Aorta
h. Trachea
i, i'. Left and right main bronchus
k. Middle tracheobronchial lymph node
l. Left tracheobronchial lymph node
m. Cranial mediastinal lymph node
n, n'. Coronary sulcus
o. Left longitudinal sulcus
p. Cranial vena cava

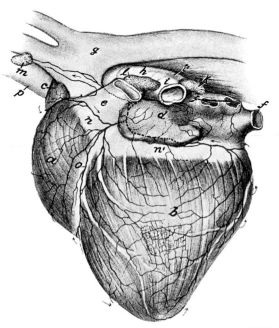

along the dorsolateral surface of the trachea, between the costocervical and azygos veins. A node may lie between the cranial vena cava and the brachiocephalic artery ventral to the trachea.

The afferent lymphatics, according to Baum (1918), come from the muscles of the neck, thorax, and abdomen, the scapula, the last six cervical vertebrae, the thoracic vertebrae, ribs, trachea, esophagus, thyroid, thymus, mediastinum, costal pleura, heart, and aorta. Lymphatics do not invade the central nervous system or the contractile elements of skeletal muscles. The mediastinal nodes receive efferent vessels from the intercostal, sternal, and middle and caudal deep cervical, tracheobronchial, and pulmonary nodes. The efferent lymphatics of all nodes caudal to the relatively constant node located in front of the costocervical vein on each side drain into it. From this node on the left side arise efferent lymphatics which empty into either the thoracic duct or the left tracheal trunk, or into both. On the right side similar efferent vessels go to the right lymphatic duct or the right tracheal trunk, or to both.

The **tracheobronchial lymph nodes** (*lymphonodi tracheobronchiales*) (Figs. 13–5, 13–6, 13–8, 13–9) include all nodes which lie on the initial parts of the bronchi at the bifurcation of the trachea. The nodes, which are constantly present, are known as the **right, left,** and **middle tracheobronchial lymph nodes.**

The **right** and **left tracheobronchial lymph nodes** are similar in size and location. Each lies on the lateral side of its respective bronchus, but also on the trachea, to a small extent. Dorsally, the right node is located ventral to the azygos vein; the left node has a similar relation to the beginning of the thoracic aorta. These nodes are 0.5 to 3 cm. long and are ellipsoidal in shape, with truncated caudal extremities. The left node is more angular, as it is wedged in the space bounded medially by the left primary bronchus and trachea, dorsally by the aorta, and ventrally by the pulmonary vein from the left apical and cardiac lobes of the lungs.

The **middle tracheobronchial lymph node** is always the largest node of this group. It is in the form of a V as it lies in

the angle formed by the origin of the primary bronchi from the trachea. The left limb of the V lies on the dorsal surface of the right pulmonary vein, from the diaphragmatic lobe; the right limb lies along the right cranial surface of this vein. The node is related to the esophagus dorsally. Its apex fits snugly into the angle formed by the bifurcation of the trachea. The vagi lie in contact with the limbs of the node as they become intimately related to the ventrolateral surfaces of the esophagus. Sometimes the tracheobronchial group includes a fourth node just cranial to the right node in the angle formed by the azygos vein at its entrance into the cranial vena cava. In other specimens the middle node and either the right or left node form one confluent mass.

The **pulmonary lymph nodes** (*lymphonodi pulmonales*) are often absent. They were present on one side in only 14 of 41 specimens examined by Baum (1918), and were never present on both sides. They are small nodes which lie on the dorsal surfaces of the primary bronchi between the peripheral ends of the right and left tracheobronchial nodes and the parenchyma of the lungs. They receive drainage from the lungs, and their efferent vessels go to the tracheobronchial nodes.

The afferent vessels to the tracheobronchial lymph nodes come from the lungs and bronchi primarily, but also from the thoracic parts of the aorta, esophagus, and trachea, and from the heart, mediastinum, and diaphragm. The two to four efferent lymphatics from each tracheobronchial node go partly to another node of this group and partly to the mediastinal nodes. Dogs which have lived in dusty or smoky environments have deeply pigmented tracheobronchial nodes. Inhaled foreign materials, regardless of their nature, are filtered out of the lymph as it passes through these nodes; this accounts for the markedly dark color of both the tracheobronchial and the pulmonary nodes in some specimens.

Kubik, Vizkelety, and Bálint (1956) have described the lymphatic drainage into each of the several lymph nodes located near the hilus of the lung. Kubik and Tömböl (1958) have investigated the tracheobronchial nodes through which the lymph from the lung passes in its course to the venous system. Miller (1937) has also described the lymphatics of the dog's lung.

LYMPH NODES AND VESSELS OF THE ABDOMINAL AND PELVIC WALLS

These nodes, like those of the thorax, can be divided into parietal and visceral groups. The parietal group includes the lumbar, aortic, medial iliac, hypogastric, sacral, and deep inguinal, or iliofemoral. The visceral group is divided largely into subgroups which serve specific organs.

The **lumbar aortic lymph nodes** (*lymphonodi lumbales aortici*) (Figs. 13–10, 13–12 to 13–14) are small nodes which lie along the aorta and caudal vena cava from the diaphragm to the deep circumflex iliac arteries. Except for a paired node near the diaphragm, they are erratic in their development and are often absent. Baum (1918) was able to demonstrate as many as 17 individual nodes in some specimens. Because of their small size and their similarity in color to the fat in which they are embedded, they are easily overlooked. The most constant in size and position is the paired lumbar node, which has somewhat different relations on either side. The left node is occasionally double. It is 1 to 2 cm. long and lies between the left crus of the diaphragm and the left sublumbar muscles, dorsal or caudal to the left renal artery. Its cranial pole usually touches or extends dorsal to the left phrenicoabdominal artery. The right member of this pair, usually smaller than the left, may have its cranial pole located dorsal to the right phrenicoabdominal artery and vein just after these vessels have crossed the respective dorsal and ventral surfaces of the right adrenal gland.

The afferent lymphatics to the lumbar nodes come from the lumbar vertebrae, the adrenal glands, and the abdominal portions of the urogenital system, including the testis of the male. They receive efferent vessels from more caudally located nodes. Efferent vessels from these nodes also empty directly into the lumbar lymphatic trunks; those from the more constant, cranially located nodes drain directly into the cisterna chyli. **Renal lymph nodes** associated with the renal vessels are included in lumbar lymphocentra.

The **medial iliac lymph node** (*lympho-nodus iliacus mediale*) (Fig. 13–13) was formerly known as the external iliac lymph node in this text. It is a large, constant node located between the deep circumflex iliac and the external iliac artery. This is usually single, but it may be double on one or both sides. It is about 4 cm. long, 1 cm. wide, and 0.5 cm. thick. It is irregular in outline, and more often its cranial end extends under or over the deep circumflex iliac vessels rather than behind the external iliac artery. It is bounded deeply on the right side by the caudal vena cava, which lies dorsal and to the right of the aorta. Each node lies in the furrow between the psoas major and the aorta and caudal vena cava, ventral to the bodies of the fifth and sixth lumbar vertebrae. Caudally, it more often bends laterally and follows along the cranial border of the external iliac artery rather than running under it.

The medial iliac lymph node (or nodes) receives afferent lymph vessels from all parts of the dorsal half of the abdomen, the pelvis, and the pelvic limb. It also receives afferent vessels from the genital system and the caudal part of the digestive and the urinary system, as well as efferent vessels from the deep and superficial inguinal, left colic, sacral, and hypogastric nodes. The efferent vessels from the medial iliac lymph node or nodes drain forward to form the lumbar lymph trunk or drain into the caudal members of the lumbar aortic lymph nodes if these nodes are present.

The **hypogastric lymph nodes** (*lymphonodi hypogastrici*) (Fig. 13–13) were formerly called the **internal iliac lymph nodes** in this text. They are usually small paired nodes which lie in the angle between the internal iliac and the median sacral artery, ventral to the body of the seventh lumbar vertebra on the ventral sacrococcaudal muscle. There may be three nodes, one behind another, on one side, or the nodes may be double on each side. In six of Baum's specimens only a single node served both sides.

The afferent lymph vessels come from the thigh, pelvis, pelvic viscera, tail, and a portion of the lumbar region; the efferent lymph vessels drain into the sacral nodes.

The **sacral lymph nodes** (*lymphonodi sacrales*) (Figs. 13–10, 13–13, 13–15, 13–17) are lacking about half of the time. They are not

Text continued on page 830

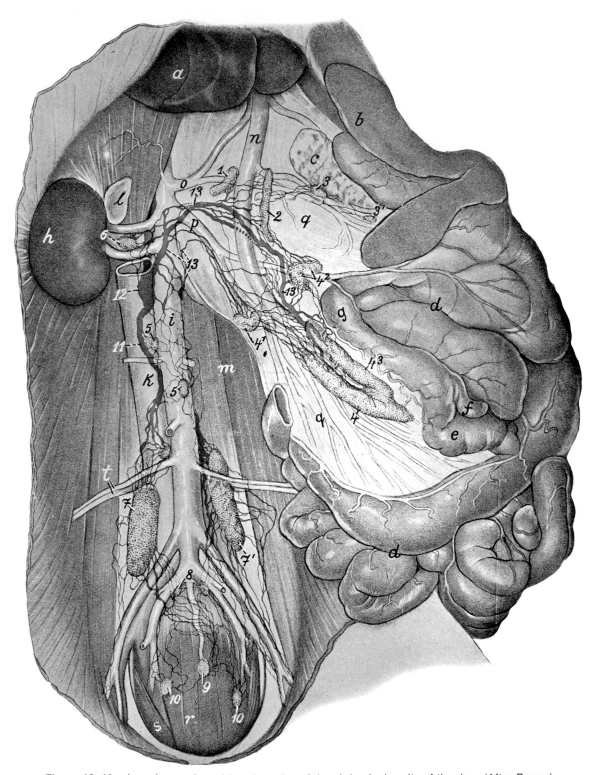

Figure 13–10. Lymph vessels and lymph nodes of the abdominal cavity of the dog. (After Baum.)

a. Liver	i. Aorta	r, s. Muscles of the tail	6. Right cranial lumbar lymph node
b. Spleen	k. Caudal vena cava	t. Deep circumflex iliac artery and vein	7, 7'. External iliac lymph nodes
c. Pancreas	l. Right adrenal gland		8. Internal lymph nodes
d. Jejunum	m. Psoas muscles	1. Right hepatic lymph node	9. Medial sacral lymph nodes
e. Ileum	n. Portal vein	2. Left hepatic lymph node	10. Lateral sacral lymph nodes
f. Cecum	o. Celiac artery	3, 3'. Splenic lymph nodes	11. Lumbar trunk
g. Colon	p. Cranial mesenteric artery	4 to 4³. Mesenteric lymph nodes	12. Cisternal chyli
h. Right kidney	q. Mesentery with blood vessels	5. Lumbar lymph nodes	13. Intestinal lymph trunk

Illustration continued on opposite page

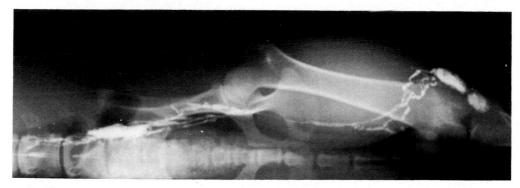

Figure 13-10. Radiograph 1. Dorsoventral radiograph of the stifle and sacral regions showing the popliteal and iliac lymph nodes.

Figure 13-10. Radiograph 2. Lateral radiograph of the popliteal and sacral lymph nodes.

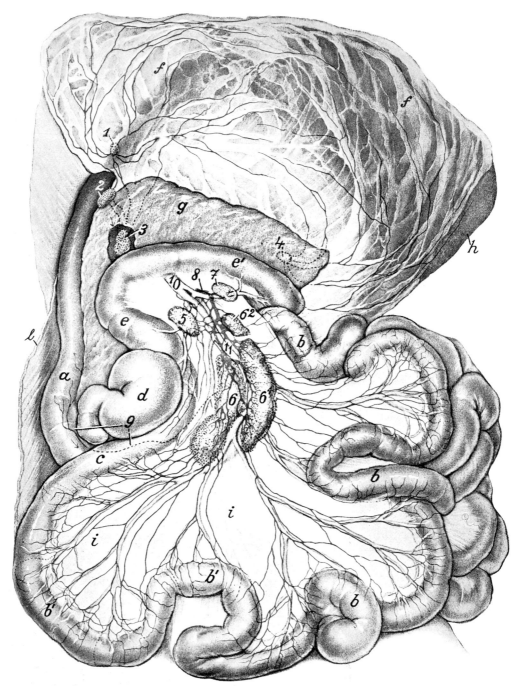

Figure 13–11.   Lymph vessels of the small intestine and omentum of a dog lying on its back. (After Baum.)

a. Duodenum
b, b'. Jejunum
c. Ileum
d. Cecum
e, e'. Colon
 f. Dorsal wall of omentum through which
    the stomach can be seen
g. Pancreas

h. Spleen (covered in part by omentum)
 i. Intestinal mesentery
 l. Cut edge of abdominal wall

1. Omental lymph node
2. Duodenal lymph node
3. Right hepatic lymph node
4. Splenic lymph node

5. Right colic lymph node
6, 6 ², 6⁴. Mesenteric lymph nodes
7. Middle colic lymph node
8. Intestinal trunk
9. Lymph vessel of the duodenum which
   goes to the right jejunal lymph node (6)
10. Cranial mesenteric artery
11. Jejunal lymph trunk

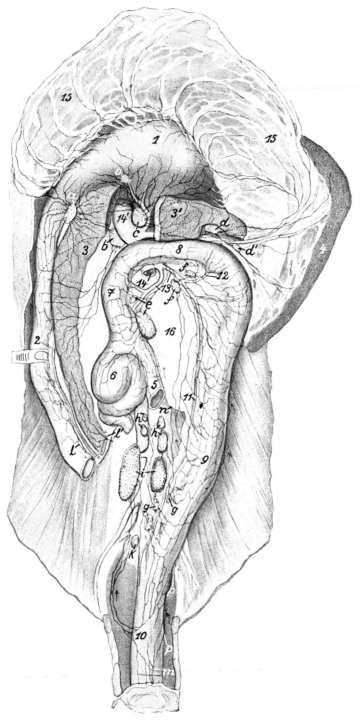

Figure 13–12. Lymph vessels and lymph nodes of the stomach, spleen, pancreas, duodenum, and large intestine of the dog. (After Baum.)

a. Duodenal lymph node
b. Right hepatic lymph node
c. Left hepatic lymph node
d, d'. Splenic lymph nodes
e. Right colic lymph node
f. Middle colic lymph nodes
g. Left colic lymph nodes
h. Lumbar lymph nodes
i. External iliac lymph nodes
k. Internal iliac lymph nodes
l,l'. Lymph vessel of the duodenum and lymph vessel of the pancreas which go

to the mesenteric lymph nodes (removed)
m. Lymph vessels of the anus and rectum
n. Lymph vessels which go directly to the lumbar cistern
o. Gastric lymph node
p. Lymph vessels of the rectum which course over the dorsal surface of the rectum to the internal and external iliac lymph nodes
1. Stomach
2. Duodenum (cut)

3,3'. Pancreas
4. Spleen (with splenic veins displaced)
5. Ileum (cut)
6. Cecum
7, 8, 9. Colon
10. Rectum
11. Left colic vein
12. Middle colic vein
13. Ileocecocolic vein
14, 14'. Portal vein
15. Ventral wall of the omental sac (reflected)
16. Mesentery of the colon

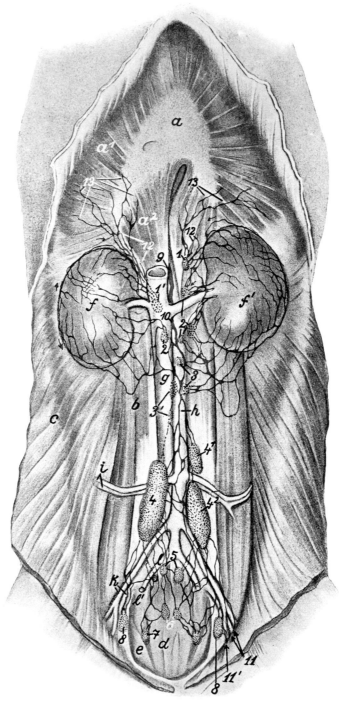

Figure 13–13. Lymph vessels of the kidney of the dog; the lymph nodes lying ventral to the aorta and its terminal branches. (The right kidney [f] is displaced toward the pelvis.) (After Baum.)

a, a¹, a². Diaphragm
 b. Psoas muscles
 c. Lateral abdominal wall
d, e. Muscles of the tail
f, f'. Kidneys
 g. Caudal vena cava
 h. Abdominal aorta
 i. Right deep circumflex iliac artery
    and vein
 k. Right external iliac artery and vein
l, l'. Right internal iliac artery and vein
1, 1'. Left and right cranial lumbar lymph

nodes (the right node is covered by
   the caudal vena cava)
 2. Lumbar lymph nodes which are lo-
    cated near the renal artery and vein
3, 3'. Lumbar lymph nodes (those la-
    beled 3 are covered by the caudal
    vena cava)
4, 4¹, 4². Medial iliac lymph nodes
 5. Hypogastric lymph nodes
 6. Middle sacral lymph nodes
 7. Lateral sacral lymph node
 8. Deep inguinal lymph nodes

 9. Cisterna chyli
10. Lumbar trunk
11. Efferent vessels of the superficial in-
    guinal and medial femoral lymph
    nodes (from them a part [11'] enters
    the deep inguinal lymph nodes [8])
12. Lymph vessels which enter the ab-
    dominal cavity from the thorax with
    the major splanchnic nerve and
    sympathetic trunk
13. Lymph vessels of the diaphragm

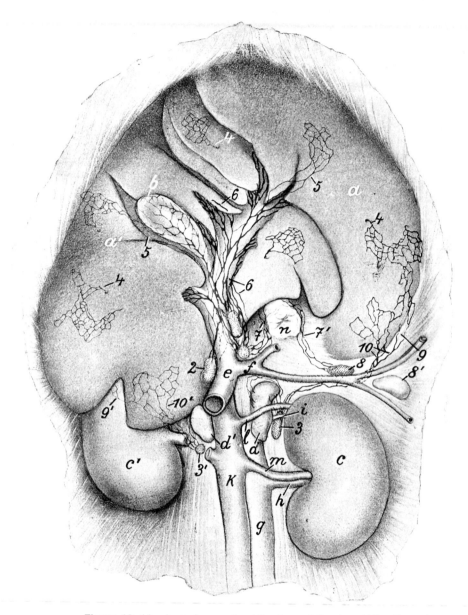

Figure 13–14. Lymph vessels of the liver of the dog. (After Baum.)

1, 2. Left and right hepatic lymph nodes
3, 3'. Cranial lumbar lymph nodes
4. Subserosal lymph vessels which course deeply
5. Subserosal lymph vessels which can be followed to the hepatic lymph nodes
6. Deep lymph vessels of the liver
7, 7'. Lymph vessels which leave the distal end of the esophagus
8, 8'. Splenic lymph nodes
9, 9'. Subserosal lymph vessels which

originate from the parietal surface of the liver
10, 10'. Subserosal lymph vessels from the visceral surface of the liver which course to the cranial lumbar lymph nodes (3, 3')

a, a'. Liver
b. Gallbladder
c, c'. Left and right kidneys
d, d'. Left and right adrenal glands

e. Portal vein
f. Splenic vein
g. Aorta
h. Renal artery
i. Phrenicoabdominal artery
k. Caudal vena cava
l. Phrenicoabdominal vein
m. Renal vein
n. Esophagus (cut)

sharply differentiated from the hypogastric nodes when more than one hypogastric node is present. They lie ventral to the body of the sacrum or ventral to the ventral sacrocaudal muscle. Small nodes, when present, lie on each side of the median sacral artery. Occasionally, there is a single small node, located in the fat ventral to the artery; in other dogs one or two nodes are present on one side. Baum (1918) occasionally found a small sacral node located between the piriformis and the sacrocaudalis ventralis muscles, closely associated with the internal iliac artery and vein.

The afferent lymph vessels come from the adjacent musculature and viscera; the efferent vessels go as a plexus to the hypogastric nodes.

The **liofemoral,** or **deep inguinal, lymph node** (*lymphonodus iliofemoralis [inguinalis profundus]*) (Fig. 13–13) was found by Baum (1918) in 18 of 50 dogs examined. A node was present on each side in 5, on the left side in 1, and on the right side in 12. It lies on the ventral surface of the tendon of the psoas minor at its insertion and is flanked by the internal and external iliac veins as these join to form the common iliac vein.

Afferent lymph vessels drain into it from the pelvic limb, and it lies in the course of the efferent lymphatics from the popliteal and superficial inguinal nodes. These lymph vessels, according to Baum, may bypass the node.

LYMPH NODES AND VESSELS OF THE GENITAL ORGANS

The lymphatics of the female genital organs (Fig. 13–15) empty into the lumbar, medial (external) iliac, hypogastric (internal) iliac, superficial inguinal, and sacral lymph nodes. A fine lymphatic network in the mesosalpinx and fat surrounding the ovary drains into lumbar lymph nodes in the region of the renal artery and vein. The lym-

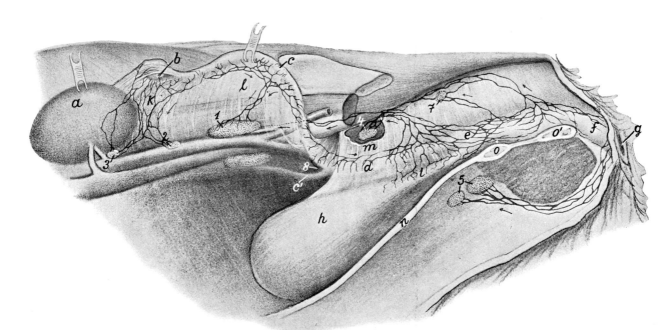

Figure 13–15. Lymph vessels of the female genital organs of the dog. (After Baum.)

1. Medial iliac lymph node	a. Left kidney (reflected)	h. Bladder
2, 3. Lumbar lymph nodes	b. Left ovary (reflected)	i. Urethra
4. Hypogastric lymph node	c. Left uterine horn (reflected)	k. Mesosalpinx
5. Superficial inguinal lymph nodes	c'. Right uterine horn	l. Broad ligament of the uterus
7. Lateral sacral lymph node	d. Body of the uterus	m. Lateral ligament of the bladder
8. Lymph vessel which courses to the other surface of the uterine horn	e. Vagina	n. Ventral abdominal wall
	f. Vestibule	o, o'. Pelvis, transected
	g. Vulva	

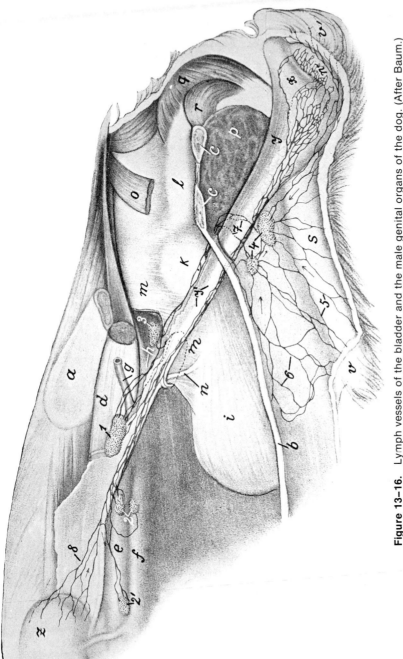

Figure 13–16. Lymph vessels of the bladder and the male genital organs of the dog. (After Baum.)

1. Medial iliac lymph node
2, 2'. Lumbar lymph nodes
3. Hypogastric lymph node
4. Superficial inguinal lymph nodes
5, 6. Lymph vessels of the preputial fold
7. Efferent vessels of the superficial inguinal
   lymph nodes
8. Lymph vessels of the testicles, which can be
   followed to the capsule of the kidney
a. Ilium
b. Ventral abdominal wall

c. Pelvis, transected
d. Lumbar muscles
e. Aorta
f. Caudal vena cava
g. Left external iliac artery
h. Internal iliac artery
i. Bladder
k. Prostate
l. Urethra
m. Lateral ligament of the bladder
n. Ureter

o. M. coccygeus (cut)
p. Cut surface of M. adductor
q. M. bulbospongiosus
r. M. ischiocavernosus
s. Penis
u. Prepuce (cut)
v'. Scrotum
w. Testicle
x. Epididymis
y. Spermatic cord with the ductus deferens (y')
z. Left kidney

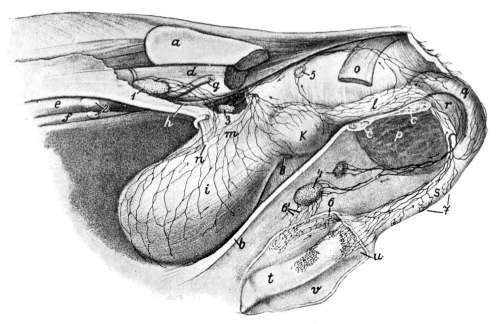

Figure 13–17. Lymph vessels of the urinary and male genital organs of the dog. (After Baum.)

1. Medial iliac lymph node
2. Lumbar lymph nodes
3. Hypogastric lymph node
4. Superficial inguinal lymph nodes
5. Lateral sacral lymph node
6, 6′. Lymph vessels of the preputial folds
7. Lymph vessels of the urethra
8. Lymph vessel of the bladder

a. Ilium

b. Ventral abdominal wall (cut)
c. Pelvis, transected
d. Lumbar muscles
e. Aorta
f. Caudal vena cava
g. Left external iliac artery
h. Internal iliac artery
i. Bladder
k. Prostate
l. Urethra
m. Lateral ligament of the bladder

n. Ureter
o. M. coccygeus (cut)
p. Cut surface of M. adductor
q. M. bulbospongiosus
r. M. ischiocavernosus
s. Penis
t. Glans penis
u. Bulbus glandis
v. Prepuce (opened and reflected)

phatics of the cranial half of the uterus empty into the lumbar and medial iliac lymph nodes, while vessels of the caudal half of the uterus drain into hypogastric iliac and sacral lymph nodes. Lymphatics of the vagina enter the hypogastric iliac lymph nodes, while those from the vestibulum vaginae in addition to entering the hypogastric iliac and sacral nodes also pass to the superficial inguinal node. Occasionally lymphatics from the vagina or vestibulum vaginae bypass the hypogastric iliac and sacral lymph nodes and empty directly into the medial iliac lymph node. The lymphatics of the vulva and clitoris pass for the most part into the superficial inguinal lymph node.

The lymphatics of the male genital organs (Fig. 13–16) empty into the same lymph nodes as do those of the female. The scrotum is drained by a coarse network which enters the superficial inguinal node. The lymphatics of the testis and epididymis extend in the spermatic cord into the abdominal cavity to enter the medial iliac and lumbar aortic lymph nodes. For the most part they accompany the blood vessels of the spermatic cord. Lymphatics of the prostate gland form a coarse network on the surface of the gland from which several vessels on each side drain into the medial iliac and hypogastric lymph nodes. Lymphatics of the prepuce and penis enter the superficial inguinal lymph node. The lymphatic system of the bladder (Fig. 13–17) was studied by Milroy and Cockett (1973). Drainage is into the lumbar and hypogastric lymph nodes.

LYMPH NODES AND VESSELS OF THE ABDOMINAL VISCERA

The lymph nodes of the abdominal viscera of the dog are not numerous, in comparison with those of other species. The following node groups serve the abdominal viscera: (1)

hepatic, (2) splenic, (3) cranial mesenteric, and (4) colic. To these may be added the inconstant gastric, duodenal, and omental nodes. Nylander and Tjernberg (1969) described lymphatics of the greater omentum.

The **hepatic lymph nodes** (*lymphonodi hepatici* [*portales*]) (portal of Baum) (Figs. 13–10 to 13–12, 13–14) usually consist of right and left nodes, lying one on each side of the portal vein, 1 or 2 cm. from the hilus of the liver. They may, however, vary greatly in number, form, and size. The node on the left is longer and larger than that on the right. It lies in the lesser omentum, dorsal to the common bile duct. It is about 3 cm. long and irregular in form; the caudal part which reaches to, extends along, or even extends beyond the splenic vein may be separated from the main lymphoid mass and form one or two additional nodes. When single, the left node is about 3 cm. long. On the right there may be one to five nodes, of various sizes and forms. They lie on the right side of the portal vein opposite the left node and dextral to the splenic vein at its termination. They are closely related to the intermediate part of the pancreas and may be flattened as they lie between the layers of peritoneum. Occasionally, the two nodal masses are joined cranially.

The afferent lymph vessels of these nodes come from the stomach, duodenum, pancreas, and liver. The several nodes are connected by lymph vessels. The efferent vessels from the right node unite into four to eight vessels which run over both surfaces of the portal vein to the cranial mesenteric artery, where they help form the intestinal trunk or the network of lymphatics which represents this trunk.

The **splenic lymph nodes** (*lymphonodi lienales*) (Figs. 13–10 to 13–12) are a group of three to five nodes that lie along the course of the splenic artery and vein and their terminal branches in the dorsal wall of the greater omentum. Most are small nodes which can be more easily palpated than seen in obese specimens. Usually the largest node lies on the cranial side of the splenic vessels or in the angle of their division, about 2 cm. from the termination of the splenic vein. The length of this node may be as much as 4 cm., but it is more commonly 1.5 cm.

The afferent vessels to the splenic lymph nodes come from the esophagus, stomach, pancreas, spleen, liver, omentum, and diaphragm. Their efferent vessels help to form the intestinal trunk or the lymphatic plexus which frequently replaces it.

The **cranial mesenteric lymph nodes** (*lymphonodi mesenterici craniales*) (Figs. 13–10, 13–11) are the largest lymph nodes of the abdomen. They are present along both sides of the vascular "tree" of the great mesentery and cover the roots of intestinal, cecal, and colic vessels. In medium-sized dogs they average 6 cm. long, 2 cm. wide, and 0.5 cm. thick. They are irregular in form. Frequently, their distal ends are knobbed or even lobated. Their middle parts may be roughly triangular in cross section. They lie between the leaves of the long jejunal mesentery, along the cranial mesenteric artery and vein. They extend from the root of the mesentery to the terminal ileal arteries of the cranial mesenteric artery. The initial parts of some 12 jejunal arteries are sandwiched between the two nodes. The right mesenteric lymph node lies between the cranial mesenteric vein and the ileum, and is seen more easily from the dorsal aspect. This node may be double or even triple. When it is double, the distal member is the larger. In 20 of Baum's (1918) 25 specimens the node on the right was single, and in 21 out of 25 the node on the left was single. As many as five nodes were present on the left in one specimen. There was no constant grouping of the smaller accessory nodes.

The afferent lymphatics to the mesenteric lymph nodes come from the jejunum, ileum, and pancreas. The efferent vessels from these nodes are the chief formative tributaries of the intestinal trunk or of the lymphatic plexus which replaces it.

The **colic lymph nodes** (*lymphonodi colici*) are found between the peritoneal laminae which form the mesocolon. They usually lie close to the gut.

Sterns and Vaughan (1970) studied the lymph drainage patterns of the dog's colon under normal and abnormal conditions by means of indirect lymphography. On the mucosal surface of the cecum, there are smooth, rounded elevations approximately 3 mm. in diameter that can be seen on the folds and between them. These structures have been described histologically by Atkins and Schofield (1972) as the sites of lymphoglandular complexes that penetrate the muscularis mucosae. There are about three per square centimeter. Similar structures were seen in the colon adjacent to the cecum but not elsewhere in the colon. These submucosal lymphoid nodules are invaginated by extensions of the overlying intestinal glands and thus are called lymphoglandular complexes. The structure and development of these complexes in the dog indicate that they have a secondary rather than a primary role as a lymphoid organ.

The **right colic lymph node** (Fig. 13–11) is disc-shaped, usually over 1 cm. in diameter, and lies dorsomedial to the right colon at the ileocolic junction. It is usually single, but as many as five nodes may be present. The largest node is located in the angle formed by the converging veins which form the v. ileocecocolica.

The **middle colic lymph node** (Figs. 13–11, 13–12) is a spherical or oval node, usually less than 1 cm. long, which lies 5 to 7 cm. from the transverse colon, near the attachment of the transverse mesocolon to the great mesentery. It is located near or on the junction of the middle colic tributary with the caudal mesenteric vein. If two nodes are present, one lies on either side of the caudal mesenteric vein. The incidence of multiple nodes in this location is the same as for the right side. The dorsocranial pole of the left mesenteric lymph node may be in apposition to the most proximal part of the middle colic node or the most proximal node of this group.

The **left colic lymph nodes** (Fig. 13–12) number two to five and lie in the caudal part of the left mesocolon near the pelvic inlet. The largest member of the group is usually located in the angle formed by the terminal branches of the caudal mesenteric artery. The others are mostly located between the cranial rectal artery and its satellite vein which lies on the intestine. They form an interconnected chain of oval nodes which do not extend further caudally than the pelvic inlet. Each node is not more than a few millimeters long.

The afferent lymph vessels to the colic lymph nodes come from the ileum, cecum, and colon. The middle colic node receives efferent vessels from the left colic nodes. The efferent lymph vessels from the right and middle colic nodes empty into the intestinal trunks. Those from the left colic nodes go as groups of one to three efferent vessels

from each node, and either enter the medial iliac, the lumbar, or the middle colic node, or empty into the intestinal trunk directly.

At various places in the omentum and mesentery are inconstant or small nodes, as follows:

The **gastric lymph nodes** (*lymphonodi gastrici*) (Fig. 13–12) lie in the lesser omentum close to the lesser curvature of the stomach near the pylorus. Occasionally, they are absent or double.

The **pancreaticoduodenal lymph nodes** (*lymphnodi pancreaticoduodenales*) (Figs. 13–11, 13–12) lie along the initial portion of the duodenum and include the nodes formerly called duodenal and omental by Baum (1918). A small constant node which may be double (duodenal) lies between the pylorus and the right limb of the pancreas. An inconstant, even smaller node (omental) may be present in the ventral wall of the omental bursa a few centimeters from the pylorus.

The afferent vessels to the pancreaticoduodenal lymph nodes come from the duodenum, pancreas, and omentum; and those to the gastric node come from the esophagus, stomach, liver, diaphragm, mediastinum, and peritoneum. The efferent vessels from the pancreaticoduodenal node go into the right hepatic or right colic node; and those from the gastric node go to the left hepatic or splenic node, or both.

LYMPH NODES AND VESSELS OF THE PELVIC LIMB

Pflug and Calnan (1969) describe three separate systems of lymphatic vessels in the pelvic limb of the dog which they believe function independently to counter any occlusion of lymphatic drainage. Communications between these systems were located at the paw, at midcrus, and in the inguinal region. In no instance were any lymphovenous anastomoses observed at peripheral sites. The three collecting systems are: (1) a superficial lateral system, (2) a superficial medial system, and (3) a deep medial system.

The superficial lateral system begins on the dorsum of the paw, parallels the dorsal metatarsal veins, and then crosses to the caudolateral surface at midcrus. After accompanying the calcanean tendon, the superficial lateral lymphatics pass over the belly of the gastrocnemius muscle and enter the popliteal lymph node.

The superficial medial lymphatics begin above the hock and are closely associated with the skin. They continue over the gracilis muscle to the level of midthigh, where they run between the gracilis and vastus medialis muscles before passing into the inguinal lymph node.

The deep medial lymphatic afferents begin just above the level of the hock between the distal end of the tibia and the calcanean tendon. They pass toward the belly of the gastrocnemius and bypass the popliteal lymph node. At the level of the stifle, they cross the insertion of the adductor muscle and run along the gracilis muscle, where they are joined by deep lymphatics before passing to the iliac lymph nodes.

The **popliteal lymph node** (*lymphocentrum popliteum*) (see Fig. 13–1) is the largest node of the pelvic limb. It is rarely double, and is constant in location. It is an oval node about 2 cm. long, which lies in the fat depot between the medial border of the biceps femoris and the lateral border of the semitendinosus, as these muscles diverge from each other. The node and its surrounding fat are subcutaneous caudally as they lie in the popliteal space caudal to the stifle joint.

The afferent lymph vessels to the popliteal node come from all parts of the pelvic limb distal to the location of the node. The 8 to 10 efferent vessels accompany the medial saphenous vein in the popliteal space and unite to form 2 to 4 trunks proximal to the origin of the m. gastrocnemius. They continue proximally, close to the femur, where they lie between the m. semimembranosus and m. adductor. They reach the apex of the femoral triangle, traverse it and the vascular lacuna within the femoral canal, and, after crossing the medial surface of the pelvis, empty into the medial (external) iliac lymph node.

The **femoral lymph node** (*lymphonodus femoralis*) is exceedingly inconstant and small. Baum (1918) found it only five times in 40 dissections, and then only once on both sides. It is never over a few millimeters in diameter, and lies in the fat under the deep medial femoral fascia at the distal part of the femoral triangle. The femoral vessels lie in front of it, and it is intercalated in the main

lymph vessels from the popliteal node. It may receive afferent lymphatics from the medial side of the pelvic limb (Baum 1918).

The **superficial inguinal, or inguinofemoral, lymph nodes** (*lymphonodi inguinales superficiales*) (Figs. 13–15, 13–16), usually two in number, begin a few millimeters cranial to the vaginal process and lie in the fat which fills the furrow between the abdominal wall and the medial surface of the thigh. In the male, right and left nodes lie along the dorsolateral borders of the penis. When a single node is present, the external pudendal vessels lie lateral to it, but when two nodes are present these vessels usually run between the cranial and caudal poles of the nodes, or the cranial node lies medial to the vessels in a deeper location than the caudal node. Their shape and size vary between wide extremes, but the nodes are usually oval and about 2 cm. long.

The afferent vessels to the superficial inguinal lymph nodes come from the ventral half of the abdominal wall, including the caudal abdominal and inguinal mammary glands. In the male, the afferent vessels come from the penis and the skin of the prepuce and scrotum. Other afferent vessels come from the ventral part of the pelvis, the tail, and the medial side of the thigh, stifle joint, and crus. The superficial inguinal lymph nodes receive the efferent vessels from the popliteal node and thus serve as one of the nodal stations for the whole pelvic limb. The efferent vessels unite to form one or two trunks which pass through the inguinal canal with the pudendal vessels and finally break up in the medial iliac lymph nodes.

SPLEEN

The **spleen** (*lien*) (Figs. 13–10, 13–11, 13–18) is situated in the left hypogastric region, approximately parallel to the greater curvature of the stomach. It is gray-brown in color, and often has a purple cast. Although it is irregular in shape, it is considerably longer than it is wide, and slightly constricted in the middle. The right fourth to half is widest and most variable in position. The location of the organ is dependent on the size and position of the other abdominal organs, particularly

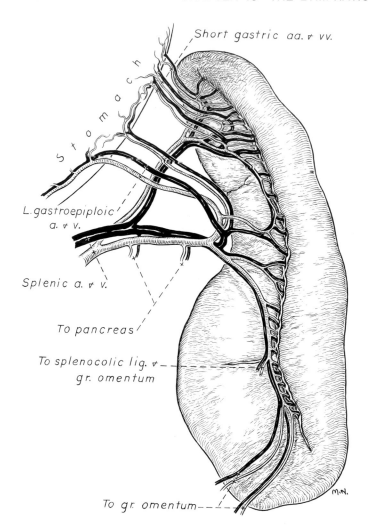

Short gastric aa. & vv.

Stomach

L.gastroepiploic
a. & v.

Splenic a. & v.

To pancreas

To splenocolic lig. &
gr. omentum

To gr. omentum

Figure 13-18. The spleen, showing its blood supply. (The cranial border is reflected laterally.)

the stomach, to which it is loosely attached. It is rather firm in consistency, especially when contracted, and has a thick trabecular framework. It is roughly tongue-shaped and presents two extremities, two surfaces, and two borders. In cross section the spleen is triangular. Its longest surface is the outer or parietal surface.

The **dorsal extremity,** rounded and wedge-shaped, lies ventral to the left crus of the diaphragm, between the fundus of the stomach and the cranial pole of the left kidney. Because of the relatively fixed position of the left kidney, the position of this part of the spleen is the least variable.

The **ventral extremity** is most variable, in both position and shape. When the spleen is maximally contracted, it is completely hidden under the middle of the caudal border of

the rib cage. When it is maximally distended, not only does most of the organ extend beyond the rib cage, but the ventral extremity moves beyond the midventral line to the right side of the floor of the abdomen. Depending on the fullness of the stomach, this may reach any level from the infrasternal fossa to a transverse plane caudal to the umbilicus. The ventral extremity may be uniformly rounded and approximately twice as wide as the dorsal half of the organ. Usually the most cranial portion of this extremity is pointed in a direction which ranges from cranioventral to craniodorsal.

The **parietal surface** faces the diaphragm and the lateral abdominal wall on the left side. It extends from the vertebral ends of the last two ribs ventrolaterally, mainly opposite the eleventh left intercostal space. As the distal portions of the last few costal cartilages bend forward and downward to form the costal arch, the spleen continues tangentially across the medial surface of the costal arch and obliquely downward, along the medial surface of the cranial part of the abdominal wall. It is at first related to the diaphragm and then to the transversus abdominis muscle. The parietal surface is slightly convex transversely. The proximal part is most convex longitudinally as it bends medially toward the median plane.

The **visceral surface** of the spleen is divided into two nearly equal longitudinal parts by the long **hilus** (*hilus lienis*) of the organ. The area cranial to the hilus is related to the greater curvature of the stomach; that caudal to the hilus is related to the left kidney proximally. It is related to the colon at its middle and to the mass of the small intestine distally. Even in hardened specimens, the organs related to the spleen do not make clear-cut impressions on it, as the fatty omentum prevents direct contact of the visceral surface with the adjacent viscera.

The **cranial and caudal borders** of the spleen are thin and irregular in contour. They may contain shallow or deep fissures. There is always a concavity of the cranial border proximal to the expanded distal end. This may involve the whole border or be replaced by an angular depression. In such instances the proximal two-thirds of the cranial border is usually sigmoid in shape. This is masked by a sweeping cranial concavity involving the middle half of the organ. The cranial border may be infolded.

The spleen is attached to the greater curvature of the stomach by the **gastrosplenic ligament** (*lig. gastrolienale*), a part of the greater omentum which runs between the stomach and the spleen. Proximally, the ligament attaches the greater omentum to the diaphragm.

The **internal structure** of the spleen consists of the parenchyma, the **red and white splenic pulp** (*pulpa lienis rubra et alba*), and a framework which consists of a capsule, which is rich in elastic and smooth muscle fibers, and **trabeculae** (*trabeculae lienis*), which are large and fibromuscular. The trabeculae form a complicated network within the organ. Some join the veins, strengthening their walls, and others are independent of them (Trautmann and Fiebiger 1952). The

larger intrasplenic arteries lie mainly in the trabeculae. The collagenous fibers of the trabeculae continue directly into the reticular fibers of the splenic pulp.

The **white pulp** in the dog consists of diffuse and nodular or follicular lymphoid tissue. The **nodules** (*folliculi lymphatici lienales*) are usually less than 1 cm. in diameter and are not grossly visible. The germinal centers of these nodules are lighter in color than the surrounding pulp. Lymphatic tissue is also elaborated along the arteries (diffuse lymphatic tissue). The **red pulp** consists of the venous sinuses and the cellular tissue filling the spaces between them. It includes many lymphocytes, free macrophages, and all the elements of the circulating blood. The non-granular leukocytes are the most numerous of these free cells (Bloom and Fawcett 1962).

The blood vessels of the spleen are the splenic artery from the celiac artery, and the splenic vein, which drains into the gastrosplenic vein. Blood enters the organ by way of about 25 **splenic branches** (*rami lienales*), which pass through the long hilus. Once within the capsule, they course in the trabeculae, branching repeatedly and becoming smaller. When they reach a diameter of 0.2 mm., they leave the trabeculae and become surrounded by lymphoid tissue (white pulp), in which they continue to divide. According to Bloom and Fawcett (1962), when they reach a caliber of 40 to 50 micrometers they leave the lymphatic tissue and enter the red pulp. Here they branch into small straight vessels, called *penicilli*. After these have divided and become smaller, their walls become greatly thickened. Further divisions reduce the arterial capillary to a caliber of not over 10 micrometers (Bloom and Fawcett 1962).

The venous side of the vascular pathway through the spleen begins in the venous sinuses. These **sinuses** (*sinus lienis*) play an active role as a part of the reticuloendothelial system. They occupy more space than does the solid part of the red pulp, among which they profusely anastomose. They range from 12 to 40 micrometers in width, and have walls which are composed of long, narrow reticuloendothelial cells. The sinuses coalesce into veins of the red pulp, and these finally merge to become the trabecular veins. The venous pathway through the splenic capsule parallels the arterial inflow.

The exact method whereby blood passes from the arterial to the venous side of the capillary bed is still in doubt (Weiss 1972). Three theories have been postulated, as follows:

1. The "open" circulation theory was held largely by Mall (1903), who did his research on the dog. According to this theory, there is no endothelial continuity between the arterial and venous sides of the capillary bed, but rather the arterial capillaries open directly into the red pulp and the blood gradually filters into the sinuses through the clefts between the reticuloendothelial cells which form their walls. Robinson (1930), working with cats, also believed in the "open" circulation theory.

2. According to the "closed" circulation theory, the arterial capillaries communicate directly with the lumen of the venous sinuses. Several studies, including those of Kniseley, indicated a direct flow from arterial terminations into sinuses.

3. The compromise theory was set forth by Kniseley (1936), who made his observations on living, unstimulated spleens. According to this theory, the walls of the venous sinuses quantitatively separate the cells of the blood from the fluid of the blood by continuous filtration processes. The salient observations of Kniseley's (1936) experiments were these: the sinuses become filled with whole blood by the closing of a physiological sphincter at the venous or efferent end of the sinus. As more blood is allowed to pass into the sinus, the fluid of the blood diffuses through the sinus wall, bringing about a great concentration of the cellular residue. Apparently, the fluid of the blood re-enters adjacent capillaries, as these have been observed to contain relatively cell-free blood. One stage merges into the next and, at the same time, other capillary-sinus units are not undergoing cyclic changes at all. In one cat the time interval for each cycle varied from a few minutes to as long as 10 hours.

True accessory spleens are rare in the dog. Armstrong (1940) described three animals having scores of accessory spleens as a result of postnatal trauma. Many small splenic grafts were dispersed in the omentum. Some spleens were divided, and others showed extensive cicatrix formation. Dogs usually do not possess hemal lymph nodes.

The nerve supply to the spleen is from the celiac plexus, and consists chiefly of nonmyelinated postganglionic sympathetic fibers. The few myelinated fibers present are probably afferent axons. The vagi also send fibers to the spleen. The nerves to the spleen form the splenic plexus of nerves which entwine the splenic artery.

Several diverse functions have been attributed to the spleen of the dog:

1. It stores and concentrates the erythrocytes and releases them during times of need.

2. It filters the blood, as the lymph nodes filter the lymph, and removes the worn-out erythrocytes from the circulation. From these cells it produces bilirubin, which is collected by the liver. From the hemoglobin it extracts the iron, which is released here, and is again used by the red bone marrow in the production of new erythrocytes.

3. It produces many of the lymphocytes and probably most of the monocytes, and has an important function in the production of antibodies.

The spleen is not essential to life or even to health, as most of its normal functions are taken over by other tissues in its absence.

THYMUS

The thymus (Fig. 13–19) is a light gray, distinctly lobulated organ with a pink tinge in fresh material. It is laterally compressed and lies in the cranial ventral part of the thoracic cavity. Because it is predominantly lymphoid in structure and hormone production is questionable, it is described under the vascular rather than the endocrine system.

The organ, relatively large at birth, grows rapidly during the first few postnatal months so that it reaches its maximum development before sexual maturity, or between the fourth and fifth postnatal months, just before the shedding of the deciduous incisor teeth. The thymus begins to involute with the changing of the teeth. Although the process is rapid at first, the organ usually does not atrophy completely even in old age. As it decreases in size and loses its lymphoid

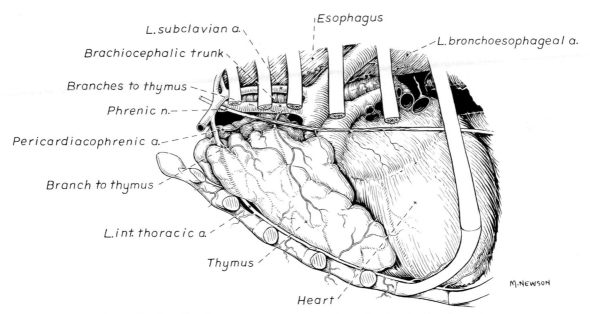

Figure 13-19. The thymus gland of a young dog, showing its blood supply.

structure, it is replaced by fat. However, evidence of a thymus can be seen in most dogs regardless of age.

The **thymus** is located almost entirely in the precardial mediastinal septum with its cranial end at the level of the thoracic inlet. The main portion of the organ extends from the thoracic inlet to a curved line opposite the left side of the heart. This place is usually located opposite the fifth costal cartilage.

Each polygonal lobule, which may measure over 1 cm. in length, is separated from adjacent lobules by a delicate but distinct connective tissue capsule. The whole organ is further divided into right and left lobes (*lobus dexter et sinister*), although the compressed lobes are difficult to separate from one another. The division between the two lobes is distinct caudally, but cranially it is not always possible to determine to which lobe the adjacent lobules belong, because the lobes are united here by essentially the same amount and kind of connective tissue that unites the lobules.

The **left lobe** extends farther caudally than the right. It occupies a special outpocketing of the precardial mediastinal portion of the pleura of the left pleural sac. Its caudal border is gently curved and thin as it lies between the left thoracic wall and the left ven-

tricle. In a small dog this portion of the left lobe may have a dorsoventral dimension of 5 cm. and may be 5 mm. thick. The left lobe becomes slightly narrower cranially as it lies in the mediastinal septum, loosely fused to the right lobe.

The **right lobe** of the thymus usually butts against the cranial surface of the pericardial sac and is expanded laterally. The thickest part of the thymus is located here. When maximally developed in the beagle, the thymus has the following measurements: 12 cm. long by 6 cm. wide by 3 cm. thick. When fully developed in this breed, it weighs about 50 gm. The right lobe accounts for 60 per cent of the weight, and the left lobe for 40 per cent. Variations in size and form are common. Latimer (1954) has studied the fetal development of the thymus in the dog.

Dorsally, the thymus is related to the phrenic nerves, cranial vena cava, and trachea. The apical lobes of the lungs produce large, smooth impressions on its lateral sides when it is maximally developed. The internal thoracic vessels form deep clefts, or even run through the cranioventral portion of the gland. The thymus receives its chief blood supply from one or two thymic branches which go to each lobe from the ipsilateral internal thoracic artery. Occasionally, an additional thymic branch leaves the brachiocephalic artery on the right side and the subclavian on the left side.

The veins from the thymus are satellites of the thymic arteries. The lymphatics from the thymus form four to six vessels which empty into the cranial mediastinal and sternal lymph nodes. Both parasympathetic (vagal) and sympathetic nerve fibers supply the organ, and are probably vasomotor. The

basic cell unit of the thymus is a small lymphocyte, sometimes referred to as a thymocyte. These cells do not differ from small lymphocytes found in other organs of the body. The thymic corpuscles (Hassall's bodies) are spherical or oval structures in the medulla of the gland, composed of concentrically arranged cells. The centers of many thymic corpuscles consist of degenerated cells which may be hyalinized, cystic, or calcified.

BIBLIOGRAPHY

Armstrong, W. H. 1940. Traumatic autotransplantation of splenic tissue with a report on three cases in the dog. Cornell Vet. 30:89–96.

Arnason, G., B. D. Janković, and B. H. Waksman. 1962. A survey of the thymus and its relation to lymphocytes and immune reaction. Blood 20:617–628.

Atkins, A. M., and G. C. Schofield. 1972. Lymphoglandular complexes in the large intestine of the dog. J. Anat. 113:169–178.

Barcroft, J., and H. W. Florey. 1928. Some factors involved in the concentration of blood by the spleen. J. Physiol. 66:231–234.

Battezzati, M., and I. Donini. 1973. The Lymphatic System. Revised edition translated from Italian by V. Cameron-Curry. 496 pp. New York, J. Wiley and Sons.

Baum, Hermann. 1918. Das Lymphgefässystem des Hundes. Arch. wiss. prakt. Tierheilk. Bd. 44:521–650. Berlin, Hirschwald.

Blalock, A., C. S. Robinson, R. S. Cunningham, and M. E. Gray. 1937. Experimental studies on lymphatic blockage. Arch. Surg. 34:1049–1071.

Bloom, W., and D. W. Fawcett. 1962. A Textbook of Histology. 8th Ed. Philadelphia, W. B. Saunders Co.

Chretien, P. B., R. J. Behar, Z. Kohn, G. Moldovanu, D. G. Miller, and W. Lawrence, Jr. 1967. The canine lymphoid system: a study of the effect of surgical excision. Anat. Rec. 159:5–15.

Clark, E. R., and E. L. Clark. 1937. Observations on living mammalian lymphatic capillaries — their relation to the blood vessels. Am. J. Anat. 60:253–298.

Drinker, C. K. 1942. The Lymphatic System. Lane Medical Lectures. Stanford Univ. Pub. Med. Sci. IV:137–235.

————.1946. Extravascular protein and the lymphatic system. Ann. N.Y. Acad. Sci. 46:807–821.

Fischer, H. W. 1959. A critique of experimental lymphography. Acta Radiol. 52:448–454.

Fischer, H. W., and G. R. Zimmerman. 1959. Roentgenographic visualization of lymph nodes and lymphatic channels. Am. J. Roentgenol. 81:517–534.

Freeman, L. W. 1942. Lymphatic pathways from the intestine in the dog. Anat. Rec. 82:543–550.

Grau, H., and R. Barone. 1970. Sur la topographie comparée et la nomenclature des nodules lymphatiques du bassin et du membre pelvien. Rev. Med. Vet. N.S. 33:649–659.

Higgins, G. M., and A. S. Graham. 1929. Lymphatic

drainage from the peritoneal cavity in the dog. Arch. Surg. 19:453–465.

Kniseley, M. H. 1936. Spleen studies: I. Microscopic observations of the circulatory system of living unstimulated mammalian spleens. Anat. Rec. 65:23–50.

Kubik, I., and T. Tömböl. 1958. Über die Abflussfolge der regionären Lymphknoten der Lunge des Hundes. Acta Anat. 33:116–121.

Kubik, I., T. Vizkelety, and J. Bálint. 1956. Die Lokalisation der Lungensegmente in den regionalen Lymphknoten. Anat. Anz. 104:104–121.

Latimer, H. B. 1954. The prenatal growth of the thymus in the dog. Growth 18:71–77.

MacKenzie, D. W., A. O. Whipple, and M. P. Wintersteiner. 1941. Studies on the microscopic anatomy and physiology of living transilluminated mammalian spleens. Am. J. Anat. 68:397–456.

Mahorner, H. R., H. D. Caylor, C. F. Schlotthauer, and J. de J. Pemberton. 1927. Observations on the lymphatic connections of the thyroid gland in man. Anat. Rec. 36:341–347.

Mall, F. P. 1903. On circulation through the pulp of the dog's spleen. Am. J. Anat. 2:315–332.

Meyer, A. W. 1906. An experimental study on the recurrence of lymphatic gland and regeneration of lymphatic vessels in the dog. Johns Hopkins Hosp. Bull. 17:185–192.

Miller, W. S. 1937. The Lung. Springfield, Ill., Charles C Thomas.

Milroy, E. J., and A. T. Cockett. 1973. Lymphatic system of the canine bladder. An anatomical study. Urology 2:375–377.

Nylander, G., and B. Tjernberg. 1969. The lymphatics of the greater omentum. An experimental study in the dog. Lymphology 2:3–7.

Pflug, J. J., and J. S. Calnan. 1969. Lymphatics: normal anatomy in the dog hind leg. J. Anat. 105:457–465.

Prier, J. E., B. Schaffer, and J. F. Skelley. 1962. Direct lymphangiography in the dog. J. Am. Vet. Med. Assoc. 140:943–947.

Reichert, F. L. 1926. The regeneration of the lymphatics. Arch. Surg. 13:871–881.

Robinson, W. L. 1930. The venous drainage of the cat spleen. Am. J. Path. 6:19–26.

Rusznyák, I., M. Földi, and G. Szabo. 1967. Lymphatics and Lymph Circulation. 2nd English Ed. by L. Youlten. New York, Pergamon Press.

Sabin, F. R. 1911. A critical study of the evidence

presented in several recent articles on the development of the lymphatic system. Anat. Rec. 5:417–446.

Skelley, J. F., J. E. Prier, and R. Koehler. 1964. Applications of direct lymphangiography in the dog. Am. J. Vet. Res. 24:747–755.

Stalker, L. K., and C. F. Schlotthauer. 1936. Neoplasms of the Mammary gland in the dog. North Am. Vet. 17:33–43.

Sterns, E. E., and G. E. Vaughan. 1970. The lymphatics of the dog colon. A study of the lymph drainage patterns by indirect lymphography in the dog under normal and abnormal conditions. Cancer 26:218–231.

Suter, P. 1969. Die Lymphographie beim Hund, eine röntgenologische Methode zur Diagnose von Veränderungen am Lymphsystem. Zurich, Juris-Verlag.

Suter, P. F., and R. W. Greene. 1971. Chylothorax in a dog with abnormal termination of the thoracic duct. J. Am. Vet. Med. Assoc. 159:302–309.

Todd, G. L., and G. R. Bernard. 1973. The sympathetic innervation of the cervical lymphatic duct of the dog. Anat. Rec. 177:303–316.

Trautmann, A., and J. Fiebiger. 1952. Fundamentals of the Histology of Domestic Animals (translated and revised from 8th and 9th German editions, 1949, by R. E. Habel and E. L. Biberstein). Ithaca, New York, Comstock Publishing Assoc.

Ullal, S. R., T. H. Kluge, W. J. Kerth, and F. Gerbode. 1972. Anatomical studies on lymph drainage of the heart in dogs. Ann. Surg. 175:305–310.

Uhley, H. N., S. E. Leeds, and M. A. Sung. 1972. The subendocardial lymphatics of the canine heart. A possible role of the lymphatics in the genesis of conduction disturbances and arrhythmias. Am. J. Cardiol. 29:367–371.

Weiss, L. 1972. The Cells and Tissues of the Immune System. Foundations of Immunology Series. Englewood Cliffs, N.J., Prentice-Hall.

Yoffey, J. M., and F. C. Courtice. 1956. Lymphatics, Lymph, and Lymphoid Tissue. Cambridge, Mass., Harvard University Press.

Yoffey, J. M., and C. K. Drinker. 1938. The lymphatic pathways from the nose and pharynx. J. Exp. Med. 68:629–640.

Zajac, S. 1972. Natural lympho-venous communication between the cisterna chyli and inferior vena cava. Pol. Med. J. 11:1271–1277.

Chapter 14

THE BRAIN

By HERMANN MEYER

The central nervous system consists of the brain and the spinal cord. The **brain** develops from the rostral cephalic portion of the neural tube. In rostrocaudal sequence it is divided into five major parts: the **telencephalon,** the **diencephalon,** the **mesencephalon,** the **metencephalon,** and the **myelencephalon** (Fig. 14–1). The pattern of this subdivision of the brain is based on the early development of the rostral part of the central nervous system, where, at first, three vesicles develop. They are called the prosencephalon, or forebrain; the mesencephalon, or midbrain; and the rhombencephalon, or hindbrain. With further development, the prosencephalon divides into the telencephalon and diencephalon. The mesencephalon does not divide. The rhombencephalon also gives rise to two parts: the metencephalon and the myelencephalon. The **spinal cord** is derived from the remainder of the neural tube. The lumen of the neural tube persists inside the central nervous system in the form of cavities and connecting canals. These spaces develop into the **central canal** of the spinal cord and the **ventricular system** of the brain and become filled with **cerebrospinal fluid.**

From a macroscopic point of view, the brain may be subdivided into three components: the cerebrum, the brain stem, and the cerebellum (Fig. 14–2). The **cerebrum** is the largest, most rostral part of the brain. It is derived from the telencephalon. The **brain stem** includes the entire diencephalon and mesencephalon, the ventral portion of the metencephalon, and the entire myelencephalon. It connects the cerebrum with the spinal cord and the cerebellum. The **cerebellum,** which develops from the dorsal part of the metencephalon, is located on the dorsal aspect of the caudal part of the brain stem.

Structurally, the central nervous system consists of gray matter *(substantia grisea),* white matter *(substantia alba),* and reticular formation *(formatio reticularis).* The **gray matter** is formed by localized aggregations of cell bodies of neurons within the central nervous system. In the spinal cord the cell bodies are arranged in centrally located, continuous **gray columns.** In the brain the spatial relation is different. The columns of gray matter, as they are found in the spinal cord, are intersected by white matter and reticular formation and form discrete entities of gray matter. Such accumulations of cell bodies within the central nervous system are called **nuclei.** In addition to these nuclear masses, gray matter is located peripherally in the cerebral hemispheres and the cerebellum as **cerebral cortex** and **cerebellar cortex,** respectively. The **white mat-**

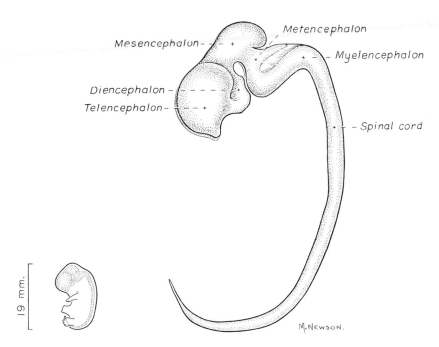

Figure 14–1. Five-vesicle stage of the dog's brain (19 mm. C-R embryo).

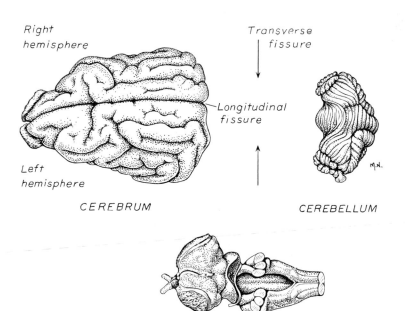

Figure 14–2. Gross subdivisions of the brain.

ter consists mainly of myelinated nerve cell processes or portions of processes inside the central nervous system. In part it is arranged in fiber groups with common connections and functions. These fiber groups or bundles are referred to as **tracts** or *fasciculi*. They connect centers in various portions of the central nervous system. The **reticular formation** extends throughout the brain stem and includes diffusely scattered

nerve cell bodies and nerve fibers which appear in a network structure, as the name implies.

TELENCEPHALON

The **telencephalon** forms the cerebrum, the largest, most rostral major subdivisions of the brain (Fig. 14–2). Caudally, it is sepa-

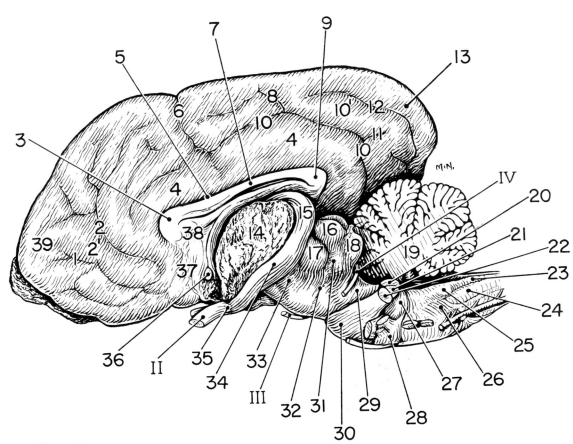

Figure 14–3.   Medial surface of the right cerebral hemisphere and lateral surface of the brain stem.

1. Ectogenual sulcus
2. Genual sulcus
2'. Genual gyrus
3. Genu of corpus callosum
4. Cingulate gyrus
5. Callosal sulcus
6. Cruciate sulcus
7. Body of corpus callosum
8. Lesser cruciate sulcus
9. Splenium of corpus callosum
10. Splenial sulcus
10'. Splenial gyrus
11. Caudal horizontal ramus of splenial sulcus
12. Suprasplenial sulcus

13. Occipital gyrus
14. Cut surface between cerebrum and brain stem
15. Lateral geniculate body
16. Rostral colliculus
17. Medial geniculate body
18. Caudal colliculus
19. Arbor vitae cerebelli
20. Rostral cerebellar peduncle
21. Caudal cerebellar peduncle
22. Middle cerebellar peduncle
23. Fasciculus cuneatus
24. Spinal tract of trigeminal nerve
25. Nucleus cuneatus lateralis
26. Superficial arcuate fibers

27. Cochlear nuclei
28. Trapezoid body
29. Lateral lemniscus
30. Transverse fibers of pons
31. Brachium of caudal colliculus
32. Transverse crural tract
33. Crus cerebri
34. Left optic tract
35. Optic chiasm
36. Rostral commissure
37. Paraterminal gyrus
38. Telencephalic septum
39. Prorean gyrus
II. Optic nerve
III. Oculomotor nerve
IV. Trochlear nerve

rated from the cerebellum and the dorsal part of the rostral brain stem by the **transverse fissure** *(fissura transversa cerebri).* Along the midline, the cerebrum is divided into two cerebral hemispheres by the **longitudinal fissure** *(fissura longitudinalis cerebri).* Each **cerebral hemisphere** *(hemispherium)* consists of four topographical areas: the frontal lobe, the parietal lobe, the occipital lobe, and the temporal lobe. The boundaries of these lobes are quite arbitrary. The **frontal lobe** comprises the rostral part of the cerebral hemisphere. The **occipi-**

tal lobe is at the caudal pole. The **temporal lobe** consists of the ventrolateral area, and the **parietal lobe** includes the remaining dorsolateral portion of the cerebrum. Structurally, the cerebrum consists of a peripheral layer of gray matter, or cerebral cortex; white matter beneath the cortex; centrally located basal nuclei; and the phylogenetically older rhinencephalon at the base of the brain. Inside each cerebral hemisphere there is a cavity, the **lateral ventricle** *(ventriculus lateralis),* which developed from the lumen of the telencephalic vesicle. The

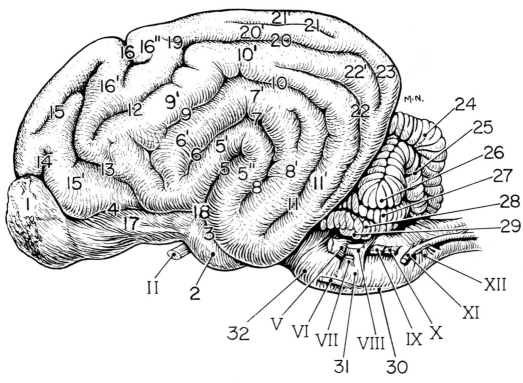

Figure 14-4. Lateral view of the brain.

1. Olfactory bulb
2. Piriform lobe
3. Caudal part of lateral rhinal sulcus
4. Rostral part of lateral rhinal sulcus
5. Pseudosylvian fissure
5'. Rostral sylvian gyrus
5". Caudal sylvian gyrus
6. Rostral ectosylvian sulcus
6'. Rostral ectosylvian gyrus
7. Middle ectosylvian sulcus
7'. Middle ectosylvian gyrus
8. Caudal ectosylvian sulcus
8'. Caudal ectosylvian gyrus
9. Rostral suprasylvian sulcus
9'. Rostral suprasylvian gyrus
10 . Middle suprasylvian sulcus
10'. Middle suprasylvian gyrus
11. Caudal suprasylvian sulcus
11'. Caudal suprasylvian gyrus

12. Coronal sulcus
13. Presylvian sulcus
14. Olfactory sulcus
15. Prorean sulcus
15'. Prorean gyrus
16. Cruciate sulcus
16'. Precruciate gyrus
16". Postcruciate gyrus
17. Olfactory peduncle
18. Insular region
19. Ansate sulcus
20. Marginal sulcus
20'. Marginal gyrus
21. Endomarginal sulcus
21'. Endomarginal gyrus
22. Ectomarginal sulcus
22'. Ectomarginal gyrus
23. Occipital gyrus
24. Vermis of cerebellum

25. Paramedian lobule
26. Ansiform lobule
27. Dorsal paraflocculus
28. Ventral paraflocculus
29. Flocculus
30. Pyramid
31. Trapezoid body
32. Pons

II. Optic nerve
V. Trigeminal nerve
VI. Abducent nerve
VII. Facial nerve
VIII. Vestibulocochlear nerve
IX. Glossopharyngeal nerve
X. Vagus nerve
XI. Accessory nerve
XII. Hypoglossal nerve

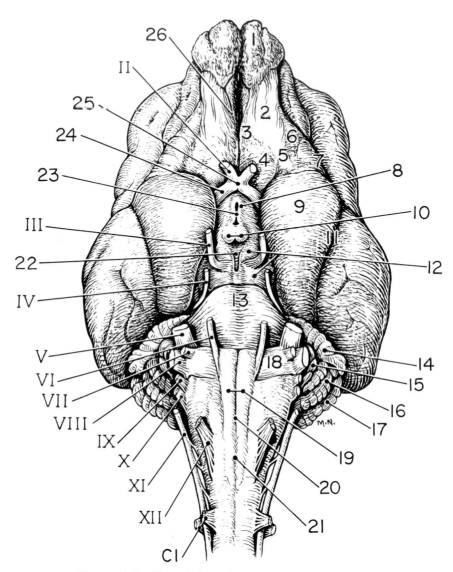

Figure 14–5.   Ventral view of the brain and cranial nerves.

1. Olfactory bulb
2. Olfactory peduncle
3. Medial olfactory tract
4. Rostral perforated substance
5. Lateral olfactory tract
6. Lateral olfactory gyrus
7. Rostral part of lateral rhinal sulcus
8. Tuber cinereum
9. Piriform lobe
10. Mamillary bodies
11. Caudal part of lateral rhinal sulcus
12. Crus cerebri
13. Transverse fibers of pons

14. Ventral paraflocculus
15. Flocculus
16. Dorsal paraflocculus
17. Ansiform lobule
18. Trapezoid body
19. Pyramids
20. Median fissure
21. Decussation of pyramids
22. Caudal perforated substance in intercrural fossa
23. Infundibulum
24. Optic tract
25. Optic chiasm

26. Medial rhinal sulcus
II. Optic nerve
III. Oculomotor nerve
IV. Trochlear nerve
V. Trigeminal nerve
VI. Abducent nerve
VII. Facial nerve
VIII. Vestibulocochlear nerve
IX. Glossopharyngeal nerve
X. Vagus nerve
XI. Accessory nerve
XII. Hypoglossal nerve
CI. First cervical nerve

first cranial nerves, or **olfactory nerves** *(nervi olfactorii),* enter the telencephalon rostroventrally by way of the rhinencephalon.

The **cerebral cortex** *(cortex cerebri)* is the outermost gray matter of the cerebrum which covers the hemispheres like a cloak or mantle. For this reason it is also referred to as the *pallium.* Its dorsal part, a phylogenetically recent acquisition, is called the *neocortex,* or *neopallium.* The olfactory cortex at the base of the brain is phylogenetically older and consists of the *paleocortex,* or *paleopallium,* and the *archicortex,* or *archipallium.* In the course of development, the cerebral cortex becomes folded, and a system of grooves and elevations is formed. The elevations or convolutions are known as **gyri** *(gyri cerebri);* the depressions between the gyri are the **sulci** *(sulci cerebri).* The sulcal and gyral pattern is typical for the species, but individual variations are relatively frequent, and the following description (Figs. 14–3 through 14–9) will be limited to structures essential for an understanding of normal conditions.

Among the sulci of the cerebral cortex, the **lateral rhinal sulcus** *(sulcus rhinalis lateralis)* is one of the most constant grooves. It is located on the ventrolateral aspect of

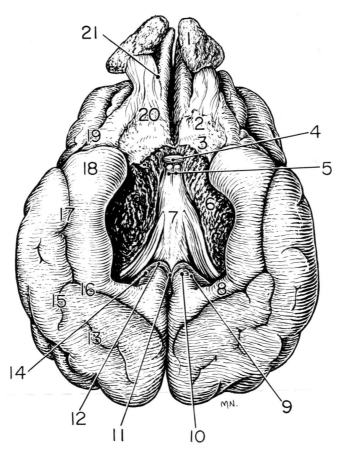

Figure 14–6. Ventral view of the cerebrum.

1. Olfactory bulb
2. Olfactory trigone
3. Diagonal gyrus
4. Rostral commissure
5. Columns of fornix
6. Cut surface between cerebrum and brain stem
7. Body of fornix
8. Parahippocampal gyrus

9. Hyppocampal sulcus
10. Callosal gyrus
11. Subsplenial flexure of dentate gyrus
12. Tubercle of dentate gyrus
13. Occipitotemporal sulcus
14. Dentate gyrus
15. Lateral limb of caudal part of lateral rhinal sulcus

16. Medial limb of caudal part of lateral rhinal sulcus
17. Caudal part of lateral rhinal sulcus
18. Piriform lobe
19. Rostral part of lateral rhinal sulcus
20. Olfactory tubercle
21. Medial rhinal sulcus

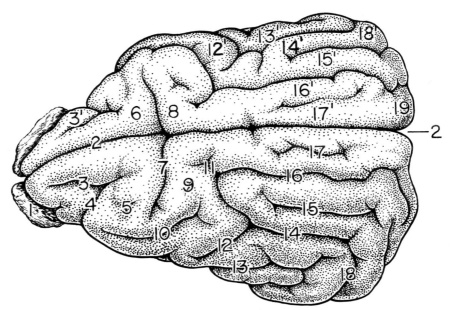

Figure 14-7.   Sulci and gyri of the cerebrum (left dorsolateral view).

1. Olfactory bulb
2. Longitudinal fissure
3. Prorean sulcus
3'. Prorean gyrus
4. Presylvian sulcus
5. Precruciate sulcus
6. Precruciate gyrus
7. Cruciate sulcus
8. Postcruciate gyrus

9. Postcruciate sulcus
10. Coronal sulcus
11. Ansate sulcus
12. Rostral suprasylvian sulcus
12'. Rostral suprasylvian gyrus
13. Ectosylvian sulcus
13'. Ectosylvian gyrus
14. Middle suprasylvian sulcus
14'. Middle suprasylvian gyrus

15. Ectomarginal sulcus
15'. Ectomarginal gyrus
16. Marginal sulcus
16'. Marginal gyrus
17. Endomarginal sulcus
17'. Endomarginal gyrus
18. Caudal suprasylvian sulcus
18'. Caudal suprasylvian gyrus
19. Occipital gyrus

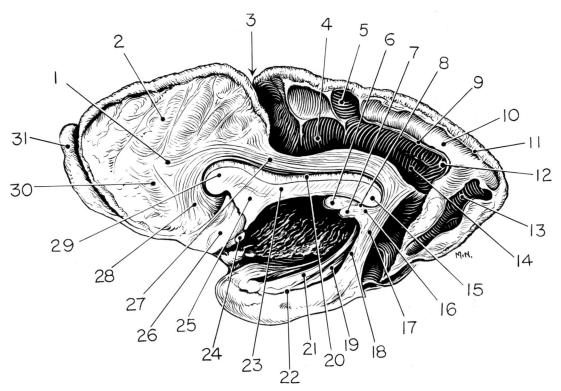

Figure 14–8. Medial view of a decorticated right cerebral hemisphere.

1. Rostral fibers of cingulum
2. Dorsal part of genual sulcus
3. Cruciate sulcus
4. Rostral part of splenial sulcus
5. Lesser cruciate sulcus
6. Tubercle of dentate gyrus
7. Callosal gyrus
8. Subsplenial flexure of dentate gyrus
9. Splenial gyrus
10. Rostral part of suprasplenial sulcus
11. Occipital gyrus
12. Caudal horizontal ramus of splenial sulcus
13. Caudal part of suprasplenial sulcus
14. Caudal part of splenial sulcus
15. Splenium of corpus callosum
16. Fibers of cingulum to callosal gyrus
17. Fibers of cingulum to parahippocampal gyrus
18. Parahippocampal gyrus
19. Dentate gyrus
20. Callosal sulcus
21. Fimbria of hippocampus
22. Hippocampal sulcus
23. Body of corpus callosum
24. Rostral commissure
25. Septum telencephali
26. Paraterminal gyrus
27. Cingulum
28. Fibers of cingulum to paraterminal gyrus
29. Genu of corpus callosum
30. Ventral part of genual sulcus
31. Olfactory bulb

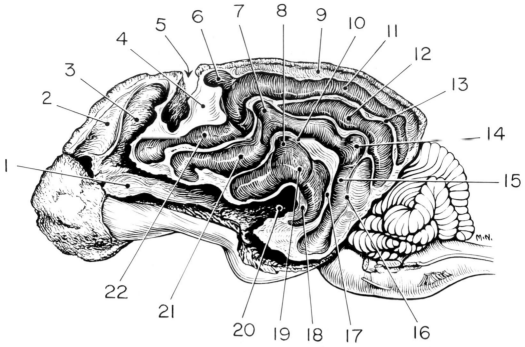

Figure 14–9. Lateral view of a decorticated left cerebral hemisphere.

1. Prorean gyrus
2. Prorean sulcus
3. Presylvian sulcus
4. Postcruciate sulcus
5. Cruciate sulcus
6. Ansate sulcus
7. Rostrodorsal branch of middle suprasylvian sulcus

8. Dorsal branch of middle ectosylvian sulcus
9. Endomarginal sulcus
10. Middle ectosylvian gyrus
11. Marginal sulcus
12. Ectomarginal sulcus
13. Ectomarginal gyrus
14. Caudodorsal branch of middle suprasylvian sulcus

15. Caudal suprasylvian sulcus
16. Caudal suprasylvian gyrus
17. Caudal ectosylvian gyrus
18. Caudal sylvian gyrus
19. Middle ectosylvian sulcus
20. Pseudosylvian fissure
21. Rostral suprasylvian sulcus
22. Coronal sulcus

the brain and extends along the entire length of the cerebrum. Together with the **medial rhinal sulcus** *(sulcus rhinalis medialis)* on the medial side of the **olfactory peduncle** *(pedunculus olfactorius)*, it separates the rhinencephalon at the base of the brain from the phylogenetically more recent neopallium. At about its middle, the lateral rhinal sulcus has a sharp flexure and is joined obliquely by another constant furrow, the relatively short **pseudosylvian fissure** *(fissura pseudosylvia)*. This junction divides the lateral rhinal sulcus into a **rostral part** *(pars rostralis)* and a **caudal part** *(pars caudalis)*. The caudal part of the lateral rhinal sulcus may bifurcate caudally and form a lateral and medial limb. From about the middle of the rostral part of the lateral rhinal sulcus the **presylvian sulcus** *(sulcus presylvius)* extends rostrodorsally toward the midline. Its dorsal end may unite with the inconstant **prorean sulcus** *(sulcus*

proreus), which, if present, appears as a sagittal sulcus and parallels the rostrodorsal margin of the frontal lobe. The **olfactory sulcus** *(sulcus olfactorius)* is the rostral continuation of the rostral part of the lateral rhinal sulcus. It runs parallel to the presylvian sulcus and is partly hidden by the **olfactory bulb** *(bulbus olfactorius)*, which belongs to the rhinencephalon.

From its junction with the lateral rhinal sulcus the pseudosylvian fissure extends for a short distance in a caudodorsal direction and is surrounded by three constant, concentric, and almost semicircular grooves. The first and second, the ectosylvian and the suprasylvian sulci, have three parts each: the **rostral, middle,** and **caudal ectosylvian** and **suprasylvian sulci** *(sulcus ectosylvius rostralis, medius, caudalis et sulcus suprasylvius rostralis, medius, caudalis)*, respectively. The third major sulcal semicircle consists of the **coronal sulcus** *(sulcus* *coronalis)* rostrally and the **marginal sulcus** *(sulcus marginalis)* caudally. In most specimens a groove is found between the marginal sulcus and the middle and caudal suprasylvian sulci. Since this groove is peripheral to the marginal sulcus, it is appropriately named the **ectomarginal sulcus** *(sulcus ectomarginalis)*. Less frequently, a depression referred to as the **endomarginal sulcus** *(sulcus endomarginalis)* is seen medial to the marginal sulcus. The **ansate sulcus** *(sulcus ansatus)* is a short dorsomedial branch of the rostral part of the marginal sulcus. Similar branches, without being specifically named, may be present along the ectosylvian and suprasylvian sulci (Fig. 14–9).

The most conspicuous sulcus on the dorsal aspect of the cerebral hemisphere is the **cruciate sulcus** *(sulcus cruciatus)*. It is very deep and runs more or less transversely. It meets the cruciate sulcus of the other side at the midline, and in this way a crosslike design is formed by the intersecting of the longitudinal fissure. Rostrally and caudally, **precruciate** and **postcruciate sulci** *(sulcus precruciatus, sulcus postcruciatus)* may be found. The postcruciate sulcus is present more frequently than is the precruciate sulcus.

On the medial surface, the *corpus callosum,* a large fiber connection between the two hemispheres, is bounded rostrally, dorsally, and caudally by the **callosal sulcus** *(sulcus corporis callosi)*. At the caudal end of the corpus callosum, the callosal sulcus blends with the **hippocampal sulcus** *(sulcus hippocampi)*, which runs ventrorostrally to the temporal lobe. In doing so, it describes a concave line and curves around the brain stem. Peripherally, about halfway between the corpus callosum and the margin of the medial side of the cerebral hemisphere, the **genual** and **splenial sulci** *(sulcus genualis, sulcus splenialis)* together describe an elongated semicircle. Rostrally, the genual sulcus is variable, but in general it parallels the course of the callosal sulcus around the rostral end or **genu of the corpus callosum** *(genu corporis callosi)*. The genual sulcus is usually accompanied peripherally by the **ectogenual sulcus** *(sulcus ectogenualis)*. Caudally, the splenial sulcus curves around the caudal end or **splenium of the corpus callosum** *(splenium corporis callosi)*. The

lesser cruciate sulcus *(sulcus cruciatus minor)* is given off from the dorsal side of the rostral part of the splenial sulcus and reaches the dorsal margin of the brain at the level of the ansate sulcus. In a substantial number of brains, the splenial sulcus gives off a **caudal horizontal ramus** which, according to Ariëns Kappers, Huber, and Crosby (1936), appears to be the forerunner of the posterior calcarine fissure of the human. (See also Cohn and Papez 1933.) The ventral end of the splenial sulcus may connect with the medial limb of the caudal part of the lateral rhinal sulcus. The **suprasplenial sulcus** *(sulcus suprasplenialis)* extends between the splenial sulcus and the dorsocaudal border of the medial surface of the hemisphere. Ventrocaudal to the ventral end of the splenial sulcus, the inconstant **occipitotemporal sulcus** *(sulcus occipitotemporalis)* may be present. Ventrally, it may join the lateral limb of the caudal part of the lateral rhinal sulcus. Dorsally, it may connect with the marginal or ectomarginal sulci.

The gyri of the cerebral cortex are separated from each other by the various sulci and with few exceptions are named for adjacent sulci. The rostral pole of the cerebrum is occupied by the **prorean gyrus** *(gyrus proreus)*. On the medial side it extends to the ectogenual sulcus. Laterally, the presylvian sulcus forms its caudal boundary. In the caudal three-fourths of the lateral aspect of the cerebrum, four major concentric arcuate convolutions are prominent. The most central one is called the sylvian gyrus *(gyrus sylvius)*. It is followed in a peripheral direction by the ectosylvian, the suprasylvian, and the marginal gyri *(gyrus ectosylvius, gyrus suprasylvius, gyrus marginalis)*, respectively.

The **sylvian gyrus** curves around the pseudosylvian fissure and may be divided into a **rostral** and a **caudal sylvian gyrus** *(gyrus sylvius rostralis, gyrus sylvius caudalis)*. These two parts cover the triangularly shaped **insular region** *(regio insularis)*, which consists of cerebral cortex in the depth of the ventral portion of the pseudosylvian fissure. The peripheral boundary of the sylvian gyrus is formed by the ectosylvian sulcus.

The **ectosylvian gyrus** lies between the ectosylvian and the suprasylvian sulci and is divided into the **rostral, middle,** and **caudal ectosylvian gyri** *(gyrus ectosylvius rostralis, medius, et caudalis).*

The third convolution, the **suprasylvian gyrus,** arches around the suprasylvian sulcus. Its rostral portion, the **rostral suprasylvian gyrus** *(gyrus suprasylvius rostralis)* lies on the rostrodorsal side of the rostral suprasylvian sulcus and caudolateral to the coronal sulcus. The **middle suprasylvian gyrus** *(gyrus suprasylvius medius)* extends between the middle suprasylvian sulcus and the rostral part of the marginal sulcus. The caudal segment of the third convolution, the **caudal suprasylvian gyrus** *(gyrus suprasylvius caudalis),* lies between the caudal suprasylvian sulcus and the ectomarginal sulcus. On the medial and caudal side of the ectomarginal sulcus, the **ectomarginal gyrus** *(gyrus ectomarginalis)* parallels the marginal sulcus, which forms its medial and caudal boundaries.

The fourth arcuate convolution, the **marginal gyrus,** extends along the dorsal and caudal margin of the cerebral hemisphere. It forms a keystone between the marginal sulcus laterally and the splenial sulcus on the medial surface of the cerebrum. If the endomarginal sulcus is present, it isolates the **endomarginal gyrus** *(gyrus endomarginalis)* from the medial side of the marginal gyrus. Rostral to the marginal gyrus, on the dorsal aspect, the cruciate sulcus separates the **precruciate gyrus** *(gyrus precruciatus)* from the **postcruciate gyrus** *(gyrus postcruciatus).*

On the medial surface, the caudodorsal part is occupied by the medial extension of the marginal gyrus. The suprasplenial sulcus divides it into dorsal and ventral parts. The **splenial gyrus** *(gyrus splenialis)* extends through the ventral area between the splenial and suprasplenial sulci. The dorsal part along the occipital pole is referred to as the **occipital gyrus** *(gyrus occipitalis).* In the frontal lobe the **genual gyrus** *(gyrus genualis)* lies between the genual sulcus and the ectogenual sulcus, which in turn, as previously mentioned, bounds the caudal margin of the prorean gyrus.

The cortical area ventral to the genu of the corpus callosum is the **paraterminal gyrus** *(gyrus paraterminalis).* It is bounded caudally by the *lamina terminalis,* which forms the rostral wall of the **third ventricle** *(ventriculus tertius).* Ventrally, the paraterminal gyrus reaches the basal border of the cerebrum. It forms part of the **subcallosal area** *(area subcallosa),* which belongs to the septal part of the rhinencephalon. The dorsal end of the paraterminal gyrus is in contact with the **cingulate gyrus** *(gyrus cinguli).* The cingulate gyrus surrounds the corpus callosum rostrally, dorsally, and caudally. It lies between the callosal sulcus and the genual and splenial sulci. The caudal end of the cingulate gyrus blends with the **parahippocampal gyrus** *(gyrus parahippocampalis).* The parahippocampal gyrus runs ventrorostrally and slightly laterad from the splenial region and lies between the medial limb of the caudal part of the lateral rhinal sulcus and the hippocampal sulcus. Ventrorostrally, it blends with the **piriform lobe** *(lobus piriformis),* which will be described with the rhinencephalon. At its dorsal end, the parahippocampal gyrus extends rostrally for a short distance on the ventral side of the splenium of the corpus callosum. This part has been referred to as the **callosal gyrus** *(gyrus callosus)* (Dexler 1932, Seiferle 1957).

The **dentate gyrus** *(gyrus dentatus)* is partly involuted in the depth of the hippocampal sulcus and extends from the temporal lobe to the splenium of the corpus callosum. Its free border is visible along the concave lateral side of the hippocampal sulcus. Ventral to the splenium it runs along the lateral side of the callosal gyrus, rostral to which it becomes prominent as the **tubercle of the dentate gyrus** *(tuberculum gyri dentati).* From the tubercle it continues mediocaudad as the **subsplenial flexure of the dentate gyrus** *(flexura subsplenialis gyri dentati)* and blends with the *gyrus fasciolaris,* a transitional area between the dentate gyrus and the *indusium griseum.* The indusium griseum consists of a thin layer of gray matter hidden in the callosal sulcus. It encircles the corpus callosum and reaches the paraterminal gyrus in the subcallosal area (Fig. 14–12).

At birth, the sulcal and gyral pattern is simple, as is to be expected in an altricial species. The lateral rhinal, suprasylvian, coronal, marginal, presylvian, cruciate, and splenial sulci usually can be recognized. The pseudosylvian fissure is an open triangular groove in the depth of which the insular region may be seen. The time of appearance of the sulci is variable, but in general all major sulci and gyri are represented at the end of the second week post partum, and after the first month the sulcal and gyral pattern is relatively well differentiated.

The **white matter** of the cerebrum basically consists of two types of fiber systems: the corticocortical fibers and the projection fibers. The **corticocortical fibers** have both their origins and their terminations in the cerebral cortex. They are subdivided into association fibers and commissural fibers. The **association fibers** connect different cortical areas in the same cerebral hemisphere. The **commissural fibers** extend between homologous areas on opposite sides of the cerebrum. The **projection fibers** either originate or terminate in the cerebral cortex. Their respective other ends are found within lower centers of the basal nuclei, the brain stem, or the spinal cord.

The **association fibers** (Figs. 14–8 through 14–11) are differentiated into short association fibers, or intralobar fibers, and long association fibers, or interlobar fibers. The **short association fibers** connect adjacent and close gyri by curving centrally around one or several intervening sulci. In doing so, they form U-shaped loops referred to as **arcuate fibers** (*fibrae arcuatae cerebri*). In addition to the fibers which unite different gyri, short association fibers can be found to connect parts within certain gyri. These fibers may be called intragyral fibers (Meyer 1957). The **long association fibers** form more or less distinct bundles which connect more distant cortical areas of the cerebrum. Among the long association fiber systems, the *cingulum* is the only long fiber bundle whose existence and associative function have never been questioned. It is the most distinct structure on the medial surface of the decorticated hemisphere (Fig. 14–8). Its fibers extend from the genual and parater-minal gyri through the cingulate gyrus to the callosal and parahippocampal gyri. Degeneration studies have shown that some of its fibers extend only partway along the bundle, while others extend along the entire length.

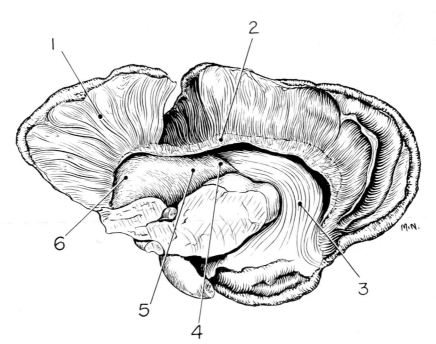

Figure 14–10. Medial view of a deep dissection of the right cerebral hemisphere.

1. Corona radiata
2. Dorsal wall of lateral ventricle
3. Subcallosal bundle
4. Tail of caudate nucleus
5. Body of caudate nucleus
6. Head of caudate nucleus

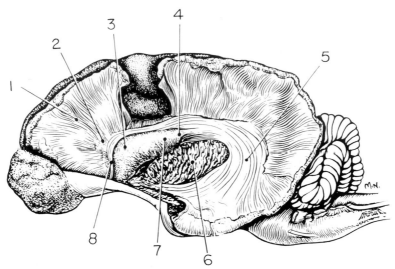

Figure 14–11. Lateral view of a deep dissection of the left cerebral hemisphere.

1. Corona radiata
2. Subcallosal bundle (rostrodorsal part)
3. Head of caudate nucleus (lateral side)
4. Tail of caudate nucleus (lateral side)

5. Subcallosal bundle (caudoventral part)
6. Cut fibers of internal capsule
7. Body of caudate nucleus
8. Subcallosal bundle (rostroventral part)

The **subcallosal bundle** *(fasciculus sub-callosus)* (Muratoff 1893a, b) can be traced as a distinct entity along the peripheral wall of the lateral ventricle. Rostrally, it lies dorsal to the caudate nucleus. The connections of the subcallosal bundle are not exclusively between cortical areas along the course of the bundle. Fibers are also received from the caudate nucleus, which would indicate that it is not a pure association system but a mixed fiber bundle (Figs. 14–10, 14–11).

The other long association bundles which are referred to in the literature—the superior longitudinal bundle *(fasciculus longitudinalis superior)*, the uncinate bundle *(fasciculus uncinatus)*, the inferior occipitofrontal bundle *(fasciculus occipito-frontalis inferior)*, and the inferior longitudinal bundle *(fasciculus longitudinalis inferior)* — are of questionable nature in the dog. A group of fibers dorsomedial to the insular region may be interpreted as the **superior longitudinal bundle.** This fiber group extends along the dorsal edge of the *claustrum*, which belongs to the basal nuclei. The **uncinate bundle** may be represented by a group of fibers at the junction of the frontal and temporal lobes. There is no proof, however, that they actually form corticocortical connections. No trace of the in-

ferior occipitofrontal bundle, which, according to Ariëns Kappers, Huber, and Crosby (1936), can be demonstrated in the dog by gross dissection, nor of the inferior longitudinal bundle, could be found by Meyer (1957).

The **commissural fibers** of the cerebrum form the corpus callosum, the rostral commissure *(commissura rostralis)*, and the dorsal and ventral commissures of the fornix *(commissura fornicis dorsalis, commissura fornicis ventralis)*. The *corpus callosum* (Figs. 14–3, 14–8, 14–12, 14–16) is the largest commissure of the brain. It connects homologous neopallial areas of the two sides. Its rostral portion is referred to as the **genu of the corpus callosum** *(genu corporis callosi)*. The **rostrum** *(rostrum corporis callosi)* is the smallest part of the corpus callosum. It tapers from the ventral side of the genu. The middle is called the **body** *(truncus corporis callosi)*. The caudal part is the **splenium** *(splenium corporis callosi)*. The corpus callosum is separated from the cingulate gyrus by the callosal sulcus. In the center, where all the fibers meet, the corpus callosum forms the roof of the lateral ventricles. On the way laterad, the fibers of the corpus callosum interdigitate with the fibers of the **internal capsule** *(capsula interna)*, which

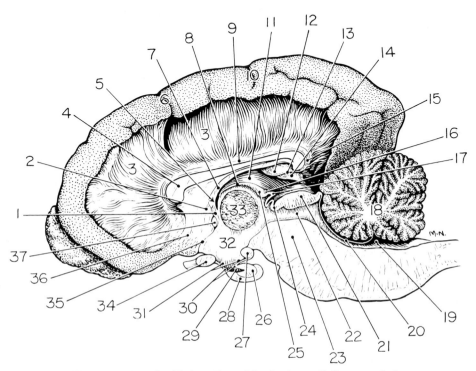

Figure 14–12. Sagittal section of the brain, partially excavated.

1. Postcommissural fornix
2. Precommissural fornix
3. Radiation of corpus callosum
4. Genu of corpus callosum
5. Interventricular foramen
6. Cruciate sulcus
7. Fornix
8. Stria habenularis thalami
9. Indusium griseum
10. Lesser cruciate sulcus
11. Dorsal aspect of thalamus
12. Tubercle of dentate gyrus

13. Subsplenial flexure of dentate gyrus
14. Callosal gyrus
15. Gyrus fasciolaris
16. Habenular commissure
17. Pineal body
18. Arbor vitae of cerebellum
19. Fourth ventricle
20. Rostral medullary velum
21. Tectum of mesencephalon
22. Mesencephalic aqueduct
23. Tegmentum of mesencephalon
24. Caudal commissure
25. Habenular nucleus

26. Neurohypophysis
27. Mamillary body
28. Adenohypophysis
29. Infundibular recess or neurohypophyseal recess of third ventricle
30. Infundibulum
31. Tuber cinereum
32. Third ventricle
33. Interthalamic adhesion
34. Optic chiasm
35. Lamina terminalis
36. Paraterminal gyrus
37. Rostral commissure

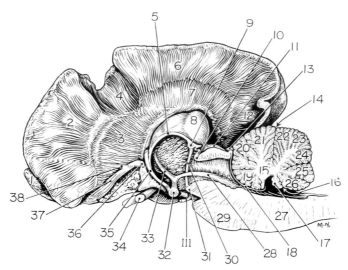

Figure 14-13. Medial view of hemisected brain with the internal capsule and diencephalon dissected.

1. Olfactory bulb
2. Corona radiata of frontal lobe
3. Rostral crus of internal capsule
4. Cruciate sulcus
5. Stria habenularis thalami
6. Corona radiata of parietal lobe
7. Fibers in corona radiata derived from caudal crus of internal capsule
8. Dorsal aspect of thalamus
9. Habenular nucleus
10. Caudal commissure
11. Rostral colliculus
12. Corona radiata of occipital lobe
13. Caudal colliculus
14. Fissura prima
15. Arbor vitae cerebelli
16. Caudolateral fissure = uvulonodular fissure
17. Nodulus
18. Rostral medullary velum
19. Lingula
20. Lobulus centralis
21. Culmen
22. Declive
23. Folium vermis
24. Tuber vermis
25. Pyramis vermis
26. Uvula vermis
27. Medulla oblongata
28. Mamillotegmental tract
29. Pons
30. Fasciculus retroflexus
31. Location of intercrural nucleus
32. Mamillary body
33. Mamillothalamic tract
34. Optic chiasm
35. Column of fornix
36. Rostral commissure (at median plane)
37. Rostral part of rostral commissure
38. Caudal part of rostral commissure
III. Oculomotor nerve

will be described with the projection fibers.

The **rostral commissure** forms connections between the olfactory areas of the cerebrum (Figs. 14-13, 14-14; Plate 10). On median sections (Figs. 14-3, 14-12) it may be seen caudal to the paraterminal gyrus and rostral to the diencephalon, about halfway between the corpus callosum and the base of the brain. It consists of a strong rostral part *(pars rostralis)* and a considerably smaller caudal part *(pars caudalis).* The **rostral part** connects the two olfactory bulbs, which are located in the rostroventral portion of the rhinencephalon. The fibers of the rostral part of the rostral commissure are embedded in the ventromedial portion of the **caudate nucleus** *(nucleus caudatus).* As these fibers run rostrad to the olfactory bulb, they travel ventral to the internal capsule. The **caudal part** of the rostral commissure extends between the piriform lobes (Fig. 14-5), which are olfactory cortical regions at the temporal poles of the cerebral hemispheres. From the midline, the fibers of the

caudal part of the rostral commissure run slightly rostrad. Then they curve laterad and become intermingled with the fibers of the internal capsule. After leaving the area of the internal capsule, they turn caudolaterad and ventrad to reach the piriform lobe.

The **dorsal commissure of the fornix** unites the hippocampi of the two sides and is located ventral to the splenium of the corpus callosum (Fig. 14-14). The **ventral commissure of the fornix** is related closely to the septal nuclei *(nuclei septi).* The hippocampus and the septal nuclei will be described with the rhinencephalon.

The **projection fibers** of the neocortical areas form the internal capsule *(capsula interna).* The fornix is a phylogenetically older projection system associated with the rhinencephalon. The **internal capsule** consists of a **rostral crus** *(crus rostrale capsulae internae),* a **caudal crus** *(crus caudale capsulae internae),* and the **genu** *(genu capsulae internae)* where the two crura meet (Figs. 14-13, 14-15; Plates 1, 10). The ros-

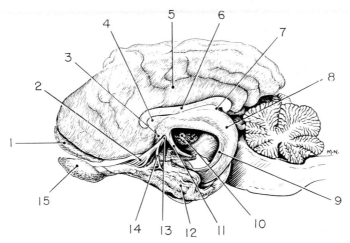

Figure 14–14.   Lateral view of the brain showing rhinencephalic structures.

1. Right olfactory bulb
2. Rostral part of rostral commissure
3. Precommissural fornix
4. Telencephalic septum
5. Medial surface of right cerebral hemisphere

6. Corpus callosum
7. Dorsal commissure of fornix
8. Alveus of hippocampus
9. Fimbria of hippocampus
10. Interthalamic adhesion

11. Column of fornix
12. Piriform lobe (from dorsal side)
13. Rostral commissure
14. Caudal part of rostral commissure
15. Left olfactory bulb

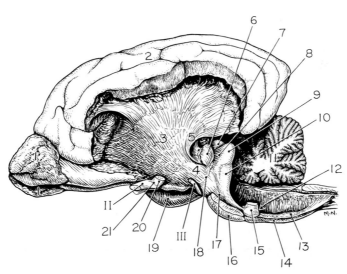

Figure 14–15.   Lateral view of the brain dissected to show projection pathways.

1. Olfactory bulb
2. Left cerebral hemisphere
3. Internal capsule (lateral view)
4. Crus cerebri
5. Acoustic radiation
6. Medial geniculate body
7. Rostral colliculus
8. Brachium of caudal colliculus

9. Caudal colliculus
10. Lateral lemniscus
11. Cerebellum
12. Location of dorsal nucleus
   of trapezoid body
13. Location of olivary nucleus
14. Pyramid
15. Trapezoid body

16. Transverse fibers of pons
17. Longitudinal fibers of pons
18. Transverse crural tract
19. Piriform lobe
20. Optic tract (cut to show internal capsule)
21. Optic chiasm
II. Optic nerve
III. Oculomotor nerve

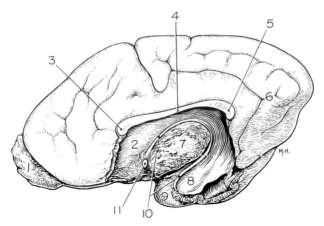

Figure 14–16. Medial view of the right cerebral hemisphere with the hippocampus and parateminal gyrus removed.

1. Olfactory bulb
2. Caudate nucleus
3. Genu of corpus callosum
4. Body of corpus callosum
5. Splenium of corpus callosum
6. Caudal horizontal ramus of splenial sulcus

7. Cut surface between cerebrum and brain stem
8. Amygdaloid body
9. Piriform lobe
10. Stria terminalis
11. Rostral commissure

tral crus passes through the corpus striatum. The caudal crus extends along the lateral aspect of the diencephalon. The fibers of the internal capsule either originate in the cerebral cortex and descend to lower centers or take origin in lower centers and carry impulses to the cerebral cortex. At the dorsolateral edge of the lateral ventricle, as previously indicated, they are intersected by the fibers of the corpus callosum. The resulting intermingled mass of white matter, the *corona radiata* (Fig. 14–13), has the appearance of a crown of rays in which it is difficult to determine the origin of individual fibers and fiber groups. The *fornix* connects the hippocampus with the septal part of the rhinencephalon and the ventral part of the diencephalon. It will be described with the structures of the rhinencephalon and the diencephalon.

The **basal nuclei** (Plate 1) consist of three nuclear masses: the corpus striatum, the claustrum, and the amygdaloid body *(corpus amygdaloideum)*. The *corpus striatum* is the largest nuclear mass among the basal nuclei. It is subdivided by the rostral crus of the internal capsule into a medial portion, the **caudate nucleus** *(nucleus caudatus)*, and a lateral portion, the **lentiform**

nucleus *(nucleus lentiformis)*. The lentiform nucleus in turn is further subdivided by the **medullary lamina** *(lamina medullaris)* into the more medially located *globus pallidus* and the more lateral *putamen*.

The caudate nucleus protrudes into the lateral ventricle from the lateral side (Figs. 14–10, 14–16). Its large rostral portion is called the **head of the caudate nucleus** *(caput nuclei caudati)*. The caudally tapering part forms the **tail** *(cauda nuclei caudati)*. The middle portion is referred to as the **body of the caudate nucleus** *(corpus nuclei caudati)*. Rostrally, across the fibers of the internal capsule, the gray matter of the caudate nucleus and of the putamen is continuous. The ventromedial portion of the head of the caudate nucleus which forms the *nucleus accumbens* is pierced ventrally by the rostral part of the rostral commissure. The tail of the caudate nucleus dwindles ventral to the beginning of the caudal fourth of the corpus callosum and becomes difficult to identify grossly.

The *globus pallidus*, the smaller medial portion of the lentiform nucleus, lies lateral to the genu of the internal capsule. Laterally, it is separated from the putamen by the medullary lamina. The *putamen* (Fig. 14–

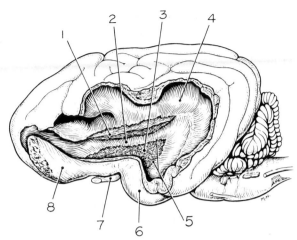

Figure 14–17. Lateral view of brain dissected to show claustrum and putamen.

1. Claustrum (ventral part removed to show external capsule)
2. External capsule
3. Putamen
4. Corona radiata
5. Cut edge of external capsule, permitting putamen to be seen
6. Piriform lobe
7. Optic chiasm
8. Olfactory bulb

17) extends beyond the lateral aspect of the globus pallidus and is in contact with the rostral and caudal crura of the internal capsule. Laterally, the lentiform nucleus is bounded by a thin layer of white matter, the **external capsule** (*capsula externa*). The *claustrum* (Fig. 14–17) is located lateral to the external capsule and extends along the lateral margin of the putamen. The very thin layer of white matter that separates the claustrum from the cortex is referred to as the **extreme capsule** (*capsula extrema*).

For morphological reasons, the **amygdaloid body** (Fig. 14–16) is customarily listed with the basal nuclei. On the basis of its connections, it may be included among the structures of the rhinencephalon. It consists of a number of nuclear masses which occupy the lateral wall and the rostral extremity of the **temporal horn** (*cornu temporale*) **of the lateral ventricle.** Ventrally, the gray matter of the amygdaloid body continues into the cerebral cortex of the piriform lobe at the temporal pole of the cerebrum.

The **rhinencephalon,** or olfactory part of the brain, is separated from the remaining telencephalon by the lateral and medial rhinal sulci. It is subdivided into a basal part (*pars basalis rhinencephali*), a septal part (*pars septalis rhinencephali*), and a limbic part (*pars limbica rhinencephali*).

The **basal part of the rhinencephalon** consists of the olfactory bulb (*bulbus olfactorius*), the olfactory peduncle (*pedunculus olfactorius*), the lateral and medial olfactory tracts (*tractus olfactorius lateralis, tractus olfactorius medialis*), the lateral olfactory gyrus (*gyrus olfactorius lateralis*), the rostral perforated substance (*substantia perforata rostralis*), the piriform lobe (*lobus piriformis*), the parahippocampal gyrus (*gyrus parahippocampalis*), and the amygdaloid body (*corpus amygdaloideum*).

The **olfactory bulbs** are rounded ventrorostral projections (Fig. 14–4) which receive special visceral afferent fibers for the sense of smell through the openings of the lamina cribrosa of the ethmoid bone. These fibers constitute the first cranial nerves, or **olfactory nerves** *(nervi olfactorii).*

The **olfactory peduncle** continues the olfactory bulb caudally (Fig. 14–5). After a short distance, it bifurcates to form the lateral and the medial olfactory tracts. The fibers of the **lateral olfactory tract** enter the piriform lobe. The olfactory peduncle and the lateral olfactory tract are accompanied by gray matter, the **lateral olfactory gyrus,** which lies between the rostral part of the lateral rhinal sulcus dorsally and the olfactory peduncle and the lateral olfactory tract ventromedially. Caudally, the lateral olfactory gyrus blends with the dorsolateral part of the piriform lobe. The fibers of the **medial olfactory tract** contribute to the rostral part of the rostral commissure and connect with the septal nuclei (nuclei septi) of the septal part of the rhinencephalon.

The **rostral perforated substance** (Fig. 14–5) is a more or less triangular area bounded by the lateral and the medial olfactory tracts, the piriform lobe, and the optic tract (tractus opticus), which will be described with the diencephalon. The rostral perforated substance is named appropriately because a large number of small blood vessels pierce the area to reach the basal nuclei. It receives fiber connections from both the medial and the lateral olfactory tracts. Along the caudal border of the rostral perforated substance, a band of lighter color, the **diagonal gyrus** *(gyrus diagonalis),* extends to the medial side, where it continues into the paraterminal gyrus in the subcallosal area. The remaining part of the rostral perforated substance is known as the **olfactory trigone** *(trigonum olfactorium).* The rostral portion of the olfactory trigone may be prominent and is customarily referred to as the **olfactory tubercle** *(tuberculum olfactorium).*

The **piriform lobe** is a conspicuous pear-shaped cortical region ventral to the middle of the lateral rhinal sulcus at the temporal pole of the cerebrum (Fig. 14–6). It belongs to the paleopallium. Rostrally, its dorsolateral portion unites with the lateral olfactory gyrus. Caudally, it joins the **parahippocampal gyrus,** which in turn connects with the cingulate gyrus dorsocaudally. Dorsally, the gray matter of the piriform lobe blends with the **amygdaloid body,** which was previously referred to in the description of the basal nuclei. Several nuclei in the amygdaloid body receive fibers from the olfactory bulbs and the rostral commissure. There are also connections with the septal part of the rhinencephalon by way of the diagonal gyrus. The main efferent pathway of the amygdaloid body is a clearly defined fiber bundle, the *stria terminalis* (Fig. 14–16), which arises from the rostromedial part of the amygdaloid body. It runs dorsad and caudad and follows the optic tract to the tail of the caudate nucleus. From there, the stria terminalis passes rostrad and ventrad in the **thalamocaudate sulcus** *(sulcus thalamocaudatus),* a groove between the caudate nucleus and the thalamus, and ends at the base of the brain in the preoptic and hypothalamic region.

The **septal part of the rhinencephalon** consists of the subcallosal area *(area subcallosa),* the septal nuclei *(nuclei septi),* and the telencephalic septum *(septum telencephali).* It is bounded by the genu of the corpus callosum, the rostral commissure, the lamina terminalis, and the olfactory tubercle.

The **subcallosal area** occupies the medial part of the cerebrum ventral to the corpus callosum and rostral to the rostral commissure. The **paraterminal gyrus,** which was described with the convolutions of the medial cerebral surface, forms part of the subcallosal area and lies immediately rostral to the lamina terminalis. The *nervus terminalis* enters the brain in this region (McCotter 1913).

The **septal nuclei** protrude into the lateral ventricle from its medial wall. They are in close spatial relation with the subcallosal area and have fiber connections with the hippocampus. These fibers will be described with the fornix. The **ventral commissure of the fornix,** as previously mentioned, is associated with the septal nuclei.

The **telencephalic septum** is a thin, small membrane which is bounded by the concave side of the genu of the corpus callosum, the fornix, and the septal nuclei. In some species the septum consists of a double layer with a cavity, the *cavum septi telencephali,* between. In the dog the cavum septi telencephali is rare or absent (Elliot Smith 1895–96; Ziehen 1906).

The **limbic part of the rhinencephalon** includes the hippocampus and the fornix. The *hippocampus* (Fig. 14–14; Plate 10) is a long, curved structure which projects into the temporal horn of the lateral ventricle. It consists of the cortical region which represents the *archipallium*, the phylogenetically oldest part of the cerebral cortex. It is involuted along the hippocampal sulcus and contacts the lateral wall of the dentate gyrus. The ventricular surface of the hippocampus is covered with a thin layer of white matter, the *alveus (alveus hippocampi)*, which consists of the efferent and commissural fibers of the hippocampus. The commissural fibers cross the midline ventral to the splenium of the corpus callosum. They constitute the **dorsal commissure of the fornix** (Fig. 14–14). The remaining major part of the fibers of the alveus gather along the concave edge of the hippocampus and form the **fimbria of the hippocampus** *(fimbria hippocampi)*. The fornix continues the fimbria at the rostrodorsal end of the hippocampus and connects it with the septal part of the rhinencephalon and with the mamillary body *(corpus mamillare)* in the hypothalamus (Fig. 14–13). This efferent hippocampal fiber system parallels the course of the stria terminalis, from which it is separated by the lateral ventricle. The part of the fornix rostral to the hippocampus is referred to as the **body of the fornix** *(corpus fornicis)*. It contributes the major portion of the longitudinal fibers along the ventral aspect of the body of the corpus callosum. On its ventrorostral course, the fornix forms on each side a roundish bundle, the **column of the fornix** *(columna fornicis)*, which becomes isolated and bypasses the rostral commissure caudally (Fig. 14–6). This portion of the fornix may be referred to as the postcommissural fornix. Under cover of the gray matter of the ventral part of the diencephalon, the column of the fornix reaches the mamillary body in the hypothalamus of the respective side. The remaining fibers of the fornix enter the septal part of the rhinencephalon, partly bypassing the rostral commissure rostrally, as the precommissural fornix (Fig. 14–12). Longitudinal fibers ventral to the body of the corpus callosum which are not included in the fornix carry impulses from the septal nuclei to the hippocampus and are called septal fibers.

On the ventral side of the corpus callosum each hemisphere contains a **lateral ventricle** *(ventriculus lateralis)* with a rostral horn *(cornu rostrale)*, a central part *(pars centralis)*, and a temporal horn *(cornu temporale)*. In the **rostral horn,** the septal nuclei and the telencephalic septum form the medial wall. The head of the caudate nucleus lies along the lateral wall. An inconstant **olfactory recess** *(recessus olfactorius)* may project into the olfactory bulb (Fizgerald 1961; Böhme 1967). In the **central part,** the fornix runs along the medial wall, and the body and the tail of the caudate nucleus extend laterally along the thalamocaudate sulcus, which is occupied by the stria terminalis. The amygdaloid body and the hippocampus and its fimbria protrude into the **temporal horn.** The **choroid plexus of the lateral ventricle** *(plexus choroideus ventriculi lateralis)* extends along the fimbria and fornix. On each side an **interventricular foramen** *(foramen interventriculare)* connects the lateral ventricle with the third ventricle *(ventriculus tertius)* in the diencephalon.

The remaining part of the brain caudal to the telencephalon, with the exception of the cerebellum, forms the **brain stem** (Fig. 14–18). It consists of nuclei and fibers of cranial nerves, fibers and associated nuclei of ascending and descending fiber tracts, and the **reticular formation** *(formatio reticularis)* of the brain stem.

DIENCEPHALON

The **diencephalon** (Plates 1, 2) is the most rostral portion of the brain stem. It lies medial to the caudal limb of the internal capsule. Along the midline it forms the lateral walls of the **third ventricle** *(ventriculus tertius)*. Rostrally, it extends to the rostral commissure and is adjacent to the caudate nucleus. Caudolaterally and dorsally, it is related to the hippocampus and the body of the fornix. Caudodorsally, its boundary with the mesencephalon is marked by the **caudal commissure** *(commissura caudalis)*. The **subcommissural organ** *(organum subcommissurale)*, a modified ependymal structure, is located ventral and rostral to the caudal commissure and protrudes into the third ventricle. The subdivisions of the diencephalon are the **epithalamus** along the dorsal midline; the **thalamus,** the largest part in the center; the **hypothalamus** at the base of the diencephalon, along the midline; and the **subthalamus** caudoventral to the thalamus and caudolateral to the hypothalamus. The second cranial nerve, or **optic nerve** (*nervus opticus*), enters the diencephalon from the ventral aspect. The nerves of both sides converge and partially decussate to form the **optic chiasm** (*chiasma opticum*), which is a landmark of the ventral surface of the diencephalon (Fig. 14–5).

The **epithalamus** is a narrow area along the dorsal midline of the diencephalon (Figs. 14–12, 14–13). It consists of the habenular stria *(stria habenularis thalami)*, the habenular nuclei *(nuclei habenulares)*, the habenular commissure *(commissura habenularum)*, and the pineal body *(corpus pineale)*. The **habenular stria** is a thin bundle of fibers which encircles the diencephalon along the dorsal midline. It connects olfactory centers at the base of the brain, mainly the rostral perforated substance, with the **habenular nuclei.** Some of its fibers from either side cross the midline and form the **habenular commissure.** The habenular nuclei are small, rounded nuclear masses at the caudal end of the epithalamus, immediately rostral to the pineal body. They serve as way stations in an olfactory reflex pathway to motor centers in the caudal brain stem. Their efferent fibers form the *fasciculus retroflexus* and connect them with the **intercrural nucleus** (*nucleus intercruralis*) in the mesencephalon, which in turn conveys impulses through the reticular formation to the motor nuclei of cranial nerves in the brain stem. The **pineal body** is very small. It projects just a few millimeters in a caudal direction from the dorsocaudal end of the epithalamus. It has some inhibitory effect on the function of the gonads.

The **thalamus** is a wedge-shaped mass of gray matter which is located ventral and lateral to the epithalamus, between the caudal limb of the internal capsule and the midline (Fig. 14–13). Along the midline, the thalami of the two sides are separated from each other by the third ventricle, except for a round area in the center of the ventricle, where the gray masses of the two sides adhere to each other. This area is called the **interthalamic adhesion** (*adhesio interthalamica*). The **third ventricle** lies in the median plane and encircles the interthalamic adhe-

sion (Fig. 14–12). It connects with each lateral ventricle in the cerebral hemispheres by ways of an **interventricular foramen** (*foramen interventriculare*) and opens caudally into the **mesencephalic aqueduct** (*aqueductus mesencephali*). The **choroid plexus of the third ventricle** (*plexus choroideus ventriculi tertii*) protrudes from the dorsal aspect of the third ventricle into its lumen. Rostrally and dorsolaterally, the thalamus is separated from the caudate nucleus by the **thalamo caudate sulcus** (*sulcus thalamocaudatus*). The stria terminalis partly encircles the thalamus along the bottom of this groove. On the concave side of the stria terminalis, the thalamus meets the internal capsule. The fibers which leave or enter the thalamus on its lateral side provide connections with the cerebral cortex by way of the internal capsule and the corona radiata (Fig. 14–13).

The subdivisions of the thalamus, the **thalamic nuclei** (*nuclei thalami*), for the most part are named for their topographical location. They consist of a number of subsidiary nuclei which perform three basic thalamic functions as relay nuclei, association nuclei, and nuclei of the diffuse projection system.

The **relay nuclei** receive sensory systems from the spinal cord, the brain stem, the cerebellum, and the eyes and ascending motor systems from the cerebellum, the brain stem, and the basal nuclei, and relay their impulses to the cerebral cortex. They occupy the ventral and caudal part of the thalamus and consist of the ventral thalamic nuclei (*nuclei ventrales thalami*) and the caudal thalamic nuclei (*nuclei caudales thalami*).

The **ventral thalamic nuclei** are subdivided into a ventral rostral nucleus (*nucleus ventralis rostralis*), a ventral lateral nucleus (*nucleus ventralis lateralis*), and ventral caudal nucleus (*nucleus ventralis caudalis*). The **ventral rostral nucleus** receives fibers from the globus pallidus of the lentiform nucleus by way of the fasciculus thalamicus. The **ventral lateral nucleus** receives fibers from the cerebellum and the red nucleus (*nucleus ruber*) in the mesencephalon by way of the rostral cerebellar peduncle (*pedunculus cerebellaris rostralis*). Both the ventral rostral and the ventral lateral nuclei relay impulses from motor systems. The ventral lateral nucleus projects primarily to the motor cortex, whereas the ventral rostral nucleus projects largely to the premotor cortex rostral to the motor area. The **ventral caudal nucleus** consists of a lateral part *(pars lateralis)* and a medial part *(pars medialis)*. The **lateral part of the ventral caudal nucleus** receives sensory impulses from the body by way of ascending fiber tracts which form or join the **medial lemniscus** *(lemniscus medialis)*. The origin of the medial lemniscus from the nucleus gracilis and cuneatus medialis will be covered in the description of the medulla oblongata. The **medial part of the ventral caudal nucleus** receives sensory impulses from the head region by way of the **trigeminal lemniscus** *(lemniscus trigeminalis)*. The efferent fibers of both the lateral and medial parts of the ventral caudal nucleus project to the sensory cortex through the somesthetic radiation.

The **caudal thalamic nuclei** consist of the lateral geniculate body *(corpus geniculatum laterale)* and the medial geniculate body *(corpus geniculatum mediale)*. The **lateral geniculate body** is located at the caudodorsolateral angle of the thalamus. It receives the fibers of the **optic tract** *(tractus opticus)*, which extends from the **optic chiasm** at the base of the diencephalon in a dorsocaudal direction. The lateral geniculate body is the thalamic relay center for visual impulses. It sends fibers by way of the **optic radiation** *(radiatio optica)* in the internal capsule and the corona radiata to the visual cortex in the occipital lobe of the cerebrum. The **medial geniculate body** lies caudoventral and somewhat medial to the lateral geniculate body and forms the caudalmost part of the diencephalon. It receives acoustic fiber connections from lower centers and serves as the thalamic relay nucleus for hearing. By way of the **acoustic radiation** *(radiatio acustica)*, it projects impulses to the acoustic cortex (Fig. 14–15), which is located in the ectosylvian gyrus (Tunturi 1948, 1970).

The **association nuclei** have exclusively diencephalic and telencephalic connections for reciprocal communications with other thalamic nuclei, the hypothalamus, and the cerebral cortex. They form the medial, lateral, and rostral thalamic nuclei *(nuclei mediales thalami, nuclei laterales thalami, nuclei rostrales thalami)* and are located dorsal to the ventral thalamic nuclei. They are separated from each other by the **internal medullary lamina** *(lamina medullaris thalami interna)*, a rostrally bifid and laterally convex sheet of white matter in the middle of each thalamus. The **medial thalamic nuclei,** which include the dorsomedial thalamic nucleus *(nucleus dorsomedialis thalami)*, are medial to the internal medullary lamina. The **lateral thalamic nuclei** lie on the lateral side of the internal medullary lamina, and the **rostral thalamic nuclei** are bounded by the two rostral prongs of the internal medullary lamina. The rostral nuclei receive fibers from the **mamillary body** in the hypothalamus (Fig. 14–13) by way of the **mamillothalamic tract** *(tractus mamillothalamicus)* and project impulses to the cingulate gyrus.

The **nuclei of the diffuse projection system** are the nuclei of the midline, or paraventricular thalamic nuclei *(nuclei paraventriculares thalami)*; the intralaminar thalamic nuclei *(nuclei intralaminares thalami)*; and the reticular nucleus of the thalamus *(nucleus reticulatus thalami)*. They convey slowly spreading impulses from the thalamus to the cerebral cortex either directly, as is the case with the reticular nucleus of the thalamus, or indirectly, by way of the thalamic association nuclei. The **paraventricular thalamic nuclei** are located along the wall of the third ventricle, medial to the medial thalamic nuclei. The **intralaminar thalamic nuclei** are associated with the internal medullary lamina of the thalamus and form part of a secondary nonspecific thalamocortical system (Crosby, Humphrey, and Lauer 1962). The **reticular nucleus of the thalamus** lies along the lateral edge of the thalamus, adjacent to the internal capsule. On its medial side, it is separated from the lateral thalamic nuclei by a sheet of white matter, the **external medullary lamina** *(lamina medullaris thalami externa)*. It belongs to the ascending reticular activating system, which is involved in arousal mechanisms of the cerebral cortex.

The **hypothalamus** lies at the base of the diencephalon on either side of the midline and is bounded externally by the rostral perforated substance, the piriform lobe, and the crura cerebri. It is divided into a **rostral hypothalamic region** *(regio hypothalamica rostralis)*, an **intermediate** or tuberal **hypothalamic region** *(regio hypothalamica intermedia)*, and a **caudal hypothalamic region** *(regio hypothalamica caudalis)*. Each of these regions is subdivided into a number of subsidiary nuclei. Major landmarks on the ventral surface (Fig. 14–5) are the optic chiasm, the tuber cinereum, and the mamillary bodies. They represent the rostral, intermediate, and caudal regions, respectively.

The **optic chiasm** belongs to the visual system, and its relation to the hypothalamus is largely topographical. It is formed by the convergence and partial decussation of the two **optic nerves,** which carry special somatic afferent impulses for vision from the retinae to the level of the optic chiasm. As a result of the partial decussation, the optic tract of each side receives through its **medial root** *(radix medialis)* fibers from the medial or nasal half of the opposite retina and through its **lateral root** *(radix lateralis)* fibers from the lateral or temporal retinal portion of the same side. The optic tract thus carries optic impulses from both sides to the lateral geniculate body, and by way of the **brachium of the rostral colliculus** *(brachium colliculi rostralis)* to the **rostral colliculus** *(colliculus rostralis)* and the **pretectal area** *(area pretectalis)* in the mesencephalon. As described with the caudal thalamic nuclei, the lateral geniculate body serves as a thalamic relay nucleus for visual impulses to the cerebral cortex. The rostral colliculus and the pretectal area function as mesencephalic optic reflex centers.

The **supraoptic nucleus** *(nucleus supraopticus)* is one of the nuclei of the rostral hypothalamic region and lies dorsal and slightly lateral to the optic chiasm.

The *tuber cinereum* is largely gray matter in the intermediate or tuberal region between the optic chiasm and the mamillary bodies. It ends ventrally in a funnel-shaped process, the *infundibulum,* by means of which the *hypophysis* is attached to the hypothalamus.

The **mamillary bodies** are two spherical eminences in the caudal hypothalamic region between the crura cerebri. They touch each other at the midline and lie rostral to the **caudal perforated substance** *(substantia perforata caudalis)* in the **intercrural fossa** *(fossa intercruralis)* of the mesencephalon. They receive impulses from the hippocampus through the fornix and convey them to the rostral thalamic nuclei as well as to the mesencephalic tegmentum by way of the **mamillothalamic** and the **mamillotegmental**

tracts *(tractus mamillothalamicus, tractus mamillotegmentalis),* respectively (Fig. 14–13).

The **subthalamus** is a transitional area between the thalamus and the midbrain and lies caudal and lateral to the hypothalamus. Laterally, it is bounded by the fibers of the internal capsule, which continue caudad into the mesencephalon as the crus cerebri. It is related to the extrapyramidal system and relays impulses from the basal nuclei to higher levels by connections with the thalamus and to lower levels by way of the midbrain. The subthalamus contains the subthalamic nucleus *(nucleus subthalamicus),* the zona incerta with its associated tegmental areas or fields of Forel, and related fiber systems like the fasciculus thalamicus, the fasciculus subthalamicus, the fasciculus lenticularis, and the ansa lenticularis.

The **subthalamic nucleus** has the shape of a biconvex lens and lies dorsomedial to the medial edge of the crus cerebri. The *zona incerta* consists of scattered cells between the subthalamic nucleus ventromedially and the thalamus dorsally. The *fasciculus thalamicus* travels along the dorsal border of the zona incerta (in field H_1 of Forel) and conveys impulses from the globus pallidus to the ventral rostral nucleus of the thalamus. The *fasciculus subthalamicus* interrelates the lentiform nucleus and the subthalamic nucleus with both efferent and afferent fibers. The subthalamic nucleus in turn has connections with the zona incerta, the substantia nigra, and the lateral part of the mesencephalic tegmentum.

The *fasciculus lenticularis* originates from the globus pallidus. Its fibers interdigitate with those of the internal capsule and reach the *zona incerta,* where some of them synapse. From the zona incerta, fibers descend to the mesencephalic tegmentum and join the **central tegmental tract** *(tractus tegmenti centralis),* which descends to the **olivary nucleus** *(nucleus olivaris)* in the myelencephalon. Most of the fibers of the fasciculus lenticularis continue medially along the dorsal side of the subthalamic nucleus (in field H_2 of Forel) and enter the prerubral nucleus (in field H of Forel) at the medial border of the zona incerta. The *ansa lenticularis* contains fibers from the lentiform nucleus; it loops around the ventral side of the internal capsule and enters the prerubral nucleus, together with the fasciculus lenticularis. After considerable, but incomplete, synapsing, both tracts descend to tegmental areas of the midbrain. The impulses from the fasciculus lenticularis reach the **interstitial nucleus** *(nucleus interstitialis)* and related tegmental nuclei of the **medial longitudinal bundle** *(fasciculus longitudinalis medialis)*, the **red nucleus** *(nucleus ruber)*, and the **deep mesencephalic nucleus** *(nucleus mesencephalicus profundus)* on the dorsolateral and lateral side of the red nucleus. Those from the ansa lenticularis contribute to the red nucleus and to more caudal portions of the lateral tegmental gray matter.

MESENCEPHALON

The **mesencephalon** (Plate 3) is the region of the brain stem between the diencephalon rostrally and the pons and the cerebellum caudally. The dorsal part of the mesencephalon is referred to as the *tectum (tectum mesencephali)*, since it provides the roof of the mesencephalon. Its rostral boundary is formed by the **caudal commissure** *(commissura caudalis)* and the **pretectal area** *(area pretectalis)*. Caudally, the tectum ends at the fourth ventricle *(ventriculus quartus)*. The ventral portion of the mesencephalon, in ventrodorsal sequence, is formed by the *crus cerebri*, the *substantia nigra*, and the *tegmentum (tegmentum mesencephali)*. The **mesencephalic aqueduct** *(aqueductus mesencephali)*, a tubelike canal which is surrounded by the **central gray matter** *(substantia grisea centralis)*, passes through the mesencephalon between the tectum and the tegmentum (Fig. 14–12). It connects the third ventricle in the diencephalon with the fourth ventricle in the metencephalon and rostral portion of the myelencephalon. The third cranial nerve, or **oculomotor nerve** *(nervus oculomotorius)*, and the fourth cranial nerve, or **trochlear nerve** *(nervus trochlearis)*, originate from the mesencephalon.

The *tectum* of the mesencephalon (Fig. 14–18) consists of the **tectal lamina** *(lamina tecti)*, which is occupied by a rostral and a caudal colliculus *(colliculus rostralis, colliculus caudalis)* on both sides. Each colliculus is connected with the brain stem by a brachium of the rostral or caudal colliculus

(brachium colliculi rostralis, brachium colliculi caudalis) respectively. The **rostral colliculi** form a pair of spherical elevations in the rostrodorsal part of the mesencephalon. Together with the pretectal area, they receive fibers from the optic tracts by way of the **brachia of the rostral colliculi** and serve as centers for visual reflexes. Within the substance of the tectum, the rostral colliculi of the two sides are connected with each other by the **commissure of the rostral colliculi** *(commissura colliculorum rostralium)*. The **caudal commissure** connects the contralateral pretectal areas and related diencephalic and mesencephalic structures. The **caudal colliculi** are bulbous protuberances in the caudolateral part of the tectum. They are somewhat smaller than the rostral colliculi and do not meet along the midline. A band of white matter, the **commissure of the caudal colliculi** *(commissura colliculorum caudalium)* forms a visible connection between the caudal colliculi immediately caudal to the rostral colliculi (Fig. 14–18). The caudal colliculi serve as centers for acoustic reflexes, and each caudal colliculus receives acoustic impulses from the ipsilateral **lateral lemniscus** *(lemniscus lateralis)*. The lateral lemniscus originates from the **dorsal nucleus of the trapezoid body** *(nucleus dorsalis corporis trapezoidei)* in the ventral metencephalon and sweeps rostrad and dorsad into the caudal colliculus (Fig. 14–15). From the lateral aspect of the caudal colliculus the **brachium of the caudal colliculus** extends rostroventrad along the ventrolateral edge of the rostral colliculus to the medial geniculate body, which relays acoustic impulses for conscious perception of sound to the cerebral cortex.

The *crus cerebri* (Fig. 14–5) is a large group of superficially located descending fibers at the base of the mesencephalon. It becomes visible as it emerges from the caudal side of the optic tract halfway between the optic chiasm and the lateral geniculate body. Rostrolaterally, it lies adjacent to the medial geniculate body (Fig. 14–3). Caudally, the crura cerebri of the two sides converge toward the midline and form the lateral boundaries of the **intercrural fossa** *(fossa intercruralis)*, which includes the **caudal perforated substance** *(substantia perforata caudalis)* and the **intercrural nucleus** *(nucleus intercruralis)*. The intercrural nucleus has previously described connections with the habenular nuclei in the epithalamus by way of the fasciculus retroflexus (Fig. 14–13) and with motor nuclei in the caudal brain stem through the reticular formation.

The crus cerebri consists of the **corticonuclear** and the **corticospinal fibers** (*fibrae corticonucleares, fibrae corticospinales*) of the **pyramidal tract** (*tractus pyramidalis*) and the **corticopontine tract** (*tractus corticopontinus*). All these fibers originate from the cerebral cortex; descend by way of the corona radiata, the internal capsule, and the crus cerebri; and continue to the pons or lower levels. The fibers of the corticopontine tract synapse with the **pontine nuclei** (*nuclei pontis*). The corticonuclear fibers are distributed bilaterally to the motor nuclei of the brain stem. The corticospinal fibers descend through the ventral metencephalon and myelencephalon and after partial decussation at the caudal end of the medulla oblongata enter the gray matter of the spinal cord at various levels. The **tranverse crural tract** (*tractus cruralis transversus*) can be seen on the lateral aspect of the crus cerebri (Fig. 14–3). It has connections with the optic tract and the oculomotor nuclei.

The *tegmentum* is separated from the crus cerebri ventrolaterally by the *substantia nigra,* a platelike nuclear mass which extends throughout the mesencephalon. The substantia nigra belongs to the extrapyramidal system and has connections with the globus pallidus, the subthalamus, the tectum, the red nucleus, the hypothalamus, the intercrural nucleus, and the reticular formation. The pigmentation which is present in the substantia nigra (Marsden 1961) is not as conspicuous as in man. The components of the tegmentum include the nuclei and fibers of the third and fourth cranial nerves, fibers and nuclear masses associated with ascending and descending fiber bundles, and the reticular formation of the mesencephalon.

The third cranial nerve, or **oculomotor nerve,** sends somatic efferent (motor) fibers to all the extrinsic muscles of the eye except those innervated by the fourth or sixth cranial nerves. The muscles supplied by the third cranial nerve are the dorsal, medial, and ventral rectus, the ventral oblique, and the levator of the upper eyelid. The dorsal oblique muscle is innervated by the fourth and the lateral rectus and the retractor bulbi by

the sixth cranial nerve. In addition to somatic efferent fibers, the oculomotor nerve provides general visceral efferent (parasympathetic) impulses to the intrinsic muscles of the eye. Its nuclei, the **motor nucleus** and the **parasympathetic nuclei of the oculomotor nerve** (*nucleus motorius nervi oculomotorii, nuclei parasympathici nervi oculomotorii*), are located ventral to the mesencephalic aqueduct at the level of the rostral colliculi. The fibers of the oculomotor nerve sweep ventrad through the tegmentum, and its lateral fibers traverse the medial portion of the red nucleus and the most medial part of the substantia nigra before they emerge from the intercrural fossa.

The fourth cranial nerve, or **trochlear nerve,** supplies somatic efferent (motor) fibers to the dorsal oblique muscle of the eye. The **motor nucleus of the trochlear nerve** (*nucleus motorius nervi trochlearis*) lies close to the caudal extremity of the oculomotor nuclei at the level of the caudal colliculi. The trochlear nerves of the two sides cross the midline (Fig. 14–18) at the **decussation of the trochlear nerves** (*decussatio nervorum trochlearium*) in the **rostral medullary velum** (*velum medullare rostrale*), which forms the roof over the caudal opening of the mesencephalic aqueduct and the rostral part of the fourth ventricle. The nerves emerge from the dorsal aspect of the mesencephalon immediately caudal to the caudal colliculi.

The **red nucleus** (*nucleus ruber*) is the most prominent of the nuclear masses of the tegmentum. It is rounded and lies in the rostral portion of the mesencephalon ventral to the rostral colliculi and ventrolateral to the nuclei of the oculomotor nerve. It receives descending fibers from the subthalamus, which were mentioned with the ansa and the fasciculus lenticularis. Ascending fibers enter the red nucleus from the opposite half of the cerebellum through the **rostral cerebellar peduncle** (*pedunculus cerebellaris rostralis*), which also transmits cerebellar fibers to the ventral lateral nucleus of the thalamus. The fibers of the rostral cerebellar peduncles form the **decussation of the rostral cerebellar peduncles** (*decussatio pedunculorum cerebellarium rostralium*) by crossing the midline in the caudal portion of the tegmentum of the mesencephalon ventromedial to the caudal colliculi. The **deep mesen-**

cephalic nucleus lies in the tegmentum, lateral and slightly dorsolateral to the red nucleus. It receives impulses from the subthalamus and gives rise to tegmentospinal and tegmentonuclear fibers.

The **medial longitudinal bundle** *(fasciculus longitudinalis medialis)* is a phylogenetically old, distinct fiber bundle next to the midline of the tegmentum of the mesencephalon immediately ventral to the central gray matter. It receives fibers from the **interstitial nucleus** *(nucleus interstitialis)* and other tegmental nuclei with connections from the subthalamus. Ascending fibers of the medial longitudinal bundle interrelate the vestibular nuclei in the caudal portion of the ventral metencephalon and the rostral myelencephalon with the nuclei of the cranial nerves which innervate the muscles of the eye. The vestibular nuclei belong to the vestibular system, which is responsible for the sensations of head position and head movement and which exerts vestibular control over eye movement.

Ascending fiber systems of the mesencephalic tegmentum include the medial lemniscus *(lemniscus medialis),* the trigeminal lemniscus *(lemniscus trigeminalis)* and the mesencephalic tract of the trigeminal nerve *(tractus mesencephalicus nervi trigemini)* with the nucleus of the mesencephalic tract of the trigeminal nerve *(nucleus tractus mesencephalici nervi trigemini).* The **medial lemniscus** and the **trigeminal lemniscus** were mentioned in the description of the thalamus as entering the lateral and medial parts of the ventral caudal nucleus, respectively. In their course through the mesencephalon, these tracts lie lateral to the decussation of the rostral cerebellar peduncles and the red nuclei. The **mesencephalic tract** and **nucleus of the mesencephalic tract of the trigeminal nerve** are associated with the fifth cranial nerve, or **trigeminal nerve** *(nervus trigeminus),* in the pons. The tract carries proprioceptive impulses from the muscles of mastication and very likely from the intrinsic muscles of the eye, by way of the oculomotor, trochlear, and abducent nerves, to be the region of the mesencephalon ventrolateral to the central gray matter, where the nucleus of the mesencephalic tract is located.

The descending fiber systems which originate or at least receive fibers from the mesencephalon are the tectospinal and the tectonuclear tracts *(tractus tectospinalis, tractus tectonuclearis),* the rubrospinal tract *(tractus rubrospinalis),* and the central tegmental tract *(tractus tegmenti centralis).* The **tectospinal** and **tectonuclear tracts** arise from the tectum and transmit impulses which involve motor neurons at lower levels in visual and acoustic reflexes. Before descending through the brain stem, the tectospinal and tectonuclear tracts cross to the opposite side through the **dorsal tegmental decussation** *(decussatio tegmenti dorsalis),* which lies in the dorsal portion of the mesencephalic raphe ventral to the medial longitudinal bundle and rostral to the decussation of the rostral cerebellar peduncles. The **rubrospinal tract** begins in the red nucleus and sends fibers to motor neurons at lower levels. It crosses the midline in the **ventral tegmental decussation** *(decussatio tegmenti ventralis)* of the mesencephalon and descends through the tegmental portion of the pons and reticular formation of the medulla oblongata to the lateral funiculus of the spinal cord. The **central tegmental tract** carries impulses from the globus pallidus, the zona incerta of the subthalamus, the central gray matter of the mesencephalon, and the red nucleus. It descends in the center of the ipsilateral half of the tegmentum and continues through the ventral metencephalon and rostral myelencephalon to the **olivary nucleus** *(nucleus olivaris)* in the medulla oblongata (Fig. 14–15).

METENCEPHALON

The **metencephalon** (Plates 4, 5) lies between the mesencephalon and the myelencephalon, or medulla oblongata. Of its two components, the **ventral metencephalon** belongs to the brain stem; the dorsal metencephalon develops into the **cerebellum.**

The **ventral metencephalon** includes the pons rostrally and extends through the trapezoid body caudally (compare Evans and deLahunta 1971). The **pons** (Figs. 14–4, 14–5; Plate 4) consists of a ventral part *(pars ventralis pontis)* and a dorsal or tegmental part *(pars dorsalis pontis).* The fifth cranial nerve, or **trigeminal nerve** *(nervus trigeminus),* takes origin from the caudoventrolateral aspect of the pons. The **trapezoid body** *(corpus trapezoideum)* is a group of transverse fibers, caudal to the ventral part of the pons, on the ventral surface of the brain stem (Fig. 14–5). The sixth cranial nerve, or **abducent nerve** *(nervus abducens)*, the seventh cranial nerve, or **facial nerve** *(nervus facialis)*, and the eighth cranial nerve, or **vestibulocochlear nerve** *(nervus vestibulocochlearis)*, arise from the caudal part of the ventral metencephalon on the ventral and lateral aspect of the trapezoid body (Fig. 14–5).

The **ventral part of the pons** is marked by **transverse fibers of the pons** *(fibrae pontis transversae)*, which bridge its ventral surface. Dorsal to the transverse fibers, longitudinal fibers continue the crus cerebri caudad (Fig. 14–20). They consist of the **corticopontine tract** *(tractus corticopontinus)* and the **pyramidal tract** *(tractus pyramidalis)*. The fibers of the corticopontine tract end in the **pontine nuclei** *(nuclei pontis)*, which consist of gray matter occupying the remainder of the ventral part of the pons. The pontine nuclei give rise to the transverse fibers of the pons, which cross the midline and ascend as the **middle cerebellar peduncle** *(pedunculus cerebellaris medius)* into the cerebellum (Figs. 14–3, 14–18). The corticonuclear fibers of the pyramidal tract convey impulses to the lower motor neurons in the brain stem. The corticospinal fibers continue through the caudal part of the ventral metencephalon and form the beginning of a **pyramid** *(pyramis)* on each side ventral to the trapezoid body. At their origins each pyramid is associated with a curved cluster of cell bodies, the **arcuate nucleus** *(nucleus arcuatus)*, which represents caudally displaced pontine nuclei. Their connections will be described with the **superficial arcuate fibers** *(fibrae arcuatae superficiales)* as they enter the caudal cerebellar peduncle.

The **dorsal part of the pons** contains nuclei and fibers of cranial nerves and of ascending and descending fiber bundles as well as the reticular formation of the pons.

The fifth cranial nerve, or **trigeminal nerve,** carries general somatic afferent (sensory) fibers from the head region and special visceral efferent (branchial motor) fibers to the muscles of mastication. It originates from the pons caudal to the middle cerebellar peduncle just ventral to the cerebellum (Figs. 14–4, 14–5). The exteroceptive sensory fibers enter the **pontine sensory nucleus** *(nucleus sensibilis pontinus nervi trigemini)*, which lies immediately medial to the superficial origin, or descend as the **spinal tract of the trigeminal nerve** *(tractus spinalis nervi trigemini)* to the **nucleus of the spinal tract of the trigeminal nerve** *(nucleus tractus spinalis nervi trigemini)*, which is located caudal to the pontine sensory nucleus and medial to the spinal tract of the trigeminal nerve. Both the spinal tract and the nucleus of the spinal tract extend caudad in the lateral part of the medulla oblongata to the beginning of the spinal cord (Fig. 14–22). The trigeminal fibers synapse with secondary sensory neurons in the pontine sensory nucleus and the nucleus of the spinal tract of the trigeminal nerve, which in turn give rise to secondary sensory fibers. These cross the midline and form the **trigeminal lemniscus** *(lemniscus trigeminalis)*, which accompanies the medial lemniscus through the pons and mesencephalon to the ventral caudal nucleus of the thalamus. As described with the thalamic relay nuclei, the medial lemniscus enters the lateral part, and the trigeminal lemniscus enters the medial part of the ventral caudal nucleus. The motor fibers of the trigeminal nerve innervate the muscles of mastication and originate from the **motor nucleus of the trigeminal nerve** *(nucleus motorius nervi trigemini)*, which lies medial to the pontine sensory nucleus (Plate 9). The proprioceptive fibers from the muscles of mastication form the **mesencephalic tract of the trigeminal nerve** *(tractus mesencephalicus nervi trigemini)* and reach the **nucleus of the mesencephalic tract of the trigeminal nerve** *(nucleus tractus mesencephalici nervi trigemini)*. Possibly, as indicated in the description of the mesencephalon, the mesencephalic nucleus also receives proprioceptive fibers from the extrinsic muscles of the eye.

The sixth cranial nerve, or **abducent nerve,** supplies the lateral rectus and the retractor bulbi muscles with somatic efferent (motor) fibers. The **nucleus of the abducent nerve** *(nucleus motorius nervi abducentis)* lies close to the dorsal midline and the medial longitudinal bundle, from which it receives vestibular impulses. The fibers of the abducent nerve run ventrad and slightly laterad through the reticular formation and along the medial border of the dorsal nucleus of the trapezoid body *(nucleus dorsalis* *corporis trapezoidei).* They emerge directly lateral to the pyramids among the fibers of the trapezoid body, which belongs to the acoustic system. The dorsal nucleus of the trapezoid body is another component of the acoustic system (Plate 5). It forms a rounded nuclear mass halfway between the midline and the lateral margin of the ventral metencephalon just dorsal to the trapezoid body (Fig. 14–15). The connections of both the trapezoid body and the dorsal nucleus of the trapezoid body will be described with the eighth cranial nerve, or vestibulocochlear nerve.

The seventh cranial nerve, or **facial nerve,** carries special visceral efferent (branchial motor) fibers to the muscles of the face, general visceral efferent (parasympathetic) fibers to the sublingual and mandibular salivary glands, nasal, palatine, and lacrimal glands, special visceral afferent fibers for taste from the rostral two-thirds of the tongue by way of the chorda tympani, and presumably general somatic afferent fibers for proprioceptive impulses from the muscles of facial expression. The **motor nucleus of the facial nerve** *(nucleus motorius nervi facialis)* lies in the rostral myelencephalon, caudal to the dorsal nucleus of the trapezoid body, and its location is marked by a slight elevation on the ventral surface of the medulla oblongata caudal to the trapezoid body (Plate 6). The motor fibers course dorsorostrad to the nucleus of the abducent nerve, where they form the **internal genu of the facial nerve** *(genu nervi facialis).* From its internal genu the facial nerve extends obliquely ventrolaterad through the reticular formation and along the lateral side of the dorsal nucleus of the trapezoid body. Its superficial origin from among the fibers of the trapezoid body is just caudal to the medial edge of the superficial origin of the trigeminal nerve (Fig. 14–5). The parasympathetic fibers come from the **parasympathetic nucleus of the facial nerve** *(nucleus parasympathicus nervi facialis),* an ill-defined, small nucleus within the reticular formation. The taste fibers contribute to the *tractus solitarius,* a distinct fiber bundle which, farther caudally, becomes associated with the glossopharyngeal (ninth cranial) and the vagus (tenth cranial) nerves in the medulla oblongata. The secondary path for taste, which extends from the nucleus of the tractus solitarius *(nucleus*

tractus solitarii) by way of the contralateral medial lemniscus to the medial part of the ventral caudal nucleus of the thalamus, will be described with the taste fibers of the myelencephalon.

The eighth cranial nerve, or **vestibulocochlear nerve,** consists of a vestibular and a cochlear part *(pars vestibularis, pars cochlearis)* which enter the central nervous system together at the ventrolateral border of the trapezoid body directly caudal to the superficial origin of the trigeminal nerve (Figs. 14–4, 14–5).

The **cochlear part** receives special somatic afferent fibers for acoustic impulses from the cochlea of the inner ear and conveys them to the dorsal and ventral cochlear nuclei *(nucleus cochlearis dorsalis, nucleus cochlearis ventralis).* The **dorsal cochlear nucleus** lies dorsolateral to the caudal cerebellar peduncle. The **ventral cochlear nucleus** is located at the lateral edge of the trapezoid body where the vestibulocochlear nerve enters the ventral metencephalon. The cochlear nuclei give rise to crossed and uncrossed secondary acoustic pathways for conscious perception of sound at the cortical level and for acoustic reflexes involving the dorsal nuclei of the trapezoid body and the caudal colliculi in the mesencephalon. Some of these secondary fibers enter the ipsilateral or the contralateral **lateral lemniscus** *(lemniscus lateralis)* and ascend to higher levels; some synapse in the reticular formation or the dorsal nucleus of the trapezoid body of the same or opposite side. The fibers from the dorsal cochlear nuclei curve dorsally around the caudal cerebellar peduncle and pass through the reticular formation of both sides. Those from the ventral cochlear nuclei form the **trapezoid body,** which crosses the midline dorsal to the pyramids (Fig. 14–5). The lateral lemniscus consists of fibers which originate from the **dorsal nucleus of the trapezoid body** and those secondary acoustic fibers which continue from the reticular formation and the trapezoid body without having synapsed at this level. At first the lateral lemniscus lies medial to the middle cerebellar peduncle, beyond which it emerges to the dorsolateral surface of the brain stem before entering the caudal colliculi. As previously stated, it carries impulses from both sides. Some of its fibers synapse in the **nucleus of the lateral lemniscus** *(nucleus* *lemnisci lateralis*), some run to the **caudal colliculi** in the tectum of the mesencephalon, and some carry impulses directly to the **medial geniculate body**, where they are relayed to the cerebral cortex. In addition to transmitting acoustic impulses to higher levels, the dorsal nuclei of the trapezoid body and the caudal colliculi serve as acoustic reflex centers. Fibers from the **dorsal nuclei of the trapezoid body**, together with secondary acoustic fibers, which synapse in the **reticular formation**, contribute to the ascending reticular activating system and to acoustic reflex arcs which influence motor nuclei of the cranial nerves either directly or by way of intercalated neurons in the reticular formation. At the level of the caudal colliculi, the acoustic impulses involve the **tectospinal** and the **tectonuclear tracts**.

The **vestibular part** of the vestibulocochlear nerve receives special somatic afferent impulses which are also referred to as special proprioceptive impulses, from receptors in the vestibular apparatus of the inner ear which register changes in head position. These changes provide the stimuli for the sense of balance or equilibrium. The vestibular fibers enter the brain stem ventral to the **caudal cerebellar peduncle** (*pedunculus cerebellaris caudalis*) and end in the **vestibular nuclei** (*nuclei vestibulares*) and the **flocculonodular lobe** (*lobus flocculonodularis*) of the cerebellum. The vestibular nuclei occupy the floor of the fourth ventricle of the caudal pons and rostral medulla oblongata and consist of the paired **rostral, caudal, medial**, and **lateral vestibular nuclei** (*nucleus vestibularis rostralis, caudalis, medialis, et lateralis*). The flocculonodular lobe will be described with the cerebellum.

Detailed information about the secondary vestibular connections is controversial. In a general way, however, the vestibular nuclei convey impulses to the cerebellum, the cerebral cortex, the motor nuclei of the cranial nerves innervating the extrinsic muscles of the eye, and the spinal cord. The secondary vestibular fibers to the **cerebellum** run in the medial portion of the **caudal cerebellar peduncle**, along with the primary vestibular fibers which reach the flocculonodular lobe directly without synapsing in the vestibular nuclei. Both crossed and uncrossed secondary vestibular fibers from the vestibular nuclei enter the **medial longitudinal bundle**

(*fasciculus longitudinalis medialis*), which is located close to the midline in the floor of the fourth ventricle. The secondary fibers to the **cerebrum** and the **nuclei of the cranial nerves** which innervate the **extrinsic muscles of the eye** ascend in the rostral portion of the medial longitudinal bundle. Fibers to the spinal cord join the caudal continuation as the **vestibulospinal part** (*pars vestibulospinalis*) of the medial longitudinal bundle. Uncrossed fibers from the lateral vestibular nucleus form the **vestibulospinal tract** (*tractus vestibulospinalis*), which descends through the lateral medulla oblongata to the ventrolateral spinal cord.

Other tracts in the dorsal part of the pons are the medial lemniscus, the rostral cerebellar peduncle, the ventral spinocerebellar tract, the tectospinal and the tectonuclear tracts, the rubrospinal tract, and the central tegmental tract. The **medial lemniscus**, with the lateral spinothalamic tract applied to its lateral border, occupies a ventromedial position in the dorsal part of the pons medial to the lateral lemniscus. It continues through the mesencephalon to the lateral part of the ventral caudal thalamic nucleus. The **rostral cerebellar peduncles** (Fig. 14–18) run dorsolaterally along the fourth ventricle throughout most of their course in the pons, except for the very rostral part, where they converge ventrad to the **decussation of the rostral cerebellar peduncles**. Their fiber components, including the ventral spinocerebellar tract, will be described in connection with the cerebellum. The **tectospinal** and the **tectonuclear tracts** descend close to the midline between the medial longitudinal bundle and the medial lemniscus. The **rubrospinal tract** is in a ventrolateral position. The **central tegmental tract** descends through the center of each half of the dorsal part of the pons on its way to the **olivary nucleus** in the medulla oblongata.

The **fourth ventricle** (*ventriculus quartus*) represents the ventricular system in the rhombencephalon (Figs. 14–12, 14–18). Rostrally, it connects with the third ventricle by way of the mesencephalic aqueduct. In the metencephalon, it is bounded ventrally by the dorsal free border of the dorsal part of the pons and laterally by the rostral cerebellar peduncles. The roof is formed by the **rostral medullary velum** (*velum medullare rostrale*) and the cerebellum. Caudally, the fourth

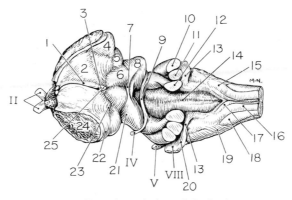

Figure 14–18. Dorsolateral view of the brain stem.

1. Stria habenularis thalami
2. Dorsal aspect of thalamus
3. Habenular commissure
4. Lateral geniculate body
5. Medial geniculate body
6. Rostral colliculus
7. Commissure of caudal colliculi
8. Caudal colliculus
9. Decussation of trochlear nerves in rostral medullary velum
10. Middle cerebellar peduncle
11. Caudal cerebellar peduncle
12. Rostral cerebellar peduncle
13. Dorsal cochlear nucleus
14. Median sulcus in fourth ventricle
15. Nucleus cuneatus lateralis
16. Fasciculus cuneatus
17. Fasciculus gracilis
18. Spinal tract of trigeminal nerve
19. Superficial arcuate fibers
20. Ventral cochlear nucleus
21. Brachium of caudal colliculus
22. Optic tract
23. Brachium of rostral colliculus
24. Cut surface between cerebrum and brain stem
25. Pineal body
  II. Optic nerve
  IV. Trochlear nerve
  V. Trigeminal nerve
VIII. Vestibulocochlear nerve

ventricle continues ventral to the **caudal medullary velum** (*velum medullare caudale*) into the myelencephalon. The **choroid plexus of the fourth ventricle** (*plexus choroideus ventriculi quarti*) protrudes from the roof into the fourth ventricle.

The **cerebellum** is derived from the dorsal portion of the metencephalon. It is a deeply fissured, more or less globular part of the brain and lies caudal to the mesencephalon and the occipital pole of the cerebrum and dorsal to the fourth ventricle. The middle portion of the ventral surface of the cerebellum forms part of the roof of the fourth ventricle between the rostral medullary velum and the caudal medullary velum. Ventrally, the cerebellum is connected to the brain stem by three pairs of cerebellar peduncles (Figs. 14–3, 14–18; Plates 4 to 7, 9). In rostrocaudal sequence, they are the **rostral cerebellar peduncle** (*pedunculus cerebellaris rostralis*), the **middle cerebellar peduncle** (*pedunculus cerebellaris medius*), and the **caudal cerebellar peduncle** (*pedunculus cerebellaris caudalis*). Like the cerebrum, the cerebellum has a peripheral cortex, white matter underneath the cortex, and centrally located nuclear masses. The cere-

bellar cortex (*cortex cerebelli*) consists of the cerebellar folia (*folia cerebelli*), which correspond to the gyri of the cerebral cortex. The white matter is called the *corpus medullare* (Fig. 14–20). It connects the cerebellar cortex and the **cerebellar nuclei** (*nuclei cerebelli*) with each other and with the brain stem by way of the cerebellar peduncles. The arrangement of gray and white matter in the cerebellar cortex results in a treelike appearance, especially in median sections, and is referred to as the *arbor vitae cerebelli* (Fig. 14–12).

In early developmental stages, the **caudolateral fissure** (*fissura caudolateralis*), also referred to as the **uvulonodular fissure** (*fissura uvulonodularis*), the first cerebellar fissure to appear in the embryo, divides the cerebellum into the **flocculonodular lobe** (*lobus flocculonodularis*) caudally and the *corpus cerebelli* rostrally (Fig. 14–13). With further development, the flocculonodular lobe stays comparatively small and is almost entirely hidden on the ventral surface of the corpus cerebelli, which becomes considerably enlarged and forms the bulk of the cerebellum. The *fissura prima*, the next cerebellar fissure to appear, divides the corpus

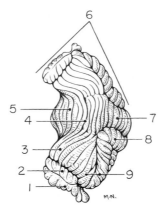

Figure 14–19. Dorsolateral view of the cerebellum.

1. Ventral paraflocculus
2. Dorsal paraflocculus
3. Lobulus simplex
4. Fissura prima
5. Vermis portion of rostral lobe
6. Right cerebellar hemisphere
7. Vermis portion of caudal lobe
8. Paramedian lobule
9. Ansiform lobule

cerebelli into the **rostral** and the **caudal lobes** (*lobus rostralis, lobus caudalis*). In spite of its considerable depth, the fissura prima is not easily detectable among the large number of similar folia unless the cerebellum is hemisected (Figs. 14–13, 14–19). The flocculonodular lobe is associated with the vestibular system. The rostral lobe receives fibers from the spinal cord, the trigeminal system, and the olivary nucleus in the medulla oblongata. The caudal lobe has largely corticopontocerebellar connections.

TABLE 14–1.   EQUIVALENTS OF DESCRIPTIVE AND NUMERICAL NOMENCLATURE OF CEREBELLAR LOBULES*

Vermis			Hemispherium	
		Corpus cerebelli		
		Lobus rostralis		
Lingula		Lob. I	Vinculum lingulae	Lob. H I
Lob. centralis }	{	Lob. II	Ala lob. centralis	{ Lob. H II
		Lob. III		Lob. H III
Culmen	{	Lob. IV	Lob. quadrangularis	{ Lob. H IV
		Lob. V		Lob. H V
		Fissura prima		
		Lobus Caudalis		
Declive		Lob. VI	Lob. simplex	Lob. H VI
Folium vermis }	{	Lob. VIIA	Lob. ansiformis	Lob. H VIIA
Tuber vermis }		Lob. VIIB		{ Lob. H VIIB
			Lob. paramedianus	
Pyramis		Lob. VIIIA		{ Lob. H VIIIA
		Lob. VIIIB	Paraflocc. dorsalis	Lob. H VIIIB
Uvula		Lob. IX	Paraflocc. ventralis	
				Lob. H IX
		Fissura caudolateralis = Fissura uvulonodularis		
		Lobus flocculonodularis		
Nodulus		Lob. X	Flocculus	Lob. H X

* From Nomina Anatomica Veterinaria, 1968 and 1973.

From a topographical point of view, the cerebellum is divided into a median portion, the *vermis*, and a **cerebellar hemisphere** (*hemispherium cerebelli*) on each side (Fig. 14–19). The vermis and the hemispheres are subdivided into a number of *lobules*, which are designated by descriptive names in the classical terminology or by Roman numerals from I to X and letters for sublobules in the nomenclature of Larsell (1953, 1954). The degree to which these classifications are interchangeable is best exemplified by a list of equivalents (Table 14–1).

The vermian lobules of the rostral lobe are numbered I to V. They are called the *lingula cerebelli*, the *lobulus centralis*, and the *culmen*. The vermis of the caudal lobe consists of the *declive*, the *folium vermis*, the *tuber vermis*, the *pyramis vermis*, and the *uvula vermis*. They correspond to lobules VI to IX. The *nodulus* is lobulus X and represents the vermian part of the flocculonodular lobe.

The extension of the rostral lobe into the hemisphere is rather small. It consists of lobules H I to H V. Their names are *vinculum lingulae, ala lobuli centralis*, and *lobulus quadrangularis*. The major portion of the cerebellar hemisphere consists of the lateral part of the caudal lobe. Its lobules are designated H VI to H IX and are named *lobulus simplex, lobulus ansiformis, lobulus paramedianus, paraflocculus dorsalis*, and *paraflocculus ventralis*. The **ventral paraflocculus** is located at the ventrolateral edge of the cerebellar hemisphere, dorsal to the flocculus. Laterally, the ventral paraflocculus projects into the **cerebellar fossa** (*fossa cerebellaris*) of the petrous part of the temporal bone. The **dorsal paraflocculus** lies dorsal to the ventral paraflocculus and reaches the **paramedian lobule** on the caudal aspect of the cerebellum. The paramedian lobule parallels the vermis on the caudal side and consists of transversely running folia. It connects with the **ansiform lobule** to form the **ansoparamedian lobule**, which rep-

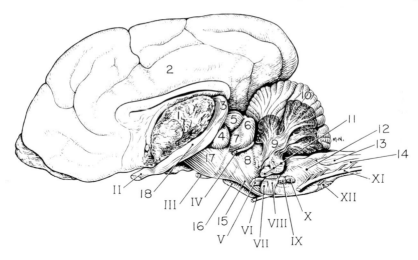

Figure 14–20. Lateral view of the brain with the left cerebral hemisphere, left transverse fibers of the pons, and left middle cerebellar peduncle removed.

1. Cut surface between cerebrum and brain stem
2. Medial surface of right cerebral hemisphere
3. Lateral geniculate body
4. Medial geniculate body
5. Rostral colliculus
6. Caudal colliculus
7. Brachium of caudal colliculus
8. Lateral lemniscus
9. Corpus medullare of cerebellum
10. Vermis of cerebellum
11. Flocculus
12. Superficial arcuate fibers
13. Spinal tract of trigeminal nerve
14. Dorsal spinocerebellar tract
15. Longitudinal fibers of pons

16. Transverse fibers of pons
17. Crus cerebri
18. Optic tract

II. Optic nerve
III. Oculomotor nerve
IV. Trochlear nerve
V. Trigeminal nerve
VI. Abducent nerve
VII. Facial nerve
VIII. Vestibulocochlear nerve
IX. Glossopharyngeal nerve
X. Vagus nerve
XI. Accessory nerve
XII. Hypoglossal nerve

resents the bulk of the cerebellar hemisphere. It is bounded rostromedially by the lobulus simplex, which lies caudal to the fissura prima in the hemisphere. The *flocculus*, lobulus H X, lies on the ventral surface of the cerebellum, dorsal to the cochlear nuclei (Fig. 14–20). It forms the lateral part of the flocculonodular lobe.

The **cerebellar nuclei** (Plates 6, 9) of each side consist of the lateral cerebellar nucleus (*nucleus lateralis cerebelli*), the nuclei interpositi cerebelli (*nucleus interpositus lateralis, nucleus interpositus medialis*), and the fastigial nucleus (*nucleus fastigii*) in lateromedial sequence. They are located in the center of the cerebellum just dorsal to the roof of the fourth ventricle. They receive fibers from the cerebellar cortex and give rise to the efferent fibers of the cerebellum. The **lateral cerebellar nucleus** is the largest of the group. The *nuclei interpositi* are two smaller, less distinct nuclear masses. The **fastigial nuclei** of the two sides are adjacent to each other at the median plane and appear roundish in cross sections.

The efferent and afferent fibers of the cerebellum pass through the cerebellar peduncles (Figs. 14–3, 14–18, 14–21, 14–22). The **rostral cerebellar peduncle** is the most medial of the three cerebellar peduncles of a given side. It carries largely efferent fibers from the lateral cerebellar nucleus and the nuclei interpositi rostrally to the opposite **red nucleus** and **ventral lateral thalamic nucleus**. The **ventral spinocerebellar tract** (*tractus spinocerebellaris ventralis*) reaches the rostral cerebellar peduncle and enters the cerebellum rostrally.

The **middle cerebellar peduncle** lies more laterally than the other two cerebellar peduncles. It consists of the continuation of the transverse fibers of the pons, which convey **corticopontocerebellar impulses** from the pontine nuclei to the cerebellar cortex.

The **caudal cerebellar peduncle** has the largest number of components. It consists of spinocerebellar fibers, trigeminocerebellar fibers, olivocerebellar fibers, superficial arcuate fibers, vestibulocerebellar fibers, and fastigiobulbar fibers. The spinocerebellar fibers come from the **dorsal spinocerebellar tract** (*tractus spinocerebellaris dorsalis*). The olivocerebellar fibers originate from the **olivary nucleus** (*nucleus olivaris*) in the medulla oblongata and form the **olivocerebellar tract** (*tractus olivocerebellaris*). The **superfi-**

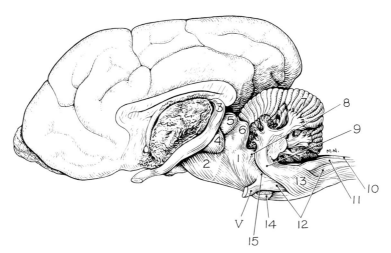

Figure 14–21.   Lateral view of the brain with the left cerebral hemisphere removed and the caudal cerebellar peduncle dissected.

1. Lateral lemniscus
2. Crus cerebri
3. Lateral geniculate body
4. Medial geniculate body
5. Rostral colliculus
6. Caudal colliculus
7. Cerebellum
8. Rostral cerebellar peduncle

9. Caudal cerebellar peduncle
10. Fasciculus gracilis
11. Fasciculus cuneatus
12. Spinal tract of trigeminal nerve
13. Superficial arcuate fibers
14. Trapezoid body (transected)
15. Location of pontine sensory nucleus of trigeminal nerve
V. Trigeminal nerve

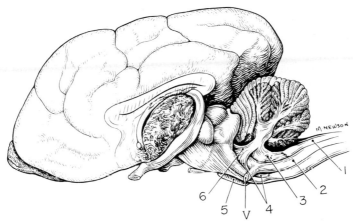

Figure 14–22. Lateral view of the brain with the left cerebral hemisphere removed and the spinal tract of trigeminal nerve dissected.

1. Fasciculus cuneatus
2. Spinal tract of trigeminal nerve
3. Location of vestibular nuclei
4. Ventral spinocerebellar tract

5. Transverse fibers of pons (transected on midline)
6. Rostral cerebellar peduncle
V. Trigeminal nerve

cial **arcuate fibers** (*fibrae arcuatae superficiales*) are subdivided into dorsal superficial arcuate fibers, derived from the nucleus cuneatus lateralis on the dorsolateral aspect of the medulla oblongata (Fig. 14–18), and ventral superficial arcuate fibers, which originate from the **arcuate nucleus** in the caudal part of the ventral metencephalon, the **nucleus of the lateral funiculus** (*nucleus funiculi lateralis*) in the caudal medulla oblongata, and disseminated cells of the reticular formation. The **vestibulocerebellar fibers** arise from vestibular nuclei and join the direct fibers from the vestibular part of the vestibulocochlear nerve. The **fastigiobulbar fibers** originate from the fastigial nucleus and, after partial crossing, pass by way of the medial part of the caudal cerebellar peduncle to vestibular nuclei, motor nuclei of the brain stem, and the reticular formation.

MYELENCEPHALON

The **myelencephalon**, or **medulla oblongata** (Plates 6 to 9), is the caudalmost portion of the brain stem and extends from the caudal edge of the trapezoid body in the ventral metencephalon to the beginning of the spinal cord. It contains nuclei and fibers of

cranial nerves, ascending and descending fiber bundles with associated nuclei, the reticular formation of the medulla oblongata, the caudal part of the fourth ventricle, and the beginning of the central canal. The cranial nerves which originate from the myelencephalon are the ninth cranial nerve, or **glossopharyngeal nerve** (*nervus glossopharyngeus*); the tenth cranial nerve, or **vagus nerve** (*nervus vagus*); the eleventh cranial nerve, or **accessory nerve** (*nervus accessorius*); and the twelfth cranial nerve, or **hypoglossal nerve** (*nervus hypoglossus*). The term "bulb," an old name for medulla oblongata, is still used in compound words, especially in clinical nomenclature.

On the ventral aspect of the medulla oblongata (Fig. 14–5), the *pyramid* (*pyramis*) on each side of the **median fissure** (*fissura mediana*) continues the corticospinal fibers from the ventral metencephalon through the medulla oblongata. The corticonuclear fibers do not enter the pyramids but reach the motor nuclei of cranial nerves on both sides through the reticular formation. At the caudal end of the medulla oblongata, about three-fourths of the fibers of the pyramids cross the midline by sweeping dorsolaterad and slightly caudad toward the **lateral funiculus** (*funiculus lateralis*) of the opposite side of the spinal cord, where they form the lateral pyramidal tract (*tractus pyramidalis lateralis*). The crossing of the fibers is referred to as the **decussation of the pyramids** (*decussatio pyramidum*). The remaining fibers do not cross at the caudal end of the medulla oblongata but continue caudad in the ipsilateral **ventral funiculus** (*funiculus ventralis*) of the spinal cord as the **ventral pyramidal tract** (*tractus pyramidalis ventralis*). Most of these fibers cross the midline farther caudally at various segmental levels of the spinal cord. A small number may remain on the same side. Crossed and uncrossed corticospinal fibers carry impulses to the lower motor neurons in the spinal cord, apparently by way of interneurons situated in the intermediary zone and in the ventral part of the dorsal column (Szentágothai-Schimmert, 1941). Lateral to the pyramids, just rostral to the decussation of the pyramids, the **olivary nucleus** (*nucleus olivaris*) forms a histological landmark of the caudal medulla oblongata (Fig. 14–15, Plate 7), which is related to the extrapyramidal system. It will be described with the central tegmental tract.

On the lateral side of the rostral part of the medulla oblongata, the **superficial arcuate fibers** (*fibrae arcuatae superficiales*) curve around the lateral and dorsolateral surface (Figs. 14–3, 14–20) and enter the caudal cerebellar peduncle, which connects the medulla oblongata with the cerebellum (Fig. 14–21). The superficial arcuate fibers consist of dorsal and ventral superficial arcuate fibers. The **dorsal superficial arcuate fibers** originate from the nucleus cuneatus lateralis on the dorsolateral aspect of the medulla oblongata. The **ventral superficial arcuate fibers** originate largely from the arcuate nucleus, the nucleus of the lateral funiculus (*nucleus funiculi lateralis*), and disseminated cells of the reticular formation. The **arcuate nuclei** are caudally displaced pontine nuclei near the beginning of the pyramids in the caudal part of the ventral metencephalon. The **nucleus of the lateral funiculus** is a fairly well-circumscribed nuclear mass lateral and caudal to the olivary nucleus, and receives impulses from the lateral funiculus of the spinal cord which it relays to the cerebellum.

In the caudal part of the lateral side of the medulla oblongata, the **spinal tract of the trigeminal nerve** (*tractus spinalis nervi trigemini*) becomes exposed. The beginning of the tract in the ventral metencephalon lies medial to the ventral cochlear nucleus. In the rostral portion of the medulla oblongata, it is hidden by the ventral superficial arcuate fibers (Figs. 14–20, 14–21, 14–22). The **nucleus of the spinal tract of the trigeminal nerve** (*nucleus tractus spinalis nervi trigemini*) accompanies the spinal tract on its medial side through the medulla oblongata. It contributes secondary sensory fibers to the trigeminal lemniscus (*lemniscus trigeminalis*), connects with motor nuclei of cranial nerves in the brain stem by way of the reticular formation, and conveys impulses through the caudal cerebellar peduncle to the cerebellum. The **trigeminal lemniscus** originates from the nucleus of the spinal tract and the pontine sensory nucleus of the trigeminal nerve. It carries secondary sensory fibers for all cutaneous sensations of the opposite side of the face to the medial part of the ventral caudal nucleus of the thalamus and is topographically related to the medial lemniscus along the dorsal aspect of the pyramid.

On the dorsal side (Fig. 14–18), the rostral part of the medulla oblongata forms the floor of the caudal portion of the **fourth ventricle** (*ventriculus quartus*). It is separated into two halves by the **median sulcus** (*sulcus medianus*). The **caudal medullary velum** (*velum medullare caudale*) forms the roof over this part of the fourth ventricle. The fourth ventricle communicates with the subarachnoid space by the **lateral apertures of the fourth ventricle** (*aperturae laterales ventriculi quarti*), and caudally it continues as the **central canal** (*canalis centralis*).

The caudodorsal portion of the medulla oblongata is occupied by the *fasciculus gracilis* and the *fasciculus cuneatus*, which carry proprioceptive and discriminating tactile impulses from the **dorsal funiculus** (*funiculus dorsalis*) of the spinal cord. The fasciculus gracilis lies along the midline. The fasciculus cuneatus runs between the fasciculus gracilis and the spinal tract of the trigeminal nerve. The fibers of the fasciculus gracilis synapse in the nucleus gracilis, those of the fasciculus cuneatus in the nucleus cuneatus medialis and lateralis (Fig. 14–3, Plate 8).

The *nucleus gracilis* and the *nucleus cuneatus medialis* give rise to **deep arcuate fibers** (*fibrae arcuatae profundae*), which curve concentrically in a ventromedial direction. At the **decussation of the medial lemnisci** (*decussatio lemniscorum medialium*), they cross the midline, form the medial lemniscus of the opposite side, and ascend through the medulla oblongata dorsal to the respective pyramid. Caudal to the olivary nucleus, the medial lemniscus receives the fibers of the **ventral spinothalamic tract** (*tractus spinothalamicus ventralis*), and rostral to the olivary nucleus, it is joined by the **lateral spinothalamic tract** (*tractus spinothalamicus lateralis*). At this point it contains all secondary fibers which carry conscious sensory impulses from the opposite side of the body: conscious proprioceptive and discriminating tactile impulses from the fasciculus gracilis and cuneatus, light touch from the ventral spinothalamic tract, and pain and temperature sensations from the lateral spinothalamic tract. From the medulla oblongata the medial lemniscus courses rostrally in the dorsal part of the ventral metencephalon and the tegmentum of the mesencephalon. In the diencephalon, its fibers synapse in the lateral part of the ventral caudal nucleus of the thalamus, which relays the impulses to the cerebral cortex.

The *nucleus cuneatus lateralis* gives rise to dorsal superficial arcuate fibers which enter the cerebellum through the caudal cerebellar peduncle. This pathway carries proprioceptive impulses, but since input to the cerebellum does not reach conscious levels, they are referred to as unconscious proprioceptive impulses. The **dorsal spinocerebellar tract** (*tractus spinocerebellaris dorsalis*) likewise carries unconscious proprioceptive impulses largely from the same side of the spinal cord to the cerebellum. In the caudal part of the medulla oblongata, before joining the ventral superficial arcuate fibers, the dorsal spinocerebellar tract runs along the ventrolateral margin of the spinal tract of the trigeminal nerve (Fig. 14–20). The **ventral spinocerebellar tract** (*tractus spinocerebellaris ventralis*) carries ipsilateral as well as contralateral unconscious proprioceptive impulses. It ascends ventral to the dorsal spinocerebellar tract and the spinal tract of the trigeminal nerve to the level of the pons, curves around the superficial origin of the trigeminal nerve, and enters the cerebellum by way of the rostral cerebellar

peduncle (Fig. 14–22). Fibers which decussate in the cord cross again within the cerebellum so that they, too, conform to the ipsilateral dominance (Jenkins 1978).

Among the fiber tracts inside the medulla oblongata, the **rubrospinal tract** is located laterally in the reticular formation ventral to the spinal nucleus of the trigeminal nerve. The **tectospinal tract** runs lateral to the midline dorsal to the medial lemniscus. The **vestibulospinal tract** lies medial to the rubrospinal tract. The **medial longitudinal bundle** has a close spatial relation to the floor of the fourth ventricle and the ventral side of the central canal. The **central tegmental tract** descends through the central portion of the ipsilateral half of the medulla oblongata and enters the olivary nucleus in the caudoventral part of the medulla oblongata (Fig. 14–15, Plate 7). Through the central tegmental tract, the olivary nucleus receives descending fibers from the globus pallidus, the zona incerta of the subthalamus, the mesencephalic central gray matter, and the red nucleus. The **spino-olivary tract** (*tractus spinoolivaris*) contributes ascending impulses from the spinal cord. The efferent fibers of the olivary nucleus form the **olivocerebellar tract** (*tractus olivocerebellaris*), which consists largely of crossed fibers and enters the cerebellum by the caudal cerebellar peduncle.

The ninth cranial nerve, or **glossopharyngeal nerve**, the tenth cranial nerve, or **vagus nerve**, and the eleventh cranial nerve, or **accessory nerve**, are closely related and may be referred to as the **vagus group**. They share some of their nuclei of origin and termination and emerge in linear fashion among the ventral superficial arcuate fibers along the ventrolateral border of the medulla oblongata just caudal to the superficial origin of the vestibulocochlear nerve. The twelfth cranial nerve, or **hypoglossal nerve**, has its superficial origin in the ventral aspect of the medulla oblongata at the level of the olivary nucleus, lateral to the pyramid.

The **glossopharyngeal nerve** carries general visceral afferent (visceral sensory) fibers from the pharynx, the caudal third of the tongue, and the carotid sinus (Adams 1958); special visceral afferent fibers for taste impulses from the caudal third of the tongue; general visceral efferent (parasympathetic) fibers to the parotid salivary gland; special visceral efferent (branchial motor) fibers to the stylopharyngeus muscle; and possibly some general somatic afferent (sensory) impulses from the skin to the ear canal.

The **vagus nerve** has its superficial origin caudal to the glossopharyngeal nerve. It consists of general visceral efferent (parasympathetic) fibers for the viscera of the neck, thorax, and abdomen; special visceral efferent (branchial motor) fibers for the striated musculature of the pharynx and larynx; general visceral afferent (visceral sensory) fibers from the pharynx, larynx, trachea, esophagus, and thoracic and abdominal viscera; special visceral afferent fibers for taste impulses from the epiglottis; and some general somatic afferent (sensory) fibers from the skin of the external ear canal.

The **accessory nerve** has spinal and cranial roots (*radices spinales, radices craniales*). The **spinal roots** originate in the cervical spinal cord and are merely topographically related to the cranial roots. The **cranial roots** leave the medulla oblongata caudal to the superficial origin of the vagus nerve. They join the fibers of the spinal roots for a short distance, but enter the vagus nerve by means of the **internal ramus** (*ramus internus*)of the accessory nerve. By this connection they convey special visceral efferent (branchial motor) impulses to the vagus nerve through which they are distributed to the pharynx and larynx. Because of this relationship, the cranial roots may be considered to belong altogether to the vagus nerve.

The **general** and **special visceral afferent fibers** of the glossopharyngeal and vagus nerves join the previously described special visceral afferent fibers of the facial nerve in forming the *tractus solitarius*. The tractus solitarius is a histological landmark of the caudodorsal part of the medulla oblongata medial to the nucleus of the spinal tract of the trigeminal nerve. Its fibers enter the **nucleus of the tractus solitarius** (*nucleus tractus solitarii*), which in turn conveys impulses to the motor nuclei of the cranial nerves by way of the reticular formation and gives origin to the secondary taste fibers. These fibers cross the midline with the deep arcuate fibers, join the medial lemniscus, and end in the medial part of the ventral caudal nucleus of the thalamus.

The **general visceral efferent (parasympathetic) fibers** of the glossopharyngeal nerve have their cell bodies in the **parasympathetic nucleus of the glossopharyngeal nerve** (*nucleus parasympathicus nervi glossopharyngei*), which lies caudal to the parasympathetic nucleus of the facial nerve in the reticular formation of the medulla oblongata. The general visceral efferent (parasympathetic) fibers of the vagus nerve originate from the **parasympathetic nucleus of the vagus nerve** (*nucleus parasympathicus nervi vagi*), which lies medial to the tractus solitarius and nucleus of the tractus solitarius. Rostrally, it protrudes into the fourth ventricle. Caudally, it becomes submerged ventral to the nucleus gracilis in the caudal medulla oblongata just lateral to the dorsal midline.

The **special visceral efferent (branchial motor) fibers** of the vagus group, which innervate the striated musculature of the pharynx and larynx, arise from the *nucleus ambiguus* in the center of the lateral part of the reticular formation. They run dorsad and then turn ventrolaterad to the root bundles of the glossopharyngeal, vagus, and accessory nerves. The fibers in the cranial roots of the accessory nerve—if not considered as part of the vagus nerve, as previously indicated — run through the internal ramus of the accessory nerve to the vagus nerve to be distributed together with the vagus fibers.

The **general somatic afferent fibers** of the glossopharyngeal and vagus nerves from the ear canal synapse in the nucleus of the spinal tract of the trigeminal nerve, which provides for the secondary connections to motor nuclei and higher levels.

The twelfth cranial nerve, or **hypoglossal nerve**, supplies the musculature of the tongue with somatic efferent (motor) fibers. Its cell bodies are located in the **hypoglossal nucleus** (*nucleus motorius nervi hypoglossi*), which lies along the midline ventromedial to the tractus solitarius and the parasympathetic nucleus of the vagus nerve. The rostral portion of the hypoglossal nucleus protrudes into the fourth ventricle medial to the parasympathetic nucleus of the vagus nerve. Its caudal part lies directly lateral to the central canal. The fibers of the hypoglossal nerve pass ventrad and slightly laterad through the reticular formation and emerge from the ventral surface of the caudal medul-

Text continued on page 890

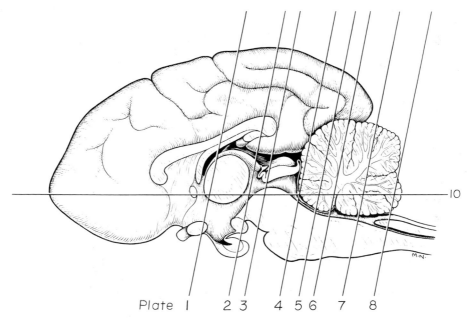

Plate I   2 3   4 5 6   7   8

Figure 14–23.   Medial view of hemisected brain, indicating planes of transverse and horizontal sections shown in plates 1 through 8 and 10. Plate 9 represents a paramedian section.

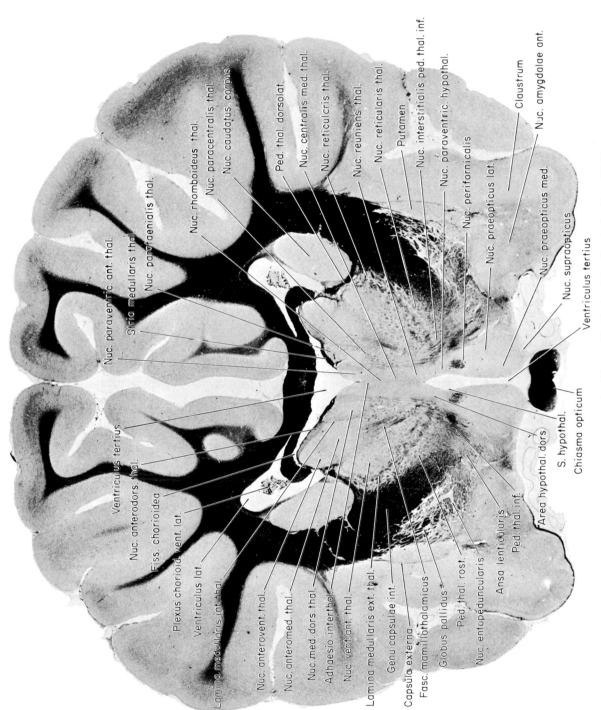

Plate 1. Transverse section of the brain. (Singer: The Brain of the Dog in Section. 1962.)

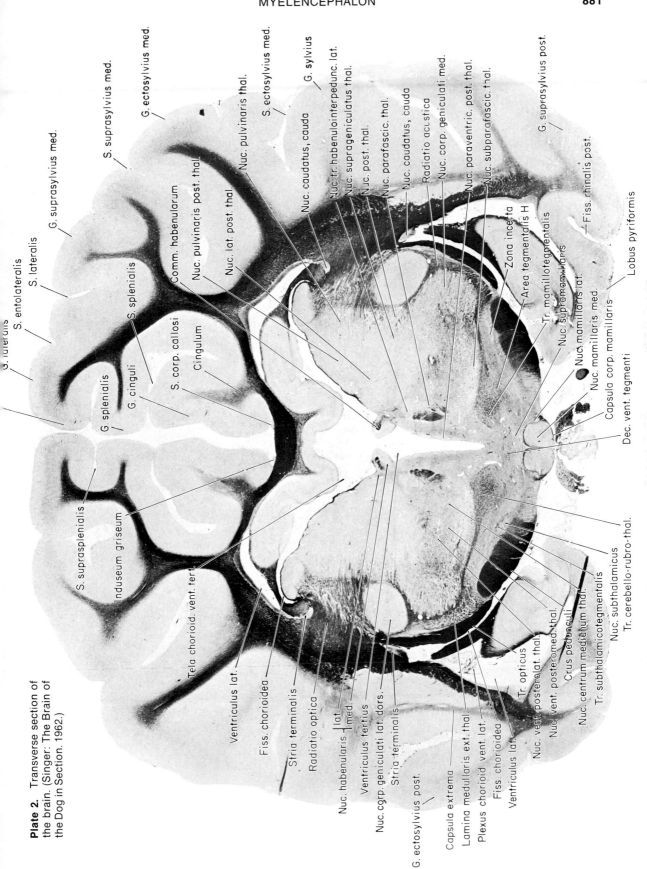

Plate 2. Transverse section of the brain. (Singer: The Brain of the Dog in Section. 1962.)

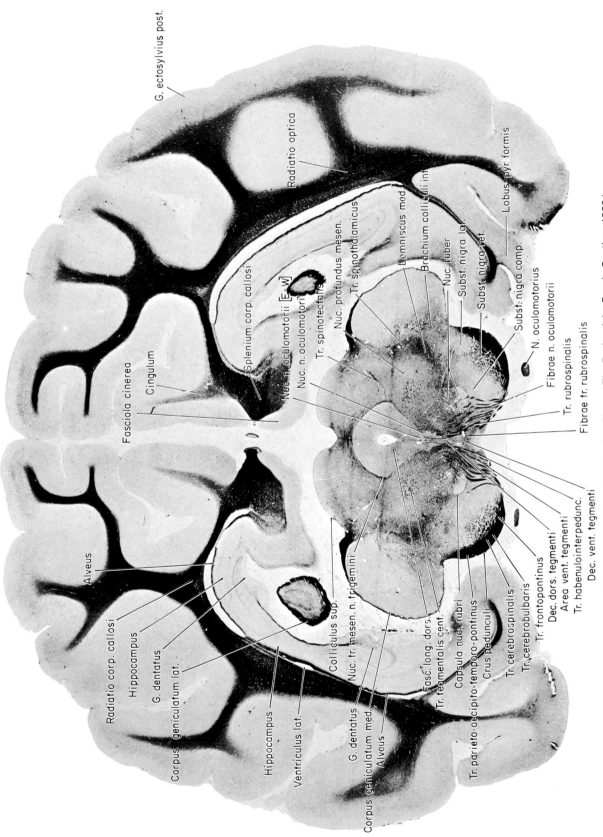

Plate 3. Transverse section of the brain. (Singer: The Brain of the Dog in Section. 1962.)

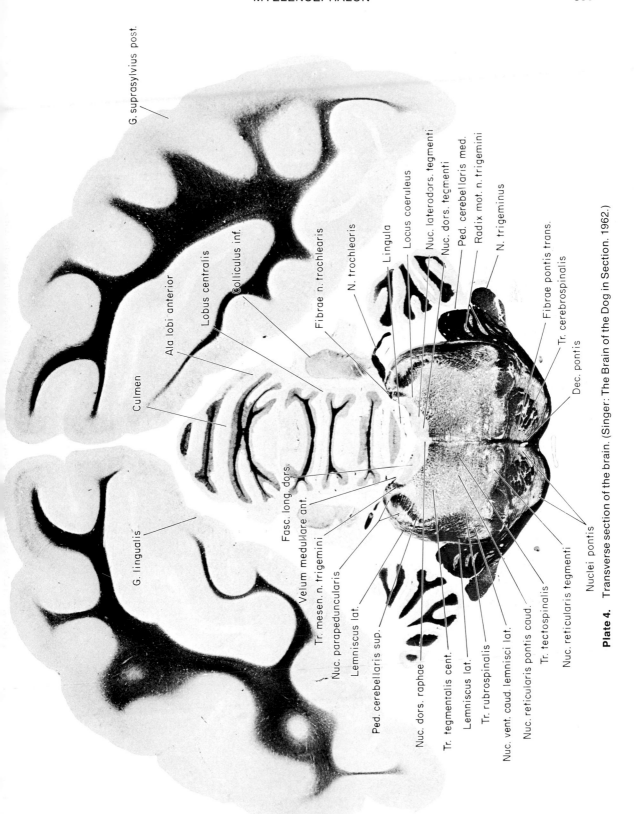

G. suprasylvius post.

Culmen

Ala lobi anterior

Lobus centralis

Colliculus inf.

Fibrae n. trochlearis

N. trochlearis

Lingula

Locus coeruleus

Nuc. laterodors. tegmenti

Nuc. dors. tegmenti

Ped. cerebellaris med.

Radix mot. n. trigemini

N. trigeminus

Fibrae pontis trans.

Tr. cerebrospinalis

Dec. pontis

G. lingualis

Fasc. long. dors.

Velum medullare ant.

Tr. mesen. n. trigemini

Nuc. parapeduncularis

Lemniscus lat.

Ped. cerebellaris sup.

Nuc. dors. raphae

Tr. tegmentalis cent.

Lemniscus lat.

Tr. rubrospinalis

Nuc. vent. caud. lemnisci lat.

Nuc. reticularis pontis caud.

Tr. tectospinalis

Nuc. reticularis tegmenti

Nuclei pontis

Plate 4. Transverse section of the brain. (Singer: The Brain of the Dog in Section. 1962.)

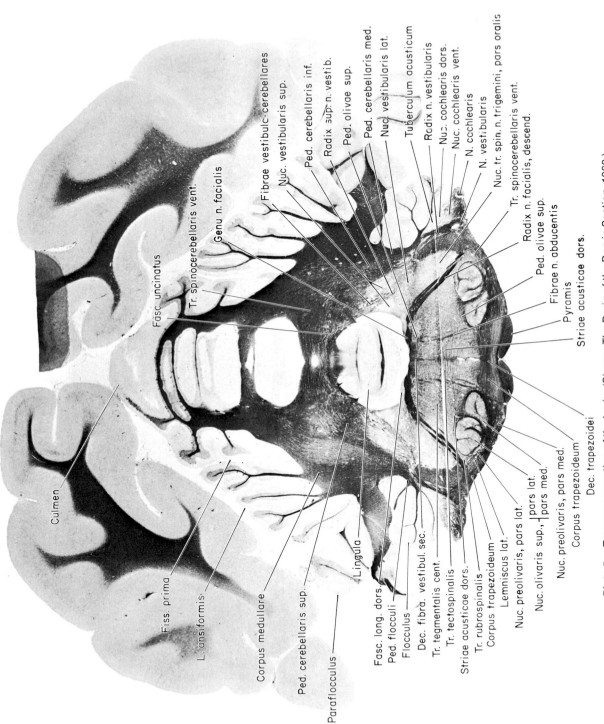

Plate 5. Transverse section of the brain. (Singer: The Brain of the Dog in Section, 1962.)

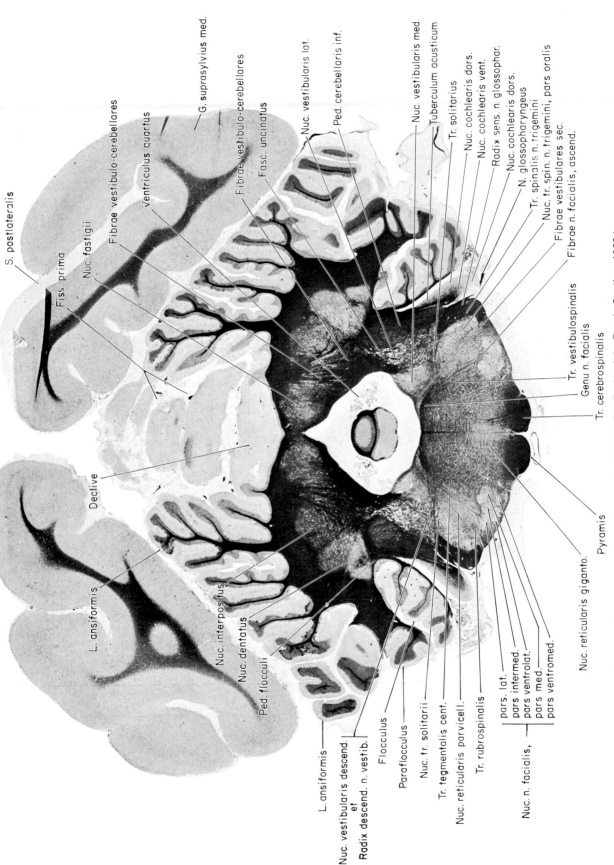

Plate 6. Transverse section of the brain. (Singer: The Brain of the Dog in Section. 1962.)

S. postlateralis

Fiss. prima

Nuc. fastigii

Fibrae vestibulo-cerebellares

Ventriculus quartus

Fibrae vestibulo-cerebellares

Fasc. uncinatus

G. suprasylvius med.

Nuc. vestibularis lat.

Ped. cerebellaris inf.

Nuc. vestibularis med.

Tuberculum acusticum

Tr. solitarius

Nuc. cochlearis dors.

Nuc. cochlearis vent.

Radix sens. n. glossophar.

Nuc. cochlearis dors.

N. glossopharyngeus

Tr. spinalis n. trigemini

Nuc. tr. spin. n. trigemini, pars oralis

Fibrae vestibulares sec.

Fibrae n. facialis, ascend.

Tr. vestibulospinalis

Genu n. facialis

Tr. cerebrospinalis

Pyramis

Nuc. reticularis giganto.

pars. lat.
pars intermed.
pars ventrolat.
pars med.
pars ventromed.

Nuc. n. facialis,

Tr. rubrospinalis

Nuc. reticularis parvicell.

Tr. tegmentalis cent.

Nuc. tr. solitarii

Paraflocculus

Flocculus

Nuc. vestibularis descend.
et
Radix descend. n. vestib.

L. ansiformis

Ped. flocculi

Nuc. dentatus

Nuc. interpositus

Declive

L. ansiformis

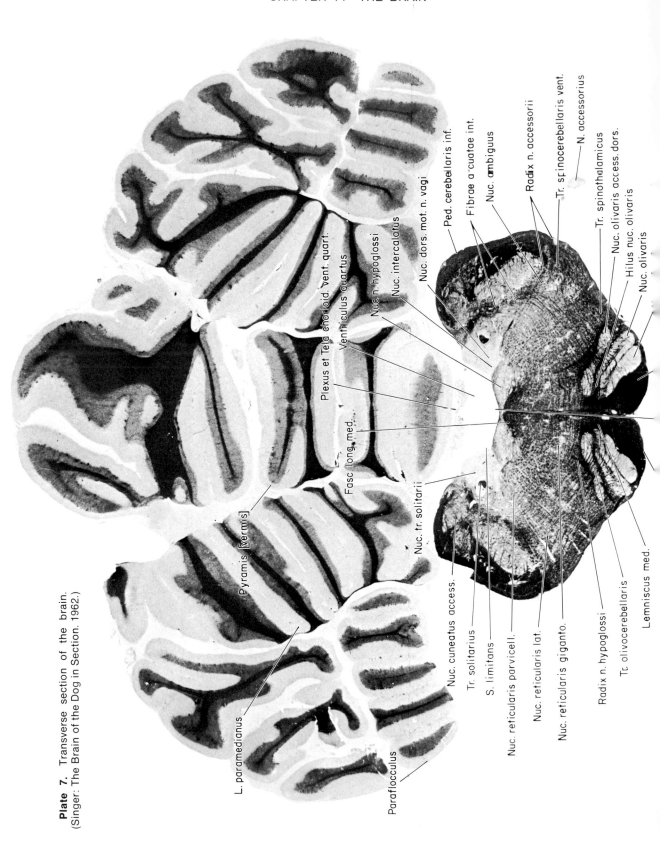

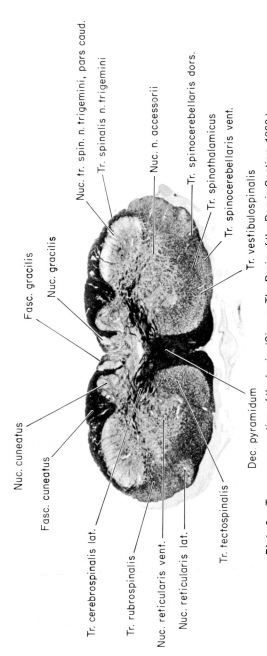

Pyramis [vermis]

Nuc. tr. spin. n. trigemini, pars caud.

Tr. spinalis n. trigemini

Nuc. n. accessorii

Tr. spinocerebellaris dors.

Tr. spinothalamicus

Tr. spinocerebellaris vent.

Tr. vestibulospinalis

Fasc. gracilis

Nuc. gracilis

Nuc. cuneatus

Fasc. cuneatus

Tr. cerebrospinalis lat.

Tr. rubrospinalis

Nuc. reticularis vent.

Nuc. reticularis lat.

Tr. tectospinalis

Dec. pyramidum

Plate 8. Transverse section of the brain. (Singer: The Brain of the Dog in Section. 1962.)

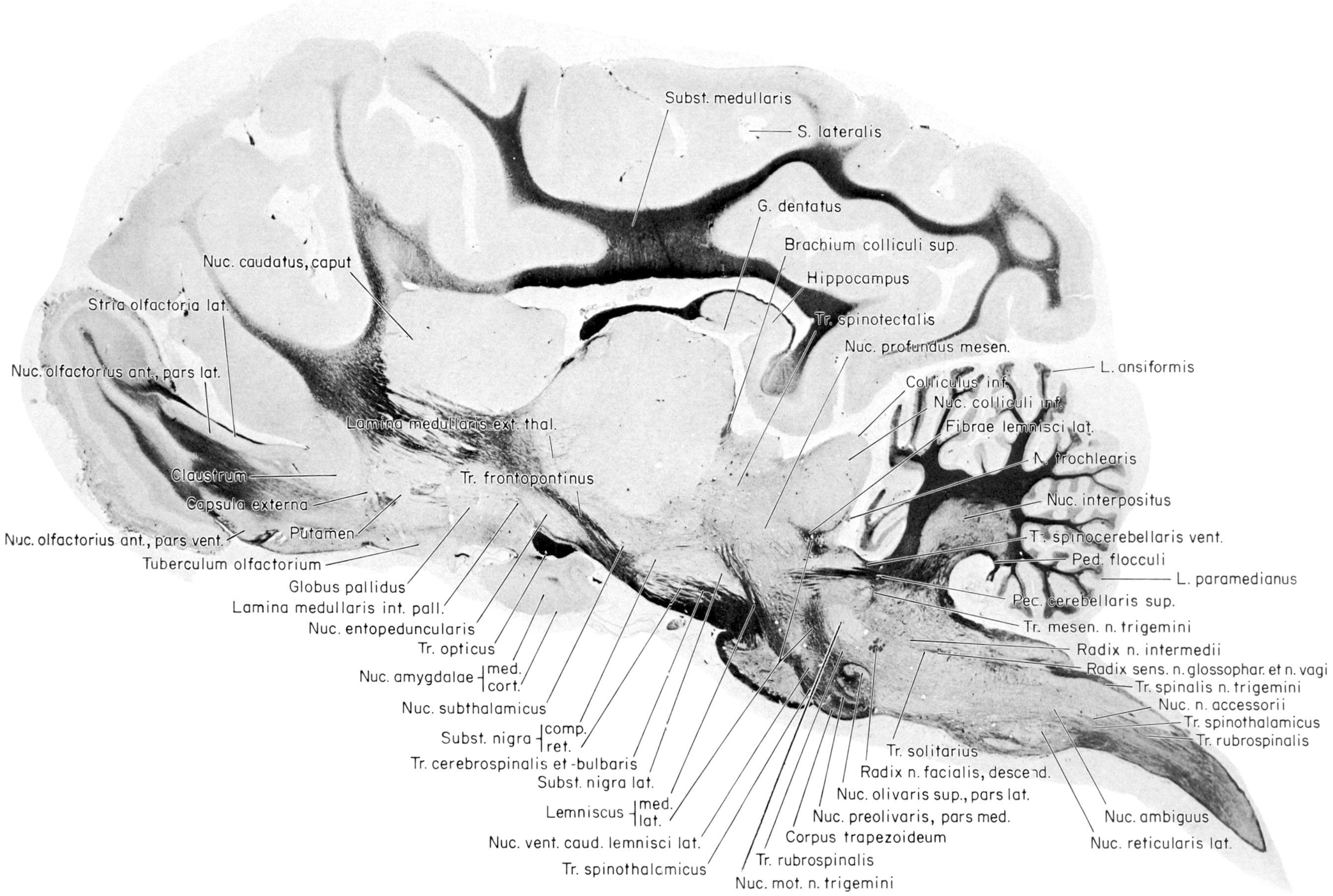

Plate 9. Sagittal section of the brain. (Singer: The Brain of the Dog in Section. 1962.)

S. lateralis

Subst. medullaris

G. dentatus

Brachium colliculi sup.

Hippocampus

Tr. spinotectalis

Nuc. profundus mesen.

Colliculus inf.

Nuc. colliculi inf.

Fibrae lemnisci lat.

N. trochlearis

L. ansiformis

Nuc. interpositus

T. spinocerebellaris vent.

Ped. flocculi

Ped. cerebellaris sup.

Tr. mesen. n. trigemini

Radix n. intermedii

L. paramedianus

Radix sens. n. glossophar. et n. vagi

Tr. spinalis n. trigemini

Nuc. n. accessorii

Tr. spinothalamicus

Tr. rubrospinalis

Nuc. ambiguus

Nuc. reticularis lat.

Tr. solitarius

Radix n. facialis, descend.

Nuc. olivaris sup., pars lat.

Nuc. preolivaris, pars med.

Corpus trapezoideum

Tr. rubrospinalis

Nuc. mot. n. trigemini

Nuc. vent. caud. lemnisci lat.

Tr. spinothalamicus

Lemniscus med. / lat.

Subst. nigra lat.

Tr. cerebrospinalis et-bulbaris

Subst. nigra comp. / ret.

Nuc. subthalamicus

Nuc. amygdalae med. / cort.

Tr. opticus

Nuc. entopeduncularis

Lamina medullaris int. pall.

Globus pallidus

Tr. frontopontinus

Lamina medullaris ext. thal.

Tuberculum olfactorium

Putamen

Capsula externa

Claustrum

Nuc. olfactorius ant., pars vent.

Stria olfactoria lat.

Nuc. caudatus, caput

Nuc. olfactorius ant., pars lat.

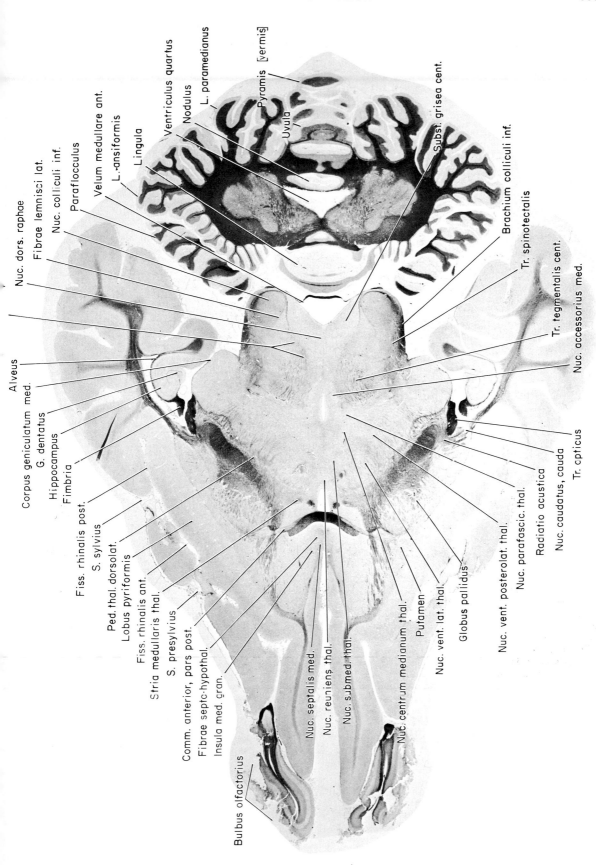

Pyramis [vermis]

Uvula

Nodulus

L. paramedianus

Ventriculus quartus

Lingula

L. ansiformis

Velum medullare ant.

Paraflocculus

Nuc. colliculi inf.

Fibrae lemnisci lat.

Nuc. dors. raphae

Subst. grisea cent.

Brachium colliculi inf.

Tr. spinotectalis

Tr. tegmentalis cent.

Nuc. accessorius med.

Tr. tegmentalis cent.

Tr. cpticus

Nuc. caudatus, cauda

Radiatio acustica

Nuc. parafascic. thal.

Nuc. vent. posterolat. thal.

Globus pallidus

Nuc. vent. lat. thal.

Putamen

Nuc. centrum medianum thal.

Nuc. submed. thal.

Nuc. reuniens thal.

Nuc. septalis med.

Bulbus olfactorius

Insula med. gran.

Fibrae septo-hypothal.

Comm. anterior, pars post.

S. presylvius

Stria medullaris thal.

Fiss. rhinalis ant.

Lobus pyriformis

Ped. thal. dorsolat.

S. sylvius

Fiss. rhinalis post.

Fimbria

Hippocampus

G. dentatus

Corpus geniculatum med.

Alveus

Plate 10. Horizontal section of the brain. (Singer: The Brain of the Dog in Section. 1962.)

la oblongata after traversing the lateral part of the olivary nucleus (Figs. 14–4, 14–5, Plate 7).

The remainder of the medulla oblongata is occupied by the **reticular formation** (*formatio reticularis*). It consists of a number of more or less well-defined reticular nuclei, including the nucleus of the **lateral funiculus**, which was referred to in connection with the ventral superficial arcuate fibers. A number of areas in the reticular formation aid in visceral functions such as cardiovascular and respiratory control, but these physiological centers are difficult to define morphologically.

At the caudal end of the medulla oblongata, the central nervous system continues as the **spinal cord** (*medulla spinalis*).

BIBLIOGRAPHY

Ackerknecht, E. 1943. Die Zentralorgane Rückenmark und Gehirn. Ellenberger-Baum, Handbuch der Vergleichenden Anatomie der Haustiere. Berlin, Springer.

Adams, W. E. 1958. The Comparative Morphology of the Carotid Body and Carotid Sinus. Springfield, Ill., Charles C Thomas.

Adrianov, O. S., and T. A. Mering. 1959. Atlas mozga sobaki. (Atlas of the Brain of the Dog). Gos. izd. med. lit. (State publishing house for medical literature). Moscow, Medgiz.

Akert, K., H. D. Potter, and J. W. Anderson, 1961. The subfornical organ in mammals; I. Comparative and topographical anatomy. J. Comp. Neurol. 116:1–13.

Allen, W. F. 1937. Olfactory and trigeminal conditioned reflexes in dogs. Am. J. Physiol. 118:532–540.

— — — — —. 1944. Degeneration in the dog's mammillary body and Ammon's horn following transection of the fornix. J. Comp. Neurol. 80:283–291.

— — — — —. 1945. Effect of destroying three localized cerebral cortical areas for sound on correct conditioned differential responses of the dog's foreleg. Am. J. Physiol. 144:415–428.

— — — — —. 1948. Fiber degeneration in Ammon's horn resulting from extirpations of piriform and other cortical areas and from transection of the horn at various levels. J. Comp. Neurol. 88:425–438.

Andres, K. H. 1965. Ependymkanälchen im Subfornikalorgan vom. Hund. Naturwiss. 52: 433.

— — — — —. 1965. Der Feinbau des Subfornikalorganes vom Hund. Z. Zellforsch. 68:445–473.

Andreyev, L. A. 1935a. Functional changes in the brain of the dog after reduction of cerebral blood supply; I. Cerebral circulation and the development of anastomosis after ligation of the arteries. Arch. Neurol. Pyschiat. (Chic.) 34:481–507.

— — — — —. 1935b. Functional changes in the brain of the dog after reduction of the cerebral blood supply; II. Disturbances of conditioned reflexes after ligation of arteries. Arch. Neurol. Psychiat. (Chic.) 34:699–713.

Arey, L. B., S. R. Bruesch, and S. Castanares. 1942. The relation between eyeball size and number of optic nerve fibers in the dog. J. Comp. Neurol. 76:417–422.

Arey, L. B., and M. Gore. 1942. The numerical relation between the ganglion cells of the retina and the fibers in the optic nerve of the dog. J. Comp. Neurol. 77:609–617.

Ariëns Kappers, C. U., G. C. Huber, and E. C. Crosby. 1936. The Comparative Anatomy of the Nervous System of Vertebrates, Including Man. New York, Macmillan Co.

Asratian, E. 1935. Motor defensive conditioned reflexes in dogs with extirpated cortical motor areas of the cerebral hemispheres. C.R. Acad. Sci. (U.S.S.R.) 1:159–164. (Quoted from Biol. Abstr. 1937/11:3517.)

Augustine, J. R., B. Vidić, and P. A. Young. 1971. The intermediate root of the trigeminal nerve in the dog (Canis familiaris). Anat. Rec. 169:697–703.

Badoni, C. T. 1973. Relationship of sensory and motor cortex to the segments of the dog spinal cord. Folia Morphol. (Praha) 21:341–344.

Bahrs, A. M. 1927. Notes on reflexes of puppies in the first six weeks after birth. Am. J. Physiol. 82:51–55.

Bailey, P., and W. Haynes. 1940. Location of inhibitory respiratory center in cerebral cortex of the dog. Proc. Soc. exp. Biol. (N.Y.) 45:686–687.

Balado, M. 1925. Anatomia externa del encephalo del perro. Bol. Inst. Clin. quir. (B. Aires) 1:128–168.

Barnhard, J. W. 1940. The hypoglossal complex of vertebrates. J. Comp. Neurol. 72:489–524.

Barone, R. 1960. La substance blanche et ses courants de fibres dans la moelle épinière des mammifères. Rev. Méd. Vét. 111:200–248.

Bartley, S. H., and E. B. Newman. 1931. Studies on the dog's cortex; I. The sensorimotor areas. Am. J. Physiol. 99:1–8.

Bary, A. 1898. Über die Entwickelung der Rindencentren. Arch. Anat. Physiol. Physiol. Abt. 1898:341–360.

Basir. M. A. 1932. The vascular supply of the pituitary body in the dog. J. Anat. (London) 66:387–398.

von Bechterew, W. 1891. Zur Frage über die äusseren Associationsfasern der Hirnrinde. Neurol. Zbl. 10:682–684.

— — — — —. 1899. Die Leitungsbahnen im Gehirn und Rückenmark. Leipzig, Georgi.

Becker, H. 1952. Zur Faseranatomie des Stamm- und Riechhirns auf Grund von Experimenten an jugendlichen Tieren. Dtsch. Z. Nervenheilk. 168:345–383.

Bekhterieff, V. 1886. De l' excitabilité des centres moteurs de l'écorce cérébrale chez les chiens nouveaunés. Arch. Slav. Biol. Mem. orig. Anat. Physiol. 2:191–198.

Bellucci, L. 1929. Sulla nuova sede del centro corticale laringeo nel cane. Am. J. Physiol. 90:279–280.

Berger, H. 1900. Experimentell-anatomische Studien über die durch den Mangel optischer Reize veranlassten Entwicklungshemmungen im Occipitallappen des Hundes und der Katze. Arch. Psychiat. Nervenkr. 33:521–567.

Berkeley, H. J. 1894. The cerebellar cortex of the dog. Johns Hopk. Hosp. Rep. 3:195–214.

Berlucci, C. 1927. Ricerche di fine anatomia sul claustrum e sull' insula del gatto. Riv. Sper. Freniat. 51:125–157.

Bertrand, I. C. Medynski, and P. Salles. 1936. Etude d'un cas d'agénésie du vermis cérébelleux chez le chien. Rev. Neurol. 66:716–733.

Bianchi, L., and G. d'Abundo. 1886a. Le degenerazioni discendenti sperimentali nel cervello e nel midollo spinale a contributo della dottrina delle localizzazioni cerebrali. Resoc. Accad. Med. Chir. Napoli 40:92–101.

— — — — —. 1886b. Die ins Gehirn und Rückenmark absteigenden experimentalen Degenerationen als Beitrag zur Lehre von den cerebralen Lokalisierungen. (abstract, transl.) Neurol. Zbl. 5:385–391.

Bikeles, G. 1894. Anatomische Befunde bei experimenteller Porencephalie am neugeborenen Hunde. Arb. Neurol. Inst. Univ. Wien 2:91–109.

Billenstien, D. C. 1953. The vascularity of the motor cortex of the dog. Anat. Rec. 117:129–144.

Bishop, G. H., and M. H. Clare. 1955. Organization and distribution of fibers in the optic tract of the cat. J. Comp. Neurol. 103:269–304.

Bleier, R. 1961. The Hypothalamus of the Cat. Baltimore, The Johns Hopkins Press.

Blinkov, S. M., and V. S. Ponomarev. 1965. Quantitative determinations of neurons and glial cells in the nuclei of the facial and vestibular nerves in man, monkey and dog. J. Comp. Neurol. 125:295–301.

Böhme, G. 1967. Unterschiede am Gehirnventrikelsystem von Hund und Katze nach Untersuchungen an Ausgusspräparaten. Berl. Münch. Tierärztl. Wschr. 80:195–196.

Bonvallet, M., P. Dell, and F. Stutinsky. 1949. Lésions hypothalamiques et comportement émotional chez le chien. C. R. Soc. Biol. (Paris) 143:80–83.

Bossy, J. 1955. Étude Topographique et Volumétrique des Noyaux d'Origine des Nerfs Craniens Chez l'Homme et Quelques Mammifères. Lyon, Bosc. Frères.

Bottazzi, F. 1893. Intorno alla corteccia cerebrale e specialmente intorno alle fibre nervose intracorticali dei vertebrati. Ric. fatte nel lab. di Anat. norm. R. Univ. Roma 3:241–316.

Bourdelle, E., and C. Bressou. 1953. Anatomie régionale des Animaux Domestiques. IV. Carnivores, Chien et Chat. Paris, Baillières et fils.

Bradley, O. C., and T. Grahame. 1948. Topographical Anatomy of the Dog. Edinburgh and London, Oliver and Boyd.

Breazile, J. E., and W. D. Thompson, 1967. Motor cortex of the dog. Am. J. Vet. Res. 28:1483–1486.

Brodal, A. 1957. The Reticular Formation of the Brain Stem. Edinburgh and London, Oliver and Boyd.

— — — — —. 1959. The Cranial Nerves; Anatomy and Anatomicoclinical Correlations. Springfield, Ill., Charles C Thomas.

Brodmann, K. 1905–06. Beiträge zur histologischen Lokalisation der Grosshirnrinde; IV. Der Riesenpyramidentypus und sein Verhalten zu den Furchen bei den Karnivoren. J. Psychol. Neurol. (Lpz.) 6:108–120.

Brown, J. O. 1943a. The nuclear pattern of the non-tectal portions of the midbrain and isthmus in the dog and cat. J. Comp. Neurol. 78:365–405.

— — — — —. 1943b. Pigmentation of the substantia nigra and the locus coeruleus in certain carnivores. J. Comp. Neurol. 79:393–405.

— — — — —. 1944. Pigmentation of certain mesencephalic tegmental nuclei in the dog and cat. J. Comp. Neurol. 81:249–257.

Bruni, A. C., and U. Zimmerl. 1951. Anatomia degli Animali Domestici. Milano, Francesco Vallardi.

Brutkowski, S., and J. Dabrowska. 1966. Prefrontal cortex control of differentiation behavior in dogs. Acta Biol. Exp. (Warsaw) 26:425–439.

Brutkowski, S., E. Fonberg, and E. Mempel. 1961. Angry behavior in dogs following bilateral lesion in the genual portion of the rostral cingulate gyrus. Acta Biol. Exp. (Warsaw) 21:199–205.

Brutkowski, S., E. Fonberg, J. Kreiner, E. Mempel, and B. Sychowa. 1962. Aphagia and adipsia in a dog with bilateral complete lesion of the amygdaloid complex. Acta Biol. Exp. (Warsaw) 22:43–50.

Burakowska, J. 1966. Extreme capsule in the dog: Myeloarchitectonics. Acta Biol. Exp. (Warsaw) 26:123–133.

Bürgi, S., and V. M. Bucher. 1960. Markhaltige Faserverbindungen im Hirnstamm der Katze. Monogr. aus dem Gesamtgebiet der Neurologie und Psychiatrie. Berlin, Springer.

Buxton, D. F. 1967. Function and anatomy of the corticospinal tracts of the dog and raccoon. Anat. Rec. 157:222–223.

Buxton, D. F., and D. C. Goodman. 1967. Motor function and the corticospinal tracts in the dog and raccoon. J. Comp. Neurol. 129:341–360.

Buytendijk, F. J. J. 1924. Über die Formwahrnehmung beim Hunde. Arch. ges. Physiol. 204:4–14.

— — — — —. 1936. The Mind of the Dog. New York, Houghton Mifflin.

Buytendijk, F. J. J., and W. Fischel. 1933. Über akustische Wahrnehmungen des Hundes. Arch. Néerl. Physiol. 18: 265–289.

Bykoff, K. 1925. Versuche an Hunden mit Durchschneiden des Corpus callosum. (abstract, transl.) Zbl. ges. Neurol. Psychiat. 39:199.

Campbell, A. E. 1904–05. Further histological studies on the localisation of cerebral function — the brains of *Felis*, *Canis* and *Sus* compared with that of *Homo*. Proc. Roy. Soc. Lond. 74:390–392.

— — — — —. 1905. Histological Studies on the Localisation of Cerebral Function. Cambridge, University Press.

Caradonna, G. B. 1902. Ricerche originali sulla forma normale del cervello del cane ed i suoi rapporti col cranio, con la età e col sesso, con la estensione della superficie cerebrale, con lo sviluppo del lobo frontale e con alcune particolarità delle scissure, solchi e circumvoluzioni cerebrali. Ann. Fac. Med. Perugia 2:5–103.

Chorazyna, H., and L. Stepien. 1961. Impairment of auditory recent memory produced by cortical lesions in dogs. Acta Biol. Exp. (Warsaw) 21:177–178.

Chusid, J. G., C. G. DeGutierrez-Mahoney, and F. Robinson. 1949. The "motor" cortex of the dog. Fed. Proc. 8:25.

Clark, R. 1968. Postnatal myelinization in the central nervous system of the beagle dog. Anat. Rec. 160:331.

Cohn, H. A., and J. W. Papez. 1933. The posterior

calcarine fissure in dog. J. Comp. Neurol. 58:593–602.

Cohrs, P. 1936. Das subfornikale Organ des 3. Ventrikels. Z. Anat. Entwickl.-Gesch. 105:491–518.

Collin, R., and P. Grognot. 1938. Sur les images alvéolaires de l'hypothalamus chez le chien. C. R. Soc. Biol. (Paris) 127:1029–1032.

Corder, R. L., and H. B. Latimer. 1947. The growth of the brain in the fetal dog. Anat. Rec. 97:383.

—————. 1949. The prenatal growth of the brain and of its parts and of the spinal cord in the dog. J. Comp. Neurol. 90:103–212.

Crosby, E. C., T. Humphrey, and E. W. Lauer. 1962. Correlative Anatomy of the Nervous System. New York, Macmillan Co.

Culler, E. A., and F. A. Mettler. 1934. Observations upon conduct of a thalamic dog; hearing and vision in decorticated animals. Proc. Soc. Exp. Biol. (N.Y.) 31:607–609.

Cummings, J. F., and J. M. Petras. 1977. The origin of spinocerebellar pathways. I. The nucleus cervicalis centralis of the cervical spinal cord. J. Comp. Neurol. 173:655–691.

Curtis, H. J. 1940. Intercortical connections of corpus callosum as indicated by evoked potentials. J. Neurophysiol. 3:407–413.

Davison, C., and W. M. Kraus. 1929. The measurement of the cerebral and cerebellar surfaces; VII. The measurement of visible and total cerebral surfaces of some vertebrates and of man. Arch. Neurol. Psychiat. (Chic.) 22:105–122.

DeGroot, J. 1958–59. A model of the rhinencephalon in the cat (Felis domestica). Acta Morphol. Neerl.-Scand. 2:140–144.

DeLahunta, A. 1977. Veterinary Neuroanatomy and Clinical Neurology. Philadelphia, W. B. Saunders Co.

Delgado, J. M. R. 1948. Movimentos de marcha provocados por excitación de un punto profundo en la corteza cerebral del perro. Rev. Esp. Fisiol. 4:173–180.

Delmas-Marsalet, P. 1932a. Le syndrome frontal de déséquilibre chez le chien. C. R. Soc. Biol. (Paris) 110:966–967.

—————. 1932b. Études sur les connexions anatomiques du lobe frontal chez le chien. C.R. Soc. Biol (Paris) 111:795–796.

Demoor, J. 1899. Les Centres Sensitivo-moteurs et les Centres d'Association Chez le Chien. Bruxelles, Hayez.

Dexler, H. 1932. Die Entwicklung und der feinere Aufbau des zentralen Nervensystems, Ellenberger-Baum, Handbuch der vergleichenden Anatomie der Haustiere. Berlin, Springer.

Dimić, J. M., and S. S. Nonin. 1954. Beitrag zur Physiologie des prefrontalen Gehirnteiles bei Haustieren. Acta Vet. (Beogr.) 4:31–36.

Döllken, A. 1898. Die Reifung der Leitungsbahnen im Thiergehirn. Vorläufige Mittheilung. Neurol. Zbl. 17:996–998.

Dow, R. S. 1940. Partial agenesis of the cerebellum in the dog. J. Comp. Neurol. 72:569–586.

Dresel, K. 1924. Die Funktion eines grosshirn- und striatum-losen Hundes. Klin. Wschr. 3:2231–2233.

Dua-Sharma, S., K. N. Sharma, and H. L. Jacobs. 1970. The canine brain in stereotaxic coordinates. Cambridge, Mass., MIT Press.

Dumenko, V. N. 1961. Changes in the electrical activity of the cerebral cortex of dogs during the formation of a stereotype of motor conditional reflexes. Zhur. Vysshei Nervnoi Deiatel' nosti im I. P. Pavlova (Transl.) 11:292–299.

Dziurdzik, B. 1965. Frontal lobe sulci in the dog. Acta Biol. Exp. (Warsaw) 25:245–261.

Edinger, L. 1897. Twelve Lectures on the Structure of the Central Nervous System. Philadelphia, F. A. Davis Co.

—————. 1911. Vorlesungen über den Bau der nervösen Zentralorgane des Menschen und der Tiere. Band 1: Das Zentralnervensystem des Menschen und der Säugetiere. Leipzig, Vogel.

Eliasson, S., P. Lindgren, and B. Uvnäs. 1952. Representation in the hypothalamus and the motor cortex in the dog of the sympathetic vasodilator outflow to the skeletal muscle. Acta Physiol. Scand. 27:18–37.

Ellenberger, W. 1889. Über die Furchen und Windungen der Grosshirnoberfläche des Hundes. Arch. wiss. prakt. Tierheilk, 15:263–282.

Ellenberger, W., and H. Baum. 1891. Systematische und topographische Anatomie des Hundes. Berlin, Parey.

—————. 1943. Handbuch der vergleichenden Anatomie der Haustiere. 18th Ed. Berlin, Springer.

Elliot Smith, G. 1895–96. Morphology of the true "limbic lobe," corpus callosum, septum pellucidum and fornix. J. Anat. (Lond.) 30:157–167; 185–205.

Evans, H. E., and A. deLahunta. 1971. Miller's Guide to the Dissection of the Dog. Philadelphia, W. B. Saunders Co.

Fankhauser, R. 1947. Makroskopische Anatomie der Stammganglien bei Pferd und Hund. Schweiz. Arch. Neurol. Psychiat. 60:400–401.

—————. 1962. The Cerebrospinal Fluid, Chapter 3, pp. 21–54, in Comparative Neuropathology, edited by J. R. M. Innes and L. Z. Saunders. New York, Academic Press.

Feliciangeli, G. 1910. Experimenteller Beitrag zur Kenntnis der Funktion des Stirnlappens des Hundehirns. Folia Neuro-Biol. (Lpz.) 4:449–466.

Ferrier, D. 1873. Experimental researches in cerebral physiology and pathology. West Riding Lunatic Asylum Medical Reports 3:30–96.

—————. 1880. The Functions of the Brain. New York, G. P. Putnam's Sons.

Filimonoff, I. N. 1928. Über die Varianten der Hirnfurchen des Hundes. J. Psychol. Neurol. (Lpz.) 36:22–43.

Fitzgerald, T. C. 1961. Anatomy of the cerebral ventricles of domestic animals. Vet. Med. 56:38–45.

Flatau, E., and L.Jacobsohn. 1899. Handbuch der Anatomie und vergleichenden Anatomie des Centralnervensystems der Säugetiere. I. Makroskopischer Teil. Berlin, Karger.

Florio, A. 1943–47. Alcuni rilievi anatomo-microscopici sul cosiddetto Gyrus sigmoideus di Canis Fam. L. Bol. Zool. 14:1–21.

Forel, A. 1872. Beiträge zur Kenntnis des Thalamus opticus und der ihn umgebenden Gebilde bei den Säugethieren. S.-B. Akad. Wiss. Wien, Math.-Nat. Kl. 66:25–58.

Fox, C. A., 1943. The stria terminalis, longitudinal association bundle and precommissural fornix fibers in the cat. J. Comp. Neurol. 79:277–295.

Fox, C. A., and J. T. Schmitz. 1943. A Marchi study of

the distribution of the anterior commissure in the cat. J. Comp. Neurol. 79:297–314.

Fox, M. W. 1963. Gross structure and development of the canine brain. Am. J. Vet. Res. 24:1240–1247.

—————. 1968. Neuronal development and ontogeny of evoked potentials in auditory and visual cortex of the dog. Electroencephalogr. Clin. Neurophysiol. 24:213–226.

Fox, M. W., and O. R. Inman 1966. Persistence of Retzius-Cajal cells in developing dog brain. Brain Res. 3:192–194.

Fox, M. W., O. R. Inman, and W. A. Himwich 1966. The postnatal development of the neocortical neurons in the dog. J. Comp. Neurol. 127:199–206.

Frandson, R. D. 1955. Herring bodies in the hypothalamic region of the Greyhound. M. S. Thesis. Colorado State University, Fort Collins, Colorado.

Frank, C. 1930. Intorno alla mia scoperta di due nuclei del mesencefalo dell' uomo ed ulteriori studi sui nuclei oculomotori dei mammiferi. Arch. Gen. Neurol. Psichiat. 11:1–41.

Frauchiger, E., and R. Fankhauser. 1949. Die Nervenkrankheiten unserer Hunde. Bern, Huber.

Friede, R. L. 1961. Surface structures of the aqueduct and the ventricular walls; a morphologic, comparative and histochemical study. J. Comp. Neurol. 116:229–247.

Fritsch, G. 1884. Demonstration des Gehirns des von Herrn Goltz auf dem III. Congress für innere Medicin vorgestellten Hundes. Dtsch. med. Wschr. 10:353–355.

Fritsch, G., and E. Hitzig. 1870. Über die elektrische Erregbarkeit des Grosshirns. Arch. Anat. Physiol. wiss. Med. 1870:330–332.

Fujita, H. 1957. Electron microscopic observation on the neurosecretory granules in the pituitary posterior lobe of the dog. Arch. Histol. Jap. 12:165–172.

Fuse, G. 1919. Experimenteller Beitrag zur Anatomie des roten Kerns bei einem Hunde (Guddensche Methode) und bei drei Kaninchen (Nisslsche Methode)...Vertikale Durchschneidung der interrubralen Region resp. des medialen Segmentes der Mittelhirnhaube. Arb. anat. Inst. Sendai 2:49–86.

—————. 1920a. Beiträge zur mikroskopischen Anatomie des Truncus cerebri. Arb. Anat. Inst. Sendai 4:1–107.

—————. 1920b. Über eine neue Endigungsstätte des v. Monakowschen rubrospinalen Bündels beim Hunde und Kaninchen. Arb. Anat. Inst. Sendai 5:83–86.

—————. 1926. Vergleichend-anatomische Beobachtungen am Hirnstamme der Säugetiere; VIII. Eine weitere Bemerkung über den Nucleus ventralis accessorius lemnisci lateralis bei einigen Karnivoren (Katze, Hund, Fuchs, Dachs, Meles anakuma). Arb. Anat. Inst. Sendai 12:39–44.

Gabrawi, A. F., and A. A. Tarkhan. 1967. On the mesencephalic nucleus and root of the fifth cranial nerve (dog, cat). Acta Anat. 67:550–560.

Gagel, O., and W. Mahoney. 1933. Zur Frage des Zwischenhirn-Hypophysensystems. Z. ges. Neurol. Psychiat. 148:272–279.

Ganser, S. 1879. Über die vordere Hirncommissur der Säugetiere. Arch. Psychiat. Nervenkr. 9:286–299.

Gantt, W. H. 1948. Cardiac reaction in partially decorticated dogs. Trans. Am. Neurol Assoc. 73:131–133.

Getty, R. 1975. Sisson and Grossman's Anatomy of the Domestic Animals. 5th Ed. Philadelphia, W. B. Saunders Co.

Gheție, V., I. T. Riga, and E. Pastea. 1956. Anatomia sistemului nervos central si neurovegetativ la animalele domestice. Bucresti, Editura Agro-Silvica de Stat.

Girden, E. 1938. Cerebral mechanisms and auditory localisation in dogs. Psychol. Bull. 35:699–700.

Glorieux, P. 1929. Anatomie et connexions thalamiques chez le chien. J. Neurol. (Brux.) 29:525–555.

Goldani, J. J., and C. H. Funke. 1963/64.Pesquisas morfologicas sobre as arquiteturas glio-vasculares do corpo estriado do cao e suas modificacoes com a idade. Univ. Rio Grande Sul. Arq. Inst. Anat. 6:97–140. (Quoted from Biol. Abstr. 48/1967: 2046.)

Goldberg, R. C., and I. L. Chaikoff. 1952. On the occurrence of six cell types in the dog anterior pituitary. Anat. Rec. 112:265–274.

Goldzbrand, M. G., S. E. Goldberg, and G. Clark. 1951. Cessation of walking elicited by stimulation of the forebrain of the unanesthetized dog. Am. J. Physiol. 167:127–133.

Goltz, F. 1884. Über die Verrichtungen des Grosshirns. Fünfte Abhandlung. Arch. ges. Physiol. 34:450–505.

—————. 1888. Über die Verrichtungen des Grosshirns. Sechste Abhandlung. Arch. ges. Physiol. 42:419–467.

—————. 1892. Der Hund ohne Grosshirn. Siebente Abhandlung über die Verrichtungen des Grosshirns. Arch. ges. Physiol. 51:570–614.

Gorska, T., and J. Czarkowska. 1978. Motor cortex development in the dog. Some cortical stimulation and behavioral data. Neurosci. Behav. Physiol. 3:129–131.

Gröschel, G. 1930. Über die Cytoarchitektonik und Histologie der Zwischenhirnbasis beim Hund. Dtsch. Z. Nervenheilk, 112:108–123.

Gross, S. W., 1939. Cerebral arteriography in the dog and in man with a rapidly excreted organic iodide. Proc. Soc. exp. Biol. (N.Y.) 42:258–259.

Grünthal, E. 1929. Der Zellaufbau des Hypothalamus beim Hunde. Z. ges. Neurol. Psychiat. 120:157–177.

Gurewitsch, M., and C. Bychowski. 1928. Zur Architektonik der Hirnrinde (Isocortex) des Hundes. J. Psychol. Neurol. 35:283–300.

Gurewitsch, M., and A. Chatschaturian. 1928. Zur Cytoarchitektonik der Grosshirnrinde der Feliden. Z. Anat. Entwickl. Gesch. 87:100–138.

Hagen, E. 1957. Morphologische Beobachtungen im Hypothalamus und in der Neurohypophyse des Hundes nach Teilläsion des Infundibulum. Acta Anat. (Basel) 31:193–219.

Hagg, S., and H. Ha. 1970. Cervicothalamic tract in the dog. J. Comp. Neurol. 139:357–373.

Haines, D. E., and T. W. Jenkins. 1968. Studies on the epithalamus. I. Morphology of post-mortem degeneration: The habenular nucleus in dog. J. Comp. Neurol. 132:405–417.

Hammouda, M. 1933. The central and the reflex mechanism of panting. J. Physiol. (Lond.) 77:319–336.

Hamuy, T. P., R. B. Bromiley, and C. N. Woolsey, 1950. Somatic afferent areas I and II of the dog's cerebral cortex. Am. J. Physiol. 163:719–720.

— — — — —. 1956. Somatic afferent areas I and II of dog's cerebral cortex. J. Neurophysiol. *19*:485–499.

Heidreich, L. 1931. Beitrag aus der Entstehung des Hydrocephalus internus und der ventrikulären Liquor-Resorptionsstellen. Bruns' Beitr. klin. Chir. *151*:607–612.

Heinbecker, P., and H. L. White. 1941. Hypothalamico-hypophysial system and its relation to water balance in the dog. Am. J. Physiol. *133*:582–593.

Herre, W., and H. Stephan. 1955. Zur postnatalen Morphogenese des Hirnes verschiedener Haushundrassen. Morph. Jb. *96*:210–264.

Himwhich, H. E., and J. F. Fazekas. 1941. Comparative studies of the metabolism of the brain of infant and adult dogs. Am. J. Physiol. *132*:454–459.

Hitzig, E. 1900a. Über das corticale Sehen des Hundes. Arch. Psychiat. Nervenkr. *33*:707–720.

— — — — —. 1900b. Über den Mechanismus gewisser cortikaler Sehstörungen des Hundes. Berl. klin. Wschr. *37*:1001–1003.

— — — — —. 1901. Alte und neue Untersuchungen über das Gehirn. Arch. Psychiat. Nervenkr. *34*:1–38.

— — — — —. 1901–02a. Alte und neue Untersuchungen über das Gehirn. Arch. Psychiat. Nervenkr. *35*:275–392.

— — — — —.1901–02b. Alte und neue Untersuchungen über das Gehrin III. Über die Beziehungen der Rinde und der subcorticalen Ganglien zum Sehact des Hundes. Arch. Psychiat. Nervenkr. *35*:585–611.

— — — — —. 1902–03a. Alte und neue Untersuchungen über das Gehirn IV. (Fortsetzung). Arch. Psychiat. Nervenkr. *36*:1–96.

— — — — —. 1902–03b. Über die Function der motorischen Region des Hundehirns und über die Polemik des Herrn H. Munk. Arch. Psychiat. Nervenkr. *36*:605–629.

— — — — —. 1903. Alte und neue Untersuchungen über das Gehirn. (Fortsetzung und Schluss). Arch. Psychiatr. Nervenkr. *37*:299–467; 849–1013.

— — — — —. 1904. Physiologische und klinische Grundlagen über das Gehirn. Berlin, Hirschwald.

Hoenig, H. 1912. Vergleichend-anatomische Untersuchung über den Hirnfurchentypus der Caniden mit besonderer Berücksichtigung des *Canis dingo*. Diss. Berlin.

Hoffmann, G. 1955. Topographischer und zytologischer Atlas der Medulla oblongata von Schwein und Hund. Berlin, Deutsche Akademie der Landwirtschaftswissenschaften.

Holbrook, J. R., H. Schapiro, 1974. The accessory optic tract in the dog; a retino-entopeduncular pathway. J. Hirnforsch. *15*:365–377.

Holl, M. 1899. Über die Insel des Carnivorengehirnes. Arch. Anat. Physiol. Anat. Abt. *1899*:217–266.

Holmes, G. M. 1901. The nervous system of the dog without a forebrain. J. Physiol. (Lond.) *27*:1–25.

Houston, M. L. 1968. The early brain development of the dog. J. Comp. Neurol. *134*:371–383.

Howard, D. R., and J. E. Breazile. 1973. Optic fiber projections to dorsal lateral geniculate nucleus in the dog. Am. J. Vet. Res. *34*:419–424.

Hukuda, S., H. D. Jameson, and C. B. Wilson, 1973. Experimental cervical myelopathy. 3. The canine corticospinal tract. Anatomy and function. Surg. Neurol. *1*:107–114.

Innes, J. R. M., and L. Z. Saunders. 1962. Comparative Neuropathology. New York, Academic Press.

Iwai, E. 1961. Visual learning and retention after ablation of inferotemporal cortex in dogs. Tohoku J. Exp. Med. 75:243–258.

Iwama, K., and C. Yamamoto, 1961. Nature of the secondary discharge of negative polarity in the cerebral cortex of cats and dogs. Tohoku J. Exp. Med. 75:43–54.

James, W. 1890. The Principles of Psychology. New York, Henry Holt.

Janischewski. 1902. Über die Technik der Durchschneidung des Corpus callosum bei experimentellen Versuchen. Neurol. Zbl. *21*:278–279.

Janklewicz, E. 1967. Habenular complex in the dog's brain. Acta Biol. Exp. (Warsaw) 27:367–387.

Jasper, H. H., and C. Ajmone-Marsan. 1954. A Stereotaxic Atlas of the Diencephalon of the Cat. Ottawa, Canada, Nat. Res. Council.

Jenkins, T. W. 1978. Functional Mammalian Neuroanatomy. Philadelphia, Lea & Febiger.

Jewell, P. A. 1953. The occurrence of vesiculated neurones in the hypothalamus of the dog. J. Physiol. (Lond.) *121*:167–181.

Johnson, H. C., and K. M. Browne, 1954. Cerebral cortical ablations in dogs with chronic renal hypertension. J. Neurophysiol. *17*:183–188.

Kaada, B. R. 1951. Somato-motor, autonomic and electrocorticographic responses to electrical stimulation of "rhinencephalic" and other structures in primates, cat and dog: A study of responses from the limbic, subcallosal, orbito-insular, piriform and temporal cortex, hippocampus-fornix and amygdala. Acta Physiol. Scand. 24(Suppl. 83): 1–285.

Kaes, T. 1891. Die Anwendung der Wolter'schen Methode auf die feinen Fasern der Hirnrinde. Neurol. Zbl. *10*:456–459.

Kalinina, T. E. 1961. The effects of olfactory stimulations on the higher nervous activity of dogs. Zhur. Vysshei Nervnoi Deiatel'nosti im I. P. Pavlova (Transl.) *11*:330–333.

Kalischer, O. 1907. Zur Funktion des Schläfenlappens des Grosshirns. Eine neue Hörprüfungsmethode bei Hunden; zugleich ein Beitrag zur Dressur als physiologischer Untersuchungsmethode. S. B. preuss. Akad. Wiss. *10*:204–216.

Katz, D. 1932. Versuche über die akustische Lokalisation beim Hunde. Z. Hundeforsch. 2:11–17.

Katzenstein, J. 1908. Über die Lautgebungsstelle in der Hirnrinde des Hundes. Arch. Laryngol. Rhin. (Berl.) 20:500–524.

Kellogg, W. N. 1949. Locomotor and other disturbances following hemidecortication in the dog. J. Comp. Physiol. Psychol. *42*:506–516.

Kirk, G. R., and J. E. Breazile. 1972. Maturation of the corticospinal tract in the dog. Exp. Neurol. *35*:394–407.

Kitchell, R. L., M. W. Stromberg, and L. H. Davis. 1977. Comparative study of the dorsal motor nucleus of the vagus nerve. Am. J. Vet. Res. *38*:37–49.

Klein, E. 1883–84. Report on the parts destroyed on the left side of the brain of the dog operated on by Prof. Goltz. J. Physiol (Lond.) *4*:310–315.

Klempin, D. 1921. Über die Architektonik der Gross-hirnrinde des Hundes. J. Psychol. Neurol. (Lpz.) 26:229–249.

Klingler, J. 1935. Erleichterung der makroskopischen Präparation des Gehirns durch den Gefrierprozess. Schweiz. Arch. Neurol. 36:247–256.

Knoche, H. 1952. Neurohistologische Untersuchungen am Hypophysenzwischenhirnsystem des Hundes. Anat. Anz. (Ergh.) 99:93.

— — — — —. 1953. Über das Vorkommen eigen-artiger Nervenfasern (Nodulus Fasern) in Hypo-physe und Zwischenhirn von Hund und Mensch. Acta Anat. (Basel) 18:208–233.

— — — — —. 1957. Über die Ausbreitung und Herkunft der Nervösen Nodulusfasern in Hypothala-mus und Retina. Z. Zellforsch. 48:602–616.

von Korányi, A. 1890. Über die Folgen der Durch-schneidung des Hirnbalkens. Arch. ges. Physiol. 47:35–42.

Kosaka, K. 1909. Über die Vaguskerne des Hundes. Neurol. Zbl. 28:406–410.

Kreiner, J. 1958. The quantitative myelinization of brains and spinal cords in dogs of various size. Acta Anat. 33:50–64.

— — — — —. 1961. The myeloarchitectonics of the frontal cortex of the dog. J. Comp. Neurol. 116:117–133.

— — — — —. 1962a. Myeloarchitectonics of the cingular cortex in dog. J. Comp. Neurol. 119:255–267.

— — — — —. 1962b. The cingular bundle of the dog brain. Acta Biol. Cracoviensia Ser. Zool. 5:253–261.

— — — — —. 1964a. Myeloarchitectonics of the sensorimotor cortex in dog. J. Comp. Neurol. 122:181–200.

— — — — —. 1964b. Myeloarchitectonics of the perisylvian cortex in dog. J. Comp. Neurol. 123:231–241.

— — — — —. 1964c. Myeloarchitectonics of the parietal cortex in dog. Acta Biol. Exp. (Warsaw) 24:195–212.

— — — — —. 1966a. Myeloarchitectonics of the oc-cipital cortex in dog and general remarks on the mye-loarchitectonics of the dog. J. Comp. Neurol. 127:531–557.

— — — — —. 1966b. Reconstruction of neocortical lesions within the dog's brain: Instructions. Acta Biol. Exp. (Warsaw) 26:221–243.

— — — — —. 1967. Myeloarchitectonic differentia-tion of the cerebellar cortex. Acta Biol. Exp. (War-saw) 27:3–9.

— — — — —. 1970. Homologies of the fissural pat-terns of the hemispheres of dog and cat. Acta Biol. Exp. (Warsaw) 30:295–305.

Kreiner, J., and K. Maksymowicz. 1962. A three-dimensional model of the striatal nuclei in the dog's brain. Acta Biol. Exp. (Warsaw) 22:69–79.

Kremer, W. F. 1947. Autonomic and somatic reactions induced by stimulation of the cingular gyrus in dogs. J. Neurophysiol. 10:371–379.

— — — — —. 1948. Blood pressure changes in re-sponse to electrical and chemical (acetyl-beta-methyl choline) stimulation of the cerebral cortex in dogs. Am. J. Physiol. 152:314–323.

Krueg, J. 1880. Über die Furchen auf der Gross-hirnrinde der zonoplacentalen Säugethiere. Z. wiss. Zool. 33:595–672.

Krüger, G. 1942. Über die absolute und die relative Querschnittgrösse der Gyri olfactorii communes und der Nervi optici beim Foxterrier. Vet. Diss. Berlin. (abstract in Jber. Vet. Med. 1942–43/70:135–136.).

Kuntz, A. 1950. A Text-book of Neuro-anatomy. Phila-delphia. Lea & Febiger.

Langley, J. N. 1883–84a. The structure of the dog's brain. J. Physiol. 4:248–285.

— — — — —. 1883–84b. Report on the parts de-stroyed on the right side of the brain of the dog operated on by Prof. Goltz. J. Physiol. (Lond.) 4:286–309.

Laqueur, G. L. 1954. Neurosecretory pathways be-tween the hypothalamic paraventricular nucleus and the neurohypophysis. J. Comp. Neurol. 101:543–563.

Larsell, O. 1953. The cerebellum of the cat and mon-key. J. Comp. Neurol. 99: 135–199.

— — — — —. 1954. The development of the cere-bellum of the pig. Anat. Rec. 118:73–107.

Lassek, A. M., L. W. Dowd, and A. Weil. 1930. The quantitative distribution of the pyramidal tract in the dog. J. Comp. Neurol. 51:153–163.

Latimer, H. B. 1942. The weights of the brain and of its parts, and the weight and length of the spinal cord in the dog. Growth 6:39–57.

— — — — —. 1946. The relative weights of the major divisions of the brain and the cord in several species of animals. Univ. Kansas Sci. Bull. 31:211–221.

— — — — —. 1954. Growth of the hypophysis, thy-roid and suprarenals in the fetal dog. Anat. Rec. 120:495–514.

Laughton, N. B. 1924. Studies on the nervous regula-tion of progression in mammals. Am. J. Physiol. 70:358–384.

— — — — —. 1925.–26. Studies on young decere-brate mammals. Am. J. Physiol. 75:339–350.

— — — — —. 1928. Studies on the occurence of extensor rigidity in mammals as a result of cortical injury. Am. J. Physiol. 85:78–90.

Lazorthes, G. 1943. Note sur le nerf terminal (d'aprés une étude sur le chien). Bull. Soc. Hist. Nat. Tou-louse 78:102–106.

Lee, J. C., M. B. Glover, D. D. Gilboe, and L. Bakay. 1968. Electron microscopy of isolated dog brain. Exp. Neurol. 20:111–119.

Lenys, R. 1966. Contribution à l'étude de la structure et du rôle de l'organe sous-commissural. Ann. Sci. Univ. Besançon Med. 3:1–147.

Leontovich, T. A., and G. P. Zhukova. 1963. The speci-ficity of the neuronal structure and topography of the reticular formation in the brain and spinal cord of carnivora. J. Comp. Neurol. 121:347–379.

Lesbre, M. X. 1884. Crâne et cerveau chez les chiens. J. Méd. Vét. Zootechn. 35:418–424.

Lim, R. K. S., Chan-Nao Liu, and R. L. Moffitt. 1960. A Stereotaxic Atlas of the Dog's Brain. Springfield, Ill., Charles C Thomas.

Lindberg, A. A. 1937. The influence of longitudinal transection of the corpus callosum upon locomotion in the dog. (In Russian.)Trud. Tsentral. Psikhon-evrol. Inst. 8:55–60. (Quoted from Biol. Abstr. 1939/13:4020.)

Lindgren, P., and U. Borje, 1953. Activation of sympathetic vasodilator and vasoconstrictor neurons by electric stimulation in the medulla of the dog and cat. Circulat. Res. 1:479–485.

Löwenthal, N. 1886. La région pyramidale de la capsule interne chez le chien et la constitution du cordon antéro-lateral de la moelle. Rev. Méd. Suisse Rom. 6:529–554.

— — — — —. 1904. Atlas zur vergleichenden Histologie der Wirbeltiere nebst erläuterndem Text. Berlin, Karger.

— — — — —, 1883. Über den Unterschied zwischen der secundären Degeneration des Seitenstrangs nach Hirnund Rückenmarksverletzungen. Arch. ges. Physiol. 31:350–354.

Lucas, F. 1939. Contribution à l'étude de la surface de l'encéphale et de la topographie cranio-encéphalique chez le chien. Diss., Paris.

Luciani, L. 1884–85. On the sensorial localization in the cortex cerebri (transl. by A. de Watteville). Brain 7:145–160.

Luciani, L. and G. Seppilli. 1885. Le locallizzazioni funzionali del cervello. Napoli, Vallardi.

Ludwig, E. 1935. Note technique sur la préparation macroscopique de l'encéphale. C. R. Ass. Anat. 30:347–348.

Ludwig, E., and J. Klingler, 1956. Atlas Cerebri Humani. The Inner Structure of the Brain Demonstrated on the Basis of Macroscopical Preparations. Basel, Karger.

Maksymowicz, K. 1963. Amygdaloid complex of the dog. Acta Biol. Exp. (Warsaw) 23:63–73.

Marburg, O. 1904. Die physiologische Funktion der Kleinhirnseitenstrangbahan (Tractus spinocerebellaris dorsalis) nach Experimenten am Hunde. Arch. Anat. Physiol. Physiol. Abt. Suppl. 1904:457–482.

— — — — —. 1934. Über Verschiedenheiten im Bau des Gehirns hochgezüchteter Hunderassen. Psychiat. Neurol. Bl. (Amst.) 38:386–395.

Marinesco, G. 1895. Des connexions du corps strié avec le lobe frontal. C.R. Soc. Biol. (Paris) 47:77–78.

Marquis, D. G. 1932a. Brightness discrimination in dogs after removal of the striate cortex. Anat. Rec. 52:67.

— — — — —. 1932b. Effects of removal of visual cortex in mammals, with observation on the retention of light discrimination in dogs. Ass. Res. nerv. Dis. Proc. 13:558–592.

Marquis, D. G., and E. R. Hilgard, 1936. Conditioned lid responses to light in dogs after removal of the visual cortex. J. Comp. Psychol. 22:157–178.

Marsden, C. D. 1961. Pigmentation in the nucleus substantiae nigrae of mammals. J. Anat. (Lond.) 95:256–261.

Martin, P. 1893. Zur Entwickelung des Gehirnbalkens bei der Katze. Anat. Anz. 9:156–162.

— — — — —. 1894a. Zur Entwickelung des Gehirnbalkens bei der Katze. Anat. Anz., 9:472–476.

— — — — —. 1894b. Bogenfurche und Balkenentwickelung bei der Katze. Jen. Z. Med. Naturw. 29:221–246.

— — — — —. 1923. Lehrbuch der Anatomie der Haustiere. IV. Band. Stuttgart, Schickhardt und Ebner.

Maspes, P. E. 1932. Studio sulla macroglia della corteccia cerebrale nel cane normale. Riv. Pat. Nerv. Ment. 40:414–439.

Massion, J. 1967. The mammalian red nucleus. Physiol. Rev. 47:383–436.

Mayser, P. 1878. Eine Erwiederung an Herrn Professor P. Flechsig in Leipzig. Arch. Psychiat. Nervenkr. 9:105–121.

McCotter, R. E. 1913. The nervus terminalis in the adult dog and cat. J. Comp. Neurol. 23:145–152.

McGrath, J. T. 1953. The neurologic examination of the dog with some clinico-pathologic observations. Univ. Penn. Bull. Vet. Ext. Quart. 132:5–47.

— — — — —. 1960. Neurologic Examination of the Dog. Philadelphia, Lea & Febiger.

Mettler, F. A., and L. J. Goss. 1946. Canine chorea due to strio-cerebellar degeneration of unknown etiology. J. Amer. Vet. Med. Assoc. 108:377–384.

Meyer, H. 1952. Zur Anatomie des Hundes in Welpenalter. 1. Beitrag: Makroskopisches. Vet. Diss., Zürich.

— — — — —. 1954. Macroscopic brain dissection in veterinary anatomy. Amer. J. Vet. Res. 15:143–146.

— — — — —. 1957. The corticocortical fiber systems of the dog's brain. Ph.D. Diss. Cornell Univ., Ithaca, N.Y.

Michaels, J. J., and C. Davison. 1930. Measurement of cerebral and cerebellar surfaces; VIII. Measurement of the motor area in some vertebrates and in man. Arch. Neurol. Psychiat. (Chic.) 23:1212–1226.

Michaels, J. J., and W. M. Kraus. 1930. Measurements of cerebral and cerebellar surfaces; IX. Measurement of cortical areas in cat, dog, and monkey. Arch. Neurol. Psychiat. (Chic.) 24:94–101.

Miller, M. E. 1958. Guide to the dissection of the dog. Ithaca, N.Y.

Mingazzini, G. 1895. Über die gekreuzte cerebro-cerebellare Bahn. Neurol. Zbl. 14:658–664.

Miodonski, A. 1962. The nucleus accumbens in the brain of the dog. Acta Biol. Cracoviensia. Ser. Zool. 5:109–115.

— — — — —. 1963. Preoptic area of the dog. Acta Biol. Exp. (Warsaw) 23:209–220.

— — — — —. 1967. Myeloarchitectonics of the septum in the brain of the dog. Acta Biol. Exp. (Warsaw) 27:11–59.

— — — — —. 1968a. The myeloarchitectonics of the nuclei of the anterior thalamic group of the dog (Canis familiaris). Acta Biol. Cracoviensia. Ser. Zool. 11:77–92.

— — — — —. 1968b. The myeloarchitectonics of the nucleus parataenialis in the brain of the dog (Canis familiaris). Acta Biol. Cracoviensia. Ser. Zool. 11:111–117.

— — — — —. 1974. The angioarchitectonics and cytoarchitectonics (impregnation modo Golgi-Cox) structure of the fissural frontal neocortex in dog. Folia Biol (Krakow) 22:237–279.

Miodoński, R. 1966. Myeloarchitectonics of the stria terminalis in the dog. Acta Biol. Exp. (Warsaw) 26:135–147.

— — — — —. 1967. Myeloarchitectonics and connections of substantia innominata in the dog brain. Acta Biol. Exp. (Warsaw) 27:61–84.

— — — — —. 1968. Bulbus olfactorius of the dog (Canis familiaris). Acta Biol. Cracoviensia. Ser. Zool. 11:65–76.

— — — — —. 1974. The structure of the posterior piriform cortex in the dog. Acta Anat. (Basel) 88:556–573.

— — — — —. 1975. The claustrum in the dog brain. Acta Anat. (Basel) 91:409–422.

Miskolczy, D. 1931. Über die Endigungsweise der spino-cerebellaren Bahnen. Z. Anat. Entwickl.-Gesch. 96:537–542.

Mittelstrass, H. 1937. Vergleichend-anatomische Untersuchungen über den Mandelkern der Säugetiere. Z. Anat. Entwickl.-Gesch. 106:717–738.

Mobilio, C. 1912. Topografia cranio-encefalica del cane, preceduta dalla descrizione del mantello cerebrale. Intern. Mschr. Anat. Physiol. 29:205–263.

Mogilnitzky, B. N. 1928a. Zur Frage über den Zusammenhang der Hypophyse mit dem Zwischenhirn. Virchows Arch. path. Anat. 267:263–268.

— — — — —. 1928b. Zur Frage über die gegenseitige Beziehung zwischen Hypophyse und Zwischenhirn. Fortschr. Röntgenstr. 37:380–392.

Molliver, M. E., and H. Van der Loos. 1969. The synaptic strata of the somesthetic cortex in neonatal dog. Anat. Rec. 163:317–318.

— — — — —. 1970. The ontogenesis of cortical circuitry: the spatial distribution of synapses in somesthetic cortex of newborn dog. Erg. Anat. Entwickl. Gesch. 42:1–54.

von Monakow, C. 1909; 1910. Der rote Kern, die Haube und die Regio hypothalamica bei einigen Säugetieren und beim Menschen. Arb. Hirnant. Inst. Zürich 3:49–267; 4:103–225.

Morawski, J. 1912. Gehirnuntersuchungen bei Katzen- und Hundefamilien (mit Berücksichtigung des Geschlechts und der Entwicklung). Jb. Psychiat. Neurol. 33:306–477.

Morgan, L. O. 1927. Symptoms and fiber degeneration following experimental lesions in the subthalamic nucleus of Luys in the dog. J. Comp. Neurol. 44:379–401.

— — — — —. 1930a. Cell groups in the tuber cinereum of the dog with a discusssion of their function. J. Comp. Neurol. 51:271–297.

— — — — —. 1930b. The rôle of the tuber cinereum and the thyroid gland in experimental fever in the dog. Anat. Rec. 45:233.

Morin, G., V. Donnet, and P. Zwirn. 1949. Nature et évolution des troubles consécutifs á la section d'une pyramide bulbaire, chez le chien. C. R. Soc. Biol. (Paris) 143:710–712.

Morin, G., Y. Poursines, and V. Donnet. 1949. Pluralité des dégénérescences produites par la cordotomie médullaire postérieure cervicale, chez le chien. C.R. Soc. Biol. (Paris) 143:1127–1129.

Morin, G., Y. Poursines, and S. Maffre. 1951. Sur l'origine de la voie pyramidale. Documents obtenus par la méthode des dégénérescences descendants, chez le chien. J. Physiol. (Paris) 43:75–96.

Mosidze, V. M. 1960. The importance of the cortical auditory area in the conditioned reflex activity of dogs. Zhur. Vysshei Nervnoi Deiatel'nosti im. I. P. Pavlova (Transl.) 10:923–928.

Munk, H. 1880. Über die Sehsphäre und die Riechsphäre der Grosshirnrinde. Arch. Physiol. Lpz. 1880:449–456.

— — — — —. 1881. Über die Functionen der Grosshirnrinde. Berlin, Hirschwald.

— — — — —. 1890. Of the visual area of the cerebral cortex, and its relation to eye movements. (transl. by F. W. Mott). Brain 13:45–70.

Muratoff, W. 1893a. Secundäre Degenerationen nach Zerstörung der motorischen Sphäre des Gehirns in Verbindung mit der Frage von der Localisation der Hirnfunctionen. Arch. Anat. Physiol. Anat. Abt. 1893:97–116.

— — — — —. 1893b. Secundäre Degeneration nach Durchschneidung des Balkens. Neurol. Zbl. 12:714–729.

Muratow, W. 1893. Secundäre Degeneration nach Durchschneidung des Corpus callosum. Neurol. Zbl. 12:316.

Nakai, Y. 1974. Comparative anatomical studies on the mamillary body of mammals. Bull. Tokyo Med. Dent. Univ. 21:189–204.

Narkiewicz, O. 1972. Frontoclaustral interrelations in cats and dogs. Acta. Neurobiol. Exp. (Warsaw) 32:141–150.

Narkiewicz, O., and S. Brutkowski. 1967. The organization of projections from the thalamic mediodorsal nucleus to the prefrontal cortex of the dog. J. Comp. Neurol. 129:361–374.

Nasedkin, V. A. 1929. Morphology and comparative anatomy of the dentate nucleus of the cerebellum in man, monkeys and other mammals. Arch. Russ. Anat. Histol. Embryol. 8:137–148. (Original in Russian, quoted from Biol. Abstr. 1932/6:20400.)

Niedzielska, B. 1966. Stria medullaris of the thalamus in the dog. Acta Biol. Exp. (Warsaw) 26:149–158.

Nigge, K.-H. 1944. Die Gewinnung und Untersuchung des Liquor cerebrospinalis beim Hund mit besonderer Berücksichtigung der Liquorbefunde bei der Hundestaupe. Dtsch. tierärztl. Wschr. tierärztl. Rdsch. 52:26–29.

Nilges, R. G. 1944. The arteries of the mammalian cornu ammonis. J. Comp. Neurol. 80:177–190.

Nowak, A. 1968a. Myeloarchitectonics of the putamen in the dog brain. Acta Biol. Cracoviensia. Ser. Zool. 11:137–147.

Novak, A. 1968b. The myeloarchitectonics of the caudate nucleus in the dog brain. Acta Biol. Cracoviensia. Ser. Zool. 11:227–241.

Obersteiner, H., and E. Redlich. 1902. Zur Kenntnis des Stratum (Fasciculus) subcallosum (Fasciculus nuclei caudati) und des Fasciculus fronto-occipitalis (reticuliertes cortico-caudales Bündel). Arb. Neurol. Inst. Univ. Wien 8:286–307.

Oboussier, H. 1949. Über Unterschiede des Hirnfurchenbildes bei Hunden. Verh. Dtsch. Zool. Mainz 1949. Leipzig 1950:109–114.

— — — — —. 1950. Zur Frage der Erblichkeit der Hirnfurchen. Untersuchungen an Kreuzungen extremer Rassetypen des Hundes. Z. menschl. Vererb.- u. Konstit.-Lehre 29:831–864.

O'Connor, W. J. 1947. Atrophy of the supraoptic and paraventricular nuclei after interruption of the pituitary stalk in dogs. Q. J. Exp. Physiol. *34*:29–42.

—————. 1952. The normal interphase in the polyuria which follows section of the supraopticohypophysial tracts in the dog. Q. J. Exp. Physiol. *37*:1–10.

Ogawa, T., and S. Mitomo. 1938. Eine experimentell-anatomische Studie über zwei merkwürdige Faserbahnen im Hirnstamm des Hundes: Tractus mesencephalo-olivaris medialis (Economo, Karplus) und Tractus tectocerebellaris. Jap. J. Med. Sci. I. Anat. *7*:77–94.

Okinaka, S., and Y. Kuroiwa. 1952. A contribution to the study of the histological relationship between hypothalamus and peripheral autonomic nervous system. Folia psychiat. Neurol. Jap. *6*:45–56.

Okinaka, S., H. Nakamura, T. Tsabaki, Y. Kuroiwa, and Y. Toyokura. 1953. Studies on the autonomic representation in the cerebral cortex. 1. On the pupillary function evokable through stimulation of the piriform region and its circumference. Folia Psychiat. Neurol. Jap. *7*:141–148.

Olivieri, L. 1946. Le arterie del mesencefalo. Arch. Ital. Anat. Embriol. *51*:243–283.

Oshinomi, T. 1930. Über die Sehbahnen, welche die primären Sehzentren mit der Sehrinde verbinden, mit besonderer Rücksicht auf die Verbindung zwischen dem Corpus geniculatum externum und dem Hinterhauptpol. Okayama Igakkwai Zasshi *42*:253–262. (Quoted from Biol. Abstr. 1934/8:12112.)

Otten, E. 1943. Umfangsmessungen an der Hypophysis cerebri und ihren Lappen beim Deutschen Schäferhund. Anat. Anz. *94*:1–25.

Palionis, T. 1950. Die Nissl-substanz in den Ganglienzellen des Riechkolbens, gyrus olfactorius, lobus piriformis und Ammonshorns des Hundes. Vet. Diss., Hannover.

Pampiglione, G. 1963. Development of Cerebral Function in the Dog. London, Butterworths.

Paneth, J. 1885. Über Lage, Ausdehnung und Bedeutung der absoluten motorischen Felder auf der Hirnoberfläche des Hundes. Arch. ges. Physiol. *37*:523–561.

Papez, J. W. 1927. Subdivisions of the facial nucleus. J. Comp. Neurol. *43*:159–191.

—————. 1929. Comparative Neurology. New York, Crowell.

—————. 1932. The nucleus of the mammillary peduncle. Anat. Rec. (Suppl.) *52*:72.

—————. 1938. Thalamic connections in a hemidecorticate dog. J. Comp. Neurol. *69*:103–120.

Papez, J. W., and R. W. Rundles. 1938. Thalamus of a dog without a hemisphere due to a unilateral congenital hydrocephalus. J. Comp. Neurol. *69*:89–102.

Parhon, C., and G. Nadejde, 1906. Nouvelle contribution à l'étude des localisations dans les noyaux des nerfs craniens et rachidiens chez l'homme et chez le chien. J. Neurol. (Brux.) *11*:129–140.

Pavlov, I. P. 1927. Conditioned Reflexes. London, Oxford University Press.

Pawlow, I. P. 1952. Sämtliche Werke, Berlin, Akademie Verlag.

Peele, T. L. 1961. The Neuroanatomic Basis for Clinical Neurology. New York, McGraw-Hill.

Perkins, L. C. 1961. The early postnatal development of the cerebellar cortex in the dog and the effect of early spinocerebellar tractotomy. Ph.D. Dissertation, Duke University. (Diss. Abstr. *22*:706.)

Petras, J. M., and J. F. Cummings. 1977. The origin of spinocerebellar pathways. II. The nucleus centrobasalis of the cervical enlargement and the nucleus dorsalis of the thoracolumbar spinal cord. J. Comp. Neurol. *173*:693–715.

Phalen, G. S., and H. A. Davenport. 1937. Pericellular endbulbs in the central nervous system of vertebrates. J. Comp. Neurol. *68*:67–76.

Phemister, R. D., and S. Young. 1968. The postnatal development of the canine cerebellar cortex. J. Comp. Neurol. *134*:243–254.

Pickford, M., and H. E. Ritchie. 1945. Experiments on the hypothalamic-pituitary control of water excretion in dogs. J. Physiol. (Lond.) *104*:105–128.

Piltz, J. 1902. Über centrale Augenmuskelnervenbahnen. Neurol. Zbl. *21*:482–487.

Pines, J. L. 1926. Über ein bisher unbeachtetes Gebilde im Gehirn einiger Säugetiere; das subfornicale Organ des III. Ventrikels. J. Psychol. Neurol. (Lpz.) *34*:186–193.

Pines, J. L., and R. Maiman. 1928. Weitere Beobachtungen über das subfornicale Organ des dritten Ventrikels der Säugetiere. Anat. Anz. *64*:424–437.

Pivetau, J. 1951. Recherches sur l'évolution de l'encéphale chez les carnivores fossiles. Ann. Paléol. *37*:133–152.

Poljak, S. 1927. An experimental study of the association, callosal, and projection fibers of the cerebral cortex of the cat. J. Comp. Neurol. *44*:197–258.

Poltyreff, S. S. 1936. Voorborgene Assoziationen des Grosshirns bei Hunden. Z. Biol. *97*:306–307.

Poltyreff, S. S., and W. A. Alexejeff. 1936. Über die Möglichkeit der Bildung bedingter Reflexe bei Hunden mit exstirpierter Hirnrinde von der der Hemisphäre gegenüberliegenden Körperoberfläche aus. Z. Biol. *97*:297–305.

Poltyrev, S. S., and G. P. Zeliony. 1929. Der Hund ohne Grosshirn. Am. J. Physiol. *90*:475–476.

—————. 1930. Grosshirnrinde und Assoziationsfunktion. Z. Biol. *90*:157–160.

Potts, D. G., M. D. F. Deck, and V. Deonarine. 1971. Measurement of the rate of cerebrospinal fluid formation in the lateral ventricles of the dog. Radiology *98*:605–610.

Preziuso, L. 1930. Sul nucleo accessorio del nervo abducente in alcuni mammiferi domestici (*Canis familiaris, Bos taurus, Equus caballus*). Pisa, Stabilimento Editoriale Vallerini.

Probst, M. 1900a. Experimentelle Untersuchungen über die Schleifenendigung, die Haubenbahnen, das dorsale Längsbündel und die hintere Commissur. Arch. Psychiat. Nervenkr. *33*:1–57.

—————. 1900b. Physiologische anatomische und pathologisch-anatomische Untersuchungen des Sehhügels. Arch. Psychiat. Nervenkr. *33*:721–817.

—————. 1901a. Zur Kenntnis des Faserverlaufes des Temporallappens, des Bulbus olfactorius, der vorderen Commissur und des Fornix nach entsprechenden Extirpations- und Durchschneidungsversuchen. Arch. Anat. Physiol. Anat. Abt. *1901*:338–356.

—————. 1910b. Über den Verlauf und die Endigung der Rinden- Sehhügelfasern des Parietallappens, sowie Bemerkungen über den Verlauf des Balkens, des Gewölbes, der Zwinge und über den Ursprung des Monakowschen Bündels. Arch. Anat. Physiol. Anat. Abt. *1901*:357–370.

— — — — —. 1902. Über den Verlauf der centralen Sehfasern (Rinden-Sehhügelfasern) und deren Endigung im Zwischen- und Mittelhirne und über die Associations- und Commissurenfasern der Sehsphäre. Arch. Psychiat. Nervenkr. *35*:22–43.

— — — — 1903a. Über die Leitungsbahnen des Grosshirns mit besonderer Berücksichtigung der Anatomie und Physiologie des Sehhügels. Jb. Psychiat. Neurol. *23*:18–106.

— — — — —. 1903b. Über die anatomischen und physiologischen Folgen der Halbseitendurchschneidung des Mittelhirns. Jb. Psychiat. Neurol. *24*:219–325.

Rademaker, G. G. J. 1926. Démonstration de deux chats d'ecérébelles, de deux chiens decérébelles et d'un chien ayant subi l'ablation, outre due cervelet, de la moitié droite du cerveau. Arch. Nérl. Physiol. *11*:445–450.

Ramirez-Corria, C.-M. 1927a. La région infundibulotubérienne du chien. C. R. Soc Biol. (Paris *97*:591–593.

— — — — —. 1927b. Étude des lésions de la région infundibulotubérienne chez des chiens polyuriques. C.R. Soc. Biol. (Paris) *97*:593–594.

Ranson, W. S., and S. L. Clark. 1959. The anatomy of the Nervous System. Philadelphia, W. B. Saunders.

Rawitz, B. 1926. Zur Kenntnis der Architektonik der Grosshirnrinde des Menschen und einiger Säugetiere; II. Die Hirnrinde von Orang, Rhesus, Lemur, Kaninchen und Hund Z. Anat. Entwickl.-Gesch. *79*:198–227.

Read, E. A. 1908. A contribution to the knowledge of the olfactory apparatus in the dog, cat and man. Am. J. Anat. *8*:17–47.

Rheingans, U. 1954. Das postnatale Oberflächenwachstum der cytoarchitektonischen Gebiete der Grosshirnrinde des Hundes (Boxer und Barsoi). Diss., Kiel.

Rioch, D. M. 1929a. Studies on the diencephalon of Carnivora; I. The nuclear configuration of the thalamus, epithalamus, and hypothalamus of the dog and cat. J. Comp. Neurol. *49*:1–119.

— — — — —. 1929b. Studies on the diencephalon of Carnivora; II. Certain nuclear configurations and fiber connections of the subthalamus and midbrain of the dog and cat. J. Comp. Neurol. *49*:121–153.

— — — — —. 1931a. Studies on the diencephalon of Carnivora; III. Certain myelinated-fiber connections of the diencephalon of the dog (*Canis familiaris*), cat (*Felis domestica*), and aevisa (*Crossarchus obscurus*). J. Comp. Neurol. *53*:319–388.

— — — — —. 1931b. Note on the center median of Luys. J. Anat. (Lond.) *65*:324–327.

Rose, J. E. 1939. The cell structure of the mammillary body in the mammals and in man. J. Anat. (Lond.) *74*:91–115.

Rosenzweig, B. M. 1935a. Der Einfluss beiderseitiger Exstirpation der Occipitallappen des Gehirns auf bedingte Gesichtsreflexe. Acta. Med. Scand. *84*:386–400.

— — — — —. 1935b. Über den Einfluss der Exstirpation der Rinde einer Grosshirnhemisphäre auf be-

dingte Augenreflexe und auf das Gesichtsfeld beim Hunde. Acta. Med. Scand. *84*:401–421.

— — — — —. 1935c. Bedingte Sehreflexe bei einem Hunde mit exstirpiertem linkem Occipitallappen. Acta. Med. Scand. *85*:169–183.

Rothmann, H. 1924. Zusammenfassender Bericht über den Rothmannschen grosshirnlosen Hund nach klinischer und anatomischer Untersuchung. Z. ges. Neurol. Psychiat. *87*:247–313.

Roussy, G., and M. Mosinger, 1933. Rapports anatomiques et physiologiques de l'hypothalamus et de l'hypophyse. Ann. Méd. *33*:301–324.

— — — — —. 1935. L'hypothalamus chez l'homme et le chien. Rev. Neurol. *63*:1–35.

Rüdinger, N. 1894a. Über die Hirne verschiedener Hunderassen. Verh. anat. Ges. (Jena) *8*:173–176.

— — — — —. 1894b. Über die Hirne verschiedener Hunderacen. S. B. bayer. Akad. Wiss. *24*:249–255.

Russel, G. V. 1954. The dorsal trigemino-thalamic tract in the cat reconsidered as a lateral reticulo-thalamic system of connections. J. Comp. Neurol. *101*:237–263.

Sander, J. 1866. Über Faserverlauf und Bedeutung der Commissura cerebri anterior bei den Säugethieren. Arch. Anat. Physiol. *1866*:750–756.

Sarkissow, S. 1929. Über die postnatale Entwicklung einzelner cytoarchitektonischer Felder beim Hund. J. Psychol. Neurol. (Lpz.) *39*:486–505.

Scharrer, E. 1954. The maturation of the hypothalamic-hypophyseal neurosecretory system in the dog. Anat. Rec. *118*:437.

Scharrer, E., and R. D. Frandson, 1954. The mode of release of neurosecretory material in the posterior pituitary of the dog. Anat. Rec. *118*:350–351.

Scharrer, E., and G. J. Wittenstein. 1952. The effect of the interruption of the hypothalamo-hypophyseal neurosecretory pathway in the dog. Anat. Rec. *112*:387.

Schneebeli, S. 1958. Zur Anatomie des Hundes im Welpenalter. 2. Beitrag. Form und Grössenverhältnisse innerer Organe. Vet. Diss. Zürich.

Schneider, A. J. 1928. The histology of the radix mesencephalica n. trigemini in the dog. Anat. Rec. *38*:321–339.

Schukowski, M. N. 1897. Über anatomische Verbindungen des Frontallappen. Neurol. Zbl. *16*:524–525.

Schüller, A. 1902. Reizversuche am Nucleus caudatus des Hundes. Arch. ges Physiol. *91*:477–508.

Schultz, O. 1939. Über die absolute und relative Ouerschnittgrösse der Gyri olfactorii communes beim Hund und bei der Katze. Arch. wiss. prakt. Tierheilk. *74*:139–200.

Schwill, A. 1951. Untersuchungen über den Feinbau einiger Stammganglien des Grosshirns beim Hund. Vet. Diss. Hannover.

Schwill, C. 1951. Untersuchungen über den Feinbau der Grosshirnrinde beim Hund. Vet. Diss. Hannover.

Seiferle, E. 1949. Über Nachtblindheit beim Hunde, Dtsch. tierärztl. Wschr. *56*:42–44.

— — — — —. 1957. Zur makroskopischen Anatomie des Pferdegehirns. Acta Anat. (Basel) *30*:775–786.

— — — — —. 1966. Zur Topographie des Gehirns

bei lang- und kurzköpfigen Hunderassen. Acta Anat. 63:346–362.

Sekita, B. 1931. Über den Faseraustausch zwischen dem nervus hypoglossus und nervus accessorius des Hundes an der Schädelbasis. Acta Sch. med. Univ. Kioto 13:239–244.

Sheiman, I. M. 1961. The formation of a conditional reflex to a moving visual stimulus in dogs. Zhur. Vysshei Nervnoi Deiatel'nosti im I. P. Pavlova (Transl.) 11:275–283.

Sheinin, J. J. 1930. Typing of the cells of the mesencephalic nucleus of the trigeminal nerve in the dog, based on Nissl-granule arrangement. J. Comp. Neurol. 50:109–131.

Simpson, R. M. 1930. Adaptive behaviour in circus movements of the dog following brain lesions. J. Comp. Psychol. 10:67–83.

Singer, M. 1962. The Brain of the Dog in Section. Philadelphia, W. B. Saunders.

Siuta, J. 1967. Preliminary allometric studies on the fissural cortex. Acta Biol. Cracoviensia. Ser. Zool. 10:227–231.

Sloper, J. C. 1955. Hypothalamic neurosecretion in the dog and cat, with particular reference to the identification of neurosecretory material with posterior lobe hormones. J. Anat. (Lond.) 89:301–316.

Śmialowski, A. 1965. The precommissural hippocampus in the dog. Acta Biol. Exp. (Warsaw) 25:289–296.

— — — — —. 1966. The myeloarchitectonics of the hypothalamus in the dog. I. The anterior nuclei. Acta Biol. Exp. (Warsaw) 26:99–122.

— — — — —. 1967. Magnocellular mammillary nucleus in the dog brain. Bull. Acad. Pol. Sci. Cl. II. Sér. Sc. Biol. 15:703–705.

— — — — —. 1968a. Studies on the hypothalamus of the dog. II. Intermediate (tuberal) part. Acta. Biol. Exp. (Warsaw) 28:121–144.

— — — — —. 1968b. Mammillary complex in the dog's brain. Acta Biol. Exp. (Warsaw) 28:225–243.

— — — — —. 1971. Subthalamus in dog brain. Acta Neurobiol. Exp. 31:203–212.

Smith, E. M., M. L. Calhoun, and E. P. Reineke. 1953. The histology of the anterior pituitary, thyroid and adrenal of thyroid stimulated purebred English bulldogs. Anat. Rec. 117:221–239.

Smith, W. K. 1933. A physiological and histological study of the motor cortex of the dog (Canis familiaris). Anat. Rec. (Suppl.)55:76.

— — — — —. 1935a. The extent and structure of the electrically excitable cerebral cortex in the frontal lobe of the dog. J. Comp. Neurol. 62:421–442.

— — — — —.1935b. Alterations of respiratory movements induced by electric stimulation of the cerebral cortex in the dog. Am. J. Physiol. 115:261–267.

Snider, R. S., and J. C. Lee. 1961. A Stereotaxic Atlas of the Cat Brain. Chicago, Univ. Chicago Press.

Sobusiak, T., R. Zimny, and Z. Matlosz. 1971. Primary glossopharyngeal and vagal afferent projection into the cerebellum in the dog. An experimental study with toluidine blue and silver impregnation methods. J. Hirnforsch. 13:117–134.

Sobusiak, T., R. Zimny, W. Silny, and K. Grottel. 1972. Primary vestibular afferents to the abducens nucleus

and accessory cuneate nucleus. An experimental study in the dog with Nauta method. Anat. Anz. 131:238–247.

Stam, F. C. 1958–59. The morphological division of the cerebellum. Acta Morphol. Neerl. -Scand. 2:97–106.

Starck, D. 1954. Die äussere Morphologie des Grosshirns zwergwüchsiger und kurzköpfiger Haushunde. Gaz. Med. Port. 7:132–146.

Starlinger, J. 1895. Die Durchschneidung beider Pyramiden beim Hunde. Neurol. Zbl. 14:390–394.

Steblow, E. M. 1933. Experimentelle Epilepsie der Hunde in atypischen Versuchsbedingungen. Gefrieren des Gehirns nach vorläufig ausgeführter Exstirpation oder Umstechen scinor verschiedenen Gebiete. Z. ges. Neurol. Psychiat. 149:255–265.

Stehr, F. 1963. Fasciculus mammillaris princeps and its branches in the dog. Acta Biol. Exp. (Warsaw) 23:221–237.

Stella, G., P. Zatti, and L. Sperti. 1955. Decerebrate rigidity in forelegs after deafferentiation and spinal transection in dogs with chronic lesions in different parts of the cerebellum. Am. J. Physiol. 181:230–234.

Stephan, H. 1954. Die Anwendung der Snellschen Formel h = k · p auf die Hirn-Körpergewichtsbeziehungen verschiedener Hunderassen. Zool. Anz. 153:15–27.

Stepien, I., L. Stepien, and J. Konorski, 1961. The effect of unilateral and bilateral ablations of sensorimotor cortex on the instrumental (type II) alimentary conditional reflexes in dogs. Acta Biol. Exp. (Warsaw) 21:121–140.

Stepien, I., L. Stepien, and B. Sychowa. 1966. Disturbances of motor conditioned behavior following bilateral ablations of the precruciate area in dogs and cats. Acta Biol. Exp. (Warsaw) 26:323–340.

Steward, A., P. R. Allott, and W. W. Mapleson. 1975. Organ weights in the dog. Res. Vet. Sci. 19:341–342.

Ström, G. 1950. Effect of hypothalamic cooling on cutaneous blood flow in the unanesthetized dog. Acta Physiol. Scand. 21:271–277.

Stutinsky, F. 1949. Sur des types cellulaires communs à l'hypothalamus et `a la neurohypophyse chez le chien. C.R. Ass. Anat. 36:652–658.

Stutinsky, F., M. Bonavallet, and P. Dell. 1949. Les modifications hypophysaires au cours du diabète insipide expérimental chez le chien. Ire partie. Ann. endocr. (Paris) 10:505–517.

— — — — —. 1950. Les modifications hypophysaires au cours du diabète insipide expérimental chez le chien. 2e partie. Ann. endocr. (Paris) 11:1–11.

Stutinsky, S. F. 1958. Contribution à l'étude du Complexe Hypothalamo-neurohypophysaire. Thèse, Paris.

Suvorov, N. F., L. K. Danilova, and S. F. Ermolenko. 1976. Direct connections of amygdaloid complex nuclei with the caudate nucleus in dogs Dokl. Akad. Nauk. SSSR. 229:1262–1265.

Świecimska, Z. 1967. The corpus callosum of the dog. Acta Biol. Exp. (Warsaw) 27:389–411.

— — — — —. 1968. Myeloarchitectonics of the medium-long and short association fibers in the frontal region of the dog brain. Acta Biol. Cracoviensia. Ser. Zool. 11:197–211.

— — — — —. 1970. Cortico-cortical connections in

the perisylvian cortex in the dog. Acta Biol. Cracoviensia. Ser. Zool. *13*:141–159.

Sych, B. 1976. Architecture of the ventral group of thalamic nuclei in the dog brain. Folia Biol. (Krakow) *24*:257–276.

Sych, L. 1960. The external capsule in the dog's brain (myeloarchitectonics and topography). Acta Biol. Exp. (Warsaw) *20*:91–101.

Sychowa, B. 1961a. Degenerations after ablations of the anterior and posterior parts of the sylvian gyrus in the dog. Bull. Acad. Pol. Sci. Cl. II. Sér. Sc. Biol. *9*:183–186.

— — — — . 1961b. The morphology and topography of the thalamic nuclei of the dog. Acta Biol. Exp. (Warsaw). *21*:101–120.

— — — — . 1962a. Medial geniculate body of the dog. J. Comp. Neurol. *118*:355–371.

— — — — . 1962b. Degeneration after ablation of the ectosylvian gyrus in dog. Bull. Acad. Pol. Sci. Cl. II. Sér. Sc. Biol. *10*:17–20.

— — — — . 1963. Degenerations of the medial geniculate body following ablations of various temporal regions in the dog. Acta Biol. Exp. (Warsaw) *23*:75–99.

Sychowa, B., L. Stepien, and I. Stepien 1968. Degeneration in the thalamus following medial frontal lesions in the dog. Acta Biol. Exp. (Warsaw) *28*:383–399.

Szentágothai, J. 1942. Die innere Gliederung des Oculomotoriuskernes. Arch. Psychiat. Nervenkr. *115*:127–135.

Szentágothai-Schimmert, J. 1941. Die Endigungsweise der absteigenden Rückenmarksbahnen. Z. Anat. Entwickl.-Gesch. *111*:322–330.

Taber, E. 1961. The cytoarchitecture of the brain stem of the cat. J. Comp. Neurol. *116*:27–69.

Takahashi, K. 1951. Experiments on the periamygdaloid cortex of cat and dog. Folia Psychiat. Neurol. Japon. *5*:147–154.

Teljatnik, T. 1897. Über Kreuzung der Opticusfasern. Neurol. Zbl. *16*:521–522.

Tenerowicz, M. 1960. The morphology and topography of the claustrum in the brain of the dog. Acta Biol. Cracoviensia Ser. zool. *3*:105–113.

Thauer, R., and F. Stuke. 1940. Über die funktionelle Bedeutung der motorischen Region der Grosshirnrinde für den Sehakt des Hundes. Arch. ges. Physiol. *243*:347–369.

Thelander, H. E. 1924. The course and distribution of the radix mesencephalica trigemini in the cat. J. Comp. Neurol. *37*:207–220.

Thompson, I. M. 1932. On the cavum septi pellucidi. J. Anat. (Lond.) *67*:59–77.

Tryhubczak, A. 1975. Myeloarchitectonics of the hippocampal formation in the dog. Folia Biol. (Krakow) *23*:177–188.

Tunturi, A. R. 1944. Audio frequency localisation in the acoustic cortex of the dog. Am. J. Physiol. *141*:397–403.

— — — — . 1945. Further afferent connections to the acoustic cortex of the dog. Am. J. Physiol. *144*:389–394.

— — — — . 1946. A study of the pathway from the medial geniculate body to the acoustic cortex in the dog. Am. J. Physiol. *147*:311–319.

— — — — . 1950. Physiological determination of the boundary of the acoustic area in the cerebral cortex of the dog. Am. J. Physiol. *160*:395–401.

— — — — . 1952. A difference in the representation of auditory signals for the left and right ears in the iso-frequency contours of the right middle ectosylvian auditory cortex of the dog. Am. J. Physiol. *168*:712–727.

Tunturi, A. R. 1970. The pathway from the medial geniculate body to the ectosylvian auditory cortex in the dog. J. Comp. Neurol. *138*:131–136.

— — — — . 1971. Classification of neurons in the ectosylvian auditory cortex of the dog. J. Comp. Neurol. *142*:153–165.

Turner, W. 1890. The convolutions of the brain, a study in comparative anatomy. J. Anat. Physiol. *25*:105–153.

Ullrich, K. 1928. Über die Gewinnung des Liquor cerebrospinalis beim Hund und seine Untersuchung. Prag. Arch. Tiermed. *8*:53–72.

Valverde, F. 1961. Reticular formation of the pons and medulla oblongata. A. Golgi study. J. Comp. Neurol. *116*:71–99.

Venzke, W. G., and J. W. Gilmore, 1940. Histological observation on the epiphysis cerebri and the choroid plexus of the third ventricle of the dog. Proc. Iowa Acad. Sci. *47*:409–413.

Verwer, M. A. J. 1952. Over punctie en onderzoek van de Liquor cerebrospinalis bij de gezonde en de zieke hond. Akad. Proefschrift, Utrecht.

Vogt, C., and O. Vogt. 1902. Zur Erforschung der Hirnfaserung. Denkschr. med. naturwiss. Ges. Jena *9*:1–145. (Neurobiol. Arb. *1*:1–145).

Volkmer, D. 1956. Cytoarchitektonische Studien an Hirnen verschieden grosser Hunde (Königspudel und Zwergpudel). Z. Mikr.-Anat. Forsch. *62*:267–315.

de Vries, E. 1910. Das Corpus striatum der Säugetiere. Anat. Anz. *37*:385–405.

Vuillaume, P. 1935. Le liquide céphalo-rachidien normal due chien. Thèse, Lyon.

Weinberg, R. 1902. Die Interzentralbrücke der Carnivoren und der Sulcus Rolandi, eine morphologische Skizze. Anat. Anz. *22*:268–280.

Whitaker, J. G., and L. Alexander. 1932. Die Verbindungen der Vestibulariskerne mit dem Mittel- und Zwischenhirn. Studien auf Grund experimenteller Verletzungen. J. Psychol. Neurol. (Lpz) *44*:253–376.

Wilder, B. G. 1873a. The outer cerebral fissures of mammalia (especially the carnivora) and the limits of their homology. Proc. Amer. Assoc. Anat. *22*:214–234.

— — — — . 1873b. Cerebral variations in domestic dogs and its bearing upon scientific phrenology. Proc. Amer. Assoc. Anat. *22*:234–249.

— — — — . 1881. The brain of the cat, *Felis domestica;* 1. Preliminary account of the gross anatomy. Proc. Amer. Philos. Soc. *19*:524–562.

Wing, K. G., and K. U. Smith. 1942. The role of the optic cortex in the dog in the determination of the functional properties of conditioned reactions to light. J. Exp. Psychol. *31*:478–496.

Woolsey, C. N. 1933. Postural relations of the frontal and motor cortex of the dog. Brain *56*:353–370.

— — — — . 1943. "Second" somatic receiving area in the cerebral cortex of cat, dog, and monkey. Fed. Proc. *2*:55–56.

Yagita, K. 1909. Weitere Untersuchungen über das Speichelzentrum. Anat. Anz. *35*:70–75.

— — — — . 1910. Experimentelle Untersuchungen über den Ursprung des Nervus facialis. Anat. Anz. 37:195–218.

Yagita, K., and S. Hayama. 1909. Über das Speichelsekretionscentrum. Neurol. Zbl. 28:738–753.

Yamagishi, Y. 1935. Über die cytoarchitektonische Gliederung des roten Kernes des Hundes. Z. mikr.-anat. Forsch. 37:659–672.

Yoda, S. 1941. Beitrag zu den Olivenkernen des Hundes. Z. mikr.-anat. Forsch. 49:516–524.

Zeliony, G. P. 1913. Observations sur les chiens auxquels on a enlevé les hémisphères cérébraux. C.R. Soc Biol. (Paris) 74:707–708.

Zernicki, B. 1961. The effect of prefrontal lobectomy on water instrumental conditional reflexes in dogs. Acta Biol. Exp. 21:157–162.

Zernicki, B., and G. Santibanez. 1961. The effects of ablation of "alimentary area" of the cerebral cortex on salivary conditional and unconditional reflexes in dogs. Acta Biol. Exp. (Warszawa) 21:163–167.

Ziehen, T. 1906. Die Histogenese von Hirn und Rückenmark. Entwickelung der Leitungsbahnen und der Nervenkerne bei den Wirbeltieren. Handbuch der vergleichenden und experimentellen Entwickelungslehre der Wirbeltiere. Zweiter Band. Dritter Teil. Herausgegeben von Oskar Hertwig. Jena, Gustav Fischer.

Zimny, R. T., Sobusiak, T., and W. Silny. 1972. The pattern of afferent projection from the 8th, 9th, and 10th cranial nerve to the inferior vestibular nucleus. An experimental study in the dog with Nauta method. Anat. Anz. 130:285–296.

Zimny, R. T., Sobusiak, K. Grottel, and J. Zabel. 1973. A bidirectional projection between the gracile nucleus and the cerebellum in the dog? An experimental study with tigrolysis and axonal degeneration methods. J. Hirnforsch. 14:89–108.

Chapter 15

THE CRANIAL NERVES

By ROBERT C. McCLURE

General Considerations

The nerves which arise from the brain are referred to as cranial nerves. The twelve pairs of cranial nerves are I, olfactory; II, optic; III, oculomotor; IV, trochlear; V, trigeminal; VI, abducent; VII, facial; VIII, vestibulocochlear; IX, glossopharyngeal; X, vagus; XI, accessory; and XII, hypoglossal.

The olfactory and optic nerves are often considered to be tracts of the central nervous system rather than peripheral nerves.

The anatomical organization of cranial nerves is similar in many ways to that of spinal nerves (see Chapter 17). Each spinal nerve is attached to the spinal cord by a ventral and a dorsal root. The fibers making up the ventral root emerge as a series of filaments along the ventrolateral surface of the spinal cord, while those forming the dorsal root emerge dorsolaterally from the spinal cord. The fibers making up the rootlets which unite to form the spinal nerve are derived from a portion of the spinal cord which is referred to as a spinal cord segment.

Fibers which conduct impulses away from the spinal cord make up the ventral roots. Their cell bodies are located in the gray matter of the spinal cord and are called efferent because they conduct impulses away from the spinal cord. The fibers which conduct impulses from the periphery of the body to the spinal cord form the dorsal roots and are referred to as afferent fibers. They have their cell bodies in the spinal, or dorsal root, ganglia. The cell bodies are considered to be unipolar or pseudounipolar in that they have a single fiber from the cell body which divides in the form of a T, one going centrally to the spinal cord and the other peripherally to a neurosensory organ. The central branch ends by synapsing with nerve cells in the gray matter of the spinal cord. Thus there are two main groups of fibers in the spinal and frequently in the cranial nerves, the efferent and afferent.

Cranial nerves can be further subdivided according to the types of structures which they supply. Somatic fibers innervate striated muscles, tendons, joints, and ligaments. Also in this group are most of the sensory fibers from nerve endings located in the skin, the somatic afferents. Those fibers which innervate the internal organs, smooth muscles, glands, and vessels are referred to as visceral fibers. There are four categories of fibers: **somatic afferent, somatic efferent, visceral afferent,** and **visceral efferent.** These principles apply to the cranial as well as to the spinal nerves.

Somatic afferent fibers transmit impulses from the skin muscles and mucous membranes in the cranial area to the brain stem.

903

TABLE 15–1. SUMMARY OF CRANIAL NERVES

Nerve	Brain Attachment	Cranial Exit	Location of Nerve Cell Bodies	Major Component Type(s)*	Function and Distribution
Olfactory (I)	Olfactory bulb	Foramina of ethmoid bone (cribriform plate)	Olfactory portion of nasal mucosa	SVA	Olfaction
Optic (II)	Ventral diencephalon	Optic canal	Ganglion cells of retina	SSA	Vision
Oculomotor (III)	Ventral mesencephalon	Orbital fissure	Motor nucleus of oculomotor n.	SE	Eyeball movement (extraocular mm.); raises dorsal eyelid (levator palpebrae m.)
			Parasympathetic nucleus of oculomotor n. (Edinger-Westphal)	GVE (ciliary ganglion)	Miosis (constricts pupil) and accommodation (rounds the lens)
Trochlear (IV)	Dorsal mesencephalon	Orbital fissure	Motor nucleus of trochlear n.	SE	Eyeball movement (dorsal oblique m.)
Trigeminal (V)	Ventral lateral metencephalon caudal to transverse fibers of pons	Orbital V₁ (Ophthalmic)— fissure / V₂ (Maxillary)— Round foramen/ alar canal	Trigeminal ganglion (semilunar or gasserian)	SA	V₁:Sensory—forehead, dorsal medial aspect of orbit / V₂:Sensory—lateral surface of head, upper jaw, and nose / V₃:Sensory—"chin," lower jaw, intermandibular surface
			Mesencephalic nucleus of trigeminal n.	SA	Proprioception of mm. of mastication and extrinsic ocular m.
		V₃ (Mandibular)— Oval foramen	Motor nucleus of trigeminal n.	SVE	Masticatory mm.

Nerve	Origin	Opening	Nucleus	Ganglion	Component	Function
Abducent (VI)	Ventral surface of myelencephalon caudal to transverse fibers of pons	Orbital fissure	Motor nucleus of abducent n.		SE	Movement of eyeball (lateral rectus and retractor bulbi mm.)
Facial (VII)	Lateroventral portion of myelencephalon caudal to transverse fibers of pons and through the rostrolateral portion of trapezoid body	Internal acoustic meatus, facial canal, stylomastoid foramen	Motor nucleus of facial n.		SVE	Cutaneous facial and colli mm., stapedius m., and digastric m. (caudal belly)
			Parasympathetic nucleus of facial n.	Pterygopalatine ganglion	GVE	Secretion of lacrimal, palatine, and nasal glands
				Mandib. and sublingual ganglia		Mandibular and sublingual salivary glands
			Geniculate ganglion		SVA	Taste, rostral two-thirds of tongue
Vestibulocochlear (VIII)	Myelencephalon, lateral end of trapezoid body	Internal acoustic meatus	Cochlear ganglion		SSA	Hearing
			Vestibular ganglion		SP (Special proprioception)	Balance, motion equilibration
Glossopharyngeal (IX)	Myelencephalon, lateral side, caudoventral to VIII	Jugular foramen	Nucleus ambiguus		SVE	Mm. of pharynx and soft palate
			Parasympathetic nucleus of glossopharyngeal n.		GVE (otic ganglion)	Salivary secretion, parotid and zygomatic salivary glands
			Distal ganglion of IX		SVA	Taste, caudal one-third of tongue
					GVA	Visceral afferent, tongue, pharynx, and carotid sinus

Table continued on following page.

TABLE 15-1. SUMMARY OF CRANIAL NERVES *Continued*

Nerve	Brain Attachment	Cranial Exit	Location of Nerve Cell Bodies	Major Component Type(s)*	Function and Distribution
Vagus (X)	Myelencephalon, lateral side caudal to IX	Jugular foramen	Parasympathetic nucleus of vagus (dorsal motor)	GVE	Parasympathetic to thoracic and abdominal viscera
			Nucleus ambiguus	SVE	Mm. of pharynx, larynx, and esophagus
			Proximal (jugular) ganglion of vagus	SA	Sensory from external ear canal via auricular br. to facial n.
			Distal (nodose) ganglion of vagus	{ VA SVA	Visceral afferent from pharynx, larynx, and thoracic and abdominal viscera
					Taste, palate and pharynx
Accessory (XI)	Lateral surface of cervical spinal cord	Jugular foramen	Motor nucleus of accessory n. (lateral "horn" in cervical spinal cord)	SVE	Mm. of neck and limb which are of branchial arch origin
Hypoglossal (XII)	Myelencephalon, ventrolateral sulcus	Hypoglossal canal	Motor nucleus of hypoglossal n.	SE	Movement of tongue; tongue mm., thyrohyoid and geniohyoid mm., and sometimes sternohyoid m.

Efferent
 SE = Somatic efferent fibers.
 GVE = General visceral efferents.
 SVE = Special visceral efferents (brachiomotor).

Afferent
 SA = Somatic afferent fibers.
 GVA = General visceral afferents.
 SVA = Special visceral afferents.
 SSA = Special somatic afferents.

Not all components are included for each of the cranial nerves, particularly those that are controversial or not well understood.

As in the spinal cord, the somatic afferent fibers have their cell bodies in sensory ganglia (for the most part), outside the central nervous system. These are associated with the roots of the cranial nerves. The cells in the ganglia are either bipolar or pseudounipolar, similar to those in the spinal ganglia, and give off one branch to the brain stem and the other to the periphery.

The **visceral afferent** fibers in the cranial nerves conduct sensory impulses from the internal organs. For example, the vagus nerve contains visceral afferent fibers from the heart, lungs, and stomach. Cell bodies of these vagal fibers are found in the distal (nodose or inferior) ganglion, separate from those containing the somatic afferent cell bodies, which are located in the proximal (jugular or superior) ganglion. In the spinal nerves, both the somatic and visceral afferent fibers have their cell bodies in the spinal ganglia.

The **visceral efferent** fibers of the cranial nerves are those fibers which have their cell bodies aggregated to form particular nuclei in the brain stem. These fibers are preganglionic. The nerve cells in these nuclei are of the same type as those in the intermediolateral column of the spinal cord, being medium-sized and multipolar. The largest of the visceral efferent cranial nerve nuclei is the parasympathetic nucleus of the vagus nerve. All the visceral efferent fibers leaving the brain in the cranial nerves belong to the craniosacral (parasympathetic) division of the autonomic nervous system. They terminate on nerve cell bodies which are in groups or scattered collections forming the visceral (autonomic) ganglia. The visceral ganglia are in or on the walls of the organs innervated. The fibers of the nerve cells in the visceral ganglia are postganglionic.

The **somatic efferent** fibers arise from nuclei composed of large multipolar nerve cell bodies which usually contain large amounts of chromophil (Nissl) substance.

In addition to the four functional types just enumerated, there are three additional types of fibers found in cranial nerves that are not found in the spinal nerves: (1) the **special visceral afferent,** (2) the **special somatic afferent,** and (3) the **special visceral efferent** fibers.

The **special visceral afferent** fibers are those which come from the visceral sensory organs (taste and olfaction). The special visceral afferent fibers are incorporated in cranial nerves I, V, VII, IX, and X, which convey impulses from the olfactory mucosa and the mucosa of the tongue.

The **special somatic afferent** fibers come from the special sensory organs which are considered to be derived from the ectodermal layer of the embryo (the eye and the ear).

The **special visceral efferent,** or branchial motor, fibers are found in cranial nerves V, VII, IX, X, and XI. These nerves supply motor fibers to striated muscle of branchial arch origin. The first branchial arch is supplied by the trigeminal (V) nerve. The musculature of the first branchial arch gives rise to the muscles of mastication, which are supplied by the mandibular division of the trigeminal nerve. The musculature of the second branchial arch develops into the muscles of facial expression and several other muscles (stapedius and caudal portion of the digastricus) which are supplied by the facial (VII) nerve. The remaining branchial arches, three through six, and sometimes seven, are supplied by cranial nerves IX and X. The musculature of the latter-named branchial arches differentiates into the muscles of the pharynx, larynx, and cranial portion of the esophagus. Some portions of the branchial musculature develop into the ventrolateral muscles of the neck and the trapezius muscle, which are supplied by cranial nerve XI, or the accessory nerve. Table 15–1 summarizes the structural and functional organization of the cranial nerves.

OLFACTORY NERVES

The **olfactory nerves** (*nn. olfactorii*) consist of many small bundles of nerve fibers which pass from the olfactory mucosa to the olfactory bulb (Fig. 15–1). They are classified as special visceral afferent fibers. The cell bodies of the nerve fibers are located in the olfactory nasal mucosa. The central processes of the neurons travel caudally in the submucosa to the cribriform plate of the ethmoid bone. The fibers pass through the many foramina of the cribriform plate to end in the olfactory bulb. The bundles of fibers are enclosed by the dura mater, arachnoid,

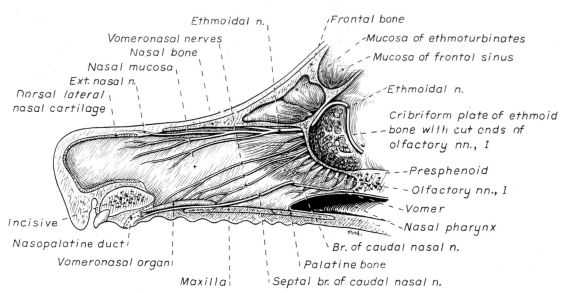

Figure 15–1.   Sagittal section of the nose with the cartilaginous and bony nasal septum removed to show distribution of nerves on the septal mucosa and to the vomeronasal organ.

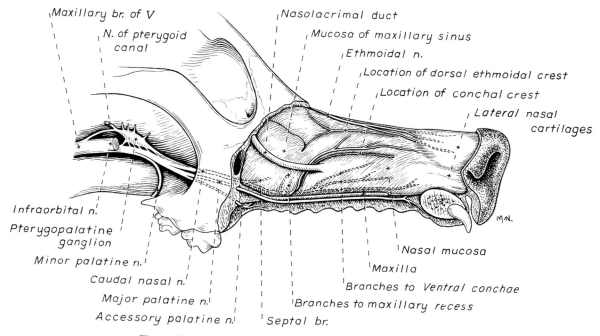

Figure 15–2.   Nerves of the lateral nasal wall and hard palate.

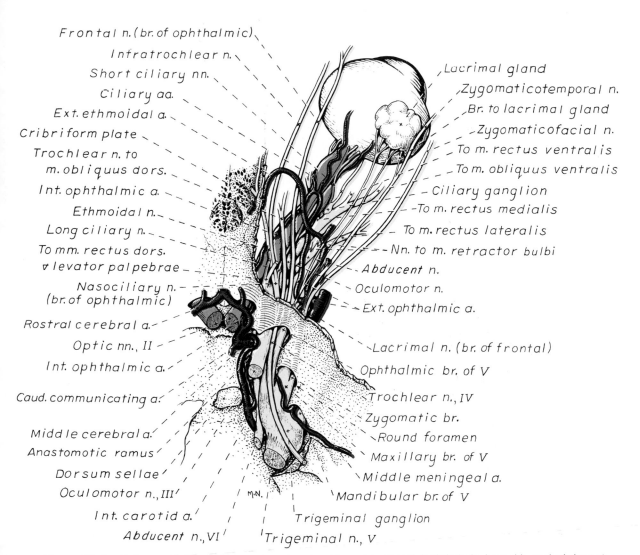

Frontal n. (br. of ophthalmic)
Intratrochlear n.
Short ciliary nn.
Ciliary aa.
Ext. ethmoidal a.
Cribriform plate
Trochlear n. to
  m. obliquus dors.
Int. ophthalmic a.
Ethmoidal n.
Long ciliary n.
To mm. rectus dors.
∇ levator palpebrae
Nasociliary n.
  (br. of ophthalmic)
Rostral cerebral a.
Optic nn., II
Int. ophthalmic a.
Caud. communicating a.
Middle cerebral a.
Anastomotic ramus
Dorsum sellae
Oculomotor n., III
Int. carotid a.
Abducent n., VI

Lacrimal gland
Zygomaticotemporal n.
Br. to lacrimal gland
Zygomaticofacial n.
To m. rectus ventralis
To m. obliquus ventralis
Ciliary ganglion
To m. rectus medialis
To m. rectus lateralis
Nn. to m. retractor bulbi
Abducent n.
Oculomotor n.
Ext. ophthalmic a.

Lacrimal n. (br. of frontal)
Ophthalmic br. of V
Trochlear n., IV
Zygomatic br.
Round foramen
Maxillary br. of V
Middle meningeal a.
Mandibular br. of V

M. N.

Trigeminal ganglion
Trigeminal n., V

Figure 15–3. Scheme of the optic, oculomotor, trochlear, trigeminal (ophthalmic branch), and abducent nerves. Dorsal aspect.

and pia mater, as they pass through the foramina.

Read (1908) described the extent of the olfactory mucosa in the dog adjacent to the cribriform plate. It occupied the surface of about half of the numerous ethmoturbinates, the caudal third to half of the nasal septum, one ethmoturbinate scroll which extended into the rostral portion of the frontal sinus, and a portion of the medial wall of the sinus. The olfactory mucosa can be distinguished in the fresh specimen by its yellow-to-brown color, which is due to the pigment of the sustentacular cells.

The **vomeronasal nerves** (*nn. vomeronasales*) arise from the dorsomedial surface of the vomeronasal organ and unite to form six to eight nerve bundles (Fig. 15–1). The bundles further unite, as they pass caudally in the submucosa of the nasal septum, to form one or two nerve trunks which pass through foramina of the ethmoid bone (cribriform plate). The nerve fibers pass caudally on the medial surface of the olfactory bulb to end on the accessory olfactory bulb, which is located on the medial surface of the olfactory tract at the caudal edge of the olfactory bulb. The vomeronasal nerves are the axis cylinder processes of the neurosensory cells in the mucosa of the vomeronasal organ. McCotter (1912) has provided a detailed description of the vomeronasal nerves and their connections.

The **terminal nerve** (*n. terminalis*) of the dog was described by McCotter (1913) as "a ganglionated nerve connected with the vomeronasal nerves on one hand and apparently with the forebrain on the other, having thereby the same morphological relations in these mammals (dog and cat) as described for the terminal nerve of lower forms." The terminal nerve is formed by several small bundles which arise from the vomeronasal nerves as they terminate on the accessory olfactory bulb. The small bundles unite to form a single trunk which runs caudoventrally on the medial surface of the olfactory tract and appears to enter the brain substance 1 or 2 cm. caudal to the olfactory bulb.

OPTIC NERVE

The second cranial nerve, or **optic nerve** (*n. opticus*), is composed of the central processes of the ganglion cells of the retina, which converge at the optic disc and pass through the choroid and scleral layers, lateroventral to the caudal pole of the eyeball. The fibers of the optic nerve are classified as special somatic afferents. The optic nerve is surrounded by extensions of the cranial meninges throughout its slightly sinuous course from the eyeball to the optic canal, where it enters the cranial cavity.

The optic nerve (and its meninges) is within a cone formed by the retractor bulbi muscle. It is related to the orbital fat and to the ciliary nerves and vessels in the rostral portion of their course (Figs. 15–3, 15–4). The ciliary ganglion is bounded ventrolaterally by the lateral rectus muscle and laterally by the ventral portion of the retractor bulbi muscle and medially by the optic nerve. Dorsally, the ophthalmic arteries and the nasociliary nerve are adjacent to the meningeal coverings in the caudal portion of the orbit. The optic nerves join at the optic chiasm shortly after entering the cranial cavity, where fibers from the medial portions of both nerves cross to the optic tract of the opposite side.

Bruesch and Arey (1942) state that the optic nerve of the dog contains 154,000 fibers, all of which are myelinated. The fibers have no neurilemma and become myelinated as soon as they leave the eyeball. The optic nerve may be considered to be a tract of the brain. Developmentally, it is the connection between the diencephalon and the retina, a derivative of the brain (Gardner et al. 1960).

OCULOMOTOR NERVE

The **oculomotor nerve** (*n. oculomotorius*) is the principal nerve to the muscles of the orbit. It contains two types of fibers: (1) general somatic efferent, which are motor to the striated ocular muscles, and (2) general visceral efferent (parasympathetic), which are destined for the ciliary ganglion. Koch (1916) studied the microscopic structure of the oculomotor nerve in the dog and found both large and small myelinated fibers. According to some authors, the oculomotor nerve conducts somatic afferent fibers concerned with proprioception for the extraocular muscles being innervated.

The oculomotor nerve leaves the brain

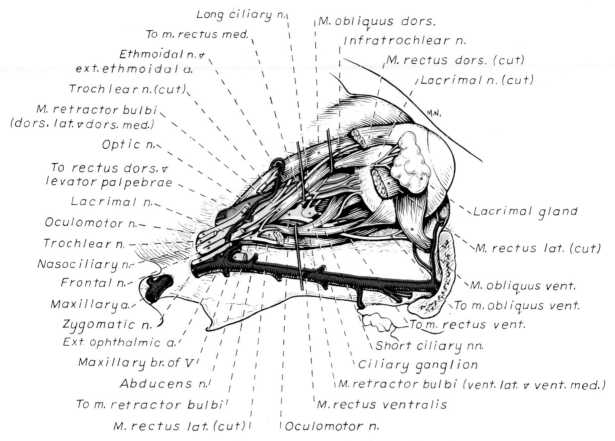

Long ciliary n.

To m.rectus med.

Ethmoidal n.v
ext.ethmoidal a.

Trochlear n.(cut)

M. retractor bulbi
(dors. lat.v dors. med.)

Optic n.

To rectus dors.v
levator palpebrae

Lacrimal n.

Oculomotor n.

Trochlear n.

Nasociliary n.

Frontal n.

Maxillary a.

Zygomatic n.

Ext. ophthalmic a.

Maxillary br. of V

Abducens n.

To m. retractor bulbi

M. rectus lat. (cut)

M. obliquus dors.

Infratrochlear n.

M. rectus dors. (cut)

Lacrimal n. (cut)

M.N.

Lacrimal gland

M. rectus lat. (cut)

M. obliquus vent.

To m. obliquus vent.

To m. rectus vent.

Short ciliary nn.

Ciliary ganglion

M. retractor bulbi (vent. lat. v vent. med.)

M. rectus ventralis

Oculomotor n.

Figure 15–4. Nerves of the eye and orbit. Lateral aspect.

stem at the medioventral surface of the mesencephalon, just medial to the cerebral peduncle. It courses rostrolaterally and enters the wall of the cavernous sinus. The oculomotor nerve emerges through the wall of the cavernous sinus and leaves the cranial cavity by way of the orbital fissure (Figs. 15–3, 15–4, 15–6). Immediately upon emergence from the orbital fissure, it divides into dorsal and ventral rami.

The **dorsal ramus,** after a short rostrodorsal course, divides into several branches which enter the muscular portion of the dorsal rectus muscle at its caudal end (Figs. 15–4, 15–6). After coursing through the muscle to approximately the junction of the distal and middle thirds, a small branch becomes superficial and enters the ventral surface of the levator palpebrae muscle, in its distal portion.

The **ventral ramus,** much the larger of the two rami, travels rostrally and gives off branches to the medial rectus, ventral rectus, and ventral oblique muscles of the eye. The parasympathetic or general visceral efferent fibers leave the ventral ramus to enter the ciliary ganglion.

The ventral ramus continues rostrally for approximately 3 to 4 cm. (Figs. 15–3, 15–4). It divides into two branches. One passes mediodorsally and ventral to the optic nerve to enter the medial rectus muscle. The other travels rostrolaterally and divides into two branches which supply the ventral rectus and ventral oblique muscles. The branch to the ventral rectus enters its dorsal surface at the junction of the proximal and middle thirds. The branch to the ventral oblique passes laterad, ventral to the ventrolateral portion of the retractor bulbi muscle, and then rostrally to enter the middle of the caudodorsal surface of the ventral oblique muscle. The ciliary ganglion is located at the point where the ventral ramus divides into branches which supply the medial rectus, the ventral rectus, and the ventral oblique muscles. The oculomotor root of the ciliary ganglion is very short and leaves the ventral ramus of the oculomotor nerve in the angle formed by the divergence of the two previously mentioned branches. The oculomotor root of the ciliary ganglion contains preganglionic parasympathetic nerve fibers.

TROCHLEAR NERVE

The fourth cranial nerve, or **trochlear nerve** (*n. trochlearis*), is the smallest of the twelve cranial nerves. It is unusual in several respects. The trochlear nerve is the only cranial nerve to emerge from the dorsal surface of the brain stem. The right and left trochlear nerves decussate (*decussatio nervorum trochlearium*) in the rostral medullary velum caudal to the caudal opening of the cerebral aqueduct. Shortly after emerging from the brain stem, the trochlear nerve pierces the dura mater and runs in the ventrolateral extension of the tentorium cerebelli along the dorsal ridge of the spine of the petrous temporal bone and passes dorsally to the trigeminal ganglion (Figs. 15–3, 15–4). The trochlear nerve remains in the dura until it passes through the orbital fissure between the ophthalmic branch of the trigeminal nerve and the oculomotor nerve. Upon emergence from the fissure, it turns dorsomediad and enters the dorsomedial surface of the dorsal oblique muscle (Figs. 15–4, 15–6). This is the only muscle supplied by the trochlear nerve. In its course from the orbital fissure to the dorsal oblique muscle, the nerve is related ventrally to the dorsal rectus and levator palpebrae muscles. Koch (1916) found both large and small myelinated fibers in the trochlear nerve of the dog.

TRIGEMINAL NERVE

The fifth cranial nerve, or **trigeminal nerve** (*n. trigeminus*), is the largest of the cranial nerves. Its sensory fibers (somatic afferents) receive impulses from the cutaneous muscles of the head, the nasal and oral cavities, and the muscles of mastication. The motor fibers (special visceral efferents) supply the muscles which are derived from the first branchial arch of the embryo, principally the muscles of mastication.

The trigeminal nerve is attached to the brain stem at the junction of the pons and trapezoid body. Its two roots, a small motor (*radix motoria*) and a large sensory (*radix sensoria*), are usually not separable. The motor and sensory roots, in their common sheath, pass through the canal of the petrous temporal bone and expand into the semilunar-shaped trigeminal ganglion (*ganglion trigeminale*), which is located in the cavum trigeminale of the dura mater lateral to the cavernous sinus at the apex of the petrous temporal bone.

The trigeminal ganglion contains most of the unipolar cell bodies of the general somatic afferent fibers which are distributed by the branches of the trigeminal nerve. The three divisions of the trigeminal nerve arise at the trigeminal ganglion. The first, the ophthalmic nerve, arises rostrally and leaves the cranial cavity through the orbital fissure. The second division, the maxillary nerve, arises from the rostrolateral side and leaves via the round foramen to enter the alar canal. The third division, the mandibular nerve, arises from the lateral side of the ganglion, caudal to the maxillary nerve, and emerges through the oval foramen. Koch (1916), in studying the intracranial portion of the trigeminal nerve in the dog, found both large and small myelinated fibers and small numbers of unmyelinated fibers, which are largely associated with the sensory portion of the nerve.

Ophthalmic Nerve

The **ophthalmic nerve** (*n. ophthalmicus*) (Fig. 15–6) is the principal sensory nerve of the orbit, the skin on the dorsum of the nose, and a portion of the mucous membranes of the nasal cavity and paranasal sinuses. It is the smallest division of the trigeminal nerve. It arises from the trigeminal ganglion and passes rostrally in the lateral wall of the cavernous sinus, ventral to the trochlear nerve. It receives filaments from the cavernous plexus of sympathetic nerves and usually connects with the oculomotor, trochlear and abducent nerves. The three primary branches — frontal, lacrimal, and nasociliary nerves — arise in or near the orbital fissure.

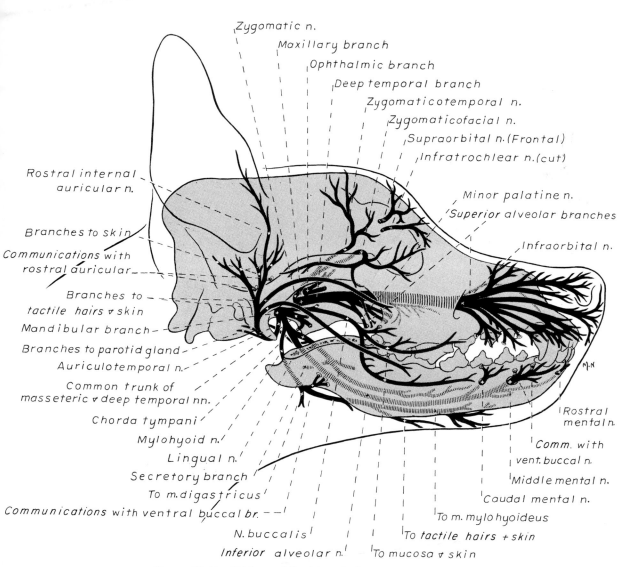

Zygomatic n.
Maxillary branch
Ophthalmic branch
Deep temporal branch
Zygomaticotemporal n.
Zygomaticofacial n.
Supraorbital n. (Frontal)
Infratrochlear n. (cut)

Minor palatine n.
Superior alveolar branches
Infraorbital n.

Rostral internal auricular n.

Branches to skin
Communications with rostral auricular
Branches to tactile hairs v skin
Mandibular branch
Branches to parotid gland
Auriculotemporal n.
Common trunk of masseteric v deep temporal nn.
Chorda tympani
Mylohyoid n.
Lingual n.
Secretory branch
To m. digastricus
Communications with ventral buccal br.
N. buccalis
Inferior alveolar n.

Rostral mental n.
Comm. with vent. buccal n.
Middle mental n.
Caudal mental n.
To m. mylohyoideus
To tactile hairs + skin
To mucosa v skin

Figure 15–5. Scheme of the trigeminal nerve. Lateral aspect.

The **frontal nerve** *(n. frontalis)* (Figs. 15–3, 15–4, 15–5, 15–6) is the sensory nerve to the middle portion of the upper eyelid and adjacent skin. It arises from the ophthalmic nerve, in the orbital fissure, in common with the lacrimal nerve. It passes rostrodorsally deep to the periorbita and dorsal to the dorsal rectus and levator palpebrae muscles. In the rostrodorsal portion of the orbit, it lies between the tendon of the dorsal oblique muscle and the supraorbital process of the frontal bone, lateral to the trochlea for the tendon of the dorsal oblique muscle.

The **lacrimal nerve** *(n. lacrimalis)* (Fig. 15–6) is a very small branch which arises from the ophthalmic division of the trigeminal nerve as stated above. The lacrimal nerve travels rostrodorsally along the lateral edge of the dorsal rectus muscle to end in the lacrimal gland.

The **nasociliary nerve** *(n. nasociliaris)* (Fig. 15–3) is the continuation of the ophthalmic division into the orbit. It passes ventral to the trochlear nerve and between the dorsal and ventral rami of the oculomotor nerve to reach the dorsal part of the optic nerve. There it gives off the **long ciliary nerve,** which travels rostrally parallel to the optic nerve. It joins the short ciliary nerves from the ciliary ganglion and enters the eyeball. The fibers in the long ciliary nerve are primarily sensory to the structures of the eyeball. In addition, it contains some postganglionic sympathetic fibers from the cavernous plexus.

A connection between the nasociliary nerve and the ciliary ganglion may be present. The nasociliary nerve, after giving off the long ciliary nerve, divides into the **infratrochlear** and **ethmoidal nerves.**

The **infratrochlear nerve** *(n. infratrochlearis)* (Figs. 15–3, 15–4, 15–6) passes rostrodorsally along the lateral edge of the dorsal oblique muscle. It passes ventral to the trochlea for the tendon of the dorsal oblique muscle and ramifies in the medial portion of the upper eyelid and adjacent skin.

The **ethmoidal nerve** *(n. ethmoidalis)* (Figs. 15–3, 15–4) may be regarded as the continuation of the nasociliary nerve. It is sometimes referred to as the external ethmoidal nerve because it accompanies the external ethmoidal artery in a portion of its course. The ethmoidal nerve runs rostrally and medially, passing between the dorsal oblique and medial rectus muscles to enter the ventral ethmoid foramen. From the ventral ethmoid foramen it enters the cranial cavity and courses dorsally and rostrally in a small groove on the lateral wall. It leaves the cranial cavity through a foramen in the dorsomedial aspect of the cribriform plate. It contributes sensory branches *(nn. nasales interni)* to the nasal conchae as it courses rostrally. The nerve terminates rostrally as the **external nasal nerve** *(n. nasalis externus),* which is distributed to the skin of the muzzle rostral to the nasal bone (Fig. 15–2).

Maxillary Nerve

The **maxillary nerve** *(n. maxillaris)* is the largest of the trigeminal divisions. It is the sensory nerve to the skin of the cheek, side of the nose, muzzle, mucous membrane of the nasopharynx, maxillary sinus, soft and hard palates, and the teeth and gingivae of the upper jaw. The nerve leaves the trigeminal ganglion and passes rostrally in the dura mater of the lateral wall of the cavernous sinus to the round foramen. It leaves the cranial cavity by way of the round foramen to enter the alar canal. It turns rostrally in the alar canal and is related to the maxillary artery, which it accompanies across the pterygopalatine fossa, after leaving the rostral alar foramen.

The **zygomatic nerve** *(n. zygomaticus)* is the first branch of the maxillary division (Figs. 15–3, 15–4). It leaves the parent trunk near the round foramen. It may accompany the parent nerve in the alar canal on its dorsal side, it may lie in the zygomatic groove on the dorsal wall of the alar canal, or it may course in the zygomatic canal of the sphenoid bone and emerge from the zygomatic foramen (Fig. 15–7) (McClure 1960). The zygomatic nerve enters the apex of the periorbita soon after emerging from the cranial cavity. It divides into two rami, the **zygomaticotemporal** and **zygomaticofacial** (Figs. 15–3, 15–5, 15–8). This division may occur before the zygomatic nerve leaves the bony canal or after it enters the periorbita.

The **zygomaticotemporal nerve** *(n. zygomaticotemporalis)* is the most dorsal of the two branches. It courses rostrad and dorsad

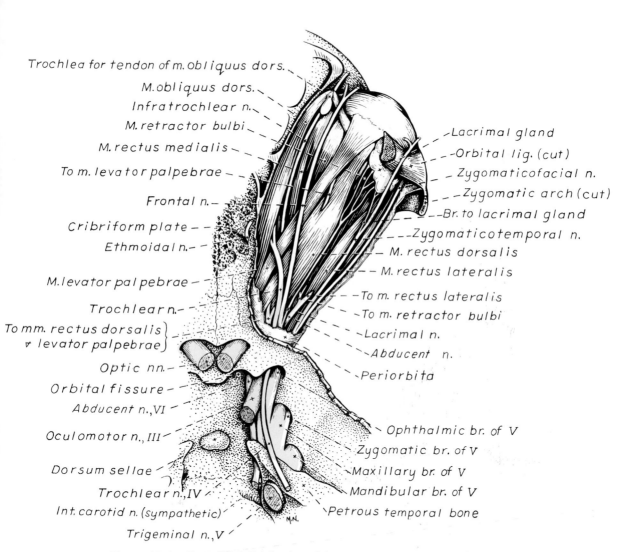

Figure 15–6. Superficial distribution of the nerves of the eye. Dorsal aspect.

immediately under the periorbita and penetrates the caudal edge of the orbital ligament to ramify in the lateral portion of the upper eyelid and skin of the rostral temporal area. In the rostrodorsal portion of the periorbita the nerve is related to the dorsolateral aspect of the lacrimal gland, to which it gives off some branches. These are believed to be the parasympathetic secretory fibers to the gland because of the communications between the pterygopalatine ganglion and the trigeminal nerve.

The **zygomaticofacial nerve** (*n. zygomaticofacialis*) parallels the zygomaticotemporal ramus within the periorbita (Figs. 15–3, 15–6, 15–8). In the rostral portion of the periorbita it deviates ventrally from the other ramus and emerges ventral to the lateral canthus of the eye. It ramifies in the lateral portion of the lower eyelid and adjacent area of skin (Fig. 15–10).

The next branches of the maxillary nerve are the **pterygopalatine nerves** (*nn. pterygopalatini*), numbering from one to three small branches, which communicate with the pterygopalatine ganglion (Figs. 15–7, 15–8). They are composed primarily of sensory fibers which pass near the ganglion to enter the maxillary nerve.

The next three nerves usually leave the ventral aspect of the maxillary nerve as a common trunk (the pterygopalatine nerve) as it courses rostrally between the medial pterygoid muscle and the ventral surface of the periorbita. They receive many branches from the pterygopalatine ganglion (Fig. 15–7).

The **minor palatine nerve** (*n. palatinus minor*) accompanies the minor palatine artery between the maxillary bone and medial pterygoid muscle to reach the soft palate (Figs. 15–2, 15–8). It contains sensory fibers from the mucosa as well as motor fibers which go from the pterygopalatine ganglion to the gland cells in the mucosa.

The **major palatine nerve** (*n. palatinus major*) accompanies the major palatine artery and enters the palatine canal of the maxillary bone (Figs. 15–2, 15–8). An **accessory palatine nerve** (*n. palatinus accessorium*) frequently leaves the ventral border of the major palatine nerve and passes through the accessory palatine canal to supply the caudal portion of the mucosa on the hard palate.

The major palatine nerve travels rostrally in the slight groove on the ventral surface of the bony hard palate, halfway between the midline and the teeth. The nerve terminates by sending a branch dorsally through the palatine fissure to communicate with the caudal nasal nerve in the nasal mucosa caudal to the incisor teeth.

The **caudal nasal nerve** (*n. nasalis caudalis*) enters the nasal cavity from the pterygopalatine fossa via the sphenopalatine foramen. In the nasal cavity it sends branches to the nasal septum and to the ventral conchae and maxillary recess (Fig. 15–2). The septal branch courses medially to the nasal septum and then rostrally on the dorsal aspect of the hard palate. It is related to the vomeronasal organ and communicates with the major palatine nerve at the incisive foramen. Many fine branches are given off to the nasal mucosa along its course. The fibers are motor (parasympathetic) from the pterygopalatine ganglion and sensory from the nasal mucosa. The branches to the maxillary recess leave the caudal nasal nerve and pass dorsally to the mucosal lining. The branches to the ventral concha lie between the bone and its mucosal covering (Fig. 15–2).

The **infraorbital nerve** (*n. infraorbitalis*) is the continuation of the maxillary nerve in the pterygopalatine fossa (Figs. 15–7, 15–8). It gives off **caudal superior alveolar branches** (*rami alveolares superiores caudales*), which supply the caudal cheek teeth (Figs. 15–5, 15–7). The infraorbital nerve enters the infraorbital canal and gives off **middle superior alveolar branches** (*rami alveolares superiores medii*) to the cheek teeth. Just before it emerges from the infraorbital canal, the infraorbital nerve gives off the **rostral superior alveolar branches** (*rami alveolares superiores rostrales*) (Fig. 15–5). The rostral maxillary alveolar nerve enters the maxilloincisive canal and supplies the upper canine and incisor teeth. The infraorbital nerve divides into a number of large fasciculi upon emerging from the infraorbital canal. These are distributed to the skin and sinus or tractile hairs of the upper lip and muzzle (Fig. 15–5). There are external and internal **nasal branches** (*rami nasales externi et interni*) and superior **labial branches** (*rami labiales superiores*).

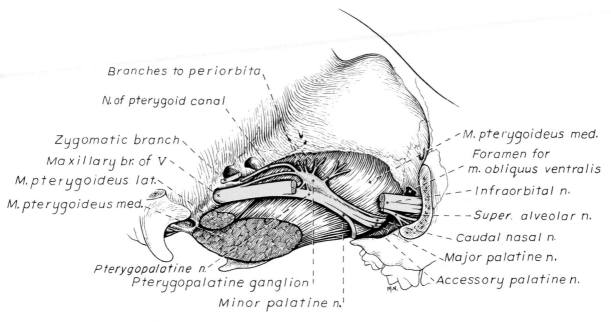

Figure 15–7. The pterygopalatine ganglion. Lateral aspect.

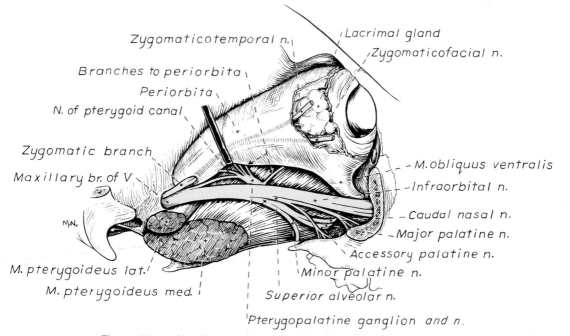

Figure 15–8. Maxillary branch of the trigeminal nerve. Lateral aspect.

Mandibular Nerve

The **mandibular nerve** (*n. mandibularis*) arises from the lateral side of the trigeminal ganglion and receives all the motor fibers (special visceral efferents) from the motor root. The nerve courses rostrolaterally and leaves the cranial cavity through the oval foramen (Figs. 15–3, 15–9). A mixed nerve, it contains both motor and sensory fibers. The trunk of the mandibular nerve is short and divides immediately upon emerging from the oval foramen. It has eight branches: the pterygoid, buccal, deep temporal, masseteric, auriculotemporal, mylohyoid, mandibular alveolar, and lingual nerves.

The **pterygoid nerves** (*nn. pterygoidei*) may arise separately or as a common trunk from the ventromedial side of the mandibular nerve. The **nerve to the lateral pterygoid muscle** (*n. pterygoideus lateralis*) is small and short (Fig. 15–9). The **nerve to the medial pterygoid muscle** (*n. pterygoideus medialis*) is larger and passes around the caudal border of the lateral pterygoid muscle to reach the medial pterygoid muscle. The nerve to the medial pterygoid muscle gives off two small branches shortly after its origin: the nerve to the tensor tympani muscle and the nerve to the tensor veli palatini muscle (Figs. 15–9, 15–16). The **nerve to the tensor tympani muscle** (*n. tensoris tympani*) passes caudally ventral to the mandibular trunk and enters the osseous auditory tube. It travels caudodorsally between the bone and the mucosal lining of the tube to reach the tensor tympani muscle. It parallels the minor petrosal nerve for a portion of its course. The **nerve to the tensor veli palatini muscle** (*n. tensoris veli palatini*) is relatively short and enters the dorsolateral side of the muscle.

The **buccal nerve** (*n. buccalis*), the masseteric nerve, and their deep temporal branches leave the mandibular nerve by a short common trunk which courses rostrally dorsal to the lateral pterygoid muscle (Fig. 15–9). The buccal nerve is the most medial, and it courses rostrally on the dorsal surface of the lateral and medial pterygoid muscles (Fig. 15–11). It emerges from the pterygopalatine fossa rostroventral to the rostral portion of the masseter muscle, and ramifies in the mucosa and skin of the cheek (Fig. 15–10). Several communications with branches of the dorsal buccal branch of the facial nerve are present. The buccal nerve receives branches from the otic ganglion. These are distributed to the zygomatic or orbital salivary gland.

The **deep temporal nerve** (*n. temporalis profundus*) is usually composed of two parts. One part arises in common with the buccal nerve and the other in common with the masseteric nerve (Fig. 15–9). The deep temporal nerve enters the temporalis muscle.

The **masseteric nerve** (*n. massetericus*) is the lateral portion of the dorsal trunk of the mandibular nerve. After giving origin to one of the deep temporal nerve branches, it passes laterad, caudal to the temporalis muscle and through the mandibular notch of the mandible to enter the masseter muscle (Figs. 15–9, 15–11). The nerve ramifies in the muscle and is its sole motor nerve supply.

The **auriculotemporal nerve** (*n. auriculotemporalis*) arises from the ventrolateral trunk of the mandibular nerve (Fig. 15–9). The fibers of the auriculotemporal nerve are sensory and are distributed to the skin in conjunction with branches of the facial nerve. It passes laterally, ventrally, and caudally to the glenoid process of the squamous temporal bone. It emerges between the caudodorsal border of the masseteric muscle and the external auditory canal (Fig. 15–11). The auricular branches supply the skin on the lateral surface of the external auditory canal and the rostral internal surface of the auricular cartilage (rostral internal auricular nerve). There are many fine communications which with the auricular branches communicate with the branches of the facial nerve and are distributed to the skin of the temporal and zygomatic areas (Fig. 15–10). One large branch goes directly to the sinus or tactile hairs on the caudolateral surface of the cheek. The parotid salivary gland also receives many small branches from the auriculotemporal nerve.

The fibers in the **parotid nerves** arise from the otic ganglion. They join the auriculotemporal nerve (Fig. 15–9) shortly after its origin from the mandibular nerve and leave it where it is related to the rostral deep aspect of the parotid gland. They supply parasympathetic fibers to the gland.

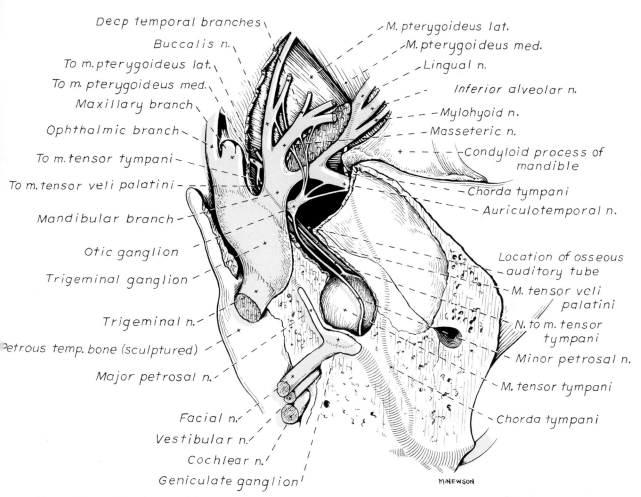

Deep temporal branches
Buccalis n.
To m. pterygoideus lat.
To m. pterygoideus med.
Maxillary branch
Ophthalmic branch
To m. tensor tympani
To m. tensor veli palatini
Mandibular branch
Otic ganglion
Trigeminal ganglion
Trigeminal n.
Petrous temp. bone (sculptured)
Major petrosal n.
Facial n.
Vestibular n.
Cochlear n.
Geniculate ganglion

M. pterygoideus lat.
M. pterygoideus med.
Lingual n.
Inferior alveolar n.
Mylohyoid n.
Masseteric n.
Condyloid process of mandible
Chorda tympani
Auriculotemporal n.
Location of osseous auditory tube
M. tensor veli palatini
N. to m. tensor tympani
Minor petrosal n.
M. tensor tympani
Chorda tympani

M·NEWSON

Figure 15–9. The temporal bone sculptured to show the otic ganglion in relation to the trigeminal nerve. Dorsal aspect.

The **mylohyoid nerve** (*n. mylohyoideus*) conducts motor and sensory impulses. It arises from the ventral lateral trunk of the mandibular nerve, rostral to the origin of the auriculotemporal nerve (Figs. 15–9, 15–11). The nerve courses ventrally over the rostral border of the medial pterygoid muscle to the ventral border of the mandible. At the ventral border of the mandible, it gives off a branch which supplies motor fibers to the rostral belly of the digastricus muscle (Fig 15–5). Another branch passes rostrally on the lateral side of the mandible dorsal to the digastricus muscle and deep to the masseteric muscle to communicate with the ventral buccal nerve (Figs. 15–5, 15–10).

The mylohyoid nerve passes rostrally on the ventral surface of the mylohyoid muscle and supplies it with many fine motor rami. It also supplies small rami to the skin of the intermandibular area. A larger ramus goes to the sinus or tactile hairs.

The **inferior alveolar nerve** (*n. alveolaris inferior*) leaves the ventral lateral trunk of the mandibular division of the trigeminal nerve and enters the mandibular canal through the mandibular foramen (Figs. 15–5, 15–9). The inferior alveolar nerve accompanies the inferior alveolar artery and gives off sensory branches to the teeth of the mandible. Several branches (mental nerves) leave the nerve rostrally and pass out through the mental foramina (Figs. 15–5, 15–10). The mental nerves are distributed to the skin ventral to the lower incisor teeth.

The **lingual nerve** (*n. lingualis*) passes rostrally on the dorsal surface of the medial pterygoid muscle and turns ventrally over its rostral border. Near its origin from the mandibular nerve, it receives the chorda tympani branch of the facial nerve (Figs. 15–5, 15–9, 15–11, 15–16). The lingual nerve gives off one or two small branches to the mucosa of the isthmus of the fauces.

The **sublingual nerve** (*n. sublingualis*) leaves the rostral surface of the lingual nerve and is distributed to the oral mucosa between the mandible and the tongue (Fig. 15–11). The **ramus communicans to the mandibular ganglion** leaves the caudal surface of the lingual nerve and courses alongside the mandibular duct to the **mandibular ganglion** located in the hilus of the mandibular salivary gland. The fibers in this ramus are para-

sympathetics derived from the chorda tympani nerve. Some branches are distributed to the sublingual salivary gland. Usually there is a small ganglion (sublingual ganglion) situated in the angle between the ramus communicans and the lingual nerve. It contains the nerve cell bodies which are secretory to the sublingual salivary gland.

The main trunk of the lingual nerve continues into the musculature of the tongue, where it has several communications with the hypoglossal nerve (Fig. 15–13) (Fitz-Gerald and Law 1958). The fibers are distributed to the dorsal mucosa of the tongue rostral to the circumvallate papillae. It conducts somatic afferent impulses to the brain stem via the trigeminal nerve. Special visceral afferent fibers from the taste buds leave the lingual nerve, via the chorda tympani nerve, and enter the brain stem with the facial nerve. (See The Tongue, Chapter 7.)

ABDUCENT NERVE

The sixth cranial nerve, or **abducent nerve** (*n. abducens*), leaves the ventral surface of the medulla oblongata just caudal to the pons. The abducent nerve conveys motor impulses to the retractor bulbi and the lateral rectus muscles. Koch (1916) states that both the large and small myelinated fibers of the abducent nerve are joined by a large bundle of unmyelinated fibers in the cavernous sinus.

The nerve courses rostrally in the subarachnoid cavity and enters the wall of the cavernous sinus. It passes rostrally, medial to the trigeminal ganglion of the fifth cranial nerve (Figs. 15–3, 15–4, 15–6), and leaves both the cavernous sinus wall and the cranial cavity by way of the orbital fissure. It is the most ventral and medial of the cranial nerves which emerge from the orbital fissure. It is related laterally to the ophthalmic division of the fifth cranial nerve and to the anastomotic branch of the external ophthalmic artery.

Approximately 1 to 2 cm. rostral to the orbital fissure, the abducent nerve gives off a branch to the retractor bulbi muscle. This short branch divides into smaller branches which supply both the dorsal and ventral parts of the muscle.

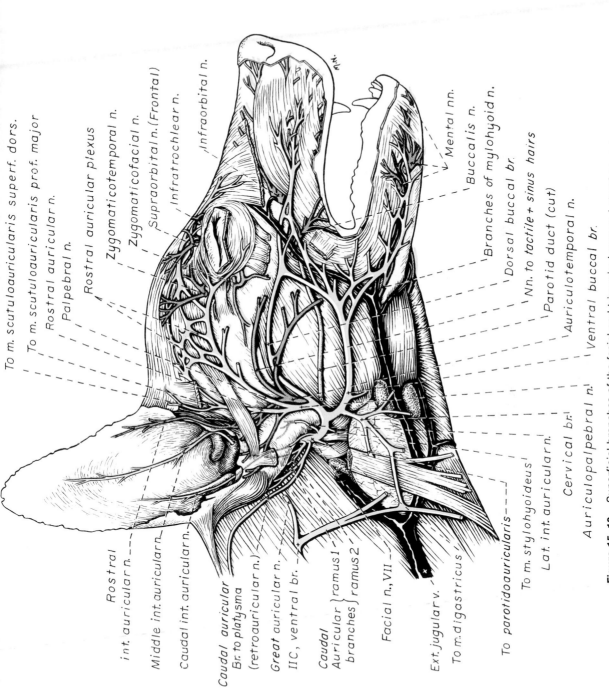

To m. scutuloauricularis superf. dors.
To m. scutuloauricularis prof. major
Rostral auricular n.
Palpebral n.
Rostral auricular plexus
Zygomaticotemporal n.
Zygomaticofacial n.
Supraorbital n. (Frontal)
Infratrochlear n.
Infraorbital n.

Mental nn.
Buccalis n.
Branches of mylohyoid n.
Dorsal buccal br.
Nn. to tactile + sinus hairs
Parotid duct (cut)
Auriculotemporal n.
Ventral buccal br.
Auriculopalpebral n.
Cervical br.
Lat. int. auricular n.
To m. stylohyoideus
To parotidoauricularis
To m. digastricus
Ext. jugular v.
Facial n., VII
Caudal Auricular ramus 1 branches ramus 2
IIC, ventral br.
Great auricular n.
Caudal auricular Br. to platysma (retroauricular n.)
Caudal int. auricular n.
Middle int. auricular n.
Rostral int. auricular n.

Figure 15–10. Superficial branches of the facial and trigeminal nerves. Lateral aspect.

FACIAL NERVE

GENERAL CONSIDERATIONS

The **facial nerve** (*n. facialis*) is a mixed nerve containing special visceral efferent (branchial motor) and afferent fibers and visceral efferent (parasympathetic) and afferent fibers.

The cell bodies of the special visceral efferent fibers are located in the **facial nucleus** (*nucleus motorius n. facialis*) in the rostroventral portion of the medulla oblongata. The special visceral efferent fibers make up the larger portion of the facial nerve. They are distributed peripherally to the auricular, facial, and other musculature derived from the second branchial arch, via the caudal auricular, digastric, stylohyoid, cutaneous colli, auriculopalpebral, stapedial, and dorsal and ventral buccal nerves.

The special visceral afferent and visceral efferent and afferent fibers make up a part of the facial nerve which is often referred to as the **nervus intermedius**. It is not grossly separable from the rest of the nerve in the dog. The visceral efferent fibers (preganglionic parasympathetic) are motor to the postganglionic nerve cells supplying the glandular cells of the nasal cavity, the mandibular and sublingual salivary glands, and the lacrimal gland. The cell bodies of the general visceral efferent fibers are located in the **parasympathetic nucleus of the facial nerve** (rostral salivatory nucleus) near the facial nucleus in the rostral portion of the medulla oblongata. The fibers are distributed by the major petrosal and chorda tympani nerve branches to the pterygopalatine, mandibular, and sublingual ganglia.

The sensory nerve fibers, both visceral afferent and special visceral afferent, have their cell bodies in the geniculate ganglion. The peripheral processes of the sensory nerves are distributed via the chorda tympani and major petrosal nerves, primarily to the taste buds in the rostral two-thirds of the tongue and to other visceral receptors in the epithelium covering the soft palate, nasopharynx, and nasal cavity.

COURSE OF THE FACIAL NERVE

The central course of the facial nerve fibers within the brain stem is described in Chapter 14. The facial nerve emerges from the brain stem at the rostral edge of the trapezoid body lateral to the origin of the abducent nerve. After a short course laterad, it becomes closely associated with the vestibulocochlear nerve and accompanies it into the internal auditory meatus. The facial nerve diverges from the vestibulocochlear nerve and enters the facial canal. Upon reaching the genu of the facial canal, it turns sharply (about 90 degrees), forming the **genu of the facial nerve** (*geniculum n. facialis*), and courses caudally.

The **geniculate ganglion** (*ganglion geniculi*) (Fig. 15–15) is located on the dorsorostral border of the facial nerve at the genu (Fig. 15–14). The cell bodies of the visceral and special visceral afferent fibers of the facial nerve are located in the ganglion.

The **major petrosal nerve** (*n. petrosus major*) arises from the geniculate ganglion and facial nerve at the genu Figs. 15–9, 15–15). The nerve passes rostroventrad in the petrous temporal bone, emerges, and becomes associated with the auditory tube (Fig. 15–16). The deep petrosal nerve (sympathetic fibers) joins with the major petrosal nerve to form the **nerve of the pterygoid canal**. The nerve of the pterygoid canal enters the caudal end of the pterygoid canal and emerges from the canal in the pterygopalatine fossa and ends in the pterygopalatine ganglion (Figs. 15–7, 15–8, 15–15).

The major petrosal nerve and the nerve of the pterygoid canal contain visceral efferent (preganglionic parasympathetic) fibers destined for the pterygopalatine ganglion. The nerve cells in the pterygopalatine ganglion give rise to the postganglionic fibers to the nasal glands and lacrimal gland. Some visceral afferents are contained in the nerves. Their distribution and function are uncertain; they may subserve general sensation from the nasal mucosa.

The **nerve to the stapedial muscle** (*n. stapedius*) (Fig. 15–15) leaves the dorsal medial surface of the facial nerve just before the latter turns ventrolaterad in the facial canal. The facial nerve also receives the auricular branch of the vagus nerve and gives rise to the chorda tympani nerve, which enters the cavity of the middle ear. In its ventrolaterad course the facial nerve receives the auricular branch of the vagus, prior to emerging from the facial canal at the stylomastoid foramen (Fig. 15–15). The fibers in the branch are

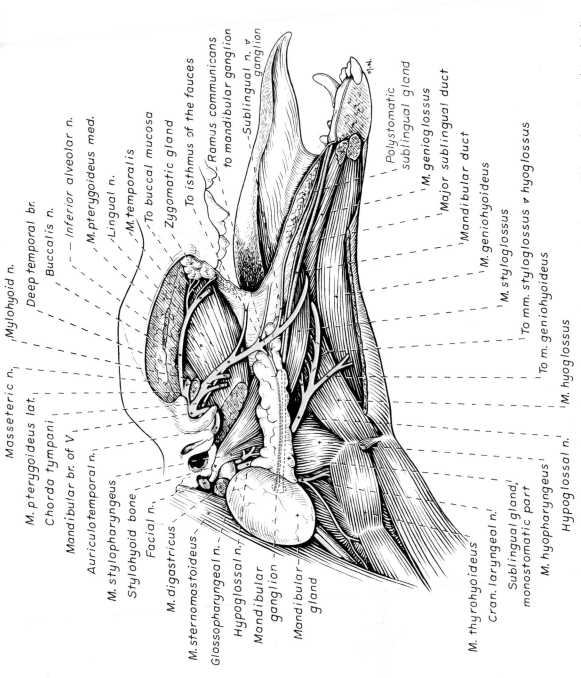

Figure 15-11. Nerve distribution medial to the mandible. (The digastricus muscles, the mandible, and structures lateral to it have been removed.)

somatic afferent, which are distributed by the internal auricular nerves to the skin on the concave surface of the auricular cartilage of the ear. Miller and Witter (1942) state that the cell bodies of the fibers in the auricular branch of the vagus are probably located in the proximal (jugular or superior) ganglion of the vagus nerve.

The **chorda tympani nerve** arises from the cranial surface of the facial nerve opposite the junction of the auricular branch of the vagus with the facial (Fig. 15–15). The chorda tympani nerve enters the cavity of the middle ear.

Just prior to emerging from the stylomastoid foramen, two or three auricular branches are given off the caudal surface. These branches accompany the main portion of the facial nerve to emerge from the stylomastoid foramen (Huber 1923).

The **caudal auricular nerves** *(nn. auriculares caudales)* (Fig. 15–10) are paired and supply the caudal or retroauricular musculature. They leave the facial nerve as it emerges from the stylomastoid foramen. They travel caudodorsally from the stylomastoid foramen in company with the great auricular artery. The most caudal of the two branches *(ramus I)* (Fig. 15–10) gives off a branch to the cervical portion of the platysma muscle. It then passes caudodorsally, parallel and deep to the muscle 2 or 3 cm. from the midline. The branch to the platysma gradually diminishes in size as it gives off fine twigs to the muscle. Grossly, it cannot be traced much farther caudally than the middle of the neck. Occasionally, this branch divides at the cranial edge of the muscle. The two resulting branches course caudally parallel to each other and to the mid line. Ramus I then gives off branches to the platysma muscle, the cervicoauricularis profundus muscle, and the musculi obliquii and transversi auriculae.

The other caudal auricular nerve *(ramus II)* may arise in common with ramus I or separately from the facial nerve at the stylomastoid foramen. It is larger than ramus I and courses dorsocaudally for 3 to 5 cm. on the caudoventral surface of the temporalis muscle. It turns dorsally and divides into branches which end in the following muscles: cervicoauricularis profundus (medial portion of the major part), cervicoauricularis superficialis, cervicoscutularis, interparie-

toauricularis, interparietoscutularis, occipitalis, interscutularis, scutuloauricularis superficialis accessorius, scutuloauricularis superficialis medius, scutuloauricularis profundus major and minor, helicis, and mandibuloauricularis. The branch to the mandibuloauricularis muscle courses dorsorostrally around the medial surface of the external ear canal to enter the dorsal portion of the muscle.

The **digastric nerve** *(n. digastricus)* innervates the caudal belly of the digastricus muscle. It leaves the caudoventral surface of the facial nerve immediately after the latter emerges from the stylomastoid foramen.

The **caudal internal auricular** and the **lateral internal auricular nerves** (Fig. 15–10) arise from the dorsal surface of the facial nerve and pass through the cartilage to ramify in the skin of the ear canal and in the auricular cartilage of the ear.

The **stylohyoid nerve** *(n. stylohyoideus)* is a small nerve which leaves the facial trunk and supplies the stylohyoid muscle (Fig. 15–10).

TERMINAL BRANCHES OF THE FACIAL NERVE

The facial trunk terminates as the auriculopalpebral, the dorsal buccal, and the ventral buccal nerve.

The **auriculopalpebral nerve** *(n. auriculopalpebralis)* (Fig. 15–10) is distributed to the rostral auricular muscles and the muscles of the eyelids. Auricular and palpebral branches arise approximately 1 to 2 cm. from the origin of the parent trunk. The auricular branches enter the following muscles: scutuloauricularis superficialis dorsalis, scutuloauricularis profundus major, and frontalis. The rostral auricular nerve has several communications with the auricular branches of the auriculotemporal nerve of the trigeminal nerve. The **palpebral branch** of the auriculopalpebral nerve passes dorsorostrally and enters into the formation of the extensive rostral auricular plexus. It communicates with the most rostral branch of the rostral auricular nerve. The palpebral nerve supplies the orbicularis oculi and corrugator supercilii muscles and terminates in the levator nasolabialis and maxillonasolabialis muscles.

The **dorsal buccal branch** *(r. buccalis dor-*

salis) (Fig. 15–10), the second terminal branch of the facial nerve, also has several communications with branches of the auriculotemporal nerve and the trigeminal nerve (source of sensory fibers). The dorsal buccal nerve passes rostrodorsally, forming an arc which roughly parallels the zygomatic arch, and communicates with the ventral buccal nerve caudodorsal to the commissure of the mouth. It passes through the orbicularis oris muscle and terminates in the maxillonasolabialis muscle. In its course through the orbicularis oris muscle it also communicates with the infraorbital and buccal nerves of the trigeminal nerve.

The **ventral buccal branch** *(r. buccalis ventralis)* (Fig. 15–10) immediately passes ventrally and slightly rostrally from the termination of the facial nerve. After coursing 1 to 2 cm., it gives off, caudoventrally, the **cervical branch,** which innervates the musculi depressor auriculae and the sphincter colli primitivus. There are several branches which join the cervical branches of the second cervical nerve. The ventral buccal nerve then passes rostrally on the lateral surface of the masseter muscle. Approximately 2 cm. caudal to the commissure of the mouth, it divides into numerous branches which enter the ventral portion of the orbicularis oris muscle. The ventral buccal nerve, at this point, also receives communicating branches from the mylohyoid nerve of the trigeminal. The terminal branches of the ventral buccal nerve also join the mental nerves, which are terminal branches of the mandibular alveolar nerve.

VESTIBULOCOCHLEAR NERVE

The eighth cranial nerve, or **vestibulocochlear nerve** *(n. vestibulocochlearis),* formerly referred to as the auditory, acoustic, or statoacoustic nerve, is composed of afferent fibers from the internal ear. The fibers are arranged in two bundles, the vestibular and cochlear parts *(nerves)* (Figs. 15–9, 15–15). The **vestibular nerve** *(pars vestibularis)* is involved with the sense of balance and is distributed to the cristae ampullares of the semicircular canals and the maculae of the saccule and utricle. The **cochlear nerve** *(pars cochlearis)* is distributed to the hair cells in the spiral organ of the cochlea and is concerned with the sense of hearing. Their origins are described in the section on the vestibulocochlear organ.

The two parts of the vestibulocochlear nerve are enclosed in a common dural sheath (with the facial nerve) as they pass through the internal acoustic meatus (Figs. 15–9, 15–15). They enter the brain stem caudal to the pons at the lateral end of the trapezoid body.

The cell bodies of the fibers in the vestibular nerve are bipolar and located in the vestibular ganglion in the vestibular portion of the inner ear. The cell bodies of the fibers in the cochlear nerve are bipolar and located in the spiral ganglion in the cochlear portion of the inner ear.

GLOSSOPHARYNGEAL NERVE

The ninth cranial nerve, or **glossopharyngeal nerve** *(n. glossopharyngeus),* is a mixed nerve. It is motor to the stylopharyngeus muscle and to the parotid and zygomatic salivary glands, and sensory to the pharynx, a portion of the tongue, and the carotid sinus. The nerve cells of origin for the fibers (special visceral efferents) to the stylopharyngeus muscle are located in the nucleus ambiguus of the medulla. The nerve cells of origin for the preganglionic parasympathetic fibers (general visceral efferents) for the glands are located in the parasympathetic nucleus of the glossopharyngeal nerve (caudal salivatory nucleus). The sensory cell bodies are in the proximal ganglion. Their peripheral processes are distributed to the pharyngeal and lingual mucosa and to the carotid sinus. Their central processes end in the nucleus of the solitary tract in the cat (Kerr 1962). Kozma and Gellert (1959) report that there are numerous nerve cells in the terminal trunk and the lingual ramus.

The glossopharyngeal nerve arises from the rostral part of the medulla oblongata by several small rootlets and leaves the cranial cavity through the jugular foramen (Figs. 15–15, 15–16). The **proximal (petrosal or superior) ganglion** is located in the jugular foramen. It is small and produces only a slight enlargement of the glossopharyngeal nerve (Fig. 15–16). Grossly, no separate distal ganglion is present in the dog.

The glossopharyngeal nerve gives rise to the following nerves.

The **tympanic nerve** (*n. tympanicus*) leaves the glossopharyngeal nerve at the level of the proximal ganglion and enters the cavity of the middle ear. It divides into several branches to form the **tympanic plexus** on the promontory of the petrous temporal bone. The **minor petrosal nerve** (*n. petrosus minor*) arises from the tympanic plexus and passes dorsally to gain the dorsal aspect of the tensor tympani muscle (Figs. 15–15, 15–16). The nerve then passes rostrally on the dorsolateral aspect of the auditory tube to the **otic ganglion** (Fig. 15–9). The preganglionic parasympathetic fibers in the nerve synapse with the postganglionic cells in the otic ganglion which are distributed to the parotid and zygomatic salivary glands.

The **pharyngeal ramus** (*r. pharyngeus*) of

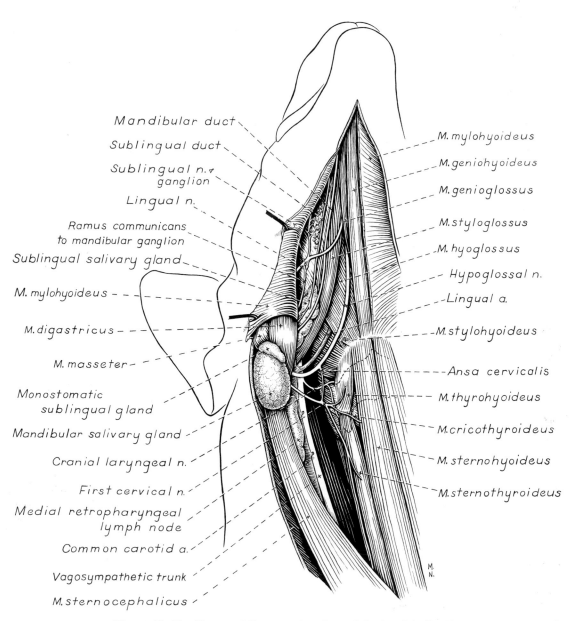

Mandibular duct

Sublingual duct

Sublingual n. ♥ ganglion

Lingual n.

Ramus communicans to mandibular ganglion

Sublingual salivary gland

M. mylohyoideus

M. digastricus

M. masseter

Monostomatic sublingual gland

Mandibular salivary gland

Cranial laryngeal n.

First cervical n.

Medial retropharyngeal lymph node

Common carotid a.

Vagosympathetic trunk

M. sternocephalicus

M. mylohyoideus

M. geniohyoideus

M. genioglossus

M. styloglossus

M. hyoglossus

Hypoglossal n.

Lingual a.

M. stylohyoideus

Ansa cervicalis

M. thyrohyoideus

M. cricothyroideus

M. sternohyoideus

M. sternothyroideus

Figure 15–12. Nerves of the ventral surface of the head and neck.

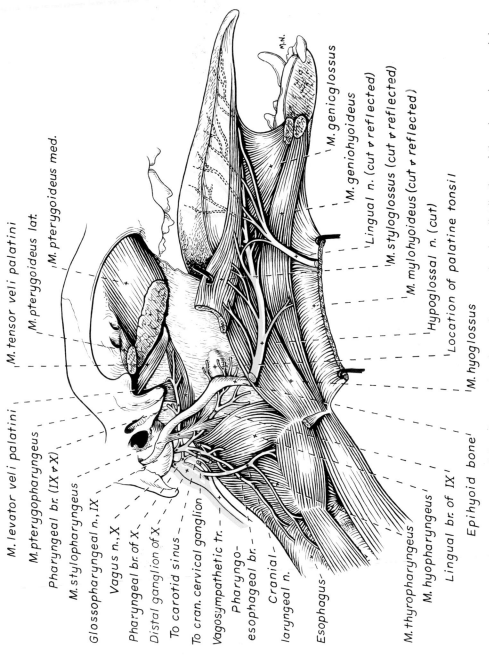

Figure 15-13. Deep dissection of the pharyngeal region and tongue showing distribution of the glossopharyngeal, hypoglossal, and lingual nerves.

M. levator veli palatini

M. tensor veli palatini

M. pterygopharyngeus

M. pterygoideus lat.

M. pterygoideus med.

M. stylopharyngeus

Glossopharyngeal n., IX

Vagus n., X

Pharyngeal br. of X

Pharyngeal br. (IX ʌ X)

Distal ganglion of X

To carotid sinus

To cran. cervical ganglion

Vagosympathetic tr.

Pharyngo-esophageal br.

Cranial-laryngeal n.

Esophagus

M. thyropharyngeus

M. hyopharyngeus

Lingual br. of IX

Epihyoid bone

M. hyoglossus

Location of palatine tonsil

Hypoglossal n. (cut)

M. mylohyoideus (cut ʌ reflected)

M. styloglossus (cut ʌ reflected)

Lingual n. (cut ʌ reflected)

M. geniohyoideus

M. genioglossus

the glossopharyngeal nerve leaves the parent nerve ventral to the petro-occipital fissure at the ventral border of the tympanic bulla (Fig. 15–12). The glossopharyngeal nerve and the pharyngeal ramus are lateral to the cranial cervical ganglion (Fig. 15–14). The pharyngeal ramus passes rostrally, medial to the stylopharyngeus muscle and the stylohyoid bone, on the dorsolateral aspect of the pterygopharyngeus muscle. The ramus courses around the rostral edge of the levator veli palatini muscle and ramifies in the mucosa of the dorsal aspect of the pharynx (Fig. 15–13). The pharyngeal ramus frequently receives a branch from the pharyngeal ramus of the vagus nerve and sometimes a branch from the cranial cervical ganglion.

The **lingual ramus** (*r. lingualis*) leaves the glossopharyngeal nerve a short distance distal to the pharyngeal ramus (Fig. 15–17). It passes through the stylopharyngeus muscle, to which motor branches are given off (special visceral efferents), and ramifies in the mucosa of the lateral pharyngeal wall and the palatine tonsil (Fig. 15–13).

The **ramus to the carotid sinus** (*r. sinus carotici*) leaves the glossopharyngeal nerve and parallels the internal carotid artery to the carotid sinus. The ramus is joined by a branch from the cranial cervical ganglion and often forms a plexus of several nerves which end at the carotid sinus and around the arteries at the termination of the common carotid artery. The glossopharyngeal nerve occasionally sends a branch to the pharyngoesophageal nerve of the vagus (Fig. 15–13).

VAGUS NERVE

The tenth cranial nerve, or **vagus nerve** (*n. vagus*), is the longest of the cranial nerves. It traverses the neck, thorax, and abdomen. A listing of its components indicates its distribution and functions.

General Visceral Efferent. These are the preganglionic parasympathetic nerve fibers that supply the muscle of the heart and the smooth muscle and glands of other thoracic and abdominal viscera.

Special Visceral Efferent. These supply motor innervation to the musculature derived from the last three branchial arches

of the embryo and to the esophageal musculature. These include the muscles of the pharynx (except the stylopharyngeus) and the larynx.

General Visceral Afferent. These visceral sensory nerves transmit impulses from the base of the tongue, pharynx, esophagus, stomach, intestines, larynx, trachea, bronchi, lungs, heart, and other viscera.

Special Visceral Afferent. A few sensory fibers come from the small number of taste buds on the epiglottis and pharyngeal wall.

General Somatic Afferent. These are the fibers which come from the skin of the external ear canal. They form the auricular branch of the vagus, which joins the facial.

The vagus nerve arises from the medulla oblongata of the brain stem by a series of fine rootlets. It passes through the middle portion of the jugular foramen (Fig. 15–15). The **proximal (jugular or superior) ganglion,** which contains the unipolar nerve cells of the general somatic afferent fibers of the vagus nerve, lies in the jugular foramen.

The vagus nerve leaves the jugular foramen and enters the petro-occipital fissure. It lies caudal to the glossopharyngeal nerve and the internal carotid artery and rostral to the accessory nerve (Figs. 15–14, 15–16). Ventral and medial to the tympanic bulla, it presents the **distal (nodose or inferior) ganglion,** which is prominent. Distal to the distal ganglion it is joined by the sympathetic trunk and becomes the vagal part of the vagosympathetic trunk, which travels in the carotid sheath (with the common carotid artery) to the thoracic inlet.

Branches in the Head and Cranial Cervical Region. The **auricular ramus** (*r. auricularis*) leaves the vagus nerve at the level of the jugular ganglion, and travels laterad through the caudal edge of the petrous temporal bone to the facial canal (Figs. 15–15, 15–16). The fibers of the auricular ramus join with the facial nerve and are distributed to the external acoustic meatus by way of the auricular branches of the facial nerve (Fig. 15–10).

The **pharyngeal ramus** (*r. pharyngeus*) leaves the vagus nerve at the proximal end of the distal ganglion. It sends a branch rostrally to join with the pharyngeal ramus of the glossopharyngeal nerve. The main trunk of the pharyngeal ramus is short and divides into many smaller branches to form the pha-

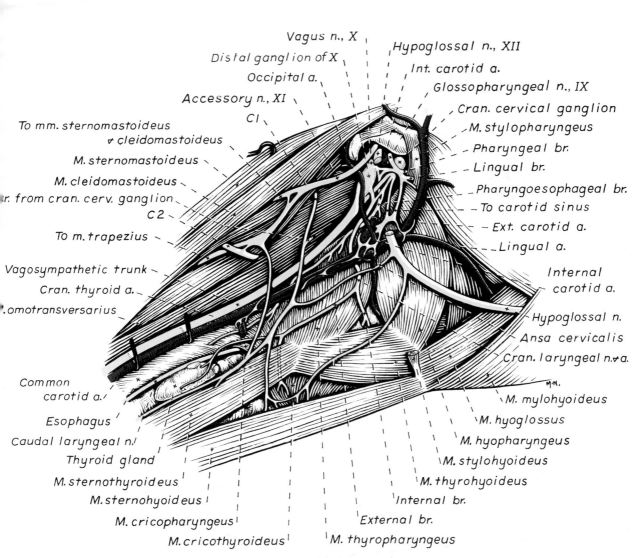

Vagus n., X
Distal ganglion of X
Occipital a.
Accessory n., XI
C1
Hypoglossal n., XII
Int. carotid a.
Glossopharyngeal n., IX
Cran. cervical ganglion
M. stylopharyngeus
Pharyngeal br.
Lingual br.
Pharyngoesophageal br.
To carotid sinus
Ext. carotid a.
Lingual a.
To mm. sternomastoideus
v cleidomastoideus
M. sternomastoideus
M. cleidomastoideus
r. from cran. cerv. ganglion
C2
To m. trapezius
Vagosympathetic trunk
Cran. thyroid a.
.omotransversarius
Internal
carotid a.
Hypoglossal n.
Ansa cervicalis
Cran. laryngeal n. v a.
M.N.
Common
carotid a.
Esophagus
Caudal laryngeal n.
Thyroid gland
M. sternothyroideus
M. sternohyoideus
M. cricopharyngeus
M. cricothyroideus
M. mylohyoideus
M. hyoglossus
M. hyopharyngeus
M. stylohyoideus
M. thyrohyoideus
Internal br.
External br.
M. thyropharyngeus

Figure 15–14. Nerves of the pharyngeal region. Lateral aspect. (The digastricus muscle and superficial structures have been removed.)

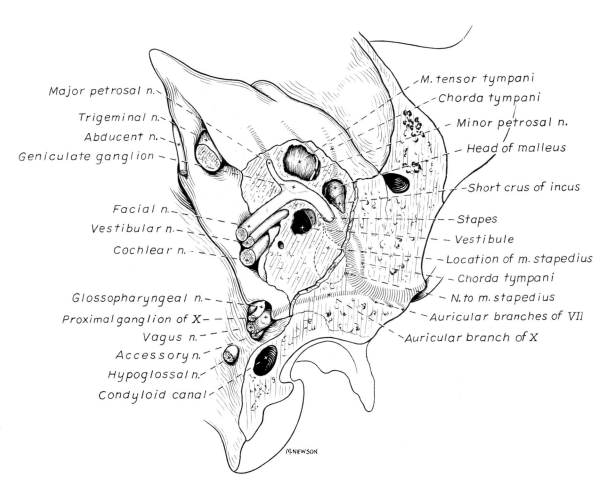

Figure 15–15. The petrous temporal bone sculptured to show the path of the facial nerve. Dorsal aspect.

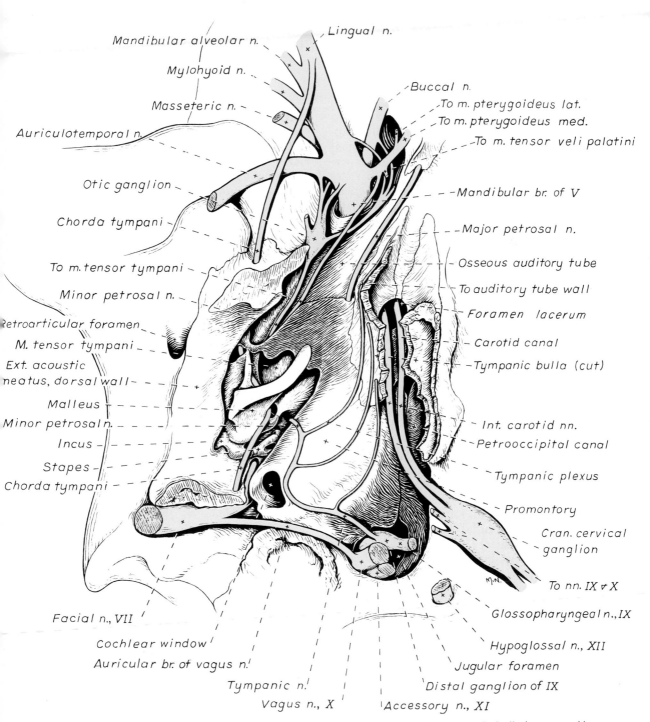

Figure 15–16. Nerves in the region of the middle ear. Ventral aspect. (The tympanic bulla is removed.)

ryngeal plexus *(plexus pharyngeus)* (Figs. 15–13, 15–14). The cranial cervical ganglion also contributes several branches to the pharyngeal plexus. A larger branch in the plexus, the **pharyngoesophageal branch** *(n. pharyngoesophageus),* usually receives branches from the glossopharyngeal nerve and the cranial cervical ganglion and is distributed to the caudal pharyngeal muscles (cricopharyngeus and thyropharyngeus) and to the esophagus. Hwang et al. (1948) report that the pharyngoesophageal branch supplies the cranial two-thirds of the cervical esophagus and, in some dogs, the entire cervical portion of the esophagus.

The **cranial laryngeal nerve** *(n. laryngeus cranialis)* leaves the vagus nerve at the distal ganglion (Figs. 15–13, 15–14). It passes ventrally, medial to the occipital and common carotid arteries, and divides into an **external branch** *(r. externus)* and an **internal branch** *(r. internus)* on the cranial portion of the thyropharyngeus muscle (Fig. 15–14).

The external branch travels caudally on the ventral portion of the thyropharyngeus muscle. A branch leaves the external branch ventrally and passes deep to the insertion of the sternothyroideus muscle, to supply the cricothyroideus muscle. The external branch continues caudad and ends in the area of the thyroid gland (Fig. 15–14).

The internal branch passes between the thyropharyngeus and hyopharyngeus muscles cranial to the edge of the thyroid cartilage to enter the larynx. It ramifies in the mucosal lining of the larynx. Before entering the cavity of the larynx, the internal branch gives off a branch caudally which communicates with the **caudal laryngeal nerve** *(n. laryngeus caudalis)* (Fig. 15–17).

The **recurrent laryngeal nerve** *(n. laryngeus recurrens)* arises from the vagus nerve at the thoracic inlet. The right recurrent nerve leaves the right vagus nerve and passes caudal to the right subclavian artery and then cranially on the right dorsolateral aspect of the trachea. The left recurrent nerve leaves the left vagus nerve and passes caudally around the ligamentum arteriosum and then cranially between the esophagus and the trachea. Both recurrent laryngeal nerves give off rami to the esophagus *(r. esophagei)* and the trachea *(r. tracheales)* as they run cranially.

The recurrent laryngeal nerve terminates as the caudal laryngeal nerve (Fig. 15–17). It also communicates with the nerves supplying the esophagus at its cranial end. The caudal laryngeal nerve is the motor nerve to all the intrinsic muscles of the larynx, except the cricothyroid muscle.

The remaining branches of the vagus nerve are described in Chapter 18.

ACCESSORY NERVE

The eleventh cranial nerve, or **accessory nerve** *(n. accessorius),* has a spinal origin in addition to several small rootlets of origin *(radices craniales)* from the medulla oblongata of the brain stem. The spinal origin *(radices spinales)* arises from cervical segments one to seven of the spinal cord (see Fig. 16–3). The spinal rootlets of origin are located between the dorsal and ventral roots of the cervical spinal nerves and form the spinal root, which is located on the dorsal side of the denticulate ligament. The spinal root enters the cranial cavity through the foramen magnum and is joined by several rootlets from the medulla oblongata and enters the jugular foramen caudal to the vagus nerve (Fig. 15–15). The cranial rootlets form a branch *(r. internus)* which leaves the accessory nerve in the jugular foramen and joins the vagus nerve distal to the proximal ganglion via common epineural sheaths. DuBois and Foley (1936) found in the cat that the branch from the accessory to the vagus nerve contained fibers from the medulla oblongata and that these conveyed motor impulses in the caudal laryngeal nerve.

The nerve cell bodies for the spinal root (motor nucleus of the accessory nerve) are located in the dorsolateral part of the ventral gray column in the spinal cord. The fibers of the cranial portion originate from nerve cells in the caudal portion of the nucleus ambiguus.

The trunk of the accessory nerve *(ramus externus)* leaves the jugular foramen and turns caudally in the petro-occipital fissure (Fig. 15–16). It passes lateral to the hypoglossal nerve and enters the cranial portion of the combined cleidomastoideus and sternomastoideus muscles. The nerve then divides into two parts, a **ventral ramus** and a **dorsal ramus.**

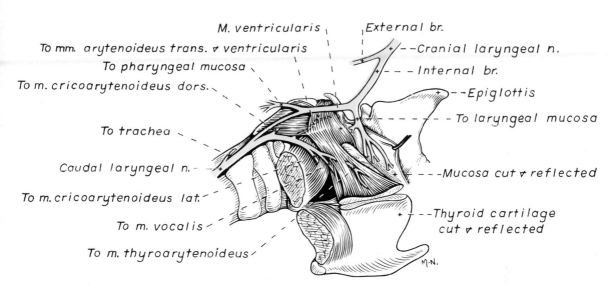

Figure 15–17. Distribution of the laryngeal nerves. Lateral aspect.

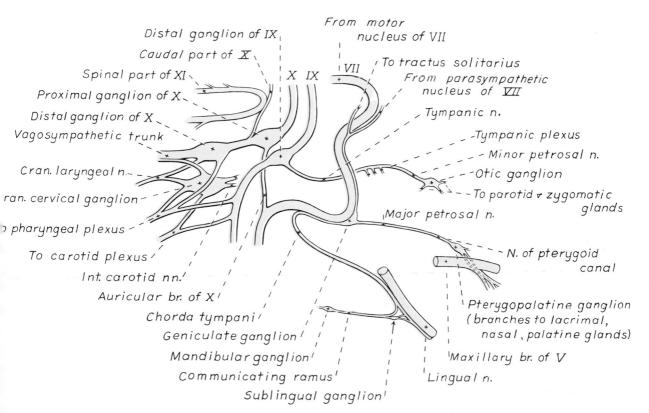

Figure 15–18. Schema of cranial nerves VII, IX, X, and XI, and autonomic interconnections.

The **ventral ramus** supplies the cleidomastoideus and sternomastoideus muscles. It gives off many branches as it passes caudally and ventrally on the medial side of the muscle. It is frequently embedded in the muscle tissue (Fig. 15–14).

The **dorsal ramus** penetrates the dorsal aspect of the cleidomastoideus muscle under the wing of the atlas and passes caudally deep to the cleidocervicalis muscle. It gives off several branches to the muscle.

The dorsal ramus has a number of communications with the second, third, and fourth cervical nerves. The dorsal ramus parallels the dorsal border of the omotransversarius muscle. At the shoulder it travels between the supraspinatus muscle and the cranial portion of the trapezius muscles. It gives off branches to the cranial or cervical portion of the trapezius muscle and terminates in its thoracic or caudal portion. These branches to the trapezius muscle constitute its sole motor supply.

HYPOGLOSSAL NERVE

The twelfth cranial nerve, or **hypoglossal nerve** (*n. hypoglossus*), arises as a series of rootlets from the ventrolateral sulcus of the medulla oblongata. The nerve leaves the cranial cavity through the hypoglossal canal (Figs. 15–15, 15–16). It passes medial to the accessory nerve and lateral to the vagus nerve, sympathetic trunk, and internal carotid artery (Fig. 15–14).

The hypoglossal nerve passes ventrally and gives rise to a descending branch which communicates with the ventral branch of the first cervical nerve, to form a part of the **ansa cervicalis** (Fig. 15–14). Another descending branch sometimes leaves the caudal side of the hypoglossal nerve about 2 cm. distal to the first and joins the ventral branch of the first cervical nerve on the dorsal aspect of the sternohyoideus muscle. It is thought to supply the thyrohyoid and, in some animals, the sternohyoid muscle.

The hypoglossal nerve curves rostrally on the lateral surface of the hyopharyngeus and hyoglossus muscles (Figs. 15–12, 15–14). Deep to the mylohyoideus it gives rise to rami which supply the styloglossus, hyoglossus, genioglossus, and geniohyoideus muscles as well as the intrinsic muscle fibers of the tongue (Fig. 15–13). Several communications occur between the rami of the hypoglossal nerve and the rami of the lingual nerve (Fig. 15–13) (FitzGerald and Law 1958).

BIBLIOGRAPHY

Bruesch, S. R., and L. B. Arey. 1942. The number of myelinated and unmyelinated fibers in the optic nerve of vertebrates. J. Comp. Neurol. 77:631–665.

DuBois, F. S., and J. O. Foley. 1936. Experimental studies on the vagus and spinal accessory nerves in the cat. Anat. Rec. 64:285–307.

FitzGerald, M. J. T., and M. E. Law. 1958. The peripheral connexions between the lingual and hypoglossal nerves. J. Anat. (Lond.) 92:178–188.

Gardner, E., D. J. Gray, and R. O'Rahilly. 1960. Anatomy: A Regional Study of Human Structure. Philadelphia, W. B. Saunders Co.

Huber, E. 1923. Über das Muskelgebiet des Nervus facialis beim Hund, nebst allgemeinen Beotrachtungen über die Facialis-muskulatur. Morph. Jb. II Teil. 52:353–414.

Hwang, K., M. I. Grossman, and A. C. Ivy. 1948. Nervous control of the cervical portion of the esophagus. Am. J. Physiol. 154:343–357.

Kerr, F. W. L. 1962. Facial, vagal and glossopharyngeal nerves in the cat. Arch. Neurol. (Chic.) 6:264–281.

Koch, S. L. 1916. Structure of the third, fourth, fifth, sixth, ninth, eleventh and twelfth cranial nerves. J. Comp. Neurol. 26:541–552.

Kozma, A., and A. Gellert. 1959. Vergleichende histologische Untersuchungen über die mikroskopischen Ganglien und Nervenzellen des Nervus glossopharyngeus. Anat. Anz. 106:38–49.

McClure, R. C. 1960. Occurrence of the zygomatic groove and canal in the sphenoid bone of the dog skull (*Canis familiaris*). Abstract, Anat. Rec. 138:366.

McCotter, R. E. 1912. The connection of the vomeronasal nerves with the accessory olfactory bulb in the opossum and other mammals. Anat. Rec. 6:299–317.

– – – – –, 1913. The nervus terminalis in the adult dog and cat. J. Comp. Neurol. 23:145–152.

Miller, M. E., and R. E. Witter. 1942. Applied anatomy of the external ear of the dog. Cornell Vet. 32:65–86.

Read, E. A. 1908. A contribution to the knowledge of the olfactory apparatus in dog, cat and man. Am. J. Anat. 8:17–48 (Plates 1–17).

Chapter 16

SPINAL CORD AND MENINGES

By THOMAS F. FLETCHER

THE SPINAL CORD

Introduction

The spinal cord is the division of the central nervous system enclosed within the vertebral column. Spinal nerves, which innervate the trunk, limbs, tail, and dorsum of the head, are attached to the spinal cord. The spinal cord performs three general operations (1) Through spinal nerves, it monitors receptors from skin, muscles, joints, and viscera, and it discharges impulses that control muscles and glands. (2) The spinal cord is a reflex center, containing neurons which continuously integrate afferent information and initiate appropriate responses in muscles and glands. (3) The spinal cord conducts information to and from the brain through an elaborate system of fiber tracts, by which the central nervous system regulates posture, movement, secretion, and afferent activity.

Spinal nerves are attached to the spinal cord by dorsal and ventral spinal roots. Dorsal roots are composed of afferent fibers which transmit information from receptors in the body. Ventral roots are bundles of efferent fibers which conduct nerve impulses to effector organs. Spinal roots and nerves constitute a basis for dividing the spinal cord into segments.

Grossly, the mammalian spinal cord appears as a continuous mass of nervous tissue; even histologically, columns of cells are distributed with little regard for the defined segmental boundaries of the spinal cord. Nevertheless, a segmental innervation pattern persists in the dog. Segmental innervation serves as a basis for understanding spinal cord organization and the manifestations of spinal disease.

Morphological Features of the Spinal Cord

The spinal cord and spinal roots are located within the **vertebral canal** (Fig. 16–1). They are surrounded and protected by three covering layers termed meninges. *Dura mater*, the most superficial meningeal coat, forms a tough, fibrous sheath, enclosing the spinal cord and roots (Fig. 16–2). **Arachnoid membrane** lies in contact with the inner surface of the dura mater. Deep to the arachnoid membrane is the subarachnoid space, which contains **cerebrospinal fluid.** *Pia mater*, the deepest,

935

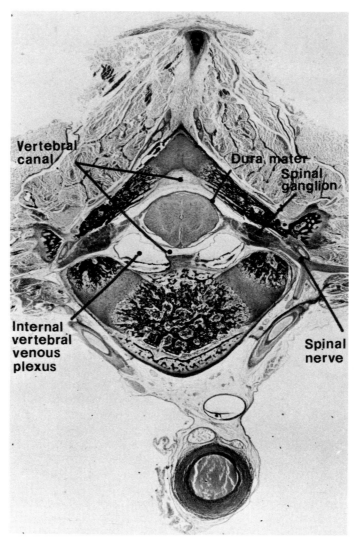

Figure 16–1.   Transverse section through the vertebral column in a dog fetus. The vertebral canal contains a paired internal vertebral venous plexus in addition to spinal cord and roots. The spinal cord, surrounded by dura mater, is connected bilaterally to a spinal nerve by means of dorsal and ventral spinal roots. A spinal ganglion is located on each dorsal root where dorsal and ventral roots combine to form a spinal nerve at an intervertebral foramen.

most vascular meninx, is fused to glial cells concentrated along the spinal cord surface (superficial glial limiting membrane). Pia mater is thickened (bilaterally) along the lateral margin of the spinal cord, forming the **denticulate ligaments**. Each denticulate ligament has attachments to dura mater, thereby suspending the spinal cord within the dura mater (Figs. 16–2, 16–4).

The center of the spinal cord is designated **gray matter** because of its appearance (Fig. 16–2). In this region, concentrations of nerve cell bodies and dense ramifications of dendrites and axons are found in a matrix of glial cells. Myelinated axons are also found in the gray matter, entering from or exiting to surrounding white matter. Myelinated nerve fibers are sparse, and capillaries are abundant in gray matter, compared with white matter.

In transverse sections (Fig. 16–2), gray matter displays a butterfly shape — wings connected by central gray substance. The connecting gray matter is called **central intermediate substance**. Located in its center is the **central canal**, lined by ependymal cells. The central canal contains cerebrospinal fluid and a proteinaceous material (Reissner's fiber). An area of concentrated glial cells immediately surrounding the central canal is designated *substantia gelatinosa centralis*. Gray matter dorsal and ventral to the central canal is the **gray commissure**.

Each wing of gray matter is divided into **dorsal** and **ventral horns** (or columns) and an intermediate zone, the junctional region between dorsal and ventral horns. The intermediate zone is referred to as **lateral intermediate substance**, since it is lateral to central intermediate substance. At thoracic and cranial lumbar levels of the spinal cord, a **lateral horn** projects into white matter from the lateral intermediate substance.

White matter surrounds the gray matter in the spinal cord. White matter is characterized by densely packed axons having myelin sheaths; lipid-rich myelin appears white when unstained. The white matter is composed of entering and exiting nerve fibers of dorsal and ventral rootlets, and of fiber tracts conveying information to and from the brain or traveling entirely within the spinal cord. Unmyelinated axons are also present among the conspicuous myelinated fibers of white matter.

Spinal cord white matter is divided into regions called funiculi (Fig. 16–2). The **dorsal funiculus** lies medial to entering dorsal rootlets. The **ventral funiculus** is located medial to exiting ventral rootlets. The remaining white matter, located between dorsal and ventral rootlets, is termed **lateral funiculus**. White matter ventral to the gray commissure is termed the (ventral) **white commissure**. It contains nerve fibers crossing from one side of the cord to the other.

The spinal cord has fissures and sulci which are used as landmarks (Fig. 16–2). Sulci are grooves. Fissures are clefts lined by pia mater. The **ventral median fissure**, into which central vessels project, and a **dorsal median fissure** divide the cord into symmetrical halves. Particularly in cervical and thoracic regions, the cleft of the dorsal median fissure may be partly obliterated, leaving a condensation of pia mater and underlying glial limiting membrane which is termed **dorsal median septum**. A **dorsolateral sulcus** is evident where dorsal rootlets enter the spinal cord. The sulcus marks the junction of the dorsal and lateral funiculi. A **ventrolateral sulcus** is sometimes discernible where ventral rootlets leave the cord, at the juncture of lateral and ventral funiculi. A **dorsal intermediate sulcus** is distinguishable in the cervical region, and a **dorsal intermediate septum** is variably developed in the cranial half of the spinal cord.

Thousands of nerve fibers compose each **dorsal root** and each **ventral root**. The nerve fibers of a root are bound together within a dural sheath near the spinal nerve; but near the spinal cord in the subarachnoid space, nerve fibers are grouped in separate bundles called rootlets. Rootlets attach serially along each spinal cord segment (Fig. 16–3). The typical number of rootlets per root is seven. Caudal (coccygeal) roots may have only 1 or 2 rootlets, while segments innervating the limbs may have 12 or more rootlets per dorsal or ventral root. Dorsal and ventral roots separately traverse the vertebral canal and unite to form a spinal nerve at each intervertebral foramen. In addition to dorsal and ventral roots, the first seven or eight cervical segments have laterally attached rootlets,

Text continued on page 945.

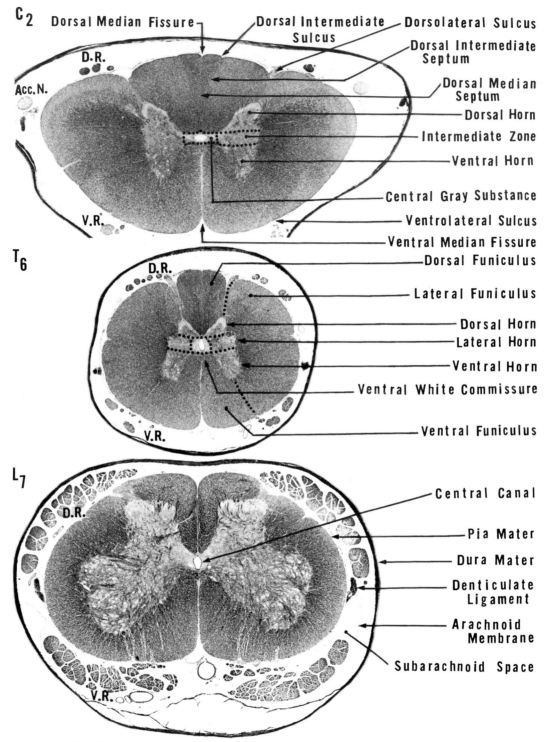

Figure 16–2. Morphological features of the canine spinal cord are illustrated in transverse sections through cervical (C2), thoracic (T6), and lumbar (L7) levels of the cord (luxol fast blue, hematoxylin stain). Centrally located gray matter is surrounded by myelinated white matter which has been stained dark. Myelinated fibers can be seen coursing into and out of the gray matter. Meninges are labeled in L7; a process of the denticulate ligament can be seen extending toward the dura mater. The escape of cerebrospinal fluid has allowed the arachnoid membrane to collapse away from the dura mater. The split ventrally in the dura mater is artifact. (D.R. = dorsal rootlets; V.R. = ventral rootlets; Acc. N. = spinal root of accessory nerve).

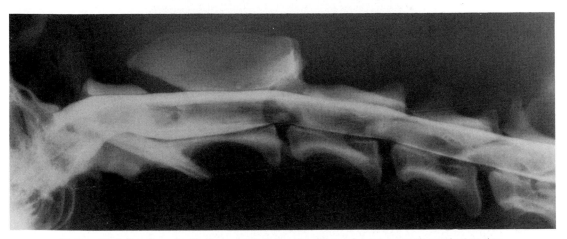

Figure 16–2 *Continued*   Radiograph 1. Lateral radiograph of the craniocervical region.

Illustration continued on following page.

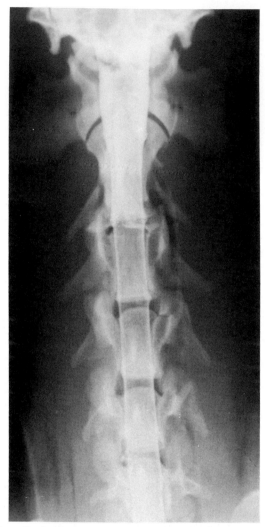

Figure 16–2 *Continued.*   Radiograph 2. Dorsoventral radiograph of the cervical region.

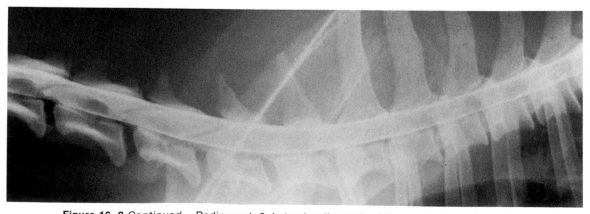

Figure 16–2 *Continued.*   Radiograph 3. Lateral radiograph of the cervicothoracic region.

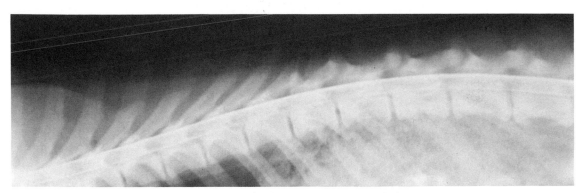

Figure 16–2 *Continued.* Radiograph 4. Lateral radiograph of the thoracic region.

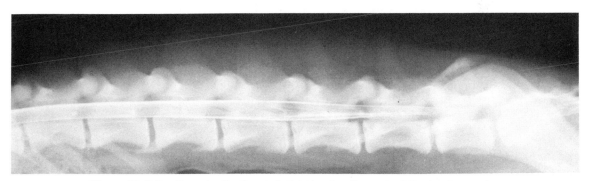

Figure 16–2 *Continued.* Radiograph 5. Lateral radiograph of the lumbar region.

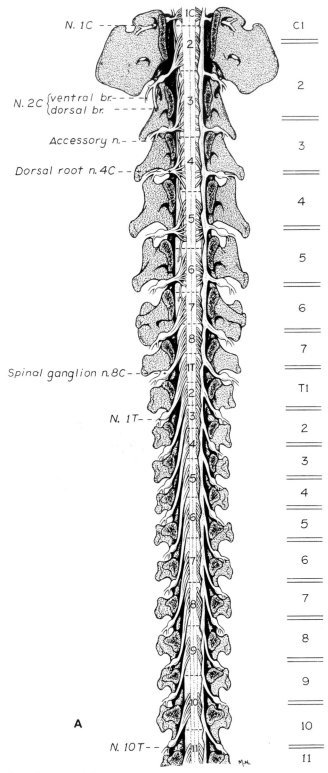

Figure 16–3. *A,* Spinal cord segmental relationship to vertebral bodies. From C1 to T11 the spinal cord, roots, ganglia, and nerves have been exposed by removal of the vertebral arches. The dura mater has been removed except on the right side. The numbers on the right represent the levels of the vertebral bodies.

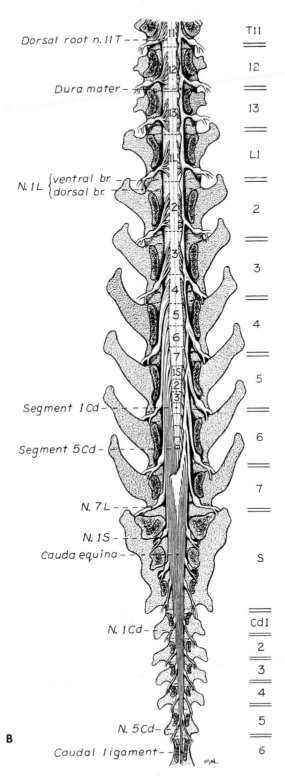

Dorsal root n. 11 T - -

Dura mater -

N. 1 L { ventral br. -
         { dorsal br. -

Segment 1 Cd -

Segment 5 Cd -

N. 7 L - - -

N. 1 S - - -
Cauda equina - -

N. 1 Cd -

N. 5 Cd -

B

Caudal ligament - -

T11

12

13

L1

2

3

4

5

6

7

S

Cd1

2

3

4

5

6

Figure 16–3 *Continued.* *B,* Spinal cord segmental relationship to vertebral bodies. From T11 through the caudal segments the spinal cord, roots, ganglia, and nerves have been exposed by removal of the vertebral arches. The dura mater has been removed except on the right side. The numbers on the right represent the levels of the vertebral bodies.

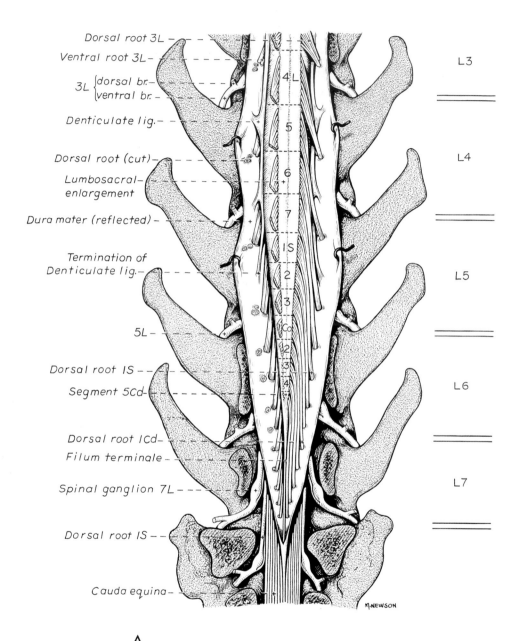

Dorsal root 3L
Ventral root 3L
3L {dorsal br.
{ventral br.
Denticulate lig.
Dorsal root (cut)
Lumbosacral-enlargement
Dura mater (reflected)
Termination of Denticulate lig.
5L
Dorsal root IS
Segment 5Cd
Dorsal root ICd
Filum terminale
Spinal ganglion 7L
Dorsal root IS
Cauda equina

L3
L4
L5
L6
L7

4L
5
6
7
IS
2
3
Co
2
3
4

M. NEWSON

A

Figure 16–4. *A,* Enlarged view of the terminal region of the spinal cord shown in Figure 16–3. Dorsal roots are removed on the left side to show the denticulate ligament, ending in a process that joins the dura mater between the entrances of L5 and L6 roots into dural sheaths. (Fletcher and Kitchell 1966a).

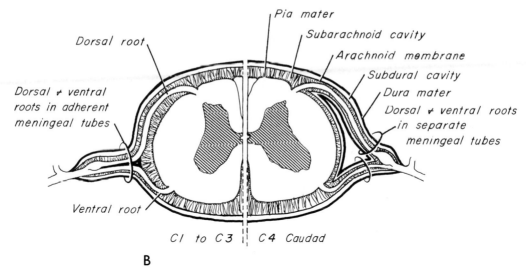

Figure 16-4 *Continued. B,* Schema of spinal meninges.

which join to form the **spinal root of the accessory nerve** (Fig. 16–3). Distally on each dorsal root, proximate to the spinal nerve, afferent neuron cell bodies are collected in a **spinal ganglion**.

The spinal cord is divided into segments, based on the attachments of dorsal and ventral roots. Boundaries between **spinal cord segments** are arbitrarily placed midway between the caudal and cranial rootlets of adjacent dorsal roots (Fig. 16–3). Spinal cord segments, roots, and spinal nerves are designated numerically according to region: cervical, 1 to 8; thoracic, 1 to 13; lumbar, 1 to 7; sacral, 1 to 3; and caudal, or coccygeal, 1 to 5.

At two locations where nerves to the limbs arise, the relative diameter of the spinal cord is increased. The **cervical enlargement**, where thoracic limbs are innervated, involves part of segment C6, segments C7 and C8, and part of segment T1. The **lumbar (lumbosacral) enlargement**, which supplies pelvic limbs, comprises part of segment L5, segments L6 and L7, and part of segment S1.

Caudal to the lumbar enlargement, the cord tapers to an elongate cone (*conus medullaris*). In this region, the remaining sacral and caudal segments are found. The segments appear successively smaller and are surrounded by caudally directed spinal roots (Fig. 16–4). Approximately 1 cm. caudal to the fifth caudal segment, the spinal cord terminates as the *filum terminale*, a uniform band of glial and ependymal cells. A dural sac and the subarachnoid space extend about 2 cm. beyond the end of the spinal cord. The caudal extension of dura mater which envelops the filum terminale is called the caudal ligament (ligament of the spinal dura mater).

Within the vertebral canal, sacral and caudal spinal roots stream caudally toward intervertebral foramina, where the roots exit. Collectively, these roots are known as the *cauda equina*. Most of the cauda equina lies caudal to the dural sac in the dog (Figs. 16–3, 16–4).

Vertebral Relationships

At birth, the canine spinal cord extends through the vertebral canal of the sacrum. As a consequence of postnatal development, the spinal cord terminates at caudal lumbar or cranial sacral levels of the vertebral column. In experimental and clinical situations, it is often useful to locate spinal cord segments with respect to more easily identifiable vertebrae.

With respect to the vertebral column, the spinal cord may be divided into four regions: (1) a cranial cervical region, where at least the first cervical segment lies within its corresponding vertebrae

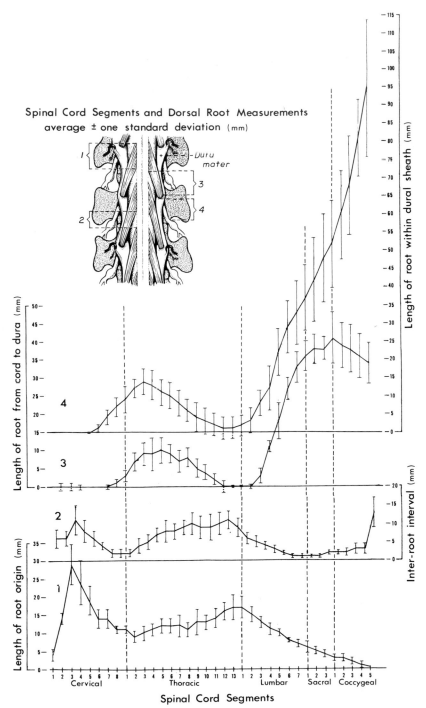

Figure 16–5. Graphs of four measurements made unilaterally on each spinal segment and dorsal root in 20 dogs. The spinal cord insert illustrates the measurements taken. The mean values for each measurement are connected by lines; vertical bars represent two standard deviations, one above and one below the average. Length of root origin (1) and inter-root interval (2) are measures of segment length, while (3) and (4) give a measure of root length. Notice that regional segments alternately lengthen, shorten, lengthen, and then progressively shorten along the spinal cord. Notice that segment shortening (1 and 2) is followed by root lengthening (3 and 4) and vice versa. This occurs because short segments shift craniad with respect to the vertebral column, while long segments shift caudad. (Fletcher and Kitchell 1966a).

(Fig. 16–3); (2) a caudal cervical through cranial thoracic region, where segments are positioned cranial to their respective vertebrae; (3) a thoracolumbar junction, where segments lie within their numerically corresponding vertebrae; and (4) a caudal lumbar through caudal region, where segments lie progressively cranial to their numerically respective vertebrae.

Regional segments in the spinal cord alternately lengthen, shorten, lengthen, and shorten again along the extent of the spinal cord (Fig. 16–5). Spinal cord segments display considerable differential growth, relative to subtle changes in vertebral length. The longest spinal cord segment is C3; caudal to it, segment length declines sharply, reaching a minimum at the level of segment T2. Caudal to this, thoracic segments lengthen, particularly by increased distance between adjacent roots (Figs. 16–3, 16–5). Segments of the thoracolumbar junction are relatively long. Caudal to these, segments become progressively shorter (Figs. 16–3, 16–5).

The location of spinal segments with respect to vertebrae may vary by half a vertebral length cranial or caudal to the usual relationship depicted for medium and large dogs in Figure 16–3. Small dogs (under 7 kg.) have relatively longer spinal cords. In these dogs, caudal displacement of the spinal cord is particularly evident in the lumbosacral and caudal regions, where segments may be one vertebra caudal to the relationship illustrated in Figure 16–4. In larger dogs, the spinal cord ends in filum terminale generally at the cranial border of the L7 vertebra; in small dogs, the spinal cord terminates at the lumbosacral vertebral junction.

The location of a spinal segment relative to its respective vertebra is indicated by spinal root length, since spinal roots travel to intervertebral foramina formed by numerically corresponding vertebrae. Spinal roots are short in cranial cervical and thoracolumbar junction regions of the cord, where segments lie near or within their corresponding vertebra. Roots are longer where segments are displaced cranial to corresponding vertebrae, as a result of segment shortening (Figs. 16–3, 16–5).

The first spinal nerve passes through a foramen in the dorsal arch of the atlas.

Spinal nerves C2 through C7 exit through intervertebral foramina formed by cranial margins of numerically corresponding vertebrae. Because of the presence of C8 spinal nerves, which leave the vertebral canal cranial to the T1 vertebra, remaining spinal nerves (T1 through Cd5) exit through intervertebral foramina formed at caudal margins of numerically respective vertebrae (Fig. 16–3).

Segmental Innervation

Segmentation is generally obscured in the dog. The spinal cord does not appear segmented, except for attachments of spinal roots. Nevertheless, innervation to the body is organized in a segmental pattern. Individual organs and structures are supplied by multiple spinal cord segments, a sometimes useful safety feature.

Segmental innervation originates in embryo. Each embryonic metamere (segment) gives rise to a segment of spinal cord, a pair of spinal nerves, somites, and mesenchyme. The spinal segment and spinal nerves of a metamere innervate somites and mesenchyme only of that metamere. In subsequent development — wherever somites and mesenchyme migrate, forming dermis, muscle, and viscera — the nerve supply is carried along, maintaining segmental innervation. Since body structures are formed by the merger of several somites, they receive innervation from multiple spinal cord segments.

The cutaneous region innervated by a single spinal nerve (or dorsal root) is termed a **dermatome**. By tactile stimulation, lumbar, sacral, and caudal dermatomes have been delineated in the dog (Fig. 16–6). Dermatomes form continuous skin fields and are arranged serially in an overlapping fashion. Dermatomes overlap to the extent that most skin loci are innervated by receptors from three spinal nerves. Thus, migration and merger of somites in skin deposition provide multiple innervation, a safety feature against sensory loss.

The musculature innervated by a single spinal nerve (or ventral root) is termed a **myotome**. Since individual muscles are formed by somite merger, muscles are in-

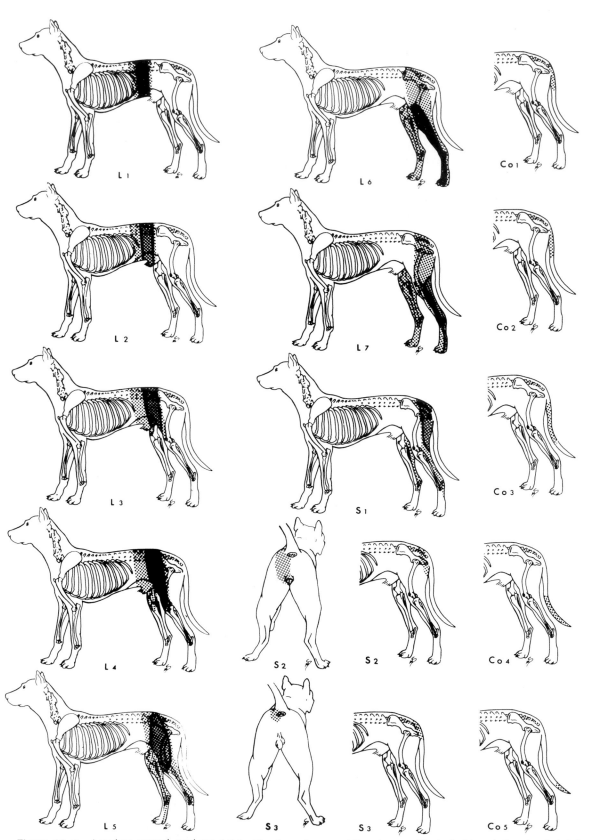

Figure 16–6. Lumbar, sacral, and caudal tactile dermatomes of the dog are illustrated in relation to skeletal landmarks. Each dotted area represents the entire distribution of a dermatome. The darker zones, shown within L1 to S1, represent regions of denser innervation. (Fletcher and Kitchell 1966b).

948

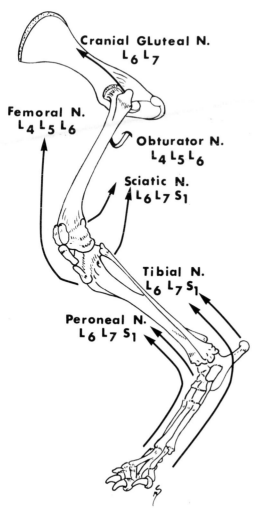

Figure 16–7. Segmental innervation to major muscle groups in the canine pelvic limb. Muscle groups are indicated by regional nerve supply and action (*arrows*). In general, each muscle within a group is innervated by all spinal segments listed for a regional nerve. Myotomes were determined by stimulating ventral roots and recording from individual muscles in dogs with median fixed plexuses. (Fletcher 1970).

nervated by multiple spinal segments. The segmental nature of muscle innervation is readily apparent in the case of expansive trunk muscles, where ventral rami from several spinal nerves can be seen entering one muscle. In the limbs, where muscles are supplied by single regional nerves, segmental nerve branches have to be exchanged at brachial or lumbosacral plexuses, where regional nerves arise. In the canine pelvic limb, most muscles are innervated by three spinal nerves; no muscle is supplied by less than two spinal cord segments (Fig. 16–7).

Regional nerves (e.g., femoral, tibial, pudendal, etc.) are consistent in their distribution to certain muscles, joints, skin, and viscera. However, regional nerves vary from dog to dog in their origin from spinal cord segments and spinal nerves. Variability in regional nerve segmental origin is most evident in the lumbosacral plexus (Fig. 16–8). A plexus is said to be **prefixed** when regional nerves are formed by more cranial spinal rootlets than is usually the case. In a **postfixed** plexus, nerve fibers forming regional nerves originate more caudally from spinal cord segments, compared to a **median fixed** plexus, which is the most common type. Examples of plexus fixation may be seen in Figure 16–8. Notice the origin of the femoral nerve and the distribution of the S1 ventral ramus and other nerve origins as well. The relative frequency of prefixed, median, and postfixed plexus types is 1:3:1 in the dog.

Organization of the Spinal Cord *

Afferent nerve fibers travel in dorsal rootlets, enter the spinal cord, and bifurcate into cranial and caudal branches in the dorsal funiculus (Fig. 16–9). The cranial and caudal branches give off collateral branches which synapse on neurons in spinal cord gray matter. The spinal cord neurons upon which collateral fibers synapse either participate in spinal reflexes or project their axons to the brain. In some cases, the cranial branches of afferent fibers continue in the dorsal funiculus to terminate directly in the brain.

In **spinal reflexes**, the excitable pathway, starting with a receptor and peripheral afferent neuron, continued by neurons in the spinal cord, and ending with an efferent neuron that leaves the cord to activate muscle or gland, is described as a **reflex arc**. The simplest reflex arc (monosynaptic) involves two neurons (Fig.

---

*In this section and in subsequent sections on gray matter and white matter of the spinal cord, much of the information presented has been obtained from experimental literature dealing with the cat. In view of the anatomical similarity of canine and feline spinal cords, no distinction has been made between these species.

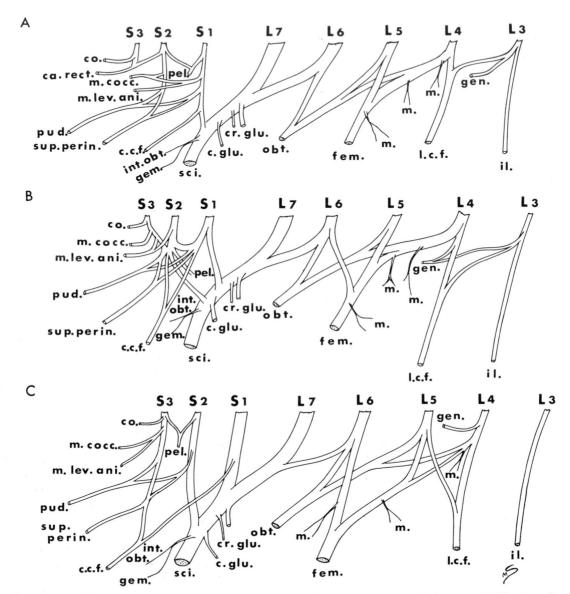

Figure 16–8. Lateral views of canine lumbosacral plexuses illustrate spectrum of plexus variability. *A,* prefixed type; *B,* median type; *C,* postfixed type. Abbreviations: L3 to S3 designate ventral rami; il., ilioinguinal; l.c.f., lateral cutaneous femoral; gen., genitofemoral; m., muscular branches to iliopsoas muscle; fem., femoral; obt., obturator; cr. glu. and c. glu., cranial and caudal gluteal; sci., ischiatic (sciatic); int. obt. and gem., branches to internal obturator, gemelli and quadratus femoris muscles; c.c.f., caudal cutaneous femoral; sup. perin., superficial perineal; pud., pudendal; m. lev. ani and cocc., branches to levator ani and coccygeus muscles; ca. rect., caudal rectal nerve; co., branch to ventral coccygeal (caudal) nerve trunk; pel., pelvic nerve. (Fletcher, 1970.)

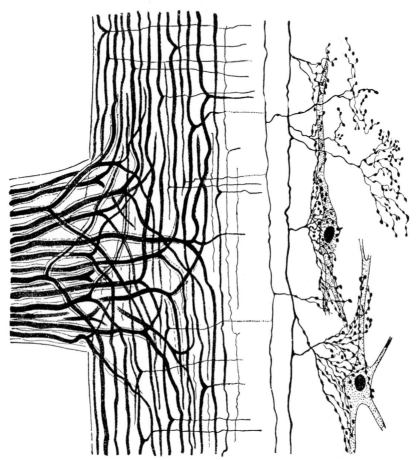

Figure 16–9. Afferent nerve fibers of a dorsal rootlet bifurcate into cranial and caudal branches as they enter the dorsal funiculus. The cranial and caudal branches give off collateral branches which terminate in spinal cord gray matter. At the right, collateral branches are shown as they terminate in synapses on neurons of the dorsal horn. (Cajal, Edinger from Ranson and Clark, 1959.)

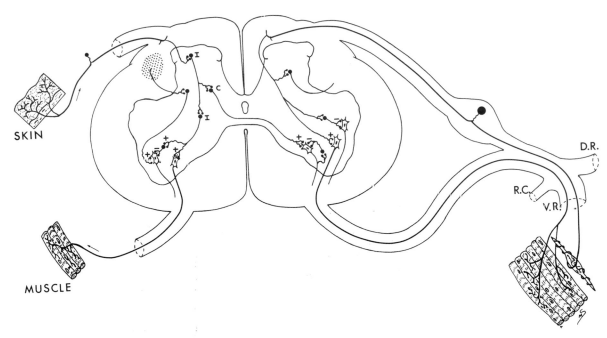

Figure 16–10. Schematic representation of spinal reflex arcs. On the right, an afferent neuron from a muscle spindle enters the cord, bifurcates, and gives off a collateral branch which synapses on a ventral horn motor neuron (monosynaptic reflex). The ventral horn motor neuron and the muscle fibers it innervates constitute a motor unit. The collateral branch may inhibit an antagonist motor neuron through an interneuron. The collateral fiber may also activate neurons in the dorsal horn, which could project to the brain or participate in intersegmental spinal reflexes. On the left, a polysynaptic reflex having two interneurons (I) is depicted. A commissural neuron (C) is seen crossing the cord to synapse on a motor neuron which activates its own inhibitory interneuron. At the upper left, a fiber from a descending tract synapses on an interneuron, through which it competes with spinal reflexes for control of a motor neuron (final common pathway).

16–10). Peripherally, the afferent neuron begins as an annulospiral receptor (in a muscle spindle), and centrally it ends by synapsing on an efferent neuron in the spinal cord. The efferent axon joins a ventral root and terminates peripherally on skeletal muscle fibers.

Most reflex arcs involve three or more neurons (polysynaptic). In these spinal reflexes, peripheral afferent neurons synapse on **interneurons**, so named because they are interposed between the afferent and efferent neurons (Fig. 16–10). The interneurons may terminate on efferent neurons directly or indirectly through other interneurons. While synapses made by afferent and efferent neurons have an excitatory effect, some interneurons inhibit neurons upon which they synapse. Interneurons which cross from one side of the spinal cord to synapse in the other side are called **commissural neurons**.

Spinal reflexes may be **segmental**, i.e.,

localized to one segment of the spinal cord, or **intersegmental**, in which case afferent activity in one segment influences efferent neurons at other spinal cord segments. Some interneurons send their axons into the white matter, where they travel to distant regions of the spinal cord to participate in intersegmental reflexes.

Efferent neurons to skeletal muscle are located in the ventral horn of the spinal cord. These **ventral horn motor neurons** are often called **final common pathway motor neurons** because they are controlled by subconscious and conscious directives from the brain, as well as by spinal reflex arcs. Competitive excitatory and inhibitory influences from varied sources impinge on a final common pathway motor neuron. Whenever the net excitatory input reaches a threshold level, the motor neuron fires, causing skeletal muscle fibers to contract.

A ventral horn motor neuron may innervate several or several hundred muscle fibers, depending on the size of the muscle it supplies. The ventral horn motor neuron and the muscle fibers it innervates are called a **motor unit** because these elements are activated as a unit and every muscle must contract in degrees of these units.

In addition to interneurons and motor neurons, **projection neurons** are located in spinal cord gray matter. Afferent fibers synapse on projection neurons, which send their axons to the brain. The axons gather according to modality in spinal cord white matter and pass to the brain as ascending spinal tracts.

Spinal cord white matter consists of ascending tracts, descending tracts from the brain, nerve fibers involved in intersegmental spinal reflexes, afferent fibers entering from dorsal roots, and efferent fibers leaving by ventral roots.

Gray Matter of the Spinal Cord

Spinal cord gray matter exhibits bilateral expansions, each divided into dorsal horn, lateral intermediate substance, and ventral horn. The gray expansions are joined across the midline by central intermediate substance (Fig. 16–11, C2), which includes: substantia gelatinosa centralis, a region of glial density surrounding the central canal; gray commissure, decussating axons and dendrites located dorsal and ventral to the central canal; and neurons of undetermined function, located at the junction of central intermediate substance and lateral intermediate substance.

The **dorsal gray horn** is often subdivided into an apex (the substantia gelatinosa), a head, a narrower neck, and a base which blends with the intermediate zone (Fig. 16–11, L3). Most dorsal root afferent fibers synapse in the dorsal horn. Dorsal horn neurons are interneurons for spinal reflexes or projection neurons for white matter pathways to the brain.

The **lateral intermediate substance (intermediate zone)** is recognized for its role in visceral (autonomic) innervation. Cell bodies and dendritic zones of preganglionic autonomic neurons, whose axons leave the spinal cord through ventral roots, are located in the intermediate zone. Sympa-

thetic preganglionic cell bodies are found in the thoracic and in the cranial four or five lumbar spinal cord segments. The cell bodies are arranged in medial and lateral cell columns, the lateral cell column projecting into white matter as the **lateral gray horn** (Fig. 16–11, T2, T4, L1). Parasympathetic preganglionic cell bodies are located in sacral spinal cord segments (Fig. 16–11, S2). Interneurons and projection neurons of unresolved significance are also located in the intermediate zone.

In the **ventral horn** are found conspicuous ventral horn motor neurons (final common pathway) which innervate somatic skeletal musculature. Large ventral horn motor neurons innervate motor units with many muscle fibers; medium-sized motor neurons supply smaller motor units. Small neurons in the ventral horn (fusimotor neurons) innervate intrafusal muscle fibers in muscle spindles. Axons of all these neurons leave the spinal cord through ventral roots. In addition, many neurons in the ventral horn serve as interneurons, and some project to the brain (ventral spinocerebellar tract).

Spinal cord neurons are multipolar in type. Each is identified morphologically by its cell body, which marks the location (though not the extent) of the receptive dendritic zone in a multipolar neuron. Neurons of similar function tend to have cell bodies (and receptive zones) distributed in a column, positioned similarly at various levels of the spinal cord. Columns of cell bodies are usually termed **nuclei** when viewed in transverse sections of the spinal cord.

Some neuronal cell columns are restricted to certain regions of the spinal cord; other columns (nuclei) may be sparsely populated and not very evident in a given section of the cord. Many neurons are scattered diffusely throughout the gray matter; at present, these cannot be differentiated into distinct groups. In order to understand the morphological basis of spinal cord organization, neuroanatomists attempt to resolve spinal cord neurons into functionally significant groups.

Two methods are presently utilized for subdividing spinal cord neurons into functionally significant groups (Fig. 16–11). One method divides the spinal cord into

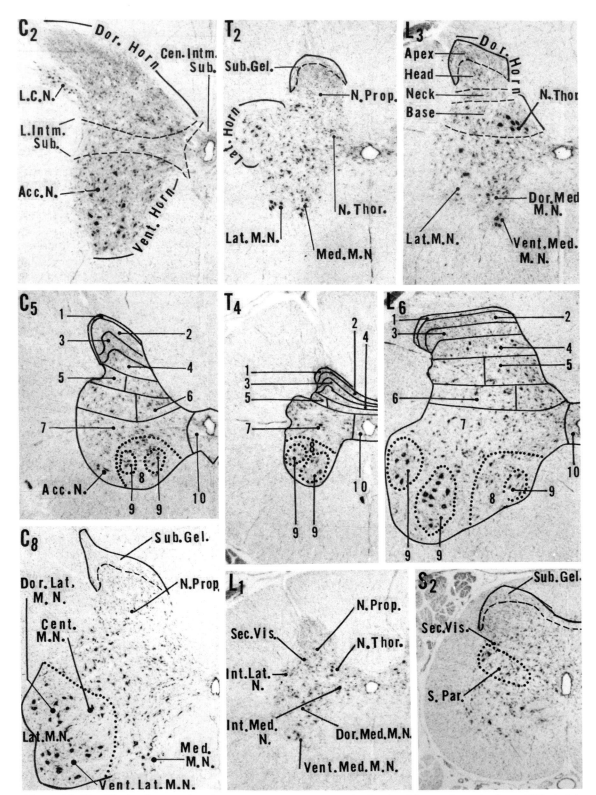

Figure 16–11. Photomicrographs of gray matter at representative levels of the canine spinal cord (Nissl stain. 12-μm. sections). Segments C5, T4, and L6 illustrate laminar divisions of gray matter applied to the dog (based on Rexed 1954). Remaining sections depict gray matter regions and nuclei (see text). Abbreviations: L.C.N., lateral cervical nucleus; L. Intm. Sub., lateral intermediate substance; Acc.N., accessory nerve motor nucleus; Cen. Intm. Sub., central intermediate substance; Sub. Gel., substantia gelatinosa; N. Prop., nucleus proprius; M.N., motor nucleus (medial, dorsomedial, ventromedial, lateral, dorsolateral, ventrolateral, central); N. Thor., nucleus thoracicus; Sec. Vis., secondary visceral cell column; Int. Lat. and Int. Med. N., intermediolateral and intermediomedial nuclei; S. Par., sacral parasympathetic nucleus.

10 laminar divisions (Rexed 1952). This classification accounts for all spinal neurons, but it is often difficult to distinguish the laminae, many of which seem inappropriate as discrete functional units. The other approach to categorizing spinal cord neurons identifies similar neurons as cell body groups which are then designated nuclei or cell columns. This method is specific for certain functional units, but it omits many neuronal units which are morphologically diffuse. In addition, conflicts in terminology and functional designation detract from the usefulness of nuclear grouping. Both the laminar and the nuclear methods of grouping spinal cord neurons are generally used, depending on which applies best to a specific situation.

DORSAL HORN NUCLEAR GROUPS

Substantia Gelatinosa. The concentration of small cells crowning the dorsum of the dorsal horn is called the substantia gelatinosa (Fig. 16–11). The nucleus extends the length of the spinal cord, varying in shape at different regions and blending cranially with the nucleus of the spinal tract of V. The substantia gelatinosa receives input from small, slowly conducting afferent fibers, particularly of pain and temperature modalities. Most cells of the substantia gelatinosa are interneurons which project to the nucleus proprius. Some of the larger neurons of substantia gelatinosa project to the brain in the lateral spinothalamic tract.

A row of larger, scattered cells, representing the **dorsomarginal nucleus**, is located along the outer surface of the substantia gelatinosa (Fig. 16–11, lamina 1 of C5, T4, L6). These cells participate in intersegmental reflexes.

Nucleus Proprius. Dorsal horn cells deep to substantia gelatinosa are collectively designated nucleus proprius of the dorsal horn (Fig. 16–11, C8, T2, L1). Nucleus proprius extends the length of the spinal cord and receives input from dorsal root afferent fibers, substantia gelatinosa, and descending spinal tracts. Many neurons of the nucleus proprius participate as interneurons in spinal reflexes. Larger neurons project to the brain in the lateral spinothalamic tract (pain, tempera-

ture), the ventral spinothalamic tract (touch), or the spinotectal tract.

Gray and white matter intermix at the lateral margin of the dorsal horn (Fig. 16–2). This region is designated (spinal) **reticular formation**. It varies in prominence but extends the length of the spinal cord.

Secondary Visceral Nucleus. The secondary visceral cell column is a group of cell bodies located laterally at the base of the dorsal horn (Fig. 16–11, L1, S2). It extends through the thoracic to the sacral regions of the cord, except for its absence in the caudal lumbar region. The small neurons of this cell column receive input from visceral afferent fibers (interoceptors) and send axons to the brain in ascending visceral tracts.

Nucleus Thoracicus. (Formerly called the nucleus dorsalis or dorsal spinocerebellar tract.) This nucleus consists of large cell bodies located medially at the base of the dorsal horn at the level of C8 through L3 spinal segments (Fig. 16–11, T2, L1, L3). The nucleus thoracicus receives input principally from muscle spindle afferents. The neurons project axons to the cerebellum in the dorsal spinocerebellar tract.

Lateral Cervical Nucleus. This nucleus appears as a peninsula or an island of scattered cells located in white matter just lateral to the dorsal gray horn in the first two cervical segments (Fig. 16–11, C2). The lateral cervical nucleus receives input from the nucleus thoracicus, from the ventral spinocerebellar tract, and from large neurons in the dorsal horn of caudal lumbar and sacral spinal cord segments. The neurons that project to the lateral cervical nucleus are activated by touch, pressure, and joint movement. The lateral cervical nucleus sends axons contralaterally to the thalamus by way of the medial lemniscus.

LATERAL INTERMEDIATE SUBSTANCE NUCLEAR GROUPS

Intermediomedial and Intermediolateral Cell Columns. These cell columns, which contain preganglionic sympathetic neurons, are recognized at spinal cord levels T1 through L4 (Fig. 16–11, L1). Embryologically, the neurons occupy a medial position, but subsequent migration positions the majority of cell bodies laterally. The intermediolateral nucleus, or cell column, extends into white matter as the lateral gray horn (Fig. 16–11, T2, T4, L1).

Sacral Parasympathetic Nucleus. This nucleus is found in the sacral spinal cord, chiefly segments S2 and S3. The parasympathetic preganglionic neurons form a mediolateral band in the intermediate zone (Fig. 16–11, S2).

In addition to autonomic preganglionic neurons, the intermediate zone contains projection neurons and many interneurons of unresolved functional significance, which are continuous in distribution with related interneurons in dorsal and ventral horns.

VENTRAL HORN NUCLEAR GROUPS

The prominent ventral horn motor neurons, which innervate skeletal muscles, are divided into as many as seven nuclear

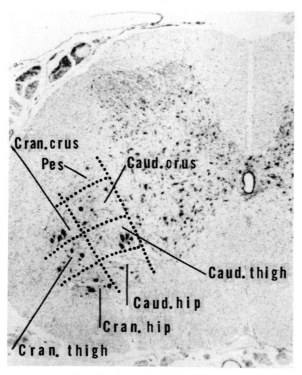

Figure 16–12. Transverse section through gray matter of L7 segment in the dog, depicting somatotopic arrangement of lateral motor neurons in the ventral horn. Because cranial thigh muscle motor neurons do not extend as far caudad as L7, neurons to muscles of the cranial crus have been appropriated to show the relative position of motor neurons which supply the cranial thigh. (Based on Romanes 1951.) Motor neurons innervating muscles of the pes constitute the retrodorsolateral motor nucleus.

groups in various segments of the spinal cord (Fig. 16–11, C8, T2, L1, L3). A **medial motor nucleus** is recognized at all levels of the spinal cord except for segments L7 and S1 (Rexed 1954). These neurons innervate musculature of the trunk. At various levels of the spinal cord, the medial nuclear group appears divisible into **dorsomedial** and **ventromedial motor nuclei** (a separation of disputed significance).

Throughout the spinal cord, cell bodies positioned laterally in the ventral horn may be considered a **lateral motor nucleus**, for innervation of extreme ventral hypaxial musculature. The lateral motor nucleus is very much expanded in regions of the cervical and lumbar enlargements, from which muscles of the limbs are innervated. In portions of the enlargements, it is possible to divide lateral motor neurons into **ventrolateral, dorsolateral, central,** and **retrodorsolateral motor nuclei** (Fig. 16–11, C8; Fig. 16–12).

Motor neurons of lateral motor nuclei are somatotopically arranged, in terms of relative locations of the limb muscles they innervate. In the lumbosacral enlargement, the proximal to distal muscles of the pelvic limb are innervated by ventral to dorsal groups of motor neurons; the more lateral motor neuron groups supply cranial limb muscles, while more medial motor neuron groups innervate caudal muscles of the limb (Fig. 16–12).

White Matter of the Spinal Cord

Spinal cord white matter consists of dorsal root afferent fibers entering the cord, ventral root efferent fibers leaving the cord, fibers conducting between segments of the spinal cord, and fiber tracts running to and from the brain. White matter of each half of the spinal cord is divided into dorsal, lateral, and ventral funiculi, and is continuous as (ventral) white commissure, composed of myelinated axons crossing from one half of the cord to the other.

Axons serve to conduct excitation from one location to another. Thus, axons which originate from a common receptor type or nucleus and terminate in a common location have a similar function. Functionally similar axons generally travel together in white matter, and such bundles of nerve fibers are called tracts or fasciculi. Fasciculi, found principally in the dorsal funiculus, are named for location or termination. Tracts are named according to origin and destination (e.g., a spinocerebellar tract runs from the spinal cord to the cerebellum; a corticospinal tract travels from the cerebral cortex to the spinal cord). In deference to their vertical direction in man, spinal tracts conducting toward the brain are referred to as ascending, and tracts conducting from the brain to the spinal cord are called descending tracts.

Dorsal root afferent fibers enter the dorsal funiculus of the spinal cord at the dorsolateral sulcus. Within the dorsal funiculus, afferent fibers bifurcate into longer cranial and shorter caudal branches. These in turn give off collateral branches that enter gray matter to synapse on projection neurons, interneurons, or ventral horn motor neurons (Fig. 16–9). Cranial and caudal branches of afferent fibers give off collaterals over a distance of from one to eight segments, depending on the receptive field of the afferent fiber. Afferent fibers having receptors clustered in small areas give collaterals only to localized levels of the cord, while afferent fibers which distribute collateral branches to a number of spinal cord segments have diffuse receptive fields.

Afferent fibers of dorsal rootlets segregate according to size upon entering the spinal cord. Larger myelinated fibers, which conduct from proprioceptive and tactile receptors, are located medially as a rootlet enters the cord. Unmyelinated and small myelinated fibers collect as a lateral division of the entering dorsal rootlet; these fibers conduct excitability from pain and temperature receptors.

The cranial and caudal branches of lateral division, small, afferent fibers overlap in adjacent spinal cord segments. The band of overlap extends from the dorsolateral sulcus of the substantia gelatinosa along the length of the spinal cord (Fig. 16–13). This bundle of axons of small diameter is called the **dorsolateral fasciculus** (Lissauer's tract); it blends with the spinal tract of the trigeminal nerve in cranial cervical segments.

Cranial and caudal branches of large, medially positioned afferent fibers form fasciculus gracilis and fasciculus cuneatus. Generally, the cranial and caudal branches

terminate as collateral branches to gray matter; however, about 25 per cent of the cranial branches continue to the brain, terminating in nucleus gracilis or nucleus cuneatus (Glees and Soler 1951). The ascending cranial branches are somatotopically arranged in the dorsal funiculus.

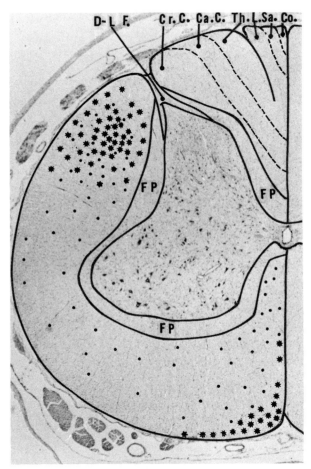

Figure 16–13. Transverse section through cervical spinal cord, schematized to illustrate features of white matter. The dorsolateral fasciculus (D–L F.) forms a band from the substantia gelatinosa to the dorsolateral sulcus. Within the somatotopically arranged dorsal funiculus, cranial ascending branches from more caudal rootlets are located dorsal and medial to those from more cranial rootlets. (Cr.C., cranial cervical; Ca.C., caudal cervical; Th., thoracic, etc.) Fasciculus proprius (FP) is found adjacent to gray matter in all three funiculi. The lateral corticospinal tract (stars in lateral funiculus) and a vestibulospinal tract (stars in ventral funiculus) represent relatively localized tracts. They have a center of high fiber density, but fibers are scattered at the tract periphery. In contrast, reticulospinal fibers (dots) are widely dispersed, having no area of high fiber density.

At each level, ascending fibers which originated more caudally are displaced medially and dorsally by entering afferent fibers (Fig. 16–13). It is generally true in all ascending and descending tracts that shorter fibers (e.g., to and from the cervical region) lie closer to the gray matter, while longer fibers (e.g., to and from the sacral region) lie closer to the exterior margin of the spinal cord.

Spinal tracts that travel to or from the brain are located more peripherally in white matter; shorter nerve fibers, originating at one spinal segment and terminating at another level of the cord, are deeply positioned, proximate to gray matter. Such intersegmental spinal nerve fibers are known collectively as *fasciculus proprius*. The fasciculus proprius lies adjacent to gray matter in all three funiculi (Fig. 16–13). Multisynaptic pathways to and from the brain (spino-bulbo-spinal fibers) also run in fasciculus proprius.

In the lateral funiculus, generally, spinal tracts which ascend to the brain are found along the margin of the spinal cord, while descending tracts from the brain are positioned between fasciculus proprius and the ascending tracts. In the ventral funiculus, which contains mainly descending fibers, tracts are smaller and more dispersed, and overlap with one another more than in the other funiculi.

Fiber tracts are never as circumscribed or localized as they are generally illustrated. A relatively localized tract may have fibers densely grouped at its center, but fibers become dispersed at the tract margin, mingling with margins and centers of other tracts. Some functionally recognized tracts have fibers scattered widely over the white matter with no apparent center of high fiber density (Fig. 16–13).

ASCENDING SPINAL PATHWAYS

Discriminative tactile and proprioceptive sensibility is conveyed to the brain in fasciculus gracilis and fasciculus cuneatus by cranial branches of large dorsal root afferent fibers (Fig. 16–14). The *fasciculus gracilis* is composed of proprioceptive and discriminative (small receptive field) tactile afferent fibers from caudal thoracic, lumbar, sacral, and caudal regions. These afferent fibers send cranial branches of the nucleus gracilis. The *fasciculus cuneatus* is composed of the same nerve fiber types. It originates from cranial thoracic and cervical segments, lies lateral to the fasciculus gracilis, and terminates in the nucleus cuneatus medialis. The dorsal intermediate sulcus and the variable dorsal intermediate septum tend to separate fasciculus gracilis from fasciculus cuneatus.

Crude tactile sensibility is carried by myelinated dorsal root fibers having diffuse receptive fields. The axons travel in the medial division of the dorsal root and bifurcate into cranial and caudal branches which give off collaterals over six to eight spinal cord segments (Crosby, Humphrey, and Lauer 1962). The pathway to the brain is continued by projection neurons in the nucleus proprius. Projected axons cross in the white commissure (some remain ipsilateral) and ascend to the thalamus in the ventrolateral margin of the ventral funiculus as the **ventral spinothalamic tract** (Fig. 16–14).

Pain and temperature sensibility is conveyed by small (thin) afferent fibers in the lateral division of the dorsal root. Projection neurons in substantia gelatinosa and nucleus proprius continue the conscious pathway. Their axons decussate in the white commissure and form the **lateral spinothalamic tract**, located ventrally in the lateral funiculus (Fig. 16–14). The tract terminates on neurons in the thalamus.

Ventral and lateral spinothalamic tracts intermingle to the extent that sometimes they are considered a combined ventrolateral ascending system. Some fibers in the system reach the thalamus directly, while other terminate in the brain stem on relay neurons which continue the pathway to the thalamus. A few spinal neurons have been found to project directly to the cerebral cortex by joining the descending corticospinal tracts (Crosby, Humphrey, and Lauer 1962).

Pain also reaches conscious centers by traveling in bilateral, multisynaptic pathways in the dorsal half of the spinal cord (Kennard 1954). These pathways probably involve dorsolateral fasciculus and fasciculus proprius. Visceral pain reaches conscious centers through bilateral multisynaptic pathways in fasciculus proprius of the lateral funiculus (Crosby, Humphrey, and Lauer 1962).

Non-conscious pathways conduct from the spinal cord to various brain stem locations. In the brain stem, information from several sources is integrated to produce somatic and visceral responses of an automatic nature. In many cases, afferents to brain stem centers (e.g., to brain stem reticular formation) are

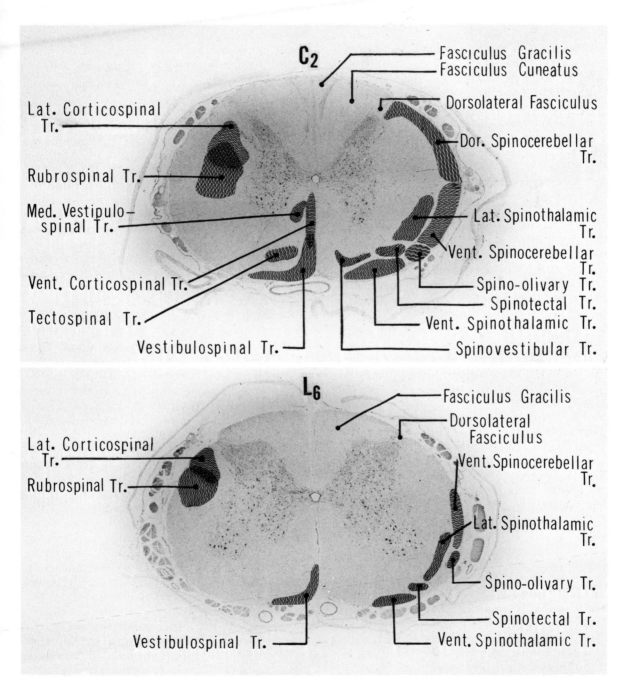

Figure 16–14. Transverse sections through cervical and lumbar levels of the spinal cord. Approximate locations of major spinal tracts are represented schematically. Descending tracts are depicted on the left, ascending on the right. (Lat. = lateral; Med. = medial; Dor. = dorsal; Vent. = ventral; Tr. = tract.)

collaterals of axons in the spinothalamic conscious pathways. The **spinotectal tract** consists of fibers in the ventrolateral ascending system which terminate in the roof of the midbrain (tectum) at the level of the rostral colliculus (Fig. 16–14). Input of these fibers is integrated with visual and other afferent information in the tectum.

Pathways to the cerebellum and associated brain stem nuclei convey proprioceptive and exteroceptive activity required for postural and movement adjustments. Proprioceptive and some tactile afferent fibers from thoracic, lumbar, sacral, and caudal regions of the body enter the cord in the medial division of the dorsal root and give collaterals to the nucleus thoracicus, which is located at thoracolumbar spinal levels. Axons of neurons in nucleus thoracicus collect ipsilaterally at the lateral margin of the dorsal half of the lateral funiculus and project to the cerebellum as **dorsal spinocerebellar tract** (Fig. 16–14). From cervical and cranial thoracic spinal nerves similar afferent fibers enter the dorsal funiculus and bifurcate. Their cranial branches ascend in fasciculus cuneatus to synapse in the lateral cuneate nucleus of the medulla, which projects to the cerebellum.

The **ventral spinocerebellar tract** is located at the lateral margin of the ventral half of the lateral funiculus (Fig. 16–14). The tract arises from projection neurons located in a lateral zone of gray matter from the neck of the dorsal horn to the middle of the ventral horn (Ha and Liu 1968). The projection neurons are activated by proprioceptive and tactile dorsal root afferents; their axons ascend in both the ipsilateral and contralateral ventral spinocerebellar tracts.

The **spino-olivary tract** is formed by ascending nerve fibers at the ventral extent of the ventral spinocerebellar tract (Fig. 16–14). The tract originates from projection neurons activated by proprioceptive and tactile inputs at all levels of the spinal cord. The axons initially decussate in the white commissure and ultimately terminate in the olivary nuclear complex of the medulla, which projects to the contralateral cerebellum.

Spinovestibular fibers originate from projection neurons located medially in the dorsal horn, mainly in the cervical region. Axons of these neurons travel in the ventral funiculus to the descending vestibular nucleus (Fig. 16–14). Fibers originating from all levels of spinal cord travel with the dorsal spinocerebellar tract and terminate in the lateral and other vestibular nuclei (Pompeiano and Brodal 1957). **Spinoreticular fibers** are diffused in the lateral funiculus; they go to the lateral reticular nucleus, which projects to the cerebellum. Spinopontine fibers to the pontine nuclei have been described, but the nature of this pathway is not understood.

DESCENDING SPINAL PATHWAYS

Neurons in the brain dispatch axons to the spinal cord in descending tracts. These tracts effect voluntary behavior, and they regulate movement, posture, and secretion of an automatic nature. Particularly in the context of movement volitionally initiated, descending tracts are often divided into pyramidal and extrapyramidal systems. The pyramidal system represents a direct corticospinal pathway: constituent neurons are situated in the cerebral cortex; their axons do not terminate until they reach spinal cord gray matter. The axons travel in the pyramids of the medulla oblongata, hence the term pyramidal system.

Other descending pathways from the brain originate or relay in the brain stem. These are termed extrapyramidal pathways, since their nerve fibers are located outside the medullary pyramids.

The cerebral cortex can initiate voluntary movements through both pyramidal and extrapyramidal pathways. Functionally, the canine pyramidal pathway is associated with fine degrees of muscle control in the extremities (Buxton and Goodman 1967), while the extrapyramidal system is involved in gross movements, posture, and muscle tone regulation.

Corticospinal fibers principally originate from the motor area (90 per cent) and the frontal region of the cerebral cortex (van Crevel and Verhaart 1963). The fibers travel in the ipsilateral pyramid of the medulla oblongata. At the junction of the medulla oblongata and spinal cord, the great majority of pyramidal tract fibers turn dorsad and decussate, forming the lateral pyramidal (corticospinal) tract. The uncrossed fibers (and some crossed axons) continue along the medial margin of the ventral funiculus as the **ventral pyramidal (corticospinal) tract** (Fig.

16–14). This tract is distributed to cervical segments, particularly those innervating the thoracic limbs; its fibers cross in the white commissure before terminating on interneurons medially in the ventral gray horn (Chambers and Liu 1957, Buxton and Goodman 1967).

The **lateral pyramidal (corticospinal) tract** is positioned between fasciculus proprius and dorsal spinocerebellar tract in the lateral funiculus (Fig. 16–14); it lies at the surface of the lateral funiculus caudal to the dorsal spinocerebellar tract (caudal lumbar, sacral, caudal segments). Most lateral pyramidal tract fibers are contralateral as a result of the pyramidal decussation, but a minority of its fibers are from the ipsilateral pyramid. Corticospinal fibers in the lateral pyramidal tract terminate on interneurons located mainly in lateral gray matter at the base of the dorsal horn, the lateral intermediate substance, and the base of the ventral horn (Buxon and Goodman 1967). The lateral pyramidal tract is distributed to all levels of the cord: 55 per cent to the cervical region, 20 per cent to the thoracic region, and 25 per cent to the lumbosacral region (Lassek, Dowd, and Weil 1930). Pyramidal tract fibers also synapse on sensory pathway neurons in gracilis and cuneatus nuclei and the nucleus thoracicus (Chambers and Liu 1957). Through these synapses, the cerebral cortex can regulate the flow of sensory information to the brain.

The **rubrospinal tract** originates from neurons in the red nucleus of the midbrain. The rubrospinal tract decussates just caudal to the red nucleus and descends in the contralateral lateral funiculus, ventral to and overlapping with the lateral pyramidal tract (Fig. 16–14). Fibers of the rubrospinal tract terminate on interneurons located mainly in the lateral gray matter of the dorsal horn base and the lateral intermediate substance, at all levels of the spinal cord (Brodal 1965).

The rubrospinal tract is the major pathway of the extrapyramidal system in terms of fiber quantity and extent of fiber distribution to the spinal cord. Correlated with this anatomical prominence is experimental evidence that the rubrospinal tract is the predominant volitional pathway from the brain to the spinal cord: experimental frontal cortex lesions, causing massive destruction of corticospinal fibers, produce minimal interference with usual voluntary movements in the dog (Buxton and Goodman 1967). The cerebral cortex and basal ganglia project heavily in an organized manner to the red nucleus (Brodal 1965). Chronic decerebration experiments (Bard and Macht 1958) indicate that complex, organized movements — i.e., spontaneous righting, crouching, sitting, standing, walking, and climbing — originate from the level of the red nucleus and are abolished by transection at the level of the pons.

The **tectospinal tract,** another component of the extrapyramidal system, originates from neurons in the rostral colliculus. The axons decussate in the midbrain and descend ventromedially in the ventral funiculus (Fig. 16–14). They are distributed primarily to cervical levels of the spinal cord (Petras 1967). The axons terminate on interneurons in the ventral horn, intermediate zone, and base of the dorsal horn. The spinotectal tract represents a pathway for turning the head (Crosby, Humphrey, and Lauer 1962).

Fibers responsible for dilating the pupils originate from the midbrain and travel ventral to the rubrospinal tract to terminate in the cranial thoracic segments of the cord (lateral tectotegmentospinal tract of Crosby, Humphrey, and Lauer).

Reticulospinal tracts originate from the reticular formation of the pons and medulla oblongata (Nyberg-Hansen 1964). Collectively, the tracts represent pathways for extrapyramidal influences, for vestibular and cerebellar control of posture and movement, and for the regulation of visceral musculature and glands. The origins of reticulospinal tracts are varied. In the spinal cord, the fiber tracts collectively are dispersed widely in the ipsilateral and contralateral lateral and ventral funiculi (Fig. 16–13). The tracts terminate on interneurons at all levels of the spinal cord. Functionally, a number of reticulospinal pathways have been demonstrated by stimulating loci in the brain stem. Several morphologically distinct reticulospinal tracts have been described by various authors (e.g., Petras 1967, Staal and Verhaart 1963, and Crosby, Humphrey, and Lauer 1962). However, terminology and anatomical descriptions of the reticulospinal tracts

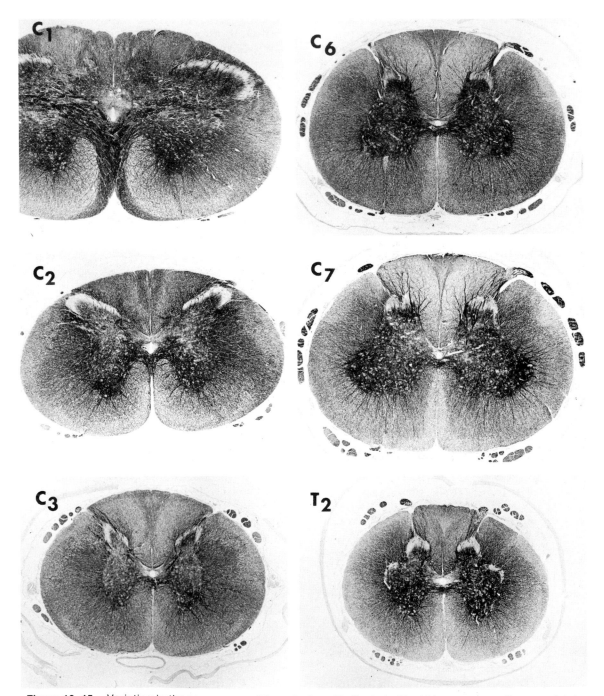

Figure 16–15. Variation in the appearance of the spinal cord is illustrated by 14 transverse sections, each taken through the middle of the segment designated (modified Holmes silver stain, all magnified 6.5 ×). Segments caudal to an illustrated segment were omitted if they appeared similar, with the exception that the omitted L6 section was similar to L7 rather than L4.

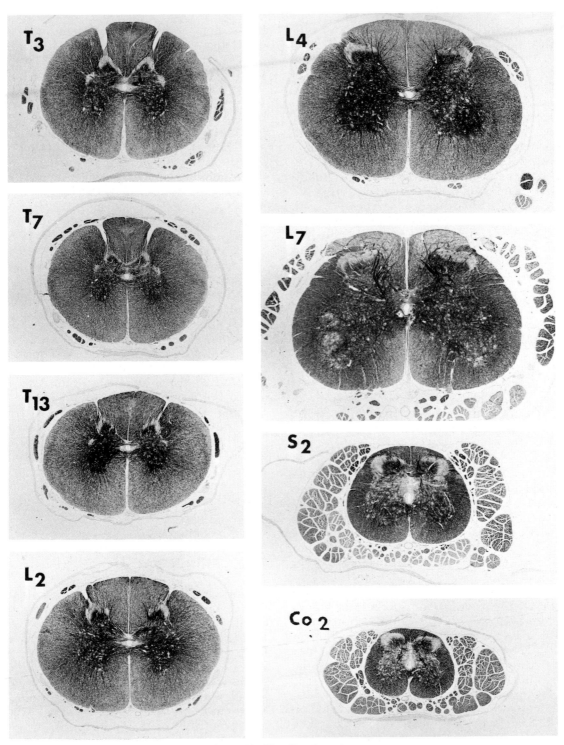

Figure 16–15. *Continued.*

are conflicting and not well correlated with pathways determined by stimulation experiments.

The **vestibulospinal tract** originates from the lateral vestibular nucleus and descends on the ventromedial margins of the ipsilateral ventral funiculus (Erulkar et al. 1966) (Fig. 16–14). Its nerve fibers terminate ipsilaterally on interneurons mainly in the medial portion of the ventral horn, at all levels of the spinal cord. Destruction of the tract causes the body to stumble toward the side of the lesion.

The **medial longitudinal fasciculus** (medial vestibulospinal tract) originates from the medial vestibular nucleus (Nyberg-Hansen 1964). The fibers descend dorsomedially in the ventral funiculus (Fig. 16–14), for the most part ipsilaterally. The tract extends to the midthoracic region, terminating on interneurons mainly medial in the ventral gray horn.

Transverse Sections of the Spinal Cord

The internal structure of the spinal cord varies at different levels. It is possible to examine a transverse section of the cord and determine the approximate segment from which the section was taken. The 14 sections illustrated in Figure 16–15 depict the variety of spinal cord appearance. Spinal cord segments omitted from Figure 16–15 appeared similar to the preceding illustrated segments; i.e., C4 and C5 were similar to C3, C8 and T1 were similar to C7, and so on. The one exception to this pattern is that the omitted L6 segment appeared similar to L7 rather than to L4.

Segments are represented by sections through the middle of dorsal root attachment regions. However, it should be appreciated that at levels of abrupt transition (e.g., near an enlargement) the center of a segment may not be representative of its cranial or caudal margins.

At the level of first cervical spinal roots, the transition from medulla oblongata to spinal cord is evident (Fig. 16–15, C1). The medullary pyramidal tract is completing its decussation to form lateral pyramidal tracts,

and gray matter is in the process of assuming the typical butterfly shape. In the cranial cervical region, substantia gelatinosa extends close to the external margin of the cord, and in transverse section, the spinal cord is strikingly oval (Fig. 16–15, C2). By the thoracic region, the cord appears circular in transverse section.

The quantity of white matter is greatly increased relative to gray matter at cervical levels of the spinal cord. The ratio of white to gray matter gradually declines along the length of the spinal cord, reflecting the fact that ascending and descending tracts of more cranial regions are subtracted from white matter at more caudal levels.

In the cranial half of the spinal cord, a dorsal intermediate sulcus or septum is generally evident in the dorsal funiculus. The dorsolateral sulcus is very prominent in the thoracic region. In lumbar and sacral regions, the dorsal median fissure is quite prominent.

The apex of the dorsal horn tends to be pointed in the cervical region, blunted in the thoracic region, and squared in the lumbosacral region. The dorsal gray horn is much reduced in the midthoracic region, where dorsal horns of each side blend across the midline (Fig. 16–15, T7).

A lateral gray horn is evident in thoracic and cranial lumbar segments of the spinal cord. The ventral gray horn is considerably enlarged by expansion of lateral motor nuclei in segments which innervate the limbs, i.e., at the cervical and lumbosacral enlargements (C6 to T1 and L4 to S1).

THE MENINGES

The central nervous system is well protected within the cranium and vertebral canal. It is enveloped in protective membranes, termed **meninges** (singular = meninx), which differentiate as dura mater, pia mater, and arachnoid membranes.

The *dura mater,* or pachymeninx, is composed mainly of collagen bundles and is tough and fibrous. It is the most superficial and protective of the meningeal layers. It encloses the brain, spinal cord, and leptomeninges.

The *pia mater* and **arachnoid membrane,** or leptomeninges, comprise two layers which are predominantly cellular in composition. The arachnoid membrane lies deep to the dura mater and is continuous with pia mater by means of trabeculae. The pia mater is attached to the nervous tissue surface by glial processes. The arachnoid membrane and pia mater are separated by a subarachnoid space which is filled with **cerebrospinal fluid.**

The brain contains cavities, **ventricles,** which are remnants of the embryonic neural tube. The ventricles are lined by ependymal epithelium. Vascular proliferations and ependyma form choroid plexuses which secrete cerebrospinal fluid into the ventricles. The cerebrospinal fluid flows from the ventricular system into the subarachnoid space and surrounds the brain and spinal cord.

Thus, the central nervous system is surrounded by fluid contained between leptomeninges, encased within dura mater, and protected by bone. Without these precautions, many of the traumas sustained daily by the average dog would be fatal.

Dura Mater

The dura mater is composed mainly of collagen bundles with a few fibrocytes, some elastic fibers, occasional blood vessels, and some sensory nerve endings.

In the vertebral canal, the **spinal dura mater** is composed of longitudinally directed collagen bundles. The dura mater is separated from the periosteum lining the vertebral canal by a wide **epidural space** which contains fat. In the cranial cavity there is no epidural space. Spinal dura mater is continuous with the **cranial dura mater** covering the brain. At the foramen magnum and on the floor of the vertebral canal in C1 and C2, the spinal dura mater fuses with the periosteum.

The dura mater which envelops the spinal cord and its leptomeninges resembles a long tube. Cranially, the spinal dura mater is attached to the foramen magnum; caudally, it tapers to a slender sheath, the *filum durae matris spinalis,* around the *filum terminale* formed by pia. The narrowed tube of dura mater is sometimes called the ligament of the spinal dura mater (Fig. 16–4A). It can be traced in the vertebral canal to the middle of the tail.

As the spinal roots traverse the vertebral canal, they are enclosed with lateral extensions of dura mater called **dural sheaths.** Generally, each dorsal and ventral root is enclosed in a separate dural sheath (Fig. 16–4B). The dorsal and ventral roots of C1, however, are enveloped by a common dural sheath, and dorsal and ventral roots of C2 and C3 have sheaths which are often attached superficially by connective tissue. In the caudal lumbar region, two separate dural sheaths are provided for each dorsal and each ventral root. Dural sheath fibrous tissue becomes continuous with the epineurium and perineurium of the spinal nerves.

In the lumbosacral region, fibrous bands may be found extending from the spinal dura mater to the floor of the vertebral canal. The bands often accompany the spinal roots and attach near the intervertebral foramina. Similar fibrous bands are given off by the filum durae matris spinalis. The fibrous bands are variably developed, and they probably augment the dural sheaths in anchoring the spinal dura mater to the vertebral canal. Dural sheaths are connected by fascia to the periosteum of intervertebral foramina.

Leptomeninges

Embryologically, the arachnoid membrane and pia mater are derived from a common membrane, and histologically the tissues are composed of a common cell type. Even grossly, the arachnoid membrane and pia mater are joined across the subarachnoid space by numerous trabecular continuities. The leptomeninges are composed of pale, watery cells, each having processes that interdigitate with neighboring cells. Bundles of collagen fibers weave between the cell processes.

Overlapping cell processes make the arachnoid membrane several layers thick. The pia mater is generally one cell layer in thickness, and its cells rest on a basement membrane of collagen fibers located outside of astrocyte foot processes that line the outer surface of the brain and cord. The combined tissues are often designated the **pia-glia membrane.** Throughout the central nervous system, astrocyte processes abut against pia mater, ependyma, and blood vessels, forming a glial barrier (Fig. 16–3). Astrocyte processes also contact neurons. The astrocytes concentrated along the outer margin of the central nervous tissue are called the **glia limiting membrane.**

Grossly, the **arachnoid membrane** appears delicate, translucent, and avascular. The arachnoid membrane forms the outer wall of the **subarachnoid space,** which is filled with cerebrospinal fluid. The superficial surface of the arachnoid membrane is in contact with dura mater, but potentially the two may be separated by a **subdural space** (see Fig. 16–3, L7). The arachnoid membrane is held against the dura mater by cerebrospinal fluid pressure and by the surface tension of a fluid film in the subdural space. The surface tension retards separation of the dural and arachnoid layers but permits them to slide against one another with minimal resistance.

The arachnoid membrane is joined to the pia mater by numerous **arachnoid trabeculae** which traverse the subarachnoid space (Allen and Low 1975). The depth of the subarachnoid space is variable, since the arachnoid membrane contacts the dura mater while the pia mater follows every irregularity of the brain surface. At certain

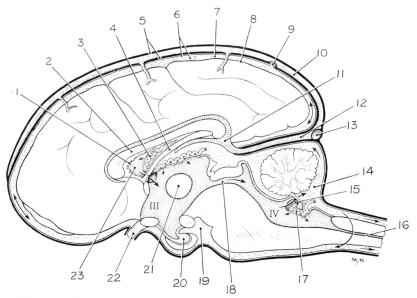

Figure 16–16. Schema of meninges and ventricles. (From Evans and deLahunta 1974.)

1. Cut edge of septum pellucidum
2. Corpus callosum
3. Choroid plexus
4. Fornix of hippocampus
5. Dura
6. Arachnoid membrane and trabeculae
7. Subarachnoid space
8. Pia
9. Arachnoid villus
10. Dorsal sagittal sinus
11. Great cerebral vein

12. Straight sinus
13. Transverse sinus
14. Cerebellomedullary cistern
15. Lateral aperture of fourth ventricle
16. Central canal
17. Choroid plexus
18. Mesencephalic aqueduct
19. Intercrural cistern
20. Hypophysis
21. Interthalamic adhesion
22. Optic nerve
23. Lateral ventricle

sites, crevices of the brain surface establish subarachnoid space enlargements known as *cisternae*. A number of cisternae may be recognized in the cranial subarachnoid space.

The *cisterna cerebellomedullaris*, also known as the cisterna magna, is located between the caudal surface of the cerebellum and the dorsal surface of the medulla oblongata (Fig. 16–16). It is the largest cisterna and the most common site for clinically tapping cerebrospinal fluid.

The *pia mater* is held tightly to the central nervous tissue surface by virtue of the glia limiting membrane. The pia mater covers the entire external surface of the central nervous system and continues into the depth of various sulci, fissures, and perivascular channels. Pia mater is very vascular, since all of the vascular pathways to and from the central nervous system must travel in the pia mater. The vessels are covered by leptomeningeal cells derived from the arachnoid trabeculae and pia mater. Fixed macrophages in the pia mater are generally adjacent to vessels. Nerve plexuses are present in the vessel adventitia.

Larger vessels penetrating into nervous tissue may be surrounded for a short distance by extensions of the subarachnoid space called perivascular spaces. The **perivascular space** is located between leptomeningeal cells covering the vessel surface and typical pia-glia membrane lining the outer wall of the space. The perivascular space is obliterated as the vessel proceeds, and smaller vessels in the central nervous system are tightly invested by a pia-glia membrane. Pia mater does not extend to the capillary level, where astrocyte processes are directly applied to a basement membrane which they share in common with capillary endothelium.

The pia mater and arachnoid membrane covering the spinal cord are reflected onto the spinal roots. Within dural sheaths, the leptomeninges envelop spinal roots distally to the level of the formation of spinal nerves.

Pia mater collagen is bilaterally thickened along the lateral surface of the spinal cord, forming denticulate ligaments. **Denticulate ligaments** have lateral extensions which traverse the subarachnoid space and attach to the dura mater between the entrances of successive roots into dural sheaths (Fig. 16–4). Caudally, each denticulate ligament terminates in a process which connects to the dura mater between the entrances of the L5 and L6 spinal roots into dural sheaths; in 25 per cent of dogs, the terminal process connects to the dura mater between the entrances of L6 and L7 roots into dural sheaths.

THE VENTRICULAR SYSTEM

The lumen of the embryonic neural tube persists as the ventricular system of the brain and the central canal of the spinal cord (Fig. 16–17). These cavities within the central nervous system are filled with cerebrospinal fluid and lined by ependymal epithelium. By means of foramina and canals, the chambers of the ventricular system are in communication with one another, with the central canal, and with the subarachnoid space.

The brain contains two **lateral ventricles,** one within each of the cerebral hemispheres. The **third ventricle** is a perpendicular chamber surrounding the interthalamic adhesion and separating the diencephalon into symmetrical halves. Each lateral ventricle communicates with the third ventricle through an **interventricular foramen.** The **mesencephalic aqueduct** of the midbrain is a canal which connects the third and fourth ventricles. The **fourth ventricle** is located in the hindbrain. It communicates with the central canal and with the subarachnoid space by means of paired **lateral recesses** and **apertures** (Fig. 16–16).

Along one region in each of the brain ventricles, nervous tissue is absent from the wall. Pia mater comes in direct contact with ependyma, and this combined tissue is called *tela choroidea*. Thus the tela choroidea forms a portion of the ventricular wall in each ventricle, namely, the medial wall of each lateral ventricle and the roof of the third and fourth ventricles.

In each ventricle, the tela choroidea is elaborated to form a choroid plexus. The **choroid plexus** projects into the ventricle and appears as a tufted band of delicate, clustered villi. A choroid plexus is extremely vascular, and its expansive surface is covered by cuboidal ependymal cells called

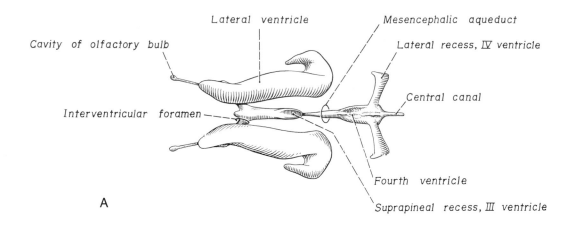

A

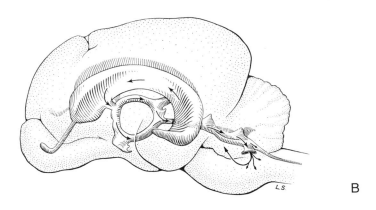

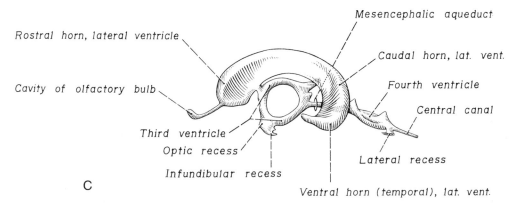

B

C

Figure 16–17. The ventricles of the brain in the dog. (From deLahunta 1977.)
A, Dorsal view.
B and C, Lateral view.

choroidal epithelium. The choroidal epithelium secretes cerebrospinal fluid.

The choroid plexus in each lateral ventricle continues into the third ventricle through the interventricular foramen. Consequently, two choroid plexuses are found in the roof of the third ventricle. The roof of the fourth ventricle also contains two choroid plexuses, and each plexus sends a short extension out through the lateral aperture.

CEREBROSPINAL FLUID

Cerebrospinal fluid is normally a clear, colorless, slightly alkaline liquid that resembles the aqueous humor of the eye. It contains inorganic ions, protein, sugar, and a few cells. Cerebrospinal fluid is a few millivolts positive with respect to the extracellular fluid of the body.

Production. Cerebrospinal fluid is actively secreted by the choroid plexuses of the brain ventricles. Its production is evidently an active process because the fluid can be produced against a hydrostatic gradient (e.g., in hydrocephalus), the distribution of ions is contrary to passive diffusion, and chemicals which inhibit active transport decrease cerebrospinal fluid production. The rate of cerebrospinal fluid formation is about 0.05 ml. per minute (3 ml./hr.) in the dog.

Cerebrospinal fluid is in contact with the interstitial fluid of the central nervous system across the pia-glia and ependyma-glia membranes. Solutes can be exchanged between these fluid compartments, and cerebrospinal fluid is augmented by a flow of fluid from the extracellular space of the nervous tissue. Large molecules can leave the neural interstitial fluid by this route, and the cerebrospinal fluid compartment can thereby function as a lymphatic drainage system for the central nervous system, which lacks lymph vessels.

Flow. There is a flow of cerebrospinal fluid from the lateral ventricles into the third ventricle and then through the mesencephalic aqueduct into the fourth ventricle. Fluid flows from the fourth ventricle into the subarachnoid space (Fig. 16–16). Within the subarachnoid space, fluid movements are generally haphazard.

The dura mater, particularly in the cranial cavity, tends to form a rigid chamber which is completely filled with nervous tissue, cerebrospinal fluid, and blood within vessels. Any volume change in one intradural component must be opposed and balanced by reciprocal volume changes in one or both of the other intradural components. During breathing, the alternate changes in thoracocranial and abdominal venous blood pressure (and volume) cause cerebrospinal fluid to shift cranially during inspiration and caudally during expiration. Cerebrospinal fluid currents are generated by coughing, straining, lying down, sitting up, running, jumping, stopping, and so on, owing to regional differences in blood pressure and pressure differentials in cerebrospinal fluid caused by gravity and other accelerating influences. Cerebrospinal arteries are protected under conditions of high blood pressure by a corresponding increase in cerebrospinal fluid pressure.

Drainage. Cerebrospinal fluid, which is derived from blood circulating through capillaries of principally the choroid plexuses, must ultimately be returned to the blood circulation, because fluid is continuously being produced, and yet the volume of cerebrospinal fluid remains approximately constant. A major drainage route for cerebrospinal fluid to blood is by means of arachnoid villi.

Arachnoid villi are extensions of leptomeninges that project into dural venous sinuses (Fig. 16–18). The cerebrospinal fluid in a villus is separated from the blood in a venous sinus by a wall of arachnoid and endothelial cells. Each arachnoid villus functions as a valve regulating the flow of cerebrospinal fluid into the venous sinus.

When the cerebrospinal fluid pressure exceeds venous blood pressure in the dural sinuses, the arachnoid villi expand. The space between arachnoid cell processes increases, and fluid flows from the subarachnoid space into the venous sinus. When blood pressure in the venous sinus exceeds cerebrospinal fluid pressure, the arachnoid villi collapse, effectively blocking reflux flow from the venous sinus to the subarachnoid space. In man, arachnoid villi often enlarge with age and may undergo calcification. Grossly visible arachnoid villi are called arachnoid granulations (pacchionian granulations). These are rare in the dog.

Cerebrospinal fluid is also absorbed from

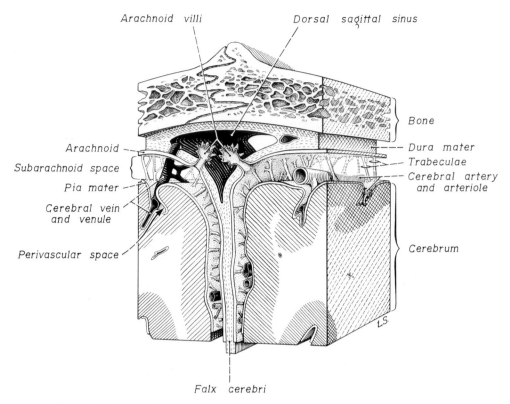

Figure 16–18. Cerebral meninges and arachnoid villi. (From deLahunta 1977.)

extensions of the spinal subarachnoid space within the dural sheaths. From distal recesses of the subarachnoid space, where spinal roots join to form spinal nerves, cerebrospinal fluid drains across the arachnoid membrane into peripheral nerve lymphatics. Arachnoid villi have been reported in association with veins in the regions of the intervertebral foramina. There is also considerable absorption around the optic and olfactory nerves.

The drainage sites previously mentioned constitute routes of massive efflux, through which large solute particles can enter the blood stream. Water and small solutes of the cerebrospinal fluid can also interchange with the blood compartment across the pia-covered walls of small vessels on the surface of the brain and spinal cord. Water and small solutes can interchange with blood in nervous tissue capillaries, through the brain extracellular space which interacts with cerebrospinal fluid across ependyma-glia and pia-glia membranes. Cerebrospinal fluid water and appropriate solutes will follow hydrostatic and electrochemical gradients in entering or leaving the blood stream by the above routes. In addition, some molecules are actively transported into and out of the cerebrospinal fluid.

BIBLIOGRAPHY

Allen, D. J., and F. N. Low. 1975. Scanning electron microscopy of the subarachnoid space in the dog. III. Cranial levels. J. Comp. Neurol. *161*:515–540.

Bard, P., and M. B. Macht. 1958. The behavior of chronically decerebrate cats. Ciba Foundation Symposium on the Neurological Basis of Behavior. London, J. and A. Churchill Ltd.

Brodal, A. 1965. Experimental anatomical studies of the corticospinal and cortico-rubrospinal connections in the cat. Symp. Biol. Hung. 5:207–217.

Brodal, A., and B. Rexed. 1953. Spinal afferents to the lateral cervical nucleus in the cat. J. Comp. Neurol. 98:179–211.

Buxton, D. F., and D. C. Goodman. 1967. Motor function and the corticospinal tracts in the dog and raccoon. J. Comp. Neurol. 129:341–360.

Chambers, W. W., and C. Liu. 1957. Cortico-spinal tracts of the cat. J. Comp. Neurol. 108:23–55.

van Crevel, H., and W. J. C. Verhaart. 1963. The "exact" origin of the pyramidal tract. A quantitative study in the cat. J. Anat. 97:495–515.

Crosby, E. C., T. Humphrey, and E. W. Lauer. 1962. Correlative Anatomy of the Nervous System. New York, Macmillan Co.

deLahunta, A. 1977. Veterinary Neuroanatomy and Clinical Neurology. Philadelphia, W. B. Saunders Co.

Erulkar, S. D., J. M. Sprague, B. L. Whitsel, S. Dogan, and P. J. Jannetta. 1966. Organization of the vestibular projection to the spinal cord of the cat. J. Neurophysiol. 29:626–664.

Evans, H. E., and A. deLahunta. 1974. Miller's Guide to Dissection of the Dog. Philadelphia, W. B. Saunders Co.

Fletcher, T. F. 1970. Lumbosacral plexus and pelvic limb myotomes of the dog. Am. J. Vet. Res. 31:35–41.

Fletcher, T. F., and R. L. Kitchell. 1966a. Anatomical studies on the spinal cord segments of the dog. Am. J. Vet. Res. 27:1759–1767.

— — — — —. 1966b. The lumbar, sacral, and coccygeal tactile dermatomes of the dog. J. Comp. Neurol. 128:171–180.

Glees, P., and J. Soler. 1951. Fiber content of the posterior column and synaptic connections of nucleus gracilis. Z. Zellforsch. 36:381–400.

Ha, H., and C. Liu. 1966. Organization of the spinocervico-thalamic system. J. Comp. Neurol. 127:445–470.

— — — — —. 1968. Cell origin of the ventral spinocerebellar tract. J. Comp. Neurol. 133:185–206.

Henneman, E., G. Somjen, and D. O. Carpenter. 1965. Functional significance of cell size in spinal motorneurons. J. Neurophysiol. 28:560–580.

Kennard, M. A. 1954. The course of ascending fibers in the spinal cord of the cat essential to the recognition of painful stimuli. J. Comp. Neurol., 100:511–524.

Lassek, A. M., L. W. Dowd, and A. Weil. 1930. The quantitative distribution of the pyramidal tract in the dog. J. Comp. Neurol. 51:153–163.

Nyberg-Hansen, R. 1964. Origin and termination of fibers from the vestibular nuclei descending in the medial longitudinal fasciculus. J. Comp. Neurol. 124:71–100.

Oliver, J. E., Jr., W. E. Bradley, and T. F. Fletcher. 1969. Identification of preganglionic parasympathetic neurons in the sacral spinal cord of the cat. J. Comp. Neurol. 137:321–328.

Palmer, A. C. 1965. Introduction to Animal Neurology. Oxford, Eng., Blackwell Publications.

Petras, J. M. 1967. Cortical, tectal, and tegmental fiber connections in the spinal cord of the cat. Brain Res. 6:275–324.

Pompeiano, O., and A. Brodal. 1957. Spino-vestibular fibers in the cat, an experimental study. J. Comp. Neurol. 108:353–381.

Ranson, S. W., and S. I. Clark. 1959. The Anatomy of the Nervous System. 10th Ed. Philadelphia, W. B. Saunders Co.

Rexed, B. 1952. The cytoarchitectonic organization of the spinal cord in the cat. J. Comp. Neurol. 96:415–495.

— — — — —. 1954. A cytoarchitectonic atlas of the spinal cord in the cat. J. Comp. Neurol. 100:297–379.

Romanes, G. J. 1951. The motor cell columns of the lumbosacral spinal cord of the cat. J. Comp. Neurol. 94:313–363.

Staal, A., and W. J. C. Verhaart. 1963. Subcortical projections on the spinal grey matter of the cat. Acta Anat. 52:235–243.

Thomas, C. E., and C. M. Combs. 1962. Spinal cord segments. A. Gross structure in the adult cat. Am. J. Anat. 110:37–48.

Truex, R. C. 1959. Human Neuroanatomy. 4th Ed. Baltimore, The Williams & Wilkins Co.

Verhaart, W. J. C. 1962. The pyramidal tract. World Neurol. 3:43–53.

Whitlock, D. G., and E. R. Perl. 1959. Afferent projections through ventrolateral funiculi to thalamus of cat. J. Neurophysiol. 22:133–148.

Wislocki, G. B., E. H. Leduc, and A. J. Mitchell. 1956. On the ending of Reissner's fiber in the filum terminale of the spinal cord. J. Comp. Neurol. 104:493–517.

Chapter 17

THE SPINAL NERVES

The **spinal nerves** *(nervi spinales)* (Figs. 17–1, 17–2) usually number 36 pairs in the dog. They have retained their embryological segmental characteristics, although the spinal cord from which they arise and the structures which they supply have largely lost this feature. Each spinal nerve is attached to its corresponding segment of the spinal cord by the dorsal and ventral rootlets or **root filaments** *(fila radicularia)*. The dorsal filaments carry sensory or afferent impulses; the ventral filaments carry motor or efferent impulses. The afferent filaments collectively are called the **dorsal root** *(radix dorsalis)*, and the efferent filaments collectively are called the **ventral root** *(radix ventralis)* of the spinal nerve. The afferent filaments enter the dorsolateral sulcus of the spinal cord. The efferent rootlets leave the ventrolateral funiculus over an area which is about 2 mm. wide. Neither the dorsal nor the ventral root filaments are compact units. They consist of loosely united bundles of nerve fibers which are difficult to differentiate from each other because of the transparency of the covering arachnoid. The number of dorsal root filaments agrees closely with the number of ventral root filaments for each spinal nerve. The number of dorsal and ventral root filaments averages six each for the first five cervical nerves. They increase in size and in number to an average of seven dorsal and seven ventral filaments from the fifth cervical segment as far caudad as the second thoracic segment. From the second thoracic segment through the thirteenth thoracic segments there are two dorsal and two ventral filaments which form each thoracic nerve root.

The filaments which merge to compose the nerves of the lumbosacral plexus are large and constantly double. There are usually two dorsal and two ventral root filaments for each of the three pairs of sacral and five pairs of caudal spinal nerves.

Because the vertebral column continues to grow after the spinal cord has ceased growing, the last several lumbar, the sacral, and the caudal nerves have to pass increasingly longer distances before they reach the corresponding intervertebral foramina. Therefore, lumbar, sacral, and caudal nerves leave the caudal part of the spinal cord and lie within the dural and arachnoid membranes. Since the caudal part of the spinal cord and the nerves which leave it resemble a horse's tail, it is called the *cauda equina.* (See Chapter 16 for illustrations.)

The **spinal ganglia** *(ganglia spinales)* are aggregations of unipolar nerve cells which are located in the dorsal root within (rarely external) the corresponding intervertebral foramen. The axons of the unipolar cells divide into central and peripheral processes. The central processes form the dorsal root filaments, whereas the peripheral processes intermingle with the axons of the ventral root filaments in forming the mixed (sensory and motor) spinal nerve.

Branches of a Typical Spinal Nerve

Just peripheral to the intervertebral foramen each spinal nerve trifurcates into dorsal, ventral, and visceral, or communicating, branches (Fig. 17–1).

The **dorsal branch** *(ramus dorsalis)* ex-

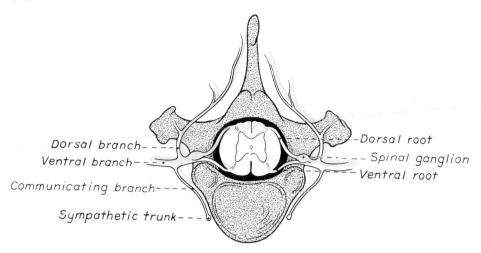

Figure 17-1. Diagram of a spinal nerve.

Dorsal branch

Ventral branch

Communicating branch

Sympathetic trunk

Dorsal root

Spinal ganglion

Ventral root

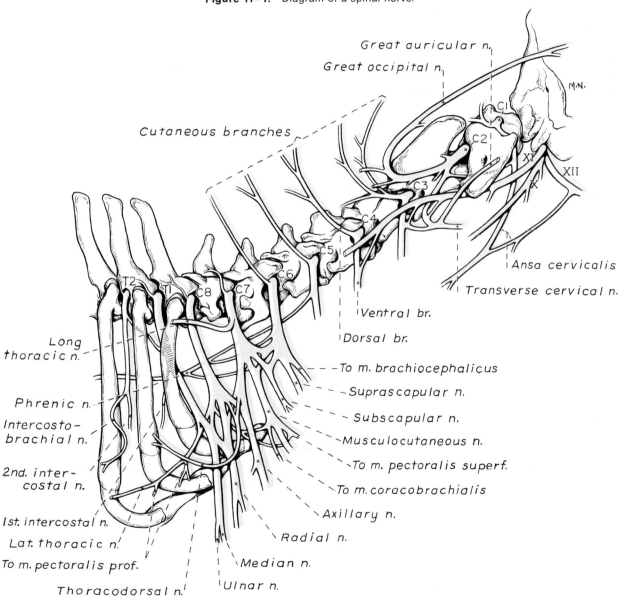

Great auricular n.

Great occipital n.

M.N.

C1

C2

XI

XII

X

Cutaneous branches

C3

C4

C5

C6

Ansa cervicalis

Transverse cervical n.

T2

T1

C8

C7

Ventral br.

Dorsal br.

Long thoracic n.

To m. brachiocephalicus

Suprascapular n.

Phrenic n.

Subscapular n.

Intercosto-brachial n.

Musculocutaneous n.

To m. pectoralis superf.

2nd. intercostal n.

To m. coracobrachialis

Axillary n.

1st. intercostal n.

Radial n.

Lat. thoracic n.

To m. pectoralis prof.

Median n.

Thoracodorsal n.

Ulnar n.

Figure 17-2. Schema of the cervical nerves and brachial plexus.

973

tends dorsad and usually subsequently divides into medial and lateral parts.

The **ventral branch** *(ramus ventralis)* supplies all hypaxial structures, including the limbs. It divides into medial and lateral parts except where it is specialized to supply the extremities or the tail.

The **communicating branch** *(ramus communicans),* also called visceral branch, differs from the dorsal and ventral branches in that it carries only motor and sensory fibers to and from visceral structures (gland tissue and smooth muscle). Some communicating branches also contain fibers which come from and go to the heart. Consult Chapter 18 for further information. The communicating branch is connected to the ventral branch of the spinal nerve.

The spinal nerves usually leave the vertebral canal by means of spaces between adjacent vertebrae, the **intervertebral foramina.** Certain discrepancies exist because the number of vertebrae and the number of spinal nerves for each of the several regions are not always the same. There are eight pairs of cervical nerves, but only seven cervical vertebrae. In most dogs there are 20 caudal vertebrae, but only the first five pairs of caudal nerves usually develop. Since the three sacral vertebrae are fused to form the sacrum, there are two dorsal and two ventral pairs of sacral foramina for the passage of the dorsal and ventral branches of the first two pairs of sacral nerves. The third pair of sacral nerves pass through the intervertebral foramina located between the sacrum and the first caudal vertebra.

CERVICAL NERVES

There are eight pairs of **cervical nerves** *(nn. cervicales)* (Fig. 17–2), although there are only seven cervical vertebrae. The intervertebral foramina, through which the first cervical nerves pass, are located not between the skull and the atlas, but in the craniodorsal portion of the arch of the atlas. The second pair of cervical nerves leave the spinal canal through the second pair of intervertebral foramina, which are located between the atlas and the axis. The last, or eighth, pair of cervical nerves pass through the eighth intervertebral foramina, which

are located between the seventh cervical and first thoracic vertebrae. Therefore, the cervical nerves leave the vertebral canal cranial to the vertebra of the same number with the exception of the first and last pairs.

Many branches of cervical nerves communicate with one another and run variable distances in common before joining still other branches. This results in the formation of a **cervical plexus** *(plexus cervicalis)* that can include axons of all cervical nerves.

The **first cervical nerve** *(n. cervicalis I)* arises from the first segment of the spinal cord, which is located just caudal to the foramen magnum, and is surrounded by the cranial portion of the atlas. Both its dorsal and ventral formative root filaments number from three to five and are approximately equal in size. In most specimens a barely distinguishable dorsal root ganglion is present, while in others it may be 1 mm. in diameter. Upon emerging through the intervertebral foramen of the atlas, the first cervical nerve divides into dorsal and ventral branches of equal size, measuring about 2 mm. in diameter.

The **dorsal branch of the first cervical nerve** *(ramus dorsalis n. cervicalis I),* or the **suboccipital nerve** *(n. suboccipitalis),* does not divide into medial and lateral branches, and no part of it supplies the skin. It lies initially under the cranial part of the large obliquus capitis caudalis muscle. It arborizes in the muscles of the cranial portion of the neck. These include the obliquus capitis cranialis; obliquus capitis caudalis; rectus capitis dorsalis, intermedius, and minor; and the cranial ends of the semispinalis capitis and splenius.

The **ventral branch of the first cervical nerve** *(ramus ventralis n. cervicalis I)* initially lies in the osseous groove of the atlas which runs transversely to the alar notch from the intervertebral foramen. The ventral branch passes through the alar notch and continues in a ventrocaudal direction by passing between the rectus capitis lateralis and the rectus capitis ventralis muscles. It is initially covered by the medial retropharyngeal lymph node. After appearing at the caudal border of this node, it continues its course down the neck in close

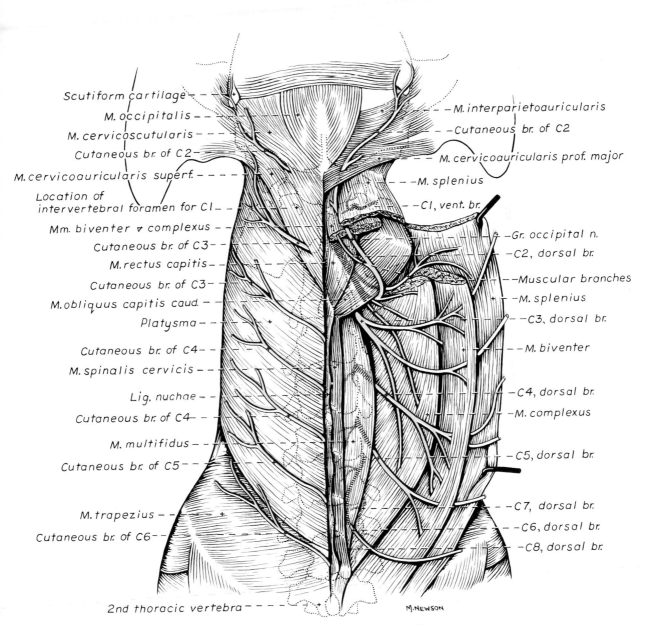

Scutiform cartilage
M. occipitalis
M. cervicoscutularis
Cutaneous br. of C2
M. cervicoauricularis superf.
Location of intervertebral foramen for CI
Mm. biventer & complexus
Cutaneous br. of C3
M. rectus capitis
Cutaneous br. of C3
M. obliquus capitis caud.
Platysma
Cutaneous br. of C4
M. spinalis cervicis
Lig. nuchae
Cutaneous br. of C4
M. multifidus
Cutaneous br. of C5
M. trapezius
Cutaneous br. of C6
2nd thoracic vertebra

M. interparietoauricularis
Cutaneous br. of C2
M. cervicoauricularis prof. major
M. splenius
CI, vent. br.
Gr. occipital n.
C2, dorsal br.
Muscular branches
M. splenius
C3, dorsal br.
M. biventer
C4, dorsal br.
M. complexus
C5, dorsal br.
C7, dorsal br.
C6, dorsal br.
C8, dorsal br.

M. NEWSON

Figure 17–3. Dorsal branches of the cervical nerves, dorsal aspect. (The muscles on the right side are reflected.)

relation to the vagosympathetic nerve trunk. It usually communicates with the smaller descending branch of the hypoglossal nerve to form the **cervical loop** *(ansa cervicalis).* (See Benson and Fletcher 1971 for variability of the ansa cervicalis.) Variations in the formation of the cervical loop are common. In about 3 per cent of dogs it fails to develop. It may be a long loop measuring 10 cm. Usually the cervical loop extends to a level through the third cervical vertebra, but in some dogs it is short, lying on the carotid sheath about 2 cm. caudal to the jugular process. Several branches arise from the cervical loop. As the sternothyroid and sternohyoid muscles cross the larynx, they each receive a branch from the loop. Another branch runs caudad on the trachea and bifurcates near the middle of the neck. The shorter of these branches becomes related to the middle of the lateral border of the sternothyroideus before entering its distal portion. The longer branch follows the lateral border of the sternohyoideus caudally and enters the muscle approximately 4 cm. cranial to the manubrium of the sternum.

The **second cervical nerve** *(n. cervicalis II)* differs from other typical spinal nerves in three respects. (1) Its afferent component is larger than that of any of the other cervical or thoracic nerves. (2) The dorsal and ventral roots fuse peripheral to the second intervertebral foramen. (3) The large dorsal root ganglion lies completely outside the vertebral canal.

The **dorsal branch of the second cervical nerve** *(ramus dorsalis n. cervicalis II),* or the **greater occipital nerve** *(n. occipitalis major),* is approximately the same size as the ventral branch (3 mm. in diameter) in a large dog. It runs caudodorsally, where it is deeply located under the obliquus capitis caudalis muscle. Emerging between this muscle and the spine of the axis, it sends muscular branches into the semispinalis capitis and the splenius muscles. It then turns cranially, perforates the overlying muscles, and supplies the skin which covers the caudal portion of the temporal muscle and the base of the pinna.

The **ventral branch of the second cervical nerve** *(ramus ventralis n. cervicalis II)* runs caudoventrad on the lateral surface of the cleidomastoideus muscle for one or two centimeters and divides into the **transverse cervical** and **great auricular nerves.**

The **transverse cervical nerve** *(n. transversus colli),* called in some texts the n. cutaneus colli, runs cranioventrad from under the platysma and crosses the external and internal maxillary veins just before they unite to form the external jugular vein. The nerve may branch before crossing these veins. The branches of this nerve arborize in the skin of the mandibular space.

The **great auricular nerve** *(n. auricularis magnus)* is the larger of the two terminal branches of the ventral branch of the second cervical nerve. It runs dorsocranially to the base of the pinna of the ear and divides into at least two branches which run toward the apex of the ear. Each of these nerves runs about midway between the intermediate auricular artery and the peripheral arteries — the lateral and the medial auricular arteries as they arborize in their course toward the apex of the ear. Variations in both the numbers and the distribution of the arteries and nerves to the pinna are common.

The **dorsal branches of cervical nerves III through VII** *(rami dorsales nn. cervicales III–VII)* (Fig. 17–3) vary in both their distributions and their form. Each of these branches sends a small branch medially into the multifidus cervicis muscle. Only their peripheral portions definitely divide into medial (cutaneous) and lateral (muscular) branches. The dorsal branches of cervical nerves III through VII gradually decrease in size caudally. The seventh dorsal cervical branch may be reduced to a muscular twig which innervates only the deep muscle fibers which lie adjacent to it. The dorsal branch of the eighth cervical nerve may be absent. The dorsal branches of the middle cervical nerves perforate the lateral portion of the multifidus cervicis muscle and run dorsally with a slight caudal inclination. Upon reaching the ventral portion of the biventer muscle, they usually bifurcate into medial and lateral branches. The third dorsal branch divides near its origin, and the fourth dorsal branch may also divide deeply.

The **lateral branches** of the ramus dorsalis of cervical nerve III supply the middle portion of the complexus muscle. The main lateral branches of the rami dorsales of cervi-

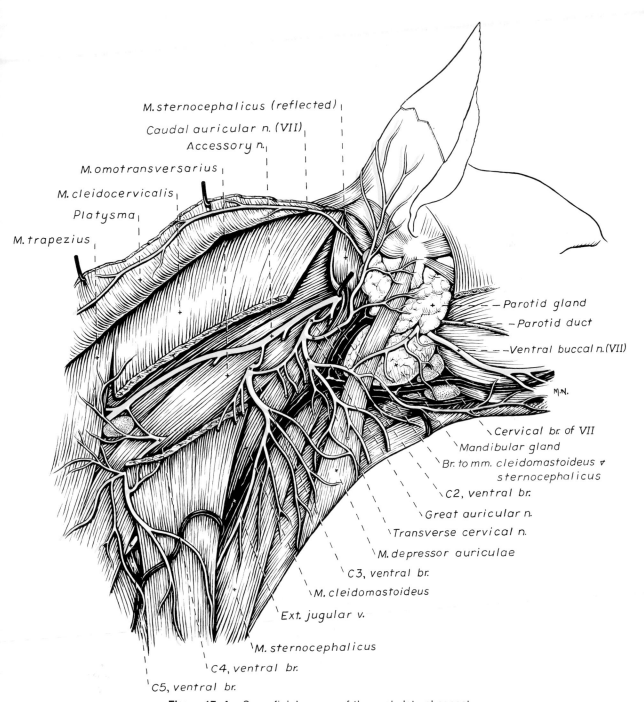

M.sternocephalicus (reflected)
Caudal auricular n. (VII)
Accessory n.
M.omotransversarius
M.cleidocervicalis
Platysma
M.trapezius

Parotid gland
Parotid duct
Ventral buccal n.(VII)

M.N.

Cervical br. of VII
Mandibular gland
Br. to mm. cleidomastoideus v
   sternocephalicus
C2, ventral br.
Great auricular n.
Transverse cervical n.
M.depressor auriculae
C3, ventral br.
M.cleidomastoideus
Ext. jugular v.
M. sternocephalicus
C4, ventral br.
C5, ventral br.

Figure 17–4. Superficial nerves of the neck, lateral aspect.

cal nerves IV and V innervate the middle portion of the biventer muscle. The splenius muscle is innervated by branches which come from the rami dorsales of cervical nerves III and IV, which leave near their origins. The branches enter the deep surface of the middle and caudal part of the muscle. The cranial part of the splenius muscle is supplied by large muscular branches from the dorsal branch of cervical nerve II.

The **medial branches** of the rami dorsales of cervical nerves III through VII perforate the intertransversarius dorsalis muscle and run almost directly dorsad. In their dorsal courses they lie between the multifidus cervicis and the spinalis cervicis muscles located medially, and the complexus and biventer muscles (the two portions of the semispinalis capitis) located laterally. Finally they cross the lateral side of the ligamentum nuchae to reach the skin. In most specimens the third and fourth medial dorsal cervical branches further divide into cranial and caudal parts. They thus appear by their spacing in the subcutaneous fascia as if they were branches of separate cervical nerves. Bilaterally, these nerves innervate the loose, thick skin of the dorsal and adjacent sides of the neck.

The **ventral branches** of cervical nerves II through V (Fig. 17–4) pass between the muscle bundles of the intertransversarius cervicis to reach the medial surface of the omotransversarius muscle. The ventral branches of cervical nerves II, III, and IV regularly communicate with the accessory nerve, however, and a connection between cervical nerves II and III is not uncommon. The ventral branch of the large, second cervical nerve has been described. The ventral branches of the third and fourth cervical nerves divide into large **lateral (cutaneous) branches** (*rami laterales*), which supply the skin of the ventrolateral part of the neck, and smaller **medial (muscular) branches** (*rami mediales*), which supply the longus capitis, longus colli, intertransversarius cervicis, omotransversarius, and brachiocephalicus muscles. The medial branches appear as loose clusters of nerves which arise just peripheral to the intervertebral foramina and, after short caudoventral courses, enter the several muscles. The size of the ventral branches of the second to the fifth cervical nerves progressively decreases. The ventral branch of the second cervical nerve is about three times larger than that of the fifth.

NERVES TO THE DIAPHRAGM

The **phrenic nerve** (*n. phrenicus*) (Fig. 17–2) reflects the fact that the diaphragm which it supplies has a cervical origin. The phrenic nerve regularly arises from the fifth, sixth, and seventh cervical nerves, and occasionally a small twig comes from the fourth.

The branches of origin of the phrenic nerve run caudally, medial to the brachial plexus. While running in the fascia adjacent to the external jugular vein, the nerve branches converge and unite to form the phrenic nerve just cranial to the thoracic inlet. The nerve on each side then passes through the thoracic inlet ventral to the subclavian artery and dorsal to the omocervical artery. At this site it is joined by a fine branch from the caudal cervical ganglion or the sympathetic trunk adjacent to the ganglion. Within the thorax the right phrenic nerve lies in a narrow plica of pleura from the right lamina of the precardial and cardiac mediastinum and the plica venae cavae. The left phrenic nerve lies in a similar plica from the left pleural sheet of the mediastinum. Each phrenic nerve spreads out on its respective half of the diaphragm, where it supplies this muscle with motor and sensory fibers. Upon reaching the diaphragm, each phrenic nerve divides into three main branches: ventral, lateral, and dorsal. This splitting takes place lateral to the middle portion of the tendinous center. Each nerve division supplies its appropriate third of its half of the diaphragm. Each dorsal branch therefore supplies the crus of its side. The dog, like man, usually has a connection between each phrenic nerve and the sympathetic system at the celiac plexus. In 300 dissections an accessory phrenic nerve in the dog has not been seen. The periphery of the diaphragm also receives sensory fibers from the last several intercostal nerves (Lemon 1928). It is generally agreed that the phrenic nerves are the only motor nerves to the diaphragm. Investigators also agree that bilateral phrenicotomy does not seriously interfere with respiration in the dog, even under moderate exercise.

BRACHIAL PLEXUS

The **brachial plexus** (*plexus brachialis*) (Figs. 17–5, 17–6, 17–7) consists of the large somatic nerve plexus which gives origin to the nerves which supply the thoracic limb. It is usually formed by the ventral branches of the sixth, seventh, and eighth cervical and the first and second thoracic spinal nerves. Occasionally, the ventral branch of the fifth cervical nerve also contributes to its formation; frequently, the second thoracic contribution is lacking. When either or both the fifth cervical and the second thoracic spinal nerves send branches which enter into the formation of the brachial plexus, they are exceedingly small compared with the other ventral branches which compose the plexus. In over 250 dissections neither one of these nerves was found to be over 1 mm. in diameter. Allam et al. (1952) found in 58 dissections that the fifth cervical and the second thoracic nerve contributions to the brachial plexus are more often absent than present.

Ventral branches of the cervical (C) and thoracic (T) spinal nerves which form the brachial plexus are distributed in a variable manner. They were grouped as follows in dogs studied by Allam et al. (1952):

58.62% formed by C 6, 7, 8, and T 1
20.69% formed by C 5, 6, 7, 8, and T 1
17.24% formed by C 6, 7, 8, and T 1 and 2
3.4% formed by C 5, 6, 7, 8, and T 1 and 2

After the ventral branches of the last three cervical and the first and second thoracic spinal nerves have passed through the intervertebral foramina and the intertransverse musculature, they cross the ventral border of the scalenus muscle and extend to the thoracic limb by traversing the axillary space. In this course, parts of these nerves unite with each other to form the various specific nerves which supply the structures of the thoracic limb and adjacent muscles and skin. The axillary artery and vein lie ventromedial to the caudal portion of the brachial plexus. The external jugular vein, after it has been augmented by the proximal tributary of the cephalic vein, crosses the ventral surfaces of the seventh and eighth cervical nerves, from which it is separated by the superficial cervical artery. The axillary artery, after having crossed the cranial margin of the first rib, lies closely applied to the ventral margin of the scalenus primae costae muscle and later follows along the craniomedial margin of the radial nerve as both the artery and nerve run distad in the brachium. They are crossed ventrally at the first rib by a muscular nerve branch which goes to the deep pectoral muscle.

Allam et al. (1952) recognize three cords in the brachial plexus of the dog to assist the exploring surgeon by establishing suitable landmarks for electrical stimulation. These lie as intermediate nerve trunks between the ventral branches of the spinal nerves which form the plexus and the named nerves which innervate structures of the limb. Most of these vary. For further information on the morphology of the brachial plexus of the dog, refer to Russell (1893), Reimers (1925), Miller (1934), and Bowne (1959).

The nerves which are branches of the brachial plexus or are direct continuations of the formative ventral branches include the suprascapular, subscapular, axillary, musculocutaneous, radial, median, ulnar, dorsal thoracic, lateral thoracic, long thoracic, pectoral, and muscular branches. The term *radices plexus* refers to the contributions of the ventral spinal nerves which form the brachial plexus. Some of the radices may form trunks (*trunci plexus*) prior to forming the definitive brachial plexus.

The basic plan of the brachial plexus appears as a variable "anastomosis" of the last three cervical and first two thoracic spinal nerves whose fibers run in common for short distances and then segregate in variable combinations to form the extrinsic and intrinsic named nerves of the thoracic limb.

The **suprascapular nerve** (*n. suprascapularis*) arises primarily and occasionally entirely from the sixth cervical nerve. It often has a contribution from the seventh, but rarely from the fifth, cervical nerve. It enters the distal end of the intermuscular space between the supraspinatus and the subscapularis muscles from the medial side. It is accompanied by the supraspinous artery and vein. The suprascapular nerve is pri-

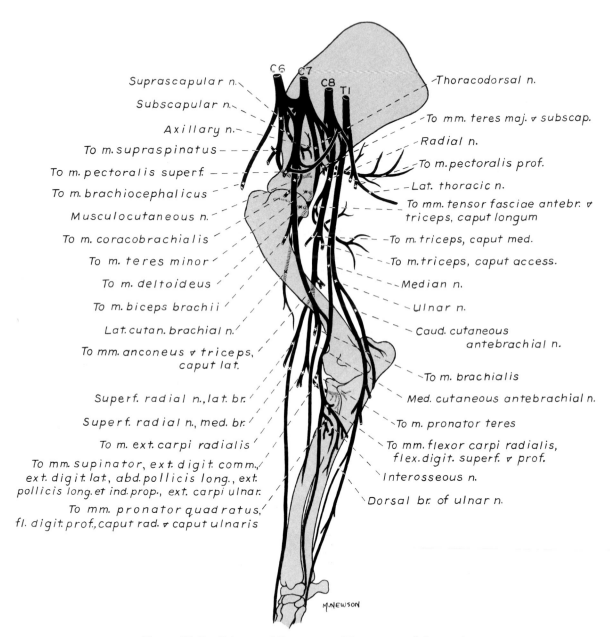

Figure 17–5. Schema of the nerves of the arm, medial aspect.

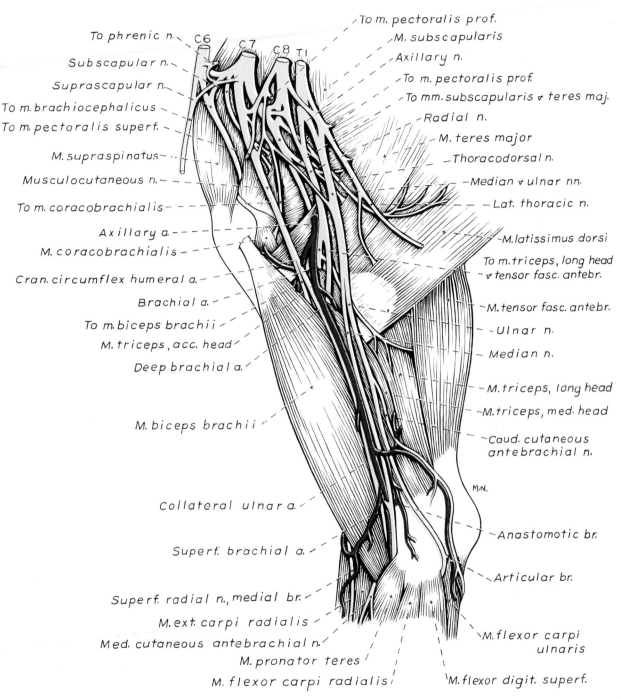

Figure 17–6. The brachial plexus, medial aspect of the right pectoral limb.

marily a motor nerve to the supraspinatus and infraspinatus muscles. Prior to crossing the neck of the scapula distal to the end of the spine, the nerve sends a delicate twig to the lateral part of the shoulder joint (Fig. 17–8).

The **subscapular nerve** (*n. subscapularis*) is usually a single, but occasionally double, nerve which arises from the union of a branch from the sixth and seventh cervical nerves, or if the nerve is double, one part usually arises from the seventh cervical nerve directly. It usually divides into cranial and caudal parts upon entering the medial surface of the distal fifth of the subscapularis muscle. The subscapular nerve is about 5 cm. long in a medium-sized dog. This permits the extensive sliding movement of the shoulder on the thorax during locomotion without nerve injury.

The **axillary nerve** (*n. axillaris*), like the subscapular nerve, is much longer than the distance between its origin and its peripheral fixed end. It arises as a branch from the combined seventh and eighth cervical nerves. A contribution from the sixth cervical nerve may also be present. It may arise completely or nearly completely from either the seventh or the eighth cervical nerve (Allam et al. 1952). It supplies mainly the muscles of the shoulder as it curves around the caudal border of the subscapularis muscle near its distal end. In its intermuscular course proximocaudal to the shoulder joint, it divides basically into two portions; one part sends twigs to the subscapularis muscle and completely supplies the teres major muscle. The other portion, accompanied by the caudal circumflex humeral vessels, runs laterally to supply the laterally lying teres minor and deltoideus muscles. Before entering the teres minor muscle, a branch enters the caudal part of the shoulder joint capsule (Fig. 17–8).

The **cranial lateral cutaneous brachial nerve** (*n. cutaneus brachii lateralis cranialis*) leaves the axillary nerve just prior to the entry of this nerve into the deltoideus muscle. It arises, therefore, lateral to the space between the origins of the lateral and long heads of the triceps muscle. It runs distally on the lateral head of the triceps muscle, where it is covered by the deltoideus muscle. It appears subcutaneously caudal to the main portion of the cephalic

vein, where it is associated with the cutaneous branches of the caudal circumflex humeral artery and vein. It supplies the skin of the lateral surface of the brachium, lapping in its distribution the area supplied by the intercostobrachial nerves caudally and the cutaneous branches of the cervical nerves cranially. It terminates in the skin of the proximocraniolateral aspect of the forearm. At the elbow joint or just distal to it, it joins the medial branch of the superficial radial nerve, and by means of this nerve its fibers are carried to the skin of the dorsum of the antebrachium and possibly the dorsum of the paw also.

The **musculocutaneous nerve** (*n. musculocutaneus*) gives muscular branches to the coracobrachialis, biceps brachii, and brachialis muscles. It continues in the forearm as the medial cutaneous antebrachial nerve. The musculocutaneous nerve arises mainly from the seventh cervical nerve. It is irregular in its formation, occasionally receiving a branch from the sixth cervical, but more frequently from the eighth cervical and even from the first thoracic nerve in rare instances. Throughout its course in the brachium it lies between or under the cranially lying biceps brachii muscle and the axillary vessels caudally. The muscular branches are three in number. Proximally, a small twig goes to the coracobrachialis muscle. This branch is small, and instead of arising from the musculocutaneous nerve directly it may exist as a separate twig which comes from the eighth cervical or first thoracic nerve, or both. In reaching the coracobrachialis, it follows the cranial circumflex humeral vessels over a portion of its course. A large branch, the **proximal muscular branch** (*ramus muscularis proximalis*), enters the deep surface of the biceps brachii muscle about 4 cm. from its origin and near its caudomedial border. In the distal third of the brachium an **anastomotic branch** (*ramus anastomaticus*) passes distocaudad usually medial to the brachial vessels and joins the median nerve, which with the ulnar nerve lies caudal to the brachial vessels. (This communication between the musculocutaneous and median nerves in the dog is not homologous to the ansa axillaris, according to N.A.V. 1973.) As the musculocutaneous nerve winds under the terminal part of the biceps brachii muscle from the medial side,

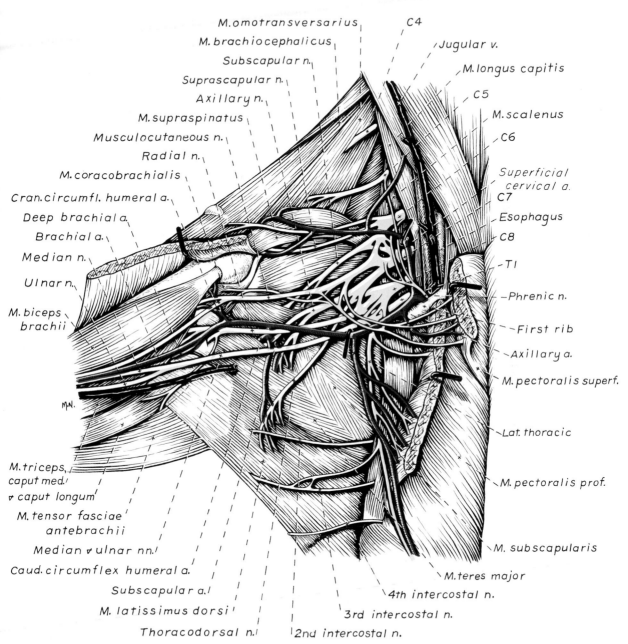

Figure 17–7. The right brachial plexus, ventral aspect.

it terminates in the **distal muscular branch** (*ramus muscularis distalis*), which enters the distal medial portion of the brachialis muscle, and the small **medial cutaneous antebrachial nerve** (*n. cutaneus antebrachii medialis*). This cutaneous branch crosses the lateral side of the tendon of the biceps brachii muscle and enters the cranial surface of the antebrachium from the flexor angle of the elbow joint. As the nerve crosses the cranial surface of the elbow joint, it sends a small branch to the craniolateral part of it. It freely branches in its course distad in the forearm as it supplies the skin of the craniomedial portion of the antebrachium. It ends at the carpus. The cutaneous area which it supplies is overlapped by the area supplied by the cutaneous branches of the medial branch of the superficial radial nerve cranially and the caudal cutaneous antebrachial nerve from the ulnar caudally.

The **radial nerve** (*n. radialis*) (Figs. 17–9, 17–10) arises from the seventh and eighth cervical and the first and second thoracic nerves. It is the largest nerve of the brachial plexus. It supplies all the extensor muscles of the elbow, carpal, and digital joints and also the supinator, the brachioradialis, and the abductor pollicis longus muscles. The skin on the cranial portion of the antebrachium and paw is also supplied by fibers of radial nerve origin via the caudal lateral cutaneous brachial nerve. As the radial nerve approaches the brachium by traversing the axillary space, it lies lateral to the axillary vein and medial to the axillary artery. Upon crossing the medial surface of the conjoined tendons of the teres major and lattissimus dorsi muscles, it lies caudal to the brachial vessels which are the continuation of the axillary vessels after these have crossed the conjoined tendon. Upon entering the interval between the medial and long heads of the triceps muscle, the radial nerve divides into a branch which runs proximolaterad and is distributed to the long head of the triceps. The second branch runs distolaterad and represents the main continuation of the radial nerve. It supplies a branch to the accessory and medial heads before it makes contact with the brachial muscle. It follows this muscle, where it lies related to the nutrient artery of the humerus, in a spiral manner around the humerus. Upon

contacting the lateral head of the triceps, it sends a branch to it, and shortly thereafter it bifurcates into deep and superficial branches. The deep branch runs under the proximocranial border of the extensor carpi radialis muscle. At the place of bifurcation of the radial nerve, a minute twig runs to the deep surface of the brachioradialis muscle. The superficial branch pursues a more cranial course and becomes superficial between the distocranial border of the lateral head of the triceps and the lateral surface of the deeply lying brachial muscle.

The **deep branch** (*ramus profundus*) of the antebrachial part of the radial nerve at first passes under the extensor carpi radialis muscle near its origin from the lateral supracondyloid crest and sends a branch into it. As the deep branch crosses the flexor surface of the elbow joint, it sends an articular branch to the craniolateral part of it. The remaining part of the deep branch then passes under the supinator muscle, which it supplies. Upon emerging from under this muscle, it immediately divides into branches which supply the common and lateral digital extensors and a small branch which closely follows the lateral border of the radius and runs distad to innervate the abductor pollicis longus and extensor pollicis longus et indicis proprius muscles.

The **superficial branch** (*ramus superficialis*) of the radial nerve is its more cranial branch. Upon emerging from under the cranial part of the distal border of the lateral head of the triceps muscle, it runs obliquely craniodistad on the brachial muscle, where it is covered by the heavy intermuscular fascia. After running about 1 cm. in this location, it perforates the heavy fascia and divides unevenly into a larger **lateral branch** (*ramus lateralis*) and a smaller **medial branch** (*ramus medialis*). These branches continue to the carpus in relation to the lateral and medial branches of the superficial brachial arteries respectively. They thus closely flank the medial and lateral sides of the cephalic vein as they traverse the antebrachium. From the lateral branch of the superficial portion of the radial nerve the usually double **lateral cutaneous antebrachial nerve** (*n. cutaneus antebrachii lateralis*) arises. Since the nerves to the skin of the lateral side of the antebrachium are appreciably larger and longer than those which

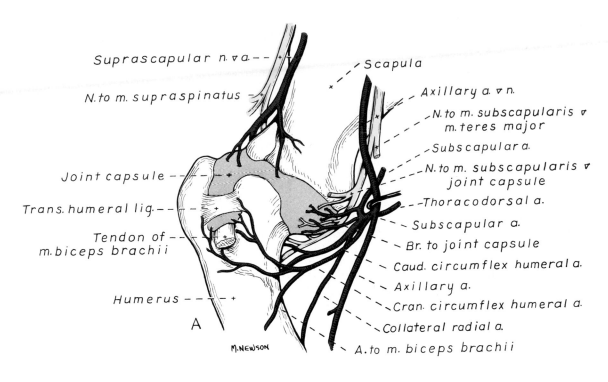

Suprascapular n. ▽ a.
N. to m. supraspinatus
Joint capsule
Trans. humeral lig.
Tendon of m. biceps brachii
Humerus
A
M. NEWSON

Scapula
Axillary a. ▽ n.
N. to m. subscapularis ▽ m. teres major
Subscapular a.
N. to m. subscapularis ▽ joint capsule
Thoracodorsal a.
Subscapular a.
Br. to joint capsule
Caud. circumflex humeral a.
Axillary a.
Cran. circumflex humeral a.
Collateral radial a.
A. to m. biceps brachii

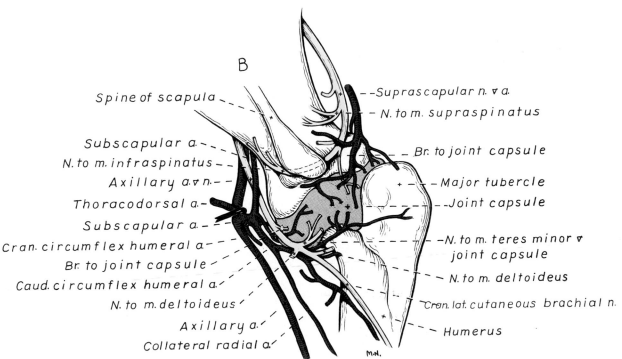

B

Spine of scapula
Subscapular a.
N. to m. infraspinatus
Axillary a. ▽ n.
Thoracodorsal a.
Subscapular a.
Cran. circumflex humeral a.
Br. to joint capsule
Caud. circumflex humeral a.
N. to m. deltoideus
Axillary a.
Collateral radial a.

Suprascapular n. ▽ a.
N. to m. supraspinatus
Br. to joint capsule
Major tubercle
Joint capsule
N. to m. teres minor ▽ joint capsule
N. to m. deltoideus
Cran. lat. cutaneous brachial n.
Humerus

M. N.

Figure 17–8.   A. Nerves and arteries of the right shoulder joint, medial aspect.
B. Nerves and arteries of the right shoulder joint, lateral aspect.

supply the skin on the dorsum of the ante-brachium, it seems feasible to designate them by distinctive names. The more proximal nerve is the larger and is the branch which is more constantly present. It arises just distal to the flexor surface of the elbow joint, and, associated with relatively large cutaneous branches of the lateral branch of the superficial brachial vessels, it supplies the skin of the proximal half to two-thirds of the lateral surface of the antebrachium. The more distally located nerve to the skin of the lateral side of the antebrachium, smaller than the more proximally located nerve, also is accompanied by a cutaneous artery and vein which serve the cutaneous area of the region. Occasionally, more than two lateral cutaneous antebrachial nerves are present. Small branches leave both the medial and lateral branches of the superficial radial nerves and innervate the skin on the cranial surface of the antebrachium as the **cranial cutaneous antebrachial branches** (*rami cutanei antebrachiales craniales*). The medial and lateral branches of the superficial radial nerves as they innervate the dorsum of the forepaw are described under the description of the nerves to the forepaw.

Because it supplies all the extensor muscles of the thoracic limb, except those of the shoulder, injury to the proximal part of the nerve results in a grave syndrome. Uncomplicated radial paralysis is rare. Surely what was diagnosed as radial paralysis a generation ago was probably an avulsion of the brachial plexus in most cases. Clifford et al. (1958) describe such a case. Bowne (1959) has demonstrated that the radial nerve can be interrupted just distal to the point where the last branch goes to the triceps brachii with no permanent impairment of locomotion. He observed, however, that his experimentally induced radial paralysis did not produce the same syndrome as that seen in typical clinical cases. Worthman (1957) has reported upon many neurectomies of the nerves of both the fore and hind limbs in the dog.

The **median nerve** (*n. medianus*) (Figs. 17–10, 17–11) arises primarily from the eighth cervical and the first and second thoracic spinal nerves. Reimers (1925) does not regard the nerve to be formed until it has received the communicating branch from the musculocutaneous nerve in the distal part of the brachium. Through this communication the median nerve is augmented by fibers from the sixth and seventh cervical nerves. The loosely joined median and ulnar nerves lie caudal to the brachial artery and lateral to the brachial vein. The median nerve is cranial in relation to the ulnar nerve. At the flexor surface of the elbow joint, the median nerve dips laterally under the pronator teres muscle and enters the large caudal group of flexor muscles of the antebrachium. It supplies the pronator teres, pronator quadratus, flexor carpi radialis, and flexor digitorum superficialis muscles, and the radial head of the flexor digitorum profundus muscle. It also sends fibers to the ulnar and humeral heads of the flexor digitorum profundus muscle, and a small articular branch to the medial aspect of the elbow joint.

Upon emerging from under the pronator teres, to which it sends a small branch, several **muscular branches** (*rami musculares*) leave the caudal portion of the nerve. The shortest and most proximal of these nerves enters the flexor carpi radialis muscle close to its humeral origin. The remaining flattened bundle of muscular branches crosses the medial surface of the median (brachial) vessels at the place where the common interosseous artery arises and, after running under the flexor carpi radialis and through the humeral head of the flexor digitorum profundus, most of them end in the superficially lying, flattened flexor digitorum superficialis muscle. In their path to this muscle, they lie about 1 cm. proximal and parallel to the deep antebrachial vessels. In this deep location a branch is sent to the radial head of the flexor digitorum profundus muscle, which it completely innervates, and smaller twigs enter the humeral and ulnar heads of this muscle, which are also supplied by the ulnar nerve. The small **interosseous nerve** of the antebrachium (*n. interosseus antebrachii anterior*) first runs on the proximal part of the delicate interosseous membrane. It then perforates this membrane and runs distally on about the proximal half of the pronator quadratus muscle, where it appears as a fine white streak. It enters this muscle in its distal half and innervates it.

The portion of the median nerve which continues distad in the antebrachium, after

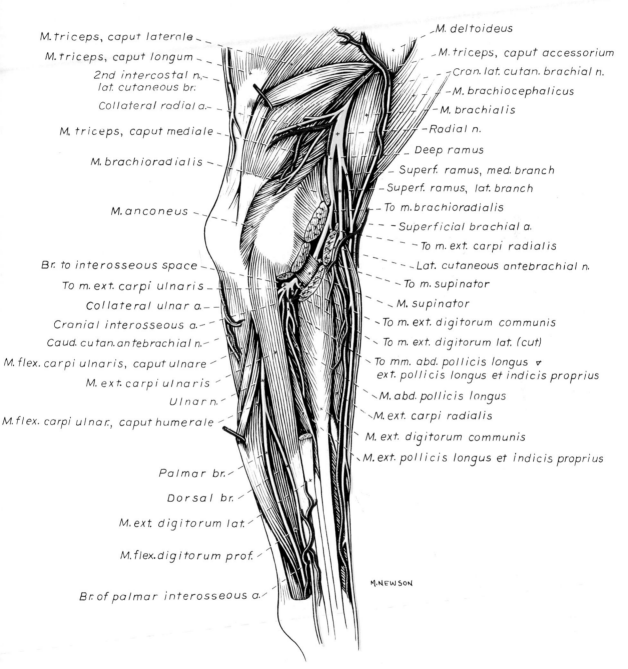

M. triceps, caput laterale

M. triceps, caput longum

2nd intercostal n.,
lat. cutaneous br.

Collateral radial a.

M. triceps, caput mediale

M. brachioradialis

M. anconeus

Br. to interosseous space

To m. ext. carpi ulnaris

Collateral ulnar a.

Cranial interosseous a.

Caud. cutan. antebrachial n.

M. flex. carpi ulnaris, caput ulnare

M. ext. carpi ulnaris

Ulnar n.

M. flex. carpi ulnar., caput humerale

Palmar br.

Dorsal br.

M. ext. digitorum lat.

M. flex. digitorum prof.

Br. of palmar interosseous a.

M. deltoideus

M. triceps, caput accessorium

Cran. lat. cutan. brachial n.

M. brachiocephalicus

M. brachialis

Radial n.

Deep ramus

Superf. ramus, med. branch

Superf. ramus, lat. branch

To m. brachioradialis

Superficial brachial a.

To m. ext. carpi radialis

Lat. cutaneous antebrachial n.

To m. supinator

M. supinator

To m. ext. digitorum communis

To m. ext. digitorum lat. (cut)

To mm. abd. pollicis longus ⱴ
ext. pollicis longus et indicis proprius

M. abd. pollicis longus

M. ext. carpi radialis

M. ext. digitorum communis

M. ext. pollicis longus et indicis proprius

M. NEWSON

Figure 17–9. Nerves of the right pectoral limb, lateral aspect.

the muscular branches have arisen, is at first related to the median artery and vein. When the median (brachial) artery terminates by dividing into the radial and ulnar arteries at about the middle of the antebrachium, the median nerve continues distad in relation to the larger ulnar artery. This portion of the median nerve is small, measuring about 0.5 mm. in diameter.

The **ulnar nerve** (*n. ulnaris*) (Figs. 17–10, 17–12) arises in close association with the radial and median nerves from the eighth cervical and the first and second thoracic nerves. After leaving the caudal part of the brachial plexus, the median and ulnar nerves are flanked by the brachial artery cranially and the brachial vein caudally. They are bound to each other by areolar tissue until they reach the middle of the brachium, where they diverge. The ulnar nerve, which measures about 3 mm. in diameter, runs distad along the cranial border of the medial head of the triceps brachii muscle and adjacent to the caudal border of the biceps brachii. Upon entering the caudomedial part of the antebrachium, the ulnar nerve runs under the heavy antebrachial fascia. After crossing the medial epicondyle of the humerus just proximal to the origin of the humeral head of the superficial digital flexor, it runs under the ulnar head of the flexor carpi ulnaris muscle. As with the median nerve, no muscular branches leave the ulnar nerve as it traverses the brachium.

The **caudal cutaneous antebrachial nerve** (*n. cutaneous antebrachii caudalis*) leaves the caudal part of the ulnar nerve near the beginning of the distal third of the brachium and passes over the medial surface of the olecranon process into the caudomedial part of the antebrachium. In its subcutaneous course throughout most of the area it supplies, it is accompanied by the collateral ulnar artery and vein. It freely sends branches to the skin as it winds across the proximal portion of the antebrachium from the medial to the caudolateral aspects. Ascending branches arborize in the skin of the distal part of the brachium. It supplies the proximal two-thirds of the skin of the caudolateral aspect of the antebrachium. The cranial cutaneous branches of the superficial radial nerve overlap its caudolateral branches, and the medial cutaneous

branches of the musculocutaneous nerve overlap its caudomedial branches.

The **muscular branches** (*rami musculares*) of the ulnar nerve, which supply the muscles of the antebrachium, arise as a short, stout trunk which leaves the caudal side of the ulnar nerve as it passes over the medial epicondyle of the humerus and plunges into the deep surface of the thin, wide ulnar head of the flexor carpi ulnaris muscle. The ulnar nerve, upon entering the septum between the ulnar and humeral heads of the flexor carpi ulnaris, sends a branch about 1 mm. in diameter and 1.5 cm. long distally into the caudal border of the humeral head of the deep digital flexor. In the proximal fifth of the antebrachium, as the ulnar nerve curves around the caudal border of the humeral head of the flexor carpi ulnaris, it sends a stout branch into its lateral surface. Throughout the middle third of the antebrachium the ulnar nerve lies on the caudal border of the deep digital flexor, where it is covered by the humeral head of the flexor carpi ulnaris muscle. At about the middle of the antebrachium, the small, cutaneous dorsal branch of the ulnar nerve arises. This branch and the palmar branch arise as terminal branches of the ulnar nerve. Both these branches are distributed to the structures of the forepaw.

Nerves of the Forepaw (Manus)

Like the vessels which serve the forepaw, the nerves may be divided into dorsal and palmar sets. Kopp (1901) described the morphology of the nerves of the forepaw of the dog. The radial nerve nearly totally supplies the dorsum of the forepaw, where it forms a single set of dorsal common digital and dorsal digital nerves. The median and ulnar nerves supply the palmar aspect of the forepaw and all other parts which are not supplied by the radial nerve. In the palmar part of the metacarpus they form the palmar common digital nerves, which are derived largely from the median nerve, and the palmar metacarpal nerves from the ulnar nerve.

In accordance with Nomina Anatomica and Nomina Anatomica Veterinaria, the superficial nerves of the metapodium are designated nn. digitales communes, whereas

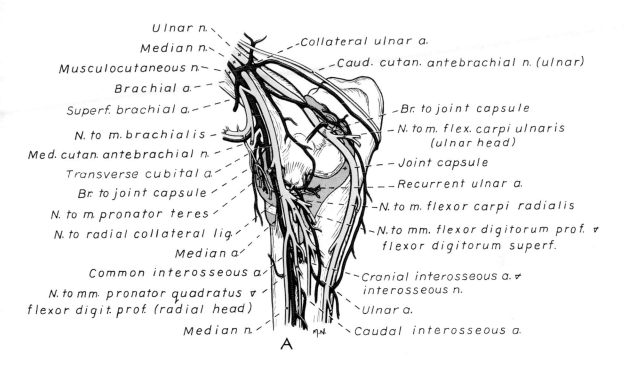

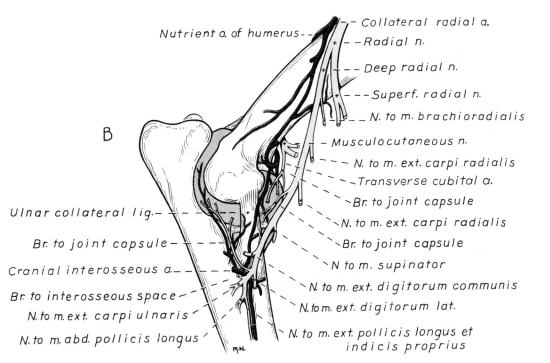

Figure 17–10. A. Nerves and arteries of the right elbow joint, medial aspect.
B. Nerves and arteries of the right elbow joint, lateral aspect.

the deep nerves are called nn. metacarpei and metatarsei. Digital nerves that originate from the bifurcation of nn. digitales communes are called nn. digitales proprii. Those that originate from some other source are simply nn. digitales. The dog has nn. digitales communes I, II, III, and IV.

The nn. digitales palmares communes I, II, III are the terminal branches of the median nerve in the dog. They receive the nn. metacarpei palmares from the deep branch of the ulnar nerve just before they divide into nn. digitales proprii. The n. digitalis palmaris communis IV is formed by the superficial branch of the ulnar nerve and is joined by the n. metacarpeus palmaris IV from the deep branch of the ulnar nerve.

The **radial nerve of the forepaw** (*n. radialis manus*) (Fig. 17–13) is represented by the terminal portions into which the medial and lateral branches of the superficial radial nerve divide.

The **medial branch of the superficial radial nerve** (*ramus medialis n. radialis superficialis*) continues into the proximal part of the metacarpus at the **dorsal common digital nerve I**. It divides early into medial and lateral branches. The medial branch arborizes in the skin on the dorsum of the small first digit as the **dorsal digital nerve I** (*n. digitalis dorsalis I*). The lateral branch of the first dorsal common digital nerve supplies the skin on the dorsomedial sides of the second metacarpal bone and second digit as the **abaxial dorsal proper digital nerve II**.

The **lateral branch of the superficial radial nerve** (*ramus lateralis n. radialis superficialis*) trifurcates at about the carpometacarpal junction into the **dorsal common digital nerves II, III, and IV**. Each of the dorsal common digital nerves II, III, and IV bifurcates before reaching the clefts which separate the four main digits. The resultant branches are either **axial or abaxial dorsal digital nerves II, III, and IV**, depending upon which side of the digit they lie. The axis of the limb passes between the third and fourth digits. Thus, the dorsal common digital nerve III divides into the axial dorsal proper digital nerve III and the axial dorsal proper digital nerve IV.

The **median nerve of the forepaw** (*n. medianus*) (Fig. 17–14) near the proximal end of the carpus divides into medial and lateral branches. The medial branch, upon reaching the metacarpus, divides again to form the **palmar metacarpal nerves I and II**.

The palmar metacarpal nerve I runs to the web of skin of the first interdigital space and bifurcates into palmar digital nerve I and abaxial palmar digital nerve II, which supply part of the skin of the adjacent palmar sides of the first two digits. These nerves will be described with the portion of the ulnar nerve which innervates the forepaw. The lateral branch into which the median nerve divides becomes the **palmar common digital nerve III** (*n. metacarpalis palmaris superficialis III*). The main palmar common digital nerves are formed as follows: nerves II and III from the median; nerve IV from the ulnar. All three of these nerves communicate with the comparable deep branches of the ulnar to form the palmar common digital nerves. These nerves and certain irregularities concerning them are presented after the description of those portions of the ulnar nerve which are located in the paw.

The **ulnar nerve of the forepaw** (*n. ulnaris*) (Fig. 17–14) is represented by the dorsal and palmar branches, which arise as terminal branches of the ulnar nerve at the junction of the proximal and middle thirds of the antebrachium.

The **dorsal branch** (*ramus dorsalis*) passes distad toward the lateral aspect of the carpus by passing obliquely laterad between the caudally lying flexor carpi ulnaris and the cranially lying extensor carpi ulnaris muscles. It perforates the heavy deep antebrachial fascia from 3 to 8 cm. proximocaudal to the styloid process of the ulna. Closely applied to the skin and in association with a cutaneous branch of the caudal interosseous artery, it obliquely crosses the lateral side of the carpus. In its subcutaneous course on the dorsolateral surface of the forepaw, it is called the **abaxial dorsal digital nerve V** (*n. digitorum dorsalis abaxialis V*), where it supplies the skin of the dorsolateral surfaces of the metacarpus and the fifth digit.

The **palmar branch** (*ramus palmaris*) of the forepaw is the main continuation of the ulnar nerve after the dorsal branch has arisen. As it passes through the distal portion of the antebrachium, it lies on the cau-

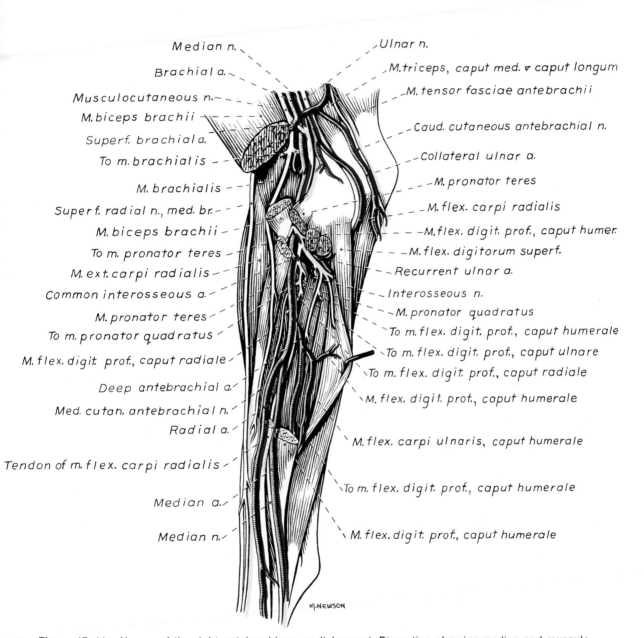

Figure 17–11. Nerves of the right antebrachium, medial aspect. Dissection showing median and musculo-cutaneous nerves.

domedial surface of the deep digital flexor muscle about 1 cm. lateral to the origin of the ulnar artery. It converges toward the artery distally in the antebrachium. The vessel and nerve pass through the deep portion of the carpal canal together. Lying medial to the accessory carpal bone, the palmar branch issues a small branch to the carpal pad. A larger branch runs almost directly medially at the distal end of the palmar carpal fibrocartilage and innervates the special muscles of the fifth digit.

The **palmar common digital nerve IV** arises from the deep branch of the ulnar nerve as it traverses the carpus. It passes laterally across the proximal end of the fifth metacarpal bone and sends a branch to the skin on the lateral sides of the fifth metacarpal bone and fifth digit as the **abaxial palmar digital nerve V** (*n. digitalis palmaris abaxialis V*).

At the proximal end of the metacarpus, the deep branch of the ulnar nerve divides into two series of branches. One set is short and is composed of the muscular branches which frequently arise from the second set, or the palmar metacarpal nerves.

The **palmar metacarpal nerves I, II, III, and IV** (*nn. metacarpales palmares I, II, III, et IV*) are usually the terminal branches of the deep branch of the ulnar nerve. The deep branch passes through the lateral portion of the carpal canal, where it lies in close association with the relatively large terminal portion of the caudal interosseous artery. On the deep surfaces of the abductors of the second and fifth digits at their origins, the deep branch of the ulnar nerve divides into the palmar metacarpal nerves I, II, III, and IV and into muscular branches.

The **muscular branches** (*rami musculares*) arise directly from the deep branch of the ulnar or individually from the several palmar metacarpal nerves. They innervate the four interosseous muscles; the three lumbricales; the special muscles of the first, second, and fifth digits; and the single, weak flexor digitorum brevis muscle.

At first the palmar metacarpal nerves lie on the palmar surfaces of the proximal ends of the interosseous muscles. As they run distad toward the digits, they lie between the main interosseous muscles in relatively superficial positions. Communications occur between the palmar common digital and palmar metacarpal nerves near or at the distal ends of the metacarpal bones.

By these unions the **palmar common digital nerves II, III, and IV** (*nn. digitales palmares communes II, III, et IV*) are formed (Fig. 17–15). Communications between the members of the two sets — superficial and deep — are irregular and in some instances multiple. Occasionally, the deep members send slender anastomotic branches to the palmar digital nerves into which the palmar common digital nerves terminally divide. The common digital nerves are therefore short as they cross the contact surfaces of adjacent metacarpophalangeal joints. From the common digital nerves II, III, and IV, or occasionally proximal or distal to these nerves, the three sensory nerves arise which innervate the large metacarpal foot pad. Minute twigs from the palmar common digital nerves supply the structures of the metacarpophalangeal joints of the four main digits.

The **axial and abaxial palmar digital nerves II, III, and IV** (*nn. digitales proprii palmares laterales et mediales*) are the terminal branches into which each of the three main common digital nerves divides. They lie on the contact sides of the four main digits, dorsal to the comparable vessels, and like the vessels the proper nerves which face the axis through the paw (axial branches) are larger and longer than are the abaxial proper nerves. They supply the skin under which they lie and the digital joints which they cross. At the distal interphalangeal joints, sensory branches are supplied to the digital foot pads. Axial and abaxial digital nerves enter the palmar vascular canal of the distal phalanges, and thereby nerve fibers reach the corium of the horny claw.

Nerves of the Brachial Plexus Which Supply Extrinsic Muscles of the Thoracic Limb

The nerves in this group are smaller than the nerves which supply the intrinsic structures of the thoracic limb. They consist of a nerve to the brachiocephalicus muscle, cranial pectoral nerves, long thoracic nerve, dorsal thoracic nerve, lateral thoracic nerve, and caudal pectoral nerves.

The **nerve to the m. brachiocephalicus**

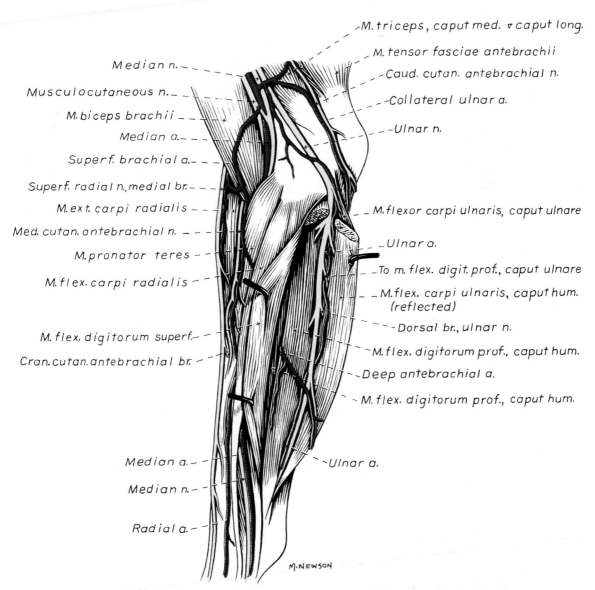

Median n.

Musculocutaneous n.

M. biceps brachii

Median a.

Superf. brachial a.

Superf. radial n., medial br.

M. ext. carpi radialis

Med. cutan. antebrachial n.

M. pronator teres

M. flex. carpi radialis

M. flex. digitorum superf.

Cran. cutan. antebrachial br.

Median a.

Median n.

Radial a.

M. triceps, caput med. ʊ caput long.

M. tensor fasciae antebrachii

Caud. cutan. antebrachial n.

Collateral ulnar a.

Ulnar n.

M. flexor carpi ulnaris, caput ulnare

Ulnar a.

To m. flex. digit. prof., caput ulnare

M. flex. carpi ulnaris, caput hum. (reflected)

Dorsal br., ulnar n.

M. flex. digitorum prof., caput hum.

Deep antebrachial a.

M. flex. digitorum prof., caput hum.

Ulnar a.

M. NEWSON

Figure 17–12. Nerves of the right antebrachium, medial aspect. Dissection showing the ulnar nerve.

(Allam et al. 1952) arises mainly from the sixth cervical nerve, but it may be joined by a branch from the fifth cervical nerve. It passes directly laterad into the brachiocephalicus muscle cranial to the shoulder joint. Branches from the cervical plexus also supply the brachiocephalicus.

The **cranial pectoral nerves** (*nn. pectorales craniales*) supply the superficial pectoral muscles. They are irregular in number and origin but usually arise as two branches from the sixth, seventh, and eighth cervical nerves. They have formerly been included under the term "ventral thoracic nerves" (Miller, Christensen, and Evans 1964).

The **long thoracic nerve** (*n. thoracicus longus*) (Fig. 17–2) usually arises from the seventh cervical nerve before it branches to aid in forming the brachial plexus. It runs largely horizontally on the superficial surface of the thoracic portion of the serratus ventralis muscle, which it supplies. Miller (1934) regarded a small branch from the fifth cervical nerve as also belonging to the long thoracic nerve.

The **dorsal thoracic nerve** (*n. thoracodorsalis*) (Fig. 17–6) arises primarily from the eighth cervical nerve with contributions from the first thoracic or the seventh cervical nerve or both. It is the motor nerve to the latissimus dorsi muscle. It runs caudodorsad in close relation to the thoracodorsal vessels on the medial surface of the muscle.

The **lateral thoracic nerve** (*n. thoracicus lateralis*) is composed of fibers which come primarily from the eighth cervical and the first thoracic nerves. This long nerve gives rise to branches which supply the deep pectoral muscle and is the sole motor supply to the cutaneous trunci muscle. At first the nerve accompanies the lateral thoracic artery and vein. It lies between the adjacent borders of the latissimus dorsi and deep pectoral muscles after passing medial to the axillary lymph nodes.

The **caudal pectoral nerves** (*nn. pectorales caudales*) are represented by three or four branches which innervate the deep pectoral muscle. They originate from the eighth cervical and the first two thoracic nerves. Some of the branches may appear to originate from the proximal portion of the lateral thoracic nerve.

THORACIC NERVES

The **thoracic nerves** (*nn. thoracici*) (Fig. 17–16) number 13 pairs in the dog, and as a group they retain the simplest segmental form of all the spinal nerves. Each pair of thoracic nerves has the same serial number as the vertebra which lies in front of their intervertebral foramina of exit. All the ventral branches with the exception of the first three or four send twigs at their distal ends into the rectus abdominis muscle, since this muscle traverses both the thorax and the abdomen. Unlike typical spinal nerves, the ventral branches do not divide into medial and lateral branches.

The **medial branches** (*rami mediales*) of the dorsal branches of the thoracic nerves run essentially parallel and caudal to the first 10 caudally sloping thoracic vertebral spines, with which they correspond in number. Caudal to the eleventh thoracic spine, the thoracic spines slope cranially, and the corresponding nerves cross them obliquely, since they run caudodorsally. The medial branches supply the multifidus thoracis, the rotatores, longissimus dorsi, and the spinalis et semispinalis thoracis et cervicis muscles. It is probable that the vertebrae, ligaments, and dura receive branches from these nerves. The medial branches do not end in cutaneous branches.

The **lateral branches** (*rami laterales*) of the dorsal branches of the thoracic nerves run caudolaterally at about a 45° angle to a sagittal plane. They course between the longissimus dorsi muscle dorsally and the iliocostalis ventrally to reach the medial surfaces of the segments of the serratus dorsalis muscles where these are present. They usually perforate these segments as well as the iliocostalis dorsi and cutaneus trunci muscles to reach the skin, where they become the proximal **lateral cutaneous branches** (*rami cutanei laterales*). These nerves divide in the superficial fascia into short medial branches and longer ventrolateral branches which supply the skin of approximately the dorsal third of the thorax. As these nerves cross the medial border of the iliocostalis dorsi muscle, they send branches to it and to the levator costae muscle segments.

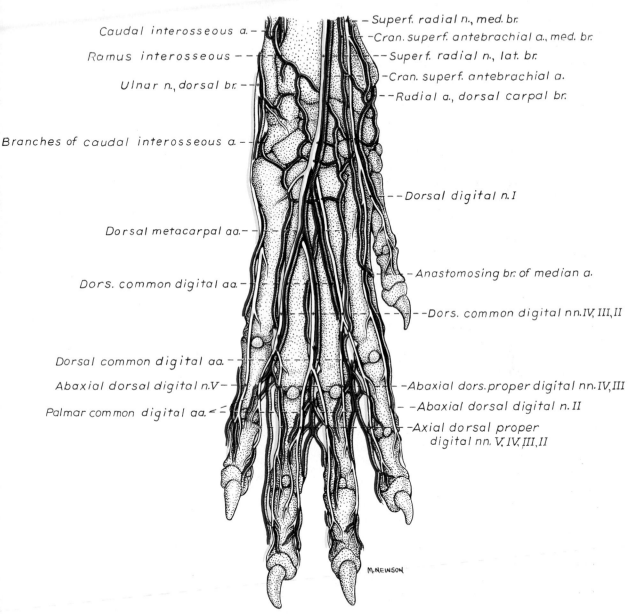

Figure 17–13. Nerves and arteries of the right forepaw, dorsal aspect.

The **ventral branches** (*rami ventrales*) of the thoracic nerves, with the exception of most of the first and the thirteenth, are more commonly known as the **intercostal nerves** (*nn. intercostales*).

The major portion of the first thoracic nerve (ventral branch) passes forward medial to the neck of the first rib and contributes appreciably to the formation of the brachial plexus. The intercostal portion of the first thoracic nerve is but a delicate twig which enters the muscles of the first intercostal space.

The **second intercostal nerve** (*n. intercostalis II*) differs from the intercostal nerves which lie caudal to it in two respects. It usually sends a communicating branch to the first thoracic nerve, and secondly its distal lateral cutaneous branch is the largest of all these branches; furthermore, this branch runs to the thoracic limb and supplies a patch of skin in the region of the elbow joint. It can be called the **second intercostobrachial nerve**. Usually the **distal lateral cutaneous branch of the third intercostal nerve** (*n. intercostobrachialis III*) also innervates a portion of the thoracic limb proximal to that of the second nerve in most specimens.

The **costoabdominal nerve** (*n. costoabdominalis*) is the ventral branch of the last or thirteenth thoracic nerve. It supplies a band of the abdominal wall which lies adjacent to the caudal border of the last rib and then continues tangentially to the last rib and costal arch in the abdominal wall. In its area of distribution it lies cranial but parallel to the bands of the abdominal wall which are supplied by the ventral branches of the first three lumbar nerves. It divides into lateral and medial branches which resemble those of the lumbar nerves lying caudal to it. In man this nerve is called the subcostal nerve because of the upright position of the body.

A typical **intercostal nerve** (*n. intercostalis*) (Fig. 17–17) is disposed as follows: each begins where the dorsal branch of the particular thoracic nerve arises. For about the first centimeter it lies embedded in the dorsal border of the cranioventrally running internal intercostal muscle. It then turns distad on the medial surface of the internal intercostal muscle, where it is separated from the caudal border of the corresponding rib by the intercostal vein and the intercostal artery. Variations in this order of arrangement occur most frequently in the cranial part of the series. This triad of structures is surrounded by a variable quantity of fat. The nerve lies adjacent to or among the fiber strands of the internal intercostal muscle. In the caudal part of the thorax, fleshy sheets of the internal intercostal muscle from the intercostal space in front of it extend over the ribs medially and cover the intercostal vessels and nerves which lie caudal to the rib. In most intercostal spaces, the intercostal vessels and nerves are covered medially only by the pleura.

A typical intercostal nerve has the following branches:

1. The **communicating branch** (*ramus communicans*) which contains efferent sympathetic fibers with which are intermingled afferent fibers from visceral structures. These branches are connections between the initial part of the intercostal nerve and sympathetic trunk, usually at a ganglion. This branch is about 1 mm. in diameter and 3 mm. long.

2. The **proximal muscular branch** (*ramus muscularis proximalis*) leaves the dorsal side of the intercostal nerve 1 or 2 cm. from its origin. Before it perforates the external intercostal muscle, it sends a long, slender branch distally on the deep (medial) surface of the external intercostal muscle, which lies only a few millimeters caudal to the corresponding rib. It sends delicate twigs to the external intercostal muscle throughout its length. Upon contact with the external intercostal muscle, a branch leaves the nerve which runs laterally and supplies the serratus dorsalis muscle.

3. The **distal lateral cutaneous branch** (*ramus cutaneus lateralis distalis*) (Fig. 17–18) passes through the midlateral portion of the thoracic wall and runs distad in the superficial fascia with the like-named artery. The aggregate of these branches supplies all the skin on the ventrolateral half of the thoracic wall except a narrow, longitudinal ventral strip. The first two of these nerves supply skin on the thorax and a portion of the thoracic limb. The limb portions of these nerves (intercostal brachial nerves) supply zones of skin covering the triceps muscle, proximal to and over the elbow joint. In the regions of the thoracic mam-

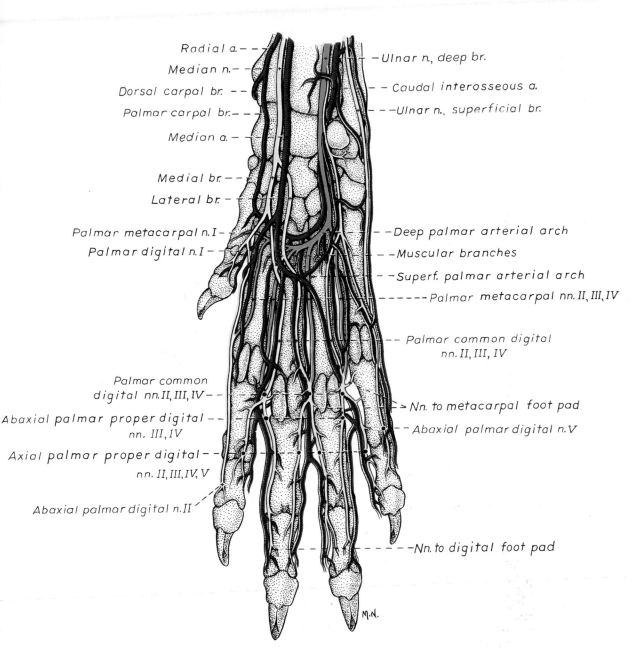

Radial a.
Median n.
Dorsal carpal br.
Palmar carpal br.
Median a.
Medial br.
Lateral br.
Palmar metacarpal n. I
Palmar digital n. I

Ulnar n., deep br.
Caudal interosseous a.
Ulnar n., superficial br.
Deep palmar arterial arch
Muscular branches
Superf. palmar arterial arch
Palmar metacarpal nn. II, III, IV

Palmar common digital
nn. II, III, IV

Palmar common
digital nn. II, III, IV
Abaxial palmar proper digital
nn. III, IV
Axial palmar proper digital
nn. II, III, IV, V
Abaxial palmar digital n. II

Nn. to metacarpal foot pad
Abaxial palmar digital n. V

Nn. to digital foot pad

M. N.

Figure 17–14. Nerves and arteries of the right forepaw, palmar aspect.

mary glands, branches from the distal lateral cutaneous branches (**lateral mammary branches**) ramify under the skin.

4. The **distal muscular branch** (*ramus muscularis distalis*) consists of two branches. One is short and enters the transversus thoracis muscle; the other branch passes to the outside of the rib cage and enters the rectus abdominis muscle.

5. The **ventral cutaneous branch** (*ramus cutaneus ventralis*) is the terminal part of each typical intercostal nerve. It runs to the skin by crossing the lateral surface of the sternum. In the superficial thoracic fascia, which lies on the medial portion of the deep pectoral muscle, it is closely bound to the only slightly larger ventral cutaneous artery. The aggregate of these nerves supplies a zone of skin about 2.5 cm. wide which lies adjacent to the midventral line. The terminal twigs of the ventral cutaneous branches of the fifth and seventh intercostal nerves ramify in the skin covering the medial portions of the two thoracic mammary glands when these glands are functional. They are named the **medial mammary branches** (*rami mammarii mediales*).

The ventral extensions of the last three intercostal nerves leave the intercostal spaces medial to the costal arch and extend toward the linea alba on the superficial surface of the transversus abdominis muscle, which they supply before terminating in the rectus abdominis. These nerves fail to send ventral cutaneous branches to the skin.

LUMBAR NERVES

The **lumbar nerves** (*nn. lumbales*) (Figs. 17–19, 17–20) are seven in number on each side. Each member of a pair has the same number as its intervertebral foramen of exit and the vertebra which lies cranial to it. The middle of the first lumbar segment of the spinal cord lies opposite the fibrocartilage which connects the first two lumbar vertebrae. The nerve roots, therefore, of the first pair of lumbar nerves lie essentially in the same transverse area as the foramina of exit of these nerves. As traced caudally, the segments of the spinal cord are shorter than the vertebral segments, so that the spinal cord ends dorsal to the fibrocartilage between the sixth and seventh lumbar vertebrae. Be-

cause of this disproportionate length, the last several pairs of spinal nerves run increasingly longer distances within the spinal canal before leaving their osseous confines by means of the intervertebral foramina than do the spinal nerves cranial to them. The leash of nerves thus formed is called the *cauda equina*. According to Hopkins (1935), in a 40- to 50-pound dog the intraspinal extent of the lumbar nerves varies from 0.6 cm. for the first pair to 3.5 cm. for the seventh. This author was undoubtedly measuring the nerve roots rather than the spinal nerves which they form. The reason for this spatial disparity between the length of the spinal cord and the vertebral column is a continuation of the growth of the vertebral column after the spinal cord has stopped growing. The lumbar spinal nerves are formed by the merging of the dorsal and ventral roots at the intervertebral foramina. As the roots of the several lumbar spinal nerves run caudally, their proximal halves lie within the dural covering of the spinal cord, and their distal halves lie in dural sheaths in soft epidural fat. Like the typical spinal nerves of the preceding regions, each lumbar spinal nerve divides upon leaving the intervertebral foramen into small dorsal and larger ventral branches. The actual length of each lumbar spinal nerve is only a few millimeters, and it lies largely in and just lateral to the intervertebral foramen through which it passes.

The **dorsal branches** (*rami dorsales*) of the lumbar nerves are similar throughout most of the region. Each typically divides into medial and lateral branches. The **medial branches** (*rami mediales*) arborize in the longissimus lumborum muscle, which they supply, and send terminal twigs to the multifidus lumborum and the interspinales lumborum muscles. They run caudodorsad obliquely across the lateral surface of the cranially inclined spinous processes of the vertebrae which follow them. They are separated from the ventral borders of the tendons of the longissimus lumborum muscle which go to the accessory processes of the lumbar vertebrae by the large branches of the dorsal branches of the lumbar segmental arteries. Only an occasional medial branch runs far enough peripherally to reach the skin. Pederson et al. (1956) found in the human being that small branches enter the

Text continued on page 1005

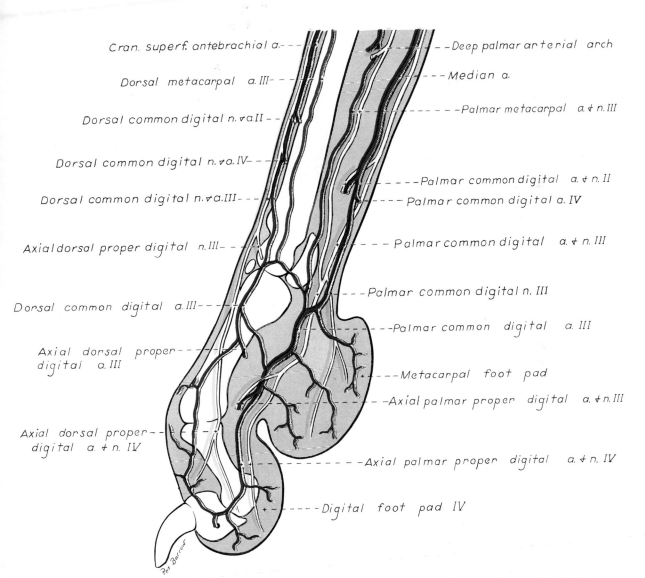

Cran. superf. antebrachial a.

Dorsal metacarpal a. III

Dorsal common digital n. v a. II

Dorsal common digital n. v a. IV

Dorsal common digital n. v a. III

Axial dorsal proper digital n. III

Dorsal common digital a. III

Axial dorsal proper digital a. III

Axial dorsal proper digital a. v n. IV

Deep palmar arterial arch

Median a.

Palmar metacarpal a. v n. III

Palmar common digital a. v n. II

Palmar common digital a. IV

Palmar common digital a. v n. III

Palmar common digital n. III

Palmar common digital a. III

Metacarpal foot pad

Axial palmar proper digital a. v n. III

Axial palmar proper digital a. v n. IV

Digital foot pad IV

Figure 17–15. Arteries and nerves of the fourth digit and metacarpus, medial aspect. (From Miller 1958.)

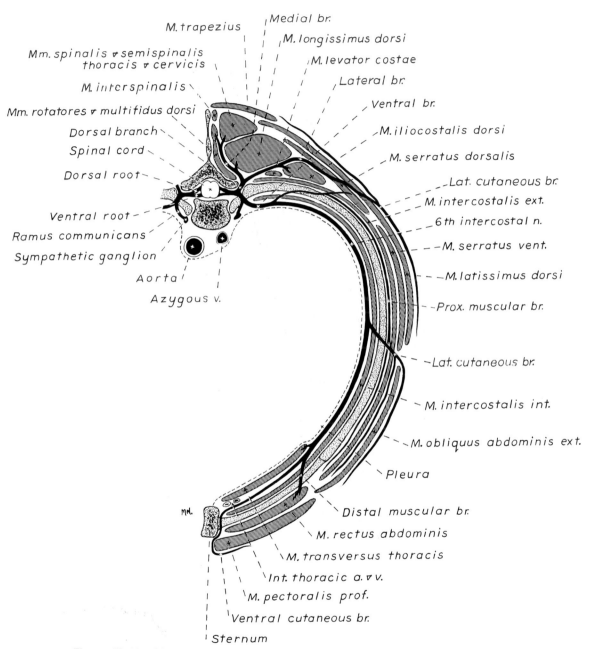

Figure 17–16. Diagram of the sixth thoracic nerve. (Section caudal to the sixth rib.)

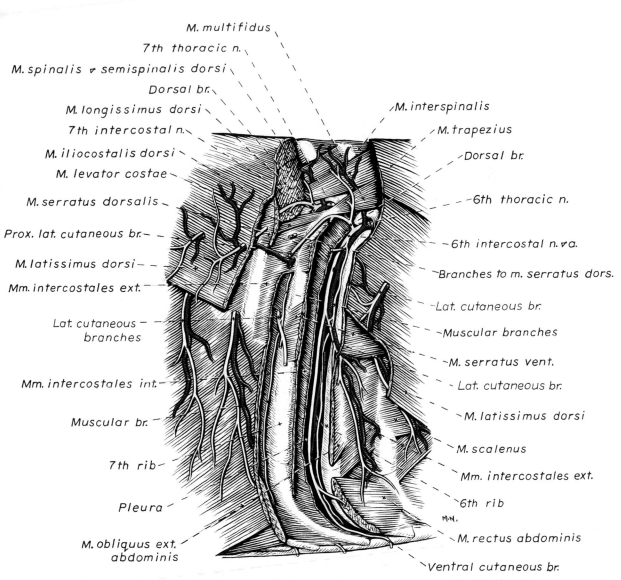

M. multifidus
7th thoracic n.
M. spinalis ♂ semispinalis dorsi
Dorsal br.
M. longissimus dorsi
7th intercostal n.
M. iliocostalis dorsi
M. levator costae
M. serratus dorsalis
Prox. lat. cutaneous br.
M. latissimus dorsi
Mm. intercostales ext.
Lat. cutaneous branches
Mm. intercostales int.
Muscular br.
7th rib
Pleura
M. obliquus ext. abdominis

M. interspinalis
M. trapezius
Dorsal br.
6th thoracic n.
6th intercostal n. ♂ a.
Branches to m. serratus dors.
Lat. cutaneous br.
Muscular branches
M. serratus vent.
Lat. cutaneous br.
M. latissimus dorsi
M. scalenus
Mm. intercostales ext.
6th rib
M. rectus abdominis
Ventral cutaneous br.

Figure 17-17. Dissection showing distribution of the thoracic nerves, right lateral aspect.

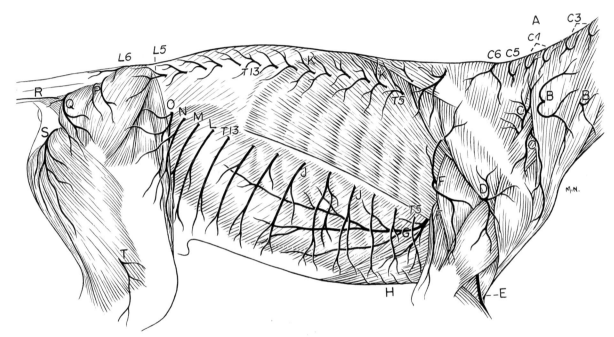

Figure 17–18. Cutaneous nerves of the trunk, lateral aspect.

A. Dorsal branches of cervical nerves
B. Ventral branches of C 3
C. Ventral branches of C 4
D. Lateral cutaneous brachial nerve (from axillary)
E. Radial nerve, superficial branch
F. Intercostobrachial nerve (from T 2)
G. Lateral thoracic nerve (from C 8 and T 1)
H. Ventral cutaneous branches of intercostal nerves
J. Distal lateral cutaneous branches of intercostal nerves
K. Proximal lateral cutaneous branches of thoracic nerves

L. Cranial iliohypogastric nerve (L 1)
M. Caudal iliohypogastric nerve (L 2)
N. Ilioinguinal nerve (L 3)
O. Lateral cutaneous femoral nerve (L 4)
P. From dorsal branch of S 1
Q. From ventral branch of S 1 and S 2
R. From ventral branch of S 3
S. Caudal cutaneous femoral nerve
T. Lateral cutaneous sural nerve

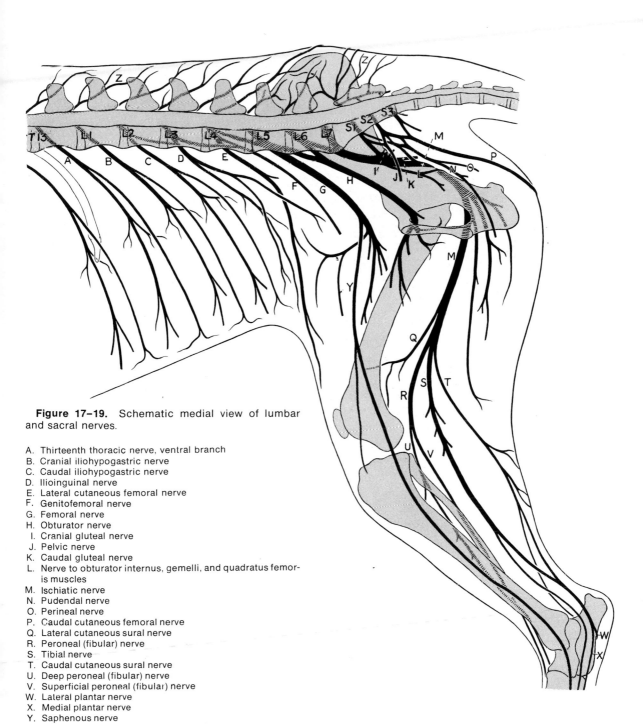

Figure 17–19. Schematic medial view of lumbar and sacral nerves.

A. Thirteenth thoracic nerve, ventral branch
B. Cranial iliohypogastric nerve
C. Caudal iliohypogastric nerve
D. Ilioinguinal nerve
E. Lateral cutaneous femoral nerve
F. Genitofemoral nerve
G. Femoral nerve
H. Obturator nerve
I. Cranial gluteal nerve
J. Pelvic nerve
K. Caudal gluteal nerve
L. Nerve to obturator internus, gemelli, and quadratus femoris muscles
M. Ischiatic nerve
N. Pudendal nerve
O. Perineal nerve
P. Caudal cutaneous femoral nerve
Q. Lateral cutaneous sural nerve
R. Peroneal (fibular) nerve
S. Tibial nerve
T. Caudal cutaneous sural nerve
U. Deep peroneal (fibular) nerve
V. Superficial peroneal (fibular) nerve
W. Lateral plantar nerve
X. Medial plantar nerve
Y. Saphenous nerve
Z. Dorsal branches of lumbar and sacral nerves

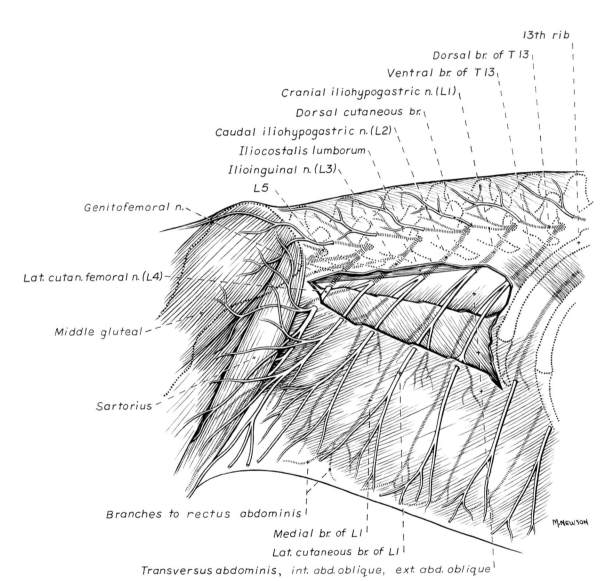

Figure 17–20.   Dissection showing the arrangement of the first four lumbar nerves, lateral aspect.

spinal canal and contain sympathetic and sensory fibers which communicate with each other and supply the dorsal longitudinal ligament, dura mater, periosteum, and blood vessels. The **lateral branches** *(rami laterales)* of the dorsal branches of the first three or four lumbar nerves are clearly separated from the medial branches. The dorsal branches of the last three or four lumbar nerves do not clearly divide into medial and lateral portions, but they arborize in the epaxial muscles of the loin. The lateral branches of the dorsal branches of the first three or four lumbar nerves run caudolaterad through the longissimus and iliocostalis muscles and perforate the iliocostalis midlaterally in a segmental manner. After continuing in the intramuscular caudolateral direction in the areolar tissue under the lumbodorsal fascia one or more centimeters, they perforate the lumbodorsal fascia and arborize in the skin of the dorsolateral parts of the lumbar and sacral regions as the **dorsal cutaneous branches** *(rami cutanei dorsales)* *(nn. clunium superiores* of N.A. terminology).

The lateral branches of the dorsal branches also supply the lumbar portion of the iliocostalis muscle. The cutaneous branches of the dorsal branches are unusually variable. Occasionally, all the lumbar nerves have these branches. In other specimens a single dorsal branch bifurcates deeply within the epaxial musculature, resulting in two cutaneous nerves which supply the skin of each side of the dorsum of the loin. This variation is similar to those encountered in the cervical region. In some specimens the dorsal cutaneous branch of either the fifth, sixth, or seventh lumbar nerve supplies the skin lying adjacent to it as well as much of that over the rump. In these specimens there is an absence of some dorsal cutaneous branches. The lumbar nerves caudal to a hyperdeveloped dorsal cutaneous branch do have dorsal muscular branches which are dissipated in the epaxial musculature without first dividing into medial and lateral parts.

The **ventral branches** *(rami ventrales)* of the seven pairs of lumbar nerves are variable, but less so than the dorsal branches. They are usually described as lumbar nerves without specifically referring to them as ventral branches. Each lumbar nerve is connected to the sympathetic trunk by a *ramus communicans,* with rare exceptions. The connections are exceedingly variable, as Mehler et al. (1952) have pointed out. In 100 dogs dissected by these investigators only 23 specimens had symmetrically located bilateral trunk ganglia at every lumbar segment. These 23 specimens also had at least two paired sacral ganglia. The rami communicantes may be double, or the rami from two adjacent nerves may go to the same ganglion. The first four or five rami contain preganglionic as well as postganglionic fibers. The remaining communicating rami contain only postganglionic and afferent fibers. The communicating rami, as they run between the lumbar nerves and the sympathetic trunk, lie largely under the psoas minor muscle. They are less than 1 mm. in diameter and about 5 mm. long. Like the brachial plexus, which gives origin to the nerves which innervate the thoracic limb, the last five lumbar nerves and all the sacral nerves are joined together to form the lumbosacral plexus *(plexus lumbosacralis),* which issues the nerves to the pelvic limb. This plexus, therefore, is divided into lumbar and sacral portions. The first two lumbar nerves are usually not joined to each other or to adjoining nerves, but run caudolaterally in the abdominal wall in series with the last several caudal thoracic nerves and are therefore not included in the lumbosacral plexus.

The **cranial and caudal iliohypogastric nerves** *(nn. iliohypogastrici craniales et caudales)* (Fig. 17–20) represent the ventral branches of the first and second lumbar nerves respectively. Each nerve is connected with the sympathetic trunk by a single ramus communicans which contains both preganglionic and postganglionic fibers. Both nerves send branches to the quadratus lumborum and the psoas minor muscles.

After leaving the caudal thoracic portion of the hypaxial musculature by passing between the two segments of the quadratus lumborum muscle, the cranial iliohypogastric nerve lies in the subserous endothoracic fascia at its origin. It then passes into the subserous transversalis fascia of the abdomen by passing dorsal to the lumbocostal arch. About 4 cm. ventrolateral to a plane through the free end of the lumbar transverse processes, the nerve, after having passed through the aponeurosis of origin of the transversus abdominis muscle, divides into a lateral and a medial branch. Before dividng, it gives branches to the serosa and to the segments of the quadratus lumborum muscle, against which it lies. It then runs between two adjacent fleshy bundles of the transversus abdominis muscle as these bundles arise from a narrow aponeurosis which attaches to the ends of the transverse processes of the lumbar vertebrae. Shortly after entering the fascia which separates the transversus abdominis from the internal abdominal oblique muscle, the iliohypogastric nerve divides into lateral and medial branches.

The **lateral branch** (*ramus lateralis*) passes through the internal abdominal oblique to run in the septum between the two abdominal oblique muscles. In its course ventrocaudally it sends most of its branches to these muscles, and near the middle of the abdomen it perforates the external abdominal oblique to become subcutaneous as the **lateral cutaneous branch** (*ramus cutaneous lateralis*). It is accompanied by a twig of the cranial abdominal artery and vein. The lateral cutaneous branch is distributed to a ventrolaterally running band of skin which crosses the junction of the cranial and middle thirds of the abdomen caudal to the ribs.

The **medial branch** (*ramus medialis*) lies closely applied to the lateral surface of the transversus abdominis muscle, where it appears in series with the last five thoracic and the second and third lumbar nerves. Like the lateral branch, it is also accompanied by a small branch of the cranial abdominal artery and vein. It supplies a band of the transversus abdominis muscle and peritoneum along its course. It ends in the first lumbar segment of the rectus abdominis muscle as well as the skin and peritoneum covering it. The ventral branch of the **second lumbar nerve** is in all respects similar to the first lumbar nerve except that it supplies the abdominal wall caudal to it and appears at the lateral border of the hypaxial musculature after having passed between the quadratus lumborum and the iliopsoas muscles at the lumbocostal arch. Occasionally, one of the iliohypogastric nerves is double with a reciprocal diminution in size of the nerve caudal to it. Rarely is a single nerve formed

by the fusion of the first and second or the second and third lumbar nerves.

The **lumbosacral plexus** (*plexus lumbosacralis*) (Fig. 17–21) consists of the intercommunicating ventral branches of the last five lumbar and the three sacral nerves. It may be divided into lumbar and sacral plexuses, although the two always have communications. There is an overlapping of contributed axons to the named nerves of the plexus so that most pelvic muscles and structures are innervated from more than one level of the spinal cord. Thus lumbar nerves 3, 4, and 5 contribute to the femoral nerve; L3 and 4 to the genitofemoral; L4, 5, and 6 to the obturator; L6 and 7 and S1 to the cranial gluteal; L6 and 7 and S1 and 2 to the sciatic and its branches; and S1, 2, and 3 to the pudendal nerves (deLahunta 1977).

Bennett (1976) studied the gross and histological features of the sciatic nerve and its branches in the cat and dog and provided a table showing details of the internal funicular structure of the peroneal and tibial nerves. His findings agree with those of Havelka (1928) and others regarding the components of the sciatic nerve.

Fletcher and Kitchell (1966) and Fletcher (1970) studied the cutaneous innervation of the pelvic limb and found variation in the spinal segments involved. Mehler et al. (1952) found that the communicating rami and sympathetic roots were highly variable, particularly in the lumbar segments. In 65 per cent of the 100 animals they dissected, L4 was the lowest from which communicating rami originated. The remaining 35 per cent of animals had communicating rami that arose from spinal nerves L5 and L6.

The **lumbar plexus** (*plexus lumbalis*) is a term ordinarily restricted to the interconnected third, fourth, and fifth lumbar nerves (Havelka 1928). In some specimens the third lumbar nerve is connected to the second by a fine branch (Bradley and Grahame 1959), and usually the fifth lumbar nerve is connected to the sixth. The division of the lumbosacral plexus into its two components is made primarily because of the location of the plexuses and not its origin. The connection between the third and fourth lumbar nerves gives origin to the genitofemoral branch whereas the connection between the fourth and fifth is usually devoid of branches. The morphology of the lumbar

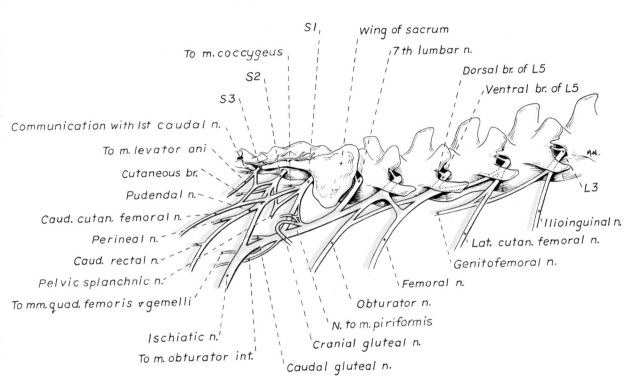

Figure 17–21. Diagram of the lumbosacral plexus, lateral aspect.

nerves is particularly variable. The lumbar plexus may be moved cranially one segment from the normal (prefixed) or caudally one segment from the normal (postfixed).

The **ilioinguinal nerve** (*n. ilioinguinalis*) (Fig. 17–20) is the direct ventrolateral continuation of the third lumbar nerve. Its ramus communicans contains both preganglionic and postganglionic fibers. It communicates with the fourth lumbar nerve and divides into medial and lateral branches which resemble those of the lumbar nerves which precede it. Cutaneous branches from its lateral portion extend caudoventrally superficial to the lateral cutaneous femoral nerve and ramify in the skin of the craniolateral surface of the thigh. The medial branch of the ilioinguinal nerve is small. It runs more caudally than ventrolaterally. It is accompanied by a cranial branch of the ascending branch of the deep circumflex iliac vessels. It usually has no grossly demonstrable cutaneous branches.

The **lateral cutaneous femoral nerve** (*n. cutaneus femoris lateralis*) is formed primarily by the ventral branch of the fourth lumbar nerve, although there are connections with both the third and fifth lumbar nerves. It is larger than the lumbar nerves which precede it, but smaller than those which follow it. As it runs caudolaterad through the substance of the psoas minor muscle, it sends branches to it and to the other hypaxial muscles of the region. According to Mehler et al. (1952), the fourth lumbar nerve is the lowest nerve from which a communicating ramus arises. The main portion of the nerve passes through the abdominal wall, lying between the deep circumflex iliac artery cranially and the satellite vein caudally. It passes between the lumbar and inguinal portions of the internal abdominal oblique and over the dorsal margin of the external abdominal oblique muscle. Its terminal cutaneous branches are variable, but in general they follow the accompanying vessels. A branch ramifies in the skin in the region of the tuber coxae and the adjacent cranial part of the rump; other branches supply the skin over the cranial portion of the thigh, while its longest branch runs distally, supplying the skin over the thigh and the lateral surface of the stifle joint. Kunzel (1957) illustrates this

nerve in his work on the topography of the hip joint.

The **genitofemoral nerve** (*n. genitofemoralis*) (Fig. 17–22) is called the external spermatic nerve in many veterinary publications. The dog lacks the separate femoral branch found in man, but does have a cutaneous ramus from the genital branch which supplies the skin of the pudendal region. The genitofemoral nerve arises from the third and fourth lumbar nerves, the root from the third being larger than that from the fourth. Bradley and Grahame (1959) state that the contribution from the fourth may be absent. Ellenberger and Baum (1891) stated that occasionally the nerve is double, and in some specimens the ilioinguinal nerve joins the genital and is distributed with it. Usually the nerve is single, small, and long. It is formed in the substance of the medial portion of the iliopsoas muscle near the body of the fourth lumbar vertebra. As it runs caudally, it leaves the substance of the medial part of the muscle and continues caudally in the fat which fills the irregularities around the caudal vena cava and aorta. After passing dorsal to the external iliac lymph node to which it sends a small branch, it becomes related to the distal portion of the external iliac artery, which it crosses medially. It leaves the abdomen by passing through the inguinal canal, where it lies medial to the spermatic cord in the male or associated with the round ligament of the uterus in the female. Minute muscular branches are given off to the cremaster muscle and the spermatic cord. Upon passing through the external inguinal ring, it goes to the skin of the scrotum and prepuce in the male or to the inguinal mammary gland and its covering skin in the female. In both sexes the terminal branch of the nerve runs caudolaterally and distally and supplies a zone of skin on the caudomedial surface of the thigh.

The **femoral nerve** (*n. femoralis*) (Fig. 17–19) arises primarily from the fifth segment of the lumbar plexus with a strong root of origin also coming from the fourth. A smaller branch of origin may come from the third, but rarely does one come from the sixth segment. After being formed in the substance of the iliopsoas muscle, it continues caudally in the substance of this muscle and

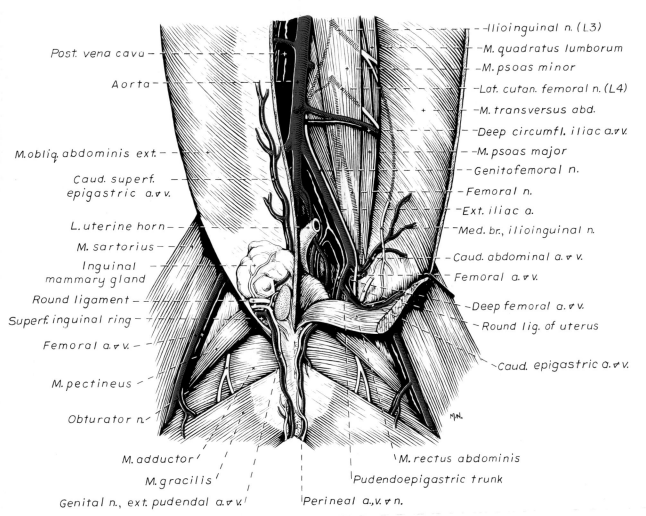

Post. vena cava —

Aorta —

M. obliq. abdominis ext. —

Caud. superf.
epigastric a. v. v. —

L. uterine horn —

M. sartorius —

Inguinal
mammary gland —

Round ligament —

Superf. inguinal ring —

Femoral a. v v. —

M. pectineus —

Obturator n. —

— Ilioinguinal n. (L3)

— M. quadratus lumborum

— M. psoas minor

— Lat. cutan. femoral n. (L4)

— M. transversus abd.

— Deep circumfl. iliac a. v. v.

— M. psoas major

— Genitofemoral n.

— Femoral n.

— Ext. iliac a.

— Med. br., ilioinguinal n.

— Caud. abdominal a. v v.

— Femoral a. v v.

— Deep femoral a. v v.

— Round lig. of uterus

— Caud. epigastric a. v v.

M. adductor

M. gracilis

Genital n., ext. pudendal a. v v.

M. rectus abdominis

Pudendoepigastric trunk

Perineal a., v. v n.

Figure 17–22.   Dissection showing course of the genital nerve in the female, ventral aspect. (The left abdominal wall is reflected.)

leaves the abdomen along with the muscle. During its intra-abdominal course, it sends muscular branches to the iliopsoas. The prominent saphenous nerve arises from its cranial side. Shortly thereafter the femoral nerve enters the quadriceps femoris muscle by passing between the rectus femoris and the vastus medialis muscles at the proximal end of the cleft which separates these heads. It supplies all four heads of the quadriceps muscle (rectus femoris, vastus medialis, vastus intermedius, and vastus lateralis) and also sends a small twig to the articularis coxae muscle. The branches to the various parts of the quadriceps largely accompany the cranial femoral artery, which is its chief source of blood supply proximally. No cutaneous branches arise from the femoral nerve, nor can a branch be found going to the hip joint.

The **saphenous nerve** *(n. saphenus)* (Figs. 17–23, 17–30) is the only superficial branch of the femoral nerve. Arising from the femoral before this nerve leaves the iliopsoas, the saphenous becomes related to the medial surface of the tensor fasciae latae muscle and immediately divides into muscular and cutaneous branches. This division is lacking in many specimens, since the muscular branch arises from the femoral nerve either proximal or distal to the origin of that part of the saphenous nerve which is cutaneous. The **muscular branch** *(ramus muscularis)* bifurcates, one twig going to the cranial belly of the sartorius muscle and the other to the caudal belly. These nerve branches accompany the blood vessels serving the proximal part of the sartorius muscle. The **cutaneous branch** *(ramus cutaneus)* of the saphenous nerve is long and slender. It lies in apposition to the cranial surface of the femoral artery as it runs distally across the medial surface of the quadriceps muscle. It sends branches to the skin of the medial surface of the thigh. Proximal to the stifle, a small nerve accompanies the descending genicular vessels to the deep structures of the medial surface of the stifle, and a cutaneous branch accompanies the medial genicular vessels to the skin supplied by this vessel. Distal to the stifle joint, the saphenous nerve continues with the dorsal branch of the saphenous vessels and supplies twigs to the skin along its course to the paw. A small cutaneous branch also follows the plantar branch of the saphenous artery and vein and supplies the skin which overlies them. In the paw it supplies the skin of the dorsomedial part of the tarsus and metatarsus and ends in the skin over the second and first digits when a first digit is present.

The **obturator nerve** *(n. obturatorius)* (Figs. 17–23, 17–24) arises from the fourth, fifth, and sixth lumbar nerves (Langley and Anderson 1896). The sixth root of origin is usually heaviest, and the fourth smallest or even absent. The roots of origin do not arise close to the intervertebral foramina, but from the main nerve trunks (sciatic for the sixth and the femoral for the fifth) which are the main continuations of the fifth and sixth lumbar nerves. The obturator nerve is formed within the caudomedial portion of the iliopsoas muscle. It leaves this muscle dorsomedially and, after crossing the ventrally lying common iliac vein, enters the subserosa of the pelvis. After running obliquely caudoventrad across the laterally lying shaft of the ilium, it disappears from view by first passing between the iliopubic and ischial portions of the levator ani muscle. The obturator nerve leaves the pelvis by passing through the cranial part of the obturator foramen, where it lies in relation to the small obturator ramus of the deep femoral artery which ascends through the opening. It sends branches into the external obturator muscle and continues through the cranial part of the obturator foramen. Thereafter it sends branches into the pectineus, gracilis, and adductor muscles, innervating all the muscles which primarily adduct the pelvic limb.

SACRAL NERVES

The **sacral nerves** *(nn. sacrales)* (Figs. 17–21, 17–23) leave the three sacral segments of the spinal cord by means of long dorsal and ventral roots, since these segments form that part of the conus medullaris which lies opposite the sixth lumbar vertebra. The roots merge to form the sacral nerves within the sacral canal prior to their intervertebral foramina of exit. Each of the first two sacral nerves then divides into dorsal and ventral branches which leave the sacrum through the first two dorsal and pelvic sacral foramina, respectively. The third pair of sacral

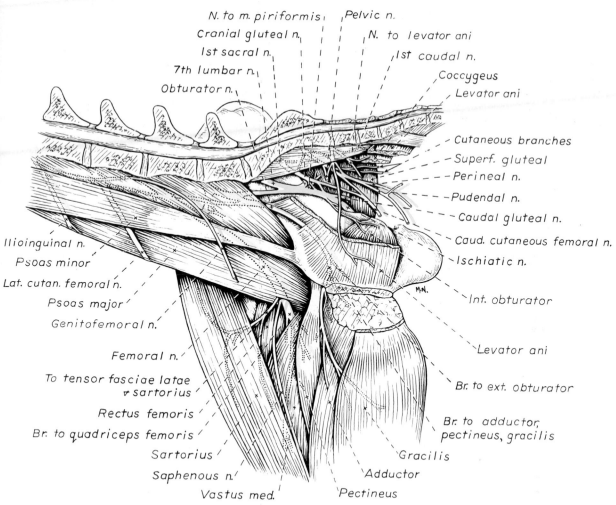

N. to m. piriformis
Cranial gluteal n.
1st sacral n.
7th lumbar n.
Obturator n.

Pelvic n.
N. to levator ani
1st caudal n.
Coccygeus
Levator ani

Cutaneous branches
Superf. gluteal
Perineal n.
Pudendal n.
Caudal gluteal n.
Caud. cutaneous femoral n.
Ischiatic n.

Ilioinguinal n.
Psoas minor
Lat. cutan. femoral n.
Psoas major
Genitofemoral n.

Femoral n.
To tensor fasciae latae v sartorius
Rectus femoris
Br. to quadriceps femoris
Sartorius
Saphenous n.
Vastus med.

Int. obturator
Levator ani
Br. to ext. obturator
Br. to adductor, pectineus, gracilis
Gracilis
Adductor
Pectineus

Figure 17–23. Dissection showing distribution of the femoral and obturator nerves, medial aspect.

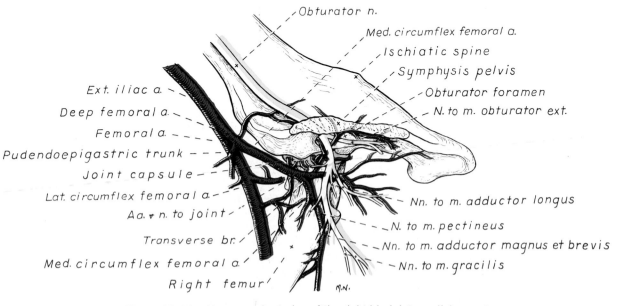

Obturator n.
Med. circumflex femoral a.
Ischiatic spine
Symphysis pelvis
Obturator foramen
N. to m. obturator ext.

Ext. iliac a.
Deep femoral a.
Femoral a.
Pudendoepigastric trunk
Joint capsule
Lat. circumflex femoral a.
Aa. v n. to joint
Transverse br.
Med. circumflex femoral a.
Right femur

Nn. to m. adductor longus
N. to m. pectineus
Nn. to m. adductor magnus et brevis
Nn. to m. gracilis

Figure 17–24. Nerves and arteries of the right hip joint, medial aspect.

nerves leaves the spinal canal by passing through the intervertebral foramina between the sacrum and first caudal vertebra like the other typical spinal nerves.

The larger **ventral branches** (*rami ventrales*), after leaving the confines of the sacral canal, are connected to the sympathetic trunk by single rami communicantes. They then continue on the medial wall of the pelvis.

The **dorsal branches** (*rami dorsales*) of the three sacral nerves leave the two dorsal sacral foramina and the intervertebral foramen between the sacrum and the first caudal vertebra. The dorsal branches are connected to each other by communicating strands forming a small **dorsal sacral trunk** or **plexus**. This trunk is usually joined to the dorsal branches of the last lumbar and first caudal nerves. They decrease in size from first to third and like the dorsal branches of the lumbar nerves are divided basically into medial muscular branches and lateral cutaneous branches. The short medial branches supply the lateral and medial dorsal sacrocaudal muscles. The lateral branches are longer. They run caudodorsolaterad between the dorsal lateral sacrocaudal and dorsal intertransverse coccygeal muscles to reach the deep gluteal fascia through which they pass, usually accompanied by the corresponding dorsal sacral vessels. The single lateral branch of the dorsal branch of the first sacral nerve bifurcates upon passing through the gluteal fascia so that two cutaneous nerves run caudoventrad on the superficial gluteal muscle and down the thigh. The second and third sacral nerves are similarly disposed, supplying bands of skin caudal to the first. These three nerves, the **middle clunial nerves** (*nn. clunii medii*) of Ellenberger and Baum (1943), supply a zone of skin which reaches to the middle of the lateral surface of the thigh before ending in a tapering cutaneous zone to communicate with each other and with the last two lumbar nerves to form the sacral plexus. The middle clunial nerves also supply the piriformis muscle (Henning 1965).

The **sacral plexus** (*plexus sacralis*) is formed by the large interconnected ventral branches of the last two lumbar nerves and the ventral branches of the three small sacral nerves which divide and redivide shortly after leaving the spinal canal, with the resultant branches uniting with adjacent nerves to

form the sacral plexus. The last two lumbar nerves run caudoventrolaterad, converge, and blend under cover of the lateral ventral sacrocaudal muscle on the pelvic surface of the sacrum medial to the sacroiliac joint. Two slender branches leave the cranial surface of the sixth nerve before it joins the seventh. The first branch joins the fifth lumbar nerve and augments the size of its continuation, the femoral nerve. The second branch unites with a like branch of the fifth to form the obturator nerve.

The **lumbosacral trunk** (*truncus lumbosacralis*) (Fig. 17–25) is the largest and most important part of the lumbosacral plexus, since it is continued outside the pelvis as the sciatic nerve. It arises primarily from the sixth and seventh lumbar nerves with a small contribution from the first and occasionally the second sacral nerves (Havelka 1928), and it has a root of origin from the fifth lumbar, according to Ellenberger and Baum (1891). The lumbosacral trunk has two medium-sized branches, the cranial and caudal gluteal nerves. The trunk lies in the pelvic fascia while crossing the shaft of the ilium in nearly a frontal plane where it is sandwiched in the space between the origin of the ventral lateral sacrocaudalis muscle medially and the thin levator ani muscle laterally. It is covered by peritoneum and is crossed obliquely on its ventral aspect by the parietal branches of the internal iliac vessels. It becomes the sciatic nerve after the last sacral branch enters it at the greater ischiatic foramen. Each of the five constant, spinal roots of the sacral plexus is usually united to the sympathetic trunk by a single ramus communicans.

The **cranial gluteal nerve** (*n. gluteus cranialis*) arises from the lumbosacral trunk or its roots mainly from the sixth and seventh lumbar nerves and from the first sacral nerves. It leaves the pelvis by passing immediately through the greater sciatic foramen and plunges into the muscles of the rump. It is accompanied by the cranial gluteal artery and vein. It circles craniad across the lateral aspect of the shaft of the ilium at the origin of the caudal bundle of the deep gluteal muscle. The cranial gluteal nerve continues cranioventral between the middle and deep gluteal muscles, usually perforates the cranial edge of the deep gluteal, and terminates in the tensor fasciae latae muscle. It supplies

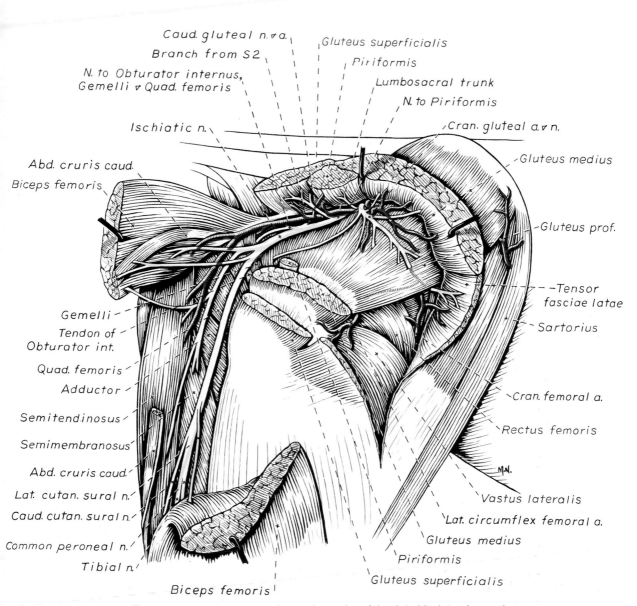

Caud. gluteal n. ᵥ a.
Branch from S2
N. to Obturator internus,
Gemelli ᵥ Quad. femoris
Gluteus superficialis
Piriformis
Lumbosacral trunk
N. to Piriformis
Ischiatic n.
Cran. gluteal a. ᵥ n.
Abd. cruris caud.
Biceps femoris
Gluteus medius
Gluteus prof.
Gemelli
Tendon of
Obturator int.
Quad. femoris
Adductor
Tensor
fasciae latae
Sartorius
Semitendinosus
Semimembranosus
Abd. cruris caud.
Lat. cutan. sural n.
Caud. cutan. sural n.
Common peroneal n.
Tibial n.
Cran. femoral a.
Rectus femoris
M.N.
Vastus lateralis
Lat. circumflex femoral a.
Gluteus medius
Piriformis
Gluteus superficialis
Biceps femoris

Figure 17–25. Nerves, arteries, and muscles of the right hip, lateral aspect.

the deep and middle gluteal and tensor fasciae latae muscles. It has no cutaneous branches.

The **caudal gluteal nerve** (*n. gluteus caudalis*) is a small nerve which may be double. It usually arises from the cranial margin of the lumbosacral trunk or from its seventh lumbar root. Occasionally the cranial and caudal gluteal nerves arise in common; at the other extreme the caudal gluteal nerve may arise only from the first and second sacral nerves independently of the lumbosacral trunk, as Bradley and Grahame (1959) illustrate. The caudal gluteal nerve runs parallel to the ventrocranial border of the lumbosacral trunk on the medial surface of the shaft of the ilium, and after passing through the greater sciatic foramen, it crosses the caudal border of the piriformis muscle or passes between the piriformis and middle gluteal muscles to enter the medial surface of the superficial gluteal muscle. It may also send a branch to the middle gluteal muscle. In its course of about 2.5 cm., it lies between the sacrotuberous ligament medially and the large caudal gluteal artery and vein laterally. It is distributed to the superficial gluteal muscle. It also supplies the piriformis muscle by a delicate branch which enters the proximal third of the muscle.

The **caudal cutaneous femoral nerve** (*n. cutaneus femoris caudalis*) (Figs. 17–26, 17–27) is nearly as large as the pudendal nerve, to which it is united for most of its intrapelvic course. It arises from the first and second sacral nerves and seldom from the third sacral (Havelka 1928). Bradley and Grahame (1959) state that it arises from the seventh lumbar and first sacral nerves with a possible addition from the sixth lumbar. As it passes out of the pelvis dorsal to the ischial arch, it divides into perineal branches and the caudal clunial nerves. The **perineal branches** (*rami perineales*) accompany the perineal vessels to the skin around the anus as several small twigs. The **caudal clunial nerves** (*nn. clunii caudales*) stream out of the ischiorectal fossa in the fat covering the dorsal surface of the internal obturator muscle, usually as three nerves, one adjacent to the proximal part of the penis or labia, one over the semitendinosus and semimembranosus muscles, and one in the furrow between the biceps femoris and semimembranosus muscles. These nerves run distally and supply the

skin of the proximal half of the caudal and the adjacent medial and lateral surfaces of the thigh.

The **pudendal nerve** (*n. pudendus*) (Fig. 17–23) usually arises from all three sacral nerves. A prefixed origin of the pudendal by one segment is described by Havelka (1928). Hummel (1965) says it may originate as far cranially as L5. As the pudendal nerve runs obliquely caudoventrad to the pelvic outlet, it lies lateral to the coccygeus muscle and appears superficially medial to the superficial gluteal muscle. It lies dorsal to the accompanying internal pudendal vessels. At the pelvic outlet the pudendal nerve gives rise to the caudal rectal and perineal nerves and the nerves to the external genital organs: the dorsal nerve of the penis of the male and the nerve of the clitoris in the female. The perineal nerve of both sexes continues ventrocranially and ends as the caudal scrotal nerves of the male or the caudal labial nerves of the female.

The **caudal rectal nerve** (*n. rectalis caudalis*) is a short nerve which leaves the pudendal at the caudal border of the levator ani muscle and enters the external anal sphincter muscle a little below the middle. It is accompanied by the more deeply lying caudal rectal artery and vein. It, and its companion nerve of the opposite side, are apparently the sole supply to the external sphincter muscle of the anus. Blakely (1957) found that unilateral severance of this nerve in perineal hernia operations could be performed without producing incontinence, but that bilateral severance of the caudal rectal nerve resulted in incontinence of feces.

The **perineal nerves** (*nn. perinei*) (Fig. 17–27) are represented by several long twigs which supply the skin and the mucous membrane in the region of the anus. The first branch to leave the pudendal nerve goes to the mucosa of the anal canal and the skin at the anus. It either crosses the external anal sphincter superficially or runs deeply between the dorsal portion of the anal sac and the internal sphincter muscle of the anus to reach the mucosa. The main portion of the perineal nerve leaves the pudendal near or in common with the nerve going to the mucosa and sends several branches to the skin of the perineum as it runs distally in the furrow located between the proximal part of the penis and the buttocks. One of these

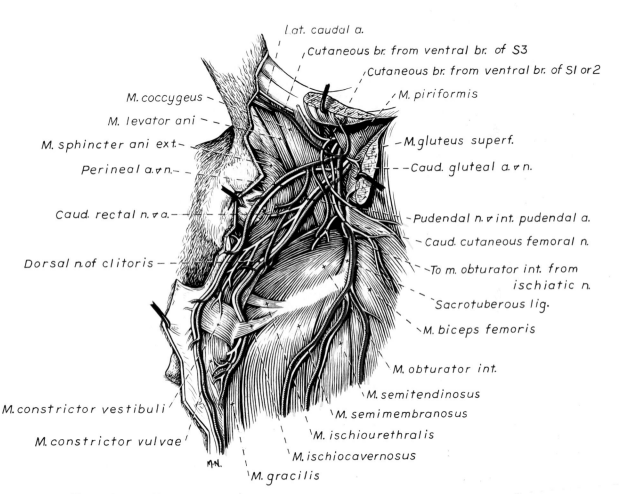

Lat. caudal a.

Cutaneous br. from ventral br. of S3

Cutaneous br. from ventral br. of SI or 2

M. piriformis

M. coccygeus

M. levator ani

M. sphincter ani ext.

Perineal a. & n.

M. gluteus superf.

Caud. gluteal a. & n.

Caud. rectal n. & a.

Pudendal n. & int. pudendal a.

Caud. cutaneous femoral n.

To m. obturator int. from ischiatic n.

Dorsal n. of clitoris

Sacrotuberous lig.

M. biceps femoris

M. obturator int.

M. semitendinosus

M. constrictor vestibuli

M. semimembranosus

M. ischiourethralis

M. constrictor vulvae

M. ischiocavernosus

M. gracilis

Figure 17–26. Nerves, arteries, and muscles of the female perineum, caudolateral aspect.

branches, larger than the others, becomes related to the root of the penis and as a branched nerve runs distally on the caudolateral surface of the proximal portion of the penis. The main perineal nerve, however, continues distally and supplies the skin of the escutcheon with its fellow. The terminal portion of this nerve supplies the skin of the scrotum as the **caudal scrotal nerve** (*n. scrotalis caudalis*). In the female the comparable nerve supplies the skin of the labia as the **caudal labial nerve** (*n. labialis caudalis*). The terminal portion of this nerve is not confined to innervating the skin of the caudal portions of these external genital parts, but continues cranial to them and is finally dissipated in the skin of the inguinal region, where its area of skin supply is overlapped by the branches from the genitofemoral nerve, which passes through the inguinal canal.

The **dorsal nerve of the penis** (*n. dorsalis penis*) in the male is the main extrapelvic continuation of the pudendal nerve. It curves around the pelvic outlet near the symphysis ischii, where it is separated from the opposite nerve by the paired artery and vein of the penis. At this place it issues a thin but long branch which inclines medially where it comes to lie between the dorsal vein and the tunica albuginea of the penis. It usually communicates with the larger nerve just caudal to the bulbus glandis. Upon reaching the dorsal surface of the penis, the dorsal nerve of the penis runs cranially on the organ, sending branches to it, and then enters the caudal part of the bulbus glandis. It continues through the glans along the dorsal surface of the os penis and finally ends in the mucosa of the apex of the glans. It is the chief sensory nerve to the penis, mediating afferent impulses which result in orgasm. The **dorsal nerve of the clitoris** (*n. dorsalis clitoridis*) in the female is the homologue of the dorsal nerve of the penis of the male and mediates similar impulses. It is much smaller than the comparable nerve of the male and runs to the ventral commissure of the vulva, where it ends in the clitoris.

The **muscular branches** of ventral sacral nerves (*rami musculares*) are usually two in number. One supplies the levator ani and coccygeus muscles, and the second, larger nerve innervates the rotators of the hip joint. Coming mainly from the second sacral

nerve, but also receiving a small twig from the third, usually, is a small single or double nerve which crosses the lateral surface of the lateral ventral sacrocaudal muscle. Its medial branch arborizes largely on the medial, subperitoneal surface of the levator ani muscle. Its lateral branch enters the cranial, medial surface of the much narrower, more caudally located coccygeus muscle. The muscular branch from the lumbosacral trunk to the muscles which rotate the hip joint was called the **rotator nerve** (*n. rotatorius*) by Schmaltz (1914). It is short. As it leaves the caudal margin of the lumbosacral trunk just prior to the passing of this trunk through the caudal part of the greater ischiatic foramen to be continued as the sciatic nerve, a twig runs caudally and arborizes in the dorsal surface of the internal obturator muscle as it lies on the ischiatic table caudomedial to the lesser ischiatic foramen. When the rotator nerve is double, the second branch leaves the caudal border of the lumbosacral trunk or its continuation, the sciatic nerve, about 1 cm. distal to the origin of the branch to the internal obturator, and curves around the caudal border of the deep gluteal muscle and sinks into the fascia between the deep gluteal and the cranial gemellus muscles. Sometimes the two parts of the rotator nerve arise in common. After crossing the lateral surface of the shaft of the ischium the second branch supplies both parts of the gemelli and after passing ventral to the cranial gemellus terminates in the larger, obliquely running quadratus femoris muscle.

The **sciatic** or **ischiatic nerve** (*n. ischiadicus*) (Figs. 17–25, 17–28) is the largest nerve in the body. It is a continuation of the lumbosacral trunk. The division between the two is marked by the second sacral nerve, contributing to this nerve complex. Since this contribution is located at the greater ischiatic foramen, the extrapelvic part of the trunk is regarded as the sciatic nerve. It consists of two nerves, the tibial and fibular, which are so closely bound together, proximally, that they appear as one. The two parts, however, can be forcefully separated back to their origins from the spinal nerves. The normal division of the sciatic nerve is variable. Occasionally it is located as far proximally as the hip joint, while at other times it may be as far distally as the popliteal space. Upon leaving the pelvis, the sciatic nerve first lies on the

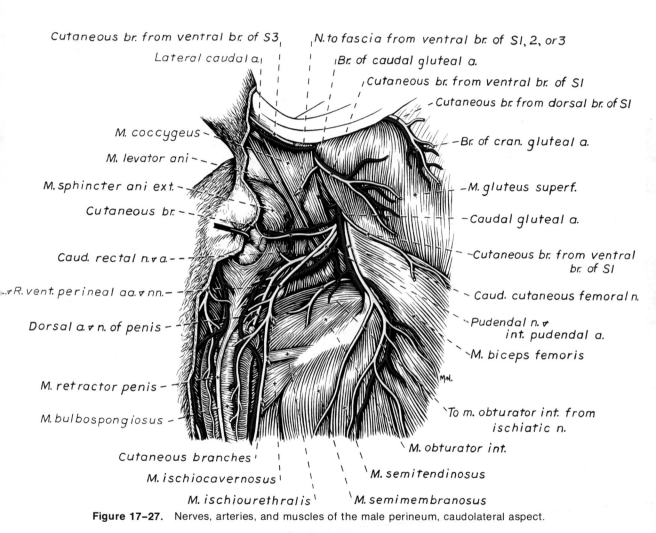

Cutaneous br. from ventral br. of S3

Lateral caudal a.

N. to fascia from ventral br. of SI, 2, or 3

Br. of caudal gluteal a.

Cutaneous br. from ventral br. of SI

Cutaneous br. from dorsal br. of SI

M. coccygeus

M. levator ani

M. sphincter ani ext.

Cutaneous br.

Caud. rectal n. v a.

R. vent. perineal aa. v nn.

Dorsal a. v n. of penis

M. retractor penis

M. bulbospongiosus

Cutaneous branches

M. ischiocavernosus

M. ischiourethralis

Br. of cran. gluteal a.

M. gluteus superf.

Caudal gluteal a.

Cutaneous br. from ventral br. of SI

Caud. cutaneous femoral n.

Pudendal n. v int. pudendal a.

M. biceps femoris

To m. obturator int. from ischiatic n.

M. obturator int.

M. semitendinosus

M. semimembranosus

Figure 17–27. Nerves, arteries, and muscles of the male perineum, caudolateral aspect.

gemelli and the tendon of the internal obturator. As it passes down the thigh, it lies in succession on the quadratus femoris, adductor, and semimembranosus muscles. It is covered first by the superficial gluteal muscle and then by the biceps femoris muscle and lies in close association with the small abductor cruris caudalis muscle, which crosses it obliquely in the proximal third of the thigh. It proximal portion is accompanied by the caudal gluteal vessels which lie caudal to it, and it is nourished by the small sciatic artery, which is partly embedded in its caudolateral surface. In addition to the rotator nerve, which is usually double and arises from the lumbosacral trunk, the true sciatic nerve which continues this trunk outside the pelvis has a single, stout, muscular branch to the hamstring muscles and a muscular branch to the abductor cruris caudalis. Other main branches of the sciatic nerve include the lateral cutaneous sural nerve, the caudal cutaneous sural nerve, and the terminal peroneal and tibial nerves. (See Bennett 1976 for origins of the peroneal and tibial nerves and details of their internal funicular structure.)

The *ramus muscularis* of the sciatic nerve is about 1 cm. in length and 3 mm. in diameter. It leaves the caudomedial border or peroneal portion of the sciatic nerve opposite the space between the middle gluteal and the cranial gemellus muscles. It lies cranial and parallel to the sacrotuberous ligament and caudal gluteal artery and vein. After running about a centimeter in the furrow (trochanteric fossa) medial to the trochanter major, it sends a branch into the larger superficial portion of the biceps femoris muscle about 3 cm. from its main origin on the tuber ischiadicum. The remaining portion of the nerve, which is approximately the same size as the branch to the superficial portion of the biceps muscle, runs distally and at the distal border of the quadratus femoris muscle branches into usually four parts which arise variably. The most caudal portion regularly bifurcates into a smaller, distolaterally running branch which supplies the smaller, deeper portion of the biceps femoris muscle and by a shorter, larger branch which innervates the proximal part of the semitendinosus muscle. The main portion of the muscular branch then continues distally and divides into two or three

branches. The more caudal branch enters the middle portion of the semitendinosus muscle distal to the tendinous intersection which partly divides the muscle. The more cranial part of the nerve runs distally and again bifurcates; one branch goes to the caudal belly and the other to the cranial belly of the semimembranosus. A long, slender, mixed nerve arises from the fibular portion of the sciatic nerve opposite or distal to the trochanteric fossa. After obliquely crossing the caudal surface of the tibial part, it becomes associated with the medial border of the abductor cruris caudalis muscle. In this region it sends a muscular branch to this small abductor. At about the middle of the thigh, it leaves the medial border of the abductor cruris caudalis muscle and obliquely crosses the caudal surface of the muscle as it runs distally. It becomes subcutaneous in the proximal part of the popliteal region where it supplies a small area of skin.

The **pelvic nerves** (*nn. pelvini*) (Figs. 17–21, 17–23) are the pelvic part of the parasympathetic portion of the autonomic nervous system. They usually arise from the first and second sacral nerves. These two nerves may anastomose to form a single pelvic splanchnic nerve or may run independently to the pelvic plexus.

The **lateral cutaneous sural nerve** (*n. cutaneus surae lateralis*) arises from the lateral surface of the fibular portion of the sciatic nerve at about the junction of the middle and distal thirds of the thigh. After running a few centimeters distally under the biceps femoris, it enters the muscle between its smaller, deep head and its larger, more cranially lying, superficial head. The nerve passes through the biceps femoris muscle without contributing to its supply and appears subcutaneously with cutaneous twigs of the caudal femoral vessels in the proximal, lateral portion of the crus. It supplies most of the skin on the lateral aspect of the crus as it ends near the tarsus.

The **caudal cutaneous sural nerve** (*n. cutaneus surae caudalis*) (Fig. 17–29) is known as the n. suralis s. communicans tibialis s. cutan. fem. et tibiae post. long. by Ellenberger and Baum (1891). It is called the n. cutaneus surae medialis by Bradley and Grahame (1959). It resembles this nerve of man more closely than any other nerve, but

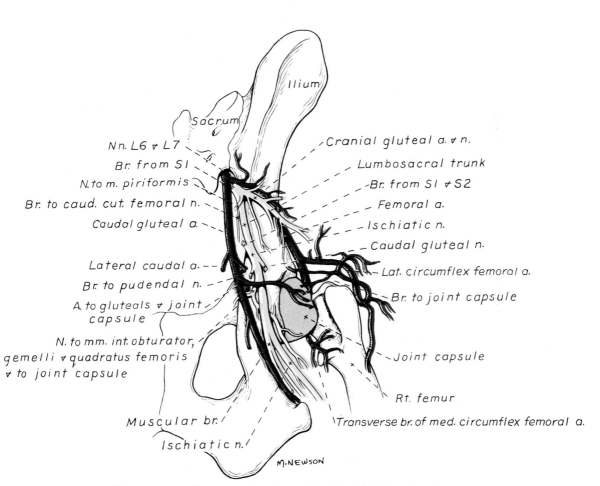

Nn. L6 & L7
Br. from S1
N. to m. piriformis
Br. to caud. cut. femoral n.
Caudal gluteal a.
Lateral caudal a.
Br. to pudendal n.
A. to gluteals & joint capsule
N. to mm. int. obturator, gemelli & quadratus femoris & to joint capsule
Muscular br.
Ischiatic n.

Ilium
Sacrum
Cranial gluteal a. & n.
Lumbosacral trunk
Br. from S1 & S2
Femoral a.
Ischiatic n.
Caudal gluteal n.
Lat. circumflex femoral a.
Br. to joint capsule
Joint capsule
Rt. femur
Transverse br. of med. circumflex femoral a.

M. NEWSON

Figure 17–28. Nerves and arteries of the right hip joint, dorsal aspect.

the differences in location, area of supply, and communication warrant the name given it. It is a long, slender nerve which arises from the caudal border of the tibial portion of the sciatic nerve or from the tibial nerve directly, usually about a centimeter distal to the origin of the lateral cutaneous sural nerve. It runs a short distance distally, bounded laterally by the biceps femoris muscle and medially by the semimembranosus muscle. Upon entering the popliteal region, it becomes associated with the plantar surface of the gastrocnemius muscle, running distally on the muscle near the fusion of its two heads and in association with the lateral saphenous vein. Throughout its course it sends branches to the skin of the caudal part of the crus. Upon reaching the calcanean tendon, it usually divides into two branches of unequal size. The smaller branch extends distally. Upon reaching the tarsus, it usually bifurcates into a small **articular branch** (*ramus articularis*) which runs over the caudal part of the lateral malleolus and supplies the lateral portion of the tarsal joint. The **lateral calcaneal branch** (*ramus calcanei lateralis*) crosses the distolateral surface of the tuber calcanei and ends in the skin of this region. A twig may innervate the tibiotarsal joint capsule. The larger branch of the caudal cutaneous sural nerve runs mediad between the calcanean tendon and the tibia and joins the tibial nerve 1 to 3 cm. proximal to the tarsal canal.

The **common peroneal (fibular) nerve** (*n. peroneus [fibularis] communis*) (Figs. 17–29, 17–30) is the smaller of the two terminal branches of the sciatic. It lies under the thin terminal part of the deep portion of the biceps femoris muscle. The nerve runs almost directly distad, obliquely crossing the lateral head of the gastrocnemius muscle. At the level of the stifle joint, it sends an articular branch to the lateral collateral ligament. Upon reaching the thin lateral border of the flexor hallucis longus muscle about 1.5 cm. distal to the stifle joint, it dips between this muscle and the lateral digital extensor on the caudal side, and the peroneus longus which lies cranial to it, and enters the muscles of the cranial part of the crus. The common peroneal nerve supplies a small branch to the peroneus longus muscle before dividing into superficial and deep peroneal nerves.

The **superficial peroneal (fibular) nerve** (*n. peroneus [fibularis] superficialis*) leaves the lateral portion of the parent nerve about 3 cm. distal to the stifle joint, where it lies in the intermuscular septum between the flexor hallucis longus muscle caudally and the peroneus longus muscle cranially. It supplies the peroneus brevis and the lateral digital extensor. As it extends distally, it curves under the distal part of the belly of the peroneus longus muscle to enter the septum between this muscle and the long digital extensor. At the beginning of the distal third of the crus, it becomes subfascial, and upon approaching the flexor surface of the tarsus, it perforates the crural fascia to become subcutaneous. At or just proximal to the tarsus it becomes related to the small, dorsal division of the saphenous artery, which it follows into the dorsum of the paw. Like the artery, it bifurcates on the dorsal surface of the proximal end of the paw into medial and lateral parts, the lateral branch being the larger. The lateral branch divides again, and the medial branch may also redivide when a first digit is present. In this manner the dorsal common digital nerves are formed. (See the explanation of the naming of the superficial and deep nerves of the digits under "Nerves of the Forepaw."

The **dorsal common digital nerves II, III, and IV** (*nn. digitales dorsales communes II, III, et IV*) (Fig. 17–31) run distally in the superficial fascia and send many delicate branches to the skin of the dorsum and sides of the metatarsus. They lie over but not in the grooves between adjacent metatarsal bones, where they are related to the corresponding arteries and deep to the corresponding veins. Upon approaching the metatarsophalangeal junctions, each nerve sends many rather large cutaneous branches to the skin of the distal portion of the metatarsus and the dorsal proximal portions of the digits. Each of the three dorsal common digital nerves receives the much smaller dorsal metatarsal nerves at the distal ends of the intermetatarsal spaces to form the several dorsal proper digital nerves.

The **deep peroneal nerve** (*n. peroneus profundus*) arises as the cranial terminal branch of the sciatic nerve on the lateral head of the gastrocnemius muscle near its cranial border. It sinks in the proximal part of the crus with the distally lying superficial peroneal nerve by passing between the

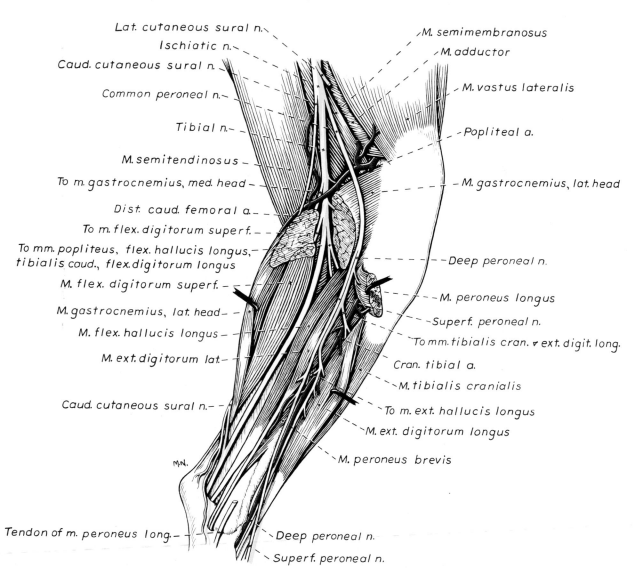

Figure 17–29. Nerves, arteries, and muscles of the right leg, lateral aspect. (The biceps femoris and abductor cruris caudalis muscles have been removed.)

deep digital flexor and lateral digital extensor muscles caudally and the peroneus longus muscle cranially. From its first 3 cm. there arise in succession usually four branches. The most proximal branch is a **muscular ramus** (ramus muscularis) to the deep face, proximal end of the peroneus longus muscle. The next branch crosses under the long digital extensor muscle and enters the medial border of the tibialis cranialis muscle. The third branch obliquely crosses the cranial surface of the delicate extensor hallucis longus muscle which it supplies and becomes related to the lateral surface of the cranial tibial artery. As it runs to the tarsus, it passes between the artery and the tibia, but remains closely united to the artery. In the proximal half of the tarsus, the deep peroneal nerve and cranial tibial artery lie in the groove formed by the tendon of the long digital extensor, laterally, and the tendon of the cranial tibial muscle, medially. At the tarsus the deep peroneal nerve sends delicate branches to the three heads of the extensor digitorum brevis muscle, which lie on the flexor surface of the tarsus. The deep peroneal nerve then divides into medial and lateral parts.

The **dorsal metatarsal nerve II** (*n. metatarsei dorsalis II*) (Fig. 17–31) is the metatarsal continuation of the medial branch of the deep peroneal nerve. The lateral branch in the proximal part of the metatarsus bifurcates to form the **dorsal metatarsal nerves III and IV** (*nn. metatarsei dorsales III et IV*). The three dorsal metatarsal nerves communicate with the corresponding superficial ones just proximal to the metatarsophalangeal joints to form the dorsal common digital nerves II, III, and IV.

The **dorsal common digital nerves II, III, and IV** (*nn. digitales dorsales communes II, III, et IV*) are formed in the distal ends of the three main intermetatarsal spaces by the union of the dorsal common digital and dorsal metatarsal nerves. They are about 1 cm. long, devoid of gross branches, and end opposite or distal to the metatarsophalangeal joints by dividing into the **axial or abaxial dorsal proper digital nerves II, III, and IV** *nn. digitales dorsales proprii axialis et abaxialis II, III, et IV*). Since the common digital nerves lie between the metatarsal bones, their terminal branches are not confined to the supply of a single digit but are distribut-

ed to the apposed sides of adjacent digits. In accordance with N.A.V., the proper digital nerves and vessels are designated as axial or abaxial, depending upon which side of the digit they lie in regard to the axis of the limb which passes between the third and fourth digits. Thus the nerves on the lateral side of the third digit and the medial side of the fourth digit are both axial digital nerves. For example, the third dorsal common digital nerve gives origin to the axial dorsal proper digital nerve IV and the abaxial dorsal proper digital nerve III, whereas the second dorsal common digital nerve provides the axial dorsal proper digital nerve II and the abaxial dorsal proper digital nerve III. The fourth dorsal common digital nerve forms the abaxial dorsal proper digital nerve IV and the axial dorsal proper digital nerve V. The dorsal proper digital nerves, as they run toward the distal ends of the digits, lie ventral to the corresponding arteries and send many branches to the skin of the dorsal and adjacent sides of the digit, finally ending in the corium of the claw.

The **tibial nerve** (*n. tibialis*) (Fig. 17–29), larger than the peroneal, is the more caudal terminal branch of the sciatic nerve. It is about 5 mm. wide at its origin and is flattened transversely as it lies between the caudal portions of the semimembranosus muscle medially and the biceps femoris muscle laterally. Bennett (1976) found it to arise from S1 and S2. It separates from the ischiatic nerve gradually in the proximal two-thirds of the thigh. It enters the crus between the two heads of the gastrocnemius muscle. The tibial nerve supplies all the muscles which lie caudal to the tibia and fibula and sends branches to the stifle, tarsal, and digital joints. Its terminal portions run to the muscles which lie in the plantar part of the hindpaw and to the skin and foot pads of the plantar surface of the hindpaw.

The **muscular branches** (*rami musculares*) (Fig. 17–30) are numerous. The first two branches arise in common closely bound to the cranial border of the caudal cutaneous sural nerve, arising about 2 cm. before the tibial nerve enters the gastrocnemius muscle. One branch enters the proximal portion of the caudal border of the lateral head, while the other branch enters the medial head of the gastrocnemius muscle at a like place. The muscular branch to the superfi-

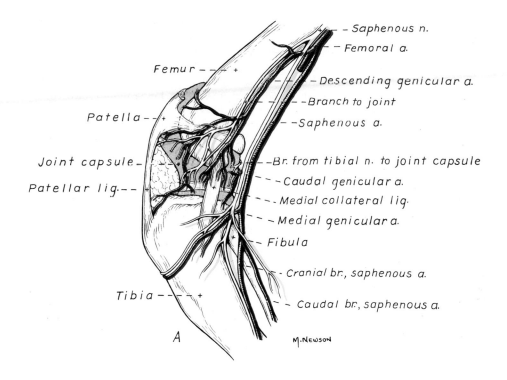

Saphenous n.
Femoral a.
Femur
Descending genicular a.
Branch to joint
Patella
Saphenous a.
Joint capsule
Br. from tibial n. to joint capsule
Patellar lig.
Caudal genicular a.
Medial collateral lig.
Medial genicular a.
Fibula
Cranial br., saphenous a.
Tibia
Caudal br., saphenous a.

A

M. NEWSON

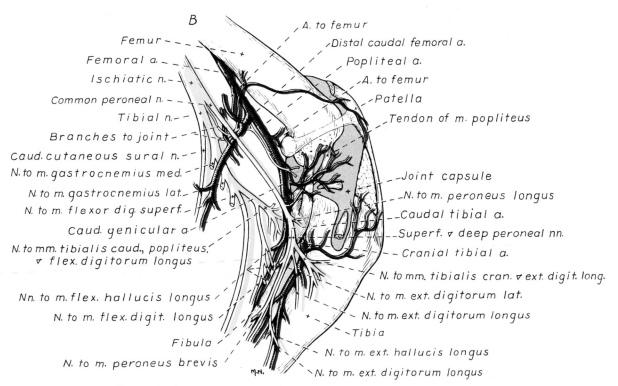

B

A. to femur
Femur
Distal caudal femoral a.
Femoral a.
Popliteal a.
Ischiatic n.
A. to femur
Common peroneal n.
Patella
Tibial n.
Branches to joint
Tendon of m. popliteus
Caud. cutaneous sural n.
N. to m. gastrocnemius med.
Joint capsule
N. to m. gastrocnemius lat.
N. to m. peroneus longus
N. to m. flexor dig. superf.
Caudal tibial a.
Caud. genicular a.
Superf. ⅋ deep peroneal nn.
N. to mm. tibialis caud., popliteus, ⅋ flex. digitorum longus
Cranial tibial a.
N. to mm. tibialis cran. ⅋ ext. digit. long.
Nn. to m. flex. hallucis longus
N. to m. ext. digitorum lat.
N. to m. flex. digit. longus
N. to m. ext. digitorum longus
Tibia
Fibula
N. to m. ext. hallucis longus
N. to m. peroneus brevis
N. to m. ext. digitorum longus

M.N.

Figure 17–30. A. Nerves and arteries of the right stifle joint, medial aspect.
B. Nerves and arteries of the right stifle joint, lateral aspect.

cial digital flexor enters the muscle at its origin. From the cranial border of the tibial nerve, as it lies between the two heads of the gastrocnemius muscle, arises the muscular branch which enters the popliteus muscle at about the middle of its obliquely running distal border. Arising from this nerve, or separately from the tibial, or from both of these, is the nerve or branches to the deep digital flexor muscle. When single, this nerve soon trifurcates into branches which supply its two constituent bellies, the flexor hallucis longus and flexor digitorum longus muscles. A small twig goes to the tibialis caudalis muscle. After these muscular branches arise, the tibial nerve then runs distad on the flexor hallucis longus muscle, where it is covered by the lateral head of the gastrocnemius muscle. About 2 cm. proximal to the tarsus, it receives the caudal cutaneous sural nerve from the lateral side cranial to the Achilles tendon. At about the middle of the crus near the medial surface, the tibial nerve comes into relationship with the plantar branch of the saphenous artery and vein. About 1 cm. proximal to the tibiotarsal (ankle) joint, the tibial nerve bifurcates into the medial and lateral plantar nerves.

Nerves of the Hindpaw (Pes)

The **medial plantar nerve** (*n. plantaris medialis*) (Fig. 17–33) is smaller and more medial and superficial than the lateral plantar nerve. It crosses the medial side of the tarsus in the superficial fascia outside the tarsal canal. At the proximal end of the metatarsus, it branches irregularly into the **plantar common digital nerves II, III, and IV** (*nn. digitales plantares communes II, III, et IV*) and the **abaxial plantar digital nerve II** (*n. digitalis plantaris abaxialis II*) or **plantar proper digital nerve I** (*n. digitalis plantaris proprius I*) if a first digit is present. The continuation of this nerve is similar to the other plantar common digital nerves. These nerves extend obliquely distolaterad in the fascia covering the four branches of the superficial flexor tendon and send minute twigs to the quadratus plantae, lumbricales, and interflexorii muscles. At or near the distal end of the metatarsus, branches are given off their plantar sides to the meta-

tarsal foot pad. Their small continuations communicate with the plantar metatarsal nerves II, III, and IV.

The **lateral plantar nerve** (*n. plantaris lateralis*) (Fig. 17–33) is larger, more lateral, and deeper than the medial plantar nerve. Caudal to the distal part of the tarsus, it enters the interval between the superficial and deep digital flexor tendons and obliquely runs distad between them a short distance before terminating as the deep plantar metatarsal nerves II, III, and IV and muscular branches. The lateral plantar nerve near its origin gives rise to the **abaxial plantar digital nerve V** (*n. digitalis plantaris abaxialis V*), which runs distad on the lateral surface of the fifth metatarsal bone and digit and terminates by dorsal and plantar branches in the terminal part of the fifth digit. The plantar branch goes to the fifth digital foot pad, and the dorsal branch goes to the corium of the claw. This nerve sends twigs all along its course to the skin of the lateral side of the paw. It issues a branch proximally which supplies the small quadratus plantae muscle.

Usually upon entering the proximal part of the metatarsus, the lateral plantar nerve divides into many branches. Three of these are the plantar metatarsal nerves II, III, and IV, and the remaining branches go to the interosseous muscles and the special muscles of the (first), second and fifth digits, and the quadratus plantae muscle. The first **muscular branch** (*ramus muscularis*) arises from the lateral plantar nerve medial to the distal portion of the fibular tarsal bone; it crosses the plantar surface of this bone and enters the small spindle-shaped belly of the abductor digiti V muscle. Another branch closely associated with it goes to the small, transversely running quadratus plantae muscle. The remaining muscular branches arise rather closely together at the proximal end of the metatarsus. The first branch leaves the lateral border of the lateral plantar nerve and enters the proximal end of the fifth interosseous muscle. A few millimeters distal to this origin, the several nerves, one to each of the interosseii (IV, III, and II), arise from the proximal portions of the deep metatarsal nerves. From the second or most medial of the plantar common digital nerves arises the branch which supplies the abductor digiti secundi muscle. When the special muscles

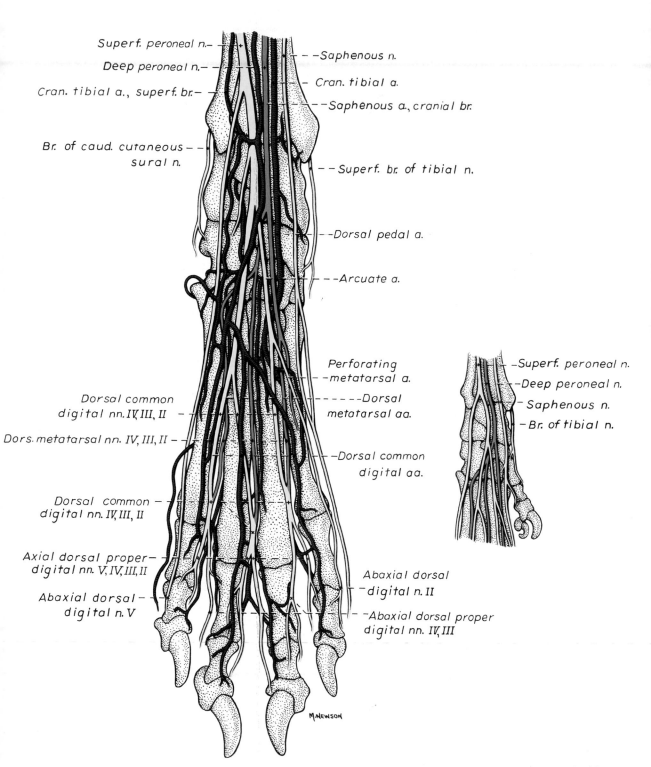

Figure 17–31. Nerves and arteries of the right hindpaw, dorsal aspect. Inset of nerve supply to a double dewclaw.

of the first digit are developed, they receive their nerve supply from this source also.

The **plantar metatarsal nerves I, II, III, and IV** (*nn. metatarsei plantares I, II, III, et IV*) are similar in location and termination to the comparable nerves of the forepaw. After running distad under the special muscles of the second and fifth digits and on the interossei, they communicate with the plantar common digital nerves near the metatarsophalangeal joints. Adjacent plantar metatarsal nerves communicate with each other, and cutaneous branches arise from their distal ends and innervate the metatarsal pad.

The plantar common digital nerves II, III, and IV are formed in the distal ends of the three main intermetatarsal spaces. After running 1 or 2 cm., each nerve bifurcates into two **plantar proper digital nerves** (*nn. digiti plantares proprii*) which supply the skin of the apposed and plantar surfaces of contiguous digits. Each nerve terminates as a **dorsal branch** which innervates the corium of the claw and the distal interphalangeal joint and a **plantar branch** which innervates a portion of the digital pad of its side.

CAUDAL NERVES

The paired **caudal nerves** (*nn. caudales [coccygei]*) (Fig. 17–32) vary in number from four (Havelka 1928) to seven (Ellenberger and Baum 1891). Hopkins (1935) found five pairs in each of nine carefully dissected mongrel specimens. Like the other spinal nerves, the caudal nerves branch immediately upon leaving their intervertebral fora-

mina into dorsal and ventral branches. Each pair of caudal nerves is numbered according to the vertebra which precedes the intervertebral foramen through which it runs. The **dorsal branch** of the first caudal nerve is joined by the dorsal branch of the third, or last, sacral nerve, and the dorsal branches of the five caudal nerves communicate with each other. In this manner a **dorsal caudal trunk** is formed. In a similar manner the **ventral caudal trunk** is formed. Baum and Zietzschmann (1936) name these trunks the *n. collector caudae dorsalis* and the *n. collector caudae ventralis*, respectively. (Some precedence for the names dorsal and ventral trunks is found in Sisson and Grossman's text [1953].) Other authors have named these caudal nerve trunks and their branches the coccygeal plexuses. Nomina Anatomica Veterinaria (1973) calls them the *plexus caudalis [coccygeus] dorsalis* and *plexus caudalis [coccygeus] ventralis*. The dorsal trunk lies directly dorsal to the transverse processes and the intertransverse muscles which extend only between several of the most proximal transverse processes. The dorsal sacrocaudal muscles lie directly dorsal to the nerve trunk which is accompanied by the dorsolateral caudal artery. The delicate **muscular branches** (*rami musculares*) which leave it extend dorsocaudally into the dorsal lateral and medial sacrocaudal muscles. In some segments there are two muscular branches leaving the trunk between the dorsal caudal nerves which contribute to its formation. In several dissections not more than two of these branches could be traced to terminations in the skin in any single speci-

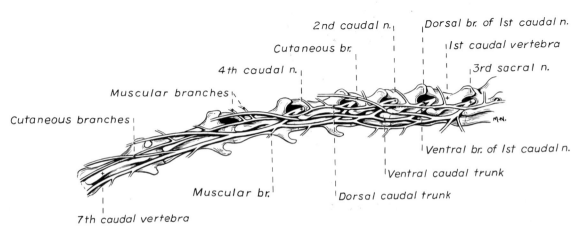

Figure 17–32. Diagram of the caudal nerves, lateral aspect.

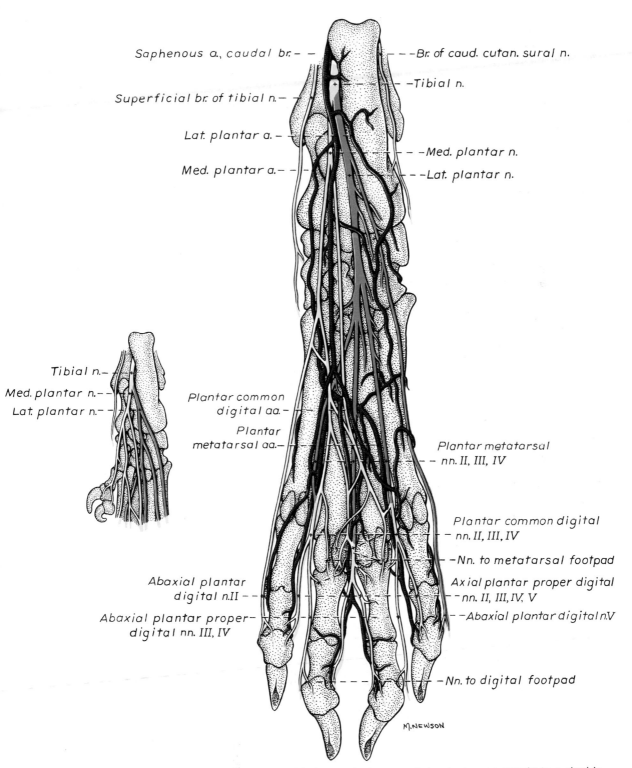

Figure 17–33. Nerves and arteries of the right hindpaw, plantar aspect. Inset of nerve supply to a double dewclaw.

men, and these were found near the root of the tail.

The **ventral trunk** is larger than the dorsal trunk. It is united with the ventral branch of the last sacral nerve at its beginning, and thereafter the trunk is augmented by the total volume of each of the ventral branches of the caudal nerves joining it. The trunk reaches its greatest width of about 2 mm. as it is flattened against the fibrocartilage located between the fifth and sixth caudal vertebrae. The ventral sacrocaudal muscles cover it ventrally. It lies ventral to the transverse processes and the intertransverse muscles. It is accompanied by the ventrolateral caudal artery; the two structures lie closely lateral to the hemal arches and more distally lateral to the hemal processes. The dorsal and ventral caudal trunks extend to the tip of the tail. *Muscular branches (rami musculares)* leave both the dorsal and ventral surfaces of the trunk. The delicate dorsal branches supply both parts of the intertransverse caudae muscle and the vertebrae. The ventrally running branches innervate the lateral and medial ventral sacrocaudal muscles. The first seven to nine ventral branches terminate in dissectable **cutaneous branches** (*rami cutanei*) which innervate the skin of the tail. Some of these fine nerves can be traced several centimeters distally in the superficial fascia. Their main trunks lie in close association with the large, superficial lateral caudal vein and its small accompanying artery.

BIBLIOGRAPHY

Allam, M. W., D. G. Lee, F. E. Nulsen, and E. A. Fortune. 1952. The anatomy of the brachial plexus of the dog. Anat. Rec. *114*:173–180.

Baum, H., and O. Zietzschmann. 1936. Handbuch der Anatomie des Hundes. Band I. Skelett- und Muskelsystem. Berlin, Paul Parey.

Bennett, D. 1976. An anatomical and histological study of the sciatic nerve, relating to peripheral nerve injuries in the dog and cat. J. Small Anim. Pract. *17*:379–386.

Benson, R. O., and T. F. Fletcher. 1971. Variability of the ansa cervicalis in dogs. Am. J. Vet. Res. *32*:1163–1168.

Blakely, C. L. 1957. Perineal hernia. Pp. 458–468 *in* Canine Surgery, edited by K. Mayer, J. V. Lacroix, and H. P. Hoskins. 4th Ed. Evanston, Ill., American Veterinary Publications, Inc.

Browne, J. G. 1959. Neuroanatomy of the brachial plexus of the dog. Thesis, Ames, Iowa.

Bradley, O. C., and T. Grahame. 1959. Topographical Anatomy of the Dog. 6th Ed. New York, Macmillan Co.

Clifford, C. H., R. L. Kitchell, and D. R. Knauff. 1958. Brachial paralysis in the dog. Am. J. Vet. Sci. *39*:49–53.

deLahunta, A. 1977. Veterinary Neuroanatomy and Clinical Neurology. Philadelphia, W. B. Saunders Co.

Ellenberger, W., and H. Baum. 1891. Systematische und topographische Anatomie des Hundes. Berlin, Paul Parey.

— — — — — 1943. Handbuch der vergleichenden Anatomie der Haustiere. 18th Ed. Berlin, Springer.

Fletcher, T. F. 1970. Lumbosacral plexus and pelvic limb myotomes of the dog. Am. J. Vet. Res. *31*:35–41.

Fletcher, T. F., and R. L. Kitchell. 1966. The lumbar, sacral, and coccygeal dermatomes of the dog. J. Comp. Neurol. *164*:140–149.

Gross, C. M. 1954. Gray's Anatomy of the Human Body. 26th ed. Philadelphia, Lea & Febiger.

Havelka, F. 1928. Plexus lumbo-sacralis u psa (in Czech with German summary: Plexus lumbo-sacralis des Hundes). Vysoká škola veterinárni Biologiché spisy 7:1–40 (Pub. biologiques de l'école des hautes etudes Vet.). Brünn, Czechoslovakia.

Henning, P. 1965. Der M. piriformis und die Nn. clunium medii des Hundes. Zbl. Vet Med (A) *12*:263–275.

Hopkins, G. S. 1935. The correlation of anatomy and epidural anesthesia in domestic animals. Cornell Vet. *25*:263–270.

Hummel, P. 1965. Die Muskel- und Hautnerven des Plexus sacralis des Hundes. Anat. Anz. *117*:385–399.

Kopp, P. 1901. Uber die Verteilung und das topographische Verhalten der Nervens an der Hand der Fleischfresser. Inaugural Diss., Bern.

Kunzel, E. 1957. Die Huftgelenkstopographie des Hundes und der Zugang zum Gelenk. Zbl. Vet.-Med. *4*:379–388.

Langley, J. N., and H. K. Anderson. 1986. The innervation of the pelvic and adjoining viscera. J. Physiol. *20*:372–406.

Langworthy, O. R. 1924. The panniculus carnosus in cat and dog and its genetical relations to the pectoral musculature. J. Mammalogy *5*:49–63.

Lemon, W. S. 1928. The function of the diaphragm. Arch. Surg. *17*:840–853.

Mehler, W. R., J. C. Fischer, and W. F. Alexander. 1952. The anatomy and variations of the lumbo-sacral sympathetic trunk in the dog. Anat. Rec. *113*:421–435.

Miller, R. A. 1934. Comparative studies upon the morphology and distribution of the brachial plexus. Am. J. Anat. *54*:143–175.

Pederson, H. E., C. F. J. Blunck, and E. Gardner. 1956. The anatomy of lumbosacral posterior rami and meningeal branches of spinal nerves (sinu-vertebral nerves) with an experimental study of their functions. J. Bone Joint Surg. *38A*:377–391.

Reimers, Hans. 1925. Der plexus brachialis der Haussäugetiere; eine vergleichend-anatomische Studie. Z. Anat. *76*:653–753.

Russell, J. S. R. 1893. An experimental investigation of the nerve roots which enter into the formation of the brachial plexus of the dog. Phil. Trans. B *184*:39–65.

Schaeffer, J. P. 1953. Morris' Human Anatomy. 11th Ed. New York, Blakiston Co.

Schmaltz, R. 1914. Atlas der Anatomie des Pferdes. Teil III. Berlin, Schoetz.

Sisson, S., and J. D. Grossman. 1953. Anatomy of the Domestic Animals. 4th Ed. Philadelphia, W. B. Saunders Co.

Worthman, R. P. 1957. Demonstrations of specific nerve paralyses in the dog. J. Am. Vet. Med. Assoc. *131*:174–178.

Chapter 18

THE AUTONOMIC NERVOUS SYSTEM

By MELVIN W. STROMBERG

If one remains aware of the artificiality of segregating any portion of the nervous system from the whole, a most useful concept can still be gained by studying the **autonomic nervous system** *(systema nervosum autonomicum)* as a more or less separate unit. Consideration must also be given to interrelations between the autonomic and the somatic nervous systems (see Gellhorn 1967).

The autonomic (general visceral efferent) system may be defined as that portion of the nervous system concerned with the motor innervation of smooth muscle, cardiac muscle, and glands. Its counterpart with respect to somatic (derived from somites of embryo) innervation is the somatic efferent system — the portion of the nervous system concerned with motor innervation of striated muscle.

The foregoing definition of autonomic purposely excludes sensory innervation of the viscera, since this is best classified separately as visceral afferent. It should be noted here that this is not a universally accepted definition of the word. Many authors prefer to include visceral afferent structures under the term "autonomic." For example, Nomina Anatomica lists under autonomic nervous system various gross structures which contain both visceral motor and visceral sensory nerves. Nevertheless, as a basis for rational consideration of function it is felt best to equate autonomic with general visceral efferent. References will be made to certain details of visceral afferent innervation when pertinent to the general plan (Fig. 18–1).

On anatomical, pharmacological, and functional bases the autonomic nervous system is further subdivided into **sympathetic** *(pars sympathica)* and **parasympathetic** *(pars parasympathica)* portions. Both are so-called two-neuron systems in the sense that two neuron chains link the central nervous system to the structure innervated. One neuron (preganglionic) of each pair is located in the central nervous system and sends its myelinated axon out as part of the peripheral nervous system. These preganglionic axons synapse with the second neuron (postganglionic) of the chain, and the nonmyelinated axons of these postganglionic neurons end among or on glandular, cardiac muscle, or smooth muscle cells. The postganglionic neurons ordinarily occur in clusters referred to as ganglia. The larger ganglia are found in specific locations and are named accordingly. Many smaller, unnamed ganglia may be found scattered

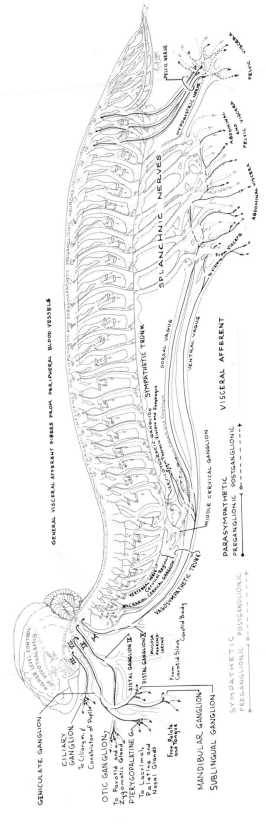

Figure 18–1. General plan (semidiagrammatic) of autonomic fiber distribution and higher level control of the peripheral portion of the autonomic system.

along the paths of the autonomic nerves (Baumann and Gajisin 1975). It must further be emphasized that variability of details of autonomic nerve distribution is common.

Autonomic innervation is known to be no longer limited to smooth muscle, cardiac muscle, and glands. Although brown fat innervation has long been known, Ballard et al. (1974) described adrenergic innervation of canine white adipose tissue. In addition to a rich vascular innervation, nerve fibers were also found between fat cells. The authors suggest that some fat cells are part of a depot that can be lysed by sympathetic activity. Kyösola et al. (1976) described both an adrenergic and a cholinergic innervation of atrial myocardial cells as well as evidence for interneuronal links between fibers at this level, ". . . a link that might function as an integrating unit at the most peripheral level." A cholinergic innervation of both canine and human cardiac conducting system which is capable of raising fibrillation threshold has been demonstrated by Kent et al. (1974). Walles et al. (1975) describe autonomic innervation of the bovine ovarian follicular wall and discuss its possible role in ovulation.

Pawlowski and Weddell (1967) describe innervation of sebaceous glands (formerly disputed) and present evidence for attempts of developing nervous tissue to wall off tumors. In some areas, growth of tumor cells appeared to be inhibited by ingrowing nerve fibers. Borodulia and Plechkowa (1977) have reported an autonomic innervation of cerebral vessels which may have a role in local cerebral blood flow regulation. A highly complex peripheral innervation of the cat colon controlled from hypothalamus, midbrain, amygdala, and cortex has been described by Rostad (1973).

Sympathetic stimulation causes a massive release of glucagon and inhibition of insulin secretion in the calf (Bloom et al. 1973). Kyösola and Rechardt (1974) have described the sphincter of Oddi in the dog and cat. It proves to be a remarkably complex apparatus for rhythmical suction-pressure pumping as well as for sensitive protective reflex functions. The complexity of innervation matches or exceeds that of its non-nervous structure.

Equally complex is a ureterovesical autonomic innervation in the cat and dog which

implies adrenergic and cholinergic interactions within ganglia as well as individual postganglionic neurons which receive both cholinergic and adrenergic synapses (Hamberger and Norberg 1965; El-Badawi and Schenk 1973). The repeated demonstration of unipolar neurons within the walls of visceral organs serves to multiply the complexity. Information on what appears to be the ultimate in specific autonomic influence has been provided by Förster et al. (1975), who describe autonomic innervation of interstitial cells in skin. These cells are able to contact neighboring cells (such as plasma and mast cells) by microvilli-like cell protrusions. The authors postulate an "electromechanical feedback system of information processing in the vegetative periphery." Vajda et al. (1973) have published beautiful pictures of an array of free and encapsulated endings in the mesentery of cats. Some are provided with separate "C" fiber innervation, which the authors feel "may control their stimulation threshold" as well as "inform on the prevailing functional state."

Results of a large series of experiments by Osterholm (1974) reveal that although initial effects of spinal cord trauma may be minimal, within one to four hours norepinephrine (via a descending system) is greatly increased locally. The result is a severe hemorrhagic necrosis and consequent loss of functional tissue. Of unique interest from a therapeutic standpoint is the fact that under the same experimental circumstances most of the lesions fail to develop if the local afferent input to the cord has been previously removed by cutting several pairs of dorsal roots. This may offer a clue to the reported efficacy of acupuncture for relief of paralysis as well as a reason for the pronounced self-immobilization practiced by animals such as the dog following severe trauma.

Gellhorn (1967) has amassed a great deal of evidence for centrally mediated "partial" responses of the autonomic system. Heart rate, force of contraction, blood pressure, pupillary diameter, and contraction of nictitating membrane are all individually separable effects. Graded reflexes such as variations in size of the low-resistance area of hand, foot, and face are also possible.

Dahlstrom and Fuxe (1965) reported a 5–HT neuronal system derived from the raphe nuclei of the medulla that projects to the sympathetic lateral column and a norepinephrine system derived from the ventrolateral reticular formation of the medulla that projects to preganglionic neurons.

Coote and Macleod (1975) have identified three areas in the medulla oblongata which inhibit sympathetic activity via two discrete pathways in the spinal cord. These could conceivably play a role in such graded responses. Gellhorn's work (based on over 1000 references) succeeds in bringing the highly complex area of somatic-autonomic interactions into excellent focus.

Gross (1974) has outlined the role of the autonomic nervous system in certain pain syndromes of man and emphasized that vascular zone topography has here the same diagnostic significance as does the segment (and dermatome) for identification of diseases of the spinal cord and its roots. That is to say, painful areas resulting from autonomic dysfunction do not match the field of distribution of most peripheral nerves, with the exception of much of the trunk region. Although primarily dealing with the results of work in rabbits, Uusitalo (1972) lends support to the premise that intraocular tension is under autonomic control. Adrenergic fibers found in the subepithelial region of the trabecular meshwork in cats may play a role in ocular drainage (Akagi et al. 1976), and sympathetic fibers to the rat cornea (Tervo and Palkama 1976) have also been demonstrated. Autonomic innervation of sensory receptors has frequently been described (for example, Klaisman 1975) and postulated to play a role in the function of these receptors. The foregoing examples are provided to illustrate the pervasiveness of the autonomic system.

Autonomic ganglia are classified as vertebral, collateral (prevertebral), and terminal (peripheral). **Vertebral ganglia** are more or less segmentally distributed along the paired sympathetic chains which in turn lie along the ventral sides of the heads of the ribs. **Collateral ganglia** are found more peripherally and in the abdominal cavity are related to some of the larger arteries such as the celiac. **Terminal ganglia** are usually small and located in various body organs.

Of interest here is the demonstration of apparently extensive sensory innervation of autonomic ganglia (both stroma and parenchyma) in the cat (Shindin 1964, Tikhonova 1968). There also seems to be considerable evidence for adrenergic varicose terminals within ganglia. These endings may modulate ganglionic activity (Jacobowitz 1974). In paravertebral ganglia these endings have been postulated to arise from dendrites of postganglionic cell bodies, from collaterals of these cells, or possibly from processes of SG (small granule–containing) cells (Matthews 1974). The last are apparently similar or identical to the SIF (small, intensely fluorescent) cells, although the morphological range appears to include cells with and without processes. A further complication has been introduced by the demonstration of both cholinoceptive and adrenoceptive sites on preganglionic terminals (Nishi 1974). Therefore, the substrate appears to exist for presynaptic modulation within the ganglion.

The earlier concept of divergence of autonomic output from small numbers of myelinated preganglionic fibers to relatively large numbers of postganglionic neurons (with unmyelinated axons) is at best inaccurate. As many as 40 preganglionic fibers may converge on a single postganglionic cell. Conduction velocity of these fibers may vary widely. These and other details point to a very precise and graded preganglionic input (Blackman 1974), a far cry from the massive and generalized activity often ascribed to the autonomic (especially sympathetic) system.

Both sympathetic and parasympathetic divisons of the autonomic system have been shown to be under direct control of the hypothalamus. For example, Beattie et al. (1930) showed that extrasystolic arrhythmia produced by chloroform anesthesia in the cat could be prevented either by the combination of sectioning sympathetic nerves to the heart and removal of the adrenal glands or by destruction of a portion of the hypothalamus. These authors were also able to trace hypothalamic efferent fibers as far as the second lumbar segment of the spinal cord (see also Katsuki et al. 1955). The hypothalamus is regulated by certain areas of the cerebral cortex (Uvnäs 1954) and in turn reciprocally influences the cortex. Certain groups of neurons in the rostral hypothalamus have been shown to exert control over the parasympathetic system, whereas neurons in the caudal and lateral hypothalamus influence sympathetic activity. Overlap of these higher autonomic centers exists, and other parts of the brain (Landau 1953, Löfving 1961, Kelts and Bignall 1973, Coote and Macleod 1975) are also involved in control of autonomic activity. Cortical neurons may even act directly on preganglionic neurons of the dorsal vagal nucleus and lateral horn of thoracic spinal cord (Gonzalo-Sanz and Ullan 1976). Therefore it may be useful to think of impulses from several parts of the brain converging upon preganglionic neurons of the autonomic system in a manner similar to the convergence of neural influences upon lower motor neurons of the somatic nervous system.

A good deal of autonomic activity is reflexly regulated and is influenced by a wide variety of sensory input.

It has long been known that a painful stimulus anywhere on the body surface will result in the typical "fight or flight" response with increased blood pressure and heart rate, piloerection, and so forth. Less well known is the fact that a depressor response (in muscle, kidney, gut, and skin), as well as decreased heart rate, follows stimulation of group III afferent nerve fibers from muscle, as well as pulling, pinching, or stretching of the viscera (Johansson 1962). These responses (pressor and depressor) are relatively generalized. Localized autonomic responses were demonstrated, for example, by Kuntz (1945). Successive warming and cooling of caudal thoracic skin left it, as well as the small intestine, hyperemic. Sato et al. (1975) found in the rat that stimulation of chest, abdominal, and inguinal skin affects heart rate, stomach motility, and bladder pressure, respectively. The general area of somatosympathetic reflexes has been reviewed by Sato and Schmidt (1973). Conversely, skeletal muscle responses to visceral irritation are well known, especially with reference to the abdominal region. Downmann (1955) found that central stimulation of splanchnic nerves in "spinal" cats produced a widespread somatic response of intercostal and lumbar nerves and sometimes of leg nerves. Ginzel et al. (1972) have pharmacologically activated receptors in the cardiopulmonary region and recorded depression of reflex response from lumbosa-

cral alpha motor neurons. Similar but independent influence on gamma motor neuron output was also demonstrated. The most universally recognized viscerosomatic effect is probably that of referred pain (Sinclair 1973). Abundant evidence appears to exist for rather localized reciprocal relations between viscera on the one hand and limbs and body wall on the other. Fields et al. (1970) have shown a remarkably specific spatial and modality convergence of somatic and visceral input on individual neurons in the spinal cord of the cat. As a matter of fact, such convergence of visceral and somatic sensory input on single neurons has also been shown for medulla, midbrain, thalamus, reticular formation, and cerebral cortex. In turn, it can be said that all major levels of the central nervous system influence autonomic function. The foregoing brief digression into autonomic function is offered to stimulate the reader to further study of an area of the greatest importance in the understanding of disease processes. One can also gain a better understanding of techniques such as acupuncture which to the casual observer appear to be without theoretical basis in terms of Western medicine.

Schofield (1961) describes enteric neurons in the wall of the gut which may be afferent in function and therefore possibly play a part in the peristaltic reflex. Such an arrangement could allow for reflex activity without mediation by the central nervous system. See also Bakeyeva (1957). Visceral afferent impulses play a large part in these reflexes, and most such impulses do not give rise to sensations in human beings. We should remain aware that no means exist for determining subjective sensations in animals, and so we sometimes resort to personal experience and feelings for interpretation. Later work (for example, Schofield 1962, Gunn 1968, Zlatitskaya 1968, Burnstock and Bell 1974) illustrates the incredible complexity of alimentary tract innervation. The discovery of intramural (in the gut wall) purinergic inhibitory neurons is but one of these developments.

Gross (1974) should be consulted concerning the role of the autonomic system in various pain syndromes. Gellhorn (1967) has published extensively on the autonomic system, and his books provide access to a vast experimental literature based largely on work in dogs and cats.

PARASYMPATHETIC DIVISION

Parasympathetic preganglionic axons leave the brain stem as part of cranial nerves III, VII, IX, and X (Figs. 18–1, 18–2, 18–3). The axons in nerves III, VII, and IX are distributed to the head region, whereas the vagus nerve distributes autonomic fibers to the cervical, thoracic, and abdominal viscera as far caudally as the left colic flexure. Parasympathetic preganglionic fibers also leave the spinal cord as part of the ventral roots of the sacral nerves and become part of the pelvic plexus. It is this brain stem and sacral spinal cord origin of parasympathetic fibers which is described by the term "craniosacral" as a synonym for parasympathetic.

OCULOMOTOR (III) (FIGS. 18–1, 18–2, 18–3)

Preganglionic neurons lie in the nucleus of Edinger-Westphal, which is unpaired in the dog. The axons run as part of the third cranial nerve as far as the level of the **ciliary ganglion** (ganglion ciliare) and leave the oculomotor nerve as the short root of the ciliary ganglion. Synapse occurs in the ciliary ganglion, and the postganglionic fibers leave as the short ciliary nerves. They supply the ciliary muscle, which regulates lens curvature, and the sphincter of the iris, which, when activated, reduces pupillary diameter.

FACIAL (VII) (FIGS. 18–1, 18–2, 18–3)

Small cells located in and about the motor nucleus of the seventh cranial nerve send their preganglionic axons out with the pars intermedia of the facial. These fibers then run in the **major petrosal nerve** (n. petrosus major) and **nerve of the pterygoid canal** (n. canalis pterygoidei) to the site of synapse in the **pterygopalatine ganglion** (ganglion pterygopalatinum). The nerve of the pterygoid canal (Wakata 1975) emerges through a small foramen rostral to the foramen rotun-

dum. The ganglion is described as lying medial to the sphenopalatine nerve on the pterygoid muscles. Postganglionic distribution to the lacrimal gland is not well defined. Apparently, some fibers may go directly as small filaments, and some may run with the lacrimal and zygomatic nerve. Other postganglionic fibers go to glands and smooth muscle of the nasal and oral cavities via branches of the fifth cranial nerve (Jung et al. 1926, Nitschke 1976).

The **rostral salivatory nucleus** is located in the dorsal part of the reticular formation dorsal to the facial nucleus and medial to the nucleus of the descending (spinal) root of the trigeminal. These preganglionic axons leave the brain stem with the pars intermedia of the facial. They join the main trunk of the facial nerve (Van Buskirk 1945) and then leave as part of the chorda tympani nerve (Foley 1945), which later joins the lingual branch of the mandibular. Synapse occurs in the **submandibular ganglion** (ganglion submandibulare), and the postganglionic axons go to the sublingual and submaxillary salivary glands. The rostral salivatory nucleus also supplies glands of the tongue (Chibuzo, The Tongue, Chapter 7).

GLOSSOPHARYNGEAL (IX) (FIGS. 18–1, 18–2, 18–3)

On a line caudal to the rostral salivatory nucleus and also in the dorsal reticular formation is found the **caudal salivatory nucleus.** The preganglionic axons leave with rootlets of the ninth cranial nerve and run as part of its **tympanic branch** (n. tympanicus), **tympanic plexus** (plexus tympanicus), and **lesser petrosal nerve** (n. petrosus minor). Synapse occurs in the **otic ganglion** (ganglion oticum) near the origin of the mandibular branch of the fifth nerve, and the postganglionic axons run with the auriculotemporal branch of the trigeminal to their destination on the gland cells of the parotid salivary gland. Holmberg (1971) concludes that a large number of these postganglionic fibers also run in the adventitia of the internal maxillary artery and a small number of fibers run in the facial nerve. Apparently the orbital salivary gland also receives parasympathetic postganglionic fibers from the otic ganglion.

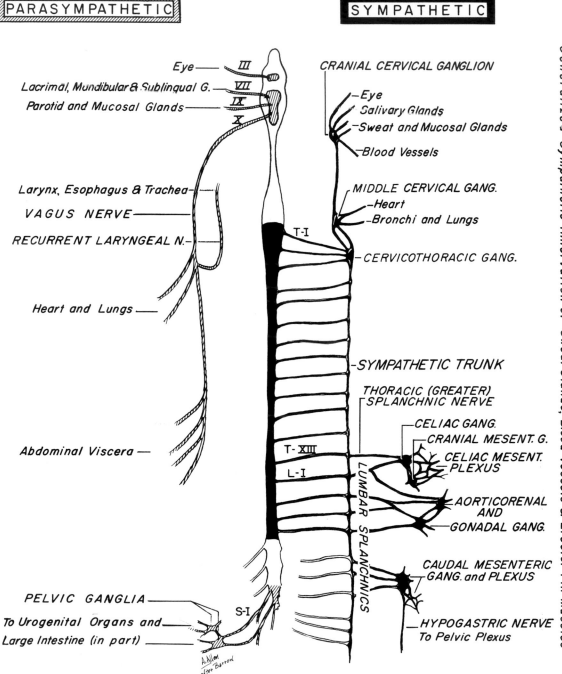

Figure 18–2. General distribution of peripheral elements of sympathetic and parasympathetic divisions.

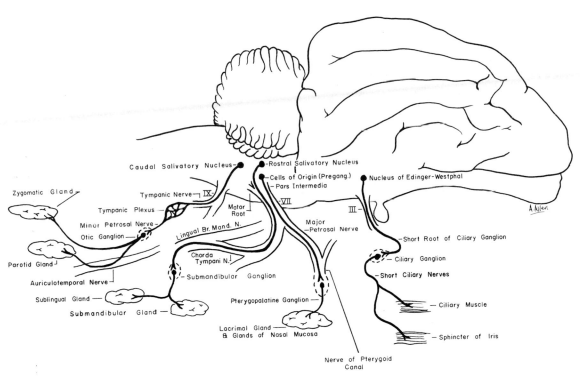

Figure 18-3.   Routes of distribution of preganglionic and postganglionic parasympathetic fibers in the head region.

VAGUS (X) (FIGS. 18–1, 18–2, 18–4, 18–5)

The **vagus nerve** contains a number of functional components which include the visceral afferent (a large number of fibers) and the general visceral efferent (parasympathetic) division. The cervical vagus is said to contain over 80 per cent "C" fibers, with the balance myelinated (Fussey et al. 1973). The majority of the "C" fibers are assumed to be afferent and the rest efferent parasympathetic fibers originating in the dorsal (motor) vagal nucleus. The paired nucleus is located in the dorsal part of the caudal medulla oblongata beneath and caudal to the small eminences referred to as the alae cinereae. Various branches of the vagus will be mentioned in this discussion, but distribution of the branches other than autonomic is described in the section on cranial nerves. The vagal rootlets leave the brain along the dorsolateral sulcus of the medulla oblongata in series with the rootlets of cranial nerves IX and XI. Close association of some of the vagal fibers with those of the

accessory (XI) nerve for a very short distance has led to the description of these vagal fibers as the bulbar or internal root of the accessory nerve in some species. Chase and Ranson (1914) concluded that a true bulbar root of the eleventh cranial nerve does not exist in the dog, a fact that can be demonstrated by careful dissection.

A few millimeters distal to the origin of the vagus is located the small **proximal vagal (jugular) ganglion** (*ganglion proximale n. vagi*). It contains unipolar sensory-type neurons whose dendrites are distributed with the **auricular branch** (*r. auricularis*) of the vagus. This branch leaves approximately at the level of the proximal ganglion and has been shown to be distributed in part via branches of the seventh cranial nerve. There are also some branches in the tongue of the dog (Chibuzo 1978) that have their cell bodies in the proximal ganglion of the vagus.

The **pharyngeal branch** (*r. pharyngeus*) of the vagus is given off between the proximal and the more distally lying **distal vagal (no-**

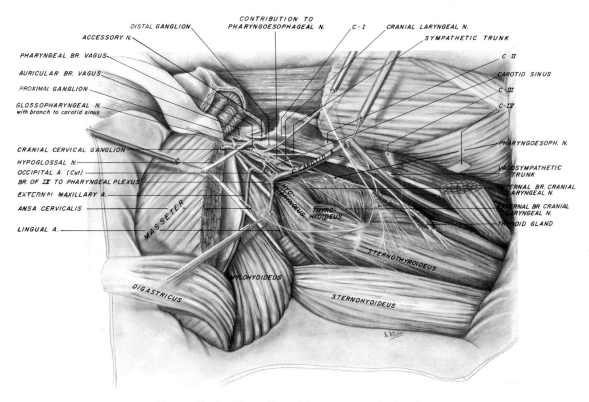

Figure 18–4. Dissection of the upper cervical region.

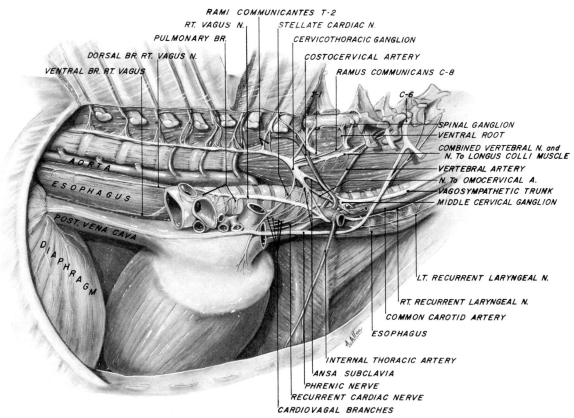

Figure 18–5. Dissection of thoracic region, right side. The right brachiocephalic (innominate) nerve(s), the sympathetic contribution to the phrenic nerve, and several of the arterial plexuses have been omitted.

dose) **ganglion** *(ganglion distale n. vagi)*. The latter ganglion is also composed mainly of unipolar sensory-type neurons which have their peripheral distribution in the viscera.

The **cranial laryngeal nerve** *(n. laryngeus superior)* leaves the vagus at the level of the distal ganglion. The cranial cervical ganglion (sympathetic) is located medial to the distal ganglion. The epineurial sheaths of the two ganglia may be separate or may show variable degrees of fusion. By careful dissection it is possible to peel the epineurial sheath away from the vagus and its ganglia, thereby clearly demonstrating the branches which leave the vagus. In some specimens a small, unnamed bundle of vagal fibers may be seen to bypass the distal ganglion.

Distal to the distal vagal ganglion, the vagus joins the sympathetic trunk, and the two are bound in a common sheath along the entire cervical region. The sympathetic trunk may retain a rounded contour (in cross section) or may assume a crescent shape where it is closely applied to the vagus. In some specimens it is nevertheless possible to separate the two easily by dissection.

At the level of the middle cervical ganglion (Fig. 18–6) the **right recurrent laryngeal nerve** *(n. laryngeus recurrens)* leaves the right vagus and passes dorsally around the caudal side of the subclavian artery. As it runs cranially on the trachea, it lies deep to the carotid artery in the angle between the longus colli muscle and the trachea. Near its origin, the right recurrent laryngeal nerve gives off a fairly large branch, the **recurrent cardiac nerve** (right middle cardiosympathetic nerve), which may receive contributions from the middle cervical ganglion, the vagal trunk (right cardiovagal nerve of Nonidez), and the left recurrent laryngeal nerve. It is distributed mainly to plexuses along the right and left coronary arteries. The nerve contains postganglionic sympathetic, preganglionic parasympathetic, and visceral afferent fibers for the heart.

Nonidez (1939) has described parasympathetic neurons of the terminal ganglia of the heart as differing markedly in size. As a consequence, the postganglionic axons also vary in diameter. These axons lie in the walls of the atria and the interatrial septum and form extensive plexuses. Smaller branches leave the plexuses and end as ring-shaped, club-shaped, or reticulated enlargements, which contact the surface of the cardiac muscle fibers. Similar enlargements may also occur along the course of the finer branchings. Terminations are especially abundant among the specialized muscle fibers of the sinoatrial and atrioventricular node. Apparently, these nodes are also richly supplied with adrenergic fibers (Dahlstrom et al. 1965). Postganglionic parasympathetic axons of the cardiac ganglia are also distributed via nerve plexuses along branches of the coronary arteries. These axons form plexuses in the adventitia and finally end in relation to the smooth muscle cells of the outer media. Primarily, these fibers innervate the arteries of the atrium and proximal part of the ventricles. Most of the parasympathetic distribution is to structures above the coronary sulcus, whereas sympathetic fibers innervate the ventricles. However, Kent et al. (1974) state that numerous cholinergic fibers were found to innervate ventricular conduction tissue and shown to mediate enhanced electrical stability (thus a reduced tendency toward arrhythmias and fibrillation). (See also Martin 1977.) Further information on intrinsic innervation of the heart may be found in: Tcheng 1950, 1951; Holmes 1956, 1957; Uchizono 1964; Hirsch et al. 1964, 1965; Napolitano et al. 1965; Ehinger et al. 1967; Denn and Stone 1976. Detailed and well-illustrated description of autonomic pathways to the atria, SA node, and AV node of the canine heart is also available in a paper by Geis et al. (1973). Further, it is of considerable interest that Kyösola et al. (1976) have shown both an adrenergic and cholinergic innervation of individual atrial myocardial cells in the human heart.

Coleridge et al. (1973) have described widely distributed high-threshold vascular receptors in the dog that are supplied by fine myelinated or unmyelinated vagal fibers. These may belong to the group which normally are not very active but can be stimulated by chemicals such as veratridine or capsaicin as well as by stretching. The other category of cardiac or vascular receptor is one which signals beat-to-beat changes in pressure or volume. These are supplied by myelinated rapidly conducting fibers.

The right vagus nerve, having usually separated from the sympathetic trunk a short distance cranial to the level of the middle cervical ganglion, runs ventral to the subclavian artery and continues caudally along the lateral surface of the trachea. At the level of the subclavian artery, the ansa subclavia may be intimately attached to the vagus for a short distance. Also in this region the vagus and recurrent laryngeal nerves may receive branches from the middle cervical ganglion or ansa subclavia. Distal to the origin of the recurrent nerve, the vagus gives off two or more fine branches called the cranial and caudal cardiovagal nerves. The **cranial cardiovagal nerves** go mainly to the pretracheal plexus. The **caudal cardiovagal nerves** distribute to the dorsal wall of the right atrium. Electrical stimulation of the right vagus nerve at various levels indicates that most of the inhibitory fibers going to the right atrium run in the cardiovagal nerves. (Results of more recent work on the effects of stimulating various canine cardiac nerves can be found in Armour and Randall 1975.) Most of the fibers in these nerves are nonmyelinated, but nevertheless are considered to be preganglionic parasympathetic (numerous exceptions exist to the "rule" of preganglionic axons being myelinated and postganglionic axons unmyelinated). Daly and Hebb (1952) have provided evidence that the cervical sympathetic trunk of the dog may contain cardioinhibitor, bronchoconstrictor, and pulmonary vasomotor (vasopressor) fibers. Such fibers could possibly be collaterals of ascending preganglionic sympathetic fibers or postganglionic fibers which originate in the cranial cervical ganglion. Furthermore, the apparent presence of cardioaccelerator fibers in the vagus has been demonstrated by Shizume (1952) and others. Pannier (1946) placed the origin of these fibers in the medulla oblongata.

The main trunk of the right vagus continues caudad dorsal to the root of the lung. At this level it gives off several prominent branches along the bronchi (Zremianski et al. 1967). Parasympathetic ganglia are found in the lung, and postganglionic parasympathetic fibers have been traced to smooth muscle and glandular structures. The lung is also richly innervated with a wide variety of receptor structures as far distally as the

alveoli. Receptors have been described in smooth muscle, tracheal and bronchial epithelium, respiratory bronchioles, alveolar ducts, and alveolar walls (Elftmann 1943). Gold et al. (1972), in a study of experimental asthma in allergic dogs, demonstrate the major role of the vagus nerve (both afferent and efferent fibers) in the bronchoconstrictor response. The localized direct response to the antigen appears to be relatively minor. The vagus supplies branches directly to the trachea and esophagus and also via the recurrent laryngeal nerve.

Immediately caudal to the root of the lung, both right and left vagi split into dorsal and ventral branches. The ventral branch of the right fuses with its left counterpart to form the ventral vagal (ventral esophageal) trunk. A similar anastomosis occurs between right and left dorsal vagal branches more distally to form the dorsal vagal (dorsal esophageal) trunk.

Both dorsal and ventral vagal trunks supply branches to the esophagus before passing through the diaphragm at the esophageal hiatus. Upon reaching the abdominal cavity, the ventral vagus becomes plexiform for a short distance. Branches from this plexus supply mainly the liver and stomach. The hepatic branches (usually two or three) run in the lesser omentum to the liver. Chiu (1943) describes three superficial hepatic branches, two of which come from the ventral vagus and one from the dorsal. These run in the lesser omentum and pass between the caudate and left lateral lobe of the liver to form a simple plexus just rostral to the porta hepatis. Chiu also describes two more plexuses in the course of distribution of these fibers. One of the plexuses receives contributions from the celiac plexus. Autonomic nerve branches are distributed from the plexuses to the cystic duct, gallbladder, left lateral lobe of the liver, and bile duct. Fiber groups are also described as passing along the right gastric and cranial pancreaticoduodenal arteries to the pancreas and along the right gastric artery to the pylorus. A small filament may go directly to the duodenum.

The second group of fibers (three or four gastric branches) from the ventral vagus supplies the ventral surface of the stomach. Mizeres (1955a) mentions the presence of a number of other filaments which join the sympathetic plexus on the branches of the left gastric artery.

The dorsal vagal trunk supplies the cardiac region of the stomach and then forms a plexus on its dorsal surface. Distribution is mainly to the lesser curvature and pyloric regions of the stomach. Stimulation experiments in unanesthetized decerebrate dogs (Okabe 1958) indicate that the vagus nerves carry motor fibers to the cardiac sphincter and that sympathetic fibers in the greater splanchnic nerve mediate inhibition of cardiac sphincter motility (Okabe 1959). Countee (1977) discusses extrinsic neural influences on gastrointestinal motility and notes, however, that the gastrointestinal tract can carry on all of its major functions after all extrinsic nerves are cut. This is due to local nervous mechanisms, inherent smooth muscle properties, and hormones. Countee further notes that all levels of the central nervous system have been shown (by stimulation and ablation studies) to influence motility of the entire gastrointestinal tract. There are many cerebral areas where stimulation causes inhibition and sometimes excitation of alimentary tract motility. These effects are mediated by both parasympathetic and sympathetic nerves, as well as by humoral agents, and are influenced by information from visceral, cranial, and somatic afferents and by psychic influences. To the foregoing concepts should be added the intricate peripheral control mechanisms referred to previously. Branches of the dorsal vagal trunk may also join the celiac, left gastric, hepatic, and cranial mesenteric nerve plexuses associated with the corresponding arteries. Mizeres (1955a) has also described filaments to the hepatic plexus. Further references on abdominal distribution of the vagus nerve are those of Obrebowski (1965), Kapeller (1965), and Kemp (1973).

Information on vagal parasympathetic distribution caudal to the stomach is sparse, but there seems to be general agreement that these fibers reach as far caudally as the left colic flexure.

Branching and distribution of the left vagus are similar to the right as far distally as the level of the middle cervical ganglion. Here a branch leaves the vagus to join a branch from the middle cervical ganglion, resulting in formation of the left brachio-cephalic (innominate) nerve (apparently left cardiovagal nerve of Nonidez). This is distributed to the base of the brachiocephalic artery and ventral surface of the aortic arch (Mizeres 1955a). A significant component of this nerve is afferent (depressor).

It must be emphasized that details of autonomic nerve distribution to the heart are extremely variable, particularly with respect to origin of fibers. Many of the nerves to the heart are mixed in the sense that they carry sympathetic, parasympathetic, and visceral afferent fibers. Terminology is varied and often confusing. In general it can be said that impulses propagated in sympathetic fibers tend to accelerate heart rate and force of contraction, whereas parasympathetic stimulation reduces heart rate.

Filaments leave the left recurrent laryngeal nerve near its origin and go to the left atrium or plexuses which supply it. Other filaments to the left atrium leave the vagus distal to the origin of the recurrent nerve. The vagus also contributes fibers to certain other cardiac nerves which will be discussed with the sympathetic system.

Fibers which make up the left recurrent laryngeal nerve leave the vagus at the level of the aortic arch and loop around the arch distal to the ligamentum arteriosum. The recurrent laryngeal nerve then proceeds cranially along the left ventrolateral surface of the trachea adjacent to the ventromedial side of the esophagus. Branches are supplied to the trachea and esophagus as the nerve passes to its termination in the larynx (see Hwang et al. 1948 and Watson 1974 for more detail on innervation of the esophagus). Gross dissection indicates that in the cervical region the left recurrent laryngeal supplies more branches to the esophagus than it does to the trachea. The opposite is true of the right recurrent laryngeal nerve. In dogs whose esophagus lies to the right of the midline in the cervical region there is a more even distribution of branches of right and left recurrent laryngeal nerves to both trachea and esophagus.

Electrical stimulation experiments conducted on the left vagus of the dog indicate that cardioinhibitory fibers leave the left recurrent laryngeal nerve as it passes around the arch of the aorta and go to the left atrial region. Other inhibitory fibers leave the left vagus distal to the origin of the recurrent nerve and are also distributed to the left atrium. Most of the fibers in these cardioinhibitory nerves are nonmyelinated.

Cranial to the root of the lung the left vagus may contribute a variable number of filaments to the cardiosympathetic nerves, as previously noted. Several branches leave the left vagus for distribution along the bronchi much as occurs on the right side. Distal to the root of the lung, the left vagus divides into dorsal and ventral branches, each of which joins its counterpart of the right side to result in dorsal and ventral vagal trunks. Their distribution has been discussed.

Both right and left thoracic vagus have been shown to contain a bundle of sympathetic fibers near the periphery of the main trunk.

Parasympathetic postganglionic neurons may be found scattered in the walls of structures which they supply or else as part of well-defined plexuses in similar regions. Most prominent of these are the myenteric and submucous plexuses of the digestive tract (Lawrentjew 1931; Filogamo 1950; Richardson 1960; Schofield 1961). The myenteric (Auerbach's) plexus (El'bert 1956; Leaming and Cauna 1961) (plexus myentericus) is located between the longitudinal and circular muscle layers, and the submucous (Meissner's) plexus (plexus submucosus) lies in the submucosa of the digestive tube. Matthews (1974) notes that in the myenteric plexus neuronal membranes often lie in apposition, even though satellite cells tend to ensheathe many of the somata and processes. Endings are not limited to cholinergic ones; adrenergic and purinergic terminals are also present. (Selective destruction of the myenteric plexus in dogs has resulted in conditions which mimic either esophageal achalasia or Hirschsprung's disease, depending upon where the destruction takes place.) Prominent ganglionated plexuses are also found in the atria of the heart. Much of their regulating effect is thought to be mediated by way of the sinoatrial and atrioventricular nodes. Rostad (1973), in a detailed and extensive paper, describes a highly complex peripheral innervation of the cat colon. This is controlled from several brain levels, including hypothalamus, midbrain, amygdala, and cerebral cortex.

The spinal parasympathetic neurons are depicted classically in the intermediolateral cell column of the sacral spinal cord. Recent experimental study in the dog (Petras and Cummings, 1978) reveals that the preganglionic cell bodies are not confined to this column, but extend medially throughout the broad zona intermedia; laterally to the adjoining part of the lateral funiculus; and ventrally among the somatic motor neurons of the ventral gray column. Preganglionic axons leave as part of the ventral roots of the sacral segments (Gaskell 1886, Mizeres 1955a, Schnitzlein et al. 1963, Wozniak and Skowronska 1967). These fibers join more distally as the **pelvic nerve** (*nervus erigens*), located on the lateral wall of the distal portion of the rectum. On both sides the pelvic nerve expands into a plexus which also receives the hypogastric (sympathetic) nerves. One or more ganglia (presumably parasympathetic) are located within the plexus (Jayle 1935). Distribution of postganglionic autonomic fibers is via branches and plexuses to the pelvic viscera and reproductive organs. Parasympathetic fibers from the sacral region have been traced through the length of the large intestine (Schmidt 1933). Iljina and Lawrentjew (1932a, 1932b) have described parasympathetic ganglia in the wall of the rectum and urinary bladder.

SYMPATHETIC DIVISION

The sympathetic preganglionic cell bodies in the dog are located in the zona intermedia of the spinal cord from C8 or T1 through the L4 or L5 segments. Most of the preganglionic neurons are contained in the triangular cell nests of the intermediolateral nucleus. However, substantial numbers also extend across the zona intermedia toward the central canal in the transverse cell bands of the intercalated nucleus (Cummings 1969, Petras and Faden 1978). There is some evidence to indicate that sympathetic preganglionic fibers may sometimes appear in the ventral roots of the more caudal cervical segments. The myelinated small-diameter axons leave in the ventral roots of the above-named segments and for a short distance are part of the corresponding spinal nerves. These fibers then leave each spinal nerve as one or more communicating rami which attach to the sympathetic trunk. The trunk is a paired strand of preganglionic and postganglionic sympathetic and visceral afferent nerve fibers located ventrolateral to the bodies of the vertebrae in the thoracic and lumbar regions. Continuations of the two trunks caudally into the sacral and cranial caudal region may be partially fused and lie ventral to the bodies of the vertebrae. Both sympathetic trunks are continued cranially into the cervical region, where each runs in a common sheath with the vagus nerve. Each trunk terminates in the cranial cervical ganglion. With certain exceptions, sympathetic chain ganglia are located at segmental intervals along the sympathetic trunk.

The ganglia are numbered according to the spinal nerve to which the postganglionic fibers attach. The part of the sympathetic trunk which joins adjacent ganglia is the interganglionic (internodal) segment. These ganglia contain the multipolar neurons which give rise to the predominantly nonmyelinated postganglionic axons. Synapses may occur, for example, between preganglionic fibers of T12 ramus communicans and cells in T12 sympathetic ganglion, or the preganglionic fibers may proceed cranially or caudally in the sympathetic trunk to synapse in ganglia at other segmental levels. Moreover, it has been demonstrated that preganglionic axons in the dog may project cranially or caudally over a number of segments within the spinal cord before exiting in the ventral roots (Faden and Petras 1978). In general, preganglionic fibers from T1 to T5 are directed cranially whereas caudal to T5 the fibers go caudally. Thus preganglionic fibers from a given segmental level of the spinal cord are distributed to vertebral ganglia at many different levels.

Preganglionic fibers may also leave the sympathetic trunk (as for example in the splanchnic nerves) and synapse in collateral ganglia such as the celiac.

Postganglionic fibers may be distributed in a number of different ways:

1. Fibers which leave the vertebral ganglia via the communicating rami to return to the spinal nerve at the same level. These are distributed to smooth muscle (vasomotor and pilomotor fibers) and glands by way of the spinal nerve.

2. Fibers which run either rostrally or caudally in the sympathetic trunk and join the adjacent spinal nerves via their respective rami.

3. Fibers which leave vertebral ganglia such as the cervicothoracic to go directly to a visceral structure such as the heart via a cardiac nerve.

4. Fibers which leave vertebral ganglia such as the cranial cervical for distribution as plexuses along arteries of the head region.

5. Fibers which leave collateral ganglia such as the celiac for distribution as plexuses along arteries to the abdominal viscera.

Other modifications of distribution route may occur.

Reference is sometimes made to gray versus white rami communicantes, the gray being made up mainly of postganglionic fibers and the white mainly of preganglionic. At most segmental levels of the thoracic and cranial lumbar regions of the dog, the rami are mixed in the sense that they contain both preganglionic and postganglionic fibers. Cranial and caudal to the thoracic and cranial lumbar levels, the communicating rami are made up mostly of nonmyelinated fibers, since usually no preganglionic contributions to the sympathetic trunk arise from the cervical, lower lumbar, and sacral regions. The comparatively few myelinated fibers which may be present in these "gray" rami are probably visceral afferent.

The small size of the sympathetic filaments which are grossly dissectable belies their extensive distribution. In general it appears that sympathetic postganglionic fibers are as widely distributed as the arterial system.

Both microscopic and macroscopic ganglia may be found more or less randomly scattered along the peripheral portions of the sympathetic system. These are for the most part unnamed ganglia either proximal or distal to ganglia such as celiac, cranial mesenteric, or those of the vertebral chain. For this reason any given segment of sympathetic nerve cannot be considered to carry all preganglionic or all postganglionic fibers. Microscopic examination of a sympathetic nerve could not be expected to enable one to identify completely the nature of the fibers, since not all preganglionic fibers are universally myelinated throughout their length. Postganglionic fibers usually have no myelin sheath, but this also is probably not a constant characteristic. Apparently, sympathetic fibers acquire myelin sheaths after birth, whereas myelinization of vagus fibers begins before birth (Diamare and deMennato 1930).

Sympathetic Distribution to Head and Neck

In the dog, preganglionic cell bodies from the C8 through the T7 spinal segments give rise to fibers that run as part of the vagosympathetic trunk to synapse in the cranial cervical ganglion (Petras and Faden 1978). The cervical sympathetic trunk of the dog has been shown to contain approximately 5000 to 13,000 fibers, of which about half appear to be myelinated (Foley and DuBois 1940). Sanbe (1961) finds less than half this number of fibers, most of which are described as myelinated. The same author states that over 40 per cent of the fibers contained in the cervical sympathetic trunk are processes of sensory neurons whose cell bodies lie in the spinal ganglia of the caudal cervical and cranial thoracic region. The cranial cervical ganglion lies deep to the tympanic bulla and the proximal portions of cranial nerves IX, X, XI, and XII. A large portion of the postganglionic fibers leaving the ganglion continue as plexuses along the arteries of the head region. For example, rather prominent bundles of fibers can be seen going to both external and internal carotid arteries. Sympathetic postganglionic fibers also follow the common carotid artery. In general, the arterially distributed sympathetic fibers supply sweat and salivary glands, nasal glands (Franke and Bramante 1964), and smooth muscle such as that of vascular walls, nictitating membrane (Esterhiuzen et al. 1967, Malmfors 1968), dilator pupillae (Ehinger 1967), and erector pili. In addition, adrenergic fibers to the eye appear to have a wide distribution, including the iris stroma, cornea, trabeculae of the iridocorneal angle, ciliary body, ciliary processes, and retina (Ehinger 1966). Some adrenergic fibers to the sphincter pupillae have also been seen, but their function appears to be unknown. A similar "double innervation" (sympathetic and parasympathetic) of the dilator pupillae of the cat has also been described by Ehinger (1967). Those sympathetic nerves which fol-

low the arteries may be fairly discrete bundles or plexiform (see Billingsley and Ranson 1918a and 1918b for distribution in the cat).

Filaments from the cranial cervical ganglion may in some specimens be seen to join the ninth, tenth, eleventh, and twelfth cranial nerves, the connection to the tenth being most common. A fairly constant branch to the first cervical nerve occurs (Fig. 18–4), and this may supply filaments to the second and third cervical nerves also. Other branches may join the pharyngeal and cranial laryngeal branches of the vagus. In some dogs one may demonstrate a pharyngeal plexus formed by contributions from cranial nerves IX and X and the cranial cervical ganglion, whereas in other specimens there is little intermingling of these fibers. Filaments also attach the cranial cervical ganglion to the region of the carotid sinus and carotid body (glomus). A larger branch (Fig. 18–4) leaves the caudal end of the cranial cervical ganglion, contributes to the pharyngeal plexus, and sends fibers (thyroid nerve of Nonidez) along the thyroid artery to the gland. The rest of the nerve continues caudally in the connective tissue membrane which joins the vagosympathetic trunk to the common carotid artery. (Intermingling of several functional types of fibers and variations in branching preclude the positive identification of some nerves in this area by other than physiological means.) It is joined by small branches of the cranial laryngeal and glossopharyngeal nerves and continues caudally as the pharyngoesophageal nerve. The thyroid gland receives sympathetic postganglionic fibers which are distributed as interfollicular nerves and plexuses along arteries. The vascular nerve plexuses lie in the tunica media and tunica adventitia. Cunliffe (1961) has found no definite histological evidence of a secretomotor nerve supply to the thyroid and suggests that rate of secretion and removal of the secretory product are indirectly controlled by rate of blood flow. Essentially the same conclusions were reached by Nonidez (1935), who also demonstrated an afferent nerve supply of the thyroid. It may be assumed that other visceral structures of the head and neck region (as well as the rest of the body) are also supplied with afferent innervation.

Norberg et al. (1975) have described

adrenergic endings close to the endocrine cells of the human parathyroid. They conclude that norepinephrine can reach a large portion of these cells and may possibly influence hormonal release. Borodulia and Plechkowa (1977) have demonstrated, on cerebral arteries of the dog, the presence of a well-developed adrenergic plexus which varies in vessels of different diameter. At sites such as branching they describe peculiar circular windings of adrenergic fibers.

The adventitia of the cervical lymphatic duct is innervated by a loose, interrupted network that appears to arise from plexuses of sympathetic fibers surrounding the vasa vasorum of the ducts. Some authors have described complex multi-layered nerve plexuses, especially in the walls of larger lymphatic vessels (Todd and Bernard 1973).

Sympathetic Distribution in Thoracic Region

Fusion of vertebral ganglia in the cranial thoracic and caudal cervical regions has resulted in formation of the **cervicothoracic** and **middle cervical ganglia.** The cervicothoracic (stellate) is the largest autonomic ganglion in the dog and is located on the lateral surface of the longus colli muscle at the level of the first intercostal space. It receives mixed rami from T1, T2, and T3 spinal nerves and sometimes from T4. Preganglionic fibers ending in the stellate ganglion may originate as far back as T5 spinal cord segment. Gray rami connect to C8 and C7 spinal nerves. Some of the rami attached to the cervicothoracic ganglion may be double. The ramus to C7 may be fused with the **vertebral nerve** *(n. vertebralis),* and both may be fused with a third nerve, the branch of the seventh cervical nerve to the longus colli muscle. For this reason the vertebral nerve may appear deceptively large. Actually, the vertebral nerve is comparatively small. Supposedly the nerve plexus which runs along the vertebral artery supplies gray rami to spinal nerves cranial to C7 as far as C3. These connections are difficult to demonstrate. In some instances the vertebral nerve (Fukuyama and Yabuki 1958) leaves the cervicothoracic ganglion as a separate entity. The nerve enters the transverse foramen of the sixth cervical vertebra along with the vertebral artery and usually continues as a plexus along the artery.

Cranioventrad to the cervicothoracic ganglion, the sympathetic trunk splits to pass around the subclavian artery as the ansa subclavia. One or both sides of this loop may be double.

At the junction of the vagosympathetic trunk with the ansa subclavia medial to the subclavian artery lies the middle cervical ganglion of the sympathetic division (Fig. 18–5). It lies in the path of the sympathetic trunk and may be variably fused with the vagal trunk.

The ansa contains largely preganglionic sympathetic fibers, but also large- and medium-diameter visceral afferent fibers. The sympathetic fibers may synapse in the middle cervical ganglion or else proceed rostrally as part of the cervical sympathetic trunk. Other preganglionic fibers may leave with cardiac branches and synapse in small unnamed ganglia along their course. The ansa subclavia may contain scattered ganglion cells, or some may be grouped as a small ganglion. On the left side, nerve filaments may join this ganglion to the subclavian artery. On the right these filaments, if present, go to the base of the cranial vena cava and the right atrium.

NERVES TO THE HEART

Autonomic nerves to the thoracic viscera may be traced from the cervicothoracic ganglion, ansa subclavia, middle cervical ganglion, recurrent laryngeal nerve, and from the vagus. (See McKibben and Getty 1968 for further information on cardiac innervation in the dog.) Pannier (1946) offers evidence to indicate that some sympathetic preganglionic and postganglionic cardioaccelerator fibers may run via the cervicothoracic ganglion, cervical sympathetic trunk, cranial cervical ganglion and back again via the cervical sympathetic trunk and middle cervical ganglion to the heart (see also Erez 1955). Since sympathetic and parasympathetic fibers appear to intermix in this region, it is difficult to designate any given grossly dissectable nerve as being one or the other (Greenberg 1954, 1956; Randall et al. 1968).

Mizeres (1955a) has named these nerves

partly according to their apparent origin and partly according to termination. This terminology is used primarily, but with some modification and the older terms of Nonidez noted in parentheses.

Right Side. RECURRENT CARDIAC NERVE. See section on Vagus.

STELLATE CARDIAC NERVES (CAUDAL OR INFERIOR CARDIOSYMPATHETIC NERVES). These filaments may arise from one or more of the following: ganglion, ansa subclavia, vagal trunk. One or more of these nerves may be present, while the nerves may be entirely absent in some animals. Most run ventral to the base of the cranial vena cava and are distributed to the right pulmonary plexus, and the dorsal and ventral walls of the right atrium.

CRANIAL CARDIOVAGAL NERVES. See section on Vagus.

CAUDAL CARDIOVAGAL NERVES. See section on Vagus.

RIGHT BRACHIOCEPHALIC (INNOMINATE) NERVES (CRANIAL OR SUPERIOR CARDIOSYMPATHETIC NERVES). These are one or more nerves arising from the middle cervical ganglion or from a fiber bundle which joins the middle cervical ganglion to the recurrent cardiac nerve. They may also include a nerve group arising via several roots from the recurrent laryngeal and recurrent cardiac nerves. Both groups go to the walls of the large arteries near the heart.

Left Side. STELLATE CARDIAC NERVES (CAUDAL OR INFERIOR CARDIOSYMPATHETIC NERVES) may be absent or, if present, may arise from either the cervicothoracic ganglion or the ansa subclavia. They supply the left pulmonary and dorsal ventricular plexuses. They send a branch to the ventrolateral cervical cardiac nerve. They convey mostly sympathetic postganglionic and afferent fibers.

VENTROLATERAL CERVICAL CARDIAC NERVE (MIDDLE CARDIOSYMPATHETIC OR LEFT ACCELERATOR NERVE). This is the largest of the cardiac nerves and the main sympathetic supply to the left ventricle. It also carries afferent cardiac fibers and preganglionic parasympathetic fibers. It arises from the middle cervical ganglion and runs caudally lateral to the vagus. More caudally it may branch, and it becomes related to the ventral surface of the arch of the aorta and the left pulmonary artery. The nerve later enters the vestigial fold of the left cranial cardinal vein (of the embryo) near the coronary sinus. It is distributed along branches of the coronary arteries, by direct branches to the dorsal atrial walls and pericardium, and via the left pulmonary plexus.

Along its course this nerve may be joined by filaments to the vagal trunk, left recurrent laryngeal nerve, and stellate cardiac nerves.

In general it appears to supply the dorsal or diaphragmatic surface of the heart.

VENTROMEDIAL CERVICAL CARDIAC NERVE (LEFT CRANIAL OR SUPERIOR CARDIOSYMPATHETIC). This nerve originates from the middle cervical ganglion and vagal trunk rostral to the ganglion. A vagal branch may also join the nerve more distally. Filaments enter the aortic arch ventrally, the pretracheal plexus, the wall of the pulmonary artery, and the pericardium.

The nerve is largely composed of sympathetic postganglionic fibers, but parasympathetic preganglionics and visceral afferents are also present.

DORSAL CERVICAL CARDIAC NERVE. This nerve originates from the middle cervical ganglion. It supplies the aortic arch, descending aorta, and pretracheal plexus. It may arise as a branch of the left brachiocephalic nerve. It may receive contributions from the ansa subclavia, cervicothoracic ganglion, and/or vagal trunk.

LEFT BRACHIOCEPHALIC (INNOMINATE NERVES (LEFT CARDIOVAGAL). These nerves originate from the middle cervical ganglion and vagus. They go to the base of the brachiocephalic artery and ventral surface of the aortic arch. Pressoreceptor afferents are carried in these nerves in addition to autonomic fibers.

OTHER SYMPATHETIC BRANCHES IN THE THORACIC REGION

Prominent branches can be seen to follow each of the arteries just rostral to the heart, such as axillary, superficial cervical, and costocervical. These may be referred to as plexuses which follow the respective arteries and are named accordingly. Contributions to these branches are mainly from the ansa subclavia and middle cervical ganglia with occasional filaments from the cervicothoracic ganglion. This ganglion may also supply a branch to the phrenic nerve.

Filaments from the rostral portions of the thoracic sympathetic trunk going directly to the heart have been reported, but apparently are not constant. Small ganglia may be located along the paths of these nerves. Mizeres (1955a) has described the autonomic plexuses of the thoracic region, and the following is based on his work.

The **pretracheal plexus** is found ventral to the tracheal bifurcation and dorsal to the right pulmonary artery. Contributions to this plexus are from the left recurrent laryngeal nerve, ventromedial cervical cardiac nerve, recurrent cardiac nerve, dorsal cervical cardiac nerve, and cranial cardiovagal nerves. This plexus supplies the dorsal atrial walls and continues as the **right and left coronary plexuses** along the coronary arteries. The latter plexuses receive the main part of the recurrent cardiac nerve. Distribution is to the ventricular walls and the ascending aorta. The **dorsal ventricular plexus** lies on the dorsal wall of the left atrium and left ventricle. The fibers, which apparently supply the ventricular wall, come mainly from the ventrolateral cervical cardiac nerve and also from the left recurrent laryngeal and the left stellate cardiac nerves.

The **right pulmonary plexus** receives branches from the vagus and stellate cardiac nerves. The **left pulmonary plexus** is supplied by the vagus and stellate cardiac nerves. It also receives filaments from the **dorsal ventricular plexus.** The pulmonary plexuses carry autonomic and sensory fibers from the parenchyma and vasculature of the lung.

Plexuses on the smaller arteries have already been discussed.

PRESSORECEPTOR AND CHEMORECEPTOR AFFERENTS

A rise in blood pressure activates pressoreceptors located at a number of different sites in the body which are primarily related to the larger vessels. As a sequel to increased pressoreceptor activity, there follows a reflex fall in blood pressure. (Löfving [1961] has delineated a cortical depressor area [sympathoinhibitory] in the rostral part of the cingulate gyrus and a pressor area [sympathoexcitatory] in the subcallosal region. These experiments were done on the cat.) Pressoreceptor cells are known to be located in the wall of the carotid sinus (innervated by the carotid sinus branch of cranial nerve IX) at the bifurcation of the internal and external carotid arteries and also in the walls of the large vessels as they enter and leave the heart. The presence of rather complex sensory spindles in the wall of the carotid sinus raises the question of possible motor control of level of sensitivity of these receptors (Rees 1967). Afferent fibers (aortic depressor fibers) which carry impulses from the aortic arch run rostrally with the vagus nerve and enter the medulla as part of the rootlets of the vagus. Cell bodies of these afferent fibers are in the distal ganglion of the vagus. Nonidez (1937) has described in considerable detail the location of the pressoreceptors and also the location of the group of chemoreceptor cells (aortic glomi or aortic bodies) in this area. The epithelioid bodies themselves are further described as being composed of one type of cell with granular cytoplasm and a round or oval nucleus. Some are closely packed, while others form irregular strands separated by capillaries. These receptor regions are abundantly supplied by nerves composed of mostly medium- and small-diameter fibers. Some fibers form "baskets" around the epithelioid cells. Finer branches end as rings or club-shaped enlargements on the cells. Garner and Duncan (1958) have reported on electron microscopic studies of the structure of the carotid body (glomus) in the cat.

ROUTES OF SYMPATHETIC FIBERS TO THE HEART

Effects of stimulating sympathetic fibers to the heart are generally those of increased rate (accelerator effect) and increased force of contraction (augmentor effect). In stimulation experiments described by Mizeres (1958), the greatest acceleratory effect was usually obtained from T2 and T3 levels of the sympathetic outflow on the right side. Stimulation at the same level on the left side was usually without effect on heart rate. An augmentor effect, however, could always be obtained by stimulating the appropriate sympathetic rami on the left side.

Thus most of the accelerator fibers come

from the right communicating rami (T1 to T4), pass through both limbs of the ansa subclavia, and proceed via the right stellate cardiac nerves to the right atrial wall. Alternate routes appear to be via the cranial and caudal cardiovagal nerves.

Augmentor fibers, which take origin mainly via the left upper thoracic rami communicantes, probably run mainly in the ventrolateral cervical cardiac nerve.

Sympathetic Distribution in the Abdominal Region (Fig. 18–6)

Branches of the thoracic and abdominal sympathetic trunk supply preganglionic fibers to the abdominal and pelvic regions. In addition, these branches also carry a few postganglionic fibers and a considerable number of visceral afferent fibers. As is true of other areas of autonomic nerve distribution in the body, the specific nerve branchings in this region may vary considerably from one animal to the next.

The **thoracic (greater) splanchnic nerve** leaves the thoracic sympathetic trunk approximately at the level of the thirteenth thoracic ganglion. The nerve is larger in diameter than the continuation of the sympathetic trunk. Both pass under the lumbocostal arch lateral to the crus of the diaphragm and enter the abdominal cavity. Other branches designated as lesser splanchnic nerves may leave the sympathetic trunk just caudal to the greater splanchnic nerve. The thoracic splanchnic nerve supplies filaments to the thoracic aorta and the adrenal gland (Marley and Prout 1968). It terminates mainly as several branches to the celiacomesenteric plexus. The adrenal ganglia may also be supplied by this nerve. The lesser splanchnic nerves, if present, distribute to the same general area. Filaments may sometimes be found joining the greater splanchnic to the first lumbar splanchnic nerve.

Immediately after the thoracic splanchnic nerve branches off, the **lumbar sympathetic trunk** is small, but becomes larger as it proceeds caudally. Ganglia may be present for each of the seven lumbar segments, or variable fusion of adjacent ganglia may occur. Preganglionic contributions to the lumbar sympathetic trunk apparently do not occur

caudal to L6. In most instances the caudal limits of the preganglionic contributions are L4 or L5.

Lumbar splanchnic nerves may arise from each of the lumbar sympathetic segmental levels, those from the first five being most constant. They are named according to the level from which they arise. Most originate as grossly dissectable single filaments, but some may be double, and anastomotic branches may join adjacent nerves. Origin may be from a lumbar ganglion or from interganglionic segments of the sympathetic trunk. As noted earlier, the first lumbar splanchnic nerve may have connections with the thoracic splanchnic nerve. In general, the first four lumbar splanchnic nerves distribute to one or more of the following: aorticorenal, cranial mesenteric, and gonadal ganglia; intermesenteric, renal, and gonadal plexuses. Fibers arising from the aorticorenal ganglion provide adrenergic terminals found on the glomerular afferent and efferent arterioles and close to the macula densa cells of the kidney. These nerves also supply the vasa recta and adjacent cortical veins (Dolezel et al. 1976). It should be remembered that veins in general are well innervated and play an active role in circulatory regulation. The fifth through seventh lumbar splanchnic nerves go to the caudal mesenteric ganglion. Filaments also go to the caudal vena caval vein and iliac arteries. Correspondingly named plexuses of variable complexity are found along the arteries of the abdominal region. Those at the root of the celiac and cranial mesenteric arteries form a dense mat or sheath called the celiacomesenteric plexus. This is continuous with the plexuses which are distributed with the branches of these two arteries. Interwoven in the celiacomesenteric plexus are the paired celiac and unpaired cranial mesenteric ganglia.

Small sympathetic ganglia containing approximately 25 to 100 neurons have been demonstrated along the course of the larger arteries of the abdominal cavity. Since these lie distal to the large ganglia such as celiac and cranial mesenteric, it follows that the nerve plexuses along the arteries can be expected to contain some sympathetic preganglionic fibers in addition to the preponderance of sympathetic postganglionics.

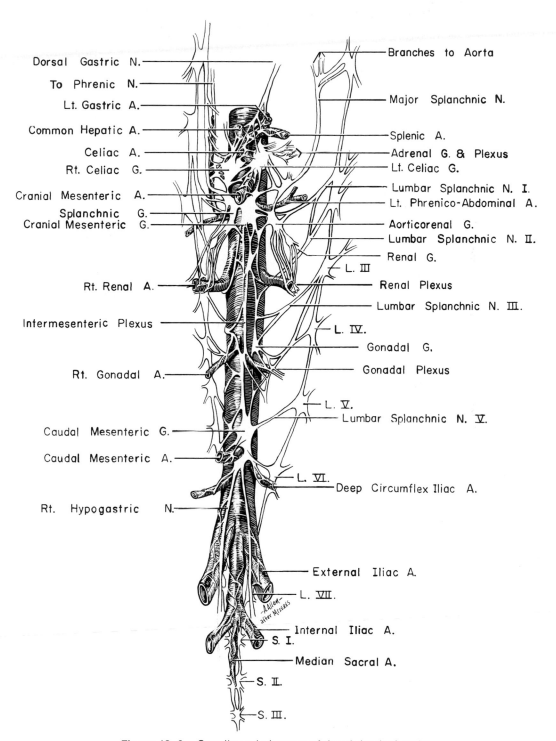

Dorsal Gastric N.
To Phrenic N.
Lt. Gastric A.
Common Hepatic A.
Celiac A.
Rt. Celiac G.
Cranial Mesenteric A.
Splanchnic G.
Cranial Mesenteric G.
Rt. Renal A.
Intermesenteric Plexus
Rt. Gonadal A.
Caudal Mesenteric G.
Caudal Mesenteric A.
Rt. Hypogastric N.

Branches to Aorta
Major Splanchnic N.
Splenic A.
Adrenal G. & Plexus
Lt. Celiac G.
Lumbar Splanchnic N. I.
Lt. Phrenico-Abdominal A.
Aorticorenal G.
Lumbar Splanchnic N. II.
Renal G.
L. III.
Renal Plexus
Lumbar Splanchnic N. III.
L. IV.
Gonadal G.
Gonadal Plexus
L. V.
Lumbar Splanchnic N. V.
L. VI.
Deep Circumflex Iliac A.
External Iliac A.
L. VII.
Internal Iliac A.
S. I.
Median Sacral A.
S. II.
S. III.

Figure 18–6. Ganglia and plexuses of the abdominal cavity.

TABLE 18–1.   AUTONOMIC PLEXUSES OF THE ABDOMINAL REGION

Plexus	Main Arterial Branches Continuing the Plexus	Grossly Dissectable Source (Need Not Imply Synapse)	Structures Supplied
Celiacomesenteric	Celiac and cranial mesenteric	Dorsal vagus, thoracic splanchnic nerves	See structures supplied by branches of celiac and cranial mesenteric arteries
Hepatic	Right gastric Proper hepatic Gastroduodenal (also direct branches to pancreas, common bile duct and pyloric region of stomach)	Right celiac ganglion, dorsal vagus	Pancreas Duodenum Stomach Cystic duct Gall bladder Liver Bile duct
Splenic	Left gastroepiploic Continuation of splenic artery to spleen and pancreas	Left celiac ganglion	Spleen, pancreas, greater curvature of stomach
Left gastric		Celiac ganglia Dorsal vagus	Lesser curvature, cardiac and fundic regions of stomach
Phrenicoabdominal	Phrenic (may arise directly from aorta) Branches to adrenal gland and abdominal wall	Celiac ganglion Adrenal ganglia	Diaphragm, part of abdominal wall Adrenal gland
Adrenal (paired)		Celiac ganglion Thoracic splanchnic nerve Adrenal ganglia Splanchnic ganglia	Adrenal gland
Renal (paired)	Phrenicoabdominal artery may arise from renal	Splanchnic ganglia Aorticorenal ganglia Renal ganglia	Kidney
Cranial mesenteric	Common colic Caudal pancreaticoduodenal Intestinal	Cranial mesenteric ganglion Celiac ganglia Dorsal vagus Lumbar splanchnic nerves	Large intestine Small intestine Cecum
Aortic (intermesenteric)		Cranial mesenteric ganglion, lumbar splanchnic nerves	Caudal mesenteric ganglion Pelvic plexus
Internal spermatic		Aortic plexus Gonadal ganglion (not constant) Lumbar splanchnic nerves	Testis Epididymis
Utero-ovarian	Cranial uterine Ovarian	Same as above	Ovary Oviduct Uterus
Caudal mesenteric	Left colic Cranial hemorrhoidal	Lumbar splanchnic nerves (caudal group) Aortic plexus	Hypogastric nerves (paired) Distal colon (left) Rectum Pelvic plexus

These plexuses also include parasympathetic preganglionic and visceral afferent fibers.

The cellular makeup of the adrenal medulla and the chemical mediators involved indicate that this portion of the adrenal glands is much like a sympathetic ganglion. The medulla is richly innervated by preganglionic fibers that originate from T4 or T5 through L1 or L2 spinal segments in the dog (Cummings 1969). This is in contrast to the sparsely innervated adrenal cortex (Teitlebaum 1942, Arikhbayev 1957, Wilkinson 1961).

Filaments which follow the external and internal iliac arteries apparently may arise from the more caudally located lumbar splanchnic nerves, the hypogastric nerves, and as continuations of the aortic plexus.

The **right and left hypogastric nerves** represent largely postganglionic connections between the caudal mesenteric ganglion and the pelvic plexuses. They are usually grossly dissectable nerves. Jaskowski and Wozniak (1968), in describing the myelinated fiber content of these nerves, note that those with diameters of 3 to 5 micrometers predominate, and none have diameters exceeding 12 micrometers.

The initial portion of the caudal mesenteric plexus has been designated as the lumbar colonic nerve.

Sympathic Distribution in the Pelvic Region (Fig. 18–6)

Both sympathetic trunks lie in close apposition in the sacral region. Variable fusion between corresponding ganglia of the two sides usually occurs. Disagreement exists (Wozniak 1966) as to whether or not a ganglion impar is present in the dog. (Fusion of right and left L7 or S1 [or both] sympathetic ganglia results in a structure which has been interpreted as an impar ganglion, but both sympathetic trunks continue caudal to this level and may contain one or more additional pair of sacral ganglia [Mizeres 1955a]. In man the impar or caudal ganglion terminates the two sympathetic trunks.) Gray rami communicantes join the sympathetic trunk to each of the sacral spinal nerves.

As noted previously, the hypogastric nerves carry sympathetic fibers to the pelvic viscera via the pelvic plexus.

The plexus is located on the lateral surface of the rectum and receives a large contribution of parasympathetic preganglionic fibers via the pelvic nerve. (See discussion on sacral portion of parasympathetic for distribution from pelvic plexus.)

An intact parasympathetic and sympathetic nerve supply is essential to normal regulation of physiological activities such as urination, defecation, erection, and ejaculation. It must be remembered that afferent pathways must also be intact for normal control of these functions.

El-Badawi and Schenk (1967) describe a vascular, interstitial, and peritubular innervation of the epididymis. This consists of both adrenergic and cholinergic fibers, with the former being dominant. The distal third of the tail of the epididymis is especially richly innervated with intricate interstitial and peritubular plexuses.

Mori (1971) reports that stimulation of the peripheral cut ends of L1 through L4 ventral roots almost always causes contraction of the uterine horn and body, followed by temporary inhibition. This effect is abolished by section of the hypogastric nerve. Similar stimulation of the ventral roots of S2, S3, and Cd1 also causes contraction of the uterine horn and body, with greater effect on the latter. The effect is eliminated by cutting the pelvic nerve.

The physiological significance of the hypogastric nerves in normal urinary bladder function has not been definitely established. Electrical stimulation of these nerves usually results in some degree of ipsilateral detrusor muscle contraction (Kuntz and Saccomano 1944). Ingersoll and Jones (1958) report that this contraction is usually followed by relaxation as the stimulus is continued. Responses varied from contraction of small areas to contraction of one entire side of the bladder. The muscle along the lateral edge of the bladder was most apt to respond. Pelvic nerve stimulation results in a more constant and generalized urinary bladder contraction. The reader is referred to papers by Nyberg-Hansen (1966), Gosling and Dixon (1975), Fletcher and Bradley (1969), and Petras and Cummings (1978) for more detailed information on innervation and control of the urinary bladder and urethra in mammals. Failure to understand clearly the comparative role of adrenergic and cholinergic fibers to the bladder most likely results from a failure to appreciate the complexity of this organ's innervation.

Sympathetic Distribution to Somatic Vasculature

Electrical stimulation experiments reported by Cloninger and Green (1955) indicate that vasomotor fibers present in the lumbar sympathetic trunk reach the vessels of the distal hind leg muscles via the sciatic and femoral nerves and also along the femoral artery. As might be expected when considering the corresponding field of somatic innervation, the sciatic nerve is the main route for these fibers. Results of lumbar sympathetic trunk stimulation were mainly vasoconstriction, but vasodilation also resulted, dependent upon exact site of stimulation and time which elapsed since the stimulation. Distribution of adrenergic fibers to blood vessels of skeletal muscle in the cat has been described by Fuxe (1965).

Vasodilator effects have been described as resulting from sympathetic nerve stimulation in the dog, and further experimentation has demonstrated that these fibers are cholinergic (Uvnäs 1954, Folkow 1955, Bolme et al. 1970). Such vasodilator effects may also be obtained by stimulating parts of the motor cortex (Lindgren et al. 1956). The descending pathway has been shown to synapse in the hypothalamus and the collicular area of the midbrain. It further descends through the pons along the dorsolateral border of the corticospinal tract. The small bundle of fibers retains the foregoing relation on its way through most of the medulla oblongata. In the caudal medulla it moves dorsolaterally to descend in the lateral funiculus of the spinal cord.

In describing the perivascular nerve plexuses of an intramuscular artery in the cat, Polley (1955) lists three components:

1. A superficial plexus of the outer adventitia and perivascular connective tissue. Here are located terminations of sensory fibers which are distributed with the segmental somatic innervation to the muscle.

2. A middle or adventitial plexus of medium-sized fibers, apparently sensory, which reaches the vessels by way of the sympathetic trunk. These fibers originate from levels rostral to those which supply the superficial plexus. They are probably the anatomical basis for persistence of some sensation in the hind limbs even after spinal cord section.

3. A deep plexus of small-diameter fibers in the outer media and inner adventitia. These fibers end as small nets and varicosities among and on the smooth muscle cells. The end-formations of each fiber tend to be discrete and do not form an anastomosing network. These are the sympathetic postganglionic endings. See Hassin (1929), White (1963), and Borodulia and Plechkowa (1977) for a description of the nerve supply to cerebral blood vessels; and Derom (1945) for information on vascular innervation. For detailed work on the veins, see Shepherd and Vanhoutte (1975).

Duncan and Shim (1977) report that intraosseous vessels of rabbits have a rich adrenergic nerve supply made up of larger nerves near the vessel surface, medium-sized adventitial nerves which spiral around the long axis of the vessel, and fine tertiary nerves which form a rich plexus at the outer area of the tunica media.

Visceral Pathways and Reflexes

Kuru et al. (1959) reported on previously demonstrated visceral afferent pathways in the spinal cord of the cat. Those which they call the sacrobulbar pathways take origin as two groups.

1. This group of fibers originates in cells of the dorsal root ganglia and runs superficially close to the midline in the dorsal funiculus of the spinal cord. The fibers end in the medulla oblongata near the dorsal nucleus of the vagus. Potentials recorded from this tract corresponded temporally to filling of the bladder.

2. The other group of fibers arises from large second-order neurons located laterally at the base of the sacral dorsal horn. In the thoracic region the fibers are found in the lateral funiculus between the lateral and ventral spinothalamic tracts. They also ter-

minate in the medulla oblongata in relation to several structures (solitary fasciculus, lateral reticular nucleus, nucleus ambiguus, and dorsal nucleus of the vagus).

Kuru et al. (1959) further describe a medullary site in the lateral reticular nucleus which upon electrical stimulation produces contraction of the urinary bladder in the cat. The pathway for these vesicoconstrictor impulses is the lateral reticulospinal tract. Some of the fibers were seen to decussate at the lumbar level. They terminate on sacral preganglionic parasympathetic neurons whose axons become part of the pelvic nerve. Kerr and Alexander (1964) have shown in the cat that autonomic fibers which control the bladder and those that mediate vasoconstriction have a subpial location in the lateral funiculus of the spinal cord. Similarly, Kerr and Brown (1964) give the location of the descending pupillomotor pathways as the most peripheral part of the ventrolateral funiculus of the cord. They further note that mydriasis may result from compression or contusion of the cervical portion of the cord.

Coote and Macleod (1975) have demonstrated sympathoinhibitory pathways which originate in the ventrolateral medulla, the caudal raphe nucleus, and the ventromedial reticular formation. Barman and Wurster (1975) refer to visceromotor pathways which lie in the dorsolateral funiculus of the spinal cord of the dog. Apparently, several descending pathways are involved in the regulation of sympathetic activity.

Miyake (1958) has demonstrated the predominantly inhibitory effect on the colon which follows experimental distention of the small intestine in the dog. Apparently, the effect is mediated through the sympathetic system, and when its influence is removed, the reflex response is reversed to one of excitation. The excitatory response was shown to be mediated by the pelvic nerves and the inhibitory response via the hypogastric colonic fibers. Distention of the urinary bladder in anesthetized dogs usually inhibits its motility of the colon (Hayasi 1959). Part of a large body of evidence relating to somatovisceral and viscerosomatic reflexes as well as to possible relations to acupuncture has been referred to earlier in this chapter. Hypothalamic and mesencephalic mic-

turition centers have been previously demonstrated.

There is a large amount of Russian literature on the autonomic system in mammals, so the reader is referred to suitable abstracting media for further information.

Autonomic End-Formations

Long-standing controversies over the morphological details of autonomic end-formations are beginning to be resolved with the aid of the electron microscope and fluorescence staining techniques. The idea of a terminal autonomic ground plexus or syncytium has been favored for many years. Although this view is still held by many, recent work indicates that what was formerly held to be a continuous network is actually a series of closely apposed Schwann cells which in turn are wrapped around the fine (down to 0.1 micrometer or less) branchings of autonomic postganglionic axons (Richardson 1960; see also Hillarp 1959). Indications are that this Schwann sheath may be lost just before the autonomic fibers terminate next to the innervated cell (Richardson 1962). Well-defined neuromuscular terminals appear to be lacking, and the usual description is that of very fine, bare axons coursing among smooth muscle fibers but separated from them by many hundreds of angstrom units. Vesicular structures of variable shape and density may be found within these terminals (Lever et al. 1967, Ono 1967). Terminal or *en passant* varicosities of 0.5 to 2.0 μm. diameter are described for both adrenergic and cholinergic fibers (Burnstock and Bell 1974, Jacobowitz 1974). Lying among the Schwann cell formations are fibroblast-like cells, the so-called interstitial cells of Cajal which were once thought to be a link in the autonomic pathway. The identification and role of these cells is also highly controversial (Jabonero 1962, DuPont and Sprinz 1964, Rogers and Burnstock 1966). Förster et al. (1975) have reintroduced the concept of interstitial cells (in the skin) which act as final links in autonomic distribution. Based on this and a variety of other evidence, it appears that the "two neuron chain" concept of the peripheral autonomic system is no longer entirely accurate.

BIBLIOGRAPHY

Akagi, Y., Y. Ibata, and Y. Sano. 1976. The sympathetic innervation of the ciliary body and trabecular meshwork of the cat. Fluorescence histochemistry and electron microscopy. Cell Tissue Res. 173:261–269.

Arikhbayev, K. P. 1957. On the problem of innervation of the adrenal glands. Tr. Stalingrad. Med. Inst. 1957(25):55–64.

Armour, J. A., and W. C. Randall. 1975. Functional anatomy of canine cardiac nerves. Acta Anat. 91:510–528.

Bakeyeva, N. A. 1957. Innervation of the ileocaecal area of the intestine in man and dog. Tr. Stalingrad. Med. Inst. 1957(25):87–135.

Ballard, K., T. Malmfors, and S. Rosell. 1974. Adrenergic innervation and vascular patterns in canine adipose tissue. Microvasc. Res. 8:164–171.

Barman, S. N., and R. D. Wurster. 1975. Visceromotor organization within descending spinal sympathetic pathways in the dog. Circ. Res. 37:209–214.

Baumann, J. A., and S. Gajisin. 1975. Multiplicity and dispersion of the parasympathetic ganglia of the head. Bull. Assoc. Anat. (Nancy) 59:329–332.

Beattie, J., G. R. Brow, and C. N. H. Long. 1930. Physiological and anatomical evidence for the existence of nerve tracts connecting the hypothalamus with spinal sympathetic centres. Proc. Roy. Soc. B 106:253–275.

Bell, M., and W. Montagna. 1972. Innervation of sweat glands in horses and dogs. Br. J. Dermatol. 86:160–163.

Billingsley, P. R., and S. W. Ranson. 1918a. On the number of nerve cells in the ganglion cervicale superius and of nerve fibers in the cephalic end of the truncus sympathicus in the cat and on the numerical relations of preganglionic and postganglionic neurones. J. Comp. Neurol. 29:359–366.

— — — —. 1918b. Branches of the ganglion cervicale superius. J. Comp. Neurol. 29:367–384.

Blackman, J. G. 1974. Function of autonomic ganglia. In The Peripheral Nervous System, edited by John I. Hubbard. New York, Plenum Press.

Bloom, S. R., A. V. Edwards, and N. J. A. Vaughan. 1973. The role of the sympathetic innervation in the control of plasma glucagon concentration in the calf. J. Physiol. 233:457–466.

Bolme, P., J. Novotny, B. Uvnäs, and P. G. Wright. 1970. Species distribution of sympathetic cholinergic vasodilator nerves in skeletal muscle. Acta. Physiol. Scand. 78:60–64.

Borodulia, A. V., and E. K. Plechkowa. 1977. Adrenergic-sympathetic nervous apparatus of the cerebral arteries and its role in regulating cerebral circulation. Zlr. Neuropathol. Psikhiati. 77:975–980.

Burnstock, G., and C. Bell. Peripheral autonomic transmission. In The Peripheral Nervous System, edited by John I. Hubbard. New York, Plenum Press.

Chase, M. R., and S. W. Ranson. 1914. The structure of the roots, trunks, and branches of the vagus nerve. J. Comp. Neurol. 24:31–60.

Chiu, S. L. 1943. The superficial hepatic branches of the vagi and their distribution to the extrahepatic biliary tract in certain mammals. Anat. Rec. 86:149–155.

Cloninger, G. L., and H. D. Green. 1955. Pathways taken by the sympathetic vasomotor nerves from the sympathetic chain to the vasculature of the hind leg muscles of the dog. Am. J. Physiol. 181:258–262.

Coleridge, H. M., J. C. G. Coleridge, A. Dangel, C. Kidd, J. C. Luck, and P. Sleight. 1973. Impulses in slowly conducting vagal fibers from afferent endings in the veins, atria and arteries of dogs and cats. Circ. Res. 33:87–97.

Coote, J. H., and V. H. Macleod. 1975. The spinal route of sympatho-inhibitory pathways descending from the medulla oblongata. Pflügers Arch. 359:335–347.

Countee, R. W. 1977. Extrinsic neural influences on gastrointestinal motility. Am. Surg. 43:621–626.

Cummings, J. F. 1969. Thoracolumbar preganglionic neurons and adrenal innervation in the dog. Acta Anat. 73:27–37.

Cunliffe, W. J. 1961. The innervation of the thyroid gland. Acta Anat. (Basel) 46:135–141.

Dahlström, A., M. Mya-Tu, and K. Fuxe. 1965. Observations on adrenergic innervation of dog heart. Am. J. Physiol. 209:689–692.

Dahlström, A., and K. Fuxe. 1965. Evidence for the existence of monamine neurons in the C.N.S. II. Experimentally induced changes in the intraneuronal amine levels of bulbospinal neuron systems. Acta Physiol. Scand. 64 (Suppl.) 247:1–36.

Daly, M. deB., and C. Hebb. 1952. Pulmonary vasomotor fibers in the cervical vagosympathetic nerve of the dog. Q. J. Exp. Physiol. 37:19–44.

Denn, M. J., and H. L. Stone. 1976. Autonomic innervation of dog coronary arteries. J. Appl. Physiol. 41:30–35.

Derom, E. 1945. Experimental researches on the vasomotor innervation of the dog's front paw. Bull. Acad. Roy. Méd. Belg. 10:427–459.

Diamare, V., and M. de Mennato. 1930. Contributo all' anatomia ed allo sviluppo del sistema nervoso simpatico. Atti R. Accad. Sci. Fis. e. Mat. (Naples) 18:1–115.

Dolezel, S., L. Edvinsson, C. Owman, and T. Owman. 1976. Fluorescence histochemistry and autoradiography of adrenergic nerves in the renal juxtaglomerular complex of mammals and man with special regard to the efferent arteriole. Cell Tissue Res. 169:211–220.

Downman, C. B. B. 1955. Skeletal muscle reflexes of splanchnic and intercostal nerve origin in acute spinal and decerebrate cats. J. Neurophysiol. 18:217–235.

Duncan, C. P., and S. S. Shim. 1977. The autonomic nerve supply of bone. An experimental study of the interosseous adrenergic nervi vasorum in the rabbit. J. Bone Joint Surg. (Br.) 59:323–330.

DuPont, J.-R., and H. Sprinz. 1964. The neurovegetative periphery of the gut. A revaluation with conventional technics in the light of modern knowledge. Am. J. Anat. 114:393–402.

Ehinger, B. 1966. Distribution of adrenergic nerves in the eye and some related structures in the cat. Acta. Physiol. Scand. 66:123–128.

— — — —. 1967. Double innervation of the feline iris dilator. Arch. Ophthalmol. 77:541–545.

Ehinger, B., B. Falck, and B. Sporrong. 1967. Adrenergic fibres to the heart and to peripheral vessels. Bibl. Anat. 8:34–45.

El-Badawi, A., and E. A. Schenk. 1967. The distribution of cholinergic and adrenergic nerves in the mammalian epididymis: a comparative histochemical study. Am. J. Anat. *121*:1–14.

————. 1973. Parasympathetic and sympathetic postganglionic synapses in ureterovesical autonomic pathways. Z. Zellforsch. *146*:147–154.

El'bert, M. E. 1956. On the problem of the cytoarchitecture of Auerbach's plexus of the small intestines in the cat and dog. Tr. Mosk. Vet. Akad. 1956(*18*):35–38.

Elftman, A. G. 1943. The afferent and parasympathetic innervation of the lungs and trachea of the dog. Am. J. Anat. *72*:1–23.

Erez, B. M. 1955. The problem of the extracardiac nerve system in dogs. Uch. Zap. Tadzh. Inst. 1955(*6*):145–153.

Esterhuizen, A. C., J. D. Graham, and J. D. Lever. 1967. The innervation of the smooth muscle of the nictitating membrane of the cat. J. Physiol. (London) *192*:41P–42P.

Faden, A. I., and Petras, J. M. 1978. An intraspinal sympathetic preganglionic pathway: anatomic evidence in the dog. Brain Res. *144*:358–362.

Fields, H. L., G. A. Meyer, and L. D. Partridge, Jr. 1970. Convergence of visceral and somatic input onto spinal neurons. Exp. Neurol. *26*:36–52.

Filogamo, G. 1950. Ricerche sul plesso mienterico. Arch. Ital. Anat. Embriol. *54*:401–412.

Fletcher, T. F., and W. F. Bradley. 1969. Comparative morphologic features of urinary bladder innervation. Am. J. Vet. Res. *30*:1655–1662.

Foley, J. O. 1945. The sensory and motor axons of the chorda tympani. Proc. Soc. Exp. Biol. (N.Y.) *60*:262–267.

Foley, J. O., and F. S. DuBois. 1940. A quantitative and experimental study of the cervical sympathetic trunk. J. Comp. Neurol. *72*:587–601.

Folkow, B. 1955. Nervous control of blood vessels. Physiol. Rev. *35*:629–663.

Förster, F. J., H. Heine, and G. Schaeg. 1975. Histophysiology of the vegetative peripheral nervous system of skin. Arch. Dermatol. Res. *254*:295–302.

Franke, F. E., and P. O. Bramante. 1964. Spinal origin of nasal vasoconstrictor innervation in the dog. Proc. Soc. Exp. Biol. Med. *117*:769–771.

Fukuyama, U., and M. Yabuki. 1958. On the vertebral nerve and the communicating rami connecting with the inferior cervical ganglion in the dog. Fukushima J. Med. Sci. *5*:63–88.

Fussy, I. F., C. Kidd, and J. G. Whitwam. 1973. Activity evoked in the brain stem by stimulation of C fibers in the cervical vagus nerve of the dog. Brain Res. *49*:436–440.

Fuxe, K. 1965. The distribution of adrenergic nerve fibers to the blood vessels in skeletal muscle. Acta. Physiol. Scand. *64*:75–86.

Garner, C. M., and D. Duncan. 1958. Observations on the fine structure of the carotid body. Anat. Rec. *130*:691–710.

Gaskell, W. H. 1886. On the distribution and function of the nerves which innervate the visceral and vascular systems. J. Physiol. *7*:1–80.

Geis, W. P., M. P. Kaye, and W. C. Randall. 1973. Major autonomic pathways to the atria and S-A and A-V nodes of the canine heart. Am. J. Physiol. *224*:202–208.

Gellhorn, E. 1967. Autonomic-Somatic Integrations. Physiological Basis and Psychological and Clinical Implications. Minneapolis, Univ. Minn. Press, 318 pp.

Ginzel, K. H., E. Eldred, and J. A. Estavillo. 1972. Depression of a motoneuron activity by excitation of visceral afferents in the cardiopulmonary region. Int. J. Neurosci. *4*:203–214.

Gold, W. M., G. F. Kessler, and D. Y. C. Yu. 1972. Role of vagus nerves in experimental asthma in allergic dogs. J. Appl. Physiol. *33*:719–725.

Gonzalo-Sanz, M., and J. Ullan. 1976. Autonomic corticofugal fibers. Bull. Assoc. Anat. (Nancy) *60*:685–693.

Gosling, J. A., and J. S. Dixon. 1975. The structure and innervation of smooth muscle in the wall of the bladder, neck and proximal urethra. Br. J. Biol. *47*:549–558.

Greenberg, S. R. 1954. Sympathetic components of the vagus cardiac nerves of the dog. Anat. Rec. *118*:304–305.

————. 1956. A fiber analysis of the vagus cardiac rami and the cervical sympathetic nerves in the dog. J. Comp. Neurol. *104*:33–48.

Gross, D. Pain and autonomic nervous system. Adv. Neurol. *4*:93–103, 1974.

Gunn, M. 1968. Histological and histochemical observations on the myenteric and submucous plexuses of mammals. J. Anat. *102*:223–239.

Hamberger, B., and K.-A. Norberg. 1965. Adrenergic synaptic terminals and nerve cells in bladder ganglia of the cat. Int. J. Neuropharmacol. *4*:41–45.

Hassin, G. B. 1929. The nerve supply of the cerebral blood vessels; a histologic study. Arch. Neurol. Psychiat. (Chic.) *22*:375–391.

Hayasi, T. 1959. On the effects of distension of the urinary bladder upon the colon. J. Physiol. Soc. Japan *21*:380–385.

Hillarp, N. A. 1959. The construction and functional organization of the autonomic innervation apparatus. Acta Physiol. Scand. *46* (Suppl. 157):1–38.

Hirsch, E. F., C. A. Nigh, M. P. Kaye, and T. Cooper. 1964. Terminal innervation of the heart II. Studies of the perimysial innervation apparatus and of sensory receptors in the rabbit and in the dog with the techniques of total extrinsic denervation, bilateral cervical vagotomy and bilateral thoracic sympathectomy. Arch. Pathol. *77*:172–187.

Hirsch, E. F., G. C. Kaiser, and T. Cooper. 1964. Experimental heart block in the dog. I. The distribution of nerves, their ganglia and terminals in the septal myocardium of the dog and human heart. Arch. Pathol. *78*:523–532.

————. 1965. Experimental heart block in the dog. III. Distribution of the vagus and sympathetic nerves in the septum. Arch. Pathol. *79*:441–451.

Holmberg, J. 1971. The secretory nerves of the parotid gland of the dog. J. Physiol. (London) *219*:463–476.

Holmberg, J. 1972. Release of acetylcholine in the parotid gland of the dog during stimulation of postganglionic nerves. Acta Physiol. Scand. *86*:115–119.

Holmes, R. L. 1956. Further observations on the nerve endings in the adult dog heart. J. Anat. (Lond.) *90*:600.

————. 1957. Structures in the atrial endocardium of the dog which stain with methylene blue, and the effects of unilateral vagotomy. J. Anat. (Lond.) *91*:259–266.

Huhtala, A., T. Tervo, and A. Palkama. 1975. Autonomic and somatic sensory nerves of the iris and cornea. Acta Opthalmol. (Kbh.) (Suppl.) 125:47–48.

Hwang, K., M. I. Grossman, and A. C. Joy. 1948. Nervous control of the cervical portion of the esophagus. Am. J. Physiol. 154:343–357.

Iljina, W. J., and B. J. Lawrentjew. 1932a. Zur Lehre von der Cytoarchitektonik des peripherischen autonomen Nervensystems; Ganglien des Rektums und ihre Beziehungen zu dem sakralen Parasympathikus. Z. mikr.-anat. Forsch. 30:530–542.

– – – – –. 1932b. Experimentell-morphologische Studien über den feineren Bau des autonomen Nervensystems; Über die Innervation der Harnblase. Z. mikr.-anat. Forsch. 30:543–550.

Ingersoll, E. H., and L. L. Jones. 1958. Effect upon the urinary bladder of unilateral stimulation of hypogastric nerves in the dog. Anat. Rec. 130:605–616.

Jabonero, V. 1962. New observations on the fine innervation of the esophagus. Trab. Inst. Cajal Invest. Biol. 54:37–92.

Jacobowitz, D. M. 1974. The peripheral autonomic system. In The Peripheral Nervous System, edited by John I. Hubbard. New York, Plenum Press.

Jaskowski, J., and W. Wozniak. 1968. Myelin fibers in the hypogastric nerves in the dog. Folia Morphol. (Warsz.) 27:387–392.

Jayle, G. E. 1935. Le centre hypogastrique du chien. Arch. Anat. (Strasbourg) 19:357–367.

Johansson, B. 1962. Circulatory response to stimulation of somatic afferents. Acta Physiol. Scand. 57 (Suppl.):198.

Jung, L., R. Tagand, and F. Chavanne. 1926. Sur l'innervation excito-sécrétoire de la muqueuse nasale. C. R. Soc. Biol. (Paris) 95:835–837.

Kapellar, K. 1965. The origin and the course of nerves leading to the liver in the dog. Folia Morph. (Prague) 13:12–14.

Katsuki, S., T. Okajima, and T. Matsumoto. 1955. Experimental studies on the autonomic function of the preoptic area; influences of electrical stimulation of the preoptic area upon blood pressure, and movement of pelvic organs (urinary bladder, rectum, and uterus). Kumamoto Med. J. 8:180–186.

Kelts, K. A., and K. E. Bignall. 1973. Spinal and cortical inhibition of intestinal motility in the squirrel monkey. Exp. Neurol. 41:387–394.

Kemp, D. R. 1973. A histological and functional study of the gastric mucosal innervation in the dog. Part I: The quantification of the fibre content of the normal supradiaphragmatic vagal trunks and their abdominal branches. Austr. N.Z. J. Surg. 43:288–294.

Kent, K. M., S. E. Epstein, T. Cooper, and D. M. Jacobowitz. 1974. Cholinergic innervation of the canine and human ventricular conducting system. Anatomic and electrophysiologic correlations. Circulation 50:948–955.

Kerr, F. W. L., and S. Alexander. 1964. Descending autonomic pathways in the spinal cord (cat, monkey). Arch. Neurol. 10:249–261.

Kerr, F. W. L., and J. A. Brown. 1964. Pupillomotor pathways in the spinal cord (cat, monkey). Arch. Neurol. 10:262–270.

Klaisman, E. B. 1975. An adrenergic component of the nervous apparatus of the aorta reflexogenic zone. Bull. Exp. Biol. Med. 77:825–828.

Kuntz, A. 1945. Anatomic and physiologic properties of cutaneo-visceral vasomotor reflex arcs. J. Neurophysiol. 8:421–430.

Kuntz, A., and G. Saccomanno. 1944. Sympathetic innervation of the detrusor muscle. J. Urol. (Baltimore) 51:535–542.

Kuru, M., T. Kurati, and Y. Koyama. 1959. The bulbar vesicoconstrictor center and the bulbo-sacral connections arising from it; a study of the function of the lateral reticulospinal tract. J. Comp. Neurol. 113:365–388.

Kyösola, S. Partanen, O. Korkala, E. Merikallio, O. Penttila, and P. Siltanen. 1976. Fluorescence histochemical and electron-microscopical observations on the innervation of the atrial myocardium of the adult human heart. Virchows Arch. A. Path. Anat. Histol. 371:101–119.

Kyösola, K., and L. Rechardt. 1974. The anatomy and innervation of the sphincter of Oddi in the dog and cat. Am. J. Anat. 140:497–522.

Landau, W. 1953. Autonomic responses mediated via the corticospinal tract. J. Neurophysiol. 16:299–311.

Lawrentjew, B. J. 1931. Zur Lehre von der Cytoarchitektonik des peripherischen autonomen Nervensystems; die Cytoarchitektonik der Ganglien des Verdauungskanals beim Hunde. Z. mikr.-anat. Forsch. 23:527–551.

Leaming, D. B., and N. Cauna. 1961. A qualitative and quantitative study of the myenteric plexus of the small intestine of the cat. J. Anat. (Lond.) 95:160–169.

Lever, J. O., J. O. Graham, and T. L. Spriggs. 1967. Electron microscopy of nerves in relation to the arteriolar wall. Bibl. Anat. 8:51–55.

Lindgren, P., A. Rosen, P. Strandberg, and B. Uvnäs. 1956. The sympathetic vasodilator outflow; a new cortico-spinal autonomic pathway. J. Comp. Neurol. 105:95–109.

Löfving, B. 1961. Cardiovascular adjustments induced from the rostral cingulate gyrus. Acta Physiol. Scand. 53 (Suppl. 184):1–82.

Malmfors, T. 1968. Histochemical studies on the adrenergic innervation of the nictitating membrane of the cat. Histochemie 13:203–206.

Marley, E., and G. I. Prout. 1968. Innervation of the cat's adrenal medulla. J. Anat. 102:257–273.

Martin, P. 1977. The influence of the parasympathetic nervous system on atrioventricular conduction. Circ. Res. 41:593–599.

Matthews, M. R. 1974. Ultrastructure of ganglionic junctions. In The Peripheral Nervous System, edited by John I. Hubbard. New York, Plenum Press.

McKibben, J. S., and R. Getty. 1968. A comparative morphologic study of the cardiac innervation of domestic animals I. The canine. Am. J. Anat. 122:533–543.

Mehler, W. R., J. C. Fischer, and W. F. Alexander. 1952. The anatomy and variations of the lumbosacral sympathetic trunk in the dog. Anat. Rec. 113:421–435.

Miyake, T. 1958. Effects of intestinal distention upon the movements of the colon. J. Physiol. Soc. Japan 20:744–751.

Mizeres, N. J. 1955a. The anatomy of the autonomic nervous system in the dog. Am. J. Anat. 96:285–318.

– – – – –. 1955b. Isolation of the cardioinhibitory branches of the right vagus nerve in the dog. Anat. Rec. 123:437–446.

– – – – –. 1957. The course of the left cardioinhibitory fibers in the dog. Anat. Rec. 127:109–115.

— — — — —. 1958. The origin and course of the cardio-accelerator fibers in the dog. Anat. Rec. *132*:261–279.

Mori, T. 1971. Origin from the spinal cord of motor innervation to uterine horn and corpus uteri in dogs. Acta Obstet. Gynaecol. Jap. *18*:101–106.

Napolitano, L. M., V. L. Willman, C. R. Hanlon, and T. Cooper. 1965. Intrinsic innervation of the heart. Am. J. Physiol. *208*:455–458.

Nishi, S. Ganglionic transmission. 1974. *In* The Peripheral Nervous System, edited by John I. Hubbard. New York, Plenum Press.

Nitschke, T. 1976. Die Rami Orbitales des Ganglion plerygapalatinum des Hundes: Zugleich ein Beitrag Über die Innervation der Tränendrüse. Anat. Anz. *139*:58–70.

Nonidez, J. F. 1935. Innervation of the thyroid gland: Distribution and termination of the nerve fibers in the dog. Am. J. Anat. *57*:135–170.

— — — — —. 1937. Distribution of the aortic nerve fibers and the epithelioid bodies (supracardial "paraganglia") in the dog. Anat. Rec. *69*:299–313.

— — — — —. 1939. Studies on the innervation of the heart; distribution of the cardiac nerves with special reference to the identification of the sympathetic and parasympathetic postganglionics. Am. J. Anat. *65*:361–407.

Norberg, K., A. B. Persson, and P. O. Granberg. 1975. Adrenergic innervation of the human parathyroid glands. Acta. Chir. Scand. *141*:319–322.

Nyberg-Hansen, R. 1966. Innervation and nervous control of the urinary bladder: anatomical aspects (man, mammals). Third Scand. Symp. on Multiple Sclerosis: The neurogenic bladder. Aarhus, Denmark, 18–19 Oct. 1965. Acta Neurol. Scand. *42* (Suppl. 20):7–24.

Obrebowski, A. 1965. Subdiaphragmatic segments of the vagus nerve in dogs. Folia Morph. (Warsz.) *24*:295–301 (Trans.).

Okabe, Y. 1958. The vagus innervation of the cardiac sphincter. J. Physiol. Soc. Jap. *20*:752–763.

— — — — —. 1959. The splanchnic innervation of the cardiac sphincter. J. Physiol. Soc. Jap. *21*:43–49.

Ono, M. 1967. Electron microscopic observations on the ganglia of Auerbach's plexus and autonomic nerve endings in muscularis externa of the mouse small intestine. Sapporo Med. J. *32*:56–74.

Osterholm, J. L. 1974. Minireview. Noradrenergic mediation of traumatic spinal cord autodestruction. Life Sciences *14*:1363–1384.

Pannier, R. 1946. Contribution a l'innervation sympathique du coeur; les nerfs cardioaccelerateurs. Arch. Int. Pharmacodyn. *73*:193–259.

Pawlowski, A., and G. Weddell. 1967. The lability of cutaneous neural elements. Br. J. Dermatol. *79*:14–19.

Petras, J. M., and J. F. Cummings. 1978. Sympathetic and parasympathetic innervation of the urinary bladder and urethra. Brain Res. *153*:363–369.

Petras, J. M., and A. I. Faden. 1978. The origin of sympathetic preganglionic neurons in the dog. Brain Res. *144*:353–357.

Polley, E. H. 1955. The innervation of blood vessels in striated muscle and skin. J. Comp. Neurol. *103*:253–268.

Randall, W. C., M. Szentivanyi, and J. B. Pace. 1968. Patterns of sympathetic nerve projections onto the canine heart. Circ. Res. *22*:315–323.

Rees, P. M. 1967. Observations on the fine structure and distribution of presumptive baroreceptor nerves at the carotid sinus. J. Comp. Neurol. *131*:517–548.

Richardson, K. C. 1960. Studies on the structure of autonomic nerves in the small intestine, correlating the silver impregnated image in light microscopy with the permanganate fixed ultrastructure in electron microscopy. J. Anat. (Lond.) *94*:457–472.

Richardson, K. C. 1962. The fine structure of autonomic nerve endings in smooth muscle of the rat vas deferens. J. Anat. (London) *96*:427–442.

Rogers, D. C., and G. Burnstock. 1966. The interstitial cell and its place in the concept of the autonomic ground plexus. J. Comp. Neurol. *126*:255–284.

Rostad, H. 1973. Central and peripheral nervous control of colonic motility in the cat. J. Oslo City Hosp. *23*:65–75.

Sanbe, S. 1961. Studies on the incoming and outgoing myelinated nerve fibers of the sympathetic superior cervical ganglion. Fukushima J. Med. Sci. 8:109–129.

Sato, A., and R. E. Schmidt. 1973. Somatosympathetic reflexes: afferent fibers, central pathways, discharge characteristics. Physiol. Rev. *54*:916–947.

Sato, A., Y. Sato, F. Shimada, and Y. Torigata. 1975. Changes in vesical function produced by cutaneous stimulation in rats. Brain Res. *94*:465–474.

Schmidt, C. A. 1933. Distribution of vagus and sacral nerves to the large intestine. Proc. Soc. Exp. Biol. (N.Y.) *30*:739–740.

Schnitzlein, H. N., H. H. Hoffman, D. M. Hamlett, and E. M. Howell. 1963. A study of the sacral parasympathetic nucleus. J. Comp. Neurol. *120*:477–493.

Schofield, G. C. 1961. Experimental studies on the innervation of the mucous membrane of the gut. Brain *83*:490–514.

— — — — —. 1962. Experimental studies on the myenteric plexus in mammals. J. Comp. Neurol. *119*:159–185.

Shepherd, J. T., and P. M. Vanhoutte. Veins and Their Control. 1975. Philadelphia, W. B. Saunders Co.

Shindin, S. M. 1964. Receptor elements in the ganglia of the autonomic part of the peripheral nervous system. Kazanskogo Med. Inst. *13*:96–97.

Shizume, K. 1952. The cardio-accelerator fibers in the vagus nerve. Nippon J. Angio-cardiol. *16*:8–14.

Sinclair, D. C. The nerves of the skin. 1973. *In* Vol. 2. The Nerves and Blood Vessels. The Physiology and Pathophysiology of Skin, edited by A. Jarret. New York, Academic Press.

Tcheng, K. T. 1950. Étude histologique de l'innervation cardiaque chez le chien. C. R. Soc. Biol. (Paris) *144*:882–883.

— — — — —. 1951. Innervation of the dog's heart. Am. Heart J. *41*:512–524.

Teitlebaum, H. 1942. The innervation of the adrenal gland. Q. Rev. Biol. *17*:135.

Tervo, T., and A. Palkama. 1976. Sympathetic nerves to the rat cornea. Acta Ophthalmol. (Kbh.) *54*:75–84.

Tikhonova, L. P. 1968. Receptor neurons of ganglia in intramural nerve plexus of the large intestine in cats. Arch. Anat. Gistol. Embriol. *54*:98–102.

Todd, G. L., and G. R. Bernard. 1973. The sympathetic innervation of the cervical lymphatic duct of the dog. Anat. Rec. *177*:303–316.

Uchizono, K. 1964. Innervation of the blood capillary in the heart of dog and rabbit. Jap. J. Physiol. *14*:587–598.

Uusitalo, R. 1972. Effect of sympathetic and parasympathetic stimulation on the secretion and outflow of aqueous humour in the rabbit eye. Acta Physiol. Scand. *86*:315–326.

Uvnäs, B. 1954. Sympathetic vasodilator outflow. Physiol. Rev. *34*:608–618.

Vajda, J., E. Feher, and K. Csanyi. 1973. The sensory nerve terminals of the mesentery. Acta Anat. *85*:514–532.

Van Buskirk, C. 1945. The seventh nerve complex. J. Comp. Neurol. *82*:303–330.

Wakata, S. 1975. Studies on the inflowing and outflowing myelinated nerve fibers of the pterygopalatine ganglion in the dog. Chiba Med. J. *51*:133–142, 1975.

Walles, B., L. Edvinsson, B. Falck, C. Owman, N. O. Sjoberg, and K. G. Svensson. 1975. Evidence for a neuromuscular mechanism involved in the contractility of the ovarian follicular wall: Fluorescence and electron microscopy and effects of tyramine on follicle strips. Biol. Reprod. *12*:239–248.

Watson, A. G. 1974. Some aspects of the vagal innervation of the canine esophagus, an anatomical study. Masters Thesis, Massey Univ., N.Z.

White, J. C. 1963. Nervous control of the cerebral vascular system. Clin. Neurosurg. *9*:67–87.

Wilkinson, I. M. S. 1961. The intrinsic innervation of the suprarenal gland. Acta Anat. (Basel) *46*:127–134.

Wirsman, G. G., D. S. Jones, and W. C. Randall. 1966. Sympathetic outflows from cervical spinal cord in the dog. Science *152*:381–382.

Wozniak, W. 1966. Sacral segments of the sympathetic trunks in the dog, cat, and man. Folia Morphol. (Warsz.) *25*:407–414.

Wozniak, W., and U. Skowronska. 1967. Comparative anatomy of pelvic plexus in cat, dog, rabbit, macaque and man. Anat. Anz. *120*:457–473.

Zlatitskaya, N. N. 1968. Sensory innervation of the small intestine in cats. Arch. Anat. Gistol. Embriol. *54*:102–105.

Zremianski, A., A. Orebowski, and A. Kampf. 1967. Innervation of the sites of division of the bronchii by the vagi nerves in the dog. Folia Morphol. (Warsz.) *26*:471–478.

Chapter 19

THE EAR[*]

The **ear,** or organ of hearing and balance *(organum vestibulocochleare),* is divided, for purposes of discussion, into three portions: the external ear, the middle ear, and the internal ear. The **external ear** *(auris externa)* consists of the auricle, or pinna, and the external auditory meatus. The **middle ear** *(auris media)* consists of the tympanic cavity, the tympanic membrane, and the three auditory ossicles with their associated ligaments and muscles. The middle ear cavity is connected with the pharynx by way of the auditory tube (formerly known as the Eustachian or pharyngotympanic tube). The **internal ear** *(auris interna)* consists of a membranous labyrinth enclosed in the petrous portion of the temporal bone (Gray 1907, Bast and Anson 1949, Anson and Donaldson 1973). The internal ear is the organ for both hearing and equilibrium, whereas the external ear and middle ear represent a sound-collecting and conducting apparatus (Honda 1908, Miller and Witter 1942, Getty et al. 1956.)

EXTERNAL EAR

The **pinna** *(auricula)* of the external ear is a funnel-like plate of cartilage which serves to receive air vibrations and transmit them via the ear canal to the tympanic membrane (eardrum) (Fig. 19–1). The tympanic membrane is enclosed in the deep portion of the external acoustic meatus and forms the lateral wall of the middle ear. The pinna is covered on both sides with skin which is tightly attached to the perichondrium. The pinnae are highly mobile and can be controlled independently. The shape of the pinna is characteristic of the breed. Some are small, erect, and V-shaped, as in toy terrier breeds; some may be slightly tipped, as in Collies and Irish terriers; others may be large and pendant, as in the hounds.

The **auricular cartilage** is pierced by many foramina which permit the passage of blood vessels. (For a discussion of the blood vessels, see Sis [1962] and Chapters 11 and 12.) The skin covering the inner or concave surface of the pinna is firmly attached, and thus, when the ear is traumatized, hemorrhage may occur between the skin and the cartilage. The auricular cartilage is attached to the external acoustic process of the tem-

---

[*]By the late Robert Getty, revised from the first edition.

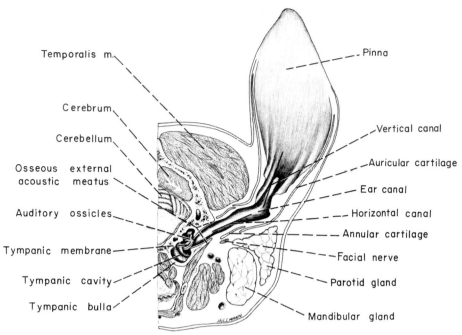

Figure 19–1. Transverse section through head showing ear canal. (Modified after Sis.)

poral bone by means of a smaller annular cartilage. This is usually about 2 cm. long, presenting a lumen of 5 to 10 mm. in diameter. The description that follows is based upon the pendant type of ear held erect. The isolated cartilages will be described separately.

The opening of the ear canal faces dorsolaterally. The apex of the pinna points dorsally, the convex or outer surface faces medially, and the concave or inner surface faces laterally. Thus the margins of the pinna are rostral and caudal in the description. The elastic cartilage is thin and pliable, and at its proximal end it thickens where it is rolled into the form of a tube. The term "helix" is applied to the slightly folded free margin of the cartilage. A low transverse ridge, the anthelix (Fig. 19–2), is present on the medial wall of the initial proximal part of the ear canal. The concave triangular area between the helix and the anthelix is the scapha. A relatively dense, irregularly quadrangular plate of cartilage known as the tragus forms the lateral boundary of the initial portion of the ear canal lying opposite the anthelix. The tragus (Fig. 19–2) curves caudomedially and with the proximal end of the antitragus completes

the caudal boundary of the opening into the ear canal. The antitragus is a thin, elongated piece of cartilage caudal to the tragus and separated from it by an important notch, the incisura intertragica.

The antitragus may be divided into two limbs: the medial cornu and the lateral cornu. They are caudally demarcated by the antitragic incisure. The apex of the lateral cornu ends in a sharp process, the styloid process of the antitragus. Just distal to this, the caudal border presents the cutaneous helicine pouch. A prominent feature of the caudal border of the auricular cartilage is the caudal process of the helix. Proximal to the caudal process in the region of the cutaneous pouch is the deep caudal incisure or antitragohelicine fissure.

The rostral border of the auricular cartilage is nearly straight. At the junction of the proximal and middle thirds, the spine of the helix or distal crus of the helix is formed by an abrupt incisure of this border. The medial crus of the helix is separated from the tragus by the tragohelicine incisure. The lateral crus of the helix arises rostral to the medial crus and extends around the medial crus to overlap the rostral border of the tragus.

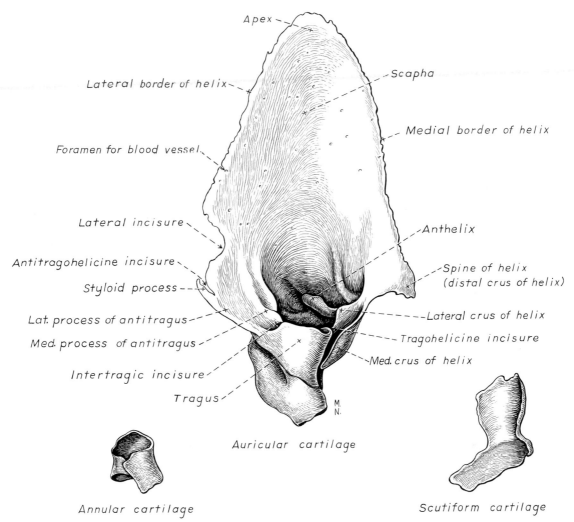

Apex

Scapha

Lateral border of helix

Medial border of helix

Foramen for blood vessel

Lateral incisure

Anthelix

Antitragohelicine incisure

Spine of helix
(distal crus of helix)

Styloid process

Lat. process of antitragus

Lateral crus of helix

Med. process of antitragus

Tragohelicine incisure

Intertragic incisure

Med. crus of helix

Tragus

M. N.

Auricular cartilage

Annular cartilage

Scutiform cartilage

Figure 19–2. Pinna and auricular cartilages. (From Getty et al. 1956.)

The groove or furrow bounded by the anthelix, tragus, and antitragus, which continues into the external auditory meatus, is termed the cavum conchae. The concha is the area of the auricular cartilage between the scapha and the cartilaginous external acoustic meatus. A shallow groove is present on the medial surface of the auricular cartilage opposite the anthelix. This is the anthelicine sulcus. The bend in the pinna of lop-eared animals occurs distal to the anthelix in the scapha. The skin lining the scapha and concha shows pigmentation characteristic of the breed. It has, in most specimens, a decreasing amount of hair from the distal to the proximal parts. A few very fine hairs are found at the entrance of the cartilaginous

external acoustic meatus. The rather prominent transverse ridges, as well as the longitudinal ridges, are simple skin folds which frequently continue toward the apex of the ear. Cartilage is not discernible in them upon histological section. The protective hair of the skin becomes fine and scanty in the conchal cavity.

Interposed between the auricular cartilage and the external acoustic process is the **annular cartilage.** This is a narrow band of cartilage rolled to form a tube (Fig. 19–2). The proximal end of the cartilage overlaps the osseous external acoustic process, with which it articulates by means of ligamentous tissue. The ear canal, therefore, is divided into lateral cartilaginous and medial osseous parts. The space between the incomplete tubes of the auricular and annular cartilages and between the annular cartilage and the external acoustic process provides for freedom of movement of the pinna.

A small cartilaginous plate, the **scutiform cartilage** *(cartilago scutiformis)* (Fig. 19–2), is located in the rostroauricular muscles medial to the ear. This cartilaginous plate is shaped somewhat like a boot with the heel directed away from the midline, thus forming no part of the external ear. Deep to the scutiform cartilage lies a fatty cushion, the corpus adiposum auriculi. This fatty pillar extends over a portion of the superficial surface of the temporal muscle and around the base of the auricular cartilage.

The bony **external auditory meatus** is lined with a thin cutaneous membrane which contains, in carnivores, large alveolar glands, according to Ellenberger and Baum (1943). The muscles of the external ear are described with the muscles of the head. The tympanic membrane (eardrum) which separates the external ear from the middle ear is a thin, semitransparent sheet, oval in shape, and concave when viewed from the external aspect. Its long axis is horizontal. The membrane is thin centrally and becomes thicker near its periphery.

GLANDS OF THE EXTERNAL AUDITORY MEATUS

The external auditory meatus presents a cutaneous lining which includes stratified squamous epithelium, sebaceous and tubular glands, and hair. The conchal cartilage is covered with skin which, according to Fraser (1961), presents fewer hair follicles on the concave inner surface than on the external surface. Both types of glands are present in the cartilaginous and bony portions of the auditory meatus of the dog.

The **sebaceous glands** form a superficial glandular bed immediately below the epithelial surface, whereas the **tubular glands** are found in the deeper connective tissue layers. The sebaceous glands are frequently associated with hair follicles, whereas the tubular ceruminous glands are located below the sebaceous glands in the deeper dermal layers. Nielsen (1953) is of the opinion that the tubular and sebaceous glands in the ear are similar in structure to those of the skin and that the normal ear secretion, the cerumen, is a product of both glandular types. Trautmann and Fiebiger (1957) also believe the secretion (ear wax) to be a mixture from both types of glands. These authors also describe the great vascularity of the subcutaneous tissue of the osseous meatus.

THE MIDDLE EAR

The **tympanic cavity** *(cavum tympani)* (Fig. 19–3) contains the auditory ossicles, the chorda tympani, the ossicular muscles, and the auditory tube, which communicates with the nasal pharynx. The middle ear is lined with a mucous membrane that is, in general, covered with a two-layered, columnar, ciliated epithelium, according to Krölling and Grau (1960). The cavum tympani is divisible into a dorsal part, the epitympanicum; a middle, mesotympanicum; and a ventral, hypotympanicum. The last corresponds to the bulla tympanica. The tympanic membrane may be divided into two parts: the pars flaccida and the pars tensa. The pars flaccida is a small, triangular portion which lies between the lateral process of the malleus and the margins of the tympanic incisure. The pars tensa constitutes the remainder of the membrane. The external aspect of the tympanic membrane is somewhat concave, owing to traction on the medial surface by the manubrium of the malleus. The most depressed point, which is opposite the distal end of the manubrium, is termed the *umbo membranae tympani.* A light-colored streak, *stria malleolaris,* may be seen running dorsocaudally from the umbo toward the pars flaccida when viewed from the external side. This is caused by the manubrium being partly visible through the tympanic membrane along its attachment. The manubrium is embedded in the tunica propria and is covered by the epithelium lining the membrane. This in turn is fastened to a collar of bone in the external acoustic meatus. This bony collar is incomplete dorsocaudally, forming the tympanic incisure.

The epitympanic recess is dorsal to a frontal plane through the osseous external

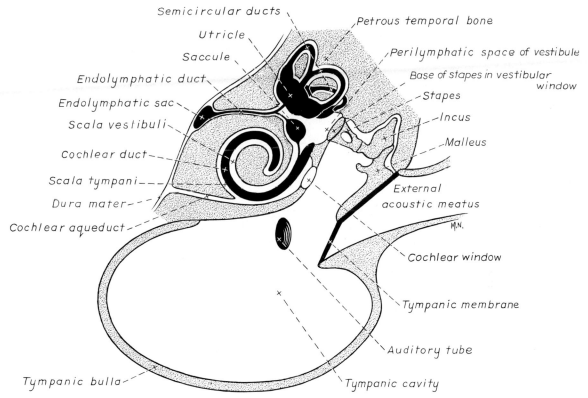

Figure 19–3. Diagram of middle ear and inner ear. (From Getty et al. 1956.)

acoustic meatus. It is the smallest of the three portions and is occupied almost entirely by the head of the malleus and the incus at their articulation.

The tympanic cavity proper is that portion adjacent to the tympanic membrane. It is irregularly quadrangular in shape, being flattened laterally by the tympanic membrane which forms its wall. In the caudal portion, but facing rostrally, is the secondary tympanic membrane closing the cochlear (round) window (*fenestra cochlea*) (Fig. 19–3).

The ventral portion, the part within the tympanic bulla, may be compared in shape to the interior of an eggshell, having an elliptical opening on the side which faces dorsally. It communicates with the tympanic cavity proper through this opening. The long axis of the tympanic cavity is about 15 mm. in length and at an angle of about 45° with the sagittal plane in a caudolateral direction. The width and depth are about equal, measuring 8 to 10 mm. The tympanic membrane is slanted ventromedially. Holz (1931) states that when viewed from the

front, a plane through the *annulus tympanicus* is in general at a 57° angle with the frontal plane in the dog.

Holz describes a membrana Shrapnelli (pars flaccida of the tympanic membrane) which helps to form the lateral boundary of an upper tympanic pouch (Prussak's pouch). This pouch lies dorsal to the lateral (short) process of the malleus and is bounded medially by the neck of the malleus. In man there are relatively small openings which communicate with the epitympanicum and the mesotympanicum, while in most animals Prussak's pouch communicates freely with the tympanic cavity, according to Holz (1931). In the dog, however, he describes a complete separation between Prussak's pouch and the tympanic cavity.

On the medial wall of the tympanic cavity is a bony eminence (*promontorium*) (Fig. 19–4), which houses the cochlea; it lies opposite the tympanic membrane medial to the epitympanic recess. The vestibular window (*fenestra vestibuli*), formerly called the oval window (Fig. 19–3), is occupied by the base of the stapes. It is located on

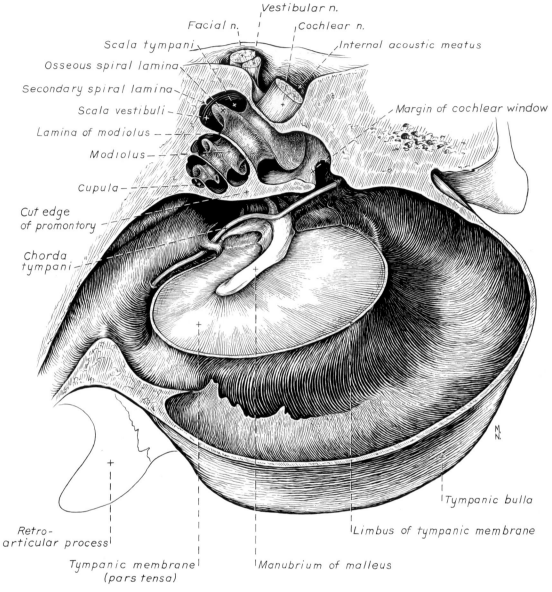

Figure 19-4. Sculptured medial view of the right middle ear and cochlea. (From Getty et al. 1956.)

the dorsolateral surface of the promontory just medial to the pars flaccida. The ostium of the auditory tube (*ostium tympanicum tubae auditivae*) is the rostral extremity of the tympanic cavity proper. The tendon of the tensor tympani muscle (Fig. 19–5) descends ventrolaterally through an arch in a thin lamina of bone which overlies the muscle. It inserts on the muscular process of the malleus. The ossicles form a short chain across the dorsal part of the tympanic cavity.

The **tympanic nerve** (*chorda tympani*) (Fig. 19–4), after leaving the facial nerve, passes through the tympanic cavity medial to the malleus to join the lingual nerve. The tympanic plexus arising from the tympanic branch of the glossopharyngeal nerve lies on the promontory and supplies the tympanic mucosa. Other nerves which contribute to this plexus are the small, superficial petrosal and the caroticotympanic nerve.

The **auditory (Eustachian) tube** (*tuba auditiva*) is a short canal which extends from

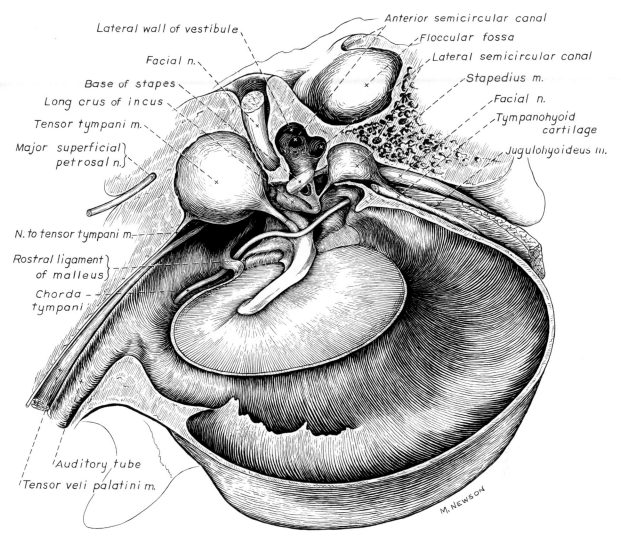

Lateral wall of vestibule

Facial n.

Base of stapes

Long crus of incus

Tensor tympani m.

Major superficial petrosal n.

N. to tensor tympani m.

Rostral ligament of malleus

Chorda tympani

Auditory tube

Tensor veli palatini m.

Anterior semicircular canal

Floccular fossa

Lateral semicircular canal

Stapedius m.

Facial n.

Tympanohyoid cartilage

Jugulohyoideus m.

M. NEWSON

Figure 19–5. Sculptured medial view of the right middle ear showing auditory ossicles and their muscles. (From Getty et al. 1956.)

the nasal pharynx to the rostral portion of the tympanic cavity proper. Its short bony wall is formed rostrally by the squamous part, and ventrally its floor is formed by the tympanic part of the temporal bone. The lateral wall, which is about 8 mm. long, is nearly twice the length of the medial wall. The tube is oval in cross section, with its greater diameter 1.5 mm. The medial wall of the membranous part of the tube is supported by a plate of hyaline cartilage, the rostral end of which curves medially to form a short hook.

The tensor veli palatini muscle (Fig. 19–5) arises in the groove of the petrous temporal bone ventrolateral to the tensor tympani muscle. It supports the lateral wall

of the auditory tube. The branch of the fifth cranial nerve which supplies the tensor tympani muscle enters the tympanic cavity in association with the tendon of origin of the tensor veli palatini muscle.

BONES OF THE MIDDLE EAR

The **auditory ossicles** (*ossicula auditus*) are three small bones which transmit air vibrations from the tympanic membrane across the cavity of the middle ear to the inner ear. The most lateral and largest of the three bones is the malleus (Fig. 19–3). The most medial is the stapes. The handle of the malleus attaches to the tympanic membrane.

The base of the stapes is attached to the margin of the vestibular window. Between the malleus and stapes is the incus.

The **malleus** consists of a head; a wide, thin neck; and a manubrium or handle. The handle is three-sided in cross section. The side embedded in the substance of the tympanic membrane is wider and smoother than the other two; it is also slightly concave longitudinally. At the base of the manubrium, extending medially and slightly rostrally, is the muscular process of the malleus. This is provided with a tiny hook at its end to which the tensor tympani muscle attaches. The rostral process (Fig. 19–6A) or long process is largely embedded in the tympanic membrane. It extends directly forward from the neck of the malleus, arising at the same level as the muscular process. Opposite the muscular process at an angle of about 90° with the rostral process is the short, lateral process. This is the most dorsal attachment of the manubrium to the tympanic membrane. The head of the malleus articulates with the body of the incus in the epitympanic recess, the most dorsal portion of the tympanic cavity.

The **incus** (Fig. 19–6C), measuring about 4 mm. long and 3 mm. high, is much smaller than the malleus. Its shape has often been likened to a human bicuspid tooth with divergent roots. The incus lies caudal to the malleus in the epitympanic recess. The crura are located on each side of a transverse ridge which forms the caudal limit of the recess. The short crus points caudally into the fossa incudis dorsal to this ridge. The long crus is also directed caudally, but presents a small bone, the os lenticularis, which extends rostrally and somewhat medially from its distal end. In some instances this connection ossifies to form the processus lenticularis.

The **stapes** (Fig. 19–6D) consists of a head, neck, two crura, a base, and a muscular process. It lies in a horizontal plane, the base facing medially. The base articulates with the cartilage which covers the edge of te vestibular (oval) window (fenestra vestibulae). The stapes is the innermost ossicle and is the smallest bone in the body, being approximately 2 mm. in length. The crura are hollowed on their concave or opposed sides. A cross section of a single crus ap-

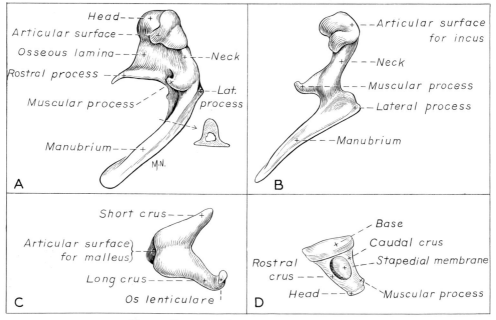

Figure 19–6. Auditory ossicles of right ear.
A. Malleus, medial aspect and cross section
B. Malleus, posterior aspect
C. Incus, medial anterior aspect
D. Stapes, medial posterior aspect

pears as a narrow semicircle of bone. There is a thin stapedial membrane (membrana stapedis), which connects one crus to the other. The rostral crus is slightly longer than the caudal crus. Arising from the caudal crus near the neck is a minute muscular process which provides attachment for the stapedius muscle (Fig. 19–6D).

Ligaments of the Ossicles

Several ligaments attach the ossicles to the wall of the tympanic cavity. A short but fairly well-defined **lateral ligament of the malleus** connects the lateral process of the malleus to the margins of the tympanic notch. The **dorsal ligament of the malleus** is a somewhat diffuse mass of ligamentous tissue which joins the head of the malleus to a small area on the roof of the epitympanic recess. The **rostral ligament of the malleus** (Fig. 19–5) is a short ligament attaching the rostral process of the malleus to the osseous tympanic ring just ventral to the canal by which the chorda tympani leaves the tympanic cavity. The body of the incus is attached to the roof of the epitympanic recess by the **dorsal ligament of the incus.** The

caudal ligament of the incus attaches the short crus of the incus to the fossa incudis. An **annular ligament** (*lig. anulare stapedis*) attaches the base of the stapes to the cartilage which lines the vestibular window.

Muscles of the Ossicles

Two tiny muscles (Fig. 19–5) are associated with two of the ossicles. The *m. tensor tympani* is spherical with its base in the fossa tensor tympani. The short tendon of insertion is attached to the hook on the apex of the muscular process of the malleus. Contraction of this muscle tends to draw the handle of the malleus medially, tensing the tympanic membrane. Innervation is by a twig from the mandibular division of the trigeminal nerve. The *m. stapedius* is the smallest skeletal muscle in the body, and its origin is in the fossa musculae stapedis. The body of the muscle lies largely medial to the facial nerve. Its tendon of insertion attaches to the muscular process of the stapes (Fig. 19–5). Contraction of the stapedius muscle moves the rostral end of the base of the stapes caudolaterally. This muscle is innervated by the stapedial branch of the facial

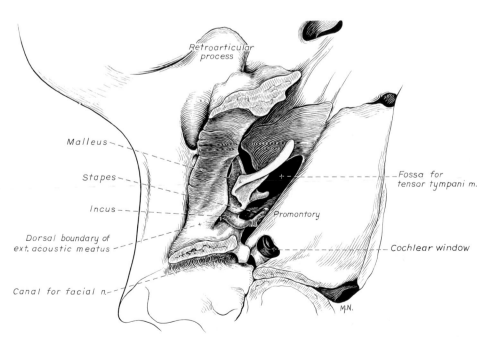

Figure 19–7. Auditory ossicles of right ear, ventral aspect. Tympani bulla removed.

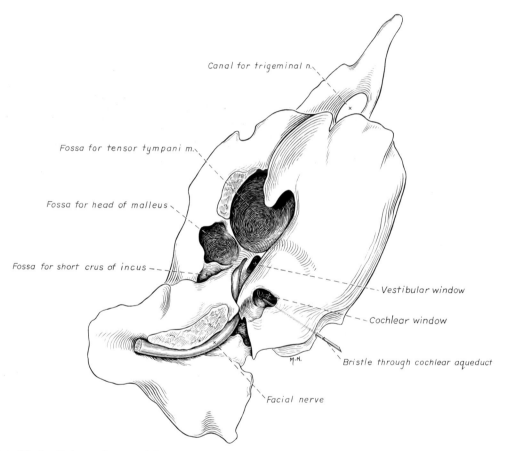

Figure 19–8. Petrous temporal bone and middle ear cavity, ventral aspect. Tympanic bulla and auditory ossicles removed.

nerve. The cavum tympani is lined with a thin mucous membrane that is partly covered with ciliated epithelium.

THE INNER EAR

The **internal ear** *(auris interna)* consists of fluid-filled ducts and sacs: a membranous labyrinth contained within an osseous labyrinth. The structures of the bony inner ear may be separated into three parts: the cochlea, the vestibule, and the semicircular canals. The rostral part is the cochlea, which contains the cochlear duct, where mechanical stimuli are converted to nerve impulses which, upon reaching the brain, result in audition. The caudal part consists of the three semicircular canals containing the semicircular ducts, each in a different plane. The third part, the osseous vestibule, contains the utricle and saccule. The semicircular ducts and utricle are directly concerned with equilibrium. Perilymph occupies a narrow space between the endolymph-filled membranous labyrinth and the osseous labyrinth.

The semicircular ducts contain the end-organs of the vestibular nerve, which conducts impulses resulting in the orientation of the body in space (equilibrium). The cochlear duct contains the end-organs (*organum spirale* or organ of Corti) of the cochlear nerve and conducts impulses concerned with hearing. For a discussion of the anatomy and pathology of hearing in dogs, see Johnson et al. (1973) and deLahunta (1977).

THE OSSEOUS COCHLEA

The **osseous cochlea** (Figs. 19–9, 19–10) is similar in shape to a snail's shell, from which it derives its name. It winds ventrally in a spiral around a hollow bony core, or modiolus, which contains the cochlear nerve. It ends blindly at the apex or cupula (Figs. 19–4, 19–10). In the dog the cochlea makes three and one quarter turns. It points ventrorostrally and slightly laterally within the promontory. The osseous spiral lamina, which winds around the modiolus much like the thread of a screw, nearly bisects the

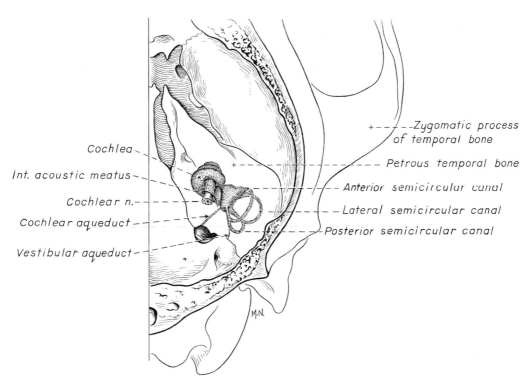

Figure 19–9. Phantom diagram of right inner ear in situ, dorsal aspect. (From Getty et al. 1956.)

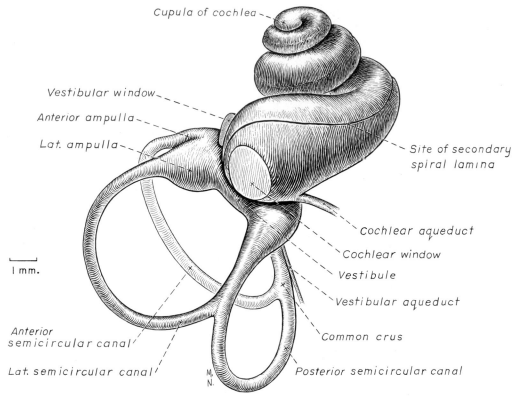

Figure 19–10. Right osseous labyrinth drawn from a latex cast, lateral aspect. (From Getty et al. 1956.)

lumen of the spiral cochlear canal into two portions called the *scala tympani* and *scala vestibuli* (Fig. 19–4). The osseous spiral lamina begins within the vestibule and ends at the apex in a free hooklike process, the hamulus. The scala vestibuli communicates with the vestibule, and hence the fluid within is acted upon by the base of the stapes in the vestibular window. The cochlear window is an opening situated near the rostral end of the vestibule by which the scala tympani communicates with the middle ear. A secondary tympanic membrane closes this cochlear window. The membranous cochlear duct completes the separation of the two scalae. The scalae communicate at the apex of the modiolus by a small opening, the helicotrema, formed at the free border of the hamulus. The basal turn of the cochlea is about 4 mm. in diameter and lies close to the medial side of the vestibule. The total height of the cochlea measures about 7 mm. Longitudinal modiolar canals and a spiral modiolar canal serve for the distribution of both blood vessels and

nerves to the cochlea. Perilymph gains access from the subarachnoid space to the vestibule, the cochlea, and the semicircular canals by means of the perilymphatic duct, which lies in a small canal. This small canal, the cochlear canaliculus (*canaliculus cochleae*), descends directly ventrad from a point on the ventral wall of the scala tympani near its origin to communicate with the arachnoid space (Fig. 19–3).

THE OSSEOUS VESTIBULE

This is an irregular oval space about 3 mm. in diameter which communicates with the cochlea rostrally and with the semicircular canal caudally. The walls of the vestibule are marked by depressions and ridges which correspond to the various portions of the membranous labyrinth. The medial wall contains two depressions. The caudodorsal one is the elliptical recess, which contains the utricle. Rostroventral to it is the spherical recess for the saccule. The vestibular crus separates the two recesses. Several groups of small openings which accommodate the nerves of this region occur near the recesses. These tiny groups of foramina are called maculae cribrosae.

The lateral wall contains the vestibular window, which is closed by the base of the stapes. Ventral to it is the cochlear window. The **vestibular aqueduct** (*aqueductus vestibuli*) descends caudoventrally from the vestibule to the caudal surface of the petrous temporal bone. The endolymphatic duct (Fig. 19–3) ends in the small endolymphatic sac just superficial to the dura. The semicircular canals open into the vestibule caudally.

THE OSSEOUS SEMICIRCULAR CANALS

There are three **semicircular canals,** an anterior, a posterior, and a lateral canal (Figs. 19–9, 19–10). They lie caudal and slightly dorsal to the vestibule. Each canal describes about two-thirds of a circle in a single plane, and each is approximately at a 90° angle to the other two. A segment of the canal proximal to the vestibule is called a crus. Each canal has two crura which communicate with the vestibule (with the exception of the common crus, to be noted later). One crus of each canal has a dilation, the osseous ampulla, near the junction with the vestibule. The lumen diameter of the canals averages roughly 0.5 mm., the ampulla being about twice as large.

The anterior canal of one ear is roughly parallel with the posterior canal of the opposite ear. The lateral canal of each side occupies a nearly horizontal plane. The anterior canal is the longest. The arc it forms measures about 6 mm. across at the widest part. The lateral canal forms an arc which measures about 4.5 mm., while the arc of the posterior semicircular canal is the smallest, measuring only 3.5 mm. in medium-sized dogs. These measurements vary with the size of the dog. The common crus is formed by the nonampullated ends of the posterior

and anterior canals. In sculptured specimens the anterior semicircular canal is seen to surround the floccular fossa, a small but deep depression on the medial side of the petrous temporal bone. This depression is occupied by the paraflocculus of the cerebellum.

The ampullated end of the posterior canal and the nonampullated end of the lateral canal are united for a short distance caudal to the vestibule.

MEMBRANOUS LABYRINTH

The **membranous labyrinth** (*labyrinthus membranaceus*) does not completely fill the hollow system within the osseous labyrinth. Thus it is slightly smaller, but similar in shape. The fluid perilymph surrounds it, and connective tissue trabeculae support and attach it to the osseous wall. Spaces comparable to those in the subarachnoid space exist among the trabeculae. There are three regions which correspond to those of the osseous labyrinth. They are the membranous cochlear duct, the membranous semicircular ducts, and the membranous vestibule (Fig. 19–3). The last, however, differs from the osseous vestibule, since the membranous part is composed of two saclike structures, the utriculus and sacculus, which occupy the lumen of the osseous part. The cochlear duct is united to the sacculus by the ductus reuniens. The cochlear duct is roughly triangular in cross section. The basilar membrane (*lamina basalis*) forms the floor of the cochlear duct, separating the endolymph of this duct from the perilymph of the scala tympani. The very thin vestibular or Reissner's membrane (*paries vestibularis ductus cochlearis*) forms the roof of the cochlear duct, separating its cavity from that of the scala vestibuli. The spiral organ (of Corti), the organ of hearing, lies upon the basilar membrane and consists of a thickened, specialized epithelium.

BIBLIOGRAPHY

Anson, B. J., and J. A. Donaldson. 1973. The Surgical Anatomy of the Temporal Bone and Ear. 2nd Ed. Philadelphia, W. B. Saunders Co.

Bast, T. H., and B. J. Anson. 1949. The Temporal Bone and the Ear. Springfield, Ill., Charles C Thomas.

de Lahunta, A. 1977. Veterinary Neuroanatomy and

Clinical Neurology. Philadelphia, W. B. Saunders Co.

Ellenberger, W., and H. Baum. 1943. Handbuch der vergleichenden Anatomie der Haustiere. Berlin, Springer.

Fraser, G. 1961. The histopathology of the external

auditory meatus of the dog. J. Comp. Pathol. *71*:253–258.

Getty, R., H. L. Foust, E. T. Presley, and M. E. Miller. 1956. Macroscopic anatomy of the ear of the dog. Am. J. Vet. Res. *17*:364–375.

Gray, A. A. 1907. The Labyrinth of Animals, Vol. I. London, I. and A. Churchill.

Holz, K. 1931. Vergleichende anatomische und topographische Studien über das Mittelohr der Säugetiere. Z. Anat. Entwickl-Gesch. *94*:757–791.

Honda, Y. 1908. Gehörorgan des Hundes. Inaugural Dissertation. Erlangen, Junge und Sohn.

Johnsson, L. G., J. E. Hawkins, Jr., A. A. Muraski, and R. E. Preston. 1973. Vascular anatomy and pathology of the cochlea in Dalmatian Dogs. pp. 249–295 *In* Vascular Disorders and Hearing Defects, A. J. D. De Lorenzo. Baltimore, University Park Press.

Krölling, O., and H. Grau. 1960. Lehrbuch der Histologie und vergleichenden mikroskopischen Anatomie der Haustiere. Berlin, Paul Parey.

Miller, M. E., and R. Witter. 1942. Applied anatomy of the external ear of the dog. Cornell Vet. *32*:64–86.

Nielsen, S. W. 1953. Glands of the canine skin—morphology and distribution. Am. J. Vet. Res. *14*:448–454.

Shambaugh, G. E. 1923. Blood stream in the labyrinth of the ear of dog and man. Am. J. Anat. *32*:189–198.

Sis, R. F. 1962. Polytetrafluoroethylene in Reconstructive Surgery of the Canine External Acoustic Meatus. M.S. Thesis, Iowa State University, Ames.

Trautmann, A., and J. Fiebiger. 1957. The Histology of Domestic Animals (translated and revised by R. Habel and E. Biberstein). Ithaca, N.Y., Comstock Pub. Assoc.

Chapter 20

THE EYE

By ROY V. H. POLLOCK

The **eye** (*organum visus*) (Fig. 20–1) develops as a neurectodermal outgrowth of the embryonic prosencephalon which contacts surface ectoderm and is enveloped by induced mesodermal derivatives. The definitive eye and its adnexa are contained within an orbit that is only partly bony (Fig. 20–5). Associated with the bulb of the eye are extraocular muscles which move it, periorbital fascia and fat which surround and cushion it, eyelids and conjunctivae which protect it, and a lacrimal apparatus which keeps its surface moist and helps to nourish the cornea.

As a consequence of its dual origin, the eye has both central and peripheral types of neural elements. The optic nerve is of the central type with myelin formed by oligodendroglial cells, whereas the nerves of the extraocular muscles and iris are of the peripheral type with lemmocyte (Schwann cell) sheaths. The vascular and fibrous tunics which surround the bulb of the eye and optic nerve are homologous to the meninges surrounding the brain and spinal cord. The intervaginal space of the optic nerve is continuous with the subarachnoid space of the brain and contains cerebrospinal fluid.

There is considerable variation between breeds in regard to the position of the eyes, the size of the orbit, and the size and shape of the palpebral opening (see Figs. 20–16, 20–20).

DEVELOPMENT

Aguirre, Rubin, and Bistner (1972) have studied the early development of the dog's eye and its adnexa from day 15 to functional maturity using serially sectioned embryos in the Cornell University Collection (Evans and Sack 1973). The approximate time periods given in the following account are from these studies.

The first indication of the formation of the eye is seen as an **optic sulcus** on the neural fold rostral to the notochord on each side. The neuroectoderm surrounding this sulcus proliferates rostrolaterally to form a hollow diverticulum of the prosencephalon, the optic vesicle (Fig. 20–2A). The anterior margin of the vesicle enlarges and folds into a hemisphere, the **optic cup**. The **optic fissure** is the ventral meridian along which the lateral and medial folds of the optic cup meet and fuse (Fig. 20–2B). Incomplete fusion results in defects of one or more of the tunics of the eye (colobomata). Such defects are common in the Collie breed as part of the heritable Collie eye syndrome.

The connection of the optic cup to the brain stem lengthens and attenuates as growth proceeds, forming the **optic stalk,** the future optic nerve. Anteriorly, the **optic vesicle** induces the overlying surface ectoderm to proliferate, forming the **lens placode,** which is present by 15 days of gesta-

1073

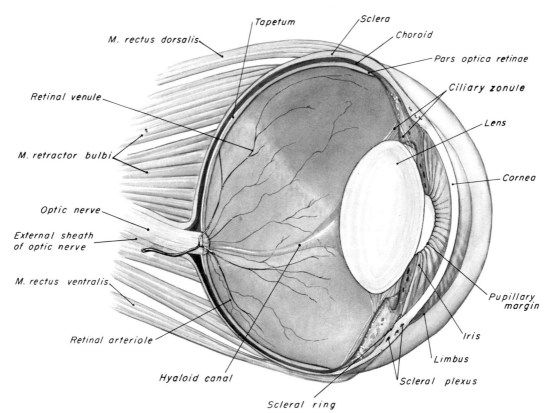

Figure 20–1. Bulbus oculi.

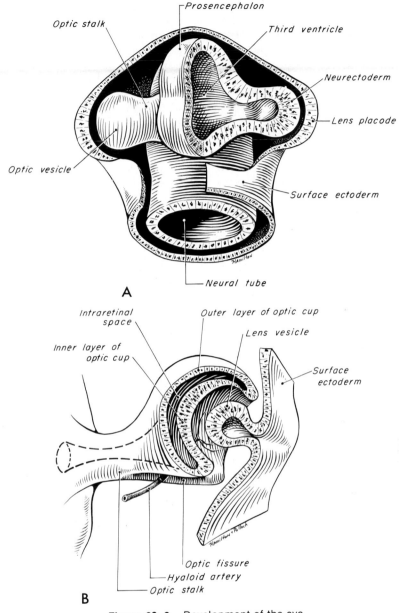

Figure 20–2.   Development of the eye.
A. Optic vesicle (4-mm. embryo).
B. Optic cup (7-mm. embryo).

tion. The placode invaginates into the optic cup and by day 25 pinches off to form the **lens vesicle**, the anlage of the crystalline lens (Fig. 20–2B). The lens induces the overlying ectoderm and mesoderm to form the cornea.

The **hyaloid artery** develops in the mesenchyme surrounding the optic cup. It enters the posterior end of the optic fissure to supply the inner surface of the cup and

the mesenchyme filling the cup, the primary vitreous (Fig. 20–2B). The hyaloid artery grows forward and reaches the posterior lens surface by day 25, where it branches extensively to form the posterior portion of the *tunica vasculosa lentis.* Secondary vitreous, secreted by the inner layer of the optic cup, surrounds the primary vitreous beginning about gestational day 26. Continued elaboration of the secondary vitreous and the growth of the eyeball reduces the primary vitreous to a narrow funnel between the optic nerve and posterior lens surface (Fig. 20–1). The hyaloid vessels between the optic disc and lens begin to atrophy at about day 45, but remnants are commonly present until 10 or 11 days after birth. The retinal arteries of the adult are derived from that portion of the hyaloid vasculature which supplied the inner layer of the optic cup.

The anterior face of the lens and primordial iris are supplied by the anterior portion of the tunica vasculosa lentis, a plexus of vessels derived from the anterior ciliary vessels. This portion of the vascular tunic supplying the lens also atrophies late in gestation so that the lens is normally avascular at birth and in adult life.

The cavity of the optic vesicle, originally continuous with the third ventricle of the brain, is obliterated on about day 25, when the inner and outer walls of the optic vesicle fuse. The outer layer becomes the retinal pigment epithelium. The inner layer proliferates to form all the other layers of the retina. Axons from the ganglion cells of the innermost layer arc toward the optic stalk, which they invade about gestational day 30, and follow toward the brain stem. Maturation of the retina proceeds from central to peripheral and is not complete until six weeks after birth (Shively et al. 1971).

The epithelia of the ciliary body and the posterior surface of the iris are also derived from the neurectoderm of the optic cup, but are non-visual (*pars ceca retinae*). The richly vascular mesoderm surrounding the anterior tunica vasculosa lentis forms the stroma and anterior epithelium of the iris. The central area of the iris (pupillary membrane) is thin and normally completely atrophies by 14 days after birth, forming the pupil. Incomplete atrophy, resulting in persistent pupillary membranes, has been observed as a heritable defect in Basenji dogs (Roberts and Bistner 1968). The dilator and sphincter muscles of the iris differentiate from the neurectodermal pars iridica retinae.

Rarefaction of the mesenchyme between that which forms the stroma of the iris and that which forms the substantia propria and posterior epithelium of the cornea is evident by day 45. Progressive rarefaction forms a cavity which fills with aqueous humor to become the **anterior chamber** of the adult eye. A continuation of this cavitation between the lens and iris forms the posterior chamber (Fig. 20–3).

The mesenchyme adjacent to the optic cup is induced by the outer layer of the cup to form the richly vascular tunica vasculosa bulbi, which is homologous to the pia-arachnoid of the brain. The vascular tunic comprises the iris, ciliary body, and choroid. At the periphery of the lens, the vascular mesenchyme proliferates into the folds of the ciliary body. Fibers form from the ciliary body to the lens equator. These elongate as the globe increases in size to form the definitive *zonula ciliaris*, the suspensory ligament of the lens.

External to the vascular tunic, the mesenchyme condenses to form the fibrous sclera, which is homologous to the dura mater of the brain. The extraocular muscles are derived from mesoderm behind the de-

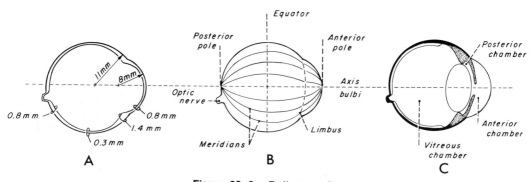

Figure 20–3. Bulbus oculi.
A. Average dimensions of sclera and cornea.
B. Directional terminology.
C. Chambers of the eyeball.

veloping eye vesicle in the somite stage. Gilbert (1947) demonstrated that the extrinsic ocular muscles of the cat arise from three distinct but closely approximated anlagen which are homologous with the premandibular, mandibular, and hyoid head cavities of lower vertebrates. The same is probably true for the dog. Gilbert noted that Bonnet (1901) described a pair of lateral mesodermal buds in contact with the cranial end of the notochord in a 16-somite dog. These were most likely the premandibular condensations arising from the prechordal plate, although Bonnet did not determine their fate.

The **eyelids** appear as folds above and below the eye on about day 25 of gestation. These enlarge, grow over the cornea, and narrow the palpebral fissure by about half on day 35. The eyelids completely cover the cornea and fuse by day 40 (Evans 1974).

The superficial musculature of the lids, the m. orbicularis oculi, forms from the platysma sheet and is recognizable by about day 40. Huber (1922) has described the postnatal development of the platysma derivatives.

Dogs are born with the lid margins still adherent to one another. Final maturation of the eye occurs after birth; most notable are changes in the retina, iridocorneal angle, corneal epithelium, and tapetum. The fused lids normally separate about two weeks post partum, and the eyes are able to open. Premature separation of the lids results in severe ophthalmitis, apparently as a result of the immaturity of the lacrimal apparatus (Aguirre and Rubin 1970).

THE EYEBALL

The **eyeball** (*bulbus oculi*) is formed by three concentric coats: the **fibrous tunic** (*tunica fibrosa bulbi*), the **middle vascular tunic** (*tunica vasculosa bulbi*) and the **inner nervous tunic** (*tunica interna bulbi*). In the dog the eyeball is nearly spherical, differing little in its sagittal, transverse, and vertical diameters. The size of the eyeball varies among breeds, but the radius is usually about 11 mm.

The transparent cornea forms the anterior one-fourth of the eyeball, and since it has a smaller radius of curvature (approximately 8 mm.) than the rest of the eye, it bulges forward (Fig. 20–3). The vertex of the cornea is designated the **anterior pole** of the eye (*polus anterior*). The point directly opposite this is the **posterior pole.** The latter is a geometric point and does not correspond to the exit point of the optic nerve, which lies ventrolateral to the posterior pole. The line connecting the anterior and posterior poles and passing through the center of the lens is the *axis bulbi* (Fig. 20–3). In the mesaticephalic dog the axis forms an angle of approximately 30° with the median plane. The angle is greater in the brachycephalic breeds (see Fig. 20–16).

Lines which connect the anterior and posterior poles of the eye on the surface of the globe are designated **meridians.** The **equator** of the globe is its maximum circumference located midway between the poles (Fig. 20–3). Because the eye is spherical, the common anatomical terms of direction are not applicable for certain structures, such as the retinal layers. In such cases, the terms "inner" and "outer" are used with reference to the center of the bulb.

Fibrous Tunic

The **fibrous outer tunic** of the eye is divided into two parts: the opaque sclera, which encloses approximately the posterior three-fourths of the globe, and the transparent cornea anteriorly. The junction of the cornea and the sclera is designated the *limbus corneae.* The fibrous coat is the strongest of the three tunics of the eyeball. It acts like a tough balloon, which, when tensed by the intraocular pressure, gives the eye its shape.

SCLERA

The **sclera** consists of a dense network of collagen and elastic fibers and their attendant fibrocytes. It varies in thickness, being greatest in the region just posterior to the corneoscleral junction, where it receives the insertions of the rectus and oblique muscles and contains the scleral venous plexus. The ciliary muscle is attached to a small ridge which forms a ring (*anulus sclerae*) on the inner surface of the sclera posterior to the iridocorneal angle (Fig. 20–9).

Near the equator, where the retractor bulbi muscle inserts, the sclera is much thinner. The sclera is thicker again on the posterior aspect of the globe (Fig. 20–3).

Where the optic nerve leaves the eyeball, the sclera is sievelike (*area cribrosa sclerae*). Here the collagen, elastic, and reticular fiber bundles of the sclera form a net through the interstices of which the optic nerve fascicles pass. The trabeculae of the area cribrosa continue caudally as the prominent connective tissue septae of the optic nerve. The dura mater surrounding the optic nerve (*vagina externa n. optici*) is continuous with the outer layers of the sclera at the periphery of the area cribrosa, and with the periorbita and dura mater encephali at the optic canal.

The ciliary nerves and the short posterior ciliary vessels enter the eyeball through foramina in the sclera at the periphery of the area cribrosa.

CORNEA

The **cornea** forms the anterior segment of the fibrous tunic. It is normally transparent in life and slightly less than 1 mm. thick. Although earlier studies reported that the cornea of the dog is thicker at the vertex, there appears to be considerable individual variability. Diesem (1975) found that the cornea may be thicker at the center or at the periphery, or be of uniform thickness in a given dog. The cornea of the dog is very slightly oval; the mediolateral dimension is usually about 10 per cent greater than the dorsoventral dimension, which is about 14 mm. in an average-sized dog.

Classically, the cornea is described as consisting of five layers: the anterior epithelium, the anterior limiting lamina, the substantia propria, the posterior limiting lamina, and the posterior epithelium. Shively and Epling (1970), however, in a study of the fine structure of the canine cornea, were unable to demonstrate a distinct layer comparable to the anterior limiting lamina (Bowman's membrane) of man.

The anterior epithelium of the cornea has a microplicated surface. The rugae of this surface are thought to be important in anchoring the precorneal tear film. Both the anterior and posterior corneal epithelia, but especially the latter, regulate the degree of hydration of the substantia propria by an active transport mechanism. Disruption of either epithelium results in corneal edema.

The transparency of the cornea is apparently due to the highly ordered arrangement of the collagen fibers which form the bulk of the substantia propria. These fibers are arranged in distinct laminae. The fibers in each lamina run parallel to one another and cross the fibers of the apposing lamina at an oblique angle. The fibrocytes of the cornea (keratocytes) are flattened between the laminae. The cornea is easily dissected along these lamellar planes. Any disruption of the highly ordered structure of the collagen fibers, either by edema or by scar tissue, results in a loss of transparency.

The cornea is normally avascular. It is nourished by the capillary loops at the corneoscleral junction (limbus), the precorneal tear film, and the aqueous humor. The cornea is innervated by branches of the ciliary nerves which arise from the ophthalmic nerve, a branch of the trigeminal nerve. It is extremely sensitive. Branches to the cornea enter the anterior layers of the stroma at the limbus and soon lose their myelin sheaths as they converge toward the vertex. This trigeminal innervation of the cornea is essential to its normal state. Loss of sensory innervation apparently disrupts a trophic influence normally supplied by the ciliary nerves. Corneal denervation results in corneal ulceration, edema, and loss of stromal tissue (neurotrophic keratitis), even though eyelid function is unimpaired (Scott and Bistner 1973).

At the corneoscleral junction, the anterior corneal epithelium is continuous with the bulbar conjunctiva (see "Conjunctiva" further on). The collagen fibers of the substantia propria become abruptly less ordered as they approach the sclera, with a resulting loss of transparency. The corneoscleral junction is oblique; the posterior aspect is more peripheral than the anterior (Fig. 20–4B). Immediately peripheral to the limbus, the posterior corneal epithelium reflects onto the anterior face of the iris, forming the iridocorneal (filtration) angle (Fig. 20–4B).

In the embryo the **iridocorneal angle** is a smooth, unfenestrated fornix. Late in gesta-

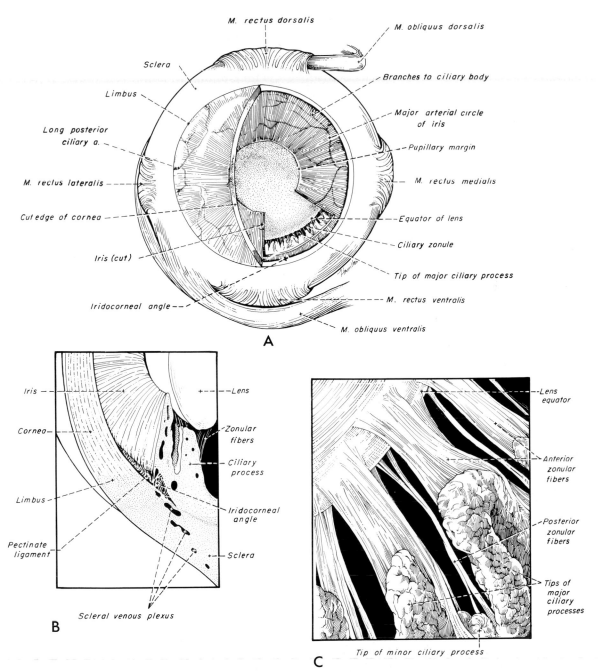

Figure 20–4. Bulbus oculi.
A. Anterior aspect of right eyeball, cornea and iris partially removed.
B. Detail of limbal region.
C. Detail of anterior aspect of ciliary zonule.

tion and continuing in the early postnatal period, the tissue in this area undergoes progressive rarefaction until a long cleft extends the anterior chamber posteriorly between the base of the iris and the sclera (Aguirre et al. 1972, Martin 1975). This cleft is bridged by a network of fine trabeculae which together form the **pectinate ligament** (*lig. pectinatum anguli iridocornealis*). The iridocorneal angle has been studied by Martin (1975), who further divides the trabeculae into primary and secondary iridocorneal trabeculae and deeper uveal trabeculae. Bedford (1977) has described the clinical appearance of the iridocorneal angle as seen with a special goniolens.

The aqueous humor leaves the eye by filtering through the spaces between these trabeculae to aqueous collector vessels that join the scleral venous plexus, which, in turn, is drained by the anterior ciliary and vorticose veins (Van Buskirk 1979). Dilation of the pupil (mydriasis) impedes the outflow of aqueous humor; the iridocorneal angle is narrowed by the increased thickness of the peripheral iris. Constriction of the pupil (miosis) opens the spaces of the iridocorneal angle and facilitates drainage.

In certain breeds, notably the Basset hound, the corneoscleral angle is dysplastic. The pectinate ligament is sheetlike with few openings (Martin 1975), which impedes the outflow of the aqueous humor. Animals with such a condition are predisposed to glaucoma.

Vascular Tunic

The **vascular tunic** (*tunica vasculosa bulbi*) is the thick middle coat of the eye, interposed between the retina and the sclera. It is commonly referred to as the uvea or uveal tract. The vascular tunic includes three contiguous parts, which, from posterior to anterior, are the choroid, the ciliary body, and the iris (Fig. 20–1).

CHOROID

The **choroid** is a pigmented vascular layer. It is continuous with the ciliary body anteriorly and completely envelops the posterior hemisphere of the eyeball, except in the region of the area cribrosa, where it is absent.

The choroid is further divided into layers which, from the outermost inward, are the **suprachoroid** (*lamina suprachoroidea*), the **perichoroidal lymphatic space** (*spatium perichoroideale*), the **vascular layer** (*lamina vasculosa*), the **reflective layer** (*tapetum lu-*

cidum), the **choroidocapillary layer** (*lamina choroidocapillaris*), and the **basal lamina** (*lamina basalis*). The last is poorly developed in the dog.

The **tapetum lucidum** is a specialized reflective layer of the choroid (Fig. 20–1). It is present in about one-third of the area of the choroid. When present it forms a distinct layer 8 to 12 cells thick between the choroidocapillary layer and the vascular layer.

In dogs, the tapetum lucidum is in roughly the shape of a right triangle with the hypotenuse resting on a dorsal plane and the right angle situated dorsally (see Fig. 20–11). The medial angle is more acute than the lateral. In large breeds of dogs, the hypotenuse (ventral border) is usually ventral to the optic disc. In small breeds of dogs, the tapetum is relatively smaller, and does not extend ventrally to include the optic disc (Habel 1975). In some toy breeds, the tapetum may be greatly diminished in area or may be entirely absent.

The tapetum of the dog is cellular. The tapetal cells contain rodlike structures which are oriented parallel to the retina and which are thought to be responsible for its high degree of reflectivity (Hebel 1969, 1971). The tapetum lucidum develops after birth. As the dog matures, the color of the tapetum changes from a slate gray to violet to red-orange at about four months of age (Rubin 1974). The color is generally uniform except at the junction of the tapetal and non-tapetal choroid, which may be quite irregular and demonstrate considerable pleochroism.

The **vascular layer** of the choroid is a homologue of the pia-arachnoid. It is a plexus of choroidal arteries and venules supported by a collagenous and elastic stroma. The choroidal vessels are the terminal branches of the ciliary arteries and the vorticose veins. Most of the choroidal arterioles run parallel to the meridians and to one another (Fig. 20–5). Radicles to the choroidocapillary layer nourish the outer layers of the retina. The choroidal venules fan outward from the point at which the vorticose veins penetrate the sclera (Fig. 20–6).

In the majority of dogs, the vessel layer of the choroid and the retinal pigment layer are darkly pigmented. The non-tapetal region in these dogs appears dark brown or

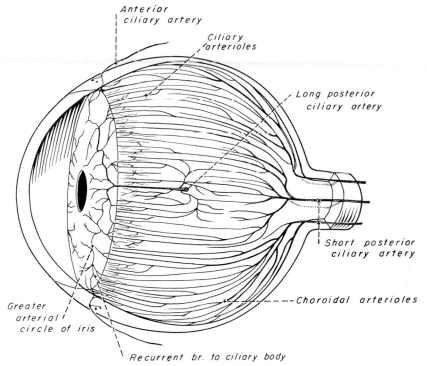

Figure 20–5. Arterial supply of the vascular tunic.

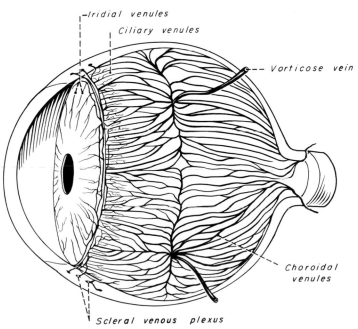

Figure 20–6. Veins of the vascular tunic.

black. In dogs with amber, blue, or he-
terochromic irides, however, the choroid
and retinal pigment layers are unpigmented
or nearly so. In these animals, the vessels of
the choroid can be visualized with the
ophthalmoscope, and the non-tapetal
fundus appears red or striped (a so-called
"tigroid" fundus). The absence of a tapetum
lucidum or choroidal pigment or both is
considered a normal variation which apparently does not affect the dog's vision.
Focal areas of choroidal hypoplasia, however, are serious ocular defects. They are a
common manifestation of the Collie eye
syndrome, a serious and widespread heritable defect in the Collie and Sheltie breeds.

CILIARY BODY

The **ciliary body** (*corpus ciliare*) is the
thickened middle segment of the vascular
tunic, between the iris and choroid (Fig.
20–1). The *ora serrata* is the slightly undulating line which demarcates the visual
retina, which rests on the ciliary body (*pars
ciliaris retinae*). Anterior to the ora serrata,
the vascular coat is elevated into several
hundred small, parallel, regular folds.
These increase rapidly in height and coalesce as they pass anteriorly along the
meridians, forming tall, thin folds rising up
from the base plate of the ciliary body (Fig.
20–7). Seventy to 80 major ciliary plicae are
thus formed. At the root of the iris (*margo
ciliaris*), these folds lose their outer attachment to the sclera and arc centrally toward
the lens as the short, blunt, free **ciliary processes.** Minor ciliary folds and processes are
interposed between the major processes, although this relationship is not constant; two
major processes may occur together without
an intervening minor process. The minor
processes are neither as tall nor do they
approximate the lens as closely as the major
processes (Fig. 20–4C). The posterior surface of the ciliary processes is almost entirely covered by fibers of the ciliary zonule
(Fig. 20–7). The major and minor processes
together form the **ciliary crown** (*corona ciliaris*). The distance between the tips of the
ciliary processes and the ora serrata is
greater on the lateral aspect of the ciliary
body.

Like the choroid, the ciliary body is highly vascular. Each ciliary fold is supplied by
a large branch from the choroidal arteries.
The greater arterial circle of the iris contributes many recurrent branches to the ciliary
body which anastomose with the choroidal
vessels (Fig. 20–5). The ciliary body is
drained by the choroidal and vorticose
veins. The overlying pars ciliaris retinae
produces the aqueous humor.

The **ciliary muscle** (*m. ciliaris*) consists of
numerous smooth muscle fascicles located
in the outer portions of the ciliary body.
There are both **circumferential fibers** (*fibrae circulares*) and **meridional fibers** (*fibrae meridionales*). The meridional fibers
predominate in the dog, although neither
group is prominent. The meridional fibers
originate from the **scleral ring** (*anulus
sclerae*) on the inner surface of the sclera
posterior to the iridocorneal angle (see Fig.
20–9). The muscle fibers form the outer
layer of the uveal trabeculae of the ciliary
cleft. The meridional fibers insert in the
stroma of the ciliary body near the ora serrata. When the fibers of the ciliary muscle
contract upon parasympathetic stimulation,
they decrease tension on the zonular fibers
supporting the lens. With the release in tension, the lens becomes more spherical as a
result of the inherent elasticity of its capsule. The more spherical lens has a shorter
focal distance, and close objects are now
brought into critical focus on the retina (accommodation). It is not known whether the
ora serrata actually moves anteriorly during
this process, or whether the zonule fibers
are relaxed because of a sphincter-like action of the ciliary muscle which decreases
the diameter of the ciliary body. The extent
of the accommodative ability of the dog is
also not known with certainty; it is pre-
sumed to be inferior to that of primates, in
which the ciliary muscle is far better developed, and probably does not exceed 1 to 3
diopters (Duke-Elder 1958).

ZONULE

The lens is fixed in position by a delicate
suspensory apparatus, the *zonula ciliaris.*
The zonule is composed of a highly ordered
array of zonular fibers which are aggregates
of fibrils of 10-nm. diameter, similar to
those described in elastic tissue by Greenlee et al. (1966). The zonule separates the

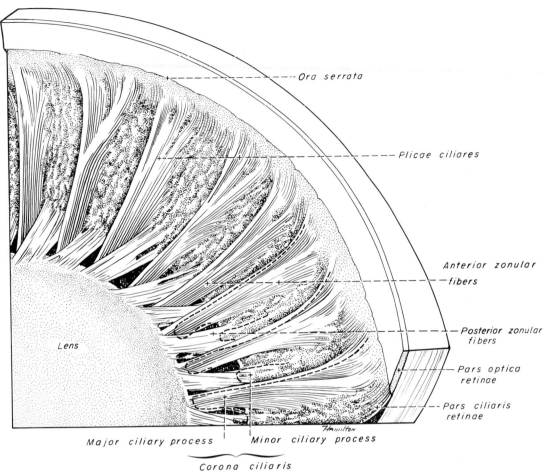

Ora serrata

Plicae ciliares

Anterior zonular fibers

Posterior zonular fibers

Pars optica retinae

Pars ciliaris retinae

Lens

Major ciliary process

Minor ciliary process

Corona ciliaris

Figure 20–7. Posterior aspect of ciliary body, ciliary zonule, and lens.

posterior chamber from the vitreous body. The zonule lies posterior to the iris and ciliary body and is not visible in the intact eye unless the iris is widely dilated or the lens is subluxated.

The apices of the ciliary processes may be in contact with the equator of the lens, but they are not attached to it by zonular fibers, as is commonly illustrated (Fig. 20–4C). The majority of the zonular fibers, in fact, originate from the pars ciliaris retinae just anterior to the ora serrata. These fibers pass anteriorly closely adherent to the surface of the ciliary body. Small auxiliary fibers arise from the epithelium of the ciliary folds and serve to anchor the main fiber bundles. As the small folds of the ciliary body unite to form the major ciliary folds, the zonule fibers also converge until they completely

cover the sides of each major ciliary process (Fig. 20–7). These fibers continue centrally beyond the apices of the processes, span the circumlental space, and insert primarily on the anterior lens capsule near the equator (Fig. 20–8). They are thus designated the **anterior zonular fibers** (Pollock 1978).

The zonular attachments to the posterior lens capsule are not as well developed. Two subsets of fibers compose the **posterior fiber group.** Where a minor ciliary fold is present, fibers arise from its surface and the surrounding epithelium and insert on the posterior lens capsule. Fibers also originate from the valleys between ciliary folds. These arc toward the major ciliary processes, cross the face of the anterior fibers obliquely, and insert on the posterior lens capsule.

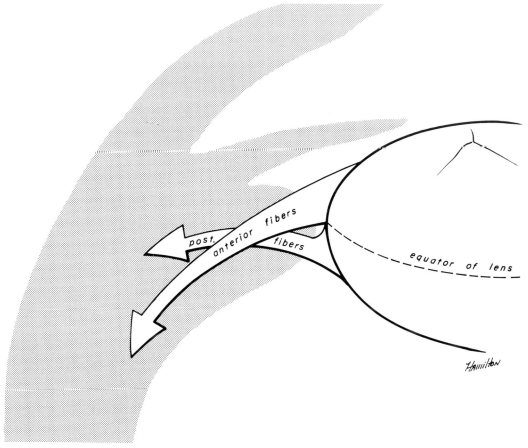

Figure 20–8. Scheme of zonular fibers.

The anterior face of the vitreous body bulges forward between the ciliary folds and into the spaces between the zonular fiber bundles (*spatia zonularia*) (see Fig. 20–14). The zonular fibers are under tension when the ciliary muscle is relaxed.

IRIS

The **iris** is the most anterior segment of the vascular tunic. It is a thin circular diaphragm which rests against the anterior surface of the lens (Fig. 20–4). The central opening in the iris, the **pupil** (*pupilla*), is circular in the dog. The size of the pupil is variable and serves to regulate the amount of light which is allowed to reach the retina. The diameter of the pupil is smallest when the intensity of illumination is greatest. The periphery of the iris (*margo ciliaris*) is continuous with the ciliary body and trabeculae of the iridocorneal angle (Fig. 20–4B).

In the fetus, the iris is not fenestrated and thus covers the anterior surface of the lens. The central portion of the iridial anlage (*membrana pupillaris*) contains vessels which nourish the growing lens (see "Development"). Normally the **pupillary membrane** atrophies so that only remnants of the vessels are present when the eyelids open at about two weeks of age. However, remnants commonly persist, especially on the dorsal pupillary margin, until the age of four to five weeks. Abnormal persistence of the pupillary membrane into adult life is a heritable condition in Basenji dogs (Roberts and Bistner 1968).

Two antagonistic muscles regulate the diameter of the pupil: the *m. sphincter pupillae* and the *m. dilator pupillae* (Fig. 20–9). Both are derived from the outer layer of neurepithelium of the pars iridica retinae. The **sphincter muscle** is a sheet of circumferentially arranged smooth muscle fibers

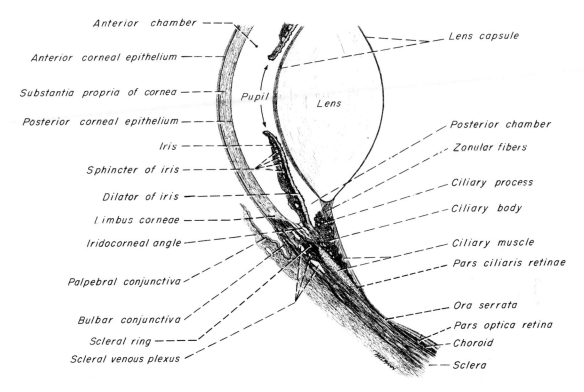

Figure 20–9. Anterior segment of eyeball.

near the pupillary margin. It is the stronger of the two muscles. The **dilator of the pupil** is composed of radially arranged smooth muscle fibers which form a meshwork through which the collagen bundles of the iris stroma are looped. The dilator is posteror to the sphincter muscle, closely applied to the pars iridica retinae.

The sphincter is innervated by parasympathetic nerve fibers whose preganglionic cell bodies are located in the parasympathetic nucleus of the oculomotor nerve in the brain stem. Axons of the preganglionic neurons follow the oculomotor nerve to the ciliary ganglion, where they synapse (Fig. 20–10). Postganglionic fibers from the ciliary ganglion reach the iris via the communicating branch with the nasociliary nerve.

The **dilator muscle** is innervated by sympathetic nerve fibers. The preganglionic neurons are located in the intermediate grey column of the first three thoracic spinal cord segments (Fig. 20–10). Their axons course rostrally in the vagosympathetic trunk. Postganglionic neurons are located in the cranial cervical ganglion. From the gan-glion the postganglionic fibers pass rostrally through the tympano-occipital fissure to join the ophthalmic nerve on the ventral surface of the trigeminal ganglion. They are distributed with the ciliary branches of the ophthalmic nerve.

The blood supply of the iris arises primarily from the long posterior ciliary arteries (Figs. 20–4A, 20–5). These arteries follow the medial and lateral meridians from the area cribrosa to the iris. They are visible in the episcleral space from the optic nerve to about the equator of the eyeball and are useful landmarks for orienting an isolated eyeball. At the equator, the long posterior ciliary arteries pass deep to the sclera and continue anteriorly to the base of the iris. Some branches may be given to the choroid, where they anastomose with the choroidal arterioles. In the ciliary margin of the iris, the medial and lateral posterior arteries each divide into dorsal and ventral branches. These branches run circumferentially in the iris, forming the **greater arterial circle of the iris** (*circulus arteriosus iridis major*) (Fig. 20–4A). The position of the circle approximates that of the equator of the

Pathway for Pupillary Control

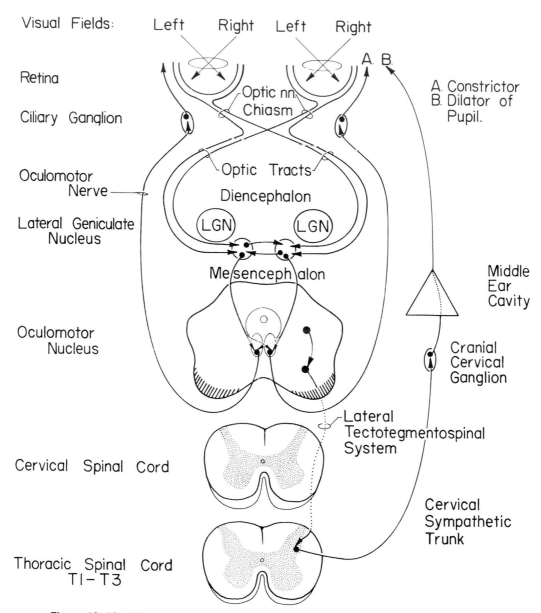

Figure 20–10. Neuroanatomic pathway for pupillary control. (From deLahunta 1977.)

lens. The circle is not complete; the medial and lateral branches do not anastomose (Purtscher 1961). The formation of the greater arterial circle is similar to that in man but differs significantly in that, in man, the greater arterial circle of the iris is actually within the ciliary body. In the dog the vasculature of the iris is often visible in the undissected eye and commonly protrudes slightly from the anterior iridial surface (Bedford 1977).

Fine radial branches originate from the arterial circle. These run either centrally toward the pupil to supply the iris stroma and musculature or peripherally to anastomose with the ciliary vessels which supply the ciliary processes (Anderson and Anderson 1977) (Fig. 20–5). A lesser arterial circle near the pupil is not present. Incision of the iris in the dog results in profuse hemorrhage.

Retina

The innermost tunic of the eye is the nervous coat, or **retina**. The retina develops from an outgrowth of the diencephalon, the optic vesicle (Fig. 20–2). The optic nerve, therefore, is actually a tract of the central nervous system and has the characteristics of central nervous tissue. The meninges of the brain are continued along the optic nerve to the eyeball as the **internal and external sheaths of the optic nerve** (*vaginae n. optici interna et externa.*).

At about 17 days of gestation, the optic vesicle invaginates to form the optic cup, which folds into a hemisphere (Fig. 20–2B). The inner and outer walls of the cup fuse and together form the definitive retina. The outer wall of the cup becomes a single layer of retinal pigment epithelium which is directly apposed to the choroid. The inner layer of the cup differentiates into the photoreceptor and synaptic layers of the pars optica retinae. In traumatic retinal detachment, or in postmortem material, the retina usually separates along the line of the embryonic intraretinal space, leaving the pigment epithelium adhering to the choroid.

There are three distinct areas of the retina which, from anterior to posterior, are the *pars iridica retinae*, the *pars ciliaris retinae*, and the *pars optica retinae*. Only the last is photosensitive (visual retina). The pars iridica and pars ciliaris together constitute the non-visual, or blind, retina (*pars ceca retinae*).

The *pars iridica retinae* is a bilayered epithelium which covers the posterior surface of the iris. The pupillary margin is the anterior limit of tissue of neurectodermal origin in the eye. The anterior face of the iris is covered by epithelium derived from mesoderm. The outer layer of the pars iridica retinae gives rise to the iridial musculature. The inner layer, in most dogs, is heavily pigmented and has prominent radial striations.

At the ciliary margin of the iris, the pars iridica retinae reflects onto the ciliary body to become the *pars ciliaris retinae.* This is a bilayered cuboidal epithelium. The outer layer is heavily pigmented. The inner layer is not pigmented. The aqueous humor which fills the anterior and posterior chambers of the eye is produced by the pars ciliaris retinae. The zonular fibers which support the lens appear to originate in the interstices between the cells of the epithelium (Raviola 1971).

The *ora serrata* is the demarcation between the visual and non-visual retina (Fig. 20–7). The pars optica retinae, posterior to the ora, is three to four times as thick as the pars ciliaris. The change is sufficiently marked and abrupt as to cast a shadow when illuminated from behind (Fig. 20–9). Intraretinal cysts adjacent to the ora serrata are a common finding in older dogs (Heywood et al. 1976, Rubin 1974). The *pars optica retinae* is classically described as having 10 layers. The photoreceptor cells form the outermost layer, which developed from the inner wall of the optic cup. The photoreceptor segments of these cells (rods and cones) are situated outward, adjacent to the pigment epithelium, which, in turn, rests on the basal lamina of the choroid. Light must pass through the retinal nerve fibers, the various synaptic cells of the retina, and the cell bodies of the rods and cones themselves before it reaches the photosensitive layer. Light passing through the photoreceptors is absorbed by the pigment of the pigment epithelium and choroid. In the area of the tapetum lucidum, the pigment epithelium of the retina lacks melanosomes, and the light is reflected by the tapetum back through the photoreceptor layer. This is presumably an adaptation for improved vision in low levels of illumination.

The rods and cones are not evenly distributed throughout the retina (Koch and Rubin 1972). In the dog about 95 per cent of the photoreceptors are rods, and, presumably, the dog has minimal color vision. The photoreceptor layer is about one-half as thick peripherally as it is centrally, where it measures about 0.24 mm. thick. Approximately 3 mm. lateral to the optic disc is the **macula** (Fig. 20–11). This area of most acute vision is poorly developed in the dog as compared to man and higher primates. There is no thinning of the inner retinal layers and, therefore, no fovea. The area is not clearly defined either grossly or by

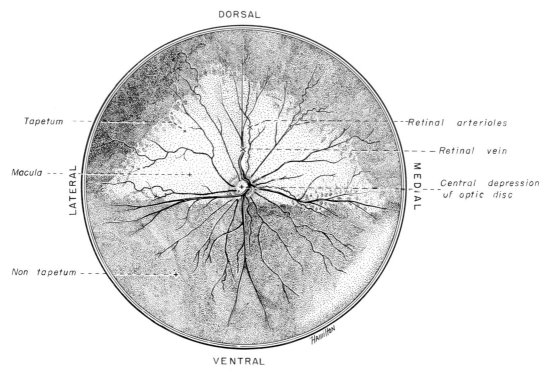

DORSAL

Tapetum - - -
Macula - - -
LATERAL
Non tapetum - - -

- - Retinal arterioles
- - - Retinal vein
MEDIAL
- - Central depression of optic disc

Hamilton

VENTRAL

Figure 20–11.   Ocular fundus of right eye.

means of the ophthalmoscope. Histological-ly, an increase in the number of cones and ganglion cells is noted, although the central area of the macula in the dog does not consist exclusively of cones, as is the case in man. Major retinal blood vessels do not cross the macula but curve dorsal or ventral to it (Fig. 20–11).

The synaptic pathway of the visual impulse initially passes inward toward the vitreous. The photoreceptors synapse with the overlying horizontal and bipolar cells. These synapse with the still more interior amacrine and ganglion cells. The axons of the ganglion cell layer arc toward the optic disc. Here they become myelinated, turn, and pass through the area cribrosa of the sclera to form the optic nerve. The intraocular myelinated portion of the nerve forms the **optic disc** (Fig. 20–1). The morphology of the disc varies greatly among dogs. It may be round to oval or triangular to quadrilateral. It is commonly pink or white but may contain pigment. A ring of pigmentation or hyper-reflectivity may surround the disc. Myelin may extend along the nerve fibers onto the retina for a variable distance, giv-

ing the disc a large and ragged appearance. The optic disc commonly contains a **central depression,** the *excavatio disci,* although in some dogs the disc is level with the surrounding retina or slightly elevated above it. There are no photoreceptors overlying the disc, and hence the optic disc represents a "blind spot" in the eye. The normal disc is 1 to 2 mm. in diameter (Wyman and Donovan 1965). It is located ventrolateral to the posterior pole of the eyeball (Fig. 20–3).

The retinal blood vessels are readily visualized with the ophthalmoscope, and their examination represents an important part of the ophthalmic examination (Magrane 1977). The part of the retina that is visible with the ophthalmoscope is referred to as the "ocular fundus" clinically.

The **retinal arteries** originate from the short posterior ciliary arteries, where the latter penetrate the sclera at the periphery of the area cribrosa. The retinal arteries pass into the eyeball closely applied to the surface of the optic nerve. There is no central artery or vein in the dog (Wyman and Donovan 1965, Engermann et al. 1966). The retinal arteries appear at the periphery of the optic disc and branch into 15 to 20 arterioles which radiate toward the periphery (Fig. 20–11). The macular area is relatively avascular; the retinal vessels curve around it. Except for the macula, the retina of the dog is uniformly vascularized and, as such, is said to be a holoangiotic retina.

The primary **retinal veins** are three or four and occasionally five in number. Veins directed dorsally, ventromedially, and ventrolaterally are constant. A fourth vein, running ventrally, is present in about 80 per cent of dogs. The retinal veins form a variable anastomosis on, or just below, the surface of the optic disc. This anastomosis may be in the form of a complete circle or any portion thereof. The veins are easily recognized ophthalmoscopically, being two to four times the diameter of the arterioles and a darker red in color. They project very slightly from the retinal surface and indent the vitreous body (see Fig. 20–14).

Lens

The **lens** of the eye is a soft, transparent, nearly spherical structure suspended in contact with the posterior face of the iris (Fig. 20–1). Its function is to bring images into critical focus on the photoreceptor layer of the retina.

Images are focused on the retina by the combined refraction of the cornea, aqueous, lens, and vitreous. Most refraction occurs at the anterior face of the cornea. The lens actually alters the path of the light rays only slightly, but is unique in that the degree of refraction is variable. The focal length of the lens is altered by changes in its shape which are brought about by the action of the ciliary muscle, zonular fibers, and lens capsule.

An anterior and posterior pole and equator are described for the lens in a manner analogous to that for the eyeball as a whole. The equator demarcates the anterior and posterior faces.

The lens is circular in transverse section but slightly ellipsoidal in sagittal or dorsal section. The dorsoventral and mediolateral diameters average about 10 mm., whereas the anteroposterior length along the axis bulbi is about 3 mm. less. The posterior face is more convex than the anterior, but the overall deviation from spherical is much less pronounced than in man. The posterior pole of the lens approximates the equatorial plane of the eyeball as a whole.

The lens is ectodermal in origin. It forms from an invagination of the surface epithelium overlying the optic cup (Fig. 20–2B). This pinches off to form the lens vesicle by about 27 days of gestation (Andersen and Shultz 1958). The cells forming the posterior wall elongate until they reach the anterior epithelium obliterating the cavity of the vesicle. These cells then lose their nuclei to become the primary lens fibers of the embryonal nucleus. As a result, in the adult, there is a cuboidal lenticular epithelium only on the anterior face of the lens (Fig. 20–12).

Throughout life the epithelium continues to proliferate slowly. Cells at the equator elongate along the meridians until their apices approach the poles. In section, the nuclei of these elongating cells form an arc, the so-called lens bow, from the equator toward the deeper portions of the lens (Fig. 20–12). As successive layers of cells accumulate, the deeper cells lose their nuclei, but remain viable as secondary **lens fibers.** This manner of growth results in a lens which is distinctly lamellar, resembling an onion in cross section. In dogs over one year of age, an embryonal, fetal, and adult nucleus and an adult cortex are recognized (Martin 1969).

Because the continued epithelial proliferation is unaccompanied by cell loss, the weight of the lens increases with age. The increase in weight is most rapid during the first few months of life but continues at a slow rate in the adult. Since individual variation of the growth rate of the lens within a species is usually slight, there is a direct correlation between age and dry lens weight which can be used to estimate the age of wild canids and other species (Lord 1961).

The nuclear portion of the lens undergoes progressive dehydration and condensation (nuclear sclerosis). In adults, this shrinkage occurs at a rate roughly equal to the rate of growth at the periphery, and thus the lens does not continually increase in size. The progressive nuclear sclerosis results in a much firmer and less elastic lens in older individuals. The deeper portions of the lens

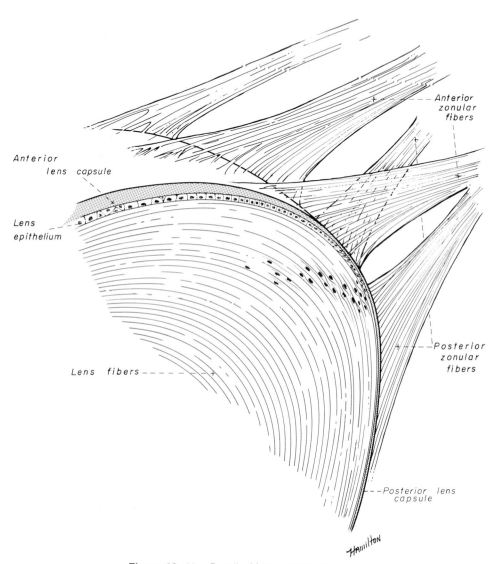

Figure 20–12. Detail of lens equator in section.

Figure 20–13. Scheme of lens fibers, posterior view.

appear blue-gray and hazy. This normal age change should not be confused with cataracts, which are pathological lenticular opacities.

The apices of the lens fibers do not all meet at a single point at each pole, like sections of an orange. Rather, the junctions form distinct linear markings, the **lens sutures** (Fig. 20–13). On the anterior face, the lens sutures form an upright letter Y. On the posterior face the Y is inverted. Lens fibers which begin at the tip of one of the arms of

the Y on the anterior face end in the crotch between the arms of the Y on the posterior face (Fig. 20–13). This pattern is most obvious in the adult nuclear region. The prominence of the sutures increases with age, beginning about the third year (Heywood et al. 1976). In the cortex the lens sutures are more stellate in form. The lens sutures are easily visualized in the living dog with a slit lamp and are frequently the site of cataracts. With suitable illumination, a small white dot, a remnant of the hyaloid artery, can be

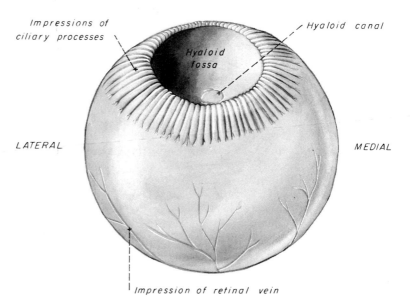

Impressions of ciliary processes

Hyaloid canal

Hyaloid fossa

LATERAL

MEDIAL

Figure 20–14. Dorsal view of isolated vitreous of left eye.

Impression of retinal vein

seen in the center of the posterior lens face (Martin 1969).

The entire lens is enveloped by the **lens capsule** (Fig. 20–12). The capsule is a basement membrane secreted by the cells of the lenticular epithelium. It is highly refractile and elastic. Because the posterior epithelium is present only in the early fetal period, the posterior capsule is much thinner than the anterior capsule, the latter being about 15 times thicker. The anterior capsule in the dog is of relatively constant thickness across the anterior face.

The zonular fibers, which suspend the lens, insert into the superficial layers of the lens capsule. A larger number of fibers that are also stronger insert in the thicker anterior capsule (see "Zonule"). The lens rests in a depression in the vitreous, the **hyaloid fossa** (Fig. 20–14). The vitreous is tightly adherent to the posterior capsule and greatly complicates surgical removal of the lens (see "Vitreous," page 1093).

In the adult, the lens, although active metabolically, is avascular. Nutrition is received from, and wastes are eliminated into, the aqueous and vitreous humors. Disease processes — for example, diabetes mellitus — which affect lenticular metabolism result in a loss of transparency (cataract).

Chambers of the Eye

Within the eyeball are three chambers: the **anterior chamber** (camera anterior bulbi), the **posterior chamber** (camera posterior bulbi), and the **vitreous chamber** (camera vitrea bulbi) (Fig. 20–3C).

The anterior chamber is the space bounded by the cornea anteriorly and the iris posteriorly. It is filled with the aqueous humor. The anterior chamber is in direct communication with the posterior chamber through the variably sized aperture of the pupil. The periphery of the chamber is continuous with the spaces of the iridocorneal angle.

The posterior chamber is smaller than the anterior chamber. It is bounded anteriorly by the posterior face of the iris, posteriorly by the anterior lens capsule, and peripherally by the zonule and anterior face of the vitreous.

Aqueous humor fills the anterior and posterior chambers. The humor is produced by an active secretory process from the epithelium (pars ciliaris retinae) of the richly vascular ciliary body. The aqueous is normally clear and colorless. It is very low in protein compared to blood plasma and closely resembles the cerebrospinal fluid in composition. Normally the aqueous humor is maintained at an intraocular pressure of approximately 15 to 30 mm. mercury in the dog (Magrane 1977). This pressure is essential to maintain the normal shape and firmness of the eyeball. When the intraocular pressure is lost post mortem or as a result of the escape of aqueous humor through a corneal laceration, the eye becomes soft and deformed.

The aqueous outflow pathway in the dog has been described by Van Buskirk (1979). The aqueous humor flows from its site of production, the epithelium of the ciliary processes, into the posterior chamber, through the pupil into the anterior chamber and peripherally to the spaces of the iridocorneal angle. Here it is resorbed into the blood stream by the scleral venous plexus. A delicate balance between production and resorption maintains the normal intraocular pressure. The production of aqueous humor is not regulated by the intraocular pressure, however. Occlusion of the primary outflow pathway, either at the pupil or iridocorneal angle, results in increased intraocular pressure (glaucoma).

VITREOUS

The **vitreous chamber** is the largest of the three chambers of the eye, accounting for approximately 60 per cent of the volume of the eyeball. The zonule and posterior lens capsule form the anterior limit of the chamber. The retina encloses the remainder.

The **vitreous body** occupies the vitreous chamber. The vitreous body is a soft, clear gel which conforms to the shape of the cavity it occupies (Fig. 20–14). Thus, the anterior face of the vitreous is indented (fossa hyaloidea) where the posterior face of the lens bulges caudally into the vitreous chamber. Similarly, the surface of the vitreous is fluted where it projects between the ciliary processes.

The vitreous body is almost entirely extracellular. The bulk of the vitreous is formed by the liquid component *(humor vitreus)*, a solution of mucopolysaccharides rich in hyaluronic acid (Balazs 1973). The vitreous body is greater than 98 per cent water. The structure of the vitreous is reinforced by fibers of vitrein *(stroma vitreum)* which are similar to collagen and are essential to the gel characteristics of the vitreous (Balazs 1973). These are especially numerous near the ora serrata, but do not lend sufficient rigidity to maintain the shape of the vitreous after its removal from the eye. Fine and Yanoff (1972) have aptly described the vitreous as the "most delicate of all the connective tissue in the body."

The anterior face of the vitreous is limited by the *membrana vitrea*. This is not a discretely demonstrable membrane in the ordinary sense, but rather a local condensation of the filamentous framework found throughout the vitreous (Fine and Yanoff 1972). In man it is of sufficient strength to contain the vitreous after removal of the lens and lens capsule. In the dog, intracapsular lens removal is impractical because the vitreous membrane is thin and tightly adherent to the posterior lens capsule. Attempts to remove the lens within its capsule usually result in tearing of the vitreous face with subsequent loss of the vitreous body (Bistner et al. 1977).

The vitreous body is also tightly adherent to the pars ciliaris retinae at the ora serrata and to the optic disc, as well as to the posterior lens capsule. It is easily separated from the retina except at these points.

The **hyaloid canal** *(canalis hyaloideus)* traverses the vitreous from the optic disc to the posterior face of the lens (Fig. 20–1). It is the funnel-shaped remnant of the primary (embryological) vitreous. As the secondary or definitive vitreous is elaborated by the retina, the primary vitreous is compressed centrally. At the same time, the primary vitreous is attenuated by being stretched as the globe increases in size. The hyaloid canal is broadest anteriorly, where it is attached to the posterior surface of the lens. The attachment to the lens can be visualized with the biomicroscope as a thin white circle about 3 mm. in diameter surrounding the posterior pole of the lens (Martin 1969). The canal tapers posteriorly toward the optic disc and usually exhibits a ventral sag between its points of attachment (Fig. 20–1); the hyaloid

canal is concave dorsally in most dogs. The primary vitreous is normally of the same optical clarity as the rest of the vitreous body and can be distinguished only by a slight difference in optical refractivity resulting from local differences in the vitreous stroma (Balazs 1973). With the appropriate illumination, a line of demarcation can be detected at the interface of the primary and secondary vitreous (the "wall" of the canal). Occasionally, hemorrhage into the vitreous will spread along the interface, clearly outlining the canal (Fine and Yanoff 1972).

In the embryo, the **hyaloid artery** courses through the hyaloid canal from the optic disc to the lens to supply the posterior surface of the growing lens. This artery normally atrophies completely by the time the eyelids open; a small white dot on the posterior pole of the lens marks its site of attachment (Martin 1969). The multilayered fenestrated sheaths peculiar to the fine structure of the primary vitreous probably derive from the tunica media of the hyaloid vessels (Balazs 1973). Remnants of the artery itself are occasionally seen ophthalmoscopically, especially in very young dogs. Rarely, persistent hyaloid arteries occur and result in posterior lenticular cataracts (Rebhun 1976).

ORBIT

The **orbit** is the conical cavity which contains the eyeball and the ocular adnexa. The orbital margin outlines the base of the cone, which is directed anterolaterally. The shape of the base approximates a rounded trapezoid more than a true circle (Fig. 20–15). The axis of the orbit, a line passing from the center of the base to the apex at the optic canal, is directed obliquely caudal and ventral. In mesaticephalic dogs, the axis of the orbit forms an angle of approximately 30° with a median plane and 30° with a dorsal plane. In the foreshortened skulls of the brachycephalic breeds, the axis of the orbit deviates as much as 50° from the median plane (Fig. 20–16). The eyeball occupies the base of the orbit and projects a variable distance anterior to the orbital margin. The average dog's visual field encompasses about 240° of arc. Binocular vision is confined to the central 60° (Sherman et al. 1975).

The orbital margin is bony for approxi-

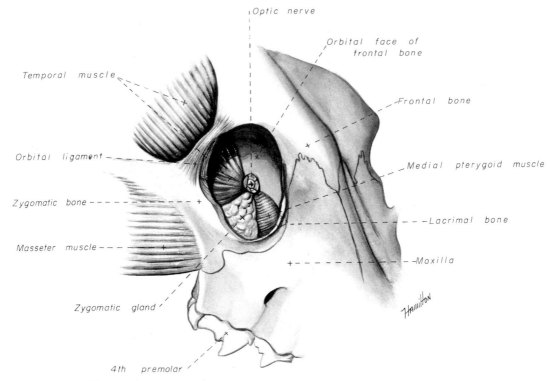

Figure 20–15. The right orbit, orbital contents removed, viewed along orbital axis.

mately four-fifths of its circumference. The posterolateral one-fifth of the margin is completed by the **orbital ligament** *(ligamentum orbitale)*. In the brachycephalic dog the liga-

ment forms a larger proportion of the circumference. The ligament is a thick fibrous band which unites the zygomatic process of the frontal bone and the frontal process of the

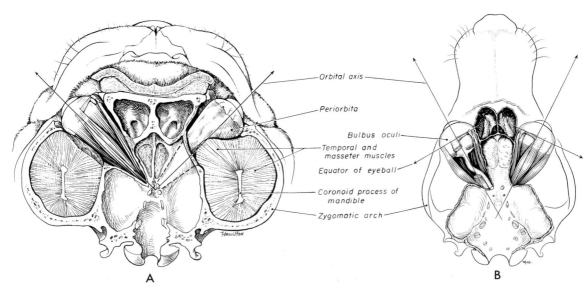

Figure 20–16. Orbital axes of a brachycephalic *(A)* versus mesocephalic *(B)* dog.

zygomatic bone (Fig. 20–15). The ligament serves as the lateral attachment of the m. orbicularis oculi and the lateral palpebral ligament. The dorsal and medial segments of the orbital margin are formed by the frontal bone. In most dogs, the lacrimal bone forms a small portion of the ventromedial orbital margin. In some brachycephalic skulls, however, the lacrimal bone is confined to the medial orbital wall and does not contribute to the orbital margin. In these cases only, the medioventral orbital margin is formed by the maxillary bone. The ventrolateral orbital margin is the orbital border of the zygomatic bone.

In man the confines of the orbit are entirely bony and readily discernible from a study of the skull. In the dog only the medial wall and part of the roof of the orbit are osseous. The lateral wall and floor are formed by soft tissue (Fig. 20–15). Consequently, the orbit of the dog cannot be properly appreciated from study of a skeletal preparation. Furthermore, this anatomical difference between the orbit of the dog and that of man is manifested by marked differences in diseases which affect the orbit and in surgical approaches to this area.

The medial wall of the orbit is osseous (Fig. 20–15). It is formed primarily by the orbital part of the frontal bone. The orbital wing of the presphenoid forms the caudal part of the medial wall and contains the optic canal. The lacrimal bone contributes to a small portion of the rostroventral medial wall and contains the fossa for the lacrimal sac and the caudal orifice of the nasolacrimal canal (see Fig. 4–32).

Five foramina are found in the medial wall of the orbit (see Fig. 4–38). At the apex of the orbit are the optic canal and orbital fissure. The optic nerve and internal ophthalmic artery leave the cranial cavity through the optic canal. The orbital fissure between the basisphenoid and presphenoid bones gives passage to the oculomotor, trochlear, abducent and ophthalmic nerves, the anastomotic branch of the external ophthalmic artery, and the orbital venous plexus. The retractor bulbi muscle originates within the orbital fissure. Rostrodorsal to the orbital fissure are the two small ethmoidal foramina which transmit the external ethmoidal artery and the ethmoidal nerve. Cranially, the fossa for the lacrimal sac occupies the center of the

orbital face of the lacrimal bone. This fossa is continued rostromedially as the lacrimal canal, which contains the nasolacrimal duct (see Fig. 20–21).

The dorsally convex **ventral orbital crest** *(crista orbitalis ventralis)* of the frontal bone demarcates the boundary between the orbit dorsally and the more ventral pterygopalatine fossa (see Fig. 4–38). The crest is not prominent, and in unfleshed skulls the orbital fossa appears to extend much farther ventrally than is actually the case. The ventral orbital crest is the dorsal boundary of the origin of the medial pterygoid muscle, which forms the medial third of the orbital floor (Fig. 20–15). The zygomatic salivary gland rests on the dorsolateral surface of the medial pterygoid muscle. Its dorsal surface forms most of the floor of the orbit, from the orbital margin to nearly the optic canal (Fig. 20–15). The maxillary artery and nerve cross the floor of the orbit near its apex.

The medial aspect of the roof of the orbit is formed by the zygomatic process of the frontal bone. A very small foramen is often found in the midorbital face of the process, through which a small artery passes dorsally. In some skulls a palpable depression, the **fossa for the lacrimal gland,** is present in the ventral surface of the zygomatic process of the frontal bone at the origin of the orbital ligament.

The orbit is bounded dorsolaterally and laterally by the medial surface of the temporalis muscle and the orbital ligament (Fig. 20–15). The ramus of the mandible is embedded in the masseter and temporal muscles immediately caudal to the orbit (Fig. 20–16). When the mouth is opened, the dorsal aspect of the ramus of the mandible moves rostrally, compressing the orbital contents. Pain on opening the mouth is a cardinal sign of retrobulbar abscesses (Severin 1976). Denervation atrophy of the masseter and temporal muscles effectively enlarges the orbit, and a sinking of the eye into the orbit (enophthalmos) results. Conversely, swelling of the muscles of mastication, as in eosinophilic myositis, results in exophthalmos.

Because the floor of the orbit is composed entirely of soft tissue, a retro-orbital abscess can be drained into the oral cavity by blunt dissection behind the last molar tooth (Magrane 1977). Similarly, the lateral aspect of the orbit can be explored surgically without requiring osseous resection.

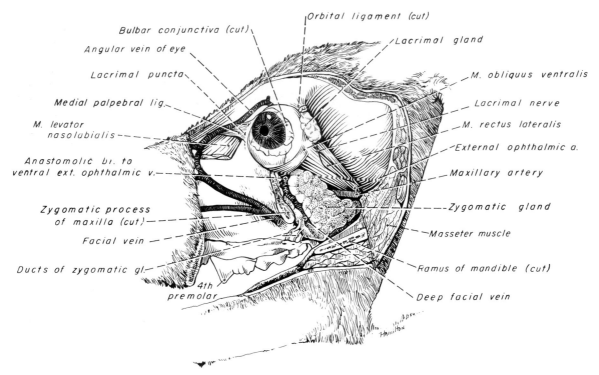

Bulbar conjunctiva (cut)
Angular vein of eye
Lacrimal puncta
Medial palpebral lig.
M. levator nasolabialis
Anastomotic br. to ventral ext. ophthalmic v.
Zygomatic process of maxilla (cut)
Facial vein
Ducts of zygomatic gl.
4th premolar

Orbital ligament (cut)
Lacrimal gland
M. obliquus ventralis
Lacrimal nerve
M. rectus lateralis
External ophthalmic a.
Maxillary artery
Zygomatic gland
Masseter muscle
Ramus of mandible (cut)
Deep facial vein

Figure 20–17. Lateral aspect of orbit; zygomatic arch, ramus of mandible, and temporal and masseter muscles removed.

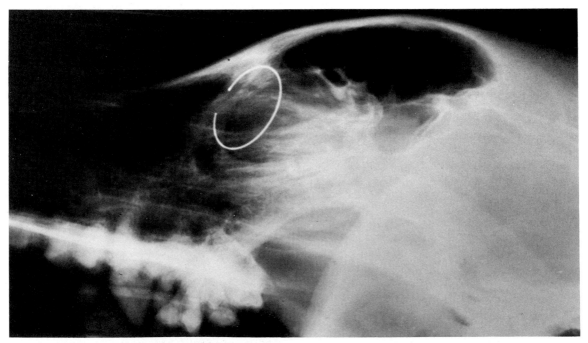

Figure 20–17 Radiograph. Lateral radiograph with a ring to delineate the orbital margin.

The relationships of the orbit to surrounding structures is well illustrated in Hamon's *Atlas of the Head of the Dog* (1977).

Zygomatic Gland

The zygomatic salivary gland forms the lateral two-thirds of the floor of the orbit (Fig. 20-15). Although the gland discharges its secretions into the vestibule and has no function in the visual system, it is described here because it lies adjacent to the orbital contents and because it produces ocular manifestations when diseased (Knecht 1970). This gland has also been called the orbital or dorsal buccal gland.

The zygomatic gland lies dorsal and lateral to the pterygoid muscles and ventral to the periorbita and the ventral orbital margin. Anteriorly and laterally it is bounded by the orbital surface of the zygomatic bone. The gland extends caudally almost to the optic canal (Fig. 20-17). It is roughly pyramidal, tapering caudomedially. The surface of the gland is lobulated and covered by a thin capsule and a layer of fat. Histologically the gland consists almost entirely of mucous acini.

The glandular secretion reaches the oral cavity through a number of ducts which leave the ventrolateral apex of the gland (Fig. 20-17). The major duct opens into the vestibule on the serosal ridge lateral to the last upper molar tooth and about 1 cm. caudal to the opening of the parotid duct. Two to four minor ducts also open onto this ridge caudal to the major duct.

The gland is usually supplied by the first branch of the infraorbital artery. Blood leaves the gland through a branch of the deep facial vein which lies in a groove on the ventral lateral surface.

Orbital Fasciae

There are two important fascial structures of the orbit: the **bulbar sheath** (*vagina bulbi*) and the **periorbita.**

The **bulbar sheath** is a thin fibrous capsule which envelops the eyeball from the limbus to the optic nerve (Fig. 20-18). It is separated from the sclera by the episcleral space. Delicate fibrous trabeculae bridge this space and anchor the sheath to the eyeball. Anteriorly, the sheath ends in the subconjunctival and episcleral tissues at the limbus. It is well developed in the dog and complicates certain ophthalmological surgical procedures. Medicants may be injected either superficial or deep to the bulbar sheath into the subconjunctival or episcleral tissues. The pharmacokinetics of the spaces are distinctly different (Severin 1976).

Near the equator of the globe the sheath reflects around the tendons of insertion of the extraocular muscles as they pass through it to their insertions on the sclera (Fig. 20-19). The bulbar sheath is continuous with the muscular fasciae at these points. Posteriorly, the sheath is continuous with the fascia binding the ciliary vessels and nerves to the dura mater of the optic nerve.

The stronger and more easily demonstrated fascial structure of the orbit is the **periorbita,** which is the conical fibrous sheet surrounding the eyeball and its associated

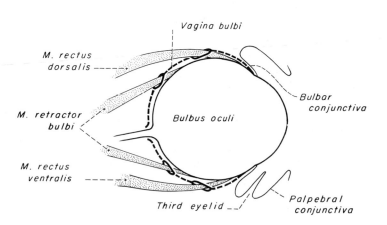

Figure 20-18. Scheme of vagina bulbi in section.

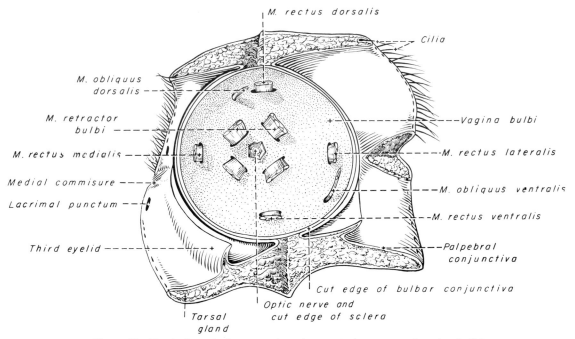

Figure 20–19. Left eyeball removed to show anterior aspect of vagina bulbi.

muscles, nerves, and vessels. In man, in whom the orbital confines are entirely osseous, the periorbita and periosteum of the orbit are one and the same. In the dog, this is only true medially and dorsally. Laterally and ventrally the periorbita is a tough fibrous sheet which separates the eye and its adnexa from the other structures of the orbital and pterygoid fossae.

The apex of the periorbita is firmly attached to the margins of the orbital fissure and the optic canal. At these sites the periorbita is continuous with the dura mater encephali and the external sheath of the optic nerve. The extraocular muscles originate from the periorbita adjacent to the orbital fissure. The cranial nerves to the eye and the internal and external ophthalmic vessels enter the periorbital cone at its apex. The base of the periorbital cone is the orbital margin. At the orbital margin the periorbita reflects onto the face to become the periosteum of the facial bones. A sheet of connective tissue, the **orbital septum,** extends from the periorbita at the orbital margin to blend with the tarsi of the lids. The orbital septum is the anterior limit of the orbit.

In the dog, the periorbita contains numerous circular smooth muscle fibers which normally exhibit a degree of tonus. This tonus acts to squeeze the eyeball out of the orbit and causes the normal prominence of the globe as it projects beyond the orbital margin. Sympathetic stimulation causes marked exophthalmos (Code and Essex 1935). The **superior and inferior tarsal muscles** *(mm. tarsalis superior et inferior)* are smooth muscle fibers also derived from the periorbita which insert in the eyelids and which help to maintain their retracted positions when the palpebral fissure is open (Nicholas 1914). Smooth muscle fibers from the periorbita also insert on the base of the cartilage of the third eyelid and help to maintain it in its retracted position. Innervation of these smooth muscle cells is derived from the cranial cervical ganglion. Lesions resulting in denervation cause a sinking of the globe into the orbit (enophthalmos), protrusion of the third eyelid, and a dropping of the upper eyelid (ptosis) (de Lahunta 1977).

The **orbital fat body** *(corpus adiposum orbitae)* cushions the contents of the orbit and, being easily deformable, permits the rotation and retraction of the eyeball. The orbital fat is found both within the periorbita *(corpus adiposum intraperiorbitale)* and between the periorbita and the surrounding structures (*corpus adiposum extraperiorbitale.*)

A well-developed intraorbital fat pad is present at the posterior pole of the eyeball. It is in the shape of a hollow cone, surrounding the optic nerve and filling the space between it and the diverging extraocular muscles. Thin sheets of intraperiorbital fat are present between the retractor bulbi and the rectus muscles. Extraorbital fat deposits˙ may be extensive ventral and lateral to the periorbita, especially in obese specimens. A prominent fat body is a constant finding caudal to the ventral orbital margin. There is no extraperiorbital fat on the medial and mediodorsal aspects of the orbit, where the periorbita is the periosteum of the frontal and presphenoid bones.

The periorbita is tightly adherent to the osseous orbital wall only at the apex, the orbital margin, the lacrimal fossa, the origin of the ventral oblique muscle, and the frontosphenoid suture. Dissection or effusive disease processes easily separate the periorbita from the medial orbital wall.

EYELIDS

The **eyelids** (*palpebrae*) are mobile folds of skin which can be drawn over the anterior aspect of the eyeball to occlude light and protect the cornea. Their development has been summarized (p. 1077).

The opening between the lids (*rima palpebrarum*) is variable in size. The width of the opening is controlled by opposing groups of muscles. The upper lid (*palpebra superior*) is slightly greater in extent and somewhat more mobile than the lower (*palpebra inferior*). The m. orbicularis oculi (see "Muscles," page 1111) acts to close the palpebral fissure. The levator palpebrae superioris, the pars palpebralis of the m. sphincter colli profundus, and smooth muscles derived from the periorbita (*m. tarsalis*) act to widen the fissure.

The upper and lower lids join at their medial and lateral aspects (*commissura palpebrarum medialis et lateralis*) (Fig. 20–20). The angles formed by the lids at the commissures are the medial and lateral angles of the eye (*anguli oculi medialis et lateralis*).

The lateral is slightly more acute (Fig. 20–20). A triangular prominence, the lacrimal caruncle, lies in the medial angle (Figs. 20–20, 20–21). Small, fine hairs project from the lacrimal caruncle, and sebaceous glands, similar to the tarsal glands but smaller, are present (*glandulae carunculae lacrimalis*). The caruncle may or may not be pigmented. The upper and lower lacrimal puncta open onto the bulbar surfaces of the lid margins 2 to 5 mm. from the medial commissure (Fig. 20–21).

The skin of the face continues onto the anterior surface of the lids with little alteration. The typical hair and glandular structure of the skin can be identified in sections of the lids (Fig. 20–21).

Long hairs (*cilia*) project from the upper lid margin. Cilia are not present on the lower lid in dogs. At the level of the dorsal medial orbital margin, there is a tuft of long tactile hairs (*pili supraorbitales*) which corresponds to the eyebrows in man (Fig. 20–20). In many dogs, the region of the pili supraorbitales contrasts in color to the rest of the face. Such dogs achieve particularly animated facial expressions.

Although similar to glands found elsewhere in the skin, the glands of the lid margin have received special designations. **Sebaceous glands** (*glandulae sebaceae*) open into the follicles of the cilia on the upper lid. **Ciliary glands** (*glandulae ciliares*) are coiled, tubular, apocrine sweat glands which secrete into hair follicles or sebaceous glands or directly onto the lid margin (Riis 1976). At the palpebral margin, the epidermis changes abruptly from a pigmented, keratinized, stratified squamous epithelium typical of skin to the unpigmented, non-keratinized, stratified squamous epithelium of the conjunctiva (see "Conjunctiva," page 1101).

Specially modified sebaceous glands, the **tarsal glands** (*glandulae tarsales*), are present in both eyelids. The openings of the ducts of the tarsal glands lie in a shallow furrow immediately posterior to the mucocutaneous junction of the palpebral margin of each lid (Fig. 20–21). They are easily visualized when the lid is everted slightly. The glands themselves are usually visible through the conjunctiva as white or yellow columnar structures (3 mm. long) which run at right angles to the palpebral margin (Fig.

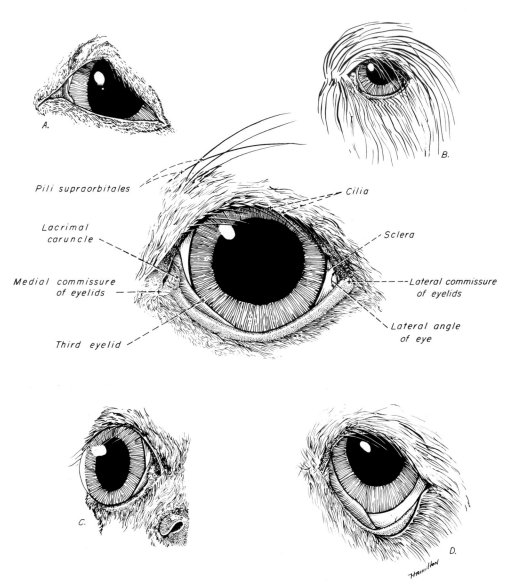

Figure 20–20. The eyelids.

A. Bull terrier.
B. English sheepdog.
C. Boston terrier.
D. St. Bernard.

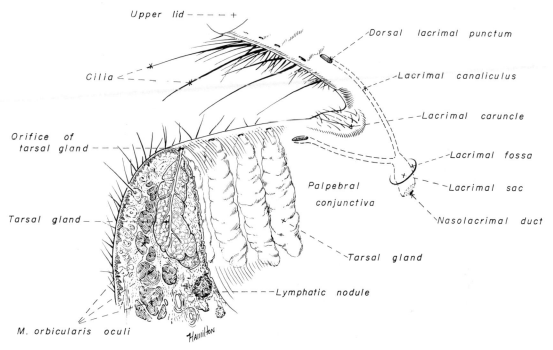

Figure 20–21. Posterior aspect of medial commissure of eyelids.

20–21). There are 20 to 40 glands in each eyelid. They are usually better developed on the upper lid. The oily superficial layer of the tear film is produced by the tarsal glands. It is common to find very fine hairs originating from some of these glands. This condition (distichiasis), if severe, may lead to corneal irritation and ulceration (Bedford 1971).

Surrounding the tarsal glands is a prominent thickening of the palpebral fibrous tissue, the **tarsus.** The tarsus stiffens the lid margin. The fibers of the tarsus course generally parallel to the margin. The tarsus is not as well developed in the dog as in man and some of the other domestic species.

The commissures of the lids are stabilized by the medial and lateral palpebral ligaments. The **lateral palpebral ligament** *(ligamentum palpebrale laterale)* is a poorly defined thickening of the orbital septum deep to the retractor anguli oculi lateralis muscle. The lateral ligament originates from the zygomatic arch and ventral end of the orbital ligament and inserts by blending with the fibers of the upper and lower tarsi.

The medial commissure is much more firmly anchored than the lateral. The **medial palpebral ligament** is a distinct fibrous band originating from the periosteum of the frontal bone near the nasomaxillary suture (Fig. 20–17). A small oval area of roughening at the site of origin is observed on most skulls. From its origin the ligament passes laterally deep to the angular vein of the eye, then superficial to the origin of the levator nasolabialis to blend with the tarsi at the medial commissure. The orbicularis oculi muscle both originates and inserts on medial palpebral ligament (see "Muscles," page 1111). The palpebral ligaments and retractor anguli oculi lateralis muscle prevent the palpebral fissure from becoming circular when the sphincter-like orbicularis oculi muscle contracts.

Conjunctiva

The inner aspect of the eyelids is lined by a special mucous membrane, the **palpebral conjunctiva.** At the level of the orbital rim, the palpebral conjunctiva reflects onto the surface of the globe to become the **bulbar**

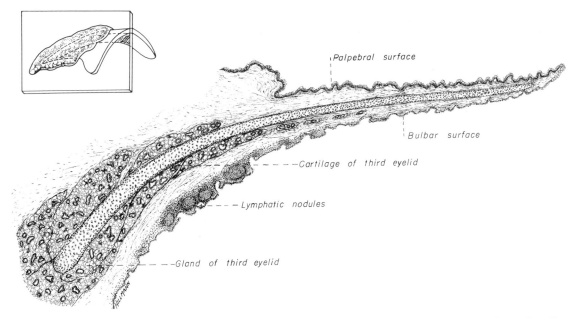

Figure 20–22. Section of the third eyelid. Inset: cartilage and gland of third eyelid showing plane of section.

conjunctiva (Fig. 20–9). The point of reflection is the **conjunctival fornix.** The **conjunctival sac** is the potential space between the lid and the eyeball which normally contains a thread of mucus and fluid tears.

The normal conjunctiva of the dog has been described by Riis (1976). Near the palpebral margin the conjunctiva is a stratified squamous epithelium. Toward the fornices the epithelium thins, and the surface cells become cuboidal. Goblet cells are present throughout most of the conjunctival sac but are extremely numerous in the fornices. The bulbar conjunctiva is very thin and is continuous with the anterior epithelium of the cornea at the limbus (Fig. 20–9). Goblet cells are absent in the perilimbal region, where the epithelium again becomes squamous. There are no perilimbal glands (so-called glands of Manz) in the dog.

The conjunctiva rests on a loose connective tissue stroma rich in fibrocytes, mast cells, plasma cells, lymphocytes, and macrophages. It is very mobile, permitting extensive excursions of the eyeball and eyelids. Multiple small folds of conjunctiva are formed in the fornix when the lids are open.

The conjunctiva is richly vascular. It is supplied by branches of the dorsal and ventral palpebral and malar arteries as well as by terminal twigs from the anterior ciliary arteries (see page 1122). Deep and superficial components of the conjunctival vasculature have been recognized. These react differently to inflammation of various segments of the eyeball and may be distinguished by biomicroscopy or topically applied vasoconstrictor pharmaceuticals (Riis 1976). The conjunctiva is well innervated by branches of the long ciliary, zygomaticofacial, zygomaticotemporal, infratrochlear, and frontal nerves (see "Innervation," page 1114). Corneal or conjunctival irritation results in reflex lacrimation.

Numerous **lymphatic nodules** are found throughout the conjunctiva (Figs. 20–21, 20–22). Their number, size, and location vary with the age of the dog and the degree of antigenic stimulation. They are especially prominent on the bulbar surface of the third eyelid (Fig. 20–22), where they may obtain a diameter of ½ cm. or more in chronically infected eyes. When enlarged, they may protrude from behind the third eyelid into the palpebral fissure. These lymphatic nodules are distinct from, and more superficial than, the superficial gland of the third eyelid, with which they have often been confused.

Third Eyelid

The **third eyelid** *(palpebra tertia)*, or **semilunar fold of the conjunctiva** *(plica semilunaris conjunctivae)*, is well developed in the dog. The third eyelid arises as a fold from the ventromedial aspect of the conjunctiva. The free edge of the fold is concave ventromedially and is usually darkly pigmented in contrast to the rest of the conjunctiva. The third eyelid is highly mobile and sufficient in extent to cover the entire anterior face of the cornea. When the eye is in its normal position in the live dog, the bulk of the third eyelid is hidden within the orbit; only the free edge is visible in the ventromedial aspect of the palpebral fissure (Fig. 20–20).

The third eyelid of the dog is reinforced by a T-shaped hyaline cartilage plate *(cartilago plica semilunaris conjunctivae)* (Fig. 20–22). The base of the T curves around the ventromedial aspect of the globe. The concave crossbar of the T stiffens the free edge of the fold. The base of the cartilage is surrounded by the superficial gland of the third eyelid (Fig. 20–22). The gland is a pink, tear drop–shaped, mixed seromucous gland. Numerous microscopic ducts empty the secretion of the gland into the inferior conjunctival fornix. The gland contributes significantly to the production of the tear film (see "Lacrimal Apparatus").

When the eyeball is retracted into the orbit, the base of the cartilage, surrounded by its bulky gland, is displaced anteriorly, and the third eyelid sweeps across the cornea from ventromedial to dorsolateral. The motion of the third eyelid is passive, the result of displacement by the globe's being pulled into the orbit by the retractor bulbi and rectus muscles (Habel 1975). Consequently, the third eyelid may be exposed for examination by manual displacement of the globe into the orbit. The dog has no specific muscle which draws the third eyelid across the globe such as is present in the cat. Smooth muscle cells derived from the periorbita attach to the fascia of the base of the third eyelid and help to maintain its retracted position. Loss of sympathetic innervation to the periorbita results in partial protrusion of the third eyelid.

LACRIMAL APPARATUS

The **precorneal tear film** is essential to maintain the normal transparent state of the cornea. Insufficient tear production results in keratinization and opacification of the cornea (keratoconjunctivitis sicca). The tear film consists of a superficial oily layer, a central aqueous layer, and a thin glycoprotein layer covering the cornea. The lacrimal apparatus includes those structures responsible for the production, dispersal, and disposal of the tears. The lacrimal gland, tarsal glands, conjunctival goblet cells, and the superficial gland of the third eyelid contribute to the tear film. Lacrimal fluid flows across the cornea, aided by blinking, to the lacrimal puncta at the medial commissure of the lids and thence through the lacrimal canaliculi and nasolacrimal duct to the nasal vestibule.

Lacrimal Gland

The **lacrimal gland** *(glandula lacrimalis)* is a pink, oval, lobated gland which lies under the periorbita on the dorsolateral aspect of the eyeball (Fig. 20–17). The gland is flattened between the eyeball and the orbital ligament and zygomatic process of the frontal bone. A shallow fossa for the lacrimal gland may be present in the orbital face of the frontal bone. A thin sheet of fascia separates the gland from the underlying dorsal and lateral rectus muscles. The seromucous secretion of the gland empties into the dorsolateral conjunctival fornix through three to five microscopic secretory ducts (Michel 1955). Removal of the lacrimal gland results in only a minor decrease in tear production, probably as a result of a compensatory increase in production by the superficial gland of the third eyelid (Helper 1970).

The lacrimal gland is innervated by the lacrimal nerve, a small branch of the ophthalmic nerve from the trigeminal nerve (Fig. 20–32). Parasympathetic postganglionic fibers from the pterygopalatine ganglion are distributed to the gland with the lacrimal nerve. Parasympathomimetic drugs increase the rate of glandular secretion. The blood supply is derived from the dorsal muscular branches of the external ophthalmic artery (Fig. 20–33). The lacrimal vein drains the gland into the dorsal external ophthalmic vein (Fig. 20–35).

Superficial Gland of the Third Eyelid

The **superficial gland of the third eyelid** *(glandula superficialis plica semilunaris conjunctivae)* in the dog is an accessory lacrimal gland which normally produces a significant proportion of the tear film (Helper et al. 1974). The gland surrounds the base of the cartilage of the third eyelid in the ventromedial aspect of the eyeball (Fig. 20–22). The gland is mixed in the dog; both mucous and serous acini are seen in section. The secretions of this gland flow through numerous microscopic ductules into the conjunctival sac. There is no deep (harderian) gland of the third eyelid in the dog.

Conjunctiva

The glycoprotein component of the tear film is produced by the goblet cells of the conjunctivae. These cells are especially numerous in the regions of the fornices and contribute significantly to the tears. The dog does not have isolated accessory lacrimal glands throughout the conjunctiva corresponding to the several types bearing eponyms in man. The tarsal glands (see "Eyelids") produce the oily superficial layer of the tear film. This oily layer is essential to prevent excessive evaporation from the surface of the cornea (Mishima and Maurice 1961).

Tears accumulate in the **lacrimal lake** *(lacus lacrimalis),* which is the shallow cleft between the third eyelid and inferior palpebral conjunctiva just lateral to the medial commissure. Blinking spreads the tear film over the cornea; this action is essential to its health. Inability to close the lids, as a result of facial nerve lesions, allows the cornea to dry and results in marked pathological alterations (neuroparalytic keratitis) (Bistner 1978). The lacrimal fluid produced in excess of that which evaporates from the surface of the cornea drains into the nasal cavity through the nasolacrimal duct system.

Nasolacrimal Duct System

The outflow pathway of excess lacrimation includes the puncta, the lacrimal canaliculi, the lacrimal sac, and the nasolacrimal duct. The **lacrimal puncta** *(puncta lacrimalia)* are located on the inner surface of the upper and lower lid margins 2 to 5 mm. from the medial commissure (Fig. 20–21). The puncta are the oval openings of the **lacrimal canaliculi** *(canaliculi lacrimale)* and measure about 0.7 mm. by 0.3 mm. The long axis is parallel to the lid margin. The puncta are distinguished from the openings of the tarsal glands by their larger size and greater distance from the lid margin (Fig. 20–21). In some dogs, one or both puncta may be smaller than normal or absent. Spilling of tears onto the face (epiphora) results.

The dorsal lacrimal canaliculus runs medially parallel to the lid margin for 3 to 7 mm. from the dorsal punctum, and then turns ventrally medial to the commissure of the lids to enter the **lacrimal sac** *(saccus lacrimalis)* (Fig. 20–21). The ventral lacrimal canaliculus arcs ventromedially from its punctum to join the dorsal duct at the lacrimal sac. The canaliculi are easily cannulated if the instrument is directed medially, parallel to the lid margin. Radiopaque contrast material can be introduced into the nasolacrimal duct system through the canaliculi to outline the duct system radiographically (Fig. 20–23) (Yakely and Alexander 1971).

The lacrimal sac is the dilated origin of the nasolacrimal duct. It is not as prominent in the dog as in man, leading some authors to conclude that it is not present (Yakely and Alexander 1971). The sac occupies a depression *(fossa sacci lacrimalis)* in the center of the orbital surface of the lacrimal bone, medial and ventral to the medial commissure (see Fig. 4–38).

The **nasolacrimal duct** *(ductus nasolacrimalis)* forms an arch which is concave dorsally as it passes rostrally from the lacrimal sac through the lacrimal canal of the lacrimal bone and maxilla (see Figs. 4–29, 4–32). Rostral to the conchal crest, the duct is no longer covered by bone but continues anteriorly

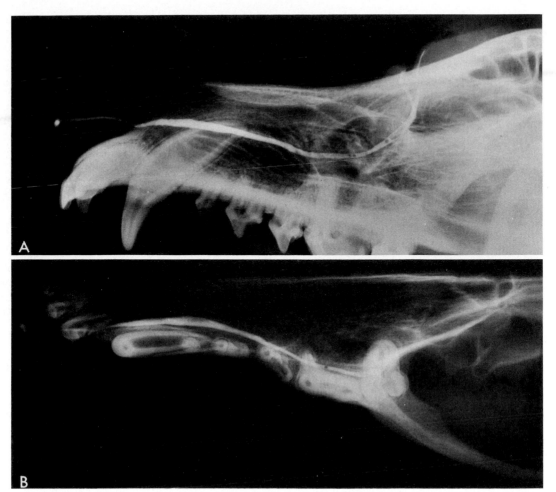

Figure 20–23. A. Lateral radiograph showing the nasolacrimal duct. (Courtesy of V. Rendano.) B. Dorso-ventral radiograph showing the course of the nasolacrimal duct. (Courtesy of V. Rendano.)

deep to the nasal mucosa on the nasal face of the maxilla. In about 50 per cent of dogs, the nasolacrimal duct has two openings. At the level of the root of the canine tooth, there is an inconstant communication of the duct with the nasal cavity below the ventral nasal concha (Michel 1955). Where present, the opening is a slit ¼ to ½ cm. long in the medial wall of the duct. Rostrally, the duct passes medial to the ventral lateral nasal cartilage and ends by opening onto the ventrolateral floor of the nasal vestibule below the alar fold (Fig. 8–8). The anterior opening cannot be visualized without a speculum in the living dog.

The nasolacrimal duct is supplied by a small branch of the malar artery.

MUSCLES

The muscles important to the function of the visual apparatus may be considered to constitute three groups: the intraocular muscles, the extraocular muscles, and the palpebral muscles. The intraocular muscles are those which lie entirely within the sclera and act to regulate pupillary diameter and the shape of the lens. The extraocular muscles insert on the sclera and effect rotation and retraction of the eyeball as a whole. The palpebral muscle group includes a number of muscles of the lids and head which regulate the shape and position of the palpebral fissure.

Intraocular Muscles

The **dilator and sphincter muscles of the iris** and the **ciliary muscles** lie entirely within the eyeball (Fig. 20–9). They are composed of smooth muscle fibers. The iris musculature acts reflexly to regulate the amount of light which reaches the retina. The ciliary muscle accomplishes visual accommodation (focusing) by altering the tension of the zonular fibers. These muscles are described in detail under the headings "Iris" and "Ciliary Body," of which they are integral parts.

Extraocular Muscles

The **extraocular muscles,** or *musculi bulbi,* are striated muscles which are seven in number: the dorsal, medial, ventral, and lateral rectus muscles; the dorsal and ventral

oblique muscles; and the retractor bulbi muscle.

The extraocular muscles rotate the globe around three mutually perpendicular axes passing through the center of the globe (Fig. 20–24). The dorsal and ventral rectus muscles rotate the globe around a medial to lateral axis. The medial and lateral rectus muscles rotate the globe about a dorsal-ventral axis, and the oblique muscles rotate the eyeball around the axis bulbi. In addition, the eyeball can be retracted into the orbit along the optic axis by the retractor bulbi muscle. Contraction of two or more muscles accomplishes oblique movements. The dog is able to rotate the eye through about 90° of arc in the dorsal plane and 60° in a sagittal plane. The rotation produced by the oblique muscles is more limited, amounting to only about 30°. The oblique muscles also help to fix the eye against the backward pull of the rectus muscles, which they accomplish because the pull of their tendons has an anterior component.

The **m. retractor bulbi** is a striated muscle, derived from the lateral rectus, which originates from the periosteum within the orbital fissure, lateral to the optic nerve (Fig. 20–25).

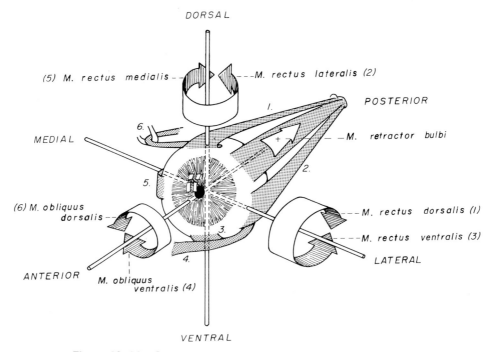

Figure 20–24.　Scheme of motion produced by extraocular muscles.

The retractor bulbi passes laterally between the dorsal and lateral rectus muscles and divides into dorsal and ventral components which come to lie above and below the optic nerve. Each component again bifurcates, forming four flat fasciculi. These diverge as they run anteriorly deep to the rectus muscles to insert on the equator of the eyeball about 1 cm. posterior to the corneoscleral junction (Fig. 20–25). The fasciculi are very broad and thin at their insertions, and in some specimens, adjacent fasciculi nearly meet to form a complete muscular cone around the posterior aspect of the eyeball. The conical space between the diverging fasciculi of the retractor bulbi and the optic nerve is filled with intraperiorbital fat.

The primary action of the retractor bulbi muscle is to pull the eyeball deeper into the orbit (Fig. 20–24). This displaces the base of the cartilage of the third eyelid with its encircling gland, causing the free edge of the third eyelid to sweep across the cornea (see "Third Eyelid"). The retractor bulbi may also play a part in the rotatory movements of the eye. Watrous and Olmsted (1941) reported that following excision of all other extrinsic muscles in the dog, the retractor bulbi was eventually capable of moving the eye in all directions. Blood supply to the retractor muscle is derived from the dorsal and ventral muscular branches of the external ophthalmic artery. The abducent nerve supplies general somatic efferent fibers to the retractor muscle.

The four **mm. recti** are named for the position of their insertion on the globe. They originate in proximity to one another at the apex of the periorbital cone (Fig. 20–25). The dorsal rectus originates between the optic canal and the orbital fissure. The lateral, medial, and ventral recti originate ventral to the orbital fissure in the stated order from dorsal to ventral. From their origin they arc to their final positions—dorsal, lateral, medial, and ventral—and diverge toward the eyeball. Over most of their course the rectus muscles are deep to the periorbita and superficial to the fascicles of the retractor bulbi. Dorsally, the m. levator palpebrae superioris is interposed between the dorsal rectus and the periorbita (Fig. 20–27).

The muscle bellies of the rectus muscles are oval in transverse section, measuring about 9 mm. wide by 2 mm. thick at their broadest point. The medial rectus is slightly larger than the other recti, which are comparable in size. At about the equator of the globe, the muscles form flat tendons which insert on the sclera anterior to the insertion of the m. retractor bulbi, 3 to 7 mm. posterior to the corneoscleral junction. Anterior to the equator of the lens, the dorsal rectus passes over the tendon of insertion of the dorsal oblique muscle (Fig. 20–4). The tendon of the ventral rectus passes deep to the ventral oblique muscle.

The action of the rectus muscles has been described (Fig. 20–24). The oculomotor nerve is motor to the dorsal, medial, and ventral recti. The abducent nerve innervates the lateral rectus. The rectus muscles are supplied by the muscular branches of the external ophthalmic artery. Venous return is by means of muscular branches of the external ophthalmic veins.

The **m. obliquus dorsalis** arises at the dorsomedial margin of the optic canal closely associated with the origin of the other extraocular muscles (Fig. 20–25). It runs anteriorly within the periorbita between the dorsal and medial recti. At about the posterior pole of the eyeball, the muscle gives rise to a thin, round tendon which passes over a small cartilaginous trochlea located at the phylogenetic origin of the muscle on the medial wall of the orbit near the medial angle of the eye (Fig. 20–26).

The **trochlea** is a small oval plate of hyaline cartilage in the periorbita. It is firmly anchored to the medial orbital wall by three ligamentous thickenings of the periorbita. The longest of these runs from the anterior end of the trochlea to the periosteum at the medial commissure of the eyelids. A short ligament anchors the trochlea to the dorsal orbital margin, and a third runs from the posterior aspect to the periosteum of the zygomatic process of the frontal bone.

The tendon of the dorsal oblique muscle runs through a groove in the trochlea formed by a prominence on its medial face near the anterior end. A synovial sheath (*vagina synovialis m. obliqui dorsalis*) is present at this point. As the tendon passes over the trochlea, it turns through an angle of approximately 135° to the muscle belly to run obliquely caudodorsolaterally to its insertion on the sclera deep to the tendon of the dorsal rectus muscle (Fig. 20–26).

Text continued on page 1111

Figure 20–25. Scheme of muscle attachments in orbital fissure and relationship of dura to periorbita.
A. Dorsal aspect. (Dotted lines are internal borders of optic canal and orbital fissure.)
    1. M. óbliquus dorsalis
    2. M. rectus medialis
    3. Optic nerve
    4. Periorbita
    5. Frontal bone
    6. Dura mater
    7. Jugum of presphenoid
    8. Rostral clinoid process
    9. Orbital fissure
    10. Internal carotid artery
    11. Oculomotor nerve
    12. Tendons of mm rectus lateralis, rectus medialis, and rectus ventralis originating ventrally in orbital fissure.
    13. M. retractor bulbi
    14. M. rectus lateralis
    15. M. rectus dorsalis
B. Origin of extraocular muscles. Rostral lateral aspect.
    1. Optic nerve
    2. M. levator palpebrae superioris
    3. M. rectus dorsalis
    4. M. retractor bulbi
    5. Orbital fissure
    6. Cut edge of periorbita
    7. Tendons of mm. rectus lateralis, rectus medialis, and rectus ventralis originating ventrally in orbital fissure.
    8. M. rectus lateralis
    9. M. rectus ventralis
    10. Periorbita
    11. M. obliquus dorsalis
    12. M. rectus medialis
C. Apex of orbit.
    1. Optic canal
    2. Orbital fissure
    3. Rostral alar foramen
    4. Zygomatic arch (cut)
D. Schematic transection of structures within the apex of the periorbita. Rostral aspect.
    1. Optic nerve in optic canal
    2. M. Levator palpebrae superioris
    3. Internal ophthalmic artery and vein
    4. Cut edge of dura lining orbital fisure
    5. Oculomotor nerve in orbital fissure
    6. Trochlear nerve
    7. Cut edge of periorbita (dura)
    8. Frontal nerve (V)
    9. Nasociliary nerve (V)
    10. Abducent nerve
    11. Anastomotic artery
    12. Emissary vein of orbital fissure
    13. Tendons of mm rectus lateralis, rectus medialis, and rectus ventralis
    14. M. retractor bulbi in orbital fissure
    15. M. rectus dorsalis
    16. M. rectus lateralis
    17. M. rectus medialis
    18. M. rectus ventralis
    19. Fibrous ring of periorbita
    20. Cut edge of periorbita
    21. M. obliquus dorsalis

See illustration on opposite page

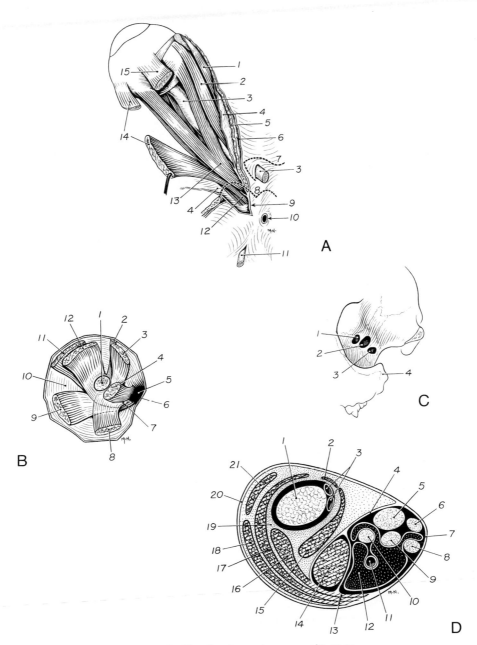

Figure 20–25.   *See legend on opposite page.*

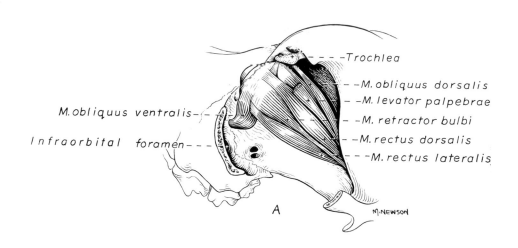

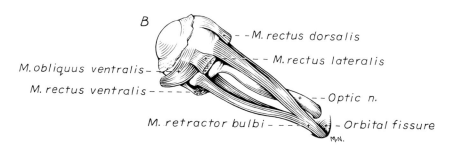

Figure 20–26. Extrinsic muscles of the eye.

A. Caudolateral aspect. (The eye is displaced slightly lateral.)
B. The m. retractor bulbi, lateral aspect.

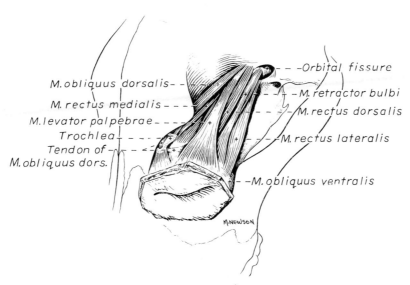

M.obliquus dorsalis- -

M. rectus medialis- - -

M.levator palpebrae- - -

Trochlea- - - -

Tendon of - -

M.obliquus dors.

Orbital fissure

M. retractor bulbi

M. rectus dorsalis

M. rectus lateralis

M. obliquus ventralis

M·NEWSON

Figure 20–27. Extrinsic muscles of the eye, dorsolateral aspect.

The dorsal oblique muscle is the only structure innervated by the fourth cranial nerve *(n. trochlearis).*

The *m. obliquus ventralis* is the only extraocular muscle which arises from a site remote to the apex of the orbit (Fig. 20–26). The ventral oblique originates from a small depression in the palatine bone near the junction of the palatomaxillary and palatolacrimal sutures. In prepared skulls this site may appear as a foramen, since the attachment plate is thin and easily lost. The muscle courses dorsolaterally, passing ventral to the insertion of the ventral rectus. It is fusiform in shape and roughly circular in transverse section. At the ventrolateral aspect of the orbit, it gives rise to two short tendons. The shorter tendon inserts deep to the insertion of the lateral rectus. The superficial portion passes over the lateral rectus to insert on the dorsolateral aspect of the eyeball (Figs. 20–4, 20–26). The ventral oblique is supplied by branches of the malar artery. The muscle is innervated by the oculomotor nerve.

Palpebral Muscles

The muscles which alter the size or position of the palpebral fissure include the m. orbicularis oculi, the m. levator palpebrae superioris, m. levator anguli oculi medialis, m. rectractor anguli oculi lateralis, and the pars palpebralis of the m. sphincter colli profundus (Figs. 20–28, 20–29).

The *m. orbicularis oculi* is the most important muscle which acts to close the eyelids. It is composed of two parts: the pars orbitalis and the pars palpebralis. The division between the parts is not distinct in the dog. The *pars palpebralis* is composed of fibers which run in the substance of the lids themselves. These fibers originate from the medial palpebral ligament, encircle the palpebral fissure, and insert again on the ligament. The muscle is wedge-shaped in transverse section, tapering toward the lid margin. Fibers of the pars palpebralis lie anterior to the tarsus and tarsal glands and closely approach the lid margin; some fibers are found almost to the level of the opening of the tarsal glands (Fig. 20–21). The pars palpebralis is better developed in the upper lid.

Constant corneal irritation results in hypertrophy of the orbicularis oculi muscle, which, if pronounced, will actually roll the haired anterior surface of the lid inward against the cornea (spastic entropion).

The *pars orbitalis* of the orbicularis oculi surrounds the pars palpebralis. It consists of dorsal and ventral components which originate from the medial palpebral ligament and follow the curve of the orbital margin laterally. At the lateral commissure of the lids, some of the peripheral fibers of the ventral component fan out caudally and dorsally on

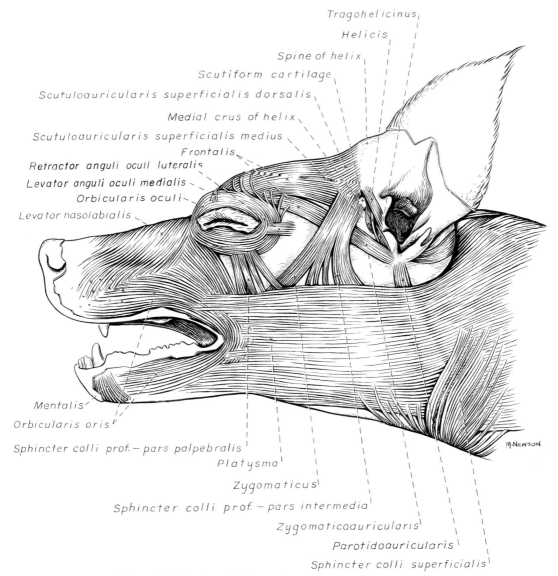

Tragohelicinus

Helicis

Spine of helix

Scutiform cartilage

Scutuloauricularis superficialis dorsalis

Medial crus of helix

Scutuloauricularis superficialis medius

Frontalis

Retractor anguli ocull luteralis

Levator anguli oculi medialis

Orbicularis oculi

Levator nasolabialis

Mentalis

Orbicularis oris

Sphincter colli prof. — pars palpebralis

Platysma

Zygomaticus

Sphincter colli prof. — pars intermedia

Zygomaticoauricularis

Parotidoauricularis

Sphincter colli superficialis

M. NEWSON

Figure 20-28. Superficial muscles of the head, lateral aspect.

the superficial surface of the frontalis muscle (Fig. 20–28). Fibers of the dorsal pars orbitalis decussate with the ventral fibers caudal to the lateral commissure, forming the **lateral palpebral raphe** *(raphe palpebralis lateralis)*. No pars lacrimalis could be identified in the dog.

The orbicularis oculi muscle exerts a sphincter-like action to close the palpebral fissure. The medial and lateral palpebral ligaments and the retractor anguli oculi lateralis muscle stabilize the commissures of the lids and prevent the fissure from becoming circular when the muscle contracts. Motor innervation is supplied by the palpebral branches of the auriculopalpebral nerve, a branch of the facial nerve. A prominent sign of facial nerve palsy, therefore, is the inability to close the palpebral fissure (de Lahunta 1977). Blood supply to the orbicularis oculi is derived from the malar artery medially and the superior and inferior lateral palpebral arteries laterally (see Fig. 11–14).

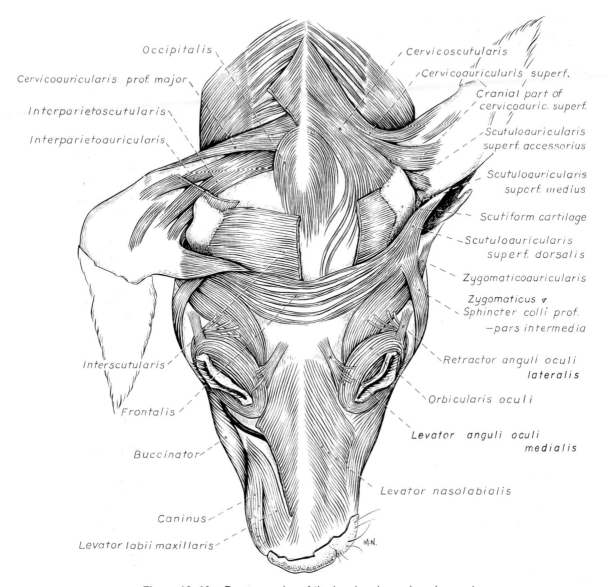

Occipitalis

Cervicoauricularis prof. major

Interparietoscutularis

Interparietoauricularis

Cervicoscutularis

Cervicoauricularis superf.

Cranial part of
cervicoauric. superf.

Scutuloauricularis
superf. accessorius

Scutuloauricularis
superf. medius

Scutiform cartilage

Scutuloauricularis
superf. dorsalis

Zygomaticoauricularis

Zygomaticus v
Sphincter colli prof.
—pars intermedia

Retractor anguli oculi
lateralis

Orbicularis oculi

Levator anguli oculi
medialis

Levator nasolabialis

Interscutularis

Frontalis

Buccinator

Caninus

Levator labii maxillaris

M·N.

Figure 20–29. Deep muscles of the head and ear, dorsal aspect.

The *m. levator palpebrae superioris* (Fig. 20–31) is the most important muscle which acts to retract the upper eyelid. The muscle originates deep within the orbit, dorsal to the optic canal between the origins of the dorsal rectus and dorsal oblique muscles. The levator courses rostrally deep to the periorbita on the dorsomedial aspect of the dorsal rectus muscle toward its insertion in the upper eyelid. The muscle becomes pro-gressively wider and flatter anteriorly. Anterior to the equator of the eyeball, a broad aponeurosis continues the muscle into the upper lid, where it inserts on the palpebral connective tissue among the fibers of the orbicularis oculi muscle. The m. levator palpebrae superioris is inner-vated by the oculomotor nerve. Blood is supplied by the dorsal muscular branch of the external ophthalmic artery.

The *m. levator anguli oculi medialis* is a
small, straplike muscle which arises caudodorsal to the medial commissure from the
nasofrontal fascia with the frontalis muscle
(Fig. 20–29). The muscle runs ventrolaterally to insert on the medial half of the upper lid
superficial to the orbicularis oculi muscle.
Contraction of the levator raises the medial
portion of the upper lid and the long tactile
hairs (pili supraorbitales) which correspond
to the eyebrows of man (Fig. 20–20). The
auriculopalpebral nerve (VII) is the motor
innervation of the levator. Blood is supplied
by branches of the superior lateral palpebral
artery.

The *m. retractor anguli oculi lateralis* is a
small, flat muscle which arises from the temporal fascia near the temporozygomatic suture (Fig. 20–28). The muscle is parallel and
superficial to the lateral palpebral ligament.
It passes rostrally superficial to the orbital
part of the orbicularis oculi muscle and inserts by blending with fascicles of the palpebral part at the lateral commissure of the lid.
The lateral retractor draws the lateral
canthus posteriorly and thus has some action
in closing the eye. The retractor is innervated by the zygomatic branch of the auro-
culopalpebral nerve, and is supplied by
branches of the lateral ventral palpebral
artery.

The *pars palpebralis* of the *m. sphincter
colli profundus* acts as a depressor of the
lower lid (Fig. 20–28). It consists of several
delicate straps of muscle which originate
near the ventral midline. These course dorsally caudal to the angle of the mouth to
insert on the lower tarsus. The ventral portion of these muscular straps is deep to the
platysma. The dorsal portion is subcutaneous and closely applied to the skin. The
muscle is innervated by the buccal branches
of the facial nerve.

INNERVATION

The eye and its adnexa are innervated by
cranial nerves II, III, IV, V, VI, and VII. The
general anatomy of these nerves is described
elsewhere (see Chapter 15, The Cranial
Nerves). Only the course of fibers which
innervate structures of the eye and its
adnexa are described in detail here.

Optic Nerve

The **optic nerve** *(n. opticus)*, or cranial
nerve II, is a component of the special somatic afferent system. It resembles a central
nervous sytem tract because it is formed by
the centripetal growth of axons of the
neurons in the ganglion cell layer of the
retina. The axons invade the hollow stalk of
the original neurectodermal outpouching
which forms the optic vesicle. Centripetal
growth of the ganglion cell axons begins at
about day 32 of gestation and is not completed until after birth.

Because the eye forms as an outgrowth of
the brain stem (see "Development"), the
myelin of the optic nerve is of the central
nervous system type, being formed by oligo-
dendrogliocytes. Ganglion cell axons are unmyelinated in the nerve fiber layer of the
retina; they become myelinated as they turn
from the nerve fiber layer into the optic
nerve. The intraocular myelinated portion is
the white or gray optic disc seen grossly.
Myelin may extend a variable distance from
the optic nerve head onto the retina, giving
the disc a large and ragged appearance. The
size and shape of the optic disc varies widely
among dogs (see "Retina").

The optic nerve is surrounded by outer
and inner sheaths *(vagina externa et vagina
interna n. optici)*, which are continuations
of the dura mater and pia mater of the brain,
respectively. The space between the sheaths
(spatia intervaginalia) is continuous with
the subarachnoid space and contains cerebrospinal fluid. Optic neuritis may thus result from a direct extension of a meningitis
(Hoerlein 1978). The ciliary vessels and
nerves are closely applied to the outer surface of the external sheath. Radicles of the
internal ophthalmic and ciliary vessels supply the intraorbital portion of the optic
nerve.

The optic nerve follows an undulating
course from its origin lateral and ventral to
the posterior pole of the eyeball to the optic
canal of the presphenoid bone (Fig. 20–16).
The nerve traverses the optic canal to the
rostroventral floor of the braincase, where
fibers are exchanged with the opposing optic
nerve at the **optic chiasm.** In the dog, approximately 75 per cent of the optic nerve fibers
cross to join the contralateral optic tract cau-

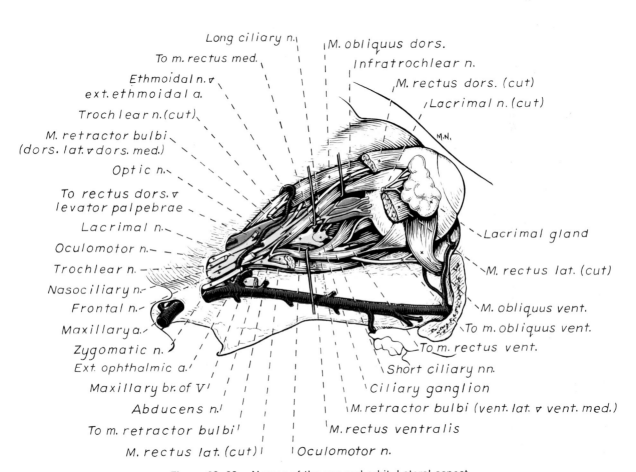

Figure 20–30. Nerves of the eye and orbit. Lateral aspect.

dal to the chiasm (de Lahunta and Cummings 1967). The crossing fibers are derived from the central and medial retinal areas; axons from the lateral retina join the ipsilateral optic tract (Fig. 20–10). Each optic nerve is composed of approximately 150,000 axons, all of which are myelinated (Bruesch and Arey 1942). There is a correlation between the size of the eyeball and the number of axons, but larger eyes have fewer axons relative to retinal surface area. (Arey et al. 1942).

In man and the cat, centrifugal fibers, as well as the centripetal ganglion cell axons, have been identified in the optic nerve. It is probable that they also occur in the dog. Their function has not yet been elucidated.

Within the orbit, the optic nerve occupies the center of the cone formed by the extraocular muscles. A well-developed intraperiorbital fat body lies between the nerve and the muscle cone.

The optic nerve is longer than the straight-line distance from the posterior pole of the eyeball to the optic canal of the presphenoid bone to allow for the rotation of the eyeball. The degree of undulation in the nerve depends upon the position of the eyeball and the degree of retraction. In dolichocephalic and mesaticephalic dogs, the eyeball can be retracted into the orbit. In the dog, the optic nerve is sufficiently long that proptosis can occur without rupture of the nerve.

Oculomotor Nerve

The **oculomotor nerve** (n. oculomotorius), or cranial nerve III, as its name implies, is the primary general somatic efferent innervation to the eye. The oculomotor nerve is motor to the dorsal, medial, and ventral rectus muscles, the ventral oblique, and the levator palpebrae superioris muscle. It also contains preganglionic general visceral efferent fibers which synapse in the ciliary ganglion (Figs. 20–10, 20–30). Postganglionic fibers from the ciliary ganglion innervate the m. sphincter pupillae of the iris and the ciliary muscle. The parasympathetic fibers are located medially at the origin of the nerve and, as such, are vulnerable to pressure damage when there is increased intracranial pressure. Thus dilation of the pupil

(mydriasis) is a common finding in dogs with cerebral edema.

The oculomotor nerve first becomes visible grossly where it leaves the ventral aspect of the mesencephalon on the medial aspect of the crus cerebri (see Fig. 14–5). The nerve runs laterally a short distance, then turns rostrally to pass on either side of the hypophysial stalk. It then enters a sulcus in the wall of the cavernous sinus which it follows rostrally a short distance. It emerges from the sinus to leave the cranial cavity through the orbital fissure (Fig. 20–25). Within the orbit the nerve divides into a small dorsal and larger ventral branch. The dorsal branch innervates the dorsal rectus and levator palpebrae muscles. The ventral ramus travels rostrally deep to the lateral rectus, sending branches to the medial and ventral rectus and to the ventral oblique. A short branch containing the parasympathetic fibers enters the **ciliary ganglion**. The ganglion is located midway between the eyeball and the orbital fissure, closely applied to the ventrolateral surface of the optic nerve. **Short ciliary nerves** (nn. ciliares breves) leave the rostral surface of the ganglion and are distributed to the eyeball. A communicating branch from the ganglion often joins the nasociliary nerve or long ciliary nerve (Fig. 20–30). Lesions of the oculomotor nerve result in a ventrolateral strabismus because of the unopposed tension in the lateral rectus. The eye cannot be retracted, and the pupil is dilated. There is ptosis due to paralysis of the levator palpebrae superioris muscle.

Trochlear Nerve

The **trochlear nerve** (n. trochlearis), or cranial nerve IV, is unique among the cranial nerves in that it leaves the dorsal surface of the brain stem. It is the smallest of the cranial nerves and difficult to preserve in anatomical preparations. Nerve fibers leave the brain stem immediately caudal to the caudal colliculi. The fibers cross in the rostral medullary velum. The nerves arc ventrally between the cerebrum and cerebellum along the tentorium cerebelli and spine of the petrous temporal bone. The nerve reaches the orbit by passing over the dorsal surface of the trigeminal ganglion and out the orbital fissure lateral to the oculomotor nerve (Fig.

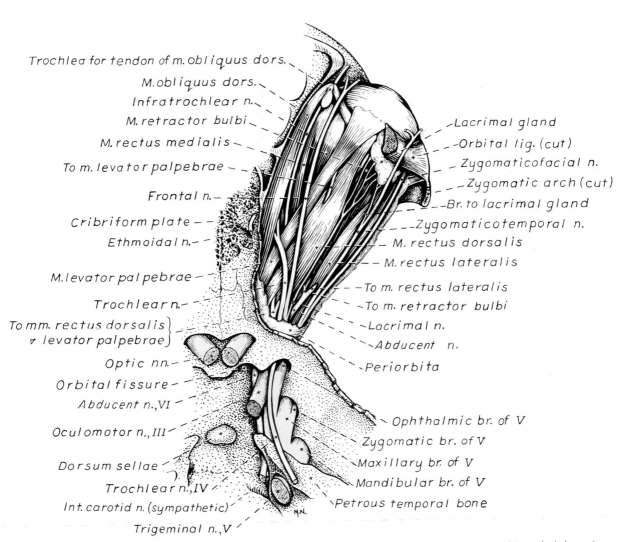

Trochlea for tendon of m. obliquus dors.
M. obliquus dors.
Infratrochlear n.
M. retractor bulbi
M. rectus medialis
To m. levator palpebrae
Frontal n.
Cribriform plate
Ethmoidal n.
M. levator palpebrae
Trochlear n.
To mm. rectus dorsalis ∿ levator palpebrae
Optic nn.
Orbital fissure
Abducent n., VI
Oculomotor n., III
Dorsum sellae
Trochlear n., IV
Int. carotid n. (sympathetic)
Trigeminal n., V

Lacrimal gland
Orbital lig. (cut)
Zygomaticofacial n.
Zygomatic arch (cut)
Br. to lacrimal gland
Zygomaticotemporal n.
M. rectus dorsalis
M. rectus lateralis
To m. rectus lateralis
To m. retractor bulbi
Lacrimal n.
Abducent n.
Periorbita
Ophthalmic br. of V
Zygomatic br. of V
Maxillary br. of V
Mandibular br. of V
Petrous temporal bone

Figure 20–31. Scheme of the optic, oculomotor, trochlear, trigeminal (ophthalmic branch), and abducent nerves. Dorsal aspect.

20–32). Upon emerging from the fissure, the nerve turns dorsomedially to enter the dorsal oblique muscle, the only structure it innervates.

Trigeminal Nerve

The eye and orbit are richly supplied by branches of the **trigeminal nerve** (*n. trigeminus*), or cranial nerve V. Although the nerve contains both general somatic afferent and special visceral efferent fibers, all of the fibers distributed to the eye are general somatic afferent in nature.

The trigeminal nerve is the largest of the cranial nerves. The nerve leaves the brain at the juncture of the pons and trapezoid body (Fig. 10–5). It passes through a canal in the petrous temporal bone. Here the **trigeminal ganglion** rests in the *cavum trigeminale* of the dura mater. The ganglion contains the cell bodies of the general somatic afferent neurons of the trigeminal nerve. Immediately distal to the ganglion the nerve divides into its three major divisions, the ophthalmic, the maxillary, and the mandibular nerves (Fig. 20–31). The ophthalmic nerve is the principal sensory innervation of the eye and orbit. Branches of the maxillary nerve innervate a part of the superficial structure of the eyelids. The mandibular nerve innervates structures of the lower face and jaw and does not play a role in the innervation of ocular structures.

OPHTHALMIC NERVE

The **ophthalmic nerve** (*n. ophthalmicus*) arises from the rostromedial aspect of the trigeminal ganglion and arcs rostromedially into the orbital fissure to join cranial nerves III, IV, and VI. Sympathetic postganglionic fibers join the ophthalmic from the cavernous plexus. Within, or immediately rostral to, the orbital fissure, the ophthalmic divides into three branches, the frontal, lacrimal, and nasociliary nerves (Fig. 20–30).

The **frontal nerve** (*n. frontalis*) is a small nerve which is sensory to the middle portion of the upper eyelid. From the orbital fissure it passes rostrodorsally between the periorbita and dorsal rectus muscles to the upper lid (Fig. 20–32).

The **lacrimal nerve** (*n. lacrimalis*) is a very small branch of the ophthalmic which travels along the lateral edge of the dorsal rectus to innervate the lacrimal gland (Fig. 20–17). Diesem (1975) found the lacrimal nerve occasionally originating from the maxillary nerve. He also was able to trace branches of the lacrimal, presumably sensory, to the lateral portion of the upper eyelid. The lacrimal nerve supplies postganglionic sympathetic fibers to the lacrimal gland.

The **nasociliary nerve** (*n. nasociliaris*) continues the ophthalmic into the orbit. It passes rostromedially between the dorsal and ventral rami of the oculomotor nerve to the dorsal surface of the optic nerve. Here it divides into the long ciliary nerves and the infratrochlear and ethmoidal nerves (Fig. 20–30).

The **long ciliary nerves** (*nn. ciliares longi*) continue rostrally closely applied to the optic nerve. Variable communications with the short ciliary nerves are observed, and a communicating branch to the ciliary ganglion is usually present (*ramus communicans cum ganglio ciliari*). The long and short ciliary nerves enter the globe adjacent to the optic nerve. According to Prince et al. (1960), the ciliary nerves continue anteriorly in the suprachoroidea, supplying sensory innervation to the choroid, ciliary body, iris, cornea, and bulbar conjunctiva. Nerves enter the cornea throughout its circumference at the limbus and run toward the center, branching dichotomously. They may be visualized in the live dog with the biomicroscope (Martin 1969). The sensory innervation of the cornea apparently exerts a trophic influence which is essential to its normal state (see "Cornea"). Sympathetic postganglionic fibers in the long ciliary nerves are motor to the dilator muscle of the pupil.

The **infratrochlear nerve** (*n. infratrochlearis*) passes rostrodorsally along the medial edge of the dorsal rectus (Fig. 20–31). The nerve passes ventral to the trochlea, as its name implies, and ramifies in the tissues surrounding the medial commissure of the lids.

The **ethmoidal nerve** (*n. ethmoidalis*) accompanies the external ethmoidal artery as it curves rostrally and medially over the extraocular muscles to leave the orbit through the ventral ethmoidal foramen. It innervates part of the nasal mucosa and skin of the muzzle.

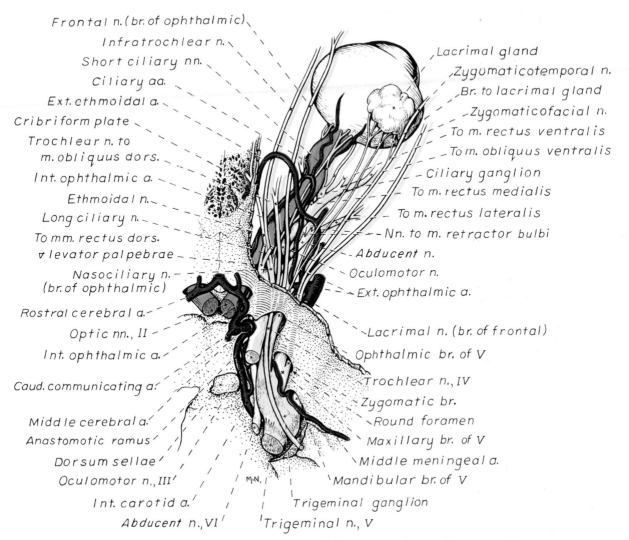

Frontal n. (br. of ophthalmic)
Infratrochlear n.
Short ciliary nn.
Ciliary aa.
Ext. ethmoidal a.
Cribriform plate
Trochlear n. to
m. obliquus dors.
Int. ophthalmic a.
Ethmoidal n.
Long ciliary n.
To mm. rectus dors.
v levator palpebrae
Nasociliary n.
(br. of ophthalmic)
Rostral cerebral a.
Optic nn., II
Int. ophthalmic a.
Caud. communicating a.
Middle cerebral a.
Anastomotic ramus
Dorsum sellae
Oculomotor n., III
Int. carotid a.
Abducent n., VI

Lacrimal gland
Zygomaticotemporal n.
Br. to lacrimal gland
Zygomaticofacial n.
To m. rectus ventralis
To m. obliquus ventralis
Ciliary ganglion
To m. rectus medialis
To m. rectus lateralis
Nn. to m. retractor bulbi
Abducent n.
Oculomotor n.
Ext. ophthalmic a.
Lacrimal n. (br. of frontal)
Ophthalmic br. of V
Trochlear n., IV
Zygomatic br.
Round foramen
Maxillary br. of V
Middle meningeal a.
Mandibular br. of V
Trigeminal ganglion
Trigeminal n., V

M.N.

Figure 20–32. Superficial distribution of the nerves of the eye. Dorsal aspect.

MAXILLARY NERVE

The **zygomatic nerve** (*n. zygomaticus*), a branch of the maxillary nerve, may enter the orbit either through the rostral alar foramen with the maxillary nerve or through a separate zygomatic foramen (McClure 1960). The nerve enters the periorbita at its apex and divides into the zygomaticofacial and zygomaticotemporal nerves (Fig. 20–30). The **zygomaticotemporal nerve** courses rostrodorsally deep to the lateral aspect of the periorbita to the region of the orbital ligament, where it ramifies in the skin and conjunctiva of the upper eyelid. As the nerve passes the lacrimal gland, branches may communicate with the lacrimal nerve, possibly supplying it with parasympathetic innervation derived from the pterygopalatine ganglion (Fig. 20–32). Parasympathomimetic pharmaceuticals are used to increase glandular secretion.

The **zygomaticofacial nerve** is ventral to the zygomaticotemporal and parallel to it over most of its course within the periorbita (Fig. 20–31). Near the orbital margin it turns ventrally and ramifies in the skin and conjunctiva of the lower lid adjacent to the lateral commissure.

Abducent Nerve

The **abducent nerve** *(n. abducens)*, or cranial nerve VI, supplies general somatic efferent (GSE) axons to the lateral rectus and retractor bulbi muscles. The GSE neuronal cell bodies are located in the rostrodorsal medulla adjacent to the midline. The abducent axons pass ventral to emerge through the trapezoid body immediately lateral to the pyramids of the medulla (see Fig. 14–5). The nerve runs rostrally in the subarachnoid space medial to the trigeminal ganglion and leaves the cranial cavity through the orbital fissure, closely applied to its medial wall (Fig. 20–32). One to two centimeters rostral to the orbital fissure, the abducent nerve gives a branch to the retractor bulbi. This branch further divides to supply the four fasciculi of the muscle. The abducent continues rostrally and laterally to reach the deep surface of the lateral rectus muscle. Lesions of the abducent nucleus or nerve result in medial strabismus because of paralysis of the lateral rectus muscle.

Facial Nerve

The **facial nerve** *(n. facialis)*, or cranial nerve VII, supplies special visceral efferent innervation to the muscles of the eyelids and parasympathetic innervation to the lacrimal gland by way of the chorda tympani and pterygopalatine ganglion.

The facial nerve emerges from the brain stem lateral to the abducent nerve and rostral to the fibers of the trapezoid body (see Fig. 14–5). The nerve courses laterally through the internal acoustic meatus with the vestibulocochlear nerve. The facial nerve enters the facial canal and turns caudally, forming the *geniculum n. facialis*. The cell bodies of the afferent axons of the facial nerve form the **geniculate ganglion** *(ganglion geniculi)*, where the nerve makes its turn.

Preganglionic parasympathetic fibers leave the facial nerve just distal to the genu as the deep petrosal nerve. This joins the major petrosal nerve to form the **nerve of the pterygoid canal,** which enters the caudal end of this small canal of the basisphenoid bone. The nerve ends at the **pterygopalatine ganglion,** which lies ventral to the periorbita on the dorsal surface of the medial pterygoid muscle. Postganglionic fibers from the pterygopalatine ganglion innervate the nasal and lacrimal glands.

The facial nerve leaves the facial canal and emerges on the caudolateral aspect of the skull through the stylomastoid foramen, behind the external acoustic meatus. The nerve branches to give rise to the caudal auricular nerves, the digastricus branch, the internal auricular nerves, and the stylohyoid nerve. The continuation of the facial nerve curves ventrally around the annular auricular cartilage, gives origin to the dorsal and ventral buccal branches, and continues as the **auriculopalpebral nerve** *(n. auriculopalpebralis)*. The auriculopalpebral nerve turns dorsally along the rostral aspect of the auricular cartilage, where it is especially liable to damage in surgical manipulations for diseases of the external ear canal (Fig. 15–10).

At the level of the origin of the zygomatic process from the temporal bone, the auriculopalpebral nerve divides to form the rostral auricular branches and the zygomatic branch. The **zygomatic branch** *(ramus zygomaticus)* curves rostrally along the dorsal margin of the zygomatic arch. Numerous branches join in the formation of the extensive rostral auricular plexus (see Fig. 15–10). Dorsocaudal to the lateral commissure, the zygomatic branch gives rise to the dorsal and ventral **palpebral branches** *(rami palpebrales)*, which innervate the dorsal and ventral portions of the orbicularis oculi muscle. The retractor anguli oculi lateralis and the levator anguli oculi medialis are palpebral muscles which are also supplied by rami of the zygomatic and palpebral branches.

Injury to the facial nerve or to its branches to the palpebral muscles paralyzes the orbicularis oculi muscle and is manifested as an inability to close the eyelids. Such paralysis prevents distribution of the tear film, which normally occurs during blinking. Marked ocular changes soon result from desiccation of the cornea. Severing only the ventral palpebral branch, as in lateral exploration of the orbit, does not interfere with normal eyelid function (Bistner, Aguirre, and Batik 1977).

Roberts et al. (1974) have described a technique for anesthetizing the auriculopalpebral nerve, where it crosses the zygomatic arch, to paralyze the orbicularis oculi muscle and facilitate ocular examination or the replacement of a proptosed eyeball.

VASCULATURE

Branches of the external carotid artery are the primary source of blood supply to the eye and its adnexa. Venous blood leaves the orbit through the angular vein of the eye, the deep facial vein, and the ophthalmic veins.

Arteries

Rostral to the base of the auricular cartilage, the external carotid terminates by branching to become the superficial temporal and maxillary arteries. The **superficial temporal artery** *(a. temporalis superficialis)* courses dorsally, supplying branches to adjacent structures, and terminates as the superior and inferior lateral palpebral arteries, which supply the lateral aspect of the eyelids and conjunctiva (see Fig. 11–14).

The corresponding superior and inferior medial palpebral arteries, which supply the conjunctiva and eyelids adjacent to the medial commissure, arise from the **malar artery** *(a. malaris),* a branch of the infraorbital artery. The malar also sends branches to the third eyelid *(a. palpebrae tertiae),* to the ventral oblique muscle, and to the nasolacrimal duct.

The continuation of the external carotid, the **maxillary artery** *(a. maxillaris),* gives rise to the inferior alveolar, caudal deep temporal, rostral tympanic, pterygoid, and middle meningeal arteries before entering the caudal alar foramen. The maxillary artery traverses the alar canal and emerges through the rostral alar foramen on the lateral aspect of the maxillary nerve. A few millimeters rostral to the alar foramen, the maxillary artery gives rise to the **external ophthalmic artery** *(a. ophthalmica externa),* which passes dorsally to enter the apex of the periorbita (Fig. 20–33). Within the periorbita an anastomotic branch leaves the external ophthalmic and passes caudally through the orbital fissure to the internal carotid artery *(ramus anastomoticus cum a. carotide interna)* at the level of the sella turcica. A similar branch anastomoses with the middle meningeal ar-

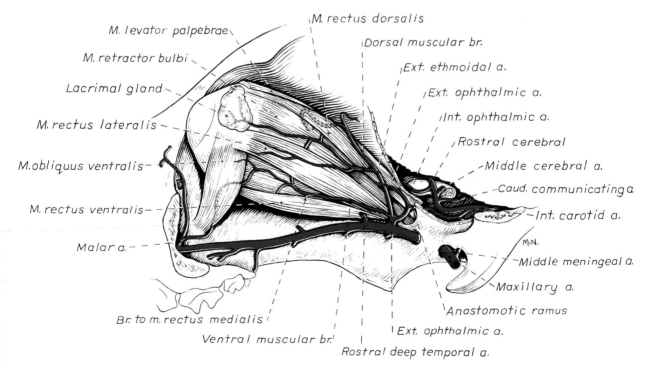

Figure 20–33. Arteries of the orbit and extrinsic ocular muscles, lateral aspect.

tery *(ramus anastomoticus cum a. meningea media)* (Fig. 20–34). These anastomotic branches may arise independently from the external ophthalmic, but more commonly arise from a single trunk which divides within the orbital fissure.

The **external ethmoidal artery** *(a. ethmoidalis externa)* arises from the external ophthalmic distal to these anastomotic branches and curves dorsomedially over the extraocular muscles to enter the dorsal ethmoidal foramen on the medial orbital wall (Fig. 20–34). There are usually two muscular branches of the external ethmoidal artery, although they may arise from a single trunk or from the external ophthalmic directly. The **ventral muscular branch** runs rostrally toward the eyeball between the ventral and lateral rectus muscles, supplying these muscles as well as the medial rectus, the ventral fasciculi of the retractor bulbi, and the gland of the third eyelid (Fig. 20–33). The muscular branches are continued as the **anterior ciliary arteries** *(aa. ciliares anteriores)*, which follow the tendons of the rectus muscles to their insertions anterior to the equator of the eyeball. Here the arteries pass into the sclera to ramify in and around the scleral venous plexus. From this arterial net fine arterioles pass inward to anastomose with terminal branches of the posterior ciliary arteries supplying the ciliary body and iris (Fig. 20–5).

Terminal twigs of the branches also contribute to the capillary loops of the bulbar conjunctiva *(aa. conjunctivales posteriores)* at the limbus and to the deeper episcleral vessels *(aa. episclerales)*. A distinct **lacrimal artery** *(a. lacrimalis)* usually arises from the dorsal muscular branch but may originate independently from the external ethmoidal or external ophthalmic arteries. It runs rostrally on the lateral edge of the dorsal rectus to supply the lacrimal gland (Fig. 20–33).

The **dorsal muscular branch** arises near or in common with the ventral branch and crosses over the lateral rectus to run rostrally between the lateral and dorsal recti (Fig. 20–33). It supplies branches to the lateral and dorsal rectus muscles, the dorsal fascicles of the retractor bulbi, and the dorsal oblique and the dorsal levator palpebral muscles. At the equator of the eyeball, the dorsal muscular branch terminates as an-terior ciliary, episcleral, and posterior conjunctival vessels analogous to those of the ventral branch.

The external ophthalmic artery continues rostrally and medially to the center of the periorbital cone, where it comes to lie on the external sheath of the optic nerve. The course of the artery along the optic nerve is quite sinuous. About midway between the optic canal and the posterior pole of the eyeball, there is a large anastomosis between the external and internal ophthalmic arteries *(ramus anastomoticus cum a. ophthalmica interna)* (Fig. 20–34). The **internal ophthalmic artery** *(a. ophthalmica interna)* is a small artery which arises from the anterior cerebral artery at the level of the optic chiasm (Fig. 20–34). The internal ophthalmic artery passes through the optic canal on the dorsal surface of the optic nerve and runs rostrally on the nerve to anastomose with the external ophthalmic. The internal ophthalmic artery is smaller than the external ophthalmic and is distributed almost exclusively to the eyeball itself. This is quite distinct from human anatomy, in which the homologous artery, the ophthalmic, is the primary source of supply for all the orbital structures as well as the eyeball itself (Last 1968). Therefore, in man the primary blood supply to the eye is derived from the internal carotid circulation, whereas the primary source of blood for the eye in the dog is the external carotid.

From the anastomosis between the internal and external ophthalmic arteries, two **long posterior ciliary arteries** *(aa. ciliares posteriores longae)* arise which run rostrally closely applied to the optic nerve (Fig. 20–34). At the posterior aspect of the eyeball, the long posterior ciliary arteries give rise to a variable number of **short posterior ciliary arteries** *(aa. ciliares posteriores breves)* which pass through the sclera adjacent to the optic nerve and ramify in the choroid. These choroidal arterioles follow primarily in a meridional course to the ciliary body and ciliary margin of the iris (Fig. 20–5). Here they form variable anastomoses with branches of the anterior and long posterior ciliary arteries. In dogs with poorly pigmented ocular fundi, the course of the choroidal vessels may be visible with the ophthalmoscope. A defect in the choroidal vasculature

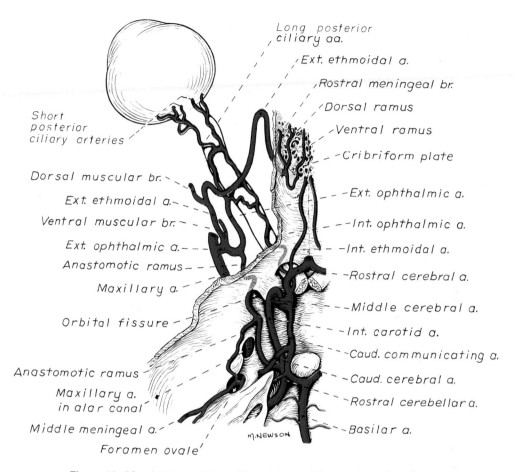

Long posterior ciliary aa.

Ext. ethmoidal a.

Rostral meningeal br.

Dorsal ramus

Ventral ramus

Cribriform plate

Short posterior ciliary arteries

Dorsal muscular br.

Ext. ethmoidal a.

Ventral muscular br.

Ext. ophthalmic a.

Anastomotic ramus

Maxillary a.

Orbital fissure

Anastomotic ramus

Maxillary a. in alar canal

Middle meningeal a.

Foramen ovale

Ext. ophthalmic a.

Int. ophthalmic a.

Int. ethmoidal a.

Rostral cerebral a.

Middle cerebral a.

Int. carotid a.

Caud. communicating a.

Caud. cerebral a.

Rostral cerebellar a.

Basilar a.

M. NEWSON

Figure 20–34. Arteries of the orbit and base of the cranium, dorsal aspect.

in the lateral quadrant of the fundus is the most common manifestation of Collie eye syndrome (Latshaw et al. 1969).

The **retinal arteries** are derived from the short posterior ciliary arteries as they pass through the sclera at the periphery of the optic nerve (Hetkamp 1972). The retinal arteries continue anteriorly through the choroid and retina and emerge in the periphery of the optic disc, where they are visible ophthalmoscopically (Fig. 20–11). The number of retinal arterioles is variable, but there are usually about 15 where they first become visible at the periphery of the optic disc. They divide repeatedly as they wind toward the periphery; secondary and tertiary branches are visible in the normal eye. The retinal arterioles are distinguished from the retinal veins by their greater tortuosity, finer caliber, and brighter red color (Wyman and Donovan 1965). Contrary to previous re-

ports, a central artery of the retina, comparable to that found in man, does not occur in the dog (see "Retina").

The long posterior ciliary arteries continue anteriorly in the episcleral tissues along the medial and lateral meridians to the equator of the eyeball. Here they disappear from view, passing deep to the sclera into the suprachoroidea. They continue anteriorly in the suprachoroidea to the ciliary margin of the iris. A few branches may be given off which anastomose with choroidal arterioles from the short posterior ciliary arteries (Fig. 20–5). In the periphery of the iris, each long posterior ciliary artery divides into a dorsal and ventral branch. These run circumferentially, forming the incomplete **major arterial circle of the iris** (Fig. 20–5). Small branches leave the circle to supply the pupillary region or to anastomose with arteries in the ciliary region (see "Vascular Tunic").

Veins

Blood leaves the orbit through one of three routes: (1) by way of the angular vein of the eye to the facial vein; (2) from the ophthalmic plexus to the cavernous sinus and maxillary vein; or (3) through an anastomosis of the ventral external ophthalmic vein with the deep facial vein (Fig. 20–35). These three drainage pathways are interconnected and are all filled in orbital venography.

The **angular vein of the eye** (*v. angularis oculi*) continues the facial vein dorsally and caudally. It originates where the dorsal nasal vein joins the facial rostral to the medial commissure of the eyelids. The angular vein passes caudally over the superficial surface of the medial palpebral ligament and passes caudodorsally medial to the commissure of the lids (Fig. 20–17). From the medial commissure the angular vein follows the dorsal orbital margin for about one-half the length of the zygomatic process of the frontal bone. It then turns caudally and enters the orbit. A small vein from the medial upper eyelid (*v. palpebralis superior medialis*) joins the angular vein where it turns to enter the orbit. At about the level of the equator of the eyeball, the angular vein passes into the periorbita to anastomose with the dorsal external ophthalmic vein. As the angular vein lacks valves,

blood may flow through it either from the orbit into the facial vein or from the facial vein into the ophthalmic vessels.

The **dorsal external ophthalmic vein** (*v. ophthalmica externa dorsalis*) is the largest intraorbital vein. After its anastomosis with the angular vein of the eye, the dorsal ophthalmic courses caudally along the dorsomedial orbital wall. At the posterior aspect of the eyeball, a large **anastomotic branch** (*ramus anastomoticus cum v. ophthalmica externa ventrali*) passes medially deep to the dorsal oblique muscle and winds down the medial orbital wall to join the ventral external ophthalmic vein (Fig. 20–35).

The dorsal vorticose veins usually join the dorsal external ophthalmic near this anastomosis, or they may enter the anastomotic branch itself. Caudal to the anastomosis, the dorsal external ophthalmic begins to dilate markedly and assumes a more dorsal position. Numerous muscular branches and the external ethmoidal vein enter the dilated caudal portion of the dorsal ophthalmic (Fig. 20–35). The **lacrimal vein** (*v. lacrimalis*), which drains the lacrimal gland, joins the dorsal external ophthalmic near the apex of the orbit. At the apex of the orbit, the dorsal external ophthalmic is massively dilated and envelops the other vessels and nerves entering the orbit, forming the so-called **ophthal-**

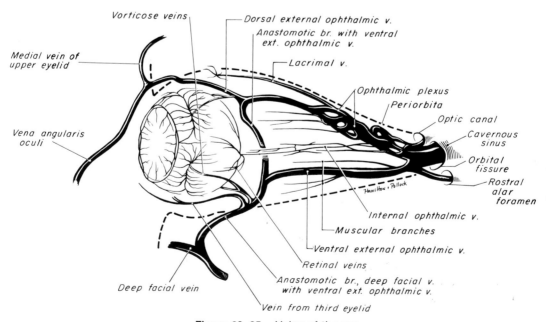

Figure 20–35. Veins of the eye.

mic plexus *(plexus ophthalmicus)*. From the ophthalmic plexus the small internal ophthalmic vein enters the optic canal with the optic nerve to anastomose with its fellow from the contralateral side on the floor of the braincase (see Fig. 12–23).

The ventral external ophthalmic vein joins the dorsal external ophthalmic ventral to the optic canal. The bulk of the blood from the union of these vessels leaves the orbit as the **emissary vein of the orbital fissure** (Fig. 20–25, 20–35). One or more small branches enter the rostral alar foramen to anastomose with the maxillary vein.

The **ventral external ophthalmic vein** *(v. ophthalmica externa ventralis)* lies within the periorbita on the floor of the orbit between the extraocular muscles and the medial pterygoid muscle. Caudally, it receives numerous muscular branches and joins the dorsal external ophthalmic in the ophthalmic plexus. Rostrally, it receives the large anastomotic branch from the dorsal ophthalmic and drainage from the ventral pair of vorticose veins. A large branch draining the base of the third eyelid joins the ventral external ophthalmic near or in common with the vorticose veins (Fig. 20–35). Caudal to the ventral orbital margin, a large anastomotic branch unites the ventral external ophthalmic to the deep facial vein *(ramus anastomoticus cum v. ophthalmica externa ventrali)* on the lateral surface of the zygomatic glands (Figs. 20–17, 20–35).

The eyeball drains into the ophthalmic vessels through the retinal, ciliary, and vorticose veins. The intraocular course of the retinal veins has already been described (see "Retina"). **Ciliary veins** *(vv. ciliares)* accompany the several ciliary arteries as satellite veins. **Retinal veins** join the posterior ciliary arteries at the periphery of the area cribrosa to form the internal ophthalmic vein. There is no central retinal vein in the dog. The **internal ophthalmic vein** comprises a number of interanastomosing vessels which are closely adherent to the external sheath of the optic nerve and which receive fine radicles from the internal sheath. The internal ophthalmic runs caudally to join in the formation of the ophthalmic plexus.

The major drainage of the vascular tunic, however, is via the **vorticose veins** *(vv. vorticosae)*, which have no accompanying arteries (Fig. 20–35). The vorticose veins are usually four in number and penetrate the sclera near the equator between the insertions of the four rectus muscles. The intraocular branches of the vorticose veins, the choroidal venules, radiate outward from the point of scleral penetration (Fig. 20–6). Anastomotic branches unite the vorticose veins with terminal branches of the ciliary veins and the scleral venous plexus (Van Buskirk 1979).

COMPARATIVE OPHTHALMOLOGY

Since publication of the first edition of *Anatomy of the Dog* (1964), major advances have been made in canine and comparative ophthalmology. Models for human ocular disease have been identified in the dog as well as a number of conditions peculiar to the dog. Current textbooks which discuss diseases of the canine eye include Magrane (1977), Severin (1976), Startup (1969), and Kómár and Szutter (1968). A manual of veterinary ophthalmological surgical procedures has recently been completed (Bistner et al. 1977), and an atlas of veterinary ocular pathology has been published by Saunders and Rubin (1975).

Major works in comparative ophthalmology include Walls (1942), Duke-Elder (1958), Prince (1956), Prince, Diesem, Eglitis, and Ruskell (1960), and Smythe (1961). A full-color atlas of comparative ophthalmoscopy is also available (Rubin 1974).

Several excellent texts of human ocular anatomy are available, including Last (1968) and Duke-Elder and Wybar (1961).

Ocular histology and fine structure are superbly treated in the text by Hogan, Alvarado, and Weddel (1971).

Standard works on ocular physiology include Davson and Graham (1974) and Moses (1970).

BIBLIOGRAPHY

Aguirre, G. D., and L. F. Rubin. 1970. Ophthalmitis secondary to congenitally open eyelids in a dog. J. Am. Vet. Med. Assoc. *156*:70–72.

Aguirre, G. D., L. F. Rubin, and S. I. Bistner, 1972. Development of the canine eye. Am. J. Vet. Res. *33*:2399–2414.

Andersen, A. C., and F. T. Shultz. 1958. Inherited (congenital) cataract in the dog. Am. J. Pathol. *34*: 965–975.

Anderson, B. G., and W. D. Anderson. 1977. Vasculature of the equine and canine iris. Am. J. Vet Res. 38:1791–1799.

Arey, L. B., S. R. Bruesch, and S. Castanares, 1942. The relation between eyeball size and the number of optic nerve fibers in the dog. J. Comp. Neurol. 76:417–422.

Balazs, E. A. 1973. The vitreous. Int. Ophthalmol. Clin. *13*:169–187.

Bedford, P. G. C. 1971. Eyelashes and adventitious cilia as causes of corneal irritation. J. Small Anim. Pract. *12*:11–17.

————. 1977. Gonioscopy in the dog. J. Small Anim. Pract. *18*:615–629.

Bistner, S. I. 1978. Neuro-ophthalmology. *In* Canine Neurology: Diagnosis and Treatment by B. F. Hoerlein. 3rd Ed. Philadelphia, W. B. Saunders Co.

Bistner, S. I., G. D. Aguirre, and G. Batik. 1977. Atlas of Veterinary Ophthalmic Surgery. Philadelphia, W. B. Saunders Co.

Bonnet, R. 1901. Beiträge zur Embryologie des Hundes. Erste Fortsetzung. Anat. Hefte *16*:231–332.

Bruesch, S. R., and L. B. Arey. 1942. The number of myelinated and unmyelinated fibers in the optic nerve of vertebrates. J. Comp. Neurol. 77:631–665.

Code, C. F., and H. E. Essex. 1935. The mechanism involved in the production of exophthalmos in the dog by vago-sympathetic stimulation. Am. J. Physiol. *113*:29.

Davson, H., and L. T. Graham. 1974. The Eye. Vol. 5, Comparative Physiology. New York, Academic Press.

deLahunta, A. 1977. Veterinary Neuroanatomy and Clinical Neurology. Philadelphia, W. B. Saunders Co.

deLahunta, A., and J. Cummings. 1967. Neuroophthalmologic lesions as a cause of visual deficit in dogs and horses. J. Am. Vet. Med. Assoc. *150*:994–1011.

Diesem, C. 1975. Organ of vision. *In* Sisson and Grossman's The Anatomy of the Domestic Animals, by Robert Getty. Philadelphia, W. B. Saunders Co.

Duke-Elder, W. S. 1958. System of Ophthalmology. Vol I, The Eye in Evolution. London, Henry Kimpton.

Duke-Elder, W. S., and K. C. Wybar. 1961. System of Ophthalmology. Vol II, The Anatomy of the Visual System. St. Louis, C. V. Mosby Co.

Engerman, R. L., D. L. Molitor, and J. M. B. Bloodworth. 1966. Vascular system of the dog retina: light and electron microscopic studies. Exp. Eye Res. 5:296–301.

Evans, H. E. 1974. Prenatal development of the dog. Publication of the Twenty-fourth Gaines Veterinary Symposium, Ithaca, N. Y., pp. 18–28.

Evans, H. E., and W. Sack. 1973. Prenatal development of domestic and laboratory mammals. Anat. Histol. Embryol. 2:11–45.

Fine, B. S., and Yanoff, M. 1972. Ocular Histology: A Text and Atlas. New York, Harper & Row.

Gilbert, P. W. 1947. The origin and development of the extrinsic ocular muscles in the domestic cat. J. Morphol. *81*:151–194.

Greenlee, T. K., R. Ross, and J. L. Hartman. 1966. The fine structure of elastic fibers. J. Cell Biol. *30*:59–71.

Habel, R. E. 1975. Applied Veterinary Anatomy. Ithaca, N. Y., published by the author.

Hamon, M. A. 1977. Atlas de la Tête du Chien. Thèse, L. Université Paul Sabatier de Toulouse.

Hebel, R. 1969. Licht- und elektronen-mikroskopische Untersuchungen an den Zellen des Tapetum lucidum des Hundes. Z. Anat. Entwickl.-Gesch. *129*:274–284.

————. 1971. Entwicklung und Struktur der Retina und des Tapetum lucidum des Hundes. Ergeb. Anat. Entwicklungsgesch. *45*.

Helper, L. C. 1970. The effect of lacrimal gland removal on the conjunctiva and cornea of the dog. J. Am. Vet. Med. Assoc. *157*:72–75.

Helper, L. C., W. G. Magrane, J. Koehm, and R. Johnson. 1974. Surgical induction of keratoconjunctivitis sicca in the dog. J. Am. Vet. Med. Assoc. *165*:172–174.

Hetkamp, D. 1972. Korrosionsanatomische Untersuchungen der Blutgefässe des Auges des Haushundes (*Canis fam. L.*) unter besonderer Berücksichtigung des Kapillarsystems. Vet. Diss. Giessen.

Heywood, R., P. L. Hepworth, and N. J. van Abbe. 1976. Age changes in the eyes of the Beagle dog. J. Small Anim. Pract. *17*:171–177.

Hoerlein, B. F. 1978. Canine Neurology: Diagnosis and Treatment. 3rd Ed. Philadelphia, W. B. Saunders Co.

Hogan, M. J., J. A. Alvarado, and J. E. Weddel, 1971. Histology of the Human Eye: An Atlas and Textbook. Philadelphia, W. B. Saunders Co.

Huber, E. 1922, 1923. Über das Muskelgebiet des N. facialis beim Hund, nebst allgemeinen Betrachtungen über die Fascialismuskulatur. Morph. Jahrb. 52:1–110, 354–414.

Knecht, C. D. 1970. Treatment of diseases of the zygomatic salivary gland. J. Am. Anim. Hosp. Assoc. 6:13–19.

Koch, S. A., and L. F. Rubin. 1972. Distribution of cones in retina of the normal dog. Am. J. Vet. Res. 33:361–363.

Kómár, G., and L. Szutter, 1968. Tierärztliche Augenheilkunde. Berlin, Verlag Paul Parey.

Last, R. J. 1968. Wolff's Anatomy of the Eye and Orbit. Philadelphia, W. B. Saunders Co.

Latshaw, W. K., M. Wyman, and W. G. Venzke. 1969. Embryologic development of an anomaly of the ocular fundus in the Collie dog. Am. J. Vet. Res. 30:211–217.

Lord, R. D. 1961. The lens as an indication of age in the grey fox. J. Mammol. *42*:109–110.

Magrane, W. G. 1977. Canine Ophthalmology. 3rd Ed. Philadelphia, Lea & Febiger.

Martin, C. L. 1969. Slit lamp examination of the normal canine anterior ocular segment: Part II, Description. J. Small Anim. Pract. *10*:151–162.

Martin, C. L. 1975. Scanning electron microscopic examination of selected canine iridocorneal angle abnormalities. J. Am. Anim. Hosp. Assoc. *11*:300–306.

McClure, R. C. 1960. Occurrence of the zygomatic groove and canal in the sphenoid bone of the dog skull (*Canis familiaris*). Anat. Rec. (Abstr.) *138*:136.

Michel, G. 1955. Beitrag zur Anatomie der Tränenorgane von Hund und Katze. Dtsch. Tierärztl. Wochenschr. *62*:347–349.

Mishima, S., and D. M. Maurice. 1961. The oily layer of

the tear film and evaporation from the corneal surface. Exp. Eye Res. *1*:39–45.

Moses, R. A. 1970. Adler's Physiology of the Eye. 5th Ed. St. Louis, C. V. Mosby Co.

Nicholas, E. 1914. Veterinary and Comparative Ophthalmology, translated by H. Gray. London, H. & W. Brown.

Pollock, R. V. H. 1978. The zonula ciliaris of the dog. Proc. Can. Assoc. of Vet. Anat. 3:11.

Prince, J. H. 1956. Comparative Anatomy of the Eye. Springfield, Ill., Charles C Thomas.

Prince, J. H., C. D. Diesem, I. Eglitis, and G. L. Ruskell, 1960. Anatomy and Histology of the Eye and Orbit in Domestic Animals. Springfield, Ill., Charles C Thomas.

Purtscher, E. 1961. Die grossen Irisarterien beim Hunde. Berlin and München. Tierärztl. Wschr. 74:436–438.

Raviola, G. 1971. The fine structure of the ciliary zonule and ciliary epithelium. Invest. Ophthalmol. *10*:851–869.

Rebhun, W. C. 1976. Persistent hyperplastic primary vitreous in a dog. J. Am. Vet. Med. Assoc. *169*:620–622.

Riis, R. C. 1976. The normal canine conjunctiva. Thesis, Cornell Univ. Ithaca, N. Y.

Roberts, S. R., and S. I. Bistner. 1968. Persistent pupillary membrane in Basenji dogs. J. Am. Vet. Med. Assoc. *153*:533–542.

Roberts, S. R., R. C. Vierheller, and W. J. Lennox. 1974. Eyes. *In* Canine Surgery, edited by J. Archibald. Santa Barbara, Calif., American Veterinary Publications.

Rubin, L. F. 1974. Atlas of Veterinary Ophthalmoscopy. Philadelphia, Lea & Febiger.

Saunders, L. Z., and L. F. Rubin. 1975. Ophthalmic Pathology of Animals: An Atlas and Reference Book. New York, S. Karger.

Scott, D. W., and S. I. Bistner. 1973. Neurotrophic keratitis in a dog. Vet. Med. Small Anim. Clin. 68:1120–1122.

Severin, G. A. 1976. Veterinary Ophthalmology Notes. 2nd Ed. Fort Collins, Col. Colorado State Univ.

Sherman, S. M., and J. R. Wilson. 1975. Behavioral and morphological evidence for binocular competition in the postnatal development of the dog's visual system. J. Comp. Neurol. *161*:183–196.

Shively, J. N., and G. P. Epling. 1970. Fine structure of the canine eye: cornea. Am. J. Vet. Res. *31*:713–722.

Shively, J., G. Epling, and R. Jensen. 1971. Fine structure of the postnatal development of the canine retina. Am. J. Vet Res. 32:383–392.

Smythe, R. H. 1961. Animal Vision: What Animals See. Springfield, Ill. Charles C Thomas.

Startup, F. G. 1969. Diseases of the Canine Eye. Baltimore, The Williams & Wilkins Co.

Van Buskirk, E. M. 1979. The canine eye: The vessels of aqueous drainage. Invest. Ophthalmol. Vis. Sci. *18*:223–230.

Walls, G. L. 1942. The Vertebrate Eye and its Adaptive Radiation. Bloomfield Hills, Mich., Cranbrook Institute of Science.

Watrous, W. G., and J. M. D. Olmstead. 1941. Reflex studies after muscle transplantation. Am. J. Physiol. *132*:607–611.

Wyman, M., and E. F. Donovan. 1965. The ocular fundus of the normal dog. J. Am. Vet. Med. Assoc. *147*:17–26.

Yakely, W. L., and J. E. Alexander. 1971. Dacryocystorhinography in the dog. J. Am. Vet. Med. Assoc. *159*:1417–1421.

INDEX

Page numbers in italics indicate illustrations. Page numbers followed by (t) indicate tables.

Tuberosity (*Continued*)
  maxillary, 141
  of fourth tarsal bone, 221
  radial, 186
  supracondylar, 209
  supraglenoid, 181, 182
  tibial, 210
  ulnar, 190
Tubular glands, 1062
Tubules, renal, 549
  seminiferous, 558
  straight, 549, 558
Tubuli seminiferi contorti, 558
Tubuli seminiferi recti, 558
Tubulus renalis rectus, 549
Tunic(s), fibrous, of eyeball,
    1077–1080
  nervous, of eyeball, 1087–1089,
    1088
  uterine, 589, 590
  vaginal, 558, 564
    of scrotum, 556
  vascular, of eyeball, 1080, *1081,*
    1082–1087
Tunica adventitia, 456
Tunica albuginea, 558
  female, 584
Tunica fibrosa bulbi, 1077
Tunica interna bulbi, 1077
Tunica mucosa. See *Mucous coat.*
Tunica mucosa linguae, 425
Tunica muscularis. See *Muscular
  coat.*
Tunica serosa. See *Serous coat.*
Tunica vaginalis parietalis, 564
Tunica vasculosa bulbi, 1080
Tunica vasculosa lentis, 1075
Turbinate(s). See *Concha(e).*
Turkish saddle, 128
Tylotrich hair, 92, 93
Tympanic artery, 666
Tympanic bullae, 151
Tympanic cavity, 132, 134,
    1062–1064
Tympanic membrane, 134, 1059,
    1060, 1062
Tympanic nerve, 926, *931,* 1035,
    1064
Tympanic part of temporal bone,
    133
Tympanic plexus, 926, 1035, 1064
Tympanic ring, 134
Tympanic surface, of pyramid, 132
  of temporal bone, 132
Tympanicum, 133
Tympanohyoid cartilage, 52, 148,
    149
Tympano-occipital fissure, 135

Ulna, *187, 188,* 189, 190
Ulna, body of, 190
  development of, 63, *64, 65*
  muscle attachments of, *351, 352*
  nutrient artery of, 702
Ulnar arteries, 696, 698

Ulnar collateral ligament, short,
    249
Ulnar head. See *Caput ulnare.*
Ulnar nerves, 988, 990
Ulnar notch, 189
Ulnar tuberosity, 190
Ulnar veins, 774, 775
Ultimobranchial bodies, 613
Umbilical artery, 746
Umbilical fold, middle, 466
Umbilical ligament, lateral, 746
Umbilical region of abdomen, 463
Umbilicus, 79, 331
Uncinate bundle, 854
Uncinate notch, 137
Uncinate process, 137, 138
Ungual crest, 197
Unguicula, 78, 99
Unipennate muscle, 271
Unit, motor, 953
Urachus, 553
Ureter(s), 551, 552, *552*
  anomalies of, 552
  blood supply of, 552
  nerves of, 552
  structure of, 552
Ureteric veins, 779, 788
Urethra, female, 594, 595
  anomalies of, 595
  blood vessels of, 594, 595
  male, 578, 579
  blood vessels of, 579
  parts of, 578, 579
  variations in, 579
Urethra masculina, 578
Urethral arteries, 579, 748
Urethral bulb, 567
  artery of, 579
Urethral crest, 578
Urethral groove, 223
Urethral orifice, external, female,
    592
Urethral tubercle, female, 592
Urethral veins, 566, 579
Urethral venous plexus, 785, 788
Urinary bladder. See *Bladder,
  urinary.*
Urinary organs, 544–554, *545–548,
    550, 552.* See also *Urogenital
    apparatus.*
  kidneys, 544–551, *545–548, 550*
  ureters, 551, 552, *552*
  urethra, female, 594, 595
    male, 578, 579
  urinary bladder, 552–554, *552*
Urogenital apparatus, 544–601.
    See also *Reproductive organs
    and Urinary organs.*
  embryology of, 595–598, *597*
  female, *545, 580–595, 597, 581*
  lymph nodes and vessels of,
    830, *830–832, 832*
  male, *597*
Urogenital ligaments of male, *560*
Urogenital vein, 788
Uterine artery, 748
Uterine cervix, 587

Uterine growth, 29, *31*
Uterine horns, 587
Uterine orifices, 588
Uterine tube, 586, 587
  anomalies of, 587
  blood vessels of, 587
  lymphatics of, 587
  nerves of, 587
  structure of, 586
Uterine vein, 788
Uterus, 16–18, 22, 587–590, *588*
  blood circulation in during
    pregnancy, 590
  blood vessels of, 590
  in labor, 589
  ligaments of, 589
    broad, 466, 553, 581–584
  lymphatics of, 590
  nerves of, 590
  relation of to other organs, 589
  structure of, 589, 590
  tunics of, 589, 590
Uterus masculinus, 578
Utricle, prostatic, 578
Utriculus prostaticus, 578
Uvea, 1080
Uvula vermis, 873
Uvulonodular fissure, 871

Vagal ganglion, 1037, 1039
Vagina, 590, 591
  anomalies of, 591
  blood vessels of, 591
  lymphatics of, 591
  nerves of, 591
  relation of to other organs, 591
  structure of, 591
Vagina bulbi, 1097, *1097, 1098*
Vagina externa n. optici, 1078
Vaginae n. optici interna et
    externa, 1087
Vaginal artery, 594, 748
Vaginal process, female, 582
  of scrotum, 556, *561*
Vaginal ring, male, 564
Vaginal tunic, 558, 564
  of scrotum, 556
Vaginal vein, 788
Vaginal venous plexus, 788
Vagus group of nerves, 877
Vagus nerve. See *Nerve, vagus.*
Vallate papillae, 427, *428–430,* 430
Vallecula, 518
Vallum, 99
Valva aortae, 646
Valva trunci pulmonalis, 646
Valvae atrioventriculares, 645
Valve(s), aortic, 646
  atrioventricular, 645, 646
  bicuspid, 645
  mitral, 645
  of foramen ovale, 640
  of pulmonary trunk, 646
  tricuspid, 645
Valvula foraminis ovalis, 640